W9-AWL-824

COLONIZATION OF MUCOSAL SURFACES

COLONIZATION OF MUCOSAL SURFACES

EDITED BY

JAMES P. NATARO
Center for Vaccine Development
Departments of Pediatrics, Medicine, and Microbiology & Immunology
University of Maryland School of Medicine
Baltimore, Maryland

PAUL S. COHEN
Department of Cell and Molecular Biology
University of Rhode Island
Kingston, Rhode Island

HARRY L. T. MOBLEY
Department of Microbiology and Immunology
University of Michigan Medical School
Ann Arbor, Michigan

AND

JEFFREY N. WEISER
Department of Microbiology and Pediatrics
University of Pennsylvania
Philadelphia, Pennsylvania

ASM PRESS
Washington, D.C.

Address editorial correspondence to ASM Press, 1752 N St. NW, Washington, DC 20036-2904, USA

Send orders to ASM Press, P.O. Box 605, Herndon, VA 20172, USA
Phone: (800) 546-2416 or (703) 661-1593
Fax: (703) 661-1501
E-mail: books@asmusa.org
Online: www.asmpress.org

Library of Congress Cataloging-in-Publication Data

Colonization of mucosal surfaces / edited by James P. Nataro ... [et al.].
p. ; cm.
Includes bibliographical references and index.
ISBN 1-55581-323-2
1. Mucous membrane—Microbiology. 2. Mucous membrane—Immunology.
[DNLM: 1. Mucous Membrane—microbiology. 2. Immunity, Mucosal—physiology. QS 532.5.M8 C719 2005] I. Nataro, James P.

QR185.9.M83C655 2005
616.9′201—dc22

2004023760

10 9 8 7 6 5 4 3 2 1

Cover photo: Salmonella enterica producing multiple peritrichous flagella (see Color Plate 4). Photo courtesy of Jorge Girón.

CONTENTS

CONTRIBUTORS

Michael A. Apicella
Department of Microbiology, University of Iowa, Iowa City, IA 52242

Andreas J. Bäumler
Department of Medical Microbiology and Immunology, College of Medicine, Texas A&M University System Health Science Center, College Station, TX 77843

Göran Bergsten
Section for Microbiology, Immunology and Glycobiology, Institute for Laboratory Medicine, Lund University, Lund Sweden

Iain A. Brownlee
School of Cell and Molecular Biosciences, The Medical School, University of Newcastle upon Tyne, Newcastle upon Tyne NE2 4HH, England

Susan M. Butler
Department of Molecular Biology and Microbiology, Tufts University School of Medicine, Boston, MA 02111

Andrew Camilli
Department of Molecular Biology and Microbiology, Tufts University School of Medicine, Boston, MA 02111

Paul S. Cohen
Department of Cell and Molecular Biology, University of Rhode Island, Kingston, RI 02881

Alexander M. Cole
Department of Molecular Biology and Microbiology, Biomolecular Science Center, University of Central Florida, Orlando, FL 32816

Tyrrell Conway
Department of Botany and Microbiology, The University of Oklahoma, Norman, OK 73019

Peggy A. Cotter
Department of Molecular, Cellular, and Developmental Biology, University of California, Santa Barbara, Santa Barbara, CA 93106

Caleb W. Dorsey
Department of Medical Microbiology and Immunology, College of Medicine, Texas A&M University System Health Science Center, College Station, TX 77843

Christopher G. Dowson
Biomedical Research Institute, Biological Sciences, University of Warwick, Coventry CV4 7AL, United Kingdom

Daniel E. Dykuizen
Department of Ecology and Evolution, State University of New York, Stony Brook, NY 11794

Leo Eberl
Department of Microbiology, Institute of Plant Biology, University of Zurich, CH-8008 Zurich, Switzerland

Hans Fischer
Section for Microbiology, Immunology and Glycobiology, Institute for Laboratory Medicine, Lund University, Lund Sweden

Gad Frankel
Centre for Molecular Microbiology and Infection, Department of Biological Sciences, Imperial College London, London SW7 2AZ, United Kingdom

Tomas Ganz
Division of Pulmonary and Critical Care Medicine and the Will Rogers Institute Pulmonary Research Laboratory, and Department of Medicine, David Geffen School of Medicine at UCLA, Los Angeles, CA 90095

Jorge A. Girón
Department of Microbiology and Immunology, College of Medicine, University of Arizona, Tucson, AZ 85743

Michael Givskov
Centre for Biomedical Microbiology, BioCentrum-DTU, Technical University of Denmark, 2800 Lyngby, Denmark

Gabriela Godaly
Section for Microbiology, Immunology and Glycobiology, Institute for Laboratory Medicine, Lund University, Lund Sweden

Elizabeth L. Hartland
Department of Microbiology, School of Biomedical Sciences, Monash University, Victoria 3800, Australia

David L. Hasty
Department of Anatomy and Neurobiology, University of Tennessee Health Science Center, and Research Service, Veterans Affairs Medical Center, Memphis, TN 38104

Heikke Irjala
Section for Microbiology, Immunology and Glycobiology, Institute for Laboratory Medicine, Lund University, Lund Sweden

Angela Jansen
Department of Microbiology and Immunology, University of Maryland School of Medicine, Baltimore, MD 21201

Ann E. Jerse
Department of Microbiology and Immunology, Uniformed Services University of the Health Services, F. Edward Hèbert School of Medicine, Bethesda, MD 20814

Paul A. Jones
Department of Microbiology, University of Iowa, Iowa City, IA 52242

James B. Kaper
Center for Vaccine Development and Department of Microbiology and Immunology, University of Maryland School of Medicine, Baltimore, MD 21201

Mogens Kilian
Department of Medical Microbiology and Immunology, University of Aarhus, Aarhus C, DK-8000, Denmark

Robert A. Kingsley
Department of Medical Microbiology and Immunology, College of Medicine, Texas A&M University System Health Science Center, College Station, TX 77843

Karen A. Krogfelt
Department of Gastrointestinal Infections, Statens Seruminstitut, 2300 Copenhagen S, Denmark

David C. Laux
Department of Cell and Molecular Biology, University of Rhode Island, Kingston, RI 02881

Ann Charlotte Lundstedt
Section for Microbiology, Immunology and Glycobiology, Institute for Laboratory Medicine, Lund University, Lund Sweden

Beth A. McCormick
Department of Pediatric Gastroenterology and Nutrition, Massachusetts General Hospital, and Department of Microbiology and Molecular Genetics, Harvard Medical School, Boston, MA 02129

Jane Michalski
Department of Microbiology and Immunology, University of Maryland School of Medicine, Baltimore, MD 21201

Harry L. T. Mobley
Department of Microbiology and Immunology, University of Michigan, Ann Arbor, MI 48109

Søren Molin
Centre for Biomedical Microbiology, BioCentrum-DTU, Technical University of Denmark, 2800 Lyngby, Denmark

James P. Nataro
Center for Vaccine Development and Departments of Pediatrics, Medicine, and Microbiology and Immunology, University of Maryland School of Medicine, Baltimore, MD 21201

Marcela F. Pasetti
Center for Vaccine Development and Department of Pediatrics, University of Maryland, Baltimore, MD 21201

Jeff P. Pearson
School of Cell and Molecular Biosciences, The Medical School, University of Newcastle upon Tyne, Newcastle upon Tyne NE2 4HH, England

Kristine M. Peterson
Division of Infectious Diseases and International Health, Department of Medicine, University of Virginia School of Medicine, Charlottesville, VA 22908

William A. Petri, Jr.
Division of Infectious Diseases and International Health, Department of Medicine, University of Virginia School of Medicine, Charlottesville, VA 22908

Alan D. Philips
Centre for Paediatric Gastroenterology, Department of Paediatrics and Child Health, Royal Free Hospital, London NW3 2QG, United Kingdom

Christopher Pritchett
Department of Microbiology and Immunology, University of Maryland School of Medicine, Baltimore, MD 21201

Manuella Raffatellu
Department of Medical Microbiology and Immunology, College of Medicine, Texas A&M University System Health Science Center, College Station, TX 77843

Bryndis Ragnarsdottir
Institute for Laboratory Medicine, Section for Microbiology, Immunology and Glycobiology, Lund University, Lund, Sweden

Adam J. Ratner
Department of Microbiology, University of Pennsylvania, Philadelphia, PA 19104

Gregor Reid
Canadian Research and Development Centre for Probiotics. Lawson Health Research Institute, and Departments of Microbiology & Immunology and Surgery, University of Western Ontario, London, Ontario N6A 4V2, Canada

Jesper Reinholdt
Department of Oral Biology, University of Aarhus, Aarhus C, DK-8000, Denmark

Roy M. Robins-Browne
Department of Microbiology and Immunology, University of Melbourne, Victoria 3010, Australia

Niamh Roche
Section for Microbiology, Immunology and Glycobiology, Institute for Laboratory Medicine, Lund University, Lund Sweden

Rosangela Salerno-Gonçalves
Center for Vaccine Development and Department of Pediatrics, University of Maryland, Baltimore, MD 21201

Patrik Samuelsson
Section for Microbiology, Immunology and Glycobiology, Institute for Laboratory Medicine, Lund University, Lund Sweden

Amy N. Simms
Department of Microbiology and Immunology, Uniformed Services University of the Health Services, F. Edward Hèbert School of Medicine, Bethesda, MD 20814

Evgeni V. Sokurenko
Department of Microbiology, University of Washington, Seattle, WA 98195

Claus Sternberg
Centre for Biomedical Microbiology, BioCentrum-DTU, Technical University of Denmark, 2800 Lyngby, Denmark

David J. Stickler
Cardiff School of Biosciences, Cardiff University, Cardiff CF10 3TL, Wales, United Kingdom

Catharina Svanborg
Section for Microbiology, Immunology and Glycobiology, Institute for Laboratory Medicine, Lund University, Lund Sweden

Majlis Svensson
Section for Microbiology, Immunology and Glycobiology, Institute for Laboratory Medicine, Lund University, Lund Sweden

Marcelo B. Sztein
Center for Vaccine Development and Departments of Pediatrics and Medicine, University of Maryland, Baltimore, MD 21201

Gerald W. Tannock
Department of Microbiology, University of Otago, Dunedin, New Zealand, and Agricultural, Food and Nutritional Science, University of Alberta, Edmonton, Alberta, Canada

Anna D. Tischler
Department of Molecular Biology and Microbiology, Tufts University School of Medicine, Boston, MA 02111

Mumtaz Virji
Department of Pathology and Microbiology, School of Medical Sciences, University of Bristol, Bristol BS8 1TD, United Kingdom

Jeffrey N. Weiser
Department of Medicine, University of Pennsylvania, Philadelphia, PA 19104

Xue-Ru Wu
Departments of Urology and Microbiology, Kaplan Comprehensive Cancer Center, New York University School of Medicine, New York, NY 10016, and Veterans Affairs Medical Center in Manhattan, New York, NY 10010

PREFACE

Multicellular organisms have established and must defend barriers that protect them from a hostile environment. Large organisms, including all mammals, store an enticing accumulation of biological energy substrates and nutrients, naturally attractive to assault by microorganisms. Thus, mammalian barriers are a battleground on which this struggle continuously occurs. The simple fact that mammals continue to exist on earth implies that they have successfully fortified and defended their barriers against microbial attack. Conversely, the existence of microbial pathogens affirms that many species have evolved mechanisms to circumvent host defenses, starting at the level of the mucosal barriers.

Importantly, however, microorganisms may be better able to utilize the nutrients of the host by negotiating symbiotic arrangements, thereby assuring that the host will continue to provide nutrients for many generations of microbial progeny. Perhaps the most secure way to assure a warm welcome for the microbe is to come bearing gifts; we see increasing evidence that mammalian hosts are provided prodigious advantages upon welcoming certain microbial guests.

Thus, the mammalian mucosa has evolved numerous types of defenses to protect itself from microbial invaders and simultaneously to provide sustenance for an abundant commensal microflora. In this book, we consider the complex ecosystems which are the mammalian mucosa and examine mechanisms adapted by microorganisms to colonize these surfaces effectively. An understanding of the biology at these sites is critical to understanding both health and disease and to developing effective means to prevent infection. In this volume, we consider many different organisms present at the major mucosal surfaces. Much excellent work has not been given its deserved attention, but we feel that essential themes of mucosal colonization are replayed at multiple sites. This book is intended to provide an overview and introduction to this vital field of microbiology.

Jim Nataro

Paul Cohen

Harry Mobley

Jeff Weiser

August 2004

I. GENERAL CONSIDERATIONS

Colonization of Mucosal Surfaces
Edited by James P. Nataro et al.

Chapter 1

Structure and Function of Mucosal Surfaces

JEFF P. PEARSON AND IAIN A. BROWNLEE

MUCOSAL SURFACES

The mucosal surfaces of the body are the areas where important absorptive and excretive functions occur. The three systems where these exchange functions take place are the gastrointestinal (GI), respiratory, and urinogenital tracts. Consequently, these surfaces are exposed to the external environment and the cells present in the mucosa, along with their secretions, form a barrier between the nonsterile external environment and the essentially sterile internal environment of the body. As a result of this exposure, the mucosal surfaces are the primary locus of attack by microorganisms.

What Is a Mucosa?

Mucosa forms the inner lining of the respiratory, GI, and urinogenital tracts. It consists of three layers. The first is made up of the epithelial cells, which can be a single layer, as in the upper airways, bronchi, and the GI tract, or three to seven cells thick, as in the human urinary bladder. These cells are attached to a basement membrane overlying the second layer, the lamina propria, which consists of subepithelial connective tissue and lymph nodes, underneath which is the third layer, a thin layer of smooth muscle called the muscularis mucosae in the GI tract. Below this is the submucosa, which differs in the different tracts. The epithelial cells vary depending on the tissue. In the upper airways and bronchi they form a columnar epithelium with goblet cells which synthesize, store, and secrete mucus and ciliated cells that move the mucus blanket (Fig. 1A). This mucosa is essentially a secretory physical barrier. The epithelial cells of the GI tract are squamous in the mouth and the esophagus and do not produce an adherent mucus layer (16). The epithelial cells of the esophageal mucosa consist of three layers. The deepest layer is the basal layer, consisting of one to three cells actively involved in mitosis. Above this is the prickle cell layer, making up 10 to 15% of the epithelium, where the cells differentiate and flatten. The space between the prickle cells is filled with glycoconjugate material, which is thought to form a physical barrier in the epithelium (26). The outermost layer is the functional layer, which is one or two cells thick. Here the cells become leaky and die before being shed into the lumen (Fig. 1B). This cell shedding (desquamation) is an important mechanism of preventing microorganism invasion since microbes attached to the surface cells are shed along with the cells into the lumen, from where they pass into the stomach to be destroyed by gastric acid. In addition, the surface squamous epithelial cells have some protection from salivary secretions and the secretions of the sparse esophageal mucus-secreting submucosal glands, which together wet the surface and lubricate the tract to aid the passage of food. This squamous epithelium is a nonabsorptive barrier.

In the rest of the GI tract the epithelial cells are columnar. In the stomach the surface mucosal cells secrete mucus and gastric pits formed from several specialized gland structures secrete pepsinogen, the inactive zymogen of pepsin from peptic cells, and hydrochloric acid and intrinsic factor from parietal cells. In addition, the glands contain several types of endocrine and neuroendocrine cells (Fig. 2). Within the gastric mucosa, stem cells are dividing and migrating up the pits to replace lost surface epithelial cells. This means that the whole exposed epithelial surface of the stomach is replaced every 72 to 96 h. Some stem cell progeny migrate in the opposite direction down the pits and become acid- and pepsinogen-secreting cells. The glands do not

Jeff P. Pearson and Iain A. Brownlee • School of Cell and Molecular Biosciences, The Medical School, University of Newcastle upon Tyne, Newcastle upon Tyne, NE2 4HH, United Kingdom.

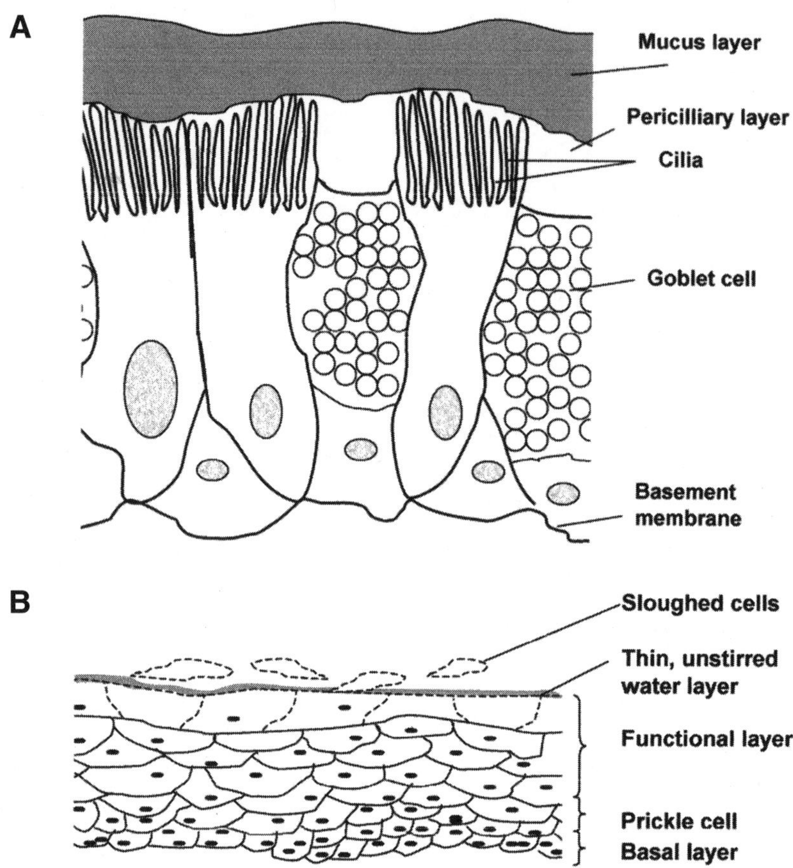

Figure 1. The diverse natures of mucosal surfaces. (A) Airway epithelium. The airway epithelium consists mainly of two cell types, ciliated cells that move the mucus blanket and goblet cells that secrete the mucus layer. Submucosal glands (not shown) also contribute to the mucus blanket. Above the cell surface is a low-viscosity pericilliary layer, which allows the cilia to beat effectively with the tips of the cilia just engaged into the mucus blanket. (B) Esophageal epithelium. The basal layer is where active cell division takes place. The prickle cell layer is where the cells differentiate into functional cells, and the functional layer is where the cells start to die and are shed into the lumen. The functional layer is covered by a thin unstirred water layer containing some salivary mucus and some mucus secreted by the esophageal submucosal glands. (C) Intestinal epithelium. The small intestinal epithelium consists of small finger-like extensions called villi, and the absorptive cells also have microvilli on their apical surface, creating the brush border. Interspaced among the absorptive cells are the mucus-secreting goblet cells. This epithelium is covered by a mucus bilayer of variable thickness. The layer is thickest in the ileum; the total thickness of 480 ± 47 μm is made up of 447 ± 47 μm of sloppy mucus and 29 ± 8 μm of firm mucus layers in the rat. (D) Urinary bladder epithelium. The urinary bladder epithelium consists of three layers: the basal, intermediate, and superficial layers. The surface is not covered by a secreted mucus gel but does have a surface protection by MUC 1 and 4 mucins extending 0.7 μm from the cell membrane.

turn over as rapidly as the surface epithelium; parietal cells have a life span of about 200 days. The gastric mucosa is therefore secretory and essentially nonadsorptive except for compounds that are lipid soluble at the low pH of the stomach, e.g., aspirin and alcohol.

In the intestines, columnar epithelium mucus is also secreted by goblet cells interspersed among absorptive cells, the enterocytes (Fig. 1C). The intestinal epithelium also contains M cells, which are present in Peyer's patches and are part of the gut-associated lymphoid tissue. The M cells are specialized epithelial cells that transport antigens and microorganisms from their apical surface through the cytoplasm to the basolateral surface by using transcytosis. Immune cells such as macrophages and lymphocytes are located in the extracellular compartment underneath these cells, waiting for antigen presentation. M cells are also present in the bronchi and upper airways, forming the bronchial-associated lymphoid tissues.

The epithelial cells of the urinary bladder and the conducting passages form a transitional epithelium, so called because its histologic appearance is somewhere between that of pseudostratified columnar and nonkeratinizing squamous epithelium (Fig.

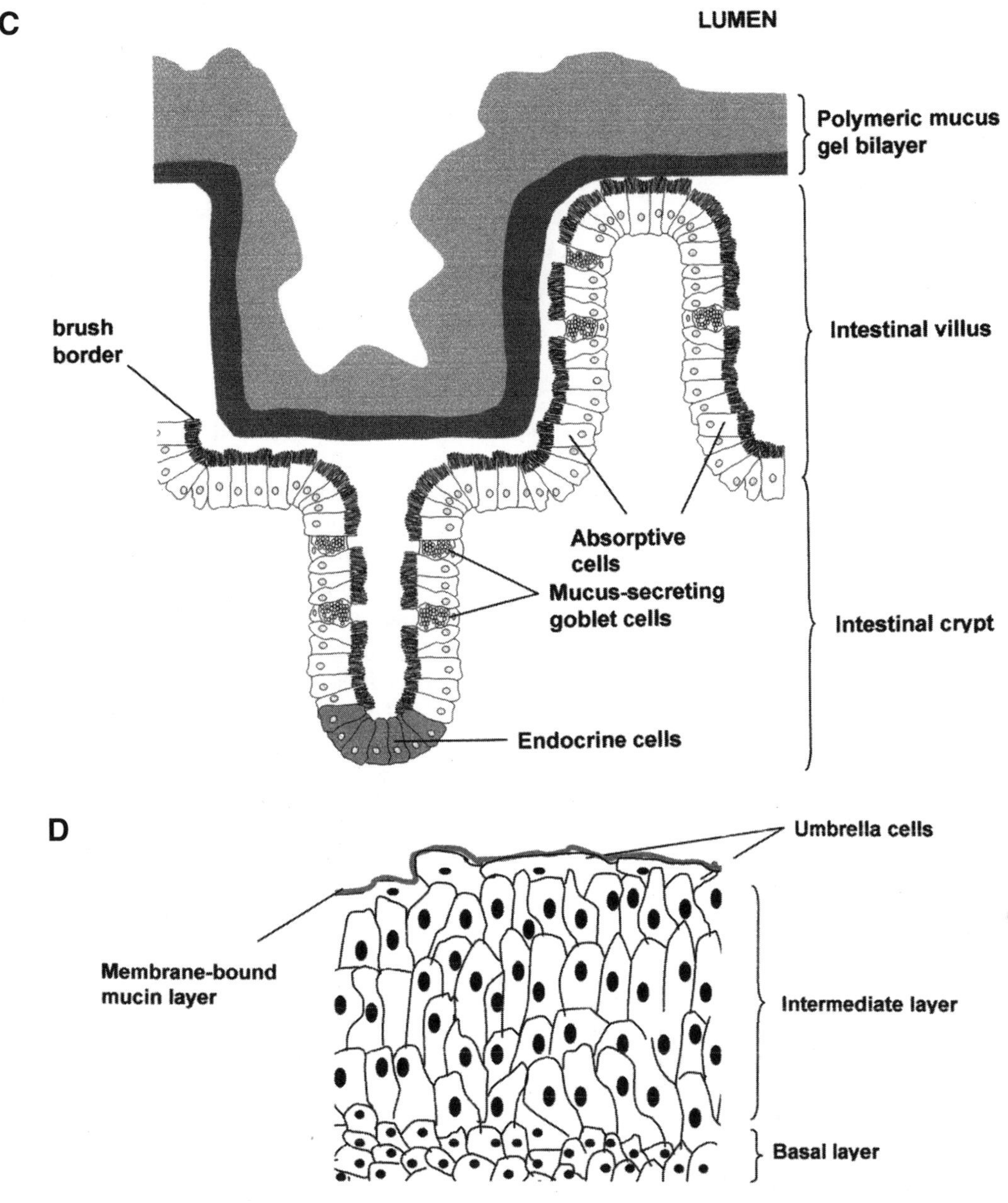

Figure 1. *Continued.*

1D). This urothelium is three to seven cells thick and is composed of a basal layer, one or more intermediate layers, and a superficial layer of umbrella cells. The thickness of these layers changes when the bladder expands as it fills with urine. The cells flatten along their long axis parallel to the basement membrane, and the epithelium is now only two or three cells thick. The normal urothelium does not contain mucus-secreting goblet cells. The umbrella cells have a specialized luminal membrane, with tight lateral junctions forming an impermeable layer to urine and also preventing water movement across the epithelium into the hypertonic urine.

WHAT DEFENSES ARE IN PLACE AT THE MUCOSAL LEVEL TO PREVENT MICROORGANISM INVASION?

The innate defense system consists of three components: mechanical, chemical, and cellular. The mechanical aspect is the barrier provided by the epithelial cells and the junctions between them and their secretions, e.g., mucus and motility, ciliary function, and desquamation. The epithelial cells are not just a physical barrier but also contribute to the chemical and cellular defenses. The chemical aspect of the defenses comes from (i) antimicrobial

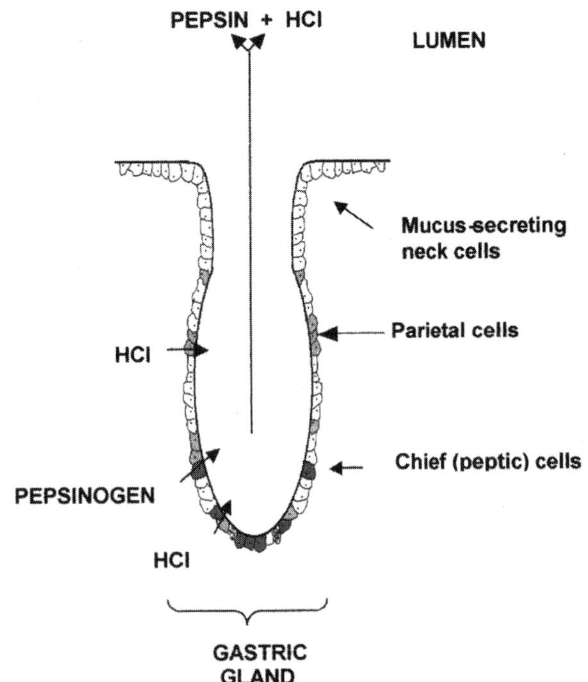

Figure 2. Structure of a gastric gland. Mucus is secreted by the neck cells and the surface mucosal cells. Acid and pepsinogen are secreted from cells deeper within the gland.

proteins, peptides, and glycoproteins; (ii) cytokines, e.g., tumor necrosis factor alpha (TNF-α) and chemokines, e.g., interleukin-8 (IL-8), that conduct the immune response via intercellular communication; and (iii) pattern recognition molecules, which are membrane bound and soluble (7). The cellular component of the innate defense system is composed of M cells, dendritic cells, phagocytic cells, mast cells, lymphocytes, and epithelial cells.

The Preepithelial Barrier

The first defense that an invading pathogen would encounter is the preepithelial barrier, consisting of a secreted mucus gel in the airways and GI tract (2). There is currently no evidence of a secreted mucus gel in the urinogenital tract. To function effectively, the gel must be able to flow and maintain a barrier while the mucosa moves during contraction of the underlying smooth muscle. This gel, as well as flowing, must be able to reanneal if fractured. Protein gels such as gelatin or carbohydrate gels such as agar, although strong, could not function as a mucosal protective because once they are broken the damage is permanent and they are too rigid to maintain a continuous layer over the mucosa during motility (58). Mucus is therefore a unique physical gel that has both flow and rigidity properties. It is somewhere between a rigid gel and an entangled system, as demonstrated by the fact that the rheological modulus G′ (the storage modulus) is always greater than G″ (the loss modulus). These moduli indicate solid- and liquid-like behavior, respectively, and the fact that mucus has a small but significant G″ value (i.e., it has some liquid-like behavior) explains to some extent its ability to flow slowly (58).

Mucins

The secreted mucins are the principal viscous and gel-forming components of mucus gel secretions (9). Mucins are high-molecular-weight glycoproteins. Currently 20 human mucin genes (the *MUC* genes) have been identified, and their protein products have been partially characterized (Table 1). It is becoming clear that mucins can be divided into two main families: (i) the membrane-tethered, transmembrane mucins, including MUC 1, 3A, 3B, 4, 11, 12, 13, 15, 16, 17, and 18, and (ii) the secreted gel-forming mucins, MUC 2, 5AC, 5B, 6, and 19 (14, 42). The *MUC 19* gene has been identified only from a genome-wide search, and although it codes for a mucin with the characteristics of a gel-forming mucin, the properties of the gene product have not been directly characterized. Therefore, its assignment to the secreted gel-forming mucins is at this stage only preliminary (12). The *MUC 8* gene is still too poorly characterized to assign to either of the groups (51). *MUC 9* appears to code for a membrane tethered mucin; however, the existence of a secreted form has been suggested (33). MUC 7 is a secreted but non-gel-forming mucin that has bacterium-binding activity in saliva.

Consequently, mucins form two lines of preepithelial defense, such that the secreted gel overlies the mucins forming part of the glycocalyx on the apical surface of the epithelial cells; e.g., in the normal colon, MUC 2 forms a gel and MUC 3 and to a lesser extent MUCs 1, 4, 11, 12, 15, and 17 form a barrier at the apical surface. At least 14 and 15 of these *MUC* are expressed in the GI tract and airways, respectively, both membrane bound and secreted. The normal urinary bladder does not express *MUC 5AC, 5B,* or *6* and has been reported to express *MUC 2* at very low levels, if at all (39, 62). Therefore, there is no evidence for a secreted gel layer protecting the urothelium. *MUC 1* and *4* are expressed, and *MUC 3* is expressed sparsely, so that the main preepithelial barrier in the bladder is mucin bound to the apical surface of the mucosa (39). There are two gene clusters associated with mucins. The genes expressing all but one of the secreted gel-forming mucins, i.e., *MUC 2, 5AC, 5B,* and *6,* have a chromosomal loca-

Table 1. Chromosomal location of *MUC* genes

Gene	Location of gene product in body	Location on chromosome
MUC 1	All epithelia, breast, pancreas, small intestine, urinary bladder	1q21
MUC 2	Colon, small intestine, airways	11p15.5
MUC 3B	Colon, small intestine, gallbladder, urinary bladder	7q22
MUC 3A	As MUC 3B, also heart, liver, thymus, pancreas	
MUC 4	Airways, colon, small intestine, stomach, cervix, urinary bladder	3q29
MUC 5AC	Airways, stomach, cervix, middle ear	11p15.5
MUC 5B	Airways, submaxillary gland, cervix, gallbladder, middle ear	11p15.5
MUC 6	Stomach, gallbladder, cervix	11p15.5
MUC 7	Salivary glands, airways	4q13-21
MUC 8	Airways	12q24.3
MUC 9	Oviduct	1p13
MUC 10		
MUC 11	Colon	7q22
MUC 12	Colon	7q22
MUC 13	GI tract, colon, airways (columnar and goblet cells)	3q13.3
MUC 15	Colon, airways, small intestine, spleen, prostate, breast, etc.	11p14.3
MUC 16	Ovarian epithelial cells	19p13.3
MUC 17	Membrane tethered; duodenum, stomach, colon	7q22
MUC 18	Melanoma cell adhesion molecule (Ig) gene superfamily [CD146], normal lung and breast	11q23
MUC 19	Salivary glands and tracheal submucosal glands	12

tion of 11p15.5, and those expressing several of the membrane-bound mucins, i.e., *MUC 3A* and *B, 11, 12,* and *17,* are clustered at 7q22 (Table 1).

Mucin Structure

Mucins are glycoproteins with a central protein core to which are attached many carbohydrate side chains via O-glycosidic links between serine and/or threonine and *N*-acetylgalactosamine; up to 80% of the molecule consists of carbohydrate. The sugars within these carbohydrate chains are galactose, fucose, *N*-acetylgalactosamine, *N*-acetylglucosamine, sialic acid, and mannose (3). These carbohydrate chains express blood group activity (1). As well as the O-linked chains, all mucins contain potential N-glycosylation sites with the amino acid sequence Asn-X-Ser/Thr, where X is any amino acid except proline. These N-linked carbohydrate chains are joined to the protein core via *N*-acetylgalactosamine, and the chains contain mannose as well as the other sugars present in the O-linked chains. Consequently, a major function of mucins is to present recognition molecules similar to the epithelial cell surface in order to provide bacteria with pseudo-cell surface attachment sites, thereby preventing true epithelial cell surface attachment.

All mucins contain a region within the protein core that is rich in the amino acids serine and or threonine and proline. This region contains a repetitive sequence of amino acids unique to each mucin gene, which can repeat a variable number of times. Consequently, it is called the variable-number-of-tandem-repeats (VNTR) region (Fig. 3). The presence of these repeating sequences means that this portion of the protein core is highly immunogenic. In addition, mRNA for the individual mucins can be easily detected by in situ hybridization because the cDNA probes made to the repeating sequence bind multiple times to the mucin mRNA and amplify the signal (50). The possession of a VNTR region is common to all mucins; however, the structure outside of the VNTR region differs significantly between the secreted gel-forming mucins and the membrane-bound mucins. The membrane-bound mucins have at least three characteristics in common: a highly glycosylated extracellular domain (the VNTR region); a transmembrane domain; and a short cytoplasmic tail, which may contain potential serine/tyrosine phosphorylation sites (as seen in the *MUC 17* gene product). In addition, the *MUC 1, 3A, 3B, 12, 13,* and *17* genes all code for a SEA domain (a sea urchin sperm protein-enterokinase-agrin domain), whose role is to regulate and/or assist binding to neighboring carbohydrate moieties (22).

All the membrane-bound mucins localized at chromosomal position 7q22 have two disulfide bridges containing epidermal growth factor (EGF)-like domains between the VNTR and the cell membrane. The *MUC 4* gene product also contains two EGF-like domains, and the *MUC 13* product has three (14, 63). The role of these EGF-like domains is probably in cell adhesion, modulation, chemotaxis, and cell signaling, and these domains may be important in mucin-growth factor interactions during wound healing. The *MUC 4* gene codes for two

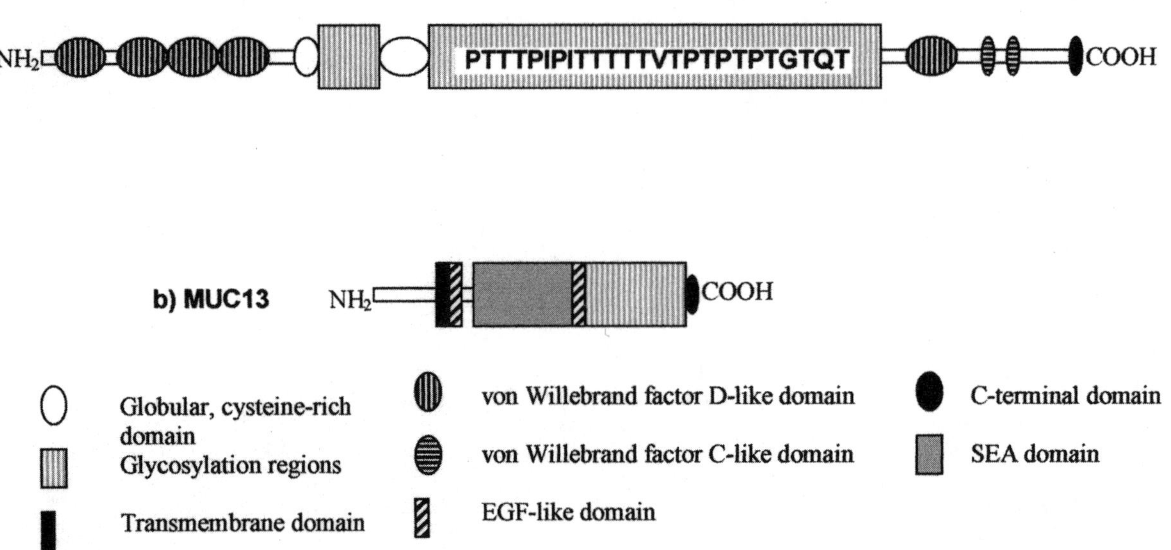

Figure 3. Structures of MUC 2 (secreted mucin) and MUC 13 (membrane-tethered mucin) proteins. (a) In the VNTR in MUC 2, there are many tandem repeats of the amino acid sequence. (b) In MUC 13 there are 10 degenerative tandem repeats rich in serine and threonine.

domains not present on any of the other membrane-bound mucins, a nidogen domain probably involved in interactions between the cell and the extracellular matrix, and a so-called AMOP domain, thought to play a role in cell adhesion (10, 14). All these domains exposed on the cell surface are potential recognition sites for bacterial binding.

The four secreted mucins whose genes are located within the 11p15.5 gene locus form gels at a mucin concentration between 30 and 100 mg/ml. MUC 5AC and 6 form the gel layer in the stomach, and MUC2 forms the gel layer in the intestines. MUC 5B and 5AC form the gel protecting the respiratory tract. These gel-forming mucins also exist in secretions at concentrations below the concentration required for gel formation, e.g., MUC 5B in saliva and bile. The gel-forming mucins contain domains outside the VNTR region that are rich in cysteine, the so-called D domains, which are homologous to the D domains of von Willebrand factor (vWF), a blood-clotting factor. These mucins polymerize end to end via these domains, using disulfide bridges to produce polymers with molecular weights of around 10^7, and this polymerization is essential for gel formation. The *MUC 2, 5AC,* and *5B* genes all have D_1, D_2, D′, and D_3 domains upstream of the VNTR region and a D_4 domain downstream. As well as the D_4 domain, there are vWF-like B, C, and CK domains. It would appear that these secreted gel-forming mucins have evolved from a common ancestor of the vWF gene (15), where the VNTR region replaces the three A domains present in vWF and the C and N terminals are partially conserved, although not completely, since the C terminal of vWF contains three B domains and two C domains compared to one of each in the mucins. The last mucin gene of the 11p15.5 cluster, *MUC 6,* differs from the others in that it has lost the exons coding for the C-terminal D_4, B, and C domains, expressing only the CK domain.

The Secreted Mucus Barrier Consists of Two Layers

The surface mucus gel forms a continuous layer in the stomach and large intestine; however, the thickness of the layer depends on the method used to measure it. The mucus layer in the human colon has been reported as 155 μm thick in the rectum, 134 μm thick in the right colon, and 107 μm thick in the left colon when unfixed sections are used. If classical histologic methods of fixation, dehydration, and staining are applied to tissues with a mucus layer, no obvious mucus layer can be discerned in the colon or the stomach. This is because mucus is 95% water and is completely dehydrated and effectively lost during this procedure. Even using a very mild fixation technique developed in our laboratory, mucus thickness in the human rectum was only 29.5 ± 5.3 μm (56). These results gave the first evidence that there

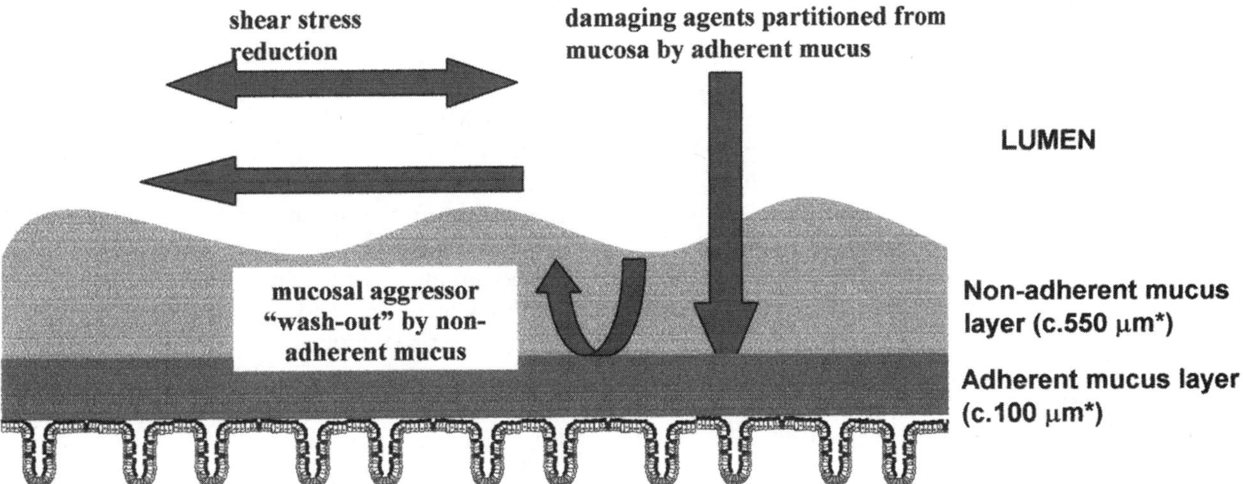

Figure 4. Colonic mucus bilayer. *, mucus layer thicknesses shown are from in vivo measurements made in the rat colon.

may be two layers within the mucus gel. Using a novel in vivo method, the presence of two layers was demonstrated in the rat stomach and colon. In the fasted-rat colon, the complete bilayer had a thickness of 642 ± 55 μm and consisted of a loosely adherent sloppy layer (which can be removed by suction) overlying a firm layer that was 101 ± 6.2 μm thick (56) (Fig. 4). It has become clear that the difference in reported thickness can be explained by the fact that even under mild fixation conditions the sloppy layer is lost and only the firmly adherent layer is retained, so, e.g., in the rat stomach the bilayer is 221 to 284 μm thick with a firm layer of 154 μm, which agrees closely with a thickness of 176 μm measured using the mild fixation technique (4, 29).

What Is the Function of the Two Layers?

In experiments with the firm and sloppy layers from pig stomachs, rheological measurements demonstrated that they were both gels with G′ dominant over G″ across the frequency range of 0.1 to 3 Hz (Fig. 5). The shear-resistant (firm) gel layer had a phase angle of 5 to 10° (n = 7), and the shear-compliant (sloppy) layer had a phase angle of 15 to 20°. This demonstrates that the shear-resistant gel is a stronger gel (the lower the phase angle, the stronger the gel). Further experiments have shown that the sloppy gel is not a pepsin digestion product of the firm gel. In breakdown studies, as the shear stress applied to the gel was increased, G′ decreased, and at the point of breakdown G″ became dominant. This occurred at ~150 Pa for the firm gel and 1 to 2 Pa for the sloppy gel (58). Both gels re-formed when the shear stress was removed. These data show two important properties of these gels: (i) mucus gels have the ability to re-form when broken, a property essential when considering that they line tissues subject to varying levels of motility and (ii) the sloppy gel layer has a significantly lower resistance to flow than does the firm mucus layer; the shear-resistant (firm) gel provides the mucus barrier in vivo, and the shear compliant (sloppy) gel acts primarily as a lubricant. It is therefore advantageous to the host to trap microorganisms in the sloppy layer, where they can easily be removed.

Can Microorganisms Modulate the Mucus Layers?

Goblet cells in the airways and digestive tracts respond to bacterial signals by increasing mucus secretion to enhance the removal of microorganisms

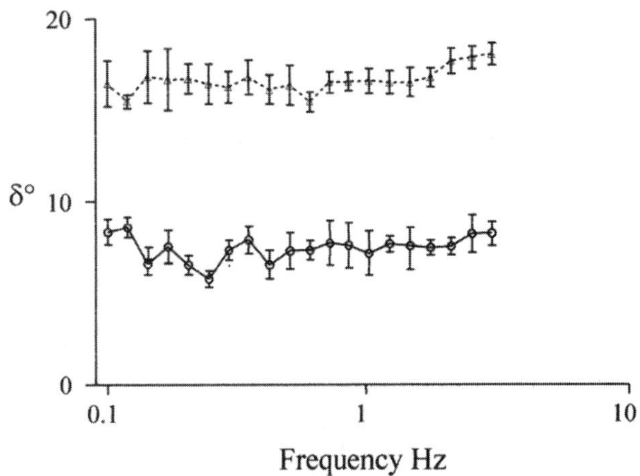

Figure 5. Frequency sweep of pig gastric mucus gels. Shear-compliant (Δ) and shear-resistant (○) gels were subjected to frequency sweeps between 0.1 and 3 Hz by using a Bohlin CV050 rheometer at 25°C. Results are shown as δ (the phase angle), a measure of gel strength. The lower the δ, the stronger the gel.

trapped in the mucus (25). Lipopolysaccharide (LPS) stimulates mucus secretion in the rat intestine (19). This newly secreted mucus is capable of binding *Escherichia coli* and clearing it from the gut. In addition, in response to the presence of microbes the epithelium can release chemokines and cytokines from the epithelial cells and immune cells in the mucosa. The proinflammatory (acute-phase) cytokines TNF-α, IL-1, and IL-6 all increase mucin synthesis and release from goblet cell lines in culture, and this up-regulation is not the same for all the mucin genes. For example, 20 ng of TNF-α per ml caused a 56% increase in secretion of the MUC 5AC mucin after 7 h, whereas the secretion of MUC 5B mucin increased by only 15% during the same period (54). The most important chemokine released from the epithelial cells is IL-8, which, as well as being an effective chemoattractant for granulocytes, stimulates mucin secretion by goblet cells. Interestingly, LPS, in addition to stimulating mucin release from goblet cells, up-regulates IL-8 production in the goblet cells (55). This consequently generates a second self-stimulus for mucin production and release. The proinflammatory cytokines and IL-8 also induce the expression of adhesion molecules. In particular, IL-8 increases the binding of lymphocyte function-associated antigen 1 (LFA-1) to intercellular cell adhesion molecule 1 (ICAM-1), and these effects enhance the process of diapedesis of phagocytes (7). Other chemokines released by the epithelium involved in the immune response and attraction of macrophages and natural killer cells are macrophage inflammatory protein 1α (MIP-1α), MIP-1β, RANTES, and GRO-α, -β, and -γ.

In the intestines, mucins inhibit viral replication and bacterial adherence. Consider the situation in the intestines where many bacterial species are competing for available space. If a bacterial species can modulate mucin production, this may provide it with a survival aid and reduce the survival of other species. For example, the so-called probiotic *Lactobacillus* strains have been demonstrated to adhere to intestinal cells via a mannose-dependent ligand (37). The response of the epithelia to this adherence is to up-regulate *MUC 3* transcription and translation (37). Only *Lactobacillus* strains that can adhere to epithelial cells stimulate this up-regulation. As well as producing the membrane-bound form of MUC 3, intestinal cells can, via an alternative splicing mechanism, produce variants of MUC 3 with the C terminus deleted (13). This protein product would have the two EGF-like domains, the transmembrane domain, and the cytoplasmic tail absent. This variant without the ability to anchor to the cell membrane is thought to be a secreted product. Consequently, *Lactobacillus* binding stimulates extracellular release of MUC 3 mucin, and in coincubation experiments *Lactobacillus* adherence reduced the adherence of enteropathogenic *E. coli*. It is tempting to speculate that *E. coli* recognizes MUC 3 on the epithelial membrane as an attachment site and the release of soluble MUC 3 blocks this site and that therefore the bacterium is unable to adhere to the cell and is washed away.

As well as stimulating the synthesis of membrane-bound mucins and their secreted variants, *Lactobacillus* strains have no degradative effect on the secreted intestinal mucus gel barrier (mainly the *MUC 2* gene product) (47), thereby retaining a mucus gel barrier to protect the *Lactobacillus* strains and the mucosa.

Some microorganisms can utilize the mucus layer for protection, e.g., *Helicobacter pylori,* which, due to its shape and possession of multiple flagella, can bore its way through the mucus layer. Once under the mucus blanket, it is protected from the harsh acidic environment of the gastric lumen. In fact, culture experiments have demonstrated that *H. pylori* is rapidly killed at pH 2 or below (32). However, it would probably enter the stomach along with food, which would buffer the acid to some extent, allowing the bacterium to survive long enough to penetrate the mucus layer and adhere to the underlying mucosal cells. A further survival aid for *H. pylori* is its urease, which converts urea found in the gastric juice to ammonia. This can then produce a neutralization barrier around the organism via the reaction of ammonia with hydrogen ions to form ammonium ions. Once under the mucus layer, this urease has a further role. The high pH generated, i.e., 9.5 and above, can cause local degradation of the mucus gel, creating a local pool of liquid mucus surrounding the bacterium and hence allowing it easier access to nutrients (40). The main body of the mucus gel is unaffected and remains intact. Therefore, the pH gradient within the gastric mucus layers, i.e., pH 7 at the mucosal surface and pH 2 at the luminal mucus surface, remains, effectively protecting and maintaining the bacteria in a neutral environment (45). *H. pylori* has therefore developed a niche environment which uses the host defenses to protect it and remove competition by other bacterial species. *H. pylori* has a tropism for the gastric mucosa. The stomach secretes two mucin gene products. MUC 5AC mucin is expressed in the surface epithelial cells and gastric glands, and MUC 6 mucin is expressed in the deeper antral-gland epithelium. Recent studies have demonstrated that MUC 5AC mucin is the major receptor for *H. pylori* (60), and this bacterium can modulate mucin secretion in the stomach (61). *H. pylori* possesses an adhesin (BabA) which interacts with the blood group antigen Lewis B (LeB). Lewis antigens are carbohydrate structures carried by glycoproteins and glycolipids in secretions and on the cell surfaces of secretors. Nonsecretors also express this antigen at low levels. MUC 5AC mucin is the main carrier of this antigen in gastric mucosa; consequently, *H. pylori* colocalizes with LeB/MUC 5AC-producing gastric cells.

What Else Is Secreted by the Mucosa to Protect Itself?

Secreted by the epithelial cells are a wide range of molecules with antimicrobial activity, contained within the secreted mucus layer, if one exists, or at the mucosal surface glycocalyx if no secreted mucus layer is present (e.g., the urinary bladder) (Table 2).

IgA and IgM

At the mucosal surface, both transudated monomeric immunoglobulin A (IgA) and locally produced actively secreted dimeric IgA are present. Only dimeric IgA, in which the IgA molecules are joined by a J chain via a covalent link and which is associated with secretory component (SC), is fully active in protecting the mucosa (43, 46). SC is provided by the epithelial cells and is derived from the polymeric-Ig receptor (pIgR), which facilitates the transcytosis of IgA from the basolateral membrane to the apical surface of epithelial cells.

Current research using recombinant SC suggests a role for SC in immobilizing IgA in the mucus layer (43). Human SC is a glycoprotein with seven carbohydrate chains which can be removed by deglycosylation with *N*-glycosidase F (57). The removal of these carbohydrate chains did not affect the ability of SC to bind to IgA, the ability of SC and dimeric IgA to bind to bacteria, or the stability of the IgA-SC complex. It did, however, prevent the complex from interacting with the mucus layer of the airways, i.e., if applied to the mucus layer, it did not bind but diffused out of the layer. In in vivo mouse experiments, in which IgA complexed with deglycosylated SC was given intranasally, the deglycosylated complex could not be retained in the mucus layers of the nose or the lungs. IgM is also present in the external secretions such as mucus and tends to be of limited antigen specificity (natural antibody).

Antimicrobial proteins and peptides

There are three families of antimicrobial peptides: defensins, cathelicidins, and histatins (7, 35). Defensins are positively charged peptides, relatively rich in arginine, with three disulfide bridges. There are three groups of defensins: α, β, and θ. There are six α defensins: 1 to 4 are human neutrophil peptides (HNP1 to HNP4), and 5 and 6 (HD5 and HD6) are produced by epithelial cells of the female urinogenial tract and Paneth cells of the small intestine. The β defensins consist of β defensins 1 and 2 (also known as HD-1 and HD-2) and are expressed by epithelial cells of the GI, urinogenital, and respiratory tracts. β defensin 1 is constitutively expressed in all these tissues (6), whereas β defensin 2 is induced in response to infection and/or inflammation; e.g., it is up-regulated

Table 2. Epithelial cell protective secretions

Secretion	Function
IgA, IgM	Opsoninize microorganisms; inactivate bacterial toxins
Antimicrobial peptides and proteins	
Defensins	Bind to and destabilize bacterial membranes
Cathelicidins	Bacteriocidal against gram-negative and gram-positive bacteria; neutralize LPS; chemotactic for white blood cells
Histatins	Antimicrobial via ROS generation, particularly against fungi, e.g., *Candida albicans*
Lactoferrin	Deprives microbes of the essential nutrient iron; binds LPS and disrupts bacterial membrane; has antibiofilm activity
Secretory proteinase inhibitors	
SLP1	Inhibits serine-dependent proteinases
Elafin	Bacteriocidal, but mode of antimicrobial action not fully characterized
Secreted enzymes	
Lysozyme, secretory phospholipase A_2	Damages bacterial cell walls
Lactoperoxidase/dual oxidase	Microbiocidal via the generation of ROS and free radicals
Antiadherence molecules	
Surfactant	Inhibits mucosal adhesion; enhances phagocytosis
Tamm-Horsfall glycoprotein and related molecules	Inhibit mucosal adhesion; stimulate polymorphonuclear leukocytes

in gastric epithelial cells by the presence of *H. pylori* (5), and several studies have shown increased expression of β defensin 2 in the presence of proinflammatory cytokines (41, 48, 59). The mode of action of defensins depends on their cationic nature, which means that they can interact with the negatively charged surface of microorganisms, resulting in pores forming directly in the membrane or producing a general destabilization of the membrane.

The cathelicidin gene is expressed in epithelial cells of the respiratory, urinogenital, and GI tracts; in human neutrophils; and in other leucocytes and keratinocytes. The product of this gene is cathelicidin LL-37/human cationic antimicrobial protein 18 (hCAP18) and is expressed within the surface and upper crypt epithelial cells of the normal human colon. There is little or no expression in the deeper crypts or in the small intestine. This suggests, because the cells differentiate as they move up the crypts, that cell differentiation is a key expression determinant for cathelicidin (24). This means that the level of protection provided by this peptide increases as the cells become functional. Unlike the defensins, the expression of cathelicidin does not appear to be up-regulated by inflammatory mediators. Cathelicidins have a conserved cathelin domain and a variable C-terminal domain. Human cathelicidin LL-37 is synthesized as a precursor called human cationic antimicrobial protein 18. This precursor is cleaved after exocytosis to the mature peptide LL-37 by a serine protease (protease 3 in neutrophils). LL-37 contains a linear amphipathic α-helical structure important for its activity (28). This cathelicidin has a broad range of bactericidal activities against both gram-negative and gram-positive bacteria. It can bind and neutralize LPS, and it is chemotactic for human peripheral monocytes, CD4 T lymphocytes, and neutrophils.

Human saliva also contains histatins, which are antimicrobial in general but are very effective against fungal infections. Histatins do not seem to interact with the microbial cell membranes but are taken up by metabolically active cells, inhibit mitochondrial respiration, and damage and kill the cell via the generation of reactive oxygen species (ROS).

Lactoferrin is a protein that can inhibit the adhesion of microbes to epithelial cells. Its primary bacteriostatic action is achieved by chelating iron and making it inaccessible to the invading organism (17). This makes lactoferrin a nutrient-depriving host defense molecule, along with calprotectin, which binds zinc, and transcobalamins, which bind vitamin B_{12}. Lactoferrin also has a direct antimicrobial action residing in a region of the molecule near the N terminus, distinct from the iron binding domain. Lactoferrin can bind LPS and disrupt bacterial membranes. The importance of lactoferrin as a mucosal protectant is demonstrated by its high concentration in external secretions, i.e., 0.4 to 1 mg/ml in the airway surface liquid and mucus layer, 1 to 4 mg/ml in tears, and 3 to 7 mg/ml in breast milk, where it enhances the protection of the developing GI tract of the infant. These properties of lactoferrin require high levels; e.g., the growth rate of free-swimming *Pseudomonas aeruginosa* was not inhibited until lactoferrin concentrations reached above 50 μg/ml. Interestingly, lactoferrin appears to play a role at lower concentrations, such as in preventing biofilm formation when present at 20 μg/ml (52). In chronic infections, bacteria live in biofilms in a sessile state surrounded by a glycosubstance which forms an extracellular matrix. As such, these bacteria are notorious for their resistance to antibiotics and host defenses and therefore present a disastrous situation for the host. Therefore, the innate mucosal defense must have a mechanism to prevent biofilm formation. Recent studies (52) have shown that in the presence of lactoferrin, i.e., reduced iron levels, *P. aeruginosa* moved across the mucosa rather than forming clusters and biofilms. Also, any biofilms that did form were less well protected from antibiotics. However, if the biofilms were formed in the absence of lactoferrin, addition of lactoferrin at this stage had no effect on biofilm stability or resistance to antibiotics. These experiments indicate that higher levels of iron are required for biofilm formation than for growth and that iron binding molecules, e.g., lactoferrin and mucin, can prevent biofilm formation. If the iron levels at the mucosal surface are too low, the bacterium continues moving and does not form microcolonies, presumably because iron is a critical nutrient, until they reach an area of the mucosa depleted in mucus and lactoferrin.

Secretory Proteinase Inhibitors

Several proteinase inhibitors are produced by epithelial cells and, where present, submucosal glands, e.g., in the airways. These inhibitors form an important part of the preepithelial defenses and the innate immune system. Secretory leukocyte proteinase inhibitor (SLPI) is one such epithelial secretion, along with elafin (SKALP), which is an elastase inhibitor with 42% sequence homology to SLPI. Both these molecules inhibit serine-dependent proteinases. SLPI is a small nonglycosylated 11.7-kDa protein with an N-terminal domain possessing antibacterial activity and a C-terminal domain possessing enzyme-inhibiting activity. Each domain contains two disulfide bridges. As well as killing bacteria, SLPI has antifungal properties and is a potent inhibitor of human immunodeficiency virus type 1. Some bacteria have developed mechanisms to overcome the actions of SLPI. Some group A streptococci, e.g., *Streptococcus pyogenes,* particularly the M1 strains, produce a 31-kDa extracellular protein called streptococcal inhibitor of complement (SIC). This protein binds to SLPI and lysozyme but not lactoferrin. SIC binds to SLPI via hydrophobic interactions with a 1:2 stoichiometry, whereas it binds lysozyme via an ionic interaction with a 1:4 stoichiometry. SIC is very effective in preventing SLPI from killing bacteria, but it does not affect its antiproteinase activity. These effects of SIC explain the virulent nature of some group A streptococcal infections (20).

Secreted Enzymes

Several enzymes are secreted by the epithelial cells into the external secretions. Lysozyme and secretory phospholipase A_2 have a direct effect on cell membranes. Lysozyme, a hydrolytic enzyme, acts by cleaving the glycosidic bonds of *N*-acetylmuramic acid, damaging the bacterial cell wall and eventually killing the bacteria by lysis.

Lactoperoxidase (LPO) is an enzyme with antimicrobial properties; it is secreted by epithelial cells and is present in saliva, breast milk, tears, and mucus secretions of the airways. It has been demonstrated in vitro (23, 30) to have antimicrobial activities against a wide range of gram-positive and gram-negative organisms. LPO generates ROS by utilizing hydrogen peroxide, the source of which has only recently been discovered (21). Mucosal surface enzymes, analogues of NADPH oxidase, dual oxidase 1 and dual oxidase 2 (Duox 1 and 2), are expressed on the mucosal surfaces of the salivary glands, colon, rectum, trachea, and bronchi. Their expression is tissue specific, with Duox 1 being expressed in the airways and Duox 2 being expressed in the salivary ducts and the rectal glands. These dual oxidases release H_2O_2 into the external secretion, where it can be acted on by LPO to generate reactive oxygen and bacteriocidal species. Therefore, the two enzyme systems, along with a sodium/iodide symporter, form part of the preepithelial host defense mechanism (Fig. 6).

LPO is secreted by the epithelial cells onto the mucosal surface, where it catalyzes the conversion of iodide and thiocyanate, transported by the mucosal sodium/iodide symporter and hydrogen peroxide released into the extracellular medium from the action of Duox into hypothiocyanate and OI^-. These ROS produced by LPO are well suited to mucosal defense because they are much less damaging to the mucosa (53) than are the reactive species produced by neutrophil myeloperoxidase, i.e., hypochlorite, and the H_2O_2 generated and released into the mucosal milieu is prevented from reacting with free iron and producing the very damaging hydroxyl free radical (Fenton reaction) by the chelation of iron by mucin and lactoferrin.

Antiadherence Molecules

A key role for the preepithelial barrier is to prevent microbial adherence by interfering with microbial adhesins and toxins. In the lungs, surfactant, a mixture containing glycoproteins and lipids such as dipalmitoylphophatidylcholine, limits microbial adhesion to mucosal surfaces and enhances macrophage phagocytosis as well as reducing the fluid surface tension in the lungs and thereby reducing the work of breathing (44).

Tamm-Horsfall glycoprotein (THP) is synthesized by the renal tubular cells in the thick ascending limb of the loop of Henle. It is the most abundant glycoprotein found in urine, and the amount excreted daily in urine is 20 to 200 mg (27). Consequently, THP coats most of the mucosal surface of the urinary tract and is very important in protecting the urinary

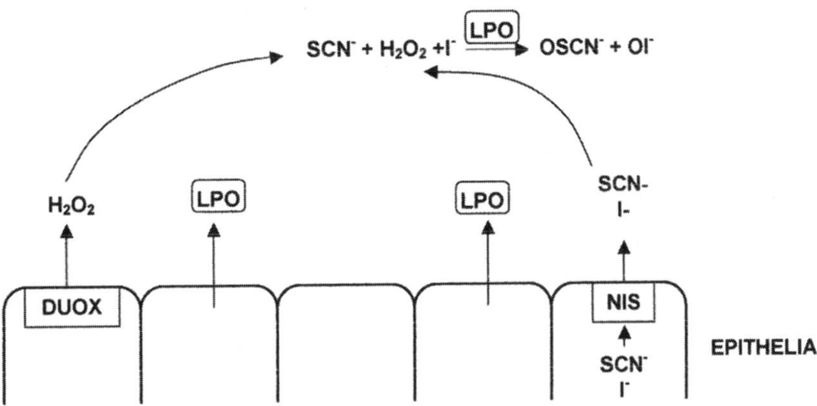

Figure 6. The LPO/Duox mucosal protective system. NIS, sodium/iodine symporter (substrates iodide and thiocyanate).

bladder, which lacks a secreted mucus barrier. Although THP is found only in the urinary tract, THP-related glycoproteins are secreted by other mucosal surfaces, e.g., the squamous epithelium of the oral mucosa, and there is some evidence that THP-related glycoproteins are secreted in the airways (38). Therefore, THP-like glycoproteins may well form a general antiadherence shield on the mucosal surfaces of the body. THP is a 90-kDa glycoprotein, with 30% of the weight of the molecule being due to N-linked sugar chains along with some O-linked sugar chains. The N-linked sugars are necessary for the cytoadherence inhibitory effects (31). *E. coli* expressing type 1 fimbriae can be prevented from binding to mucosal surfaces by THP, and this effect is mediated by the mannose content of the N-linked sugars on THP. In this situation, THP is protecting the mucosa by providing bacterial binding sites, i.e., mimicking those present on the epithelial cells. THP also inhibits the binding of S- and P-fimbriated *E. coli* (34). It can also prevent the adherence of virulent *Bordetella pertussis* to pulmonary epithelial cells. This inhibitory effect relies on THP binding to the S_2 subunit of the pertussis toxin. It is thought that THP is providing a copy of the epithelial cell surface receptor used for pertussis toxin binding. The receptor has a sialic acid moiety on an N-linked oligosaccharide, and THP contains a copy of this, so that the bacterium will bind to THP and be washed away, thereby preventing bacterial adherence to epithelial cells (38).

Epithelial Cells

As well as cell membrane-associated mucin, the apical surface of the epithelial cells has other molecules presented to approaching microbes. Many epithelial surfaces are essentially sterile environments and can therefore respond to bacterial adhesion and LPS with a full-blown immune response. The situation in the intestine is, however, different, and the epithelial cells must be tolerant. In the colon and to some extent the small intestine, the microbial flora is necessary and beneficial. Therefore, some degree of selectivity is necessary for homeostasis. Human intestinal cells express a protein of 55 to 60 kDa, which is bound to the apical surface; this protein is called bactericidal permeability-increasing protein (BPI), and it can impart such selectivity. BPI was first detected in neutrophils as an antibacterial and endotoxin-neutralizing molecule (18, 36). It has since been shown to also be expressed in oral, respiratory and GI tract mucosa. It acts as a molecular shield to protect mucosal surfaces from gram-negative bacteria and their endotoxins. It is cytotoxic at nanomolar levels via damage to the inner and outer bacterial membranes. It can neutralize LPS and serve as an opsonin for phagocytosis of gram-negative bacteria. BPI has two domains: the N-terminal domain is cationic and has the antibacterial and endotoxin-neutralizing activity, and the C-terminal domain is involved in oponization (8). This ability to bind endotoxin can explain the selective tolerance of the intestinal mucosa to gram-negative bacteria (11). As a consequence of the presence of commensal bacteria, there is a high level of endotoxin at the mucosal surface. The epithelial cells respond to LPS via binding it to CD14 and Toll-like receptor 4 (TLR4) on the cell surface, which, unchecked, would generate an immune response to the commensal bacteria. This response therefore needs to be damped. BPI, because of its endotoxin binding capacity, can compete with CD14 and TLR4 for endotoxin and thereby reduce the immune response. TLRs are an important part of the luminal sampling process. TLR4 and TLR2 bind bacterial products at the apical surface and translocate them to the basolateral surface. In a further mechanism to down-regulate an inappropriate response to commensal bacteria, TLRs are expressed at only a low level at the apical surface of epithelial cells. This situation changes in chronic inflammation, when they are strongly expressed at the apical surface.

Studies of mice that have a TLR4 mutation producing nonfunctional receptors on epithelial cells and/or hematopoietic cells have shown that when these mice are challenged with uropathagenic strains of *E. coli*, both cell types must have functional TLR4 for an acute inflammatory response to happen. $TLR4^+$ bladder epithelial cells alone were not sufficient to activate a full immune response (49). This suggests that the role of TLR4 and epithelial cells in bacterial invasion is to augment rather than initiate the immune response via the production and release of cytokines, chemokines, and adhesion molecules and to augment bacterial killing and clearance by the release of local inflammatory mediators, NO, and degradative enzymes.

CONCLUSION

The mucosal surface possesses a diverse array of antimicrobial defenses, including a secreted mucus bilayer providing a physical barrier, which can also bind and cause microorganisms to be washed away, nutrient-depriving molecules such as lactoferrin, degradative enzymes, antimicrobial peptides, proteins and glycoproteins, epithelial cell surface mucin and glycoproteins that prevent adhesion, and cell surface receptors that generate a local immune response.

REFERENCES

1. **Allen, A.** 1989. Gastrointestinal mucus, p. 359–382. *In* J. G. Forte (ed.), *Handbook of Physiology—The Gastrointestinal System,* vol. 3. American Physiological Society, Bethesda, Md.
2. **Allen, A., D. A. Hutton, A. J. Leonard, J. P. Pearson, and L. A. Sellers.** 1986. The role of mucus in the protection of the gastroduodenal mucosa. *Scand. J. Gastroenterol.* **21:**71–78.
3. **Allen, A., D. A. Hutton, and J. P. Pearson.** 1998. The MUC2 gene product: a human intestinal mucin. *Int. J. Biochem. Cell Biol.* **30:**797–801.
4. **Atuma, C., V. Strugala, A. Allen, and L. Holm.** 2001. The adherent gastrointestinal mucus gel layer: thickness and physical state in vivo. *Am. J. Physiol. Ser. G* **280:**G922–G929.
5. **Bajaj-Elliott, M., P. Fedeli, G. V. Smith, P. Domizio, L. Maher, R. S. Ali, A. G. Quinn, and M. J. G. Farthing.** 2002. Modulation of host antimicrobial peptide (beta-defensins 1 and 2) expression during gastritis. *Gut* **51:**356–361.
6. **Bals, R.** 2000. Epithelial antimicrobial peptides in host defense against infection. *Respir. Res.* **1:**141–150.
7. **Basset, C., J. Holton, R. O'Mahony, and I. Roitt.** 2003. Innate immunity and pathogen-host interaction. *Vaccine* **21:**S12–S23.
8. **Beamer, L. J., S. F. Carroll, and D. Eisenberg.** 1997. Crystal structure of human BPI and two bound phospholipids at 2.4 angstrom resolution. *Science* **276:**1861–1864.
9. **Bell, A. E., A. Allen, E. R. Morris, and S. B. Rossmurphy.** 1984. Functional interactions of gastric mucus glycoprotein. *Int. J. Biol. Macromol.* **6:**309–315.
10. **Bork, P., and L. Patthy.** 1995. The sea module—a new extracellular domain associated with O-glycosylation. *Protein Sci.* **4:**1421–1425.
11. **Canny, G., O. Levy, G. T. Furuta, S. Narravula-Alipati, R. B. Sisson, C. N. Serhan, and S. P. Colgan.** 2002. Lipid mediator-induced expression of bactericidal/permeability-increasing protein (BPI) in human mucosal epithelia. *Proc. Nat. Acad. Sci. USA* **99:**3902–3907.
12. **Chen, Y., Y. H. Zhao, T. B. Kalaslavadi, E. Hamati, K. Nehrke, A. D. Le, D. K. Ann, and R. Wu.** 2003. Genome-wide search and identification of a novel gel-forming mucin MUC19/Muc19 in glandular tissues. *Am. J. Respir. Cell Mol. Biol.* **30:**155–165.
13. **Crawley, S. C., J. R. Gum, J. W. Hicks, W. S. Pratt, J. P. Aubert, D. M. Swallow, and Y. S. Kim.** 1999. Genomic organization and structure of the 3′ region of human MUC3: alternative splicing predicts membrane-bound and soluble forms of the mucin. *Biochem. Biophys. Res. Communi.* **263:**728–736.
14. **Dekker, J., J. W. A. Rossen, H. A. Buller, and A. W. C. Einerhand.** 2002. The MUC family: an obituary. *Trends Biochem. Sci.* **27:**126–131.
15. **Desseyn, J. L., J. P. Aubert, N. Porchet, and A. Laine.** 2000. Evolution of the large secreted gel-forming mucins. *Mol. Biol. Evol.* **17:**1175–1184.
16. **Dixon, J., V. Strugala, S. M. Griffin, M. R. Welfare, P. W. Dettmar, A. Allen, and J. P. Pearson.** 2001. Esophageal mucin: an adherent mucus gel barrier is absent in the normal esophagus but present in columnar-lined Barrett's esophagus. *Am. J. Gastroenterol.* **96:**2575–2583.
17. **Ellison, R. T.** 1994. The effects of lactoferrin on Gram-negative bacteria. *Adv. Exp. Med. Biol.* **357:**71–90.
18. **Elsbach, P., and J. Weiss.** 1998. Role of the bactericidal/permeability-increasing protein in host defense. *Curr. Opini. Immunol.* **10:**45–49.
19. **Enss, M. L., H. Muller, U. Schmidt-Wittig, R. Kownatzki, M. Coenen, and H. J. Hedrich.** 1996. Effects of perorally applied endotoxin on colonic mucins of germfree rats. *Scand. J. Gastroenterol.* **31:**868–874.
20. **Fernie-King, B. A., D. J. Seilly, A. Davies, and P. J. Lachmann.** 2002. Streptococcal inhibitor of complement inhibits two additional components of the mucosal innate immune system: secretory leukocyte proteinase inhibitor and lysozyme. *Infect. Immun.* **70:**4908–4916.
21. **Geiszt, M., J. Witta, J. Baffi, K. Lekstrom, and T. L. Leto.** 2003. Dual oxidases represent novel hydrogen peroxide sources supporting mucosal surface host defense *FASEB J.* **17:**U362–U375.
22. **Gendler, S. J., and A. P. Spicer.** 1995. Epithelial mucin genes. *Annu. Rev. Physiol.* **57:**607–634.
23. **Gerson, C., J. Sabater, M. Scuri, A. Torbati, R. Coffey, J. W. Abraham, I. Lauredo, R. Forteza, A. Wanner, M. Salathe, W. M. Abraham, and G. E. Conner.** 2000. The lactoperoxidase system functions in bacterial clearance of airways. *Am. J. Respir. Cell Mol. Biol.* **22:**665–671.
24. **Hase, K., L. Eckmann, J. D. Leopard, N. Varki, and M. F. Kagnoff.** 2002. Cell differentiation is a key determinant of cathelicidin LL-37/human cationic antimicrobial protein 18 expression by human colon epithelium. *Infect. Immun.* **70:**953–963.
25. **Hecht, G.** 1999. Innate mechanisms of epithelial host defense: spotlight on intestine. *Am. J. Physiol. Ser. C* **277:**C351–C358.
26. **Hopwood, D.** 1997. Oesophageal damage and defense in reflux oesophagitis: pathophysiological and cell biological mechanisms. *Prog. Histochem. Cytochem.* **32:**1–42.
27. **Hunt, J. S., A. R. McGiven, A. Groufsky, K. L. Lynn, and M. C. Taylor.** 1985. Affinity-purified antibodies of defined specificity for use in a solid-phase microplate radioimmunoassay of human Tamm-Horsfall glycoprotein in urine. *Biochem. J.* **227:**957–963.
28. **Johansson, J., G. H. Gudmundsson, M. E. Rottenberg, K. D. Berndt, and B. Agerberth.** 1998. Conformation-dependent antibacterial activity of the naturally occurring human peptide LL-37. *J. Biol. Chem.* **273:**3718–3724.
29. **Jordan, N., J. Newton, J. Pearson, and A. Allen.** 1998. A novel method for the visualization of the in situ mucus layer in rat and man. *Clin. Sci.* **95:**97–106.
30. **Klebanoff, S. J., W. H. Clem, and R. G. Luebke.** 1966. The peroxidase-thiocyanate-hydrogen peroxide antimicrobial system. *Biochim. Biophys. Acta* **117:**63–72.
31. **Kuriyama, S. M., and F. J. Silverblatt.** 1986. Effect of Tamm-Horsfall urinary glycoprotein on phagocytosis and killing of type I-fimbriated *Escherichia coli. Infect. Immun.* **51:**193–198.
32. **Lambert, J. R., S. K. Lin, and J. Arandamichel.** 1995. *Helicobacter pylori. Scand. J. Gastroenterol.* **30:**33–46.
33. **Lapensee, L., Y. Paquette, and G. Bleau.** 1997. Allelic polymorphism and chromosomal localization of the human oviductin gene (MUC9). *Fertil. Steril.* **68:**702–708.
34. **Leeker, A., B. Kreft, J. Sandmann, J. Bates, G. Wasenauer, H. Muller, K. Sack, and S. Kumar.** 1997. Tamm-Horsfall protein inhibits binding of S- and P-fimbriated *Escherichia coli* to human renal tubular epithelial cells. *Exp. Nephrol.* **5:**38–46.
35. **Lehrer, R. I., and T. Ganz.** 2002. Defensins of vertebrate animals. *Curr. Opini. Immunol.* **14:**96–102.
36. **Levy, O.** 2000. A neutrophil-derived anti-infective molecule: bactericidal/permeability-increasing protein. *Antimicrob. Agents Chemother.* **44:**2925–2931.
37. **Mack, D. R., S. Ahrne, L. Hyde, S. Wei, and M. A. Hollingsworth.** 2003. Extracellular MUC3 mucin secretion follows adherence of *Lactobacillus* strains to intestinal epithelial cells in vitro. *Gut* **52:**827–833.
38. **Menozzi, F. D., A. S. Debrie, J. P. Tissier, C. Locht, K. Pethe, and D. Raze.** 2002. Interaction of human Tamm-Horsfall glycoprotein with *Bordetella pertussis* toxin. *Microbiology* **148:** 1193–1201.

39. **N'Dow, J., C. N. Robson, J. N. S. Matthews, D. E. Neal, and J. P. Pearson.** 2001. Reducing mucus production after urinary reconstruction: a prospective randomized trial. *J. Urol.* **165:** 1433–1440.
40. **Oliver, L., J. Newton, P. Dettmar, J. Pearson, and A. Allen.** 1996. Effects of *Helicobacter pylori* colonisation on the adherent gastric mucus barrier. *Immunology* **89:**OG415–OG415.
41. **O'Neil, D. A., E. M. Porter, D. Elewaut, G. M. Anderson, L. Eckmann, T. Ganz, and M. F. Kagnoff.** 1999. Expression and regulation of the human beta-defensins hBD-1 and hBD-2 in intestinal epithelium. *J. Immunol.* **163:**6718–6724.
42. **Pearson, J. P., I. A. Brownlee, and C. Taylor.** Mucin genes in the GI tract. *In* P. A. Williams, and G. O. Phillips (ed.), *Gums and Stabilisers for the Food Industry,* in press. Royal Society of Chemistry, Cambridge, United Kingdom.
43. **Phalipon, A., A. Cardona, J. P. Kraehenbuhl, L. Edelman, P. J. Sansonetti, and B. Corthesy.** 2002. Secretory component: a new role in secretory IgA-mediated immune exclusion in vivo. *Immunity* **17:**107–115.
44. **Pison, U., M. Max, A. Neuendank, S. Weissbach, and S. Pietschmann.** 1994. Host-defense capacities of pulmonary surfactant—evidence for nonsurfactant functions of the surfactant system. *Eur. J. Clini. Investig.* **24:**586–599.
45. **Quigley, E. M. M., and L. A. Turnberg.** 1987. pH of the microclimate lining human gastric and duodenal mucosa in vivo—studies in control subjects and in duodenal-ulcer patients. *Gastroenterology* **92:**1876–1884.
46. **Rindisbacher, L., S. Cottet, R. Wittek, J. P. Kraehenbuhl, and B. Corthesy.** 1995. Production of human secretory component with dimeric IgA binding-capacity using viral expression systems. *J. Biol. Chem.* **270:**14220–14228.
47. **Ruselervanembden, J. G. H., L. M. C. Vanlieshout, M. J. Gosselink, and P. Marteau.** 1995. Inability of *Lactobacillus casei* strain Gg, *L. acidophilus*, and *Bifidobacterium bifidum* to degrade intestinal mucus glycoproteins. *Scand. J. Gastroenterol.* **30:**675–680.
48. **Russell, J. P., G. Diamond, A. P. Tarver, T. F. Scanlin, and C. L. Bevins.** 1996. Coordinate induction of two antibiotic genes in tracheal epithelial cells exposed to the inflammatory mediators lipopolysaccharide and tumor necrosis factor alpha. *Infect. Immun.* **64:**1565–1568.
49. **Schilling, J. D., S. M. Martin, C. S. Hung, R. G. Lorenz, and S. J. Hultgren.** 2003. Toll-like receptor 4 on stromal and hematopoietic cells mediates innate resistance to uropathogenic *Escherichia coli. Proc. Nat. Acad. Sci. USA* **100:**4203–4208.
50. **Severn, T. L., D. A. Hutton, A. Sama, J. A. Wilson, J. P. Birchall, and J. P. Pearson.** 1999. Is hypertrophic nasal mucosa a good model system for studying mucus production in OME? p. 503–507. *In* M. Tos, J. Thomsen, and V. Balle (ed.), *Otitis Media Today.* Kugler Publications, The Hague, The Netherlands.
51. **Shankar, V., M. S. Gilmore, R. C. Elkins, and G. P. Sachdev.** 1994. A novel human airway mucin cDNA encodes a protein with unique tandem-repeat organization. *Biochem. J.* **300:** 295–298.
52. **Singh, P. K., M. R. Parsek, E. P. Greenberg, and M. J. Welsh.** 2002. A component of innate immunity prevents bacterial biofilm development. *Nature* **417:**552–555.
53. **Slungaard, A., and J. R. Mahoney.** 1991. Thiocyanate is the major substrate for eosinophil peroxidase in physiological fluids—implications for cytotoxicity. *J. Biol. Chem.* **266:**4903–4910.
54. **Smirnova, M. G., J. P. Birchall, and J. P. Pearson.** 2000. TNF-alpha in the regulation of MUC5AC secretion: some aspects of cytokine-induced mucin hypersecretion on the in vitro model. *Cytokine* **12:**1732–1736.
55. **Smirnova, M. G., L. Guo, J. P. Birchall, and J. P. Pearson.** 2003. LPS up-regulates mucin and cytokine mRNA expression and stimulates mucin and cytokine secretion in goblet cells. *Cell. Immunol.* **221:**42–49.
56. **Strugala, V., A. Allen, P. W. Dettmar, and J. P. Pearson.** 2003. Colonic mucin: methods of measuring mucus thickness. *Proc. Nutri. Soc.* **62:**237–243.
57. **Stubbe, H., J. Berdoz, J. P. Kraehenbuhl, and B. Corthesy.** 2000. Polymeric IgA is superior to monomeric IgA and IgG carrying the same variable domain in preventing *Clostridium difficile* toxin A damaging of T84 monolayers. *J. Immunol.* **164:**1952–1960.
58. **Taylor, C., A. Allen, P. W. Dettmar, and J. P. Pearson.** 2003. The gel matrix of gastric mucus is maintained by a complex interplay of transient and nontransient associations. *Biomacromolecules* **4:**922–927.
59. **Valore, E. V., C. H. Park, A. J. Quayle, K. R. Wiles, P. B. McCray, and T. Ganz.** 1998. Human beta-defensin-1: an antimicrobial peptide of urogenital tissues. *J. Clin. Investig.* **101:** 1633–1642.
60. **Van de Bovenkamp, J. H. B., J. Mahdavi, A. M. Korteland-Van Male, H. A. Buller, A. W. C. Einerhand, T. Boren, and J. Dekker.** 2003. The MUC5AC glycoprotein is the primary receptor for *Helicobacter pylori* in the human stomach. *Helicobacter* **8:**521–532.
61. **Van den Brink, G. R., K. Tytgat, R. W. M. Van der Hulst, C. M. Van der Loos, A. W. C. Einerhand, H. A. Buller, and J. Dekker.** 2000. *H. pylori* colocalises with MUC5AC in the human stomach. *Gut* **46:**601–607.
62. **Walsh, M. D., B. G. Hohn, W. Thong, P. L. Devine, R. A. Gardiner, M. Samaratunga, and M. A. McGuckin.** 1994. Mucin expression by transitional-cell carcinomas of the bladder. *Br. J. Urol.* **73:**256–262.
63. **Williams, S. J., D. H. Wreschner, M. Tran, H. J. Eyre, G. R. Sutherland, and M. A. McGuckin.** 2001. MUC13, a novel human cell surface mucin expressed by epithelial and hemopoietic cells. *J. Biol. Chem.* **276:**18327–18336.

Colonization of Mucosal Surfaces
Edited by James P. Nataro et al.

Chapter 2

Defensins and Other Antimicrobial Peptides: Innate Defense of Mucosal Surfaces

Alexander M. Cole and Tomas Ganz

Mucosal secretions form a mechanical and chemical barrier to microbes and constitute a first-line host defense against microbial invaders. Among the components of the secretions, mediators of adaptive immunity include immunoglobulin A (IgA) and IgG, produced by plasma cells adjacent to submucosal glands (137). Immunoglobulins are thought to act in part by preventing the attachment and invasion of pathogenic organisms and to act as beacons for the recruitment and activation of innate effector cells. Mediators of innate mucosal host defense are also found in mucosal secretions; they include substances that sequester microbial nutrients, selectively disrupt bacterial cell walls and membranes, or act as decoys for the attachment of microbes. The antimicrobial properties of mucosal secretions are largely due to the contributions of antimicrobial proteins and peptides.

In 1922, Alexander Fleming attributed the antimicrobial properties of human nasal secretions to an enzyme he named "lysozyme" (56). Since the discovery of lysozyme, additional antimicrobial components of nasal and other mucosal secretions have been identified, including defensins, lactoferrin, secretory leukocyte protease inhibitor (SLPI), cathelicidins, calprotectin, chemokines, and histatins (37, 40, 77, 104, 105, 209, 210). Lysozyme and lactoferrin, two polypeptides expressed at high concentrations in the mucosa, are stored in and secreted from serous cells in submucosal glands, are found in inflamed secretions, and are also released from neutrophils. SLPI, a serine protease inhibitor, is also present at potentially antimicrobial concentrations (86). Human β-defensin 1 (HBD-1) is an antimicrobial peptide constitutively expressed at low levels in both the lower and upper airways, while a structurally similar peptide, human β-defensin 2 (HBD-2), is induced predominantly at sites of inflammation (136, 188). Human β-defensin 3 is induced in keratinocytes during wounding and healing (194). The α-defensin, HNP-1 (65), has also been found at different concentrations in bronchoalveolar lavage (BAL) fluid (177) and is probably a degranulation product of neutrophils that extravasate into tissues during infection. Other antimicrobial (poly)peptides that are present in mucosal secretions include the cathelicidin LL-37 (13, 14) and secretory phospholipase A_2 (4, 37). Anionic antimicrobial (poly)peptides probably also contribute to the antimicrobial activity of the airways and have been detected in ovine and human BAL fluid (27, 28).

Mucosal surfaces are blanketed by a mixture of antimicrobial peptides and proteins, perhaps because such a mixture provides a broad antimicrobial spectrum and decreases the likelihood that microbial resistance will emerge. By interacting in an additive or synergistic manner, individual peptides and proteins with modest intrinsic antimicrobial activity could produce considerably greater effects when combined. For example, lysozyme alone is bactericidal for gram-positive and in concert with lactoferrin is bactericidal for some gram-negative bacteria (51, 168). Other studies have revealed a synergistic microbicidal effect of lysozyme with lactoferrin and SLPI (51, 172, 189) and anionic peptides (103) and an additive killing with β-defensins (189). Nevertheless, the abundance and multiplicity of antimicrobial peptides and proteins in vivo suggests that overall protection is best achieved from the combined effects of multiple host-derived molecules.

How do these substances kill microbes or inhibit their multiplication? Some are enzymes that disrupt

Alexander M. Cole • Department of Molecular Biology and Microbiology, Biomolecular Science Center, University of Central Florida, Orlando, FL 32816-2364. **Tomas Ganz** • Division of Pulmonary and Critical Care Medicine and the Will Rogers Institute Pulmonary Research Laboratory, Department of Medicine, David Geffen School of Medicine, University of California Los Angeles, Los Angeles, CA 90095-1690.

essential microbial structures, as exemplified by lysozyme and its ability to hydrolyze peptidoglycan, a key structural component of the bacterial cell wall. Others bind essential nutrients, denying them to microbes. Lactoferrin, an iron-binding protein, is representative of this host defense strategy. Others, especially small antimicrobial proteins and peptides, act by disrupting microbial membranes. These antimicrobial peptides are phylogenetically ancient molecules whose manufactured counterparts comprise a new class of antibiotics (reviewed in references 25, 75, and 76). Many show activity against a broad range of bacteria and fungi. In evolutionary terms, antimicrobial peptides and proteins are among the earliest molecular effectors of innate immunity. Although antimicrobial peptides can be evolutionarily and structurally diverse, common properties typically include amphipathicity (spatial separation of polar and nonpolar residues) and cationicity (a net positive charge at physiological pH). Many families of peptides display similar mechanisms of action against microbes: most are broad-spectrum microbicides that target gram-positive and gram-negative bacteria as well as fungi and some enveloped viruses (59, 115, 178).

Antimicrobial peptides are cationic and hydrophobic at physiologic pH, attributes that assist peptide binding and insertion into microbial membranes. Some peptides then aggregate to form pores, and the microbe dies once a threshold number of pores have formed (126, 129, 134, 184). Indeed, biophysical studies have revealed that antimicrobial peptides have two binding states in lipid bilayers. As peptide molecules enter the membrane, they form a carpet (67), also described as the surface state "S." As the concentration of peptide is increased, the membrane is progressively strained due to membrane thinning (33). The phase transition into the pore-forming state "I" partially resolves the strain and makes the membrane permeable. Exceptions to the two-phase model have been observed, suggesting that the mechanism of action of certain peptides may be different from the standard pore-forming mechanism (223).

This chapter gives an in-depth account of defensins, the primary topic, and ties together the roles of other antimicrobial peptides and proteins that together contribute to mucosal innate host defense. While the focus is on defensins and antimicrobial peptides from humans, peptides from other species are also included where appropriate.

DEFENSINS

Defensins are one of the most widely studied families of antimicrobial peptides. Over 80 different defensins are expressed by the leukocytes and epithelial cells of various birds and mammals (59, 60, 95, 115, 117, 131, 154). These peptides can be divided into three subfamilies: α-defensins, β-defensins, and θ-defensins (201). All of these defensins derive from an ancestral gene (124) that existed before the evolutionary divergence of reptiles and birds (237), contain six cysteines, and have largely β-sheet structures that are stabilized by three intramolecular disulfide bonds. Humans produce at least 10 different defensin molecules: 6 α-defensins and 4 β-defensins. Although α- and β-defensins differ in the spacing and connectivity of their cysteines, they have similar topology (240). By contrast, despite originating from the same ancestral precursor as other defensins, θ-defensins are structurally unique. The following sections detail the structural and functional relationships between these three classes of defensins.

α-Defensins

In 1966, Zeya and Spitznagel described the antibacterial activities of "cationic proteins of polymorphomuclear leukocyte lysosomes" of guinea pigs and rabbits (233, 234). Nearly two decades passed before α-defensins were sequenced and described as one of the principal antibacterial components of neutrophils from rabbits (182), guinea pigs (181), and humans (65). The α-defensins have 29- to 35-residue β-sheet structures with six cysteines that form three intramolecular disulfide bonds by pairing Cys1-Cys6, Cys2-Cys4, and Cys3-Cys5 (65). α-Defensins are effective at low micromolar concentrations against a wide spectrum of gram-positive and gram-negative bacteria, fungi, yeast, and some enveloped viruses (44, 65, 114, 139, 149). Human α-defensins are found in neutrophils as well as in Paneth cells of the small intestine. Four α-defensins (HNP1 to HNP4) are stored in azurophilic granules of the neutrophil, where they constitute at least 30% of the granules' total protein (63). Neutrophil defensins are synthesized in bone marrow promyelocytes as 93- to 94-amino-acid preprodefensins that undergo proteolytic processing by sequential removal of the signal sequence and propiece, so that the granules contain almost exclusively the mature active peptides (81, 214). Before its removal, the anionic propiece functions effectively to inactivate defensins (215). During phagocytosis of bacteria by neutrophils, defensins are discharged into phagocytic vacuoles of neutrophils, where they reach very high concentrations (in the milligrams-per-milliliter range), become a predominant protein constituent (100), and presumably contribute to the microbicidal milieu in the phagocytic vacuoles. At much lower concentrations, HNP1 to HNP3 have also been identified as components of primary T lymphocytes, natural killer (NK) cells, and monocytic cell lines (1).

Intestinal Paneth cells are host defense cells that occupy the basal portions of the small intestinal crypts adjacent to the zone that contains the small intestinal stem cells. While at least 20 α-defensins are found in Paneth cells of mice (155), human Paneth cells express only two α-defensins. Human defensin 5 (HD-5) and HD-6 are stored as inactive precursor peptides in secretory granules of the Paneth cells (101, 102). Paneth cell-derived trypsin activates HD-5 in the small intestine by proteolytic cleavage (68). This contrasts with the mouse Paneth cell defensins, which are stored in active form after posttranslational processing that involves the matrix metalloproteinase matrilysin (224). Paneth cell granules and their defensins are discharged into the narrow crypt lumen by cholinergic stimuli associated with digestion and by the presence of microbes (12, 165, 173, 174). High concentrations of defensins (in the milligrams-per-milliliter range) are attained in the crypt, but even in the intestinal lumen the defensin concentrations are sufficient to exert antimicrobial effects (12).

β-Defensins

β-Defensins were first identified in the tracheal epithelium (45) and granulocytes (183) of cows over a decade ago. Although β-defensins are slightly larger (up to ~45 residues) than their α-defensin counterparts and have a different disulfide connectivity (Cys1-Cys5, Cys2-Cys4, and Cys3-Cys6 pairing), the shape of their peptide backbones is very similar to that in α-defensins (240). Human β-defensins are produced predominantly by epithelia (19, 77, 78, 117). Human β-defensin-1 (HBD-1) is expressed constitutively by barrier epithelial cells and keratinocytes and may provide a ubiquitous protective blanket at those sites. Recently, several groups have shown that HBD-1 is also upregulated by proinflammatory processes in monocytes/macrophages and dendritic cells (48). Expression of HBD-2, HBD-3, and HBD-4 is induced on stimulation by inflammatory mediators by mast cells, epithelial cells, keratinocytes, monocytes, and dendritic cells (48, 146). Using genomic techniques, 28 additional β-defensins were discovered (179, 226), most of which were potentially transcribed (179). Novel human β-defensin genes were also discovered by another group, who reported that the rapid sequence divergence of β-defensins is a result of adaptive evolution, as judged by the ratio of nonsynonymous to synonymous substitutions (135). Additional research is required to establish how this cornucopia of defensins collectively functions in host defense.

Although some defensins are expressed constitutively, others are produced in response to infection. Identification of microbial products by receptors of innate immunity is one mechanism to initiate the host immune response to infection. Among other molecules that function as detectors of infection, Toll-like receptors (TLRs) recognize conserved molecular patterns produced by microbial pathogens. Examples of these pathogen recognition receptors that recognize pathogen-associated molecular patterns include TLR4, which recognizes lipopolysaccharide (LPS) from gram-negative bacteria, and TLR2, which is triggered by peptidoglycan of gram-positive bacteria. Pathogen-associated molecular pattern recognition is a component of the innate ability of the host to discriminate between self and nonself. Activated TLRs induce the transcription of genes of immunity and inflammation (98).

Early studies of LPS induction of HBD-2 in human tracheobronchial epithelium revealed that HBD-2 utilized CD14, which may have subsequently formed a complex with a TLR to induce NF-κB nuclear translocation and antimicrobial peptide gene activation (17). Similarly, mouse β-defensin 2, a peptide with similarities to human β-defensins, was also found to be LPS inducible (141). In other studies, the expression of HBD-2 in airway epithelia was found to be mediated through TLR2, activated by either lipoteichoic acid, a component of peptidoglycan (219), or lipopeptide derived from mycobacteria (85). Likewise, HEK293 cells transfected with TLR2 also produced HBD-2 in response to bacterial lipoproteins (24). Live rhinovirus-16, a respiratory virus for the common cold, but not UV-inactivated virus, increased the expression of HBD-2 and HBD-3 in bronchial epithelial cells (47). The investigators reported that these cells expressed TLR3, a receptor that recognizes double-stranded RNA, and that double-stranded RNA alone was sufficient to cause an upregulation of β-defensins, implicating TLR3 in rhinoviral infection. Conversely, β-defensins can also act as endogenous ligands for TLRs. In mice, murine β-defensin 2 (MBD-2) acts as a ligand for TLR4 on immature dendritic cells, inducing dendritic cell maturation and the upregulation of costimulatory molecules (22).

The connectivities of the disulfide bonds formed between pairs of cysteine residues in the mature oxidized peptide are unique for each class of defensins. Two examples of the detailed structure of the α- and β-defensins were determined by X-ray diffraction of the α-defensin HNP-3 (89) and high-resolution nuclear magnetic resonance spectroscopy for the β-defensin HBD-2 (176). These examples revealed how disulfide pairing defines the alignment of the β-sheet structures in determining specific types of molecular folds. In addition, one structural feature that distinguishes the β-defensins from the α-defensins is the presence of a short α-helical segment near the N terminus (93, 94, 176). Similar α-helical segments have been found in a β-defensin-like peptide, DLP-2, from platypus venom (205) but not in bovine β-defensin BNBD-12 (240). Although the function of the N-terminal helical segment is currently unknown, it may be a (co)factor that contributes to the self-association of the peptide in microbial membranes.

θ-Defensins

Three members of a new defensin subfamily, the θ-defensins, were isolated recently from the leukocytes and bone marrow of the rhesus monkey, *Macacca mulatta* (120, 201, 208). These arginine-rich, circular 18-amino-acid peptides arose from two precursor peptides, each of which contributed nine residues to the mature θ-defensin (120, 201, 208). They have a β-sheet conformation (207) and represent the first truly circular peptides of vertebrate origin. Their intracellular cyclization occurs via two posttranslational head-to-tail ligations of the backbone of two 9-residue precursors to make the resulting peptide circular. Oxidation of the six cysteines forms an internal tridisulfide ladder. Because the incorporated 9-residue segments can be derived from identical or different genes, the θ-defensin family can generate molecular diversity through a unique posttranslational mechanism (120, 208). In principle, n θ-defensin genes could produce $(n/2) \times (n + 1)$ peptides (120). Therefore, just as two different θ-defensin (DEFT) genes in the rhesus monkey actually produced three distinct θ-defensins, four genes could produce 10 distinct peptides and five genes could result in 15 distinct peptides.

Whereas α- and β-defensin peptides are produced by both human and nonhuman primates (201), only certain nonhuman primates produce mature θ-defensin peptides. Although humans express mRNA encoded by DEFT genes, a stop codon within the signal sequence prematurely arrests its translation (39). Consequently, human cells express θ-defensin mRNA but not θ-defensin precursor peptides. It should be noted that the inability to express mature retrocyclin molecules in humans does not result from the inability to perform posttranslational processing—it is known that the cellular machinery responsible for processing θ-defensin precursors remains operational in human leukocytes (201).

Having cloned the human θ-defensin cDNA and also identified several DEFT genes by surveying the human genome database, it was a logical step to deduce the sequences of the 9-residue precursor elements that were incorporated into the θ-defensins produced by our primate ancestors before the mutational event silenced translation of the precursors. These ancestral peptides were then recreated by solid-phase synthesis. The name "retrocyclin" resulted from splicing the terms "retro-" from the Latin word for back or backward, and "-cyclin," referring to the circular nature of the peptide. One might notice that it was fortuitous that the "retro-" portion of the name foretold the peptide's activity against the human immunodeficiency virus type (HIV-1) "retro"virus (39). Solid-phase FMoc synthesis of retrocyclin-1 has been described previously (39, 201). Rather than attempting to synthesize and ligate two 9-residue peptides, a linear 18-residue peptide was prepared and subjected to disulfide oxidation and backbone cyclization. The antibacterial and antifungal activities of the resulting retrocyclin were modest and were similar to those of synthetic and native rhesus θ-defensins (RTD-1, RTD-2, and RTD-3) (39, 120, 201, 208).

It was later found that several retrocyclin pseudogenes cluster on chromosome 8p23 (145), a locus that also contains the human α- and β-defensin genes (124). This and other pieces of evidence indicate that θ-defensin genes are the evolutionary descendants of α-defensin genes (146, 201). Human retrocyclins were silenced after the orangutan and hominid lineages had diverged, approximately 7.5 million to 10 million years ago (146). Given the antiviral properties of retrocyclin against HIV-1, the evolutionary loss of retrocyclin may have contributed to HIV-1 susceptibility in modern humans. If so, then the administration of retrocyclins to humans, either systemically or as a topical microbicide, would restore effector molecules that were never relinquished by many of our nonhuman primate relatives.

Activities

Antibacterial, antifungal, and antiprotozoal activities

Under optimal environmental conditions, most defensins at micromolar concentrations display a remarkably broad spectrum of antimicrobial activity, including activity against gram-positive and gram-negative bacteria, fungi, and yeast (6, 65, 181–183), mycobacteria (149), and protozoa (7). In the presence of plasma components, including salt, other monovalent and especially divalent cations (118, 119), and defensin-binding proteins (156, 158, 159), higher concentrations of defensins are required for microbial killing. Microbes also become more resistant to defensins when microbial metabolism is inhibited by the lack of nutrients or the presence of metabolic toxins (118). These observations are consistent with the proposed mechanism of electrostatic attraction of defensins to negatively charged microbial surfaces, followed by insertion of defensins into the cell membrane of microbes under the influence of local electric fields and the transmembrane potential. In this model, the presence of salts would shield the electrostatic attraction of peptides to microbial surfaces. Calcium and other divalent cations may also stabilize the outer membrane of gram-negative bacteria and make it less permeable to defensins. The loss of transmembrane potential from metabolic inhibitors would diminish the force that drives cationic defensins to enter into the membrane. Countering these inhibitory influences, defensins reach high concentrations in several common biological contexts, including the phagocytic vacuoles of neutrophils (62, 100), small intestinal crypts (12), and intercellular spaces between keratinocytes (152, 186). Nevertheless, even lower concentrations of defensins, such as those that prevail in the lumen of the small intestine, appear to have significant effects on the intestinal flora and pathogens (171), as illustrated by the ability of transgenic HD-5 to protect mice against intestinal infection with *Salmonella enterica* serovar Typhimurium.

Antiviral activities

Whereas the ability of defensins to kill bacteria and fungi is well documented, relatively few studies have examined their antiviral effects. In retrospect, this should not be taken to reflect the inactivity of antimicrobial peptides against viruses—only that this possibility until recently received relatively little attention in the field. Certain enveloped viruses can be inactivated by several α-defensins purified from rabbit (116) or human (44) leukocytes. These viruses included herpes simplex virus types 1 and 2 (HSV-1 and HSV-2), cytomegalovirus, vesicular stomatitis virus, and influenza virus A/WSN. In contrast, two nonenveloped viruses, echovirus type 11 and reovirus type 3, were not inactivated. The inhibitory effect of human α-defensin HNP-1 on HSV-1 was temperature dependent and was inhibited by the presence of serum. Rabbit α-defensin NP-1 was recently shown to be an entry inhibitor, preventing an early step in HSV-1 infection—either viral adherence or subsequent membrane fusion (190).

Stimulated by the global epidemic of AIDS, recent studies have shown that several antimicrobial peptides are active against HIV. Evidence that defensins are active against HIV-1 has been available for nearly 10 years. Nakashima and colleagues reported that rabbit, rat, and guinea pig defensins can inhibit HIV-1-induced cytopathogenicity of a $CD4^+$ human T-cell line (H. Nakashima, N. Yamamoto, M. Masuda, and N. Fujii, Letter, *AIDS* 7:1129, 1993). This short report stimulated further investigation by another group that noted structural and functional similarities between the looped motifs of α-defensins and peptides derived from HIV-1 gp41 (140). It was postulated that these loops may be required for viral fusion and infectivity. Following an 8-year lag in research on defensin anti-HIV activity, a persuasive and controversial study implicated human α-defensins as a component of the $CD8^+$ T-cell antiviral factor (CAF) (235). CAF has been an elusive low-molecular-weight, noncytolytic antiviral factor, or group of factors, secreted by stimulated $CD8^+$ lymphocytes in a small subset of individuals who have immunological control of HIV replication (217). However, two recent papers (31, 127) have disputed this claim, asserting that although α-defensins can inhibit HIV infection, the mechanism is distinct from the transcriptional inhibition mechanism described for CAF (121). Several other molecules have been reported as likely CAF candidates, and it is unlikely that disagreement regarding the finding that α-defensins are a component of CAF will be fully resolved in the near future (36).

β-Defensins were also recently shown to inhibit HIV-1 infection. Quinones-Mateu et al. (166) revealed that HIV-1 induced HBD-2 and HBD-3 expression in human oral epithelial cells. Moreover, expression of β-defensins downmodulated the expression of CXCR4, a chemokine (co)receptor required for HIV-1 entry. These results may help describe why infection by HIV-1 via the oral route is uncommon compared with infection via the rectal or vaginal route (170).

θ-Defensin peptides, including retrocyclin and RTD-1, can protect primary and transformed T lymphocytes from in vitro infection by both X4 and R5 strains of HIV-1 (39) and are much more active in vitro than their α-defensin brethren (A. M. Cole and R. I. Lehrer, unpublished data). Studies with luciferase reporter viruses indicated that retrocyclins act by preventing the entry of HIV-1 (142). Furthermore, the antiviral properties of retrocyclin and other θ-defensins might be intimately linked to an ability to bind carbohydrate epitopes displayed by viral and cell surface glycoproteins involved in viral entry (218). Together, these studies suggest that retrocyclins inhibit HIV-1 entry by (i) blocking a postbinding conformational change in the viral envelope glycoproteins gp120 or gp41 that occurs on the pathway to fusion, (ii) preventing the insertion of the gp41 amino-terminal fusion peptide into the target cell membrane, or (iii) by analogy to the fusion peptide inhibitor T20 (108), preventing prehairpin-to-hairpin conversion of gp41. RTD-1, the rhesus monkey homolog of retrocyclin, binds parallel to lipid membrane surfaces (223). Similar binding by retrocyclin may alter cell surface properties in a manner that prevents the fusion of cellular and viral membranes. X4 strains utilize the chemokine coreceptor CXCR4 along with the primary receptor CD4 to aid entry into receptive cells, while R5 strains instead utilize the coreceptor CCR5. Since retrocyclins protect against both tropisms, they may be better therapeutic candidates than peptides that protect only against one. Indeed, since R5 strains are purported to be the primary tropism involved during sexual transmission of HIV-1, it is important to design and develop compounds that are active against this HIV-1 tropism.

Other functions

It is curious that humans have retained multiple versions of defensin molecules. One speculation is that defensin sequences have diverged and generated different, and at times partially overlapping, spectra of activity. Thus, microbes that are resistant to one type of defensin might still preserve sensitivity to another. Because of overlapping functions of defensins, microbes encountering a number of defensin types would be less likely to subvert the host defenses. Molecular diversity is well demonstrated by the antimicrobial activity of β-defensins. *Staphylococcus aureus* is markedly insensitive to HBD-1 and HBD-2, yet even multidrug-resistant strains are susceptible to HBD-3 (77). Interestingly, low concentrations of HBD-3 are also active at physiological concentrations of salt, a property that is not held by HBD-1 and HBD-2. However, there may be other explanations for the multitude and apparent duplication of defensins in humans, including alternative activities for certain subgroups of molecules.

Studies implicating defensins as chemotactic agents for cells mediating the adaptive immune response have partially bridged the gap between innate and adaptive immunity. Human α-defensins and β-defensins, as well as the cathelicidin LL-37, have been reported to be chemotactic for leukocytes at subnanomolar concentrations (1, 34, 66, 202, 227, 229). The chemotactic activity of HBD-2 for T cells and dendritic cells is mediated by the CC chemokine receptor CCR6 (229). HBD-2 and MIP-3α/CCL20, the cognate ligand for CCR6 in mice and humans, are structurally similar (160). Not surprisingly, murine β-defensins 2 and 3 bind murine CCR6 and chemoattract immature dendritic cells but not mature dendritic cells (23). Intranasal delivery of α-defensins plus the antigen ovalbumin (OVA) enhanced OVA-specific systemic IgG, but not IgA, antibody responses through help provided by cytokines produced in Th1 and Th2 responses (123). Additionally, α-defensins in conjunction with OVA helped to mediate B- and T-cell interactions, thus providing another example of a link between innate and adaptive immunity (123).

Several studies have also suggested additional roles of defensins aside from their microbistatic/microbicidal and chemotactic functions. Some defensins suppress the activity of adrenocorticotropin on the adrenal cortex by antagonizing the adrenocorticotropin receptor (193, 204, 238, 239). Others act as inhibitors of tissue-type plasminogen activator-mediated fibrinolysis (87, 88) and promote wound healing (9, 74, 143). Defensins can also bind and inactivate serine proteinase inhibitors (serpins) (158), α_2-macroglobulin (156), and C1 complement (159). Defensins are mitogenic for epithelial cells and fibroblasts (143). Certain β-defensins degranulate mast cells and thus release histamine, as well as inducing the production of prostaglandin D_2 (147). Mouse intestinal α-defensins (cryptdins 2 and 3) are able to activate nifedipine-sensitive calcium channels (16, 128). In view of the many reported activities of defensins in vitro, it will be necessary to determine which are biologically relevant.

OTHER ANTIMICROBIAL PEPTIDES AND PROTEINS OF MUCOSAL SURFACES

Lysozyme

Lysozyme (muramidase) is a 14.6-kDa enzyme whose bacteriolytic properties derive from cleavage of the bond linking *N*-acetylmuramic acid and *N*-acetyl-D-glucosamine residues in peptidoglycan, the macromolecule that confers shape and rigidity on cell walls of bacteria. Lysozyme also has nonenzymatic antimicrobial activity that stems from its highly cationic nature, which presumably allows it to disrupt microbial membranes or activate autolytic bacterial enzymes (96, 109, 220). However, lysozyme alone is inactive against many pathogens, either because it cannot reach the peptidoglycan layer or because the structure of peptidoglycan is modified to make it less sensitive to the enzymatic action of lysozyme. Such resistance may be overcome by synergy with other antimicrobial components of epithelial secretions (51, 189). Mice deficient in lysozyme M, the form normally expressed in neutrophils and macrophages, have only a mild deficit in microbial killing but show a greatly increased inflammatory reaction when challenged with *Micrococcus luteus* or its peptidoglycan (61). Thus, lysozyme-mediated degradation of peptidoglycan could also function to limit and terminate the inflammatory reaction to this pathogen-associated macromolecule. Lee-Huang et al. reported that lysozyme purified from human milk and neutrophils reduced HIV-1 production in chronically infected T lymphocytes and monocytes (113). The activity reported was weak and has not since been confirmed by other laboratories.

Lactoferrin

Lactoferrin, an 80-kDa homolog of the iron carrier protein transferrin, is an abundant constituent of the specific granules of neutrophils and of epithelial secretions, notably including milk and tears, where it is present at concentrations similar to those of lysozyme, reaching the milligrams-per-milliliter range. This cationic protein exerts antibacterial activity through at least two distinct mechanisms (10, 29): by sequestration of iron, an essential element for microbes, and directly, independently of iron. Unlike transferrin, lactoferrin binds iron under acidic conditions in areas of microbial infection and inflammation. The iron-independent activity may be due to a highly cationic segment of lactoferrin (32) that may disrupt microbial membranes. There is evidence that lysozyme and lactoferrin act synergistically against gram-negative bacteria (51). In the digestive tract, lactoferrin from milk may be cleaved by pepsin or other proteases to free an antimicrobial peptide segment called lactoferricin (18). Lactoferrin was reported to directly and indirectly inhibit several viruses, including HIV-1, HSV, and human cytomegalovirus (8, 79, 83, 200). The mechanism of its action against HIV-1 occurs in an early phase of infection, probably during adsorption of virus to the target cell (164). In vitro, lactoferrin binds strongly to the V3 loop of the gp120 receptor on HIV-1 and HIV-2, causing inhibition of virus-cell fusion and entry of the virus into cells (200).

Secretory Leukocyte Protease Inhibitor

SLPI is a 12-kDa protein found in epithelial secretions but also produced by macrophages. It is especially abundant in cervical mucus plugs of pregnancy (84), where it reaches concentrations in the range of 100 to 1,000 μg/ml. SLPI is approximately 10- to 100-fold less abundant in normal cervical secretions, nasal fluid, and other respiratory secretions. It is weakly antibacterial and antifungal (86, 203), but adequate SLPI concentrations and appropriate conditions for its activity may not be present in most secretions except perhaps in cervical mucus plugs. The observation that HIV-1 is rarely transmitted through salivary secretions provided impetus to study endogenous inhibitors of HIV-1 produced by the oral mucosa. Several laboratories have reported an inhibitory activity of SLPI against HIV-1 (reviewed in reference 187). However, other studies failed to confirm SLPI-mediated activity against HIV-1 (213), leaving its antiretroviral activity uncertain.

Cathelicidin (LL-37)

Cathelicidins are a family of antimicrobial peptides with a common precursor "cathelin" sequence similar to that of the thiol protease inhibitor cystatin but with the C-terminal addition of very diverse peptide sequences. While animals such as pigs and cows are endowed with multiple cathelicidin genes (230), humans have only one, called hCAP-18/LL-37, so named because its precursor is a cationic protein of 18 kDa whose mature form has 37 amino acids with two leucines at the N terminus (2, 43, 110). The sole murine cathelicidin, cathelin-related antimicrobial peptide (CRAMP) (58), is similar to LL-37 in its structure and antimicrobial properties. Despite the conservation of the precursor sequence, the cathelicidin family includes antimicrobial peptides that are α-helical (e.g., LL-37) (99), form β-sheets (e.g., porcine protegrins) (11, 54), or have repeating motifs (e.g., PR-39) (3). It appears that the diversity of peptide sequences attached to the same cathelin precursor was generated by repeated evolutionary shuffling of the last exon into a primordial cathelicidin gene. The cathelicidins are abundant components of mammalian neutrophils but are also found in semen and other secretions. Human cathelicidin LL-37 has a broad spectrum of activity against bacteria and fungi (212) even in the presence of salt in complex media. Cathelicidins are found in neutrophil granule types that are predominantly destined for extracellular release (43), suggesting that this peptide family may play an important role in host defense of extracellular spaces.

Calprotectin (Calgranulins)

Calprotectin, also called leukocyte protein L1, MRP-8/14, and S100A8/9, is an abundant cytoplasmic protein complex of neutrophils, monocytes, and keratinocytes, accounting for about 40% of the cytoplasmic protein content of human neutrophils (26). It is found in inflammatory fluids, where it may be released from neutrophils by an as yet incompletely characterized secretory mechanism. At realistic concentrations, calprotectin inhibits the growth of fungi and bacteria (196), in part by sequestering zinc (35, 192). Recently developed knockout mice lacking MRP-14, one of the components of this complex, lack an obvious phenotype, suggesting that the contribution of this protein complex to host defense may be subtle (91, 130).

Histatins

Histatins are histidine-rich salivary peptides (151), 3 to 5 kDa in size, synthesized from two human genes and posttranslationally cleaved to yield diverse peptide species. They are especially active against yeast. Their distribution in nonsalivary material has not been reported. Two nice reviews are available for those interested in the details of histatin structure and function (50, 211).

Secretory Phospholipase A_2

The large family of phospholipases includes group IIA secretory phospholipase A_2 ($sPLA_2$), which, distinct from other family members, apparently evolved to target microbial phospholipids (222). It is abundant in human tears (144), intestinal Paneth cells (4, 82, 165), and inflammatory exudates (225), and it may contribute to the infection-inducible antibacterial activity of human serum (73). It is active against many gram-positive bacteria by itself and against gram-negative bacteria in the presence of bacterial permeability-increasing protein (57). Overexpression of human $sPLA_2$ partially protects transgenic mice against *S. aureus* sepsis.

Bactericidal Permeability-Increasing Protein

BPI, a 56-kDa cationic LPS-binding protein, was formerly thought to be expressed solely in neutrophils but was recently also found in lipoxin-stimulated epithelia (30). The protein is highly active against some gram-negative bacteria (52) and can neutralize LPS by competing for it with its physiologic carrier, LPS-binding protein. A related family of proteins, exemplified by PLUNC (palate, lung, and nasal epithelium clone), is found in nasal fluid (41) and other respiratory fluids (21), but little is known about the function of these proteins.

Chemokines

Several chemokines have similar characteristics to defensins, including size, disulfide bonding, interferon (IFN) inducibility, and cationic charge at neutral pH. Indeed, MIG/CXCL9, IP-10/CXCL10, and I-TAC/CXCL11, members of the IFN-γ-inducible ELR^- CXC chemokines, were antimicrobial against *Escherichia coli* and *Listeria monocytogenes* (38). The noticeable salt sensitivity of ELR^- CXC chemokines was similar to that of defensins and many other antimicrobial peptides (59). Additionally, the amounts of chemokines released per monocyte suggest that, in tissues with mononuclear cell infiltration, IFN-γ-inducible chemokines may reach concentrations necessary for microbicidal activity (38, 111, 175). IFN-γ administered therapeutically to patients with chronic granulomatous disease improved the ability of phagocytes to clear bacterial infections (reviewed in reference 20). Although there is controversy, it is plausible that the increase in antimicrobial activity results from oxygen-independent bactericidal mechanisms, including the IFN-γ-inducible ELR^- CXC chemokines. Collectively, chemokines and chemotactic defensins released by sentinel cells could directly target and inactivate microbial pathogens while also serving as signals for leukocyte recruitment to infected tissues.

A report published after the identification of antimicrobial IFN-γ-inducible chemokines indicated that there are other chemokines, including CCL20/MIP3α, which can display antimicrobial activity (228). The report also suggested that activity for these chemokines may be a result of a cationic patch serving to separate charged from uncharged residues on the molecular surface, not unlike the amphipathic nature of defensins (228). While it is interesting that HBD-2 and CCL20 can both bind and activate the chemokine receptor CCR6 (229), there is no apparent primary sequence similarity between HBD-2 and CCL20/MIP-3α, except for disulfide bridges and a large number of cationic residues. However, X-ray crystallography revealed structural similarities between these two peptides that map help explain the overlap of chemotactic and antimicrobial activities (92).

IN VIVO EVIDENCE FOR A HOST DEFENSE ROLE OF DEFENSINS AND OTHER ANTIMICROBIAL (POLY)PEPTIDES

Clinical Disorders

The abundance of defensins and other antimicrobial peptides at sites of infection and inflammation lend one strand of evidence that these molecules may play an important role in host defense. High concentrations of α-defensins are frequently observed in chronically inflamed tissues. HBD-2 concentrations in plasma have been reported to increase fourfold in patients with bacterial pneumonia (90). Inflamed intestinal (150) and gingival (132) epithelia also contain high concentrations of HBD-2 compared with uninflamed tissue, and lesional psoriatic scales contain large amounts of both HBD-2 and HBD-3 (77, 78). Likewise, the absence of HBD-2 in epidermal and pulmonary burns has been correlated with increased susceptibility to infection (138, 153; M. R. Ortega, A. M. Cole, T. Ganz, M. H. Bakir, and S. M. Milner, submitted for publication).

Antimicrobial peptides and proteins have been implicated in the pathogenesis of cystic fibrosis (CF). Most of the morbidity and mortality of CF is the result of progressive respiratory damage due to uncontrolled airway infection and inflammation. The genetic cause of CF has been localized to a defective CF transmembrane regulator, resulting in a failure of epithelial chloride channel function (71). The search for the link between the CF gene defect and bacterial colonization of the airways was given new momentum by the proposal that the bacterial colonization in CF is due to the abnormal composition of airway surface fluid (69, 191). Certain factors within airway fluid, including electrolytes and anionic macromolecules, decrease the activity of many antimicrobial peptides, including defensins. Initial studies of cultured airway epithelial cells revealed that common CF pathogens were killed when added to the apical fluid of cultured normal airway epithelia but proliferated in the fluid of cultured CF airway epithelia (191). Additionally, the concentration of NaCl in apical fluid was elevated in CF airway secretions and the difference in antimicrobial activity between normal and CF epithelia was in part attributed to salt-sensitive, low-molecular-weight factors such as defensins. However, in a slightly different system of cultured epithelia, salt concentrations did not differ between cultures of CF and normal epithelia (106, 133). Rather, in this system, CF airway epithelia absorbed airway surface liquid at an elevated rate, which depleted the periciliary liquid layer and subsequently reduced mucus transport (133). Using a bronchial xenograft model, Bals et al. also demonstrated a salt-independent defect in bacterial killing in CF airway surface fluid that could be corrected by adenovirus gene transfer of CFTR (15). Alternative hypotheses for the role of airway surface fluid electrolyte concentrations and defective gland functions are reviewed by Verkman et al. (219).

The activity of antimicrobial peptides could be adversely affected regardless of which of the above-mentioned models applies in vivo. While increased salt concentrations would interfere with the function of many antimicrobial polypeptides, diminished mucociliary clearance would prolong the time for which inhaled microbes were exposed to antimicrobial substances and could permit microbial adaptation or the emergence of resistant variants. There are several specific questions that remain to be answered about CF airway fluid. Are antimicrobial polypeptides produced in normal or sufficient quantities in the CF airway, and which, if any, of the polypeptides are defective? Are inhibitory substances, for example polyanionic macromolecules, interfering with effective antimicrobial activity? Does the increased production or retention of antimicrobial peptides contribute to destructive inflammation in the airways of CF patients?

There are other situations in which the concentrations of defensins and other antimicrobial polypeptides are known to be decreased. Neutrophils from neonates contain three- to fourfold less BPI than do adult neutrophils, and this reduction in the amount of BPI was directly correlated with a reduction antibacterial activity of the neutrophil extracts (122). Neutrophils from patients with neutrophil-specific granule deficiency have severely reduced concentrations of defensins and lactoferrin yet maintain normal concentrations of the proteases cathepsin G and elastase (64, 125, 167). These patients suffer from recurrent bacterial infections. Furthermore, levels of the mRNA of BPI and the cathelicidin hCAP18/LL-37 were shown to be significantly decreased in patients with neutrophil-specific granule deficiency (70). In separate studies, slow resolution of hyaline membrane disease-induced bronchopulmonary dysplasia in infants has been found to be associated with reduced production of lysozyme and lactoferrin (169), suggesting a biological role for these molecules in the airways.

In children and adults with enteric infections, the production of HBD-1 and LL-37 by colonic enterocytes is downregulated during infection, and the levels normalize following the resolution of disease (97). Microarray analyses revealed that HD-5 and HD-6 were particularly overexpressed in inflamed colonic tissue from patients with Crohn's disease, a form of chronic inflammatory bowel disease with unclear etiology (112). However, other studies reveal that Crohn's disease is characterized by impaired upregulation of HBD-2 and HBD-3 (55, 221). Whether the expression of HD-5 and HD-6 compensates for the lack for HBD-2 and HBD-3 production has not been investigated.

Because of the complexity of these disorders, it is often difficult to attribute the clinical picture to the deficiency of antimicrobial polypeptides. In this regard, investigators have applied defined animal models to study the biological role of antimicrobial peptides.

Evidence from In Vivo Animal Models

The importance of defensins and other antimicrobial peptides in combating infections has been best demonstrated by experiments utilizing transgenic mice engineered to produce reduced or increased amounts of peptides. Three influential studies are discussed here, involving either (i) the targeted disruption or "knocking out" of an antimicrobial peptide gene, (ii) the disruption of a processing enzyme required for the production of active antimicrobial peptides, or (iii) the transgenic expression or "knocking in" of a gene to overexpress an antimicrobial protein. Additionally, one study with pigs revealed that the in vivo chemical inhibition of neutrophil elastase, a processing enzyme required for the production of certain active forms of antimicrobial peptides, increases the susceptibility of skin wounds to infection. In each case, described in detail below, the presence of active antimicrobial peptides resulted in decreased host susceptibility to infection and the reduction or inactivation of antimicrobial peptides contributed to an increased susceptibility to infection and in some cases to decreased host survival.

To investigate the function of the mouse cathelicidin CRAMP, Nizet et al. (148) and Dörschner and colleagues (46) utilized a CRAMP knockout mouse model of cutaneous infection and showed that this mouse was more susceptible to group A *Streptococcus* than were wild-type control mice. Thus, cathelicidins are a native component of innate host defense in mice, providing protection against skin infections caused by group A *Streptococcus*. Additionally, these studies revealed that mouse (CRAMP) and human (LL-37) cathelicidins are similarly expressed during cutaneous injury (46). These important studies lend insight into how the human counterpart to CRAMP may function during infection, since the LL-37 and CRAMP peptides have similar structures, spectra of antimicrobial activity, and tissue distribution (163).

The precursors of cathelicidins, abundant antimicrobial peptides of mammalian neutrophils, are activated by catalytic amounts of neutrophil elastase and proteinase 3 (157, 180, 185, 195, 231, 232). Although the neutrophils of many if not all mammalian species contain cathelicidins, porcine neutrophils are unusually well endowed with these peptides, including protegrins, prophenins, PR-39, and porcine myeloid antimicrobial peptides (3, 80, 107, 197–199, 206, 236); however, porcine neutrophils have not been reported to contain any peptides of the defensin family. Porcine neutrophils are the sole known sources of protegrins, potently and broadly microbicidal peptides of 16 to 18 residues that are structurally related to defensins (107). One study explored the effects of a specific neutrophil elastase inhibitor (NEI) on innate host defense in porcine skin wounds (42). The authors demonstrated that wound fluid generated in the presence of NEI was deficient in mature protegrin content, protegrin-associated antibacterial activity, and total antibacterial activity against a wide range of microbes. Moreover, NEI-treated wounds showed diminished clearance of fungi and gram-positive and gram-negative bacteria. The reduction of antimicrobial activity caused by NEI could be corrected by replacing the mature protegrin, confirming the important contribution of elastase-processed protegrin to the antimicrobial activity of wound fluid. Together, these findings support the proposed role of antimicrobial peptides as host defense molecules in mammals (42).

Although there is only one reported mouse cathelicidin gene, Paneth cells of the mouse small intestine express over 20 α-defensin genes called cryptdins, with high sequence and structural similarity to human neutrophil defensins. However, the sheer number of cryptdin genes makes it difficult to disrupt or delete them all in order to see a pathophysiological phenotype. However, both mouse and human defensin peptides are synthesized as larger, inactive precursor molecules that require proteolytic activation. For cryptdins, the matrix metalloproteinase matrilysin was found to activate cryptdin precursors in vitro through the removal of an amino-terminal anionic peptide segment (224). Mice with a homozygous disruption of their matrilysin genes lacked mature cryptdins and showed increased susceptibility to orally administered *S. enterica* serovar Typhimurium (224). Although other potential roles of matrilysin in host defense cannot be ruled out, matrilysin functions in intestinal mucosal defense at least through the regulation of defensin activity. In humans, matrilysin is constitutively expressed in the epithelium of peribronchial glands and conducting airways in the normal lungs, and expression is increased in airway epithelial cells of CF patients (49). Even though the processing of human neutrophil α-defensins occurs during neutrophil maturation prior to cellular release from the bone marrow, matrilysin may participate in the activation of β-defensins or other antimicrobial peptides in the human airways.

Lysozyme has long been presumed to play a significant role in mucosal host defense due to its in vitro microbicidal activity and its ubiquitous as well as abundant epithelial expression. Until recently, direct evidence of the biological role of lysozyme has been limited. Akinbi et al. reported that bacterial killing was enhanced in the lungs of mice that transgenically overexpressed rat lysozyme in distal respiratory epithelial cells (5). The concentration of lysozyme in BAL fluid from transgenic mice was two- to fourfold greater than that in BAL fluid from wild-type mice. Mice that overexpressed lysozyme were more efficient than wild-type mice in killing group B streptococci and mucoid *Pseudomonas aeruginosa* and had an increased survival rate following infection.

Another group examined transgenic mice deficient in lysozyme M but sufficient in lysozyme P after challenge by the normally nonpathogenic and highly lysozyme-sensitive bacterium *M. luteus* (61). Despite partial compensation by newly expressed lysozyme P in macrophages, lysozyme M-deficient mice developed much more severe lesions than wild-type mice did. The tissue injury was due to the failure of lysozyme M-deficient mice to inactivate peptidoglycan, resulting in an intense and prolonged inflammatory response. These data suggest that tissue injury is normally limited by prompt degradation of bacterial macromolecules that trigger innate immunity and inflammation (61).

Microbial Evasion of Antimicrobial Peptides

Several mechanisms of bacterial resistance to cationic antimicrobial (poly)peptides reflect the same molecular strategy even when the target molecules and genes might be unrelated. The common theme for this resistance involves diminution of negative surface charge, which in turn reduces the electrostatic attraction of cationic antimicrobials such as defensins. Antimicrobial peptides that cannot bind to membrane surfaces have little chance of exerting their action on target microbes. Gram-positive bacteria, such as *S. aureus,* have found efficient ways to counteract antimicrobial mechanisms of human leukocytes and epithelia. One mechanism of staphylococcal resistance to cationic antimicrobial peptides, including defensins and cathelicidins, is achieved by modification of teichoic acid polymers in staphylococcal cell walls with D-alanine (162) and of phospholipids in their membrane with L-lysine (161). The net effect results in a electrostatic repulsion of defensins and cathelicidins (and possibly other antimicrobial peptides) on the bacterial surface, which enables *S. aureus* to more effectively colonize host mucosal surfaces such as those in the nose. Pathogenesis of the gram-negative bacterium *S. enterica* serovar Typhimurium in the alimentary canal can be intimately linked to its ability to evade the innate host defense. A two-component signal transduction system, PhoP-PhoQ, is necessary for *S. enterica* serovar Typhimurium resistance to antimicrobial peptides in mice and humans (reviewed in reference 53). PhoP is an environmental response regulator that is phosphorylated by the histidine kinase sensor PhoQ, and the combination of the two controls the expression of more than 40 proteins. A major function of the PhoP-PhoQ system involves remodeling of the bacterial envelope during host colonization. Proteins activated by PhoP include enzymes that modify the lipid A structure in the bacterial outer membrane, rendering the surface less favorable to binding by cationic peptides. Details of outer membrane remodeling in *S. enterica* serovar Typhimurium are summarized in two elegant reviews (53, 72).

SUMMARY AND CONCLUSIONS

Defensins and other antimicrobial peptides and proteins coat mucosal surfaces and are among the primary early mediators of host defense against colonization and tissue invasion by pathogenic microbes. In vitro and in vivo studies indicate that most antimicrobial peptides probably act as endogenous antibiotics. However many, if not most, antimicrobial peptides also have alternative modes of action. Certain peptides initially regarded as antimicrobials with effective concentrations in the micromolar range are chemotactic at (sub)nanomolar concentrations for cells involved in the adaptive immune response. In addition to their well-recognized chemotactic roles, several chemokines, whose cationic surface charge distribution resembles the amphipathic nature of antimicrobial (poly)peptides, are indeed microbicidal at micromolar concentrations. These often redundant roles further blur the division between innate and adaptive immunity and underscore the notion that adaptive responses emerged as an extension of the more ancient innate immune system.

Acknowledgments. This work was supported by grants HL70876 and AI52017 (to A.M.C.) and AI46514 and HL 46809 (to T.G.) from the National Institutes of Health.

REFERENCES

1. **Agerberth, B., J. Charo, J. Werr, B. Olsson, F. Idali, L. Lindbom, R. Kiessling, H. Jornvall, H. Wigzell, and G. H. Gudmundsson.** 2000. The human antimicrobial and chemotactic peptides LL-37 and alpha-defensins are expressed by specific lymphocyte and monocyte populations. *Blood* **96:**3086–3093.
2. **Agerberth, B., H. Gunne, J. Odeberg, P. Kogner, H. G. Boman, and G. H. Gudmundsson.** 1995. FALL-39, a putative human peptide antibiotic, is cysteine-free and expressed in bone marrow and testis. *Proc. Natl. Acad. Sci. USA* **92:**195–199.
3. **Agerberth, B., J. Y. Lee, T. Bergman, M. Carlquist, H. G. Boman, V. Mutt, and H. Jornvall.** 1991. Amino acid sequence of PR-39. Isolation from pig intestine of a new member of the family of proline-arginine-rich antibacterial peptides. *Eur. J. Biochem.* **202:**849–854.
4. **Aho, H. J., R. Grenman, J. Sipila, H. Peuravuori, J. Hartikainen, and T. J. Nevalainen.** 1997. Group II phospholipase A2 in nasal fluid, mucosa and paranasal sinuses. *Acta Otolaryngol.* **117:**860–863.
5. **Akinbi, H. T., R. Epaud, H. Bhatt, and T. E. Weaver.** 2000. Bacterial killing is enhanced by expression of lysozyme in the lungs of transgenic mice. *J. Immunol.* **165:**5760–5766.
6. **Alcouloumre, M. S., M. A. Ghannoum, A. S. Ibrahim, M. E. Selsted, and J. E. J. Edwards.** 1993. Fungicidal properties of defensin NP-1 and activity against *Cryptococcus neoformans* in vitro. *Antimicrob. Agents Chemother.* **37:**2628–2632.
7. **Aley, S. B., M. Zimmerman, M. Hetsko, M. E. Selsted, and F. D. Gillin.** 1994. Killing of *Giardia lamblia* by cryptdins and cationic neutrophil peptides. *Infect. Immun.* **62:**5397–5403.

8. **Andersen, J. H., S. A. Osbakk, L. H. Vorland, T. Traavik, and T. J. Gutteberg.** 2001. Lactoferrin and cyclic lactoferricin inhibit the entry of human cytomegalovirus into human fibroblasts. *Antiviral Res.* **51:**141–149.
9. **Andreu, D., and L. Rivas.** 1998. Animal antimicrobial peptides: an overview. *Biopolymers* **47:**415–433.
10. **Arnold, R. R., J. E. Russell, W. J. Champion, M. Brewer, and J. J. Gauthier.** 1982. Bactericidal activity of human lactoferrin: differentiation from the stasis of iron deprivation. *Infect. Immun.* **35:**792–799.
11. **Aumelas, A., M. Mangoni, C. Roumestand, L. Chiche, E. Despaux, G. Grassy, B. Calas, and A. Chavanieu.** 1996. Synthesis and solution structure of the antimicrobial peptide protegrin-1. *Eur. J. Biochem.* **237:**575–583.
12. **Ayabe, T., D. P. Satchell, C. L. Wilson, W. C. Parks, M. E. Selsted, and A. J. Ouellette.** 2000. Secretion of microbicidal α-defensins by intestinal Paneth cells in response to bacteria. *Nat. Immunol.* **1:**113–118.
13. **Bals, R.** 2000. Epithelial antimicrobial peptides in host defense against infection. *Respir. Res.* **1:**141–150.
14. **Bals, R., X. Wang, M. Zasloff, and J. M. Wilson.** 1998. The peptide antibiotic LL-37/hCAP-18 is expressed in epithelia of the human lung where it has broad antimicrobial activity at the airway surface. *Proc. Natl. Acad. Sci. USA* **95:**9541–9546.
15. **Bals, R., D. J. Weiner, R. L. Meegalla, F. Accurso, and J. M. Wilson.** 2001. Salt-independent abnormality of antimicrobial activity in cystic fibrosis airway surface fluid. *Am. J. Respir. Cell Mol. Biol.* **25:**21–25.
16. **Bateman, A., R. J. MacLeod, P. Lembessis, J. Hu, F. Esch, and S. Solomon.** 1996. The isolation and characterization of a novel corticostatin/defensin-like peptide from the kidney. *J. Biol. Chem.* **271:**10654–10659.
17. **Becker, M. N., G. Diamond, M. W. Verghese, and S. H. Randell.** 2000. CD14-dependent lipopolysaccharide-induced beta-defensin-2 expression in human tracheobronchial epithelium. *J. Biol. Chem.* **275:**29731–29736.
18. **Bellamy, W., M. Takase, H. Wakabayashi, K. Kawase, and M. Tomita.** 1992. Antibacterial spectrum of lactoferricin B, a potent bactericidal peptide derived from the N-terminal region of bovine lactoferrin. *J. Appl. Bacteriol.* **73:**472–479.
19. **Bensch, K. W., M. Raida, H. J. Magert, P. Schulz-Knappe, and W. G. Forssmann.** 1995. hBD-1: a novel beta-defensin from human plasma. *FEBS Lett.* **368:**331–335.
20. **Billiau, A. and K. Vandenbroeck.** 2000. IFN-γ, p. 641–688. *In* J. J. Oppenheim, M. Feldmann, S. K. Duram, T. Hirano, J. Vilcek, and N. A. Nicola (ed.), *Cytokine Reference. A Compendium of Cytokines and Other Mediators of Host Defense.* Academic Press, Inc., New York, N.Y.
21. **Bingle, C. D., and C. J. Craven.** 2002. PLUNC: a novel family of candidate host defence proteins expressed in the upper airways and nasopharynx. *Hum. Mol. Genet.* **11:**937–943.
22. **Biragyn, A., P. A. Ruffini, C. A. Leifer, E. Klyushnenkova, A. Shakhov, O. Chertov, A. K. Shirakawa, J. M. Farber, D. M. Segal, J. J. Oppenheim, and L. W. Kwak.** 2002. Toll-like receptor 4-dependent activation of dendritic cells by beta-defensin 2. *Science* **298:**1025–1029.
23. **Biragyn, A., M. Surenhu, D. Yang, P. A. Ruffini, B. A. Haines, E. Klyushnenkova, J. J. Oppenheim, and L. W. Kwak.** 2001. Mediators of innate immunity that target immature, but not mature, dendritic cells induce antitumor immunity when genetically fused with nonimmunogenic tumor antigens. *J. Immunol.* **167:**6644–6653.
24. **Birchler, T., R. Seibl, K. Buchner, S. Loeliger, R. Seger, J. P. Hossle, A. Aguzzi, and R. P. Lauener.** 2001. Human Toll-like receptor 2 mediates induction of the antimicrobial peptide human beta-defensin 2 in response to bacterial lipoprotein. *Eur. J. Immunol.* **31:**3131–3137.
25. **Boman, H. G.** 1995. Peptide antibiotics and their role in innate immunity. *Annu. Rev. Immunol.* **13:**61–92.
26. **Brandtzaeg, P., T. O. Gabrielsen, I. Dale, F. Muller, M. Steinbakk, and M. K. Fagerhol.** 1995. The leucocyte protein L1 (calprotectin): a putative nonspecific defence factor at epithelial surfaces. *Adv. Exp. Med. Biol.* **371A:**201–206.
27. **Brogden, K. A., M. Ackermann, and K. M. Huttner.** 1998. Detection of anionic antimicrobial peptides in ovine bronchoalveolar lavage fluid and respiratory epithelium. *Infect. Immun.* **66:**5948–5954.
28. **Brogden, K. A., M. R. Ackermann, P. B. McCray, Jr., and K. M. Huttner.** 1999. Differences in the concentrations of small, anionic, antimicrobial peptides in bronchoalveolar lavage fluid and in respiratory epithelia of patients with and without cystic fibrosis. *Infect. Immun.* **67:**4256–4259.
29. **Bullen, J. J., and J. A. Armstrong.** 1979. The role of lactoferrin in the bactericidal function of polymorphonuclear leucocytes. *Immunology* **36:**781–791.
30. **Canny, G., O. Levy, G. T. Furuta, S. Narravula-Alipati, R. B. Sisson, C. N. Serhan, and S. P. Colgan.** 2002. Lipid mediator-induced expression of bactericidal/permeability-increasing protein (BPI) in human mucosal epithelia. *Proc. Natl. Acad. Sci. USA* **99:**3902–3907.
31. **Chang, T. L., F. Francois, A. Mosoian, and M. E. Klotman.** 2003. CAF-mediated human immunodeficiency virus (HIV) type 1 transcriptional inhibition is distinct from alpha-defensin-1 HIV inhibition. *J. Virol.* **77:**6777–6784.
32. **Chapple, D. S., C. L. Joannou, D. J. Mason, J. K. Shergill, E. W. Odell, V. Gant, and R. W. Evans.** 1998. A helical region on human lactoferrin. Its role in antibacterial pathogenesis. *Adv. Exp. Med. Biol.* **443:**215–220.
33. **Chen, F. Y., M. T. Lee, and H. W. Huang.** 2003. Evidence for membrane thinning effect as the mechanism for peptide-induced pore formation. *Biophys. J.* **84:**3751–3758.
34. **Chertov, O., D. F. Michiel, L. Xu, J. M. Wang, K. Tani, W. J. Murphy, D. L. Longo, D. D. Taub, and J. J. Oppenheim.** 1996. Identification of defensin-1, defensin-2, and CAP37/azurocidin as T-cell chemoattractant proteins released from interleukin-8-stimulated neutrophils. *J. Biol. Chem.* **271:**2935–2940.
35. **Clohessy, P. A., and B. E. Golden.** 1996. The mechanism of calprotectin's candidastatic activity appears to involve zinc chelation. *Biochem. Soc. Trans.* **24:**309S.
36. **Cohen, J.** 2002. AIDS Research: mystery anti-HIV factor unmasked? *Science* **297:**2188.
37. **Cole, A. M., P. Dewan, and T. Ganz.** 1999. Innate antimicrobial activity of nasal secretions. *Infect. Immun.* **67:**3267–3275.
38. **Cole, A. M., T. Ganz, A. M. Liese, M. D. Burdick, L. Liu, and R. M. Strieter.** 2001. Cutting edge: IFN-inducible ELR$^-$ CXC chemokines display defensin-like antimicrobial activity. *J. Immunol.* **167:**623–627.
39. **Cole, A. M., T. Hong, L. M. Boo, T. Nguyen, C. Zhao, G. Bristol, J. A. Zack, A. J. Waring, O. O. Yang, and R. I. Lehrer.** 2002. Retrocyclin: a primate peptide that protects cells from infection by T- and M-tropic strains of HIV-1. *Proc. Natl. Acad. Sci. USA* **99:**1813–1818.
40. **Cole, A. M., Y. H. Kim, S. Tahk, T. Hong, P. Weis, A. J. Waring, and T. Ganz.** 2001. Calcitermin, a novel antimicrobial peptide isolated from human airway secretions. *FEBS Lett.* **504:**5–10.
41. **Cole, A. M., H. I. Liao, O. Stuchlik, J. Tilan, J. Pohl, and T. Ganz.** 2002. Cationic polypeptides are required for antibacterial activity of human airway fluid. *J. Immunol.* **169:** 6985–6991.
42. **Cole, A. M., J. Shi, A. Ceccarelli, Y.-H. Kim, A. Park, and T. Ganz.** 2001. Inhibition of neutrophil elastase prevents

cathelicidin activation and impairs clearance of bacteria from wounds. *Blood* **97**:297–304.

43. **Cowland, J. B., A. H. Johnsen, and N. Borregaard.** 1995. hCAP-18, a cathelin/pro-bactenecin-like protein of human neutrophil specific granules. *FEBS Lett.* **368**:173–176.
44. **Daher, K. A., M. E. Selsted, and R. I. Lehrer.** 1986. Direct inactivation of viruses by human granulocyte defensins. *J. Virol.* **60**:1068–1074.
45. **Diamond, G., M. Zasloff, H. Eck, M. Brasseur, W. L. Maloy, and C. L. Bevins.** 1991. Tracheal antimicrobial peptide, a cysteine-rich peptide from mammalian tracheal mucosa: peptide isolation and cloning of a cDNA. *Proc. Natl. Acad. Sci. USA* **88**:3952–3956.
46. **Dorschner, R. A., V. K. Pestonjamasp, S. Tamakuwala, T. Ohtake, J. Rudisill, V. Nizet, B. Agerberth, G. H. Gudmundsson, and R. L. Gallo.** 2001. Cutaneous injury induces the release of cathelicidin anti-microbial peptides active against group A *Streptococcus*. *J. Investig. Dermatol.* **117**: 91–97.
47. **Duits, L. A., P. H. Nibbering, E. van Strijen, J. B. Vos, S. P. Mannesse-Lazeroms, M. A. van Sterkenburg, and P. S. Hiemstra.** 2003. Rhinovirus increases human beta-defensin-2 and -3 mRNA expression in cultured bronchial epithelial cells. *FEMS Immunol. Med. Microbiol.* **38**:59–64.
48. **Duits, L. A., B. Ravensbergen, M. Rademaker, P. S. Hiemstra, and P. H. Nibbering.** 2002. Expression of beta-defensin 1 and 2 mRNA by human monocytes, macrophages and dendritic cells. *Immunology* **106**:517–525.
49. **Dunsmore, S. E., U. K. Saarialho-Kere, J. D. Roby, C. L. Wilson, L. M. Matrisian, H. G. Welgus, and W. C. Parks.** 1998. Matrilysin expression and function in airway epithelium. *J. Clin. Investig.* **102**:1321–1331.
50. **Edgerton, M., and S. E. Koshlukova.** 2000. Salivary histatin 5 and its similarities to the other antimicrobial proteins in human saliva. *Adv. Dent. Res.* **14**:16–21.
51. **Ellison, R. T., III, and T. J. Giehl.** 1991. Killing of Gram-negative bacteria by lactoferrin and lysozyme. *J. Clin. Investig.* **88**:1080–1091.
52. **Elsbach, P., and J. Weiss.** 1993. The bactericidal/permeability-increasing protein (BPI), a potent element in host-defense against gram-negative bacteria and lipopolysaccharide. *Immunobiology* **187**:417–429.
53. **Ernst, R. K., T. Guina, and S. I. Miller.** 2001. *Salmonella typhimurium* outer membrane remodeling: role in resistance to host innate immunity. *Microbes Infect.* **3**:1327–1334.
54. **Fahrner, R. l., T. Dieckmann, S. S. Harwig, R. I. Lehrer, D. Eisenberg, and J. Feigon.** 1996. Solution structure of protegrin-1, a broad-spectrum antimicrobial peptide from porcine leukocytes. *Chem. Biol.* **3**:543–550.
55. **Fellermann, K., J. Wehkamp, K. R. Herrlinger, and E. F. Stange.** 2003. Crohn's disease: a defensin deficiency syndrome? *Eur. J. Gastroenterol. Hepatol.* **15**:627–634.
56. **Fleming, A.** 1922. On a remarkable bacteriolytic element found in tissues and secretions. *Proc. R. Soc. Lond. Ser. B* **93**:306–317.
57. **Forst, S., J. Weiss, J. M. Maraganore, R. L. Heinrikson, and P. Elsbach.** 1987. Relation between binding and the action of phospholipases A2 on *Escherichia coli* exposed to the bactericidal/permeability-increasing protein of neutrophils. *Biochim. Biophys. Acta* **920**:221–225.
58. **Gallo, R. L., K. J. Kim, M. Bernfield, C. A. Kozak, M. Zanetti, L. Merluzzi, and R. Gennaro.** 1997. Identification of CRAMP, a cathelin-related antimicrobial peptide expressed in the embryonic and adult mouse. *J. Biol. Chem.* **272**: 13088–13093.
59. **Ganz, T.** 1999. Defensins and host defense. *Science* **286**:420–421.
60. **Ganz, T.** 2001. Defensins in the urinary tract and other tissues. *J. Infect. Dis.* **183** (Suppl. 1):S41–S42.
61. **Ganz, T., V. Gabayan, H. I. Liao, L. Liu, A. Oren, T. Graf, and A. M. Cole.** 2002. Increased inflammation in lysozyme M-deficient mice in response to *Micrococcus luteus* and its peptidoglycan. *Blood* **101**:2388–2392.
62. **Ganz, T., and R. I. Lehrer.** 1994. Defensins. *Curr. Opin. Immunol.* **6**:584–589.
63. **Ganz, T., and R. I. Lehrer.** 1997. Antimicrobial peptides of leukocytes. *Curr. Opin. Hematol.* **4**:53–58.
64. **Ganz, T., J. A. Metcalf, J. I. Gallin, L. A. Boxer, and R. I. Lehrer.** 1988. Microbicidal/cytotoxic proteins of neutrophils are deficient in two disorders: Chediak-Higashi syndrome and "specific" granule deficiency. *J. Clin. Investig.* **82**:552–556.
65. **Ganz, T., M. E. Selsted, D. Szklarek, S. S. Harwig, K. Daher, D. F. Bainton, and R. I. Lehrer.** 1985. Defensins. Natural peptide antibiotics of human neutrophils. *J. Clin. Investig.* **76**:1427–1435.
66. **Garcia, J. R., F. Jaumann, S. Schulz, A. Krause, J. Rodriguez-Jimenez, U. Forssmann, K. Adermann, E. Kluver, C. Vogelmeier, D. Becker, R. Hedrich, W. G. Forssmann, and R. Bals.** 2001. Identification of a novel, multifunctional beta-defensin (human beta-defensin 3) with specific antimicrobial activity. Its interaction with plasma membranes of *Xenopus* oocytes and the induction of macrophage chemoattraction. *Cell Tissue Res.* **306**:257–264.
67. **Gazit, E., I. R. Miller, P. C. Biggin, M. S. Sansom, and Y. Shai.** 1996. Structure and orientation of the mammalian antibacterial peptide cecropin P1 within phospholipid membranes. *J. Mol. Biol.* **258**:860–870.
68. **Ghosh, D., E. Porter, B. Shen, S. K. Lee, D. Wilk, J. Drazba, S. P. Yadav, J. W. Crabb, T. Ganz, and C. L. Bevins.** 2002. Paneth cell trypsin is the processing enzyme for human defensin-5. *Nat. Immunol.* **3**:583–590.
69. **Goldman, M., G. Anderson, E. D. Stolzenberg, U. P. Kari, M. Zasloff, and J. M. Wilson.** 1997. Human β-defensin-1 is a salt-sensitive antibiotic in lung that is inactivated in cystic fibrosis. *Cell* **88**:553–560.
70. **Gombart, A. F., M. Shiohara, S. H. Kwok, K. Agematsu, A. Komiyama, and H. P. Koeffler.** 2001. Neutrophil-specific granule deficiency: homozygous recessive inheritance of a frameshift mutation in the gene encoding transcription factor CCAAT/enhancer binding protein–epsilon. *Blood* **97**:2561–2567.
71. **Gregory, R. J., S. H. Cheng, D. P. Rich, J. Marshall, S. Paul, K. Hehir, L. Ostedgaard, K. W. Klinger, M. J. Welsh, and A. E. Smith.** 1990. Expression and characterization of the cystic fibrosis transmembrane conductance regulator. *Nature* **347**:382–386.
72. **Groisman, E. A.** 2001. The pleiotropic two-component regulatory system PhoP-PhoQ. *J. Bacteriol.* **183**:1835–1842.
73. **Gronroos, J. O., V. J. Laine, and T. J. Nevalainen.** 2002. Bactericidal group IIA phospholipase A2 in serum of patients with bacterial infections. *J. Infect. Dis.* **185**:1767–1772.
74. **Gudmundsson, G. H., and B. Agerberth.** 1999. Neutrophil antibacterial peptides, multifunctional effector molecules in the mammalian immune system. *J. Immunol. Methods* **232**: 45–54.
75. **Hancock, R. E.** 1997. Peptide antibiotics. *Lancet* **349**:418–422.
76. **Hancock, R. E., and A. Patrzykat.** 2002. Clinical development of cationic antimicrobial peptides: from natural to novel antibiotics. *Curr. Drug Targets Infect. Disord.* **2**:79–83.
77. **Harder, J., J. Bartels, E. Christophers, and J. M. Schroder.** 2001. Isolation and characterization of human beta-defensin-3, a novel human inducible peptide antibiotic. *J. Biol. Chem.* **276**:5707–5713.

78. **Harder, J., J. Bartels, E. Christophers, and J.-M. Schroeder.** 1997. A peptide antibiotic from human skin. *Nature* **387:** 861–862.
79. **Harmsen, M. C., P. J. Swart, M. P. de Bethune, R. Pauwels, E. De Clercq, T. H. The, and D. K. Meijer.** 1995. Antiviral effects of plasma and milk proteins: lactoferrin shows potent activity against both human immunodeficiency virus and human cytomegalovirus replication in vitro. *J. Infect. Dis.* **172:** 380–388.
80. **Harwig, S. S., V. N. Kokryakov, K. M. Swiderek, G. M. Aleshina, C. Zhao, and R. I. Lehrer.** 1995. Prophenin-1, an exceptionally proline-rich antimicrobial peptide from porcine leukocytes. *FEBS Lett.* **362:**65–69.
81. **Harwig, S. S., A. S. Park, and R. I. Lehrer.** 1992. Characterization of defensin precursors in mature human neutrophils. *Blood* **79:**1532–1537.
82. **Harwig, S. S. L., X.-D. Qu, L. Tan, and R. I. Lehrer.** 1994. Small intestinal Paneth cells contain a potently antimicrobial group II phospholipase A-2. *FASEB J.* **8:**A516.
83. **Hasegawa, K., W. Motsuchi, S. Tanaka, and S. Dosako.** 1994. Inhibition with lactoferrin of in vitro infection with human herpes virus. *Jpn. J. Med. Sci. Biol.* **47:**73–85.
84. **Hein, M., E. V. Valore, R. B. Helmig, N. Uldbjerg, and T. Ganz.** 2002. Antimicrobial factors in the cervical mucus plug. *Am. J. Obstet. Gynecol.* **187:**137–144.
85. **Hertz, C. J., Q. Wu, E. M. Porter, Y. J. Zhang, K. H. Weismuller, P. J. Godowski, T. Ganz, S. H. Randell, and R. L. Modlin.** 2003. Activation of Toll-like receptor 2 on human tracheobronchial epithelial cells induces the antimicrobial peptide human beta defensin-2. *J. Immunol.* **171:**6820–6826.
86. **Hiemstra, P. S., R. J. Maassen, J. Stolk, R. Heinzel-Wieland, G. J. Steffens, and J. H. Dijkman.** 1996. Antibacterial activity of antileukoprotease. *Infect. Immun.* **64:**4520–4524.
87. **Higazi, A. A., I. I. Barghouti, and R. Abu-Much.** 1995. Identification of an inhibitor of tissue-type plasminogen activator-mediated fibrinolysis in human neutrophils. A role for defensin. *J. Biol. Chem.* **270:**9472–9477.
88. **Higazi, A. A. R., T. Ganz, K. Kariko, and D. B. Cines.** 1996. Defensin modulates tissue-type plasminogen activator and plasminogen binding to fibrin and endothelial cells. *J. Biol. Chem.* **271:**17650–17655.
89. **Hill, C. P., J. Yee, M. E. Selsted, and D. Eisenberg.** 1991. Crystal structure of defensin HNP-3, an amphiphilic dimer: mechanisms of membrane permeabilization. *Science* **251:** 1481–1485.
90. **Hiratsuka, T., M. Nakazato, Y. Date, J. Ashitani, T. Minematsu, N. Chino, and S. Matsukura.** 1998. Identification of human beta-defensin-2 in respiratory tract and plasma and its increase in bacterial pneumonia. *Biochem. Biophys. Res. Commun.* **249:**943–947.
91. **Hobbs, J. A., R. May, K. Tanousis, E. McNeill, M. Mathies, C. Gebhardt, R. Henderson, M. J. Robinson, and N. Hogg.** 2003. Myeloid cell function in MRP-14 (S100A9) null mice. *Mol. Cell. Biol.* **23:**2564–2576.
92. **Hoover, D. M., C. Boulegue, D. Yang, J. J. Oppenheim, K. Tucker, W. Lu, and J. Lubkowski.** 2002. The structure of human macrophage inflammatory protein-3alpha /CCL20. Linking antimicrobial and CC chemokine receptor-6-binding activities with human beta-defensins. *J. Biol. Chem.* **277:** 37647–37654.
93. **Hoover, D. M., O. Chertov, and J. Lubkowski.** 2001. The structure of human beta-defensin-1: new insights into structural properties of beta-defensins. *J. Biol. Chem.* **276:**39021–39026.
94. **Hoover, D. M., K. R. Rajashankar, R. Blumenthal, A. Puri, J. J. Oppenheim, O. Chertov, and J. Lubkowski.** 2000. The structure of human beta-defensin-2 shows evidence of higher order oligomerization. *J. Biol. Chem.* **275:**32911–32918.
95. **Hughes, A. L.** 1999. Evolutionary diversification of the mammalian defensins. *Cell Mol. Life Sci.* **56:**94–103.
96. **Ibrahim, H. R., T. Matsuzaki, and T. Aoki.** 2001. Genetic evidence that antibacterial activity of lysozyme is independent of its catalytic function. *FEBS Lett.* **506:**27–32.
97. **Islam, D., L. Bandholtz, J. Nilsson, H. Wigzell, B. Christensson, B. Agerberth, and G. Gudmundsson.** 2001. Downregulation of bactericidal peptides in enteric infections: a novel immune escape mechanism with bacterial DNA as a potential regulator. *Nat. Med.* **7:**180–185.
98. **Janeway, C. A., Jr., and R. Medzhitov.** 2002. Innate immune recognition. *Annu. Rev. Immunol.* **20:**197–216.
99. **Johansson, J., G. H. Gudmundsson, M. E. Rottenberg, K. D. Berndt, and B. Agerberth.** 1998. Conformation-dependent antibacterial activity of the naturally occurring human peptide LL-37. *J. Biol. Chem.* **273:**3718–3724.
100. **Joiner, K. A., T. Ganz, J. Albert, and D. Rotrosen.** 1989. The opsonizing ligand on *Salmonella typhimurium* influences incorporation of specific, but not azurophil, granule constituents into neutrophil phagosomes. *J. Cell Biol.* **109:**2771–2782.
101. **Jones, D. E., and C. L. Bevins.** 1992. Paneth cells of the human small intestine express an antimicrobial peptide gene. *J. Biol. Chem.* **267:**23216–23225.
102. **Jones, D. E., and C. L. Bevins.** 1993. Defensin-6 mRNA in human Paneth cells: implications for antimicrobial peptides in host defense of the human bowel. *FEBS Lett.* **315:**187–192.
103. **Kalfa, V. C., and K. A. Brogden.** 1999. Anionic antimicrobial peptide-lysozyme interactions in innate pulmonary immunity. *Int. J. Antimicrob. Agents* **13:**47–51.
104. **Kaliner, M. A.** 1991. Human nasal respiratory secretions and host defense. *Am. Rev. Respir. Dis.* **144:**S52–S56.
105. **Kaliner, M. A.** 1992. Human nasal host defense and sinusitis. *J. Allergy Clin. Immunol.* **90:**424–430.
106. **Knowles, M. R., J. M. Robinson, R. E. Wood, C. A. Pue, W. M. Mentz, G. C. Wager, J. T. Gatzy, and R. C. Boucher.** 1997. Ion composition of airway surface liquid of patients with cystic fibrosis as compared with normal and disease-control subjects. *J. Clin. Investig.* **100:**2588–2595.
107. **Kokryakov, V. N., S. S. Harwig, E. A. Panyutich, A. A. Shevchenko, G. M. Aleshina, O. V. Shamova, H. A. Korneva, and R. I. Lehrer.** 1993. Protegrins: leukocyte antimicrobial peptides that combine features of corticostatic defensins and tachyplesins. *FEBS Lett.* **327:**231–236.
108. **LaBranche, C. C., G. Galasso, J. P. Moore, D. P. Bolognesi, M. S. Hirsch, and S. M. Hammer.** 2001. HIV fusion and its inhibition. *Antiviral Res.* **50:**95–115.
109. **Laible, N. J., and G. R. Germaine.** 1985. Bactericidal activity of human lysozyme, muramidase-inactive lysozyme, and cationic polypeptides against *Streptococcus sanguis* and *Streptococcus faecalis:* inhibition by chitin oligosaccharides. *Infect. Immun.* **48:**720–728.
110. **Larrick, J. W., M. Hirata, R. F. Balint, J. Lee, J. Zhong, and S. C. Wright.** 1995. Human CAP18: a novel antimicrobial lipopolysaccharide-binding protein. *Infect. Immun.* **63:**1291–1297.
111. **Lauw, F. N., A. J. H. Simpson, J. M. Prins, S. J. H. van Deventer, W. Chaowagul, N. J. White, and T. van der Poll.** 2000. The CXC chemokines gamma interferon (IFN-γ)-inducible protein 10 and monokine induced by IFN-γ are released during severe melioidosis. *Infect. Immun.* **68:**3888–3893.
112. **Lawrance, I. C., C. Fiocchi, and S. Chakravarti.** 2001. Ulcerative colitis and Crohn's disease: distinctive gene expression

profiles and novel susceptibility candidate genes. *Hum. Mol. Genet.* **10:**445–456.

113. **Lee-Huang, S., P. L. Huang, Y. Sun, P. L. Huang, H. F. Kung, D. L. Blithe, and H. C. Chen.** 1999. Lysozyme and RNases as anti-HIV components in beta-core preparations of human chorionic gonadotropin. *Proc. Natl. Acad. Sci. USA* **96:** 2678–2681.
114. **Lehrer, R. I., A. Barton, K. A. Daher, S. S. Harwig, T. Ganz, and M. E. Selsted.** 1989. Interaction of human defensins with *Escherichia coli.* Mechanism of bactericidal activity. *J. Clin. Investig.* **84:**553–561.
115. **Lehrer, R. I., C. L. Bevins, and T. Ganz.** 1999. Defensins and other antimicrobial peptides, p. 89–99. *In* P. L. Ogra, J. Mestecky, M. E. Lamm, W. Strober, J. Bienenstock, and J. R. McGhee (ed.), *Mucosal Immunology* Academic Press, Inc., San Diego, Calif.
116. **Lehrer, R. I., K. Daher, T. Ganz, and M. E. Selsted.** 1985. Direct inactivation of viruses by MCP-1 and MCP-2, natural peptide antibiotics from rabbit leukocytes. *J. Virol.* **54:**467–472.
117. **Lehrer, R. I., and T. Ganz.** 2002. Defensins of vertebrate animals. *Curr. Opin. Immunol.* **14:**96–102.
118. **Lehrer, R. I., T. Ganz, D. Szklarek, and M. E. Selsted.** 1988. Modulation of the in vitro candidacidal activity of human neutrophil defensins by target cell metabolism and divalent cations. *J. Clin. Investig.* **81:**1829–1835.
119. **Lehrer, R. I., D. Szklarek, T. Ganz, and M. E. Selsted.** 1985. Correlation of binding of rabbit granulocyte peptides to *Candida albicans* with candidacidal activity. *Infect. Immun.* **49:**207–211.
120. **Leonova, L., V. N. Kokryakov, G. M. Aleshina, T. Hong, T. Nguyen, C. Zhao, A. J. Waring, and R. I. Lehrer.** 2001. Circular minidefensins and posttranslational generation of molecular diversity. *J. Leukoc. Biol.* **70:**461–464.
121. **Levy, J. A.** 2001. The importance of the innate immune system in controlling HIV infection and disease. *Trends Immunol.* **22:**312–316.
122. **Levy, O., S. Martin, E. Eichenwald, T. Ganz, E. Valore, S. F. Carroll, K. Lee, D. Goldmann, and G. M. Thorne.** 1999. Impaired innate immunity in the newborn: newborn neutrophils are deficient in bactericidal/permeability-increasing protein. *Pediatrics* **104:**1327–1333.
123. **Lillard, J. W., Jr., P. N. Boyaka, O. Chertov, J. J. Oppenheim, and J. R. McGhee.** 1999. Mechanisms for induction of acquired host immunity by neutrophil peptide defensins. *Proc. Natl. Acad. Sci. USA* **96:**651–656.
124. **Liu, L., C. Zhao, H. H. Q. Heng, and T. Ganz.** 1997. The human β-defensin-1 and α-defensins are encoded by adjacent genes: two peptide families with differing disulfide topology share a common ancestry. *Genomics* **43:**316–320.
125. **Lomax, K. J., J. I. Gallin, D. Rotrosen, G. D. Raphael, M. A. Kaliner, E. J. Jr. Benz, L. A. Boxer, and H. L. Malech.** 1989. Selective defect in myeloid cell lactoferrin gene expression in neutrophil specific granule deficiency. *J. Clin. Investig.* **83:**514–519.
126. **Ludtke, S. J., K. He, W. T. Heller, T. A. Harroun, L. Yang, and H. W. Huang.** 1996. Membrane pores induced by magainin. *Biochemistry* **35:**13723–13728.
127. **Mackewicz, C. E., J. Yuan, P. Tran, L. Diaz, E. Mack, M. E. Selsted, and J. A. Levy.** 2003. α-Defensins can have anti-HIV activity but are not CD8 cell anti-HIV factors. *AIDS* **17:**F23–F32.
128. **MacLeod, R. J., J. R. Hamilton, A. Bateman, D. Belcourt, J. Hu, H. P. Bennett, and S. Solomon.** 1991. Corticostatic peptides cause nifedipine-sensitive volume reduction in jejunal villus enterocytes. *Proc. Natl. Acad. Sci. USA* **88:**552–556.
129. **Mangoni, M. E., A. Aumelas, P. Charnet, C. Roumestand, L. Chiche, E. Despaux, G. Grassy, B. Calas, and A. Chavanieu.** 1996. Change in membrane permeability induced by protegrin 1: implication of disulphide bridges for pore formation. *FEBS Lett.* **383:**93–98.
130. **Manitz, M. P., B. Horst, S. Seeliger, A. Strey, B. V. Skryabin, M. Gunzer, W. Frings, F. Schonlau, J. Roth, C. Sorg, and W. Nacken.** 2003. Loss of S100A9 (MRP14) results in reduced interleukin-8-induced CD11b surface expression, a polarized microfilament system, and diminished responsiveness to chemoattractants in vitro. *Mol. Cell. Biol.* **23:**1034–1043.
131. **Martin, E., T. Ganz, and R. I. Lehrer.** 1995. Defensins and other endogenous peptide antibiotics of vertebrates. *J. Leukoc. Biol.* **58:**128–136.
132. **Mathews, M., H. P. Jia, J. M. Guthmiller, G. Losh, S. Graham, G. K. Johnson, B. F. Tack, and P. B. McCray, Jr.** 1999. Production of b-defensin antimicrobial peptides by the oral mucosa and salivary glands. *Infect. Immun.* **67:**2740–2745.
133. **Matsui, H., B. R. Grubb, R. Tarran, S. H. Randell, J. T. Gatzy, C. W. Davis, and R. C. Boucher.** 1998. Evidence for periciliary liquid layer depletion, not abnormal ion composition, in the pathogenesis of cystic fibrosis airways disease. *Cell* **95:**1005–1015.
134. **Matsuzaki, K.** 2001. Why and how are peptide-lipid interactions utilized for self defence? *Biochem. Soc. Trans.* **29:**598–601.
135. **Maxwell, A. I., G. M. Morrison, and J. R. Dorin.** 2003. Rapid sequence divergence in mammalian beta-defensins by adaptive evolution. *Mol. Immunol.* **40:**413–421.
136. **McCray, P. B., and L. Bentley.** 1997. Human airway epithelia express a β-defensin. *Am. J. Respir. Cell Mol. Biol.* **16:**343–349.
137. **Meredith, S. D., G. D. Raphael, J. N. Baraniuk, S. M. Banks, and M. A. Kaliner.** 1989. The pathophysiology of rhinitis. III. The control of IgG secretion. *J. Allergy Clin. Immunol.* **84:**920–930.
138. **Milner, S. M., and M. R. Ortega.** 1999. Reduced antimicrobial peptide expression in human burn wounds. *Burns* **25:**411–413.
139. **Miyasaki, K. T., A. L. Bodeau, T. Ganz, M. E. Selsted, and R. I. Lehrer.** 1990. In vitro sensitivity of oral, gram-negative, facultative bacteria to the bactericidal activity of human neutrophil defensins. *Infect. Immun.* **58:**3934–3940.
140. **Monell, C. R., and M. Strand.** 1994. Structural and functional similarities between synthetic HIV gp41 peptides and defensins. *Clin. Immunol. Immunopathol.* **71:**315–324.
141. **Morrison, G. M., D. J. Davidson, and J. R. Dorin.** 1999. A novel mouse beta defensin, Defb2, which is upregulated in the airways by lipopolysaccharide. *FEBS Lett.* **442:**112–116.
142. **Münk, C., G. Wei, O. O. Yang, A. J. Waring, W. Wang, T. Hong, R. I. Lehrer, N. R. Landau, and A. M. Cole.** 2003. The theta-defensin, retrocyclin, inhibits HIV-1 entry. *AIDS Res. Hum. Retroviruses* **19:**875–881.
143. **Murphy, C. J., B. A. Foster, M. J. Mannis, M. E. Selsted, and T. W. Reid.** 1993. Defensins are mitogenic for epithelial cells and fibroblasts. *J. Cell. Physiol.* **155:**408–413.
144. **Nevalainen, T. J., H. J. Aho, and H. Peuravuori.** 1994. Secretion of group 2 phospholipase A2 by lacrimal glands. *Investig. Ophthalmol. Visual Sci.* **35:**417–421.
145. **Nguyen, T. X., A. M. Cole, and R. I. Lehrer.** 2003. Evolution of primate theta-defensins: a serpentine path to a sweet tooth. *Peptides* **24:**1647–1654.
146. **Niyonsaba, F., K. Iwabuchi, H. Matsuda, H. Ogawa, and I. Nagaoka.** 2002. Epithelial cell-derived human beta-defensin-2 acts as a chemotaxin for mast cells through a

pertussis toxin-sensitive and phospholipase C-dependent pathway. *Int. Immunol.* **14**:421–426.

147. **Niyonsaba, F., A. Someya, M. Hirata, H. Ogawa, and I. Nagaoka.** 2001. Evaluation of the effects of peptide antibiotics human beta-defensins-1/-2 and LL-37 on histamine release and prostaglandin D(2) production from mast cells. *Eur. J. Immunol.* **31**:1066–1075.
148. **Nizet, V., T. Ohtake, X. Lauth, J. Trowbridge, J. Rudisill, R. A. Dorschner, V. Pestonjamasp, J. Piraino, K. Huttner, and R. L. Gallo.** 2001. Innate antimicrobial peptide protects the skin from invasive bacterial infection. *Nature* **414**:454–457.
149. **Ogata, K., B. A. Linzer, R. I. Zuberi, T. Ganz, R. I. Lehrer, and A. Catanzaro.** 1992. Activity of defensins from human neutrophilic granulocytes against *Mycobacterium avium-Mycobacterium intracellulare. Infect. Immun.* **60**:4720–4725.
150. **O'Neil, D. A., E. M. Porter, D. Elewaut, G. M. Anderson, L. Eckmann, T. Ganz, and M. F. Kagnoff.** 1999. Expression and regulation of the human beta-defensins hBD-1 and hBD-2 in intestinal epithelium. *J. Immunol.* **163**:6718–6724.
151. **Oppenheim, F. G., T. Xu, F. M. McMillian, S. M. Levitz, R. D. Diamond, G. D. Offner, and R. F. Troxler.** 1988. Histatins, a novel family of histidine-rich proteins in human parotid secretion. Isolation, characterization, primary structure, and fungistatic effects on *Candida albicans. J. Biol. Chem.* **263**:7472–7477.
152. **Oren, A., T. Ganz, L. Liu, and T. Meerloo.** 2003. In human epidermis, [beta]-defensin 2 is packaged in lamellar bodies. *Exp. Mol. Pathol.* **74**:180–182.
153. **Ortega, M. R., T. Ganz, and S. M. Milner.** 2000. Human beta defensin is absent in burn blister fluid. *Burns* **26**:724–726.
154. **Ouellette, A. J., and C. L. Bevins.** 2001. Paneth cell defensins and innate immunity of the small bowel. *Inflamm. Bowel Dis.* **7**:43–50.
155. **Ouellette, A. J., and M. E. Selsted.** 1996. Paneth cell defensins: endogenous peptide components of intestinal host defense. *FASEB J.* **10**:1280–1289.
156. **Panyutich, A., and T. Ganz.** 1991. Activated alpha 2-macroglobulin is a principal defensin-binding protein. *Am. J. Respir. Cell Mol. Biol.* **5**:101–106.
157. **Panyutich, A., J. Shi, P. L. Boutz, C. Zhao, and T. Ganz.** 1997. Porcine polymorphonuclear leukocytes generate extracellular microbicidal activity by elastase-mediated activation of secreted proprotegrins. *Infect. Immun.* **65**:978–985.
158. **Panyutich, A. V., P. S. Hiemstra, S. Van Wetering, and T. Ganz.** 1995. Human neutrophil defensin and serpins form complexes and inactivate each other. *Am. J. Respir. Cell Mol. Biol.* **12**:351–357.
159. **Panyutich, A. V., O. Szold, P. H. Poon, Y. Tseng, and T. Ganz.** 1994. Identification of defensin binding to C1 complement. *FEBS Lett.* **356**:169–173.
160. **Perez-Canadillas, J. M., A. Zaballos, J. Gutierrez, R. Varona, F. Roncal, J. P. Albar, G. Marquez, and M. Bruix.** 2001. NMR solution structure of murine CCL20/MIP-3alpha, a chemokine that specifically chemoattracts immature dendritic cells and lymphocytes through its highly specific interaction with the beta-chemokine receptor CCR6. *J. Biol. Chem.* **276**:28372–28379.
161. **Peschel, A., R. W. Jack, M. Otto, L. V. Collins, P. Staubitz, G. Nicholson, H. Kalbacher, W. F. Nieuwenhuizen, G. Jung, A. Tarkowski, K. P. van Kessel, and J. A. van Strijp.** 2001. *Staphylococcus aureus* resistance to human defensins and evasion of neutrophil killing via the novel virulence factor MprF is based on modification of membrane lipids with l-lysine. *J. Exp. Med.* **193**:1067–1076.
162. **Peschel, A., M. Otto, R. W. Jack, H. Kalbacher, G. Jung, and F. Gotz.** 1999. Inactivation of the *dlt* operon in *Staphylococcus aureus* confers sensitivity to defensins, protegrins, and other antimicrobial peptides. *J. Biol. Chem.* **274**:8405–8410.
163. **Pestonjamasp, V. K., K. H. Huttner, and R. L. Gallo.** 2001. Processing site and gene structure for the murine antimicrobial peptide CRAMP. *Peptides* **22**:1643–1650.
164. **Puddu, P., P. Borghi, S. Gessani, P. Valenti, F. Belardelli, and L. Seganti.** 1998. Antiviral effect of bovine lactoferrin saturated with metal ions on early steps of human immunodeficiency virus type 1 infection. *Int. J. Biochem. Cell Biol.* **30**:1055–1062.
165. **Qu, X. D., K. C. Lloyd, J. H. Walsh, and R. I. Lehrer.** 1996. Secretion of type II phospholipase A2 and cryptdin by rat small intestinal Paneth cells. *Infect. Immun.* **64**:5161–5165.
166. **Quinones-Mateu, M. E., M. M. Lederman, Z. Feng, B. Chakraborty, J. Weber, H. R. Rangel, M. L. Marotta, M. Mirza, B. Jiang, P. Kiser, K. Medvik, S. F. Sieg, and A. Weinberg.** 2003. Human epithelial beta-defensins 2 and 3 inhibit HIV-1 replication. *AIDS* **17**:F39–F48.
167. **Raphael, G. D., J. L. Davis, P. C. Fox, H. L. Malech, J. I. Gallin, J. N. Baraniuk, and M. A. Kaliner.** 1989. Glandular secretion of lactoferrin in a patient with neutrophil lactoferrin deficiency. *J. Allergy Clin. Immunol.* **84**:914–919.
168. **Raphael, G. D., E. V. Jeney, J. N. Baraniuk, I. Kim, S. D. Meredith, and M. A. Kaliner.** 1989. Pathophysiology of rhinitis. Lactoferrin and lysozyme in nasal secretions. *J. Clin. Investig.* **84**:1528–1535.
169. **Revenis, M. E., and M. A. Kaliner.** 1992. Lactoferrin and lysozyme deficiency in airway secretions: association with the development of bronchopulmonary dysplasia. *J. Pediatr.* **121**:262–270.
170. **Rothenberg, R. B., M. Scarlett, C. del Rio, D. Reznik, and C. O'Daniels.** 1998. Oral transmission of HIV. *AIDS* **12**: 2095–2105.
171. **Salzman, N. H., D. Ghosh, K. M. Huttner, Y. Paterson, and C. L. Bevins.** 2003. Protection against enteric salmonellosis in transgenic mice expressing a human intestinal defensin. *Nature* **422**:522–526.
172. **Samaranayake, Y. H., L. P. Samaranayake, P. C. Wu, and M. So.** 1997. The antifungal effect of lactoferrin and lysozyme on *Candida krusei* and *Candida albicans. APMIS* **105**:875–883.
173. **Satoh, Y.** 1988. Effect of live and heat-killed bacteria on the secretory activity of Paneth cells in germ-free mice. *Cell Tissue Res.* **251**:87–93.
174. **Satoh, Y., K. Ishikawa, Y. Oomori, S. Takeda, and K. Ono.** 1992. Bethanechol and a G-protein activator, NaF/AlCl3, induce secretory response in Paneth cells of mouse intestine. *Cell Tissue Res.* **269**:213–220.
175. **Sauty, A., M. Dziejman, R. A. Taha, A. S. Iarossi, K. Neote, E. A. Garcia-Zepeda, Q. Hamid, and A. D. Luster.** 1999. The T cell-specific CXC chemokines IP-10, Mig, and I-TAC are expressed by activated human bronchial epithelial cells. *J. Immunol.* **162**:3549–3558.
176. **Sawai, M. V., H. P. Jia, L. Liu, V. Aseyev, J. M. Wiencek, P. B. McCray, Jr., T. Ganz, W. R. Kearney, and B. F. Tack.** 2001. The NMR structure of human beta-defensin-2 reveals a novel alpha-helical segment. *Biochemistry* **40**:3810–3816.
177. **Schnapp, D., and A. Harris.** 1998. Antibacterial peptides in bronchoalveolar lavage fluid. *Am. J. Respir. Cell Mol. Biol.* **19**:352–356.
178. **Schonwetter, B. S., E. D. Stolzenberg, and M. A. Zasloff.** 1995. Epithelial antibiotics induced at sites of inflammation. *Science* **267**:1645–1648.

179. **Schutte, B. C., J. P. Mitros, J. A. Bartlettt, J. D. Walters, H. P. Jia, M. J. Welsh, T. L. Casavant, and P. B. McCray.** 2002. Discovery of five conserved beta-defensin gene clusters using a computational search strategy. *Proc. Nat. Acad. Sci. USA* **99:**2129–2133.

180. **Scocchi, M., B. Skerlavaj, D. Romeo, and R. Gennaro.** 1992. Proteolytic cleavage by neutrophil elastase converts inactive storage proforms to antibacterial bactenecins. *Eur. J. Biochem.* **209:**589–595.

181. **Selsted, M. E., and S. S. Harwig.** 1987. Purification, primary structure, and antimicrobial activities of a guinea pig neutrophil defensin. *Infect. Immun.* **55:**2281–2286.

182. **Selsted, M. E., D. Szklarek, and R. I. Lehrer.** 1984. Purification and antibacterial activity of antimicrobial peptides of rabbit granulocytes. *Infect. Immun.* **45:**150–154.

183. **Selsted, M. E., Y. Q. Tang, W. L. Morris, P. A. McGuire, M. J. Novotny, W. Smith, A. H. Henschen, and J. S. Cullor.** 1993. Purification, primary structures, and antibacterial activities of beta-defensins, a new family of antimicrobial peptides from bovine neutrophils. *J. Biol. Chem.* **268:**6641–6648.

184. **Shai, Y.** 1999. Mechanism of the binding, insertion and destabilization of phospholipid bilayer membranes by alpha-helical antimicrobial and cell non-selective membrane-lytic peptides. *Biochim. Biophys. Acta* **1462:**55–70.

185. **Shi, J., and T. Ganz.** 1998. The role of protegrins and other elastase-activated polypeptides in the bactericidal properties of porcine inflammatory fluids. *Infect. Immun.* **66:**3611–3617.

186. **Shi, J., G. Zhang, H. Wu, C. Ross, F. Blecha, and T. Ganz.** 1999. Porcine epithelial beta-defensin 1 is expressed in the dorsal tongue at antimicrobial concentrations. *Infect. Immun.* **67:**3121–3127.

187. **Shugars, D. C., A. L. Alexander, K. Fu, and S. A. Freel.** 1999. Endogenous salivary inhibitors of human immunodeficiency virus. *Arch. Oral Biol.* **44:**445–453.

188. **Singh, P. K., H. P. Jia, K. Wiles, J. Hesselberth, L. Liu, B. D. Conway, E. P. Greenberg, E. V. Valore, M. J. Welsh, T. Ganz, B. F. Tack, and P. B. J. McCray.** 1998. Production of β-defensins by human airway epithelia. *Proc. Natl. Acad. Sci. USA* **95:**14961–14966.

189. **Singh, P. K., B. F. Tack, P. B. McCray, Jr., and M. J. Welsh.** 2000. Synergistic and additive killing by antimicrobial factors found in human airway surface liquid. *Am. J. Physiol. Ser. L* **279:**L799–L805.

190. **Sinha, S., N. Cheshenko, R. I. Lehrer, and B. C. Herold.** 2003. NP-1, a rabbit alpha-defensin, prevents the entry and intercellular spread of herpes simplex virus type 2. *Antimicrob. Agents Chemother.* **47:**494–500.

191. **Smith, J. J., S. M. Travis, E. P. Greenberg, and M. J. Welsh.** 1996. Cystic fibrosis airway epithelia fail to kill bacteria because of abnormal airway surface fluid. *Cell* **85:**229–236.

192. **Sohnle, P. G., M. J. Hunter, B. Hahn, and W. J. Chazin.** 2000. Zinc-reversible antimicrobial activity of recombinant calprotectin (migration inhibitory factor-related proteins 8 and 14). *J. Infect. Dis.* **182:**1272–1275.

193. **Solomon, S., J. Hu, Q. Zhu, D. Belcourt, H. P. Bennett, A. Bateman, and T. Antakly.** 1991. Corticostatic peptides. *J. Steroid Biochem. Mol. Biol.* **40:**391–398.

194. **Sorensen, O. E., J. B. Cowland, K. Theilgaard-Monch, L. Liu, T. Ganz, and N. Borregaard.** 2003. Wound healing and expression of antimicrobial peptides/polypeptides in human keratinocytes, a consequence of common growth factors. *J. Immunol.* **170:**5583–5589.

195. **Sorensen, O. E., P. Follin, A. H. Johnsen, J. Calafat, G. S. Tjabringa, P. S. Hiemstra, and N. Borregaard.** 2001. Human cathelicidin, hCAP-18, is processed to the antimicrobial peptide LL-37 by extracellular cleavage with proteinase 3. *Blood* **97:**3951–3959.

196. **Steinbakk, M., C. F. Naess-Andresen, E. Lingaas, I. Dale, P. Brandtzaeg, and M. K. Fagerhol.** 1990. Antimicrobial actions of calcium binding leucocyte L1 protein, calprotectin. *Lancet* **336:**763–765.

197. **Storici, P., and M. Zanetti.** 1993. A cDNA derived from pig bone marrow cells predicts a sequence identical to the intestinal antibacterial peptide PR-39. *Biochem. Biophys. Res. Commun.* **196:**1058–1065.

198. **Storici, P., and M. Zanetti.** 1993. A novel cDNA sequence encoding a pig leukocyte antimicrobial peptide with a cathelin-like pro-sequence. *Biochem. Biophys. Res. Commun.* **196:** 1363–1368.

199. **Strukelj, B., J. Pungercar, G. Kopitar, M. Renko, B. Lenarcic, S. Berbic, and V. Turk.** 1995. Molecular cloning and identification of a novel porcine cathelin- like antibacterial peptide precursor. *Biol. Chem. Hoppe-Seyler* **376:**507–510.

200. **Swart, P. J., E. M. Kuipers, C. Smit, B. W. Van Der Strate, M. C. Harmsen, and D. K. Meijer.** 1998. Lactoferrin. Antiviral activity of lactoferrin. *Adv. Exp. Med. Biol.* **443:**205–213.

201. **Tang, Y. Q., J. Yuan, G. Osapay, K. Osapay, D. Tran, C. J. Miller, A. J. Ouellette, and M. E. Selsted.** 1999. A cyclic antimicrobial peptide produced in primate leukocytes by the ligation of two truncated α-defensins. *Science* **286:**498–502.

202. **Territo, M. C., T. Ganz, M. E. Selsted, and R. Lehrer.** 1989. Monocyte-chemotactic activity of defensins from human neutrophils. *J. Clin. Investig.* **84:**2017–2020.

203. **Tomee, J. F., G. H. Koeter, P. S. Hiemstra, and H. F. Kauffman.** 1998. Secretory leukoprotease inhibitor: a native antimicrobial protein presenting a new therapeutic option? *Thorax* **53:**114–116.

204. **Tominaga, T., J. Fukata, Y. Naito, Y. Nakai, S. Funakoshi, N. Fujii, and H. Imura.** 1990. Effects of corticostatin-I on rat adrenal cells in vitro. *J. Endocrinol.* **125:**287–292.

205. **Torres, A. M., G. M. de Plater, M. Doverskog, L. C. Birinyi-Strachan, G. M. Nicholson, C. H. Gallagher, and P. W. Kuchel.** 2000. Defensin-like peptide-2 from platypus venom: member of a class of peptides with a distinct structural fold. *Biochem. J.* **348:**649–656.

206. **Tossi, A., M. Scocchi, M. Zanetti, P. Storici, and R. Gennaro.** 1995. PMAP-37, a novel antibacterial peptide from pig myeloid cells. cDNA cloning, chemical synthesis and activity. *Eur. J. Biochem.* **228:**941–946.

207. **Trabi, M., H. J. Schirra, and D. J. Craik.** 2001. Three-dimensional structure of RTD-1, a cyclic antimicrobial defensin from *Rhesus* macaque leukocytes. *Biochemistry* **40:**4211–4221.

208. **Tran, D., P. A. Tran, Y. Q. Tang, J. Yuan, T. Cole, and M. E. Selsted.** 2002. Homodimeric theta-defensins from *Rhesus* macaque leukocytes—isolation, synthesis, antimicrobial activities, and bacterial binding properties of the cyclic peptides. *J. Biol. Chem.* **277:**3079–3084.

209. **Travis, S. M., B. A. Conway, J. Zabner, J. J. Smith, N. N. Anderson, P. K. Singh, E. P. Greenberg, and M. J. Welsh.** 1999. Activity of abundant antimicrobials of the human airway. *Am. J. Respir. Cell Mol. Biol.* **20:**872–879.

210. **Travis, S. M., P. K. Singh, and M. J. Welsh.** 2001. Antimicrobial peptides and proteins in the innate defense of the airway surface. *Curr. Opin. Immunol.* **13:**89–95.

211. **Tsai, H., and L. A. Bobek.** 1998. Human salivary histatins: promising anti-fungal therapeutic agents. *Crit. Rev. Oral Biol. Med.* **9:**480–497.

212. **Turner, J., Y. Cho, N. N. Dinh, A. J. Waring, and R. I. Lehrer.** 1998. Activities of LL-37, a cathelin-associated antimicrobial

peptide of human neutrophils. *Antimicrob. Agents Chemother.* **42:**2206–2214.

213. **Turpin, J. A., C. A. Schaeffer, M. Bu, L. Graham, R. W. Buckheit, Jr., D. Clanton, and W. G. Rice.** 1996. Human immunodeficiency virus type-1 (HIV-1) replication is unaffected by human secretory leukocyte protease inhibitor. *Antiviral Res.* **29:**269–277.
214. **Valore, E. V., and T. Ganz.** 1992. Posttranslational processing of defensins in immature human myeloid cells. *Blood* **79:** 1538–1544.
215. **Valore, E. V., E. Martin, S. S. Harwig, and T. Ganz.** 1996. Intramolecular inhibition of human defensin HNP-1 by its propiece. *J. Clin. Investig.* **97:**1624–1629.
216. **Verkman, A. S., Y. Song, and J. R. Thiagarajah.** 2003. Role of airway surface liquid and submucosal glands in cystic fibrosis lung disease. *Am. J. Physiol. Ser. C* **284:**C2–C15.
217. **Walker, C. M., D. J. Moody, D. P. Stites, and J. A. Levy.** 1986. $CD8^+$ lymphocytes can control HIV infection in vitro by suppressing virus replication. *Science* **234:**1563–1566.
218. **Wang, W., A. M. Cole, T. Hong, A. J. Waring, and R. I. Lehrer.** 2003. Retrocyclin, an antiretroviral θ-defensin, is a lectin. *J. Immunol.* **170:**4708–4716.
219. **Wang, X., Z. Zhang, J. P. Louboutin, C. Moser, D. J. Weiner, and J. M. Wilson.** 2003. Airway epithelia regulate expression of human beta-defensin 2 through Toll-like receptor 2. *FASEB J.* **17:**1727–1729.
220. **Wecke, J., M. Lahav, I. Ginsburg, and P. Giesbrecht.** 1982. Cell wall degradation of *Staphylococcus aureus* by lysozyme. *Arch. Microbiol.* **131:**116–123.
221. **Wehkamp, J., J. Harder, M. Weichenthal, O. Mueller, K. R. Herrlinger, K. Fellermann, J. M. Schroeder, and E. F. Stange.** 2003. Inducible and constitutive beta-defensins are differentially expressed in Crohn's disease and ulcerative colitis. *Inflamm. Bowel Dis.* **9:**215–223.
222. **Weiss, J., M. Inada, P. Elsbach, and R. M. Crowl.** 1994. Structural determinants of the action against *Escherichia coli* of a human inflammatory fluid phospholipase A_2 in concert with polymorphonuclear leukocytes. *J. Biol. Chem.* **269:** 26331–26337.
223. **Weiss, T. M., L. Yang, L. Ding, W. C. Wang, A. J. Waring, R. I. Lehrer, and H. W. Huang.** 2002. Two states of a cyclic antimicrobial peptide theta-defensin in lipid bilayers. *Biophys. J.* **82:**7A.
224. **Wilson, C. L., A. J. Ouellette, D. P. Satchell, T. Ayabe, Y. S. Lopez-Boado, J. L. Stratman, S. J. Hultgren, L. M. Matrisian, and W. C. Parks.** 1999. Regulation of intestinal alpha-defensin activation by the metalloproteinase matrilysin in innate host defense. *Science* **286:**113–117.
225. **Wright, G. W., C. E. Ooi, J. Weiss, and P. Elsbach.** 1990. Purification of a cellular (granulocyte) and an extracellular (serum) phospholipase A_2 that participate in the destruction of *Escherichia coli* in a rabbit inflammatory exudate. *J. Biol. Chem.* **265:**6675–6681.
226. **Yamaguchi, Y., T. Nagase, R. Makita, S. Fukuhara, T. Tomita, T. Tominaga, H. Kurihara, and Y. Ouchi.** 2002. Identification of multiple novel epididymis-specific beta-defensin isoforms in humans and mice. *J. Immunol.* **169:** 2516–2523.
227. **Yang, D., Q. Chen, O. Chertov, and J. J. Oppenheim.** 2000. Human neutrophil defensins selectively chemoattract naive T and immature dendritic cells. *J. Leukoc. Biol.* **68:**9–14.
228. **Yang, D., Q. Chen, D. M. Hoover, P. Staley, K. D. Tucker, J. Lubkowski, and J. J. Oppenheim.** 2003. Many chemokines including CCL20/MIP-3alpha display antimicrobial activity. *J. Leukoc. Biol.* **74:**448–455.
229. **Yang, D., O. Chertov, S. N. Bykovskaia, Q. Chen, M. J. Buffo, J. Shogan, M. Anderson, J. M. Schroder, J. M. Wang, O. M. Howard, and J. J. Oppenheim.** 1999. β-Defensins: linking innate and adaptive immunity through dendritic and T cell CCR6. *Science* **286:**525–528.
230. **Zanetti, M., R. Gennaro, and D. Romeo.** 1995. Cathelicidins: a novel protein family with a common proregion and a variable C-terminal antimicrobial domain. *FEBS Lett.* **374:** 1–5.
231. **Zanetti, M., L. Litteri, R. Gennaro, H. Horstmann, and D. Romeo.** 1990. Bactenecins, defense polypeptides of bovine neutrophils, are generated from precursor molecules stored in the large granules. *J. Cell Biol.* **111:**1363–1371.
232. **Zanetti, M., L. Litteri, G. Griffiths, R. Gennaro, and D. Romeo.** 1991. Stimulus-induced maturation of probactenecins, precursors of neutrophil antimicrobial polypeptides. *J. Immunol.* **146:**4295–4300.
233. **Zeya, H. I., and J. K. Spitznagel.** 1966. Cationic proteins of polymorphonuclear leukocyte lysosomes. I. Resolution of antibacterial and enzymatic activities. *J. Bacteriol.* **91:**750–754.
234. **Zeya, H. I., and J. K. Spitznagel.** 1966. Cationic proteins of polymorphonuclear leukocyte lysosomes. II. Composition, properties and mechanism of antibacterial action. *J. Bacteriol.* **91:**755–762.
235. **Zhang, L., W. Yu, T. He, J. Yu, R. E. Caffrey, E. A. Dalmasso, S. Fu, T. Pham, J. Mei, J. J. Ho, W. Zhang, P. Lopez, and D. D. Ho.** 2002. Contribution of human α-defensin-1, -2 and -3 to the anti-HIV-1 activity of CD8 antiviral factor. *Science* **298:**995–1000.
236. **Zhao, C., L. Liu, and R. I. Lehrer.** 1994. Identification of a new member of the protegrin family by cDNA cloning. *FEBS Lett.* **346:**285–288.
237. **Zhao, C., T. Nguyen, L. Liu, R. E. Sacco, K. A. Brogden, and R. I. Lehrer.** 2001. Gallinacin-3, an inducible epithelial β-defensin in the chicken. *Infect. Immun.* **69:**2684–2691.
238. **Zhu, Q., A. Bateman, A. Singh, and S. Solomon.** 1989. Isolation and biological activity of corticostatic peptides (anti-ACTH). *Endocr. Res.* **15:**129–149.
239. **Zhu, Q. Z., A. V. Singh, A. Bateman, F. Esch, and S. Solomon.** 1987. The corticostatic (anti-ACTH) and cytotoxic activity of peptides isolated from fetal, adult and tumor-bearing lung. *J. Steroid Biochem.* **27:**1017–1022.
240. **Zimmermann, G. R., P. Legault, M. E. Selsted, and A. Pardi.** 1995. Solution structure of bovine neutrophil beta-defensin-12: the peptide fold of the beta-defensins is identical to that of the classical defensins. *Biochemistry* **34:**13663–13671.

Colonization of Mucosal Surfaces
Edited by James P. Nataro et al.

Chapter 3

Mechanisms of Adaptive Immunity That Prevent Colonization at Mucosal Surfaces

MARCELA F. PASETTI, ROSANGELA SALERNO-GONÇALVES, AND MARCELO B. SZTEIN

THE ROLE OF ADAPTIVE IMMUNITY IN PROTECTION OF MUCOSAL SURFACES

The mucosal surfaces are covered by tight barriers of epithelial cells which separate the highly regulated internal compartment from the external environment, a compartment laden with microbes and other potentially harmful agents. Specialized innate and adaptive host defense mechanisms, the latter providing specific antigen recognition and immunologic memory, play an important role in maintaining the integrity of the mucosal barrier. The adaptive immune responses at the mucosal surfaces originate from complex associations between epithelial cells, the mucosa-associated lymphoid tissue (MALT), and the systemic immune cells.

The MALT is anatomically and functionally distinct from the systemic immune system, having developed distinct processes for antigen uptake, transport, processing, and presentation, as well as specialized immune effector mechanisms. Mucosal tissues can act as primary lymphoid organs, wherein B and T lymphocytes originating from immature precursors can differentiate into effector cells. Moreover, immune effector cells primed in the mucosa against antigens and pathogens acquire specific migration (or "homing") patterns that allow them to travel to distant mucosal sites. The mucosal innate immune mechanisms are addressed in detail in chapter 2. In this chapter, we discuss the chief mechanisms of adaptive immunity in mucosal tissues with particular emphasis on the gut, the best understood of the mucosal immune organs. We describe the organization and main components of the mucosal immune system and the adaptive host defense mechanisms that prevent microbial infection.

STRUCTURE AND CELLS OF THE GUT-ASSOCIATED LYMPHOID TISSUES

The mucosa of the gastrointestinal tract contains large numbers of lymphocytes organized in several compartments that are morphologically and functionally distinct. These compartments can be broadly divided into immune inductive and effector sites. Inductive sites include Peyer's patches (PP) and single isolated lymphoid follicles beneath the epithelial layer and mesenteric lymph nodes (MLN). Effector sites include less highly organized lymphoid regions in the mucosa containing the lamina propria lymphoid cells (LPL) and intraepithelial lymphocytes (IEL). Color Plate 1 shows a schematic representation of the organizational structure of the gut-associated lymphoid tissue (GALT) and the key elements involved in mucosal immunity.

Mucosal Lymphoid Follicles and Peyer's Patches

PP are organized mucosal lymphoid follicles situated beneath a thin layer of epithelial cells in the small intestine; they serve as key inductive sites of the mucosal immune responses. In contrast to peripheral lymphoid tissues, mucosal isolated lymphoid follicles and PP lack afferent lymphatics, which transport antigens and antigen-loaded antigen-presenting cells (APC) for immunologic priming. Instead, they possess a specialized dome epithelium containing microfold-epithelial (M) cells, which sample antigens from the intestinal lumen (described in detail below). The basolateral membrane of M cells folds into a pocket-like structure which contains dendritic cells (DC), macrophages, and B and T cells, each of which plays a key role in antigen

Marcela F. Pasetti, Rosangela Salerno-Gonçalves, and Marcelo B. Sztein • Center for Vaccine Development and Department of Pediatrics and Medicine, University of Maryland, 685 West Baltimore St., Room 480, Baltimore, MD 21201.

presentation and immune priming. The B cells in the pocket express naive (secretory immunoglobulin D-containing [sIgD$^+$]) and memory (sIgD$^-$) markers (78), suggesting that some of them serve as antigen-presenting B cells; these B cells interact with T cells (105) to promote antibody diversification and immunologic memory (6). Most T cells contained in the pocket are CD4$^+$ T helper (Th) lymphocytes that express the αβ T-cell receptor (TCRαβ), in contrast to the T cells that reside within the epithelial layer, which are mostly CD8$^+$ and express TCRγδ. Most of the pocket T cells express CD69 and CD45RO surface molecules, denoting early activation and memory, respectively. Interestingly, the development of B- and T-cell clusters associated with M cells depends on microbial colonization; these cells are absent in germ-free mice (6, 104).

Beneath the dome, PP are organized in germinal centers, each comprising an assembly of mostly naive B cells capable of intense clonal expansion. PP are supported by a network of interfollicular dendritic cells. Very few plasma cells reside in the PP. Differentiating B and T cells appear to leave the PP, migrating to MLN and other lymphoid tissues and effector sites in the gut (19).

The GALT also contains abundant isolated organized lymphoid follicles which are dispersed throughout the small and large intestine and are morphologically and functionally similar to the PP. The primary role of these follicles is antigen sampling and priming of the antigen-specific effector mechanism(s). As such, they play a key role in the decision between induction of active immunity and tolerance. Each mucosal B follicle is flanked by T-cell-rich areas in which high endothelial venules and lymphatic vessels are embedded, serving as entry and exit points for migrating cells (77). Homing of immune cells to the intestine is mediated by high endothelial venue expression of mucosal vascular addressin cell adhesion molecule 1 (MAdCAM-1), which is recognized by lymphocytes bearing either integrin-$\alpha_4\beta_7$ (primed lymphocytes) or L-selectin (naive lymphocytes) (2), as described in detail below.

Lymphocytes and Other Immune Cells Present in the Lamina Propria

The human intestinal lamina propria, a layer of connective tissue underlying the mucosal epithelium, contains an abundance of primarily activated IgA$^+$ B lymphocytes that are capable of differentiating into IgA-secreting plasma cells. IgM and IgG plasma cells are also present, albeit in smaller numbers. The lamina propria also contains T cells, including CD4$^+$ Th cells (60 to 70%) and CD8$^+$ (~30%) cells expressing mostly the TCRαβ (19, 44). Although the CD4/CD8 ratio in LPL and peripheral tissues is similar, mucosal T cells are phenotypically and functionally different from peripheral T cells. Most of the LPL T cells display a memory phenotype (CD45RO$^+$ CD45RA$^-$) and a higher state of activation (based on the expression of interleukin-2 receptor alpha [IL-2Rα] chain, HLA-DR molecules, and transferrin receptors) when compared to peripheral T cells (19). On activation, these cells produce larger amounts of cytokines, consistent with their increased helper activity for B-cell responses (as described in detail below). Naive LPL are primed in the mucosal lymphoid follicles and, following differentiation within the regional lymph nodes, seed the lamina propria as specific effector and memory cells. DC, macrophages, natural killer cells (NK), eosinophils, and mast cells are also found in this tissue (23).

Intraepithelial Lymphocytes

IEL, composed mostly of CD8$^+$ T cells, are found between epithelial cells on the luminal side of the lamina propria basement membrane (44). Their development appears to depend on IL-7 and IL-15 produced by the enterocytes themselves (57, 63). There are several populations of IEL. While some bear TCRγδ, composed of γ (Vγ) and δ (Vδ) chains, others express TCRαβ, composed of α (Vα) and β (Vβ) chains. Functionally, the Vβ chain has been associated with the antigen recognition process, resulting in clonal expansion of particular Vβ antigen-specific T cells (16). Studies addressing the TCR Vβ usage in the human mucosa have shown that only a few Vβ families predominate among human IEL (5). The most frequent Vβ families observed in IEL are Vβ1, Vβ2, Vβ3, and Vβ6 (102).

The expression of CD8α, CD8β, and CD4 permits the additional subdivision of IEL into four main subsets: TCRγδ$^+$ CD4$^-$ CD8α$^-$ CD8β$^-$ (γδDN), TCRγδ$^+$ CD4$^-$ CD8α$^+$ CD8β$^-$ (γδ CD8αα), TCRαβ$^+$ CD4$^-$ CD8α$^+$ CD8β$^-$ (αβ CD8αα), and TCRαβ$^+$ CD4$^-$ CD8α$^+$ CD8β$^+$ (αβ CD8αβ) (93). Populations expressing TCRαβ$^+$ CD4$^-$ CD8$^-$ (αβ DN), TCRαβ$^+$ CD4$^+$ CD8$^-$ (αβ CD4$^+$), TCRαβ$^+$ CD4$^+$ CD8$^+$ (αβ CD8αα DP), and TCRαβ$^+$ CD8αβ$^+$ CD8αα$^+$ have also been described (14). The proportion of the various subpopulations varies among species. In mice a significant proportion (10 to 50%) of IEL are TCRγδ$^+$, while in humans the proportion is ~10%. This is in contrast to other tissues, including peripheral blood, where <5% of T cells are TCRγδ$^+$(22). Of note, a significant proportion of TCRγδ$^+$ and TCRαβ$^+$ are CD8αα rather than TCRαβ CD8αβ$^+$, which are the predominant T cells present in peripheral blood and spleen (74). There is evidence suggesting that TCRαβ$^+$ CD8αβ$^+$

cells may enter the PP or other organized sites in the gut and recirculate to populate the gut via homing mechanisms following systemic activation, whereas the other subsets may develop extrathymically and associate with the epithelium directly (89).

INDUCTION OF ADAPTIVE IMMUNITY AT MUCOSAL SURFACES

Antigen Sampling across M Cells

Adaptive immune responses in the mucosal tissues require close collaboration between epithelial cells, APC, and lymphoid cells. Antigens are sampled from the gut lumen across the epithelial barrier by M cells in the follicle-associated epithelium (FAE). Endocytic or phagocytic uptake of foreign antigens or particles is followed by rapid transcytosis via an active transepithelial vesicular transport system that releases them into the intraepithelial pocket, with little or no retention in M-cell lysosomes (79). Antigens and microorganisms are generally delivered intact across M cells (76); however, it is not known whether they can be altered within vesicles while in transit (79). In the intraepithelial pocket, antigens interact with professional APC (i.e., DC and macrophages), as well as CD4 Th and B cells, all of which are involved in immune priming (78).

M-cell adherence and internalization of pathogens is often the first step in the initiation of mucosal and/or systemic infections. Several gram-negative bacteria, such as *Vibrio cholerae,* enteropathogenic *Escherichia coli, Salmonella enterica* serovars Typhi and Typhimurium, *Shigella flexneri,* enteropathogenic *Yersinia* spp., and *Campylobacter jejuni,* cause disease by colonizing the intestinal mucosa following binding to M cells (79). M-cell surface glycoconjugates serve as receptors for bacterial adhesins, promoting efficient internalization (79). Entry of poliovirus via M cells in human intestinal explants has also been demonstrated (95).

Role of Dendritic Cells in Antigen Uptake, T-Cell Priming, and B-Cell Differentiation

DC are present in PP, beneath the epithelial dome, as well as in the interfollicular T-cell areas (49). DC underneath the FAE take up luminal antigens transported by M cells and present them to naive T cells. At least three distinct DC subsets (in the unperturbed state) have been found in mouse PP: (i) myeloid ($CD11c^+$ $CD11b^+$ $CD8\alpha^-$) DC in the subepithelial dome, (ii) lymphoid ($CD11c^+$ $CD11b^-$ $CD8\alpha^+$) DC in the interfollicular T-cell area, and (iii) a novel population of $CD11c^+$ $CD11b^-$ $CD8\alpha^-$ DC at both sites (42, 43).

Both immature myeloid and lymphoid DC are recruited to the PP via chemokine-receptor interactions (42). Once within the PP, myeloid DC migrate to the subepithelial region, where they remain until they encounter antigens, transported via M cells. If the antigen encountered is an innocuous protein antigen (e.g., food), DC mediate the development of Th2/Th3 cells by secreting high levels of IL-10 and possibly transforming growth factor β (TGF-β). If the antigen is a microbial component, myeloid $CD11b^+$ DC undergo maturation and differentiate into APC, which stimulate Th1 responses by secreting high levels of IL-12. Indeed, gastrointestinal pathogens such as *S. enterica* serovar Typhimurium are known to induce strong Th1 responses at mucosal sites, a process that is initiated by IL-12 secretion from DC (98).

DC are also present in the gut epithelium and the lamina propria (85). DC in the lamina propria of the small intestine form a tight network similar to epidermal Langerhans' cells, whereas colonic DC are found mainly in lymphoid follicles and subepithelial regions (3). It has been shown that DC recruited to the epithelium or lamina propria can extend their dendrites between epithelial tight junctions and can sample intestinal microbes and antigens directly (85). During this process, DC-expressed tight-junction proteins, such as occludin, claudin 1, zonula occludens 1, and junctional adhesion molecule, facilitate the interjection of DC dendrites between epithelial cells without breaching the cell barrier (27).

Inflammatory stimuli from mucosal pathogens, inflammatory cytokines, or necrotic cells induce a rapid influx of DC or precursors to mucosal tissues. Following antigen stimulation, DC are activated and undergo a maturation process while migrating to regional draining lymph nodes, where they activate naive T cells. Pathogenic bacteria, but not commensal organisms, are thought to induce DC maturation and migration through Toll-like receptors (TLR) and cytokines, shaping the type of response induced (e.g., tolerance or immune priming). It has been proposed that lamina propria DC primed by pathogenic bacteria might be prompted to migrate whereas DC that have encountered commensals remain in the lamina propria, activating local B and T cells (85).

Role of Intestinal Epithelial Cells in Antigen Presentation

In vitro studies have demonstrated that intestinal epithelial cells (IEC) can present antigens to T cells (36). Intestinal cells in the small intestine express functional major histocompatibility complex (MHC) class II molecules, lack classical costimulatory molecules (e.g., CD80), but express other costimulatory molecules such as CD58 (signaling via CD2), CD86, and gp189. IEC also express nonclassical class I MHC molecules (CD1) that contribute to resistance to bacterial, viral, parasitic, and fungal infections (e.g., by *Pseudomonas aeruginosa* and *Mycobacterium tuberculosis*). These receptors enhance innate and adaptive immunity, including antibody production, through presentation of microbial lipids and lipopeptide antigens to CD1-restricted T cells (10). Interestingly, the phenotype and function of IEC change while they are migrating from crypt to villi. Crypt epithelial cells express polymeric immunoglobulin receptor (pIgR) and efficiently transport sIgA and IgM but do not express MHC class II antigens, whereas villus cells no longer express pIgR but express class II antigens (36). This distinct phenotype of IEC suggests that their role in T-cell activation is substantially different from that of traditional APC and is likely to be unique to the mucosal environment. The precise role of IEC-induced immune priming in vivo is unknown.

ANTIBODY- AND B-CELL-MEDIATED IMMUNITY IN MUCOSAL TISSUES

Secretory IgA: Structure and Transepithelial Transport

sIgA is the most abundant immunoglobulin on the intestinal mucosal surface, and it is known to play a major role as a first line of defense against adherence and invasion by enteric pathogens (58). It is produced mainly by IgA^+ plasma cells in the intestinal lamina propria. In contrast to systemic IgA, which circulates as a monomer, sIgA is typically secreted as a dimer, consisting of at least two monomeric IgA units and two additional polypeptide chains: the J chain, a 15-kDa glycoprotein produced in the plasma cell and covalently linked to polymeric IgA or IgM before secretion, and the secretory component, a ~100-kDa transmembrane glycoprotein contributed by epithelial cells and exocrine glands (8). The polymeric structure enhances the avidity of mucosal antibodies, and their distinct carbohydrate composition enhances antibody binding to mucus components and prevents proteolysis in the luminal environment. Because of its high stability, sIgA maintains antibody activity for remarkably long periods in the gastrointestinal and oral mucosa (6).

The translocation of plasma cell-secreted IgA (or IgM, although in much smaller amounts) from the lamina propria to the luminal compartment occurs through dimeric IgA (dIgA) (or pentameric IgM) binding to the pIgR expressed at the basolateral surface of secretory epithelial cells. The complex is internalized into the basolateral endosomal compartment and is sorted into transcytotic vesicles, which reach the apical endosome; here the ligand-receptor complex recycles transiently. At the apical surface, the pIgR is proteolytically cleaved and the extracellular fragment, known as secretory component, is released, either bound or not bound to pIgA (72). The expression of pIgR in mucosal epithelia is regulated by a complex interplay among lymphocytes, macrophages, and epithelial cells. It is up-regulated in response to Th1 (gamma interferon [IFN-γ]), Th2 (IL-4), and proinflammatory (tumor necrosis factor alpha [TNF-α] and IL-1) cytokines (8), which in turn stimulate the transcytosis of IgA antibodies.

Humans have two subclasses of monomeric and polymeric IgA: IgA1 and IgA2. IgA1, but not IgA2, can be cleaved by IgA-specific bacterial proteases (50). The proportion of IgA1 and IgA2 varies in individual secretions: IgA1 prevails in nasal and bronchial mucosa, while IgA2 prevails in the large intestine (8). Mucosal secretions contain larger amounts of IgA2 than peripheral tissue does, an advantage because of the presence of microbial proteases in this environment (58). Protein antigens elicit mainly IgA1, whereas lipopolysaccharide (LPS) induces IgA2. IgA does not fix complement (like IgG does), providing a highly effective yet noninflammatory immune effector mechanism (58).

Priming and Maturation of Mucosal B Cells

Most IgA-producing cells originate in organized inductive lymphoid follicles in the MALT, such as PP in the intestine, where they encounter the antigen. Following antigenic stimulation, naive B cells migrate to the follicle-associated germinal center (GC), where they clonally expand and differentiate. During this clonal expansion, B cells undergo affinity maturation by somatic hypermutation and selection of the highest-affinity clones. Cells with high-affinity Ig receptor are maintained, whereas low-affinity clones are eliminated by apoptosis. Antigen-specific B cells can either become memory cells ($sIgD^-$ IgM^+ $CD38^-$ $B7^+$ B cells) or initiate isotype switching. In the MALT GCs, B lymphocytes receive costimulation by CD4 Th cells, undergo isotype switching, and differentiate further into B cells that express high-affinity IgA receptor. TGF-β and IL-10 are crucial in triggering IgA switching, thereby inducing antigen-specific B cells to become predominantly IgA-committed plasmablasts (8). Ligation of costimulatory molecules, such as CD40, provides an important signal for switch induction. Bacterial LPS, cholera toxin, and *E. coli* heat-labile toxin are known to favor the mucosal Ig switch. Subsequently, B lymphocytes differentiate into effector or memory cells following contact with Th lymphocytes and CD40-CD40L interactions. IgA^+ B cells originating in PP and mucosal follicles migrate to the draining MLN, where they proliferate and further differentiate. While isotype switching and differentiation occur sequentially in mucosal inductive sites (i.e., the PP) and regional lymph nodes, IgA production by plasma cells takes place in mucosal effector sites (i.e., the lamina propria). Terminal differentiation of effector B cells into polymeric IgA-secreting plasma cells occurs in the lamina propria in a process regulated by cytokines derived from activated Th2 CD4 Th cells (IL-2, IL-5, and IL-10), DC, and IEC (IL-6). Plasmablasts can differentiate locally into plasma cells remaining in the secondary lymphoid tissue of origin, or they can traffic back though the efferent lymph to the blood to populate distant sites.

In situ class switching and differentiation from IgM to IgA has recently been detected in lamina propria B cells (21). In contrast to the GC-supported class switch, under T-cell control, there is evidence that activated lamina propria DC might be sufficient to activate B cells independently of T-cell costimulation (20). Early studies suggest a similar role in isotype switching for PP DC (97), showing that polyclonal activation of sIgM B cells, in the presence of mixtures of PP DC and T cells, enhance IgA secretion and that this effect was mediated by DC rather than T cells. Moreover, polyclonal activation of naive, $sIgM^+$ human tonsillar B cells in the presence of CD40 ligand and blood-derived DC led to skewed IgA responses (24). The mechanisms of DC-mediated B-cell isotype switching are unknown, yet evidence indicates that it might result from direct DC–B-cell contact before or independent of T-cell involvement (103).

B1 cells from the peritoneal and pleural cavities also appear to contribute mucosal IgA^+ plasma cells, at least in mice (53). Different from the IgA responses generated in the GCs within the mucosal follicle, which require T-cell collaboration, IgA production by B1 cells in the lamina propria appear to be T-cell independent (21, 65).

Homing Pattern of Mucosal IgA-Secreting Cells

Stimulated MALT B and T cells acquire a mucosal homing program. The effector and memory lymphocytes lose their adhesion to stromal cells, leave organized MALT structures, and disseminate via draining lymph and blood circulation to mucosal effector sites, where IgA production occurs, or they return to the MALT. The tissue specificity of IgA^+ B-cell homing is determined by complex interactions between lymphocyte homing receptors, which are differentially expressed depending on the priming site, and ligands expressed on the vascular endothelium of target tissues. Virtually all IgA and even IgG antibody-secreting cells detected after peroral and rectal immunization express integrin-α_4/β_7 receptors, while only a minority express the peripheral lymph node CD62L (L-selectin) receptor, which is induced mainly after systemic immunization (84). In contrast, circulating B cells, induced by intranasal immunization, coexpress L-selectin and integrin-α_4/β_7 (84), possibly explaining the compartmentalized responses.

Lymphocytes expressing integrin-α_4/β_7 interact with MAdCAM-1 that mediate selective binding of lymphocytes to the postcapillary blood vessels of the lamina propria endothelial cells. Extravasation of lymphocytes either in the lymph nodes or at effector sites requires chemoattractant signals. Splenic, MLN, and PP IgA antibody secreting cells (ASC) express chemokine receptor CCR9 and migrate efficiently to its ligand, CCL25 (expressed mainly by intestinal epithelium), whereas IgG ASC are responsive to other chemokines including CXCL12 and CXCL9, ligands for CXCR4 and CXCR3, respectively, and most express integrin-$\alpha_4\beta_1$ (55). Thus, selectivity in isotype-specific ASC chemokine receptor expression is probably a major determinant of IgA localization to mucosal gut epithelial surfaces and IgG localization to systemic sites of chronic inflammation (55). It has been suggested that antigen-specific plasmablasts become locally enriched in mucosal sites via retention at sites of antigen deposition.

Mechanisms of sIgA-Mediated Protection

The exact mechanisms involved in IgA-mediated protection of the mucosal surfaces are only partly understood. One of the main roles of sIgA, in cooperation with innate mucosal defense mechanisms, is to perform immune exclusion, preventing the attachment of pathogens to the mucosal surfaces (6). Antibodies bound to surface antigens form large immune complexes that prevent colonization and invasion by microbes, facilitating their entrapment in the mucus and subsequent peristaltic or ciliary clearance. IgA may also specifically block and sterically hinder the microbial surface proteins that mediate cell attachment. IgA also seizes incoming pathogens within the epithelial cell vesicular compartments and exports antigens back into the lumen (40). Additional adaptive defense functions include the ability of sIgA to shuttle antigens to the lumen by using trancytosis mediated by pIgR and the potential enhancement of mucosal immunity by favoring the uptake of luminal antigens via the IgA binding receptor in M cells (6, 58, 79). Descriptions of these protective mechanisms are given in the following sections.

Immune exclusion

IgA in the secretions protect the mucosal surfaces by cross-linking microorganisms or macromolecules, thereby facilitating their elimination and preventing their attachment to the mucosal surfaces. IgA can also interfere with the function of microbial adhesins or cause steric hindrance that prevents interaction between pathogens and mucosal cell receptors. These processes are collectively known as immune exclusion (58). Immune exclusion provides an immunological barrier that prevents microbial attachment and diffusion through the glycocalyx covering the epithelial cells. It also facilitates the entrapment of microorganisms and macromolecules in the mucus, blocking their binding to epithelial surface receptors. Neutralizing antibodies are important in protection from viral infection by preventing virus attachment and subsequent internalization. sIgA antibodies cooperate with the innate defense system and enhance bacteriostatic activity. Additionally, sIgA can induce the loss of bacterial plasmids encoding adherence factors and antibiotic resistance and can interfere with enzymes and bacterial nutrients, crippling their growth (6). These mechanisms apply to both pathogenic and commensal organisms, and they are thought to attenuate the chronic stimulation of GC reactions by bacteria that persist in the GALT. Mucosal IgA antibodies cannot function systemically; they do not neutralize microorganisms as does IgG, and they do not activate the classical complement pathway.

Intracellular antigen neutralization

IgA antibodies neutralize viral pathogens within epithelial cells (40, 90, 92). Soluble dIgA binding to newly synthesized viral protein within infected cells was found to block the proliferation and assembly of virus during transcytosis via the polymeric Ig receptor (92). Following infection, intracellular virus neutralization reduces the levels of virus measured in cell lysates and apical fluids, decreasing virus spread along the mucosal surface (69). sIgA within IEC was also reported to neutralize LPS molecules after gram-negative bacterial infections. Specific dIgA, colocalizing with LPS in the apical recycling endosome compartment, prevented LPS-induced NF-κB translocation and subsequent proinflammatory responses (25).

sIgA-mediated shuttle/excretion of extracellular antigen

A large amount of IgA secreted by plasma cells ends up in the extracellular space of the lamina propria, within the layer of loose connective tissue underlying the epithelial basement membrane. Before reaching the epithelium, the IgA contained in this compartment can bind (and clear) antigens that have crossed the mucosal lining. The immune complexes formed could be absorbed into the circulation or be actively transported through the mucosal epithelium from the basal to the apical surface by using the pIgR, the same mechanism that transports free IgA into the mucosal secretions (88). Thus, through this "shuttle service," the IgA in the lamina propria can excrete antigens, providing a second layer of immune exclusion to support the primary immune exclusion barrier in the lumen.

Interaction of sIgA with M Cells and Leukocytes Bearing Fc-α Rc

sIgA produced by lamina propria plasma cells and transported to the lumen by crypt epithelial cells does not adhere to the apical surfaces of human enterocytes or follicle-associated epithelial cells but selectively adheres to the apical surface of PP M cells (66). This binding process requires the Cα1 and Cα2 domains of IgA (66), but not the associated SC as for mucus anchoring (81), and is likely to be mediated by an IgA-specific receptor, which enables sIgA transport from the intestinal lumen to underlying mucosal tissue (66). sIgA taken up from the apical surfaces of M cells accumulates within vesicles clustered near the intraepithelial pocket membranes. It has recently been shown that following M-cell transport, exogenous sIgA-antigen complexes are internalized by pocket DC that selectively associate with subepithelial $CD4^+$ T and B lymphocytes (86). DC in the M-cell pocket present luminal antigens to neighboring resident T cells. It has been speculated that IgA transport by M cells might promote antigen presentation under neutralizing conditions and act as an immunomodulator to control mucosal homeostasis (86). sIgA-immune complexes could also interact with the DC network below the FAE, thought to migrate to the M-cell pocket.

Monocytes, macrophages, neutrophils, and eosinophils recruited into mucosal tissues following microbe-epithelial cell interactions preferentially express Fcα receptor for IgA in mice and humans. These Fc-α receptor-bearing cells may facilitate the uptake of antigen-antibody complexes or IgA-opsonized organisms and may modulate immune functions through the release of specific cytokines (e.g., IFN-γ, IL-1β, and IL-6) (82).

Role of Other Antibodies in Mucosal and Glandular Secretions

Although IgA is the main isotype in the mucosal secretions, significant quantities of IgG are also present. It has long been understood that IgG reaches luminal secretions through epithelial transudation of systemic IgG and (less frequently) through IgG synthesized locally. The intestine of suckling rodents and the human placenta express an MHC class I-related neonatal Fc receptor (FcRn) for IgG, which mediates its transport across epithelial barriers by transcytosis. After weaning, however, epithelial cells of the rodent intestine downregulate the expression of FcRn to nearly undetectable levels. In contrast, absorptive epithelial cells lining the intestine of humans continue to express FcRn in adult life (41) and exhibit in vitro FcRn-dependent transcytosis of IgG in both directions across the epithelial monolayer (18). It has recently been confirmed that FcRn is the vehicle that transports IgG across the intestinal epithelial barrier into the lumen. The FcRn can also recycle the IgG-antigen complex back across the intestinal barrier into the lamina propria for processing by DC cells and presentation to $CD4^+$ T cells in regional organized lymphoid structures (106).

IgE is also produced by lamina propria plasma cells, which may mediate protection against parasites through activation of local mast cells, which in turn release TNF-α. This, however, may be considered "pathologic enhancement" of local defenses (80). Monomeric Igs might be cotransported via pIgR, concomitantly with an immune complex involving at least one pIg. Mucosal pIgM is less abundant but is also present in secretions (6). In the mucosal environment, all plasma cells, irrespective of their Ig isotype, express J chain, the small polypeptide required for IgA polymerization.

CELL-MEDIATED IMMUNITY IN MUCOSAL TISSUES

Antigen Presentation

As outlined above, pathogens transported by M cells may be endocytosed by immature DC in the subepithelial dome region of the M-cell pocket and ferried to adjacent interfollicular T-cell zones and/or regional lymph nodes, where DC maturation and antigen presentation to naive T cells occurs (77). Naive T cells continuously recirculate though secondary lymphoid tissue (e.g., spleen, lymph nodes, and PP), increasing the likelihood that the naive T cells will encounter their cognate antigen (11). Naive-T-cell priming requires sustained TCR stimulation, which is achieved by the formation of an immunological synapse, a specialized area of contact between T cells and DC (60). Immunological synapses enable costimulation and the sustained relatively low-affinity interaction between TCR and specific peptide-MHC complexes (28). The synapse is a dynamic structure, as evidenced by the ability of T cells that are engaged with APC to form new synapses within minutes after encountering APC displaying larger amounts of antigen (101).

Antigen-specific responses require that DC provide at least two signals: antigenic peptide presented in the context of MHC class I or II molecules and costimulatory signals (74). The specificity of T-cell recognition depends on interactions between the peptide-MHC complex and the TCR. In general, peptide antigens from pathogens replicating inside APC (endogenous antigens), such as viruses, bind to MHC class I, leading to the induction of antigen-specific $CD8^+$ T lymphocytes. $CD8^+$ cells are likely to play an important role in host defenses by several effector mechanisms, including lysis of infected cells (cytotoxicity; mediated by cytotoxic T lymphocytes [CTL]) and secretion of IFN-γ and other cytokines. Mucosal CTL play a role in protective immunity against intracellular pathogens by both of these mechanisms (14). In contrast, antigens from extracellular pathogens are usually processed differently from endogenous antigens; they bind to MHC class II and stimulate $CD4^+$ Th cells. Th cells are composed of two main subpopulations, Th1 and Th2, which primarily support the production of specific antibodies or elicit inflammatory responses and differentiation of $CD8^+$ cells, respectively. However, these antigen presentation rules are not absolute. Some peptides from extracellular pathogens can also be presented on MHC class I molecules, in a process known as cross-priming (17, 32, 38). Of note, while DC are required for the efficient priming of antigen-specific lymphocyte responses, T cells in turn contribute to DC maturation (87, 94). For example, DC mature in response to CD40L, which is expressed at high levels on activated memory T cells (60).

The ability of DC to discriminate between pathogens and commensals might be related, in part, to differential signaling by TLR and other pattern recognition receptors that recognize invariant molecules present in microbes but not in the host (74). Different organisms activate DC by interacting with discrete TLR and pattern recognition receptors. For example, LPS from gram-negative bacteria and CpG DNA motifs activate DC mostly via TLR4 and TLR9, respectively (46). On the other hand, lipoproteins from gram-positive bacteria activate the IL-12p40 promoter mostly though a TLR2-dependent mechanism. Signaling through IL-12, a key cytokine involved in promoting Th1-type cell-mediated immunity (9), is dependent on the expression of the high-affinity IL-12 receptors (IL-12R), which are composed of two subunits, IL-12Rβ1 and IL-12Rβ2, the signal-transducing subunit (100). Th1 cells express both the IL-12Rβ1 and IL-12Rβ2 subunits, whereas Th2 cells express only the IL-12Rβ1 chain. Signaling through TLR also induces the up-regulation of costimulatory molecules that provide the requisite "second signal" for the activation of naive T cells (74). Among costimulatory molecules, the principal ligand-receptor pair is the association of CD28 on T cells with either of two ligands, CD80 and CD86 molecules, on DC. These interactions play a key role in TCR-dependent proliferation, IL-2 production, cell survival, and the production of effector cytokines, such as IFN-γ, IL-4, and IL-5. Following antigen presentation and activation, T cells proliferate vigorously and partially differentiate in PP and regional lymph nodes, generating memory and effector T cells.

Migration and Differentiation of T Cells

T-cell activation is followed by cell surface expression of high-affinity IL-2 receptor complexes, composed of α (CD25)-β (CD122)-γ (CD132) chains. The interaction of high-affinity IL-2 receptors with IL-2 is a critical event leading to T-cell proliferation, differentiation, and function (30). Precursor memory T cells elicited in PP rapidly migrate to MLN. After further differentiation, these cells enter the systemic circulation via the efferent lymph (7) and then enter the diffuse lamina propria compartment of the mucosa, where they accumulate, awaiting the next antigenic challenge (67, 68).

Memory T cells have a unique phenotypic profile that includes changes in chemokine receptor and cell adhesion molecule expression (59). They can be subdivided into two categories: (i) central memory T cells, which express high levels of chemokine receptor CCR7 and the adhesion molecule CD62L, and (ii) effector memory T cells, which express low levels of CCR7 and CD62L or do not express them at all. The memory cells that accumulate in the lamina propria are mostly "effector memory" cells (91).

Effector memory T cells migrate preferentially to nonlymphoid tissues that are connected to the secondary lymphoid organs where the antigen was first encountered (13). This migration is a multistep process, and the central event is the adhesion of lymphocyte surface homing receptors to their counterparts, addresins, on endothelial cells (7). In the gut, the selective homing of effector memory cells to the lamina propria of the small intestine is driven, to a large extent, by the expression of the integrin-$\alpha_4\beta_7$ ligand MAdCAM-1, which is selectively expressed by venules in normal and inflamed intestinal endothelium (12). Other molecules have recently been shown to also participate in this process. For example, expression of CCR9 in subsets of circulating integrin-$\alpha_4\beta_7^{hi}$ lymphocytes play an important role in homing of these cells to the small intestine (33, 47, 99). TECK/CCL25, present on the crypt epithelium in the jejunum and ileum, mediates the chemotaxis of memory integrin-$\alpha_4\beta_7^{hi}$ $CD4^+$ and $CD8^+$ T lymphocytes expressing CCR9 receptors into the lamina propria. Virtually all T cells in the small intestine express CCR9 (56). Interestingly, TECK/CCL25 is absent or only weakly expressed in other segments of gastrointestinal tract, suggesting that even organs thought to be part of a common mucosal immune system may have different lymphocyte homing pathways (54).

Regulatory and Effector Functions of T Lymphocytes

T cells exert their effector and regulatory functions by two main mechanisms, i.e., cytokine and chemokine production and cytotoxic activity. Notable differences have been observed between regulatory and effector functions of the unique T-cell populations in MALT and those in other tissues. A description of the T-cell regulatory and effector mechanisms shown to be operational in mucosal tissues follows.

Cytokines and chemokines as key regulatory and effector molecules secreted by $CD4^+$ T cells

Consensus is emerging that mucosal T cells may be substantially different from T cells in others sites in regard to their potential for modulating immune responses. Induction of protective immunity requires efficient antigen presentation to T lymphocytes. Production of inflammatory cytokines, in particular IL-12, and increased expression of costimulatory molecules on the surface of APC are essential for effective antigen presentation. Whereas IL-12-producing DC prime predominantly Th1 responses, DC that fail to produce IL-12 preferentially prime Th2 responses, a process driven by IL-4 produced by activated $CD4^+$ T cells (60). Th1 cells secrete proinflammatory cytokines (e.g., IFN-γ and TNF-α), resulting in inflammatory responses characterized by the activation of macrophages and NK cells (which play a key role in eliminating commensal or pathogenic bacteria that penetrate the mucosa) and induction of CTL. In contrast, Th2, Th3, and T regulatory 1 (Tr1) cells exhibit mostly immunoregulatory functions. Th2 cells secrete IL-4, IL-5, IL-10, and IL-13 and provide critical help in the generation of antibody responses, whereas Th3 and Tr1 secrete TGF-β and IL-10 respectively, which contribute to immune homeostasis by down regulating immune responses (64) (see below).

Results from in vitro stimulation of isolated DC suggest that different populations of murine DC may dramatically affect T-cell priming by preferentially producing IL-12 or IL-10 (48). For example, a recent study showed that lamina propria and PP contain a subset of CD8α^- CD11b$^+$ DC with immunomodulatory or suppressor properties that produces primarily IL-10 rather than IL-12. Of note, T-cell responses in the gastrointestinal tract have generally been considered to be regulatory, since it would seem counterintuitive to mount a Th1-cell response to the normal or food antigens (64). However, there are many examples, involving rodents, that show Th1-biased immune responses to fed antigens, and this Th1 bias is even more pronounced in humans (64). It has been shown that freshly isolated human LPL contain 50 to 100 times more IFN-γ than IL-4, IL-5, and IL-10 (26, 34). Other cytokines also play important roles in T-cell activation and differentiation in the MALT. For example, the development of IEL and LPL appears to depend on IL-7 and IL-15, which are produced by the gastrointestinal epithelium (57, 62). In addition, proliferation induced by IL-7 appears to depend on IL-2, which emphasizes the critical role of IL-2 in lamina propria T-cell function.

Regulatory and effector mechanisms of CD8$^+$ T cells

Effector immune responses mediated by CD8$^+$ cells in mucosal tissues include immunoregulation and activation of innate immune responses, which are dependent largely on cytokine production, as well as cytotoxicity mechanisms. Of the various CD8$^+$ populations found in MALT (e.g., IEL, LPL, and PP), acquired effector/memory TCRαβ expressing CD8αβ or CD8αβ plus CD8αα have mostly CTL activity while natural memory TCRαβ double negative or expressing only CD8αα have predominantly immunoregulatory activity. On the other hand, TCRγδ DN or expressing CD8αα have predominantly immunoregulatory and tissue repair activities (14). The generation of mucosal effector/memory T cells has been observed in bacterial, viral and parasitic infections (51, 83). Protection against some pathogens that enter the host via mucosal surfaces has been shown to be mediated mostly by CD8αβ$^+$ TCRαβ$^+$ IEL and LPL, which exhibit potent CTL activity against specific-pathogen-infected target cells, e.g., during lymphocytic choriomeningitis virus infection (14,73). Other examples include strong, long-lasting CTL responses to target cells expressing HIV antigens observed in PP and MLN of mice immunized with attenuated *Listeria monocytogenes* encoding HIV gag (61) and following transcutaneous immunization with an HIV peptide vaccine (4). Of note, effector memory CD8$^+$ T cells in mucosal tissues, as in other nonlymphoid tissues, have been shown to exert immediate cytolytic activity. Concerning immunoregulatory properties, IFN-γ, TNF-α, and other cytokines secreted by CD8$^+$ cells following an encounter with specific antigens play a key role in inducing inflammatory responses that include the activation of APC, particularly macrophages present in the lamina propria, enabling them to effectively kill microorganisms that have breached the mucosal barrier.

The function of IEL, which, as described above, are composed largely of CD8$^+$ T cells, a sizable proportion of which express TCRγδ, remains elusive. These cells have been implicated in regulating the development of epithelial cells (52) and in killing both infected and malignant cells (35). Human TCRγδ T cells recognize nonclassical HLA class I molecules, such as MICA and MICB; these molecules are upregulated in IEC following adhesion of *E. coli,* through NKG2D receptors and probably also TCR γδ (31). Moreover, in a *Yersinia pseudotuberculosis* infection model, γδ IEL were found to constitutively express cytotoxic genes, including granzymes A and B, and the inflammatory chemokine RANTES (CCL5) (22). These observations suggest that, in contrast to lymph node CD8αβ$^+$ cells, which must be activated to become cytotoxic effectors, γδ IEL might become rapidly effective against a wide range of pathogens without transcriptional delays.

Regulatory T cells and GALT immune homeostasis

The risk of widespread inflammation of intestinal mucosal tissues is constant. Subsets of suppressor or regulatory T cells (Tregs) play important roles in both immune suppression and the maintenance of immune homeostasis in the potentially inflammatory gut microenvironment. It is now generally accepted that Tregs control inflammatory responses to commensal bacteria and pathogens (70, 96).

Several subsets of Tregs have been identified based on their expression of CD4 and CD8. Among the CD4$^+$ Tregs, two main subsets have been described: naturally occurring Tregs and induced Tregs (71). Naturally occurring Tregs consist of CD4$^+$ T cells that mature in the thymus (71). These cells are prevalent in the naturally activated lymphocytes (CD45RB^{low}) and CD25-expressing subsets. Additional surface molecules that have been associated with some subpopulations of naturally occurring Tregs express GITR (TNFRSF18), CD152 (CTLA-4), and integrin-$\alpha_E\beta_7$ (CD103) (71). Naturally occurring Tregs need TCR signals for activation but, in contrast to conventional effector T cells, show only marginal proliferation and production of mitogenic cytokines following TCR stimulation in vitro (71). There is evidence that $CD4^+$ $CD25^+$ Tregs also exist in mice. A unique subset of murine $CD4^+$ $CD25^+$ Tregs identified by the presence of integrin-$\alpha_E\beta_7$ (CD103) was shown to have a marked capacity to prevent inflammatory bowel disease (45). This subset of Tregs suppresses in vitro conventional $CD4^+$ T cells in a contact-dependent manner and induces IL-10 secretion by cocultured $CD25^-$ $CD4^+$ T cells (45).

The second subtype of $CD4^+$ Tregs, designated "induced Tregs," in contrast to naturally occurring $CD4^+$ $CD25^+$ Tregs, are cell contact independent and mediate their activity mainly though soluble suppressive cytokines such as IL-10 and TGF-β. These cells are secondary suppressor T cells, and they develop from conventional $CD4^+$ $CD25^-$ Tregs at the periphery (45). Two types of induced Tregs have been described: regulatory type 1 (Tr1) and Th type 3 (Th3). IL-10-secreting Tr1 cells were shown to be effective in preventing chronic intestinal inflammation, and TGF-β-secreting Th3 cells were postulated to be involved in the maintenance of tolerance to dietary antigen (75). In the absence of regulatory T cells, alymphoid animals reconstituted with CD4 cells isolated from normal mice develop severe inflammatory bowel disease if colonized by enteric bacteria (96) or develop lethal pneumonia if infected by *Pneumocystis carinii* (37).

Expansion of $CD8^+$ T cells with regulatory function after interaction with IEC has also been described (1). Suppressive activity was found to be mediated by a $CD101^+$ $CD103^+$ subset of IEC-activated $CD8^+$ LPL and appeared to require cell contact. Moreover, it has been found that deficiency of $CD8^+$ T cells, although not affecting systemic tolerance to fed antigen, did abolish the suppression of local gut IgA that is specific for antigen, indicating that tolerance to oral antigen is compartmentalized and requires $CD8^+$ T cells for local suppression of IgA responses (29). The mechanisms by which these regulatory $CD8^+$ T cells exert their activity are still unknown. However, a recent study suggests that a subpopulation of $CD8^+$ T cells might suppress the response of successfully activated $CD4^+$ T cells and B cells through an interaction that depends on expression of the MHC class Ib molecule Qa-1 on activated target cells (39). It was shown that $CD8^+$ Tregs suppress $CD4^+$ T cells expressing the same target Qa-1–self peptide complex. Preliminary evidence for the existence of this mechanism in humans is supported by experiments in vitro showing that human $CD8^+$ T cells can be induced to differentiate into regulatory cells whose function is dependent on HLA-E, the human homologue of Qa-1 (15).

CONCLUDING REMARKS

Mucosal surfaces are continually exposed to massive numbers of commensal and potentially pathogenic organisms, as well as many environmental antigens. The immune system of mucosal surfaces has evolved a wide array of mechanisms to protect the host from pathogens while maintaining a relatively "peaceful coexistence" with commensal organisms, avoiding exaggerated responses to food and other environmental antigens and insults, and avoiding the generation of autoimmune disease. In recent years, our understanding of these mechanisms has grown exponentially in many areas, including the interactions between innate and adaptive immune responses, the role and transport of Igs across the epithelial layer, the immunoregulatory effects of cytokines and chemokines, the expression of key molecules directing lymphoid cell homing to mucosal surfaces, and the modulatory mechanisms that down-regulate immune responses. However, despite this remarkable progress in our knowledge of the immune mechanisms underlying protective immunity at mucosal surfaces, our understanding is far from complete. For example, it is well established that "uncontrolled" inflammatory responses to pathogenic or commensal organisms and food antigens lead to many of the pathological conditions observed in the gastrointestinal tract. However, very little is known about the events leading to these disease states. One of the biggest challenges that lies ahead is the unraveling of the extraordinarily complex mechanisms that underlie the generation of effective immune responses to potentially pathogenic organisms while controlling inflammatory responses to commensal organisms and food antigens. This knowledge will greatly enhance our ability to prevent inflammatory diseases in organs lined by large mucosal surfaces. Moreover, this information will also be invaluable in designing new generations of vaccines that can be administered via mucosal surfaces, such as attenuated live vectors, which have the potential of inducing strong mucosal and systemic immune responses.

REFERENCES

1. **Allez, M., J. Brimnes, I. Dotan, and L. Mayer.** 2002. Expansion of $CD8^+$ T cells with regulatory function after interaction with intestinal epithelial cells. *Gastroenterology* **123:** 1516–1526.
2. **Bargatze, R. F., M. A. Jutila, and E. C. Butcher.** 1995. Distinct roles of L-selectin and integrins alpha 4 beta 7 and LFA-1 in lymphocyte homing to Peyer's patch-HEV in situ: the multistep model confirmed and refined. *Immunity* **3:**99–108.
3. **Becker, C., S. Wirtz, M. Blessing, J. Pirhonen, D. Strand, O. Bechthold, J. Frick, P. R. Galle, I. Autenrieth, and M. F.**

Neurath. 2003. Constitutive p40 promoter activation and IL-23 production in the terminal ileum mediated by dendritic cells. *J. Clin. Investig.* **112:**693–706.

4. **Belyakov, I. M., S. A. Hammond, J. D. Ahlers, G. M. Glenn, and J. A. Berzofsky.** 2004. Transcutaneous immunization induces mucosal CTLs and protective immunity by migration of primed skin dendritic cells. *J. Clin. Investig.* **113:**998–1007.
5. **Blumberg, R. S., C. E. Yockey, G. G. Gross, E. C. Ebert, and S. P. Balk.** 1993. Human intestinal intraepithelial lymphocytes are derived from a limited number of T cell clones that utilize multiple V beta T cell receptor genes. *J. Immunol.* **150:**5144–5153.
6. **Brandtzaeg, P.** 2003. Role of secretory antibodies in the defence against infections. *Int. J. Med. Microbiol.* **293:**3–15.
7. **Brandtzaeg, P., I. N. Farstad, and G. Haraldsen.** 1999. Regional specialization in the mucosal immune system: primed cells do not always home along the same track. *Immunol. Today* **20:**267–277.
8. **Brandtzaeg, P., I. N. Farstad, F. E. Johansen, H. C. Morton, I. N. Norderhaug, and T. Yamanaka.** 1999. The B-cell system of human mucosae and exocrine glands. *Immunol. Rev.* **171:**45–87.
9. **Brightbill, H. D., and R. L. Modlin.** 2000. Toll-like receptors: molecular mechanisms of the mammalian immune response. *Immunology* **101:**1–10.
10. **Brigl, M., and M. B. Brenner.** 2004. CD1: Antigen presentation and T cell function. *Annu. Rev. Immunol.* **22:**817–890.
11. **Butcher, E. C., and L. J. Picker.** 1996. Lymphocyte homing and homeostasis. *Science* **272:**60–66.
12. **Butcher, E. C., M. Williams, K. Youngman, L. Rott, and M. Briskin.** 1999. Lymphocyte trafficking and regional immunity. *Adv. Immunol.* **72:**209–253.
13. **Campbell, D. J., and E. C. Butcher.** 2002. Rapid acquisition of tissue-specific homing phenotypes by $CD4^+$ T cells activated in cutaneous or mucosal lymphoid tissues. *J. Exp. Med.* **195:**135–141.
14. **Cheroutre, H., and L. Madakamutil.** 2004. Acquired and natural memory T cells join forces at the mucosal front line. *Nat. Rev. Immunol.* **4:**290–300.
15. **Chess, L., and H. Jiang.** 2004. Resurrecting $CD8^+$ suppressor T cells. *Nat. Immunol.* **5:**469–471.
16. **Davis, M. M., and P. J. Bjorkman.** 1988. T-cell antigen receptor genes and T-cell recognition. *Nature* **334:**395–402.
17. **den Haan, J. M., and M. J. Bevan.** 2001. Antigen presentation to $CD8^+$ T cells: cross-priming in infectious diseases. *Curr. Opin. Immunol.* **13:**437–441.
18. **Dickinson, B. L., K. Badizadegan, Z. Wu, J. C. Ahouse, X. Zhu, N. E. Simister, R. S. Blumberg, and W. I. Lencer.** 1999. Bidirectional FcRn-dependent IgG transport in a polarized human intestinal epithelial cell line. *J. Clin. Investig.* **104:**903–911.
19. **Elson, C. O., and J. Mestecky.** 2002. The mucosal immune system, p. 127–139. *In* M. J. Blaser, P. D. Smith, H. B. Greenberg, J. I. Ravdin, and R. L. Guerrant (ed.), *Infections of the Gastrointestinal Tract.* Lippincott Williams & Wilkins, Philadelphia, Pa.
20. **Fagarasan, S., and T. Honjo.** 2004. Regulation of IgA synthesis at mucosal surfaces. *Curr. Opin. Immunol.* **16:**277–283.
21. **Fagarasan, S., K. Kinoshita, M. Muramatsu, K. Ikuta, and T. Honjo.** 2001. In situ class switching and differentiation to IgA-producing cells in the gut lamina propria. *Nature* **413:**639–643.
22. **Fahrer, A. M., Y. Konigshofer, E. M. Kerr, G. Ghandour, D. H. Mack, M. M. Davis, and Y. H. Chien.** 2001. Attributes of gammadelta intraepithelial lymphocytes as suggested by their transcriptional profile. *Proc. Natl. Acad. Sci. USA* **98:**10261–10266.
23. **Farstad, I. N., H. Carlsen, H. C. Morton, and P. Brandtzaeg.** 2000. Immunoglobulin A cell distribution in the human small intestine: phenotypic and functional characteristics. *Immunology* **101:**354–363.
24. **Fayette, J., B. Dubois, S. Vandenabeele, J. M. Bridon, B. Vanbervliet, I. Durand, J. Banchereau, C. Caux, and F. Briere.** 1997. Human dendritic cells skew isotype switching of CD40-activated naive B cells towards IgA1 and IgA2. *J. Exp. Med.* **185:**1909–1918.
25. **Fernandez, M. I., T. Pedron, R. Tournebize, J. C. Olivo-Marin, P. J. Sansonetti, and A. Phalipon.** 2003. Anti-inflammatory role for intracellular dimeric immunoglobulin a by neutralization of lipopolysaccharide in epithelial cells. *Immunity* **18:**739–749.
26. **Fuss, I. J., M. Neurath, M. Boirivant, J. S. Klein, C. de la Motte, S. A. Strong, C. Fiocchi, and W. Strober.** 1996. Disparate $CD4^+$ lamina propria (LP) lymphokine secretion profiles in inflammatory bowel disease. Crohn's disease LP cells manifest increased secretion of IFN-gamma, whereas ulcerative colitis LP cells manifest increased secretion of IL-5. *J. Immunol.* **157:**1261–1270.
27. **Granucci, F., and P. Ricciardi-Castagnoli.** 2003. Interactions of bacterial pathogens with dendritic cells during invasion of mucosal surfaces. *Curr. Opin. Microbiol.* **6:**72–76.
28. **Granucci, F., I. Zanoni, S. Feau, and P. Ricciardi-Castagnoli.** 2003. Dendritic cell regulation of immune responses: a new role for interleukin 2 at the intersection of innate and adaptive immunity. *EMBO J.* **22:**2546–2551.
29. **Grdic, D., E. Hornquist, M. Kjerrulf, and N. Y. Lycke.** 1998. Lack of local suppression in orally tolerant CD8-deficient mice reveals a critical regulatory role of $CD8^+$ T cells in the normal gut mucosa. *J. Immunol.* **160:**754–762.
30. **Greene, W. C., W. J. Leonard, and J. M. Depper.** 1986. Growth of human T lymphocytes: an analysis of interleukin 2 and its cellular receptor. *Prog. Hematol.* **14:**283–301.
31. **Groh, V., A. Steinle, S. Bauer, and T. Spies.** 1998. Recognition of stress-induced MHC molecules by intestinal epithelial gammadelta T cells. *Science* **279:**1737–1740.
32. **Guermonprez, P., L. Saveanu, M. Kleijmeer, J. Davoust, P. Van Endert, and S. Amigorena.** 2003. ER-phagosome fusion defines an MHC class I cross-presentation compartment in dendritic cells. *Nature* **425:**397–402.
33. **Hamann, A., D. P. Andrew, D. Jablonski-Westrich, B. Holzmann, and E. C. Butcher.** 1994. Role of alpha 4-integrins in lymphocyte homing to mucosal tissues in vivo. *J. Immunol.* **152:**3282–3293.
34. **Hauer, A. C., E. J. Breese, J. A. Walker-Smith, and T. T. MacDonald.** 1997. The frequency of cells secreting interferon-gamma and interleukin-4, -5, and -10 in the blood and duodenal mucosa of children with cow's milk hypersensitivity. *Pediatr. Res.* **42:**629–638.
35. **Hayday, A., E. Theodoridis, E. Ramsburg, and J. Shires.** 2001. Intraepithelial lymphocytes: exploring the Third Way in immunology. *Nat. Immunol.* **2:**997–1003.
36. **Hershberg, R. M., and L. F. Mayer.** 2000. Antigen processing and presentation by intestinal epithelial cells—polarity and complexity. *Immunol. Today* **21:**123–128.
37. **Hori, S., T. L. Carvalho, and J. Demengeot.** 2002. $CD25^+CD4^+$ regulatory T cells suppress $CD4^+$ T cell-mediated pulmonary hyperinflammation driven by *Pneumocystis carinii* in immunodeficient mice. *Eur. J. Immunol.* **32:**1282–1291.
38. **Houde, M., S. Bertholet, E. Gagnon, S. Brunet, G. Goyette, A. Laplante, M. F. Princiotta, P. Thibault, D. Sacks, and**

M. Desjardins. 2003. Phagosomes are competent organelles for antigen cross-presentation. *Nature* **425:**402–406.

39. **Hu, D., K. Ikizawa, L. Lu, M. E. Sanchirico, M. L. Shinohara, and H. Cantor.** 2004. Analysis of regulatory CD8 T cells in Qa-1-deficient mice. *Nat. Immunol.* **5:**516–523.
40. **Hutchings, A. B., A. Helander, K. J. Silvey, K. Chandran, W. T. Lucas, M. L. Nibert, and M. R. Neutra.** 2004. Secretory immunoglobulin A antibodies against the sigma1 outer capsid protein of reovirus type 1 Lang prevent infection of mouse Peyer's patches. *J. Virol.* **78:**947–957.
41. **Israel, E. J., S. Taylor, Z. Wu, E. Mizoguchi, R. S. Blumberg, A. Bhan, and N. E. Simister.** 1997. Expression of the neonatal Fc receptor, FcRn, on human intestinal epithelial cells. *Immunology* **92:**69–74.
42. **Iwasaki, A., and B. L. Kelsall.** 2000. Localization of distinct Peyer's patch dendritic cell subsets and their recruitment by chemokines macrophage inflammatory protein (MIP)-3alpha, MIP-3beta, and secondary lymphoid organ chemokine. *J. Exp. Med.* **191:**1381–1394.
43. **Iwasaki, A., and B. L. Kelsall.** 2001. Unique functions of $CD11b^+$, CD8 $alpha^+$, and double-negative Peyer's patch dendritic cells. *J. Immunol.* **166:**4884–4890.
44. **James, S. P.** 2002. Cellular immune mechanisms of defense in the gastrointestinal tract (including mast cells), p. 183–198. *In* M. J. Blaser, P. D. Smith, H. B. Greenberg, J. I. Ravdin, and R. L. Guerrant (ed.), *Infections of the Gastrointestinal Tract.* Lippincott Williams & Wilkins, Philadelphia, Pa.
45. **Jonuleit, H., and E. Schmitt.** 2003. The regulatory T cell family: distinct subsets and their interrelations. *J. Immunol.* **171:**6323–6327.
46. **Kaisho, T., and S. Akira.** 2001. Dendritic-cell function in Toll-like receptor- and MyD88-knockout mice. *Trends Immunol.* **22:**78–83.
47. **Kantele, A., J. Zivny, M. Hakkinen, C. O. Elson, and J. Mestecky.** 1999. Differential homing commitments of antigen-specific T cells after oral or parenteral immunization in humans. *J. Immunol.* **162:**5173–5177.
48. **Kelsall, B. L., C. A. Biron, O. Sharma, and P. M. Kaye.** 2002. Dendritic cells at the host-pathogen interface. *Nat. Immunol.* **3:**699–702.
49. **Kelsall, B. L., and W. Strober.** 1996. Distinct populations of dendritic cells are present in the subepithelial dome and T cell regions of the murine Peyer's patch. *J. Exp. Med.* **183:**237–247.
50. **Kilian, M., J. Reinholdt, H. Lomholt, K. Poulsen, and E. V. Frandsen.** 1996. Biological significance of IgA1 proteases in bacterial colonization and pathogenesis: critical evaluation of experimental evidence. *APMIS* **104:**321–338.
51. **Kim, S. K., K. S. Schluns, and L. Lefrancois.** 1999. Induction and visualization of mucosal memory CD8 T cells following systemic virus infection. *J. Immunol.* **163:**4125–4132.
52. **Komano, H., Y. Fujiura, M. Kawaguchi, S. Matsumoto, Y. Hashimoto, S. Obana, P. Mombaerts, S. Tonegawa, H. Yamamoto, S. Itohara, et al.** 1995. Homeostatic regulation of intestinal epithelia by intraepithelial gamma delta T cells. *Proc. Natl. Acad. Sci. USA* **92:**6147–6151.
53. **Kroese, F. G., and N. A. Bos.** 1999. Peritoneal B-1 cells switch in vivo to IgA and these IgA antibodies can bind to bacteria of the normal intestinal microflora. *Curr. Top. Microbiol. Immunol.* **246:**343–349.
54. **Kunkel, E. J., and E. C. Butcher.** 2002. Chemokines and the tissue-specific migration of lymphocytes. *Immunity* **16:**1–4.
55. **Kunkel, E. J., and E. C. Butcher.** 2003. Plasma-cell homing. *Nat. Rev. Immunol.* **3:**822–829.
56. **Kunkel, E. J., J. J. Campbell, G. Haraldsen, J. Pan, J. Boisvert, A. I. Roberts, E. C. Ebert, M. A. Vierra, S. B. Goodman, M. C. Genovese, A. J. Wardlaw, H. B. Greenberg, C. M. Parker, E. C. Butcher, D. P. Andrew, and W. W. Agace.** 2000. Lymphocyte CC chemokine receptor 9 and epithelial thymus-expressed chemokine (TECK) expression distinguish the small intestinal immune compartment: epithelial expression of tissue-specific chemokines as an organizing principle in regional immunity. *J. Exp. Med.* **192:**761–768.
57. **Laky, K., L. Lefrancois, U. Freeden-Jeffry, R. Murray, and L. Puddington.** 1998. The role of IL-7 in thymic and extrathymic development of TCR gamma delta cells. *J. Immunol.* **161:**707–713.
58. **Lamm, M. E.** 2002. Structure and function of mucosal immunoglobulin A, p. 157–163. *In* M. J. Blaser, P. D. Smith, H. B. Greenberg, J. I. Ravdin, and R. L. Guerrant (ed.), *Infections of the Gastrointestinal Tract.* Lippincott Williams & Wilkins, Philadelphia, Pa.
59. **Lanzavecchia, A., and F. Sallusto.** 2000. Dynamics of T lymphocyte responses: intermediates, effectors, and memory cells. *Science* **290:**92–97.
60. **Lanzavecchia, A., and F. Sallusto.** 2001. The instructive role of dendritic cells on T cell responses: lineages, plasticity and kinetics. *Curr. Opin. Immunol.* **13:**291–298.
61. **Lieberman, J., and F. R. Frankel.** 2002. Engineered *Listeria monocytogenes* as an AIDS vaccine. *Vaccine* **20:**2007–2010.
62. **Lodolce, J. P., D. L. Boone, S. Chai, R. E. Swain, T. Dassopoulos, S. Trettin, and A. Ma.** 1998. IL-15 receptor maintains lymphoid homeostasis by supporting lymphocyte homing and proliferation. *Immunity* **9:**669–676.
63. **Lodolce, J. P., D. L. Boone, S. Chai, R. E. Swain, T. Dassopoulos, S. Trettin, and A. Ma.** 1998. IL-15 receptor maintains lymphoid homeostasis by supporting lymphocyte homing and proliferation. *Immunity* **9:**669–676.
64. **MacDonald, T. T., and G. Monteleone.** 2001. IL-12 and Th1 immune responses in human Peyer's patches. *Trends Immunol.* **22:**244–247.
65. **Macpherson, A. J., D. Gatto, E. Sainsbury, G. R. Harriman, H. Hengartner, and R. M. Zinkernagel.** 2000. A primitive T cell-independent mechanism of intestinal mucosal IgA responses to commensal bacteria. *Science* **288:**2222–2226.
66. **Mantis, N. J., M. C. Cheung, K. R. Chintalacharuvu, J. Rey, B. Corthesy, and M. R. Neutra.** 2002. Selective adherence of IgA to murine Peyer's patch M cells: evidence for a novel IgA receptor. *J. Immunol.* **169:**1844–1851.
67. **Masopust, D., J. Jiang, H. Shen, and L. Lefrancois.** 2001. Direct analysis of the dynamics of the intestinal mucosa CD8 T cell response to systemic virus infection. *J. Immunol.* **166:**2348–2356.
68. **Masopust, D., V. Vezys, A. L. Marzo, and L. Lefrancois.** 2001. Preferential localization of effector memory cells in nonlymphoid tissue. *Science* **291:**2413–2417.
69. **Mazanec, M. B., C. S. Kaetzel, M. E. Lamm, D. Fletcher, and J. G. Nedrud.** 1992. Intracellular neutralization of virus by immunoglobulin A antibodies. *Proc. Natl. Acad. Sci. USA* **89:**6901–6905.
70. **McGuirk, P., C. McCann, and K. H. Mills.** 2002. Pathogen-specific T regulatory 1 cells induced in the respiratory tract by a bacterial molecule that stimulates interleukin 10 production by dendritic cells: a novel strategy for evasion of protective T helper type 1 responses by *Bordetella pertussis. J. Exp. Med.* **195:**221–231.
71. **Mittrucker, H. W., and S. H. Kaufmann.** 2004. Mini-review: regulatory T cells and infection: suppression revisited. *Eur. J. Immunol.* **34:**306–312.
72. **Mostov, K. E.** 1994. Transepithelial transport of immunoglobulins. *Annu. Rev. Immunol.* **12:**63–84.
73. **Muller, S., M. Buhler-Jungo, and C. Mueller.** 2000. Intestinal intraepithelial lymphocytes exert potent protective cytotoxic

activity during an acute virus infection. *J. Immunol.* **164:** 1986–1994.
74. **Nagler-Anderson, C.** 2001. Man the barrier! Strategic defences in the intestinal mucosa. *Nat. Rev. Immunol.* **1:**59–67.
75. **Nagler-Anderson, C., A. K. Bhan, D. K. Podolsky, and C. Terhorst.** 2004. Control freaks: immune regulatory cells. *Nat. Immunol.* **5:**119–122.
76. **Neutra, M. R., A. Frey, and J. P. Kraehenbuhl.** 1996. Epithelial M cells: gateways for mucosal infection and immunization. *Cell* **86:**345–348.
77. **Neutra, M. R., N. J. Mantis, and J. P. Kraehenbuhl.** 2001. Collaboration of epithelial cells with organized mucosal lymphoid tissues. *Nat. Immunol.* **2:**1004–1009.
78. **Neutra, M. R., E. Pringault, and J. P. Kraehenbuhl.** 1996. Antigen sampling across epithelial barriers and induction of mucosal immune responses. *Annu. Rev. Immunol.* **14:**275–300.
79. **Neutra, M. R., P. Sansonetti, K, and J. P. Kraehenbuhl.** 2002. Interactions of microbial pathogens with intestinal M cells, p. 141–156. *In* M. J. Blaser, P. D. Smith, H. B. Greenberg, J. I. Ravdin, and R. L. Guerrant (ed.), *Infections of the Gastrointestinal Tract.* Lippincott Williams & Wilkins, Philadelphia, Pa.
80. **Ogra, P. L., H. Faden, and R. C. Welliver.** 2001. Vaccination strategies for mucosal immune responses. *Clin. Microbiol. Rev.* **14:**430–445.
81. **Phalipon, A., A. Cardona, J. P. Kraehenbuhl, L. Edelman, P. J. Sansonetti, and B. Corthesy.** 2002. Secretory component: a new role in secretory IgA-mediated immune exclusion in vivo. *Immunity* **17:**107–115.
82. **Pilette, C., Y. Ouadrhiri, V. Godding, J. P. Vaerman, and Y. Sibille.** 2001. Lung mucosal immunity: immunoglobulin-A revisited. *Eur. Respir. J.* **18:**571–588.
83. **Pope, C., S. K. Kim, A. Marzo, D. Masopust, K. Williams, J. Jiang, H. Shen, and L. Lefrancois.** 2001. Organ-specific regulation of the CD8 T cell response to *Listeria monocytogenes* infection. *J. Immunol.* **166:**3402–3409.
84. **Quiding-Jarbrink, M., I. Nordstrom, G. Granstrom, A. Kilander, M. Jertborn, E. C. Butcher, A. I. Lazarovits, J. Holmgren, and C. Czerkinsky.** 1997. Differential expression of tissue-specific adhesion molecules on human circulating antibody-forming cells after systemic, enteric, and nasal immunizations. A molecular basis for the compartmentalization of effector B cell responses. *J. Clin. Investig.* **99:**1281–1286.
85. **Rescigno, M., M. Urbano, B. Valzasina, M. Francolini, G. Rotta, R. Bonasio, F. Granucci, J. P. Kraehenbuhl, and P. Ricciardi-Castagnoli.** 2001. Dendritic cells express tight junction proteins and penetrate gut epithelial monolayers to sample bacteria. *Nat. Immunol.* **2:**361–367.
86. **Rey, J., N. Garin, F. Spertini, and B. Corthesy.** 2004. Targeting of secretory IgA to Peyer's patch dendritic and T cells after transport by intestinal M cells. *J. Immunol.* **172:**3026–3033.
87. **Rissoan, M. C., V. Soumelis, N. Kadowaki, G. Grouard, F. Briere, M. R. de Waal, and Y. J. Liu.** 1999. Reciprocal control of T helper cell and dendritic cell differentiation. *Science* **283:**1183–1186.
88. **Robinson, J. K., T. G. Blanchard, A. D. Levine, S. N. Emancipator, and M. E. Lamm.** 2001. A mucosal IgA-mediated excretory immune system in vivo. *J. Immunol.* **166:**3688–3692.
89. **Rocha, B., P. Vassalli, and D. Guy-Grand.** 1994. Thymic and extrathymic origins of gut intraepithelial lymphocyte populations in mice. *J. Exp. Med.* **180:**681–686.
90. **Rojas, R., and G. Apodaca.** 2002. Immunoglobulin transport across polarized epithelial cells. *Nat. Rev. Mol. Cell Biol.* **3:**944–955.
91. **Sallusto, F., D. Lenig, R. Forster, M. Lipp, and A. Lanzavecchia.** 1999. Two subsets of memory T lymphocytes with distinct homing potentials and effector functions. *Nature* **401:** 708–712.
92. **Schwartz-Cornil, I., Y. Benureau, H. Greenberg, B. A. Hendrickson, and J. Cohen.** 2002. Heterologous protection induced by the inner capsid proteins of rotavirus requires transcytosis of mucosal immunoglobulins. *J. Virol.* **76:**8110–8117.
93. **Shires, J., E. Theodoridis, and A. C. Hayday.** 2001. Biological insights into TCRgammadelta$^+$ and TCRalphabeta$^+$ intraepithelial lymphocytes provided by serial analysis of gene expression (SAGE). *Immunity* **15:**419–434.
94. **Shreedhar, V., A. M. Moodycliffe, S. E. Ullrich, C. Bucana, M. L. Kripke, and L. Flores-Romo.** 1999. Dendritic cells require T cells for functional maturation in vivo. *Immunity* **11:**625–636.
95. **Sicinski, P., J. Rowinski, J. B. Warchol, Z. Jarzabek, W. Gut, B. Szczygiel, K. Bielecki, and G. Koch.** 1990. Poliovirus type 1 enters the human host through intestinal M cells. *Gastroenterology* **98:**56–58.
96. **Singh, B., S. Read, C. Asseman, V. Malmstrom, C. Mottet, L. A. Stephens, R. Stepankova, H. Tlaskalova, and F. Powrie.** 2001. Control of intestinal inflammation by regulatory T cells. *Immunol. Rev.* **182:**190–200.
97. **Spalding, D. M., S. I. Williamson, W. J. Koopman, and J. R. McGhee.** 1984. Preferential induction of polyclonal IgA secretion by murine Peyer's patch dendritic cell-T cell mixtures. *J. Exp. Med.* **160:**941–946.
98. **Sundquist, M., C. Johansson, and M. J. Wick.** 2003. Dendritic cells as inducers of antimicrobial immunity in vivo. *APMIS* **111:**715–724.
99. **Svensson, M., J. Marsal, A. Ericsson, L. Carramolino, T. Broden, G. Marquez, and W. W. Agace.** 2002. CCL25 mediates the localization of recently activated CD8alphabeta$^+$ lymphocytes to the small-intestinal mucosa. *J. Clin. Investig.* **110:**1113–1121.
100. **Szabo, S. J., A. S. Dighe, U. Gubler, and K. M. Murphy.** 1997. Regulation of the interleukin (IL)-12R beta 2 subunit expression in developing T helper 1 (Th1) and Th2 cells. *J. Exp. Med.* **185:**817–824.
101. **Valitutti, S., S. Muller, M. Dessing, and A. Lanzavecchia.** 1996. Signal extinction and T cell repolarization in T helper cell-antigen-presenting cell conjugates. *Eur. J. Immunol.* **26:**2012–2016.
102. **Van Kerckhove, C., G. J. Russell, K. Deusch, K. Reich, A. K. Bhan, H. DerSimonian, and M. B. Brenner.** 1992. Oligoclonality of human intestinal intraepithelial T cells. *J. Exp. Med.* **175:**57–63.
103. **Wykes, M., A. Pombo, C. Jenkins, and G. G. MacPherson.** 1998. Dendritic cells interact directly with naive B lymphocytes to transfer antigen and initiate class switching in a primary T-dependent response. *J. Immunol.* **161:**1313–1319.
104. **Yamanaka, T., L. Helgeland, I. N. Farstad, H. Fukushima, T. Midtvedt, and P. Brandtzaeg.** 2003. Microbial colonization drives lymphocyte accumulation and differentiation in the follicle-associated epithelium of Peyer's patches. *J. Immunol.* **170:**816–822.
105. **Yamanaka, T., A. Straumfors, H. Morton, O. Fausa, P. Brandtzaeg, and I. Farstad.** 2001. M cell pockets of human Peyer's patches are specialized extensions of germinal centers. *Eur. J. Immunol.* **31:**107–117.
106. **Yoshida, M., S. M. Claypool, J. S. Wagner, E. Mizoguchi, A. Mizoguchi, D. C. Roopenian, W. I. Lencer, and R. S. Blumberg.** 2004. Human neonatal fc receptor mediates transport of IgG into luminal secretions for delivery of antigens to mucosal dendritic cells. *Immunity* **20:**769–783.

Colonization of Mucosal Surfaces
Edited by James P. Nataro et al.

Chapter 4

In Situ Monitoring of Bacterial Presence and Activity

CLAUS STERNBERG, MICHAEL GIVSKOV, LEO EBERL, KAREN A. KROGFELT, AND SØREN MOLIN

Molecular microbial ecology has developed rapidly into a major discipline during the last decade. The transfer of molecular biology methods and concepts to environmental microbiology and microbial ecology has provided new insights into microbial complexity and activity in many different types of natural settings, and the rapidly expanding genomic databases have further accelerated the development. Comparative genomics, meta-genomes, techniques for in situ metabolic activity monitoring, and DNA chips for rapid identification of hundreds of species as well as for transcriptomic investigations are tools which have recently been added to those of fluorescent in situ hybridization (FISH) and reporter gene techniques. Although most of these methods have been used mainly in the context of environmental microbiology, they are also highly relevant in connection with investigation of the interactions between the human commensal microflora and pathogens. The possibilities of performing in situ examinations of both the composition and the physiology of these bacterial populations open up ecological approaches to the understanding of the coinhabitants of our bodies, which will complement the traditional clinical and molecular microbiological investigations of bacterial infections and the role of our normal flora as both a source of pathogens and a barrier against infection. Here we review the application of some of these in situ methods and tools in the context of mucosal colonization, after which we present one specific example—monitoring of quorum sensing-based cell-cell communication in colonized lungs of cystic fibrosis (CF) animal models.

FLUORESCENT IN SITU HYBRIDIZATION

Detection and identification of many of the organisms in the microflora of the intestinal tract have been significantly hampered by the incomplete knowledge of their culture conditions. Use of rRNA and its encoding genes as phylogenetic markers has led to characterization of the interactions between members of the same bacterial species and between different species in such communities and how such interactions can be affected by diet and health (85). The protocol described for rRNA FISH (68) is generally applicable to cells located in different natural environments. A natural limitation is the sample size and the bacterial load. As is the case with conventional microscopy, a large number of bacteria (10^5 to 10^6 per g of feces) must be present in order to be visible in intestinal sections. Frozen tissue sections (65) or paraffin-embedded tissue sections can be used (47). Both types of section can be made by standard histological methods, and the sections can subsequently be immobilized on coated microscope slides.

Identifying a specific pathogen based on FISH requires knowledge of that pathogen. Nevertheless, in cases where an infectious agent is a fastidious grower, FISH is an excellent choice. For example, human spirochetosis has been diagnosed by using a *Brachyspira* genus-specific probe and probes targeting 16S or 23S rRNA of *Brachyspira aalborgi* and *B. pilosicoli*. However, only 55% of the samples were identified as a species, in this case *B. aalborgi*. Spirochetes in biopsy specimens from the other patients hybridized only with the *Brachyspira* probe, possibly

Claus Sternberg, Michael Givskov, and Søren Molin • Centre for Biomedical Microbiology, BioCentrum-DTU, Technical University of Denmark, Building 301, 2800 Lyngby, Denmark. **Leo Eberl** • Department of Microbiology, Institute of Plant Biology, University of Zürich, CH-8008 Zürich, Switzerland. **Karen A. Krogfelt** • Department of Gastrointestinal Infections, Statens Serumsinstitut, 2300 Copenhagen S, Denmark.

demonstrating the involvement of as yet uncharacterized *Brachyspira* spirochetes in human intestinal spirochetosis (42). FISH targeting 18S rRNA was also shown to be a sensitive method for the specific diagnosis of *Pneumocystis carinii* pneumonia in foals and pigs, compared to the conventional staining techniques (41).

FISH has been successfully used in the diagnosis of *Helicobacter pylori* infections. *H. pylori* is one of the major infectious gastric agents. About 60% of the human population is infected worldwide. FISH can be performed on formalin-fixed and paraffin-embedded gastric biopsy specimens that have been prepared for pathological examination (84). Besides being used for *H. pylori* identification in gastric specimens, FISH has been used for genotypic determination of clarithomycin resistance in *H. pylori*. Whole-cell hybridization of rRNA holds great promise for cultivation-independent, reliable, and rapid genotypic determination of *H. pylori*. Compared with PCR techniques, FISH does not require nucleic acid preparations, is not prone to inhibition, and allows semiquantitative visualization of the bacteria within intact tissue samples. However, in the microbiology routine diagnostic laboratory, the combination of FISH and conventional culturing significantly increases the sensitivity in detection of *H. pylori* (71, 72).

Besides identification of the numerous species present in the intestine (2, 8, 9, 37, 83, 95, 96), ribosomal probes have been used extensively to show the effect of diet and of probiotics and prebiotics in the composition of the intestinal flora (6, 26, 52). Furthermore, by using a combination of specific and general probes, at least two-thirds of the members of the fecal flora have been detected (23, 27). In situ hybridization was also found to be a powerful tool in studying the distribution of *Escherichia coli* in different parts of the mouse gastrointestinal tract (65). In growing bacteria, the cellular content of ribosomes reflects their growth rate (19); i.e., the faster the cells grow, the more ribosomes are in the cells. In special quantitative applications of FISH, this particular relationship has been used to obtain in situ average estimates of the growth rates of *E. coli* cells in the large intestines of mice (54, 66). In a subsequent investigation, it was shown that essentially all *E. coli* growth was localized to the mucus; i.e., *E. coli* cells did not grow at all in the lumen of the intestine (48).

To determine whether specific bacterial virulence factors play a role in the mammalian host, it is important to know whether these factors are actually produced during infection. This has been achieved by combining molecular and fluorescence-based techniques for visualizing bacterial cells in mammalian tissues. The adhesive type 1 fimbriae expressed by virtually all *E. coli* strains have been studied extensively and were found to be produced during infection in experimental-animal models (13, 78). *E. coli* surface components and the genes that regulate their synthesis are important for colonization of the gut and for the spatial distribution of the bacteria in the intestine (55, 86). The role of *Salmonella enterica* serovar Typhimurium lipopolysaccharide in the mouse intestine has been studied; it was shown that the O side chain is important for penetration through the mucus layer and attachment to the gut epithelium (i.e., an O-side-chain mutant failed to penetrate deeply and became "stuck" in the mucus layer) (47). Although it was found that the presence or absence of the *Klebsiella pneumoniae* capsule had no apparent effect on its colonization of the gastrointestinal tract, the *K. pneumoniae* capsule was identified as an important virulence factor in the urinary tract (77). In contrast, in vitro expression of the capsule dramatically reduced the ability of *K. pneumoniae* to bind to epithelial cells compared to its noncapsulated variant (22, 77). These results demonstrate the great caution needed when extrapolating from results of in vitro studies to the in vivo situation and emphasize the necessity for in vivo models in studies of bacterial virulence.

Fluorescent Proteins as Cellular Markers

Direct-labeling in situ methods (such as FISH) provides insight into the identity and location of bacterial cells in natural systems and is frequently the only available option. However, when the species under investigation is known and genetic manipulation is possible, additional options are at hand. Apart from knowing exactly where a given cell is positioned, we often would like to know what it is actually doing. Fluorescent proteins facilitate the monitoring of gene expression in situ, since these components can be expressed in a very wide range of organisms and usually add little metabolic load to their host cells. In general, fluorescent proteins are inherently fluorescent; i.e., there is no need for the addition of external factors to initiate fluorescence (other than low concentrations of oxygen), thereby facilitating observations of live systems.

The first fluorescent protein to be useful as a tool for gene expression was the green fluorescent protein (GFP) from the jellyfish *Aequorea victoria*. GFP had been studied for several decades (57, 73) before the gene was cloned in 1994 (11). Almost immediately thereafter, the *gfp* gene became the marker of choice in many molecular microbiology and cell biology studies. Wild-type GFP consists of 238 amino acids

(27 kDa) and is excited by long-wave UV to blue light. The excitation spectrum has a major peak at 395 nm and a minor one at 470 nm. Emission is at 507 nm, with a shoulder at 540 nm (56, 73). The maturation time of the newly synthesized protein is several hours, effectively incapacitating the protein as a marker in dynamic gene expression studies (30). Consequently, GFP has been extensively modified to optimize several of its features. The excitation optimum was changed toward the red end (red shifted), while the emission wavelength was left largely unaltered. The first such mutant, S65T, was characterized by a single-amino-acid exchange (serine 65 was replaced by threonine) (28). This mutant protein has an excitation maximum at 488 nm and an emission maximum at 509 nm. Also, the very long maturation time of GFP was shortened by genetic modification of the gene. Cormack et al. (14) used fluorescence-activated cell sorting to obtain brighter mutants, which turned out to also have notably shorter maturation times. The mut1 to mut3 mutants showed 50% maturation of the protein 30 min after synthesis, whereas the wild type required 2 h.

As the name implies, the *A. victoria* GFP is green, but the need for other colors was soon recognized. Several variants with other emission wavelengths were constructed by random mutagenesis. Due to the chemical nature of the protein, the range in which the wavelengths could vary was somewhat limited, i.e., from bluish to greenish yellow. The classical variants include BFP (blue), CFP (cyan), and YFP (yellow) (29, 30, 59, 74). UV light of 380 nm is required to excite BFP, which emits blue light at 440 nm. CFP is excited by light at 433 and 453 nm and emits at 475 and 501 nm. YFP is optimally excited by green light (513 nm) and emits at 527 nm (greenish yellow).

The quest for variants of *A. victoria* GFP with true red emission turned out to be futile, but it was found that GFP synthesized in the absence of oxygen produced a deep red fluorescent protein, which rapidly turned green on exposure to oxygen (30). True red fluorescent proteins (RFPs) were only found when the source protein came from organisms other than *A. victoria*. The gene encoding the first RFP from the coral *Discomonas* sp. was cloned and expressed by Matz et al. (51). The protein, designated DsRed, fluoresces red at 583 nm. The excitation maximum is 558 nm. The DsRed protein has a very long maturation time, even compared to wild-type GFP. Detection of the protein in cells expressing DsRed can be achieved in 24 h, although the intensity remains relatively low. As was the case with GFP, DsRed has been mutagenized extensively, and as a result, variants with shorter maturation times and decreased tendencies to form aggregates have been identified, e.g., DsRed2. The same is true for DsRed-Express (7), which has six amino acid changes compared to the wild-type DsRed.

Other commercially available fluorescent proteins include AmCyan from the sea anemone *Anemonea majano* (458-nm excitation; 489-nm emission), (51), AsRed2 from *Anemonia sulcata* (576-nm excitation; 592-nm emission) (49), HcRed from *Heterectis crispa* (588-nm excitation; 618-nm emission), (25), ZsGreen from *Zoanthus* sp. (493-nm excitation; 505-nm emission), and ZsYellow (529-nm excitation; 539-nm emission (51).

The majority of the modifications mentioned above aim at altering the maturation times, excitation or emission wavelengths, or aggregation tendency. However, modifications have been made which alter other properties, in particular the stability of the proteins. Since most fluorescent proteins are very stable, only dilution by cell division will lead to reduced cellular fluorescence signals. For gene expression studies, the extreme stability of the protein reduces the usefulness of using GFP as a gene activity marker. Two methods have been used to destabilize the proteins. For use in eukaryotic cells, part of the mouse ornithine decarboxylase was fused to the C-terminal end of the GFP. The motif from the mouse gene encodes a so-called PEST domain, which is the target for indigenous protein degradation in higher organisms (46). For destabilizing GFP when expressed in prokaryotes, a short oligopeptide (3 to 10 amino acids) containing the target sequence for the ClpP/X protease was added to the C-terminal end of the protein (3). In both systems, protein degradation is greatly accelerated, making dynamic investigations possible.

To investigate cell or organelle movements, photobleaching techniques can be used. For instance, fluorescence recovery after photobleaching can be used to monitor particle movements in a defined area. In brief, a small area of a cell containing fluorescently labeled particles (e.g., organelles or macromolecules) is illuminated with a light source that bleaches the fluorophore. Subsequently, the bleached area is observed over time for recovery of fluorescence due to the migration of fluorescent particles into this area. This gives information on motion dynamics or solute dissipation in the cell. For reviews, see, e.g., Meyvis et al. (53) or, with GFP as fluorophore, Wouters et al. (90). Photoactivated GFP is a novel variant which is essentially nonfluorescent until illuminated with light of a certain wavelength. For example, the variant PA-GFP can be observed using 488 nm and virtually no signal is detected. However, after photoactivation using 413-nm light, the protein increases in fluorescence more than 100-fold (61). The activation occurs within seconds, and the activated proteins remain fluorescent for days. Photoactivation enables the observation of diffusion or active movements inside a cell much like the fluorescence recovery after photobleaching method. Other photoactivated proteins with similar properties have been described, e.g., the Keade protein from the stony coral *Trachypyllia geoffroyi* (5) and the KFP1 protein, a modified form of the sea anemone *Anemona sulcata* asCP fluorescent protein (12). Both of these proteins change color from green to red on illumination.

At the time of its discovery, it was suggested that DsRed required an additional autocatalytic reaction before red fluorescence could be achieved (51). During the protein maturation process, the occurrence of an intermediate, possibly fluorescent species was therefore expected. Using error-prone PCR cloning, Terskikh et al. (82) generated a range of mutants of DsRed, which had in common an altered fluorescence. Some of these mutants fluoresced red, some fluoresced green, and some fluoresced yellow or orange, indicating a mixed emission of red and green fluorescence. One mutant, E5, is particularly interesting since it showed a temporal dependency on emission wavelengths. Shortly after synthesis, the protein emits green fluorescence and then changes conformation after some time and switches to red fluorescence. This "fluorescent timer" has been used for studying newly activated genes (green) in regions of *Caenorhabditis elegans* embryos, regions with constant gene expression (yellow), and regions with little remaining gene expression following a a period of a higher level of expression (red).

Relatively little work has been done with fluorescent proteins used as markers in animal gut systems. GFP and its relatives require oxygen for activation, and the gut is known as a generally anaerobic environment. This poses limitations on possible applications, since the proteins cannot fold correctly in the absence of oxygen. Rather, the proteins must be fluorescent on introduction to the gut or visualized by exposure to air after completion of the experiment.

Gene fusions to monitor gene activity in situ have been used in animal model systems. For example, two *Yersinia enterocolitica* genes involved in iron scavenging, *fyuA* and *hemR,* were fused to *gfp,* and recombinant *Y. enterocolitica* cells were used to infect mice. Subsequently, fluorescence was detected from affected organs of the infected mice. While quantitative measurements could be obtained from cell homogenates, localization of the bacteria was verified by confocal microscopy of the affected organs (40).

Using *Lactococcus lactis* transformed with pGFPuv, a plasmid expressing a UV-light-excitable variant of GFP, the degree of cell lysis in different parts of the rat digestive tract was examined (20). The animals were given *L. lactis* cells containing pGFPuv in their food. When the rats were sacrificed, cells isolated from various compartments of the digestive tract were analyzed for green fluorescence and viable counts. Since only about 10% of the *L. lactis* cells survive in the intestine, the parallel loss of fluorescence and viable counts was taken as an indication of cell lysis. These data suggest the possibility that *L. lactis* might be used as a vehicle for delivering, e.g., digestive enzymes to treat patients with pancreatic deficiencies.

In a very interesting study by Zhao et al. (93), GFP was visualized inside intact animals. Mice were fed with brightly green fluorescent *E. coli,* and the fate of the inoculum was monitored from the outside by illuminating the entire animal and recording color photographs of the resulting fluorescence. The digestive system could be readily identified, and passage of the bacteria could be timed. Some animals were treated with antibiotics, and the effect was demonstrated as a reduction and eventual vanishing of fluorescence. The same procedure was used for experimental peritoneal and colonic infections, and in these cases whole-body visualization of the tagged bacteria was also achieved. Methods like these open up the possibility of new and highly informative investigations of bacterial interactions with animal host models. Obviously the efficacy and perseverance of antimicrobial drugs may be assessed directly in intact animals by using this technique. In fact, the methods of in situ monitoring of cell-cell communication described in the following section may soon be complemented by intact-animal inspection focusing on the lungs of *Pseudomonas aeruginosa*-infected normal mice and those with CF.

USE OF GREEN FLUORESCENT PROTEIN REPORTERS FOR IN SITU STUDIES OF CELL-CELL COMMUNICATION

In most pathogenic bacteria, expression of virulence factors is not constitutive but is regulated in a cell density-dependent manner. In this way, the organisms ensure that pathogenic traits are expressed only when the bacterial population density is high enough to overwhelm the host before it is able to mount an efficient response. To monitor the size of the population, many gram-negtive bacteria utilize a cell-cell communication system that relies on diffusible *N*-acylhomoserine lactone (AHL) signal molecules in a process known as quorum sensing (for recent reviews, see references 10, 18, 87, and 88).

These communication systems depend on an AHL synthase, usually a member of the LuxI family of proteins, and an AHL receptor protein, which belongs to the LuxR family of transcriptional regulators. At low population densities, cells produce a basal level of AHL via the activity of the AHL synthase. As the cell density increases, the diffusible AHL signal molecule accumulates in the growth medium. On reaching a critical threshold concentration, the AHL binds to the cognate LuxR-type receptor protein, which in turn leads to the induction or repression of target genes. Since quorum sensing is a regulatory principle that coordinates the activity of individual cells within a bacterial population, it obviously represents a particularly worthwhile target for the application of GFP single-cell technology.

The best-investigated example of an organism utilizing AHLs to control expression of virulence factors is the opportunistic human pathogen *P. aeruginosa*. This ubiquitous environmental organism represents an increasingly prevalent opportunistic human pathogen and is the most common gram-negative bacterium found in nosocomial and life-threatening infections of immunocompromised patients (1, 87). Patients with CF are especially disposed to *P. aeruginosa* infections, and for these persons the bacterium is responsible for high rates of morbidity and mortality (35, 50). Two quorum-sensing systems have been identified in *P. aeruginosa* that orchestrate the expression of virulence factors and participate in the development of biofilms: the *las* system, consisting of the transcriptional activator LasR and the AHL synthase LasI, which directs the synthesis of *N*-3-oxo-dodecanoylhomoserine lactone (3-oxo-C_{12}-HSL), and the *rhl* system, consisting of RhlR and RhlI, which directs the synthesis of *N*-butanoylhomoserine lactone (C_4-HSL). The two systems do not operate independently, since the *las* system positively regulates the expression of both *rhlR* and *rhlI*. Thus, the two quorum-sensing systems of *P. aeruginosa* are hierarchically arranged, with the *las* system being on top of the signaling cascade (44, 64). In complex interplays with additional regulators, including Vfr, GacA, RsaL, MvaT, and RpoS, the quorum-sensing cascade regulates the expression of a battery of extracellular virulence factors such as exoenzymes (elastase and alkaline protease), secondary metabolites (pyocyanin, hydrogen cyanide, and pyoverdin), and toxins (exotoxin A) (60, 89) and the development of biofilms (17, 31, 92). The pivotal role of quorum sensing for the pathogenicity of *P. aeruginosa* has been demonstrated in experiments with a number of animal models, including the neonatal-mouse model of pneumonia (81), the burned-mouse model (70), the mouse agar beat model (31, 91), and a *C. elegans* model (79, 80).

During chronic infection of the lungs of CF patients, *P. aeruginosa* produces copious amounts of alginate, which forms a matrix completely embedding the cells and becomes highly resistant to antibiotic treatment. These observations led to the suggestion that *P. aeruginosa* may exist as a biofilm in the lungs of CF patients (15, 43). This hypothesis was recently corroborated through profiling of AHL signal molecules (75). An involvement of the quorum-sensing circuitry in biofilm development by *P. aeruginosa* was originally reported by Davis et al. (16). We recently initiated a detailed analysis of the role of quorum sensing in biofilm formation in flowthrough cells (34). These investigations revealed that the primary attachment of cells to the surface and the early stages of biofilm development do not differ significantly between the wild-type and quorum-sensing-deficient mutants. However, after 5 to 7 days of growth, structural differences were noticeable, and after 10 days the biofilm formed by a *lasI rhlI* double mutant contained 30% less biomass relative to the wild-type biofilm. Even more pronounced effects were observed with a *lasR rhlR* double mutant. In conclusion, our results suggest that both quorum-sensing systems of *P. aeruginosa* participate in the regulation of biofilm development.

To visualize AHL-mediated communication between bacteria at the single-cell level, we devised three GFP-based AHL sensor cassettes, which, depending on the components used for their construction, respond to different spectra of AHL molecules. The first cassette constructed was based on components of the *lux* quorum-sensing system of *Vibrio fischeri* and contains a P_{luxI}-*gfp*(ASV) transcriptional fusion together with a constitutively transcribed *luxR* gene. The *lux* quorum-sensing system utilizes *N*-3-oxo-hexanoylhomoserine lactone (3-oxo-C_6-HSL) and the sensor cassette consequently exhibits the highest sensitivity for this signal molecule. However, it also responds to various related AHL molecules, albeit with decreased sensitivity (4). For the construction of this sensor cassette (as well as for the AHL sensors described below), an unstable variant of GFPmut3*, namely, GFP(ASV), was used. Colonies of *E. coli* harboring the *luxR*-P_{luxI}-RBSII-*gfp*(ASV)-cassette on a high-copy-number pUC-derived plasmid appeared completely dark under the epifluorescence microscope following 20 h of incubation at 30°C on Luria-Bertani plates. In striking contrast, the same strain produced bright green fluorescent colonies when cultivated under identical conditions in the presence of 1 nM 3-oxo-C_6-HSL. The most important lesson we learned from constructing various versions of this GFP-based AHL sensor cassette was that single-cell analysis is possible only when the *gfp* mRNA is efficiently translated and *luxR* is constitutively expressed at a high level. For this purpose, we inserted an optimized ribosomal binding site (RBSII) upstream of *gfp*(ASV) and placed *luxR* downstream of P_{lac}. A consequence of these modifications was that the specificity of the sensor was broadened, with single-cell detection limits in the range of 1 nM for 3-oxo-C_6-HSL, 10 nM for C_6-, C_8-, and 3-oxo-C_{12}-HSL, and 1,000 nM for C_4-HSL within a response time of 15 min. As with TraR, LuxR overexpression appears to increase nonspecific agonistic activities of AHL analogs (94).

For analyzing AHL-mediated communication between cells of *P. aeruginosa,* we constructed an AHL sensor cassette, which is based on components of the *las* quorum-sensing system. Specifically, this cassette consists of a P_{lasB}-*gfp*(ASV) translational fusion together with the *lasR* gene placed under control of P_{lac}. As expected, since the cognate AHL of the *las* system is 3-oxo-C_{12}-HSL, this system is most sensitive for this and other long-chain AHL molecules (32). Finally, for detection of AHL molecules with medium-lengths acyl side chains, we constructed a sensor cassette from components of the *cep* system of *Burkholderia cepacia,* which utilizes the signal molecule *N*-octanoyl-L-homoserine lactone (C_8-HSL) (38, 45) and contains a P_{cepI}-*gfp*(ASV) translational fusion together with the *cepR* gene transcribed from the P_{lac} promoter. This sensor responds most efficiently to C_8-HSL and with a lower efficiency to related AHL molecules (69).

The various sensor cassettes were cloned on broad-host-range vectors and/or mini-Tn*5* transposons. These can therefore be readily transferred to a wide range of bacteria. However, we noticed that the functionality of the sensor plasmids is strongly influenced by the choice of the background strain. For example, while both the *las*- and *cep*-based sensors worked very well in *Pseudomonas putida,* they did not work in *Serratia liquefaciens.* Conversely, the *lux*-based sensor cassette functioned well only in the latter strain (76). The reasons for these effects are at present unclear. However, given that cells are freely permeable for short-chain AHLs but that long-chain AHLs are actively transported (62), it is tempting to speculate that differences in the presence and/or specificity of long-chain AHL transporters in the two strains may account for the observed differences.

These GFP-based AHL biosensors have opened unprecedented possibilities for in situ studies of quorum sensing. For example, the biosensors were used for detection of AHL-mediated cell-to-cell communication in a swarming colony of *S. liquefaciens* (4, 21) and for the visualization of intergeneric communication in the rhizosphere of tomatoes (76). Besides these ecological applications, we were interested mainly in utilizing the sensors for investigation of the role of quorum sensing in *P. aeruginosa* pathogenesis in clinically relevant scenarios. In fact, the initial motivation for the construction of GFP-based AHL biosensors was to investigate whether AHL signal molecules are indeed produced during lung infection. To this end, we established a chronic lung infection with a mixture of signal-producing and signal-receiving bacteria carrying a GFP-based AHL monitor system in a mouse agar bead model (91). In this infection model, the bacteria are entrapped in alginate before they are intratracheally introduced into the lungs of mice (58, 63). Mice were challenged with alginate beads containing *P. aeruginosa* and *E. coli* harboring the *lux*-based AHL biosensor and sacrificed on different days after the challenge. At 1 to 3 days postinfection, green fluorescent cells were detected when frozen sections of the lung tissue were observed under the confocal microscope, indicating that the production of AHLs by *P. aeruginosa* occurred during the infection process (Color Plate 2). Interestingly, the histopathological changes in the mouse lung tissues showed similarities to those in the lungs of CF patients; i.e., inflammation was dominated by the presence of polymorphonuclear leukocytes (36). The use of GFP-based AHL biosensors allowed for the first time a direct visualization of AHL-mediated cell-to-cell communication during lung infection with *P. aeruginosa.* One problem that became apparent in these sets of experiments was that it was difficult to track cells of the biosensor when little or no AHL stimulated GFP expression. The next generation of AHL biosensors was therefore modified in a way so that the strains constitutively expressed RFP for easy detection and localization in the lung tissue and additionally expressed GFP in response to the presence of AHLs. This modified sensor strain was used to show that introduction of 3-oxo-C_6 into the mouse blood circulation through the tail vein activates the LuxR-controlled P_{luxI}-*gfp*(ASV) fusion in a concentration-dependent manner. These data imply that 3-oxo-C_6 is transported by the blood and is capable of penetrating the lung tissue.

B. cepacia is another important opportunistic pathogen in patients with CF. Infection with *B. cepacia* normally occurs in patients who are already colonized with *P. aeruginosa.* Cocolonization can result in three clinical outcomes: asymptomatic carriage, slow and continuous decline in lung function, or, for approximately 20% of the patients, fulminant and fatal pneumonia, the so-called cepacia syndrome (39). Like *P. aeruginosa, B. cepacia* controls the expression of various extracellular factors and biofilm formation by an AHL-dependent quorum-sensing system (24, 38, 45). The major signal molecule utilized by the organism is C_8-HSL. During chronic coinfection, *P. aeruginosa* and *B. cepacia* form mixed biofilms in the lungs of CF patients. Given that both bacteria utilize the same chemical language, it appeared likely that the two strains are capable of communicating with each and that these interactions may be of profound importance for the virulence of the mixed consortium. Direct visual evidence for AHL-mediated cross-communication between the two bacteria in mouse lung tissue was obtained by employing the *cep*- and *las*-based biosensors (69). These experiments demonstrated that *B. cepacia* is capable of perceiving the AHL signals produced by *P. aeruginosa* while the latter strain does not respond to the molecules produced by *B. cepacia*. Hence, cross-communication does occur but is unidirectional.

Recent work has provided evidence that blockade of quorum sensing by means of AHL antagonists represents a very attractive option to combat *P. aeruginosa* infections (33). A key issue in the development of suitable compounds was to test their functionality in vivo (31). To this end, we employed the mouse agar bead infection model in combination with an RFP-labeled *P. aeruginosa* wild-type strain harboring the *las*-based AHL sensor cassette. The infection was allowed to establish for 2 days before test compounds were introduced through the tail vein. When one of our lead compounds, furanone C-30 or C-56, was coadministered intravenously with 3-oxo-C_6-HSL, we observed that over a period of 4 to 6 h after administration, the GFP signal from the dually labeled biosensor was significantly reduced. After 8 h, however, the GFP signal reappeared, indicating that the furanone was cleared from the mouse blood circulation and thus de novo GFP synthesis had recommenced. The inhibitory effect of the furanones was reversed by increasing the dosage of 3-oxo-C_6-HSL. These experiments revealed important information about the mode of action, the concentration for functionality in vivo, and, importantly, the time for which a single injection was effective in the animal.

CONCLUDING COMMENTS

The last 10 years of development of new molecular tools for microscopic investigations, along with the rapid development and dispersal of advanced fluorescence microscopy methods, have resulted in greatly improved techniques for studies of microbial performance in very complex settings, including those found in connection with both commensal and pathogenic bacteria in animals and even human patients. This has opened up new possibilities of comparing in vitro results with reference strains and clinical isolates with direct observations in vivo of these bacteria as they live in their host organisms. Confirmation of extrapolations from laboratory experiments is necessary before we may claim to understand some of the mechanisms behind specific cases of pathogenicity, and likewise it is very important to understand the life of commensal organisms in order to interpret data obtained from related pathogens. The great progress that has been made in the in situ techniques permits us to ask the same questions about single bacterial cells in their complex environments in an animal or human as we have been asking for many years about the same bacteria growing as monospecies cultures in our laboratory test tubes.

REFERENCES

1. **Albus, A. M., E. C. Pesci, L. J. Runyen-Janecky, S. E. West, and B. H. Iglewski.** 1997. Vfr controls quorum sensing in *Pseudomonas aeruginosa*. *J. Bacteriol.* **179:**3928–3935.
2. **Amann, R. I., L. Krumholz, and D. A. Stahl.** 1990. Fluorescent-oligonucleotide probing of whole cells for determinative, phylogenetic, and environmental studies in microbiology. *J. Bacteriol.* **172:**762–770.
3. **Andersen, J. B., C. Sternberg, L. K. Poulsen, S. P. Bjørn, M. Givskov, and S. Molin.** 1998. New unstable variants of green fluorescent protein for studies of transient gene expression in bacteria. *Appl. Environ. Microbiol.* **64:**2240–2246.
4. **Andersen, J. B., A. Heydorn, M. Hentzer, L. Eberl, O. Geisenberger, B. B. Christensen, S. Molin, and M. Givskov.** 2001. *gfp* based *N*-acyl-homoserine-lactone monitors for detection of bacterial communication *Appl. Environ. Microbiol.* **67:**575–585.
5. **Ando, R., H. Hama, M. Yamamoto-Hino, H. Mizuno, and A. Miyawaki.** 2002. An optical marker based on the UV-induced green-to-red photoconversion of a fluorescent protein. *Proc. Natl. Acad. Sci. USA* **99:**12651–12656.
6. **Apostolou, E., L. Pelto, P. V. Kirjavainen, E. Isolauri, S. J. Salminen, and G. R. Gibson.** 2001. Differences in the gut bacterial flora of healthy and milk-hypersensitive adults, as measured by fluorescence in situ hybridization. *FEMS Immunol. Med. Microbiol.* **30:**217–221.
7. **Bevis, B. J., and B. S. Glick.** 2002. Rapidly maturing variants of the *Discosoma* red fluorescent protein (DsRed). *Nat. Biotechnol.* **20:**1159–1159.
8. **Boye, M., T. K. Jensen, K. Moller, T. D. Leser, and S. E. Jorsal.** 1998. Specific detection of the genus *Serpulina, S. hyodysenteriae* and *S. pilosicoliin* porcine intestines by fluorescent rRNA in situ hybridization. *Mol. Cell. Probes* **12:**323–330.
9. **Boye, M., T. K. Jensen, P. Ahrens, T. Hagedorn-Olsen, and N. F. Friis.** 2001. In situ hybridization for identification and differentiation of *Mycoplasma hyopneumoniae, Mycoplasma hyosynoviae* and *Mycoplasma hyorhinis* in formalin-fixed porcine tissue sections. *APMIS* **109:**656–664.
10. **Camara, M., P. Williams, and A. Hardman.** 2002. Controlling infection by tuning in and turning down the volume of bacterial small-talk *Lancet Infect. Dis.* **2:**667–676.
11. **Chalfie, M., Y. Tu, G. Euskirchen, W. W. Ward, and D. C. Prasher.** 1994. Green fluorescent protein as a marker for gene expression. *Science* **263:**802–805.

12. **Chudakov, D. M., V. V. Belousov, A. G. Zaraisky, V. N. Novoselov, D. B. Staroverov, D. B. Zorov, S. Lukyanov, and K. A. Lukyanov.** 2003. Kindling fluorescent proteins for precise in vivo photolabeling. *Nat. Biotechnol.* **91:**191–194.
13. **Connell, H., L. K. Poulsen, and P. Klemm.** 2000. Expression of type 1 and P fimbriae in situ and localisation of a uropathogenic *Escherichia coli* strain in the murine bladder and kidney. *Int. J. Med. Microbiol.* **290:**587–597.
14. **Cormack, B., R. H. Valdivia, and S. Falkow.** 1996. FACS-optimized mutants of green fluorescent protein (GFP). *Gene* **173:**33–38.
15. **Costerton, J. W., P. S. Stewart, and E. P. Greenberg.** 1999. Bacterial biofilms: a common cause of persistent infections. *Science* **284:**318–1322.
16. **Davies, D. G., M. R. Parsek, J. P. Pearson, B. H. Iglewski, J. W. Costerton, and E. P. Greenberg.** 1998. The involvement of cell-to-cell signals in the development of a bacterial biofilm. *Science* **280:**295–298.
17. **Davies, J. C., D. M. Geddes, and E. W. Alton.** 1998. Prospects for gene therapy for cystic fibrosis *Mol. Med. Today* **4:**292–299.
18. **De Kievit, T. R., and B. H. Iglewski.** 2000. Bacterial quorum sensing in pathogenic relationships. *Infect. Immun.* **68:**4839–4849.
19. **Dennis, P. P., and H. Bremer.** 1973. Regulation of ribonucleic acid synthesis in *Escherichia coli* B-r: an analysis of a shift-up. *J. Mol. Biol.* **75:**145–159.
20. **Drouault, S., G. Corthier, D. Ehrlich, and P. Renault.** 1999. Survival, phyisology, and lysis of *Lactococcus lactis* in the digestive tract. *Appl. Environ. Microbiol.* **65:**4881–4886.
21. **Eberl, L., S. Molin, and M. Givskov.** 1999. Surface motility of *Serratia liquefaciens* MG1. *J. Bacteriol.* **181:**1703–1712.
22. **Favre-Bonté, S., T. R. Licht, C. Forestier, and K. A. Krogfelt.** 1999. *Klebsiella pneumoniae* capsule expression is necessary for colonization of large intestines of streptomycin-treated mice. *Infect. Immun.* **67:**6152–6156.
23. **Franks, A. H., H. J. Harmsen, G. C. Raangs, G. J. Jansen, F. Schut, and G. W. Welling.** 1998. Variations of bacterial populations in human feces measured by fluorescent in situ hybridization with group-specific 16S rRNA-targeted oligonucleotide probes. *Appl. Environ. Microbiol.* **64:**3336–3345.
24. **Gotschlich, A., B. Huber, O. Geisenberger, A. Togl, A. Steidle, K. Riedel, P. Hill, B. Tummler, P. Vandamme, B. Middleton, M. Camara, P. Williams, A. Hardman, and L. Eberl.** 2001. Synthesis of multiple *N*-acylhomoserine lactones is wide-spread among the members of the *Burkholderia cepacia* complex. *Syst. Appl. Microbiol.* **24:**1–14.
25. **Gurskaya, N. G., A. F. Fradkov, A. Terskikh, M. V. Matz, Y. A. Labas, V. I. Martynov, Y. G. Yanushevich, K. A. Lukyanov, and S. A. Lukyanov.** 2001. GFP-like chromoproteins as a source of far-red fluorescent proteins. *FEBS Lett.* **507:**16–20.
26. **Harmsen, H. J., A. C. Wildeboer-Veloo, G. C. Raangs, A. A. Wagendorp, N. Klijn, J. G. Bindels, and G. W. Welling.** 2000. Analysis of intestinal flora development in breast-fed and formula-fed infants by using molecular identification and detection methods. *J. Pediatr. Gastroenterol. Nutr.* **30:**61–67.
27. **Harmsen, H. J., G. R. Gibson, P. Elfferich, G. C. Raangs, A. C. Wildeboer-Veloo, A. Argaiz, M. B. Roberfroid, and G. W. Welling.** 2000. Comparison of viable cell counts and fluorescence in situ hybridization using specific rRNA-based probes for the quantification of human fecal bacteria. *FEMS Microbiol. Lett.* **183:**125–129.
28. **Heim, R., A. B. Cubitt, and R. Y. Tsien.** 1995. Improved green fluorescence. *Nature* **373:**663–664.
29. **Heim, R., and R. Y. Tsien.** 1996. Engeneering green fluorescent protein for improved brightness, longer wavelengths and fluorescence resonance energy transfer. *Curr. Biol.* **6:**178–182.
30. **Heim, R., D. C. Prasher, and R. Y. Tsien.** 1994. Wavelength mutations and posttranslational autooxidation of green fluorescent protein. *Proc. Natl. Acad. Sci. USA* **91:**12501–12504.
31. **Hentzer, M., H. Wu, J. B. Andersen, K. Riedel, T. B. Rasmussen, N. Bagge, N. Kumar, M. Schembri, Z. Song, P. Kristoffersen, M. Manefield, J. W. Costerton, S. Molin, L. Eberl, P. Steinberg, S. Kjelleberg, N. Høiby, and M. Givskov.** 2003. Attenuation of *Pseudomonas aeruginosa* virulence by quorum sensing inhibitors. *EMBO J.* **22:**1–13.
32. **Hentzer, M., K. Riedel, T. B. Rasmussen, A. Heydorn, J. B. Andersen, M. R. Parsek, S. A. Rice, L. Eberl, S. Molin, and M. Givskov.** 2002. Inhibition of quorum sensing in *Pseudomonas aeruginosa* biofilm bacteria by a halogenated furanone compound. *Microbiology* **148:**87–102.
33. **Hentzer, M., L. Eberl, J. Nielsen, and M. Givskov.** 2003. Quorum sensing: a novel target for the treatment of biofilm infections. *BioDrugs* **17:**241–250.
34. **Hentzer, M., M. Givskov, and L. Eberl.** 2004. Quorum sensing in biofilms: gossip in slime city, p. 118–140. *In* M. A. Ghannoum and G. O'Toole (ed.), *Microbial Biofilms.* ASM Press, Washington, D.C.
35. **Høiby, N., and B Frederiksen.** 2000. Microbiology of cystic fibrosis, p. 83–107. *In* M. E. Hodson and D. M. Geddes (ed.), *Cystic Fibrosis.* Arnold, London, United Kingdom.
36. **Høiby, N., B. Giwercman, T. Jensen, H. K. Johansen, G. Kronborg, T. Pressler, and A. Kharazmi.** 1993. Immune response in cystic fibrosis—helpful or harmful? p. 133–144. *In* H. Scobar (ed.), *Clinical Ecology of Cystic Fibrosis.* Exerpta Medica, Amsterdam, The Netherlands.
37. **Hold, G. L., A. Schwiertz, R. I. Aminov, M. Blaut, and H. J. Flint.** 2003. Oligonucleotide probes that detect quantitatively significant groups of butyrate-producing bacteria in human feces. *Appl. Environ. Microbiol.* **69:**4320–4324.
38. **Huber, B., K. Riedel, M. Hentzer, A. Heydorn, A. Gotschlich, M. Givskov, S. Molin, and L. Eberl.** 2001 The *cep* quorum-sensing system of *Burkholderia cepacia* H111 controls biofilm formation and swarming motility. *Microbiology* **147:**2517–2528.
39. **Isles, A., I. Maclusky, M. Corey, R. Gold, C. Prober, P. Fleming, and H Levison.** 1984. *Pseudomonas cepacia* infection in cystic fibrosis: an emerging problem. *J. Pediatr.* **104:**206–210.
40. **Jacobi, C. A., S. Gregor, A. Rakin, and J. Heesemann.** 2001. Expression analysis of the yersiniabactin receptor gene *fyuA* and the heme receptor *hemA* of *Yersinia entericolitica* in vitro and in vivo using the reporter genes for green fluorescent protein and luciferase. *Infect. Immun.* **69:**7782–7772.
41. **Jensen, T. K., K. Moller, M. Boye, T. D. Leser, and S. E. Jorsal.** 2000. Scanning electron microscopy and fluorescent in situ hybridization of experimental *Brachyspira (Serpulina) pilosicoli* infection in growing pigs. *Vet. Pathol.* **37:**22–32.
42. **Jensen, T. K., M. Boye, P. Ahrens, B. Korsager, P. S. Teglbjaerg, C. F. Lindboe, and K. Moller.** 2001. Diagnostic examination of human intestinal spirochetosis by fluorescent in situ hybridization for *Brachyspira aalborgi, Brachyspira pilosicoli,* and other species of the genus *Brachyspira (Serpulina). J. Clin. Microbiol.* **39:**4111–4118.
43. **Lam, J., R. Chan, K. Lam, and J. W. Costerton.** 1980. Production of mucoid microcolonies by *Pseudomonas aeruginosa* within infected lungs in cystic fibrosis. *Infect. Immun.* **28:**546–556.
44. **Latifi, A., M. Foglino, K. Tanaka, P. Williams, and A. Lazdunski.** 1996. A hierarchical quorum-sensing cascade in

Pseudomonas aeruginosa links the transcriptional activators LasR and RhlR (VsmR) to expression of the stationary-phase sigma factor RpoS. *Mol. Microbiol.* **21**:1137–1146.

45. **Lewenza, S., B. Conway, E. P. Greenberg, and P. A. Sokol.** 1999 Quorum sensing in *Burkholderia cepacia:* identification of the LuxRI homologs CepRI. *J. Bacteriol.* **181**:748–756.
46. **Li, X., X. N. Xiao, Y. Fang, X. Jiang, T. Doung, C. Fan, C. C. Huang, and S. R. Kain.** 1998. Generation of destabilized green fluorescent protein transcription reporter. *J. Biol. Chem.* **273**:34970–34975.
47. **Licht, T. R., K. A. Krogfelt, P. S. Cohen, L. K. Poulsen, J. Urbace, and S. Molin.** 1996. Role of lipopolysaccharide in colonization of the mouse intestine by *Salmonella typhimurium* studied by in situ hybridization. *Infect. Immun.* **64**:3811–3817.
48. **Licht, T. R., T. Tolker-Nielsen, K. Holmstrøm, K. A. Krogfelt, and S. Molin.** 1999. Inhibition of *Escherichia coli* precursor-16S rRNA processing by mouse intestinal contents. *Environ. Microbiol.* **1**:23–32.
49. **Lukayanov, K. A., A. F. Fradkov, N. G. Gurskaya, M. V. Matz, Y. A. Labas, A. P. Savitsky, M. L. Markelov, A. G. Zaraisky, X. Zhao, Y. Fang, W. Tan, and S. A. Lukyanov.** 2000. Natural animal coloration can be determined by a nonfluorescent green fluorescent protein homolog. *J. Biol. Chem.* **275**:25879–25882.
50. **Lyczak, J. B., C. L. Cannon, and G. B. Pier.** 2002. Lung infections associated with cystic fibrosis. *Clin. Microbiol. Rev.* **15**:194–222.
51. **Matz, M. V., A. F. Fradkov, Y. A. Labas, A. P. Savitsky, A. G. Zaraisky, M. L. Markelov, and S. A. Lukyanov.** 1999. Fluorescent proteins from nonbioluminescent *Anthozoa* species *Nat. Biotechnol.* **17**:969–973.
52. **McCartney, A. L.** 2002. Application of molecular biological methods for studying probiotics and the gut flora. *Br. J. Nutr.* **88**(Suppl. 1):S29–S37.
53. **Meyvis, T. K. L., S. C. De Smedt, P. van Oostveldt, and J. Demeester.** 1999. Fluorescence recovery after photobleaching: a versatile tool for mobility and interaction measurements in pharmaceutical research. *Pharmacol. Res.* **16**:1153–1162.
54. **Molin, S., and M. Givskov.** 1999. Application of molecular tools for *in situ* monitoring of bacterial growth activity. *Environ. Microbiol.* **1**:383–391.
55. **Møller, A. K., M. P. Leatham, T. Conway, P. J. M. Nuijten, L. A. M. de Haan, K. A. Krogfelt, and P. S. Cohen.** 2003. An *Escherichia coli* MG1655 lipopolysaccharide deep-rough core mutant grows and survives in mouse cecal mucus but fails to colonize the mouse large intestine. *Infect. Immun.* **71**:2142–2152.
56. **Morin, J. G., and J. W. Hastings.** 1971. Energy transfer in a bioluminescent system. *J. Cell. Physiol.* **77**:313–318.
57. **Morise, H., P. Shimomura, F. H. Johnson, and J. Winant.** 1974. Intramolecular energy-transfer in bioluminescent system of *Aequorea. Biochemistry* **13**:2656–2662.
58. **Moser, C., H. K. Johansen, Z. J. Song, H. P. Hougen, J. Rygaard, and N. Hoiby.** 1997. Chronic *Pseudomonas aeruginosa* lung infection is more severe in Th2 responding BALB/c mice compared to Th1 responding C3H/HeN mice. *APMIS* **105**:838–842.
59. **Ormö, M., A. B. Cubitt, K. Kallio, L. A. Gross, R. Y. Tsien, and S. J. Remington.** 1996. Crystal structure of the *Aequorea victoria* green fluorescent protein. *Science* **273**:1392–1395.
60. **Passador, L., J. M. Cook, M. J. Gambello, L. Rust, and B. H. Iglewski.** 1993. Expression of *Pseudomonas aeruginosa* virulence genes requires cell-to-cell communication. *Science* **260**:1127–1130.
61. **Patterson, G. H., and J. Lippincott-Schwartz.** 2002. A photoactviatable GFP for selective photolabeling of proteins and cells. *Science* **297**:1873–1877.
62. **Pearson, J. P., C. Van Delden, and B. H. Iglewski.** 1999. Active efflux and diffusion are involved in transport of *Pseudomonas aeruginosa* cell-to-cell signals. *J. Bacteriol.* **181**:1203–1210.
63. **Pedersen, S. S., G. H. Shand, B. L. Hansen, and G. N. Hansen.** 1990. Induction of experimental chronic *Pseudomonas aeruginosa* lung infection with *P. aeruginosa* entrapped in alginate microspheres. *APMIS* **98**:203–211.
64. **Pesci, E. C., J. P. Pearson, P. C. Seed, and B. H. Iglewski.** 1997. Regulation of *las* and *rhl* quorum sensing in *Pseudomonas aeruginosa. J. Bacteriol.* **179**:3127–3132.
65. **Poulsen, L. K., F. Lan, C. S. Kristensen, S. Molin, and K. A. Krogfelt.** 1994. Spatial distribution of *Escherichia coli* in the mouse large intestine inferred by rRNA in situ hybridization. *Infect. Immun.* **62**:5191–5194.
66. **Poulsen, L. K., T. R. Licht, C. Rang, K. A. Krogfelt, and S. Molin.** 1995. The physiological state of *Escherichia coli* BJ4 growing in the large intestine of streptomycin treated mice. *J. Bacteriol.* **177**:5840–5845.
67. **Purevdorj, B., J. W. Costerton, and P. Stoodley.** 2002. Influence of hydrodynamics and cell signaling on the structure and behavior of *Pseudomonas aeruginosa* biofilms. *Appl. Environ. Microbiol.* **68**:4457–4464.
68. **Ramos, C., C. Sternberg, T. R. Licht, K. A. Krogfelt, and S. Molin.** 2001. Monitoring bacterial growth activity in biofilms from laboratory flow-chambers, plant rhizosphere and animal intestine. *Methods Enzymol.* **337B**:21–42.
69. **Riedel, K., M. Hentzer, O. Geisenberger, B. Huber, A. Steidle, H. Wu, N. Hoiby, M. Givskov, and L. Eberl.** 2001. *N*-Acylhomoserine-lactone-mediated communication between *Pseudomonas aeruginosa* and *Burkholderia cepacia* in mixed biofilms. *Microbiology* **147**:3249–3262.
70. **Rumbaugh, K. P., J. A. Griswold, B. H. Iglewski, and A. N. Hamood.** 1999. Contribution of quorum sensing to the virulence of *Pseudomonas aeruginosa* in burn wound infections. *Infect. Immun.* **67**:5854–5862.
71. **Russmann, H., A. Feydt-Schmidt, K. Adler, D. Aust, A. Fischer, and S. Koletzko.** 2003. Detection of *Helicobacter pylori* in paraffin-embedded and in shock-frozen gastric biopsy samples by fluorescent in situ hybridization. *J. Clin. Microbiol.* **41**:813–815.
72. **Russmann, H., V. A. Kempf, S. Koletzko, J. Heesemann, and I. B. Autenrieth.** 2001. Comparison of fluorescent in situ hybridization and conventional culturing for detection of *Helicobacter pylori* in gastric biopsy specimens. *J. Clin. Microbiol.* **39**:304–308.
73. **Shimomura, O., F. H. Johnson, and Y. Saiga.** 1962 Extraction, purification and properties of Aequorin, a bioluminescent protein from the luminous hydromedusan *Aequorea. J. Cell. Comp. Physiol.* **59**:223–239.
74. **Siemering, K. R., R. Golbik, R. Sever, and J. Haselhoff.** 1996. Mutants that suppress the thermosensitivity of green fluorescent protein. *Curr. Biol.* **6**:1653–1663.
75. **Singh, P. K., A. L. Schaefer, M. R. Parsek, T. O. Moninger, M. J. Welsh, and E. P. Greenberg.** 2000. Quorum-sensing signals indicate that cystic fibrosis lungs are infected with bacterial biofilms. *Nature* **407**:762–764.
76. **Steidle, A., K. Sigl, R. Schuhegger, A. Ihring, M. Schmid, S. Gantner, M. Stoffels, K. Riedel, M. Givskov, A. Hartmann, C. Langebartels, and L. Eberl.** 2001. Visualization of *N*-acylhomoserine lactone-mediated cell-cell communication between bacteria colonizing the tomato rhizosphere. *Appl. Environ. Microbiol.* **67**:5761–5770.

77. **Struve, C., and K. A. Krogfelt.** 2003. Role of capsule in *Klebsiella pneumoniae* virulence: lack of correlation between in vitro and in vivo studies. *FEMS Microbiol. Lett.* **218:**149–154.

78. **Struve, C., and K. A. Krogfelt.** 1999. In vivo detection of *Escherichia coli* type 1 fimbrial expression and phase variation during experimental urinary tract infection. *Microbiology* **145:**2683–2690.

79. **Tan, M. W., L. G. Rahme, J. A. Sternberg, R. G. Tompkins, and F. M- Ausubel.** 1999. *Pseudomonas aeruginosa* killing of *Caenorhabditis elegans* used to identify *P. aeruginosa* virulence factors. *Proc. Natl. Acad. Sci. USA* **96:**2408–2413.

80. **Tan, M. W., S. Mahajan-Miklos, and F. M. Ausubel.** 1999. Killing of *Caenorhabditis elegans* by *Pseudomonas aeruginosa* used to model mammalian bacterial pathogenesis. *Proc. Natl. Acad. Sci. USA* **96:**715–720.

81. **Tang, H. B., E. DiMango, R. Bryan, M. Gambello, B. H. Iglewski, J. B. Goldberg, and A. Prince.** 1996. Contribution of specific *Pseudomonas aeruginosa* virulence factors to pathogenesis of pneumonia in a neonatal mouse model of infection. *Infect. Immun.* **64:**37–43.

82. **Terskikh, A., A. Fradkov, G. Ermakova, A. Zaraisky, P. Tan, A. V. Kajava, X. Zhao, S. Lukyanov, M. Matz, S. Kim, I. Weissman, and P. Siebert.** 2000. "Fluorecent timer": protein that changes color with time. *Science* **290:**1585–1588.

83. **Trebesius, K., K. Adler, M. Vieth, M. Stolte, and R. Haas.** 2001. Specific detection and prevalence of Helicobacter heilmannii-like organisms in the human gastric mucosa by fluorescent in situ hybridization and partial 16S ribosomal DNA sequencing. *J. Clin. Microbiol.* **39:**1510–1516.

84. **Trebesius, K., K. Panthel, S. Strobel, K. Vogt, G. Faller, T. Kirchner, M. Kist, J. Heesemann, and R. Haas.** 2000. Rapid and specific detection of *Helicobacter pylori* macrolide resistance in gastric tissue by fluorescent in situ hybridization. *Gut* **46:**608–614.

85. **Urdaci, M. C., B. Regnault, and P. A. Grimont.** 2001. Identification by in situ hybridization of segmented filamentous bacteria in the intestine of diarrheic rainbow trout (*Oncorhynchus mykiss*). *Res. Microbiol.* **152:**67–73.

86. **Van Bost, S., S. Roels, and J. Mainil.** 2001. Necrotoxigenic *Escherichia coli* type 2 invade and cause diarrhoea during experimental infection in colostrum-restricted newborn calves. *Vet. Microbiol.* **81:**315–329.

87. **Van Delden, C., and B. H. Iglewski.** 1998. Cell-to-cell signaling and *Pseudomonas aeruginosa* infections. *Emerg. Infect. Dis.* **4:**551–560.

88. **Williams, P., M. Camara, A. Hardman, S. Swift, D. Milton, V. J. Hope, K. Winzer, B. Middleton, D. I. Pritchard, and B. W. Bycroft.** 2000. Quorum sensing and the population-dependent control of virulence. *Philos. Trans. R. Soc. Lond. Ser. B* **355:**667–680.

89. **Winson, M. K., M. Camara, A. Latifi, M. Foglino, S. R. Chhabra, M. Daykin, M. Bally, V. Chapon, G. P. Salmond, B. W. Bycroft, A. Lazdunski, G. S. A. B. Stewart, and P. Williams.** 1995. Multiple *N*-acyl-L-homoserine lactone signal molecules regulate production of virulence determinants and secondary metabolites in *Pseudomonas aeruginosa. Proc. Natl. Acad. Sci. USA* **92:**9427–9431.

90. **Wouters, F. S., P. J. Verveer, and P. I. H. Bastiaens.** 2001. Imaging biochemistry inside cells. *Trends Cell. Biol.* **11:**203–211.

91. **Wu, H., Z. Song, M. Hentzer, J. B. Andersen, A. Heydorn, K. Mathee, C. Moser, L. Eberl, S. Molin, N. Høiby, and M. Givskov.** 2000. Detection of *N*-acylhomoserine lactones in lung tissues of mice infected with *Pseudomonas aeruginosa. Microbiology* **146:**2481–2493.

92. **Yoon, S. S., R. F. Hennigan, G. M. Hilliard, U. A. Ochsner, K. Parvatiyar, M. C. Kamani, H. L. Allen, T. R. DeKievit, P. R. Gardner, U. Schwab, J. J. Rowe, B. H. Iglewski, T. R. McDermott, R. P. Mason, D. J. Wozniak, R. E. Hancock, M. R. Parsek, T. L. Noah, R. C. Boucher, and D. J. Hassett.** 2002. *Pseudomonas aeruginosa* anaerobic respiration in biofilms. Relationships to cystic fibrosis pathogenesis. *Dev. Cell* **3:**593–603.

93. **Zhao, M., M. Yang, E. Baranov, X. Wang, S. Penman, A. R. Moosa, and R. M. Hoffman.** 2001. Spatial-temporal imaging of bacterial infection and antibiotic response in intact animals. *Proc. Natl. Acad. Sci. USA* **98:**9814–9818.

94. **Zhu, J., J. W. Beaber, M. I. More, C. Fuqua, A. Eberhard, and S. C. Winans.** 1998. Analogs of the autoinducer 3-oxo-octanoyl-homoserine lactone strongly inhibit activity of the TraR protein of *Agrobacterium tumefaciens. J. Bacteriol.* **180:** 5398–5405.

95. **Zhu, X. Y., and R. D. Joerger.** 2003. Composition of microbiota in content and mucus from cecae of broiler chickens as measured by fluorescent in situ hybridization with group-specific, 16S rRNA-targeted oligonucleotide probes. *Poult. Sci.* **82:**1242–1249.

96. **Zoetendal, E. G., K. Ben-Amor, H. J. Harmsen, F. Schut, A. D. Akkermans, and W. M. de Vos.** 2002. Quantification of uncultured *Ruminococcus obeum*-like bacteria in human fecal samples by fluorescent in situ hybridization and flow cytometry using 16S rRNA-targeted probes. *Appl. Environ. Microbiol.* **68:**4225–4232.

II. COLONIZATION OF THE RESPIRATORY TRACT

Colonization of Mucosal Surfaces
Edited by James P. Nataro et al.

Chapter 5

Role of Phosphorylcholine in Respiratory Tract Colonization

JEFFREY N. WEISER

GENETIC APPROACH TO DEFINING MOLECULAR MECHANISMS FOR COLONIZATION OF THE RESPIRATORY TRACT

The human oro- and nasopharynx is colonized with hundreds of different bacterial species (26). Are there common features among this group that determine the ability to exist in this particular niche? To address this question, a genetic approach was taken to identifying genes common to species residing in the respiratory tract (58). The whole genomes of *Streptococcus pneumoniae* (the pneumococcus) and *Haemophilus influenzae,* two otherwise distantly related species (gram-positive and gram-negative bacteria, respectively) that both reside exclusively in the human nasopharynx, were systematically compared by searching all *H. influenzae* open reading frames for homologues in *S. pneumoniae.* The majority of the homologous sequences were "housekeeping genes" that are widely distributed among both respiratory and nonrespiratory tract bacterial species. The data set obtained was then made more specific by subtracting homologous genes and sequences found in two representative species that are not found in the human respiratory tract, the gram-positive bacterium *Bacillus subtilis* and the gram-negative bacterium *Escherichia coli.* The use of a BLAST cutoff score of $P < 1.0e^{-20}$ as the criteria for homology revealed a set of only 12 genes common to *H. influenzae* and *S. pneumoniae,* without homologs in either *B. subtilis* or *E. coli,* as candidate sequences that could encode important functions for survival in the respiratory tract. When these genes were then compared to the entire set of publically available microbial genomes by using PSI-BLAST, only a hypothetical protein (HI660) and products of a single operon (HI1537, HI1538, and HI1540) had homologs only in other organisms found predominantly or exclusively in the respiratory tract (based on the gene designation and annotation of *H. influenzae* KW20 by the Institute for Genomic Research [TIGR]). This operon, designated *lic,* consists of genes involved in choline uptake and incorporation, suggesting that the ability to utilize choline may contribute to colonization of the airway.

PHOSPHORYLCHOLINE: A COMMON SURFACE STRUCTURE AMONG BACTERIAL SPECIES IN THE RESPIRATORY TRACT

It is a generally accepted concept that commensal and pathogenic species express different surface antigens to optimize immune evasion. It is now clear that an exception to this dogma is the expression of choline phosphate [$(CH_3)_3N^+CH_2CH_2PO_4^-$] (ChoP). ChoP was first described as a constituent of the C-polysaccharide (teichoic acid) of *S. pneumoniae* decades ago, but until recently it was thought to be a highly unusual structural feature in prokaryotes (6). In contrast, it is a major component of the host cell surface in the form of phosphatidylcholine, the most abundant eukaryotic membrane phospholipid, and is found on other host molecules such as sphingomyelin, platelet-activating factor (PAF) (1-O-alkyl-2-acetyl-*sn*-glcero-3-phosphocholine) and lung surfactant (dipalmitoyl phosphatidylcholine). In 1996, this laboratory detected ChoP on the exposed surface of the lipopolysaccharide (LPS) or lipooligosaccharide of *H. influenzae* and identified the genes (*lic1A–D*) involved in choline utilization for synthesis of LPS-ChoP (65). Subsequently, operons with *licA* to *licD* homologs were described in *S. pneumoniae* and commensal species of the genus *Neisseria* (46, 70). The availability of a well-characterized monoclonal antibody MAb (the "natural" myeloma line expressing

Jeffrey N. Weiser • Departments of Microbiology and Pediatrics, University of Pennsylvania, 402A Johnson Pavilion, Philadelphia, PA 19104-6076.

the T-15 idiotype, TEPC-15) recognizing this structure and whole bacterial genome databases then facilitated the identification of ChoP as a structural feature of many other species of bacteria (Table 1; Fig. 1) (13).

The presence of the same "unusual" bacterial structure on so many diverse species, including gram-positive and gram-negative bacteria, mollicutes, and spirochetes, that reside predominantly in the oro- or nasopharynx provided further evidence that this structure may be advantageous for organisms in this host environment. In fact, it appears that the expression of ChoP among respiratory tract organisms is an example of convergent evolution since ChoP is present on distinct bacterial structures (LPS, teichoic acid, and fimbriae) in different species. An additional consideration is that in each of these species, ChoP is exposed on the bacterial cell surface, suggesting that it plays a role in host-bacterium interaction.

VARIABILITY OF PHOSPHORYLCHOLINE EXPRESSION

Another striking characteristic common to the diverse species with cell surface ChoP is that its expression is in general not constitutive (Fig. 1). The expression of a surface characteristic common to so many species inhabiting the respiratory tract indicates that the presence of ChoP confers important advantages in this environment. The variable expression of ChoP (so far described in *Haemophilus, Streptococcus, Neisseria, Actinobacillus,* and *Pseudomonas*) suggests that the molecular mimicry of host structures conferred by ChoP is not always advantageous for the organism (60–62). On-off switching or phase variation in the surface expression of ChoP allows for selection of subpopulations with or without this antigenic structure, depending on the immune response or at different times or sites in the host.

BACTERIAL ACQUISITION OF CHOLINE

Bacteria generally obtain choline from environmental sources since only a few species, such as *Pseudomonas aeruginosa, Treponema denticola,* and *Rhodobacter sphaeroides,* are capable of de novo synthesis (1, 47). Species that synthesize choline generally use it solely to generate the membrane phospholipid phosphatidylcholine. In contrast, species such as *Haemophilus, Streptococcus,* commensal *Neisseria,* and *Mycoplasma,* which utilize choline for other types of structural synthesis, lack membranes containing phosphatidylcholine. These species acquire choline exclusively from potentially abundant sources in their host by using a choline kinase pathway that in many aspects resembles that for choline utilization in eukaryotes (Fig. 2) (45, 65, 70). For streptococci with ChoP-teichoic acid, choline is a required nutrient (although some strains exhibit limited growth in ethanolamine, a structural analog) (54, 69). Members of the other ChoP-expressing genera grow normally in chemically defined media lacking choline and express it on their cell surface only when it is available in their environment. Many other bacterial species not shown in Table 1 take up choline as a precursor of the synthesis of glycine betaine, which functions as an osmolyte under conditions of osmotic stress, but do not use choline in any type of structural synthesis.

Mechanism of Choline Uptake and Utilization

Two genes (*licB* and *betT*) revealed by whole-genomic analysis as encoding potential choline transporters have been examined for their role in LPS-ChoP synthesis in *H. influenzae* (12). The *betT* gene in *H. influenzae* is similar to *betT* in *E. coli,* which functions in choline transport for the generation of glycine betaine in osmoprotection (28). The *licB* gene has homology to genes encoding bacterial permeases, including limited similarity to *betT,* and is found in

Table 1. Bacteria expressing the ChoP antigen/structure[a]

Genus	Species[b]	Cell structure with ChoP
Streptococcus	*pneumoniae,* oralis,* mitis,* sanguis**	Teichoic acid
Haemophilus	*influenzae,* somnus,* aphrophilus**	LPS
Neisseria (pathogenic)	*meningitidis,* gonorrheae*	Fimbriae
Neisseria (commensal)	*lactamica,* subflava**	LPS
Actinobacillus	*actinomycetemcomitans**	LPS
Pseudomonas	*aeruginosa*	Cell envelope protein
Actinomyces	*naeslundii**	ND[c]
Fusobacterium	*nucleatum**	ND
Treponema	*denticola**	Membrane phospholipid
Mycoplasma	*pneumoniae,* fermentans*	Polar membrane lipid

[a]Based on a survey of >45 "common" species (61).
[b]Asterisks indicate species generally found in the upper airway (others may be found in the respiratory tract).
[c]ND, not determined.

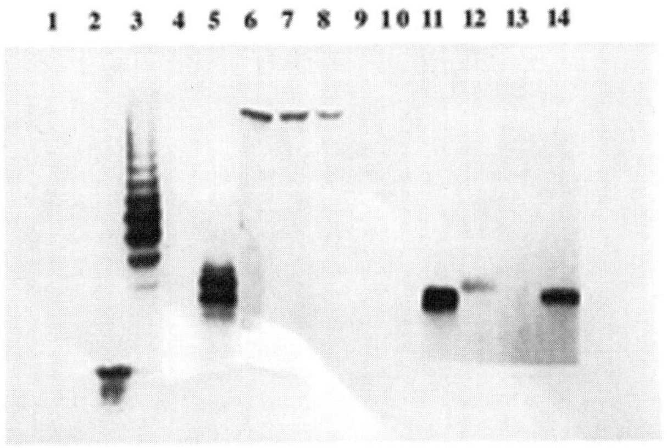

Figure 1. Composite Western immunoblot with MAb TEPC-15, summarizing the common phosphorylcholine epitope on six species of mucosal pathogens and examples of variation in its expression. Lanes: 1 and 2, nontypeable *H. influenzae* strain H233 phase variants with ChoP$^-$ and ChoP$^+$ LPS, respectively; 3, *S. pneumoniae* strain R6 lipoteichoic acid; 4 and 5, *A. actinomycetemcomitans* phase variants of strain Jp-2 without and with ChoP LPS, respectively; 6 to 10, *P. aeruginosa* grown at 26.5, 30.0, 33.5, 37.0, and 39.5°C, respectively; 11, purified *N. meningitidis* pilin; 12 to 14, piliated *N. gonorrhoeae* phase variants of strain MS11 without and with the ChoP-LPS. Reprinted from reference 58 with permission from Elsevier.

the *lic* locus, which is required for the expression of LPS-ChoP. A homolog of *licB* is also found in ChoP-expressing species of *Streptococcus* and *Neisseria*. However, in the presence of high concentrations of free choline, neither *licB* nor *betT* was necessary for expression of LPS-ChoP in *H. influenzae*, raising the possibility that other, unidentified choline uptake mechanisms and/or sources of choline may exist for this species. Further analysis revealed that under choline-limiting conditions, including growth in human nasal airway surface fluid, the *licB*, but not the *betT*, gene was required for the transport of free choline and subsequent synthesis of LPS-ChoP, confirming that LicB functions as a high-affinity choline permease. The *betT*, but not *licB*, gene was shown to function in osmoprotection in *H. influenzae*, similar to the role of *betT* in *E. coli*. Further analysis demonstrated that the transcription of the *licB* and *betT* genes is regulated differently under normal growth conditions. These observations suggest that *H. influenzae* has multiple mechanisms for choline uptake and distinct pathways for choline utilization in LPS-ChoP biosynthesis and osmoregulation.

Acquisition of Choline from Host Sources

Thus, the availability of free choline on the mucosal surface of the airway appears to be limiting. This raises the question whether other choline-containing molecules associated with eukaryotic membranes could provide an alternative source of free choline (11). In fact, *H. influenzae* is able to use glycerophosphorylcholine, an abundant, soluble degradation

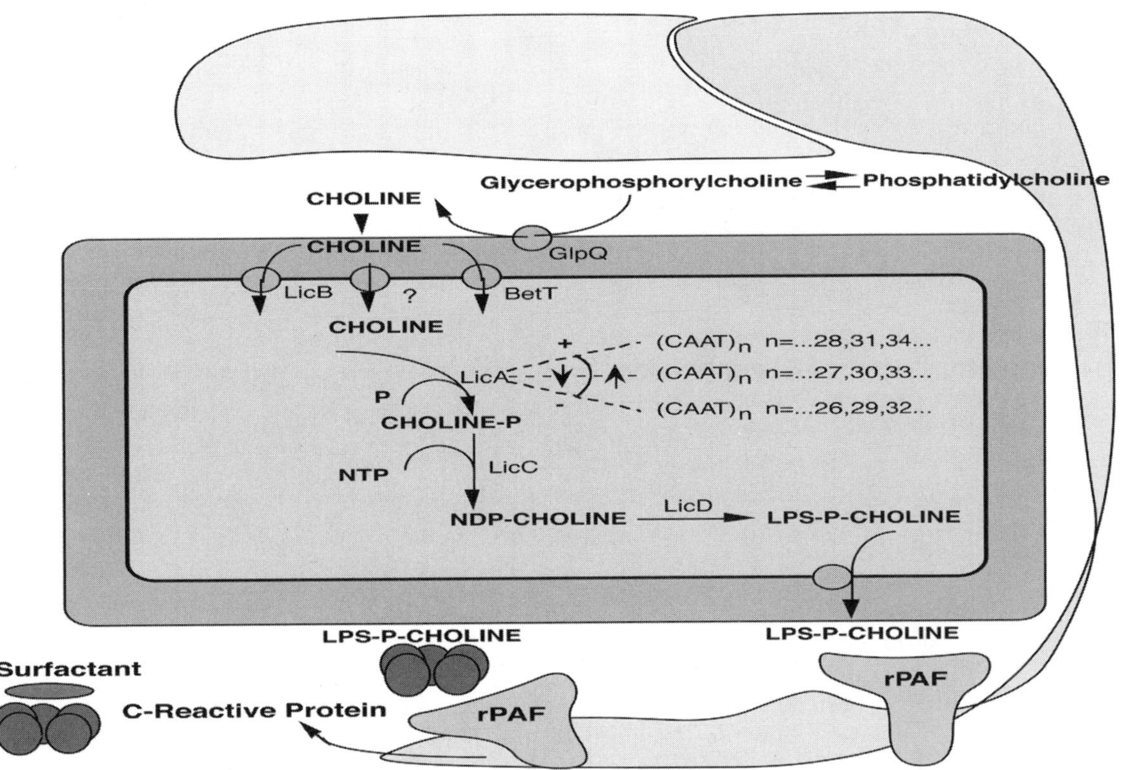

Figure 2. Current model of the pathway for expression of ChoP in *H. influenzae* and its proposed interactions within the airway.

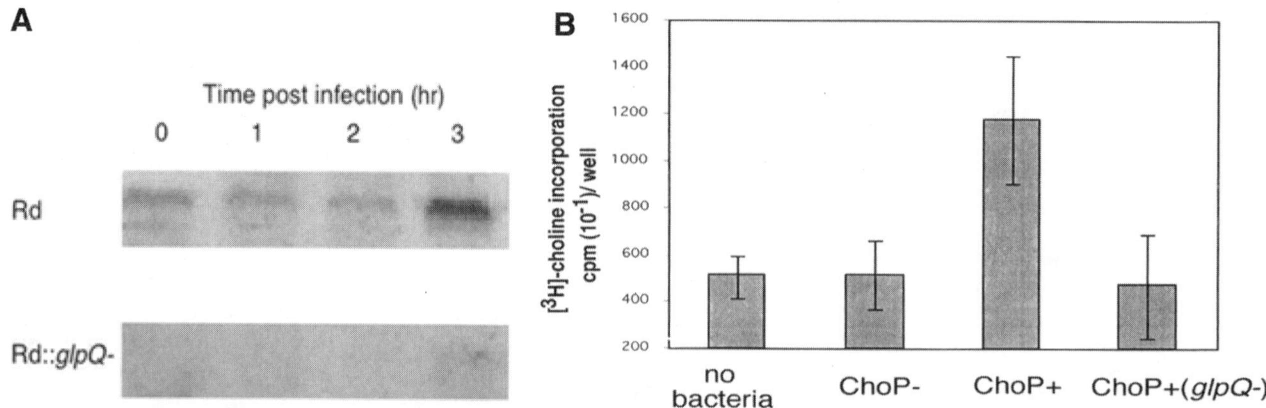

Figure 3. When host cells are the sole source of choline, *H. influenzae* is able to acquire choline by a *glpQ* dependent mechanism. (A) Time course of expression of ChoP on the LPS in bacteria adherent to D562 cells. Strains Rd ($ChoP^+$) and Rd::$glpQ^-$ ($ChoP^+$) were allowed to adhere to host epithelial cells in a medium lacking free choline for the period indicted, and the presence of ChoP on the LPS was detected by Western blotting of whole-cell lysates using MAb TEPC-15 to ChoP. (B) GlpQ is required for *H. influenzae* to transfer choline from host epithelial cells. D562 cells were grown in medium containing [^{3}H]choline and infected with the strain indicated (or control without bacteria) for 4 h, and incorporation of the radiolabel was detected in the bacterial fraction as the mean of counts per minute per well ± standard deviation (n = 3). Panels A and B reprinted from reference 11 with permission from Blackwell Publishing.

product of phosphatidylcholine, as efficiently as free choline. Utilization of GPC requires *glpQ,* which expresses an enzyme linked to the outer membrane with glycerophosphodiester phosphodiesterase activity (35). In the absence of free choline, this gene was shown to be required for adherent *H. influenzae* to obtain choline directly from epithelial cells in culture when these cells are the only available source of choline. GlpQ therefore allows choline to be transferred from the host to the bacterial cell surface (molecular mimicry through molecular thievery) (Fig. 3). GlpQ has also been reported to contribute to the virulence of *H. influenzae* in a rat model of otitis media (20). Furthermore, GlpQ facilitates damage to cilia in tissue explants infected with *H. influenzae,* although it is unknown whether this cytotoxic effect is due to its enzymatic activity in removing and possibly depleting choline from the host cell membrane lipid pool (19).

PHOSPHORYLCHOLINE EXPRESSION

Expression in *H. influenzae*

A survey of encapsulated and nontypeable strains indicated that the *lic1* locus is common to the species and that the ability to express ChoP is ubiquitous among *H. influenzae* strains (63). In this species, ChoP appears to be a phase-variable "decoration" on the exposed surface of LPS, where it may be covalently linked to different outer core oligosaccharide hexose residues in different *H. influenzae* strains (30, 65). This species does not have a nutritional requirement for choline, and in vitro growth of variants with and without ChoP is indistinguishable. In *H. influenzae,* high-frequency ($\sim10^{-3}$ to 10^{-4}/cell/generation) reversible on-off switching of ChoP expression is determined by a translational switch through slip-strand mispairing involving multiple tandem repeats of 5′-CAAT-3′ within the *licA* gene, encoding a putative choline kinase (62, 65). Thus, phase variation in this species appears to be a completely stochastic process, with high-frequency variation resulting from mutation in a hypermutable gene or "contingency locus" independently of any known environmental signals (33, 34)

Detailed structural analysis of the oligosaccharide region of the *H. influenzae* LPS showed that ChoP is a terminal structure linked to different hexose residues on different chain extensions in the unrelated strains Rd and Eagan (30). Differences in the molecular environment of ChoP affect the epitope defined by MAb 12D9 and were associated with polymorphisms within LicD, a putative diphosphonucleoside choline transferase. Exchanging the *licD* genes between the two strains with ChoP on different chain extensions was sufficient to switch its position. Therefore, in addition to the mechanism involving phase variation, structural rearrangements within the oligosaccharide have the potential to aid the evasion of an immune response targeting ChoP (see "Innate immunity targeting phosphorylcholine" below).

Further structural analysis of the oligosaccharide of additional nontypeable *H. influenzae* isolates carried out in collaboration with J. Richards, (National Research Council, Ottawa, Canada) revealed additional oligosaccharide structural possibilities (8,

42, 43). Beyond the variation in the arrangement of hexose residues, ChoP may be positioned as a terminal structure on a chain extension of each of the three heptoses or as an internal structure in different strains. These observations point out the inherent difficulties in comparing different isolates and emphasize the need for well-characterized constitutive mutants for meaningful functional analyses. To address this problem, Lysenko et al. have taken advantage of this feature to genetically modify the organism to excise the repetitive feature, leaving *licA* in frame (phase-on, ChoP$^+$) or out of frame (phase-off, ChoP$^-$) with unmarked, nonpolar mutations (30). This manipulation has eliminated the heterogeneity that otherwise characterizes this system and has enabled well-controlled comparisons of ChoP$^+$ and ChoP$^-$ phenotypes in an isogenic background.

Expression in *S. pneumoniae*

In the pneumococcus, there is variation between phenotypes with larger (transparent colonies) and smaller (opaque colonies) amounts of ChoP-teichoic acid through an unknown genetic mechanism (60). Differences of two- to fivefold in the total amount of ChoP per cell has secondary effects on other surface components, including its major virulence determinant, the capsular polysaccharide, which is covalently linked to teichoic acid, as well as a family of surface proteins referred to as choline-binding proteins (14, 48, 59, 60). Clinical isolates with the more transparent colony phenotype are more likely to be found in the human nasopharynx and are more efficient at colonization in an infant-rat model of carriage compared to the opaque variant of the same strain (56). The more efficient colonization of transparent pneumococci correlates with increased adherence to cytokine-stimulated A549 type II pneumocytes in cell culture (10). When the requirements of colonization were bypassed by intraperitoneal challenge of mice, only the opaque phenotype was able to cause a sustained bacteremia and sepsis, indicating that the biological role of opaque pneumococci appears to be in events following colonization of the mucosal surface (24). Thus, the greater quantity of capsular polysaccharide and the smaller amount of ChoP-teichoic acid associated with the opaque variants may enhance virulence by inhibiting opsonophagocytosis, whereas the greater quantity of ChoP-teichoic acid may render the transparent phenotype better adapted for colonization by enhancing adhesive interactions with host cells (25).

Thus, the prospects for using the pneumococcus to analyze ChoP-specific effects have been complicated by the pleotropic effects of opacity variation and the nutritional requirement for choline. Among the pneumococcal *lic* genes (*licA, licB, licC, licD1,* and *licD2*), mutagenesis has been has been successful only in *licD2*. The *licD2* mutant, however, retains a functional *licD1* encoding a second phosphocholine transferase and still expresses one of two ChoP residues per teichoic acid repeat unit (70).

Expression by Other Pathogens

ChoP expression on the LPS of *Actinobacillus* and the commensal *Neisseria* species and its effects on host-bacterium interaction appear to closely resemble those in *H. influenzae* (41, 45). In contrast, the pathogenic neisseriae lack *licA* to *licD* and express ChoP on their glycosylated pili in a phase-variable pattern that is independent of phase variation of fimbriae themselves (61). The dichotomy in ChoP expression between commensal and pathogenic species of the same genera would seem to be of biological significance. The very recent identification of an ortholog of a phosphoethanolamine transferase that is required for ChoP expression on menginococcal pili should facilitate a better understanding of the precise pilin linkage of ChoP (to the carbohydrate or protein backbone) and the mechanism of choline acquisition (55), in particular whether ChoP expression influences adherence mediated by fimbriae and/or allows direct binding of the meningococcus or gonococcus via ChoP. For *P. aeruginosa,* a cell envelope protein containing the ChoP epitope expressed in a temperature-dependent manner has been described (61). The nature of this protein and the functional significance of this observation remain unknown, although a pathway for de novo ChoP synthesis has recently been found in this opportunistic pathogen of the airway (66).

The following sections discuss the role of ChoP in the host-bacterium interaction and the advantages and disadvantages that expression of this host-like structure may confer on bacterial survival.

CONTRIBUTIONS OF PHOSPHORYLCHOLINE TO HOST-BACTERIUM INTERACTIONS

A series of experiments examined whether bacteria are predominantly ChoP phase-on or phase-off during colonization and systemic infection. These studies relied on *H. influenzae* because of the "on" or "off" expression of this structure. Following challenge with a mixed inoculum, a gradual selection for phase-on variants over many days was observed during nasopharyngeal carriage in an infant-rat model (64). Moreover, the phenotype of nontypeable *H. influenzae* in the human airway was determined in clinical specimens by direct genotypic analysis without in vitro culture, which could alter results because of phase variation occurring in vitro. By chance alone, one in three PCR products would have a number of CAAT repeats generating an in-frame *licA* gene. However, PCR analysis of the 5′ end of *licA* in clinical specimens from the upper airway revealed that the number of repeats was in frame in >90% of cases. This indicates that there is a selection for the ChoP-expressing phenotype within the respiratory tract of the natural human host (64). Similar direct genotypic analysis of the molecular switch that controls phase variation carried out with blood from patients with clinical cases of *H. influenzae* bacteremia without in vitro culture showed that beyond the mucosal surface of the respiratory tract, the organism is predominantly (>80%) ChoP nonexpressing (unpublished data). Findings with *H. influenzae* were similar to those with *S. pneumoniae:* transparent variants expressing large amounts of ChoP colonized more efficiently and opaque variants with less ChoP predominated during invasive disease. Together, these in vivo observations were supportive of a role for ChoP in bacterial colonization of the airway and a selection against its expression at other sites or stages in the pathogenesis of infection.

Biological Role of the Phosphorylcholine Phase-Off Phenotype

Innate immunity targeting phosphorylcholine

Why, then, is a selection against phase variants expressing ChoP observed during an inflammatory response or in invasive infection? One factor in negative selection acting on ChoP expression is C-reactive protein (CRP), named for its ability to bind to ChoP-containing cell wall carbohydrate or C-polysaccharide of the pneumococcus (51). CRP, originally thought to be only a serum component, is an acute-phase protein whose levels rise by up to 1,000-fold in response to an inflammatory stimulus. The protective effect of CRP is mediated by its ability to act as an opsonin and, when bound to ChoP, to activate the complement pathway through interaction with complement component C1q (21). Indirect evidence for a role for CRP in ChoP-specific innate immunity comes from experiments with transgenic mice carrying the human CRP gene (52). Mice, which have a constitutively low level of CRP expression, are more resistant to experimental pneumococcal sepsis when carrying the human CRP transgene conferring inducible high-level expression as in humans. Since CRP transgenic mice are also more resistant to invasive pathogens lacking ChoP, it has been unclear whether the protective role CRP in pneumococcal sepsis was mediated by its interaction with ChoP (50).

More recently, ChoP-expressing variants of *H. influenzae* were shown to be more sensitive to the bactericidal activity of human serum independently to the presence of naturally acquired antibody to ChoP. This bactericidal activity required the binding of human CRP, with subsequent activation of complement through the classical pathway. Purified human serum CRP killed ChoP^{+} *H. influenzae* at concentrations of >10 ng/ml, well below the amounts in serum in the absence of an inflammatory stimulus, and had no effect on variants of the same strain lacking its ChoP target. This was the first in vitro demonstration of any antimicrobial effect of CRP and its dependence on the presence of ChoP. Further analysis revealed that allelic variants with ChoP positioned on a hexose extending from heptose III rather than heptose I were sensitive to CRP-mediated serum bactericidal activity regardless of the genetic background. These differences in CRP-mediated killing correlated with differences in the binding of CRP from human serum to whole bacteria, suggesting that the accessibility of this structure on the variable oligosaccharide is an additional factor in sensitivity to the antibacterial effects of CRP. Innate immunity due to CRP could therefore be a factor in the selection of phase variants lacking ChoP during bacteremic infection.

CRP is known to be synthesized by hepatocytes and induced by proinflammatory cytokines. The presence of CRP on mucosal surfaces and its role in the innate immunity of the human respiratory tract where ChoP-containing organisms reside had not been previously reported. Recently, this was addressed by using a monoclonal antibody (MAb) to human CRP to show that CRP is in fact present in both infected (0.17 to 42 mg/ml) and uninfected (<0.05 to 0.88 mg/ml) secretions from the human respiratory tract in sufficient quantities to yield an antimicrobial effect (15). In addition, the CRP gene was shown, by using in situ hybridization on human nasal polyps and reverse transcription-PCR of pharyngeal cells in culture, to be transcribed in epithelial cells of the respiratory tract. Moreover, the complement-dependent bactericidal activity of normal nasal airway surface fluid and sputum against ChoP-expressing *H. influenzae* was abolished when the secretions were pretreated to remove CRP. These results indicated that (i) CRP is present in secretions of the human respiratory tract, (ii) human respiratory tract epithelial cells are capable of CRP expression, and (iii) this protein may contribute to bacterial clearance in the human respiratory tract. The absence of CRP in the airways of normal mice or mice carrying the human CRP transgene has limited our ability to test the role of innate immunity based on CRP in nasopharyngeal colonization in any relevant animal model (unpublished data).

Humoral immune response to phosphorylcholine

Despite its presence on host tissues, ChoP is antigenic (a type I thymus-independent antigen) on carriers other than phospholipids. Sufficient levels of Ab could promote the selection of a subpopulation of bacteria lacking this antigen/structure. The effects of anti-ChoP Ab have not been studied in colonization with any bacterial species and have been assessed only in the context of invasive disease for pneumococcus. It has also been suggested that ChoP has immunomodulatory properties (17). At high concentrations (25 to 50 μg/ml) of ChoP-containing molecules, a polyclonal stimulation of B cells is noted, whereas smaller amounts (0.2 to 2 μg/ml) can partially inhibit the B-cell receptor-dependent activation and proliferation of B cells (67).

Some strains of mice produce a germ line Ab, encoded by a specific combination of the V_H1 immunoglobulin heavy and Vk22 light chains, that recognizes ChoP and is referred to as the T15 idiotype. *xid* mice, which are deficient at generating Abs to polysaccharide antigens and ChoP, are more susceptible to invasive pneumococcal infection and have been used to investigate the protective role of anti-ChoP Abs (4, 5). Immunoglobulin G2a (IgG2a), IgG3, and, to a lesser extent, IgM MAbs of the T15 idiotype are variably protective in *xid* mice against lethal infection with some mouse-virulent pneumococcal types when given passively (3). Moreover, $V1^{-/-}$ knockout mice, lacking a T15 response, are also more sensitive to invasive pneumococcal infection (31). Both normal and *xid* mice are more resistant to fatal pneumococcal infection following immunization with ChoP coupled to a protein carrier (22). In a separate study, a C-polysaccharide conjugate was immunogenic when injected in both rabbits and mice and gave opsonophagocytic titers that correlated with ChoP-specific Ab only in mice (27, 53). This suggests that the induction of anti-ChoP Abs might be species dependent.

Humans produce Ab reactive with ChoP but lack the equivalent of the murine V1 heavy-chain gene and therefore do not express immunoglobulins of the T15 idiotype (31). The majority of immunocompetent adults have measurable titers of naturally occurring Ab to ChoP, possibly as a result of previous exposure to ChoP-expressing microbes, with titers of both IgM and IgG averaging 30 to 40 μg/ml (36, 38). It has been assumed that the human immune response to this antigen is not protective since levels of anti-ChoP or anti-cell wall polysaccharide Ab tend to be higher in individuals with pneumococcal disease than in healthy controls (37). Human anti-ChoP IgM purified from normal donors was protective against invasive pneumococcal infection when delivered passively to mice (20 μg/animal), but in another study it failed to demonstrate opsonic activity (4, 36, 44). Thus, the effectiveness of the human immune response to ChoP remains unclear and has not been examined in reference to IgG or mucosal immunity.

Subsequently, the issue of whether human antibody to ChoP has protective activity was addressed by using affinity-purified antibodies from the total gamma globulin fraction of pooled normal human serum. The purified Ab, which was >99% IgG2, was specific to ChoP, as shown by its binding only to ChoP-expressing *H. influenzae* and to *S. pneumoniae* in whole-cell lysates. The efficacy of the human anti-ChoP IgG was assessed by using in vitro assays that correlate with protection against *H. influenzae* and *S. pneumoniae* infection. The purified Ab showed complement-dependent bactericidal activity against ChoP-on phase variants of nontypeable *H. influenzae* (and no killing of the corresponding ChoP-off variant), with a 50% lethal dose of 3.5 mg/ml. Killing of *S. pneumoniae* requires phagocytosis. Anti-ChoP IgG, together with human neutrophils and complement, resulted in >50% killing of four representative isolates of the five most common pneumococcal serotypes at a dose of 20 to 60 μg/ml. For a serotype 6A strain, the 50% lethal dose was 8.5 μg/ml. A mouse model of fatal pneumococcal infection following intraperitoneal challenge was then used to examine protection by passively administered human anti-ChoP antibody. A single dose of 40 μg/mouse yielded a concentration of ~20 μg/ml in serum, similar to average levels in humans, at 24 h after intraperitoneal administration. *xid* mice were challenged with transparent or opaque variants of the serotype 6A clinical isolate that was sensitive to the anti-ChoP antibody in opsonophagocytic assays. In contrast to the more virulent opaque variant, there was complete protection against lethal doses of transparent pneumococci. Mice, however, lack FcγRIIa, the only Fcγ receptor that binds human IgG2, and since available evidence points to the importance of the Fcγ receptor in IgG2-mediated protection of humans, protective levels in the natural host might be far lower (18, 68). In summary, available evidence suggests that both mouse and human antibody to ChoP may be protective and therefore may be a factor in the selection of $ChoP^-$ variants in situations when there are sufficient amounts of this antibody. The effect of mucosal antibody to ChoP on colonization has not yet been investigated.

Biological Role of the Phosphorylcholine Phase-On Phenotype

Effect on susceptibility to antimicrobial peptides in the respiratory tract

Antimicrobial peptides are thought to be effectors of innate immunity through their antibiotic activity and direct killing of microorganisms and may be a determining factor in the characteristics of the flora colonizing mucosal surfaces (2). Successful commensals would probably have evolved mechanisms to evade the peptides expressed in their niche. This laboratory tested the hypothesis that bacterial ChoP, which mimics the head group on phosphatidylcholine on eukaryotic cells, acts to decrease killing by antimicrobial peptides that target differences between bacterial and host membranes (29). There was a bactericidal effect of the peptide LL-37/hCAP18 on a nontypeable *H. influenzae* strain, with an increasing selection for the ChoP$^+$ phase as the concentration of the peptide was raised from 0 to 10 μg/ml. Moreover, a constitutive ChoP-expressing mutant showed up to 1,000-fold-greater survival in the presence of this peptide than did its isogenic mutant without ChoP. The effect of ChoP on resistance to killing by LL-37/hCAP18 was dependent on the salt concentration and was observed only when bacteria were grown in the presence of environmental choline, a requirement for the expression of ChoP on the LPS. Although ChoP is zwitterionic, the positively charged quaternary amine on choline would be predicted to orient toward the exterior of the cell. Since mutants or phase variants lacking ChoP have no charged residue replacing this structure, it is plausible that the decreased killing by cationic antimicrobial peptides is primarily due the effect of ChoP on the surface charge. In situ hybridization studies of nasal polyps established that there is transcription of the LL-37/hCAP18 gene on the epithelial surface of the human nasopharynx. Highly variable amounts of LL-37/hCAP18 in normal nasal secretions (<1.2 to >80 μg/ml) were demonstrated with an Ab against this peptide. It was concluded that ChoP alters the bacterial cell surface so as to mimic host membrane lipids and decrease killing by LL-37/hCAP18, an antimicrobial peptide that is expressed on the mucosal surface of the nasopharynx in potentially bactericidal concentrations.

Several other antimicrobial peptides, including β-defensins, which are structurally unrelated to LL-37/hCAP18, are now known to be expressed in the human respiratory tract (7). Unlike LL-37/hCAP18, which may be synthesized in an active form, β-defensins are highly folded and difficult to synthesize in quantities sufficient to test their function in similar assays. To investigate the function of β-defensins, mice deficient in an antimicrobial peptide, mouse β-defensin 1 (mBD-1), were tested for their ability to clear isogenic constitutive ChoP-expressing and ChoP-nonexpressing *H. influenzae*. mBD-1$^{-/-}$ mice showed delayed clearance of *H. influenzae* from the lungs (32). Prior to this in vivo demonstration of an antibiotic effect of β-defensins, evidence to support the hypothesis of the antimicrobial effect of antimicrobial peptides in vertebrates was indirect, based on expression profiles and in vitro assays using purified peptides. These data gave a direct demonstration that antimicrobial peptides of vertebrates provide an initial block to bacteria at epithelial surfaces.

In contrast to in vitro results with LL-37/hCAP18, there was no difference between mBD-1$^{-/-}$ and mBD-1$^{+/+}$ mice in the clearance of ChoP-expressing compared to ChoP-nonexpressing bacteria. This correlated with a lack of an effect of bacterial ChoP on sensitivity to killing by mBD-1 in vitro. It was concluded that there is heterogeneity among antimicrobial peptides that impacts their spectrum of activity between species as well as within a species when there are structural variations that alter susceptibility, as demonstrated by the presence of ChoP. This could account for the need for multiple peptides of different types in the human airway as well as other sites.

ChoP promotes colonization by functioning as a bacterial ligand

It appears that multiple pathogens, including members of at least four genera, utilize a common mechanism involving ChoP binding to the receptor for PAF (rPAF) to adhere to and invade host cells or tissues (Fig. 4). The observation that pneumococcal ChoP mimics choline phosphate on PAF, allowing for adherence via the G-protein-coupled rPAF, was first made by E. Tuomanen (9). This was confirmed by showing that COS cells acquire the capacity to bind pneumococci when transfected to express rPAF and that this interaction is diminished when pneumococci are grown under conditions such that they express ethanolamine in lieu of choline. Moreover, the interaction of ChoP-expressing pneumococci and rPAF-expressing cells is blocked by rPAF antagonists. In addition, in a model of early events in meningitis, invasion of human brain microvascular endothelial cells by pneumococci required ChoP and was partially inhibited by blockage of rPAF trafficking (40).

There are now data showing that nontypeable *H. influenzae* is among the growing list of species dependent on the same receptor-ligand interaction. Multiple adhesive mechanisms have been described for *H. influenzae* (39). Earlier studies of this subject had not taken LPS phase variation into considera-

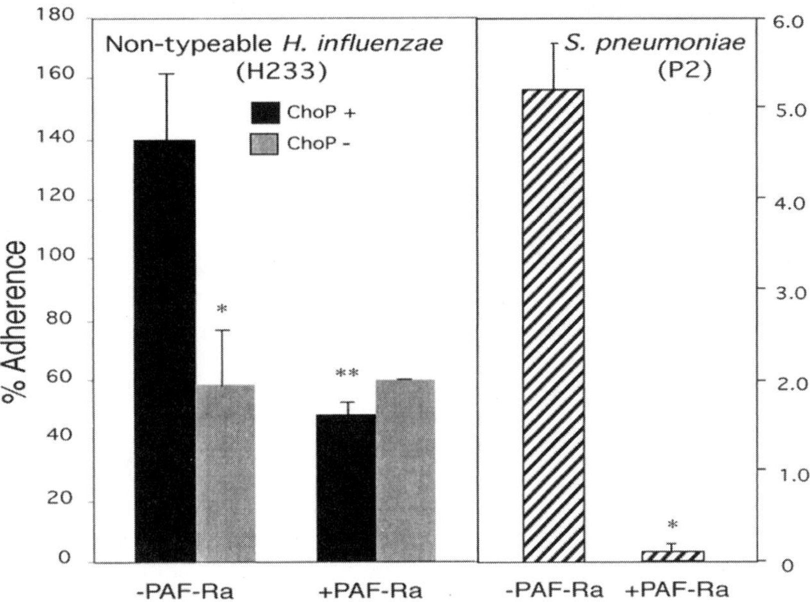

Figure 4. Effect of an rPAF antagonist (PAF-Ra) on bacterial adherence. Adherence of nontypeable *H. influenzae* or *S. pneumoniae* to Detroit 562 pharyngeal epithelial cells in the presence or absence of 1 μM PAF-Ra [1-O-hexadecyl-2-acetyl-*sn*-glycerol-3-phospho-(*N*,*N*,*N*-trimethyl)hexanolamine] is expressed as a percentage of the inoculum binding to epithelial cells after 60 min and values represent the means of at least three determinations in duplicate. Error bars indicate standard deviation. Asterisks indicate $P < 0.05$ compared to first bar in each panel.

tion. When adherence is examined with a population of nontypeable *H. influenzae* variants expressing ChoP, ChoP-mediated adherence (and invasion) is found (49). In these studies, adherence was assessed by monitoring the binding of *H. influenzae* LPS-coated polystyrene beads to 16HBE14o$^-$ human bronchial epithelial cells. Competitive-inhibition studies with a panel of compounds containing structures found within nontypeable *H. influenzae* LPS suggested that the ChoP moiety is involved in adherence. Further experiments revealed that mutations affecting the incorporation of ChoP were associated with a decrease in adherence to and invasion of primary human bronchial cells. Analysis of primary bronchial epithelial cells in culture revealed that ChoP$^+$ but not ChoP$^-$ mutants colocalized with the rPAF (49). Pretreatment of bronchial cells with an rPAF antagonist inhibited the invasion by nontypeable *H. influenzae* expressing ChoP$^+$ LPS glycoforms, suggesting that this organism invades bronchial cells by means of interaction with the rPAF. As is the case for *H. influenzae,* ChoP on commensal *Neisseria* species enhances adherence to and invasion of human epithelial cells via rPAF (45). Another report described rPAF-mediated invasion of vascular endothelial cells by *Actinobacillus actinomycetemcomitans,* the etiologic agent of juvenile peridontitis (41).

The distribution of the rPAF in the respiratory tract was not well described. In situ hybridization of samples of human nasal polyps showed the expression of rPAF on the epithelial surface of the human nasopharynx, where many of the commensal bacteria expressing the ChoP ligand reside (16). In addition, D562 human epithelial cells in culture express the receptor for bacterial ChoP as shown by reverse transcription-PCR for rPAF.

Other factors affecting phosphorylcholine-mediated adherence

Many of the major bacterial pathogens of the human respiratory tract express ChoP, a ligand for the receptor for rPAF and the target of CRP, raising the possibility that mucosal CRP interferes with or contributes to adherence. In fact, CRP can effectively block the attachment of bacteria expressing cell surface ChoP to host cells (16). Concentrations of CRP equivalent to those on the mucosal surface of the inflamed human airway (2 to 4 μg/ml) blocked >50% of the adherence of both *S. pneumoniae* and *H. influenzae* to human pharyngeal epithelial cells. Moreover, the antiadhesive effects of the rPAF antagonist and CRP were not additive, suggesting that CRP activity is specific to the portion of adherence mediated by the rPAF. The antiadhesive effect of CRP requires the presence of bacterial ChoP but, unlike its antimicrobial activity, is independent of complement, whose activity on mucosal surfaces remains poorly described. Thus, mucosal CRP could serve as a barrier that limits or controls the numbers of colonizing $ChoP^+$ bacteria.

In a related study, the binding of CRP to ChoP and the effect of CRP on adherence were inhibited by human surfactant, which is composed predominantly (~70%) of choline phosphate (16). Thus, the antiadhesive effect of CRP may be diminished in the terminal airway, where surfactant is abundant. The inhibition of CRP-mediated innate immunity by surfactant could provide an explanation for why the same ChoP-expressing species acts as commensals when resident in the nasopharynx, where there is no surfactant, and as pathogens after gaining access to the lower airway.

Many ChoP-expressing species are encapsulated, and the expression of capsular polysaccharide by bacteria inhibits adherence to an epithelial surface. In addition, IgA, the major class of immunoglobulin in secretions, classically functions by interfering with microbial attachment to host tissues. Many mucosal pathogens, including *S. pneumoniae, H. influenzae,* and the neisseriae, express an IgA1 protease that is highly specific for human IgA1 and may circumvent the protective effects of this immunoglobulin subclass (23). However, rather than inhibiting adherence, human serotype-specific IgA1 MAb markedly enhanced bacterial attachment to host cells, but only when cleaved by IgA1 protease (57). Neither Abs of protease-insensitive subclasses (IgA2 and IgG) nor those directed against heterologous capsules had such activity. The adherence-promoting properties of cleaved Abs correlated with the cationic characteristics of their variable segments, suggesting that bound Fab fragments generated by IgA1 protease cleavage may neutralize the inhibitory effect of negatively charged capsules on adhesive interaction with host cells. Coating of pneumococci with anticapsular polysaccharide Ab unmasked the bacterial ChoP ligand, as shown by its accessibility to CRP, allowing for increased adherence mediated by binding to the rPAF on epithelial cells. IgA1 proteases may therefore enable pathogens to subvert the antigen specificity of the humoral immune response to facilitate otherwise inefficient adhesive interactions and promote persistence on the mucosal surface.

CONCLUSION

The surface expression of ChoP appears to be particularly common among species that colonize predominantly the upper respiratory tract. This allows for bacterial mimicry of the host cell surface since choline is a prominent feature of the host cell membrane. In addition, the expression of ChoP by bacteria enables their utilization of host choline-dependent pathways and cellular processes. The ChoP phase-on phenotype facilitates binding to a common receptor (rPAF) and decreases bacterial susceptibility to some types of antimicrobial peptides. Since the host appears to specifically target microbial ChoP by using both innate (CRP) and adaptive (anti-ChoP antibody) immunity, successful pathogens have mechanisms for turning off or down regulating its expression. Colonization, therefore, may be a balance between factors selecting for and against this conserved structure/antigen.

REFERENCES

1. **Arondel, V., C. Benning, and C. Somerville.** 1993. Isolation and functional expression in *Escherichia coli* of a gene encoding phosphatidylethanolamine methyltransferase (EC 2.1.1.17) from *Rhodobacter sphaeroides. J. Biol. Chem.* **268:**16002–16008.
2. **Boman, H.** 1991. Antibacterial peptides: key components needed in immunity. *Cell* **65:**205–207.
3. **Briles, D., J. Claflin, K. Schroer, and C. Forman.** 1981. Mouse IgG3 antibodies are highly protective against infection with *Streptococcus pneumoniae. Nature* **294:**88–90.
4. **Briles, D. E., C. Forman, J. C. Horowitz, J. E. Volanakis, W. J. Benjamin, L. S. McDaniel, J. Eldridge, and J. Brooks.** 1989. Antipneumococcal effects of C-reactive protein and monoclonal antibodies to pneumococcal cell wall and capsular antigens. *Infect. Immun.* **57:**1457–1464.
5. **Briles, D. E., M. Nahm, K. Schroer, J. Davie, P. Baker, J. Kearney, and R. Barletta.** 1981. Antiphosphocholine antibodies found in normal mouse serum are protective against intravenous infection with type 3 *Streptococcus pneumoniae. J. Exp. Med.* **153:**694–705.
6. **Brundish, D. E., and J. Baddiley.** 1968. Pneumococcal C-substance, a ribitol teichoic acid containing choline phosphate. *Biochem. J.* **110:**573.
7. **Cole, A., H. Liao, O. Stuchlik, J. Tilan, J. Pohl, and T. Ganz.** 2002. Cationic polypeptides are required for antibacterial activity of human airway fluid. *J. Immunol.* **169:**6985–6991.
8. **Cox, A., H. Masoud, P. Thibault, J. Brisson, M. van der Zwan, M. Perry, and J. Richards.** 2001. Structural analysis of the lipopolysaccharide from the nontypable *Haemophilus influenzae* strain SB 33. *Eur. J. Biochem.* **268:**5278–5286.
9. **Cundell, D. R., N. P. Gerard, C. Gerard, I. Idanpaan-Heikkila, and E. I. Tuomanen.** 1995. *Streptococcus pneumoniae* anchor to activated human cells by the receptor for platelet-activating factor. *Nature* **377:**435–438.
10. **Cundell, D. R., J. N. Weiser, J. Shen, A. Young, and E. I. Tuomanen.** 1995. Relationship between colonial morphology and adherence of *Streptococcus pneumoniae. Infect. Immun.* **63:** 757–761.
11. **Fan, X., H. Goldfine, E. Lysenko, and J. Weiser.** 2001. The transfer of choline from the host to the bacterial cell surface requires glpQ in *Haemophilus influenzae. Mol. Microbiol.* **41:** 1029–1036.
12. **Fan, X., C. Pericone, E. Lysenko, H. Goldfine, and J. Weiser.** 2003. Multiple mechanisms for choline transport and utilization in *Haemophilus influenzae. Mol. Microbiol.* **50:**537–548.
13. **Gillespie, S. H., S. Ainscough, A. Dickens, and J. Lewin.** 1996. Phosphorylcholine-containing antigens in bacteria from the mouth and respiratory tract. *J. Med. Microbiol.* **44:**35–40.
14. **Gosink, K., E. Mann, C. Guglielmo, E. Tuomanen, and H. Masure.** 2000. Role of novel choline binding proteins in

virulence of *Streptococcus pneumoniae*. *Infect. Immun.* **68**: 5690–5695.

15. **Gould, J., and J. Weiser.** 2001. Expression of C-reactive protein in the human respiratory tract. *Infect. Immun.* **69**:1747–1754.
16. **Gould, J., and J. Weiser.** 2002. The inhibitory effect of C-reactive protein on bacterial phosphorylcholine-platelet activating factor receptor mediated adherence is blocked by surfactant. *J. Infect. Dis.* **186**:361–371.
17. **Harnett, W., and M. Harnett.** 1999. Phosphorylcholine: friend or foe of the immune system? *Immunol. Today* **20**:125–129.
18. **Hulett, M., and P. Hogarth.** 1994. Molecular basis of Fc receptor function. *Adv. Immunol.* **57**:1–127.
19. **Janson, H., B. Carlen, A. Cervin, A. Forsgren, A. Magnusdottir, S. Lindberg, and T. Runer.** 1999. Effects on the ciliated epithelium of protein D-producing and -nonproducing nontypeable *Haemophilus influenzae* in nasopharyngeal tissue cultures. *J. Infect. Dis.* **180**:737–746.
20. **Janson, H., A. Melhus, A. Hermansson, and A. Forsgren.** 1994. Protein D, the glycerophosphodiester phosphodiesterase from *Haemophilus influenzae* with affinity for human immunoglobulin D, influenzae virulence in a rat otitis model. *Infect. Immun.* **62**:4848–4854.
21. **Kaplan, M. H., and J. E. Volankis.** 1974. Interaction of C-reactive protein complexes with the complement system. I. Consumption of human complement associated with the reaction of CRP with pneumococcal C-polysaccharide and with choline phosphates, lecthin and sphingomyelin. *J. Immunol.* **112**:2135–2147.
22. **Kenny, J., G. Guelde, R. Fischer, and D. Longo.** 1994. Induction of phosphocholine-specific antibodies in X-linked immune deficient mice: in vivo protection against a *Streptococcus pneumoniae* challenge. *Int. Immunol.* **6**:561–568.
23. **Kilian, M., J. Reinholdt, H. Lomholt, K. Poulsen, and E. Frandsen.** 1996. Biological significance of IgA1 proteases in bacterial colonization and pathogenesis: critical evaluation of experimental evidence. *APMIS* **104**:321–338.
24. **Kim, J., and J. Weiser.** 1998. Association of intrastrain phase variation in quantity of capsular polysaccharide and teichoic acid with the virulence of *Streptococcus pneumoniae*. *J. Infect. Dis.* **177**:368–377.
25. **Kim, J. O., S. Romero-Steiner, U. Sörensen, J. Blom, M. Carvalho, S. Barnardi, G. Carlone, and J. N. Weiser.** 1999. Relationship between cell-surface carbohydrates and intrastrain variation on opsonophagocytosis of *Streptococcus pneumoniae*. *Infect. Immun.* **67**:2327–2333.
26. **Kroes, I., P. Lepp, and D. Relman.** 1999. Bacterial diversity within the human subgingival crevice. *Proc. Natl. Acad. Sci. USA* **96**:14547–14552.
27. **Laferriere, C. A., R. K. Soo, J. M. de Muys, F. Michon, and H. J. Jennings.** 1997. The synthesis of *Streptococcus pneumoniae* polysaccharide-tetanus toxoid conjugates and the effect of chain length on immunogenicity. *Vaccine* **15**:179–186.
28. **Lamark, T., I. Kaasen, M. W. Eshoo, P. Falkenberg, J. McDougall, and A. R. Strom.** 1991. DNA sequence and analysis of the bet genes encoding the osmoregulatory choline-glycine betaine pathway of *Escherichia coli*. *Mol. Microbiol.* **5**:1049–1064.
29. **Lysenko, E., J. Gould, R. Bals, J. Wilson, and J. Weiser.** 2000. Bacterial phosphorylcholine decreases susceptibility to the antimicrobial peptide LL-37/hCAP18 expressed in the upper respiratory tract. *Infect. Immun.* **68**:1664–1671.
30. **Lysenko, E., J. Richards, A. Cox, A. Stewart, A. Martin, M. Kapoor, and J. Weiser.** 2000. The position of phosphorylcholine on the lipopolysaccharide of *Haemophilus influenzae* affects binding and sensitivity to C-reactive protein mediated killing. *Mol. Microbiol.* **35**:234–245.
31. **Mi, Q., L. Zhou, D. Schulze, R. Fischer, A. Lustig, L. Rezanka, D. Donovan, D. Longo, and J. Kenny.** 2000. Highly reduced protection against *Streptococcus pneumoniae* after deletion of a single heavy chain gene in mouse. *Proc. Natl. Acad. Sci. USA* **97**:6031–6036.
32. **Moser, C., D. Weiner, E. Lysenko, R. Bals, J. Weiser, and J. Wilson.** 2002. β-Defensin 1 contributes to pulmonary innate immunity in mice. *Infect. Immun.* **70**:3068–3072.
33. **Moxon, E., R. Lenski, and P. Rainey.** 1998. Adaptive evolution of highly mutable loci in pathogenic bacteria. *Perspect. Biol. Med.* **42**:154–155.
34. **Moxon, E. R., and C. Wills.** 1999. DNA microsatellites: agents of evolution? *Sci. Am.* **280**:94–99.
35. **Munson, R. S. J., and K. Sasaki.** 1993. Protein D, a putative immunoglobulin D-binding protein produced by *Haemophilus influenzae*, is a glycerophosphodiester phophodiesterase. *Infect. Immun.* **175**:4569–4571.
36. **Musher, D. M., A. J. Chapman, A. Goree, S. Jonsson, D. Briles, and R. E. Baughn.** 1986. Natural and vaccine-related immunity to *Streptococcus pneumoniae*. *J. Infect. Dis.* **154**:245–256.
37. **Nordenstam, G., B. Andersson, D. Briles, J. W. J. Brooks, A. Oden, A. Svanborg, and C. S. Eden.** 1990. High anti-phosphorylcholine antibody levels and mortality associated with pneumonia. *Scand. J. Infect. Dis.* **22**:187–195.
38. **Purkall, D., J. Tew, and H. Schenkein.** 2002. Opsonization of *Actinobacillus actinomycetemcomitans* by immunoglobulin G antibody reactive with phosphorylcholine. *Infect. Immun.* **70**: 6485–6488.
39. **Rao, V., G. Krasan, D. Hendrixson, S. Dawid, and J. W. St. Geme III.** 1999. Molecular determinants of the pathogenesis of disease due to non-typable *Haemophilus influenzae*. *FEMS Microbiol. Rev.* **23**:99–129.
40. **Ring, A., J. N. Weiser, and E. I. Tuomanen.** 1998. Pneumococcal penetration of the blood-brain barrier: molecular analysis of a novel re-entry path. *J. Clin. Investig.* **102**:347–360.
41. **Schenkein, H., S. Barbour, C. Berry, B. Kipps, and J. Tew.** 2000. Invasion of human vascular endothelial cells by *Actinobacillus actinomycetemcomitans* via the receptor for platelet-activating factor. *Infect. Immun.* **68**:5416–5419.
42. **Schweda, E., J. Li, E. Moxon, and J. Richards.** 2002. Structural analysis of lipopolysaccharide oligosaccharide epitopes expressed by non-typeable *Haemophilus influenzae* strain 176. *Carbohydr. Res.* **337**:409–420.
43. **Schweda, E. K. H., H. Masoud, A. Martin, A. Risberg, D. W. Hood, E. R. Moxon, J. N. Weiser, and J. C. Richards.** 1997. Phase variable expression and characterization of phosphorylcholine oligosaccharide epitopes in *Haemophilus influenzae* lipopolysaccharides. *Glycoconj. J.* **14**:S23.
44. **Scott, M. G., D. E. Briles, P. G. Shackelford, D. S. Smith, and M. H. Nahm.** 1987. Human antibodies to phosphocholine. IgG anti-PC antibodies express restricted numbers of V and C regions. *J. Immunol.* **138**:3325–3331.
45. **Serino, L., and M. Virji.** 2002. Genetic and functional analysis of the phosphorylcholine moiety of commensal *Neisseria* lipopolysaccharide. *Mol. Microbiol.* **43**:437–448.
46. **Serino, L., and M. Virji.** 2000. Phosphorylcholine decoration of lipopolysaccharide differentiates commensal neisseriae from pathogenic strains: identification of *licA*-type genes in commensal neisseriae. *Mol. Microbiol.* **35**:1550–1559.
47. **Sohlenkamp, C., I. Lopez-Lara, and O. Geiger.** 2003. Biosynthesis of phosphatidylcholine in bacteria. *Prog. Lipid Res.* **42**: 115–162.
48. **Sorenson, U. B. S., J. Henrichsen, H.-C. Chen, and S. C. Szu.** 1990. Covalent linkage between the capsular polysaccharide and cell wall peptidoglycan of *Streptococcus pneumoniae* revealed by immunochemical methods. *Microb. Pathog.* **8**:325–334.

49. **Swords, W., B. Buscher, K. Ver Steeg Ii, A. Preston, W. Nichols, J. Weiser, B. Gibson, and M. Apicella.** 2000. Nontypeable *Haemophilus influenzae* adhere to and invade human bronchial epithelial cells via an interaction of lipooligosaccharide with the PAF receptor. *Mol. Microbiol.* **37:**13–27.

50. **Szalai, A., J. VanCott, J. McGhee, J. Volanakis, and W. J. Benjamin.** 2000. Human C-reactive protein is protective against fatal *Salmonella enterica* serovar Typhimurium infection in transgenic mice. *Infect. Immun.* **68:**5652–5656.

51. **Szalai, A. J., A. Agrawal, T. J. Greenhough, and J. E. Volanakis.** 1997. C-reactive protein. *Immunol. Res.* **16:**127–136.

52. **Szalai, A. J., D. E. Briles, and J. E. Volanakis.** 1995. Human C-reactive protein is protective against fatal *Streptococcus pneumoniae* infection in transgenic mice. *J. Immunol.* **155:** 2557–2563.

53. **Szu, S., R. Schneerson, and J. Robbins.** 1986. Rabbit antibodies to the cell wall polysaccharide of *Streptococcus pneumoniae* fail to protect mice from lethal challenge with encapsulated pneumococci. *Infect. Immun.* **54:**448–455.

54. **Thomas, A. M., P. A. Lambert, and I. R. Poxton.** 1978. The uptake of choline by *Streptococcus pneumoniae. J. Gen. Microbiol.* **109:**313–317.

55. **Warren, M. J., and M. P. Jennings.** 2003. Identification and characterization of *pptA:* a gene involved in the phase-variable expression of phosphorylcholine on pili of *Neisseria meningitidis. Infect. Immun.* **71:**6892–6898.

56. **Weiser, J., D. Bae, H. Epino, S. Gordon, M. Kapoor, L. Zenewicz, and M. Shchepetov.** 2001. Changes in availability of oxygen accentuate differences in capsular polysaccharide expression by phenotypic variants and clinical isolates of *Streptococcus pneumoniae. Infect. Immun.* **69:**5430–5439.

57. **Weiser, J., D. Bae, C. Fasching, R. Scamurra, A. Ratner, and E. Janoff.** 2003. Antibody-enhanced pneumococcal adherence requires IgA1 protease. *Proc. Natl. Acad. Sci. USA* **100:**415–420.

58. **Weiser, J., and E. Tuomanen.** 2002. A disease-oriented approach to the discovery of novel vaccines, p. 139–148. *In* B. Bloom and P.-H. Lambert (ed.), *The Vaccine Book.* Academic Press, Inc., New York, N.Y.

59. **Weiser, J. N.** 1999. Phase-variation of *Streptococcus pneumoniae,* p. 225–231. *In* V. Fischetti (ed.), *Gram-Positive Pathogens.* ASM Press, Washington, D.C.

60. **Weiser, J. N., R. Austrian, P. K. Sreenivasan, and H. R. Masure.** 1994. Phase variation in pneumococcal opacity: relationship between colonial morphology and nasopharyngeal colonization. *Infect. Immun.* **62:**2582–2589.

61. **Weiser, J. N., J. B. Goldberg, N. Pan, L. Wilson, and M. Virji.** 1998. The phosphorylcholine epitope undergoes phase variation on a 43-kDa protein in *Pseudomonas aeruginosa* and on pili of pathogenic *Neisseria. Infect. Immun.* **66:** 4263–4267.

62. **Weiser, J. N., J. M. Love, and E. R. Moxon.** 1989. The molecular mechanism of phase variation of *H. influenzae* lipopolysaccharide. *Cell* **59:**657–665.

63. **Weiser, J. N., D. J. Maskell, P. D. Butler, A. A. Lindberg, and E. R. Moxon.** 1990. Characterization of repetitive sequences controlling phase variation of *Haemophilus influenzae* lipopolysaccharide. *J. Bacteriol.* **172:**3304–3309.

64. **Weiser, J. N., N. Pan, K. L. McGowan, D. Musher, A. Martin, and J. C. Richards.** 1998. Phosphorylcholine on the lipopolysaccharide of *Haemophilus influenzae* contributes to persistence in the respiratory tract and sensitivity to serum killing mediated by C-reactive protein. *J. Exp. Med.* **187:**631–640.

65. **Weiser, J. N., M. Shchepetov, and S. T. H. Chong.** 1997. Decoration of lipopolysaccharide with phosphorylcholine: a phase-variable characteristic of *Haemophilus influenzae. Infect. Immun.* **65:**943–950.

66. **Wilderman, P., A. Vasil, W. Martin, R. Murphy, and M. Vasil.** 2002. *Pseudomonas aeruginosa* synthesizes phosphatidylcholine by use of the phosphatidylcholine synthase pathway. *J. Bacteriol.* **184:**4792–4799.

67. **Wilson, E., M. Deehan, E. Katz, K. Brown, K. Houston, J. O'Grady, M. Harnett, and W. Harnett.** 2003. Hyporesponsiveness of murine B lymphocytes exposed to the filarial nematode secreted product ES-62 in vivo. *Immunology* **109:**238–245.

68. **Yee, A., H. Phan, R. Zuniga, J. Salmon, and D. Musher.** 2000. Association between FcgammaRIIa-R131 allotype and bacteremic pneumococcal pneumonia. *Clin. Infect. Dis.* **30:** 25–28.

69. **Yother, J., K. Leopold, J. White, and W. Fischer.** 1998. Generation and properties of a *Streptococcus pneumoniae* mutant which does not require choline or analogs for growth. *J. Bacteriol.* **180:**2093–2101.

70. **Zhang, J.-R., I. Idanpaan-Heikkila, W. Fischer, and E. Tuomanen.** 1999. Pneumococcal *licD2* gene is involved in phosphorylcholine metabolism. *Mol. Microbiol.* **31:**1477–1488.

Colonization of Mucosal Surfaces
Edited by James P. Nataro et al.

Chapter 6

Sialylation of the Gram-Negative Bacterial Cell Surface

Michael A. Apicella and Paul A. Jones

Sialic acids (NeuAc) are a family of nine-carbon acidic sugars, of which there are around 40 derivatives known to date (90). The most common sialic acid is *N*-acetylneuraminic acid, or Neu5Ac (Fig. 1). This sugar is formed by the condensation of N-acetylmannosamine-6-phosphate and pyruvate (Fig. 2). The most distinctive features of the molecule are the carboxyl group at C-1 and the amino group at C-5. The carboxyl group at C-1 gives the molecule a net negative charge and is responsible for its acidic nature. The negative charge and relatively large size of this molecule are two key features in how it plays a role in biology. Neu5Ac is usually the terminal sugar on carbohydrate structures (oligo- and polysaccharides), glycoproteins, and glycolipids of mammals, birds, fungi, bacteria metazoans, and archaea.

NeuAc and its derivatives are found on cell membranes and in body fluids in all mammals and many higher-order animals, as well as pathogenic microorganisms. Typically, only a few derivatives of Neu5Ac are seen in each individual species. In addition, the derivatives of Neu5Ac can be tissue or developmentally specific. The molecule is typically modified at positions C-1, C-4, C-5, C-7, C-8, or C-9. The most common sialic acids of mammalian cells are the *N*-acetylneuraminic acids and *N*-glycolylneuraminic acid (Neu5Gc) (108). The addition of a hydroxyl group to the amino moiety at C-5 forms *N*-glycolylneuraminic acid (NeuGc). NeuGc is fairly common in animal species other than humans but is found in humans only in association with certain forms of cancer (72, 97). NeuGc functions in a number of important activities from recognition to antirecognition. Normal humans are the exception, because of a mutation in CMP-sialic acid hydrolase. This enzyme is responsible for the addition of a single oxygen atom to Neu5Ac, converting it to Neu5Gc (10). In humans, this enzyme has no hydroxylase activity because of a large deletion caused by a deletion/frameshift mutation in the human gene. In contrast, the great apes, including the chimpanzee, bonobo, gorilla, and orangutan, our closest animal relatives, have a functional CMP-Neu5Ac hydrolase and can express Neu5Gc in large amounts. In some tissues, this is the predominant NeuAc.

Other common forms of modification are O methylation and O acetylation. These modifications can have a profound impact on the properties of sialic acid. For example, acetylation at C-9 can make the sialic acid resistant to cleavage by sialidases, which catalyze the removal of sialic acid from glycoproteins and glycolipids. O acetylation and *N*-acetyl hydroxylation modifications of sialic acid can have an antirecognition effect on the binding of influenza A and B viruses (35). In addition to NeuAc, two other derivatives are typically found in healthy humans: *N*-acetyl-9-*O*-acetylneuraminic acid and 2-deoxy-2,3-didehydro-*N*-acetylnuraminic acid.

The metabolism and nucleotide derivatization of Neu5Ac are well known (Fig. 2). Metabolites can enter the pathway at multiple points. Fructose and D-glucosamine can be an entry point for the pathway. Neu5Ac and CMP-Neu5Ac can enter the pathway in bacteria from the external environment through dedicated cell membrane transporters (61).

Many pathogenic bacteria have evolved mechanisms to evade human defenses based on the surface exposure of Neu5Ac. A common theme in this mimicry is the incorporation of Neu5Ac as a component of either the capsular polysaccharide or the lipooligosaccharide (LOS) or both. Based on an understanding of the biosynthesis of sialic acid and the evolving elucidation of the genomes of multiple microbes, Vimr and Lichtensteiger have described at least four mechanisms of microbial surface sialylation (112). These include de novo synthesis, donor scavenging, *trans*-sialylation and precursor scavenging. *Neisseria meningitidis* serogroups B, C, Y, and

Michael A. Apicella and Paul A. Jones • Department of Microbiology, The University of Iowa, Iowa City, Iowa 52242.

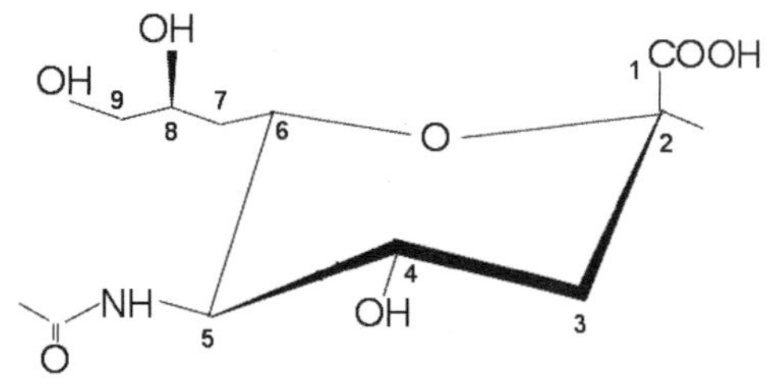

Figure 1. The nine-carbon structure of *N*-acetylneuraminic acid. Natural substitutions of the basic structure result in over 40 different compounds. Recently, studies of unnatural substitutions have suggested possible innovative approaches to the treatment of a number of human diseases (11, 12, 53, 88).

W-135 can synthesize sialic acid for incorporation into capsular polysaccharides and LOS. *Neisseria gonorrhoeae* cannot synthesize or convert Neu5Ac to the nucleotide sugar form and must scavange the donor, CMP-Neu5Ac, to sialylate its LOS structure. *Trypanosoma cruzi* does not synthesize or catabolize free sialic acid. It expresses a developmentally regulated sialidase, which it uses for surface sialylation by a *trans*-sialidase mechanism. The sialyltransferase activity of the sialidase is greater than its hydrolytic activity as long as the appropriate galactosyl acceptor is available. *Haemophilus influenzae* scavenges the precursor, Neu5Ac, to sialylate its LOS and biofilm. It must rely on scavenging of Neu5Ac from the environment because it cannot synthesize Neu5Ac. In addition, it has the means to use Neu5Ac as a source of carbon and nitrogen through a complex Neu5Ac catabolic system.

SIALIC ACIDS IN EUKARYOTIC BIOLOGY

To understand the role of sialic acids in pathogenesis, it is important to examine their role in eukaryotic systems. Sialic acids in eukaryotes can be grouped into three categories based on their functions, as compiled by Schauer et al. (90). The first deals with the general properties of sialic acid, such as its size and its hydrophilic and acidic properties. The net negative charge on the molecule at physiological pH provides a repulsive force that helps stabilize the conformation of glycoproteins and glycolipids in

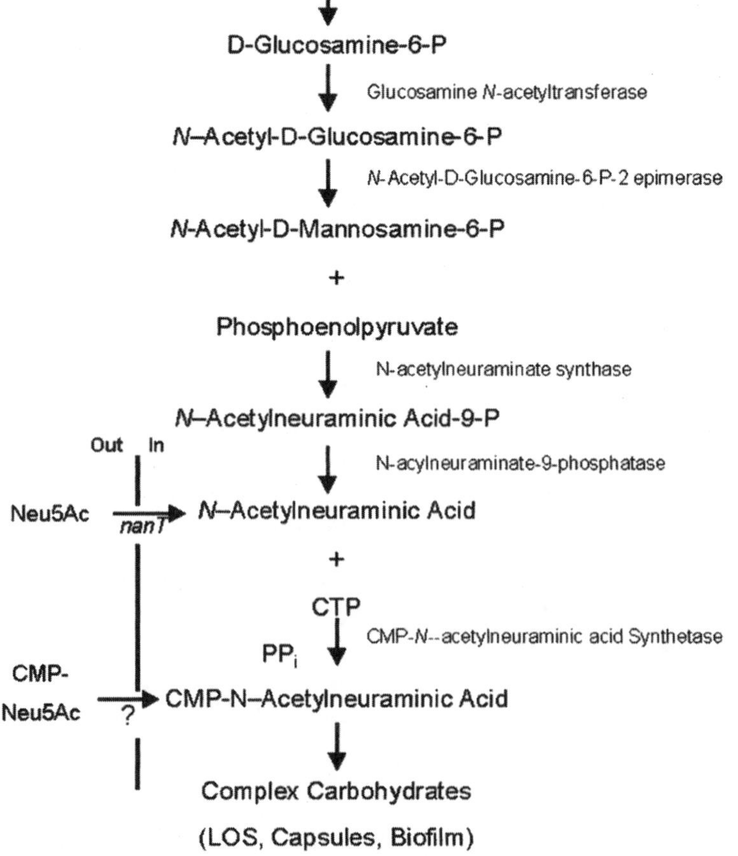

Figure 2. Biosynthesis of sialic acid.

the cell membrane. Sialic acid also increases the thermal stability and proteolytic stability of glycoproteins. The hydration and viscosity of mucins can to a large degree be attributed to the large amount of sialic acid incorporation.

The second function is the ability of sialic acid to mask carbohydrate recognition sites (47, 89). Usually this is a subterminal galactose residue. Once sialic acid is removed, the structure is recognized by a specific receptor on macrophages or hepatocytes and the molecule or cell is taken up and degraded. This is an important process in the clearance of erythrocytes, leukocytes, and thrombocytes. Sialic acid can also block the recognition of antibodies (20). This effect can be due to steric hindrance, negative charge, or hydrophilicity.

The third function is diametrically opposed to its second function. Sialic acid can itself act as a recognition determinant. Some examples of this are the specific interactions of bacteria, viruses, and toxins with sialylated structures on cells and extracellular proteins. Sialic acid is also a crucial component of ligands used by the immune system and is recognized by the selectin and sialoadhesin families (47).

MAMMALIAN SIALIC ACID-DEPENDENT RECEPTORS

Sialic acid-dependent receptors play an important role in adhesion to mammalian cells. A great deal of evidence for this comes from the study of receptor-ligand interactions of the immune system. Two examples are discussed here: the selectin and sialoadhesin families.

Selectins

Selectins are a family of protein molecules expressed on leukocytes and vascular endothelial cells (47, 50, 109, 110). They are associated with the homing of leukocytes and lymphocytes to the sites of tissue injury and inflammation. There are three main members, designated L-selectin (formerly MEL14 antigen), P-selectin (formerly gmp140/PADGEM), and E-selectin (formerly ELAM-1) (6). All of these members recognize glycans terminating in sialylated Lewis antigens (sLe^x) and sialyl-α-2,3-Lewis^a (sLe^a). These molecules have common structural features. They are all type I transmembrane proteins with an N-terminal carbohydrate recognition domain and high amino acid sequence homology. The interaction with sialic acid is not very specific, and elimination of the C-7 to C-9 side chain or replacement with sulfate does not have a drastic effect on binding.

L-selectin is expressed on lymphocytes and leukocytes. They are involved in the homing of lymphocytes to the high endothelial venules of peripheral lymph nodes and the recruitment of neutrophils to the site of acute inflammation. One of the receptors recognized by L-selectin is GlyCAM-1. Sialic acid present on this receptor is required for recognition (40).

E- and P-selectins are expressed on endothelial cells, and P-selectin is also found on platelets. Both of these receptors are activated by cytokines and are involved in the rolling event of leukocytes targeting to sites of tissue injury. E-selectin has a lag time in expression, due to its requirement for transcription and translation after induction. In contrast, P-selectin is produced almost immediately, being stored in vesicles below the cell surface.

Sialoadhesins

Another group of proteins, the sialoadhesin family, demonstrates a true sialic acid-dependent interaction with sialylated glycoproteins or glycolipids (48, 110). This is in contrast to the weak affinity of the selectins for sialic acid. In fact, modifications of sialic acid result in decreased binding efficiency for many members of this family (95). All of the members of this family belong to the immunoglobulin superfamily. Representative members of the sialoadhesin family from mammals are sialoadhesin (Sn), CD22, CD33, and myelin-associated glycoprotein.

The sialoadhesin family members are involved in a much more diverse set of biological processes, such as hematopoiesis, neuronal development, and immunity (48). Sialoadhesin is found on macrophage subpopulations and probably functions in the development of myeloid cells in bone marrow and in the trafficking of leukocytes in lymphatic organs. CD22 is found exclusively on B cells and is involved in signaling with T cells and other B cells. The minimal binding structure for its cognate ligand is NeuAcα2, 6Hex(NAc), where Hex is glucose or galactose. The presentation of the ligand seems to be important, since many glycoproteins and glycolipids have this linkage but do not bind CD22. CD33 is found exclusively on myelomonocytic progenitors, monocytes, and tissue macrophages. No biological function is known to date, but the presence of a predicted tyrosine phosphorylation site indicates that it may be involved in cell signaling. Myelin-associated glycoprotein is expressed on myelinating oligodendrocytes and Schwann cells. It functions in myelination, axonal growth and regulation, and signal transduction. The use of sialic acid in such a variety of cellular adhesin and signaling systems highlights its importance in eukaryotic biology.

MAMMALIAN SIALYLTRANSFERASES

Sialyltransferase enzymes catalyze the transfer of NeuAc from CMP-NeuAc to an acceptor structure (glycoprotein, glycolipid, LOS lipopolysaccharide [LPS], etc.). Mammalian sialyltransferases have a high degree of specificity with regard to the acceptor sugar or other structure to which the sialic acid is added (4). This degree of specificity has led to the identification of a family of 15 mammalian sialyltransferases that add sialic acid to an acceptor sugar with α2,3, α2,6, α2,8, or α2,9 linkages (102). There is more ambiguity in the derivatives of sialic acid that can be transferred, with slight modifications of NeuAc being tolerated by most of the sialyltransferases. Mammalian sialyltransferases are all type II membrane proteins containing a N-terminal cytoplasmic domain, a hydrophobic signal anchor sequence, and a short "stem" region that separates the catalytic domain from the transmembrane region (17). The expression of the sialyltransferases is regulated and differentially expressed during development and in different tissue types (103). Transformation to malignancy also increases sialyltransferase activity and the transfer of NeuGc, which is not normally found in glycoproteins and glycolipids of healthy cells (97).

The most distinctive feature found in all mammalian sialyltransferase to date is referred to as the sialylmotif. Comparison of the first three members of this family revealed a stretch of 55 amino acids with high homology. The identification of this region allowed the cloning of more mammalian sialyltransferase genes by PCR (52, 103). After the discovery of another, smaller sialylmotif, the larger one was referred to as the L-sialylmotif while the smaller one was designated the S-sialylmotif (17).

There is evidence that these sialylmotifs play a functional role. Using site-directed mutagenesis, Datta et al. demonstrated that the L-sialylmotif is responsible for binding of the CMP-NeuAc substrate (17). Site-directed mutagenesis of the S-sialylmotif revealed that it is involved in binding to both the acceptor and donor substrates (17). Besides the two sialylmotifs, there is no other known amino acid homology between members of this group (103).

SIALIC ACID AND CANCER BIOLOGY

There is a correlation between sialic acid levels and the development of cancer. A tumor cell has an increased amount of sialylation and sialyltransferase activity (118). Tumor cell metastasis has also been correlated with an increase in sialylation (124). The heavily sialylated surface of the tumor cell may assist in preventing surveillance recognition and clearance by the immune system (89). Metastasis is further increased by the ability of sialic acid on the tumor cell surface to facilitate adherence to the vascular endothelium and relocation to secondary sites (124). In addition, different forms of sialic acids, such as NeuGc, are detected on transformed cells but not on healthy cells (97).

Efforts have been made to find a specific correlation between increased sialylation and certain forms of cancer. If this were accomplished, it could be used as a quick diagnostic tool. There are problems with this approach, however, because an increase in sialic acid detection is not specific for cancer. Larger amounts of sialic acid are also detected in inflammatory disorders and diseases, cardiovascular disease, and diabetes. Another problem with using sialic acid as a diagnostic tool is that it is not specific for each different form of cancer. There is also a correlation between a decrease in sialic acid levels and remission and recovery of cancer patients. With this in mind, sialic acid levels can be used as a means of monitoring the progress of a patient recovering from cancer.

SIALIC ACID ACTIVATION IN BACTERIA

Prior to its transfer to glycolipids and glycoproteins, sialic acid has to be activated by the addition of a CMP group (Fig. 2). This activation of sialic acid is essential for sialylation to occur. A unique feature of this activation is the presence of only one phosphate in the nucleotide-sugar linkage (104) as opposed to the two phosphates present in most other sugar nucleotides (46). A gene responsible for this activity has been identified in a number of bacterial organisms, including *E. coli, Streptococcus agalactiae, N. meningitidis, Haemophilus ducreyi,* and *H. influenzae* (26, 32, 39, 104, 125). The gene, called *neuA* in *E. coli* and *N. meningitides,* is part of the capsular loci in these bacteria. The *neuA* gene in *H. ducreyi* has homology to other known CMP-NeuAc synthetases and is transcriptionally coupled to its sialyltransferase gene (8, 104). This gene in *H. influenzae* (designated *siaB*) is not in close proximity to any other known LOS, capsule, or sialylation genes (39). In all cases, a deletion in the CMP-NeuAc synthetase results in no detectable sialic acid in the LOS, LPS, or capsules of the organisms.

BACTERIAL SIALYLTRANSFERASES

The sialyltransferases from bacteria catalyze the same basic reaction as the mammalian sialyltransferases. There are key differences in how they are isolated, their structures, and the specificity for acceptor structures, however. Bacteria contain sialylated structures in their LOS or LPS or in their capsule. A great deal of work has been done on *Neisseria* sialyltransferases from both *N. gonorrhoeae* and *N. meningitidis.* To date, no sialyltransferases have been identified that have homology to mammalian sialyltransferases. Most notably, there is no homology to the "classic" sialylmotifs of these mammalian enzymes. For this reason, identification by PCR or Southern hybridization using mammalian sequences has not been possible. Sialyltransferases from both *N. meningitidis* and *N. gonorrhoeae* LOS were first cloned and characterized by Gilbert et al. (28). They were cloned by screening a DNA library made from a strain of *N. meningitidis.* Using a sensitive enzyme assay based on the transfer of sialic acid to synthetic fluorescent acceptors, this group was able to identify a sialyltransferase capable of transferring NeuAc with an α2,3-linkage to a *N*-acetyllactosamine acceptor. Strains of *N. meningitidis* capable of producing a sialic acid capsule contain a α2,8-polysialyltransferase in addition to the α2,3-sialyltransferase (19).

Once the gene from *N. meningitidis* was cloned, probes were made to amplify a similar enzyme from *N. gonorrhoeae.* There was a high degree of homology between the two sialyltransferases, with only 16 to 17-amino-acid differences sporadically located along the length of the 371-amino-acid proteins (28). In addition to not containing any homology to the sialylmotifs of the mammalian sialyltransferases, these bacterial enzymes had more ambiguity in the acceptor structures that they could recognize. Although there was a preference for the *N*-acetyllactosamine structure, sialic acid could also be transferred to lactose or galactose. In a more detailed study of the enzymatic properties of this enzyme, it was shown that NeuAc could be transferred to α-terminal or β-terminal galactose residues and β1,4-linked or β1,3-linked galactose (28). Ambiguity was also observed in the type of sialic acid able to be transferred, since *N*-propionylneuraminic acid and NeuGc could also be used with reduced efficiency.

Campylobacter jejuni is able to synthesize a number of sialylated LPS structures, some of which mimic human gangliosides (3, 27, 31, 68, 81). *C. jejuni* strain OH4384 produces an LPS similar to the ganglioside GD1a [NeuAcα2,8NeuAcα2,3Galβ1, 3GalNAcβ1,4Gal(NeuAcα2,3)β1,3Hep]. Using the same strategy used to clone the *Neisseria* sialyltransferases, two genes were identified that encoded sialyltransferase activity and were designated *cstI* and *cstII* (27). There was a great deal of homology between the two genes, and it was hypothesized that *cstI* occurred through a duplication event and subsequent genetic drift. CstI transfers NeuAc α2,3 linked to lactose, while CstII was shown to be bifunctional, capable of transferring sialic acid α2,3 linked to lactose or α2,8 linked to another NeuAc.

A sialyltransferase has also been identified in *H. ducreyi* by genetic methods. The LOS structure sialylated in *H. ducreyi* contains a terminal lactosamine structure (64, 65). A gene was identified in this organism, designated *lst,* which was located downstream of *neuA* (a CMP-NeuAc synthetase) (8). Mutational analysis of the two genes indicated that they are transcriptionally linked. The *lst* gene is thought to represent a novel class of bacterial sialyltransferases, because it does not contain strong homology to any known bacterial or mammalian sialyltransferases and contains only weak homology to the *E. coli* capsular polysialyltransferase (8, 113).

Another novel class of bacterial sialyltransferases was identified in the marine bacterium *Photobacterium damsela* (123). The protein was purified using a detergent extract, which was also required for activity, indicating that it was a membrane-associated protein. Further analysis of the cloned gene (*bst*) revealed a predicted α-helical structure in the C terminus that could function in membrane insertion (123). Deletion of this C-terminal sequence resulted in detection of the soluble enzyme in the supernatant. Enzyme assay analysis showed that both lactose and *N*-acetyllactosamine were the preferred acceptor structures; however, NeuAc could be transferred to either a galactose or an *N*-acetylgalactosamine (122). The most surprising function of this enzyme was its linkage specificity, transferring NeuAc α2,6 linked to the terminal sugar. This was the first identified bacterial sialyltransferase to preferentially add sialic acid with this linkage. Wakarchuk et al. identified an α2,6-linked sialic acid attached to a P^k antigen structure (Galα1,4Galβ1, 4Glc) on the LOS of *N. meningitidis* (116). It is not known if there is a second sialyltransferase involved in this linkage or if the α2,3-sialyltransferase already identified in this organism can account for this activity.

Studies of *H. influenzae* LOS have indicated the presence of sialic acid (39, 59). The sequencing of the genome of *H. influenzae* strain Rd has greatly increased the possibility of identifying genes involved in LOS biosynthesis, including those encoding sialyltransferases (25). There are three genes in the Rd database with homology to those encoding known sialyltransferases. The first is *lsgB,* which has 27% identity and 46% similarity to the *Neisseria* sialyltransferase genes (80). This gene is the second sialyltransferase gene in *H. influenzae* (43). A third sialyltransferase gene in *H. influenzae* has 48% identity and 59% similarity (at the protein level) to the *lst* gene from *H. ducreyi* over the entire length of the proteins (8, 43). Findings indicate that this protein adds sialic acid to a terminal *N*-acetyllactosamine structure on the LOS. Finally, *lic3A* has around 40% identity (at the protein level) to *cstII* from *C. jejuni* (38). Results from our laboratory and that of Hood et al. (38) indicate that Lic3A is capable of transferring sialic acid to a terminal lactose. It is not clear whether Lic3A contains bifunctional activity similar to that of CstII.

trans-SIALIDASES

trans-Sialidase (TS) is a bifunctional enzyme found only in the genus *Trypanosoma*. The enzyme catalyzes two reactions. The first is a sialidase reaction, in which sialic acid is cleaved from glycoconjugates free in solution or attached to a host membrane. The sialic acid can then either be hydrolyzed to form free sialic acid or be transferred to a mucin-like glycoprotein found on the *Trypanosoma* cell surface (91). The enzyme has been identified in two pathogenic forms of trypanosomes: *T. brucei,* the causative agent of African sleeping sickness, and *T. cruzi,* the causative agent of Chagas' disease.

The biological properties of TS have been studied by a number of investigators (13, 15, 91). TS has a very high specificity for α2,3-linked sialic acid attached to a penultimate galactose residue on host glycoconjugates. The enzyme is not able to transfer either free sialic acid or CMP-NeuAc. The best acceptor structure is lactose, but maltose (Glcα1,4Glc) or cellobiose (Glcβ1,4Glc) can be used with lower efficiency. The pH optimum for TS is around 7.0, which is different from that for the other sialidase enzymes, which have pH optimua between 5.0 and 5.5.

There are differences in the expression patterns of TS in *T. brucei* and *T. cruzi* (91). *T. brucei* expresses TS only in the insect vector, while *T. cruzi* expresses TS in both the insect and human hosts. The expression of TS is developmentally regulated during the life stages of *T. cruzi* as well. The first stage of development is the flagellate epimastigote, when the organism resides in the insect gut and TS expression is low. TS activity increases as the organism enters the infective metacyclic trypomastigote stage. At this point, the organism enters human cells and forms a phagolysosome. After the formation of the phagolysosome, the amastigote stage begins and TS activity is absent. The tropomastigote stage is marked by release of the organism from infected cells, and there is a return of TS activity.

The role of sialic acid on the surface of the organism and the TS in the pathogenesis of Chagas' disease is not completely clear and may be multifactorial. The presence of sialic acid on their surface protects trypomastigotes from complement-mediated lysis, but this protection is only about 24% effective, indicating that other factors add to the serum resistance (101). Desialylation of the organism increases its uptake into phagocytes, but sialic acid is required for uptake in nonphagocytic cells. It is possible that the TS of the organism is binding sialic acid on these cells, so that cleavage and transfer could expose other ligands needed for uptake of the organism, or the cleavage event itself could signal for internalization (13, 91). A study by Pereira-Chioccola and Schenkman looked at the possible effects of sialic acid on survival inside the harsh environment of the phagolysosome (79, 91). Their results showed that sialylation did not have an effect on low pH or H_2O_2 production but did protect against nitric oxide killing, although the mechanism of how this occurs is unclear. These results demonstrate that sialic acid can aid the intracellular survival of the organism.

SIALIC ACID IN LIPOOLIGOSACCHARIDE, CAPSULES, AND BIOFILMS

Sialic acid can be found in the LOS or LPS and the capsules of a number of different bacterial species. The capsules of most pathogenic *N. meningitidis* strains as well as *E. coli* K-12 strains contain polysialic acid (45). The similarity of the polysialic acid capsule in *N. meningitidis* serogroup B to human structures is thought to be the reason why a vaccine is not protective for this serogroup (7, 87). It is interesting that almost all of the microorganisms containing sialic acid are pathogenic. An example of this can been seen in the examination of pathogenic and nonpathogenic strains of *Neisseria*. Mandrell et al. tested a number of pathogenic and nonpathogenic strains of *Neisseria* by using an assay to measure sialyltransferase activity (57). They found that all pathogenic strains of *Neisseria* and some strains of the nonpathogenic *N. lactamica* tested had sialyltransferase activity while the majority of the nonpathogenic strains lacked sialyltransferase activity (56). When sialic acid is present on the LOS, it can endow the pathogenic bacteria with certain survival advantages. The best-studied example of this comes from work done with *N. gonorrhoeae*. Early studies found that *N. gonorrhoeae* organisms taken directly from urethral or cervical exudates were resistant to the killing action of normal human serum but that this capacity was lost on subculture of the bacterium (98, 117). The factor responsible for this sialylation was found to be CMP-NeuAc, acquired exogenously by the organism from its host (71, 76, 77). *N. gonorrhoeae* strains grown with sialic acid become resistant to the effects of normal human serum when grown in the presence of CMP-NeuAc (75).

One of the reasons for this decreased serum sensitivity could be the blocking of host antibody molecules. Parsons et al. demonstrated that sialylation of gonococci prevented the organisms from absorbing bactericidal antibody from normal human serum (75). The ability of sialic acid present on LOS to block antibody binding to epitopes other than LOS has also been studied. Wetzler et al. observed that antibodies to porin protein I bound the bacteria but were not bactericidal when sialic was present on the LOS (120). Other investigators have observed variable results with binding of PI antibodies and protection from bactericidal effects when the LOS is sialylated, indicating that the epitope presentation on the surface may affect the protective properties of sialic acid (18, 20).

Sialic acid also has an effect on opsonized phagocytosis by human neutrophils (49, 120). *N. gonorrhoeae* can also be phagocytosed by human neutrophils via a nonopsonic event mediated by opacity-associated outer membrane proteins. This event is also decreased in frequency when the LOS of the organism is sialylated (85). Similar results are seen with *N. meningitidis* serogroup C (22). The function of sialic acid in this inhibition is not completely clear but may involve blocking of antibody or receptor recognition; alternatively, its presence could be altering the membrane conformation.

The presence of sialic acid is also able to mitigate the effects of the complement cascade. Reactions of both the classical and alternative complement cascades converge at the interaction of complement component C3 with the bacterial membrane (69). Fragmentation of C3 to C3a and C3b and subsequent binding of Bb, along with an additional C3b, forms a C3 convertase complex that is able to amplify C3b and Bb deposition on the membrane. This eventually leads to the formation of a membrane attack complex and lysis of the cell (69). This process is negatively regulated by factors I (fI) and H (fH) (111). C3b can be inactivated by digestion by fI, forming a truncated iC3b. fH both aids in the prevention of the C3b-Bb interaction and is able to recruit fI. Studies with *N. gonorrhoeae* have demonstrated that sialylated bacteria convert C3b bound to the surface to iC3b at a greater rate than do bacteria lacking sialylated LOS (63).

The reason for this conversion seems to be the ability of sialic acid to bind fH. Studies using sialylated and nonsialylated erythrocytes have demonstrated that sialic acid on the cell surface was responsible for binding fH (24, 73). Ram et al., using flow cytometry, showed that sialylated *N. gonorrhoeae* organisms bound fH to their surface and that this binding was specific for sialic acid (83, 84). Characterization of this binding indicated that the C terminus of the elongated and flexible fH molecule was the region that bound sialic acid.

Sialylation also has an effect on the interaction of *N. gonorrhoeae* with host cells. The presence of sialic acid on the LOS makes the organisms much less invasive than nonsialylated organisms (106, 107). In human volunteer studies, fully sialylated *N. gonorrhoeae* organisms were less infective than desialylated bacteria. It appears that the ability to phase vary sialylation is important in allowing the organism to be internalized by cells while still allowing it to evade the immune system. Sialylation of the LOS is also an in vivo event. Immunoelectron microscopy studies indicate that the organisms can be sialylated when incubated with human neutrophils (58) or isolated from human volunteers (1).

Studies of sialylation with *N. meningitidis* are complicated by the fact that sialic acid can be found in both the capsule and the LOS of the organism. Furthermore, *N. meningitis* differs from *N. gonorrhoeae* in that it is able to endogenously sialylate its LOS (57). Investigators using capsule and LOS sialylation mutants have shown that both LOS and capsule can mediate serum resistance, but the LOS sialylation plays a much less pivotal role than in *N. gonorrhoeae* (21, 41, 44, 115)

Studies have also focused on the role of LOS sialylation in *C. jejuni*. This organism produces a number of sialylated LPS glycoforms, some of which mimic human gangliosides (31, 67, 81). Since previous *C. jejuni* infections have been implicated in Guillain-Barré syndrome, it is thought that molecular mimicry of these gangliosides may play a role in this disease. The most recent evidence shows that lack of LOS sialylation decreases serum resistance (31) but that the presence of NeuAc does not guarantee association with the disease (81).

The role played by sialylated LOS in *H. influenzae* biology and pathogenicity is starting to be addressed. *H. influenzae* is capable of activating both the classical and alternative complement cascades (121). Since *H. influenzae* type b is associated with systemic infections, studies of complement-mediated killing of these strains in the infant-rat model have been undertaken. When complement-depleted rats were used, encapsulated *H. influenzae* caused a greater level of bacteremia than when rats with a normal complement cascade were used (14, 16, 126). The role of sialic acid in complement-mediated killing has not been studied using this model.

In a study by Hood et al., sialylated nontypeable *H. influenzae* (NTHi) organisms differed slightly from nonsialylated organisms in their adherence to and invasion of cultured epithelial cell lines, but this difference was not statistically significant (39). This group did note a difference in the ability of the sialylated bacteria to survive serum-mediated killing. Since NTHi is not normally associated with invasive disease, the significance of these findings is unclear.

The biological effects of sialylation, which mediate antiphagocytosis, anticomplement activity, and protection against bactericidal killing, have the potential to act with sialic acid binding immunoglobulin-like lectins (siglecs) on the surface of hematopoietic and immune system cells. Jones et al. have shown that sialylated but not nonsialylated meningococcal strains were specifically recognized by two siglecs (siglec-1 and siglec-5). Macrophages expressing siglec-sn phagocytosed sialylated bacteria in a siglec- and sialic acid-dependent manner (42).

Many bacteria produce neuraminidases, which can modify the sialylation of microbial and human tissues. Peltola and McCullers discuss the modification of surface sialylated airway epithelial cell structures by the neuraminidase of *Streptococcus pneumoniae* enhancing pneumococcal invasion (78). Conversely, bacterial neuraminidases produced by one species in a population can effect sialylation of other members of that population. *S. pneumoniae, N. meningitidis,* and *H. influenzae* can be concomitant members of the resident flora of the upper respiratory tract. Shakhnovich et al. have shown that a neuraminidase expressed by *S. pneumoniae* (NanA) is able to desialylate the cell surface LOS of both *N. meningitidis* and *H. influenzae* (94).

Bacterial biofilms and their role in pathogenicity have generated considerable interest because of their role in antimicrobial resistance and pathogenesis. Bacterial biofilms are complex communities of microorganisms that develop on surfaces in diverse environments (33). Biofilms are dynamic structures, whose formation starts when bacteria attach to a surface; this is followed by the formation of microcolonies and, finally, the development of the mature, structurally complex biofilm. Bacteria eventually detach from the mature biofilm to enter the surrounding fluid phase, becoming planktonic organisms that can initiate biofilm development on other parts of the surface (33). Some NTH: strains are capable of forming biofilms (70). Swords et al. have shown that the formation of the biofilm of NTH: strain 2019 is inhibited when *siaB* (the gene for CMP-Neu5Ac synthetase) is mutated, indicating that the sialic acid is a component of this biofilm (100). Other studies have shown that the biofilm of nontypeable *H. influenzae* strain 2019 may mimic the sialylated O-linked dextrans of mucins (30).

MICROBIAL MOLECULAR MIMICRY

The LOS structures of *H. influenzae* and other pathogenic bacteria are similar to those of known host structures. This has been defined as molecular mimicry, and it is thought that similarity to host structures provides a selective advantage for the bacteria. This selective advantage could be a camouflage mechanism to avoid the immune system, or it could play a more active role by taking advantage of preexisting host receptor-ligand interactions. This molecular mimicry has been studied most extensively in *N. gonorrhoeae, N. meningitidis, C. jejuni, Helicobacter pylori,* and *H. influenzae* (2, 3, 55, 66, 81, 82, 112). Both *Neisseria* and *Haemophilus* LOSs have similarity to the P^k glycosphingolipid oligosaccharide moiety (Galα1,4Galβ1,4Glc) (114). In a study by Mandrell, 51% of *N. gonorrhoeae* and 9% of *N. meningitidis* stains tested positive with an anti-P^k monoclonal antibody specific for this epitope (54). *H. influenzae* strains are also recognized by anti-P^k antibodies (114), and in a similar study, 18 of 19 *H. influenzae* type b and 7 of 17 NTHi strains contained this structure (59). The P^k epitope is a phase-variable epitope in *H. influenzae* (114). The first structural evidence that this epitope is present on *H. influenzae* LOS came from studies by Masoud et al. (62).

Structures in *H. influenzae* LOS have been found that are also similar to the lacto-*N*-neotetranose structure (Galβ1,4GlcNAcβ1,3Galβ1,4Glc) present on paragloboside, which is a precursor of the ABH blood group glycolipid antigens present on human erythrocytes (67). The terminal lactose (Galβ1,4GlcNAc) was shown to be present on 13 of 19 Hib and 10 of 16 NTHi strains (59). In addition, neuraminidase treatment of these strains increased the binding of anti-P^k more than twofold in 47% of the strains tested. This indicates that this epitope is a site for the addition of sialic acid. Phillips et al. observed this structure when *H. influenzae* genes were expressed in *E. coli,* producing a chimeric LPS, indicating that *H. influenzae* has the ability to assemble this epitope (80).

SIALIC ACID-DEPENDENT RECEPTORS AND MICROORGANISMS

Sialic acid is also used as a receptor ligand by microorganisms. One of the best-studied examples of this is the binding of influenza virus to sialylated glycans (99). There are three main types of influenza viruses (A, B, and C), and they differ in their major internal protein antigens (99). Influenza viruses, being myxoviruses, must perform three basic functions in order to invade a cell (47). First, they must possess a hemagglutinin that allows the virus to bind to a receptor on the cell. Second, they must have a receptor-destroying enzyme that allows the virion to be released from the cell. Finally, they must have a factor involved in fusion with the host cell membrane. For influenza A virus, the hemagglutination and fusion properties are on the same protein, designated HA. This protein is the major glycoprotein present on the virion. The receptor for HA is sialic acid present on the terminal sugars of various gangliosides. Enzyme crystallography using HA has demonstrated a binding site on the globular head (119). The binding of influenza A virus to sialic acid has historical significance as well, since it was the first functional role reported for this acidic sugar (37). The binding to sialic acid is very specific and is also dependent on the presentation of the sialic acid and subterminal sugars (99). Of the three main HA types (HA1, HA2, and HA3) known to infect humans, α2,6-linked NeuAc and, in some instances, α2,3-linked NeuAc are the most common forms recognized (35, 99). The specificity for α2,6-linked NeuAc correlates very well with its presence on the ciliated human bronchial epithelium, the primary site of influenza virus colonization and infection (5). In contrast to influenza A and B viruses, the sialic acid derivative recognized by influenza C virus is $Neu5,9Ac_2$ (86). This is an example of how a slight modification of the sialic acid molecule can have a dramatic effect on receptor recognition.

A number of bacterial species also contain adherence factors that bind to sialic acid-containing ligands on the cell surface. One example of this is provided by the S-fimbriae of *E. coli.* Strains expressing S-fimbriae cause severe meningitis in newborn infants. The fimbriae preferentially bind to NeuAc α2,3-linked to Gal or NeuAc α2,8-linked to another NeuAc molecule (34, 74, 93). Consistent with the situation for many other NeuAc-dependent binding events, such as sialoadhesin and influenza A and B viruses, modification of C-8 and C-9 groups has an effect on S-fimbria binding (34, 93). A minor fimbrial subunit, SfaS, mediates the binding of the sialic acid ligand (92). Although this is not universally accepted among scientists (51), one of the receptors for *H. influenzae,* probably mediated by pili, is a sialylated structure (23, 29, 105). *H. pylori* is a human pathogen associated with gastritis and peptic ulcers. Hirmo et al., using 14 strains of *H. pylori,* showed that they could be grouped into sialic acid binding and nonbinding groups (36). The sialic acid binding strains were specific for α2,3-linked NeuAc, in contrast to α2,6-linked NeuAc. There was much less specificity when modifications of the C-5 amino group were compared. For example, modifications of hydroxyl (forming NeuGc) groups and formyl groups did not inhibit binding (36). Bacterial toxins, such as pertussis toxin, also recognize sialylated ligands on the cell surface (9).

Bacteria pathogenic for humans have very successfully incorporated sialylation into their repertoire of virulence factors. Sialylation of surface structure provides antiphagocytic, anticomplement, and antibactericidal properties. These are all advantageous for survival on cell surfaces and protection in the systemic environment. Recent studies demonstrating the potential for cooperative behavior between bacteria suggest that in complex communities, the disadvantages of surface sialylation may be obviated by neuraminidase production by a neighboring microbial partner. As we learn more about human receptors and bacterium-cell and bacterium-bacterium interactions, this will become an increasingly important area for research in bacterial pathogenesis.

REFERENCES

1. **Apicella, M. A., R. E. Mandrell, M. Shero, M. Wilson, J. M. Griffiss, G. F. Brooks, C. Fenner, J. F. Breen, and P. A. Rice.** 1990. Modification of sialic acid of *Neisseria gonorrhoeae* lipooligosaccharide epitope expression in human urethral exudates: an immunoelectron microscopic analysis. *J. Infect. Dis.* **162:**506–512.
2. **Appelmelk, B., R. Negrini, A. Moran, and E. Kuipers.** 1997. Molecular mimicry between *Helicobacter pylori* and the host. *Trends Microbiol.* **5:**70–73.
3. **Aspinall, G. O., A. G. McDonald, T. S. Raju, H. Pang, L. A. Kurjanczyk, J. L. Penner, and A. P. Moran.** 1993. Chemical structure of the core region of *Campylobacter jejuni* serotype O:2 lipopolysaccharide. *Eur. J. Biochem.* **213:**1029–1037.
4. **Basu, S. S., M. Basu, Z. Li, and S. Basu.** 1996. Characterization of two glycolipid: alpha 2-3sialyltransferases, SAT-3 (CMP-NeuAc:nLcOse4Cer alpha 2-3sialyltransferase) and SAT-4 (CMP-NeuAc:GgOse4Cer alpha 2-3sialyltransferase), from human colon carcinoma (Colo 205) cell line. *Biochemistry* **35:**5166–5174.
5. **Baum, L. G., and J. C. Paulson.** 1990. Sialyloligosaccharides of the respiratory epithelium in the selection of human influenza virus receptor specificity. *Acta Histochem. Suppl.* **40:** 35–38.
6. **Bevilacqua, M., E. Butcher, B. Furie, M. Gallatin, M. Gimbrone, J. Harlan, K. Kishimoto, L. Lasky, R. McEver, J. Paulson, S. Rosen, B. Seed, M. Siegalman, T. Springer, L. Stoolman, T. Tedder, A. Varki, D. Wagner, I. Weissman, and G. Zimmerman.** 1991. Selectins: a family of adhesion receptors. *Cell* **67:**233.
7. **Bitter-Suermann, D., and J. Roth.** 1987. Monoclonal antibodies to polysialic acid reveal epitope sharing between invasive pathogenic bacteria, differentiating cells and tumor cells. *Immunol. Res.* **6:**225–237.

8. **Bozue, J. A., M. V. Tullius, J. Wang, B. W. Gibson, and R. S. Munson, Jr.** 1999. *Haemophilus ducreyi* produces a novel sialyltransferase. Identification of the sialyltransferase gene and construction of mutants deficient in the production of the sialic acid-containing glycoform of the lipooligosaccharide. *J. Biol. Chem.* **274:**4106–4114.
9. **Brennan, M. J., J. L. David, J. G. Kenimer, and C. R. Manclark.** 1988. Lectin-like binding of pertussis toxin to a 165-kilodalton Chinese hamster ovary cell glycoprotein. *J. Biol. Chem.* **263:**4895–4899.
10. **Brinkman-Van der Linden, E. C., E. R. Sjoberg, L. R. Juneja, P. R. Crocker, N. Varki, and A. Varki.** 2000. Loss of *N*-glycolylneuraminic acid in human evolution. Implications for sialic acid recognition by siglecs. *J. Biol. Chem.* **275:**8633–8640.
11. **Charter, N. W., L. K. Mahal, D. E. Koshland, Jr., and C. R. Bertozzi.** 2000. Biosynthetic incorporation of unnatural sialic acids into polysialic acid on neural cells. *Glycobiology* **10:** 1049–1056.
12. **Charter, N. W., L. K. Mahal, D. E. Koshland, Jr., and C. R. Bertozzi.** 2002. Differential effects of unnatural sialic acids on the polysialylation of the neural cell adhesion molecule and neuronal behavior. *J. Biol. Chem.* 277:9255–9261.
13. **Colli, W.** 1993. *trans*-Sialidase: a unique enzyme activity discovered in the protozoan *Trypanosoma cruzi. FASEB J.* **7:** 1257–1264.
14. **Corrall, C. J., J. A. Winkelstein, and E. R. Moxon.** 1982. Participation of complement in host defense against encapsulated *Haemophilus influenzae* types a, c, and d. *Infect. Immun.* **35:**759–763.
15. **Cross, G. A., and G. B. Takle.** 1993. The surface *trans*-sialidase family of *Trypanosoma cruzi. Annu. Rev. Microbiol.* **47:**385–411.
16. **Crosson, F. J., Jr., J. A. Winkelstein, and E. R. Moxon.** 1976. Participation of complement in the nonimmune host defense against experimental *Haemophilus influenzae* type b septicemia and meningitis. *Infect. Immun.* **14:**882–887.
17. **Datta, A. K., A. Sinha, and J. C. Paulson.** 1998. Mutation of the sialyltransferase *S*-sialyl motif alters the kinetics of the donor and acceptor substrates. *J. Biol. Chem.* **273:**9608–9614.
18. **de la Paz, H., S. J. Cooke, and J. E. Heckels.** 1995. Effect of sialylation of lipopolysaccharide of *Neisseria gonorrhoeae* on recognition and complement-mediated killing by monoclonal antibodies directed against different outer-membrane antigens. *Microbiology* **141:**913–920.
19. **Edwards, U., A. Muller, S. Hammerschmidt, R. Gerardy-Schahn, and M. Frosch.** 1994. Molecular analysis of the biosynthesis pathway of the alpha-2,8-polysialic acid capsule by *Neisseria meningitidis* serogroup B. *Mol. Microbiol.* **14:** 141–149.
20. **Elkins, C., N. H. Carbonetti, V. A. Varela, D. Stirewalt, D. G. Klapper, and P. F. Sparling.** 1992. Antibodies to N-terminal peptides of gonococcal porin are bactericidal when gonococcal lipopolysaccharide is not sialylated. *Mol. Microbiol.* **6:** 2617–2628.
21. **Estabrook, M. M., J. M. Griffiss, and G. A. Jarvis.** 1997. Sialylation of *Neisseria meningitidis* lipooligosaccharide inhibits serum bactericidal activity by masking lacto-*N*-neotetraose. *Infect. Immun.* **65:**4436–4444.
22. **Estabrook, M. M., D. Zhou, and M. A. Apicella.** 1998. Nonopsonic phagocytosis of group C *Neisseria meningitidis* by human neutrophils. *Infect. Immun.* **66:**1028–1036.
23. **Fakih, M. G., T. F. Murphy, M. A. Pattoli, and C. S. Berenson.** 1997. Specific binding of *Haemophilus influenzae* to minor gangliosides of human respiratory epithelial cells. *Infect. Immun.* **65:**1695–1700.
24. **Fearon, D. T.** 1978. Regulation by membrane sialic acid of beta1H-dependent decay-dissociation of amplification C3 convertase of the alternative complement pathway. *Proc. Natl. Acad. Sci. USA* **75:**1971–1975.
25. **Fleischmann, R. D., M. D. Adams, O. White, R. A. Clayton, E. F. Kirkness, A. R. Kerlavage, C. J. Bult, J. F. Tomb, B. A. Dougherty, J. M. Merrick, K. McKenney, G. Sutton, W. Fitzhugh, C. Fields, J. D. Gocayne, J. Scott, R. Shirley L.-I. Liu, A. Glodek, J. M. Kelley, J. F. Weidman, C. A. Phillips, T. Spriggs, E. Hedblom, M. D. Cotton, T. R. Utterback, M. C. Hanna, D. T. Nguyen, D. M. Saudek, R. C. Brandon, L. D. Fine, J. L. Fritchman, J. L. Furhmann, N. S. M. Geoghagen, C. L. Gnehm, L. A. McDonald, K. V. Small, and C. M. Fraser.** 1995. Whole-genome random sequencing and assembly of *Haemophilus influenzae* Rd. *Science* **269:**496–512.
26. **Ganguli, S., G. Zapata, T. Wallis, C. Reid, G. Boulnois, W. F. Vann, and I. S. Roberts.** 1994. Molecular cloning and analysis of genes for sialic acid synthesis in *Neisseria meningitidis* group B and purification of the meningococcal CMP-NeuNAc synthetase enzyme. *J. Bacteriol.* **176:**4583–4589.
27. **Gilbert, M., J. R. Brisson, M. F. Karwaski, J. Michniewicz, A. M. Cunningham, Y. Wu, N. M. Young, and W. W. Wakarchuk.** 2000. Biosynthesis of ganglioside mimics in *Campylobacter jejuni* OH4384. Identification of the glycosyltransferase genes, enzymatic synthesis of model compounds, and characterization of nanomole amounts by 600-mhz (1)H and (13)C NMR analysis. *J. Biol. Chem.* **275:**3896–3906.
28. **Gilbert, M., D. Watson, A.-M. Cunningham, M. Jennings, N. Young, and W. Wakarchuk.** 1996. Cloning of the lipooligosaccharide alpha-2,3-sialyltransferase from the bacterial pathogens *Neisseria meningitidis* and *Neisseria gonorrhoeae. J. Biol. Chem.* **273:**28271–28276.
29. **Gilsdorf, J. R., M. Tucci, and C. F. Marrs.** 1996. Role of pili in *Haemophilus influenzae* adherence to, and internalization by, respiratory cells. *Pediat. Res.* **39:**343–348.
30. **Greiner, L. L., H. Watanabe, N. J. Phillips, J. Shao, A. Morgan, A. Zaleski, B. W. Gibson, and M. A. Apicella.** 2004. Nontypeable *Haemophilus influenzae* strain 2019 produces a biofilm containing *N*-acetylneuraminic acid that may mimic sialylated O-linked glycans. *Infect. Immun.* **72:**4249–4260.
31. **Guerry, P., C. P. Ewing, T. E. Hickey, M. M. Prendergast, and A. P. Moran.** 2000. Sialylation of lipooligosaccharide cores affects immunogenicity and serum resistance of *Campylobacter jejuni. Infect. Immun.* **68:**6656–6662.
32. **Haft, R. F., M. R. Wessels, M. F. Mebane, N. Conaty, and C. E. Rubens.** 1996. Characterization of cpsF and its product CMP-*N*-acetylneuraminic acid synthetase, a group B streptococcal enzyme that can function in K1 capsular polysaccharide biosynthesis in *Escherichia coli. Mol. Microbiol.* **19:**555–563.
33. **Hall-Stoodley, L., and P. Stoodley.** 2002. Developmental regulation of microbial biofilms. *Curr. Opin. Biotechnol.* **13:**228–233.
34. **Hanisch, F. G., J. Hacker, and H. Schroten.** 1993. Specificity of S fimbriae on recombinant *Escherichia coli:* preferential binding to gangliosides expressing NeuGc alpha (2-3)Gal and NeuAc alpha (2-8)NeuAc. *Infect. Immun.* **61:**2108–2115.
35. **Higa, H. H., G. N. Rogers, and J. C. Paulson.** 1985. Influenza virus hemagglutinins differentiate between receptor determinants bearing *N*-acetyl-, *N*-glycolyl-, and *N,O*-diacetylneuraminic acids. *Virology* **144:**279–282.
36. **Hirmo, S., S. Kelm, R. Schauer, B. Nilsson, and T. Wadstrom.** 1996. Adhesion of *Helicobacter pylori* strains to alpha-2,3-linked sialic acids. *Glycoconj. J.* **13:**1005–1011.
37. **Hirst, G. K.** 1941. Agglutinatin of red cells by allantonic fluid of chick embryos infected with influenzae virus. *Science* **94:** 22–23.

38. **Hood, D. W., A. D. Cox, M. Gilbert, K. Makepeace, S. Walsh, M. E. Deadman, A. Cody, A. Martin, M. Mansson, E. K. Schweda, J. R. Brisson, J. C. Richards, E. R. Moxon, and W. W. Wakarchuk.** 2001. Identification of a lipopolysaccharide alpha-2,3-sialyltransferase from *Haemophilus influenzae*. *Mol. Microbiol.* **39:**341–350.
39. **Hood, D. W., K. Makepeace, M. E. Deadman, R. F. Rest, P. Thibault, A. Martin, J. C. Richards, and E. R. Moxon.** 1999. Sialic acid in the lipopolysaccharide of *Haemophilus influenzae:* strain distribution, influence on serum resistance and structural characterization. *Mol. Microbiol.* **33:**679–692.
40. **Imai, Y., L. A. Lasky, and S. D. Rosen.** 1993. Sulphation requirement for GlyCAM-1, an endothelial ligand for L-selectin. *Nature* **361:**555–557.
41. **Jarvis, G. A., and N. A. Vedros.** 1987. Sialic acid of group B *Neisseria meningitidis* regulates alternative complement pathway activation. *Infect. Immun.* **55:**174–180.
42. **Jones, C., M. Virji, and P. R. Crocker.** 2003. Recognition of sialylated meningococcal lipopolysaccharide by siglecs expressed on myeloid cells leads to enhanced bacterial uptake. *Mol. Microbiol.* **49:**1213–1225.
43. **Jones, P. A., N. M. Samuels, N. J. Phillips, R. S. Munson, Jr., J. A. Bozue, J. A. Arseneau, W. A. Nichols, A. Zaleski, B. W. Gibson, and M. A. Apicella.** 2002. *Haemophilus influenzae* type b strain A2 has multiple sialyltransferases involved in lipooligosaccharide sialylation. *J. Biol. Chem.* **277:**14598–14611.
44. **Kahler, C. M., L. E. Martin, G. C. Shih, M. M. Rahman, R. W. Carlson, and D. S. Stephens.** 1998. The ($\alpha 2 \rightarrow 8$)-linked polysialic acid capsule and lipooligosaccharide structure both contribute to the ability of serogroup B *Neisseria meningitidis* to resist the bactericidal activity of normal human serum. *Infect. Immun.* **66:**5939–5947.
45. **Kasper, D. J., J. L. Winkelhake, B. L. Brandt, and M. S. Artenstein.** 1973. Antigenic specificity of bactericidal antibodies in antisera to *Neisseria meningitidis*. *J. Infect. Dis.* **127:**378–387.
46. **Kean, E. L.** 1991. Sialic acid activation. *Glycobiology* **1:**441–447.
47. **Kelm, S., and R. Schauer.** 1997. Sialic acids in molecular and cellular interactions. *Int. Rev. Cytol.* **175:**137–240.
48. **Kelm, S., R. Schauer, and P. R. Crocker.** 1996. The sialoadhesins—a family of sialic acid-dependent cellular recognition molecules within the immunoglobulin superfamily. *Glycoconj. J.* **13:**913–926.
49. **Kim, J. J., D. Zhou, R. E. Mandrell, and J. M. Griffiss.** 1992. Effect of exogenous sialylation of the lipooligosaccharide of *Neisseria gonorrhoeae* on opsonophagocytosis. *Infect. Immun.* **60:**4439–4442.
50. **Koenig, A., R. Jain, R. Vig, K. E. Norgard-Sumnicht, K. L. Matta, and A. Varki.** 1997. Selectin inhibition: synthesis and evaluation of novel sialylated, sulfated and fucosylated oligosaccharides, including the major capping group of GlyCAM-1. *Glycobiology* **7:**79–93.
51. **Krivan, H. C., D. D. Roberts, and V. Ginsburg.** 1988. Many pulmonary pathogenic bacteria bind specifically to the carbohydrate sequence GalNAc beta 1-4Gal found in some glycolipids. *Proc. Natl. Acad. Sci. USA* **85:**6157–6161.
52. **Livingston, B. D., and J. C. Paulson.** 1993. Polymerase chain reaction cloning of a developmentally regulated member of the sialyltransferase gene family. *J. Biol. Chem.* **268:**11504–11507.
53. **Luchansky, S. J., S. Goon, and C. R. Bertozzi.** 2004. Expanding the diversity of unnatural cell-surface sialic acids. *Chembiochem* **5:**371–374.
54. **Mandrell, R. E.** 1992. Further antigenic similarities of *Neisseria gonorrhoeae* lipooligosaccharides and human glycosphingolipids. *Infect. Immun.* **60:**3017–3020.
55. **Mandrell, R. E., and M. A. Apicella.** 1993. Lipo-oligosaccharides (LOS) of mucosal pathogens: molecular mimicry and host-modification of LOS. *Immunobiology* **187:**382–402.
56. **Mandrell, R. E., J. M. Griffiss, H. Smith, and J. A. Cole.** 1993. Distribution of a lipooligosaccharide-specific sialytransferase in pathogenic and non-pathogenic *Neisseria*. *Microb. Pathog.* **14:**315–327.
57. **Mandrell, R. E., J. J. Kim, C. M. John, B. W. Gibson, J. V. Sugal, M. A. Apicella, J. M. Griffiss, and R. Yamasaki.** 1991. endogenous sialylation of the lipooligosaccharides of *Neisseria meningitidis*. *J. Bacteriol.* **173:**2823–2832.
58. **Mandrell, R. E., A. J. Lesse, J. V. Sugai, M. Shero, J. M. Griffiss, J. A. Cole, N. J. Parsons, H. Smith, S. A. Morse, and M. A. Apicella.** 1990. In vitro and in vivo modification of *Neisseria gonorrhoeae* lipooligosaccharide epitope structure by sialylation. *J. Exp. Med.* **171:**1649–1664.
59. **Mandrell, R. E., R. McLaughlin, Y. Aba Kwaik, A. Lesse, R. Yamasaki, B. Gibson, S. M. Spinola, and M. A. Apicella.** 1992. Lipooligosaccharides (LOS) of some *Haemophilus* species mimic human glycosphingolipids, and some LOS are sialylated. *Infect. Immun.* **60:**1322–1328.
60. Reference deleted.
61. **Martinez, J., S. Steenbergen, and E. Vimr.** 1995. Derived structure of the putative sialic acid transporter from *Escherichia coli* predicts a novel sugar permease domain. *J. Bacteriol.* **177:**6005–6010.
62. **Masoud, H., E. Moxon, A. Martin, D. Krajcarski, and J. Richards.** 1997. Structure of the variable and conserved lipopolysaccharide oligosaccharide epitopes expressed by *Haemophilus influenzae* serotype b strain Eagan. *Biochemistry* **36:**2091–2103.
63. **McQuillen, D. P., S. Gulati, S. Ram, A. K. Turner, D. B. Jani, T. C. Heeren, and P. A. Rice.** 1999. Complement processing and immunoglobulin binding to *Neisseria gonorrhoeae* determined in vitro simulates in vivo effects. *J. Infect. Dis.* **179:** 124–135.
64. **Melaugh, W., N. J. Phillips, A. A. Campagnari, R. Karalus, and B. W. Gibson.** 1992. Partial characterization of the major lipooligosaccharide from a strain of *Haemophilus ducreyi*, the causative agent of chancroid, a genital ulcer disease. *J. Biol. Chem.* **267:**13434–13439.
65. **Melaugh, W., N. J. Phillips, A. A. Campagnari, M. V. Tullius, and B. W. Gibson.** 1994. Structure of the major oligosaccharide from the lipooligosaccharide of *Haemophilus ducreyi* strain 35000 and evidence for additional glycoforms. *Biochemistry* **33:**13070–13078.
66. **Moran, A. P.** 1999. Helicobacter pylori lipopolysaccharide-mediated gastric and extragastric pathology. *J. Physiol. Pharmacol.* **50:**787–805.
67. **Moran, A. P., M. M. Prendergast, and B. J. Appelmelk.** 1996. Molecular mimicry of host structures by bacterial lipopolysaccharides and its contribution to disease. *FEMS Immunol. Med. Microbiol.* **16:**105–115.
68. **Moran, A. P., U. Zahringer, U. Seydel, D. Scholz, P. Stutz, and E. T. Rietschel.** 1991. Structural analysis of the lipid A component of *Campylobacter jejuni* CCUG 10936 (serotype O:2) lipopolysaccharide. *Eur. J. Biochem.* **198:**459–469.
69. **Muller-Eberhard, H. J.** 1988. The molecular basis of target cell killing by human lymphocytes and of killer cell self-protection. *Immunol. Rev.* **103:**87–98.
70. **Murphy, T. F., and C. Kirkham.** 2002. Biofilm formation by nontypeable *Haemophilus influenzae:* strain variability, outer membrane antigen expression and role of pili. *BMC Microbiol.* **2:**7.
71. **Nairn, C. A., J. A. Cole, P. V. Patel, N. J. Parsons, J. E. Fox, and H. Smith.** 1988. Cytidine 5′-monophospho-*N*-

acetylneuraminic acid or a related compound is the low M_r factor from human red blood cells which induces gonococcal resistance to killing by human serum. *J. Gen. Microbiol.* **134:**3295–3306.

72. **Narayanan, S.** 1994. Sialic acid as a tumor marker. *Ann. Clin. Lab. Sci.* **24:**376–384.
73. **Pangburn, M. K., and H. J. Muller-Eberhard.** 1978. Complement C3 convertase: cell surface restriction of beta1H control and generation of restriction on neuraminidase-treated cells. *Proc. Natl. Acad. Sci. USA* **75:**2416–2420.
74. **Parkkinen, J., G. N. Rogers, T. Korhonen, W. Dahr, and J. Finne.** 1986. Identification of the O-linked sialyloligosaccharides of glycophorin A as the erythrocyte receptors for S-fimbriated *Escherichia coli. Infect. Immun.* **54:**37–42.
75. **Parsons, N. J., J. R. Andrade, P. V. Patel, J. A. Cole, and H. Smith.** 1989. Sialylation of lipopolysaccharide and loss of absorption of bactericidal antibody during conversion of gonococci to serum resistance by cytidine 5′-monophospho-*N*-acetyl neuraminic acid. *Microb. Pathog.* **7:**63–72.
76. **Parsons, N. J., A. A. Kwaasi, P. V. Patel, C. A. Nairn, and H. Smith.** 1986. A determinant of resistance of *Neisseria gonorrhoeae* to killing by human phagocytes: an outer membrane lipoprotein of about 20 kDa with a high content of glutamic acid. *J. Gen. Microbiol.* **132:**3277–3287.
77. **Patel, P. V., P. M. Martin, E. L. Tan, C. A. Nairn, N. J. Parsons, M. Goldner, and H. Smith.** 1988. Protein changes associated with induced resistance of *Neisseria gonorrhoeae* to killing by human serum are relatively minor. *J. Gen. Microbiol.* **134:**499–507.
78. **Peltola, V. T., and J. A. McCullers.** 2004. Respiratory viruses predisposing to bacterial infections: role of neuraminidase. *Pediatr. Infect. Dis. J.* **23:**S87–S97.
79. **Pereira-Chioccola, V. L., and S. Schenkman.** 1999. Biological role of *Trypanosoma cruzi trans*-sialidase. *Biochem. Soc. Trans.* **27:**516–518.
80. **Phillips, N. J., T. J. Miller, J. J. Engstrom, W. Melaugh, R. McLaughlin, M. A. Apicella, and B. W. Gibson.** 2000. Characterization of chimeric lipopolysaccharides from *Escherichia coli* strain JM109 transformed with lipooligosaccharide synthesis genes (*lsg*) from *Haemophilus influenzae. J. Biol. Chem.* **275:**4747–4758.
81. **Prendergast, M. M., A. J. Lastovica, and A. P. Moran.** 1998. Lipopolysaccharides from *Campylobacter jejuni* O:41 strains associated with Guillain-Barré syndrome exhibit mimicry of GM_1 ganglioside. *Infect. Immun.* **66:**3649–3655.
82. **Preston, A., R. E. Mandrell, B. W. Gibson, and M. A. Apicella.** 1996. The lipooligosaccharides of pathogenic gram-negative bacteria. *Crit. Rev. Microbiol.* **22:**139–180.
83. **Ram, S., D. P. McQuillen, S. Gulati, C. Elkins, M. K. Pangburn, and P. A. Rice.** 1998. Binding of complement factor H to loop 5 of porin protein 1A: a molecular mechanism of serum resistance of nonsialylated *Neisseria gonorrhoeae. J. Exp. Med.* **188:**671–680.
84. **Ram, S., A. K. Sharma, S. D. Simpson, S. Gulati, D. P. McQuillen, M. K. Pangburn, and P. A. Rice.** 1998. A novel sialic acid binding site on factor H mediates serum resistance of sialylated *Neisseria gonorrhoeae. J. Exp. Med.* **187:**743–752.
85. **Rest, R. F., and J. V. Frangipane.** 1992. Growth of *Neisseria gonorrhoeae* in CMP-*N*-acetylneuraminic acid inhibits nonopsonic (opacity-associated outer membrane protein-mediated) interactions with human neutrophils. *Infect. Immun.* **60:**989–997.
86. **Rogers, G. N., and B. L. D'Souza.** 1989. Receptor binding properties of human and animal H1 influenza virus isolates. *Virology* **173:**317–322.
87. **Roth, J., D. J. Taatjes, D. Bitter-Suermann, and J. Finne.** 1987. Polysialic acid units are spatially and temporally expressed in developing postnatal rat kidney. *Proc. Natl. Acad. Sci. USA* **84:**1969–1973.
88. **Saxon, E., S. J. Luchansky, H. C. Hang, C. Yu, S. C. Lee, and C. R. Bertozzi.** 2002. Investigating cellular metabolism of synthetic azidosugars with the Staudinger ligation. *J. Am. Chem. Soc.* **124:**14893–14902.
89. **Schauer, R.** 1985. Sialic acids and their role as biological masks. *Trends Biochem. Sci.* **10:**357–360.
90. **Schauer, R., S. Kelm, G. Reuter, P. Roggentin, and L. Shaw.** 1995. Biochemistry and role of sialic acids, p. 7–67. *In* A. Rosenberg (ed.), *Biology of the Sialic Acids.* Plenum Press, New York, N.Y.
91. **Schenkman, S., D. Eichinger, M. E. Pereira, and V. Nussenzweig.** 1994. Structural and functional properties of *Trypanosoma trans*-sialidase. *Annu. Rev. Microbiol.* **48:**499–523.
92. **Schmoll, T., H. Hoschutzky, J. Morschhauser, F. Lottspeich, K. Jann, and J. Hacker.** 1989. Analysis of genes coding for the sialic acid-binding adhesin and two other minor fimbrial subunits of the S-fimbrial adhesin determinant of *Escherichia coli. Mol. Microbiol.* **3:**1735–1744.
93. **Schroten, H., R. Plogmann, F. G. Hanisch, J. Hacker, R. Nobis-Bosch, and V. Wahn.** 1993. Inhibition of adhesion of S-fimbriated *E. coli* to buccal epithelial cells by human skim milk is predominantly mediated by mucins and depends on the period of lactation. *Acta Paediatr.* **82:**6–11.
94. **Shakhnovich, E. A., S. J. King, and J. N. Weiser.** 2002. Neuraminidase expressed by *Streptococcus pneumoniae* desialylates the lipopolysaccharide of *Neisseria meningitidis* and *Haemophilus influenzae:* a paradigm for interbacterial competition among pathogens of the human respiratory tract. *Infect. Immun.* **70:**7161–7164.
95. **Shi, W. X., R. Chammas, N. M. Varki, L. Powell, and A. Varki.** 1996. Sialic acid 9-O-acetylation on murine erythroleukemia cells affects complement activation, binding to I-type lectins, and tissue homing. *J. Biol. Chem.* **271:**31526–31532.
96. Reference deleted.
97. **Sillanaukee, P., M. Ponnio, and I. P. Jaaskelainen.** 1999. Occurrence of sialic acids in healthy humans and different disorders. *Eur. J. Clin. Investig.* **29:**413–425.
98. **Smith, H., J. A. Cole, and N. J. Parsons.** 1992. The sialylation of gonococcal lipopolysaccharide by host factors: a major impact on pathogenicity. *FEMS Microbiol. Lett.* **100:** 287–292.
99. **Suzuki, H., S. Eiumtrakul, T. Ariya, J. Supawadee, N. Maneekarn, M. Tanaka, M. Ueda, K. Kadoi, and S. Takahashi.** 1997. Antigenic analysis of influenza viruses isolated in Thailand between 1991 and 1994. *New Microbiol.* **20:**207–214.
100. **Swords, W. E., M. L. Moore, L. Godzicki, G. Bukofzer, M. J. Mitten, and J. VonCannon.** 2004. Sialylation of lipooligosaccharides promotes biofilm formation by nontypeable *Haemophilus influenzae. Infect. Immun.* **72:**106–113.
101. **Tomlinson, S., L. C. Pontes de Carvalho, F. Vandekerckhove, and V. Nussenzweig.** 1994. Role of sialic acid in the resistance of *Trypanosoma cruzi* trypomastigotes to complement. *J. Immunol.* **153:**3141–3147.
102. **Traving, C., and R. Schauer.** 1998. Structure, function and metabolism of sialic acids. *Cell. Mol. Life Sci.* **54:**1330–1349.
103. **Tsuji, S.** 1996. Molecular cloning and functional analysis of sialyltransferases. *J. Biochem.* (Tokyo) **120:**1–13.
104. **Tullius, M. V., R. S. Munson, Jr., J. Wang, and B. W. Gibson.** 1996. Purification, cloning, and expression of a cytidine 5′-

monophosphate *N*-acetylneuraminic acid synthetase from *Haemophilus ducreyi*. *J. Biol. Chem.* **271:**15373–15380.

105. **van Alphen, L., L. Geelen-van den Broek, L. Blaas, M. van Ham, and J. Dankert.** 1991. Blocking of fimbria-mediated adherence of *Haemophilus influenzae* by sialyl gangliosides. *Infect. Immun.* **59:**4473–4477.

106. **van Putten, J. P., H. U. Grassme, B. D. Robertson, and E. T. Schwan.** 1995. Function of lipopolysaccharide in the invasion of *Neisseria gonorrhoeae* into human mucosal cells. *Prog. Clin. Biol. Res.* **392:**49–58.

107. **van Putten, J. P. M., T. D. Duensing, and J. Carlson.** 1998. Gonococcal invasion of epithelial cells driven by P.IA, a bacterial ion channel with GTP binding properties. *J. Exp. Med.* **188:**941–952.

108. **Varki, A.** 2001. *N*-Glycolylneuraminic acid deficiency in humans. *Biochimie* **83:**615–622.

109. **Varki, A.** 1997. Selectin ligands: will the real ones please stand up? *J. Clin. Investig.* **100:**S31–S35.

110. **Varki, A.** 1997. Sialic acids as ligands in recognition phenomena. *FASEB J.* **11:**248–255.

111. **Vik, D. P., P. Munoz-Canoves, D. D. Chaplin, and B. F. Tack.** 1990. Factor H. *Curr. Top. Microbiol. Immunol.* **153:**147–162.

112. **Vimr, E., and C. Lichtensteiger.** 2002. To sialylate, or not to sialylate: that is the question. *Trends Microbiol.* **10:**254–257.

113. **Vimr, E. R., R. Bergstrom, S. M. Steenbergen, G. Boulnois, and I. Roberts.** 1992. Homology among *Escherichia coli* K1 and K92 polysialytransferases. *J. Bacteriol.* **174:**5127–5131.

114. **Virji, M., J. N. Weiser, A. A. Lindberg, and E. R. Moxon.** 1990. Antigenic similarities in lipopolysaccharides of *Haemophilus* and *Neisseria* and expression of a diagalacatoside structure also present on human cells. *Microb. Pathog.* **9:**441–450.

115. **Vogel, U., A. Weinberger, R. Frank, A. Muller, J. Kohl, J. P. Atkinson, and M. Frosch.** 1997. Complement factor C3 deposition and serum resistance in isogenic capsule and lipooligosaccharide sialic acid mutants of serogroup B *Neisseria meningitidis*. *Infect. Immun.* **65:**4022–4029.

116. **Wakarchuk, W. W., M. Gilbert, A. Martin, Y. Wu, J. R. Brisson, P. Thibault, and J. C. Richards.** 1998. Structure of an alpha-2,6-sialylated lipooligosaccharide from *Neisseria meningitidis* immunotype L1. *Eur. J. Biochem.* **254:**626–633.

117. **Ward, M. E., P. J. Watt, and A. A. Glynn.** 1970. Gonococci in urethral exudates possess a virulence factor lost on subculture. *Nature* **227:**382–384.

118. **Warren, L., J. P. Fuhrer, and C. A. Buck.** 1972. Surface glycoproteins of normal and transformed cells: a difference determined by sialic acid and a growth-dependent sialyl transferase. *Proc. Natl. Acad. Sci. USA* **69:**1838–1842.

119. **Weis, W., J. H. Brown, S. Cusack, J. C. Paulson, J. J. Skehel, and D. C. Wiley.** 1988. Structure of the influenza virus haemagglutinin complexed with its receptor, sialic acid. *Nature* **333:**426–431.

120. **Wetzler, L. M., K. Barry, M. S. Blake, and E. C. Gotschlich.** 1992. Gonococcal lipooligosaccharide sialylation prevents complement-dependent killing by immune sera. *Infect. Immun.* **60:**39–43.

121. **Winkelstein, J. A., and E. R. Moxon.** 1992. The role of complement in the host's defense against *Haemophilus influenzae*. *J. Infect. Dis.* **165** (Suppl. 1):S62–S65.

122. **Yamamoto, T., M. Nakashizuka, H. Kodama, Y. Kajihara, and I. Terada.** 1996. Purification and characterization of a marine bacterial beta-galactoside alpha-2,6-sialyltransferase from *Photobacterium damsela* JT0160. *J. Biochem.* (Tokyo) **120:**104–110.

123. **Yamamoto, T., M. Nakashizuka, and I. Terada.** 1998. Cloning and expression of a marine bacterial beta-galactoside alpha-2,6-sialyltransferase gene from *Photobacterium damsela* JT0160. *J. Biochem.* (Tokyo) **123:**94–100.

124. **Yogeeswaran, G., and P. L. Salk.** 1981. Metastatic potential is positively correlated with cell surface sialylation of cultured murine tumor cell lines. *Science* **212:**1514–1516.

125. **Zapata, G., W. F. Vann, W. Aaronson, M. S. Lewis, and M. Moos.** 1989. Sequence of the cloned *Escherichia coli* K1 CMP-*N*-acetylneuraminic acid synthetase gene. *J. Biol. Chem.* **264:**14769–14774.

126. **Zwahlen, A., J. A. Winkelstein, and E. R. Moxon.** 1983. Participation of complement in host defense against capsule-deficient *Haemophilus influenzae*. *Infect. Immun.* **42:**708–715.

Colonization of Mucosal Surfaces
Edited by James P. Nataro et al.

Chapter 7

Competitive and Cooperative Interactions in the Respiratory Microflora

Adam J. Ratner

Rupture of the amniotic membranes marks the transition from a sterile prenatal environment to one of constant contact with microorganisms. The establishment of the complex ecosystem of the upper respiratory mucosa and its subsequent maintenance are rapid and dynamic processes (65). Over the course of a lifetime, people are exposed to a vast number of bacterial species that have the potential to colonize the upper respiratory tract, yet only a small number ever do so. Of these, many are rapidly cleared, but some may persist for months or years by either joining or displacing existing organisms at a mucosal surface. A myriad of factors may influence changes in the upper respiratory flora, but the overriding principle is that of continuous interaction among the species present and the extensive host defense mechanisms of the respiratory mucosa. Some members of the bacterial flora have the ability to prevent subsequent colonization by other species ("bacterial interference"), and there have been attempts to harness this for therapeutic or prophylactic means. Conversely, there may be cooperation among species to maintain colonization through effects on the host immune system, the epithelial surface, or the ability of bacteria to withstand environmental challenges (including antibiotic therapy).

There are examples of both cooperative and competitive interactions among respiratory organisms in in vitro and in vivo models. Observation of clinical phenomena such as bacterial superinfection in the setting of respiratory viral infection has suggested the potential importance of these processes. In addition, alteration of the colonizing flora of humans as the result of medical interventions such as antibiotic therapy or vaccination provides both a means and a compelling justification for the study of interactions among the members of the upper respiratory flora. Hospitalization (especially in intensive care units) and antibiotic use can predispose to colonization and overgrowth with gram-negative organisms and fungi that may go on to cause life-threatening disease (75). Artificial colonization with avirulent bacteria in the hope of excluding more harmful organisms has been tried in this setting, with some encouraging results (discussed below). In addition, the successes of vaccination against some upper respiratory organisms (including *Haemophilus influenzae* type b [Hib] and *Streptococcus pneumoniae*) has led to changes not only in the rates of invasive disease due to these organisms but also to a decline in their carriage (13, 16, 17, 38, 52, 54, 84). This has raised the question whether new niches are created by vaccination that might be filled by other potentially virulent organisms (41). One mechanism to explain this phenomenon is the removal of competitors via drug therapy or vaccination, which highlights the potential role of competition in determining the makeup of the indigenous flora of humans.

In this chapter, I review aspects of interaction among the members of the respiratory microflora by considering relevant laboratory models and examples of both cooperation and competition. In general, I focus on the anatomic area between the nasal mucosa and the epiglottis, where several bacterial species coexist and may cocolonize for extended periods. The lower airway, below the epiglottis, is a sterile environment under normal conditions. In contrast, another neighbor of the nasopharynx, the oral cavity, is home to literally hundreds of distinct species of bacteria (57). These exist in microenvironments such as the subgingival crevice, where cooperative interactions, including the formation of highly complex multispecies biofilms, are crucial elements of their pathogenesis (62). The species that colonize the dental surface are distinct from those that inhabit

Adam J. Ratner • Department of Microbiology, University of Pennsylvania, Philadelphia, PA 19104.

the respiratory mucosal surface and are not considered here. Specific data regarding the anatomic localization of colonizing bacteria in the nasopharynx and the relative burdens of each species are limited; however, evolving concepts of interaction among the members of the commensal flora are furthering our understanding of carriage.

EXPERIMENTAL MODELS

Questions regarding bacterial colonization of the respiratory system are best answered by in vivo studies of a natural host; much of what we know about carriage comes from studies of humans. As might be expected, these investigations have been largely observational, rather than experimental, in nature. Because respiratory pathogens often colonize prior to causing invasive disease, a good deal of attention has been paid to the natural history of carriage of these organisms (reviewed in reference 26). Children acquire potentially pathogenic respiratory organisms shortly after birth and may carry them for prolonged periods (3). They may harbor more than one of these species or more than one serotype of a particular species. Spread of colonizing bacteria from child to child takes place efficiently in the winter months and in crowded situations such as day care centers. However, our understanding of the nature of cooperative and/or competitive interactions within the microflora that influence carriage is limited. The few studies of human colonization with relevance to competitive or cooperative interactions have tended to involve attempts to use "bacterial interference"—instilling a nonpathogenic microbe in an attempt to prevent colonization or infection with more virulent organisms. In recent years, a model of experimental human carriage of *S. pneumoniae* has been established (49, 50), although experiments regarding cocolonization with distinct species have not been reported. In addition, some inferences regarding competition among nasopharyngeal colonizers may be drawn from vaccination studies. These sources of data are discussed in more detail below.

Mouse models of carriage of microorganisms that usually have tropism for human tissues are more convenient but may be problematic. Colonization of mice with members of the human flora tends to require high intranasal doses of bacteria for establishment, to be transient, and to necessarily involve alteration of some important interactions between the colonizing species and the host. For example, immunoglobulin A (IgA) protease expressed by both *S. pneumoniae* and *H. influenzae* is not active against murine IgA, making it difficult to assess the importance of that system in a mouse colonization model. However, short-term murine colonization has been used to assess some specific questions about the dynamics of pathogen interaction. One study assessed competition among serotypes of *S. pneumoniae* (42). Mice colonized with a streptomycin-resistant type 6B pneumococcus resisted challenge with either an optochin-resistant type 6B strain or an unrelated type 23F strain. However, animals that were first colonized with the optochin-resistant type 6B could still acquire the streptomycin-resistant variant, showing that inhibition was not always reciprocal. More subtle effects of competition or cooperation may require the use of genetically altered mice expressing human receptors or lacking immune system components responsible for clearance of organisms.

In vitro modeling of microbial interactions has been the most widely employed means of investigating competition and cooperation among species. Early studies of bacterial antagonism extended known microbiological techniques to the study of multiple organisms. Topley and Fielden (86) demonstrated an ordered progression of quantitatively dominant species in liquid culture of organisms from stool, and Colebrook (10) showed that when grown on solid media, a streak of pneumococcus could inhibit the growth of an intersecting streak of meningococcus. These techniques are still useful for the generation of novel data (58), but better-controlled systems such as continuous-culture chemostats (19) are becoming more widely used. The chemostat allows populations of bacteria to be kept in relatively static conditions for long periods. Nutrient levels, drug concentrations, and the presence of other species can each be altered, and the effects on the organisms and on the media can be measured. The concept of a chemostat has also been used as a starting point for mathematical modeling of competition and resistance to new colonizing strains among bacteria (77). The ability to incubate bacteria and/or viruses with cultured mammalian cells, epithelial and other types, has spawned a new discipline, cellular microbiology (11), and has facilitated investigations of the mechanisms of microbial interactions. For example, much of what is known about viral-bacterial synergy in the respiratory tract, discussed in the next section, has been defined in these in vitro systems.

COOPERATIVE INTERACTIONS AMONG RESPIRATORY MICROORGANISMS

Cultures of the upper respiratory tract rarely reveal a single organism, and studies of colonization by potential bacterial pathogens show that simultaneous carriage of more than one species is common (6). Colonization of the nasopharynx by one species may alter the dynamics of that niche, changing the probability that another species may succeed at entry. One possible result of this is bacterial interference, or exclusion of subsequent species from the niche. This is discussed in detail below. However, another possibility is that colonization by one species may change the microenvironment in a way that makes it more hospitable to another. I refer to this as bacterial cooperation.

The mere description of multiple species in a single anatomic site does not imply interaction. Smith (71) described cooperative interactions among bacterial and viral species in the pathogenesis of respiratory disease and laid out a model of synergy among several weakly pathogenic species in areas such as periodontal disease, but he did not address the question of the impact of these interactions on asymptomatic carriage. The first in vitro descriptions of cooperative interactions among bacterial species date to the late 19th century (cited in reference 64). More recently, enhancement of the growth of *Moraxella catarrhalis* by oral streptococci was noted by Rosebury et al. (64). In that study, various pairs of microorganisms were tested on solid media for either inhibitory or cooperative interactions. When changes in growth were noted, they were mostly inhibitory. The exception was that several species enhanced the growth of *M. catarrhalis* and one (*Candida albicans*) appeared to increase the growth of *Escherichia coli*.

Some clinical aspects of microbial synergism in both infection and, to a lesser extent, colonization have been reviewed (45, 46). Brogden (8) described a framework for consideration of potential mechanisms of polymicrobial infections, including those involving the respiratory tract. (Table 1). Despite the inherent differences between colonization and disease states, the concepts are useful in investigating the ecosystem of the upper respiratory flora. Alterations in the host immune system may be important in determining the composition of the respiratory flora. Even in the absence of "obliteration" of the immune system, subtle changes may be enough to affect colonization. For example, alteration of ciliary motility in the setting of viral infection can impair the clearance of bacterial pathogens, setting the stage for prolonged colonization or disease such as otitis media (4, 56). In addition, people with IgA deficiency are at increased risk of respiratory infections, possibly as a result of increased colonization.

Alteration in the mucosa by one organism may favor colonization by another. The best described examples of this involve the interplay among respiratory viruses and members of the bacterial flora. It has been known for decades that viral infections can increase both colonization (29) and invasive disease such as pneumonia due to bacterial pathogens (31). Correspondingly, in vitro models demonstrate that infection with several different types of viruses (influenza virus, respiratory syncytial virus [RSV], and rhinovirus) can enhance the attachment of bacteria to respiratory epithelial cells. This may occur through damage to host cells and exposure of receptors for bacteria that are normally sequestered. For example, infection of respiratory epithelial cells with influenza virus (51) or rhinovirus (35) can increase interactions between the phosphorylcholine moiety on the surface of *S. pneumoniae* and the receptor for platelet-activating factor on the epithelial surface. A similar mechanism appears to exist for *Neisseria meningitidis* binding during RSV infection. RSV infection of epithelial cells in vitro leads to upregulation of several host cell molecules, including CD14 and CD18, both of which contribute to meningococcal adherence (60). Similar upregulation of CD14 and CD18 has been reported after epithelial cell infection with influenza virus (22). Adenovirus infection may also increase the epithelial attachment of

Table 1. Mechanisms of pathogenesis in polymicrobial disease[a]

Stress, physiologic abnormalities, or metabolic disease favors the colonization of multiple organisms

- Stress and acute respiratory infection (28)

Alterations in the mucosa by one organism favor the colonization of another

- Increased exposure of bacterial receptors during viral infection
  - Influenza virus/platelet-activating factor receptor (51)
  - Influenza virus RSV/CD14, CD18 (22, 60)
- Bacterial binding to viral proteins in the host membrane
  - RSV glycoprotein G/*N. meningitidis* (59)
  - Influenza virus proteins/group B *Streptococcus* (67)

Synergistic triggering of proinflammatory cytokines increases severity of disease, reactivates latent infections, or favors the colonization of another organism

- Enhanced TNF-α response of influenza virus-infected macrophages to *H. influenzae* lipopolysaccharide (55)
- Synergistic epithelial cell response to muramyl peptide and lipopolysaccharide (24)

Sharing of determinants among organisms allows activities that neither organism possesses individually

- "Piracy of adhesins" *S. pneumoniae, S. aureus,* and *H. influenzae* bind to secreted proteins of *B. pertussis* (87)
- IgA protease-expressing *H. influenzae* enhances the adherence of IgA protease-insufficient *S. pneumoniae* by cleaving specific antibody (93)

Obliteration of the immune system by one organism allows colonization by others

- Overall immune suppression in the setting of viral illness
- Alteration of ciliary motility by viruses leading to increased bacterial colonization (4, 56)

[a]Modified from reference 8 with permission.

S. pneumoniae, although the mechanism is unclear (30). In addition, viral proteins that are inserted into host cell membranes during viral replication can act as direct receptors for bacterial binding. This has been described for glycoprotein G of RSV, which, when present on host epithelial cells, enhances the adherence of meningococci (59). A similar mechanism may exist for influenza virus, since an infection of epithelial cells increases the adherence of group B streptococcus, an effect that can be blocked by antibodies to influenza A virus proteins (67).

Another potential mechanism for bacterial competition, the synergistic triggering of a cytokine response by the host, is of unclear importance in human upper respiratory colonization. However, expression of some bacterial receptors, such as the receptor for platelet-activating factor, is upregulated in the presence of proinflammatory cytokines including tumor necrosis factor alpha (TNF-α) (12). Synergistic production of TNF-α from influenza virus-infected macrophages has been described following stimulation with *H. influenzae* lipopolysaccharide (55). In addition, exposure of epithelial cells to muramyl peptide (a component of peptidoglycan) and lipopolysaccharide together induces a synergistic cytokine response (24), raising the possibility that different species of bacteria acting in concert might alter the immune response of the respiratory epithelium in vivo.

Sharing of determinants among diverse species in the ecosystem of the upper respiratory tract is another means for the promotion of colonization. Cohabitation of an environmental niche suggests that the activities of one group of organisms, especially those that operate through secreted factors, might affect other residents of the niche. For example, one mechanism postulated to explain episodes of failure of penicillin therapy in the treatment of group A beta-hemolytic streptococcal pharyngitis is the presence of colonizing bacteria that elaborate β-lactamases (20); however, the clinical significance of that interaction has been questioned (27). The ability of many diverse species of bacteria to cleave human IgA in the hinge region is thought to facilitate colonization of the nasopharynx. Cleavage of type-specific IgA by *S. pneumoniae* facilitates the adherence to respiratory epithelial cells, and while an IgA protease-deficient pneumococcus does not manifest this increase in adherence, it can be complemented by coincubation with an IgA protease-producing strain of *H. influenzae* (93). Thus, an enzyme produced by a distinct species of bacteria is able to cleave IgA specific for the pneumococcus and enhance its adherence, demonstrating a potential means for cooperation among species in colonization of the nasopharynx. A cooperative model involving shared determinants has also been proposed to explain the increased rate of colonization and infection with other respiratory bacteria during infection with *Bordetella pertussis. B. pertussis* secretes both filamentous hemagglutinin and pertussis toxin into the environment. These then mediate bacterial binding to the cilia of respiratory epithelial cells. However, in an example of molecular cooperation (or "piracy"), other respiratory pathogens including *S. pneumoniae, Staphylococcus aureus,* and *H. influenzae* are also able to bind these proteins and thus enhance their own adherence manyfold during coinfection with *B. pertussis* (87).

It is likely that cooperative interactions that influence the initiation and maintenance of upper respiratory colonization are both more common and more diverse than we currently appreciate. Accumulating evidence suggests that cross-species quorum sensing and multispecies coaggregation and biofilm formation (62) may be common in other niches, and it is possible that these play a role in the respiratory tree as well. In addition, the role of anaerobic and difficult-to-culture pathogens in respiratory colonization is an area awaiting further study.

COMPETITION AND "BACTERIAL INTERFERENCE" IN THE RESPIRATORY SYSTEM

The study of bacterial competition or, as it is more commonly called, bacterial interference is nearly as old as bacteriology itself. It has been the subject of numerous review articles spanning decades and of at least one book (2, 9, 25, 61, 64, 71, 73). The first report of therapy using a microorganism involved an attempt to utilize yeast as a treatment for furunculosis (53). Pasteur and Joubert described inhibition of the growth of *Bacillus anthracis* in urine during coculture with the "common bacterium" (described in reference 25). They extended this result to an in vivo model, showing decreased disease in animals that received a mixture of the two bacteria compared to those inoculated with *B. anthracis* alone. The ensuing years saw other attempts at both in vivo therapy and in vitro models of bacterial antagonism, with variable results. Two particular studies merit mention here, since they involve alteration of the respiratory tract flora. In the early 1900s, a patient with a staphylococcal pharyngitis was mistakenly hospitalized in a ward for diphtheria patients but did not develop diphtheria despite prolonged contact with other patients on the ward. An attempt was then made to use staphylococci to replace *Corynebacterium diphtheriae* in the throats of carriers, with some success (cited in reference 25). Having described in vitro inhibition of *N. meningitidis* by *S. pneumoniae,* Colebrook (10) attempted to use live pneumococci to displace meningococci in asymptomatic carriers. The effect, when noted, was short-lived.

Other epidemiologic observations have suggested that bacterial interference may be important in the establishment of the upper respiratory flora. Carriage of bacteria that cause in vitro inhibition of other respiratory organisms is common among both children and adults (36, 66). May noted an inverse relationship in the frequency of isolation of *S. pneumoniae* and *H. influenzae* from the sputa of patients with bronchitis (48). In a study of the bacteriology of middle ear fluid and nasal colonization in children with otitis media, *S. pneumoniae* and *H. influenzae* were found frequently but were isolated simultaneously less often than would be predicted mathematically (43). Similar findings of inverse relationships among species of bacteria have been found in investigations of the carrier state (18, 39, 72). Bakir et al. studied nearly 1,400 healthy children and noted a positive correlation between carriage of either *S. pneumoniae* or *H. influenzae* and carriage of *N. meningitidis*. However, they also described a protective effect of colonization with the nonpathogenic *N. lactamica* against meningococcal carriage (5). Similarly, overgrowth of antibiotic-resistant bacteria or fungi may follow alteration of the upper respiratory flora by antibiotic therapy. Many such "overgrowers" are environmental, and people are constantly exposed to them. However, they are able to take up residence in the upper respiratory tract only following elimination of the normal flora, implying that the normal flora plays an inhibitory role.

More recently, the lessons of these early studies in bacterial interference have been applied in medical settings. Probiotics are frequently used to alter the intestinal flora and to treat gastrointestinal diseases (44), but experience of using bacterial interference as therapy in the respiratory tract has been more limited. In the early 1960s, the seminal work of Shinefield et al. on artificial colonization and bacterial interference using *S. aureus* demonstrated the feasibility of the clinical use of this phenomenon (70). Following an outbreak of staphylococcal disease among newborns in a nursery, an epidemiologic investigation implicated a nurse who was a nasal carrier of the virulent 80/81 *S. aureus* strain in the spread of this organism to the infants in her care. However, the investigators made the striking observation that while 22% of infants younger than 24 h old who were cared for by that nurse became colonized with the 80/81 strain, none of those infants who were older than 24 h of age at the time that they were in her care did so. This raised the possibility that there was an age-related susceptibility to colonization with this virulent *S. aureus* strain or that prior colonization with other *S. aureus* strains inhibited colonization with the 80/81 strain. Evidence from colonization rates of children transferred from other nurseries suggested that the latter was the more likely possibility, so the investigators took the striking approach of directly testing the phenomenon of bacterial interference. After demonstrating that a relatively avirulent *S. aureus* strain (strain 502A) could colonize the nasopharynx of infants following deliberate inoculation, they effectively used this artificial colonization to prevent the acquisition of (and development of disease from) the more virulent 80/81. This approach was subsequently used in other outbreak settings (40, 69), as well as in treatment of adults with chronic furunculosis (47, 78). Unfortunately, reports of serious infections due to strain 502A have limited its further investigation (7, 34).

In another study, artificial colonization of the nasopharynx with diphtheroids protected infants against subsequent colonization with gram-negative bacilli but not pneumococci or streptococci (21). Sprunt and Redman examined the presence of alpha-hemolytic streptococci in the throats of children before and after antibiotic therapy prior to surgery. They found that elimination of these potentially inhibitory organisms from the nasopharynx predisposed patients to developing bacterial overgrowth with gram-negative organisms, a phenomenon that was not observed if the patients carried alpha-hemolytic streptococci that were resistant to the antibiotics used (76). Pretreatment of patients with oral penicillin for a period of days or weeks prior to surgery selected for colonization with resistant alpha-hemolytic streptococci and prevented overgrowth following subsequent antibiotic therapy (75). Artificial implantation of alpha-hemolytic streptococci into newborns with "abnormal" colonization (with gram-negative organisms or *S. aureus*) led to rapid replacement (74).

In some instances, one bacterial species or serotype may compete so effectively with another that it essentially excludes it from the nasopharynx. In this case, the interaction may become apparent only if the more effective organism is removed from its niche by vaccination or by antibiotic therapy. Vaccination against a bacterial pathogen can reduce both asymptomatic carriage and invasive disease due to that organism, as has been shown for both Hib (1, 83) and *S. pneumoniae* (15, 94). However, decreased carriage of one species may have effects on other members of the nasopharyngeal flora. A new niche may be created that allows the carriage of a species or serotype not seen prior to vaccination, and competitive or cooperative systems may be disrupted, with consequences for other members of the ecosystem. The primary concern with vaccination against Hib and *S. pneumoniae* has been the potential for serotype replacement, where other, potentially more virulent serotypes of bacteria fill the niche left by those removed after immunization. This has not been a major problem with the Hib vaccine to date, but increases in the carriage of nonvaccine serotypes of *S. pneumoniae* have been documented in studies of pneumococcal vaccination (14, 15). In addition, at least one group has described an increase in otitis media due to *H. influenzae* in a cohort of children immunized with a pneumococcal conjugate vaccine (23). This 11% increase was not statistically significant but is suggestive that removal of certain members of the upper respiratory flora may predispose to increased carriage of and possibly invasive disease from other species.

There are at least two mechanisms of interaction that may be important in interspecies competition in the respiratory tract: direct interbacterial interactions and competition for nutrients or receptors in a particular niche. There are numerous examples of direct interactions among bacterial species, although there is a lack of in vivo data to assess the relative importance of these. *S. pneumoniae* produces a neuraminidase that is capable of desialylation of the lipooligosaccharide of *H. influenzae* or *N. meningitidis,* two species that inhabit the same environmental niche (68). Sialylation is important in the protection of these bacteria from the host immune system (33, 88) as well as in biofilm formation in *H. influenzae* (80); therefore, desialylation may represent a means for one species to expose another to potential harm, thereby potentially gaining a competitive advantage.

Inhibition of growth or direct killing among upper respiratory bacteria has been described in vitro. Early studies (10) showed antagonism of *N. meningitidis* and *S. pneumoniae* under laboratory conditions. Other groups have shown in vitro inhibition among various members of the nasopharyngeal flora (64). The best-described mechanism for this interspecies inhibition in the production of (and resistance to) high concentrations of hydrogen peroxide by *S. pneumoniae* (58). The pneumococcus generates concentrations of hydrogen peroxide sufficient to efficiently kill *H. influenzae* and *N. meningitidis* but not *M. catarrhalis* in vitro. This mechanism may also be active in other inhibitory interactions in the nasopharynx (90). Production of bacteriocins may represent another means by which bacteria compete with rivals in a particular niche. Bacteriocins are antimicrobial molecules which can be produced by a wide variety of bacterial species and which target related bacteria in the environment (63). Bacteriocins are generally produced under conditions of stress and nutrient limitation. Important respiratory organisms, including group A *Streptococcus,* produce and are acted on by them (81). A bacteriocin produced by *S. salivarius* with action against group A *Streptococcus* may reduce the rate of streptococcal pharyngitis (82). In addition, a bacteriocin-like inhibitory substance produced by both *S. pneumoniae* and *S. pyogenes* may play a role in protection against otitis media in children (92). The relative importance of each of these mechanisms in determining the composition of the upper respiratory flora in vivo remains to be investigated.

Competition among bacterial species in the respiratory tract may also take place at the level of epithelial attachment. Organisms that fail to adhere to the mucosal surface risk removal by the mucociliary apparatus, and many distinct bacterial receptors have been described (89). However, there is overlap in receptor tropism among species that inhabit the nasopharynx, and competition for limited binding sites may occur. Both commensal and pathogenic neisseriae can adhere to the cell surface antigen CEACAM1 (CD66a) (85), and competition between these species and between *Neisseria* and *Haemophilus* species which also bind CEACAM1 has been detected in vitro (91). Krivan et al. described the ability of many respiratory pathogens to bind to a common carbohydrate sequence, Gal-NAc-β(1-4)-Gal, which is expressed on epithelial cells (37). In addition, several inhabitants of the nasopharynx, including *S. pneumoniae* and *H. influenzae,* display the phosphorylcholine moiety that allows interaction with the eukaryotic receptor for platelet-activating factor (12, 79), but the presence of competition among any of these species has not been well documented in vivo. Competition for receptor availability among upper respiratory tract species is in contrast to the increased availability of bacterial receptors accounting for synergy in mixed bacterial-viral infections as described above.

In addition, many of the members of the upper respiratory flora are auxotrophs. Competition for nutrients among members of mixed bacterial populations has been proposed (32). Since the airway surface fluid is not a rich growth medium, differential access to limited resources may be an important determinant of the composition of the flora, although this area has not been investigated in vivo.

FUTURE DIRECTIONS

There is ample evidence that competitive and cooperative interactions among species can play important roles in shaping the resident flora of the upper respiratory tract and in influencing the course of respiratory infections. Colonization is a dynamic process that involves a complex interplay among microbial species and host defense mechanisms. Numerous species coinhabit the human nasopharynx, and both in vitro and in vivo studies suggest that there is significant potential for interaction. Overlap in receptor tropism, competition for growth on solid or in liquid medium, and evidence of synergy in disease processes in humans all indicate that both cooperation and antagonism may be in constant play at the respiratory epithelial surface. There are significant technical challenges involved in accurately modeling this system, although progress in understanding the factors involved in the establishment and maintenance of carriage has been made. Artificial alteration of colonization through vaccination or antibiotic therapy may disrupt the ecological balance in this microenvironment and have unintended consequences. Further research in this area, especially if relevant in vivo models are employed, will contribute to our evolving understanding of the ecology of the upper respiratory tract.

REFERENCES

1. **Adams, W. G., K. A. Deaver, S. L. Cochi, B. D. Plikaytis, E. R. Zell, C. V. Broome, and J. D. Wenger.** 1993. Decline of childhood *Haemophilus influenzae* type b (Hib) disease in the Hib vaccine era. *JAMA* **269:**221–226.
2. **Aly, R., and H. R. Shinefield.** 1982. *Bacterial Interference.* CRC Press, Inc., Boca Raton, Fla.
3. **Aniansson, G., B. Alm, B. Andersson, P. Larsson, O. Nylen, H. Peterson, P. Rigner, M. Svanborg, and C. Svanborg.** 1992. Nasopharyngeal colonization during the first year of life. *J. Infect. Dis.* **165**(Suppl. 1):S38–S42.
4. **Bakaletz, L. O., R. L. Daniels, and D. J. Lim.** 1993. Modeling adenovirus type 1-induced otitis media in the chinchilla: effect on ciliary activity and fluid transport function of eustachian tube mucosal epithelium. *J. Infect. Dis.* **168:**865–872.
5. **Bakir, M., A. Yagci, N. Ulger, C. Akbenlioglu, A. Ilki, and G. Soyletir.** 2001. Asymtomatic carriage of *Neisseria meningitidis* and *Neisseria lactamica* in relation to *Streptococcus pneumoniae* and *Haemophilus influenzae* colonization in healthy children: apropos of 1400 children sampled. *Eur. J. Epidemiol.* **17:** 1015–1018.
6. **Berkovitch, M., M. Bulkowstein, D. Zhovtis, R. Greenberg, Y. Nitzan, B. Barzilay, and I. Boldur.** 2002. Colonization rate of bacteria in the throat of healthy infants. *Int. J. Pediatr. Otorhinolaryngol.* **63:**19–24.
7. **Blair, E. B., and A. H. Tull.** 1969. Multiple infections among newborns resulting from colonization with *Staphylococcus aureus* 502A. *Am. J. Clin. Pathol.* **52:**42–49.
8. **Brogden, K. A.** 2002. Polymicrobial diseases of animals and humans, p. 3–20. *In* K. A. Brogden and J. M. Guthmiller (ed.), *Polymicrobial Infections.* ASM Press, Washington, D.C.
9. **Brook, I.** 1999. Bacterial interference. *Crit. Rev. Microbiol.* **25:**155–172.
10. **Colebrook, L.** 1915. Bacterial antagonism, with particular reference to meningococcus. *Lancet* **ii:**1136–1138.
11. **Cossart, P.** 2000. *Cellular Microbiology.* ASM Press, Washington, D.C.
12. **Cundell, D. R., N. P. Gerard, C. Gerard, I. Idanpaan-Heikkila, and E. I. Tuomanen.** 1995. *Streptococcus pneumoniae* anchor to activated human cells by the receptor for platelet-activating factor. *Nature* **377:**435–438.
13. **Dagan, R., D. Fraser, N. Givon, and P. Yagupsky.** 1999. Carriage of resistant pneumococci by children in southern Israel and impact of conjugate vaccines on carriage. *Clin. Microbiol. Infect.* **5**(Suppl. 4):S29–S37.
14. **Dagan, R., N. Givon-Lavi, O. Zamir, and D. Fraser.** 2003. Effect of a nonavalent conjugate vaccine on carriage of antibiotic-resistant *Streptococcus pneumoniae* in day-care centers. *Pediatr. Infect. Dis. J.* **22:**532–540.
15. **Dagan, R., N. Givon-Lavi, O. Zamir, M. Sikuler-Cohen, L. Guy, J. Janco, P. Yagupsky, and D. Fraser.** 2002. Reduction of nasopharyngeal carriage of *Streptococcus pneumoniae* after administration of a 9-valent pneumococcal conjugate vaccine to toddlers attending day care centers. *J. Infect. Dis.* **185:**927–936.
16. **Dagan, R., R. Melamed, M. Muallem, L. Piglansky, and P. Yagupsky.** 1996. Nasopharyngeal colonization in southern Israel with antibiotic-resistant pneumococci during the first 2 years of life: relation to serotypes likely to be included in pneumococcal conjugate vaccines. *J. Infect. Dis.* **174:**1352–1355.
17. **Dagan, R., M. Muallem, R. Melamed, O. Leroy, and P. Yagupsky.** 1997. Reduction of pneumococcal nasopharyngeal carriage in early infancy after immunization with tetravalent pneumococcal vaccines conjugated to either tetanus toxoid or diphtheria toxoid. *Pediatr. Infect. Dis. J.* **16:**1060–1064.
18. **Davis, N. A., and G. H. Davis.** 1965. Ecology of nasal staphylococci. *J. Bacteriol.* **89:**1163–1168.
19. **Drake, D. R., and K. A. Brogden.** 2002. Continuous-culture chemostat systems and flowcells as methods to investigate microbial interactions, p. 21–30. *In* K. A. Brogden and J. M. Guthmiller (ed.), *Polymicrobial Infections.* ASM Press, Washington, D.C.
20. **Dykhuizen, R. S., D. Golder, T. M. Reid, and I. M. Gould.** 1996. Phenoxymethyl penicillin versus co-amoxiclav in the treatment of acute streptococcal pharyngitis, and the role of beta-lactamase activity in saliva. *J. Antimicrob. Chemother.* **37:**133–138.
21. **Ehrenkranz, N. J.** 1970. Bacterial colonization of newborn infants and subsequent acquisition of hospital bacteria. *J. Pediatr.* **76:**839–847.
22. **El Ahmer, O. R., M. W. Raza, M. M. Ogilvie, D. M. Weir, and C. C. Blackwell.** 1999. Binding of bacteria to HEp-2 cells infected with influenza A virus. *FEMS Immunol. Med. Microbiol.* **23:**331–341.
23. **Eskola, J., T. Kilpi, A. Palmu, J. Jokinen, J. Haapakoski, E. Herva, A. Takala, H. Kayhty, P. Karma, R. Kohberger, G. Siber, and P. H. Makela.** 2001. Efficacy of a pneumococcal conjugate vaccine against acute otitis media. *N. Engl. J. Med.* **344:**403–409.
24. **Flak, T. A., L. N. Heiss, J. T. Engle, and W. E. Goldman.** 2000. Synergistic epithelial responses to endotoxin and a naturally occurring muramyl peptide. *Infect. Immun.* **68:**1235–1242.
25. **Florey, H. W.** 1946. The use of micro-organisms for therapeutic purposes. *Yale J. Biol. Med.* **19:**101–117.
26. **Garcia-Rodriguez, J. A., and M. J. Fresnadillo Martinez.** 2002. Dynamics of nasopharyngeal colonization by potential respiratory pathogens. *J. Antimicrob. Chemother.* **50**(Suppl. S2):59–73.
27. **Gerber, M. A., R. R. Tanz, W. Kabat, G. L. Bell, B. Siddiqui, T. J. Lerer, M. L. Lepow, E. L. Kaplan, and S. T. Shulman.** 1999. Potential mechanisms for failure to eradicate group A streptococci from the pharynx. *Pediatrics* **104:**911–917.
28. **Graham, N. M., R. M. Douglas, and P. Ryan.** 1986. Stress and acute respiratory infection. *Am. J. Epidemiol.* **124:**389–401.

29. **Gwaltney, J. M., Jr., M. A. Sande, R. Austrian, and J. O. Hendley.** 1975. Spread of *Streptococcus pneumoniae* in families. II. Relation of transfer of *S. pneumoniae* to incidence of colds and serum antibody. *J. Infect. Dis.* **132:**62–68.
30. **Hakansson, A., A. Kidd, G. Wadell, H. Sabharwal, and C. Svanborg.** 1994. Adenovirus infection enhances in vitro adherence of *Streptococcus pneumoniae. Infect. Immun.* **62:** 2707–2714.
31. **Hament, J. M., J. L. Kimpen, A. Fleer, and T. F. Wolfs.** 1999. Respiratory viral infection predisposing for bacterial disease: a concise review. *FEMS Immunol. Med. Microbiol.* **26:**189–195.
32. **Harder, W., and L. Dijkhuizen.** 1982. Strategies of mixed substrate utilization in microorganisms. *Philos. Trans. R. Soc. Lond. Ser. B* **297:**459–480.
33. **Hood, D. W., K. Makepeace, M. E. Deadman, R. F. Rest, P. Thibault, A. Martin, J. C. Richards, and E. R. Moxon.** 1999. Sialic acid in the lipopolysaccharide of *Haemophilus influenzae:* strain distribution, influence on serum resistance and structural characterization. *Mol. Microbiol.* **33:**679–692.
34. **Houck, P. W., J. D. Nelson, and J. L. Kay.** 1972. Fatal septicemia due to *Staphylococcus aureus* 502A. Report of a case and review of the infectious complications of bacterial interference programs. *Am. J. Dis. Child.* **123:**45–48.
35. **Ishizuka, S., M. Yamaya, T. Suzuki, H. Takahashi, S. Ida, T. Sasaki, D. Inoue, K. Sekizawa, H. Nishimura, and H. Sasaki.** 2003. Effects of rhinovirus infection on the adherence of *Streptococcus pneumoniae* to cultured human airway epithelial cells. *J. Infect. Dis.* **188:**1929–1940.
36. **Johanson, W. G., Jr., R. Blackstock, A. K. Pierce, and J. P. Sanford.** 1970. The role of bacterial antagonism in pneumococcal colonization of the human pharynx. *J. Lab. Clin. Med.* **75:** 946–952.
37. **Krivan, H. C., D. D. Roberts, and V. Ginsburg.** 1988. Many pulmonary pathogenic bacteria bind specifically to the carbohydrate sequence GalNAc beta 1-4Gal found in some glycolipids. *Proc. Natl. Acad. Sci. USA* **85:**6157–6161.
38. **Lakshman, R., C. Murdoch, G. Race, R. Burkinshaw, L. Shaw, and A. Finn.** 2003. Pneumococcal nasopharyngeal carriage in children following heptavalent pneumococcal conjugate vaccination in infancy. *Arch. Dis. Child.* **88:**211–214.
39. **Light, I. J., J. M. Sutherland, M. L. Cochran, and J. Sutorius.** 1968. Ecologic relation between *Staphylococcus aureus* and *Pseudomonas* in a nursery population. Another example of bacterial interference. *N. Engl. J. Med.* **278:**1243–1247.
40. **Light, I. J., R. L. Walton, J. M. Sutherland, H. R. Shinefield, and V. Brackvogel.** 1967. Use of bacterial interference to control a staphylococcal nursery outbreak. Deliberate colonization of all infants with the 502A strain of *Staphylococcus aureus. Am. J. Dis. Child.* **113:**291–300.
41. **Lipsitch, M.** 1999. Bacterial vaccines and serotype replacement: lessons from *Haemophilus influenzae* and prospects for *Streptococcus pneumoniae. Emerg. Infect. Dis.* **5:**336–345.
42. **Lipsitch, M., J. K. Dykes, S. E. Johnson, E. W. Ades, J. King, D. E. Briles, and G. M. Carlone.** 2000. Competition among *Streptococcus pneumoniae* for intranasal colonization in a mouse model. *Vaccine* **18:**2895–2901.
43. **Luotonen, J.** 1982. Streptococcus pneumoniae and Haemophilus influenzae in nasal cultures during acute otitis media. *Acta Otolaryngol.* **93:**295–299.
44. **Macfarlane, G. T., and J. H. Cummings.** 2002. Probiotics, infection and immunity. *Curr. Opin. Infect. Dis.* **15:**501–506.
45. **Mackowiak, P. A.** 1978. Microbial synergism in human infections (first of two parts). *N. Engl. J. Med.* **298:**21–26.
46. **Mackowiak, P. A.** 1978. Microbial synergism in human infections (second of two parts). *N. Engl. J. Med.* **298:**83–87.
47. **Maibach, H. I., W. G. Strauss, and H. R. Shinefield.** 1969. Bacterial interference: relating to chronic furunculosis in man. *Br. J. Dermatol.* **81**(Suppl. 1):69–76.
48. **May, J. R.** 1954. Pathogenic bacteria in chronic bronchitis. *Lancet* **267:**839–842.
49. **McCool, T. L., T. R. Cate, G. Moy, and J. N. Weiser.** 2002. The immune response to pneumococcal proteins during experimental human carriage. *J. Exp. Med.* **195:**359–365.
50. **McCool, T. L., T. R. Cate, E. I. Tuomanen, P. Adrian, T. J. Mitchell, and J. N. Weiser.** 2003. Serum immunoglobulin G response to candidate vaccine antigens during experimental human pneumococcal colonization. *Infect. Immun.* **71:**5724–5732.
51. **McCullers, J. A., and J. E. Rehg.** 2002. Lethal synergism between influenza virus and *Streptococcus pneumoniae:* characterization of a mouse model and the role of platelet-activating factor receptor. *J. Infect. Dis.* **186:**341–350.
52. **Mohle-Boetani, J. C., G. Ajello, E. Breneman, K. A. Deaver, C. Harvey, B. D. Plikaytis, M. M. Farley, D. S. Stephens, and J. D. Wenger.** 1993. Carriage of *Haemophilus influenzae* type b in children after widespread vaccination with conjugate *Haemophilus influenzae* type b vaccines. *Pediatr. Infect. Dis. J.* **12:**589–593.
53. **Mosse, J. R.** 1852. Of the use of yeast in the treatment of boils. *Lancet* **ii:**113.
54. **Murphy, T. V., P. Pastor, F. Medley, M. T. Osterholm, and D. M. Granoff.** 1993. Decreased *Haemophilus* colonization in children vaccinated with *Haemophilus influenzae* type b conjugate vaccine. *J. Pediatr.* **122:**517–523.
55. **Nain, M., F. Hinder, J. H. Gong, A. Schmidt, A. Bender, H. Sprenger, and D. Gemsa.** 1990. Tumor necrosis factor-alpha production of influenza A virus-infected macrophages and potentiating effect of lipopolysaccharides. *J. Immunol.* **145:**1921–1928.
56. **Ohashi, Y., Y. Nakai, Y. Esaki, Y. Ohno, Y. Sugiura, and H. Okamoto.** 1991. Influenza A virus-induced otitis media and mucociliary dysfunction in the guinea pig. *Acta Otolaryngol. Suppl.* **486:**135–148.
57. **Paster, B. J., S. K. Boches, J. L. Galvin, R. E. Ericson, C. N. Lau, V. A. Levanos, A. Sahasrabudhe, and F. E. Dewhirst.** 2001. Bacterial diversity in human subgingival plaque. *J. Bacteriol.* **183:**3770–3783.
58. **Pericone, C. D., K. Overweg, P. W. Hermans, and J. N. Weiser.** 2000. Inhibitory and bactericidal effects of hydrogen peroxide production by *Streptococcus pneumoniae* on other inhabitants of the upper respiratory tract. *Infect. Immun.* **68:**3990–3997.
59. **Raza, M. W., C. C. Blackwell, M. M. Ogilvie, A. T. Saadi, J. Stewart, R. A. Elton, and D. M. Weir.** 1994. Evidence for the role of glycoprotein G of respiratory syncytial virus in binding of *Neisseria meningitidis* to HEp-2 cells. *FEMS Immunol. Med. Microbiol.* **10:**25–30.
60. **Raza, M. W., O. R. El Ahmer, M. M. Ogilvie, C. C. Blackwell, A. T. Saadi, R. A. Elton, and D. M. Weir.** 1999. Infection with respiratory syncytial virus enhances expression of native receptors for non-pilate *Neisseria meningitidis* on HEp-2 cells. *FEMS Immunol. Med. Microbiol.* **23:**115–124.
61. **Reid, G., J. Howard, and B. S. Gan.** 2001. Can bacterial interference prevent infection? *Trends Microbiol.* **9:**424–428.
62. **Rickard, A. H., P. Gilbert, N. J. High, P. E. Kolenbrander, and P. S. Handley.** 2003. Bacterial coaggregation: an integral process in the development of multi-species biofilms. *Trends Microbiol.* **11:**94–100.
63. **Riley, M. A., and J. E. Wertz.** 2002. Bacteriocins: evolution, ecology, and application. *Annu. Rev. Microbiol.* **56:**117–137.

64. **Rosebury, T., D. Gale, and D. F. Taylor.** 1954. An approach to the study of interactive phenomena among microorganisms indigenous to man. *J. Bacteriol.* **67:**135–152.
65. **Rotimi, V. O., and B. I. Duerden.** 1981. The development of the bacterial flora in normal neonates. *J. Med. Microbiol.* **14:** 51–62.
66. **Sanders, E.** 1969. Bacterial interference. I. Its occurrence among the respiratory tract flora and characterization of inhibition of group A streptococci by viridans streptococci. *J. Infect. Dis.* **120:**698–707.
67. **Sanford, B. A., A. Shelokov, and M. A. Ramsay.** 1978. Bacterial adherence to virus-infected cells: a cell culture model of bacterial superinfection. *J. Infect. Dis.* **137:**176–181.
68. **Shakhnovich, E. A., S. J. King, and J. N. Weiser.** 2002. Neuraminidase expressed by *Streptococcus pneumoniae* desialylates the lipopolysaccharide of *Neisseria meningitidis* and *Haemophilus influenzae:* a paradigm for interbacterial competition among pathogens of the human respiratory tract. *Infect. Immun.* **70:**7161–7164.
69. **Shinefield, H. R., M. Boris, J. C. Ribble, E. F. Cale, and H. F. Eichenwald.** 1963. Bacterial interference: its effect on nursery-acquired infection with *Staphylococcus aureus.* III. The Georgia epidemic. *Am. J. Dis. Child.* **105:**663–673.
70. **Shinefield, H. R., J. C. Ribble, M. Boris, and H. F. Eichenwald.** 1963. Bacterial interference: its effect on nursery-acquired infection with *Staphylococcus aureus.* I. Preliminary observations on artificial colonzation of newborns. *Am. J. Dis. Child.* **105:**646–654.
71. **Smith, H.** 1982. The role of microbial interactions in infectious disease. *Philos. Trans. R. Soc. Lond. Ser. B* **297:**551–561.
72. **Speck, W. T., J. M. Driscoll, R. A. Polin, and H. S. Rosenkranz.** 1978. Effect of bacterial flora on staphylococcal colonisation of the newborn. *J. Clin. Pathol.* **31:**153–155.
73. **Sprunt, K., and G. Leidy.** 1988. The use of bacterial interference to prevent infection. *Can. J. Microbiol.* **34:**332–338.
74. **Sprunt, K., G. Leidy, and W. Redman.** 1980. Abnormal colonization of neonates in an ICU: conversion to normal colonization by pharyngeal implantation of alpha hemolytic streptococcus strain 215. *Pediatr. Res.* **14:**308–313.
75. **Sprunt, K., G. A. Leidy, and W. Redman.** 1971. Prevention of bacterial overgrowth. *J. Infect. Dis.* **123:**1–10.
76. **Sprunt, K., and W. Redman.** 1968. Evidence suggesting importance of role of interbacterial inhibition in maintaining balance of normal flora. *Ann. Intern. Med.* **68:**579–590.
77. **Stemmons, E. D., and H. L. Smith.** 2000. Competition in a chemostat with wall attachment. *SIAM J. Appl. Math.* **61:**567–595.
78. **Strauss, W. G., H. I. Maibach, and H. R. Shinefield.** 1969. Bacterial interference treatment of recurrent furunculosis. 2. Demonstration of the relationship of strain to pathogeneicity. *JAMA* **208:**861–863.
79. **Swords, W. E., B. A. Buscher, K. Ver Steeg Ii, A. Preston, W. A. Nichols, J. N. Weiser, B. W. Gibson, and M. A. Apicella.** 2000. Non-typeable *Haemophilus influenzae* adhere to and invade human bronchial epithelial cells via an interaction of lipooligosaccharide with the PAF receptor. *Mol. Microbiol.* **37:**13–27.
80. **Swords, W. E., M. L. Moore, L. Godzicki, G. Bukofzer, M. J. Mitten, and J. VonCannon.** 2004. Sialylation of lipooligosaccharides promotes biofilm formation by nontypeable *Haemophilus influenzae. Infect. Immun.* **72:**106–113.
81. **Tagg, J. R.** 1984. Production of bacteriocin-like inhibitors by group A streptococci of nephritogenic M types. *J. Clin. Microbiol.* **19:**884–887.
82. **Tagg, J. R., and K. P. Dierksen.** 2003. Bacterial replacement therapy: adapting 'germ warfare' to infection prevention. *Trends Biotechnol.* **21:**217–223.
83. **Takala, A. K., J. Eskola, M. Leinonen, H. Kayhty, A. Nissinen, E. Pekkanen, and P. H. Makela.** 1991. Reduction of oropharyngeal carriage of *Haemophilus influenzae* type b (Hib) in children immunized with an Hib conjugate vaccine. *J. Infect. Dis.* **164:**982–986.
84. **Takala, A. K., M. Santosham, J. Almeido-Hill, M. Wolff, W. Newcomer, R. Reid, H. Kayhty, E. Esko, and P. H. Makela.** 1993. Vaccination with *Haemophilus influenzae* type b meningococcal protein conjugate vaccine reduces oropharyngeal carriage of *Haemophilus influenzae* type b among American Indian children. *Pediatr. Infect. Dis. J.* **12:**593–599.
85. **Toleman, M., E. Aho, and M. Virji.** 2001. Expression of pathogen-like Opa adhesins in commensal *Neisseria:* genetic and functional analysis. *Cell Microbiol.* **3:**33–44.
86. **Topley, W. W. C., and H. A. Fielden.** 1922. The succession of dominant species in a mixed bacterial culture in a fluid medium. *Lancet* **iii:**1164–1165.
87. **Tuomanen, E.** 1986. Piracy of adhesins: attachment of superinfecting pathogens to respiratory cilia by secreted adhesins of *Bordetella pertussis. Infect. Immun.* **54:**905–908.
88. **Unkmeir, A., U. Kammerer, A. Stade, C. Hubner, S. Haller, A. Kolb-Maurer, M. Frosch, and G. Dietrich.** 2002. Lipooligosaccharide and polysaccharide capsule: virulence factors of *Neisseria meningitidis* that determine meningococcal interaction with human dendritic cells. *Infect. Immun.* **70:**2454–2462.
89. **Van Nhieu, G. T., and P. Sansonetti.** 2000. Cell adhesion molecules and bacterial pathogens, p. 97–111. *In* P. Cossart, P. Boquet, S. Normark, and R. Rappuoli (ed.), *Cellular Microbiology.* ASM Press, Washington, D.C.
90. **Vernazza, T. R., and T. H. Melville.** 1979. Inhibitory activity of *Streptococcus mitis* against oral bacteria. *Microbios* **26:**95–101.
91. **Virji, M., D. Evans, J. Griffith, D. Hill, L. Serino, A. Hadfield, and S. M. Watt.** 2000. Carcinoembryonic antigens are targeted by diverse strains of typable and non-typable *Haemophilus influenzae. Mol. Microbiol.* **36:**784–795.
92. **Walls, T., D. Power, and J. Tagg.** 2003. Bacteriocin-like inhibitory substance (BLIS) production by the normal flora of the nasopharynx: potential to protect against otitis media? *J. Med. Microbiol.* **52:**829–833.
93. **Weiser, J. N., D. Bae, C. Fasching, R. W. Scamurra, A. J. Ratner, and E. N. Janoff.** 2003. Antibody-enhanced pneumococcal adherence requires IgA1 protease. *Proc. Natl. Acad. Sci. USA* **100:**4215–4220.
94. **Whitney, C. G., M. M. Farley, J. Hadler, L. H. Harrison, N. M. Bennett, R. Lynfield, A. Reingold, P. R. Cieslak, T. Pilishvili, D. Jackson, R. R. Facklam, J. H. Jorgensen, and A. Schuchat.** 2003. Decline in invasive pneumococcal disease after the introduction of protein-polysaccharide conjugate vaccine. *N. Engl. J. Med.* **348:**1737–1746.

Colonization of Mucosal Surfaces
Edited by James P. Nataro et al.

Chapter 8

Bacterial Adherence and Tropism in the Human Respiratory Tract

MUMTAZ VIRJI

Adhesion to target molecules by microbial ligands in general constitutes the primary and principal event in the colonization of mucosal tissues and as such also represents the first step in pathogenesis. Despite the presence of antimicrobial factors and the mucociliary escalator that operate to deter microbes from establishing in the niche, considerable numbers of microbes have evolved strategies to attach specifically to the molecules on the epithelial cells and in mucus secretions present in the human nasopharynx. The mechanisms used for this purpose are often complex, multifactorial, dynamic, and carefully orchestrated. Both redundancy of targeting schemes and surface variation by frequent antigenic/structural and phase (on/off) variation are common to respiratory bacteria. Redundancy compensates for the phase variation of adhesins, a strategy that helps avert host immune mechanisms and brings into play novel ligands that may bind to the same receptor or a distinct host molecule, resulting in an altered mode or specificity of interaction. Antigenic or structural variations of adhesins may achieve the same ultimate goal (45). Multiple adhesins may also operate simultaneously to increase the avidity of binding. This is often a prelude to internalization into epithelial cells and involves manipulation of target cytoskeleton. Entry into nonphagocytic host cells may be one of the strategies for immune evasion and access to a new source of nutrients. Once inside epithelial cells, the microbes may traverse the epithelium and emerge at the basolateral surface. In an immunocompromised host, this may be a deleterious situation leading to disease. Some microbes find temporary refuge from host immunity by sliding between epithelial cells and have been found in vivo and in vitro between or under epithelial barriers (46, 104, 122, 146, 173) (Fig. 1).

Within the respiratory tract, humans harbor a large number of species of distinct genera. Some of these are common residents that colonize a large proportion (possibly exceeding 50%) of healthy individuals, e.g., several species of *Streptococcus, Neisseria, Haemophilus,* and *Moraxella* (*Branhamella*). The carriage rate reflects their success as human commensals. Some species are occasional residents that are found in <10% of healthy individuals, e.g., *Streptococcus pneumoniae* and *Neisseria meningitidis*. Carriage of various bacteria may vary in distinct age groups and populations as well as between seasons. The asymptomatic commensals and colonizers can become pathogenic in rare situations, especially in immunocompromised hosts, to cause localized or disseminated infections. Uncommon residents include contagious bacteria such as *Bordetella pertussis,* whose entry into the respiratory tract is very frequently associated with disease in nonimmune hosts (12, 36, 67, 142, 147).

Many excellent reviews that have covered adhesion aspects of these and other respiratory bacteria are cited in the text, and the reader is referred to these articles for supplementary information. This chapter aims to present a comparative overview of several major adhesins and invasins of four human-tropic respiratory bacteria and includes examples of frequent to rare colonizers, namely, *Haemophilus influenzae, Moraxella catarrhalis, N. meningitidis,* and *B. pertussis,* and their extensively studied host tissue-targeting mechanisms. It considers our current understanding of the bacterial adhesion factors that determine their tropism in the human respiratory tract, although it has to be noted that other factors important for bacterial survival (e.g., Fe acquisition mechanisms and immunoglobulin A1 [IgA1] protease) also constitute determinants of host and tissue tropism.

Mumtaz Virji • Department of Pathology and Microbiology, School of Medical Sciences, University of Bristol, Bristol BS8 1TD, United Kingdom.

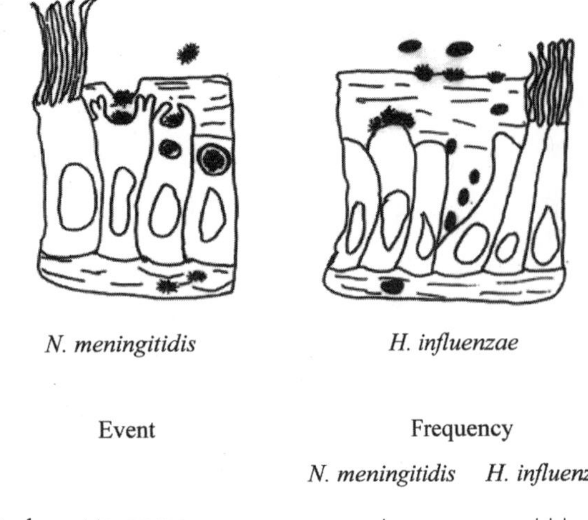

Event	Frequency	
	N. meningitidis	*H. influenzae*
Attachment to mucus	+	+++
Binding to ciliated cells	-	-
Binding to non-ciliated cells	++	++
Cellular entry	++	-
Paracytosis	-	++
Transcytosis	+	-

Figure 1. Observations of the modes of interaction of *N. meningitidis* and *H. influenzae* with human nasopharyngeal organ cultures (122). Both bacteria attach to nonciliated cells; whereas *N. meningitidis* enters cells, *H. influenzae* appears to transmigrate between cells. Both bacteria were found beneath epithelial cells. *H. influenzae* had a much greater tendency to bind to mucus than did *N. meningitidis*. Adapted from reference 122 with permission from the publisher.

TARGETS IN THE RESPIRATORY TRACT

The respiratory tract of a healthy individual presents an intact epithelium composed of distinct epithelial cells with specialized functions and may bear distinct receptors. The intact polarized epithelium is also a selective permeability barrier with intercellular occluding junctions that separate the apical and basal surfaces. These domains have a unique macromolecular composition (66). Overlying this are the mucus and associated molecules that are products of epithelial and immune cells, including secretory IgA, particularly IgA1 (75, 167, 168). Many respiratory bacteria produce IgA1 protease. Cleavage of IgA1 by *S. pneumoniae* was shown to result in the deposition of Fab fragments on bacteria, leading to unmasking of adhesive ligands on the bacterial surface and increased adherence (164). Interactions with some host components enable bacteria to target multiple sites. For example, targeting fibronectin (Fn) can be expected to enable bacteria to attach to cell surface-located Fn as well as to the basement membrane that contains Fn, and to its integrin receptors. In addition, some microbes such as *Neisseria* spp. have tropism for receptors on nonciliated epithelial cells (122) (Fig. 1), while others, e.g., *Bordetella*, specifically target only ciliated cells (114). The mechanisms of such cell tropism are not fully resolved, but many receptors targeted by distinct microbes have been discovered by in vitro investigations of cultured human cells that are providing molecular explanations for tropism.

The respiratory tract of an individual with compromised defenses, such as damaged airway mucosa, presents a novel environment to bacteria. Some bacteria produce cytotoxins that destroy epithelial cells. In addition, viral infections or smoking may damage the epithelial barrier. Such damage uncovers a novel set of receptors not only in the exposed extracellular matrix (ECM) but also on actively proliferating and unpolarized cells that may express novel molecules and present their basolateral receptors to the external/luminal milieu. Furthermore, products of inflammation, such as neutrophil defensins, may directly tether bacteria to target cells (35). Plasminogen is also proposed to be a receptor for bacteria and may mediate adhesion to human respiratory epithelial cells as well as aiding in the translocation of bacteria through the ECM (26, 93). Moreover, inflammatory cytokines, via upregulation of certain cellular receptors, may augment bacterial interactions with epithelia. In some pathogenic conditions, where mucus clearance is impeded, bacteria attaching to mucus can persist for long periods and contiguous spread to new sites of colonization may occur. Common residents of the upper respiratory tract often become opportunistic pathogens in such situations; in patients with lung diseases such as chronic obstructive pulmonary disease and cystic fibrosis, persistent infections of the lungs are often caused by upper respiratory tract residents, including *H. influenzae* and *M. catarrhalis*.

RECURRENT THEMES AND DIVERSITY OF MICROBIAL TARGETING MECHANISMS

Choice of Receptor

Bacteria may attach to a wide variety of molecules and include adhesion receptors belonging to the Ig superfamily, the selectins, the cadherins, and the proteoglycans such as syndecans, glypicans, and epiglycans. Targeting these allows bacteria to manipulate the host cell function. The initiation of signaling frequently involves lateral movement of the receptors within the membrane, or clustering. Bacteria, by virtue of their polyvalency, are effective mediators of receptor clustering. Such binding may be expected not only to successfully compete with the natural ligands but also to deliver effective signals (Color Plate 3). In terms of bacterial survival strategies, a notable choice includes the complement receptor CR3, the $\alpha_m\beta_2$-integrin, on macrophages. This nonopsonic interaction leads to bacterial entry into phagocytic cells without eliciting the respiratory burst, thus aiding bacterial survival within phagocytes (114).

Diverse Mechanisms of Host Mimicry

Molecular mimicry is probably one of the most striking, if not surprising, recurrent themes in bacterial strategy for targeting specific receptors, especially integrins. Bacterial ligands may bind to natural ligands of integrins present in host secretions (as exemplified by Fn targeting [discussed above]). By such surrogate means, microbes exhibit "indirect" or "assisted" mimicry, which is exemplified by *N. meningitidis* Opc protein binding to Fn or vitronectin (Vn) and via these to cellular integrins (140, 154). Studies of *Streptococcus pyogenes* have also shown that a number of distinct ECM molecules can mediate bacterial binding to and invasion of human lung epithelial cells (18). Another example of mimicry involves the motif RGD (Arg-Gly-Asp), which is central to the recognition of their ligands by a subfamily of integrins such as the vitronectin receptor ($\alpha_v\beta_3$-integrin) and the Fn receptor ($\alpha_5\beta_1$-integrin). This motif is present on a number of microbial structures exemplified by *B. pertussis* filamentous haemagglutinin (FHA) and pertactin (Prn) (67, 114) and is associated with direct interactions with integrins. Since RGD is a recognition sequence for several receptors (α_4- α_5- α_8-, α_{IIb}-, and α_v-containing integrins), the possession of this sequence has the potential to mediate interactions with multiple receptors and cell types (144). Thus, FHA targets both epithelial and macrophage integrins via the RGD motif. However, the interaction does not yield host specificity since RGD recognition is not unique to human integrins, and the RGD sequence of pertactin has been shown to mediate binding to Chinese hamster ovary cells (67, 114).

Numerous microbes have also developed another exquisite mechanism of mimicry, the expression of phosphorylcholine on their surfaces. The ChoP moiety also occurs in platelet-activating factor (PAF) and is involved in recognition by the PAF receptor on distinct cell types. Accordingly, bacteria utilize this adhesion mechanism to adhere to and invade target tissues (see chapter 5).

Common Receptors

Several ligands of the organisms described in this chapter target ECM proteins and directly or indirectly target the RGD-binding integrins. Multiple mechanisms have evolved that target specifically the nasopharyngeal or tracheal mucins. In addition, heparan sulfate proteoglycans (HSPGs) and members of Ig superfamily, especially the carcinoembryonic antigen-related cell adhesion molecules (CEACAMs), are targeted by multiple mucosal bacteria and are specifically discussed below (see "Some common receptor-targeting mechanisms").

Surface Charge and Adherence

Epithelial cells and bacteria carry overall negative charge and may be expected to repel each other (DLVO theory). However, positively charged domains, termed microplicea, occur on epithelial cells, as shown by atomic force microscopy and electron microscopy (EM), and interactions between these and the negatively charged surface of *M. catarrhalis* have been reported (3).

MECHANISMS OF TISSUE ADHESION VIA PERICELLULAR APPENDAGES

Types of Pericellular Appendages

Bacterial protein adhesins can be expressed as monomers, simple oligomeric structures, or polymeric fibers, the pili or fimbriae. The adhesins that form layers above the bacterial outer membrane due to their extended morphology are categorized in this section and include polymeric as well as monomeric or small multimeric filamentous or protruding structures visible by EM. The nomenclature for the polymeric structures is based on the prevalent nomenclature used in the respective literature, i.e., pili for *N. meningitidis, H. influenzae,* and *M. catarrhalis* and fimbriae for *B. pertussis.*

One of the most intensively studied mechanisms of adherence is attachment via pili or fimbriae, which enables the initial localization at the respiratory tract, particularly of capsulate bacteria. *N. meningitidis* pili belong to the type 4 pili, are ca. 6 nm in diameter, and extend several micrometers from the bacterial surface. They consist of identical pilin monomers arranged in a helical form and are expressed by a number of respiratory pathogens, including *N. meningitidis, M. catarrhalis,* and *Pseudomonas aeruginosa* (40, 79, 89). However, different type 4 piliated bacteria use different mechanisms for host and tissue targeting. For example, it appears that the epithelial binding site on *P. aeruginosa* pili lies in the major pilin subunit (40), whereas *N. meningitidis* pili may use a tip-located adhesin for host targeting (89, 112). However, other data also suggest that the *N. meningitidis* pilin subunit structure also affects cellular targeting (78, 86, 95, 119, 149, 156). Interestingly, in *P. aeruginosa*, the receptor-binding site is exposed only in the tip-located pilin subunit, which therefore participates primarily in adhesion (40).

Pili of *H. influenzae* are like type I and pap pili of *Escherichia coli* described as LKP (for "long, thick, and hemagglutination positive") or HA (for "hemagglutinating") (142). In this case also, the pilus is composed of a polymer of identical repeating subunit forming a fiber 5 nm in diameter and 450 nm long (124), with a tip-located protein, HifE; both the fiber and HifE act as adhesins (69, 80). In contrast to *N. meningitidis, H. influenzae* possesses a second filamentous adhesive structure called Hsf (for "*Haemophilus* surface fibrils") of 240 kDa, a monomeric extended fine hair-like structure visible by EM (125, 127). In addition, the P5 protein of nontypeable *H. influenzae* (NTHi), with a molecular mass of ca. 30 kDa, has been proposed as a fimbrial protein composed of a coiled-coil structure (121). However, more recent studies have provided evidence of the presence of extensive β-sheets, and P5 is predicted to form an *E. coli* OmpA-like β-barrel structure (161, 162). As such, it is discussed in "Nonfilamentous adhesins of respiratory bacteria" (below).

Only relatively limited information is available regarding the pili of *M. catarrhalis*. Although they are expressed by some strains and are similar to type 4 pili of *N. meningitidis* (79), *M. catarrhalis* isolates also produce high-molecular-weight adhesins, the ubiquitous surface proteins UspA1 and UspA2 and the haemagglutinin Hag. From sequence and structural analysis, UspA proteins are proposed to form a tripartite structure, with their C-terminal ends traversing the outer membrane and their N-terminal ends forming a globular structure extended out from the bacterial surface by a coiled-coil stalk (50). The Hag protein also forms a pericellular fiber-like structure (97).

B. pertussis fimbriae are structurally related to *H. influenzae* pili; two different types of subunits may form the shaft (Fim2 or Fim3) and have adhesive functions. In addition, the tip-located minor subunit FimD serves as an adhesin (170). *B. pertussis* FHA has a 19-residue repeat motif and is rich in β-strands and turns. It folds into a monomeric tapered rigid rod ca. 4 nm wide and 50 nm long whose appearance by EM has been likened to that of a horseshoe nail (63, 76).

Pili of *N. meningitidis*

Although *N. meningitidis* strains may frequently colonize the human respiratory tract asymptomatically, in some cases they disseminate from this primary and specific site of colonization to cause septicemia and meningitis. *N. meningitidis* characteristically is a capsulate bacterium whose signature property is the frequency of antigenic variation of its surface structures. Distinct strains of *N. meningitidis* may elaborate one of a dozen different capsules, but only a few (serogroups A, B, C, Y, and W135) are associated with serious infections (12). The molecular bases of *N. meningitidis* colonization and pathogenesis have been a subject of intense research over the past two decades.

Pili are generally regarded as the most important adhesins in capsulate phenotypes of *N. meningitidis* and are thought to determine host and tissue tropism mediating primary interactions with human epithelial and endothelial cells (86). They are required for twitching motility and partake in DNA uptake and thus facilitate natural competence. *N. meningitidis* and the related *N. gonorrhoeae* produce homologous pili termed class I pili. In addition, approximately 50% of *N. meningitidis* strains may produce somewhat structurally variant pili, termed class II pili (Fig. 2). The major class I subunit, PilE (pilin), is generally larger than that of class II (ca. 20 and 15 kDa, respectively) (24). Within each class, further pilin structural and antigenic variations also occur at high frequencies (10^{-2} to 10^{-3} per generation) via RecA-dependent recombinational events between silent (*pilS*) and expression (*pilE*) loci. *N. meningitidis* and *N. gonorrhoeae* appear to use similar mechanisms of adhesion via pili and other adhesins (86).

Both classes of *N. meningitidis* pili and those of *N. gonorrhoeae* are glycosylated and may carry a phosphorylcholine moiety (62, 94, 156, 166). In one class I piliated strain of *N. meningitidis*, several post-translational modifications have been identified and include a trisaccharide moiety (digalactosyl-2,4-diacetamido-2,4,6-trideoxyhexose) and α-glycerophosphate in addition to phosphorylcholine (129, 130, 166). As noted above, phosphorylcholine has the potential to influence bacterial adhesion (see chapter 5). However, the precise roles of these modifications of pili are not clearly defined, although glycosylation does not appear to influence adhesion in dramatic manner (86, 130). In vitro studies have demonstrated that both classes of pili mediate binding primarily to human epithelial cells, and generally no binding to human phagocytic cells (83, 84) or cells of nonhuman origin has been reported (156). However, Knepper et al. reported a role for pili in human monocyte interactions (68), and early reports suggested the presence of binding of *N. gonorrhoeae* pili

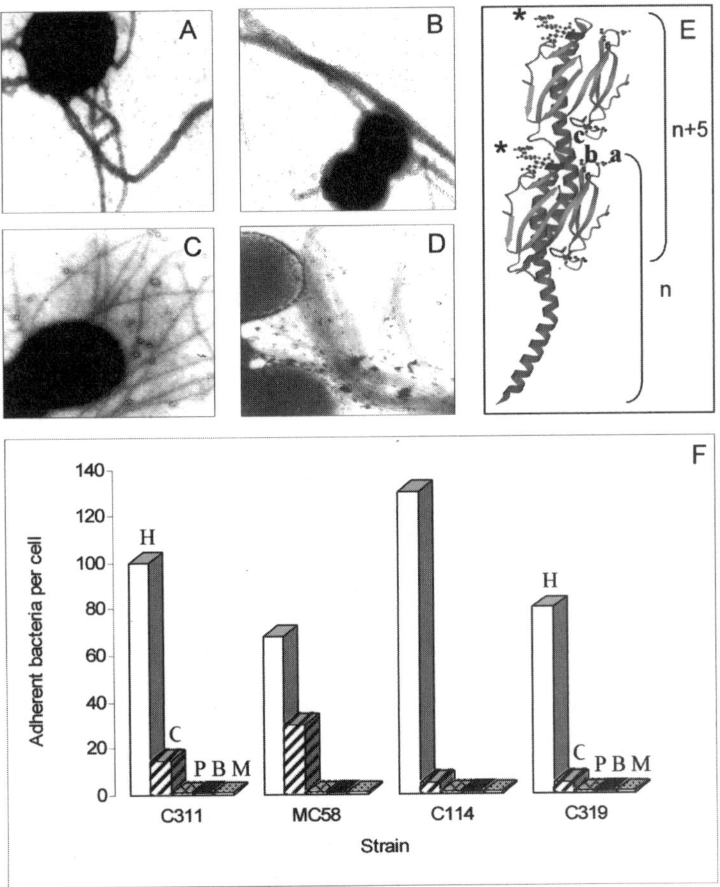

Figure 2. Pilus morphology, structure, and function in *Neisseria* species. (A to D) Negative-stain transmission electron micrographs of a class II piliated *N. meningitidis* isolate (A) and a class I piliated *N. gonorrhoeae* isolate (B), showing long filamentous pili forming rope-like bundles, which occur in both classes of pili. Pili were labeled with antibody against pilin conjugated to gold. In *N. meningitidis* and *N. gonorrhoeae,* pilin structural variants within a single strain may produce morphologically distinct pili. For example, in class I piliated strain C311, variant 3 (C) elaborates individual pili, whereas variant 16 (D) produces laterally aggregated bundles. Pilus aggregation in this case is not related to glycosylation status since removal of glycans does not affect aggregation (130, 156). (E) A three-dimensional molecular model of a pilin of strain C311 was based on that of *N. gonorrhoeae* MS11 pilin, which was determined by X-ray crystallography (94). The model was built with the help of structural databases and minimized using the program X-plor. Positions a, b, and c mark the positions of sequence differences between the pilins of variants 3 and 16. In a fiber model of the variant 16 pilus (constructed using transformations suggested by Forest and Tainer [29], pilin n of one helical turn and n+5 of the next are so juxtapositioned as to bring loops a and b of pilin n very close to loop c of pilin n+5 (only pilins n and n+5 of a variant 16 pilus model are shown). Therefore, the three loops may present a single epitope on the surface of the fiber, which is repeated many times along its length. Single amino acid changes introduced into any one of the sites affect both the lateral aggregation and adherence properties of the pili (L. Serino, A. Hadfield, and M. Virji, unpublished studies). Asterisks show the positions of glycans. (F) Adherence of distinct piliated *N. meningitidis* isolates to human cells and cells of animal origin: human umbilical vein endothelial cells (H), human conjunctiva epithelial cell line (C), human polymorphonuclear phagocytes and monocytes (P), bovine endothelial cells (B), and Madin-Darby canine kidney cells (M). In general, piliated *N. meningitidis* organisms adhere specifically to human endothelial and epithelial cells with greater variation observed in binding of variant pili to human epithelial cell types (83, 84, 149). C311 and MC58 are class I piliated strains; C114 and C319 are class II piliated strains.

to CHO cells (39). Pili were shown to bind to a number of epithelial cells, e.g., Chang conjunctiva and A549 lung carcinoma cells, but less well to HEp-2 laryngeal epithelial cells, although class I and class II pili exhibit some variations in targeting of distinct epithelial cell lines (149). The lack of binding of either class of pili to some respiratory epithelial cell types may be due to down modulation of pilus receptors or coreceptors in immortalized cell lines and may not reflect the true in vivo tropism for the site of their origin. That immortalized cells may not always present the repertoire of receptors that may be expressed in vivo is becoming increasingly recognized as a potential limitation of in vitro model systems. Besides these, *N. meningitidis* pili bind to Hec1-B human endometrial carcinoma epithelial cells and T84 colonic carcinoma cells (86, 88). The related class I pili of *N. gonorrhoeae* were shown to bind to a wide range of human tissues, including the intestines and stomach but not the kidneys. They did not bind to equivalent rat or mouse cells, and binding to green monkey intestine and stomach but not vagina was reported (61).

Pilin versus pilus-associated adhesins

N. meningitidis and *N. gonorrhoeae* express two PilC proteins, PilC1 and PilC2 (ca. 110 kDa), as minor components of the pili. PilC molecules are involved in pilus biogenesis and may also carry a cell-binding domain. In addition to being located in the outer membrane of the bacteria, PilC may be located at the tip of the pilus fibers (103, 112). It appears that PilC1 and PilC2 can function interchangeably in *N. meningitidis* and *N. gonorrhoeae* pilus biogenesis, but in *N. meningitidis* only PilC1 affects epithelial adhesion directly by targeting host receptors (86, 89). Recent studies of *N. gonorrhoeae* have shown that another pilin-like protein, PilV, may also be required for efficient adhesion via pili, perhaps by promoting the functional display of PilC on the fiber, although the mechanism in not known (171).

The numbers of bacteria associated with target cells may also be affected by bacterial aggregation. Primary sequence changes of pilins in naturally occurring variants of *N. meningitidis* or in mutants in which changes have been introduced by site-directed mutagenesis exhibit variations in bundling characteristics of pili which are often but not always associated with increased bacterial clumping (78, 89, 95, 130, 156) (Fig. 2). The bundle-forming pili tend to increase bacterium-host cell association. Nonetheless, other observations show that over and above this, pilin structural changes affect the binding to distinct target cells independently of pilus aggregation. Thus, it is possible that PilC targets the primary receptor and pilins may target coreceptors on target cells, which may vary between cell types, and primary sequence changes may alter the affinity of pili for such coreceptors (61, 78, 86, 119, 156). Neisserial pili also cause hemagglutination, a property dependent on pilin primary structure and independent of PilC (24, 98, 119).

Pilus receptors

A human cell surface molecule that may engage with *N. gonorrhoeae* and *N. meningitidis* pili has been identified as the membrane cofactor protein (CD46), a complement-regulatory protein (58, 65, 86). Interestingly, both pilin and PilC appear to interact with CD46 (113). This molecule is widely expressed in locations other than on erythrocytes, but CD46-expressing cells of many distinct origins do not support the attachment of piliated bacteria. It is possible that, in part, distinct CD46 isoform expression or its glycosylation status may determine tropism (64, 113). Other studies have reported a negative correlation of CD46 expression and pilus-mediated binding to human cells (135). Furthermore, a recent study has also found no interactions of *N. gonorrhoeae* pili with CD46 (32). Nonetheless, several reports suggest that CD46-mediated signaling occurs on piliated bacterial engagement with human target cells (32, 86). Pili may also bind to C4b-binding protein (C4BP), an important inhibitor of the classical complement pathway (9). According to Blom et al., the N-terminal part of PilC apparently engages with C4BP, whereas the C terminus binds CD46 (9). C4BP is also targeted by other respiratory bacterial ligands, including streptococcal M protein and *B. pertussis* FHA (7, 59). Pilus binding to CR3 in *N. gonorrhoeae* has also been reported (27).

Coordinated modulation of surface ligands during host targeting

Recently, studies by Taha and colleagues (133) have shown that pilus-mediated interactions of *N. meningitidis* with human epithelial cells result in upregulation of *crg,* a contact-regulated gene, whose product negatively controls the expression of pili and capsule, thus leading to more intimate interactions of bacteria with target cells via outer membrane adhesins.

Filamentous Adhesins of *H. influenzae*

H. influenzae is classified on the basis of the presence or absence of one of six distinct capsule types (serotypes a to f) (142). The capsulate or typeable *H. influenzae* (THi), especially type b, is responsible for the majority of disseminated outbreaks, although recently other typeable strains have been recognized as causes of disseminated infections (110). Localized mucosal infections such as pneumonia, bronchitis, sinusitis, and otitis media are caused by noncapsulate strains of *H. influenzae* (NTHi), which are a distinct group of bacteria with a close evolutionary relationship to THi (128, 142).

Pili

Pili are found both on THi and a subset of NTHi and are encoded by the *hif* gene cluster. The shaft of the *H. influenzae* pilus is composed of a helix of repeating subunits of HifA (ca. 25 kDa), which exhibits interstrain variation. The HifA subunits of THi and NTHi are ca. 70% identical. THi and NTHi from sputum or from the nasopharynx are piliated, whereas blood isolates tend to be nonpiliated (16, 34, 122, 142). *H. influenzae* pilus expression is subject to phase variation at a frequency of ca. 10^{-4}. This is achieved by transcriptional regulation with

Table 1. Adhesins of *H. influenzae*

Protein and properties	Expression	Functional characteristics	Adherence to cell types or receptor	Reference(s)
Pili (related to *E. coli* type 1/Pap pili; similar to *B. pertussis* fimbriae)	THi, NTHi	Lectin-like	Buccal cells, erythrocytes, mucus, ECM, AnWj, gangliosides	33, 69, 141, 158
Hsf[a] (240 kDa) (fibrillar monomeric structures)	THi	Binding pattern distinct from pili	Chang conjunctival, KB oropharyngeal, HEp-2 laryngeal	124, 127
Hia[a] (115 kDa) (homologous to Hsf)	NTHi in 25% of HMW⁻ strains	Binding similar to Hsf	As Hsf	124, 126
HMW1[a] (125 kDa) (loosely associated with OM; glycosylated protein; phase variable in degrees)	NTHi in 75% of Hia⁻ strains	Lectin-like; adhesion to various cell types	Chang conjunctival, KB oropharyngeal, Hep-2 laryngeal, HSPG, α2-3 sialylated glycoproteins on Chang cells	21, 37, 90, 125
HMW2[a] (120 kDa) (loosely associated with OM; 71% identity to HMW1)	NTHi	Adhesion complementary to HMW1		21, 37, 54
Hap[a] (160 kDa) (Hap_s [110 kDa] + Hap_β [45 kDa] serine protease)	THi, NTHi	Adhesion to cells via Hap_s	Chang conjunctival, A549 lung carcinoma, Fn, Lm, Col-IV[b]	28, 124, 126
OapA (c. 80 kDa) (expression correlates negatively with colony opacity; conserved)	THi, NTHi	Adhesion	Chang conjunctival, rat nasopharyngeal colonization	101, 165
LPD (lipoprotein D) (42 kDa) (conserved)	THi, NTHi	Adhesion, toxicity for ciliated cells	Monocytes	4, 56
P2 (39–42 kDa)	THi, NTHi	Lectin-like	Mucin sialic acids	8, 106
P5 (37–39 kDa) (heat modifiable)	THi, NTHi	Lectin-like, protein-protein interactions	Mucin sialic acids, CEACAM1-Fc	8, 47, 106
ChoP on LPS (LOS)	THi, NTHi	PAF mimicry via ChoP	PAFR on target cells	132

[a]Autotransporters (44, 124).
[b]Lm, laminin; Col-IV, collagen type IV.

the aid of AT repeats in the clustered promoters of HifA and HifB to HifE (142). HifD is a fibrillar structure located at the tip of the fiber together with HifE; the latter contains the erythrocyte- and epithelial cell-binding domain (34, 80). As with *N. meningitidis,* capsulate *H. influenzae* strains adhere more efficiently to a number of mucosal epithelial cells via pili (Table 1) (124).

Pilus receptors and tropism

Studies by van Alphen and colleagues (141, 142) have shown that *H. influenzae* pili show a low-affinity lactosyl-ceramide-binding property and that this specificity allows binding to similar structures on Anton (AnWj)-positive erythrocytes and buccal epithelial cells. The likely carriers of these structures are the epiglycan proteoglycan CD44 molecules (33). *H. influenzae* pili show tropism for some epithelial cells of the respiratory tract such as buccal cells and bronchial epithelial cells but not human nasal epithelial cells or tracheal fibroblasts (34). In studies of nasopharyngeal organ cultures, piliated *H. influenzae* strains bound mainly to nonciliated cells and preferentially to damaged epithelial cells (33, 104, 127). In another study, using human tissue sections, pilus-mediated binding to ciliated cells of the rhinopharynx and nasopharynx, adenoids, and bronchi, as well as pseudostratified columnar cells of the epiglottis but not to squamous or cubic cells of pharynx, was observed—an observation that was correlated with the ability of *H. influenzae* to colonize or infect these particular tissues (123). The host specificity of *H. influenzae* pilus binding has been studied, and no binding to a range of animal cells tested was observed, including buccal or nasal epithelial cells from rats, Buffalo green monkey cells, mink epithelial cells, and Madin-Darby canine kidney cells (33, 124). However, pili apparently help the colonization of rat and rhesus monkey nasopharynx (5, 163). In addition to cellular receptors, *H. influenzae* pili facilitate adherence to human respiratory mucins. The major subunit, HifA, appears to determine the specificity of the strain for airway mucin (6, 69). Piliated and nonpiliated *H. influenzae* isolates also bind to the mammalian ECM (158, 159). In summary, it appears that *H. influenzae* pili initiate the colonization of the human oropharynx by adhesion to oropharyngeal epithelial cells via lectin-like interactions and may also bind to mucin. They are of particular importance in capsulate bacteria. They also appear to target damaged tissue, perhaps via ECM proteins.

Fibrils

Fibrils (Hsf) are expressed primarily by THi and are relatively short structures (50 nm long) that do not traverse the capsule and require its down modulation to function (127). Based on studies of a monkey model of colonization and from in vitro studies of human oropharyngeal cells, it has been proposed that after the early phase of colonization, pili become dispensable and fibrils participate in cellular interactions. The capsule expression of *H. influenzae* can be modulated in response to the growth phase or environmental signals, suggesting that coordinate regulation of pilus and capsule expression occurs after initial contact with oropharyngeal cells (127), as has been proposed for *N. meningitidis* (133). Pili and fibrils possess distinct cell binding specificities and hence must recognize distinct receptors (Table 1).

M. catarrhalis Pericellular Proteins

M. catarrhalis, an acapsulate organism, is often isolated together with NTHi from patients with localized infections. It is the third most common cause of otitis media in children and causes persistent lower respiratory tract infections in adults with lung diseases (147).

Pili

Type 4 pili of *M. catarrhalis* may be present in a subset of strains. It was proposed that the gangliosides on human pharyngeal cells may serve as the receptor for *M. catarrhalis* pili (2, 79). Pili appeared to mediate the attachment of *M. catarrhalis* strains to cilia and basal parts of primary human bronchial cells, although not to oropharyngeal cells (109).

UspA

Most *M. catarrhalis* isolates express high-molecular-mass outer membrane proteins (HMW-OMP) or ubiquitous surface proteins (UspA). Two related UspA proteins, UspA1 and UspA2 (ca. 88 and c. 62 kDa, respectively), have been identified. They show sequence homology to *H. influenzae* Hsf and *Yersinia* YadA. YadA, UspA1, and UspA2 form similar extended "lollipop-like" peritrichously located structures (50, 81).

UspA1 appears to be an adhesin for human epithelial cells such as Chang conjunctiva and HEp-2 laryngeal carcinoma cell lines and binds to Fn. UspA2 is associated with resistance to human serum bactericidal activity and binds to Vn, a regulator of the complement system (1, 70, 81). A second type of hybrid UspA protein, UspA2H, has been found in some strains and contains both UspA1 and UspA2 sequences. UspA1 and UspA2H have identity in their N-terminal regions and are both able to bind to Chang epithelial cells (70). These proteins may be subject to phase variation via slipped-strand mispairing, and high- and low-UspA1-expressing variants have been isolated (71). Our investigations show that *M. catarrhalis* UspA1 targets CEACAMs and that anti-CEACAM antibodies inhibit the binding of *M. catarrhalis* strain MX1 to the A549 lung carcinoma epithelial cell line, which is known to express CEACAMs, but not to Chang or HEp-2 cells, which express very low levels of the receptor (48) (Fig. 3) (also see "Some common receptor-targeting mechanisms" below).

Hag and MID

M. catarrhalis hemagglutinin (Hag) (M_r, ca. 200) has C-terminal homology to *H. influenzae* Hia (Table 1) and forms fibrillar structures visible by EM (97). It binds to some human IgDs and mediates hemagglutination as well as autoagglutination. This protein may be similar to the *M. catarrhalis* IgD-binding protein, MID (30). It is suggested that IgD, which is often present in mucus secretions, may aid bacterial attachment (97). UspA1 and MID of an *M. catarrhalis* isolate were shown to collectively target A549 (30).

B. pertussis Filamentous Adhesins

B. pertussis is the causative agent of pertussis (whooping cough), an exclusively human and highly contagious infection. Effective vaccines have been used in developed countries for over 50 years, with successful reduction of the *B. pertussis* disease burden; nonetheless, it remains an endemic disease in adolescents and adults, who are thus reservoirs for the organism (36). *B. pertussis* remains localized to the respiratory tract by binding to ciliated cells of the mucosa. It causes local tissue damage, increased mucus secretion, and inflammatory cell infiltration; it enters macrophages and inhibits macrophage diapedesis. It expresses several virulence factors under the control of the *bvg* locus, including several adhesins that mediate binding to human and nonhuman cells (67, 114, 169).

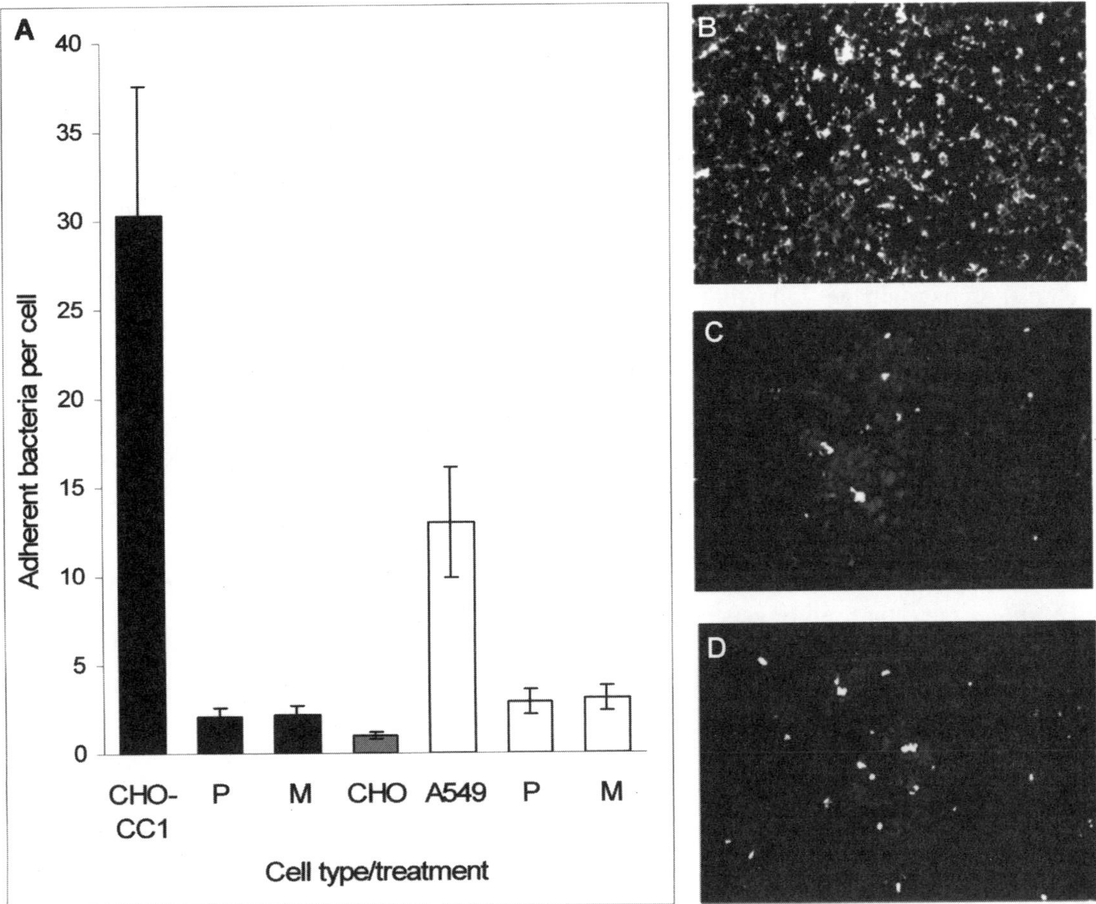

Figure 3. Adherence of *M. catarrhalis* MX1 to CEACAMs. (A) Relative adherence of bacteria (determined by a viable-count assay) to Chinese hamster ovary cell transfectants expressing CEACAM1 (black columns) without or with anti-CEACAM-specific polyclonal (P) or monoclonal (M) blocking antibodies. The CHO column shows adherence to sham-transfected CHO cells. Blank columns show quantification of the binding of MX1 to A549 cells and inhibition in the presence of the antibodies, demonstrating that CEACAMs are the primary targets of the adhesins of this isolate on A549 cells. (B to D) Immunofluorescence analysis showing the tropism of *M. catarrhalis* MX1 for human epithelial cell lines. (B) A549; (C) HEp-2; (D) Chang. Adherent bacteria were labeled using rhodamine-conjugated antibacterial antibodies. Based on published information, the lack of adherence to Chang and HEp-2 cells suggests that MX1 does not express adhesins previously implicated in binding to these cell lines (Table 2) (48).

Fimbriae

The two serotypes, 2 and 3, of *B. pertussis* major fimbrial subunits (22.5 and 22 kDa, respectively) are encoded by the *fim2* and *fim3* genes; *fimD* encodes the minor (40-kDa) adhesin, which is also required for fimbrial biogenesis (67, 170). Expression of the *fim2* and *fim3* genes is regulated by a poly(C) region in their promoter, which results in switching between high and low levels of expression of a particular *fim* gene (169). *B. pertussis* fimbriae can bind to sulfated glycosaminoglycans on the surface of epithelial cells through the major subunits (31). FimD also binds to heparin and to the very late antigen 5 $\alpha_5\beta_1$-integrin of macrophages as well as epithelial cells (67). Notably, *B. pertussis* fimbriae appear to mediate attachment to laryngeal epithelial cells but not bronchial cell lines (143).

FHA

FHA is a 220-kDa protein with multiple adhesion functions and is the dominant adhesin for bronchial epithelial cells (114, 143). It is produced as a 370-kDa precursor, FhaB, which undergoes N- and C-terminal processing, resulting in the mature FHA, which can be both surface expressed and secreted (17, 74). From a mouse model of colonization, FHA release from bacteria may be necessary to facilitate the dispersal of bacteria for colonization to new sites (17). Three distinct domains of the mature FHA are responsible for the differential recognition of receptors on host cells, a feature similar to that of eukaryotic Fn. One domain binds to sulfated glycosaminoglycans on ECM and epithelial cells, while the other sites are responsible for binding to lactosylceramides on macrophages and epithelial cells, as well as an

RGD sequence that binds to integrins on macrophages and ciliated epithelial cells (67, 85, 114). *B. pertussis* may also induce an inflammatory response via the action of FHA. A recent study has suggested that ligation of FHA RGD with $\alpha_5\beta_1$-integrin on bronchial epithelial cells leads to activation of NF-κB and upregulation of intercellular cell adhesion molecule (ICAM1), a receptor critical in recruitment of inflammatory cells (55). FHA can then bind to CR3 ($\alpha_m\beta_2$-integrin) of macrophages following its upregulation by pertactin (see "Nonfilamentous adhesins of respiratory bacteria" below) and can enter the phagocytic cells, an entry pathway that does not trigger a respiratory burst (67, 114).

Requirement for Pericellular Appendages

The obvious advantage of the long filamentous adhesins in capsulate bacteria such as *N. meningitidis* and *H. influenzae* is that pili traverse the capsule and enable capsulate bacteria to colonize host cells, since the adhesion efficacy of outer membrane proteins is reduced by both steric hindrance and the negative charge of the capsule. In addition, in acapsulate bacteria, pili mediate long-range interactions and thus are less strongly affected by electrostatic repulsion due to negative charges on bacteria and target cells. Indeed, *M. catarrhalis* pili were shown by EM to bind to raised, positively charged areas on epithelial cells (3); the binding may involve specific molecules, although this is not known. Pili thus allow the bacteria to make an initial contact with a host cell surface, leading to more intimate interactions via nonfilamentous adhesins. It is therefore also credible that the primary adhesins such as pili in general determine host and tissue tropism.

Fibrils, however, are too short to protrude from bacterial surfaces in fully capsulate bacteria; in capsulate *H. influenzae,* the roles of fibrils (Hsf) could not be observed unless the bacteria had reached the late stationary phase, when capsule production is reduced. In such circumstances, with a diminished capsule, fibrils may protrude and become functional. Indeed, all distinct capsulate lineages of *H. influenzae* contain the larger Hsf homolog (110), whereas NTHi strains contain Hia, which has functional similarity to Hsf but is smaller (see "Nonfilamentous adhesins of respiratory bacteria" below). It has been proposed that since NTHi strains do not express any level of capsulation, the smaller Hia protein can function effectively (125).

Surface Sugars: Capsule and LPS and Their Roles

Some gram-positive capsules are directly involved in cellular interactions; e.g., the hyaluronic acid capsule of group A streptococci interacts with CD44 on target tissues (19). In contrast, capsular polysaccharides of *N. meningitidis* and *H. influenzae* are generally anti-interactive, as already described above. Lipopolysaccharide (LPS, as well as lipooligosaccharide), however, appears to be both adhesive and antiadhesive. Sialylation of LPS interferes with *N. meningitidis* Opa- and Opc-mediated interactions (23, 83, 155). However, specific interactions of *N. meningitidis* asialylated LPS to asialoglycoprotein receptor and of sialylated LPS to sialic acid-binding Ig-like lectins (Siglecs) have been described previously (42, 60). *M. catarrhalis* LPS has also been implicated in adhesion (53); in addition, *H. influenzae* LPS mediates binding to PAFR via its ChoP moiety (132) (see chapter 5).

NONFILAMENTOUS ADHESINS OF RESPIRATORY BACTERIA

A wide variety of nonfilamentous structures embedded in or associated with bacterial outer membranes, ranging in molecular mass from approximately 25 to 200 kDa, have been implicated in host cell targeting. Many relatively smaller molecules are predicted to be β-barrel proteins that span the bacterial outer membranes several times with surface-exposed loops. This class of proteins includes the *N. meningitidis* opacity proteins Opa and Opc and the *H. influenzae* proteins P2 and P5 (77, 102, 162) Others are autotransporters that may remain cell associated as complete molecules (e.g., *H. influenzae* Hia) or are partly lost by autolytic cleavage (e.g., *H. influenzae* Hap and *B. pertussis* pertactin).

Opacity Proteins of *N. meningitidis*

The expression of opacity proteins Opa and Opc in *N. meningitidis* gives rise to opaque colonies only discernible in those of acapsulate phenotypes. The Opa proteins (initially termed PII or class 5 proteins [49]) are expressed by both pathogenic *Neisseria* spp., *N. meningitidis* and *N. gonorrhoeae,* as well as by some commensal *Neisseria* strains (43, 136). Opc expression is unique to *N. meningitidis.* Opa and Opc are of similar size (24 to 35 kDa) and share some physicochemical properties but are structurally distinct. In *N. meningitidis* there may be up to four complete *opa* genes encoding distinct proteins with homology in the transmembrane and intracellular regions but with diverse sequences in three of the four surface exposed loops. Opc is antigenically relatively invariant and is encoded by a single gene (92). In the absence of capsule, Opa or Opc expression confers adhesion and, under some situations, invasion property to these phenotypes (22, 140, 152, 155). However, under some conditions, interactions of capsulate meningococci via Opa and Opc do occur (140, 157). Opa and Opc proteins undergo phase variation via distinct mechanisms. Opa expression is under translational regulation, and one or more Opa proteins may be in frame at any one time. Opc is transcriptionally controlled, and the efficiency of its transcription may vary, giving rise to high, low, or no expression (43, 92, 115, 155); Opc-mediated cellular interactions can be demonstrated only when it is expressed at high levels. Additionally, there is considerable interplay between adhesins of *N. meningitidis;* for example, pili augment cellular invasion mediated by the Opc protein (155).

Opc targets multiple ECM proteins and, via these, endothelial integrins (140, 154). However, Opc may bind directly to epithelial cells via HSPGs (22). The availability of integrins and HSPGs on the luminal surfaces of target epithelial cells in the respiratory tract is unclear; however, as described above, during mucosal damage and repair these receptors could become available to bacteria in effective numbers. The Opa proteins facilitate the interaction of gonococci and meningococci with epithelial, endothelial, and phagocytic cells (43, 83, 84, 108, 131, 152). The antigenic and structural diversity of the surface-exposed loops of distinct Opa proteins of *N. gonorrhoeae* and *N. meningitidis* isolates affects their ability to interact with distinct receptors and may direct bacterial tropism. Opa proteins target some extracellular matrix proteins, HSPGs, and the CEACAMs (86, 145, 157) (see "Some common receptor-targeting mechanisms" below).

Other *N. meningitidis* adhesins

More recently, other outer membrane adhesins of *N. meningitidis* have been identified; they include the *N. meningitidis* adhesion and penetration protein (App), which belongs to the autotransporter family of adhesins (120).

Adhesins of *H. influenzae*

Numerous outer membrane adhesins have been described in THi and NTHi and are summarized in Table 1. The high-molecular-weight proteins, HMW-1 and HMW-2, are frequently present in nontypeable strains; approximately 75% of NTHi strains express these proteins. Recently, their expression has been demonstrated in a small number (ca. 10%) of THi (non-type b) strains, although these strains appear to be closely related to NTHi (110). HMW1 and HMW2 are autotransporters and are loosely associated with the outer membrane (37). HMW1 must be glycosylated for effective tethering to the bacterial surface (37). NTHi isolates that lack these genes (ca. 25% of NTHi strains) usually possess another adhesin gene, *hia* (126). Hia, with significant homology to the Hsf of THi, shares with it the specificity of tissue targeting. HMW1 also binds to a similar range of epithelial cells. Hap is another member of the autotransporter family with sequence similarity to *H. influenzae* IgA protease. Under certain conditions, it undergoes autoproteolytic cleavage and release of the adhesive 110-kDa passenger domain Hap_s. Inhibition of proteolysis increases its interactions with target cells. It has been suggested that the protease inhibitors released from respiratory cells may thus increase Hap adhesive activity (126). Hap also mediates binding to Fn, laminin, and collagen type IV via the 45-kDa gelatin-binding domain (28). Other adhesins of *H. influenzae* that are ubiquitously expressed include OapA, lipoprotein D (LPD), P2, and P5 (99). OapA can mediate adhesion to human conjunctival epithelial cells in vitro and can mediate rat nasopharyngeal colonization (101, 165). LPD has glycerophosphodiester phosphodiesterase activity, is ciliotoxic in human nasopharyngeal organ cultures, and promotes adhesion and uptake into human monocytic cell lines (4, 56). The outer membrane proteins, P2 and P5, are thought to mediate adherence to respiratory mucin via sialic acid residues (106). A potential role of P5 in mediating interactions with CEACAMs has been demonstrated by mutation of the gene in THi and NTHi strains (47) (see "Some common receptor-targeting mechanisms" below).

Outer Membrane Adhesins of *M. catarrhalis*

Outer membrane protein CD

The major outer membrane proteins of *M. catarrhalis,* ranging in molecular mass from 21 to 98 kDa (designated A to H), have been described previously (147). Outer membrane proteins C and D are identical proteins and constitute the heat-modifiable protein CD (ca. 45 kDa). Based on its homology to the OprF protein of *P. aeruginosa,* CD may be a porin, but it is also a mucin-binding protein (106, 107). Recently, CD has been shown to target the lung epithelial cell line, A549 (51).

McaP

A recent study has reported that *M. catarrhalis* adherence protein (McaP), with structural similarities to autotransporter proteins, may be involved in adherence to a number of epithelial cells, including A549

Table 2. Adhesins of *M. catarrhalis*

Protein and properties	Expression	Other functions[b]	Adhesion to	Reference(s)
Pili (type 4)	Some isolates		Human pharyngeal cell gangliosides, human bronchial cells	2, 79, 109
UspA1[a] (ca. 88 kDa) (multimeric in membrane)	Universal phase variable		Chang, HEp-2, Fn, A549, CEACAMs on A549	30, 48, 50, 70, 82
UspA2[a] (ca. 62 kDa) (multimeric in membrane)	Universal	SR	Vn	1, 50, 70, 82
UspA2H[a] (ca. 75 kDa) (hybrid of UspA1 and UspA2)	ca. 20% of isolates	SR	Chang	70
Hag/MID (200 kDa)	In majority, phase variable		A549, primary human middle ear cells, erythrocytes, Some human IgD, B-cell receptor surface IgD	30, 52, 97
OMPCD (ca. 45/60 kDa [heat modifiable] (conserved; similarity to porins)	Virtually universal	SR ?	Mucin, A549	8, 51, 107
McaP[a] (possibly surface exposed, autotransporter-like sequence, phospholipase B activity)			Chang, A549, 16HBE14o−	134
LOS		Adhesion?, SR	Antibody inhibits binding to Chang	53, 172

[a]Belong to the autotransporter family (44).
[b]?, property not fully defined; SR, serum resistance.

(134). Thus, at least four *M. catarrhalis* proteins have been implicated in binding to A549 cells; these include UspA1, Hag, outer membrane protein CD, and Mcap (Table 2). Our studies indicate that CEACAMs are targets for UspA1 (48). Other molecular targets on A549 for various ligands remain to be shown.

B. pertussis Adhesins

Pertactin

Prn (P69) is a product of the *prn* gene, which encodes a 93-kDa precursor. Prn is the cleaved 60-kDa N-terminal external domain (molecular mass of 69 kDa on sodium dodecyl sulfate-polyacrylamide gel electrophoresis), with the C-terminal 30-kDa domain remaining cell associated and determining the correct localization of Prn. Prn has two RGD sequences in the mature protein. Studies with CHO and HeLa cells suggest that RGD motifs of Prn are essential for cell adhesion via Prn (72). However, other studies have found no evidence for the contribution of Prn to adherence to either bronchial or laryngeal cells (143).

Pertussis toxin

B. pertussis, unlike other bacteria discussed here, is unique in producing an exotoxin. Pertussis toxin (PT) is a 105-kDa hexameric exotoxin with a typical AB_5 architecture. In addition to being released into the extracellular environment, it may remain cell bound. Its subunit A (subunit S1), has ADP-ribosyltransferase activity and is delivered to the target cell by the B pentamer (subunits S2 to S5), which is the adhesin. It interacts with glycoproteins and glycolipids on many eukaryotic cells, including erythrocytes, leukocytes, alveolar macrophages, and ciliated epithelial cells (67, 114). The differential targeting of ciliated epithelial cells and macrophages resides in the distinct subunits of the pentamer, with S2 and S3 exhibiting tropism for cilia and macrophages (67, 114). Since PT may be cell bound as well as released, the former may increase bacterial adhesion to target cells. However, in the presence of soluble PT, the cell adhesion may be reduced by competitive inhibition (143). PT exhibits the selectin-like property of upregulation of CR3 on macrophages (67, 114, 116).

SOME COMMON RECEPTOR-TARGETING MECHANISMS

Adhesion to Mucus

The upper respiratory mucus blanket is a 2- to 5-μm-thick viscous gel that physically protects the underlying epithelium. It contains mucin glycoproteins (apomucin peptide core linked to multiple glycan chains of 1 to 20 residues), together with a variety of products of the epithelial cells and local immune factors, e.g., secretory IgA and, depending on the inflammation state, various plasma factors (75, 122, 167, 168). Under conditions when mucociliary clearance is impaired, e.g., chronic obstructive pulmonary disease and cystic fibrosis, the occurrence of a thickened mucus blanket and impaired ciliary action in the bronchial tree encourage bacteria that can target mucus-associated molecules. A number of respiratory bacteria, including *H. influenzae and M. catarrhalis,* as well as the opportunistic *P. aeruginosa,* attach to mucus (105, 106). In addition, these bacteria secrete factors that are ciliostatic or toxic to ciliated cells, thus exacerbating the mucociliary stasis. Under such conditions, bacteria are able to multiply and escape clearance. It does not seem coincidental that bacteria that cause persistent ear and lung infections have the ability to bind to mucus. Bacterial binding to many soluble factors in the respiratory secretions may occur, as mentioned in the introduction to this chapter; however, several bacteria target specific mucins.

H. influenzae has particular affinity for mucus, and several factors are involved in mucus targeting, including both pilus (HifA) and nonpilus adhesins (P2 and P5) (6, 13, 69, 106). Binding of the latter to purified nasopharyngeal mucin glycoprotein appears to involve interaction with the sialic acid-containing oligosaccharides of the mucin (106). Bernstein and Reddy (8) reported specific and strong targeting of P2 and P5 of *H. influenzae* to purified human mucins from the NP and middle ear and weak binding to tracheobronchial mucin but not to mucins MG1 and MG2 of saliva. A similar targeting of the *M. catarrhalis* protein CD was observed. Furthermore, *S. pneumoniae* bound only to nasopharyngeal and middle ear mucin, whereas *P. aeruginosa* bound to all mucins examined via pili or outer membrane proteins (8). For *H. influenzae,* it has been suggested that in situations where mucus might be slow-moving or static, pili could confer a colonization advantage and that in healthy individuals such areas occur in the respiratory tract over adenoids and tonsils, where pools of mucus may form. It has also been suggested that piliated bacteria may be confined to mucus and may be cleared, whereas nonpiliated bacteria, which show greater affinity for epithelial cells, may be selected for (6).

Binding to ECM Proteins and Proteoglycans

Microbial surface components recognizing adhesive matrix molecules (MSCRAMMs) of a large number of gram-positive and gram-negative bacteria have been investigated (96). Binding to ECM proteins may also bring into play proteoglycans (PGs) since many ECM proteins bind to heparin or HSPG. Bacteria may also target ECM and PGs simultaneously.

PGs are an important class of adhesion receptors that decorate a wide range of mammalian cells. They are present both on surfaces of cells and in the ECM. Expression of distinct proteoglycans and apical versus basal location is tissue specific and temporally regulated during development (85, 111, 160). The transmembrane syndecans are decorated with three to five heterogeneous heparan sulfate or chondroitin sulfate glycosaminoglycan (GAG) chains. The patterns of sugars in the GAGs define the binding site for a ligand. Heparan sulfate and heparin are related structures; heparin, produced by mast cells, may contain subpopulations that bind specifically to Fn, type 1 collagen, laminin, or growth factors (111). The primary targeting of GAGs by the GAG-binding proteins is determined by defined clustered sequences of basic amino acids (111). Several bacterial ligands that target PGs also contain binding domains similar to the natural GAG-binding proteins. For bacterial ligands also, the specificity is directed by the GAG heterogeneity, which thus determines bacterial tissue tropism. Interactions of bacteria with PGs may be of low affinity, and coligation of secondary receptors appears to be required for uptake of microbes, as reported for *Neisseria* spp.

Opc and certain Opa proteins of *Neisseria* spp. bind to heparin and HSPGs (14, 22, 145, 151). The mechanisms of Opa-HSPG interactions have been studied in greater detail in *N. gonorrhoeae* and it has been reported that the overexpression of Syndecan-4 in HeLa cells increases Vn-triggered uptake of the Opa-expressing bacteria, whereas only basal levels of invasion occur in the absence of Vn. This secondary recruitment of Vn and Fn by Opa-expressing *N. gonorrhoeae* triggers efficient uptake in an integrin-dependent manner. In addition, *N. gonorrhoeae* porins may facilitate cellular invasion following Opa-HSPG interactions (25, 43, 86).

The HMW proteins of NTHi have homology to *B. pertussis* FHA, and both interact with cellular PGs (85, 90). In addition, *B. pertussis* can bind to sulfated GAGs through the major and the minor fimbrial subunits. The extent and pattern of sulfation determine the affinity of *H. influenzae* and *B. pertussis* interactions with PGs. The molecular mimicry of the *B. pertussis* major subunit Fim2 is noteworthy. It contains two regions, H1 and H2, that have sequence similarity to the heparin-binding regions of Fn (31, 111). *H. influenzae* pili and Hap protein bind to multiple heparin-binding ECM proteins (28, 158).

Binding to CEACAMs

The CEACAM family comprises the cell surface-expressed structurally related receptors, whose functions may be divergent (Fig. 4). CEACAM1 is a transmembrane signaling molecule, has the broadest tissue distribution, and is expressed on the apical surfaces of epithelial cells of human mucosa, cells of the myeloid lineage, and some endothelial cells (41, 91). In considering respiratory pathogens, it is important to note that CEACAM expression on normal epithelial cells in oral, tonsillar, and lung tissues has been reported (10, 100, 139). We have demonstrated the expression of the receptor on the apical surfaces of tonsillar epithelium and on buccal cells (148). Targeting of the CEACAM1 subgroup provides the organism with a wide tissue range for colonization and a means of host cell manipulation.

CEACAMs, especially CEA, have long been associated with carcinomas, especially those of the gastric epithelial cells, and have been used as markers of tumor load (41, 91). In addition, CEACAMs have been reported to bind via their mannosyl residues to mannose-sensitive pili of intestinal mucosal pathogens (73, 117, 118). However, within the respiratory bacteria, the first ligands to be reported to bind to CEACAMs were those of *N. meningitidis* (157). The Opa proteins of *N. meningitidis* as well as *N. gonorrhoeae* were shown to bind to several members of the family via exposed protein regions of their highly homologous N-terminal domains (11, 15, 38, 151, 153, 157). Since some commensal *Neisseria* strains were shown to express Opa-like proteins, their ability to target CEACAMs was investigated. Indeed, a small number of commensal *Neisseria* strains also targeted CEACAMs (136). More recently, CEACAM-binding ligands of *H. influenzae* and *M. catarrhalis* have been identified that are not structurally related to Opa proteins (47, 48). In *H. influenzae,* mutation of the outer membrane protein, P5, abrogated binding to the soluble receptor construct containing the receptor ectodomains linked to human Fc. Interestingly, these mutants continued to bind to the cell-expressed receptor, suggesting the presence of a second CEACAM-binding ligand on *H. influenzae* strains with somewhat different requirements for targeting CEACAMs (47). Studies of *M. catarrhalis* showed specific binding of all *M. catarrhalis* strains but not of the other *Moraxella* spp. examined. The ligand responsible in this case is the UspA1 protein (48).

Whereas *N. meningitidis* and *N. gonorrhoeae* strains have the capacity to target multiple members of the CEACAM family (11, 15, 38, 157), our recent studies have shown that THi, NTHi, and *M. catarrhalis* strains tend to target CEACAM1 as their principal receptor (unpublished data). All these bacteria bind to overlapping regions of the N domain of CEACAMs, and monoclonal antibodies to this region of the N-domain abrogate bacterial interactions. Such unique targeting may also suggest that bacteria have the potential to compete with each other in targeting the receptor in their common niche. Indeed, this can be demonstrated in vitro by using CEACAM1-transfected cells (150) (Fig. 4). Bacterial interactions with epithelial cells expressing CEACAMs also results in transmigration across monolayers (Color Plate 3).

CEACAMs as determinants of tropism

Since CEACAMs are expressed in various human tissues including respiratory epithelia, the targeting of CEACAMs by multiple respiratory pathogens implies the importance of these interactions in colonization and survival of the bacteria in this niche. CEACAM targeting by mucosal pathogens appears to be specific for human CEACAMs and is focused primarily on CEACAM1, a molecule widely expressed on epithelial cells. The diversity of adhesins involved in the targeting of CEACAMs suggests that the property has been acquired by convergent evolution. This, together with the fact that CEACAMs may be upregulated in a variety of situations (below), also suggests the clinical importance of the receptor and the need for greater understanding of the structural basis of the interaction of diverse microbial ligands with CEACAMs.

CONCLUSION

Receptor Modulation and Implications

Receptor expression on target cells is a dynamic process controlled by hormones and cytokines. In addition, certain microbes induce their own receptors by supplementary mechanisms, as has been documented for the interplay between *B. pertussis* PT and FHA, where PT, via its selectin-like action, upregulates CR3, which in turn is targeted by FHA (114, 116). In *Neisseria* spp., LPS may be involved in inducing the upregulation of CEACAM expression in some cell types (43, 87). Many receptors may also be upregulated on cells exposed to inflammatory cytokines and other factors, and CEACAMs are upregulated on distinct cell types by gamma interferon and tumor necrosis factor alpha (20, 87; C. Bradley, R. Heyderman, and M. Virji, unpublished observations). Additionally, viral infections may alter receptor profiles on target tissues. For example, respiratory syncytial virus affects some host cells such that

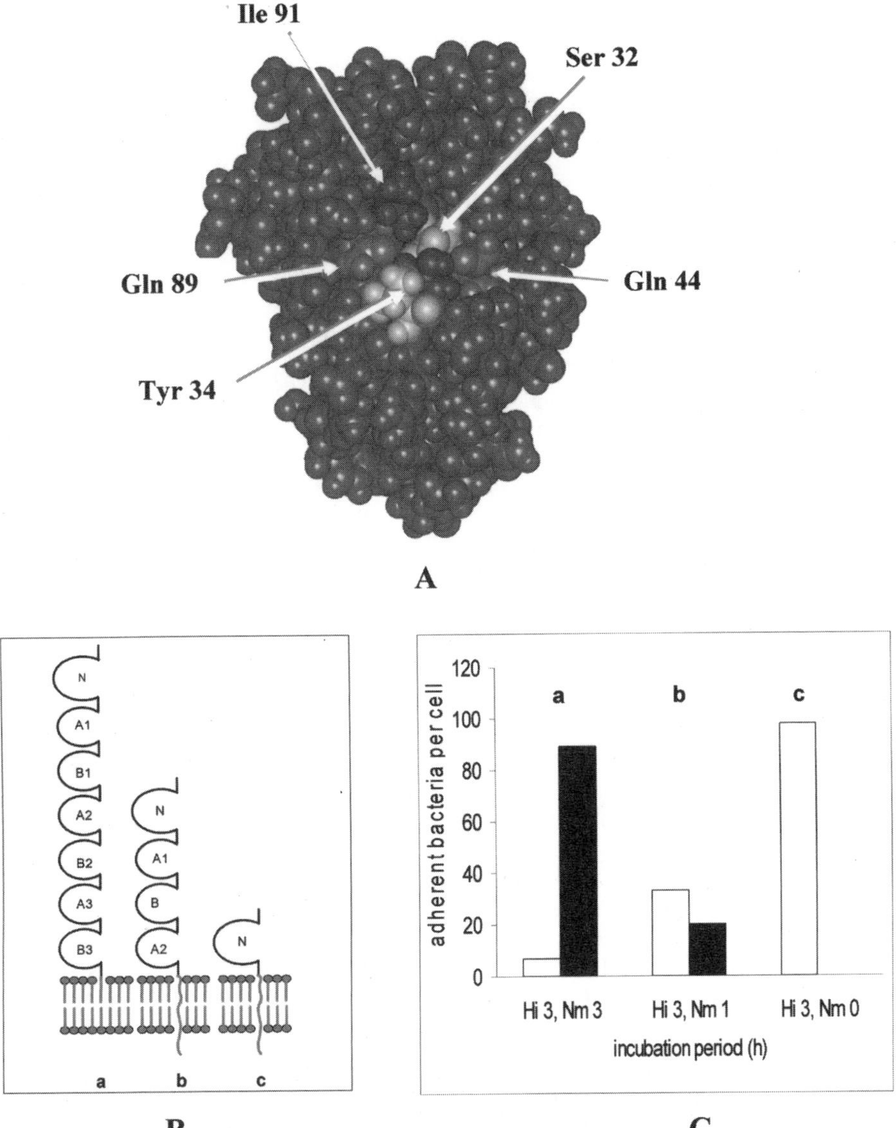

Figure 4. (A) Van der Waals' surface representation of the model (front views) of the N domain of CEACAM1, showing the critical common residues that are required for the interactions of *N. meningitidis* and *H. influenzae* with the receptor. The amino acids exerting the greatest influence on binding of *N. meningitidis* and *H. influenzae* are Ile 91 and Tyr 34. Ile 91 lies in close proximity to Tyr 34, which lies in the center of the illustrated face of the N-terminal domain, which is devoid of carbohydrate. The other amino acids shown, which appear to affect the binding of distinct strains and variants to variable extents, also lie in close proximity. The binding regions are overlapping, such that *N. meningitidis* and *H. influenzae* may compete for binding to the receptor (150, 151). (B) Domain structure of several CEA family members of the Ig superfamily (adapted from the CEA website, http://cea.klinikum.uni-muenchen.de/). The family characteristically contains a single N-terminal 1gV-like domain, and in addition, most members contain several 1gC2-like domains (A1, B1, etc.). CEA (a) is anchored via a glycosylphosphatidylinositol extension, whereas CEACAM1 (b) and CEACAM3 (c) are transmembrane molecules. (C) Competition between *N. meningitidis* and *H. influenzae* strains in targeting CEACAM1 receptors on CHO transfectants. Adhesion of *H. influenzae* (Hi) (blank columns) and *N. meningitidis* (Nm) (black columns) after 3 h of incubation in the absence of *N. meningitidis* (c) and when the *N. meningitidis* strain was added at $t = 0$ (a) or at 2 h (b) after inoculation with *H. influenzae*. Within 1 h, *N. meningitidis* inhibited the binding of *H. influenzae* that had been inoculated 2 h prior to *N. meningitidis* addition (b). In each case, the number of bacteria adhering per cell is shown (150).

they downregulate some adhesion receptors. However, respiratory syncytial virus infection of A549 cells has been shown to increase *H. influenzae* interactions, and thus it upregulates receptors on A549 recognized by *H. influenzae* (57). Parainfluenza virus type 2 also upregulates several receptors on human tracheal epithelial cells (137). Thus, bacterial localization in the respiratory tract may be determined by the state of the host. In addition, studies of *Yersinia pseudotuberculosis* demonstrated that receptor density as well as ligand affinity may play a role in microbial location in or out of the host cell (138). Therefore, receptor density that may alter as a result of prior viral or other infections may be of major importance in determining both host colonization levels and susceptibility to further infection. Indeed, our investigations have shown that at high receptor density the inhibitory action of the *N. meningitidis* capsule can be overcome. Invasion of capsulate bacteria, albeit at low levels, can be demonstrated in vitro by using target cells with high receptor densities, as can be achieved using cytokine-stimulated respiratory epithelial cells or transfected cell lines (reference 157 and our unpublished studies). Epidemiological investigations also suggest that host susceptibility to *N. meningitidis* increases following certain viral infections. Therefore, it is entirely possible that circulating cytokines upregulate microbial receptors and promote invasion rather than colonization of the mucosa. As noted above, if this were to occur with capsulate bacteria that can survive in the blood, the outcome would be serious disseminated infection.

Determinants of Tropism and Limitations of Current Understanding

Most of our current knowledge of the molecular basis of bacterium-host interactions comes from in vitro studies that have used immortalized cell lines or animal models that are not entirely satisfactory for human-tropic bacteria. However, studies of more relevant human polarized epithelial and primary cells have been relatively sparse. While studies of transformed and transfected cells have undoubtedly expanded our base of knowledge of the potential molecular participants at the bacterium-host interface, it is largely unknown whether these contribute to tissue tropism in vivo. The progress in this regard is hampered by the lack of knowledge of whether the potential target molecules are indeed expressed at the site of colonization. More investigations of the receptor profiles of potential sites of colonization in health and disease would be valuable in this regard. Complete knowledge of the identity of bacterial ligands and host receptors central to colonization will undoubtedly also aid in the development of antiadhesive strategies for prevention of diseases caused by both highly contagious and occasionally pathogenic bacteria.

REFERENCES

1. **Aebi, C., E. R. Lafontaine, L. D. Cope, J. L. Latimer, S. L. Lumbley, G. H. McCracken, Jr., and E. J. Hansen.** 1998. Phenotypic effect of isogenic *uspA1* and *uspA2* mutations on *Moraxella catarrhalis* O35E. *Infect. Immun.* **66:**3113–3119.
2. **Ahmed, K., K. Matsumoto, N. Rikitomi, and T. Nagatake.** 1996. Attachment of *Moraxella catarrhalis* to pharyngeal epithelial cells is mediated by a glycosphingolipid receptor. *FEMS Microbiol. Lett.* **135:**305–309.
3. **Ahmed, K., T. Nakagawa, Y. Nakano, G. Martinez, A. Ichinose, C. H. Zheng, M. Akaki, M. Aikawa, and T. Nagatake.** 2000. Attachment of *Moraxella catarrhalis* occurs to the positively charged domains of pharyngeal epithelial cells. *Microb. Pathog.* **28:**203–209.
4. **Ahren, I. L., H. Janson, A. Forsgren, and K. Riesbeck.** 2001. Protein D expression promotes the adherence and internalization of non-typeable *Haemophilus influenzae* into human monocytic cells. *Microb. Pathog.* **31:**151–158.
5. **Anderson, P. W., M. E. Pichichero, and E. M. Connor.** 1985. Enhanced nasopharyngeal colonization of rats by piliated *Haemophilus influenzae* type b. *Infect. Immun.* **48:** 565–568.
6. **Barsum, W., R. Wilson, R. C. Read, A. Rutman, H. C. Todd, N. Houdret, P. Roussel, and P. J. Cole.** 1995. Interaction of fimbriated and nonfimbriated strains of unencapsulated *Haemophilus influenzae* with Human respiratory-tract mucus in vitro. *Eur. Respir. J.* **8:**709–714.
7. **Berggard, K., E. Johnsson, F. R. Mooi, and G. Lindahl.** 1997. *Bordetella pertussis* binds the human complement regulator C4BP: role of filamentous hemagglutinin. *Infect. Immun.* **65:** 3638–3643.
8. **Bernstein, J. M., and M. Reddy.** 2000. Bacteria-mucin interaction in the upper aerodigestive tract shows striking heterogeneity: implications in otitis media, rhinosinusitis, and pneumonia. *Otolaryngol. Head Neck Surg.* **122:**514–520.
9. **Blom, A. M., A. Rytkonen, P. Vasquez, G. Lindahl, B. Dahlback, and A. B. Jonsson.** 2001. A novel interaction between type IV pili of *Neisseria gonorrhoeae* and the human complement regulator C4B-binding protein. *J. Immunol.* **166:**6764–6770.
10. **Bordessoule, D., M. Jones, K. C. Gatter, and D. Y. Mason.** 1993. Immunohistological patterns of myeloid antigens: tissue distribution of CD13, CD14, CD16, CD31, CD36, CD65, CD66 and CD67. *Br. J. Haematol.* **83:**370–383.
11. **Bos, M. P., F. Grunert, and R. J. Belland.** 1997. Differential recognition of members of the carcinoembryonic antigen family by Opa variants of *Neisseria gonorrhoeae. Infect. Immun.* **65:**2353–2361.
12. **Cartwright, K.** 1995. Meningococcal carriage and disease, p. 115–146. *In* K. Cartwright (ed.), *Meningococcal Disease.* John Wiley & Sons, Chichester, United Kingdom.
13. **Chance, D. L., T. J. Reilly, and A. L. Smith.** 1999. Acid phosphatase activity as a measure of *Haemophilus influenzae* adherence to mucin. *J. Microbiol. Methods* **39:**49–58.
14. **Chen, T., R. J. Belland, J. Wilson, and J. Swanson.** 1995. Adherence of pilus− Opa+ gonococci to epithelial cells in vitro involves heparan sulfate. *J. Exp. Med.* **182:**511–517.
15. **Chen, T., and E. C. Gotschlich.** 1996. CGM1a antigen of neutrophils, a receptor of gonococcal opacity proteins. *Proc. Natl. Acad. Sci. USA* **93:**14851–14856.
16. **Clemans, D. L., C. F. Marrs, M. Patel, M. Duncan, and J. R. Gilsdorf.** 1998. Comparative analysis of *Haemophilus influenzae hifA* (pilin) genes. *Infect. Immun.* **66:**656–663.
17. **Coutte, L., S. Alonso, N. Reveneau, E. Willery, B. Quatannens, C. Locht, and F. Jacob-Dubisson.** 2003. Role of adhesin release for mucosal colonization by a bacterial pathogen. *J. Exp. Med.* **197:**735–742.
18. **Cue, D., P. E. Dombek, H. Lam, and P. P. Cleary.** 1998. *Streptococcus pyogenes* serotype M1 encodes multiple pathways

for entry into human epithelial cells. *Infect. Immun.* **66:** 4593–4601.

19. **Cywes, C., and M. R. Wessels.** 2001. Group A *Streptococcus* tissue invasion by CD44-mediated cell signalling. *Nature* **414:**648–652.
20. **Dansky-Ullmann, C., M. Salgaller, S. Adams, J. Schlom, and J. W. Greiner.** 1995. Synergistic effects of IL-6 and IFN-gamma on carcinoembryonic antigen (CEA) and HLA expression by human colorectal carcinoma cells: role for endogenous IFN-beta. *Cytokine* **7:**118–129.
21. **Dawid, S., S. J. Barenkamp, and J. W. St. Geme.** 3rd. 1999. Variation in expression of the *Haemophilus influenzae* HMW adhesins: a prokaryotic system reminiscent of eukaryotes. *Proc. Natl. Acad. Sci. USA* **96:**1077–1082.
22. **de Vries, F. P., R. Cole, J. Dankert, M. Frosch, and J. P. van Putten.** 1998. *Neisseria meningitidis* producing the Opc adhesin binds epithelial cell proteoglycan receptors. *Mol. Microbiol.* **27:**1203–1212.
23. **de Vries, F. P., A. van der Ende, J. P. van Putten, and J. Dankert.** 1996. Invasion of primary nasopharyngeal epithelial cells by *Neisseria meningitidis* is controlled by phase variation of multiple surface antigens. *Infect. Immun.* **64:** 2998–3006.
24. **Diaz, J.-L., M. Virji, and J. E. Heckels.** 1984. Structural and antigenic differences between two types of meningococcal pili. *FEMS Microbiol. Lett.* **21:**181–184.
25. **Duensing, T. D., and J. P. M. van Putten.** 1998. Vitronectin binds to the gonococcal adhesin OpaA through a glycosaminoglycan molecular bridge. *Biochem. J.* **334:**133–139.
26. **Eberhard, T., G. Kronvall, and M. Ullberg.** 1999. Surface bound plasmin promotes migration of *Streptococcus pneumoniae* through reconstituted basement membranes. *Microb. Pathog.* **26:**175–181.
27. **Edwards, J. L., E. J. Brown, S. Uk-Nham, J. G. Cannon, M. S. Blake, and M. A. Apicella.** 2002. A co-operative interaction between *Neisseria gonorrhoeae* and complement receptor 3 mediates infection of primary cervical epithelial cells. *Cell. Microbiol.* **4:**571–584.
28. **Fink, D. L., B. A. Green, and J. W. St Geme III.** 2002. The *Haemophilus influenzae* Hap autotransporter binds to fibronectin, laminin, and collagen IV. *Infect. Immun.* **70:**4902–4907.
29. **Forest, K. T., and J. A. Tainer.** 1997. Type-4 pilus-structure: outside to inside and top to bottom—a minireview. *Gene* **192:**165–169.
30. **Forsgren, A., M. Brant, M. Karamehmedovic, and K. Riesbeck.** 2003. The immunoglobulin D-binding protein MID from *Moraxella catarrhalis* is also an adhesin. *Infect. Immun.* **71:**3302–3309.
31. **Geuijen, C. A. W., R. J. L. Willems, P. Hoogerhout, W. C. Puijk, R. H. Meloen, and F. R. Mooi.** 1998. Identification and characterization of heparin binding regions of the Fim2 subunit of *Bordetella pertussis. Infect. Immun.* **66:**2256–2263.
32. **Gill, D. B., M. Koomey, K. G. Cannon, and J. P. Atkinson.** 2003. Down-regulation of CD46 by piliated *Neisseria gonorrhoeae. J. Exp. Med.* **198:**1313–1322.
33. **Gilsdorf, J. R.** 2002. Role of pili in *Haemophilus influenzae* adherence, colonisation and diseases, p. 139–161. *In* M. Wilson (ed.), *Bacterial Adhesion to Host Tissues: Mechanisms and Consequences.* Cambridge University Press, Cambridge, United Kingdom.
34. **Gilsdorf, J. R., K. W. McCrea, and C. F. Marrs.** 1997. Role of pili in *Haemophilus influenzae* adherence and colonization. *Infect. Immun.* **65:**2997–3002.
35. **Gorter, A. D., P. S. Hiemstra, S. de Bentzmann, S. van Wetering, J. Dankert, and L. van Alphen.** 2000. Stimulation of bacterial adherence by neutrophil defensins varies among bacterial species but not among host cell types. *FEMS Immunol. Med. Microbiol.* **28:**105–111.
36. **Grant, C. C., and J. D. Cherry.** 2002. Keeping pace with the elusive *Bordetella pertussis. J. Infect.* **44:**7–12.
37. **Grass, S., A. Z. Buscher, W. E. Swords, M. A. Apicella, S. J. Barenkamp, N. Ozchlewski, and J. W. St Geme III.** 2003. The *Haemophilus influenzae* HMW1 adhesin is glycosylated in a process that requires HMW1C and phosphoglucomutase, an enzyme involved in lipooligosaccharide biosynthesis. *Mol. Microbiol.* **48:**737–751.
38. **Gray-Owen, S. D., D. R. Lorenzen, A. Haude, T. F. Meyer, and C. Dehio.** 1997. Differential Opa specificities for CD66 receptors influence tissue interactions and cellular response to *Neisseria gonorrhoeae. Mol. Microbiol.* **26:**971–980.
39. **Gubish, E. R. J., K. C. Chen, and T. M. Buchanan.** 1982. Attachment of gonococcal pili to lectin-resistant clones of Chinese hamster ovary cells. *Infect. Immun.* **37:**189–194.
40. **Hahn, H. P.** 1997. The type-4 pilus is the major virulence-associated adhesin of *Pseudomonas aeruginosa:* a review. *Gene* **192:**99–108.
41. **Hammarstrom, S.** 1999. The carcinoembryonic antigen (CEA) family: structures, suggested functions and expression in normal and malignant tissues. *Semin. Cancer Biol.* **9:**67–81.
42. **Harvey, H. A., M. P. Jennings, C. A. Campbell, R. Williams, and M. A. Apicella.** 2001. Receptor-mediated endocytosis of *Neisseria gonorrhoeae* into primary human urethral epithelial cells: the role of the asialoglycoprotein receptor. *Mol. Microbiol.* **42:**659–672.
43. **Hauck, C. R., and T. F. Meyer.** 2003. 'Small' talk: Opa proteins as mediators of *Neisseria*-host-cell communication. *Curr. Opin. Microbiol.* **6:**43–49.
44. **Henderson, I. R., and J. P. Nataro.** 2001. Virulence functions of autotransporter proteins. *Infect. Immun.* **69:**1231–1243.
45. **Henderson, I. R., P. Owen, and J. P. Nataro.** 1999. Molecular switches—the ON and OFF of bacterial phase variation. *Mol. Microbiol.* **33:**919–932.
46. **Hers, J. F. P., and J. Mulder.** 1953. The mucosal epithelium of the respiratory tract in muco-purulent bronchitis caused by *Haemophilus influenzae. J. Pathol. Bacteriol.* **66:**103–108.
47. **Hill, D. J., M. A. Toleman, D. J. Evans, S. Villullas, L. Van Alphen, and M. Virji.** 2001. The variable P5 proteins of typeable and non-typeable *Haemophilus influenzae* target human CEACAM1. *Mol. Microbiol.* **39:**850–862.
48. **Hill, D. J., and M. Virji.** 2003. A novel cell-binding mechanism of *Moraxella catarrhalis* ubiquitous surface protein UspA: specific targeting of the N-domain of carcinoembryonic antigen-related cell adhesion molecules by UspA1. *Mol. Microbiol.* **48:**117–129.
49. **Hitchcock, P. J.** 1989. Unified nomenclature for pathogenic *Neisseria* species. *Clin. Microbiol. Rev.* **2:**S64–S65.
50. **Hoiczyk, E., A. Roggenkamp, M. Reichenbecher, A. Lupas, and J. Heesemann.** 2000. Structure and sequence analysis of *Yersinia* YadA and *Moraxella* UspAs reveal a novel class of adhesins. *EMBO J.* **19:**5989–5999.
51. **Holm, M. M., S. L. Vanlerberg, I. M. Foley, D. D. Sledjeski, and E. R. Lafontaine.** 2004. The *Moraxella catarrhalis* porin-like outer membrane protein CD is an adhesin for human lung cells. *Infect. Immun.* **72:**1906–1913.
52. **Holm, M. M., S. L. Vanlerberg, D. D. Sledjeski, and E. R. Lafontaine.** 2003. The hag protein of *Moraxella catarrhalis* strain O35E is associated with adherence to human lung and middle ear cells. *Infect. Immun.* **71:**4977–4984.
53. **Hu, W. G., J. Chen, J. C. McMichael, and X. X. Gu.** 2001. Functional characteristics of a protective monoclonal antibody against serotype A and C lipooligosaccharides from *Moraxella catarrhalis. Infect. Immun.* **69:**1358–1363.

54. **Hultgren, S. J., S. Abraham, M. Caparon, P. Falk, J. W. St. Geme III, and S. Normark.** 1993. Pilus and nonpilus bacterial adhesins: assembly and function in cell recognition. *Cell* 73:887–901.

55. **Ishibashi, Y., and A. Nishikawa.** 2003. Role of nuclear factor-kappa B in the regulation of intercellular adhesion molecule 1 after infection of human bronchial epithelial cells by *Bordetella pertussis*. *Microb. Pathog.* **35:**169–177.

56. **Janson, H., B. Carlen, A. Cervin, A. Forsgren, A. B. Magnusdottir, S. Lindberg, and T. Runer.** 1999. Effects on the ciliated epithelium of protein D-producing and -nonproducing nontypeable *Haemophilus influenzae* in nasopharyngeal tissue cultures. *J. Infect. Dis.* **180:**737–746.

57. **Jiang, Z., N. Nagata, E. Molina, L. O. Bakaletz, H. Hawkins, and J. A. Patel.** 1999. Fimbria-mediated enhanced attachment of nontypeable *Haemophilus influenzae* to respiratory syncytial virus-infected respiratory epithelial cells. *Infect. Immun.* **67:**187–192.

58. **Johansson, L., A. Rytkonen, P. Bergman, B. Albiger, H. Kallstrom, T. Hokfelt, B. Agerberth, R. Cattaneo, and A. B. Jonsson.** 2003. CD46 in meningococcal disease. *Science* **301:** 373–375.

59. **Johnsson, E., A. Thern, B. Dahlback, L. O. Heden, M. Wikstrom, and G. Lindahl.** 1996. A highly variable region in members of the streptococcal M protein family binds the human complement regulator C4BP. *J. Immunol.* **157:**3021–3029.

60. **Jones, C., M. Virji, and P. R. Crocker.** 2003. Recognition of sialylated meningococcal lipopolysaccharide by siglecs expressed on myeloid cells leads to enhanced bacterial uptake. *Mol. Microbiol.* **49:**1213–1225.

61. **Jonsson, A. B., D. Ilver, P. Falk, J. Pepose, and S. Normark.** 1994. Sequence changes in the pilus subunit lead to tropism variation of *Neisseria gonorrhoeae* to human tissue. *Mol. Microbiol.* **13:**403–416.

62. **Kahler, C. M., L. E. Martin, Y. L. Tzeng, Y. K. Miller, K. Sharkey, D. S. Stephens, and J. K. Davies.** 2001. Polymorphisms in pilin glycosylation locus of *Neisseria meningitidis* expressing class II pili. *Infect. Immun.* **69:**3597–3604.

63. **Kajava, A. V., N. Cheng, R. Cleaver, M. Kessel, M. N. Simon, E. Willery, F. Jacob-Dubuisson, C. Locht, and A. C. Steven.** 2001. Beta-helix model for the filamentous haemagglutinin adhesin of *Bordetella pertussis* and related bacterial secretory proteins. *Mol. Microbiol.* **42:**279–292.

64. **Kallstrom, H., D. Blackmer Gill, B. Albiger, M. K. Liszewski, J. P. Atkinson, and A. B. Jonsson.** 2001. Attachment of *Neisseria gonorrhoeae* to the cellular pilus receptor CD46: identification of domains important for bacterial adherence. *Cell. Microbiol.* **3:**133–143.

65. **Kallstrom, H., M. K. Liszewski, J. P. Atkinson, and A. B. Jonsson.** 1997. Membrane cofactor protein (MCP or CD46) is a cellular pilus receptor for pathogenic *Neisseria*. *Mol. Microbiol.* **25:**639–647.

66. **Kazmierczak, B. I., K. Mostov, and J. N. Engel.** 2001. Interaction of bacterial pathogens with polarized epithelium. *Annu. Rev. Microbiol.* **55:**407–435.

67. **Kerr, J. R., and R. C. Matthews.** 2000. *Bordetella pertussis* infection: pathogenesis, diagnosis, management, and the role of protective immunity. *Eur. J. Clin. Microbiol. Infect. Dis.* **19:**77–88.

68. **Knepper, B., I. Heuer, T. F. Meyer, and J. P. van Putten.** 1997. Differential response of human monocytes to *Neisseria gonorrhoeae* variants expressing pili and opacity proteins. *Infect. Immun.* **65:**4122–4129.

69. **Kubiet, M., R. Ramphal, A. Weber, and A. Smith.** 2000. Pilus-mediated adherence of *Haemophilus influenzae* to human respiratory mucins. *Infect. Immun.* **68:**3362–3367.

70. **Lafontaine, E. R., L. D. Cope, C. Aebi, J. L. Latimer, G. H. McCracken, Jr., and E. J. Hansen.** 2000. The UspA1 protein and a second type of UspA2 protein mediate adherence of *Moraxella catarrhalis* to human epithelial cells in vitro. *J. Bacteriol.* **182:**1364–1373.

71. **Lafontaine, E. R., N. J. Wagner, and E. J. Hansen.** 2001. Expression of the *Moraxella catarrhalis* UspA1 protein undergoes phase variation and is regulated at the transcriptional level. *J. Bacteriol.* **183:**1540–1551.

72. **Leininger, E., C. A. Ewanowich, A. Bhargava, M. S. Peppler, J. G. Kenimer, and M. J. Brennan.** 1992. Comparative roles of the Arg-Gly-Asp sequence present in the *Bordetella pertussis* adhesins pertactin and filamentous hemagglutinin. *Infect. Immun.* **60:**2380–2385.

73. **Leusch, H. G., Z. Drzeniek, Z. Markos-Pusztai, and C. Wagener.** 1991. Binding of *Escherichia coli* and *Salmonella* strains to members of the carcinoembryonic antigen family: differential binding inhibition by aromatic alpha-glycosides of mannose. *Infect. Immun.* **59:**2051–2057.

74. **Locht, C., P. Bertin, F. D. Menozzi, and G. Renauld.** 1993. The filamentous haemagglutinin, a multifaceted adhesion produced by virulent *Bordetella* spp. *Mol. Microbiol.* **9:**653–660.

75. **Lopez-Vidriero, M. T.** 1981. Airway mucus; production and composition. *Chest* **80:**799–804.

76. **Makhov, A. M., J. H. Hannah, M. J. Brennan, B. L. Trus, E. Kocsis, J. F. Conway, P. T. Wingfield, M. N. Simon, and A. C. Steven.** 1994. Filamentous hemagglutinin of *Bordetella pertussis*—a bacterial adhesin formed as a 50-nm monomeric rigid-rod based on a 19-residue repeat motif rich in beta-strands and beta-turns. *J. Mol. Biol.* **241:**110–124.

77. **Malorny, B., G. Morelli, B. Kusecek, J. Kolberg, and M. Achtman.** 1998. Sequence diversity, predicted two-dimensional protein structure, and epitope mapping of neisserial Opa proteins. *J. Bacteriol.* **180:**1323–1330.

78. **Marceau, M., J. L. Beretti, and X. Nassif.** 1995. High adhesiveness of encapsulated *Neisseria meningitidis* to epithelial cells is associated with the formation of bundles of pili. *Mol. Microbiol.* **17:**855–863.

79. **Marrs, C. F., and S. Weir.** 1990. Pili (fimbriae) of *Branhamella* species. *Am. J. Med.* **88:**36S–40S.

80. **McCrea, K. W., W. J. Watson, J. R. Gilsdorf, and C. F. Marrs.** 1997. Identification of two minor subunits in the pilus of *Haemophilus influenzae*. *J. Bacteriol.* **179:**4227–4231.

81. **McMichael, J. C.** 2000. Progress toward the development of a vaccine to prevent *Moraxella* (*Branhamella*) *catarrhalis* infections. *Microb. Infect.* **2:**561–568.

82. **McMichael, J. C., M. J. Fiske, R. A. Fredenburg, D. N. Chakravarti, K. R. van der Meid, V. Barniak, J. Caplan, E. Bortell, S. Baker, R. Arumugham, and D. Chen.** 1998. Isolation and characterization of two proteins from *Moraxella catarrhalis* that bear a common epitope. *Infect. Immun.* **66:**4374–4381.

83. **McNeil, G., and M. Virji.** 1997. Phenotypic variants of meningococci and their potential in phagocytic interactions: the influence of opacity proteins, pili, PilC and surface sialic acids. *Microb. Pathog.* **22:**295–304.

84. **McNeil, G., M. Virji, and E. R. Moxon.** 1994. Interactions of *Neisseria meningitidis* with human monocytes. *Microb. Pathog.* **16:**153–163.

85. **Menozzi, F. D., K. Pethe, P. Bifani, F. Soncin, M. J. Brennan, and C. Locht.** 2002. Enhanced bacterial virulence through exploitation of host glycosaminoglycans. *Mol. Microbiol.* **43:**1379–1386.

86. **Merz, A. J., and M. So.** 2000. Interactions of pathogenic neisseriae with epithelial cell membranes. *Annu. Rev. Cell Dev. Biol.* **16:**423–457.

87. **Muenzner, P., M. Naumann, T. F. Meyer, and S. D. Gray-Owen.** 2001. Pathogenic *Neisseria* trigger expression of their carcinoembryonic antigen-related cellular adhesion molecule 1 (CEACAM1; previously CD66a) receptor on primary endothelial cells by activating the immediate early response transcription factor, nuclear factor-kappaB. *J. Biol. Chem.* **276:**24331–24340.
88. **Nassif, X., J. Lowy, P. Stenberg, P. O'Gaora, A. Ganji, and M. So.** 1993. Antigenic variation of pilin regulates adhesion of *Neisseria meningitidis* to human epithelial cells. *Mol. Microbiol.* **8:**719–725.
89. **Nassif, X., M. Marceau, C. Pujol, B. Pron, J. L. Beretti, and M. K. Taha.** 1997. Type-4 pili and meningococcal adhesiveness. *Gene* **192:**149–153.
90. **Noel, G. J., D. C. Love, and D. M. Mosser.** 1994. High-molecular-weight proteins of nontypeable *Haemophilus influenzae* mediate bacterial adhesion to cellular proteoglycans. *Infect. Immun.* **62:**4028–4033.
91. **Obrink, B.** 1997. CEA adhesion molecules: multifunctional proteins with signal-regulatory properties. *Curr. Opin. Cell Biol.* **9:**616–626.
92. **Olyhoek, A. J., J. Sarkari, M. Bopp, G. Morelli, and M. Achtman.** 1991. Cloning and expression in *Escherichia coli* of *opc,* the gene for an unusual class 5 outer membrane protein from *Neisseria meningitidis* (meningococci/surface antigen). *Microb. Pathog.* **11:**249–257.
93. **Pancholi, V., P. Fontan, and H. Jin.** 2003. Plasminogen-mediated group A streptococcal adherence to and pericellular invasion of human pharyngeal cells. *Microb. Pathog.* **35:**293–303.
94. **Parge, H. E., K. T. Forest, M. J. Hickey, D. A. Christensen, E. D. Getzoff, and J. A. Tainer.** 1995. Structure of the fibre-forming protein pilin at 2.6 A resolution. *Nature* **378:**32–38.
95. **Park, H. S. M., M. Wolfgang, J. P. M. van Putten, D. Dorward, S. F. Hayes, and M. Koomey.** 2001. Structural alterations in a type IV pilus subunit protein result in concurrent defects in multicellular behaviour and adherence to host tissue. *Mol. Microbiol.* **42:**293–307.
96. **Patti, J. M., B. L. Allen, M. J. McGavin, and M. Hook.** 1994. MSCRAMM-mediated adherence of microorganisms to host tissues. *Annu. Rev. Microbiol.* **48:**585–617.
97. **Pearson, M. M., E. R. Lafontaine, N. J. Wagner, J. W. St. Geme III, and E. J. Hansen.** 2002. A *hag* mutant of *Moraxella catarrhalis* strain O35E is deficient in hemagglutination, autoagglutination, and immunoglobulin D-binding activities. *Infect. Immun.* **70:**4523–4533.
98. **Pinner, R. W., P. A. Spellman, and D. S. Stephens.** 1991. Evidence for functionally distinct pili expressed by *Neisseria meningitidis. Infect. Immun.* **59:**3169–3175.
99. **Poolman, J. T., L. Bakaletz, A. Cripps, P. A. Denoel, A. Forsgren, J. Kyd, and Y. Lobet.** 2000. Developing a nontypeable *Haemophilus influenzae* (NTHi) vaccine. *Vaccine* **19:**S108–S115.
100. **Prall, F., P. Nollau, M. Neumaier, H. D. Haubeck, Z. Drzeniek, U. Helmchen, T. Loning, and C. Wagener.** 1996. CD66a (BGP), an adhesion molecule of the carcinoembryonic antigen family, is expressed in epithelium, endothelium, and myeloid cells in a wide range of normal human tissues. *J. Histochem. Cytochem.* **44:**35–41.
101. **Prasadarao, N. V., E. Lysenko, C. A. Wass, K. S. Kim, and J. N. Weiser.** 1999. Opacity-associated protein A contributes to the binding of *Haemophilus influenzae* to chang epithelial cells. *Infect. Immun.* **67:**4153–4160.
102. **Prince, S. M., M. Achtman, and J. P. Derrick.** 2002. Crystal structure of the OpcA integral membrane adhesin from *Neisseria meningitidis. Proc. Natl. Acad. Sci. USA* **99:**3417–3421.
103. **Rahman, M., H. Kallstrom, S. Normark, and A. B. Jonsson.** 1997. PilC of pathogenic *Neisseria* is associated with the bacterial cell surface. *Mol. Microbiol.* **25:**11–25.
104. **Read, R. C., R. Wilson, A. Rutman, V. Lund, H. C. Todd, A. P. R. Brain, P. K. Jeffery, and P. J. Cole.** 1991. Interaction of nontypable *Haemophilus influenzae* with human respiratory mucosa in vitro. *J. Infect. Dis.* **163:**549–558.
105. **Reddy, M. S.** 1992. Human tracheobronchial mucin—purification and binding to *Pseudomonas aeruginosa. Infect. Immun.* **60:**1530–1535.
106. **Reddy, M. S., J. M. Bernstein, T. F. Murphy, and H. S. Faden.** 1996. Binding between outer membrane proteins of nontypeable *Haemophilus influenzae* and human nasopharyngeal mucin. *Infect. Immun.* **64:**1477–1479.
107. **Reddy, M. S., T. F. Murphy, H. S. Faden, and J. M. Bernstein.** 1997. Middle ear mucin glycoprotein: purification and interaction with nontypable *Haemophilus influenzae* and *Moraxella catarrhalis. Otolaryngol. Head Neck Surg.* **116:**175–180.
108. **Rest, R. F., and W. M. Shafer.** 1989. Interactions of *Neisseria gonorrhoeae* with human neutrophils. *Clin. Microbiol. Rev.* **2**(Suppl.):S83–S91.
109. **Rikitomi, N., K. Ahmed, and T. Nagatake.** 1997. *Moraxella (Branhamella) catarrhalis* adherence to human bronchial and oropharyngeal cells: the role of adherence in lower respiratory tract infections. *Microbiol. Immunol.* **41:**487–494.
110. **Rodriguez, C. A., V. Avadhanula, A. Buscher, A. L. Smith, J. W. St Geme III, and E. E. Adderson.** 2003. Prevalence and distribution of adhesins in invasive non-type b encapsulated *Haemophilus influenzae. Infect. Immun.* **71:**1635–1642.
111. **Rostand, K. S., and J. D. Esko.** 1997. Microbial adherence to and invasion through proteoglycans. *Infect. Immun.* **65:**1–8.
112. **Rudel, T., I. Scheurerpflug, and T. F. Meyer.** 1995. *Neisseria* PilC protein identified as type-4 pilus tip-located adhesin. *Nature* **373:**357–359.
113. **Rytkonen, A., L. Johansson, V. Asp, B. Albiger, and A. B. Jonsson.** 2001. Soluble pilin of *Neisseria gonorrhoeae* interacts with human target cells and tissue. *Infect. Immun.* **69:**6419–6426.
114. **Sandros, J., and E. Tuomanen.** 1993. Attachment factors of *Bordetella pertussis:* mimicry of eukaryotic cell recognition molecules. *Trends Microbiol.* **1:**192–196.
115. **Sarkari, J., N. Pandit, E. R. Moxon, and M. Achtman.** 1994. Variable expression of the Opc outer membrane protein in *Neisseria meningitidis* is caused by size variation of a promoter containing poly-cytidine. *Mol. Microbiol.* **13:**207–217.
116. **Saukkonen, K., W. N. Burnette, V. L. Mar, H. R. Masure, and E. I. Tuomanen.** 1992. Pertussis toxin has eukaryotic-like carbohydrate recognition domains. *Proc. Natl. Acad. Sci. USA* **89:**118–122.
117. **Sauter, S. L., S. M. Rutherfurd, C. Wagener, J. E. Shively, and S. A. Hefta.** 1991. Binding of nonspecific cross-reacting antigen, a granulocyte membrane glycoprotein, to *Escherichia coli* expressing type 1 fimbriae. *Infect. Immun.* **59:**2485–2493.
118. **Sauter, S. L., S. M. Rutherfurd, C. Wagener, J. E. Shively, and S. A. Hefta.** 1993. Identification of the specific oligosaccharide sites recognized by type 1 fimbriae from *Escherichia coli* on nonspecific cross-reacting antigen, a CD66 cluster granulocyte glycoprotein. *J. Biol. Chem.* **268:**15510–15516.
119. **Scheuerpflug, I., T. Rudel, R. Ryll, J. Pandit, and T. F. Meyer.** 1999. Roles of PilC and PilE proteins in pilus-mediated adherence of *Neisseria gonorrhoeae* and *Neisseria meningitidis* to human erythrocytes and endothelial and epithelial cells. *Infect. Immun.* **67:**834–843.

120. **Serruto, D., J. Adu-Bobie, M. Scarselli, D. Veggi, M. Pizza, R. Rappuoli, and B. Arico.** 2003. *Neisseria meningitidis* App, a new adhesin with autocatalytic serine protease activity. *Mol. Microbiol.* **48:**323–334.
121. **Sirakova, T., P. E. Kolattukudy, D. Murwin, J. Billy, E. Leake, D. Lim, T. DeMaria, and L. Bakaletz.** 1994. Role of fimbriae expressed by nontypeable *Haemophilus influenzae* in pathogenesis of and protection against otitis media and relatedness of the fimbrin subunit to outer membrane protein A. *Infect. Immun.* **62:**2002–2020.
122. **Stephens, D. S., and M. M. Farley.** 1991. Pathogenic events during infection of the human nasopharynx with *Neisseria meningitidis* and *Haemophilus influenzae*. *Rev. Infect. Dis.* **13:**22–33.
123. **Sterk, L. M. T., L. van Alphen, L. Geelen-van den Broek, H. J. Houthoff, and J. Dankert.** 1991. Differential binding of *Haemophilus influenzae* to human tissues by fimbriae. *J. Med. Microbiol.* **35:**129–138.
124. **St. Geme, J. W., III.** 2002. Molecular and cellular determinants of non-typeable *Haemophilus influenzae* adherence and invasion. *Cell. Microbiol.* **4:**191–200.
125. **St. Geme, J. W., III.** 1996. Molecular determinants of the interaction between *Haemophilus influenzae* and human cells. *Am. J. Respir. Crit. Care Med.* **154:**S192–S196.
126. **St. Geme, J. W., III.** 2000. The pathogenesis of nontypable *Haemophilus influenzae* otitis media. *Vaccine* **19:**S41–S50.
127. **St. Geme, J. W., III, and D. Cutter.** 1996. Influence of pili, fibrils, and capsule on in vitro adherence by *Haemophilus influenzae* type b. *Mol. Microbiol.* **21:**21–31.
128. **St. Geme, J. W., III, A. Takala, E. Esko, and S. Falkow.** 1994. Evidence for capsule gene sequences among pharyngeal isolates of nontypeable *Haemophilus influenzae*. *J. Infect. Dis.* **169:**337–342.
129. **Stimson, E., M. Virji, S. Barker, M. Panico, I. Blench, J. Saunders, G. Payne, E. R. Moxon, A. Dell, and H. R. Morris.** 1996. Discovery of a novel protein modification: alpha-glycerophosphate is a substituent of meningococcal pilin. *Biochem. J.* **316:**29–33.
130. **Stimson, E., M. Virji, K. Makepeace, A. Dell, H. R. Morris, G. Payne, J. R. Saunders, M. P. Jennings, S. Barker, and M. Panico.** 1995. Meningococcal pilin: a glycoprotein substituted with digalactosyl-2,4- diacetamido-2,4,6-trideoxyhexose. *Mol. Microbiol.* **17:**1201–1214.
131. **Swanson, J., E. Sparks, D. Young, and G. King.** 1975. Studies on gonococcus infection. X. Pili and leukocyte association factor as mediators of interactions between gonococci and eukaryotic cells in vitro. *Infect. Immun.* **11:**1352–1361.
132. **Swords, W. E., B. A. Buscher, K. V. S. Li, A. Preston, W. A. Nichols, J. N. Weiser, B. W. Gibson, and M. A. Apicella.** 2000. Non-typeable *Haemophilus influenzae* adhere to and invade human bronchial epithelial cells via an interaction of lipooligosaccharide with the PAF receptor. *Mol. Microbiol.* **37:**13–27.
133. **Taha, M. K.** 2002. Transcription regulation of meningococcal gene expression upon adhesion to target cells, p. 165–182. *In* M. Wilson (ed.), *Bacterial Adhesion to Host Tissues: Mechanisms and Consequences.* Cambridge University Press, Cambridge, United Kingdom.
134. **Timpe, J. M., M. M. Holm, S. L. Vanlerberg, V. Basrur, and E. R. Lafontaine.** 2003. Identification of a *Moraxella catarrhalis* outer membrane protein exhibiting both adhesin and lipolytic activities. *Infect. Immun.* **71:**4341–4350.
135. **Tobiason, D. M., and H. S. Seifert.** 2001. Inverse relationship between pilus-mediated gonococcal adherence and surface expression of the pilus receptor, CD46. *Microbiology* **147:** 2333–2340.
136. **Toleman, M., E. Aho, and M. Virji.** 2001. Expression of pathogen-like Opa adhesins in commensal *Neisseria:* genetic and functional analysis. *Cell. Microbiol.* **3:**33–44.
137. **Tosi, M. F., J. M. Stark, A. Hamedani, C. W. Smith, D. C. Gruenert, and Y. T. Huang.** 1992. Intercellular-adhesion molecule-1 (ICAM-1)-dependent and ICAM-1-independent adhesive interactions between polymorphonuclear leukocytes and human airway epithelial cells infected with parainfluenza virus type-2. *J. Immunol.* **149:**3345–3349.
138. **Tran Van Nhieu, G., and R. R. Isberg.** 1993. Bacterial internalization mediated by beta 1 chain integrins is determined by ligand affinity and receptor density. *EMBO J.* **12:**1887–1895.
139. **Tsutsumi, Y., N. Onoda, M. Misawa, M. Kuroki, and Y. Matsuoka.** 1990. Immunohistochemical demonstration of nonspecific cross-reacting antigen in normal and neoplastic human tissues using a monoclonal-antibody: comparison with carcinoembryonic antigen localization. *Acta Pathol. Jpn.* **40:**85–97.
140. **Unkmeir, A., K. Latsch, G. Dietrich, E. Wintermeyer, B. Schinke, S. Schwender, K. S. Kim, M. Eigenthaler, and M. Frosch.** 2002. Fibronectin mediates Opc-dependent internalization of *Neisseria meningitidis* in human brain microvascular endothelial cells. *Mol. Microbiol.* **46:**933–946.
141. **van Alphen, L., L. Geelen-van den Broek, L. Blaas, M. van Ham, and J. Dankert.** 1991. Blocking of fimbria-mediated adherence of *Haemophilus influenzae* by sialyl gangliosides. *Infect. Immun.* **59:**4473–4477.
142. **van Alphen, L., and S. M. van Ham.** 1994. Adherence and invasion of *Haemophilus influenzae*. *Rev. Med. Microbiol.* **5:** 245.
143. **van den Berg, B. M., H. Beekhuizen, R. J. L. Willems, F. R. Mooi, and R. van Furth.** 1999. Role of *Bordetella pertussis* virulence factors in adherence to epithelial cell lines derived from the human respiratory tract. *Infect. Immun.* **67:**1056–1062.
144. **van der Flier, A., and A. Sonnenberg.** 2001. Function and interactions of integrins. *Cell Tissue Res.* **305:**285–298.
145. **van Putten, J. P., and S. M. Paul.** 1995. Binding of syndecan-like cell surface proteoglycan receptors is required for *Neisseria gonorrhoeae* entry into human mucosal cells. *EMBO J.* **14:**2144–2154.
146. **van Schilfgaarde, M., L. van Alphen, P. Eijk, V. Everts, and J. Dankert.** 1995. Paracytosis of *Haemophilus influenzae* through cell layers of Nci-H292 lung epithelial-cells. *Infect. Immun.* **63:**4729–4737.
147. **Verduin, C. M., C. Hol, A. Fleer, H. van Dijk, and A. van Belkum.** 2002. *Moraxella catarrhalis:* from emerging to established pathogen. *Clin. Microbiol. Rev.* **15:**125–144.
148. **Virji, M.** 2001. CEA and innate immunity. *Trends Microbiol.* **9:**258–259.
149. **Virji, M., C. Alexandrescu, D. J. Ferguson, J. R. Saunders, and E. R. Moxon.** 1992. Variations in the expression of pili: the effect on adherence of *Neisseria meningitidis* to human epithelial and endothelial cells. *Mol. Microbiol.* **6:**1271–1279.
150. **Virji, M., D. Evans, J. Griffith, D. Hill, L. Serino, A. Hadfield, and S. M. Watt.** 2000. Carcinoembryonic antigens are targeted by diverse strains of typable and non-typable *Haemophilus influenzae*. *Mol. Microbiol.* **36:**784–795.
151. **Virji, M., D. Evans, A. Hadfield, F. Grunert, A. M. Teixeira, and S. M. Watt.** 1999. Critical determinants of host receptor targeting by *Neisseria meningitidis* and *Neisseria gonorrhoeae:* identification of Opa adhesiotopes on the N-domain of CD66 molecules. *Mol. Microbiol.* **34:**538–551.
152. **Virji, M., K. Makepeace, D. J. Ferguson, M. Achtman, and E. R. Moxon.** 1993. Meningococcal Opa and Opc proteins:

their role in colonization and invasion of human epithelial and endothelial cells. *Mol. Microbiol.* **10:**499–510.

153. **Virji, M., K. Makepeace, D. J. Ferguson, and S. M. Watt.** 1996. Carcinoembryonic antigens (CD66) on epithelial cells and neutrophils are receptors for Opa proteins of pathogenic *Neisseriae. Mol. Microbiol.* **22:**941–950.

154. **Virji, M., K. Makepeace, and E. R. Moxon.** 1994. Distinct mechanisms of interactions of Opc-expressing meningococci at apical and basolateral surfaces of human endothelial cells: the role of integrins in apical interactions. *Mol. Microbiol.* **14:**173–184.

155. **Virji, M., K. Makepeace, I. R. Peak, D. J. Ferguson, M. P. Jennings, and E. R. Moxon.** 1995. Opc- and pilus-dependent interactions of meningococci with human endothelial cells: molecular mechanisms and modulation by surface polysaccharides. *Mol. Microbiol.* **18:**741–754.

156. **Virji, M., J. R. Saunders, G. Sims, K. Makepeace, D. Maskell, and D. J. Ferguson.** 1993. Pilus-facilitated adherence of *Neisseria meningitidis* to human epithelial and endothelial cells: modulation of adherence phenotype occurs concurrently with changes in primary amino acid sequence and the glycosylation status of pilin. *Mol. Microbiol.* **10:**1013–1028.

157. **Virji, M., S. M. Watt, S. Barker, K. Makepeace, and R. Doyonnas.** 1996. The N-domain of the human CD66a adhesion molecule is a target for Opa proteins of *Neisseria meningitidis* and *Neisseria gonorrhoeae. Mol. Microbiol.* **22:**929–939.

158. **Virkola, R., M. Brummer, H. Rauvala, L. van Alphen, and T. K. Korhonen.** 2000. Interaction of fimbriae of *Haemophilus influenzae* type B with heparin-binding extracellular matrix proteins. *Infect. Immun.* **68:**5696–5701.

159. **Virkola, R., K. Lahteenmaki, T. Eberhard, P. Kuusela, L. Van Alphen, M. Ullberg, and T. K. Korhonen.** 1996. Interaction of *Haemophilus influenzae* with the mammalian extracellular matrix. *J. Infect. Dis.* **173:**1137–1147.

160. **Wadstrom, T., and A. Ljungh.** 1999. Glycosaminoglycan-binding microbial proteins in tissue adhesion and invasion: key events in microbial pathogenicity. *J. Med. Microbiol.* **48:**223–233.

161. **Webb, D. C., and A. W. Cripps.** 1999. A method for the purification and refolding of a recombinant form of the nontypeable *Haemophilus influenzae* P5 outer membrane protein fused to polyhistidine. *Protein Expression Purif.* **15:**1–7.

162. **Webb, D. C., and A. W. Cripps.** 1998. Secondary structure and molecular analysis of interstrain variability in the P5 outer-membrane protein of non-typable *Haemophilus influenzae* isolated from diverse anatomical sites. *J. Med. Microbiol.* **47:**1059–1067.

163. **Weber, A., K. Harris, S. Lohrke, L. Forney, and A. L. Smith.** 1991. Inability to express fimbriae results in impaired ability of *Haemophilus influenzae* B to colonize the nasopharynx. *Infect. Immun.* **59:**4724–4728.

164. **Weiser, J. N., D. Bae, C. Fasching, R. W. Scamurra, A. J. Ratner, and E. N. Janoff.** 2003. Antibody-enhanced pneumococcal adherence requires IgA1 protease. *Proc. Natl. Acad. Sci. USA* **100:**4215–4220.

165. **Weiser, J. N., S. T. Chong, D. Greenberg, and W. Fong.** 1995. Identification and characterization of a cell envelope protein of *Haemophilus influenzae* contributing to phase variation in colony opacity and nasopharyngeal colonization. *Mol. Microbiol.* **17:**555–564.

166. **Weiser, J. N., J. B. Goldberg, N. Pan, L. Wilson, and M. Virji.** 1998. The phosphorylcholine epitope undergoes phase variation on a 43-kilodalton protein in *Pseudomonas aeruginosa* and on pili of pathogenic neisseriae. *Infect. Immun.* **66:**4263–4267.

167. **Widdicombe, J.** 1995. Relationships among the composition of mucus, epithelial lining liquid, and adhesion of microorganisms. *Am. J. Respir. Crit. Care Med.* **151:**2088–2093.

168. **Widdicombe, J. H.** 2002. Regulation of the depth and composition of airway surface liquid. *J. Anat.* **201:**313–318.

169. **Willems, R., A. Paul, H. G. J. van der Heide, A. R. Teravest, and F. R. Mooi.** 1990. Fimbrial phase variation in *Bordetella pertussis*—a novel mechanism for transcriptional regulation. *EMBO J.* **9:**2803–2809.

170. **Willems, R. J., C. Geuijen, H. G. van der Heide, M. Matheson, A. Robinson, L. F. Versluis, R. Ebberink, J. Theelen, and F. R. Mooi.** 1993. Isolation of a putative fimbrial adhesin from *Bordetella pertussis* and the identification of its gene. *Mol. Microbiol.* **9:**623–634.

171. **Winther-Larsen, H. C., F. T. Hegge, M. Wolfgang, S. F. Hayes, J. P. van Putten, and M. Koomey.** 2001. *Neisseria gonorrhoeae* PilV, a type IV pilus-associated protein essential to human epithelial cell adherence. *Proc. Natl. Acad. Sci. USA* **98:**15276–15281.

172. **Zaleski, A., N. K. Scheffler, P. Densen, F. K. Lee, A. A. Campagnari, B. W. Gibson, and M. A. Apicella.** 2000. Lipooligosaccharide P(k) (Galα1-4Galβ1-4Glc) epitope of *Moraxella catarrhalis* is a factor in resistance to bactericidal activity mediated by normal human serum. *Infect. Immun.* **68:**5261–5268.

173. **Zhang, J. R., K. E. Mostov, M. E. Lamm, M. Nanno, S. Shimida, M. Ohwaki, and E. Tuomanen.** 2000. The polymeric immunoglobulin receptor translocates pneumococci across human nasopharyngeal epithelial cells. *Cell* **102:**827–837.

Colonization of Mucosal Surfaces
Edited by James P. Nataro et al.

Chapter 9

Immunoglobulin A1 Proteases of Pathogenic and Commensal Bacteria of the Respiratory Tract

MOGENS KILIAN AND JESPER REINHOLDT

One of the recent evolutionary events affecting the immune system is the introduction of a second subclass of immunoglobulin A (IgA), denoted IgA1, in the immediate common ancestor of humans and hominoid primates (chimpanzees, gorillas, and orangutans). For reasons that remain unknown, this subclass is vastly predominant in secretions of the upper respiratory tract and in serum, while gut and genital secretions contain approximately even proportions of IgA1 and the more ancient IgA2 (36, 43, 54). An extended hinge region resulting from an insert of 13 amino acids (Fig. 1) allows the two Fab regions of IgA1 antibodies increased conformational freedom. This advantage was achieved at the expense of introducing an open stretch, which potentially is more susceptible to attack by proteolytic enzymes. However, an unusual proline-rich amino acid sequence, combined with extensive O glycosylation of several serine and threonine residues (53) (Fig. 1), renders the IgA1 hinge resistant to traditional proteolytic enzymes. Nevertheless, a group of respiratory tract pathogens and related commensal bacteria possess highly specialized IgA1 proteases that allow them to evade and take advantage of this principal adaptive immune factor of respiratory tract secretions, even in its further proteolysis-resistant secretory form, S-IgA1.

IgA1 PROTEASES: A COMMON PROPERTY OF THE PRINCIPAL CAUSES OF BACTERIAL MENINGITIS

Strikingly, all three principal causes of bacterial meningitis, *Neisseria meningitidis, Streptococcus pneumoniae,* and *Haemophilus influenzae,* produce an IgA1 protease (37, 64). Apart from occasional mutated strains, all members of these species, including noncapsulated ("nontypeable") strains, have this property. In addition, IgA1 proteases are produced by two urogenital pathogens, five species of commensal gram-positive cocci found in the pharynx and oral cavity, *H. parahaemolyticus,* and all human-associated species of the genera *Capnocytophaga* and *Prevotella* (Fig. 1). Some of the latter two genera have been implicated in the pathogenesis of periodontal diseases (for references to the individual findings, see reference 40).

The IgA1 proteases share several unique enzymatic properties. They are post-proline endopeptidases and, in most cases, attack the heavily glycosylated IgA hinge immediately adjacent to a carbohydrate side chain. As illustrated in Fig. 1, all IgA1 proteases that have been identified cleave a particular one of the several prolyl-seryl or prolyl-threonyl peptide bonds that are present in the IgA1 hinge. Recent studies using IgAs of gorillas, chimpanzees, and orangutans, which show minor sequence aberrations relative to human IgA1, and recombinant human IgA1 with specific point mutations in the hinge region demonstrate that limited amino acid substitutions are permissible (3, 75). Apart from hominoid primates, no animal species produces IgA that is susceptible to cleavage by known IgA1 proteases.

Some strains of the intestinal, anaerobic species *Clostridium ramosum* produce a unique protease that cleaves both IgA1 and the A2m(1) allotype of IgA2 at the prolyl-valyl bond just before the hinge shared by these two human α-chains (17) but lacking in the two other IgA2 allotypes (Fig. 1).

Mogens Kilian • Department of Medical Microbiology and Immunology, University of Aarhus, Aarhus C, DK-8000, Denmark.
Jesper Reinholdt • Department of Oral Biology, University of Aarhus, Aarhus C, DK-8000, Denmark.

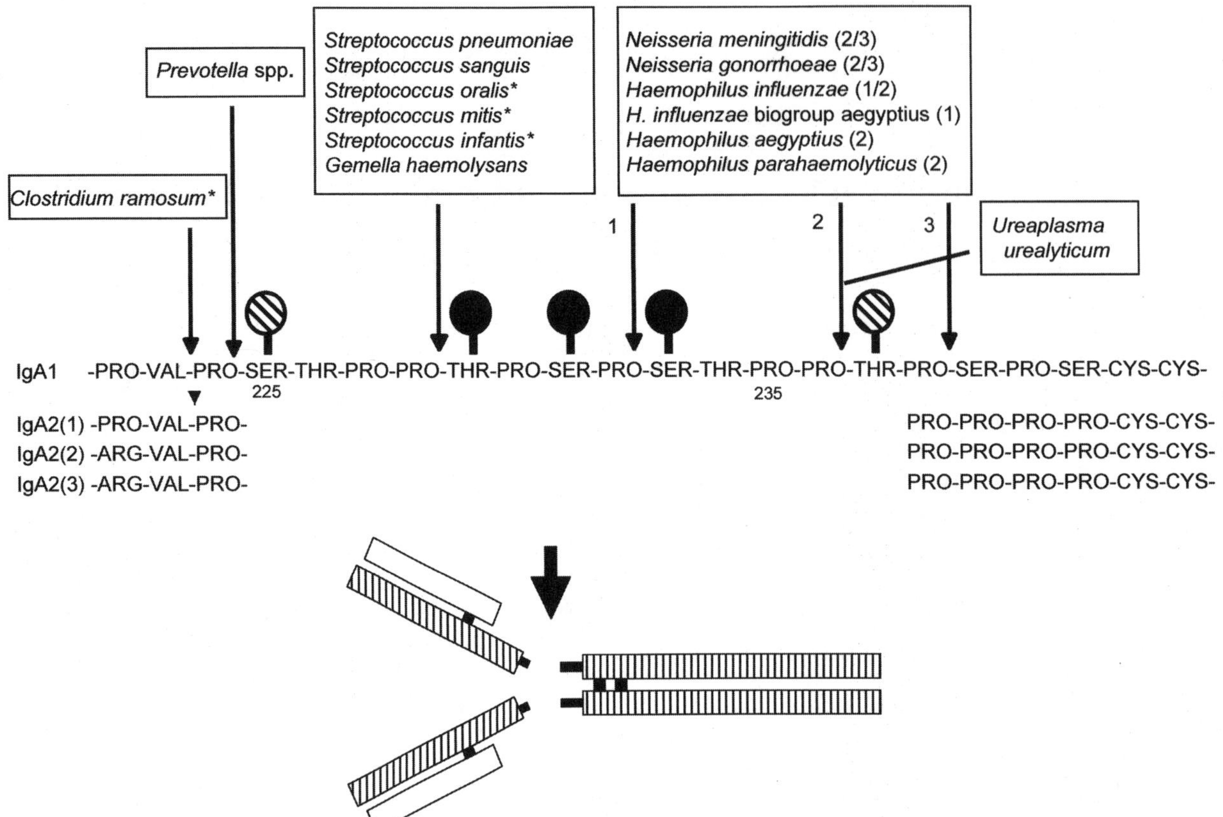

Figure 1. Sites of IgA1 protease-induced cleavage within the hinge region of human IgA. Note that strains of *H. influenzae*, *N. meningitidis*, and *N. gonorrhoeae* may cleave either a Pro-Ser peptide bond (type 1 protease) or a Pro-Thr peptide bond (type 2 protease). The *C. ramosum* IgA protease cleaves both IgA1 and the IgA2 allotype A2m(1), as indicated by the arrow. The shaded circles indicate carbohydrate side chains in the IgA1 hinge region, those that are hatched being variably present in IgA1 molecules. Species indicated by an asterisk show variable IgA1 protease activity.

A REMARKABLE EXAMPLE OF CONVERGENT EVOLUTION OF HIGHLY SPECIALIZED PROTEASES

Comparison of gene sequences and enzyme properties reveals that IgA protease activity in bacteria evolved independently along at least five lineages (Table 1). Although they share the uniquely specialized ability to cleave IgA1, they do so by several different catalytic mechanisms and, accordingly, belong to the three major groups of proteinases, i.e., serine-type proteinases, metalloproteinases, and thiol proteinases.

Table 1. Bacterial IgA proteases grouped according to enzyme characteristics and phylogenic relationships

Serine-type IgA1 proteinases
Neisseria and *Haemophilus* EC 3.4.21.72
Ureaplasma urealyticum
Metalloproteinases
Streptococcus and *Gemella* EC 3.4.24.13
Clostridium ramosum IgA protease family M64
Capnocytophaga spp.
Cysteine proteinases
Prevotella spp.

It is also remarkable that gene duplication has been involved in the evolution of IgA1 protease genes, both in *Streptococcus* species and in *H. influenzae*, meningococci, and gonococci. Several paralogous genes in these species encode proteases of the same type. However, only one of these has the ability to cleave IgA1 in spite of a significant degree of sequence homology.

The published genome sequences of *S. pneumoniae* strains show that they possess genes encoding two to four large zinc metalloproteinases (1,800 to 2,001 amino acids) associated with the bacterial surface (13, 30, 88). The gene encoding the IgA1 protease (*iga* = *zmbA*) is present in all *S. pneumoniae* strains that have been examined (73). Although the primary function of the pneumococcal IgA1 protease is assumed to be evasion of IgA1 antibodies, recent evidence suggests that it has additional yet unidentified virulence-associated properties (see "Alternative substrates and functions of IgA1 proteases" below). Currently, the only function ascribed to the second pneumococcal zinc metalloproteinase, ZmbB, is its ability to contribute to inflammation in the lower respiratory tract (7). It was initially reported that ZmbB regulates the expression of choline-binding proteins, including the major autolysin, LytA (59), but subsequent studies failed to confirm this (5). The third pneumococcal zinc metalloproteinase, ZmpC, is able to cleave human matrix metalloproteinase 9 (60) and is part of the virulence determinants important for experimental mouse models of intranasal challenge and septicemia (9, 25, 70). A fourth paralogous gene fragment is present in strain TIGR4 (88). Common to this family of proteases is a cell wall anchor motif in the N-terminal part, in contrast to all other known wall-associated proteins in gram-positive bacteria. A significant part of the IgA1 protease activity of streptococci remains associated with the bacterial surface (72, 77, 95).

In spite of these virulence associations in pneumococci, homologous proteins are expressed by several commensal *Streptococcus* species (Fig. 2a). However, the *iga* gene and IgA1 protease activity are variably present among strains of *S. mitis, S. oralis,* and *S. infantis* and are lacking in all strains of *S. gordonii* (79; M. Kilian, K. Poulsen, and U. S. Sørensen, unpublished data). In contrast, all strains of *S. sanguis* show IgA1 protease activity. This may indicate that the *iga* gene evolved in the common ancestor of these phylogenetically related streptococci and was subsequently lost within multiple lineages of commensal streptococci due to a more subtle relationship with the adaptive mucosal immune system than that of pneumococci (78). The presence of a homologous IgA1 protease in the more distantly related *Gemella haemolysans* conceivably is a result of horizontal gene transfer (51).

The serine-type IgA1 proteases of *Haemophilus* and *Neisseria* belong to a large family of genetically related, secreted proteins that may be involved in colonization and invasion in a diverse group of gram-negative pathogens, including *H. influenzae, Salmonella, Escherichia coli, Shigella, Serratia marcescens, Bordetella, Moraxella, Helicobacter,* and rickettsiae (for a review, see reference 27). The IgA1 proteases are the prototypes of this growing family of autotransporter proteins (67) and form a subfamily together with the Hap (for "*Haemophilus* adhesion and penetration") proteins and their homologs in *Neisseria* and the so-called SPATES (for "serine protease autotransporters secreted by members of the *Enterobacteriaceae*"). These proteins are synthesized as large precursors consisting of an N-terminal signal peptide, a passenger effector domain (e.g., the IgA1 protease), a linker peptide, and a carboxy-terminal β-barrel domain, which forms a pore in the outer membrane through which the effector protein is transported (67). Putative serine-type protease motifs are present in most of these proteins. In some proteins, including the IgA1 protease, this is crucial for the autoproteolytic cleavage that results in eventual release from the bacterial surface and concurrent liberation of small peptides (α-peptides) from the linker sequence (67).

The genomes of the *H. influenzae, N. meningitidis,* and *N. gonorrhoeae* strains that have been sequenced to completion contain two or three paralogous genes encoding proteins that belong in this group of serine-type proteases (Fig. 2b) (16, 61, 89; unpublished *N. gonorrhoeae* genome at http://www.genome.ou.edu.). The IgA1 protease gene (*iga*) is invariably one of these paralogs, in accordance with the observation that all isolates of these three species produce an IgA1 protease. The high sequence homology and aberrant G+C content of the *Neisseria iga* gene suggest that it may have been acquired from *H. influenzae* by the common ancestor of meningococci and gonococci (33, 49). The second protein, which appears to be expressed in the majority of strains, is the adhesion and penetration protein, of which the prototype is the Hap protein of *H. influenzae* (87). Hap has serine protease activity but does not cleave IgA1. Although it is autoproteolytically released from the cell surface like the IgA1 protease, in vitro studies indicate that it functions as an adhesin and allows intimate and concentrated contact with epithelial cells (28, 87). Similar functions are ascribed to the equivalent App protein in *N. meningitidis* (21, 25, 66, 84, 91). Surprisingly, the Hap protein is not expressed by strains of *H. aegyptius* and *H. influenzae* biogroup aegyptius, the causative agent of Brazilian purpuric fever due to several recent mutations (38). For the commensal species of *Haemophilus* and *Neisseria,* only limited information is available. None of them produce IgA1 protease (42, 57). A third paralogous gene (NMB1998) (Fig. 2b) encoding a protein with a putative serine-type protease motif is present in the genome of *N. meningitidis* group B strain MC58 but not in group A strain Z2491 (61, 89). The function of this protein is unknown. The NalP serine protease gene is another paralog of the *iga* gene present in some strains of *N. meningitidis* but is more distantly related (Fig. 2b). Interestingly, NalP cleaves both the IgA1 protease and the App protein in culture

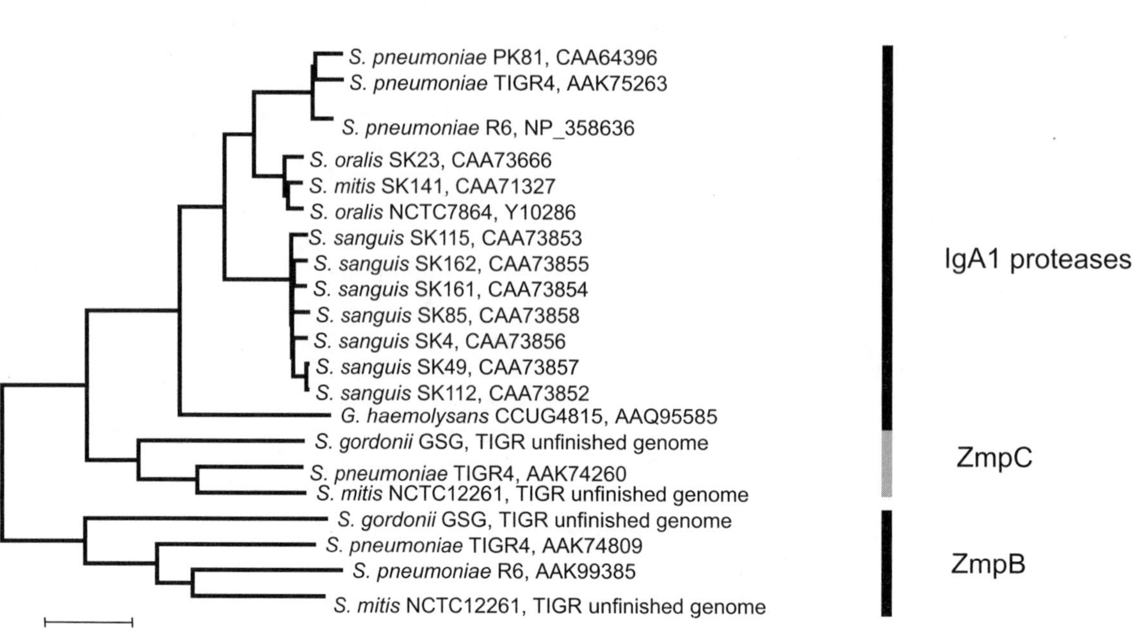

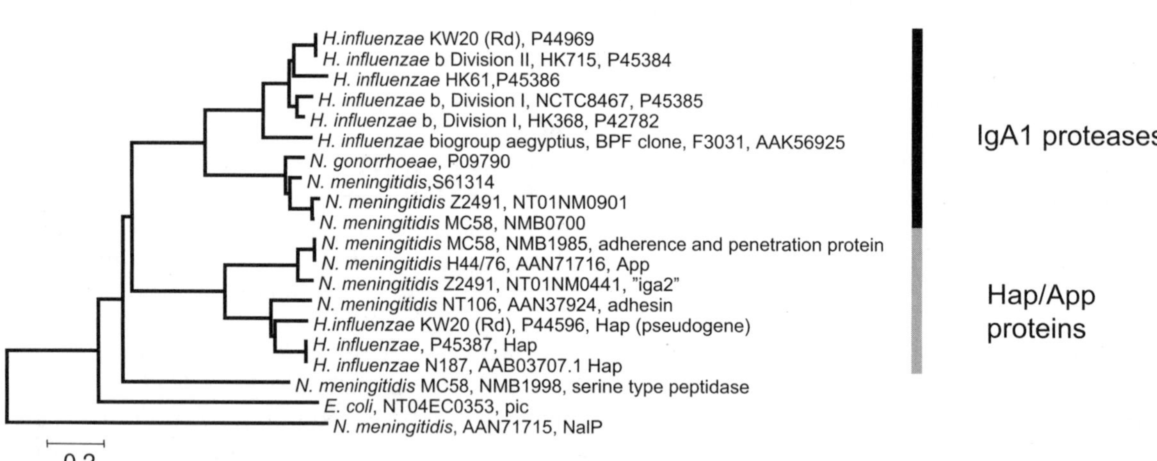

Figure 2. Phylogeny of IgA1 protease and paralogous genes of *Streptococcus* (a) and *Haemophilus* and *Neisseria* (b). Note that the *H. influenzae* strain Rd *hap* gene has a mutation resulting in a stop codon in the middle of the gene. The sequences from N. *lactamica* (AJ001739 and AJ001740) are 578-bp fragments reminiscent of the *hap/app* gene family (not shown).

supernatants. The biological significance of this processing is yet to be determined (92). *Neisseria lactamica,* which is the closest relative of the two pathogenic *Neisseria* species, contains a gene fragment reminiscent of the *hap/app* type (Fig. 2b) but has no trace of an *iga* gene (33).

THE SPECIFICITY OF IgA PROTEASES REFLECTS ADAPTATION TO HOST AND HABITAT

In accordance with the substrate specificity of IgA1 proteases, humans and hominoid primates are the exclusive hosts of the bacteria that produce these enzymes. The one partial exception is *S. pneumoniae,* which occasionally causes infections in animal species, although its primary habitat is the human respiratory tract. The specificity of IgA proteases of respiratory tract bacteria for the IgA1 subclass is, furthermore, in agreement with the predominance of that subclass in secretions of the upper respiratory tract (comprising ca. 95% of S-IgA in the nasal cavity and ca. 70% in the oral cavity). Likewise, IgA1 accounts for ca. 95% of serum IgA in humans (36, 43, 54). The ability of the IgA protease of the gut commensal *Clostridium ramosum* to cleave both IgA1 and the A2m(1) allotype of IgA2 (17) is in accordance with the even distribution of the two subclasses in that habitat. Whether the occurrence of *C. ramosum* in the population reflects the distribution of the IgA2 allotypes has not been examined. *N. gonorrhoeae* encounters IgA1 and IgA2 in approximately equal proportions in its primary habitat. A possible explanation for this seeming lack of optimal adaptation is that the recent common ancestor of the two pathogenic *Neisseria* species belonged in the upper respiratory tract, as indicated by phylogenetic evidence. The selection pressure on gonococci for further adaptation may be limited because of an abundance of other immune escape mechanisms employed by this species (55).

In spite of comprehensive screenings of animal-pathogenic bacteria for the ability to cleave IgA of their respective hosts, the only equivalent example remains that of certain *Ureaplasma* strains that are associated with infections in dogs and that cleave canine IgA in the hinge region (35).

IN VIVO ACTIVITY OF IgA1 PROTEASES

In contrast to many other microbial proteinases, IgA1 proteases are not inhibited by physiological protease inhibitors such α_2-macroglobulin, α_1-protease inhibitor, and others, a property explainable in part by the substrate specificity, which, for example, precludes activation of α_2-macroglobulin (63). However, human milk lactoferrin is reported to inactivate IgA1 protease and the related Hap protein from *H. influenzae* by cleaving these autotransporter proteins from the bacterial cell membrane (75). The full significance of this interaction in the protection against infections caused by IgA1 protease-producing bacteria is still unclear.

Under conditions that allow prolonged exposure of S-IgA1 in mucosal secretions to IgA1 protease-producing bacteria, it is possible to demonstrate characteristic fragments of S-IgA1 (1, 86). However, the failure to demonstrate characteristic cleavage fragments of IgA in vaginal secretions of women with gonococcal infection (26) and in nasal rinses from healthy human adults colonized with IgA1 protease-producing bacteria (43) suggests that cleavage, in most cases, is restricted to the microenvironment of the individual bacterial cells.

INCREASED PRODUCTION OF IgA1 PROTEASE IN VIRULENT BACTERIA

Significant inter- and intraspecies differences have been detected in the IgA1 protease activity released by bacteria in vitro (77). Although there is no overall relationship between pathogenicity of species and the IgA1 protease activity, meningococci generally show activity that may be several hundred times higher than that of some of the commensal streptococci. However, significant variations may be seen within species (77). Recent reports suggest that isolates of *N. meningitidis* and *H. influenzae* from invasive infections produce more IgA1 protease than do isolates from healthy carriers (93, 94).

IgA1 PROTEASE ACTIVITY INTERFERES WITH IMMUNE ELIMINATION AND PROMOTES MUCUS PENETRATION AND ADHERENCE IN VITRO

IgA1 protease-induced cleavage of IgA1 antibodies efficiently separates all secondary effector functions associated with Fc_α from target microorganisms bound at the antigen-binding sites. These functions include agglutination, mucus trapping, inhibition of adherence, and opsonization for interaction with various white blood cells, including phagocytes (for a review, see reference 80). We have postulated that the key to understanding why such specialized proteases are employed by several successful pathogens is that they combine the elimination of antibody effector functions with the release of monomeric Fab fragments that retain antigen-binding capacity (39). As a consequence of the latter, the IgA1 protease-producing bacteria become coated with monomeric Fab fragments when exposed to IgA1 antibodies to their surface antigens (1). Although Fab fragments of naturally induced IgA1 antibodies still may bind to potential adhesins, this does not interfere with the ability of the bacteria or purified adhesins to adhere to their target surface under in vitro conditions (22, 58, 76, 90, 96). Recent studies using pneumococci with and without IgA1 protease activity suggest that coating with hydrophobic Fab fragments of IgA1 antibodies to capsular polysaccharides may even be a prerequisite for the ability of encapsulated bacteria to adhere to epithelial cells (96).

During further advanced stages of infection, Fab_{α} fragments remaining on the bacterial surface will conceivably block access of intact antibodies of other isotypes and interfere with complement activation and complement-mediated lysis in the presence of IgM and IgG antibodies (31, 81).

Due to the specificity for human IgA1, animal studies have not been able to evaluate the contribution of these immune escape mechanisms to pathogenicity. Two studies have attempted to elucidate the biological significance of IgA1 proteases in human organ culture models. Isogenic strains of *N. gonorrhoeae* and *H. influenzae* with or without IgA1 protease activity were used to infect human fallopian tube and nasopharyngeal organ cultures. No differences in the ability to attach to or penetrate the mucosal tissues were observed (11, 15). However, these studies do not elucidate potential effects related to IgA, since the models lacked specific IgA1 antibodies. The same limitation probably applies to studies using a human infection model of initial gonococcal infection that, likewise, failed to reveal signs of reduced infectivity of IgA1 protease-negative mutants (32). The above-mentioned potentially reduced significance of the IgA1 protease to gonococcal virulence in particular may also be relevant in this context.

THE BALANCE BETWEEN CLEAVAGE-RELEVANT AND ENZYME-NEUTRALIZING ANTIBODIES MAY BE A DECISIVE FACTOR

Like most other surface-exposed or secreted bacterial proteins, IgA1 proteases of pathogenic bacteria induce significant responses of serum and secretory antibodies during both colonization and infection (8, 12, 19, 44). Neutralizing antibodies inhibit the activity of IgA1 proteases even when present as released monomeric Fab fragments (19). Antibodies may also block the release of serine-type, autotransporter IgA1 proteases of *H. influenzae,* meningococci, and gonococci due to inhibition of the autocatalytic cleavage. As a result, antibodies against IgA1 proteases may agglutinate these bacteria through interaction with the surface-bound protease, which constitutes a novel surface antigen (65).

IgA1 proteases of the pathogenic species show considerable antigenic diversity as a result of accumulated mutations combined with recombination both within and between species and affecting particular regions in *iga* genes (23, 48, 56, 74). The most extensive antigenic diversity has been detected among proteases of *H. influenzae* and *S. pneumoniae* (41, 47). Successive clones of *H. influenzae* colonizing an individual produce antigenically different IgA1 proteases that are not inhibited by antibodies induced by previously colonizing clones (50).

It is conceivable that once a human host has responded to an acquired protease-producing bacterium with IgA1 antibodies, whose cleavage would be beneficial to the bacterium, neutralizing antibodies against its protease are likely to be present too. We have presented the hypothesis that invasive infection in occasional individuals is a result of nonsynchronized induction of the two types of antibodies by successive encounters with two different microorganisms: (i) colonization in the gut or upper respiratory tract with bacteria expressing surface epitopes similar or identical to those of the respective pathogen (e.g., *E. coli* K100 in the case of *H. influenzae* type b, and *E. coli* K1 or *Moraxella nonliquefaciens* in the case of *N. meningitidis* group B), and (ii) subsequent colonization with the actual pathogen. As a result of the prior colonization with a cross-reactive microorganism, the pathogen encounters preexisting IgA1 antibodies to its surface epitopes but no antibodies that will neutralize its IgA1 protease. This situation enables the pathogen to take full advantage of IgA1 antibodies by becoming coated with Fab_{α} fragments, with the consequences discussed above (Fig. 3). In the majority of individuals, acquisition of *N. meningitidis, S. pneumoniae,* and *H. influenzae* serotype b results in the concurrent induction of S-IgA antibodies to surface antigens and the IgA1 protease of the bacteria. The protease-neutralizing antibodies prevent cleavage of IgA1 antibodies and hence prevent coating of the pathogen with Fab_{α} fragments. The concurrent induction of these antibodies results in immunity to invasive infection and subsequent attacks by bacteria expressing the same combination of surface epitopes and the IgA1 protease "inhibition type." This hypothetical model is in agreement with several hitherto unexplained observations made with both human and animal models (39).

ALTERNATIVE SUBSTRATES AND FUNCTIONS OF IgA1 PROTEASES

IgA1s of humans and hominoid primates were the only known substrates of IgA1 proteases until recently. Now other permissive substrates have been revealed. Most of these are proteins of immunological relevance, but for some of the substrates, their accessibility to IgA1 protease in vivo is a matter of speculation. Often, they have been identified as a result of

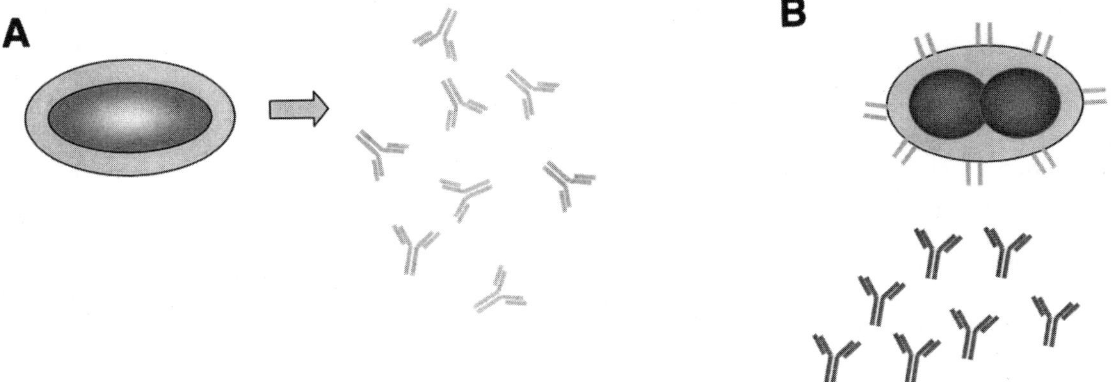

Figure 3. Hypothetical model for induction of susceptibility to invasive infection with IgA1 protease-producing pathogens. (A) Colonization with a nonpathogen with a cross-reactive surface antigen, resulting in the induction of IgA1 antibodies. (B) Colonization with an IgA1 protease-producing pathogen, resulting in elimination of IgA1-mediated protection and blockage of intact antibodies.

database searches for protein sequences homologous to that of the α1 chain hinge. Generally, cleavage has been documented only under experimental conditions, and experiments have involved only the serine-type IgA1 proteases of *N. gonorrhoeae* and *N. meningitidis*.

A non-IgA substrate was first identified through the observation by Pohlner et al. (67) that, during the secretion process, the 165-kDa precursor of the IgA1 protease of *N. gonorrhoeae* undergoes autoproteolysis at several sites within a sequence (linker) similar but not identical to that of the α1 hinge. Synthetic decapeptides based on the sequences of two of these sites were cleaved by the type 2 IgA1 protease of *N. gonorrhoeae* (97). Homology searches identified similar target sequences in the CD8 molecule and in granulocyte-macrophage colony-stimulating factor (68), but cleavage of these proteins in their natural configuration has not been demonstrated. Autoproteolysis within a sequence similar but not identical to that of the a1 hinge appears to be part of the secretion process for serine-type IgA1 proteases in general (49, 71).

Shoberg and Mulks (85) found that the type 2 IgA1 protease from *N. gonorrhoeae* degrades several unidentified proteins in preparations of cytoplasmic and outer membranes of gonococci and certain other gram-negative bacteria. This phenomenon, if it occurs in live IgA1 protease-producing bacteria, suggests that the protease may play a role in modulating proteins of the microorganism itself and possibly of nearby bacteria. Notably, the degraded proteins did not include any of the functionally characterized gonococcal major outer membrane proteins.

One of the lysosome-associated membrane proteins, LAMP-1, can be cleaved by the type 2 gonococcal IgA1 protease (24, 45). LAMP-1 contains a large N-terminal segment directed toward the lysosomal/endosomal lumen, which consists of two highly N-glycosylated domains separated by an O-glycosylated hinge similar to the hinge of the α1 chain (18). On exocytosis, LAMP-1 also occurs on the cell surface. The biological function of LAMP-1 is incompletely understood (14). Significantly reduced levels of LAMP-1 have been detected in cultured epithelial cells infected with gonococci or meningococci compared to the levels in control cells infected with IgA1 protease-deficient mutants (45), indicating that IgA1 protease-producing *Neisseria* strains can also cleave LAMP-1 in living cells. It has been doubted that cleavage takes place in the acidic milieu of the lysosome/endosome, and LAMP-1 on the cell surface may be the main target (2). However, IgA1 protease activity promotes the intracellular survival of the bacteria (45) and *iga* mutants have a significant defect in their ability to traverse monolayers of polarized epithelial cells (29). These results may be of significance to the ability of the pathogenic *Neisseria* strains to invade mucosal membranes by epithelial transcytosis (62). Notably, LAMP-1 from granulocytes, which is more densely glycosylated than its epithelial counterpart, resisted degradation by neisserial IgA1 protease (24).

Another protein shown to be cleaved by neisserial type 2 IgA1 protease is synaptobrevin-2, also known as vesicle-associated membrane protein 2 (Vamp-2). If cleaved, this protein is prevented from performing its function as a SNARE protein governing the fusion of vesicles with the plasma membrane in exocytosis (6). In view of the involvement of synaptobrevin-2 in fundamental cellular reactions, including innate defense functions (20, 46), inactivation of this protein is a conceivable means of immune escape for the pathogenic *Neisseria*. However, unlike tetanus toxin, which also cleaves and inactivates synaptobrevin-2 (82), the neisserial IgA1 protease cannot by itself penetrate eucaryotic cells. It is not known if IgA1 protease produced by endocytosed neisseriae may approach and degrade synaptobrevin-2 of the containing cell.

IgA1 proteases, at 100 to 200 kDa, are unusually large proteases. This raises the question whether these proteins have functions unrelated to their proteolytic activity. Thus, during autoproteolysis and secretion of the neisserial IgA1 protease precursors, a so-called α-protein with suspected biological activity is released (67). Interest has been paid in particular to the finding that α-proteins carry amino acid sequences demonstrated to function as nuclear location signals (69). Also, they share several features with eucaryotic transcription factors, leading to the proposal that α-proteins play a role in the regulation of host cell functions (69). Also, α-proteins carry an abundance of T-cell epitopes (34). In other studies, the native gonococcal type 2 IgA1 protease, but not the denatured protein, was found to induce the release of proinflammatory cytokines such as tumor necrosis factor alpha (TNF-α), interleukin-1β (IL-1β), IL-6, and IL-8 from peripheral blood mononuclear cells (52). Interestingly, the same IgA1 protease was subsequently shown to inhibit TNF-α-mediated apoptosis of a human monocytic cell line and to cleave the TNF-RII receptor in situ or in purified form (4). A similar effect of IgA1 protease produced by bacteria in tissues in vivo is theoretically conceivable in view of the resistance of IgA1 proteases to physiological protease inhibitors (63).

The β subunit of human chorionic gonadotropin is cleavable by the type 1 but not the type 2 gonococcal IgA1 protease, and a hypothetical role of the cleavable hormone in gonococcal virulence has been suggested (83).

The alternative proteolytic and other activities of the neisserial IgA1 proteases described above, if they occur in the infected host, would suggest that these proteases may function as virulence factors irrespective of whether specific IgA1 antibodies are present. However, studies with isogenic pairs of wild-type and *iga* mutant strains in organ culture models (11, 15) or in the initial phase of gonococcal infection in male volunteers (32) failed to support this hypothesis.

Conversely, studies with pneumococcal strains strongly support the concept that IgA1 proteases may play a role in infections, in addition to cleavage of IgA1. Thus, inactivation of IgA1 protease in *S. pneumoniae* leads to reduced virulence in mouse infection models even though the protease does not cleave murine IgA (10, 25, 70). It is not known if the streptococal enzyme exerts this effect by proteolysis or by other means. No non-IgA substrate of pneumococcal or other streptococcal IgA1 proteases has been identified. Future searches for such hypothetical substrates should take into consideration the fact that the pneumococcal IgA1 protease for cleaving IgA1 shows no absolute requirement for either proline or threonine at residues 227 to 228 (3). Besides, studies with immunoglobulin hybrid test substrates have shown, for streptococcal as well as other IgA1 proteases, that cleavability of an α1 hinge-like peptide sequence may depend on elements of the protein structure distant—usually C terminally—from the potential cleavage site (10, 49).

REFERENCES

1. **Ahl, T., and J. Reinholdt.** 1991. Detection of immunoglobulin A1 protease-induced Fab alpha fragments on dental plaque bacteria. *Infect. Immun.* **59:**563–569.
2. **Ayala, B. P., B. Vasquez, S. Clary, J. A. Tainer, K. Rodland, and M. So.** 2001. The pilus-induced Ca^{2+} flux triggers lysosome exocytosis and increases the amount of Lamp1 accessible to *Neisseria* IgA1 protease. *Cell. Microbiol.* **3:**265–275.
3. **Batten, M. R., B. W. Senior, M. Kilian, and J. M. Woof.** 2003. Amino acid sequence requirements in the hinge of human immunoglobulin A1 (IgA1) for cleavage by streptopcoccal IgA1 proteases. *Infect. Immun.* **71:**1462–1469.
4. **Beck, S. C., and T. F. Meyer.** 2000. IgA1 protease from *Neisseria gonorrhoeae* inhibits TNFalpha-mediated apoptosis of human monocytic cells. *FEBS Lett.* **472:**287–292.
5. **Bergé, M., P. Garcia, F. Iannelli, M. F. Prere, C. Granadel, A. Polissi, and J. P. Claverys.** 2001. The puzzle of *zmpB* and extensive chain formation, autolysis defect and non-translocation of choline-binding proteins in *Streptococcus pneumoniae*. *Mol. Microbiol.* **39:**1651–1660.
6. **Binscheck, T., F. Bartels, H. Bergel, H. Bigalke, S. Yamasaki, T. Hayashi, H. Niemann, and J. Pohlner.** 1995. IgA protease from *Neisseria gonorrhoeae* inhibits exocytosis in bovine chromaffin cells like tetanus toxin. *J. Biol. Chem.* **270:**1770–1774.
7. **Blue, C. E., G. K. Paterson, A. R. Kerr, M. Bergé, J. P. Claverys, and T. J. Mitchell.** 2003. ZmpB, a novel virulence factor of *Streptococcus pneumoniae* that induces tumor necrosis factor alpha production in the respiratory tract. *Infect. Immun.* **71:**4925–4935.
8. **Brooks, G. F., C. J. Lammel, M. S. Blake, B. Kusecek, and M. Achtman.** 1992. Antibodies against IgA1 protease are stimulated both by clinical disease and asymptomatic carriage of serogroup A *Neisseria meningitidis*. *J. Infect. Dis.* **166:** 1316–1321.
9. **Chiavolini, D., G. Memmi, T. Maggi, F. Iannelli, G. Pozzi, M. R. Oggioni.** 2003. The three extra-cellular zinc metalloproteinases of *Streptococcus pneumoniae* have a different impact on virulence in mice. *BMC Microbiol.* **3:**14.
10. **Chintalacharuvu, K. R., P. D. Chuang, A. Dragoman, C. Z. Fernandez, J. Qiu, A. G. Plaut, K. R. Trinh, F. A. Gala, and S. L. Morrison.** 2003. Cleavage of the human immunoglobulin A1 (IgA1) hinge region by IgA1 proteases requires structures in the Fc region of IgA. *Infect. Immun.* **71:**2563–2570.
11. **Cooper, M. D., Z. A. McGee, M. H. Mulks, J. M. Koomey, and T. L. Hindman.** 1984. Attachment to and invasion of human fallopian tube mucosa by an IgA1 protease-deficient

mutant of *Neisseria gonorrhoeae* and its wild-type parent. *J. Infect. Dis.* **150**:737–744.

12. **Devenyi, A.G., A. G. Plaut, F. J. Grundy, and A. Wright.** 1993. Post-infectious human serum antibodies inhibit IgA1 proteinases by interaction with the cleavage site specificity determinant. *Mol. Immunol.* **30**:1243–1248.
13. **Dopazo, J., A. Mendoza, J. Herrero, F. Caldara, Y. Humbert, L. Friedli, M. Guerrier, E. Grand-Schenk, C. Gandin, M. de Francesco, A. Polissi, G. Buell, G. Feger, E. Garcia, M. Peitsch, and J. F. Garcia-Bustos.** 2001. Annotated draft genomic sequence from *Streptococcus pneumoniae* type 19F clinical isolate. *Microb. Drug Resist.* **7**:99–125.
14. **Eskelinen, E.-L., T. Yoshitaka, and P. Saftig.** 2003. At the acidic edge: emerging functions for lysosomal membrane proteins. *Trends Cell Biol.* **13**:137–145.
15. **Farley, M. M., D. S. Stephens, M. H. Mulks, M. D. Cooper, J. V. Bricker, S. S. Mirra, and A. Wright.** 1986. Pathogenesis of IgA1 protease-producing and nonproducing *Haemophilus influenzae* in human naso-pharyngeal organ cultures. *J. Infect. Dis.* **154**:752–759.
16. **Fleischmann, R. D., M. D. Adams, O. White, R. A. Clayton, E. F. Kirkness, A. R. Kerlavage, C. J. Bult, J. F. Tomb, B. A. Dougherty, J. M. Merrick, B. A. Dougherty, J. M. Merrick, K. McKenney, G. Sutton, W. Fitzhugh, C. Fields, J. D. Gocayne, J. Scott, R. Shirley, L. I. Liu, A. Glodek, J. M. Kelley, J. F. Weidman, C. A. Phillips, T. Spriggs, E. Hedblom, M. D. Cotton, T. R. Utterback, M. C. Hanna, D. T. Nguyen, D. M. Saudek, R. C. Brandon, L. D. Fine, J. L. Fritchman, J. L. Fuhrmann, N. S. M. Geoghagen, C. L. Gnehm, L. A. McDonald, K. V. Small, C. M. Fraser, H. O. Smith, and J. C. Venter.** 1995. Whole-genome random sequencing and assembly of *Haemophilus influenzae* Rd. *Science* **269**:496–512.
17. **Fujiyama, Y., M. Iwaki, K. Hodohara, S. Hosoda, and K. Kobayashi.** 1986. The site of cleavage in human alpha chains of IgA1 and IgA2: A2m(1) allotype paraproteins by the clostridial IgA protease. *Mol. Immunol.* **23**:147–150.
18. **Fukuda, M., J. Viitala, J. Matteson, and S. R. Carlsson.** 1988. Cloning of cDNAs encoding human lysosomal membrane glycoproteins, h-lamp-1 and h-lamp-2. Comparison of their deduced amino acid sequences. *J. Biol. Chem.* **263**:18920–18928.
19. **Gilbert, J.V., A. G. Plaut, B. Longmaid, and M. E. Lamm.** 1983. Inhibition of microbial IgA proteases by human secretory IgA and serum. *Mol. Immunol.* **20**:1039–1049.
20. **Hackam, D. J., O. D. Rotstein, C. Sjolin, A. D. Schreiber, W. S. Trimble, and S. Grinstein.** 1998. v-SNARE-dependent secretion is required for phagocytosis. *Proc. Natl. Acad. Sci. USA* **95**:11691–11696.
21. **Hadi, H. A., K. G. Wooldridge, K. Robinson, and D. A. Ala'Aldeen.** 2001. Identification and characterization of App: an immunogenic autotransporter protein of *Neisseria meningitidis*. *Mol. Microbiol.* **41**:611–623.
22. **Hajishengallis, G., E. Nikolova, and M. W. Russell.** 1992. Inhibition of *Streptococcus mutans* adherence to saliva-coated hydroxyapatite by human secretory immunoglobulin A (SIgA) antibodies to cell surface protein antigen I/II: reversal by IgA1 protease cleavage. *Infect. Immun.* **60**:5057–5064.
23. **Halter, R., J. Pohlner, and T. F. Meyer.** 1989. Mosaic-like organization of IgA protease genes in *Neisseria gonorrhoeae* generated by horizontal genetic exchange in vivo. *EMBO J.* **8**:2737–2744.
24. **Hauck, C. R., and T. F. Meyer.** 1997. The lysosomal/phagosomal membrane protein h-lamp-1 is a target of the IgA1 protease of *Neisseria gonorrhoeae*. *FEBS Lett.* **405**:86–90.
25. **Hava, D. L., and A. Camilli.** 2002. Large-scale identification of serotype 4 *Streptococcus pneumoniae* virulence factors. *Mol. Microbiol.* **45**:1389–1405.
26. **Hedges, S. R., M. S. Mayo, L. Kallman, J. Mestecky, E. W. Hook III, and M. W. Russell.** 1998. Evaluation of immunoglobulin A1 (IgA1) protease and IgA1 protease-inhibitory activity in human female genital infection with *Neisseria gonorrhoeae. Infect. Immun.* **66**:5826–5832.
27. **Henderson, I.R., and J.P. Nataro.** 2001. Virulence functions of autotransporter proteins. *Infect. Immun.* **69**:1231-1243.
28. **Hendrixon, D. R., and J. W. St. Geme III.** 1998. The *Haemophilus influenzae* Hap serine protease promotes adherence and microcolony formation, potentiated by a soluble host protein. *Mol. Cell* **2**:841–850.
29. **Hopper, S., B. Vasquez, A. Merz, S. Clary, J. S. Wilbur, and M. So.** 2000. Effects of the immunoglobulin A1 protease on *Neisseria gonorrhoeae* trafficking across polarized T84 epithelial monolayers. *Infect. Immun.* **68**:906–911.
30. **Hoskins, J., W. E. Alborn, J. Arnold, L. C. Blaszczak, S. Burgett, B. S. Dehoff, S. T. Estrem, D. J. Fu, W. Fuller, C. Geringer, R. Gilmour, H. Khoja, A. R. Kraft, R. L. Lagace, D. J. LeBlanc, L. N. Lee, E. J. Lefkowitz, J. Lu, P. Matsushima, S. M. McAhren, M. McHenney, K. McLeaster, C. W. Mundy, T. I. Nicas, F. H. Norris, M. O'Gara, R. B. Peery, G. T. Robertson, P. Rockey, P. M. Sun, M. E. Winkler, Y. Yang, M. Young-Bellido, G. Zhao, C. A. Zook, R. H. Baltz, S. R. Jaskunas, P. R. Rosteck, P. L. Skatrud, and J. I. John.** 2001. Genome of the bacterium *Streptococcus pneumoniae* strain R6. *J. Bacteriol.* **183**:5709–5717.
31. **Jarvis, G. A., and J. M. Griffiss.** 1991. Human IgA1 blockade of IgG-initiated lysis of *Neisseria meningitidis* is a function of antigen-binding fragment binding to the polysaccharide capsule. *J. Immunol.* **147**:1962–1967.
32. **Johannsen, D.B., D. M. Johnston, H. O. Koymen, M. S. Cohen, and J. G. Cannon.** 1999. A *Neisseria gonorrhoeae* immunoglobulin A1 protease mutant is infectious in the human challenge model of urethral infection. *Infect. Immun.* **67**: 3009–3013.
33. **Jose, J., G. W. Otto, and T. F. Meyer.** 2003. The integration site of the *iga* gene in commensal *Neisseria* sp. *Mol. Gen. Genom.* 269:197-204.
34. **Jose, J., U. Wölk, D. Lorenzen, H. Wenschuh, and T. F. Meyer.** 2000. Human T-cell response to meningococcal immunoglobulin A1 protease associated α-proteins. *Scand. J. Immunol.* **51**: 176–185.
35. **Kapatais-Zoumbos, K., D. K. Chandler, and M. F. Barile.** 1985. Survey of immunoglobulin A protease activity among selected species of *Ureaplasma* and *Mycoplasma* specificity for host immunoglobulin A. *Infect. Immun.* **47**:704–709.
36. **Kett, K., P. Brandtzaeg, J. Radl, and J. T. Haaijman.** 1986. Different subclass distribution of IgA-producing cells in human lymphoid organs and various secretory tissues. *J. Immunol.* **136**:3631–3635.
37. **Kilian, M., J. Mestecky, and R. E. Schrohenloher.** 1979. Pathogenic species of *Haemophilus* and *Streptococcus pneumoniae* produce immunoglobulin A1 protease. *Infect. Immun.* **26**: 143–149.
38. **Kilian, M., K. Poulsen, and H. Lomholt.** 2002. Evolution of the paralogous *hap* and *iga* genes in *Haemophilus influenzae* and evidence for a conserved *hap* pseudogene in the recently diverged *Haemophilus aegyptius* and *H. influenzae* biogroup aegyptius. *Mol. Microbiol.* **46**:1367–1380.
39. **Kilian, M., and J. Reinholdt.** 1987. A hypothetical model for the development of invasive infection due to IgA1 protease-producing bacteria. *Adv. Exp. Med. Biol.* **216B**:1261–1269.
40. **Kilian, M., J. Reinholdt, H. Lomholt, K. Poulsen, and E. V. G. Frandsen.** 1996. Biological significance of IgA1 proteases in bacterial colonization and pathogenesis: critical evaluation of experimental evidence. *APMIS* **104**:321–338.

41. **Kilian, M., and B. Thomsen.** 1983. Antigenic heterogeneity of immunoglobulin A1 proteases from encapsulated and non-encapsulated *Haemophilus influenzae. Infect. Immun.* **42:** 126–132.
42. **Kilian, M., B. Thomsen, T. E. Petersen, and H. S. Bleeg.** 1983. Occurrence and nature of bacterial IgA proteases. *Ann. N.Y. Acad. Sci.* **409:**612–624.
43. **Kirkeby, L., T. T. Rasmussen, J. Reinholdt, and M. Kilian.** 2000. Immunoglobulins in nasal secretions of healthy humans: structural integrity of secretory immunoglobulin A1 (IgA1) and occurrence of neutralizing antibodies to IgA1 proteases of nasal bacteria. *Clin. Diagn. Lab. Immunol.* **7:**31–39.
44. **Kobayashi, K., Y. Fujiyama, K. Hagiwara, and H. Kondoh.** 1987. Resistance of normal serum IgA and secretory IgA to bacterial IgA proteases: evidence for the presence of enzyme-neutralizing antibodies in both serum and secretory IgA, and also in serum IgG. *Microbiol. Immunol.* **31:**1097–1106.
45. **Lin, L., P. Ayala, J. Larson, M. Mulks, M. Fukuda, S. R. Carlsson, C. Enns, and M. So.** 1997. The *Neisseria* type 2 IgA1 protease cleaves LAMP1 and promotes survival of bacteria within epithelial cells. *Mol. Microbiol.* **24:**1083–1094.
46. **Logan, M. R., S. O. Odemuyiwa, and R. Moqbel.** 2003. Understanding exocytosis in immune and inflammatory cells: the molecular basis of mediator secretion. *J. Allergy Clin. Immunol.* **111:**923–932.
47. **Lomholt, H.** 1995. Evidence of recombination and an antigenically diverse immunoglobulin A1 protease among strains of *Streptococcus pneumoniae. Infect. Immun.* **63:**4238–4243.
48. **Lomholt, H., K. Poulsen, D. A. Caugant, and M. Kilian.** 1992. Molecular polymorphism and epidemiology of *Neisseria meningitidis* immunoglobulin A1 proteases. *Proc. Natl. Acad. Sci. USA* **89:**2120–2124.
49. **Lomholt, H., K. Poulsen, and M. Kilian.** 1995. Comparative characterization of the *iga* gene encoding IgA1 protease in *Neisseria meningitidis, Neisseria gonorrhoeae* and *Haemophilus influenzae. Mol. Microbiol.* **15:**495–506.
50. **Lomholt, H., L. van Alphen, and M. Kilian.** 1993. Antigenic variation of irnmunoglobulin A1 proteases among sequential isolates of *Haemophilus influenzae* from healthy children and patients with chronic obstructive pulmonary disease. *Infect. Immun.* **61:**4575–4581.
51. **Lomholt, J. A., and M. Kilian.** 2000. Immunoglobulin A1 protease activity in *Gemella haemolysans. J. Clin. Microbiol.* **38:**2760–2762.
52. **Lorenzen, D. R., F. Dux, U. Wölk, A. Tsirpouchtsidis, G. Haas, and T. F. Meyer.** 1999. Immunoglobulin A1 protease, an exoenzyme of pathogenic neisseriae, is a potent inducer of proinflammatory cytokines. *J. Exp. Med.* **190:**1049–1058.
53. **Mattu, T. S., R. J. Pleass, A. C. Willis, M. Kilian, M. R. Wormald, A. C. Lellouch, P. M. Rudd, J. M. Woof, and R. A. Dwek.** 1998. The glycosylation and structure of human serum IgA1, Fab and Fc regions and the role of N-glycosylation on FcαR interactions. *J. Biol. Chem.* **273:**2260–2272.
54. **Mestecky, J., and M. W. Russell.** 1986. IgA subclasses. *Monogr. Allergy* **19:**277–301.
55. **Meyer, T. F.** 1990. Pathogenic neisseriae—a model of bacterial virulence and genetic flexibility. *Int. J. Med. Microbiol.* **274:** 135–154.
56. **Morelli, G., J. del Valle, C. J. Lammel, J. Pohlner, K. Müller, M. Blake, G. F. Brooks, T. F. Meyer, B. Koumare, N. Brieske, and M. Achtman.** 1994. Immunogenicity and evolutionary variability of epitopes within IgA1 protease from serogroup A *Neisseria meningitidis. Mol. Microbiol.* **11:**175–187.
57. **Mulks, M. H., and A. G. Plaut.** 1978. IgA protease production as a characteristic distinguishing pathogenic from harmless *Neisseriaceae. N. Engl. J. Med.* **299:**973–976.
58. **Mulks, M. H., A. G. Plaut, and M. Lamm.** 1980. Gonococcal IgA protease reduces inhibition of bacterial adherence by human secretory IgA, p. 217–220. *In* S. Normark and D. Danielsson (ed.), *Genetics and Immunobiology of Pathogenic* Neisseria. University of Umea, Umea, Sweden.
59. **Novak, R., E. Charpentier, J. S. Braun, E. Park, S. Murti, E. Tuomanen, R. Masure.** 2000. Extracellular targeting of choline-bidning proteins in *Streptococcus pneumoniae* by a zinc-metalloprotease. *Mol. Microbiol.* **36:**366–376.
60. **Oggioni, M. R, G. Memmi, T. Maggi, D. Chiavolini, F. Iannelli, and G. Pozzi.** 2003. Pneumococcal zinc metalloproteinase ZmpC cleaves human matrix metalloproteinase 9 and is a virulence factor in experimental pneumonia. *Mol. Microbiol.* **49:**795–805.
61. **Parkhill, J., M. Achtman, K. D. James, S. D. Bentley, C. Churcher, S. R. Klee, G. Morelli, D. Basham, D. Brown, T. Chillingworth, R. M. Davies, P. Davis, K. Devlin, T. Feltwell, N. Hamlin, S. Holroyd, K. Jagels, S. Leather, S. Moule, K. Mungall, M. A. Quail, M. A. Rajandream, K. M. Rutherford, M. Simmonds, J. Skelton, S. Whitehead, B. G. Spratt, and B. G. Barrell.** 2000. Complete DNA sequence of a group A strain of *Neisseria meningitidis* Z2491. *Nature* **404:**502–506.
62. **Plant, L., and A. B. Jonsson.** 2003. Contacting the host: insights and implications of pathogenic *Neisseria* cell interactions. *Scand. J. Infect. Dis.* **35:**608–613.
63. **Plaut, A. G.** 1983. The IgA proteases of pathogenic bacteria. *Annu. Rev. Microbiol.* **37:**603–622.
64. **Plaut, A. G., J. V. Gilbert, M. S. Artenstein, and J. D. Capra.** 1975. *Neisseria gonorrhoeae* and *Neisseria meningitidis:* extracellular enzyme cleaves human immunoglobulin A. *Science* **193:**1103–1105.
65. **Plaut, A. G., J. Qiu, F. Grundy, F., and A. Wright.** 1992. Growth of *Haemophilus influenzae* in human milk: synthesis, distribution and activity of IgA protease as determined by study of Iga^+ and mutant Iga^- cells. *J. Infect. Dis.* **166:**43–52.
66. **Pizza, M., V. Scarlato, V. Masignani, M. M. Giuliani, B. Arico, M. Commanduci, G. T. Jennings, L. Baldi, E. Bartolini, B. Capecchi, C. L. Galeotti, E. Luzzi, R. Manetti, E. Marchetti, M. Mora, S. Nuti, G. Ratti, L. Santini, S. Savino, M. Scarselli, E. Storni, P. Zuo, M. Broeker, E. Hundt, B. Knapp, E. Blair, T. Mason, H. Tettelin, D. W. Hood, A. C. Jeffries, N. J. Saunders, D. M. Granoff, J. C. Venter, E. R. Moxon, G. Grandi, and R. Rappuoli.** 2000. Identification of vaccine candidates against group B meningococcus by whole-genome sequencing. *Science* **287:**1816–1820.
67. **Pohlner, J., R. Halter, K. Bayreuther, and T. F. Meyer.** 1987. Gene structure and extracellular secretion of *Neisseria gonorrhoeae* IgA protease. *Nature* **325:**452–462.
68. **Pohlner, J., R. Halter, and T. F. Meyer.** 1987. *Neisseria gonorrhoeae* IgA protease. Secretion and implications for pathogenesis. *Antonie Leeuwenhoek* **53:**479–484.
69. **Pohlner, J., U. Langenberg, U. Wölk, S. C. Beck, and T. F. Meyer.** 1995. Uptake and nuclear transport of *Neisseria* IgA1 protease-associated a-proteins in human cells. *Mol. Microbiol.* **17:**1073–1083.
70. **Polissi, A., A. Pontiggia, G. Feger, M. Altieri, H. Mottl, L. Ferrari, and D. Simon.** 1998. Large-scale identification of virulence genes from *Streptococcus pneumoniae. Infect. Immun.* **66:**5620–5629.
71. **Poulsen, K., J. Brandt, J. P. Hjorth, H. C. Thøgersen, and M. Kilian.** 1989. Cloning and sequencing of the immunoglobulin A1 protease gene (*iga*) of *Haemophilus influenzae* serotype b. *Infect. Immun.* **57:**3097–3105.
72. **Poulsen, K., J. Reinholdt, and M. Kilian.** 1996. Characterization of the *Streptococcus pneumoniae* immunoglobulin A1

protease gene (*iga*) and its translation product. *Infect. Immun.* **64:**3957–3966.

73. **Poulsen, K., J. Reinholdt, C. Jespergaard, K. Boye, T. A. Brown, M. Hauge, and M. Kilian.** 1998. A comprehensive genetic study of streptococcal immunoglobulin A1 proteases: evidence for recombination within and between species. *Infect. Immun.* **66:**181–190.
74. **Poulsen, K., J. Reinholdt, and M. Kilian.** 1992. A comparative genetic study of serologically distinct *Haemophilus influenzae* type 1 immunoglobulin A1 proteases. *J. Bacteriol.* **174:**2913–2921.
75. **Qiu, J., G. P. Brackee, and A. G. Plaut.** 1996. Analysis of the specificity of bacterial immunoglobulin A (IgA) protease by a comparative study of ape serum IgAs as substrates. *Infect. Immun.* **64:**933–937.
76. **Reinholdt, J., and M. Kilian.** 1987. Interference of IgA protease with the effect of secretory IgA on adherence of oral streptococci to saliva-coated hydroxyapatite. *J. Dent. Res.* **66:** 492–497.
77. **Reinholdt, J., and M. Kilian.** 1997. Comparative analysis of immunoglobulin A1 protease activity among bacteria representing different genera, species, and strains. *Infect. Immun.* **65:**4452–4459.
78. **Reinholdt, J., V. Friman, and M. Kilian.** 1993. Similar proportions of immunoglobulin A1 (IgA1) protease-producing streptococci in initial dental plaque of selectively IgA-deficient and normal individuals. *Infect. Immun.* **61:**3998–4000.
79. **Reinholdt, J., M. Tomana, S. B. Mortensen, and M. Kilian.** 1990. Molecular aspects of immunoglobulin A1 degradation by oral streptococci. *Infect. Immun.* **58:**1186–1194.
80. **Russell, M. W., M. Kilian, and M. E. Lamm.** 1999. Biological activities of IgA, p. 225–240. *In* P. L. Ogra, J. Mestecky, M. E. Lamm, W. Strober, J. Bienenstock, and J. R. McGhee. (ed.), *Handbook of Mucosal Immunology.* Academic Press, Inc., San Diego, Calif.
81. **Russell, M.W., J. Reinholdt, and M. Kilian.** 1989. Anti-inflammatory activity of human IgA antibodies and their Fabα fragments: inhibition of IgG-mediated complement activation. *Eur. J. Immunol.* **19:**2243–2249.
82. **Schiavo, G., F. Benfenati, B. Poulain, O. Rossetto, P. Polverino de Laureto, B. R. DasGupta, and C. Montecucco.** 1992. Tetanus and botulinum-B neurotoxins block neurotransmitter release by proteolytic cleavage of synaptobrevin. *Nature* **359:** 832–835.
83. **Senior, B. W., W. W. Stewart, C. Galloway, and M. A. Kerr.** 2001. Cleavage of the hormone human chorionic gonadotropin, by the type 1 IgA1 protease of *Neisseria gonorrhoeae,* and its implications. *J. Infect. Dis.* **184:**922–925.
84. **Serruto, D., J. Adu-Bobie, M. Scarselli, D. Veggi, M. Pizza, R. Rappuoli, and B. Aricò.** 2003. *Neisseria meningitidis* App, a new adhesion with autocatalytic protease activity. *Mol. Microbiol.* **48:**323–334.
85. **Shoberg, R. J., and M. H. Mulks.** 1991. Proteolysis of bacterial membrane proteins by *Neisseria gonorrhoeae* type 2 immunoglobulin A1 protease. *Infect. Immun.* **59:**2535–2541.
86. **Sørensen, C. H., and M. Kilian.** 1984. Bacterium-induced cleavage of IgA in nasopharyngeal secretions from atopic children. *Acta Pathol. Microbiol. Immunol. Scand. Ser. C* **92:** 85–87.
87. **St. Geme, J. W., III, M. L. de la Morena, and S. Falkow.** 1994. A *Haemophilus influenzae* IgA protease-like protein promotes intimate interaction with human epithelial cells. *Mol. Microbiol.* **14:**217–233.
88. **Tettelin, H., K. E. Nelson, I. T. Paulsen, J. A. Eisen, T. D. Read, S. Peterson, J. Heidelberg, R. T. DeBoy, D. H. Haft, R. J. Dodson, A. S. Durkin, M. Gwinn, J. F. Kolonay, W. C. Nelson, J. D. Peterson, A. U. Lowell, O. White, S. L. Salzberg, M. R. Lewis, D. Radune, E. Holtzapple, H. Khouri, A. M. Wolf, T. R. Utterback, C. L. Hansen, L. A. McDonald, T. V. Feldblyum, S. Angiuoli, T. Dickinson, E. K. Hickey, E. H. Ingeborg, B. J. Loftus, F. Yang, H. O. Smith, J. C. Venter, B. A. Dougherty, D. A. Morrison, S. K. Hollingshead, and C. M. Fraser.** 2001. Complete genome sequence of a virulent isolate of *Streptococcus pneumoniae. Science* **293:**498–506.
89. **Tettelin, H., N. J. Saunders, J. Heidelberg, A. C. Jeffries, K. E. Nelson, J. A. Eisen, K. A. Ketchum, D. W. Hood, J. F. Peden, R. J. Dodson, W. C. Nelson, M. L. Gwinn, R. DeBoy, J. D. Peterson, E. K. Hickey, D. H. Haft, S. L. Salzberg, O. White, R. D. Fleischmann, B. A. Dougherty, T. Mason, A. Ciecko, D. S. Parksey, E. Blair, H. Cittone, E. B. Clark, M. D. Cotton, T. R. Utterback, H. Khouri, H. Qin, J. Vamathevan, J. Gill, V. Scarlato, V. Masignani, M. Pizza, G. Grandi, L. Sun, H. O. Smith, C. M. Fraser, E. R. Moxon, R. Rappuoli, and J. C. Venter.** 2000. Complete genome sequence of *Neisseria meningitidis* serogroup B strain MC58. *Science* **287:**1809–1815.
90. **Tyler, B. M., and M. F. Cole.** 1998. Effect of IgA1 protease on the ability of secretory IgA1 antibodies to inhibit the adherence of *Streptococcus mutans. Microbiol. Immunol.* **42:**503–508.
91. **Van Ulsen, P., L. van Alphen, C. T. Hopman, A. van der Ende, and J. Thommassen.** 2001. In vivo expression of *Neisseria meningitidis* proteins homologous to the *Haemophilus influenzae* Hap and Hia autotransporters. *FEMS Immunol. Med. Microbiol.* **32:**53–64.
92. **Van Ulsen, P., L. van Alphen, J. ten Hove, F. Fransen, P. van der Ley, and J. Thommassen.** 2003. A neisserial autotransporter NalP modulating the processing of other autotransporters. *Mol. Microbiol.* **50:**1017–1030.
93. **Vitovski, S., K. T. Dunkin, A. J. Howard, and J. R. Sayers.** 2002. Nontypeable *Haemophilus influenzae* in carriage and disease: a difference in IgA1 protease activity levels. *JAMA* **287:**1699–1705.
94. **Vitovski, S., R. C. Read, and J. R. Sayers.** 1999. Invasive isolates of *Neisseria meningitidis* possess enhanced immunoglobulin A1 protease activity compared to colonizing strains. *FASEB J.* **13:**331–337.
95. **Wani, J. H., J. V. Gilbert, A. G. Plaut, and J. N. Weiser.** 1996. Identification, cloning, and sequencing of the immunoglobulin A1 protease gene of *Streptococcus pneumoniae. Infect. Immun.* **64:**3967–3974.
96. **Weiser, J. N., D. Bae, C. Fasching, R. W. Scamurra, A. J. Ratner, and E. N. Janoff.** 2003. Antibody-enhanced pneumococcal adherence requires IgA1 protease. *Proc. Natl. Acad. Sci. USA* **100:**4215–4220.
97. **Wood, S. G., and J. Burton.** 1991. Synthetic peptide substrates for the immunoglobulin A1 protease from *Neisseria gonorrhoeae* (type 2). *Infect. Immun.* **59:**1818–1822.

Colonization of Mucosal Surfaces
Edited by James P. Nataro et al.

Chapter 10

Genetic Exchange in the Respiratory Tract

CHRISTOPHER G. DOWSON

The bacterial flora of the mouth and upper respiratory tract is legion (33, 65, 76, 108, 109). Different organisms inhabit a diverse array of niches. They may exist as part of a complex community in which there are multiple strains or species, or they may be rather more isolated, sometimes apparently in pure culture. In addition to this genomic diversity, different organisms may harbor plasmids (120), conjugative transposons (20, 118), or lysogenic phage (115) carrying novel resistance (48) or virulence (101) determinants.

Among this zoo of bacteria, chromosomal diversity, and extrachromosomal elements, there are a handful of organisms, including *Streptococcus pneumoniae, Haemophilus influenzae,* and *Neisseria meningitidis,* that for the most part exist as harmless commensals but can cause devastating morbidity and mortality as the causative agents of pneumonia, otitis media, septicemia, and meningitis (96, 97, 103, 112). Although this chapter touches on other organisms, its focus is on these three important pathogens and the role that genetic exchange has played in their continuing evolution.

Over the past six decades there have been dramatic changes to the selectives pressures imposed on bacteria within the respiratory tract, during both carriage and invasive disease. Prior to the advent of vaccines and antibiotics, selective forces would have been imposed primarily by the physical and biological environment of the niche, including competition for nutrients (9), the production of bacteriocins by other bacteria (4, 26, 128), and interactions with the human immune system (51). Since the beginning of the antimicrobial era, these organisms have been confronted by a plethora of broad-spectrum antibiotics targeting central metabolic pathways, the machinery of transcription, translation, and cell wall biosynthesis. More recently, vaccines have been introduced that specifically target individual pneumococcal, meningococcal, or *H. influenzae* polysaccharide capsular types or proteins within the cell wall (121). Given all this, there are significant pressures for these organisms to escape.

A feature shared by pneumococci, meningococci, and *H. influenzae* is that they are naturally transformable (80). They possess the ability to exchange genetic material by the direct uptake of naked DNA elaborated from other strains or even other bacterial species (125). This is in addition to the alternative mechanisms of conjugation (mating) and transduction (bacteriophage-mediated transfer of DNA). There are now numerous examples where genetic exchange has been responsible for the evolution of antibiotic resistance and virulence determinants (35) and wholesale acquisition of pathogenicity islands (53), and there is increasing evidence that recombination is important in populations of both naturally transformable and nontransformable organisms (47).

The duplicity of commensalism and pathogenicity, coexistence with related strains and species, and genetic malleability are associated with numerous potential problems and pitfalls for strategies that might be applied to the control of these organisms. To anticipate these problems, we need a clear understanding of the natural ecology of these bacteria as well as an ability to identify the important mechanisms of genetic transfer and to determine how this affects their population structure and the targets selected for antibiotic therapy or vaccination.

ROLE OF RECOMBINATION

The prevalence of genetic exchange varies among different bacteria, as does the major mechanism of exchange. If these mechanisms enable the transmission of extrachromosomal elements, plasmids

Christopher G. Dowson • Biomedical Research Institute, Biological Sciences, University of Warwick, Coventry CV4 7AL, United Kingdom.

transposons, or lysogenic bacteriophage, bacteria can acquire completely novel resistance or virulence determinants (12, 55, 63, 104). For chromosomal DNA, genetic exchange is mediated primarily by homologous recombination. Where recombination occurs, the evolutionary process is accelerated (135), not only by helping the dissemination of adaptive mutations but also by breaking up associations between deleterious mutations (93) and assisting in the process of genomic repair (16, 40, 56, 59, 129).

Illegitimate recombination, the nonhomologous "shoe-horning" of DNA into a chromosome experimentally, is a rare event but can be increased in *E. coli* by inactivating DNA gyrase (123) or overproducing the helicase DnaB (134). Given the power of natural selection, it is not surprising that even rare events become biologically significant, and analysis of sequenced genomes and virulence determinants provides compelling evidence for the role of illegitimate recombination in horizontal genetic exchange in vivo (16, 111, 119). This would include past events in certain alleles of the neuraminidase gene *nanA* of pneumococci (35). Indeed, analysis of illegitimate recombination in pneumococci in vitro, using antibiotic resistance as a driver, has shown that illegitimate recombination occurred at between 1 and 2% of the frequency of homologous recombination (114).

RECOMBINATION IN BACTERIAL POPULATIONS

The population structures of bacterial species, at the simplest level, vary from being clonal where there is widespread linkage of loci around the chromosome to being nonclonal where such linkage is not apparent. The extremes of these are usually typified by clonal organisms such as *E. coli* and *Salmonella* and the apparent absence of clonality in *N. gonorrhoeae* (89). However, the picture within a given species of bacteria is not necessarily straightforward.

The very existence of clones within bacterial populations or the observation of linkage disequilibrium between alleles at different loci, using techniques such as multilocus enzyme electrophoresis (MLEE), has in the past been used as evidence for low rates of recombination. However, clones can be present in populations where recombination is frequent, arising not only from genetic isolation but also from the rapid proliferation and spread of a successful genotype. For example, there is evidence of a stable clonal lineage of arginine-, hypoxanthine-, uracil-requiring isolates within the wider nonclonal population among strains of *N. gonorrhoeae* (52). Among *N. meningitidis,* serogroup A isolates present a clonal structure (8) compared to the epidemic population structure for representative isolates from the species as a whole (81). There is also evidence suggesting that encapsulated strains of *H. influenzae* possess a more clonal structure than nontypeable strains (21, 92, 99).

Because of the sequence-based approach of multilocus sequence typing MLST to population genetics, it has now become possible to directly compare rates of recombination in different species (47). The data for this requires large MLST studies with a significant number of strains forming clonal complexes, which are closely related strains that vary from a recent common ancestor by one or two loci of the seven or more than have been sequenced. Basically, the relative roles of recombination and mutation at these divergent loci is then examined to determine whether the divergent allele has arisen by a single alteration or multiple changes. The latter is most probably associated with a recombinational event rather than point mutation. These studies indicate that in many bacterial species, including *N. meningitidis* and *S. pneumoniae,* evolutionary change at neutral (housekeeping) loci, i.e., where there is no evidence of a positive selective pressure, is more likely to occur by recombination than mutation. The per-allele recombination/mutation (r/m) parameter is 4.75:1 for *N. meningitidis* and 8.9:1 for *S. pneumoniae* (45). The relatively high ratio of recombination to mutation in pneumococci may not be solely due to the frequency of homologous recombination but, rather, may indicate that there is no SOS-induced mutagenic response in *S. pneumoniae.* This is probably because of the lack of error-prone repair function (86), perhaps explaining why mutation makes a minor contribution to genetic plasticity in *S. pneumoniae.* There is evidence that both *H. influenzae* and *N. menigitidis* also lack an error-prone repair system that responds to DNA damage (11, 68).

In such species, recombination is therefore the important mechanism for the long-term evolution of the population, but this does not occur at such a high frequency as to prevent the emergence and spread of successful clones, which are clearly identified in both of these organisms (46, 91).

RECOMBINATION AND ANTIBIOTIC RESISTANCE

Antibiotics are generally directed to single bacterial targets, e.g., rifampin to RNA polymerase and tetracyclines to the 30S subunit of ribosomes. By contrast, β-lactam antibiotics target an array of several different penicillin binding proteins (PBPs) (54). It is therefore not surprising that there are a wide range of different resistance mechanisms. Some are effective by restricting access of the antibiotic to the cellular target by reducing permeability or increasing efflux; others modify the antibiotic so that it is no longer effective, while yet others alter the antibiotic target. Either these mechanisms require the acquisition of novel proteins, e.g., those that alter efflux (61), destroy the antibiotic (120, 122), or modify the target (15), or some of the mechanisms can be attained by mutation of the chromosomally encoded targets of the antibiotic (126) or chromosomal genes encoding outer membrane proteins affecting permeability (107).

Genetic exchange involving the acquisition of plasmids or conjugative transposons carrying novel resistance determinants is undoubtedly the most widespread route by which bacteria have acquired resistance to antibiotics (25, 132). Conjugative transposon-mediated resistance is prevalent among pneumococci and is frequently responsible for reduced susceptibility to aminoglycosides, macrolides, and tetracyclines (32, 133). Target-mediated resistance to β-lactams is now widespread among pneumococci (39), and rifampin resistance has been reported on numerous occasions (42).

It is perhaps interesting to look a little more closely into how penicillin, rifampin, and tetracycline resistance has evolved, particularly in pneumococci, given that they involve a single transposon-encoded determinant and single or multiple chromosomal targets without the added complexity of additional alterations to the outer membrane, as often found in gram-negative organisms.

Penicillin Resistance in Pneumococci

PBP-mediated resistance in pneumococci requires the alteration of three *pbp* genes, 1A, 2X, and 2B (7). There are no β-lactamases associated with pneumococcal resistance, and this organism does not possess an outer membrane to restrict access of the antibiotic. Therefore, reducing the natural affinities of these three PBPs is the route by which pneumococci acquire resistance. Comparisons of *pbp* genes from a diverse range of susceptible pneumococci revealved a high degree of sequence conservation, with about 1% sequence variation among different alleles of the same *pbp*. By contrast, *pbp* genes from penicillin resistant isolates were remarkably diverse (37, 75). The conclusion drawn was that recombination had occurred between pneumococci and chromosomal DNA elaborated from related streptococci possessing homologous *pbp* genes. Further work identified two of several different donors involved in the process as *S. mitis* (36) and *S. oralis* (124).

It is still unclear whether the original donor organisms possessed the amino acid alterations conferring a low-affinity PBP by chance or whether low-affinity forms had evolved by mutation, e.g., changing one key amino acid from Thr to Ala as required for *S. mitis* PBP2B (36). The most likely scenario is one of mutation within the oral streptococci followed by transfer to pneumococci. There are clear examples of mosaic pneumococcal *pbp* genes from penicillin-resistant strains evolving by point mutation when subject to high-level cephalosporin selection (23). Nevertheless, it is clear that in all clinical isolates of pneumococci examined to date, *pbp* genes have been modified first and foremost by single or multiple rounds of interspecies recombination.

Why is this the case? It might be the result of simple probability, since multiple mutations are required in three different PBPs. Moreover, mutations conferring penicillin resistance would have to arise first in PBP2X, giving a maximum 10-fold increase in resistance from 0.008 to 0.06 μg per ml (which clinically may well be too low). This would then most probably be followed by PBP2B (a further 5-fold increase) and finally PBP1A, permitting up to a further 10- or 20-fold increase (different β-lactams would achieve this process in a different order due to the different natural affinities of PBPs for β-lactams). Premature alterations to 1A and 2B in pneumococci would be effectively cryptic. It may be that cryptic resistant PBPs are found within the oral streptococci. Although there has not been a comprehensive search for cryptic resistance, the evidence to date would indicate that this is not the case. There is, however, a far greater pool of genetic diversity among *S. mitis* and *S. oralis* than that found among pneumococci, and examination of a handful of isolates is clearly not representative of this group of organisms as a whole (130).

Like pneumococci, the oral streptococci are naturally transformable (18). They are present in large numbers as part of a complex multistrain population (58). The oral streptococci are repeatedly subject to the selective pressure of antibiotics (100), and it is clearly possible that recombination would accelerate the evolutionary process by helping the dissemination of adaptive mutations and breaking up associations between deleterious mutations. Certainly the laboratory-derived *pbp* mutants described have been slow-growing ones (74). However, genetic scrambling would generate altered PBPs that were biologically "fit" (3). It is undoubtedly the successful fit recombinants that have since spread into pneumococci. The reverse flow, involving mosaic genes acquired by pneumococci moving back into the oral streptococci, has also been observed, indicating that there is widespread interspecies recombination (38). This phenomenon is not just important for the evolution of resistance but carries important lessons about the evolution process operating within these organisms and how that might affect novel targets for vaccination.

Interestingly, β-lactam (ampicillin) resistance in *H. influenzae* has arisen by the insertion of TEM β-lactamase into phenotypically cryptic plasmids present only rarely in *H. influenzae* but frequently isolated from *H. parainfluenzae* (13). By way of contrast, for *S. pneumoniae*, penicillin resistance is not associated with plasmids and the small, stable, cryptic plasmids found in some clinical isolates have not acquired β-lactamase (105).

Rifampin Resistance

Rifampin resistance has arisen in several different species of bacteria, either by alteration of the target, the β-subunit of RNA polymerase encoded by *rpoB* (126), or by changes in outer membrane permeability (1). It has been used as a prophylactic to eliminate carriage of both *N. meningitidis* and *H. influenzae* (64) but not pneumococci (31). This method is more effective for meningococci than for *Haemophilus* because of the relative levels of susceptibility and penetration of rifampin into the nasal cavity (29). However, even with concerns about overprescribing prophylactic treatment (110), the overall frequency of resistance to rifampin among meningococci and *Haemophilus* is low (44). This may well be due to a combination of the relatively infrequent use of the antibiotic and the initial apparent fitness cost associated with *rpoB* alterations (10, 41, 131).

It is clear that in an environment with continual selection, compensatory mutations will arise, resulting in the evolution of resistant organisms that are biologically fit (117). Obviously, organisms that are highly recombinogenic will accelerate the rate of the process. Given this fact, it was interesting to examine the relative roles of mutation and recombination in the evolution of resistance in pneumococci, where resistance to rifampin is due to alterations to *rpoB* (42).

Nucleotide sequence analysis of a 270-bp fragment of *rpoB* from 16 clinical rifampin-susceptible isolates of *S. pneumoniae*, 8 clinical rifampin-resistant isolates, and 3 spontaneous rifampin-resistant mutants revealed that, as with other species, point mutations within the cluster I region of *rpoB*, at sites encoding Asp516 and HiS526, also confer resistance to rifampin (41). Sequence analysis of *rpoB*, both from these isolates of *S. pneumoniae* and from two strains of *S. mitis*, revealed that, in a number of clinical isolates, resistance to rifampin in *S. pneumoniae* has clearly arisen by point mutation. However, the nucleotide sequence of *rpoB* from one isolate suggested that interspecies gene transfer may also have played a role in the evolution of rifampin resistance in *S. pneumoniae*. There is also evidence that mutation and recombination both play roles in the evolution of fluoroquinolone resistance in pneumococci, with *S. oralis* and *S. mitis* once again being the genetic donors for interspecies recombination (6).

Tetracycline Resistance in Pneumococci

Tetracyclines are broad-spectrum bacteriostatic drugs that are active against pneumococci. In many countries, lack of susceptibility to tetracyclines is the most frequently observed resistance phenotype in pneumococci (49). The only known mechanism of tetracycline resistance in pneumococci is the protection of the bacterial 30S ribosome subunit against antibiotic binding by the TetM (2) or TetO (50) proteins, with the *tet*(M) gene being more common than the *tet*(O) gene in pneumococci. Analysis of the nucleotide sequences of *tet*(M) genes from a diverse range of bacteria clearly reveals that *tet*(M) has evolved by recombination between what appears to be an allele of *tet*(M) first identified within pneumococci, carried on Tn*1545*, and a divergent allele of *tet*(M) initially identified from *Staphylococcus aureus* (104); however, it is still unclear what selective pressure is responsible for driving this recombination.

Further data have shown that *tet*(M) gene variation in *S. pneumoniae* can occur at the inter- and intraclone levels. Identical genomic backgrounds may possess different allelic variants of *tet*(M), or the converse may occur, with the same *tet*(M) and alleles observed in unrelated *S. pneumoniae* strains (34). This indicates active movement and evolution of these genes carried on highly promiscuous conjugative transposons (20). It is also apparent that the relative stability or instability of transposon-encoded determinants differs among different clones of *S. pneumoniae*, with some clones possessing single *tet*(M) alleles and others possessing multiple alleles (34). Little has been done to assess the relative ability of different clonal groups within pneumococci (43) to participate in conjugative transposition or transformation.

GENETIC EXCHANGE AT CAPSULE LOCI

The polysaccharide capsules, which surround *H. influenzae, S. pneumoniae,* and *N. meningitidis,* are key virulence determinants, protecting the bacteria from phagocytosis. Although purified capsular polysaccharides elicit T-independent antibody responses without a memory function, covalent linkage of the capsular polysaccharide to immunogenic carrier proteins creates glycoconjugates which are T-dependent antigens. The members of this new generation of vaccines, exemplified by type B *H. influenzae* conjugate, are highly effective in protection against and reduction in the carriage of the target organisms (2, 5, 85, 106).

For *H. influenzae* there is a strong association between the capsular serotype and genotype observed in studies using both MLST and MLEE (73, 92). This suggests that capsular genes have only very rarely become established in new lineages as a result of horizontal transfer and recombination at the capsular biosynthetic locus. The presence of the type a and b capsular genes in the distantly related lineages indicates that these genes were distributed horizontally at some time in the past (92). However, for *S. pneumoniae* and *N. meningitidis,* isolates with very different genotypes frequently have the same capsular serotype (serogroup), and variation in the capsular serotype is found even within isolates of the same clone (17, 24, 72, 98, 127).

The greatest diversity in capsular type amongst these organisms is found in *S. pneumoniae.* There are currently 90 recognized immunologically distinct pneumococcal capsules (57). Each has a structurally distinct polysaccharide, composed of repeating oligosaccharide units joined by glycosidic linkages. Apart from two notable exceptions (serotypes 3 and 37), the majority of capsule loci are very similar in structure: a central region of different numbers of serotype-specific genes is preceded by four genes thought to be involved in capsule regulation and export (50, 62, 95). Except for the single gene responsible for biosynthesis of the serotype 37 capsule (*tts*) (78), the genes responsible for capsule synthesis in all serotypes sequenced so far are located between the *dexB* and *aliA* genes. Primers directed to *dexB* and *aliA* are currently being used to amplify and sequence the capsule loci from the remaining, as yet uncharacterized, serotypes (http://www.sanger.ac.uk/Projects/S_pneumoniae/CPS/).

The diversity of capsule loci is quite remarkable, not only in the different sizes of loci but also in the diversity of glycosyltransferases being employed. For example, glycosyltransferase assays showed that among 21 pneumococcal types which contain glucose in the core of their capsule polysaccharide, 19 express glucosyl-1-phosphate transferase activity. However, not all of these types hybridized strongly with Cps14E, the type 14 glucosyl-1-phosphate transferase gene. Thus, pneumococci possess glucosyltransferase genes distinct from *cps14E* but encoding enzymes with identical activity (70). This clearly indicates diverse evolutionary origins for the component parts of different capsular types. Furthermore, the composition of these capsule loci indicates the roles of both illegitimate and homologous recombination in their formation (19, 62, 94). Current work has identified a diverse array of genes within the oral streptococci that are also found within a *dexB aliA* framework and are homologues of genes found within pneumococal cap operons (C. G. Dowson et al., unpublished data). Given the evolutionary processes already described, it is likely that these genes are intimately involved in the past and potentially ongoing evolution of the pneumococcal cap locus.

Horizontal transfer of functional capsule loci, because of their general common location between *dexB* and *aliA,* is facile. The loci possess a conserved flanking sequence (*dexB* and *aliA*) that facilitates their ready transfer between strains with differing capsule types in vitro (67). The frequency with which this happens in vivo is uncertain; however, there is clear evidence of large recombinational replacements (in the region of 22 kbp) of the whole cap locus, resulting in a serotype 9V clone acquiring, on multiple occasions, serotype 14 variants (22). There is also evidence from population genetic studies that capsule switching does occur relatively frequently in vivo. Data from two studies have revealed that serotype exchange occurs in 4 to 6% of isolates (43, 98). It is unclear whether the prevalence of serotype exchange would increase as a result of mass vaccination.

Several studies have shown that vaccination with pneumococcal conjugate vaccine reduces the carriage of vaccine serotypes (27, 28, 90, 102). While replacement of vaccine-related serotypes with non-vaccine serotypes (which does not necessarily involve recombination) may well have been shown (90, 102), in other experiments serotype replacement resulting from vaccination was not shown to occur (27, 28). However, determining whether there has been a statistically significant change in serotype postvaccination may understate the underlying changes that have occurred in small trials, which may become more apparent on mass vaccination (77).

THE BOUNDARIES OF GENETIC EXCHANGE

From what is known about the promiscuity of conjugative transposons (20) and broad-host-range plasmids (116), it might be thought that there are no absolute constraints on the evolutionary distances that can be traversed through recombination in the bacterial world. However, interspecies recombination in bacteria is reduced by a variety of factors, including ecological isolation, physical obstacles to DNA entry (e.g., the lack of pili in *Neisseria* [71]), restriction endonuclease activity (66), the divergence of nucleotide sequence hindering recombination (83), and the ability to control recombination by mismatch repair (60, 83).

It would appear that different bacterial species are genetically variable for most of these factors. For example, restriction systems which differ widely between organisms function primarily against incoming double-stranded DNA (79) but may also play a role in recombination (69, 113). Furthermore, the mismatch repair system, cited as being the major barrier to sexual isolation between *E. coli* and *Salmonella* (87, 88), is only partly responsible for sexual isolation in other species (60, 83). However, an interesting feature determined for *S. pneumoniae, Bacillus* spp., and yeast (*Saccharomyces cerevisiae*) is that an approximately log-linear relationship is observed between the recombination rate and the level of sequence divergence (30, 82, 83). Recombination becomes increasingly more difficult (by log orders of magnitude) as sequences become more divergent on a linear scale. However, it has been noted before that DNA with intermediate levels of sequence divergence (1 to 10%) from *S. pneumoniae* (60) saturates the mismatch repair system, rendering it ineffective in the correction of mismatching transforming DNA.

This ability to saturate the mismatch repair system does result in an increased ease of DNA transfer from closely related organisms. Given this, along with the fact that the oral streptococci represent an enormous and poorly characterized pool of genetic information, it will not be surprising if, over the coming decades, we see further examples of these organisms playing an important role in the continuing evolution of pneumococci. This is especially the case in the field of novel anti-infectives if product manufacturers do not fully anticipate the potential changes that lie ahead. Interspecies recombination may also foster speciation by introducing novel gene loci from divergent species, allowing the invasion of new niches. This needs some serious consideration, since more conjugate vaccines strive to eradicate the benign carrier state of target organisms from the nasopharynx.

Ever-increasing mass vaccination will cause some, sometimes very significant, ecological disturbance within the nasopharynx. The long-term outcome of such a disturbance is not known. Disturbance of bacterial populations in heterogeneous environments does result in the generation of novel variants (14). Although the future for the pathogenic commensal organisms of the upper respiratory tract is uncertain, it is clear that genetic exchange will continue to play an important role in helping to shape that future.

REFERENCES

1. **Abadi, F. J., P. E. Carter, P. Cash, and T. H. Pennington.** 1996. Rifampin resistance in *Neisseria meningitidis* due to alterations in membrane permeability. *Antimicrob. Agents Chemother.* **40:**646–651.
2. **Adegbola, R. A., E. K. Mulholland, O. Secka, S. Jaffar, and B. M. Greenwood.** 1998. Vaccination with a *Haemophilus influenzae* type b conjugate vaccine reduces oropharyngeal carriage of *H. influenzae* type b among Gambian children. *J. Infect. Dis.* **177:**1758–1761.
3. **Aita, T., and Y. Husimi.** 2000. Theory of evolutionary molecular engineering through simultaneous accumulation of advantageous mutations. *J. Theor. Biol.* **207:**543–556.
4. **Balakrishnan, M., R. S. Simmonds, and J. R. Tagg.** 2001. Diverse activity spectra of bacteriocin-like inhibitory substances having activity against mutans streptococci. *Caries Res.* **35:** 75–80.
5. **Balmer, P., R. Borrow, and E. Miller.** 2002. Impact of meningococcal C conjugate vaccine in the UK. *J. Med. Microbiol.* **51:**717–722.
6. **Balsalobre, L., M. J. Ferrandiz, J. Linares, F. Tubau, and A. G. de la Campa.** 2003. Viridans group streptococci are donors in horizontal transfer of topoisomerase IV genes to *Streptococcus pneumoniae. Antimicrob. Agents Chemother.* **47:**2072–2081.
7. **Barcus, V. A., and D. C.G.** 2001. Site-directed mutagenesis to determine structure function relationships in *Streptococcus pneumoniae* penicillin-binding protein genes, p. 245–264. *In* S. H. Gillespie (ed.), *Antibiotic Resistance: Methods and Protocols.* Humana Press, Totowa, N.J.
8. **Bart, A., C. Barnabe, M. Achtman, J. Dankert, A. van der Ende, and M. Tibayrenc.** 2001. The population structure of *Neisseria meningitidis* serogroup A fits the predictions for clonality. *Infect. Genet. Evol.* **1:**117–122.
9. **Basson, N. J.** 2000. Competition for glucose between *Candida albicans* and oral bacteria grown in mixed culture in a chemostat. *J. Med. Microbiol.* **49:**969–975.
10. **Billington, O. J., T. D. McHugh, and S. H. Gillespie.** 1999. Physiological cost of rifampin resistance induced in vitro in *Mycobacterium tuberculosis. Antimicrob. Agents Chemother.* **43:**1866–1869.
11. **Black, C. G., J. A. Fyfe, and J. K. Davies.** 1998. Absence of an SOS-like system in *Neisseria gonorrhoeae. Gene* **208:**61–66.
12. **Brunton, J., D. Clare, and M. A. Meier.** 1986. Molecular epidemiology of antibiotic resistance plasmids of *Haemophilus* species and *Neisseria gonorrhoeae. Rev. Infect. Dis.* **8:**713–724.
13. **Brunton, J. L., D. Clare, N. Ehrman, and M. A. Meier.** 1983. Evolution of antibiotic resistance plasmids in *Neisseria gonorrhoeae* and *Haemophilus* species. *Clin. Investig. Med.* **6:** 221–228.
14. **Buckling, A., R. Kassen, G. Bell, and P. B. Rainey.** 2000. Disturbance and diversity in experimental microcosms. *Nature* **408:**961–964.
15. **Burdett, V.** 1986. Streptococcal tetracycline resistance mediated at the level of protein synthesis. *J. Bacteriol.* **165:**564–569.
16. **Cavalier-Smith, T.** 2002. Origins of the machinery of recombination and sex. *Heredity* **88:**125–141.

17. **Claus, H., U. Vogel, M. Muhlenhoff, R. Gerardy-Schahn, and M. Frosch.** 1997. Molecular divergence of the sia locus in different serogroups of *Neisseria meningitidis* expressing polysialic acid capsules. *Mol. Gen. Genet.* **257:**28–34.
18. **Claverys, J. P., and B. Martin.** 2003. Bacterial "competence" genes: signatures of active transformation, or only remnants? *Trends Microbiol.* **11:**161–165.
19. **Claverys, J. P., M. Prudhomme, I. Mortier-Barriere, and B. Martin.** 2000. Adaptation to the environment: *Streptococcus pneumoniae,* a paradigm for recombination-mediated genetic plasticity? *Mol. Microbiol.* **35:**251–259.
20. **Clewell, D. B., S. E. Flannagan, and D. D. Jaworski.** 1995. Unconstrained bacterial promiscuity: the Tn*916*-Tn*1545* family of conjugative transposons. *Trends Microbiol.* **3:**229–236.
21. **Cody, A. J., D. Field, E. J. Feil, S. Stringer, M. E. Deadman, A. G. Tsolaki, B. Gratz, V. Bouchet, R. Goldstein, D. W. Hood, and E. R. Moxon.** 2003. High rates of recombination in otitis media isolates of non-typeable *Haemophilus influenzae. Infect. Genet. Evol.* **3:**57–66.
22. **Coffey, T. J., M. Daniels, M. C. Enright, and B. G. Spratt.** 1999. Serotype 14 variants of the Spanish penicillin-resistant serotype 9V clone of *Streptococcus pneumoniae* arose by large recombinational replacements of the *cpsA-pbp1a* region. *Microbiology* **145:**2023–2031.
23. **Coffey, T. J., M. Daniels, L. K. McDougal, C. G. Dowson, F. C. Tenover, and B. G. Spratt.** 1995. Genetic analysis of clinical isolates of *Streptococcus pneumoniae* with high-level resistance to expanded-spectrum cephalosporins. *Antimicrob. Agents Chemother.* **39:**1306–1313.
24. **Coffey, T. J., C. G. Dowson, M. Daniels, J. Zhou, C. Martin, B. G. Spratt, and J. M. Musser.** 1991. Horizontal transfer of multiple penicillin-binding protein genes, and capsular biosynthetic genes, in natural populations of *Streptococcus pneumoniae. Mol. Microbiol.* **5:**2255–2260.
25. **Courvalin, P.** 1994. Transfer of antibiotic resistance genes between gram-positive and gram-negative bacteria. *Antimicrob. Agents Chemother.* **38:**1447–1451.
26. **Crupper, S. S., A. J. Gies, and J. J. Iandolo.** 1997. Purification and characterization of staphylococcin BacR1, a broad-spectrum bacteriocin. *Appl. Environ. Microbiol.* **63:**4185–4190.
27. **Dagan, R., R. Melamed, M. Muallem, L. Piglansky, D. Greenberg, O. Abramson, P. M. Mendelman, N. Bohidar, and P. Yagupsky.** 1996. Reduction of nasopharyngeal carriage of pneumococci during the second year of life by a heptavalent conjugate pneumococcal vaccine. *J. Infect. Dis.* **174:**1271–1278.
28. **Dagan, R., M. Muallem, R. Melamed, O. Leroy, and P. Yagupsky.** 1997. Reduction of pneumococcal nasopharyngeal carriage in early infancy after immunization with tetravalent pneumococcal vaccines conjugated to either tetanus toxoid or diphtheria toxoid. *Pediatr. Infect. Dis. J.* **16:**1060–1064.
29. **Darouiche, R., B. Perkins, D. Musher, R. Hamill, and S. Tsai.** 1990. Levels of rifampin and ciprofloxacin in nasal secretions: correlation with MIC_{90} and eradication of nasopharyngeal carriage of bacteria. *J. Infect. Dis.* **162:**1124–1127.
30. **Datta, A., M. Hendrix, M. Lipsitch, and S. Jinks-Robertson.** 1997. Dual roles for DNA sequence identity and the mismatch repair system in the regulation of mitotic crossing-over in yeast. *Proc. Natl. Acad. Sci. USA* **94:**9757–9762.
31. **Davies, A. J., A. Dyas, and V. R. Rao.** 1985. Prophylaxis in bacterial meningitis. *J. Hosp. Infect.* **6:**281–284.
32. **de Azavedo, J. C., P. Pieroni, P. Chang, and D. E. Low.** 1997. Association of transposon Tn*1545* with multidrug resistant strains of *Streptococcus pneumoniae* isolated in Canada. *Adv. Exp. Med. Biol.* **418:**475–478.
33. **Dewhirst, F. E., M. A. Tamer, R. E. Ericson, C. N. Lau, V. A. Levanos, S. K. Boches, J. L. Galvin, and B. J. Paster.** 2000. The diversity of periodontal spirochetes by 16S rRNA analysis. *Oral Microbiol. Immunol.* **15:**196–202.
34. **Doherty, N., K. Trzcinski, P. Pickerill, P. Zawadzki, and C. G. Dowson.** 2000. Genetic diversity of the *tet*(M) gene in tetracycline-resistant clonal lineages of *Streptococcus pneumoniae. Antimicrob. Agents. Chemother.* **44:**2979–2984.
35. **Dowson, C. G., V. Barcus, S. King, P. Pickerill, A. Whatmore, and M. Yeo.** 1997. Horizontal gene transfer and the evolution of resistance and virulence determinants in *Streptococcus. Soc. Appl. Bacteriol. Symp. Ser.* **26:**42S–51S.
36. **Dowson, C. G., T. J. Coffey, C. Kell, and R. A. Whiley.** 1993. Evolution of penicillin resistance in *Streptococcus pneumoniae:* the role of *Streptococcus mitis* in the formation of a low affinity PBP2B in *S. pneumoniae. Mol. Microbiol.* **9:**635–643.
37. **Dowson, C. G., A. Hutchison, J. A. Brannigan, R. C. George, D. Hansman, J. Linares, A. Tomasz, J. M. Smith, and B. G. Spratt.** 1989. Horizontal transfer of penicillin-binding protein genes in penicillin- resistant clinical isolates of *Streptococcus pneumoniae. Proc. Natl. Acad. Sci. USA* **86:**8842–8846.
38. **Dowson, C. G., A. Hutchison, N. Woodford, A. P. Johnson, R. C. George, and B. G. Spratt.** 1990. Penicillin-resistant viridans streptococci have obtained altered penicillin-binding protein genes from penicillin-resistant strains of *Streptococcus pneumoniae. Proc. Natl. Acad. Sci. USA* **87:**5858–5862.
39. **Dowson, C. G., and K. Trzcinski (ed.).** 2001. *Evolution and Epidemiology of Antibiotic Resistant Pneumococci.* Marcel Dekker, Inc., New York, N.Y.
40. **El Karoui, M., V. Biaudet, S. Schbath, and A. Gruss.** 1999. Characteristics of Chi distribution on different bacterial genomes. *Res. Microbiol.* **150:**579–587.
41. **Enne, V. I., A. A. Delsol, J. M. Roe, and P. M. Bennett.** 2004. Rifampicin resistance and its fitness cost in *Enterococcus faecium. J. Antimicrob. Chemother.* **53:**203–207.
42. **Enright, M., P. Zawadski, P. Pickerill, and C. G. Dowson.** 1998. Molecular evolution of rifampicin resistance in *Streptococcus pneumoniae. Microb. Drug Resist.* **4:**65–70.
43. **Enright, M. C., and B. G. Spratt.** 1998. A multilocus sequence typing scheme for *Streptococcus pneumoniae:* identification of clones associated with serious invasive disease. *Microbiology* **144:**3049–3060.
44. **Enting, R. H., L. Spanjaard, D. van de Beek, E. F. Hensen, J. de Gans, and J. Dankert.** 1996. Antimicrobial susceptibility of *Haemophilus influenzae, Neisseria meningitidis,* and *Streptococcus pneumoniae* isolates causing meningitis in The Netherlands, 1993–1994. *J. Antimicrob. Chemother.* **38:** 777–786.
45. **Feil, E. J., E. C. Holmes, D. E. Bessen, M. S. Chan, N. P. Day, M. C. Enright, R. Goldstein, D. W. Hood, A. Kalia, C. E. Moore, J. Zhou, and B. G. Spratt.** 2001. Recombination within natural populations of pathogenic bacteria: short-term empirical estimates and long-term phylogenetic consequences. *Proc. Natl. Acad. Sci. USA* **98:**182–187.
46. **Feil, E. J., M. C. Maiden, M. Achtman, and B. G. Spratt.** 1999. The relative contributions of recombination and mutation to the divergence of clones of *Neisseria meningitidis. Mol. Biol. Evol.* **16:**1496–1502.
47. **Feil, E. J., and B. G. Spratt.** 2001. Recombination and the population structures of bacterial pathogens. *Annu. Rev. Microbiol.* **55:**561–590.
48. **Gomez-Lus, R.** 1998. Evolution of bacterial resistance to antibiotics during the last three decades. *Int. Microbiol.* **1:**279–284.

49. **Gruneberg, R. N.** 2002. Global surveillance through PROTEKT: the first year. *J. Chemother.* **14**(Suppl. 3):9–16.
50. **Guidolin, A., J. K. Morona, R. Morona, D. Hansman, and J. C. Paton.** 1994. Nucleotide sequence analysis of genes essential for capsular polysaccharide biosynthesis in *Streptococcus pneumoniae* type 19F. *Infect. Immun.* **62:**5384–5396.
51. **Gupta, S., and R. M. Anderson.** 1999. Population structure of pathogens: the role of immune selection. *Parasitol. Today* **15:**497–501.
52. **Gutjahr, T. S., M. O'Rourke, C. A. Ison, and B. G. Spratt.** 1997. Arginine-, hypoxanthine-, uracil-requiring isolates of *Neisseria gonorrhoeae* are a clonal lineage with a non-clonal population. *Microbiology* **143:**633–640.
53. **Hacker, J., G. Blum-Oehler, I. Muhldorfer, and H. Tschape.** 1997. Pathogenicity islands of virulent bacteria: structure, function and impact on microbial evolution. *Mol. Microbiol.* **23:**1089–1097.
54. **Hakenbeck, R., M. Tarpay, and A. Tomasz.** 1980. Multiple changes of penicillin-binding proteins in penicillin-resistant clinical isolates of *Streptococcus pneumoniae. Antimicrob. Agents Chemother.* **17:**364–371.
55. **Hayashi, T., T. Baba, H. Matsumoto, and Y. Terawaki.** 1990. Phage-conversion of cytotoxin production in *Pseudomonas aeruginosa. Mol. Microbiol.* **4:**1703–1709.
56. **Heinemann, J. A.** 1991. Genetics of gene transfer between species. *Trends Genet.* **7:**181–185.
57. **Henrichsen, J.** 1995. Six newly recognized types of *Streptococcus pneumoniae. J. Clin. Microbiol.* **33:**2759–2762.
58. **Hohwy, J., J. Reinholdt, and M. Kilian.** 2001. Population dynamics of *Streptococcus mitis* in its natural habitat. *Infect. Immun.* **69:**6055–6063.
59. **Howard-Flanders, P.** 1975. Repair by genetic recombination in bacteria: overview. *Basic Life Sci.* **5A:**265–274.
60. **Humbert, O., M. Prudhomme, R. Hakenbeck, C. G. Dowson, and J. P. Claverys.** 1995. Homeologous recombination and mismatch repair during transformation in *Streptococcus pneumoniae:* saturation of the Hex mismatch repair system. *Proc. Natl. Acad. Sci. USA* **92:**9052–9056.
61. **Jacobs, M. R., and C. E. Johnson.** 2003. Macrolide resistance: an increasing concern for treatment failure in children. *Pediatr. Infect. Dis. J.* **22:**S131–S138.
62. **Jiang, S. M., L. Wang, and P. R. Reeves.** 2001. Molecular characterization of *Streptococcus pneumoniae* type 4, 6B, 8, and 18C capsular polysaccharide gene clusters. *Infect. Immun.* **69:**1244–1255.
63. **Johnson, L. P., P. M. Schlievert, and D. W. Watson.** 1980. Transfer of group A streptococcal pyrogenic exotoxin production to nontoxigenic strains of lysogenic conversion. *Infect. Immun.* **28:**254–257.
64. **Jones, D. M.** 1991. Chemoprophylaxis of meningitis. *Trans. R. Soc. Trop. Med. Hyg.* **85**(Suppl. 1):44–45.
65. **Kazor, C. E., P. M. Mitchell, A. M. Lee, L. N. Stokes, W. J. Loesche, F. E. Dewhirst, and B. J. Paster.** 2003. Diversity of bacterial populations on the tongue dorsa of patients with halitosis and healthy patients. *J. Clin. Microbiol.* **41:**558–563.
66. **Kelleher, J. E., and E. A. Raleigh.** 1991. A novel activity in *Escherichia coli* K-12 that directs restriction of DNA modified at CG dinucleotides. *J. Bacteriol.* **173:**5220–5223.
67. **Kelly, T., J. P. Dillard, and J. Yother.** 1994. Effect of genetic switching of capsular type on virulence of *Streptococcus pneumoniae. Infect. Immun.* **62:**1813–1819.
68. **Kimball, R. F., M. E. Boling, and S. W. Perdue.** 1977. Evidence that UV-inducible error-prone repair is absent in *Haemophilus influenzae* Rd, with a discussion of the relation to error-prone repair of alkylating-agent damage. *Mutat. Res.* **44:**183–196.
69. **King, G., and N. E. Murray.** 1994. Restriction enzymes in cells, not eppendorfs. *Trends Microbiol.* **2:**465–469.
70. **Kolkman, M. A., B. A. van der Zeijst, and P. J. Nuijten.** 1998. Diversity of capsular polysaccharide synthesis gene clusters in *Streptococcus pneumoniae. J. Biochem.* (Tokyo) **123:**937–945.
71. **Koomey, M.** 1998. Competence for natural transformation in *Neisseria gonorrhoeae:* a model system for studies of horizontal gene transfer. *APMIS Suppl.* **84:**56–61.
72. **Kriz, P., D. Giorgini, M. Musilek, M. Larribe, and M. K. Taha.** 1999. Microevolution through DNA exchange among strains of *Neisseria meningitidis* isolated during an outbreak in the Czech Republic. *Res. Microbiol.* **150:**273–280.
73. **Kroll, J. S.** 1992. The genetics of encapsulation in *Haemophilus influenzae. J. Infect. Dis.* **165**(Suppl. 1):S93–S96.
74. **Laible, G., and R. Hakenbeck.** 1991. Five independent combinations of mutations can result in low-affinity penicillin-binding protein 2x of *Streptococcus pneumoniae. J. Bacteriol.* **173:**6986–6990.
75. **Laible, G., B. G. Spratt, and R. Hakenbeck.** 1991. Interspecies recombinational events during the evolution of altered PBP 2x genes in penicillin-resistant clinical isolates of *Streptococcus pneumoniae. Mol. Microbiol.* **5:**1993–2002.
76. **Linoli, O., S. Marconi, and M. Garaffa.** 1981. Quantitative bacterial ecology of normal nasal mucosa. *Ann. Sclavo* **23:**151–161. (In Italian.)
77. **Lipsitch, M.** 2001. Interpreting results from trials of pneumococcal conjugate vaccines: a statistical test for detecting vaccine-induced increases in carriage of nonvaccine serotypes. *Am. J. Epidemiol.* **154:**85–92.
78. **Llull, D., E. Garcia, and R. Lopez.** 2001. Tts, a processive beta-glucosyltransferase of *Streptococcus pneumoniae,* directs the synthesis of the branched type 37 capsular polysaccharide in pneumococcus and other gram-positive species. *J. Biol. Chem.* **276:**21053–21061.
79. **Loenen, W. A.** 2003. Tracking *Eco*KI and DNA fifty years on: a golden story full of surprises. *Nucleic Acids Res.* **31:**7059–7069.
80. **Lorenz, M. G., and W. Wackernagel.** 1994. Bacterial gene transfer by natural genetic transformation in the environment. *Microbiol. Rev.* **58:**563–602.
81. **Maiden, M. C., J. A. Bygraves, E. Feil, G. Morelli, J. E. Russell, R. Urwin, Q. Zhang, J. Zhou, K. Zurth, D. A. Caugant, I. M. Feavers, M. Achtman, and B. G. Spratt.** 1998. Multilocus sequence typing: a portable approach to the identification of clones within populations of pathogenic microorganisms. *Proc. Natl. Acad. Sci. USA* **95:**3140–3145.
82. **Majewski, J., and F. M. Cohan.** 1999. DNA sequence similarity requirements for interspecific recombination in *Bacillus. Genetics* **153:**1525–1533.
83. **Majewski, J., P. Zawadzki, P. Pickerill, F. M. Cohan, and C. G. Dowson.** 2000. Barriers to genetic exchange between bacterial species: *Streptococcus pneumoniae* transformation. *J. Bacteriol.* **182:**1016–1023.
84. Reference deleted.
85. **Makela, P. H.** 2003. Conjugate vaccines—a breakthrough in vaccine development. *Southeast Asian J. Trop. Med. Public Health* **34:**249–253.
86. **Martin, B., P. Garcia, M. P. Castanie, B. Glise, and J. P. Claverys.** 1995. The *recA* gene of *Streptococcus pneumoniae* is part of a competence-induced operon and controls an SOS regulon. *Dev. Biol. Stand.* **85:**293–300.

87. **Matic, I., F. Taddei, and M. Radman.** 1996. Genetic barriers among bacteria. *Trends Microbiol.* **4:**69–72.
88. **Matic, I., F. Taddei, and M. Radman.** 2000. No genetic barriers between *Salmonella enterica* serovar Typhimurium and *Escherichia coli* in SOS-induced mismatch repair-deficient cells. *J. Bacteriol.* **182:**5922–5924.
89. **Maynard-Smith, J., N. H. Smith, M. O'Rouke, and B. G. Spratt.** 1993. How clonal are bacteria? *Proc. Nat. Acad. Sci. USA* **90:**4384–4388.
90. **Mbelle, N., R. E. Huebner, A. D. Wasas, A. Kimura, I. Chang, and K. P. Klugman.** 1999. Immunogenicity and impact on nasopharyngeal carriage of a nonavalent pneumococcal conjugate vaccine. *J. Infect. Dis.* **180:**1171–1176.
91. **McGee, L., L. McDougal, J. Zhou, B. G. Spratt, F. C. Tenover, R. George, R. Hakenbeck, W. Hryniewicz, J. C. Lefevre, A. Tomasz, and K. P. Klugman.** 2001. Nomenclature of major antimicrobial-resistant clones of *Streptococcus pneumoniae* defined by the pneumococcal molecular epidemiology network. *J. Clin. Microbiol.* **39:**2565–2571.
92. **Meats, E., E. J. Feil, S. Stringer, A. J. Cody, R. Goldstein, J. S. Kroll, T. Popovic, and B. G. Spratt.** 2003. Characterization of encapsulated and noncapsulated *Haemophilus influenzae* and determination of phylogenetic relationships by multilocus sequence typing. *J. Clin. Microbiol.* **41:**1623–1636.
93. **Moran, N. A.** 1996. Accelerated evolution and Muller's rachet in endosymbiotic bacteria. *Proc. Natl. Acad. Sci. USA* **93:**2873–2878.
94. **Morona, J. K., R. Morona, and J. C. Paton.** 1999. Analysis of the 5′ portion of the type 19A capsule locus identifies two classes of *cpsC, cpsD,* and *cpsE* genes in *Streptococcus pneumoniae. J. Bacteriol.* **181:**3599–3605.
95. **Morona, J. K., J. C. Paton, D. C. Miller, and R. Morona.** 2000. Tyrosine phosphorylation of CpsD negatively regulates capsular polysaccharide biosynthesis in *Streptococcus pneumoniae. Mol. Microbiol.* **35:**1431–1442.
96. **Mulholland, E. K., and R. A. Adegbola.** 1998. The Gambian *Haemophilus influenzae* type b vaccine trial: what does it tell us about the burden of *Haemophilus influenzae* type b disease? *Pediatr. Infect. Dis. J.* **17:**S123–S125.
97. **Mulholland, K.** 1999. Strategies for the control of pneumococcal diseases. *Vaccine* **17**(Suppl. 1)**:**S79–S84.
98. **Muller-Graf, C. D., A. M. Whatmore, S. J. King, K. Trzcinski, A. P. Pickerill, N. Doherty, J. Paul, D. Griffiths, D. Crook, and C. G. Dowson.** 1999. Population biology of *Streptococcus pneumoniae* isolated from oropharyngeal carriage and invasive disease. *Microbiology* **145:**3283–3293.
99. **Musser, J. M., J. S. Kroll, D. M. Granoff, E. R. Moxon, B. R. Brodeur, J. Campos, H. Dabernat, W. Frederiksen, J. Hamel, G. Hammond, et al.** 1990. Global genetic structure and molecular epidemiology of encapsulated *Haemophilus influenzae. Rev. Infect. Dis.* **12:**75–111.
100. **Nakayama, A., and A. Takao.** 2003. Beta-lactam resistance in *Streptococcus mitis* isolated from saliva of healthy subjects. *J. Infect. Chemother.* **9:**321–327.
101. **Novick, R. P.** 2003. Mobile genetic elements and bacterial toxinoses: the superantigen-encoding pathogenicity islands of *Staphylococcus aureus. Plasmid* **49:**93–105.
102. **Obaro, S., and R. Adegbola.** 2002. The pneumococcus: carriage, disease and conjugate vaccines. *J. Med. Microbiol.* **51:**98–104.
103. **O'Brien, K. L., and H. Nohynek.** 2003. Report from a WHO working group: standard method for detecting upper respiratory carriage of *Streptococcus pneumoniae. Pediatr. Infect. Dis. J.* **22:**133–140.
104. **Oggioni, M. R., C. G. Dowson, J. M. Smith, R. Provvedi, and G. Pozzi.** 1996. The tetracycline resistance gene *tet*(M) exhibits mosaic structure. *Plasmid* **35:**156–163.
105. **Oggioni, M. R., F. Iannelli, and G. Pozzi.** 1999. Characterization of cryptic plasmids pDP1 and pSMB1 of *Streptococcus pneumoniae. Plasmid* **41:**70–72.
106. **Overturf, G. D.** 2000. American Academy of Pediatrics Committee on Infectious Diseases technical report. Prevention of pneumococcal infections, including the use of pneumococcal conjugate and polysaccharide vaccines and antibiotic prophylaxis. *Pediatrics* **106:**367–376.
107. **Pan, W., and B. G. Spratt.** 1994. Regulation of the permeability of the gonococcal cell envelope by the mtr system. *Mol. Microbiol.* **11:**769–775.
108. **Paster, B. J., S. K. Boches, J. L. Galvin, R. E. Ericson, C. N. Lau, V. A. Levanos, A. Sahasrabudhe, and F. E. Dewhirst.** 2001. Bacterial diversity in human subgingival plaque. *J. Bacteriol.* **183:**3770–3783.
109. **Paster, B. J., W. A. Falkler, Jr., C. O. Enwonwu, E. O. Idigbe, K. O. Savage, V. A. Levanos, M. A. Tamer, R. L. Ericson, C. N. Lau, and F. E. Dewhirst.** 2002. Prevalent bacterial species and novel phylotypes in advanced noma lesions. *J. Clin. Microbiol.* **40:**2187–2191.
110. **Pearson, N., D. J. Gunnell, C. Dunn, T. Beswick, A. Hill, and B. Ley.** 1995. Antibiotic prophylaxis for bacterial meningitis: overuse and uncertain efficacy. *J. Public Health Med.* **17:**455–458.
111. **Pedulla, M. L., M. E. Ford, J. M. Houtz, T. Karthikeyan, C. Wadsworth, J. A. Lewis, D. Jacobs-Sera, J. Falbo, J. Gross, N. R. Pannunzio, W. Brucker, V. Kumar, J. Kandasamy, L. Keenan, S. Bardarov, J. Kriakov, J. G. Lawrence, W. R. Jacobs, Jr., R. W. Hendrix, and G. F. Hatfull.** 2003. Origins of highly mosaic mycobacteriophage genomes. *Cell* **113:**171–182.
112. **Pollard, A. J., and D. Scheifele.** 2001. Meningococcal disease and vaccination in North America. *J. Paediatr. Child Health* **37:**S20–S27.
113. **Price, C., and T. A. Bickle.** 1986. A possible role for DNA restriction in bacterial evolution. *Microbiol. Sci.* **3:**296–299.
114. **Prudhomme, M., V. Libante, and J. P. Claverys.** 2002. Homologous recombination at the border: insertion-deletions and the trapping of foreign DNA in *Streptococcus pneumoniae. Proc. Natl. Acad. Sci. USA* **99:**2100–2105.
115. **Ramirez, M., E. Severina, and A. Tomasz.** 1999. A high incidence of prophage carriage among natural isolates of *Streptococcus pneumoniae. J. Bacteriol.* **181:**3618–3625.
116. **Rawlings, D. E., and E. Tietze.** 2001. Comparative biology of IncQ and IncQ-like plasmids. *Microbiol. Mol. Biol. Rev.* **65:**481–496.
117. **Reynolds, M. G.** 2000. Compensatory evolution in rifampin-resistant *Escherichia coli. Genetics* **156:**1471–1481.
118. **Roberts, A. P., G. Cheah, D. Ready, J. Pratten, M. Wilson, and P. Mullany.** 2001. Transfer of Tn*916*-like elements in microcosm dental plaques. *Antimicrob. Agents Chemother.* **45:**2943–2946.
119. **Rocha, E. P.** 2003. An appraisal of the potential for illegitimate recombination in bacterial genomes and its consequences: from duplications to genome reduction. *Genome Res.* **13:**1123–1132.
120. **Saunders, J. R., C. A. Hart, and V. A. Saunders.** 1986. Plasmid-mediated resistance to beta-lactam antibiotics in gram-negative bacteria: the role of in-vivo recyclization reactions in plasmid evolution. *J. Antimicrob. Chemother.* **18**(Suppl. C)**:**57–66.

121. **Segal, S., and A. J. Pollard.** 2003. The future of meningitis vaccines. *Hosp. Med.* **64:**161–167.
122. **Shaw, W. V.** 1983. Chloramphenicol acetyltransferase: enzymology and molecular biology. *Crit. Rev. Biochem.* **14:**1–46.
123. **Shimizu, H., H. Yamaguchi, Y. Ashizawa, Y. Kohno, M. Asami, J. Kato, and H. Ikeda.** 1997. Short-homology-independent illegitimate recombination in *Escherichia coli:* distinct mechanism from short-homology-dependent illegitimate recombination. *J. Mol. Biol.* **266:**297–305.
124. **Sibold, C., J. Henrichsen, A. Konig, C. Martin, L. Chalkley, and R. Hakenbeck.** 1994. Mosaic *pbpX* genes of major clones of penicillin-resistant *Streptococcus pneumoniae* have evolved from *pbpX* genes of a penicillin-sensitive *Streptococcus oralis. Mol. Microbiol.* **12:**1013–1023.
125. **Smith, J. M., C. G. Dowson, and B. G. Spratt.** 1991. Localized sex in bacteria. *Nature* **349:**29–31.
126. **Spratt, B. G.** 1994. Resistance to antibiotics mediated by target alterations. *Science* **264:**388–393.
127. **Swartley, J. S., A. A. Marfin, S. Edupuganti, L. J. Liu, P. Cieslak, B. Perkins, J. D. Wenger, and D. S. Stephens.** 1997. Capsule switching of *Neisseria meningitidis. Proc. Natl. Acad. Sci. USA* **94:**271–276.
128. **Upton, M., J. R. Tagg, P. Wescombe, and H. F. Jenkinson.** 2001. Intra- and interspecies signaling between *Streptococcus salivarius* and *Streptococcus pyogenes* mediated by SalA and SalA1 lantibiotic peptides. *J. Bacteriol.* **183:**3931–3938.
129. **van Ham, R. C., J. Kamerbeek, C. Palacios, C. Rausell, F. Abascal, U. Bastolla, J. M. Fernandez, L. Jimenez, M. Postigo, F. J. Silva, J. Tamames, E. Viguera, A. Latorre, A. Valencia, F. Moran, and A. Moya.** 2003. Reductive genome evolution in *Buchnera aphidicola. Proc. Natl. Acad. Sci. USA* **100:**581–586.
130. **Whatmore, A. M., A. Efstratiou, A. P. Pickerill, K. Broughton, G. Woodard, D. Sturgeon, R. George, and C. G. Dowson.** 2000. Genetic relationships between clinical isolates of *Streptococcus pneumoniae, Streptococcus oralis,* and *Streptococcus mitis:* characterization of "atypical" pneumococci and organisms allied to *S. mitis* harboring *S. pneumoniae* virulence factor-encoding genes. *Infect. Immun.* **68:**1374–1382.
131. **Wichelhaus, T. A., B. Boddinghaus, S. Besier, V. Schafer, V. Brade, and A. Ludwig.** 2002. Biological cost of rifampin resistance from the perspective of *Staphylococcus aureus. Antimicrob. Agents Chemother.* **46:**3381–3385.
132. **Williams, J. D., and F. Moosdeen.** 1986. Antibiotic resistance in *Haemophilus influenzae:* epidemiology, mechanisms, and therapeutic possibilities. *Rev. Infect. Dis.* **8**(Suppl. 5)**:**S555–S561.
133. **Woolley, R. C., A. Pennock, R. J. Ashton, A. Davies, and M. Young.** 1989. Transfer of Tn*1545* and Tn*916* to *Clostridium acetobutylicum. Plasmid* **22:**169–174.
134. **Yamashita, T., K. Hanada, M. Iwasaki, H. Yamaguchi, and H. Ikeda.** 1999. Illegitimate recombination induced by overproduction of DnaB helicase in *Escherichia coli. J. Bacteriol.* **181:**4549–4553.
135. **Zhang, Y. X., K. Perry, V. A. Vinci, K. Powell, W. P. Stemmer, and S. B. del Cardayre.** 2002. Genome shuffling leads to rapid phenotypic improvement in bacteria. *Nature* **415:**644–646.

Colonization of Mucosal Surfaces
Edited by James P. Nataro et al.

Chapter 11

Regulation in Response to Environmental Conditions

Peggy A. Cotter

In healthy immunocompetent individuals, mechanical, chemical, and cellular defenses operate effectively to keep the trachea and lungs free of biological (and nonbiological) airborne insults. Consequently, a majority of the respiratory mucosa, unlike that of the gastrointestinal tract, is maintained in a state of near sterility. The nasopharynx and nasal cavity, however, harbor a variety of commensal bacteria. Interestingly, it is a subset of these that is responsible for most of the serious respiratory illnesses that occur. As might be expected, it is this subset of respiratory mucosal colonizers that has received the most attention from the scientific and medical communities. Research has focused primarily on identifying and characterizing bacterial factors that mediate adherence, invasion, induction of cytotoxicity, and evasion of innate and adaptive immune mechanisms. It is becoming clear, though, that precise and coordinate control of virulence factor expression is as important to the infectious process as the factors themselves. This chapter discusses what is currently known regarding how and why a few of the most important human respiratory pathogens control gene expression in response to environmental conditions.

NEISSERIA MENINGITIDIS

Neisseria meningitidis (the meningococcus) is a gram-negative bacterium that usually colonizes the healthy human nasopharynx transiently (duration times are estimated at 4.1 months [73]) and asymptomatically, such that 3 to 30% of the population is colonized at any given time. Occasionally, however, these bacteria traverse the nasopharyngeal epithelium to enter the bloodstream and cause septicemia. From the bloodstream, meningococci can interact with endothelial cells of brain microvessels, cross the blood-brain barrier, and cause fulminant meningitis (146, 206). In developing countries, especially those within the African meningitis belt, epidemics of invasive meningococcal infection are major causes of morbidity and mortality, with 10,000 or more deaths resulting from a single outbreak (80). *N. meningitidis* infection remains a significant problem in developed countries as well—the annual incidence of invasive meningococcal disease in Europe is between 0.3 and 7.1 per 100,000, and in the United States *N. meningitidis* has become the leading cause of bacterial meningitis in older children and young adults.

Comparative analyses using in vitro, ex vivo, and in vivo models have provided significant insight into the complex series of events that occur during meningococcal infection, and several surface structures involved in adherence, invasion, and evasion of host immune defense mechanisms have been identified. Initial specific but relatively weak contact of meningococci with host cells occurs via type IV pili, which bind to the complement resistance protein family member CD46 (or membrane cofactor protein) (104, 183). Type IV pili also mediate twitching motility, which contributes to dispersal over the epithelial surface (163). At later times during infection, the bacteria lose their pili and become intimately and tightly attached to host cells, into which they may ultimately be engulfed (182). Intimate attachment and invasion appears to be mediated by integral membrane proteins called Opa proteins (190, 210). Identified Opa receptors include heparan sulfate proteoglycans (208), vitronectin (which binds $\alpha_v\beta_5$- or $\alpha_v\beta_3$-integrins on host cells) (213) and CD66 family members (208, 212, 216). Opc proteins, which bear some amino acid sequence similarity to Opa proteins, are also involved. They allow meningococci to adhere to and invade endothelial cells in the absence of Opa proteins and pili by binding vitronectin (210,

Peggy A. Cotter • Department of Molecular, Cellular and Developmental Biology, University of California Santa Barbara, Santa Barbara, CA 93106-9610.

211, 213, 214). Internalized bacteria grow within and traverse through epithelial cells, egressing to subepithelial stromal tissues and then to the bloodstream. During the course of invasive disease, meningococci encounter components of innate and adaptive host immune responses (117, 129, 162). Resistance to complement, opsonization, and engulfment by phagocytic cells are mediated, at least in part, by the meningococcal capsule, which is composed of long polysialic acid chains, variations in which form the basis for classifying *N. meningitidis* strains into serogroups (217). The meningococcal capsule has also been postulated to increase the survival of meningococci in the environment and to enhance transmission between hosts (38, 143).

With the exception of the group B capsule, all of the factors described above are highly immunogenic, and *N. meningitidis* has evolved complex genetic mechanisms to vary its antigenicity (antigenic variation) and/or expression (phase variation). Genes encoding the pilin subunits that compose the type IV pili undergo nonreciprocal recombination-mediated gene conversion events, resulting in the expression of pili with alterations in immunoreactivity, posttranslational modification, and adhesive function (121, 122, 146, 209, 215). Opa proteins are encoded by four genes in *N. meningitidis* that undergo phase variation via a slipped-strand mispairing (SSM) mechanism and antigenic variation via horizontal gene transfer and recombination (184, 185). Opc protein levels vary due to mutations that occur in a poly(C) tract within the promoter region of the *opc* gene (173), and capsular phase variation results from frameshift mutations within a poly(C) tract in the coding region of the polysialyltransferase gene (*siaD*) (78).

Phase and antigenic variation are so prevalent and have proven to be so important in neisserial pathogenesis that investigating these processes, from their underlying mechanisms to their roles in pathogenesis, has been the predominant focus of research aimed at understanding virulence gene regulation in these bacteria. Consequently, relatively little is known about how meningococci might control the expression of virulence genes in response to changes in environmental conditions. Since meningococci are obligate human respiratory commensals that only occasionally (perhaps accidentally) cause invasive disease, it may be that the number of different environmental conditions encountered by *N. meningitidis* is actually small, a hypothesis supported by the fact that only four two-component regulatory systems and a handful of putative regulators belonging to other families are predicted to be encoded in the sequenced *N. meningitidis* genomes (151, 194). Recent data, however, indicate that meningococci do regulate virulence gene expression in response to at least one "environmental" cue—that of host cell contact.

As mentioned above, relatively weak pilus-mediated adherence of encapsulated meningococci to epithelial and endothelial cells is followed by intimate adherence mediated by Opa and/or Opc proteins. This process also requires loss of the polysialic acid capsule from group B meningococci. Using *pilC1-lacZ* fusions, Taha et al. showed that transcription of *pilC1*, which encodes a tip-located pilus adhesin, is transiently upregulated following contact with the Hec-1-B human epithelial cell line (192). Subsequent work by this group identified a LysR-type transcriptional regulator, encoded by *crgA*, that is also transiently induced in response to epithelial cell contact. CrgA appears to function as an autorepressor as well as a repressor of *pilE*, which encodes the major pilin subunits, and the *sia* operon, which encodes genes required for synthesis of the polysialic acid capsule (48, 49). These data suggest a temporal sequence of events that begins with host cell contact, which causes the induction of *pilC* transcription (to increase pilus-mediated binding) and *crgA* transcription. Once sufficiently expressed, CrgA represses the synthesis of pili and capsule so that Opa- and Opc-mediated intimate adherence can occur. While it appears that CrgA plays a critical role in mediating the switch between initial adherence and intimate contact, it does not appear to be responsible for sensing the signal. In *E. coli*, pilus-mediated contact itself appears to generate signals that are sensed by two-component regulatory systems (150, 156). Whether a similar mechanism is operating in *N. meningitidis* remains to be determined.

HAEMOPHILUS INFLUENZAE

Like *N. meningitidis*, *Haemophilus influenzae* is a gram-negative obligate human respiratory tract commensal that only occasionally causes severe invasive disease such as meningitis. Also similar to *N. meningitidis*, the ability of *H. influenzae* to cause systemic illness in healthy individuals requires expression of a polysaccharide capsule that protects the bacteria from phagocytosis and the bactericidal activity of complement (144). Vaccines consisting of type b capsular polysaccharide conjugated to non-*Haemophilus* immunogenic carrier proteins have been hugely successful, resulting in dramatic decreases in the incidence of invasive *Haemophilus* disease. These vaccines do not, however, protect against *H. influenzae* of other capsular types or against unencapsulated (nontypeable) *H. influenzae* (NTHi) strains, which, despite their lack of capsule, are not avirulent. NTHi strains are a common cause of otitis media, sinusitis, and conjunctivitis in healthy children and adults and are frequently implicated in exacerbations of underlying lung disease, including chronic bronchitis, bronchiectasis, and cystic fibrosis. They can also cause community-acquired pneumonia and serious systemic illness in immunocompromised individuals.

As with *N. meningitidis,* several surface-associated protein factors with postulated roles in *H. influenzae* pathogenesis have been identified. Type I pili promote adherence to human oro- and nasopharyngeal epithelial cells and nasal tissue in organ culture (57, 119, 165, 205). Hap, an autotransporter protein that is expressed by all *H. influenzae* strains, promotes adherence to and invasion of cultured epithelial cells and also mediates aggregation and microcolony formation (90, 187). HMW1 and HMW2 are high-molecular-weight surface-exposed proteins of the unlinked autotransporter family that are expressed by most NTHi strains (89). Despite significant similarity between them, they have different binding specificities, suggesting that they may interact with distinct cellular receptors, perhaps at different times or locations during the infection process (47, 97). NTHi strains that do not express HMW1 and HMW2 express Hia, another large surface-exposed protein, which promotes adherence to a broad range of epithelial cell types in vitro (111, 186). An Hia homolog, called Hsf, appears to be expressed by all typeable *H. influenzae* strains (186). Two outer membrane porins have also been implicated in adherence. P2 interacts with mucin via recognition of sialic acid-containing oligosaccharides (166), while P5 interacts with human CEACAM1 (93, 207). OapA, a surface-associated lipoprotein, seems to play a minor role in attachment to epithelial cells in vitro (161) but is required for efficient colonization of the nasopharynx in an infant rat model (222).

H. influenzae lipopolysaccharide (LPS) also plays an important role in adherence and invasion. It is composed of lipid A, an inner core of singly phosphorylated 2-keto-3-deoxyoctulosonic acid (KDO) linked to three heptose molecules, and an outer core consisting of a heteropolymer of glucose and galactose. The *licA* gene product, phosphorylcholine kinase, catalyzes the substitution of the outer core with phosphorylcholine (ChoP) (223, 224). (The presence and role of ChoP on respiratory tract bacteria are discussed in detail in chapter 5.) ChoP is a component of platelet-activating factor, and ChoP-substituted LPS appears to mediate the invasion of *H. influenzae* by interacting with platelet-activating factor receptors (191).

Although not as extensive (or possibly just not as extensively characterized) as in *Neisseria,* phase and antigenic variation occur in *H. influenzae* as well. For example, phase-variable incorporation of ChoP into LPS occurs and is controlled by SSM-mediated frameshifting of *licA* (223, 224). An SSM-mediated mechanism also appears to be involved in the control of HMW1 and HMW2 expression. Dawid et al. found that the number of heptanucleotide tandem repeats in the promoters of HMW1 and HMW2 correlated inversely with transcription levels (46). Interestingly, *H. influenzae* strains isolated from the nasopharynx contained small numbers of the heptanucleotide repeats while isolates from the middle ear contained large numbers, suggesting that HMW1 and HMW2 are important for nasopharyngeal colonization but not for otitis media.

Even less well characterized is the ability of *H. influenzae* to control gene expression in response to environmental conditions. Since *H. influenzae* is an obligate human respiratory tract commensal, like *N. meningitidis,* the number of different environmental conditions it encounters may be small. Correspondingly, with only six two-component systems predicted in the sequenced genome, its capacity for signal transduction appears to be somewhat limited (63). A recent report indicates that one of the two-component systems, ArcAB, may play a role in serum resistance (51). Like its *Escherichia coli* ortholog, ArcAB of *H. influenzae* controls gene expression in response to redox conditions during growth in vitro (69). Whether these in vitro conditions reflect actual in vivo (micro)environments encountered during infection, however, is unknown. Another recent report indicates that *H. influenzae* may be able to sense when it is internalized by macrophages and to respond by activating the expression of an extracytoplasmic function sigma factor (37). It is not known how expression of the sigma factor is controlled, but it (and presumably the genes it controls) is required for intracellular survival of the bacteria (37). Thus, while studies investigating how *H. influenzae* responds to its environment have begun, there is much to be learned. With genome sequence availability and microarray technology becoming routine, it is likely that our understanding of virulence gene regulation in *H. influenzae* will expand rapidly in the coming years.

STREPTOCOCCUS PNEUMONIAE

Streptococcus pneumoniae (the pneumococcus) is also a commensal of the human nasopharynx. Although disease is rare in relation to colonization, damage occurs when pneumococci spread to the lungs to cause pneumonia, to the middle ear to cause otitis media, or to the bloodstream to cause septicemia. From the bloodstream, it can invade the meninges, causing meningitis. Worldwide, *S. pneumoniae* is the most common cause of acute respiratory infection and otitis media and is estimated to result in 3 million deaths in children every year from pneumonia, bacteremia, or meningitis. *S. pneumoniae* is the leading cause of community-acquired pneumonia and meningitis in the elderly (53, 196). The mortality rate of 40,000 per year caused by *S. pneumoniae* in the United States is higher than that caused by any other bacterial pathogen. Like *N. meningitidis* and *H. influenzae, S. pneumoniae* expresses a polysaccharide capsule that plays a major role in pathogenesis. More than 90 structurally distinct capsular polysaccharide types have been identified, which form the basis for serotyping. Anticapsular antibodies are highly protective, and following on the success of *H. influenzae* vaccines, polyvalent conjugate antipneumococcal vaccines have recently been developed that are proving to be highly efficacious (13, 14, 62).

Like *H. influenzae, S. pneumoniae* expresses ChoP on its surface, and this contributes to adherence to epithelial cells by binding the receptor for platelet-activating factor (40, 41, 107). As discussed in chapter 5, surface exposure of ChoP in this gram-positive bacterium occurs via its incorporation into cell wall teichoic acids and lipoteichoic acids (LTA). In addition to capsule and ChoP-decorated LTA, many surface-exposed protein factors with proposed roles in pneumococcal pathogenesis have been identified. Pneumococcal surface protein A (PspA) is a large, antigenically variable protein that is attached to the surface via noncovalent binding to ChoP (197). It appears to prevent complement activation, thus reducing complement-mediated clearance and phagocytosis (100). Hyaluronate lyase (Hyl), an LPXTG-containing cell wall-anchored protein, is a member of the hyaluronidase family of enzymes that breaks down extracellular matrix components (9, 160). Autolysin (LytA) is a cell wall-degrading enzyme involved in peptidoglycan turnover during cell growth that can also cause cell lysis (138, 198). Its action can result in the release of highly inflammatory cell wall components and/or intracellular enzymes such as pneumolysin (Ply), a pore-forming toxin that is cytotoxic to ciliated bronchial epithelial cells and that inhibits the microbicidal activity of polymorphonuclear leukocytes (PMNs) and macrophages (164, 172, 181). Pneumococcal surface antigen A (PsaA) appears to be involved in importing Mn^{2+} and Zn^{2+} into the cytoplasm and is required for virulence in a mouse model (10, 52). Choline binding protein A (CbpA), which is attached to the bacterial cell surface via binding to choline, promotes adherence to host cells, contributes to nasophayngeal colonization, and binds the secretory component of secretory immunoglobulin A (79, 171). *S. pneumoniae* expresses two neuraminidases, NanA and NanB, which have been hypothesized to contribute to adherence by cleaving terminal sialic acid residues from host cell surface glycans (8, 19). A recent report by Shakhnovich et al., however, suggests that these enzymes may also desialylate the LPS of *N. meningitidis* and *H. influenzae* to give *S. pneumoniae* a competitive advantage during infection (178).

With 13 two-component regulatory systems plus one "unpaired" response regulator identified in the sequenced genomes (96, 193), pneumococci appear to have considerably more potential for regulating gene expression in response to environmental conditions than meningococci and *H. influenzae*. To investigate the roles of these systems in pneumococcal pathogenesis, Throup et al. systematically mutated each locus and compared the resulting mutant strains with wild-type pneumococci in a mouse pneumonia model (195). Eight of the mutants were attenuated in this model. Using an infant-rat model, Sebert et al. found only one of those eight loci, *ciaRH,* to be important for nasopharyngeal colonization (177), suggesting that the other seven may be important under conditions encountered during colonization and/or invasion of the lower respiratory tract. Using microarray technology, Sebert et al. also identified several CiaRH-regulated genes and showed that one of them, *htrA,* which encodes a putative serine protease, is important for nasal colonization. The virulence defect displayed by the Δ*htrA* strain was not as severe as that of the Δ*ciaRH* strain, however, suggesting that additional CiaRH-regulated genes are important for pathogenesis. The CiaRH regulon appears to comprise at least 18 chromosomal regions and includes genes that are involved in the synthesis and modification of cell wall polymers and production of pheromones and bacteriocin in addition to the *htrA-spo0J* region (126). As is the case with most two-component regulatory systems, the environmental signals to which CiaRH responds are unknown, but a recent report suggests that its expression is controlled by RegR, a LacI/GalR family regulator, in response to changes in pH (23). CiaRH may therefore be part of a large regulatory hierarchy, with its expression controlled in response to one set of environmental conditions (by RegR) and its activity controlled by another (via CiaH phosphorylation).

Among the genes that appear to be regulated when pneumococci transition from nasal cavity colonization to severe, invasive disease are those encoding the polysaccharide capsule. As mentioned above,

S. pneumoniae is capable of expressing more than 90 distinct capsular types. In addition, pneumococcal isolates of the same capsular type show variation in the amount of capsular polysaccharide they produce (120). Weiser and colleagues have shown that *S. pneumoniae* strains alternate with characteristic frequencies between opaque (O) and transparent (T) forms (220). The O forms express more capsular polysaccharide and less teichoic acid and are more virulent for systemic infection than the T forms, which are better able to colonize the nasopharynx (107). More recently, this group showed that the amount of capsular polysaccharide expressed by the O variants is controlled in response to oxygen availability: 5- to 10-fold more polysaccharide is produced in cells grown under anaerobic conditions than under atmospheric growth conditions (221). Interestingly, the phosphorylation state of CpsD, an autophosphorylating tyrosine protein kinase, was found to also vary in response to oxygen concentration (221). CpsD is thought to function together with CpsC in polymerization and export of capsular polysaccharide (141). CpsB is a manganese-dependent phosphotyrosine protein phosphatase that is required to dephosphorylate CpsD. Phosphorylation of CpsD attenuates its activity, possibly by preventing it from interacting productively with CpsC. Thus, these proteins function together to control capsular polysaccharide expression via a posttranslational mechanism. How oxygen availability might alter CpsD phosphorylation/dephosphorylation kinetics is unknown. In a recent report, Kadioglu et al. demonstrated that MicAB, a two-component regulatory system that contributes to the repression of competence under oxygen limitation (55), is required for lung colonization and systemic infection (103). Although this has not been tested, it is intriguing to hypothesize that capsule expression might also fall under the control of MicAB. Another regulatory locus, *regM*, has recently been implicated in controlling capsule expression at the level of transcription of the capsular polysaccharide biosynthesis locus (*cps*) (72). Like RegR, RegM is a LacI/GalR-type transcriptional regulator. Capsular biosynthesis, therefore, may also be controlled by hierarchically arranged regulatory systems acting at the level of transcription as well as control of biosynthetic enzymes. Understanding how these regulators coordinate to mediate precise control of capsule and other virulence factors in response to specific environmental conditions awaits further study.

Several groups have now used signature-tagged mutagenesis and murine pneumonia models to perform unbiased searches for pneumococcal virulence genes (84, 113, 159). In each case, genes encoding putative two-component systems and transcriptional regulators belonging to other families were identified. Among those identified by Hava and Camilli (84) was *rlrR*, a homolog of *rofA* and *nra* of *Streptococcus pyogenes* (see below). *rlrA* resides in a pathogenicity islet that also contains genes encoding three putative sortase-anchored surface proteins and three sortase enzymes. Two of those genes (one encoding a sortase and the other encoding a surface protein) were required for lung infection (84). Subsequent analysis showed that RlrA activates transcription of all of the genes in the pathogenicity islet (85). Also identified in this study was *mgrA*, which encodes an ortholog of Mga, a transcriptional repressor of many virulence genes in *S. pyogenes* (see below). Microarray analysis showed that MgrA functions to repress the *rlrA* pathogenicity islet, but additional MgrA-regulated genes were not detected (88). Comparative analyses between *S. pneumoniae* and *S. pyogenes* should facilitate understanding the structure and function of the Mga/MgrA regulons of these two streptococcal species. Indeed, such analyses, facilitated by genome sequences and genomic techniques, are likely to yield significant insight into the basic pathogenic mechanisms used by these and other bacteria to colonize respiratory tissues and, sometimes, cause severe invasive disease.

STREPTOCOCCUS PYOGENES

S. pyogenes (group A streptococcus [GAS]) differs from the bacteria described above in its prevalence, in its ability to colonize the skin as well as the respiratory mucosa, and in the diversity of diseases it causes. It is the most common cause of bacterial pharyngitis (strep throat) worldwide and the cause of scarlet fever. Colonization of the skin can result in impetigo (a localized infection of the superficial epidermis), ecthyma (infection of the entire epidermis), erysipelas (infection that spreads laterally through the upper dermis), and cellulitis (infection that spreads through the lower dermis). GAS may also produce more invasive life-threatening illnesses such as puerperal sepsis, myositis, necrotizing fasciitis, and toxic shock (see reference 42 for a review). GAS can further elicit delayed nonsuppurative sequelae that include rheumatic fever, acute glomerulonephritis, and reactive arthritis, and it has recently been associated with Tourette's syndrome, tics, and movement and attention deficit disorders (42). Although in development, safe and effective vaccines for GAS are not currently available. Post-streptococcal rheumatic fever remains a major cause of morbidity and mortality in developing countries, and life-threatening invasive GAS infections appear to be on the rise in North America and Europe. A population-based surveillance study between 1995 and 1999 found that invasive GAS infections occurred at a rate of 3.6 per 100,000 annually in the United States, accounting for 9,600 to 9,700 cases and 1,100 to 1,300 deaths (147). Mortality rates exceeded 20% for pneumonia, necrotizing fasciitis, and central nervous system infections and 44.5% for toxic shock syndrome.

A remarkably large number of surface-exposed and secreted factors with postulated roles in GAS virulence have been identified. LTA has been proposed to provide initial relatively weak adherence by interacting with fibronectin (82). Stronger adherence may be mediated in part by M proteins; large, surface-exposed coiled-coil proteins that have their C termini anchored in the cell wall and their N termini exposed. Hypervariability at the N terminus forms the basis for the 124 Lancefield serological classification groups of GAS that have been identified so far. M proteins mediate adherence to HEp-2 cells in vitro (36, 219), but their role in adherence to the oropharyngeal cavity has not been definitively established (22, 34, 95). They promote adherence to skin keratinocytes by binding membrane cofactor CD46 (148, 149). GAS expresses a variety of fibronectin binding proteins that appear to be important mediators of adherence to both throat and skin epithelium. These include protein F1 (PrtF1), SbfII, FBP54, protein F2, and PFBP (see reference 35 for a review). After initial adherence, persistence on the mucosal surface may be facilitated by microcolony formation, which requires the expression of M protein (22). Although GAS strains are traditionally considered extracellular pathogens, recent in vitro studies indicate that they may invade a variety of respiratory epithelial cell lines (39, 64, 75, 77, 99, 112) and both M protein and fibronectin binding proteins have been implicated in mediating this process (54, 99, 139). Whether cell invasion plays a role in deep tissue invasion and systemic disease remains to be elucidated. M protein also plays an important role in resistance to phagocytosis. It binds complement control proteins, such as factor H, factor H-like protein 1 and C4b binding protein, as well as fibrinogen, to prevent activation of complement via binding by C3b (6, 102, 155, 228). In immune hosts, specific opsonic antibodies override these protective mechanisms. The GAS capsule, however, is also an important antiphagocytic factor, and since it is composed of hyaluronic acid, which is similar to that present in human connective tissue, it is poorly immunogenic. Capsule expression varies greatly among strains; in heavily encapsulated strains, it may be the primary mediator of resistance to phagocytosis (45). A variety of secreted proteins that may serve to facilitate the spread of invasive disease by lysing host cells and/or destroying connective tissue have been identified. These include streptolysin O (a pore-forming cytolysin similar to pneumolysin), streptolysin S, streptokinase, hyaluronidase, streptococcal pyrogenic exotoxin B (SpeB), C5a peptidase, and several DNases (see references 11 and 42 for reviews). GAS strains also secrete a family of at least 10 proteins (streptococcal pyrogenic exotoxins) that function as superantigens. These proteins, which are highly similar structurally to staphylococcal superantigens, simultaneously bind major histocompatibility complex class II molecules and T-cell receptors, resulting in antigen-independent activation of a large number of T cells (123). The inappropriate release of immunomodulatory cytokines by these T cells is thought to be responsible ultimately for the hypotension and multiorgan failure characteristic of streptococcal toxic shock syndrome (15, 83).

Like the pneumococci, group A streptococci appear to have considerable potential for sensing and responding to changes in their environment; 13 two-component regulatory systems and a variety of transcriptional regulatory proteins belonging to other families have been identified in the genomes of the three GAS strains for which nucleotide sequences have been determined (5, 60, 179). Given the diversity of diseases GAS is capable of causing and the multitude of virulence factors it is capable of expressing, it is perhaps surprising that there are not more. As discussed below, only a few of these systems have been characterized so far. Moreover, much of what has been learned comes from analysis of in vitro-grown cells and in vivo experiments using wound infection models. Deciphering how the information obtained from these studies relates to the events that occur during respiratory infection will require careful comparative analyses and relevant in vivo models. It is apparent from the progress that has been made so far, however, that regardless of the site of entry, the regulatory circuitry that controls virulence gene regulation in GAS is complex, multilayered, and far from being completely understood.

The first identified and most extensively characterized regulator of virulence gene expression in GAS is the multiple-gene activator Mga. Originally identified as a positive regulator of M-protein expression (21, 170), Mga was subsequently shown to activate the transcription of many GAS virulence genes including, depending on the serotype, those encoding M and M-related proteins (*emm*, *enn*, *mrp*, and *arp*), streptococcal collagen-like protein (*scl1*/*sclA*), C5a peptidase (*scpA*), fibronectin binding proteins (*sfb2* and *sof*), streptococcal pyrogenic exotoxin B (*speB*), secreted inhibitor of complement (*sic*), and *mga* itself (see references 11, 42, and 110 for reviews). In the laboratory, expression of these genes (i.e., the Mga regulon) is controlled in response to environmental and growth phase conditions; it is maximal during exponential growth or in the presence of elevated levels of CO_2 and minimal during stationary-phase growth or in high-salt, ambient O_2 and CO_2, or limiting iron or at 25°C (21, 130, 132, 157). It has been hypothesized that these in vitro conditions, especially exponential-phase growth, correspond to those encountered in vivo during the rapid growth that is associated with entry of GAS into new tissue sites, a phase for which many Mga-regulated genes are required (110). What are the specific signals to which the Mga regulon responds, and how does Mga convert those signals into changes in gene expression patterns? Although DNA binding studies have clearly demonstrated the importance of Mga in transcriptional control, its role in sensing environmental cues is less clear. Although Mga was long thought to be a response regulator of a two-component regulatory system, it now appears that it may instead represent a novel class of gram-positive bacterial regulatory factors (131). Interestingly, Mga contains four discernible helix-turn-helix DNA binding motifs at its N terminus. Two have been shown to be important for DNA binding and transcriptional activation at the promoters for *emm, scpA,* and *mga* (131). Roles for the other two, or for the C-terminal domain of the protein, remain to be determined, but there is presently no evidence that Mga protein activity is controlled in response to environmental conditions. Rather, control appears to be at the level of transcription of *mga* itself, since changes in *mga* expression are required for the observed changes in expression of Mga-regulated genes (131, 133). Mga may therefore function as an intermediate component of a regulatory cascade. With this hypothesis in mind, Ribardo and McIver sought genes required for the control of *mga* transcription using transposon mutagenesis (169). Their search identified a gene, *amrA,* predicted to encode an integral membrane sugar exporter involved in translocation of oligosaccharides. Ribardo and McIver hypothesize that rather than functioning directly as a regulator, AmrA might be involved in generating a signal (such as cell wall polyaccharide precursor pools) that links the rapid growth of the exponential phase to Mga regulon expression. It remains to be determined if and how such a signal might be sensed by Mga or as yet unidentified regulatory proteins.

Transition of GAS from an initial colonization phase to a persistence and spreading phase in vivo has been hypothesized to correlate with late-exponential/early-stationary-phase growth in vitro (110). During this phase, expression of the Mga regulon is minimal. Two regulatory protein families whose members are variably present in different GAS strains have been implicated in mediating the repression of *mga:* RofA-like proteins (RALPs) and Rgg/RopB. RofA was first identified in an M6 serotype strain as an activator of *prtF* under O_2-rich conditions (65, 66). More recently, it was shown to also be involved in the repression of *speA, sagA,* and *mga* (4). Nra, a RALP characterized in an M49 strain of GAS, mediates repression of fibronectin binding protein 2 (*prtF2*), collagen binding protein (*cpa*), and *mga,* but its activity is not influenced by O_2 concentration (158). Nra is apparently controlled at the level of gene expression, which is maximal under stationary-phase conditions, but how that control is achieved is unknown. The functions of other RALPs (identified in M1 strains) have not been investigated. Members of the Rgg/RopB family are similar to transcriptional regulators in other gram-positive bacteria that control the expression of extracellular products during stationary-phase growth (25). In GAS, Rgg/RopB activates the expression of *speB* during stationary-phase growth (24, 204) and microarray analyses revealed a role for Rgg/RopB in repression of a large number of genes both within and outside the Mga regulon (26). It is not known if and how RALPs and Rgg/RopB sense environmental cues or exactly what those cues are. It appears, though, that by mediating the activation and repression of different classes of genes within and outside the Mga regulon, RALPs and Rgg/RopB may play important roles in controlling the shift in virulence gene expression patterns required to make the transition from colonization to invasive disease.

Like *N. meningitidis, H. influenzae,* and *S. pneumoniae,* GAS produce a polysaccharide capsule that is critical for the development of infection. Acapsular mutants are impaired in their ability to colonize the upper airways of intranasally infected mice and baboons (3, 98, 226) and are relatively avirulent in experimental models of systemic disease and invasive soft tissue infection (2, 175, 227), while highly encapsulated strains have greater lethality in a mouse model of soft tissue infection (2) and have been linked to acute rheumatic fever and severe invasive infection in humans (101). The hyaluronic acid capsule has been reported to inhibit GAS invasion of HEp-2 cells and human keratinocytes (43, 176). However, it may be that deep tissue invasion is mediated by bacteria that move between cells rather than through them since Cywes et al. showed recently that capsule-mediated binding to CD44 induces cytoskeletal rearrangements that lead to disruption of tight junctions and paracellular movement of the bacteria across the epithelial layer (44). Capsule-mediated resistance to complement-mediated phagocytic killing (45, 142) is expected to be important for survival in this environment. Since the expression of this important virulence factor was known to vary but was not controlled by Mga, identification of genes involved in its regulation was a major research goal. Four groups independently and nearly simultaneously identified a two-component regulatory system, *csrRS* (also called *covRS*), responsible for repression of capsule biosynthesis (*hasAB* expression) (7, 58, 86, 116). CsrRS was subsequently shown to control several additional GAS virulence genes. It negatively regulates expression of the pyrogenic exotoxin B (*speB*), the streptolysin S-associated gene (*sagA*), streptokinase (*ska*), mitogenic factor (*speMF*), inhibitor of innate immunity (*mac*), and itself (58, 86, 114). A recent microarray analysis indicates that the total number of genes controlled by CsrRS may in fact be quite large, perhaps up to 15% of all GAS genes (74). While repression of *hasAB, speB, speMF, sagA,* and *ska* appears to be direct (137), it is likely that in most cases, control is indirect.

What role does CsrRS-mediated signal transduction play in GAS pathogenesis? Levin and Wessels showed that a Δ*csrR* strain was resistant to opsonophagocytic killing by blood leukocytes and was hypervirulent in a mouse model of systemic lethal infection (116). Similarly, Heath et al. showed that Δ*csrR* strains, but not their wild-type parents, caused necrotic lesions in a mouse soft tissue damage model (86). Some, but not all, of this hypervirulence could be attributed to increased capsule production. Interestingly, Engleberg et al. found that coinoculation of wild-type and Δ*csrRS* GAS strains into the same site resulted in enhanced in vivo growth of the wild-type strain in the necrotic lesions that formed (56). Moreover, they found that spontaneous *csrRS* mutants arose at a relatively high rate during experimental murine skin infections and that human clinical isolates obtained from invasive skin infections frequently contained mutations in *csrRS* (56). The fact that *csrRS* mutants were more virulent than wild-type GAS in all of these experimental systems indicates that the CsrRS regulatory system must be at least somewhat active during infection. The fact that *csrRS* mutants that arise during naturally occurring human invasive skin infections are hypervirulent but have not become the predominant strains circulating in the population indicates that there must be a selective advantage for maintaining an intact *csrRS* locus. When, where, and why might it be advantageous for CsrRS to repress the genes under its control?

Gryllos et al. showed recently that CsrRS responds to Mg^{2+} and that the concentration required for its full activation in vitro is ~10 mM (76). Since the concentration of Mg^{2+} in extracellular fluids is estimated to be ~1 mM, they hypothesized that CsrRS is predominantly inactive during infection and that it may become active only when the bacteria are in an ex vivo environment outside the human host. This conclusion is in apparent contrast to the observations described above. However, inspection of the data presented by Gryllos et al. reveals that CsrRS activity varies incrementally in response to the Mg^{2+} concentration and appears to be at least partially active at concentrations as low as 1 mM. Thus, while CsrRS may not be fully active during infection, it may be partially and significantly active at specific locations in vivo. Miller et al. noted differential expression of CsrRS-regulated genes and suggested that such control could result from differences in CsrRS protein concentration, the degree to which CsrR is phosphorylated (which would, according to Gryllos et al. [76], vary in response to [Mg^{2+}]), and the number and affinity of CsrR binding sites present at each promoter (137). At least two regulatory loci, *rgg* and *rocA,* were recently implicated in controlling *csrRS* transcription (12, 26). Together, these data suggest that CsrRS may repress different subsets of genes at different locations within and outside the human host, depending not only on Mg^{2+} concentration and promoter binding site differences but also on the activity of Rgg and RocA (and possibly other, as yet unidentified regulatory factors). Thus, like CiaRH of *S. pneumoniae,* CsrRS may be a part of a regulatory hierarchy in which its expression is controlled in response to one set of environmental conditions and its activity is controlled in response to another.

Investigation of just a few regulatory systems in *S. pyogenes* has already revealed considerable overlap. Investigation of two more, the Ihk/Irr two-component regulatory system and the *sil* locus, has just begun. Ihk/Irr appears to be involved in resistance to killing by PMNs and induction of host cell lysis (218), while the *sil* locus contains genes with similarity to those involved in quorum sensing and control of competence in *S. pneumoniae* and appear to be important for the development of invasive disease in GAS (92). Deciphering how these regulatory systems, and those remaining to be studied, contribute to the complex regulatory circuitry that controls virulence in *S. pyogenes* will undoubtedly be essential for understanding how these organisms colonize healthy tissue and how they cause disease.

BORDETELLA PERTUSSIS

Like the other bacteria discussed in this chapter, *Bordetella pertussis* has adapted exclusively to the human host and has no known (nonhuman) animal or environmental reservoir. Unlike the species described above, though, *B. pertussis* is not considered a commensal—its presence has always been associated with disease and is diagnostic for the illness known as whooping cough or pertussis. Pertussis is a severe and highly contagious respiratory illness of infants and young children that is characterized by intense paroxysmal coughing spasms that can last for up to 6 months (27). Also unlike the bacteria described above, *B. pertussis* does not cause severe invasive or systemic disease, although serious complications, including pneumonia, seizures, encephalopathy, and even death, sometimes occur. An efficacious pertussis vaccine has been developed, but despite its widespread use, disease persists. Pertussis cases are estimated to approach 1 million per year in the United States alone, and the overall incidence has been increasing over the last several years (230), particularly in older children and young adults, possibly due to the relatively short-lived immunity conferred by vaccination.

B. pertussis is a member of "the *Bordetella bronchiseptica* cluster" (70), a group of extremely closely related gram-negative bacteria that colonize respiratory surfaces. The progenitor organism for this group appears to have been *B. bronchiseptica* or a *B. bronchiseptica*-like bacterium. In contrast to *B. pertussis, B. bronchiseptica* has a very broad host range (it has been isolated from nearly every mammalian species) and, like the other bacteria discussed in this chapter, typically colonizes its hosts chronically and asymptomatically and only rarely causes severe disease such as bronchopneumonia. A third member of the *B. bronchiseptica* cluster, *B. parapertussis*$_{hu}$, is more closely related to *B. bronchiseptica* than it is to *B. pertussis,* but it has been isolated only from humans and it causes a disease that is nearly indistinguishable from that caused by *B. pertussis.* Comparative analyses including these three subspecies have proven to be extremely useful for determining the roles of virulence gene regulation and of regulated factors in the establishment, maintenance, and transmission of respiratory infection.

Bordetellae employ a relatively simple pathogenic strategy: adherence to respiratory epithelium followed by the secretion of toxins that contribute to persistence and/or transmission to a new host. A variety of factors with proven or postulated roles in adherence or toxicity have been identified (see references 30 and 127 for reviews). Putative adhesins include filamentous hemagglutinin (FHA), fimbriae (Fim), pertactin, BrkA, TcfA, and BipA. Like HMW1 and HMW2 of *H. influenzae,* FHA is a member of the unlinked autotransporter family. It is a large (>230-kDa), highly immunogenic, rod-shaped protein that mediates adherence to a variety of cell types in vitro, and studies with *B. bronchiseptica* and its natural hosts have shown that FHA is absolutely required for tracheal colonization in vivo (33, 167, 168). Members of the *B. bronchiseptica* cluster express at least two serologically distinct fimbrial types, Fim2 and Fim3. Although direct evidence for Fim-mediated adherence in vitro has been difficult to obtain, experiments using *B. bronchiseptica* and natural hosts have shown that fimbriae are essential for persistence in the trachea and play an important role in influencing the immune response that develops (71, 128, 140). Pertactin, BrkA, and TcfA are all autotransporter proteins. Roles in adherence have been implicated from in vitro studies, but roles in vivo have not been demonstrated (59, 61, 115). BipA is a large outer membrane protein with significant amino acid sequence similarity to intimin of enteropathogenic and enterohemorrhagic *E. coli* and invasin of *Yersinia* spp. (189). Based on this similarity and its unique expression pattern (see below), BipA has been hypothesized to function as an adhesin that is important for aerosol transmission (68, 189). Protein toxins expressed by *Bordetella* include a calmodulin-activated adenylate cyclase that possesses hemolytic activity (CyaA) (91), a dermonecrotic toxin (Dnt) (118), pertussis toxin (Ptx) (105), and those exported by a type III secretion system (231, 232). CyaA inhibits the microbicidal activity of PMNs and macrophages in vitro and is essential for overcoming PMN-mediated defenses in vivo (67, 81, 154). As its name implies, Dnt causes dermonecrotic lesions when injected intradermally (118, 153), but an in vivo role for this protein, which is not secreted by the bacteria, has not been clearly established. Ptx, which is expressed only by *B. pertussis* strains, is an AB-type toxin that ADP-ribosylates a variety of G proteins in vitro (106, 203). It is associated with leukocytotsis in vivo (87, 229). The *Bordetella* type III secretion system, encoded by genes within the *bsc* locus, is a potent inducer of cytotoxicity in epithelial cells and macrophages in vitro and appears to be involved in modulating host immunity in vivo (188, 231, 232).

The genome sequences for one strain each of *B. pertussis, B. parapertussis*$_{hu}$, and *B. bronchiseptica* have been determined (152). Their inspection reveals the potential to encode 15 (*B. pertussis*) to 21 (*B. bronchiseptica*) two-component regulatory systems and several transcriptional regulators belonging to other families, suggesting significant potential for regulating gene expression in response to variations in environmental conditions. Remarkably, one two-component regulatory system, BvgAS, is responsible for positively regulating all of the *Bordetella* virulence genes identified so far. Transposon mutagenesis and, more recently, microarray analyses have revealed that in fact BvgAS controls much more than virulence gene expression; the BvgAS regulon appears to comprise more than 150 genes, including many that encode proteins with putative roles in basic metabolic processes (109; H. J. Bootsma et al., unpublished data). For most genes encoding adhesins and toxins, transcriptional control is direct; however, control of many genes is likely to be indirect, and for some, this is known to be the case. Thus, BvgAS appears to sit at the top of a large and complex modulon, playing a central role in controlling the entire physiological state of the cell in addition to virulence genes and other factors that perform specific and potentially accessory functions.

BvgAS is a member of the phosphorelay family of signal-transducing proteins that communicate using a four-step His-Asp-His-Asp phosphotransfer mechanism (94). BvgS, the polydomain sensor, contains two periplasmic binding protein-like domains at its N terminus that are thought to be involved in signal recognition (125, 200). Although the true signals recognized by BvgS in nature are unknown, in the laboratory it is active when the bacteria are grown at 37°C in the relative absence of $MgSO_4$ and nicotinic acid. Under these conditions, BvgS autophosphorylation and subsequent phosphoryl group transfer results in phosphorylation of BvgA, rendering it more competent to bind DNA and activate or repress the transcription of target genes (17, 180, 199, 201, 202, 233). Growth at low temperature (<26°C) or at 37°C in the presence of high concentrations of $MgSO_4$ or nicotinic acid renders BvgAS inactive.

Similar to that which has been proposed for CsrRS of GAS, BvgAS activity appears to vary incrementally in response to differences in signal intensity, providing it with the ability to control potentially multiple distinct phenotypic phases (see reference 29 for a review). Three Bvg-regulated phenotypic phases have been described so far. The Bvg^- phase, expressed when BvgAS is inactive, is characterized by the expression of Bvg-repressed (*vrg*) loci. *vrg* loci expressed by *B. pertussis* and *B. parapertussis*$_{hu}$ include those that encode outer membrane and secreted proteins of unknown function, while *vrg* loci expressed by *B. bronchiseptica* include those encoding proteins involved in flagellar synthesis and motility. A phase designated the Bvg-intermediate (Bvg^i) phase is expressed during growth in the presence of 0.4 to 1.0 mM (*B. bronchiseptica, B. parapertussis*$_{hu}$, and some *B. pertussis* strains) or ~30 mM (other *B. pertussis* strains) $MgSO_4$ or nicotinic acid, conditions which presumably render BvgAS partially active. The Bvg^i phase is characterized by expression of most known and putative adhesins (such as FHA, Fim, and BipA) and BvgAS itself. The Bvg^+ phase, expressed when BvgAS is fully active, is characterized by expression of adhesins and toxins and lack of expression of BipA and other factors that, like BipA, are expressed only in the Bvg^i phase.

How does BvgAS mediate differential regulation of the various classes of genes under its control? Data from a large number of in vivo and in vitro studies indicate that BvgA controls the expression of most genes encoding adhesins and toxins directly and that it distinguishes those expressed under the different phases based on the position and affinity of BvgA binding sites present at their promoters (see references 29 and 30 for reviews). For example, the genes encoding FHA and BipA (*fhaB* and *bipA*, respectively) contain high-affinity BvgA binding sites located such that two or three dimers of BvgA-phosphate (BvgA~P) bind cooperatively with RNA polymerase to activate transcription (16, 50). The high affinity of these binding sites for BvgA~P allows for transcription of these genes under Bvg^i phase conditions when BvgA~P levels are predicted to be moderately low. Genes encoding toxins, such as *cyaA* and *ptxA*, contain low-affinity BvgA binding sites and require greater numbers of BvgA~P dimers to bind cooperatively with RNA polymerase to activate transcription (18, 233). These genes are therefore activated only under Bvg^+ phase conditions, when BvgA~P levels are expected to be maximal. Although FHA levels remain high under Bvg^+ phase conditions, BipA levels do not. The *bipA* promoter contains low-affinity BvgA binding sites located 3′ to the transcriptional start site. Binding at these sites, which requires a high concentration of BvgA~P, inhibits transcription (C. L. Williams et al., submitted for publication). Thus, *bipA* transcription is activated by low levels of BvgA~P and repressed by high levels of BvgA~P. Whether other genes that are maximally expressed in the Bvg^i phase are controlled by a similar mechanism remains to be investigated.

Control of several Bvg-regulated genes has now been shown to involve intermediate regulatory loci. For example, motility in *B. bronchiseptica* is controlled via FrlAB (1). Analogous to F1hDC of *E. coli*, FrlAB appears to form a sigma factor that sits at the top of the motility regulatory hierarchy in *Bordetella*. BvgAS represses motility by repressing transcription of *frlAB* under Bvg^i and Bvg^+ phase conditions. Bvg-repressed genes in *B. pertussis* are also controlled by an intermediate regulatory factor. In this case, expression of the factor, BvgR, is activated by BvgAS.

When expressed, BvgR mediates the repression of *vrg* loci (134, 135). It was recently discovered that the type III secretion system in *B. bronchiseptica* is controlled by intermediate loci as well. *btrS,* which appears to encode an ECF sigma factor, is positively regulated by BvgAS. Its product, BtrS, is required for transcription of the *bsc* locus (which encodes the type III secretion apparatus) and *btrU, btrV,* and *btrW. btrU, btrV,* and *btrW* encode proteins with similarity to gram-positive bacterial partner-switching regulatory proteins, and they are involved in posttranscriptional control and export of type III secretion-related proteins (128a). Inclusion of additional regulatory loci in the BvgAS regulon may provide a mechanism for increasing the number of environmental signals to which the expression of specific subsets of Bvg-regulated genes is controlled. Although an intriguing hypothesis, this possibility has not yet been explored experimentally.

Why does BvgAS differentially regulate the various classes of genes it controls, and what roles do the BvgAS-controlled phenotypic phases play in the *Bordetella* infectious cycle? Conservation and functional interchangeability of BvgAS among the bordetellae suggest that it performs the same function in each subspecies (125). Because *B. pertussis* is thought to exist nowhere other than the human respiratory tract, it was assumed that transition between the Bvg$^+$ and Bvg$^-$ phases (and all phases in between) must occur at some in vivo location during infection. Experiments with Bvg$^+$ and Bvg$^-$ phase-locked mutants of *B. pertussis* and *B. bronchiseptica,* however, demonstrated that the Bvg$^+$ phase is necessary and sufficient for respiratory infection and that the Bvg$^-$ phase of *B. bronchiseptica* is adapted for survival under conditions of extreme nutrient deprivation (1, 31, 124, 136). Experiments with ectopic expression mutants demonstrated that failure to repress the Bvg$^-$ phase genes under Bvg$^+$ phase conditions can actually be detrimental to the development of infection (1, 124, 136). These data suggest that the function of BvgAS is to sense whether the bacteria are within or outside a mammalian respiratory tract and to control the expression of the phenotypic phases required for virulence (Bvg$^+$ phase) or survival outside the host (Bvg$^-$ phase) accordingly. Since only *B. bronchiseptica* appears to be capable of extended survival outside the host, it has been hypothesized that the Bvg$^-$ phases of *B. pertussis* and *B. parapertussis*$_{hu}$ may be evolutionary remnants with no present function (30, 136). Compared with Bvg$^+$ phase bacteria, Bvg^i phase *B. bronchiseptica* displays decreased virulence (it is able to colonize the nasal cavity but not the trachea) and an increased ability to withstand nutrient deprivation (32). The Bvg^i phases of *B. pertussis, B. parapertussis*$_{hu}$, and *B. bronchiseptica* appear to be similar (68, 189), and all of these organisms are notoriously capable of transmission by the aerosol route. These observations suggest that transition to the Bvg^i phase may be important for aerosol transmission, and preliminary data showing decreased efficiency of transmission by Bvg$^+$ phase-locked strains support this hypothesis (C. L. Williams and P. A. Cotter, unpublished data). Two prominent features of the Bvg^i phase could conceivably contribute to transmission: autoaggregation, which could allow for the formation of desiccation-resistant particles of a size conducive to aerosolization, and expression of the outer membrane protein BipA, which, as an invasin/intimin family member, could mediate adherence to host tissues that specifically contribute to either dispersal or initial establishment of colonization (32, 189). The Bvg^i phase may therefore be expressed within a specific in vivo niche that promotes transmission. Alternatively, Scarlato and Rappuoli have proposed that when *Bordetella* first encounters a new host, it may be important for the development of infection that adhesins are expressed first, to mediate attachment to respiratory tissues, and that toxins are not expressed until several hours later (174). During those first few hours, the bacteria would be "primed" either to establish a full-blown infection by continuing on their temporally programmed path toward expression of the Bvg$^+$ phase or to revert quickly to the Bvg$^-$ phase if expelled back into the environment. To begin to test these hypotheses, Kinnear et al. constructed *B. pertussis* strains in which the transcription profiles of *fhaB* and *ptxA* were reversed. These mutants were defective in their ability to establish respiratory colonization in a mouse model (108). Although it was not possible to determine if the observed virulence defect resulted from aberrant regulation in response to temporal or spatial cues, the results underscore the importance of precise Bvg-mediated control of virulence gene expression in vivo. We have recently developed natural-host animal models that allow us to study the entire infectious cycle in the laboratory (Williams and Cotter, unpublished). Comparison of mutants altered in signal transduction and/or in Bvg-mediated control of specific factors in these models should allow us to gain a better understanding of the importance of precise control of gene expression in the establishment, maintenance, and transmission of *Bordetella* infection.

BvgAS has served as a regulatory paradigm. Its discovery in 1984 represented one of the first examples of coordinate control of virulence gene expression in a bacterial pathogen (225). Intense research, focused primarily on the human-adapted pathogen *B. pertussis,* provided significant insight into how the

BvgAS phosphorelay functions mechanistically to control multiple classes of genes and multiple phenotypic phases in response to subtle changes in environmental conditions (reviewed in references 28 and 29). As virulence regulators and regulons were subsequently discovered and characterized in other bacterial pathogens, many were found to fit, at least to some extent, the BvgAS paradigm. Progress toward understanding why BvgAS controls the expression of virulence genes and other factors has resulted primarily from comparative analyses that included additional members of the *B. bronchiseptica* cluster, especially *B. bronchiseptica,* because of the availability of natural-host animal models. Indeed, it is unlikely that the conclusion that the Bvg$^-$ phase is important for ex vivo survival would have been drawn from studies focused only on *B. pertussis.* Although BvgAS may be unique in its penetrance, it seems likely that functional orthologs with the ability to distinguish ex vivo and in vivo environments, as proposed for BvgAS, exist in other pathogens as well. Their discovery and/or characterization may also be facilitated by comparative analyses across related species that vary with regard to host specificity, the propensity to cause acute versus chronic disease, and/or modes of transmission.

Acknowledgments. I thank Jeff Miller, Vic DiRita, Steve Julio, Corinne Williams, and Robert Haines for insightful discussions and critical comments that improved the manuscript. Research in my laboratory is supported by the NIH (AI43986), the USDA (2003-02486), the UC SBREP, and Santa Barbara Cottage Hospital.

REFERENCES

1. **Akerley, B. J., P. A. Cotter, and J. F. Miller.** 1995. Ectopic expression of the flagellar regulon alters development of the *Bordetella*-host interaction. *Cell* **80:**611–620.
2. **Ashbaugh, C. D., S. Alberti, and M. R. Wessels.** 1998. Molecular analysis of the capsule gene region of group A *Streptococcus:* the *hasAB* genes are sufficient for capsule expression. *J. Bacteriol.* **180:**4955–4959.
3. **Ashbaugh, C. D., T. J. Moser, M. H. Shearer, G. L. White, R. C. Kennedy, and M. R. Wessels.** 2000. Bacterial determinants of persistent throat colonization and the associated immune response in a primate model of human group A streptococcal pharyngeal infection. *Cell. Microbiol.* **2:**283–292.
4. **Beckert, S., B. Kreikemeyer, and A. Podbielski.** 2001. Group A streptococcal *rofA* gene is involved in the control of several virulence genes and eukaryotic cell attachment and internalization. *Infect. Immun.* **69:**534–547.
5. **Beres, S. B., G. L. Sylva, K. D. Barbian, B. Lei, J. S. Hoff, N. D. Mammarella, M. Y. Liu, J. C. Smoot, S. F. Porcella, L. D. Parkins, D. S. Campbell, T. M. Smith, J. K. McCormick, D. Y. Leung, P. M. Schlievert, and J. M. Musser.** 2002. Genome sequence of a serotype M3 strain of group A *Streptococcus:* phage-encoded toxins, the high-virulence phenotype, and clone emergence. *Proc. Natl. Acad. Sci. USA* **99:**10078–10083.
6. **Berggard, K., E. Johnsson, E. Morfeldt, J. Persson, M. Stalhammar-Carlemalm, and G. Lindahl.** 2001. Binding of human C4BP to the hypervariable region of M protein: a molecular mechanism of phagocytosis resistance in *Streptococcus pyogenes. Mol. Microbiol.* **42:**539–551.
7. **Bernish, B., and I. van de Rijn.** 1999. Characterization of a two-component system in *Streptococcus pyogenes* which is involved in regulation of hyaluronic acid production. *J. Biol. Chem.* **274:**4786–4793.
8. **Berry, A. M., R. A. Lock, and J. C. Paton.** 1996. Cloning and characterization of *nanB,* a second *Streptococcus pneumoniae* neuraminidase gene, and purification of the NanB enzyme from recombinant *Escherichia coli. J. Bacteriol.* **178:** 4854–4860.
9. **Berry, A. M., R. A. Lock, S. M. Thomas, D. P. Rajan, D. Hansman, and J. C. Paton.** 1994. Cloning and nucleotide sequence of the *Streptococcus pneumoniae* hyaluronidase gene and purification of the enzyme from recombinant *Escherichia coli. Infect. Immun.* **62:**1101–1108.
10. **Berry, A. M., and J. C. Paton.** 1996. Sequence heterogeneity of PsaA, a 37-kilodalton putative adhesin essential for virulence of *Streptococcus pneumoniae. Infect. Immun.* **64:**5255–5262.
11. **Bisno, A. L., M. O. Brito, and C. M. Collins.** 2003. Molecular basis of group A streptococcal virulence. *Lancet Infect. Dis.* **3:**191–200.
12. **Biswas, I., and J. R. Scott.** 2003. Identification of *rocA,* a positive regulator of *covR* expression in the group A *Streptococcus. J. Bacteriol.* **185:**3081–3090.
13. **Black, S., H. Shinefield, B. Fireman, E. Lewis, P. Ray, J. R. Hansen, L. Elvin, K. M. Ensor, J. Hackell, G. Siber, F. Malinoski, D. Madore, I. Chang, R. Kohberger, W. Watson, R. Austrian, K. Edwards, and the Northern California Kaiser Permanente Vaccine Study Center Group.** 2000. Efficacy, safety and immunogenicity of heptavalent pneumococcal conjugate vaccine in children. *Pediatr. Infect. Dis. J.* **19:**187–195.
14. **Black, S. B., H. R. Shinefield, S. Ling, J. Hansen, B. Fireman, D. Spring, J. Noves, E. Lewis, P. Ray, J. Lee, and J. Hackell.** 2002. Effectiveness of heptavalent pneumococcal conjugate vaccine in children younger than five years of age for prevention of pneumonia. *Pediatr. Infect. Dis. J.* **21:**810–815.
15. **Bohach, G. A., D. J. Fast, R. D. Nelson, and P. M. Schlievert.** 1990. Staphylococcal and streptococcal pyrogenic toxins involved in toxic shock syndrome and related illnesses. *Crit. Rev. Microbiol.* **17:**251–272.
16. **Boucher, P. E., A. E. Maris, M. S. Yang, and S. Stibitz.** 2003. The response regulator BvgA and RNA polymerase alpha subunit C-terminal domain bind simultaneously to different faces of the same segment of promoter DNA. *Mol. Cell* **11:** 163–173.
17. **Boucher, P. E., K. Murakami, A. Ishihama, and S. Stibitz.** 1997. Nature of DNA binding and RNA polymerase interaction of the *Bordetella pertussis* BvgA transcriptional activator at the *fha* promoter. *J. Bacteriol.* **179:**1755–1763.
18. **Boucher, P. E., and S. Stibitz.** 1995. Synergistic binding of RNA polymerase and BvgA phosphate to the pertussis toxin promoter of *Bordetella pertussis. J. Bacteriol.* **177:**6486–6491.
19. **Camara, M., G. J. Boulnois, P. W. Andrew, and T. J. Mitchell.** 1994. A neuraminidase from *Streptococcus pneumoniae* has the features of a surface protein. *Infect. Immun.* **62:**3688–3695.
20. **Caparon, M. G., R. T. Geist, J. Perez-Casal, and J. R. Scott.** 1992. Environmental regulation of virulence in group A streptococci: transcription of the gene encoding M protein is stimulated by carbon dioxide. *J. Bacteriol.* **174:**5693–5701.

21. **Caparon, M. G., and J. R. Scott.** 1987. Identification of a gene that regulates expression of M protein, the major virulence determinant of group A streptococci. *Proc. Natl. Acad. Sci. USA* **84:**8677–8681.
22. **Caparon, M. G., D. S. Stephens, A. Olsen, and J. R. Scott.** 1991. Role of M protein in adherence of group A streptococci. *Infect. Immun.* **59:**1811–1817.
23. **Chapuy-Regaud, S., A. D. Ogunniyi, N. Diallo, Y. Huet, J. F. Desnottes, J. C. Paton, S. Escaich, and M. C. Trombe.** 2003. RegR, a global LacI/GalR family regulator, modulates virulence and competence in *Streptococcus pneumoniae*. *Infect. Immun.* **71:**2615–2625.
24. **Chaussee, M. S., D. Ajdic, and J. J. Ferretti.** 1999. The *rgg* gene of *Streptococcus pyogenes* NZ131 positively influences extracellular SPE B production. *Infect. Immun.* **67:**1715–1722.
25. **Chaussee, M. S., J. Liu, D. L. Stevens, and J. J. Ferretti.** 1997. Effects of environmental factors on streptococcal erythrogenic toxin A (SPE A) production by *Streptococcus pyogenes*. *Adv. Exp. Med. Biol.* **418:**551–554.
26. **Chaussee, M. S., G. L. Sylva, D. E. Sturdevant, L. M. Smoot, M. R. Graham, R. O. Watson, and J. M. Musser.** 2002. Rgg influences the expression of multiple regulatory loci to coregulate virulence factor expression in *Streptococcus pyogenes*. *Infect. Immun.* **70:**762–770.
27. **Cherry, J. D., and U. Heininger.** 1999. Pertussis, p. 1423–1440. *In* R. D. Feigin and J. D. Cherry (ed.), *Textbook of Pediatric Infectious Diseases,* 4th. ed. The W. B. Saunders Co., Philadelphia, Pa.
28. **Cotter, P. A., and V. J. DiRita.** 2000. Bacterial virulence gene regulation: an evolutionary perspective. *Annu. Rev. Microbiol.* **54:**519–565.
29. **Cotter, P. A., and A. M. Jones.** 2003. Phosphorelay control of virulence gene expression in *Bordetella*. *Trends Microbiol.* **11:**367–373.
30. **Cotter, P. A., and J. F. Miller.** 2001. *Bordetella,* p. 619–674. *In* E. A. Groisman (ed.), *Principles of Bacterial Pathogenesis.* Academic Press, Inc. San Diego, Calif.
31. **Cotter, P. A., and J. F. Miller.** 1994. BvgAS-mediated signal transduction: analysis of phase-locked regulatory mutants of *Bordetella bronchiseptica* in a rabbit model. *Infect. Immun.* **62:**3381–3390.
32. **Cotter, P. A., and J. F. Miller.** 1997. A mutation in the *Bordetella bronchiseptica bvgS* gene results in reduced virulence and increased resistance to starvation, and identifies a new class of Bvg-regulated antigens. *Mol. Microbiol.* **24:**671–685.
33. **Cotter, P. A., M. H. Yuk, S. Mattoo, B. J. Akerley, J. Boschwitz, D. A. Relman, and J. F. Miller.** 1998. The filamentous hemagglutinin (FHA) of *Bordetella bronchiseptica* is required for efficient establishment of tracheal colonization. *Infect. Immun.* **66:**5921–5929.
34. **Courtney, H. S., M. S. Bronze, J. B. Dale, and D. L. Hasty.** 1994. Analysis of the role of M24 protein in group A streptococcal adhesion and colonization by use of omega-interposon mutagenesis. *Infect. Immun.* **62:**4868–4873.
35. **Courtney, H. S., D. L. Hasty, and J. B. Dale.** 2002. Molecular mechanisms of adhesion, colonization, and invasion of group A streptococci. *Ann. Med.* **34:**77–87.
36. **Courtney, H. S., S. Liu, J. B. Dale, and D. L. Hasty.** 1997. Conversion of M serotype 24 of *Streptococcus pyogenes* to M serotypes 5 and 18: effect on resistance to phagocytosis and adhesion to host cells. *Infect. Immun.* **65:**2472–2474.
37. **Craig, J. E., A. Nobbs, and N. J. High.** 2002. The extracytoplasmic sigma factor, final sigma(E), is required for intracellular survival of nontypeable *Haemophilus influenzae* in J774 macrophages. *Infect. Immun.* **70:**708–715.
38. **Cross, A. S.** 1990. The biologic significance of bacterial encapsulation. *Curr. Top. Microbiol. Immunol.* **150:**87–95.
39. **Cue, D., P. E. Dombek, H. Lam, and P. P. Cleary.** 1998. *Streptococcus pyogenes* serotype M1 encodes multiple pathways for entry into human epithelial cells. *Infect. Immun.* **66:**4593–4601.
40. **Cundell, D. R., N. P. Gerard, C. Gerard, I. Idanpaan-Heikkila, and E. I. Tuomanen.** 1995. *Streptococcus pneumoniae* anchor to activated human cells by the receptor for platelet-activating factor. *Nature* **377:**435–438.
41. **Cundell, D. R., J. N. Weiser, J. Shen, A. Young, and E. I. Tuomanen.** 1995. Relationship between colonial morphology and adherence of *Streptococcus pneumoniae*. *Infect. Immun.* **63:**757–761.
42. **Cunningham, M. W.** 2000. Pathogenesis of group A streptococcal infections. *Clin. Microbiol. Rev.* **13:**470–511.
43. **Cywes, C., I. Stamenkovic, and M. R. Wessels.** 2000. CD44 as a receptor for colonization of the pharynx by group A *Streptococcus*. *J. Clin. Investig.* **106:**995–1002.
44. **Cywes, C., and M. R. Wessels.** 2001. Group A *Streptococcus* tissue invasion by CD44-mediated cell signalling. *Nature* **414:**648–652.
45. **Dale, J. B., R. G. Washburn, M. B. Marques, and M. R. Wessels.** 1996. Hyaluronate capsule and surface M protein in resistance to opsonization of group A streptococci. *Infect. Immun.* **64:**1495–1501.
46. **Dawid, S., S. J. Barenkamp, and J. W. St. Geme III.** 1999. Variation in expression of the *Haemophilus influenzae* HMW adhesins: a prokaryotic system reminiscent of eukaryotes. *Proc. Natl. Acad. Sci. USA* **96:**1077–1082.
47. **Dawid, S., S. Grass, and J. W. St. Geme III.** 2001. Mapping of binding domains of nontypeable *Haemophilus influenzae* HMW1 and HMW2 adhesins. *Infect. Immun.* **69:**307–314.
48. **Deghmane, A. E., D. Giorgini, M. Larribe, J. M. Alonso, and M. K. Taha.** 2002. Down-regulation of pili and capsule of *Neisseria meningitidis* upon contact with epithelial cells is mediated by CrgA regulatory protein. *Mol. Microbiol.* **43:**1555–1564.
49. **Deghmane, A. E., S. Petit, A. Topilko, Y. Pereira, D. Giorgini, M. Larribe, and M. K. Taha.** 2000. Intimate adhesion of *Neisseria meningitidis* to human epithelial cells is under the control of the *crgA* gene, a novel LysR-type transcriptional regulator. *EMBO J.* **19:**1068–1078.
50. **Deora, R., H. J. Bootsma, J. F. Miller, and P. A. Cotter.** 2001. Diversity in the *Bordetella* virulence regulon: transcriptional control of a Bvg-intermediate phase gene. *Mol. Microbiol.* **40:**669–683.
51. **De Souza-Hart, J. A., W. Blackstock, V. Di Modugno, I. B. Holland, and M. Kok.** 2003. Two-component systems in *Haemophilus influenzae:* a regulatory role for ArcA in serum resistance. *Infect. Immun.* **71:**163–172.
52. **Dintilhac, A., G. Alloing, C. Granadel, and J. P. Claverys.** 1997. Competence and virulence of *Streptococcus pneumoniae:* Adc and PsaA mutants exhibit a requirement for Zn and Mn resulting from inactivation of putative ABC metal permeases. *Mol. Microbiol.* **25:**727–739.
53. **Doern, G. V., A. B. Brueggemann, H. Huynh, and E. Wingert.** 1999. Antimicrobial resistance with *Streptococcus pneumoniae* in the United States, 1997–98. *Emerg. Infect. Dis.* **5:** 757–765.
54. **Dombek, P. E., D. Cue, J. Sedgewick, H. Lam, S. Ruschkowski, B. B. Finlay, and P. P. Cleary.** 1999. High-frequency intracellular invasion of epithelial cells by serotype M1 group A streptococci: M1 protein-mediated invasion and cytoskeletal rearrangements. *Mol. Microbiol.* **31:**859–870.

55. **Echenique, J. R., and M. C. Trombe.** 2001. Competence repression under oxygen limitation through the two-component MicAB signal-transducing system in *Streptococcus pneumoniae* and involvement of the PAS domain of MicB. *J. Bacteriol.* **183:**4599–4608.
56. **Engleberg, N. C., A. Heath, A. Miller, C. Rivera, and V. J. DiRita.** 2001. Spontaneous mutations in the CsrRS two-component regulatory system of *Streptococcus pyogenes* result in enhanced virulence in a murine model of skin and soft tissue infection. *J. Infect. Dis.* **183:**1043–1054.
57. **Farley, M. M., D. S. Stephens, S. L. Kaplan, and E. O. Mason, Jr.** 1990. Pilus- and non-pilus-mediated interactions of *Haemophilus influenzae* type b with human erythrocytes and human nasopharyngeal mucosa. *J. Infect. Dis.* **161:**274–280.
58. **Federle, M. J., K. S. McIver, and J. R. Scott.** 1999. A response regulator that represses transcription of several virulence operons in the group A *Streptococcus. J. Bacteriol.* **181:**3649–3657.
59. **Fernandez, R. C., and A. A. Weiss.** 1994. Cloning and sequencing of a *Bordetella pertussis* serum resistance locus. *Infect. Immun.* **62:**4727–4738.
60. **Ferretti, J. J., W. M. McShan, D. Ajdic, D. J. Savic, G. Savic, K. Lyon, C. Primeaux, S. Sezate, A. N. Suvorov, S. Kenton, H. S. Lai, S. P. Lin, Y. Qian, H. G. Jia, F. Z. Najar, Q. Ren, H. Zhu, L. Song, J. White, X. Yuan, S. W. Clifton, B. A. Roe, and R. McLaughlin.** 2001. Complete genome sequence of an M1 strain of *Streptococcus pyogenes. Proc. Natl. Acad. Sci. USA* **98:**4658–4663.
61. **Finn, T. M., and L. A. Stevens.** 1995. Tracheal colonization factor: a *Bordetella pertussis* secreted virulence determinant. *Mol. Microbiol.* **16:**625–634.
62. **Fireman, B., S. B. Black, H. R. Shinefield, J. Lee, E. Lewis, and P. Ray.** 2003. Impact of the pneumococcal conjugate vaccine on otitis media. *Pediatr. Infect. Dis. J.* **22:**10–16.
63. **Fleischmann, R. D., M. D. Adams, O. White, R. A. Clayton, E. Kirkness, A. R. Kerlavage, C. J. Bult, J. F. Tomb, B. A. Dougherty, J. M. Merrick, A. Glodek, J. L. Scott, S. M. Geoghagen, J. F. Weidman, J. L. Fuhrmann, D. Nguyen, T. R. Utterback, J. M. Kelley, J. D. Peterson, P. W. Sadow, M. C. Hanna, M. D. Cotton, K. M. Roberts, M. A. Hurst, B. P. Kaine, M. Borodovsky, H.-P. Klenk, C. M. Fraser, H. O. Smith, C. R. Woese, and J. C. Venter.** 1995. Whole-genome random sequencing and assembly of *Haemophilus influenzae* Rd. *Science* **269:**496–512.
64. **Fluckiger, U., K. F. Jones, and V. A. Fischetti.** 1998. Immunoglobulins to group A streptococcal surface molecules decrease adherence to and invasion of human pharyngeal cells. *Infect. Immun.* **66:**974–979.
65. **Fogg, G. C., and M. G. Caparon.** 1997. Constitutive expression of fibronectin binding in *Streptococcus pyogenes* as a result of anaerobic activation of *rofA. J. Bacteriol.* **179:**6172–6180.
66. **Fogg, G. C., C. M. Gibson, and M. G. Caparon.** 1994. The identification of *rofA*, a positive-acting regulatory component of *prtF* expression: use of an m gamma delta-based shuttle mutagenesis strategy in *Streptococcus pyogenes. Mol. Microbiol.* **11:**671–684.
67. **Friedman, R. L.** 1987. *Bordetella pertussis* adenylate cyclase: isolation and purification by calmodulin-sepharose 4B chromatography. *Infect. Immun.* **55:**129–134.
68. **Fuchslocher, B., L. L. Millar, and P. A. Cotter.** 2003. Comparison of *bipA* alleles within and across *Bordetella* species. *Infect. Immun.* **71:**3043–3052.
69. **Georgellis, D., O. Kwon, E. C. Lin, S. M. Wong, and B. J. Akerley.** 2001. Redox signal transduction by the ArcB sensor kinase of *Haemophilus influenzae* lacking the PAS domain. *J. Bacteriol.* **183:**7206–7212.
70. **Gerlach, G., F. von Wintzingerode, B. Middendorf, and R. Gross.** 2001. Evolutionary trends in the genus *Bordetella. Microbes Infect.* **3:**61–72.
71. **Geuijen, C. A., R. J. Willems, M. Bongaerts, J. Top, H. Gielen, and F. R. Mooi.** 1997. Role of the *Bordetella pertussis* minor fimbrial subunit, FimD, in colonization of the mouse respiratory tract. *Infect. Immun.* **65:**4222–4228.
72. **Giammarinaro, P., and J. C. Paton.** 2002. Role of RegM, a homologue of the catabolite repressor protein CcpA, in the virulence of *Streptococcus pneumoniae. Infect. Immun.* **70:**5454–5461.
73. **Gold, R., I. Goldschneider, M. L. Lepow, T. F. Draper, and M. Randolph.** 1978. Carriage of *Neisseria meningitidis* and *Neisseria lactamica* in infants and children. *J. Infect. Dis.* **137:**112–121.
74. **Graham, M. R., L. M. Smoot, C. A. Migliaccio, K. Virtaneva, D. E. Sturdevant, S. F. Porcella, M. J. Federle, G. J. Adams, J. R. Scott, and J. M. Musser.** 2002. Virulence control in group A *Streptococcus* by a two-component gene regulatory system: global expression profiling and *in vivo* infection modeling. *Proc. Natl. Acad. Sci. USA* **99:**13855–13860.
75. **Greco, R., L. De Martino, G. Donnarumma, M. P. Conte, L. Seganti, and P. Valenti.** 1995. Invasion of cultured human cells by *Streptococcus pyogenes. Res. Microbiol.* **146:**551–560.
76. **Gryllos, I., J. C. Levin, and M. R. Wessels.** 2003. The CsrR/CsrS two-component system of group A *Streptococcus* responds to environmental Mg^{2+}. *Proc. Natl. Acad. Sci. USA* **100:**4227–4232.
77. **Hagman, M. M., J. B. Dale, and D. L. Stevens.** 1999. Comparison of adherence to and penetration of a human laryngeal epithelial cell line by group A streptococci of various M protein types. *FEMS Immunol. Med. Microbiol.* **23:**195–204.
78. **Hammerschmidt, S., A. Muller, H. Sillmann, M. Muhlenhoff, R. Borrow, A. Fox, J. van Putten, W. D. Zollinger, R. Gerardy-Schahn, and M. Frosch.** 1996. Capsule phase variation in *Neisseria meningitidis* serogroup B by slipped-strand mispairing in the polysialyltransferase gene (*siaD*): correlation with bacterial invasion and the outbreak of meningococcal disease. *Mol. Microbiol.* **20:**1211–1220.
79. **Hammerschmidt, S., S. R. Talay, P. Brandtzaeg, and G. S. Chhatwal.** 1997. SpsA, a novel pneumococcal surface protein with specific binding to secretory immunoglobulin A and secretory component. *Mol. Microbiol.* **25:**1113–1124.
80. **Hart, C. A., and L. E. Cuevas.** 1997. Meningococcal disease in Africa. *Ann. Trop. Med. Parasitol.* **91:**777–785.
81. **Harvill, E. T., P. A. Cotter, and J. F. Miller.** 1998. Probing the function of a bacterial virulence factor by manipulating host immunity. *Infect. Immun.* **67:**1493–1500.
82. **Hasty, D. L., I. Ofek, H. S. Courtney, and R. J. Doyle.** 1992. Multiple adhesins of streptococci. *Infect. Immun.* **60:**2147–2152.
83. **Hauser, A. R., D. L. Stevens, E. L. Kaplan, and P. M. Schlievert.** 1991. Molecular analysis of pyrogenic exotoxins from *Streptococcus pyogenes* isolates associated with toxic shock-like syndrome. *J. Clin. Microbiol.* **29:**1562–1567.
84. **Hava, D. L., and A. Camilli.** 2002. Large-scale identification of serotype 4 *Streptococcus pneumoniae* virulence factors. *Mol. Microbiol.* **45:**1389–1406.
85. **Hava, D. L., C. J. Hemsley, and A. Camilli.** 2003. Transcriptional regulation in the *Streptococcus pneumoniae rlrA* pathogenicity islet by RlrA. *J. Bacteriol.* **185:**413–421.
86. **Heath, A., V. J. DiRita, N. L. Barg, and N. C. Engleberg.** 1999. A two-component regulatory system, CsrR-CsrS, represses

expression of three *Streptococcus pyogenes* virulence factors, hyaluronic acid capsule, streptolysin S, and pyrogenic exotoxin B. *Infect. Immun.* **67:**5298–5305.

87. **Heininger, U., K. Stehr, S. Schmitt-Grohe, C. Lorenz, R. Rost, P. D. Christenson, M. Uberall, and J. D. Cherry.** 1994. Clinical characteristics of illness caused by *Bordetella parapertussis* compared with illness caused by *Bordetella pertussis. Pediatr. Infect. Dis. J.* **13:**306–309.
88. **Hemsley, C., E. Joyce, D. L. Hava, A. Kawale, and A. Camilli.** 2003. MgrA, an orthologue of Mga, acts as a transcriptional repressor of the genes within the *rlrA* pathogenicity islet in *Streptococcus pneumoniae. J. Bacteriol.* **185:**6640–6647.
89. **Henderson, I. R., R. Cappello, and J. P. Nataro.** 2000. Autotransporter proteins, evolution and redefining protein secretion. *Trends Microbiol.* **8:**529–532.
90. **Hendrixson, D. R., and J. W. St. Geme III.** 1998. The *Haemophilus influenzae* Hap serine protease promotes adherence and microcolony formation, potentiated by a soluble host protein. *Mol. Cell* **2:**841–850.
91. **Hewlett, E. L., and N. J. Maloney.** 1994. Adenylyl cyclase toxim from *Bordetella pertussis. In* B. Iglewski, J. Moss, A. T. Tu, and M. Vaughan (ed.), *Handbook of Natural Toxins,* vol. 8. Marcel Dekker, Inc., New York, N.Y.
92. **Hidalgo-Grass, C., M. Ravins, M. Dan-Goor, J. Jaffe, A. E. Moses, and E. Hanski.** 2002. A locus of group A *Streptococcus* involved in invasive disease and DNA transfer. *Mol. Microbiol.* **46:**87–99.
93. **Hill, D. J., M. A. Toleman, D. J. Evans, S. Villullas, L. Van Alphen, and M. Virji.** 2001. The variable P5 proteins of typeable and non-typeable *Haemophilus influenzae* target human CEACAM1. *Mol. Microbiol.* **39:**850–862.
94. **Hoch, J. A.** 2000. Two-component and phosphorelay signal transduction. *Curr. Opin. Microbiol.* **3:**165–170.
95. **Hollingshead, S. K., J. W. Simecka, and S. M. Michalek.** 1993. Role of M protein in pharyngeal colonization by group A streptococci in rats. *Infect. Immun.* **61:**2277–2283.
96. **Hoskins, J., W. E. Alborn, Jr., Arnold, L. C. Blaszczak, S. Burgett, B. S. DeHoff, S. T. Estrem, L. Fritz, D. J. Fu, W. Fuller, C. Geringer, R. Gilmour, J. S. Glass, H. Khoja, A. R. Kraft, R. E. Lagace, D. J. LeBlanc, L. N. Lee, E. J. Lefkowitz, J. Lu, P. Matsushima, S. M. McAhren, M. McHenney, K. McLeaster, C. W. Mundy, T. I. Nicas, F. H. Norris, M. O'Gara, R. B. Peery, G. T. Robertson, P. Rockey, P. M. Sun, M. E. Winkler, Y. Yang, M. Young-Bellido, G. Zhao, C. A. Zook, R. H. Baltz, S. R. Jaskunas, P. R. Rosteck, Jr., P. L. Skatrud, and J. I. Glass.** 2001. Genome of the bacterium *Streptococcus pneumoniae* strain R6. *J. Bacteriol.* **183:**5709–5717.
97. **Hultgren, S. J., S. Abraham, M. Caparon, P. Falk, J. W. St. Geme III, and S. Normark.** 1993. Pilus and nonpilus bacterial adhesins: assembly and function in cell recognition. *Cell* **73:**887–901.
98. **Husmann, L. K., D. L. Yung, S. K. Hollingshead, and J. R. Scott.** 1997. Role of putative virulence factors of *Streptococcus pyogenes* in mouse models of long-term throat colonization and pneumonia. *Infect. Immun.* **65:**1422–1430.
99. **Jadoun, J., V. Ozeri, E. Burstein, E. Skutelsky, E. Hanski, and S. Sela.** 1998. Protein F1 is required for efficient entry of *Streptococcus pyogenes* into epithelial cells. *J. Infect. Dis.* **178:**147–158.
100. **Jedrzejas, M. J., S. K. Hollingshead, J. Lebowitz, L. Chantalat, D. E. Briles, and E. Lamani.** 2000. Production and characterization of the functional fragment of pneumococcal surface protein A. *Arch. Biochem. Biophys.* **373:**116–125.
101. **Johnson, D. R., D. L. Stevens, and E. L. Kaplan.** 1992. Epidemiologic analysis of group A streptococcal serotypes associated with severe systemic infections, rheumatic fever, or uncomplicated pharyngitis. *J. Infect. Dis.* **166:**374–382.
102. **Johnsson, E., K. Berggard, H. Kotarsky, J. Hellwage, P. F. Zipfel, U. Sjobring, and G. Lindahl.** 1998. Role of the hypervariable region in streptococcal M proteins: binding of a human complement inhibitor. *J. Immunol.* **161:**4894–4901.
103. **Kadioglu, A., J. Echenique, S. Manco, M. C. Trombe, and P. W. Andrew.** 2003. The MicAB two-component signaling system is involved in virulence of *Streptococcus pneumoniae. Infect. Immun.* **71:**6676–6679.
104. **Kallstrom, H., M. K. Liszewski, J. P. Atkinson, and A. B. Jonsson.** 1997. Membrane cofactor protein (MCP or CD46) is a cellular pilus receptor for pathogenic *Neisseria. Mol. Microbiol.* **25:**639–647.
105. **Kaslow, H. R., and D. L. Burns.** 1992. Pertussis toxin and target eukaryotic cells: binding, entry, and activation. *FASEB J.* **6:**2684–2690.
106. **Katada, T., and M. Ui.** 1982. Direct modification of the membrane adenylate cyclase system by islet-activating protein due to ADP-ribosylation of a membrane protein. *Proc. Natl. Acad. Sci. USA* **79:**3129–3133.
107. **Kim, J. O., and J. N. Weiser.** 1998. Association of intrastrain phase variation in quantity of capsular polysaccharide and teichoic acid with the virulence of *Streptococcus pneumoniae. J. Infect. Dis.* **177:**368–377.
108. **Kinnear, S. M., R. R. Marques, and N. H. Carbonetti.** 2001. Differential regulation of Bvg-activated virulence factors plays a role in *Bordetella pertussis* pathogenicity. *Infect. Immun.* **69:**1983–1993.
109. **Knapp, S., and J. J. Mekalanos.** 1998. Two *trans*-acting regulatory genes (*vir* and *mod*) control antigenic modulation in *Bordetella pertussis. J. Bacteriol.* **170:**5059–5066.
110. **Kreikemeyer, B., K. S. McIver, and A. Podbielski.** 2003. Virulence factor regulation and regulatory networks in *Streptococcus pyogenes* and their impact on pathogen-host interactions. *Trends Microbiol.* **11:**224–232.
111. **Laarmann, S., D. Cutter, T. Juehne, S. J. Barenkamp, and J. W. St. Geme.** 2002. The *Haemophilus influenzae* Hia autotransporter harbours two adhesive pockets that reside in the passenger domain and recognize the same host cell receptor. *Mol. Microbiol.* **46:**731–743.
112. **LaPenta, D., C. Rubens, E. Chi, and P. P. Cleary.** 1994. Group A streptococci efficiently invade human respiratory epithelial cells. *Proc. Natl. Acad. Sci. USA* **91:**12115–12119.
113. **Lau, G. W., S. Haataja, M. Lonetto, S. E. Kensit, A. Marra, A. P. Bryant, D. McDevitt, D. A. Morrison, and D. W. Holden.** 2001. A functional genomic analysis of type 3 *Streptococcus pneumoniae* virulence. *Mol. Microbiol.* **40:**555–571.
114. **Lei, B., F. R. DeLeo, N. P. Hoe, M. R. Graham, S. M. Mackie, R. L. Cole, M. Liu, H. R. Hill, D. E. Low, M. J. Federle, J. R. Scott, and J. M. Musser.** 2001. Evasion of human innate and acquired immunity by a bacterial homolog of CD11b that inhibits opsonophagocytosis. *Nat. Med.* **7:**1298–1305.
115. **Leininger, E., M. Roberts, J. G. Kenimer, I. G. Charles, N. Fairweather, P. Novotny, and M. J. Brennan.** 1991. Pertactin, an Arg-Gly-Asp-containing *Bordetella pertussis* surface protein that promotes adherence of mammalian cells. *Proc. Natl. Acad. Sci. USA* **88:**345–349.
116. **Levin, J. C., and M. R. Wessels.** 1998. Identification of *csrR/csrS*, a genetic locus that regulates hyaluronic acid capsule synthesis in group A *Streptococcus. Mol. Microbiol.* **30:** 209–219.
117. **Lin, L., P. Ayala, J. Larson, M. Mulks, M. Fukuda, S. R. Carlsson, C. Enns, and M. So.** 1997. The *Neisseria* type 2

IgA1 protease cleaves LAMP1 and promotes survival of bacteria within epithelial cells. *Mol. Microbiol.* **24**:1083–1094.

118. **Livey, I., and A. Wardlaw.** 1984. Production and properties of *Bordetella pertussis* heat-labile toxin. *J. Med. Microbiol.* **17**:91–103.
119. **Loeb, M. R., E. Connor, and D. Penney.** 1988. A comparison of the adherence of fimbriated and nonfimbriated *Haemophilus influenzae* type b to human adenoids in organ culture. *Infect. Immun.* **56**:484–489.
120. **MacLeod, C. M., and M. R. Krauss.** 1950. Relation of virulence of pneumococcal strains for mice to the quantity of capsular polysaccharide formed in vitro. *J. Exp. Med.* **92**:1–9.
121. **Marceau, M., J. L. Beretti, and X. Nassif.** 1995. High adhesiveness of encapsulated *Neisseria meningitidis* to epithelial cells is associated with the formation of bundles of pili. *Mol. Microbiol.* **17**:855–863.
122. **Marceau, M., K. Forest, J. L. Beretti, J. Tainer, and X. Nassif.** 1998. Consequences of the loss of O-linked glycosylation of meningococcal type IV pilin on piliation and pilus-mediated adhesion. *Mol. Microbiol.* **27**:705–715.
123. **Marrack, P., and J. Kappler.** 1990. The staphylococcal enterotoxins and their relatives. *Science* **248**:705–711.
124. **Martinez de Tejada, G., P. A. Cotter, U. Heininger, A. Camilli, B. J. Akerley, J. J. Mekalanos, and J. F. Miller.** 1998. Neither the Bvg$^-$ phase nor the *vrg6* locus of *Bordetella pertussis* are required for respiratory infection in mice. *Infect. Immun.* **66**:2762–2768.
125. **Martinez de Tejada, G., J. F. Miller, and P. A. Cotter.** 1996. Comparative analysis of the virulence control systems of *Bordetella pertussis* and *Bordetella bronchiseptica. Mol. Microbiol.* **22**:895–908.
126. **Mascher, T., D. Zahner, M. Merai, N. Balmelle, A. B. de Saizieu, and R. Hakenbeck.** 2003. The *Streptococcus pneumoniae cia* regulon: CiaR target sites and transcription profile analysis. *J. Bacteriol.* **185**:60–70.
127. **Mattoo, S., A. K. Foreman-Wykert, P. A. Cotter, and J. F. Miller.** 2001. Mechanisms of *Bordetella* pathogenesis. *Front. Biosci.* **6**:E168–186. [Online] http //www bioscience org/ http//www bioscience org/2001/v6/e/mattoo/fulltext htm.
128. **Mattoo, S., J. F. Miller, and P. A. Cotter.** 2000. Role of *Bordetella bronchiseptica* fimbria in tracheal colonization and development of a humoral immune response. *Infect. Immun.* **68**:2024–2033.

128a. **Mattoo, S., M. H. Yuk, L. L. Huang, and J. F. Miller.** 2004. Regulation of type III secretion in *Bordetella. Mol. Microbiol.* **52**:1201–1214.

129. **McGee, Z. A., D. S. Stephens, L. H. Hoffman, W. F. Schlech III, and R. G. Horn.** 1983. Mechanisms of mucosal invasion by pathogenic *Neisseria. Rev. Infect. Dis.* **5**(Suppl. 4):S708–S714.
130. **McIver, K. S., A. S. Heath, and J. R. Scott.** 1995. Regulation of virulence by environmental signals in group A streptococci: influence of osmolarity, temperature, gas exchange, and iron limitation on emm transcription. *Infect. Immun.* **63**:4540–4542.
131. **McIver, K. S., and R. L. Myles.** 2002. Two DNA-binding domains of Mga are required for virulence gene activation in the group A streptococcus. *Mol. Microbiol.* **43**:1591–1601.
132. **McIver, K. S., and J. R. Scott.** 1997. Role of *mga* in growth phase regulation of virulence genes of the group A *Streptococcus. J. Bacteriol.* **179**:5178–5187.
133. **McIver, K. S., A. S. Thurman, and J. R. Scott.** 1999. Regulation of *mga* transcription in the group A *Streptococcus:* specific binding of mga within its own promoter and evidence for a negative regulator. *J. Bacteriol.* **181**:5373–5383.
134. **Merkel, T. J., C. Barros, and S. Stibitz.** 1998. Characterization of the *bvgR* locus of *Bordetella pertussis. J. Bacteriol.* **180**:1682–1690.
135. **Merkel, T. J., P. E. Boucher, S. Stibitz, and V. K. Grippe.** 2003. Analysis of *bvgR* expression in *Bordetella pertussis. J. Bacteriol.* **185**:6902–6912.
136. **Merkel, T. J., S. Stibitz, J. M. Keith, M. Leef, and R. Shahin.** 1998. Contribution of regulation by the *bvg* locus to respiratory infection of mice by *Bordetella pertussis. Infect. Immun.* **66**:4367–4373.
137. **Miller, A. A., N. C. Engleberg, and V. J. DiRita.** 2001. Repression of virulence genes by phosphorylation-dependent oligomerization of CsrR at target promoters in *S. pyogenes. Mol. Microbiol.* **40**:976–990.
138. **Mitchell, T. J., J. E. Alexander, P. J. Morgan, and P. W. Andrew.** 1997. Molecular analysis of virulence factors of *Streptococcus pneumoniae. Soc. Appl. Bacteriol. Symp. Ser.* **26**:62S–71S.
139. **Molinari, G., S. R. Talay, P. Valentin-Weigand, M. Rohde, and G. S. Chhatwal.** 1997. The fibronectin-binding protein of *Streptococcus pyogenes,* SfbI, is involved in the internalization of group A streptococci by epithelial cells. *Infect. Immun.* **65**:1357–1363.
140. **Mooi, F. R., W. H. Jansen, H. Brunings, H. Gielen, H. G. J. van der Heide, H. C. Walvoort, and P. A. M. Guinee.** 1992. Construction and analysis of *Bordetella pertussis* mutants defective in the production of fimbriae. *Microb. Pathog.* **12**:127–135.
141. **Morona, J. K., J. C. Paton, D. C. Miller, and R. Morona.** 2000. Tyrosine phosphorylation of CpsD negatively regulates capsular polysaccharide biosynthesis in *Streptococcus pneumoniae. Mol. Microbiol.* **35**:1431–1442.
142. **Moses, A. E., M. R. Wessels, K. Zalcman, S. Alberti, S. Natanson-Yaron, T. Menes, and E. Hanski.** 1997. Relative contributions of hyaluronic acid capsule and M protein to virulence in a mucoid strain of the group A *Streptococcus. Infect. Immun.* **65**:64–71.
143. **Moxon, E. R., and J. S. Kroll.** 1990. The role of bacterial polysaccharide capsules as virulence factors. *Curr. Top. Microbiol. Immunol.* **150**:65–85.
144. **Moxon, E. R., and K. A. Vaughn.** 1981. The type b capsular polysaccharide as a virulence determinant of *Haemophilus influenzae:* studies using clinical isolates and laboratory transformants. *J. Infect. Dis.* **143**:517–524.
145. **Nassif, X.** 1999. Interaction mechanisms of encapsulated meningococci with eucaryotic cells: what does this tell us about the crossing of the blood-brain barrier by *Neisseria meningitidis? Curr. Opin. Microbiol.* **2**:71–77.
146. **Nassif, X., J. Lowy, P. Stenberg, P. O'Gaora, A. Ganji, and M. So.** 1993. Antigenic variation of pilin regulates adhesion of *Neisseria meningitidis* to human epithelial cells. *Mol. Microbiol.* **8**:719–725.
147. **O'Brien, K. L., B. Beall, N. L. Barrett, P. R. Cieslak, A. Reingold, M. M. Farley, R. Danila, E. R. Zell, R. Facklam, B. Schwartz, and A. Schuchat.** 2002. Epidemiology of invasive group a streptococcus disease in the United States, 1995–1999. *Clin. Infect. Dis.* **35**:268–276.
148. **Okada, N., M. K. Liszewski, J. P. Atkinson, and M. Caparon.** 1995. Membrane cofactor protein (CD46) is a keratinocyte receptor for the M protein of the group A *Streptococcus. Proc. Natl. Acad. Sci. USA* **92**:2489–2493.
149. **Okada, N., A. P. Pentland, P. Falk, and M. G. Caparon.** 1994. M protein and protein F act as important determinants of cell-specific tropism of *Streptococcus pyogenes* in skin tissue. *J. Clin. Investig.* **94**:965–977.
150. **Otto, K., and T. J. Silhavy.** 2002. Surface sensing and adhesion of *Escherichia coli* controlled by the Cpx-signaling pathway. *Proc. Natl. Acad. Sci. USA* **99**:2287–2292.

151. **Parkhill, J., M. Achtman, K. D. James, S. D. Bentley, C. Churcher, S. R. Klee, G. Morelli, D. Basham, D. Brown, T. Chillingworth, R. M. Davies, P. Davis, K. Devlin, T. Feltwell, N. Hamlin, S. Holroyd, K. Jagels, S. Leather, S. Moule, K. Mungall, M. A. Quail, M. A. Rajandream, K. M. Rutherford, M. Simmonds, J. Skelton, S. Whitehead, B. G. Spratt, and B. G. Barrell.** 2000. Complete DNA sequence of a serogroup A strain of *Neisseria meningitidis* Z2491. *Nature* **404:** 502–506.

152. **Parkhill, J., M. Sebaihia, A. Preston, L. D. Murphy, N. Thomson, D. E. Harris, M. T. Holden, C. M. Churcher, S. D. Bentley, K. L. Mungall, A. M. Cerdeno-Tarraga, L. Temple, K. James, B. Harris, M. A. Quail, M. Achtman, R. Atkin, S. Baker, D. Basham, N. Bason, I. Cherevach, T. Chillingworth, M. Collins, A. Cronin, P. Davis, J. Doggett, T. Feltwell, A. Goble, N. Hamlin, H. Hauser, S. Holroyd, K. Jagels, S. Leather, S. Moule, H. Norberczak, S. O'Neil, D. Ormond, C. Price, E. Rabbinowitsch, S. Rutter, M. Sanders, D. Saunders, K. Seeger, S. Sharp, M. Simmonds, J. Skelton, R. Squares, S. Squares, K. Stevens, L. Unwin, S. Whitehead, B. G. Barrell, and D. J. Maskell.** 2003. Comparative analysis of the genome sequences of *Bordetella pertussis, Bordetella parapertussis* and *Bordetella bronchiseptica. Nat. Genet.* **35:**32–40.

153. **Parton, R.** 1985. Effect of predisolone on the toxicity of *Bordetella pertussis* for mice. *J. Med. Microbiol.* **19:**3814–3818.

154. **Pearson, R. D., P. Symes, M. Conboy, A. A. Weiss, and E. L. Hewlett.** 1987. Inhibition of monocyte oxidative responses by *Bordetella pertussis* adenylate cyclase toxin. *J. Immunol.* **139:**2749–2754.

155. **Perez-Caballero, D., S. Alberti, F. Vivanco, P. Sanchez-Corral, and S. Rodriguez de Cordoba.** 2000. Assessment of the interaction of human complement regulatory proteins with group A *Streptococcus*. Identification of a high-affinity group A *Streptococcus* binding site in FHL-1. *Eur. J. Immunol.* **30:** 1243–1253.

156. **Pettersson, J., R. Nordfelth, E. Dubinina, T. Bergman, M. Gustafsson, K. E. Magnusson, and H. Wolf-Watz.** 1996. Modulation of virulence factor expression by pathogen target cell contact. *Science* **273:**1231–1233.

157. **Podbielski, A., J. A. Peterson, and P. Cleary.** 1992. Surface protein-CAT reporter fusions demonstrate differential gene expression in the vir regulon of *Streptococcus pyogenes. Mol. Microbiol.* **6:**2253–2265.

158. **Podbielski, A., M. Woischnik, B. A. Leonard, and K. H. Schmidt.** 1999. Characterization of *nra*, a global negative regulator gene in group A streptococci. *Mol. Microbiol.* **31:** 1051–1064.

159. **Polissi, A., A. Pontiggia, G. Feger, M. Altieri, H. Mottl, L. Ferrari, and D. Simon.** 1998. Large-scale identification of virulence genes from *Streptococcus pneumoniae. Infect. Immun.* **66:**5620–5629.

160. **Ponnuraj, K., and M. J. Jedrzejas.** 2000. Mechanism of hyaluronan binding and degradation: structure of *Streptococcus pneumoniae* hyaluronate lyase in complex with hyaluronic acid disaccharide at 1.7 A resolution. *J. Mol. Biol.* **299:** 885–895.

161. **Prasadarao, N. V., E. Lysenko, C. A. Wass, K. S. Kim, and J. N. Weiser.** 1999. Opacity-associated protein A contributes to the binding of *Haemophilus influenzae* to chang epithelial cells. *Infect. Immun.* **67:**4153–4160.

162. **Pujol, C., E. Eugene, L. de Saint Martin, and X. Nassif.** 1997. Interaction of *Neisseria meningitidis* with a polarized monolayer of epithelial cells. *Infect. Immun.* **65:**4836–4842.

163. **Pujol, C., E. Eugene, M. Marceau, and X. Nassif.** 1999. The meningococcal PilT protein is required for induction of intimate attachment to epithelial cells following pilus-mediated adhesion. *Proc. Natl. Acad. Sci. USA* **96:**4017–4022.

164. **Rayner, C. F., A. D. Jackson, A. Rutman, A. Dewar, T. J. Mitchell, P. W. Andrew, P. J. Cole, and R. Wilson.** 1995. Interaction of pneumolysin-sufficient and -deficient isogenic variants of *Streptococcus pneumoniae* with human respiratory mucosa. *Infect. Immun.* **63:**442–447.

165. **Read, R. C., A. A. Rutman, P. K. Jeffery, V. J. Lund, A. P. Brain, E. R. Moxon, P. J. Cole, and R. Wilson.** 1992. Interaction of capsulate *Haemophilus influenzae* with human airway mucosa in vitro. *Infect. Immun.* **60:**3244–3252.

166. **Reddy, M. S., J. M. Bernstein, T. F. Murphy, and H. S. Faden.** 1996. Binding between outer membrane proteins of nontypeable *Haemophilus influenzae* and human nasopharyngeal mucin. *Infect. Immun.* **64:**1477–1479.

167. **Relman, D., E. Tuomanen, S. Falkow, D. T. Golenbock, K. Saukkonen, and S. D. Wright.** 1990. Recognition of a bacterial adhesion by an integrin: macrophage CR3 (αM β2, CD11b/CD18) binds filamentous hemagglutinin of *Bordetella pertussis. Cell* **61:**1375–1382.

168. **Relman, D. A., M. Domenighini, E. T. Tuomanen, R. Rappuoli, and S. Falkow.** 1989. Filamentous hemagglutinin of *Bordetella pertussis:* nucleotide sequence and crucial role in adherence. *Proc. Natl. Acad. Sci. USA* **86:**2634–2641.

169. **Ribardo, D. A., and K. McIver.** 2003. *amrA* encodes a putative membrane protein necessary for maximal exponential phase expression of the Mga virulence regulon in *Streptococcus pyogenes. Mol. Microbiol.* **50:**673–685.

170. **Robbins, J. C., J. G. Spanier, S. J. Jones, W. J. Simpson, and P. P. Cleary.** 1987. *Streptococcus pyogenes* type 12 M protein gene regulation by upstream sequences. *J. Bacteriol.* **169:** 5633–5640.

171. **Rosenow, C., P. Ryan, J. N. Weiser, S. Johnson, P. Fontan, A. Ortqvist, and H. R. Masure.** 1997. Contribution of novel choline-binding proteins to adherence, colonization and immunogenicity of *Streptococcus pneumoniae. Mol. Microbiol.* **25:**819–829.

172. **Rubins, J. B., and E. N. Janoff.** 1998. Pneumolysin: a multifunctional pneumococcal virulence factor. *J. Lab. Clin. Med.* **131:**21–27.

173. **Sarkari, J., N. Pandit, E. R. Moxon, and M. Achtman.** 1994. Variable expression of the Opc outer membrane protein in *Neisseria meningitidis* is caused by size variation of a promoter containing poly-cytidine. *Mol. Microbiol.* **13:**207–217.

174. **Scarlato, V., and R. Rappuoli.** 1991. Differential response of the bvg virulence regulon of *Bordetella pertussis* to $MgSO_4$ modulation. *J. Bacteriol.* **173:**7401–7404.

175. **Schmidt, K. H., E. Gunther, and H. S. Courtney.** 1996. Expression of both M protein and hyaluronic acid capsule by group A streptococcal strains results in a high virulence for chicken embryos. *Med. Microbiol. Immunol.* **184:**169–173.

176. **Schrager, H. M., S. Alberti, C. Cywes, G. J. Dougherty, and M. R. Wessels.** 1998. Hyaluronic acid capsule modulates M protein-mediated adherence and acts as a ligand for attachment of group A *Streptococcus* to CD44 on human keratinocytes. *J. Clin. Invest.* **101:**1708–1716.

177. **Sebert, M. E., L. M. Palmer, M. Rosenberg, and J. N. Weiser.** 2002. Microarray-based identification of *htrA*, a *Streptococcus pneumoniae* gene that is regulated by the CiaRH two-component system and contributes to nasopharyngeal colonization. *Infect. Immun.* **70:**4059–4067.

178. **Shakhnovich, E. A., S. J. King, and J. N. Weiser.** 2002. Neuraminidase expressed by *Streptococcus pneumoniae* desialylates the lipopolysaccharide of *Neisseria meningitidis* and *Haemophilus influenzae:* a paradigm for interbacterial competition among pathogens of the human respiratory tract. *Infect. Immun.* **70:**7161–7164.

179. **Smoot, J. C., K. D. Barbian, J. J. Van Gompel, L. M. Smoot, M. S. Chaussee, G. L. Sylva, D. E. Sturdevant, S. M. Ricklefs, S. F. Porcella, L. D. Parkins, S. B. Beres, D. S. Campbell, T. M. Smith, Q. Zhang, V. Kapur, J. A. Daly, L. G. Veasy, and J. M. Musser.** 2002. Genome sequence and comparative microarray analysis of serotype M18 group A *Streptococcus* strains associated with acute rheumatic fever outbreaks. *Proc. Natl. Acad. Sci. USA* **99:**4668–4673.

180. **Steffen, P., S. Goyard, and A. Ullmann.** 1996. Phosphorylated BvgA is sufficient for transcriptional activation of virulence-regulated genes in *Bordetella pertussis. EMBO J.* **15:** 102–109.

181. **Steinfort, C., R. Wilson, T. Mitchell, C. Feldman, A. Rutman, H. Todd, D. Sykes, J. Walker, K. Saunders, P. W. Andrew, G. J. Boulnois, and P. J. Cole.** 1989. Effect of Streptococcus pneumoniae on human respiratory epithelium in vitro. *Infect. Immun.* **57:**2006–2013.

182. **Stephens, D. S., L. H. Hoffman, and Z. A. McGee.** 1983. Interaction of *Neisseria meningitidis* with human nasopharyngeal mucosa: attachment and entry into columnar epithelial cells. *J. Infect. Dis.* **148:**369–376.

183. **Stephens, D. S., and Z. A. McGee.** 1981. Attachment of *Neisseria meningitidis* to human mucosal surfaces: influence of pili and type of receptor cell. *J. Infect. Dis.* **143:**525–532.

184. **Stern, A., M. Brown, P. Nickel, and T. F. Meyer.** 1986. Opacity genes in *Neisseria gonorrhoeae:* control of phase and antigenic variation. *Cell* **47:**61–71.

185. **Stern, A., and T. F. Meyer.** 1987. Common mechanism controlling phase and antigenic variation in pathogenic neisseriae. *Mol. Microbiol.* **1:**5–12.

186. **St Geme, J. W., III, D. Cutter, and S. J. Barenkamp.** 1996. Characterization of the genetic locus encoding *Haemophilus influenzae* type b surface fibrils. *J. Bacteriol.* **178:**6281–6287.

187. **St Geme, J. W., III, M. L. de la Morena, and S. Falkow.** 1994. A *Haemophilus influenzae* IgA protease-like protein promotes intimate interaction with human epithelial cells. *Mol. Microbiol.* **14:**217–233.

188. **Stockbauer, K. E., A. K. Foreman-Wykert, and J. F. Miller.** 2003. *Bordetella* type III secretion induces caspase 1-independent necrosis. *Cell Microbiol.* **5:**123–132.

189. **Stockbauer, K. E., B. Fuchslocher, J. F. Miller, and P. A. Cotter.** 2001. Identification and characterization of BipA, a *Bordetella* Bvg-intermediate phase protein. *Mol. Microbiol.* **39:** 65–78.

190. **Swanson, J., R. J. Belland, and S. A. Hill.** 1992. Neisserial surface variation: how and why? *Curr. Opin. Genet. Dev.* **2:**805–811.

191. **Swords, W. E., B. A. Buscher, K. Ver Steeg Ii, A. Preston, W. A. Nichols, J. N. Weiser, B. W. Gibson, and M. A. Apicella.** 2000. Non-typeable *Haemophilus influenzae* adhere to and invade human bronchial epithelial cells via an interaction of lipooligosaccharide with the PAF receptor. *Mol. Microbiol.* **37:**13–27.

192. **Taha, M. K., P. C. Morand, Y. Pereira, E. Eugene, D. Giorgini, M. Larribe, and X. Nassif.** 1998. Pilus-mediated adhesion of *Neisseria meningitidis:* the essential role of cell contact-dependent transcriptional upregulation of the PilC1 protein. *Mol. Microbiol.* **28:**1153–1163.

193. **Tettelin, H., V. Masignani, M. J. Cieslewicz, J. A. Eisen, S. Peterson, M. R. Wessels, I. T. Paulsen, K. E. Nelson, I. Margarit, T. D. Read, L. C. Madoff, A. M. Wolf, M. J. Beanan, L. M. Brinkac, S. C. Daugherty, R. T. DeBoy, A. S. Durkin, J. F. Kolonay, R. Madupu, M. R. Lewis, D. Radune, N. B. Fedorova, D. Scanlan, H. Khouri, S. Mulligan, H. A. Carty, R. T. Cline, S. E. Van Aken, J. Gill, M. Scarselli, M. Mora, E. T. Iacobini, C. Brettoni, G. Galli, M. Mariani, F. Vegni, D. Maione, D. Rinaudo, R. Rappuoli, J. L. Telford, D. L. Kasper, G. Grandi, and C. M. Fraser.** 2002. Complete genome sequence and comparative genomic analysis of an emerging human pathogen, serotype V *Streptococcus agalactiae. Proc. Natl. Acad. Sci. USA* **99:**12391–12396.

194. **Tettelin, H., Saunders, J. Heidelberg, A. C. Jeffries, K. E. Nelson, J. A. Eisen, K. A. Ketchum, D. W. Hood, J. F. Peden, R. J. Dodson, W. C. Nelson, M. L. Gwinn, R. DeBoy, J. D. Peterson, E. K. Hickey, D. H. Haft, S. L. Salzberg, O. White, R. D. Fleischmann, B. A. Dougherty, T. Mason, A. Ciecko, D. S. Parksey, E. Blair, H. Cittone, E. B. Clark, M. D. Cotton, R. R. Utterback, H. Khouri, H. Qin, J. Vamathevan, J. Gill, V. Scarlato, V. Masignani, M. Pizza, G. Grandi, L. Sun, H. O. Smith, C. M. Fraser, E. R. Moxon, R. Rappuoli, and J. C. Venter.** 2000. Complete genome sequence of *Neisseria meningitidis* serogroup B strain MC58. *Science* **287:**1809–1815.

195. **Throup, J. P., K. K. Koretke, A. P. Bryant, K. A. Ingraham, A. F. Chalker, Y. Ge, A. Marra, N. G. Wallis, J. R. Brown, D. J. Holmes, M. Rosenberg, and M. K. Burnham.** 2000. A genomic analysis of two-component signal transduction in *Streptococcus pneumoniae. Mol. Microbiol.* **35:**566–576.

196. **Tomasz, A.** 1995. The pneumococcus at the gates. *N. Engl. J. Med.* **333:**514–515.

197. **Tomasz, A., M. Westphal, E. B. Briles, and P. Fletcher.** 1975. On the physiological functions of teichoic acids. *J. Supramol. Struct.* **3:**1–16.

198. **Tuomanen, E.** 1999. Molecular and cellular biology of pneumococcal infection. *Curr. Opin. Microbiol.* **2:**35–39.

199. **Uhl, M. A., and J. F. Miller.** 1994. Autophosphorylation and phosphotransfer in the *Bordetella pertussis* BvgAS signal transduction cascade. *Proc. Natl. Acad. Sci. USA* **91:**1163–1167.

200. **Uhl, M. A., and J. F. Miller.** 1995. *Bordetella pertussis* BvgAS virulence control system, p. 333–349. *In* J. A. Hoch and T. Silhavy (ed.), *Two-Component Signal Transduction.* ASM Press, Washington, D.C.

201. **Uhl, M. A., and J. F. Miller.** 1996. Central role of the BvgS receiver as a phosphorylated intermediate in a complex two-component phosphorelay. *J. Biol. Chem.* **271:**33176–33180.

202. **Uhl, M. A., and J. F. Miller.** 1996. Integration of multiple domains in a two-component sensor protein: the *Bordetella pertussis* BvgAS phosphorelay. *EMBO J.* **15:**1028–1036.

203. **Ui, J.** 1990. Pertussis toxin as a valuable probe for G-protein involvement in signal transduction. *In* J. Moss and M. Vaughan (ed.), *ADP-Ribosylating Toxins and G-Proteins: Insights into Signal Transduction.* American Society for Microbiology, Washington, D.C.

204. **Unnikrishnan, M., J. Cohen, and S. Sriskandan.** 1999. Growth-phase-dependent expression of virulence factors in an M1T1 clinical isolate of *Streptococcus pyogenes. Infect. Immun.* **67:**5495–5499.

205. **van Alphen, L., N. van den Berghe, and L. Geelen-van den Broek.** 1988. Interaction of *Haemophilus influenzae* with human erythrocytes and oropharyngeal epithelial cells is mediated by a common fimbrial epitope. *Infect. Immun.* **56:**1800–1806.

206. **van Deuren, M., P. Brandtzaeg, and J. W. van der Meer.** 2000. Update on meningococcal disease with emphasis on pathogenesis and clinical management. *Clin. Microbiol. Rev.* **13:**144–166.

207. **Virji, M., D. Evans, J. Griffith, D. Hill, L. Serino, A. Hadfield, and S. M. Watt.** 2000. Carcinoembryonic antigens are targeted by diverse strains of typable and non-typable *Haemophilus influenzae. Mol. Microbiol.* **36:**784–795.

208. **Virji, M., D. Evans, A. Hadfield, F. Grunert, A. M. Teixeira, and S. M. Watt.** 1999. Critical determinants of host receptor targeting by *Neisseria meningitidis* and *Neisseria gonorrhoeae:* identification of Opa adhesiotopes on the N-domain of CD66 molecules. *Mol. Microbiol.* **34:**538–551.

209. **Virji, M., H. Kayhty, D. J. Ferguson, C. Alexandrescu, J. E. Heckels, and E. R. Moxon.** 1991. The role of pili in the interactions of pathogenic *Neisseria* with cultured human endothelial cells. *Mol. Microbiol.* **5:**1831–1841.

210. **Virji, M., K. Makepeace, D. J. Ferguson, M. Achtman, and E. R. Moxon.** 1993. Meningococcal Opa and Opc proteins: their role in colonization and invasion of human epithelial and endothelial cells. *Mol. Microbiol.* **10:**499–510.

211. **Virji, M., K. Makepeace, D. J. Ferguson, M. Achtman, J. Sarkari, and E. R. Moxon.** 1992. Expression of the Opc protein correlates with invasion of epithelial and endothelial cells by *Neisseria meningitidis. Mol. Microbiol.* **6:**2785–2795.

212. **Virji, M., K. Makepeace, D. J. Ferguson, and S. M. Watt.** 1996. Carcinoembryonic antigens (CD66) on epithelial cells and neutrophils are receptors for Opa proteins of pathogenic neisseriae. *Mol. Microbiol.* **22:**941–950.

213. **Virji, M., K. Makepeace, and E. R. Moxon.** 1994. Distinct mechanisms of interactions of Opc-expressing meningococci at apical and basolateral surfaces of human endothelial cells; the role of integrins in apical interactions. *Mol. Microbiol.* **14:**173–184.

214. **Virji, M., K. Makepeace, I. R. Peak, D. J. Ferguson, M. P. Jennings, and E. R. Moxon.** 1995. Opc- and pilus-dependent interactions of meningococci with human endothelial cells: molecular mechanisms and modulation by surface polysaccharides. *Mol. Microbiol.* **18:**741–754.

215. **Virji, M., J. R. Saunders, G. Sims, K. Makepeace, D. Maskell, and D. J. Ferguson.** 1993. Pilus-facilitated adherence of *Neisseria meningitidis* to human epithelial and endothelial cells: modulation of adherence phenotype occurs concurrently with changes in primary amino acid sequence and the glycosylation status of pilin. *Mol. Microbiol.* **10:**1013–1028.

216. **Virji, M., S. M. Watt, S. Barker, K. Makepeace, and R. Doyonnas.** 1996. The N-domain of the human CD66a adhesion molecule is a target for Opa proteins of *Neisseria meningitidis* and *Neisseria gonorrhoeae. Mol. Microbiol.* **22:**929–939.

217. **Vogel, U., and M. Frosch.** 1999. Mechanisms of neisserial serum resistance. *Mol. Microbiol.* **32:**1133–1139.

218. **Voyich, J. M., D. E. Sturdevant, K. R. Braughton, S. D. Kobayashi, B. Lei, K. Virtaneva, D. W. Dorward, J. M. Musser, and F. R. DeLeo.** 2003. Genome-wide protective response used by group A *Streptococcus* to evade destruction by human polymorphonuclear leukocytes. *Proc. Natl. Acad. Sci. USA* **100:**1996–2001.

219. **Wang, J. R., and M. W. Stinson.** 1994. M protein mediates streptococcal adhesion to HEp-2 cells. *Infect. Immun.* **62:** 442–448.

220. **Weiser, J. N., R. Austrian, P. K. Sreenivasan, and H. R. Masure.** 1994. Phase variation in pneumococcal opacity: relationship between colonial morphology and nasopharyngeal colonization. *Infect. Immun.* **62:**2582–2589.

221. **Weiser, J. N., D. Bae, H. Epino, S. B. Gordon, M. Kapoor, L. A. Zenewicz, and M. Shchepetov.** 2001. Changes in availability of oxygen accentuate differences in capsular polysaccharide expression by phenotypic variants and clinical isolates of *Streptococcus pneumoniae. Infect. Immun.* **69:**5430–5439.

222. **Weiser, J. N., S. T. Chong, D. Greenberg, and W. Fong.** 1995. Identification and characterization of a cell envelope protein of *Haemophilus influenzae* contributing to phase variation in colony opacity and nasopharyngeal colonization. *Mol. Microbiol.* **17:**555–564.

223. **Weiser, J. N., J. M. Love, and E. R. Moxon.** 1989. The molecular mechanism of phase variation of *H. influenzae* lipopolysaccharide. *Cell* **59:**657–665.

224. **Weiser, J. N., M. Shchepetov, and S. T. Chong.** 1997. Decoration of lipopolysaccharide with phosphorylcholine: a phase-variable characteristic of *Haemophilus influenzae. Infect. Immun.* **65:**943–950.

225. **Weiss, A. A., and S. Falkow.** 1984. Genetic analysis of phase change in *Bordetella pertussis. Infect. Immun.* **43:**263–269.

226. **Wessels, M. R., and M. S. Bronze.** 1994. Critical role of the group A streptococcal capsule in pharyngeal colonization and infection in mice. *Proc. Natl. Acad. Sci. USA* **91:**12238–12242.

227. **Wessels, M. R., J. B. Goldberg, A. E. Moses, and T. J. DiCesare.** 1994. Effects on virulence of mutations in a locus essential for hyaluronic acid capsule expression in group A streptococci. *Infect. Immun.* **62:**433–441.

228. **Whitnack, E., and E. H. Beachey.** 1985. Inhibition of complement-mediated opsonization and phagocytosis of *Streptococcus pyogenes* by D fragments of fibrinogen and fibrin bound to cell surface M protein. *J. Exp. Med.* **162:**1983–1997.

229. **Wirsing von Konig, C. H., and H. Finger.** 1994. Role of pertussis toxin in causing symptoms of *Bordetella parapertussis* infection. *Eur. J. Clin. Microbiol. Infect. Dis.* **13:**455–458.

230. **Yeh, S. H.** 2003. Pertussis: persistent pathogen, imperfect vaccines. *Expert Rev. Vaccines* **2:**113–127.

231. **Yuk, M. H., E. T. Harvill, P. A. Cotter, and J. F. Miller.** 2000. Modulation of host immune responses, induction of apoptosis and inhibition of NF-κB activation by the *Bordetella* type III secretion system. *Mol. Microbiol.* **35:**991–1104.

232. **Yuk, M. H., E. T. Harvill, and J. F. Miller.** 1998. The *bvgAS* virulence control system regulates type III secretion in *Bordetella bronchiseptica. Mol. Microbiol.* **28:**945–959.

233. **Zu, T., R. Manetti, R. Rappuoli, and V. Scarlato.** 1996. Differential binding of BvgA to two classes of virulence genes of *Bordetella pertussis* directs promoter selectivity by RNA polymerase. *Mol. Microbiol.* **21:**557–565.

III. COLONIZATION OF THE GASTROINTESTINAL TRACT

Colonization of Mucosal Surfaces
Edited by James P. Nataro et al.

Chapter 12

Microbiota of Mucosal Surfaces in the Gut of Monogastric Animals

GERALD W. TANNOCK

THE GUT MICROBIOTA

The gut of monogastric animals, including that of humans, is inhabited by microbial populations, mostly bacterial, throughout the life of the host. Not all regions of the gut are suitable for microbial persistence: the acid secreted in the stomach and the swift flow of contents in the duodenum and jejunum ensure that the more proximal regions contain only transient bacterial cells in the healthy host (89). An exception to this is the proliferation of lactobacilli on the epithelial surface of the rodent forestomach, the porcine pars oesophagea, and the avian crop (3, 26, 90). *Helicobacter pylori* associates with the surface of the human gastric and duodenal mucosa, but since the presence of this bacterial species is always associated with inflammation of the tissue, it is probably best for the present to consider it a pathogen and for its presence to be termed an infection (51). The terminal ileum and the large bowel (cecum and colon) are hospitable places for bacterial proliferation, and a complex and numerous bacterial community resides in this site. This community is often referred to as the normal gut microbiota (microflora).

Many of the numerically important members of the gut microbiota have not yet been cultivated under laboratory conditions and are known and detected on the basis of their 16S rRNA gene sequences (100). As in terrestrial and aquatic microbial ecology, studies of the gut communities rely largely on nucleic acid-based methods of analysis (72). The preparation of random clone libraries of 16S rRNA sequences amplified by PCR from gut samples has been of particular importance. Clone libraries prepared from human feces and from murine, porcine, and avian gut samples have revealed a considerable complexity in composition of the respective gut communities. Moreover, a new perspective concerning the prevalence of bacterial species has been obtained: clostridia and related genera are among the predominant bacteria, a feature not easily recognised from the results of culture-based studies. Summaries of the composition of the distal gut microbiota of humans, mice, pigs, and broilers are provided in Table 1.

Monitoring the composition of the fecal community of individual humans by the use of PCR combined with temperature or denaturing gradient gel electrophoresis (TGGE, DGGE) has shown that the composition of the fecal microbiota of each person, with respect to the numerically dominant species, is unique and extremely stable over time (125). Immense variation in the proportions in which particular phylogenetic divisions of bacteria comprise the human fecal microbiota has also been detected by fluorescent in situ hybridization (FISH), in which the cells of particular bacterial groups were enumerated by fluorescently labeled DNA probes (97). Collectively, these observations indicate that the bacterial community of the large bowel is well regulated and resists minor perturbations that could be induced by a varied daily diet or minor fluctuations in host physiological parameters. Uniqueness in the composition of bacterial communities probably reflects consistent physiological and immunological idiosyncrasies of humans that are controlled by the genetic constitution of the host. Monozygotic twins, for example, have DGGE profiles that are more similar to one another than to those of unrelated subjects (126).

Although the phylogeny of individual bacterial communities seems unique among humans, the overall metabolic profile of the microbiota in terms of fermentative ability is similar. The same short-chain fatty acids in similar proportions are detected in human feces regardless of the bacterial community profile obtained by PCR/DGGE (111). This argues for considerable redundancy among the bacterial species

Gerald W. Tannock • Department of Microbiology, University of Otago, Dunedin, New Zealand, and Agricultural, Food and Nutritional Science, University of Alberta, Edmonton, Canada.

Table 1. Investigations of the phylogeny of the gut microbiota of humans, mice, chickens, and pigs by the analysis of random clone libraries of 16S rRNA genes

Investigation	Observations
Analysis of a 16S rRNA gene library derived from a human fecal sample (100)	A 284-clone 16S rRNA gene library was derived. Three phylogenetic groups comprised 95% of the clones: the *Bacteroides-Prevotella* group, the *Clostridium coccoides* group, and the *Clostridium leptum* subgroup. Most of the sequences (76%) did not correspond to bacteria that had been cultivated in the laboratory (known bacteria).
Phylogenetic analysis of the human gut microbiota by using 16S rRNA gene clone libraries (37)	Fecal samples from three adult humans were used to derive a 16S rRNA gene library of the microbiota. A total of 744 clones were partially sequenced. Of the sequences, 25% represented 31 species, including those of the *Bacteroides* group, *Streptococcus* group, *Bifidobacterium* group, and *Clostridium* RNA clusters IV, IX, XIVa, and XVIII. There were marked differences between the microbiotas of the three humans.
Molecular analysis of the fecal microbiota of a strict vegetarian (38)	The diversity of the fecal microbiota of a vegetarian was studied by deriving a 16S rRNA gene library. Among 183 clones, 29% represented 13 known species. The sequences mainly represented the *Bacteroides* group, *Bifidobacterium* group, and *Clostridium* rRNA clusters IV, XIVa, XVI, and XVIII.
Molecular analysis of the fecal microbiota of elderly humans (39)	The fecal microbiota of six elderly humans (at least 74 years old) was characterized by deriving 16S rRNA gene libraries. Among 240 clones that were partially sequenced, 46% represented 27 species. Most of the sequences represented the *Bacteroides* group, *Clostridium* rRNA clusters IV, IX, XIVa, and gamma proteobacteria.
Analysis of 16S rRNA gene libraries derived from mouse gut samples (84)	Ten independent libraries of cloned 16S rRNA gene sequences (one cecal, one fecal, five small bowel, and three large bowel) were derived. Seventy clones were used in the analysis. A total of 40 unique rRNA gene sequences were detected, of which 30 were novel. Of these, 11 belonged to a new murine *Cytophaga-Flavobacter-Bacteroides* operational taxonomic unit.
Analysis of the porcine gut microbiota by using 16S rRNA gene libraries (52)	Fifty two samples from the ileum, cecum, or colon of 24 pigs were used to generate a 4,270-clone library of 16S rRNA gene sequences. A total of 375 operational taxonomic units were detected, representing 13 major phylogenetic groups. Of the sequences, 83% did not closely resemble (<97% identity) any held in the database.
Analysis of 16S rRNA gene sequences derived from cecal contents and cecal mucosa of broiler chickens (123)	A total of 1,656 partial 16S rRNA gene sequences were obtained, of which 243 were unique. Fifty phylogenetic groups or subgroups were detected, but 89% of the sequences represented just four phylogenetic groups (*Clostridium leptum* subgroup, *Sporomusa* species, *Clostridium coccoides* group, and enterics).
Analysis of 16S rRNA gene sequences derived from ileal and cecal samples from broiler chickens (53)	Analysis of 1,230 partial 16S rRNA gene sequences showed that almost 70% of sequences derived from ileal samples were related to those of *Lactobacillus*, with the remainder being mostly related to the *Clostridiaceae*, *Streptococcus* and *Enterococcus*. The most abundant sequences detected in the ceca represented the *Clostridiaceae*, with *Fusobacterium*, *Lactobacillus*, and *Bacteroides* also being represented.
Analysis of 16S rRNA gene clones derived from ileal and cecal samples from broiler chickens (30, 31)	Ileal and cecal samples (digesta and mucosal washings) collected from 10 chickens and pooled were used to derive 16S rRNA clone libraries. Fifty-one partial sequences from ileal bacteria were analyzed: the majority of these sequences were derived from *Lactobacillus* and enterococci. In the cecum, sequences from *Fusobacterium prausnitzii* and butyrate-producing bacteria predominated among 116 clones. Of the sequences, 25% had <95% homology to those in the data bank. Many sequences were related to those of not-yet-cultivated bacteria that have been detected in human feces or the bovine rumen.
Phylogenetic analysis of the cecal microbiota of layer chickens (50)	The cecal microbiota was analyzed by deriving a 16S rRNA gene library. A total of 104 clone sequences were obtained, of which 94% represented low-G+C-content bacteria. *Clostridium* subcluster XIVa (38% of clones), *Clostridium* cluster IV (13%), *Lactobacillus* (24%), and the *Bacteroides* group (4%) were the major groups of bacteria detected.

that can inhabit the gut: probably several bacterial species can fill a given ecological niche, and each niche is filled differently from human to human. Thus, no matter what the composition of the complex microbiota, the bowel ecosystem functions in the same, predictable manner. It may be speculated, therefore, that amassing catalogues of bacterial species will not promote a much clearer understanding of the gut ecosystem than that which already exists. Mere detection of a bacterial species in a gut sample reveals little about the role of this organism in the ecosystem or even whether the bacterium is metabolically active. As recently reported, there is a striking difference in DGGE profiles of bacterial communities in human feces when bacterial RNA is used as the PCR template and when DNA is used. Profiles generated from bacterial RNA showed intensely stained fragments clustered in the middle of the denaturing gradient and were markedly different from those generated from DNA (110). RNA extracted from bacterial cells is mostly rRNA and can be used as an indicator of metabolic activity since the ribosome-per-cell ratio is roughly proportional to the growth rate of the bacteria. While DNA-based analytical procedures provide a phylogenetic picture of the community, they do not reflect metabolic activity because the DNA could originate from living active cells, living dormant cells, lysed cells, or dead cells (24). On the basis of these observations, it can be proposed that the human gut provides a habitat for a diversity of bacterial species, as seen in DNA-DGGE profiles, the composite of which varies from one human to another. The "core" or "true" gut microbiota may be composed of relatively few populations that provide the major metabolic activities. This is clearly a topic that requires further investigation, because the physiological interactions between the host and the cellular and metabolic products of the overall bacterial community may be an important aspect with regard to the gut microbiota in health and disease.

IMPACT OF THE GUT MICROBIOTA ON THE HOST

Comparisons of the characteristics of germfree and conventional animals have clearly demonstrated that the gut microbiota has considerable influences on host biochemistry, physiology, immunology, and low-level resistance to gut infections (32, 105) (Table 2). Gnotobiotic research can now be done at a sophisticated level because of the convergence of genome sequencing of animals and bacteria and the consequent manufacture of DNA microarrays that feature sequences representative of the entire genome of the animal or bacterium. It is possible, for example, to measure the transcription of mammalian genes resulting from exposure of the animal to specific bacteria and vice versa. The potential for obtaining exciting knowledge about the mechanistic influences of the microbiota on the host by using this approach has been demonstrated by the pioneering work of Hooper et al. (41). They studied the impact of colonization of ex-germfree mice by *Bacteroides thetaiotaomicron* (Table 3). In the complex, conventional ecosystem, however, the up- or down-regulation of host gene expression induced by the presence of one bacterial species could be negated by the impact of another species. Therefore, a more realistic comparison would be between germfree and conventional animals. Hooper et al. (41) observed that ileal expression of colipase (involved in lipid metabolism) and angiogenin-3 (involved in blood vessel development) was similar in conventionalized ex-germfree mice (inoculated with combined ileal and cecal contents from conventional mice) to that induced by specific bacterial species in monoassociation experiments. In contrast, Mdr1a (export of glutathione conjugates from epithelium) and glutathione *S*-transferase (glutathione conjugation to electrophiles) showed species-specific responses in monoassociation experiments, whereas conventionalized ex-germfree mice did not show a significant change. If it is accepted that the phylogeny of the gut community is largely irrelevant, gene expression studies should be planned on the basis of alterations to the functioning of the ecosystem as a whole (synecology) rather than by concentrating on the impact of individual bacterial species (autecology).

AUTOCHTHONY AND ALLOCHTHONY

Recent studies of the composition of the fecal bacterial community of humans has shown that the detection of a particular bacterial species does not necessarily mean that the bacteria are inhabitants of the gut ecosystem. *Lactobacillus* populations in human feces, for example, fluctuated in species and strain composition in about half of the group of humans who were examined in one study (111). Most probably, these *Lactobacillus* species were transient in the gut because they were present in foods and are often used as starter organisms in food production, as well as in probiotic products. Food-associated lactobacilli survive passage through the gut but do not persist there in the absence of continued consumption (15, 111, 116). Their presence in the gut is dependent on external factors (consumption of food in which they are present) and can be termed "allochthonous"

Table 2. Comparison of selected biochemical properties of the intestinal tracts of germfree and conventional animals

Property	Finding in: Conventional animals[a]	Finding in: Germfree animals[a]
Bile acid metabolism	Deconjugation, dehydrogenation, and dehydroxylation	Absence of deconjugation, dehydrogenation, and dehydroxylation
Bilirubin metabolism	Deconjugation and reduction	Little deconjugation and absence of reduction
Cholesterol	Reduction to coprostanol	Absence of coprostanol
β-Aspartylglycine	Absent	Present
Intestinal gases	Hydrogen, methane, and carbon dioxide	Absence of hydrogen and methane; less carbon dioxide
Short-chain fatty acids	Large amounts of several acids	Small amounts of a few acids
Tryptic activity	Little activity	High activity
Urease	Present	Absent
β-Glucuronidase (pH 6.5)	Present	Absent
Extent of degradation of mucins	More	Less
Serum cholesterol concn	Lower	Higher

[a]Conventional, raised in association with a normal microbiota; germfree, raised in the absence of demonstrable microbes.

(found in a place other than where they were formed). In contrast, some *Lactobacillus* species can be consistently detected in the feces of a given human host over long periods. These lactobacilli, notably *Lactobacillus ruminis,* can be referred to as being "autochthonous" to the gut (inhabiting a place or region from earliest times) (18, 88).

Discovery of the molecular traits that confer autochthony on a bacterial species should have high priority. The bacteria residing in the gut of mammals have coevolved with their host and have developed a high degree of adaptation and specialization. These bacteria must possess traits that enable them to establish and maintain themselves in a lotic and highly competitive environment. The challenge for the bacteria is to satisfy their own growth requirements as well as to cope with the hostile conditions generated by the competing members of the microflora and by the defense mechanisms of the host.

Although minor members in terms of numbers in the human bowel, lactobacilli are relatively numerous in the proximal regions of the gut of mice, rats, pigs, and chickens. Due to the presence of a nonsecretory, squamous epithelium lining the rodent forestomach, the avian crop, and the porcine pars oesophagea, lactobacilli numerically dominate these proximal regions of the gut (3, 26, 90). Since a mucus layer is absent from the surface of these specific epithelia, the lactobacilli adhere directly to epithelial cells, forming a layer of bacterial cells. Lactobacilli shed from this layer inoculate the digesta, and lactobacilli are therefore detected throughout the remainder of the gut (26). The mechanism by which *Lactobacillus* strains adhere to these epithelia has not yet been determined, but preliminary in vitro investigations have shown that both carbohydrate and protein molecules are involved (25, 101). *Lactobacillus* strains that adhere to epithelial cells, in general, show

Table 3. Examples of murine gene transcription affected by monoassociation of ex-germfree mice with *B. thetaiotaomicron* [a]

Murine gene or product	Fold change in gene expression relative to germfree mice[b]	Function
Sodium/glucose cotransporter (SGLT-1)	2.6 (0.9)	Glucose uptake
Colipase	6.6 (1.9)	Lipid metabolism
Liver fatty acid-binding protein (L-FABP)	4.4 (1.4)	Lipid metabolism
Metallothionein	−5.4 (0.7)	Copper/zinc sequestration
Polymeric immunoglobulin receptor (pIgR)	2.6 (0.7)	Transepithelial immunoglobulin A transport
Decay-accelerating factor (DAF)	5.7 (1.5)	Complement inactivation
Small proline-rich protein 2a (Sprr2a)	205 (64)	Cross-linking protein (fortification of intestinal barrier)
Glutathione *S*-transferase (GST)	−2.1 (0.1)	Glutathione conjugation to electrophiles
Multidrug resistance protein (Mdr1a)	−3.8 (1.0)	Export of glutathione conjugates from epithelium
Lactase-phlorizin hydrolase	−4.1 (0.6)	Lactose hydrolysis
Adenosine deaminase (ADA)	2.6 (0.5)	Adenosine inactivation
Angiogenin-3	9.1 (1.8)	Angiogenesis factor

[a]Data from reference 41.
[b]Real-time quantitative PCR values obtained using ileal RNA samples. Data are means of triplicate measurements with standard deviations in parentheses.

specificity for an animal host (112). Strains originating from the rodent forestomach do not adhere to crop epithelial cells, while isolates from poultry do not adhere to epithelial cells from the rodent forestomach or the pars oesophagea of pigs. However, some exceptions occur. For example, in one study a *Lactobacillus* strain isolated from calf feces adhered strongly to the squamous epithelium of the mouse stomach (94). Attachment to the epithelium must be a critical factor in the persistence of these lactobacilli in the gut: when attached and replicating on a host surface, a microbe can persist in a flowing habitat, whereas nonadherent microbes are carried away by the flow of secretions.

Lactobacillus reuteri is autochthonous to the rodent gut because it has been detected there in several studies, adheres to the nonsecretory epithelium of the forestomach, and persists at constant population levels throughout life in the gut of ex-*Lactobacillus*-free mice inoculated by mouth with a pure culture on a single occasion (40, 64, 118). *L. reuteri* and the gut ecosystem of mice therefore provide an excellent paradigm to study the molecular basis of autochthony. In the past decade, promoter-trapping technologies have been developed to overcome the limitation of in vitro models to study the traits that enhance ecological performance in complex ecosystems. For example, in vivo expression technology (IVET) was developed by Mahan et al. (55) to study gene expression by *Salmonella enterica* serovar Typhimurium during infection of mice. IVET has also been used to identify in vivo-induced (*ivi*) genes for a number of other pathogens, and mutations within a subset of these *ivi* genes resulted in a decrease in virulence (74). IVET recently identified *Lactobacillus reuteri* strain 100-23 genes that were specifically induced in the murine gut (115). A plasmid-based system was constructed containing *'ermGT* (confers lincomycin resistance) as the primary reporter gene for selection of promoters active in the gastrointestinal tract of mice treated with lincomycin. A second reporter gene, *'bglM* (β-glucanase), allowed the differentiation between constitutive and in vivo-inducible promoters. Application of the IVET system using *L. reuteri* and ex-*Lactobacillus*-free mice revealed three genes induced specifically during colonization. Sequences showing homologies to xylose isomerase (*xylA*) and methionine sulfoxide reductase (*msrB*) were detected. The third locus showed homology to a protein of unknown function. Xylose is a plant-derived sugar commonly found in straw and bran and is introduced into the gut in the food. Xylose in the gut could be derived from the hydrolysis of xylans and pectins by the gut microflora. *L. reuteri* may also utilize isoprimeverose, which is the major building block of xyloglucans, as described for *L. pentosus*. The selective expression of xylose isomerase suggests that *L. reuteri* 100-23 meets its energy requirements in the gut at least partly by the fermentation of xylose or isoprimeverose. Methionine sulfoxide reductase is a repair enzyme protecting bacteria against oxidative damage caused by reactive nitrogen and oxygen intermediates. Nitric oxide is produced by epithelial cells of the ileum and colon and possibly acts as an oxidative barrier, maintaining intestinal homeostasis, reducing bacterial translocation, and providing a means of defense against pathogens. This pioneering IVET study showed the utility of the technology in investigating the molecular basis of autochthony and identified bacterial properties that may be essential for *L. reuteri* to persist in the gut.

In the ileum of mice and some other animal species, filamentous segmented bacteria related to the clostridia attach by one end to enterocytes, particularly in the vicinity of Peyer's patches (47, 95). The attachment involves the insertion of the base of the filament into an enterocyte. Cytoskeletal rearrangements within the enterocyte form a socket by which the filament becomes permanently attached to the mucosal surface (44). The presence of filamentous segmented bacteria in the mouse intestinal tract results in the up-regulation of the murine gene that encodes a fucosyl transferase involved in fucosylation of the asialo-GM1 glycolipid that is associated with the enterocyte surface (114). This gene expression phenomenon has also been described in relation to *B. thetaiotaomicron* in the gut of monoassociated ex-germfree mice (7, 42). The bacteria utilize L-fucose, salvaged from glycoconjugates, as an energy source. The induction of fucosylation of glycoconjugates in the bowel of mice was shown to be dependent on the presence of a critical concentration of *Bacteroides* cells (10^6 to 10^7 CFU/ml) and on the ability of the bacteria to utilize L-fucose. A mutant strain, unable to utilize L-fucose, was less efficient at inducing fucosylation. The linkage between L-fucose utilization and the signaling system that induces fucosylation of glycoconjugates is mediated by a bacterial protein, FucR. It is supposed that the *Bacteroides* cells, by influencing host biochemistry, ensure that L-fucose is constantly available as an energy source in an ecosystem that is probably largely regulated by competition for nutrients in growth rate-limiting concentrations.

The ability of bacterial cells to alert their kindred to increasing cell density (quorum sensing) is well known (23). Typically, cells produce a small extracellular signal molecule, the autoinducer, and simultaneously sense it at the cell surface. If the concentration of autoinducer exceeds a threshold, gene expression is induced, often resulting in the production of other extracellular substances. Established paradigms of this phenomenon include cell-to-cell signaling in the formation of fruiting bodies by myxobacteria and luminescence by *Vibrio fischeri* when inhabiting the light organ of the squid *Euprymna scolopes* (33, 45, 58). Cell-to-cell signaling could be important in the establishment of biofilms associated with epithelia in the gut. A clue to this is provided by observations obtained with *Pseudomonas aeruginosa* biofilms on glass surfaces (16). Films formed by the wild-type strain contained mushroom- and pillar-shaped microcolonies separated by water-filled spaces. Mutant strains that could not produce a signaling molecule (an acyl-homoserine lactone) produced thin, densely packed films. Addition of the signaling molecule to the mutant biofilm resulted in a return to a film with a normal structure. For this reason, *Lactobacillus* species that form biofilms on epithelia should be investigated for cell-to-cell signaling molecules. Oligopeptides seem the most likely proposition in gram-positive bacteria (46), but the "universal" signaling molecule Autoinducer 2 (AI-2), associated with the presence of the *luxS* gene, may be worth investigating. The bioluminescent bacterium *Vibrio harveyi* produces a homoserine lactone autoinducer termed AI-1, typical of gram-negative bacterial quorum-sensing molecules. Additionally, it produces and senses a second autoinducer, AI-2 (23, 122). A wide range of gram-positive and gram-negative bacterial species produce AI-2 by a common biosynthetic pathway, and it has been proposed that AI-2 is a universal signaling molecule that functions in interspecies cell-to-cell communication. AI-2 is produced from *S*-adenosylmethionine (SAM) in three enzymatic steps. The use of SAM as a methyl donor in metabolic processes such as DNA synthesis produces a toxic intermediate, *S*-adenosylhomocysteine (an inhibitor of the SAM-dependent methyltransferases) that is hydrolyzed to *S*-riboadenosylhomocysteine (SRH) by a nucleosidase. The protein LuxS catalyzes the cleavage of SRH to form homocysteine and 4,5-dihydroxy-2,3-pentanedione (DPD). The DPD molecule cyclizes to form pro-AI-2, which is then transformed to AI-2 by the addition of boron. It can be argued that AI-2 has a cell-signaling function because extracellular accumulation would be proportional to the cell number. Additionally, the link between AI-2 production and cell syntheses (utilization of SAM) could make AI-2 accumulation a device for measuring the metabolic potential of the cell population (23, 122). If AI-2 molecules produced by every bacterial species capable of doing so are identical, then AI-2 could only inform bacterial cells that other bacterial cells were present; it would not allow cells to determine the identities of the other species that were present. Nevertheless, AI-2 has been reported to influence biofilm formation by *Streptococcus mutans* and *Streptococcus gordonii,* which are members of the microbiota of the oral cavity (5, 60, 61). The role of AI-2 as a quorum-sensing molecule is controversial, however, because as Winzer et al. (120) have pointed out, AI-2 production can equally reflect the detoxification mechanism, which is therefore closely linked to metabolic activity but not necessarily to quorum sensing. The altered expression of genes observed in strains in which the *luxS* gene has been inactivated could result from the metabolic defect: the inability to degrade SRH and not to an inability of the cells to signal to one another. It would seem that the role of AI-2 could be resolved relatively easily if purified or synthesized autoinducer was added to *luxS* mutants that had an altered phenotype. Exogenous AI-2 could not affect the cleavage of SRH since it is *luxS* that has this function. Restoration of the wild-type phenotype by the addition of AI-2 would show that AI-2 has a cell-signaling function. A *luxS* mutant of *Clostridium perfringens* produces reduced amounts of alpha-, kappa- and theta-toxins (71). The production of these toxins by the mutant was stimulated by the addition of cell-free supernatant from a wild-type culture. The supernatant could, of course, have contained an autoinducer other than AI-2. However, supernatant from a culture of *Escherichia coli* DH5α in which the *luxS* gene of wild-type *Clostridium perfringens* had been cloned, also induced expression of the theta-toxin gene. However, a supernatant from a culture of non-recombinant DH5α, which does not produce AI-2 (102), was apparently not tested; hence, this was not a conclusive experiment. Nevertheless, this study points to the need to critically examine reports of phenomena that apparently involve cell signaling. Redfield (80) has proposed that the release of autoinducers such as AI-2 by bacterial cells is not so much quorum sensing (releasing and sensing autoinducer molecules to measure cell density) as it is a means of recognizing the boundaries of the habitat (diffusion of autoinducers away from the cell is measured so as to regulate the secretion of degradative enzymes). Much needs to be clarified, and the study of AI-2 in relation to biofilm-forming lactobacilli inhabiting epithelial surfaces in the gut may be useful.

BACTERIAL SUCCESSION IN THE GUT

Relatively well-described examples of the biological successions that occur in the infant gut are provided by acquisition of the gut microbiota by mice and humans. The initial collection of bacterial species inhabiting the gut of neonatal mice is composed of environmentally derived bacteria (such as flavobacteria) and facultatively anaerobic members of the microbiota of the mother (staphylococci, enterococci, and *Escherichia coli*) as well as lactobacilli (107, 109). Environmental bacteria do not persist; by about 10 days after birth, lactobacilli, enterococci, and *E. coli* are numerous in the large-bowel content. Enterococci and *E. coli* form microcolonies in the mucus layer covering the large-bowel epithelium; these clusters of bacteria can be seen microscopically in stained histological cryosections. Microaerophilic, gram-negative, spiral-shaped bacteria also colonize the mucus layer. Obligately anaerobic bacteria do not become established in the large bowel until the infant mouse has begun to supplement its milk diet by nibbling on solid food. Many of these anaerobic species associate with the mucus layer and numerically dominate the microbiota from then on. The establishment of the obligate anaerobes results in a marked reduction in the numbers of *E. coli* and enterococcal cells due to the production of short-chain fatty acids (particularly butyric acid) by the obligate anaerobes. These short-chain fatty acids are inhibitory to the facultative anaerobes under the conditions of E_h and pH present in the large bowel. The microcolonies of enterococci and *E. coli* are obliterated in the mucus layer by a mass of fusiform-shaped cells of obligate anaerobes at this time.

The composition of the fecal microbiota of human infants differs from that of adults. Particularly noticeable are the higher proportions of facultative anaerobes (enterococci, enterobacteria) and bifidobacteria relative to the total microbiota in infants than in adults. A general decline in the numbers of enterococci and enterobacteria in the feces occurs as the intestinal microbiota of infants matures and as short-chain fatty acids increase in quantity and diversity in the intestine (107, 109). In this respect, therefore, the biological successions in the guts of mice and humans follow the same general progression. In contrast to mice, in which lactobacilli are notable members of the microbiota, bifidobacteria are predominant members of the human infant intestinal microbiota, forming between 60 and 91% of the total bacterial community in breast-fed infants and 28 to 75% (average, 50%) in formula-fed infants (34). The activities of these pioneering populations in the gut ecosystem need to be defined thoroughly, especially their interaction with the developing immune system of the host.

The influence of the gut microbiota of neonates on the immune system is of especial interest because of the observed increase in the incidence of allergies in children in affluent countries over recent decades (8, 9). Pediatricians have developed the "hygiene hypothesis" to try to explain this increased prevalence (22, 81). Strachan advanced the hypothesis in 1989, but Gerrard had published earlier work that supports it (27, 99). In brief, the hypothesis states that in affluent countries, the pattern of microbial exposure in early life has changed. Families are smaller, epidemics of childhood infectious diseases are rare, infections with helminths are rare, living standards are higher, and antibiotics are widely used to treat infants. This altered exposure to microbes, it is proposed, has predisposed the immune system of children to react to environmental antigens. A correlation between the occurrence of an atopic disease, asthma, and the frequency of treatments with antibiotics during the first year of life has been reported. Epidemiological data show that children in New Zealand are four times more likely to have asthma if they have been treated with three or four courses of antibiotics during the first year of life than those who have not (119). Examination of the fecal microbiota of antibiotic-treated babies has demonstrated that the microbiota profile changes during the treatment period, including the elimination of bifidobacteria during periods of antibiotic administration (G. W. Tannock, M. A. Simon and J. Crane, unpublished observations).

A HUMAN MUCOSA-ASSOCIATED MICROBIOTA?

The adherence of *Lactobacillus* cells to, and proliferation on, epithelial surfaces in rodents, pigs, and poultry has tempted some researchers to consider that the same phenomenon occurs in the human gut (66). This overlooks the differences in the anatomy and histology of the human gut relative to that of other monogastric animals. The formation of *Lactobacillus* biofilms in the gut is conditional on the presence of a nonsecretory squamous epithelium such as that of the rodent forestomach or avian crop. The gastrointestinal tract of humans lacks this prerequisite for biofilm formation because it is lined by an epithelium composed of columnar cells. Lactobacilli and other bacteria could, of course, become trapped and possibly multiply in the mucus layer that is continuously secreted onto the mucosal surface of the human gut. Indeed, in the proximal colon of rodents, a bacterial community distinct from that of the lumen, has been observed in the mucus layer (90, 106). The presence of a discrete community like this has not so far been demonstrated in healthy humans, but studies aimed at defining the composition of the mucsoa-associated (biopsy-associated) microbiota have been reported. Most of this work is aimed at unraveling the bacteriology of the inflammatory bowel diseases (IBD).

IBD (Crohn's disease and ulcerative colitis) result from a genetically predisposed loss of tolerance by the immune system to the presence of the gut microbiota. The latter provides a constant antigenic stimulus for the host immune system (73, 85, 86). Normally, immunological tolerance toward the intestinal microbiota prevents continuous intestinal inflammation (19). There is evidence from experiments with human patients and experimental animal models of IBD that this controlled, homeostatic response is lost in genetically susceptible hosts (dysfunctional immune system), which thus develop chronic immune-mediated colitis (85, 86).

Targeted deletion or overexpression of a variety of genes that regulate immune or mucosal barrier function in experimental animal models leads to an excessively aggressive cellular immune response confined to the intestine (21). Results from experiments with germfree rodents, in which intestinal inflammation is absent or only very mildly expressed, indicate that bacteria are indispensable contributors to the pathogenesis of chronic, immune-mediated intestinal inflammation. This concept is further supported by the rapid development of colitis when germfree HLA-B27 transgenic rats and interleukin-10-deficient mice are colonized with bacteria from healthy specific-pathogen-free SPF animals (76, 77, 79, 86, 93). Decreasing the bacterial load by the administration of broad-spectrum antibiotics reduces colitis (54, 78). In addition, transfer of subsets of normal T cells into immunodeficient mice demonstrates that different T-cell populations induce or prevent colitis (67). In most of these models, T-helper 1 $CD4^+$ lymphocytes secreting high levels of gamma interferon mediate disease (4).

From studies with human subjects, it is notable that Crohn's disease lesions are located mainly in regions of the gut colonized by large numbers of bacteria (the ileum and colon) and that diversion of the fecal stream from inflamed sites often results in clinical improvement in the patient. Moreover, infusion of intestinal contents into the excluded ileum of postoperative Crohn's disease patients induced inflammation (17). Antibiotic therapy results in transient clinical improvement of Crohn's disease patients with colonic involvement, supporting the etiological involvement of bacteria (73). In support of the hypothesis that IBD results from a loss of immunological tolerance to intestinal bacteria, Duchmann et al. (19) derived T-cell clones from peripheral blood and colonic biopsy specimens from IBD patients and controls and tested their proliferative responses after exposure to autologous and heterologous isolates of intestinal bacteria. Lymphocytes proliferated to a greater extent after exposure to heterologous isolates in controls, whereas proliferation in response to autologous and heterologous strains was equally great in IBD patients.

It remains unclear whether disease is due to the presence of specific bacterial species in the gut of IBD patients or whether it results from an excessively aggressive immune response to substances associated with many of the bacterial types that are capable of residence in the gut (92). Despite evidence from studies of patients and experimental animals that intestinal bacteria and/or their products at least fuel the chronic inflammation associated with IBD, immunological and clinical investigations have far outweighed attempts to define the gut microbiota in health and disease.

It has become clear that the composition of the bacterial community in the gut of humans is highly variable from subject to subject and that little will be gained from comparing the fecal microbiota of healthy subjects and patients; the uniqueness of human microbiotas confounds meaningful comparisons in the absence of a specific causative agent of IBD. It could be argued that microbiological information obtained from analysis of feces does not provide accurate information about the microbiota of the proximal regions of the intestine. Based on the examination of murine feces, for example, it would not have been possible to deduce the significant epithelial association of lactobacilli with the forestomach epithelium (89). Moore et al. (65) demonstrated that the composition of the human fecal microbiota reflects that of the colon, but the bacterial community is more metabolically active in the proximal colon than in the distal colon (14). This is a reflection of the relative availability of fermentable substrates in the two sites. These observations have led investigators of IBD to attempt to investigate the mucosa-associated microbiota, should one exist, of humans. Mucosa-associated bacteria would be, due to their intimate location, the cells most likely to be encountered by the gut-associated immune system and therefore the ones most likely to stimulate inflammatory processes. Research seeking to implicate overt bacterial pathogens in the pathogenesis of IBD has been performed without success (92). Summaries of other, more general studies in which biopsy specimen-associated bacteria were detected are provided in Table 4.

Accurate definition of the mucosa-associated microbiota will be difficult because mucosal biopsy specimens must be collected: an invasive procedure that is ill defined in terms of collection procedure from study to study. Prior to colonoscopy and biopsy, the patient is purged to remove colonic contents. The effect of this preparative treatment on the composition of the microbiota is unknown. Further, the bowel is not completely decontaminated by this procedure, and a fecal fluid continues to be present in the bowel and to bathe mucosal surfaces. Therefore, it is not clear what exactly is being sampled in the current studies: the mucosal surface contaminated with luminal bacteria or true mucosa-associated inhabitants. Additionally, the extent of contamination of the colonoscope with bacteria from the fecal fluid has never been determined. Therefore, reports about the mucosa-associated microbiota of humans must be treated with caution, but at least these studies have begun to relieve the previous focus on the microbiota of fecal (stool) specimens. They now seek to define the microbiology of the affected site.

If autochthonous mucosa-associated bacteria exist, it is likely that they will vary from patient to patient; therefore, comparisons of healthy subjects and patients will not be useful. Comparisons of the microbiota within an individual (inflamed and noninflamed mucosa) will be useful, however, because recognition of a microbiota of "abnormal" composition could provide a therapeutic or diagnostic target. However, these observations will not permit a causative role in IBD to be assigned to specific bacterial populations. Alterations in the mucosa-associated microbiota could occur as a result of inflammation-induced changes in the habitat. It is important to determine the spatial distribution of the microbiota in the human gut, but phylogenetic studies of the microbiota will not result in a complete understanding of IBD.

Heterogeneity of the immune system (HLA restriction), to some degree, further confounds IBD research. HLA genes determine the specificity of the immune response. Hence, HLA restriction means that the immune system of each animal is not necessarily capable of reacting to a particular antigen. In other words, the immune systems of different humans or other animals recognize different epitopes. This is apparent from experimental-animal studies because the composition of the gut microbiota of HLA-B27 rats and interleukin-10-deficient mice is different, yet colitis results in both types of animal (G. W. Tannock, M. A. Simon, R. Bibiloni, W. McBurney, H. C. Rath, and R. B. Sartor, unpublished observations). The immune systems of these two types of animal must be reacting to different bacterial antigens. A change in thinking from "what are the important microbial species" in IBD, which is surely like looking for a needle in a haystack, to "what are the microbes producing" would be a useful advance. Efforts should be made to identify antigens that are always produced in the gut ecosystem because of species redundancy inherent in the bacterial community and to which the dysfunctional immune system responds abnormally.

On a sanguine note, it is likely that there will be HLA similarities between IBD patients, which would increase the likelihood that their immune systems would react to the same antigens. A classic example of an association of an HLA with disease is provided by HLA-B27 and the spondyloarthropathies (121). The prevalence of HLA-B27 genes among ankylosing spondylitis patients is 95%, compared to 6 to 8% in the general (Caucasian) population (56). Indeed, 30% of Crohn's disease and ulcerative colitis patients have this immune cell marker. A meta-analysis of HLA association studies of IBD patients has found the following positive associations: ulcerative colitis and HLA-DR2, especially the DRB1*1502 allele, ulcerative colitis and HLA-DR9, ulcerative colitis and the DRB1*0103 allele, Crohn's disease and HLA-DR7, and Crohn's disease and the DRB3*0301 allele (98). On this basis, the HLA phenotype of patients in future studies must be determined so that the results of microbiological and immunological studies will have relevance to patients with a particular HLA polymorphism.

MODIFICATION OF THE GUT MICROBIOTA

Directed modification of the composition of the gut microbiota is a long-held desire of microbiologists (62, 63). Probiotics (dietary supplements or foods that contain living microbial cells that are believed to promote health) have enjoyed an uncritical popularity in recent decades (108). Historically, a probiotic was thought to alter the composition of the gut microbiotia, but recent studies have shown that this does not occur beyond a transient modification of a target population: ingestion of lactobacilli, for example, modifies the *Lactobacillus* population (1, 20, 87, 96, 111). Representative of these studies is the one conducted by Tannock et al. (111), in which molecular typing of *Lactobacillus* isolates was used to monitor the fate of a probiotic strain in the guts of humans. To do this, restriction endonuclease digests of DNA extracted from pure *Lactobacillus* cultures were analyzed by pulsed-field gel electrophoresis. The resulting patterns of DNA fragments were characteristic of each bacterial strain (the genetic fingerprint). This method was used in a study in which human subjects consumed a probiotic product on a daily basis for 6 months. The product contained a strain, DR20, of *Lactobacillus rhamnosus* that could be easily differentiated from other lactobacilli by genetic fingerprinting. The period of consumption (the test period) was preceded by a control period of 6 months so that the investigators could determine whether lactobacilli were already resident

Table 4. Investigations of gut tissue-associated bacteria of humans

Investigation	Observations
Scanning electron microscopic observations of spirochetes associated with the colonic mucosa (104)	Spirochetes were observed in association with the colonic mucosal surface of 3% of healthy humans. The spirochete cells were perpendicular to the mucosal surface and measured 5 by 0.5 μm.
Colonic tissue sections obtained from four sudden-death victims examined by scanning electron microscopy and by bacteriological culture (13)	The cultivable microbiota from tissue sections from each subject was different at the species level, although the genera *Bacteroides* and *Fusobacterium* were predominant. The ratio of obligate anaerobes to facultative anaerobes was variable among the subjects. Scanning electron microscopy showed the bacteria to be present in the mucus layer associated with the tissue. In two subjects, long (60-μm) helical organisms were present.
Microscopic examination and selective culture for *Enterobacteriaceae* of biopsy specimens collected from the cecum, colon, and rectum (36)	Biopsy specimens were obtained from 17 patients, 14 of whom did not have bowel diseases. Bacteria were seen adhering to epithelial cells and in the mucus. *E. coli* was cultured from the biopsy specimens, and serotyping showed that each patient harbored a unique O-antigen type.
Bacteriological culture and microscopic observations of biopsy specimens collected from various gastrointestinal sites (68)	Homogenates of biopsy specimens were cultured. Bacteria were not detected in homogenates of gastric and duodenal tissue from healthy subjects, but samples from patients generally contained bacteria. Bacteria were cultured from healthy jejunal tissue in four of seven subjects and from all appendix and colon samples. Anaerobic streptococci were most frequently detected. Bacterial cells were seen in mucus in stained histological sections of the jejunum, appendix, and colon but were most numerous in association with colonic mucus.
Microscopy and bacteriological culture to detect and enumerate bacteria associated with biopsy specimens from pouch patients (70)	Electron microscopy and bacteriological culture showed that 64 of 78 biopsy specimens contained bacteria. Facultative anaerobes were cultured from 54 biopsy specimens, and obligate anaerobes were cultured from 50. The total number of bacteria was greater for samples from patients with pouchitis than from those with normal pouches.
Clone libraries prepared from biopsy specimens collected from the terminal ileum, proximal, and distal colon of two subjects with mild diverticulosis (117)	A total of 361 16S rRNA gene sequences were analyzed. The most abundant clones represented *Clostridium* group XIVa and the *Cytophaga-Flavobacter-Bacteroides* cluster.
Scanning electron microscopy of biopsy specimens collected from the stomach and small bowel (10)	Biopsy specimens were collected from patients with rheumatic diseases and dyspepsia. Of these, 62% were associated with long helical organisms, measuring 5 to 30 μm, on the surface of the mucosa of both the stomach and small bowel.
PCR and culture-based investigation of biopsy samples collected from the colon of ulcerative colitis patients and controls (124)	Group-specific PCR and enrichment culture were used to detect sulfate-reducing bacteria. The bacteria were more prevalent in samples from ulcerative colitis patients.
FISH to detect bacteria associated with tissue from surgical resections (48)	Tissue sections (ileum and colon) obtained from 24 IBD patients and 14 controls were examined. Larger numbers of bacteria were associated with colonic tissue from ulcerative colitis patients than from controls, and bacterial invasion was evident in some of the patients but not in the controls. Phylogenetic groups detected included *Enterobacteriaceae, Clostridium histolyticum/Clostridium litusburense* group, *Clostridium coccoides/Eubacterium rectale* group, *Bacteroides-Prevotella* cluster, high-G+C-content gram-positive bacteria, and sulfate-reducing bacteria.
PCR/DGGE analysis of biopsy specimens from the ascending, transverse, and descending colon (127)	Biopsy samples were collected from 10 subjects during routine diagnostic colonoscopy. One of the patients had polyposis, and three had ulcerative colitis. Flow-cytometric measurement of labeled bacteria showed that 10^5 to 10^6 bacteria were associated with each biopsy specimen. PCR/DGGE detected unique profiles for each subject. The profiles were constant between sampling sites for each subject. Sequences having 99% identity to that of *Lactobacillus gasseri* were detected in nine of the subjects.
Bacteriological cultures of rectal biopsy specimens from Crohn's disease patients before and during antimicrobial drug administration (43)	Rectal biopsy specimens from 12 patients were investigated before and during treatment with metronidazole and co-trimoxazole. *Bacteroides* and other gram-negative bacilli were reduced in numbers, but the numbers of gram-positive cocci increased during treatment. There was clinical improvement in eight of the patients, but this could not be related to specific changes in microbiota composition.

Continued on following page

Table 4. *Continued*

Investigation	Observations
Bacteriological cultures from rectal mucosal biopsy specimens from patients with ulcerative colitis (57)	Biopsy samples were collected from inflamed rectal mucosa of 17 patients with ulcerative colitis. Samples were also collected from seven controls. Bacteria were cultured from the biopsy specimens from both groups but were more numerous in patient samples than in control samples. *Bacteroides vulgatus* cells were the most numerous.
PCR/DGGE analysis of biopsy specimens from the colon (69)	Universal bacterial and group- and genus-specific PCR primers were used to generate DGGE profiles from DNA extracted from colonic biopsy specimens collected from four subjects with a clinical history of polyposis. Unique profiles were obtained for each subject when universal primers were used. 16S rRNA sequences from bifidobacterial and lactic acid bacteria were detected.
Bacteriological culture, electron microscopy, FISH, and quantitative PCR to enumerate and locate bacterial cells associated with biopsy specimens collected from ileum and colon (103)	Biopsy specimens from 304 patients with bowel inflammation and 40 controls were examined. The total bacterial counts associated with biopsy specimens were larger in the patients than in the controls, but the kinds of bacteria present did not differ between the groups. The association of bacteria with inflamed biopsy specimens could be seen by microscopy. Evidence of bacterial invasion was reported for some patients, but this may have been a misinterpretation of secretory granules of colonic endocrine cells as "intraepithelial bacterial inclusions."
FISH to detect bacteria associated with rectal biopsy specimens (91)	Rectal biopsy specimens were collected from 19 IBD patients and 14 controls. Bacterial cells were absent from the samples from 10 controls and 6 patients. Semiquantitative measurements indicated that larger numbers of bacterial cells were associated with the intestinal mucus of IBD patients than of controls.
Histological and microbiological features of biopsy samples from patients with normal or inflamed pouches (59)	Tissue samples were obtained from 78 subjects, of whom 23 had normal pouches and 26 had pouchitis. Further groups of ileostomy patients and healthy subjects were also examined. Intramural facultatively anaerobic bacteria were more numerous in biopsy specimens collected from pouchitis patients than in those collected from the ileostomy patients or healthy controls.

in the guts of the subjects, and a posttest period. The last period was included to see if the probiotic strain persisted in the gut once probiotic consumption had ceased. Lactobacilli were enumerated on Rogosa SL agar, and a standard number of colonies were subcultured from appropriate dilution plates. These isolates were examined by pulsed-field gel electrophoresis of chromosomal DNA digests in order to distinguish between *Lactobacillus* strains on the basis of genetic fingerprints. This enabled strain DR20 to be tracked during the study. In 9 of 10 subjects, the strain of *L. rhamnosus* could be detected by culture of the feces during the period of probiotic consumption. In 8 of these 9 subjects, the ingested *Lactobacillus* strain was no longer detectable in the feces by 1 month after consumption of the probiotic ceased. DR20 was not detected by culture in the feces of the 10th subject at any stage of the study. This subject was colonized throughout the study by a strain of *Lactobacillus ruminis* that dominated the population numerically, so that colonies of DR20 were not detected by random subculture. Administration of the probiotic did not otherwise alter the composition of the fecal microbiota; indeed, as is commonly observed, lactobacilli comprised less than 1% of the total fecal microbiota even when administered in relatively large numbers (111). The maintenance of a self-regulating community structure explains the stability in fecal microbiota composition. Homeostasis is a powerful phenomenon in nature: everything changes, yet everything stays the same (2). This is the basis on which competitive exclusion operates. A mature bacterial community makes it difficult for an allochthonous organism to become established because the community is complete for that host and is homeostatic (2).

Under these conditions, it is doubtful that probiotics have any impact on the large-bowel ecosystem, but they might be effective in the small bowel. After all, consumption of a probiotic product delivers about 10^9 allochthonous bacterial cells to this site with every dose. A realistic probiotic target may not be the modification of the colonic microbiota but the stimulation of the mucosal immune system of the small intestine. Saavedra (83) has summarized studies in which lactic acid-producing bacteria have been administered to children with the aim of preventing or minimizing diarrheal diseases. He concluded that the impact of the probiotic products appeared to be most significant against rotavirus infections and suggested that an immunological mechanism was responsible for the beneficial effects.

Perhaps more attractive in an ecological sense are attempts to modify microbiota composition by the consumption of prebiotics (potential substrates for bacterial inhabitants of the intestine) (12, 29). It has the rationale that autochthonous bacterial strains could be encouraged to proliferate in the gut, hopefully with beneficial consequences to the consumer. Oligosaccharides and inulin have been reported to selectively increase the numbers of bifidobacteria, but most of the studies have involved the consumption of prebiotic powders (6, 28, 35, 49, 75, 113). The results of these studies may not be relevant to everyday life because the substrates, for commercial use, would be incorporated into a food product. In a recent study, the diets of human subjects were supplemented with biscuits that contained oligosaccharides at concentrations that produced edible products of commercial quality (110). The daily consumption of three biscuits provided a dose of oligosaccharides that did not cause flatus or abdominal pain in the subjects. The daily dosage was lower than that used in most prebiotic studies reported in the literature, yet changes to the DGGE profiles of the fecal microbiota were observed. Alterations to the bacterial community profiles in response to dietary supplementation were detected only in profiles that were generated from bacterial RNA. This means that changes in bacterial metabolic activity in response to consumption of the biscuits were detected (RNA-DGGE). Since alterations to DNA-DGGE profiles were not detected, changes to the population structure of the community did not occur. Although oligosaccharides added to the gut ecosystem are generally expected to increase the number of bifidobacterial cells in particular, changes in population sizes were not detected by selective culture or nucleic acid-based analysis (FISH) of bifidobacterial populations. Increased metabolic activity without an increase in the number of bacterial cells is not contradictory, since it has been shown in studies with rumen bacteria that "energy spillage" occurs in bacterial communities (11). As noted by Russell and Cook (82), it has generally been assumed by microbiologists that the yield of bacterial cells is directly proportional to the amount of ATP produced. The strict coupling of anabolism and catabolism has, however, been contradicted by the observation that resting-cell suspensions utilize energy sources in the complete absence of growth and by the fact that the correlation between ATP and biomass formation is often poor. Some of this variation in growth efficiency can be explained by maintenance energy expenditure, but bacteria have other mechanisms of nongrowth energy dissipation ("futile cycles"). Such energy spilling may occur when growth is limited by nutrients other than energy source. DNA fragments representing *Bifidobacterium adolescentis* and *Colinsella aerofaciens* were present in RNA-DGGE gel profiles, especially when biscuits had been consumed. These bacterial cells could have been in a metabolically quiescent state prior to dietary supplementation with biscuits. When the host diet was supplemented by oligosaccharides, substrate for metabolism could have become available for the bacteria in the large bowel but, because other nutrients were growth limiting in the highly regulated bacterial community, an increase in *Bifidobacterium* and *Colinsella* cell numbers did not occur. Therefore, *Bifidobacterium* and *Colinsella* populations remained constant in size but RNA-DGGE provided evidence of increased metabolic activity. The significance of this increased metabolism to the bacterial community and to the human host could be an important topic for future research.

REFERENCES

1. **Alander, M., R. Satokari, R. Korpela, M. Saxelin, T. Vilpponen-Samela, T. Mattila-Sandholm, and A. von Wright.** 1999. Persistence of colonization of human colonic mucosa by a probiotic strain, *Lactobacillus rhamnosus* GG, after oral consumption. *Appl. Environ. Microbiol.* **65:**351–354.
2. **Alexander, M.** 1971. *Microbial Ecology.* John Wiley & Sons, Inc., New York, N.Y.
3. **Barrow, P. A., B. E. Brooker, R. Fuller, and M. J. Newport.** 1980. The attachment of bacteria to the gastric epithelium of the pig and its importance in the microecology of the intestine. *J. Appl. Bacteriol.* **48:**147–154.
4. **Berg, D. J., N. Davidson, R. Kuhn, W. Muller, S. Menon, G. Holland, L. Thompson-Snipes, M. W. Leach, and D. Rennick.** 1996 Enterocolitis and colon cancer in interleukin-10-deficient mice are associated with aberrant cytokine production and $CD4^+$ TH1-like responses. *J. Clin. Investig.* **98:** 1010–1020.
5. **Blehert, D. S., R. J. Palmer, J. B. Xavier, J. S. Almeida, and P. E. Kolenbrander.** 2003. Autoinducer 2 production by *Streptococcus gordonii* DL1 and the biofilm phenotype of a *luxS* mutant are influenced by nutritional conditions. *J. Bacteriol.* **185:**4851–4860.
6. **Bouhnik, Y., B. Flourie, M. Riottot, N. Bisetti, M.-F. Gailing, A. Guibert, F. Bornet and J.-C. Rambaud.** 1996. Effects of fructo-oligosaccharides ingestion on fecal bifidobacteria and selected metabolic indexes of colon carcinogenesis in healthy humans. *Nutr. Cancer* **26:**21–29.
7. **Bry, L., P. G. Falk, T. Midtvedt, and J. I. Gordon.** 1996. A model of host-microbial interactions in an open mammalian ecosystem. *Science* **273:**1380–1383.
8. **Burney, P. G., S. Chinn, and R. J. Rona.** 1990. Has the prevalence of asthma increased in children? Evidence from the national study of health and growth 1973–1986. *Br. Med. J.* **300:**1306–1310.
9. **Burr, M. L., B. K. Butland, S. King, and W. E. Vaughan.** 1989. Changes in asthma prevalence: two surveys 15 years apart. *Arch. Dis. Child.* **61:**1452–1456.
10. **Collins, A. J., L. J. Notarianni, and U. J. Potter.** 1999. An unusual helical micro-organism found in the gut lumen of human subjects. *J. Med. Microbiol.* **48:**401–405.

11. **Cook, G. M. and J. B. Russell.** 1994. Energy spilling reactions of *Streptococcus bovis* and resistance of its membrane to proton conductance. *Appl. Environ. Microbiol.* **60:**1942–1948.
12. **Crittenden, R. G.** 1999. Prebiotics, p. 141–156. *In* G. W. Tannock (ed.), *Probiotics: a Critical Review.* Horizon Scientific Press, Wymondham, United Kingdom.
13. **Croucher, S. C., A. P. Houston, C. E. Bayliss, and R. J. Turner.** 1983. Bacterial populations associated with different regions of the human colon wall. *Appl. Environ. Microbiol.* **45:**1025–1033.
14. **Cummings, J. H., and G. T. Macfarlane.** 1991. The control and consequences of bacterial fermentation in the human colon. *J. Appl. Bacteriol.* **70:**443–459.
15. **Dal Bello, F., J. Walter, W. P. Hammes, and C. Hertel.** 2003. Increased complexity of the species composition of lactic acid bacteria in human feces revealed by alternative incubation condition. *Microb. Ecol.* **45:**455–463.
16. **Davies, D. G., M. R. Parsek, J. P. Pearson, B. H. Iglewski, J. W. Costerton, and E. P. Greenberg.** 1998. The involvement of cell-to-cell signals in the development of a bacterial biofilm. *Science* **280:**295–298.
17. **D'Haens, G. R., K. Geboes, M. Peeters, F Baert, F. Penninckx, and P. Rutgeerts.** 1998. Early lesions of recurrent Crohn's disease caused by infusion of intestinal contents in excluded ileum. *Gastroenterology* **114:**262–267.
18. **Dubos, R., R. W. Schaedler, R. Costello, and P. Hoet.** 1965. Indigenous, normal, and autochthonous flora of the gastrointestinal tract. *J. Exp. Med.* **122:**67–76.
19. **Duchmann, R., I. Kaiser, E. Hermann, W. Mayet, K. Ewe, and K-H. Meyer zum Buschenfelde.** 1995. Tolerance exists towards resident intestinal flora but is broken in active inflammatory bowel disease (IBD). *Clin. Exp. Immunol.* **102:** 448–455.
20. **Dunne, C., L. Murphy, S. Flynn, L. Omahory, S. O'Halloran, M. Feeney, D. Morrissey, G. Thornton, G. Fitzgerald, C. Daly, B. Keily, E. M. Quigley, G. C. O'Sullivan, F. Shanahan, and J. K. Collins.** 1999. Probiotics: from myth to reality. Demonstration of functionality in animal models of disease and in human clinical trials. *Antonie Leeuwenhoek* **76:**279–292.
21. **Elson, C. O., R. B. Sartor, G. S. Tennyson, and R. H. Riddell.** 1995. Experimental models of inflammatory bowel disease. *Gastroenterology* **109:**1344–1367.
22. **Farooqi, I. S., and J. M. Hopkin.** 1998. Early childhood infection and atopic disorder. *Thorax* **53:**927–932.
23. **Federle, M. J., and B. L. Bassler.** 2003. Interspecies communications in bacteria. *J. Clin. Investig.* **112:**1291–1299.
24. **Felske, A., H. Rheims, A. Wolerink, E. Stackebrandt, and A. D. L. Akkermans.** 1997. Ribosome analysis reveals prominent activity of an uncultured member of the class *Actinobacteria* in grassland soils. *Microbiology* **143:**2983–2989.
25. **Fuller, R.** 1975. Nature of the determinant responsible for adhesion of lactobacilli to chicken crop epithelial cells. *J. Gen. Microbiol.* **87:**245–250.
26. **Fuller, R., and B. E. Brooker.** 1974. Lactobacilli which attach to the crop epithelium of the fowl. *Am. J. Clin. Nutr.* **27:**1305–1312.
27. **Gerrard, J. W., C. A. Geddes, P. L. Reggin, C. D. Gerrard, and S. Horne.** 1976. Serum IgE levels in white and metis communities in Saskatchewan. *Ann. Allergy* **37:**91–100.
28. **Gibson, G. R., E. R. Beatty, X. Wang and J. H. Cummings.** 1995. Selective stimulation of bifidobacteria in the human colon by oligofructose and inulin. *Gastroenterology* **108:**975–982.
29. **Gibson, G. R., and M. B. Roberfroid.** 1995. Dietary modulation of the human colonic microbiota: introducing the concept of prebiotics. *J. Nutr.* **125:**1401–1412.
30. **Gong, J., R. J. Forster, H. Yu, J. R. Chambers, P. M. Sabour, R. Wheatcroft, and S. Chen.** 2002. Diversity and phylogenetic analysis of bacteria in the mucosa of chicken ceca and comparison with bacteria in the cecal lumen. *FEMS Microbiol. Lett.* **208:**1–7.
31. **Gong, J., R. J. Forster, H. Yu, J. R. Chambers, P. M. Sabour, R. Wheatcroft, and S. Chen.** 2002. Molecular analysis of bacterial populations in the ileum of broiler chickens and comparison with bacteria in the cecum. *FEMS Microbiol. Ecol.* **41:**171–179.
32. **Gordon, H. A., and L. Pesti.** 1971. The gnotobiotic animal as a tool in the study of host-microbial relationships. *Bacteriol. Rev.* **35:**390–429.
33. **Greenberg, E. P.** 1997. Quorum sensing in gram-negative bacteria. *ASM News* **63:**371–377.
34. **Harmsen, H. J. M., A. C. M. Wildeboer-Veloo, G. C. Raangs, A. A. Wagendorp, N. Klijn, J. G. Bindels, and G. W. Welling.** 2000. Analysis of intestinal flora development in breast-fed and formula-fed infants by using molecular identification and detection methods. *J. Pediatr. Gastroenterol. Nutr.* **30:**61–67.
35. **Harmsen, H. J. M., G. C. Raangs, A. H. Franks, A. C. M. Wildeboer-Veloo, and G. W. Welling.** 2002. The effect of the prebiotic inulin and the probiotic *Bifidobacterium longum* on the fecal microflora of healthy volunteers measured by FISH and DGGE. *Microb. Ecol. Health Dis.* **14:**211–219.
36. **Hartley, C. L., C. S. Neumann, and M. H. Richmond.** 1979. Adhesion of commensal bacteria to the large intestine wall in humans. *Infect. Immun.* **23:**128–132.
37. **Hayashi, H., M. Sakamoto, and Y. Benno.** 2002. Phylogenetic analysis of the human gut microbiota using 16S rDNA clone libraries and strictly anaerobic culture-based methods. *Microbiol. Immunol.* **46:**535–548.
38. **Hayashi, H., M. Sakamoto, and Y. Benno.** 2002. Fecal microbial diversity in a strict vegetarian as determined by molecular analysis and cultivation. *Microbiol. Immunol.* **46:**819–831.
39. **Hayashi, H., M. Sakamoto, M. Kitihara, and Y. Benno.** 2003. Molecular analysis of fecal microbiota in elderly individuals using 16S rDNA library and T-RFLP. *Microbiol. Immunol.* **47:**557–570.
40. **Heng, N.C.K., J. M., Bateup, D. M. Loach, X. Wu, H. F. Jenkinson, M. Morrison, and G. W. Tannock.** 1999. Influence of different functional elements of plasmid pGT232 on maintenance of recombinant plasmids in *Lactobacillus reuteri* populations in vitro and in vivo. *Appl. Environ. Microbiol.* **65:**5378–5385.
41. **Hooper, L. V., M. H. Wong, A. Thelin, L. Hansson, P. G. Falk, and J. I. Gordon.** 2001. Molecular analysis of commensal host-microbial relationships in the intestine. *Science* **291:**881–884.
42. **Hooper, L. V., J. Xu, P. G. Falk, T. Midtvedt, and J. I. Gordon.** 1999. A molecular sensor that allows a gut commensal to control its nutrient foundation in a competitive ecosystem. *Proc. Natl. Acad. Sci. USA* **96:**9833–9838.
43. **Hudson, M. J., M. J. Hill, P. R. Elliott, L. M. Berghouse, W. R., Burnham, and J. E. Lennard-Jones.** 1984. The microbial flora of the rectal mucosa and faeces of patients with Crohn's disease before and during antimicrobial chemotherapy. *J. Med. Microbiol.* **18:**335–345.
44. **Jepson, M. A., M. A. Clark, N. L. Simmons, and B. H. Hirst.** 1993. Actin accumulation at sites of attachment of indigenous apathogenic segmented filamentous bacteria to mouse ileal epithelial cells. *Infect. Immun.* **61:**4001–4004.
45. **Kaiser, D., and R. Losick.** 1993. How and why bacteria talk to each other. *Cell* **73:**873–885.
46. **Kleerebezem, M., L. E. N. Quadri, O. P. Kuipers, and W. M. de Vos.** 1997. Quorum sensing by peptide pheromones and

two-component signal transduction systems in Gram-positive bacteria. *Mol. Microbiol.* **24:**895–904.

47. **Klaasen, H. L., J. P. Koopman, F. G. Poelma, and A. C. Beynen.** 1992. Intestinal, segmented, filamentous bacteria. *FEMS Microbiol. Rev.* **8:**165–180.
48. **Kleesen, B., A. J. Kroesen, H. J. Buhr, and M. Blaut.** 2002. Mucosal and invading bacteria in patients with inflammatory bowel disease compared with controls. *Scand. J. Gastroenterol.* **37:**1034–1041.
49. **Kruse, H.-P., B. Kleesen, and M. Blaut.** 1999. Effects of inulin on faecal bifidobacteria in human subjects. *Br. J. Nutr.* **82:**375–382.
50. **Lan, P. T. N., H. Hayashi, M. Sakamoto, and Y. Benno.** 2002. Phylogenetic analysis of cecal microbiota in chicken by use of 16S rDNA clone libraries. *Microbiol. Immunol.* **46:** 371–382.
51. **Lee, A.** 1999. *Helicobacter pylori:* opportunistic member of the normal microflora or agent of communicable disease? p. 128–163. *In* G. W. Tannock (ed.), *Medical Importance of the Normal Microflora.* Kluwer Academic Publishers, Dordrecht, The Netherlands.
52. **Leser, T. D., J. Z. Amenuvor, T. K. Jensen, R. H. Lindecrona, M. Boye, and K. Moller.** 2002. Culture-independent analysis of gut bacteria: the pig gastrointestinal tract microbiota revisited. *Appl. Environ. Microbiol.* **68:**673–690.
53. **Lu, J., U. Idris, B. Harmon, C. Hofaere, J. J. Maurer, and M. D. Lee.** 2003. Diversity and succession of the intestinal bacterial community of the maturing broiler chicken. *Appl. Environ. Microbiol.* **69:**6816–6824.
54. **Madsen, K. L., J. S. Doyle, M. E. Tavernini, L. D. Jewel, R. P. Rennie, and R. N. Fedorak.** 2000. Antibiotic therapy attenuates colitis in interleukin-10 gene-deficient mice. *Gastroenterology* **118:**1094–1105.
55. **Mahan, M. J., J. M. Slauch, and J. J. Mekalanos.** 1993. Selection of bacterial virulence genes that are specifically induced in host tissues. *Science* **259:**686–688.
56. **Marker-Hermann, E., and T. Hohler.** 1998. Pathogenesis of human leucocyte antigen HLA-B27-positive arthritis. *Rheum. Dis. Clin. North Am.* **24:**865–881.
57. **Matsuda, H., Y. Fujiyama, A. Andoh, T. Ushijima, T. Kajinami, and T. Bamba.** 2000. Characterization of antibody responses against rectal mucosa-associated bacterial flora in patients with ulcerative colitis. *J. Gastroenterol. Hepatol.* **15:** 61–68.
58. **McFall-Ngai, M. J., and E. G. Ruby.** 1991. Symbiont recognition and subsequent morphogenesis as early events in an animal-bacterial mutualism. *Science* **254:**1491–1494.
59. **McLeod, R. S., D. Antonioli, J. Cullen, A. Dvorak, A. Onderdonk, W. Silen, J. E. Blair, R. Monahan-Early, R. Cisneros, and Z. Cohen.** 1994. Histologic and microbiologic features of biopsy samples from patients with normal and inflamed pouches. *Dis. Colon Rectum* **37:**26–31.
60. **McNab, R., S. K. Ford, A. El-Sabaeny, B. Barbieri, G. S. Cook, and R. J. Lamont.** 2003. LuxS-based signalling in *Streptococcus gordonii:* autoinducer 2 controls carbohydrate metabolism and biofilm formation with *Porphyromonas gingivalis. J. Bacteriol.* **185:**274–284.
61. **Merritt, J., F. Qi, S. D. Goodman, M. H. Anderson, and W. Shi.** 2003. Mutation of *luxS* affects biofilm formation in *Streptococcus mutans. Infect. Immun.* **71:**1972–1979.
62. **Metchnikoff, E.** 1907. *The Prolongation of Life. Optimistic Studies.* William Heinemann, London, United Kingdom.
63. **Metchnikoff, E.** 1908. *The Nature of Man. Studies in Optimistic Philosophy.* William Heinemann, London, United Kingdom.
64. **Mitsuoka, T.** 1992. The human gastrointestinal tract, p. 69–114. *In* B. J. B. Wood (ed.), *The Lactic Acid Bacteria,* vol. 1. *The Lactic Acid Bacteria in Health and Disease.* Elsevier Applied Science, London, United Kingdom.
65. **Moore, W. E. C., E. P. Cato and L. V. Holdeman.** 1978. Some current concepts in intestinal bacteriology. *Am. J. Clin. Nutr.* **31:**S33–S42.
66. **Morelli, L.** 2000. *In vitro* selection of probiotic bacteria: a critical appraisal. *Curr. Issues Intest. Microbiol.* **1:**59–67.
67. **Morrissey, I., K. Charrier, S. Braddy, D. Liggitt, and J. D. Watson.** 1993. $CD4^+$ T cells that express high levels of CD45RB induce wasting disease when transferred into congenic severe combined immunodeficient mice. Disease development is prevented by cotransfer of purified $CD4^+$ T cells. *J. Exp. Med.* **178:**237–244.
68. **Nelson, D. P., and L. J. Mata.** 1970. Bacterial flora associated with the human gastrointestinal mucosa. *Gastroenterology* **58:**56–61.
69. **Nielsen, D. S., P. L. Moller, V. Rosenfeldt, A. Parregaard, K. F. Michaelsen, and M. Jakobsen.** 2003. Case study of the distribution of mucosa-associated *Bifidobacterium* species, *Lactobacillus* species, and other lactic acid bacteria in the human colon. *Appl. Environ. Microbiol.* **69:**7545–7548.
70. **Onderdonk, A. B., A. M. Dvorak, R. L. Cisneros, R. S. McLeod, D. Antionoli, W. Silen, J. E. Blair, R. A. Monahan-Early, J. Cullen, and Z. Cohen.** 1992. Microbiologic assessment of tissue biopsy samples from ileal pouch patients. *J. Clin. Microbiol.* **30:**312–317.
71. **Ohtani, K., H. Hayashi, and T. Shimizu.** 2002. The *luxS* gene is involved in cell-cell signalling for toxin production in *Clostridium perfringens. Mol. Microbiol.* **44:**171–179.
72. **O'Sullivan, D. J.** 1999. Methods of analysis of the intestinal microflora, p. 23–44. *In* G. W. Tannock (ed.), *Probiotics: a Critical Review.* Horizon Scientific Press, Wymondham, United Kingdom.
73. **Podolsky, D. K.** 2002. Inflammatory bowel disease. *N. Engl. J. Med.* **347:**417–429.
74. **Rainey, P. B., and G. M. Preston.** 2000. *In vivo* expression technology strategies: valuable tools for biotechnology. *Curr. Opin. Biotechnol.* **11:**440–444.
75. **Rao, A. V.** 1999. Dose-response effects of inulin and oligofructose on intestinal bifidogenesis effects. *J. Nutr.* **129:** 1442S–1445S.
76. **Rath, H. C., D. E. Bender, L. C. Holt, T. Grenther, J. D. Taurog, R. E. Hammer, and R. B. Sartor.** 1995. Metronidazole attenuates colitis in HLA-B27/β2 microglobulin transgenic rats: a pathogenic role for anaerobic bacteria. *Clin. Immunol. Immunopathol.* **76:**S45.
77. **Rath, H. C. H. H. Herfath, J. S. Ikeda, W. B. Grenther, T. E. Hamm, E. Balish, K. H. Wilson, and R. B. Sartor.** 1996. Normal luminal bacteria, especially *Bacteroides* species, mediate chronic colitis, gastritis, and arthritis in HLA-B27/human β2 microglobulin transgenic rats. *J. Clin. Investig.* **98:** 945–953.
78. **Rath, H. C, M. Schultz, L. A. Dieleman, F. Li, H. Kobl, W. Falk, J. Scholmerich, and R. B. Sartor.** 1998. Selective vs. broad spectrum antibiotics in the prevention and treatment of experimental colitis in two rodent models. *Gastroenterology* **114:**A1067.
79. **Rath, H.C., K. H. Wilson, and R. B. Sartor.** 1999. Differential induction of colitis and gastritis in HLA-B27 transgenic rats selectively colonized with *Bacteroides vulgatus* or *Escherichia coli. Infect. Immun.* **67:**2969–2974.
80. **Redfield, R. J.** 2002. Is quorum sensing a side effect of diffusion sensing? *Trends Microbiol.* **10:**365–370.

81. **Romagnani, S.** 1997. Atopic allergy and other hypersensitivities. Interactions between genetic susceptibility, innocuous and/or microbial antigens and the immune system. *Curr. Opin. Immunol.* 9:773–775.
82. **Russell, J. B., and G. M. Cook.** 1995. Energetics of bacterial growth: balance of anabolic and catabolic reactions. *Microbiol. Rev.* **59:**48–62.
83. **Saavedra, J. M.** 2000. Probiotics and infectious diarrhea. *Am. J. Gastroenterol.* **95:**S16–S18.
84. **Salzman, N. H., H. de Jong, Y. Paterson, H. J. M. Harmsen, G. W. Welling, and N. A. Bos.** 2002. Analysis of 16S libraries of mouse gastrointestinal microflora reveals a large new group of mouse intestinal bacteria. *Microbiology* **148:**3651–3660.
85. **Sartor, R. B.** 1997. The influence of the normal microbial flora on the development of chronic mucosal inflammation. *Res. Immunol.* **148:**567–576.
86. **Sartor, R. B., H. C. Rath, S. N. Lichtman, and E. A. van Tol.** 1996. Animal models of intestinal and joint inflammation. *Bailliere's Clin. Rheumatol.* **10:**55–76.
87. **Satokari, R. M., E. E. Vaughan, A. D. L. Akkermans, M. Sareela, and W. M. De Vos.** 2001. Bifidobacterial diversity in human feces detected by genus-specific PCR and denaturing gradient gel electrophoresis. *Appl. Environ. Microbiol.* **67:** 504–513.
88. **Savage, D. C.** 1972. Associations and physiological interactions of indigenous microorganisms and gastrointestinal epithelia. *Am. J. Clin. Nutr.* **25:**1372–1379.
89. **Savage, D. C.** 1977. Microbial ecology of the gastrointestinal tract. *Annu. Rev. Microbiol.* **31:**107–133.
90. **Savage, D. C., R. Dubos, and R. W. Schaedler.** 1968. The gastrointestinal epithelium and its autochthonous bacterial flora. *J. Exp. Med.* **127:**67–76.
91. **Schultsz, C., F. M. van den Berg, F. W. Ten Kate, G. N. J. Tytgat, and J. Dankert.** 1999. The intestinal mucus layer from patients with inflammatory bowel disease harbors high numbers of bacteria compared with controls. *Gastroenterology* **117:**1089–1097.
92. **Schultz, M., and H. C. Rath.** 2002. The possible role of probiotic therapy in inflammatory bowel disease, p. 239–261. *In* G. W. Tannock (ed.), *Probiotics and Prebiotics: Where Are We Going?* Caister Academic Press, Wymondham, United Kingdom.
93. **Sellon, R. K., S. Tonkonogy, M. Schultz, L. A. Dieleman, W. Grenther, E. Balish, D. M. Rennick, and R. B. Sartor.** 1998. Resident enteric bacteria are necessary for development of spontaneous colitis and immune system activation in interleukin-10-deficient mice. *Infect. Immun.* **66:**5224–5231.
94. **Sherman, L. A., and D. C. Savage.** 1986. Lipoteichoic acids in *Lactobacillus* strains that colonize the mouse gastric epithelium. *Appl. Environ. Microbiol.* **52:**302–304.
95. **Snel, J., H. J. Blok, H. M. P. Kengen, W. Ludwig, F. G. J. Poelma, J. P. Koopman, and A. D. L. Akkermans.** 1994. Phylogenetic characterization of the *Clostridium* related segmented filamentous bacteria in mice based on 16S ribosomal RNA analysis. *Syst. Appl. Microbiol.* **17:**172–179.
96. **Spanhaak, S., R. Havenaar, and G. Schaafsma.** 1998. The effect of consumption of milk fermented by *Lactobacillus casei* strain Shirota on the intestinal microflora and immune parameters in humans. *Eur. J. Clin. Nutr.* **52:**899–907.
97. **Stebbings, S., K. Munro, M. A. Simon, G. W. Tannock, J. Highton, H. Harmsen, G. Welling, P. Seksik, J. Dore, G. Grame, and A. Tilsala-Timisjarvi.** 2002. Comparison of the faecal microflora of patients with ankylosing spondylitis and controls using molecular methods of analysis. *Rheumatology* **41:**1395–1401.
98. **Stokkers, P. C., P. H. Reitsma, G. N. Tytgat, and S. J. Deventer.** 1999. HLA-DR and -DQ phenotypes in inflammatory bowel disease: a meta-analysis. *Gut* **45:**395–401.
99. **Strachan, D. P.** 1989. Hay fever, hygiene, and household size. *Br. Med. J.* **299:**1259–1260.
100. **Suau, A., R. Bonnet, M. Sutren, J. J. Godon, G. R. Gibson, M. D. Collins, and J. Dore.** 1999. Direct analysis of genes encoding 16S rRNA from complex communities reveals many novel molecular species within the human gut. *Appl. Environ. Microbiol.* **65:**4799–4807.
101. **Suegara, N., M. Morotomi, T. Watanabe, Y. Kawai, and M. Mutai.** 1975. Behavior of microflora in the rat stomach: adhesion of lactobacilli to the keratinized epithelial cells of the rat stomach in vitro. *Infect. Immun.* **12:**173–179.
102. **Surette, M. G., and B. L. Bassler.** 1998. Quorum sensing in *Escherichia coli* and *Salmonella typhimurium*. *Proc. Natl. Acad. Sci. USA* **95:**7046–7050.
103. **Swidsinski, A., A. Ladhoff, A. Pernthaler, S. Swidsinski, V. Loening-Baucke, M. Ortner, J. Weber, U. Hoffmann, S. Schreiber, M. Dietal, and H. Lochs.** 2002. Mucosal flora in inflammatory bowel disease. *Gastroenterology* **122:**44–54.
104. **Takeuchi, A., and J. A. Zeller.** 1972. Scanning electron microscopic observations on the surface of the normal and spirochaete-infested colonic mucosa of the rhesus monkey. *J. Ultrastruct. Res.* **40:**313–324.
105. **Tannock, G. W.** 1984. Control of gastrointestinal pathogens by normal flora, p. 374 –382. *In* M. J. Klug and C. A. Reddy (ed.), *Current Perspectives in Microbial Ecology.* American Society for Microbiology, Washington, D.C.
106. **Tannock, G. W.** 1987. Demonstration of mucosa-associated microbial populations in the colons of mice. *Appl. Environ. Microbiol.* **53:**1965–1968.
107. **Tannock, G. W.** 1994. The acquisition of the normal microflora in the gastrointestinal tract, p. 1–16. *In* S. A. W. Gibson (ed.), *Human Health: the Contribution of Microorganisms.* Springer-Verlag, London, United Kingdom.
108. **Tannock, G. W.** 2003. Probiotics: time for a dose of realism. *Curr. Issues Intest. Microbiol.* **4:**33–42.
109. **Tannock, G. W., and G. Cook.** 2002. Enterococci as members of the intestinal microflora of humans, p. 101–132. *In* M. S. Gilmore, D. B. Clewell, P. Courvalin, G. M. Dunny, B. E. Murray, and L. B. Rice. (ed.), *The Enterococci: Pathogenesis, Molecular Biology, and Antibiotic Resistance.* ASM Press, Washington, D.C.
110. **Tannock, G. W., K. Munro, R. Bibiloni, M. A. Simon, P. Hargreaves, P. Gopal, H. Harmsen, and G. Welling.** 2004. Impact of the consumption of oligosaccharide-containing biscuits on the fecal microbiota of humans. *Appl. Environ. Microbiol.* **70:**2129–2136.
111. **Tannock, G. W., K. Munro, H. J. M. Harmsen, G. W. Welling, J. Smart, and P. K. Gopal.** 2000. Analysis of the fecal microflora of human subjects consuming a probiotic containing *Lactobacillus rhamnosus* DR20. *Appl. Environ. Microbiol.* **66:**2578–2588.
112. **Tannock, G. W., O. Szylit, Y. Duval, and P. Raibaud.** 1982. Colonization of tissue surfaces in the gastrointestinal tract of gnotobiotic animals by lactobacillus strains. *Can. J. Microbiol.* **28:**1196–1198.
113. **Tuohy, K. M., S. Kolida, A. M. Lustenberger and G. R. Gibson.** 2001. The prebiotic effects of biscuits containing partially hydrolysed guar gum and fructo-oligosaccharides—a human volunteer study. *Br. J. Nutr.* **86:**341–348.

114. **Umesaki, Y., Y. Okada, S. Matsumoto, A. Imaoka, and H. Setoyama.** 1995. Segmented filamentous bacteria are indigenous intestinal bacteria that activate intraepithelial lymphocytes and induce MHC class II molecules and fucosyl asialo GM_1 glycolipids on the small intestinal epithelial cells in the ex-germ-free mouse. *Microbiol. Immunol.* **39:**555–562.
115. **Walter, J., N. C. K. Heng, W. P. Hammes, D. M. Loach, G. W. Tannock, and C. Hertel.** 2003. Identification of *Lactobacillus reuteri* genes specifically induced in the mouse gastrointestinal tract. *Appl. Environ. Microbiol.* **69:**2044–2051.
116. **Walter, J., C. Hertel, G. W. Tannock, C. M. Lis, K. Munro, and W. P. Hammes.** 2001. Detection of *Lactobacillus, Pediococcus, Leuconostoc,* and *Weissella* species in human feces by using group-specific PCR primers and denaturing gradient gel electrophoresis. *Appl. Environ. Microbiol.* **67:**2578–2585.
117. **Wang, X., S. P. Heazlewood, D. O. Krause, and T. H. J. Florin.** 2003. Molecular characterization of the microbial species that colonize human ileal and colonic mucosa by using 16S rDNA sequence analysis. *J. Appl. Microbiol.* **95:**508–520.
118. **Wesney, E., and G. W. Tannock.** 1979. Association of rat, pig and fowl biotypes of lactobacilli with the stomach of gnotobiotic mice. *Microb. Ecol.* **5:**35–42.
119. **Wickens, K., N. Pearce, J. Crane, and R. Beasley.** 1999. Antibiotic use in early chidhood and the development of asthma. *Clin. Exp. Allergy* **29:**766–771.
120. **Winzer, K., K. R. Hardie, and P. Williams.** 2002. Bacterial cell-cell communication: sorry, can't talk now—gone to lunch! *Curr. Opin. Microbiol.* **5:**216–222.
121. **Wordsworth P.** 1998. Genes in the spondyloarthropathies. *Rheum. Dis. Clin. North Am.* **24:**845–863.
122. **Xavier, K. B., and B. L. Bassler.** 2003. LuxS quorum sensing: more than just a numbers game. *Curr. Opin. Microbiol.* **6:** 191–197.
123. **Zhu, Y. Y., T. Zhong, Y. Pandya, and R. D. Joerger.** 2002. 16S rRNA-based analysis of microflora from the cecum of broiler chickens. *Appl. Environ. Microbiol.* **68:**124–137.
124. **Zinkevich, V., and I. B. Beech.** 2000. Screening of sulfate-reducing bacteria in colonoscopy samples from healthy and colitic human gut mucosa. *FEMS Microb. Ecol.* **34:**147–155.
125. **Zoetendal, E. G., A. D. L. Akkermans, and W. M. de Vos.** 1998. Temperature gradient gel electrophoretic analysis of 16S rRNA from fecal samples reveals stable and host-specific communities of active bacteria. *Appl. Environ. Microbiol.* **64:**3854–3859.
126. **Zoetendal, E.G., A. D. L. Akkermans, W. M. Akkermans-van Vliet, J. A. G. M. de Viosser, and W. M. de Vos.** 2001. The host genotype affects the bacterial community in the human gastrointestinal tract. *Microb. Ecol. Health Dis.* **13:**129–134.
127. **Zoetendal, E. G., A. von Wright, T. Vilponen-Samela, K. Ben-Amor, A. D. L. Akkermans, and W. M. de Vos.** 2002. Mucosa-associated bacteria in the human gastrointestinal tract are uniformly distributed along the colon and differ from the community recovered from the feces. *Appl. Environ. Microbiol.* **68:**3401–3407.

Colonization of Mucosal Surfaces
Edited by James P. Nataro et al.

Chapter 13

Interactions of the Commensal Flora with the Human Gastrointestinal Tract

JAMES P. NATARO

Throughout this volume, we have considered the effects of natural selection on the development of mucosal virulence factors and on concurrent development of the anti-infective weapons of the host mucosal immune system. However, it must be considered that natural selection has molded all aspects of human coexistence with the commensal bacteria, both in sickness and in health. This chapter surveys recent information on the roles of the commensal intestinal flora and provides an overview of how that natural symbiosis can be enhanced.

ONTOGENY OF THE INTESTINAL FLORA

The mammalian gastrointestinal (GI) tract is sterile at birth. However, within hours after parturition, bacteria can be cultured from the GI tract (1). Infants with obstructed GI tracts yield bacterial growth proximal to the obstruction while the distal segment remains largely sterile, revealing that inoculation of the enteric flora occurs by mouth.

The first bacteria cultured from the neonatal GI tract are aerobes and facultative anaerobes, which quickly establish the low redox potential required to sustain anaerobic organisms. Clostridia are typically the first anaerobes detected, usually by 1 week of age (5); soon thereafter, especially in breast-fed infants, bifidobacteria become abundant members of the flora. The fecal flora of the formula-fed infant is typically more diverse than that of the breast-fed infant, although at weaning, a greatly increased microbial diversity is observed. Over the first 2 years of life, the fecal flora becomes increasingly adult-like and is fully constituted by 2 years of age (1, 43). By this time, anaerobes outnumber aerobes by approximately 1,000:1. The dominant microbial genera of the human GI tract include *Bacteroides, Bifidobacterium, Eubacterium, Lactobacillus, Clostridium, Fusobacterium, Peptococcus, Peptostreptococcus, Escherichia,* and *Veillonella* (41). Microbial counts are most abundant in the colon, reaching 10^{11} to 10^{12} CFU/ml, whereas counts in the small bowel rise from 10^{3}/ml near the gastric outlet to 10^{10}/ml at the ileocecal valve (5).

CONTROL OF THE INTESTINAL FLORA

The colonic environment has been modeled, both mathematically and experimentally, as an anaerobic fermentation ecosystem (9, 42, 46, 47). Within this complex environment, the relative abundance of the microflora remains generally stable over long periods in an individual, although there exist significant differences among the floras of different subjects. It is thought that the availability of specific nutrients controls the abundance of enteric species (10, 11) and that the most successful enteric commensals are those that exhibit the most efficient utilization of a particular nutritive substrate (12). *Escherichia coli,* for example, has been postulated to be the most efficient user of gluconate (73), which is uniformly available in the lumen of the human colon. Interestingly, *E. coli* pathotypes may increase their colonization efficiency by exploiting a nutritional niche distinct from that of commensal *E. coli* (51). As discussed below, certain commensals may even be able to modify the availability of luminal nutrients to maximize their chance of metabolic success (29).

The effect of the GI immune system on the control of the intestinal flora has been controversial. The ability of the host to accept the continuous presence

James P. Nataro • Center for Vaccine Development, Departments of Pediatrics, Medicine, and Microbiology and Immunology, University of Maryland School of Medicine, 685 W. Baltimore St., Baltimore, MD 21201.

of foreign species does not rely on immune tolerance in the strict sense; indeed immune responses to the enteric flora can be readily engendered. However, at least some commensal bacteria adapted to this state have developed remarkable abilities to generate antigenic diversification, which may impede the antibacterial control mechanisms of the adaptive immune system. Krinos et al. have reported that the commensal anaerobe *Bacteroides fragilis* modulates its surface antigenicity by producing at least eight distinct capsular polysaccharides and that the organism regulates their expression in an on-off manner by the reversible inversion of DNA segments containing the promoters for capsular expression (39). This kind of phase variation of surface structures is also exercised by pathogenic bacteria, as discussed in the chapter on *Salmonella* (see chapter 21).

The ability of the commensal flora to persist in the intestinal lumen stands in stark contrast to the abundance and vigor of the intestinal immune system (see chapter 3). Is this immune system present strictly to thwart the assault of mucosal pathogens? Whereas it was long assumed that the commensal flora lacked the ability to translocate across the epithelium or to intiate inflammatory responses, the identification of pathogen-associated molecular patterns on commensal organisms (e.g., lipopolysaccharide and flagellin) has called this assumption into question and has begun to blur the traditional distinctions between pathogen and commensal.

Recent observations have illuminated one essential contribution of the mucosal immune system to orchestrating interactions with the commensal flora: preventing invasion of the commensal bacteria into the bloodstream. MacPherson and Uhr have observed that members of the commensal flora do translocate across the intestinal epithelium of mice at low efficiency and that they are preferentially taken up by dendritic cells (44). Rather than being quickly killed by dendritic cells (as they are by macrophages), the bacteria survive, but do not multiply, and are transported to the mesenteric lymph nodes. At this site, the immune system generates specific secretory immunoglobulin A (sIgA)-producing B cells, which home back to the GI mucosa, resulting in the secretion of a large amount of sIgA, which effectively limits further translocation of the target microbes. Resection of mesenteric lymph nodes in experimental mice resulted in the recovery of live commensal bacteria from the spleen, suggesting that confinement of the bacteria to the lymph nodes precludes systemic immune interactions. These observations, which require confirmation from studies of other systems, would suggest that the gut has evolved specific mechanisms to ensure that the commensal flora is kept at bay.

BENEFICIAL EFFECTS OF THE INTESTINAL FLORA

Nutritional Contributions

The commensal enteric flora makes significant metabolic contributions to the health of the host (43). Production of vitamin K by the commensal flora is essential to normal blood coagulation, whereas vitamins of the B group play multiple roles in normal homeostasis (22). In addition, short-chain fatty acid production by commensal bacteria supports healthy epithelial cell growth and provides additional energy sources to the body (43). Bacterial metabolism of bile acids may be part of the normal recycling mechanism of these compounds (21). An increasing number of additional effects of the colonic flora have also been suggested. Chief among these is the relationship between enteric bacteria and carcinogenesis (23, 45, 47, 52, 55, 75–77). Interestingly, commensal bacteria have been suggested to both increase and decrease the risk of colon cancer. Some common bacterial strains are able to degrade mutagenic compounds consumed in food; other bacteria, however, express enzymes (such as azoreductase, nitroreductase, and glucuronidase) that may produce directly genotoxic products (75). The overall contribution of the commensal flora to the risk of colon cancer remains controversial.

Protection from Infection

Virtually all mammalian species studied to date harbor an abundant intestinal microflora, strongly suggesting that the presence of these bacteria confers a tangible advantage to the host. One readily observable advantage is resistance to colonization by enteric pathogens. Indeed, the study of colonization by mucosal pathogens uncovers a story of offensive and defensive coadaptation of pathogen and host. Accordingly, we observe throughout this volume the intricate nuances of adherence, immune avoidance, and metabolic fine-tuning that prepare the pathogen for the initial phase of infection. At the same time, we see equally remarkable host defenses and realize that in the natural state, encounter with pathogens may result in colonization and disease, colonization without disease, or failure to establish colonization at all.

One example of this tenuous balance between host and pathogen is illustrated by the study of colonization resistance provided by the commensal flora (6, 71). Hexaflora-associated gnotobiotic mice demonstrate significant resistance to incoming *E. coli* and *Pseudomonas* (71), and the influence of antibiotics on the enteric flora and the predisposition to

Clostridium difficile colitis are well known. The precise mechanisms of colonization resistance are not fully elucidated. Presumably, competition for nutrients plays some role, yet recent data suggest that incoming pathogens have evolved alternative metabolic pathways which permit them to exploit nutrients left available by the commensal flora. Miranda et al. have reported that *E. coli* O157:H7 is able to utilize the same glycolytic substrates metabolized by commensal *E. coli* strains (51). However, when fed to mice harboring large numbers of nonpathogenic *E. coli,* O157:H7 successfully exploits gluconeogenic pathways, thereby occupying a new metabolic niche.

Whereas the presence of an abundant commensal flora may provide some protection against incoming enteric pathogens, it has recently been suggested that some pathogens may sense the presence of resident nonpathogenic flora and thereby activate the expression of virulence-related genes (69, 70), a phenomenon that utilizes endogenous bacterial quorum-sensing pathways (see chapter 14). Alternatively, various studies have shown that specific commensal strains exert direct antagonistic effects on enteric pathogens (3). Work in my laboratory has shown that *Lactobacillus* and *Veillonella* species specifically repress the expression of AggR, the master regulator of virulence functions in the diarrheagenic pathogen enteroaggregative *E. coli* (63) (Fig. 1) Presumably, the selective pressure toward the development and maintenance of beneficial commensals in the human GI tract has been driven by the benefit of resistance to infectious diseases and the extraordinary benefit to the commensal bacteria of having plentiful and healthy hosts in which to propagate.

Maturation of the Immune System

Maturity of the infant immune system is more closely related to postnatal age than to postconceptual age. This clinical observation suggests that some acquired stimulus directs or impacts the development of the infant immune system. Growing evidence suggests that this stimulus is provided by the commensal enteric flora (74).

Germfree animals manifest a large number of immune deficits (66; reviewed in reference 60). Gut-associated lymphoid tissue in these animals is sparse and disorganized, failing to demonstrate Peyer's patches and other lymphoid follicles (20, 68); IgA production is poor; and systemic antibody responses

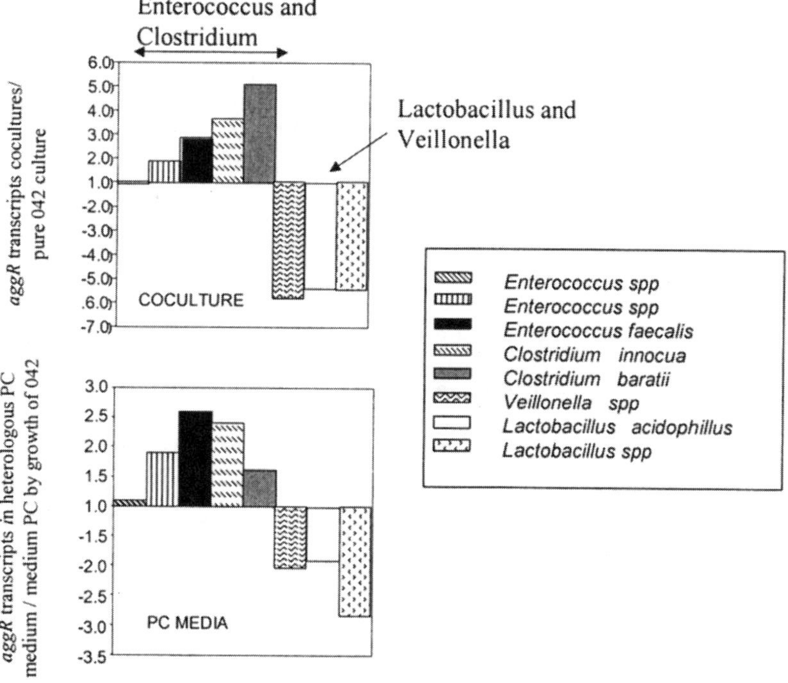

Figure 1. Effects of coculture and preconditioned media on the expression of the global regulator of enteroaggregative *E. coli* virulence, AggR. Enteroaggregative *E. coli* strain 042 was cocultivated individually with various enteric bacteria in Luria broth at 37°C to the late log phase. Conditioned media were prepared by cultivating the enteric species, filtering out bacterial growth, and correcting pH and nutrient concentration. The *aggR* transcript in both experiments was quantitated by real-time reverse transcription-PCR. Values expressed are the number of *aggR* transcripts in coculture or preconditioned media compared with *aggR* expression in pure control cultures of 042. *aggR* expression is enhanced by *Enterococcus* and *Clostridium* species and diminished by *Lactobacillus* and *Veillonella* species. Reprinted from reference 63 with permission.

are weak (74). Accumulating evidence also suggests that the antigens of the commensal flora may help to balance the Th2-dominant adaptive immune responses that are characteristic of healthy newborns (31, 33, 34, 57). Perhaps more importantly, the commensal flora appears to elicit the maturation of Tr responses, resulting in secretion of the immunomodulatory cytokines interleukin-10 and transforming growth factor β (33).

When fully constituted, the abundant leukocyte populations of the intestinal submucosa might be considered to represent a low level of constitutive inflammation. The presence of abundant proinflammatory stimuli within the GI lumen, particularly on weaning from the breast, necessitates the ability of the host to modulate the interactions of these immune cells and to control inflammatory cascades. In a landmark observation, Neish et al. showed that nonpathogenic *Salmonella* strains induced stabilization of the NF-κB complex, resulting in decreased inflammatory responses to the commensal flora and potentially to other stimuli (54). To be extrapolatable to the GI tracts of all human infants, this capability would presumably be possessed by many additional nonpathogenic bacteria.

Maturation of the Mucosa

Permeability

The intestinal mucosa must maintain a highly selective barrier function, capable of permitting the absorption of highly variable nutrients and the sampling of antigens while excluding pathogenic microorganisms. Just as the presence of the commensal flora is needed to drive the maturation of the immune system, several studies have suggested that the flora is required to establish normal epithelial barrier function (31, 33). Germfree animals possess deranged intestinal permeability, potentially absorbing large molecules excluded by conventional animals (48, 49). Using an ex vivo system, El Asmar et al. demonstrated that exposure to healthy commensal bacteria results in establishment of the normal tight-junction barrier between epithelial cells (8), the major determinant of gut permeability. The precise mechanism of this acitivity is under investigation. The importance of these studies is underscored by the observations of increased gut permeability in a growing number of noninfectious diseases, including celiac disease and type I diabetes mellitus.

Absorptive and other functions

The studies by Gordon and Hooper have illuminated dramatically the contributions of the commensal flora to ontogeny of the intestinal mucosa (4, 7, 19, 24–26, 40, 50, 53, 58, 72, 78, 79). Gordon's group first observed that monoassociation of germfree mice with the common commensal *Bacteroides thetaiotaomicron* induced the activation of a large number of genes in intestinal epithelial cells (28). Among these genes were several encoding products involved in digestive functions, including the Na^+/glucose transporter (SGLT-1), colipase, and pancreatic lipase-related protein 2. *B. thetaiotaomicron* induced a ca. 200-fold increase in the expression of small proline-rich protein 2, a contributor to epithelial barrier function. Interestingly, colonization by *B. thetaiotaomicron* also repressed the expression of several genes that participate in detoxification processes, including the glutathione *S*-transferase gene, although colonization by *E. coli* induced such genes. These data suggest temporal requirements for high levels of detoxification enzymes, which would presumably be in higher demand before the intestinal barrier function is fully established.

One particularly fascinating response to *B. thetaiotaomicron* colonization is the induction of fucosylases on the apical membrane of epithelial cells (7). Conventional mice begin life with a repertoire of plasma membrane glycoconjugates (glycans) dominated by sialic acid at the apical membrane. During weaning, however, these glycans become predominantly fucosylated, as a result of fucosyltransferase induction within the epithelial cells. Studies of germfree mice have revealed that this transition can be induced by monoassociation with *B. thetaiotaomicron*. Even more remarkably, induction of this response requires an intact fucose utilization pathway in *B. thetaiotaomicron* (29). Thus, one possible evolutionary scenario is that the early introduction of *B. thetaiotaomicron* results in the expression of cell surface glycans that can be utilized by the bacterium as a preferred nutrient source, establishing for itself a metabolic niche. In return, the host derives a large number of benefits from this developing partnership.

Perhaps principally among these benefits is the enhancement of innate and adaptive immune capability. Colonization with *B. thetaiotaomicron* and other organisms results in the induction of several genes that contribute to mucosal and systemic defense. Among the former is that encoding decay-accelerating factor (CD55), which provides resistance to endogenous complement. The polymeric immunoglobulin receptor (pIgR) is also induced, accompanied by increased numbers of sIgA-producing B cells in the lamina propria. Conventional mice possess larger numbers of intraepithelial αβ T-cell receptor-bearing intraepithelial lymphocytes, and γδ T-cell intraepithelial lymphocytes from conventional mice are dramatically more cytolytic than are those of germfree mice.

Hooper et al. have shown that the commensal flora also contributes to intestinal capillary morphogenesis, an effect possibly mediated by Paneth cells (27). Interestingly, these investigators have reported that angiogenin-4 (Ang4), a granule protein secreted by small intestinal Paneth cells, may play a role both in capillary angiogenesis and as an antibacterial factor induced during the weaning phase.

The complex panoply of developmental changes induced during weaning are intuitively the result of adaptation to the coming adult state. During weaning, the infant diet becomes more complex, the intestine becomes exposed to a far greater number of microbial invaders, both pathogenic and nonpathogenic, and maternal protective factors become increasingly unavailable to deal with these foes. The transition of the commensal flora during this critical phase may be the key signal for the intestine, and the body as a whole, to adapt to these looming challenges. If so, our understanding of these events is itself in its infancy, although molecular-biology approaches are rapidly illuminating important features of the transition.

PROBIOTICS

The most exciting implication of an expanded role for the commensal bacterial flora is the possibility that human health can be positively influenced by manipulation of the luminal bacterial populations. This stems from the growing realization that certain enteric bacterial species have adverse and/or beneficial influences on human health (Fig. 2). The study of health-promoting effects conferred by administration

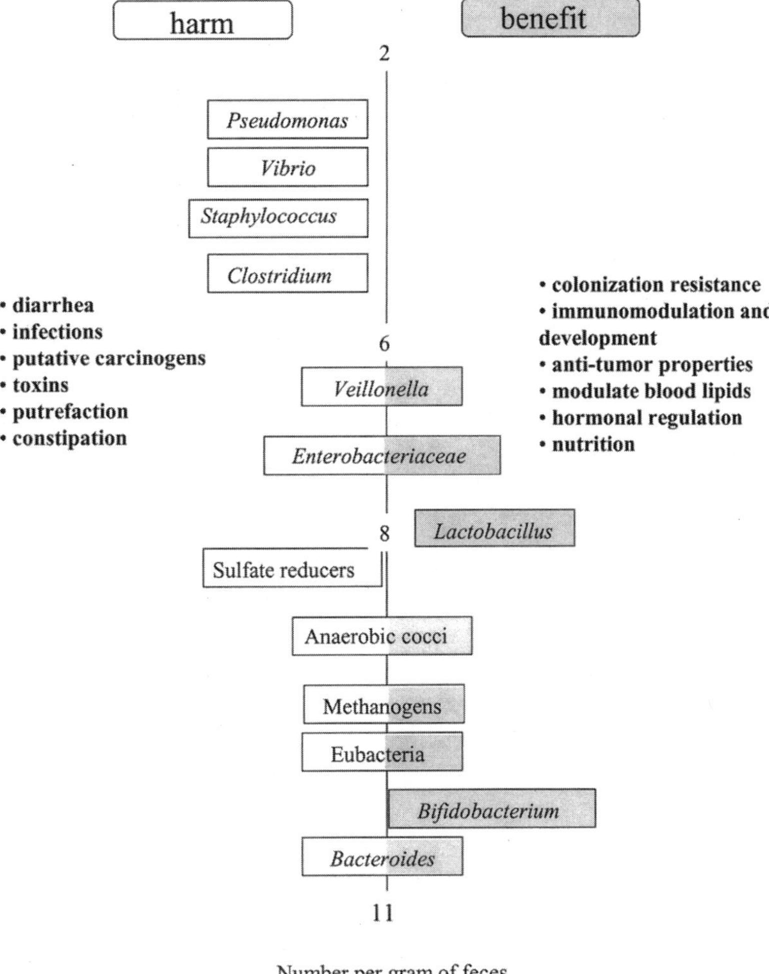

Figure 2. The commensal enteric flora can confer adverse and/or beneficial effects on human health. These effects are illustrated on a continuum representing the typical abundance of these species. Adapted from reference 15 with permission from the American Society for Nutritional Sciences.

of a live commensal flora, so called probiotic species, has a long but often confusing history. The problems derive largely from reliance on incomplete, small, or flawed clinical studies, often not directed at the most appropriate population. Moreover, interpretation of studies of probiotic effects has suffered from a lack of plausible scientific underpinning, which is only now becoming available (see above). The recognition that normal gut development is in many ways influenced by the commensal succession naturally implies that this ecological phenomenon can be influenced.

Probiotic studies have focused largely on three disease states: infectious diarrhea, inflammatory bowel disease, and allergy. Data are now fairly clear that certain probiotic strains, such as *Lactobacillus rhamnosus* GG, can both prevent and ameliorate enteric infection (2, 30, 32, 56).

The benefit of probiotic bacteria in inflammatory bowel disease has been studied intensively. The most dramatic benefit appears to derive from the administration of probiotic species to individuals after reestablishment of diverting intestinal pouches (17, 18, 65, 67). In this setting, inflamed segments of bowel are sequestered from the fecal stream and subsequently undergo amelioration of inflammation. If probiotic species are administered to the patient when the pouch is reexposed to fecal bacteria, the rate of inflammatory relapse is substantially diminished. These data suggest either that certain commensal species are less aggravating to the mucosa of inflammatory bowel disease patients or that certain strains may have active immunosuppressive effects.

In a landmark study, Kalliomaki et al. administered *L. rhamnosus* GG to 132 expectant mothers with a family history of atophy and to their infants for the first 6 months of life (35). These authors found a 50% decrease in the incidence of eczema in these infants at 2 years of age and demonstrated a persistent positive effect when the children were followed up at 4 years (36). This study has stimulated great interest in strain GG and other probiotics in the management of pediatric allergy. However, studies of probiotic efficacy in the management of established allergy have not been as positive, suggesting that the benefit of commensal bacteria may be to the ontogeny of the normal immune response.

An interesting and potentially very important twist on the use of probiotics is the concept of prebiotics, which are defined as nondigestible food ingredients that beneficially affect the host by selectively stimulating the growth and/or activity of one or a limited number of bacteria in the colon (15, 59, 64). The bacteria most commonly targeted by prebiotic supplements are lactobacilli and bifidobacteria. Thus, prebiotic substances presumably represent nutrients that are exploited by beneficial bacteria and that aim to promote a favorable microbial milieu.

The fructooligosaccharide inulin has been evaluated extensively as a prebiotic (13, 14, 16, 38, 61). Kleessen et al. fed inulin or lactose to a group of 25 elderly patients and demonstrated a significant increase in the proportion of bifidobacteria and a decrease in the proportion of fecal enterococci in the inulin group (37). The inulin group experienced a significant beneficial laxative effect. Roller et al. have reported that fructooligosaccharide in combination with the probiotics *L. rhamnosus* and *Bifidobacterium lactis* increased intestinal sIgA production and interleukin-10 secretion in rats (62). Due to its beneficial effect on enteric flora, inulin has been proposed as an intervention to decrease the risk of colon cancer (38).

CONCLUSIONS

The contribution of the enteric commensal flora to human health is only beginning to be appreciated, and many more studies are required. The availability of molecular approaches will greatly accelerate laboratory investigations, but careful clinical observations are required to ascertain the full scope of these effects.

Acknowledgments. Work in the Nataro laboratory is supported by U.S. Public Health Service grants AI33096, AI43615, and AI056578. This chapter was edited by P. Cohen.

REFERENCES

1. **Adlerberth, I.** 1999. Establishment of a normal intestinal microflora in the newborn infant, p. 63–78. *In* L. A. Hanson and R. H. Yolken (ed.), *Probiotics, Other Nutritional Factors, and Intestinal Microflora,* vol. 42. Lippincott-Raven Publishers, Philadelphia, Pa.
2. **Arvola, T., K. Laiho, S. Torkkeli, H. Mykkanen, S. Salminen, L. Maunula, and E. Isolauri.** 1999. Prophylactic *Lactobacillus* GG reduces antibiotic-associated diarrhea in children with respiratory infections: a randomized study. *Pediatrics* **104:**e64.
3. **Asahara, T., K. Shimizu, K. Nomoto, T. Hamabata, A. Ozawa, and Y. Takeda.** 2004. Probiotic bifidobacteria protect mice from lethal infection with Shiga toxin-producing *Escherichia coli* O157:H7. *Infect. Immun.* **72:**2240–2247.
4. **Bjorkholm, B. M., J. L. Guruge, J. D. Oh, A. J. Syder, N. Salama, K. Guillemin, S. Falkow, C. Nilsson, P. G. Falk, L. Engstrand, and J. I. Gordon.** 2002. Colonization of germ-free transgenic mice with genotyped *Helicobacter pylori* strains from a case-control study of gastric cancer reveals a correlation between host responses and HsdS components of type I restriction-modification systems. *J. Biol. Chem.* **277:** 34191–34197.
5. **Borriello, S. P.** 2002. The normal flora of the gastrointestinal tract, p. 3–12. *In* A. L. Hart, A. J. Stagg, H. Graffner, H. Glise, P. Falk, and M. A. Kamm (ed.), *Gut Ecology.* Martin Dunitz, Ltd., London, United Kingdom.
6. **Bourlioux, P., B. Koletzko, F. Guarner, and V. Braesco.** 2003. The intestine and its microflora are partners for the protection

of the host: report on the Danone Symposium "The Intelligent Intestine," held in Paris, June 14, 2002. *Am. J. Clin. Nutr.* **78:** 675–683.

7. **Bry, L., P. G. Falk, T. Midtvedt, and J. I. Gordon.** 1996. A model of host-microbial interactions in an open mammalian ecosystem. *Science* **273:**1380–1383.
8. **El Asmar, R., P. Panigrahi, P. Bamford, I. Berti, T. Not, G. V. Coppa, C. Catassi, and A. Fasano.** 2002. Host-dependent zonulin secretion causes the impairment of the small intestine barrier function after bacterial exposure. *Gastroenterology* **123:**1607–1615.
9. **Freter, R.** 1999. Continuous-flow culture models of intestinal microecology, p. 97–110. *In* L. Hanson and R. Yolken (ed.), *Probiotics, Other Nutritional Factors, and Intestinal Microflora,* vol. 42. Lippincott-Raven, Philadelphia, Pa.
10. **Freter, R.** 1989. Control mechanisms of the large-intestinal microflora and its influence on the host. *Acta Gastroenterol. Latinoam.* **19:**197–217.
11. **Freter, R.** 1983. Mechanisms that control the microflora in the large intestine, p. 33–54. *In* D. J. Hentges (ed.), *Human Intestinal Microflora in Health and Disease.* Academic Press, Inc., New York, N.Y.
12. **Freter, R., H. Brickner, M. Botney, D. Cleven, and A. Aranki.** 1983. Mechanisms that control bacterial populations in continuous-flow culture models of mouse large intestinal flora. *Infect. Immun.* **39:**676–685.
13. **Gibson, G. R.** 1999. Dietary modulation of the human gut microflora using the prebiotics oligofructose and inulin. *J. Nutr.* **129:**1438S–1441S.
14. **Gibson, G. R., E. R. Beatty, X. Wang, and J. H. Cummings.** 1995. Selective stimulation of bifidobacteria in the human colon by oligofructose and inulin. *Gastroenterology* **108:**975–982.
15. **Gibson, G. R., and M. B. Roberfroid.** 1995. Dietary modulation of the human colonic microbiota: introducing the concept of prebiotics. *J. Nutr.* **125:**1401–1412.
16. **Gibson, G. R., and X. Wang.** 1994. Enrichment of bifidobacteria from human gut contents by oligofructose using continuous culture. *FEMS Microbiol. Lett.* **118:**121–127.
17. **Gionchetti, P., F. Rizzello, A. Venturi, P. Brigidi, D. Matteuzzi, G. Bazzocchi, G. Poggioli, M. Miglioli, and M. Campieri.** 2000. Oral bacteriotherapy as maintenance treatment in patients with chronic pouchitis: a double-blind, placebo-controlled trial. *Gastroenterology* **119:**305–309.
18. **Gionchetti, P., F. Rizzello, A. Venturi, and M. Campieri.** 2000. Probiotics in infective diarrhoea and inflammatory bowel diseases. *J. Gastroenterol. Hepatol.* **15:**489–493.
19. **Gordon, J. I., L. V. Hooper, M. S. McNevin, M. Wong, and L. Bry.** 1997. Epithelial cell growth and differentiation. III. Promoting diversity in the intestine: conversations between the microflora, epithelium, and diffuse GALT. *Am. J. Physiol. Ser. G* **273:**G565–G570.
20. **Helgeland, L., J. T. Vaage, B. Rolstad, T. Midtvedt, and P. Brandtzaeg.** 1996. Microbial colonization influences composition and T-cell receptor V beta repertoire of intraepithelial lymphocytes in rat intestine. *Immunology* **89:**494–501.
21. **Hill, M. J.** 1991. Bile acids and colorectal cancer: hypothesis. *Eur. J. Cancer Prev.* **1**(Suppl. 2)**:**69–74.
22. **Hill, M. J.** 1997. Intestinal flora and endogenous vitamin synthesis. *Eur. J. Cancer Prev.* **6**(Suppl. 1)**:**S43–S45.
23. **Hill, M. J.** 1991. The ratio of lithocholic to deoxycholic acid in faeces: a risk factor in colorectal carcinogenesis. *Eur. J. Cancer Prev.* **1**(Suppl. 2)**:**75–78.
24. **Hooper, L. V., L. Bry, P. G. Falk, and J. I. Gordon.** 1998. Host-microbial symbiosis in the mammalian intestine: exploring an internal ecosystem. *Bioessays* **20:**336–343.
25. **Hooper, L. V., P. G. Falk, and J. I. Gordon.** 2000. Analyzing the molecular foundations of commensalism in the mouse intestine. *Curr. Opin. Microbiol.* **3:**79–85.
26. **Hooper, L. V., T. Midtvedt, and J. I. Gordon.** 2002. How host-microbial interactions shape the nutrient environment of the mammalian intestine. *Annu. Rev. Nutr.* **22:**283–307.
27. **Hooper, L. V., T. S. Stappenbeck, C. V. Hong, and J. I. Gordon.** 2003. Angiogenins: a new class of microbicidal proteins involved in innate immunity. *Nat. Immunol.* **4:**269–273.
28. **Hooper, L. V., M. H. Wong, A. Thelin, L. Hansson, P. G. Falk, and J. I. Gordon.** 2001. Molecular analysis of commensal host-microbial relationships in the intestine. *Science* **291:**881–884.
29. **Hooper, L. V., J. Xu, P. G. Falk, T. Midtvedt, and J. I. Gordon.** 1999. A molecular sensor that allows a gut commensal to control its nutrient foundation in a competitive ecosystem. *Proc. Natl. Acad. Sci. USA* **96:**9833–9838.
30. **Isolauri, E.** 2003. Probiotics for infectious diarrhoea. *Gut* **52:** 436–437.
31. **Isolauri, E.** 2001. Probiotics in human disease. *Am. J. Clin. Nutr.* **73:**1142S–1146S.
32. **Isolauri, E.** 2000. The use of probiotics in paediatrics. *Hosp. Med.* **61:**6–7.
33. **Isolauri, E., Y. Sutas, P. Kankaanpaa, H. Arvilommi, and S. Salminen.** 2001. Probiotics: effects on immunity. *Am. J. Clin. Nutr.* **73:**444S–450S.
34. **Kalliomaki, M., and E. Isolauri.** 2003. Role of intestinal flora in the development of allergy. *Curr. Opin. Allergy Clin. Immunol.* **3:**15–20.
35. **Kalliomaki, M., S. Salminen, H. Arvilommi, P. Kero, P. Koskinen, and E. Isolauri.** 2001. Probiotics in primary prevention of atopic disease: a randomised placebo-controlled trial. *Lancet* **357:**1076–1079.
36. **Kalliomaki, M., S. Salminen, T. Poussa, H. Arvilommi, and E. Isolauri.** 2003. Probiotics and prevention of atopic disease: 4-year follow-up of a randomised placebo-controlled trial. *Lancet* **361:**1869–1871.
37. **Kleessen, B., B. Sykura, H. J. Zunft, and M. Blaut.** 1997. Effects of inulin and lactose on fecal microflora, microbial activity, and bowel habit in elderly constipated persons. *Am. J. Clin. Nutr.* **65:**1397–1402.
38. **Kolida, S., K. Tuohy, and G. R. Gibson.** 2002. Prebiotic effects of inulin and oligofructose. *Br. J. Nutr.* **87**(Suppl. 2)**:**S193–S197.
39. **Krinos, C. M., M. J. Coyne, K. G. Weinacht, A. O. Tzianabos, D. L. Kasper, and L. E. Comstock.** 2001. Extensive surface diversity of a commensal microorganism by multiple DNA inversions. *Nature* **414:**555–558.
40. **Lopez-Boado, Y. S., C. L. Wilson, L. V. Hooper, J. I. Gordon, S. J. Hultgren, and W. C. Parks.** 2000. Bacterial exposure induces and activates matrilysin in mucosal epithelial cells. *J. Cell Biol.* **148:**1305–1315.
41. **MacFarlane, G. T., and S. Macfarlane.** 1997. Human colonic microbiota: ecology, physiology and metabolic potential of intestinal bacteria. *Scand. J. Gastroenterol. Suppl.* **222:**3–9.
42. **MacFarlane, G. T., S. Macfarlane, and G. R. Gibson.** 1998. Validation of a three-stage compound continuous culture system for investigating the effect of retention time on the ecology and metabolism of bacteria in the human colon. *Microb. Ecol.* **35:**180–187.
43. **MacFarlane, G. T., and A. J. McBain.** 1999. The human colonic microbiota, p. 1–25. *In* G. R. Gibson and M. Roberfroid (ed.), *Colonic Microbiota, Nutrition and Health.* Kluwer Academic Publishers, Dortrecht, The Netherlands.
44. **MacPherson, A. J., and T. Uhr.** 2004. Induction of protective IgA by intestinal dendritic cells carrying commensal bacteria. *Science* **303:**1662–1665.

45. **McBain, A. J., and G. T. Macfarlane.** 1998. Ecological and physiological studies on large intestinal bacteria in relation to production of hydrolytic and reductive enzymes involved in formation of genotoxic metabolites. *J. Med. Microbiol.* **47:** 407–416.
46. **McBain, A. J., and G. T. MacFarlane.** 1997. Investigations of bifidobacterial ecology and oligosaccharide metabolism in a three-stage compound continuous culture system. *Scand. J. Gastroenterol. Suppl.* **222:**32–40.
47. **McBain, A. J., and G. T. MacFarlane.** 2001. Modulation of genotoxic enzyme activities by non-digestible oligosaccharide metabolism in in-vitro human gut bacterial ecosystems. *J. Med. Microbiol.* **50:**833–842.
48. **Mehrazar, K., A. Gilman-Sachs, and Y. B. Kim.** 1993. Intestinal absorption of immunologically intact macromolecules in germfree colostrum-deprived piglets maintained on total parenteral nutrition. *J. Parenter. Enteral Nutr.* **17:**8–15.
49. **Mehrazar, K., and Y. B. Kim.** 1988. Total parenteral nutrition in germfree colostrum-deprived neonatal miniature piglets: a unique model to study the ontogeny of the immune system. *J. Parenter. Enteral Nutr.* **12:**563–568.
50. **Mills, J. C., N. Andersson, C. V. Hong, T. S. Stappenbeck, and J. I. Gordon.** 2002. Molecular characterization of mouse gastric epithelial progenitor cells. *Proc. Natl. Acad. Sci. USA* **99:** 14819–14824.
51. **Miranda, R. L., T. Conway, M. P. Leatham, D. E. Chang, W. E. Norris, J. H. Allen, S. J. Stevenson, D. C. Laux, and P. S. Cohen.** 2004. Glycolytic and gluconeogenic growth of *Escherichia coli* O157:H7 (EDL933) and *E. coli* K-12 (MG1655) in the mouse intestine. *Infect. Immun.* **72:**1666–1676.
52. **Moore, W. E., and L. H. Moore.** 1995. Intestinal floras of populations that have a high risk of colon cancer. *Appl. Environ. Microbiol.* **61:**3202–3207.
53. **Mysorekar, I. U., R. G. Lorenz, and J. I. Gordon.** 2002. A gnotobiotic transgenic mouse model for studying interactions between small intestinal enterocytes and intraepithelial lymphocytes. *J. Biol. Chem.* **277:**37811–37819.
54. **Neish, A. S., A. T. Gewirtz, H. Zeng, A. N. Young, M. E. Hobert, V. Karmali, A. S. Rao, and J. L. Madara.** 2000. Prokaryotic regulation of epithelial responses by inhibition of IkappaB-alpha ubiquitination. *Science* **289:**1560–1563.
55. **Pool-Zobel, B. L., A. Bub, U. M. Liegibel, S. Treptow-van Lishaut, and G. Rechkemmer.** 1998. Mechanisms by which vegetable consumption reduces genetic damage in humans. *Cancer Epidemiol. Biomarkers Prev.* **7:**891–899.
56. **Rautanen, T., E. Isolauri, E. Salo, and T. Vesikari.** 1998. Management of acute diarrhoea with low osmolarity oral rehydration solutions and *Lactobacillus* strain GG. *Arch. Dis. Child.* **79:**157–160.
57. **Rautava, S., and E. Isolauri.** 2002. The development of gut immune responses and gut microbiota: effects of probiotics in prevention and treatment of allergic disease. *Curr. Issues Intest. Microbiol.* **3:**15–22.
58. **Rawls, J. F., B. S. Samuel, and J. I. Gordon.** 2004. Gnotobiotic zebrafish reveal evolutionarily conserved responses to the gut microbiota. *Proc. Natl. Acad. Sci. USA* **101:**4596–4601.
59. **Reid, G., M. E. Sanders, H. R. Gaskins, G. R. Gibson, A. Mercenier, R. Rastall, M. Roberfroid, I. Rowland, C. Cherbut, and T. R. Klaenhammer.** 2003. New scientific paradigms for probiotics and prebiotics. *J. Clin. Gastroenterol.* **37:**105–118.
60. **Rhee, K. J., P. Sethupathi, A. Driks, D. K. Lanning, and K. L. Knight.** 2004. Role of commensal bacteria in development of gut-associated lymphoid tissues and preimmune antibody repertoire. *J. Immunol.* **172:**1118–1124.
61. **Roberfroid, M. B., J. A. Van Loo, and G. R. Gibson.** 1998. The bifidogenic nature of chicory inulin and its hydrolysis products. *J. Nutr.* **128:**11–19.
62. **Roller, M., G. Rechkemmer, and B. Watzl.** 2004. Prebiotic inulin enriched with oligofructose in combination with the probiotics *Lactobacillus rhamnosus* and *Bifidobacterium lactis* modulates intestinal immune functions in rats. *J. Nutr.* **134:** 153–156.
63. **Ruiz-Perez, F., S. Davis, and J. Nataro.** 2004. Use of a continuous-flow anaerobic culture to characterize enteric virulence gene expression. *Infect. Immun.* **72:**3793–3802.
64. **Salminen, S., C. Bouley, M. C. Boutron-Ruault, J. H. Cummings, A. Franck, G. R. Gibson, E. Isolauri, M. C. Moreau, M. Roberfroid, and I. Rowland.** 1998. Functional food science and gastrointestinal physiology and function. *Br. J. Nutr.* **80** (Suppl. 1):S147–S171.
65. **Sartor, R. B.** 2000. Probiotics in chronic pouchitis: restoring luminal microbial balance. *Gastroenterology* **119:**584–587.
66. **Schaedler, R. W., R. Dubos, and R. Costello.** 1965. The development of the bacterial flora in the gastrointestinal tract of mice. *J. Exp. Med.* **122:**59–66.
67. **Schultz, M., and R. B. Sartor.** 2000. Probiotics and inflammatory bowel diseases. *Am. J. Gastroenterol.* **95:**S19–S21.
68. **Shroff, K. E., K. Meslin, and J. J. Cebra.** 1995. Commensal enteric bacteria engender a self-limiting humoral mucosal immune response while permanently colonizing the gut. *Infect. Immun.* **63:**3904–3913.
69. **Sperandio, V., J. L. Mellies, W. Nguyen, S. Shin, and J. B. Kaper.** 1999. Quorum sensing controls expression of the type III secretion gene transcription and protein secretion in enterohemorrhagic and enteropathogenic *Escherichia coli*. *Proc. Natl. Acad. Sci. USA* **96:**15196–15201.
70. **Sperandio, V., A. G. Torres, J. A. Giron, and J. B. Kaper.** 2001. Quorum sensing is a global regulatory mechanism in enterohemorrhagic *Escherichia coli* O157:H7. *J. Bacteriol.* **183:** 5187–5197.
71. **Srivastava, K. K.** 1978. Colonization resistance against potentially pathogenic bacteria in hexaflora-associated gnotobiotic mice. *Can. J. Microbiol.* **24:**79–83.
72. **Stappenbeck, T. S., L. V. Hooper, J. K. Manchester, M. H. Wong, and J. I. Gordon.** 2002. Laser capture microdissection of mouse intestine: characterizing mRNA and protein expression, and profiling intermediary metabolism in specified cell populations. *Methods Enzymol.* **356:**167–196.
73. **Sweeney, N. J., P. Klemm, B. A. McCormick, E. Moller-Nielsen, M. Utley, M. A. Schembri, D. C. Laux, and P. S. Cohen.** 1996. The *Escherichia coli* K-12 *gntP* gene allows *E. coli* F-18 to occupy a distinct nutritional niche in the streptomycin-treated mouse large intestine. *Infect. Immun.* **64:** 3497–3503.
74. **Umesaki, Y., and H. Setoyama.** 2000. Structure of the intestinal flora responsible for development of the gut immune system in a rodent model. *Microbes Infect.* **2:**1343–1351.
75. **Van Tassell, R. L., D. G. Kingston, and T. D. Wilkins.** 1990. Dietary genotoxins and the human colonic microflora. *Prog. Clin. Biol. Res.* **340E:**149–158.
76. **Van Tassell, R. L., D. G. Kingston, and T. D. Wilkins.** 1990. Metabolism of dietary genotoxins by the human colonic microflora: the fecapentaenes and heterocyclic amines. *Mutat. Res.* **238:**209–221.
77. **Weisburger, J. H.** 2001. Antimutagenesis and anticarcinogenesis, from the past to the future. *Mutat. Res.* **480–481:**23–35.
78. **Wong, M. H., T. S. Stappenbeck, and J. I. Gordon.** 1999. Living and commuting in intestinal crypts. *Gastroenterology* **116:**208–210.
79. **Xu, J., M. K. Bjursell, J. Himrod, S. Deng, L. K. Carmichael, H. C. Chiang, L. V. Hooper, and J. I. Gordon.** 2003. A genomic view of the human-*Bacteroides thetaiotaomicron* symbiosis. *Science* **299:**2074–2076.

Colonization of Mucosal Surfaces
Edited by James P. Nataro et al.

Chapter 14

Quorum Sensing in the Gastrointestinal Tract

JAMES B. KAPER, CHRISTOPHER PRITCHETT, AND JANE MICHALSKI

THE GASTROINTESTINAL ECOSYSTEM

The gastrointestinal (GI) tract is the site of the largest and most complex environment in the mammalian host. The total microbial population in the adult human (ca. 10^{14}) is estimated to exceed our total number of somatic and germ cells by at least an order of magnitude (5). The GI tract features several diverse macroenvironments, including the oral cavity, the stomach, the small intestine (including the three major regions, i.e., the duodenum, jejunum, and ileum), and the large intestine (colon). The density of bacteria along the GI tract can vary greatly, from 10^3/ml near the gastric outlet to 10^{10}/ml at the ileocecal valve to 10^{11} to 10^{12}/ml in the colon (6). Within these macro environments are several micro environments where bacteria can live, such as the lumen of the bowel, the mucus layer overlying the epithelium, mucus within intestinal crypts, and the surface of mucosal epithelial cells. This wide diversity of environments harbors an even wider diversity of bacterial species, and an estimated 500 to 1,000 different species are present in the GI tract, with an aggregate biomass of ca. 1.5 kg (72). This immense microbial population represents an enormous genetic diversity. Assuming an average genome that is the size of *Escherichia coli* for 1,000 species, the number of genes in this "microbiome" may exceed the total number of human genes by a factor of ca. 100 (72).

A sobering fact about the microbial population of the GI tract is that the majority of the estimated 500 to 1,000 species have not yet been cultured in vitro. Molecular techniques such as microbial community genome-sequencing projects (66) are being applied to the microbiome to more fully characterize this population. Of the commensal bacteria that have been cultured from the GI tract, >99.9% are obligate anaerobes. The dominant commensal microbial genera that can be cultured from the GI tract include *Bacteroides, Bifidobacterium, Eubacterium, Lactobacillus, Clostridium, Fusobacterium, Peptococcus, Peptostreptococcus, Escherichia,* and *Veillonella* (31). As discussed in chapter 13, the commensal flora has a number of benefits to the host including nutritional contributions, protection from infection, maturation of the immune system, and maturation of the intestinal mucosa.

Given the enormous number and diversity of bacteria that comprise the GI environment, it would not be surprising if the members of this community were to somehow communicate among themselves to coordinate various processes ranging from maintenance of the commensal population to aiding or resisting infectious diseases. Quorum sensing (QS) is an important mechanism of cell-to-cell communication that involves density-dependent recognition of signaling molecules resulting in the modulation of gene expression. The first report of a potential role for QS in GI infections was published in 1999 (58) and many reports for different GI pathogens have followed. This chapter reviews the published literature concerning QS among bacteria of the GI tract, with a particular emphasis on pathogenic species that cause infection in the GI tract.

QUORUM SENSING

Overview

QS refers to the ability of bacteria to respond in a dose-dependent fashion to chemical molecules, called autoinducers, in their environment. The autoinducers can be produced from bacteria of the same species or bacteria of different genera. When an

James B. Kaper • Department of Microbiology and Immunology and Center for Vaccine Development, University of Maryland School of Medicine, Baltimore, MD 21201. **Christopher Pritchett and Jane Michalski** • Department of Microbiology and Immunology, University of Maryland School of Medicine, Baltimore, MD 21201.

autoinducer reaches a critical threshold, the bacteria detect and respond to this signal by altering their gene expression. This phenomenon allows bacteria to act as a collective unit, i.e., a multicellular entity, as opposed to individual cells all performing individual functions.

QS was first characterized in the marine bacterium *Vibrio fischeri* (19, 48). This species lives in symbiotic associations with several different marine animal hosts and can colonize the light organ of the host, in which it grows to high densities. This species produces a luciferase enzyme complex encoded by the *luxCDABEGH* genes that is responsible for light production. However, transcription of the *lux* and light production occurs only at high densities of *V. fischeri* and is repressed at low densities. A protein called LuxI is responsible for the production of a signaling molecule, i.e., an autoinducer (AI), that diffuses across the membrane into the extracellular environment or back into the cytoplasm, where it binds to a protein called LuxR. The AI of *V. fischeri* is an acylated homoserine lactone. When the bacteria are in high density, the AI is in high concentration and the LuxR-AI complex becomes an activated transcription factor that induces transcription of the *luxCDABEGH* genes. When bacteria are in low densities, the concentration of AI molecules is below the threshold required to activate transcription and the *lux* genes encoding luciferase are not expressed.

Since the initial description in *V. fischeri,* QS has been recognized to regulate a wide range of activities in diverse bacteria, including plasmid transfer and plant tumor induction by *Agrobacterium tumefaciens,* antibiotic production in *Erwinia carotovora,* biofilm production and virulence gene expression in *Pseudomonas aeruginosa,* competence and sporulation in *Bacillus subtilis,* competence for DNA uptake in *Streptococcus pneumoniae,* and virulence gene expression in numerous pathogens including *Staphylococcus aureus, Vibrio cholerae,* and diarrheagenic *E. coli* (reviewed in reference 62 and this chapter). In addition to modulating the expression of specific functions that are best achieved by a whole population rather than individual bacteria, QS may be used as a system for bacteria to prevent their population from growing to levels that are unsustainable in their environment. If all the nutrients are depleted and waste products are not removed from their environment, it will be deleterious for the community as a whole. In effect, QS is used to determine the fitness of a population (71).

Three major QS circuits have been described: one used primarily by gram-negative bacteria, one used primarily by gram-positive bacteria, and one that is universal. The gram-negative bacterial QS system involves the use of acyl homoserine lactones (AHLs) as autoinducers, which then bind to response regulators that affect gene expression. The gram-positive bacteria use oligopeptide autoinducers that are detected by two-component systems. The third QS system is a universal system that allows interspecies communication and is found in both gram-negative and gram-positive bacteria. There have been numerous recent reviews of the various QS systems, and the reader is referred to one or more of these for additional details (4, 19, 44, 48, 62, 71). The broad concepts of these three systems are briefly reviewed below.

LuxIR Systems in Gram-Negative Bacteria

The paradigm for QS in gram-negative bacteria is the LuxIR system, first described in *V. fischeri.* The LuxIR system uses the LuxI protein, or a homologue thereof, to synthesize an AI, and LuxR (or a homologue) is the regulator that binds to the AI and affects gene expression (20). This system exhibits great specificity, since the AI produced by one species of bacteria can rarely, if ever, interact with the LuxR-type regulator of another species. More than 70 LuxIR QS systems have been found in gram-negative bacteria (13, 20, 43). Interestingly, the great specificity seen with the AI-LuxR interaction is not seen at the level of binding of the activated LuxR transcriptional factor to DNA in the promoter region of the regulated gene since LuxR proteins from different species all bind to similar DNA sequences called "lux boxes" (62).

The AI molecule produced by the LuxIR systems is an AHL, in which there is a common homoserine lactone moiety but variable acyl side chains. The function of the LuxI protein is to link the side-chain group of specific acyl-acyl carrier proteins to the homocysteine moiety of *S*-adenosylmethionine (SAM). Some species may produce more than one AHL AI and have more than one LuxIR pair. For example, in *P. aeruginosa,* one pair of LuxIR homologues, called LasI-LasR, produces and responds to an AHL called 3-oxo-C_{12}-HSL, and in the same strain, the RhlI-RhlR proteins produce and respond to an AHL called C_4-HSL (13, 20, 43). The LasI-LasR system regulates exotoxin A, LasA, LasB, Xcp, and biofilm formation, while the RhlI-RhlR system regulates LasB, rhamnolipid, RpoS, and secondary metabolites. Interestingly, LasI-LasR regulates RhlI-RhlR, thereby allowing the genes controlled by the former to be expressed prior to genes controlled by the latter in a hierarchy of temporal gene expression.

Oligopeptide Systems in Gram-Positive Bacteria

Rather than AHLs, the QS system used by gram-positive bacteria utilizes peptides as AI signaling molecules. These autoinducing polypeptides (AIPs) are produced in the cytoplasm as precursor peptides and then cleaved, modified, and exported. The extracellular AIPs are detected via two-component systems whereby the external portion of a membrane-bound sensor kinase protein detects the AIP and then phosphorylates and activates a response regulator that binds to DNA and modulates transcription. *Staphyloccus aureus* has served as a prototype for the gram-positive AIP systems, and the *S. aureus* Agr (for "accessory gene regulator") QS system regulates virulence gene expression and biofilm formation (reviewed in references 33, 40, and 73). The *S. aureus* Agr system utilizes an oligopeptide produced by AgrD that is modified by AgrB. The resulting AIP is 8 or 9 amino acids long and contains thiolactone rings. Detection of the extracellular AIP and subsequent gene activation is done by the two-component system encoded by *agrAC*. The AIP of *S. aureus* is even more specific than AHLs, and there are four subgroups of this species defined by the AIP they produce. Not only will an AIP produced by one subgroup of *S. aureus* not activate gene expression in another subgroup, but also it will inhibit the QS system in another subgroup. This was demonstrated in a mouse model of infection whereby mice infected with *S. aureus* from one subgroup were protected from disease if an AIP from another subgroup was added to the inoculum (34).

The LuxS/AI-2 System

The third major QS system present in bacteria is found in a wide variety of bacteria, including both gram-negative and gram-positive species (49). This system, called the LuxS or AI-2 system, exists in at least one-third of the bacterial genomes that have been sequenced to date (71). LuxS was initially characterized in *V. harveyi,* which also has an AHL QS system. Starting from SAM, LuxS and other enzymes produce AI-2 through a series of three enzymatic steps (71). The AI-2 molecule was determined to be a furanosyl borate diester, an unusual molecule that uses boron as a cofactor (8). It is thought that AI-2 diffuses across the membrane to the extracellular environment, and, at least in *V. harveyi* and *V. cholerae,* it is detected in the extracellular environment by a two-component system called LuxP-LuxQ, and the resulting phosphorylation cascade results in activation of gene transcription (38). In *Salmonella enterica* serotype Typhimurium, AI-2 is taken up by an ABC cassette transporter system encoded by the *lsr* operon (62). The *lsrK* gene product phosphorylates AI-2 on entry into the cell, and this modification results in its sequestration within the cytoplasm. The removal of AI-2 from the extracellular environment via this mechanism could serve to modulate cell-to-cell signaling within an environment. Although the mechanisms of AI-2 recognition and the resulting signaling cascades may vary among different species, the AI-2 molecule appears to be identical in all species so far examined; thus, AI-2 produced by a gram-positive bacterium can be recognized by a gram-negative species.

The LuxS system regulates the expression of a wide range of phenotypes in different species, including virulence factors, motility, proteases, biofilms, colony morphology, cell division, DNA processing, and cell shape and morphology (71). Some have argued that AI-2 should not be considered a true quorum-signaling molecule because it is a by-product of normal cellular metabolism rather than being a signal that is dedicated solely to intercellular communication (70). While this finely drawn distinction may be valid, it seems that it is not particularly relevant if this metabolic by-product can nevertheless modulate gene transcription of other bacteria in a community, as has been repeatedly demonstrated.

Schauder et al. (48) have identified the *luxS* gene in several species of bacteria that may colonize the GI tract temporarily or in the long term; these species include *Escherichia coli, S. enterica* serotypes Typhimurium and Typhi, *Shigella flexneri, Helicobacter pylori, Campylobacter jejuni, V. cholerae, Enterococcus faecalis, S. aureus, Clostridium difficile, C. perfringens, Bacillus* species, and *Streptococcus* species. AI-2 is also produced by several species of ruminal bacteria, including *Butyrivibrio fibrisolvens, Eubacterium ruminantium, Ruminococcus flavefaciens,* and *Succinimonas amylolytica* (39). As discussed below, several important virulence factors responsible for infectious disease in the GI tract are regulated by the LuxS/AI-2 system.

QUORUM SENSING BY GASTROINTESTINAL PATHOGENS

For bacterial pathogens, QS has been proposed to aid the disease process by allowing bacteria to appropriately time the expression of virulence factors that might activate a defensive immune response before the infection has progressed. As described by de Kievit et al., through QS, "bacteria can amass a high cell density before virulence determinants are expressed, and in so doing, the bacteria are able to make a concerted attack and produce ample virulence factors to overwhelm the host defenses" (13). QS has been detected in numerous bacterial pathogens of the GI tract, and in many cases the role of QS in the pathogenesis of disease caused by these organisms is not so clear-cut. However, many examples of QS in enteric pathogens have been found that have yielded new insights into the disease process.

Enteropathogenic and Enterohemorrhagic *E. coli*

The first evidence that QS could be involved in the regulation of virulence factors of GI pathogens was found with enteropathogenic *E. coli* (EPEC), which causes nonbloody diarrhea primarily in infants in developing countries, and enterohemorrhagic *E. coli* (EHEC), which causes bloody diarrhea and hemolytic-uremic syndrome. Although an important difference between these two pathogens is the expression of Shiga toxin (Stx) by EHEC but not by EPEC, they both produce a characteristic intestinal histopathology known as attaching and effacing (AE) (28). The genes responsible for the AE phenotype are found within a pathogenicity island known as the locus of enterocyte effacement (LEE) (36). The LEE encodes a type III secretion system (T3SS) that translocates effector proteins into intestinal epithelial cells, thereby causing the dramatic cytoskeletal changes that lead to the AE lesion. Sperandio et al. (58) found that cell-free supernatants from culture media preconditioned by growth of EHEC or *E. coli* K-12 strains activated the expression of the LEE genes. When the EHEC *luxS* gene was mutated, this activation was abrogated. Because EHEC is notable for the very low infectious dose needed to cause infection (10 to 100 CFU), it is unlikely that EHEC is "talking" to itself via QS, at least in the initial stages of infection. Instead, the authors proposed a model in which the EHEC LEE genes are induced by signaling molecules produced by large numbers of commensal *E. coli* organisms and other species present in the colon. In support of this model, these authors found in a subsequent study that fecal specimens from 10 of 12 healthy individuals contained signaling molecules that were capable of inducing light in a reporter strain of *V. harveyi* BB170, which is a standard bioassay for production of AI-2 (60). Thus, it appears that EHEC uses QS to detect signaling molecules produced by large numbers of bacteria from the commensal flora to determine when it is in an appropriate environment to express genes necessary for intestinal colonization and disease.

A microarray analysis using an *E. coli* K-12 genome and mRNA or cDNA from EHEC (an EHEC array was not available at the time) found that approximately 10% of the genes in the K-12 genome were regulated fivefold by the LuxS QS system, with roughly equal numbers of genes being positively regulated and negatively regulated (58, 59). Genes involved in such fundamental properties as flagellum production, motility, and cell division were among those found to be QS regulated in this study. Other genes found to be regulated by QS included three genes whose predicted protein products were similar to regulatory proteins but for which no function had been found in K-12 or EHEC. These genes, subsequently characterized and named *qseA*, *qseB*, and *qseC*, were found to be part of a regulatory cascade in which transcription of all three was increased in response to QS signals. Increased levels of QseA in turn increased the expression of the LEE genes (57), and increased levels of QseBC, which encode a predicted two-component regulatory system, in turn increase the expression of the flagellar genes (61).

The most recent study in this series demonstrated that an EHEC *luxS* mutant, unable to produce AI-2 and unable to express the LEE-encoded T3SS at normal levels, nonetheless still produced AE lesions that were indistinguishable from those seen with the wild type on epithelial cells (60). The *luxS* mutant was still responding to a eukaryotic cell signal to activate the expression of the LEE genes. This signal was identified as the hormone epinephrine, and β- and α-adrenergic antagonists that block the effect of epinephrine can block the bacterial response to this hormone. The *luxS* mutant also responded similarly to the hormone norepinephrine. Using purified and in vitro-synthesized AI-2, it was found that AI-2 is not the AI molecule that is involved in the bacterial signaling. EHEC produces another, previously undescribed autoinducer (AI-3) whose synthesis depends on the presence of LuxS. The exact identity of AI-3 is not known, but initial purification studies show that it has a molecular mass of 213.1 Da. These results imply a potential cross-communication between the *luxS*/AI-3 bacterial QS system and the epinephrine host signaling system. The multiple signals present in the intestine, of either bacterial or host origin, could allow further fine-tuning for the expression of EHEC genes, whereby one set of genes, e.g., those encoding flagella, may be expressed at one time while another set of genes, e.g., those encoding the LEE T3SS, may be expressed at a slightly different time. Since eukaryotic cell-to-cell signaling typically occurs through hormones and bacterial cell-to-cell signaling occurs through QS, the authors speculate that QS might be a "language" by which bacteria and host cells communicate.

Although EHEC and EPEC both possess the LEE, there are some differences in QS regulation between the two pathogens (18). Unlike EHEC, EPEC contains a plasmid-encoded regulator (Per) that increases the expression of the chromosomal LEE genes. Also, EPEC produces the bundle-forming pilus (Bfp), which mediates the formation of tight microcolonies of EPEC adhering together on epithelial cells, which would allow the accumulation of locally high densities of signaling molecules. Disease due to EPEC also requires a higher infectious dose than that required for EHEC disease, and EPEC colonizes primarily the small intestine rather than the colon, the location colonized by EHEC. These differences suggest a model whereby smaller numbers of bacteria from the commensal flora in the small intestine require additional regulatory help for EPEC in the form of Per- and Bfp-mediated microcolony formation. These compensatory mechanisms then allow the EPEC LEE to be regulated by QS in the small intestine without the extremely high levels of bacteria from the commensal flora present in the colon.

Homologues of the LuxI-LuxR system have been sought in *E. coli* and although a LuxR homologue known as SdiA has been found, no obvious genes encoding LuxI homologues that could synthesize an AHL signaling molecule are present in the *E. coli* genome. Although a cloned *sdiA* gene on a multicopy plasmid can upregulate the expression of *ftsQAZ* genes, which encode proteins essential for cell division, an *sdiA* mutant has no apparent cell division defects (68). Kanamaru et al. (27) found that expression of SdiA from a high-copy-number plasmid in EHEC caused abnormal cell division, reduced adherence to cultured epithelial cells, and reduced expression of the intimin adhesin protein and the EspD protein, both of which are encoded on the LEE. However, no *sdiA* EHEC mutant was constructed and tested, and so the effects seen could be artifacts of the abnormally high expression of SdiA. Because no *E. coli* genes from either EHEC or K-12 strains have yet been demonstrated to be regulated by the single chromosomal copy of *sdiA*, Ahmer (1) recently concluded that there are no confirmed members of an SdiA regulon in this species.

Salmonella

Two QS systems have been characterized in *S. enterica* (reviewed in reference 1). The LuxR homologue, SdiA, has been characterized in *Salmonella*, but there does not appear to be a corresponding signal-generating enzyme similar to LuxI in this species. However, *Salmonella* SdiA can detect AHLs produced by a variety of bacterial species, leading to the suggestion that SdiA appears to be dedicated to detecting signals produced by other species without playing any role in autoregulation (37). These results suggest that both AHL and AI-2 can be used in interspecies communication within a mixed-species community. SdiA regulates few genes in *Salmonella*, but one gene potentially involved in resistance to human complement, *rck*, is regulated by SdiA (2). However, mutation of the *sdiA* gene had no effect on the virulence of *Salmonella* in mouse, chicken, or bovine models of disease (1).

Salmonella also produces AI-2, and the only potential virulence phenotype so far identified with a *luxS* mutant is a failure to form biofilms in an in vitro model of biofilm formation on human gallstones. LuxS does regulate the expression of an ABC transport system encoded by the *lsr* (for "LuxS-regulated") operon, which is involved in uptake and internalization of the AI-2 molecule as discussed above (63). The signaling cascade for AI-3 and epinephrine described above for EHEC is also present in *Salmonella*, and production of AI-3 by this species has been found (Sperandio et al. unpublished observations cited in reference 18).

Vibrio cholerae

In contrast to the usual paradigm of QS increasing the expression of bacterial virulence factors at high cell densities, QS in *V. cholerae* appears to act in a way that results in repression of the major virulence factors at high densities and expression at low cell densities. The major virulence factors for *V. cholerae* are cholera toxin (CT) and the toxin-coregulated pilus (TCP), both of which are regulated as part of the ToxR regulon. There appear to be three parallel QS systems that all converge at the response regulator LuxO, which is a homologue of the LuxO regulator in *V. harveyi* (22, 38). Mutation of *luxO* in *V. cholerae* results in severe intestinal colonization defects (75). System 1 has homologues to the *V. harveyi* QS system 1, which is a homoserine lactone AI system. The AI synthase for *V. cholerae* system 1 is called CqsA, and the sensor for this system is CqsS, a homologue of *V. harveyi* LuxN. The AI for this system is called CAI-1. The second QS system in *V. cholerae* is the LuxS/AI-2 system, which uses LuxQ and LuxP as sensors of AI-2. Genetic evidence suggests the existence of a third system, whose components have not yet been identified. All three systems involve a LuxR homologue called HapR that serves as a repressor of virulence genes and biofilm formation and as an activator of the Hap protease. The LuxO regulator is activated by phosphorylation and in turn activates a putative repressor (X) that represses the expression of *hapR*. As if this were not sufficiently complicated, some *V. cholerae* strains contain a frameshift in *hapR,* thereby resulting in an inactive HapR protein, yet these strains do not appear to be attenuated in human or animal infections.

Details of how QS in *V. cholerae* responds to cell density can be found in the original papers (22, 30, 38, 74, 75), but a broad outline is as follows. At low cell densities, the AI concentrations are low and phosphate is transferred from the unoccupied sensors to the LuxU protein, which in turn transfers phosphate to LuxO. The phosphorylated LuxO activates repressor X, which represses *hapR*. The low level of HapR allows the expression of CT and TCP as well as *vps* genes involved in biofilm formation but does not allow the expression of the Hap protease. At high cell densities, the flow of phosphate reverses, resulting in inactivation of LuxO and hence leading to expression of HapR. HapR then represses CT, TCP, and biofilm formation and activates the expression of the Hap protease. Hammer and Bassler (22) have proposed a model in which TCP and CT are expressed early in the infection at low densities, along with biofilm, which helps intestinal colonization. Later in the infection, when *V. cholerae* is at high densities, biofilm formation ceases and Hap protease production increases, allowing the organism to exit the host and adapt to a environmental reservoir where expression of CT and TCP is not necessary (see also chapter 18). Given the strong conservation of QS systems between *V. cholerae* and the marine organism *V. harveyi,* it is quite plausible that the QS systems would have some common functions in allowing the organism to survive in an environmental reservoir.

Enteroaggregative *E. coli*

Enteroaggregative *E. coli* (EAEC) is an increasingly recognized cause of diarrhea that is often persistent in children and adults in both developing and developed countries. Evidence for QS in the regulation of virulence genes of EAEC was recently discovered by using a continuous-flow anaerobic fecal culture system. Using this simulated model of the colonic environment, Ruiz-Perez et al. (46) found that the presence of fecal commensal bacteria increased the expression of *aggR,* which encodes a global transcriptional regulator of EAEC virulence genes (see chapter 18). By coculturing EAEC with individual strains of typical commensal bacteria, these investigators found that one or more substances produced by strains of *Enterococcus* and *Clostridium* increased the expression of *aggR* while strains of *Lactobacillus* and *Veillonella* downregulated the expression of this gene. Although the specific QS systems responsible for modulating the expression of *aggR* by these commensal species have not been identified, these results clearly show that EAEC is responding to signaling compounds produced by the colonic flora and that these signals are modulating the expression of a crucial global regulator of virulence.

Enterococcus faecalis

Enterococcus species are normal inhabitants of the human and other mammalian GI tracts, but they are also important causes of nosocomial infections such as surgical-site infections, bloodstream infections, and urinary tract infections. A major virulence factor of enterococci is a cytolysin that contributes to the pathogenesis of a variety of infections caused by *E. faecalis* (10). Not only is this cytolysin lethal for a broad range of eukaryotic cells, but also it is toxic to a number of gram-positive bacteria and serves as an autoinducer for QS induction of the cytolysin operon. The cytolysin, which can be encoded on a plasmid or in the chromosome, is unique among bacterial hemolytic toxins in having both bacteriocin and hemolytic activities in a single system. The cytolysin is made up of two subunits, $\mathrm{CylL_L}$ and $\mathrm{CylL_S}$, that are posttranslationally modified by other proteins encoded in the eight-open-reading-frame *cyl* operon to produce the active extracellular forms denoted as $\mathrm{CylL_L}''$ and $\mathrm{CylL_S}''$. Haas et al. (21) demonstrated that expression of the cytolysin is autoregulated by the presence of a threshold concentration of the mature extracellular $\mathrm{CylL_S}''$ subunit. In this case, the cytolysin subunit is acting as the autoinducer that activates transcription from the *cyl* promoter. Recent analysis of conditions for expression of the toxin have revealed that cytolysin expression is increased under anaerobic conditions (11). Anaerobic conditions are a major environmental signal in the GI tract, and so sorting out the direct regulation of enterococcal genes via anaerobiasis versus indirect regulation by anaerobiasis through QS will be a complicated endeavor.

Not only does the *E. faecalis* cytolysin serve as an autoinducer of enterococcal gene expression and as a lethal toxin active against a variety of eukaryotic cells, but also it is a bacteriocin and hence can kill other bacterial cells in the same environment. Interestingly, a recent study (42) examined 139 healthy subjects for intestinal colonization by *C. difficile,* an organism that is notorious for causing diarrhea and colitis in hospitalized individuals who have had their normal intestinal flora disrupted by treatment with broad-spectrum antibiotics. This study found that many healthy individuals without a recent history of diarrhea or antibiotic usage were persistently colonized by *C. difficile* and that the number of fecal enterococci was significantly larger in these individuals than in those who were not colonized with *C. difficile*. It is tempting to speculate that the cytolysin of *E. faecalis,* which has both AI and bacterocin activity, is responsible for the change in the intestinal flora that allows *C. difficile* to colonize these healthy individuals, but this hypothesis remains to be proven.

Another QS system in *Enterococcus* species, called Fsr (for "*E. faecalis* regulator"), is homologous in many respects to the Agr system of *S. aureus*. Enterococci lack a homologue of the staphylococcal AgrD but possess homologues of AgrABC. The lack of an AgrD homologue is consistent with the exquisite specificity seen with AIP QS systems for each species and subspecies. The Fsr QS system in *E. faecalis* activates *gelE,* encoding gelatinase, and *sprE,* encoding a serine protease, which are two virulence factors that are important in both the invertebrate *Caenorhabditis elegans* and the mammalian mouse models of infection (45, 51, 52). Finally, *E. faecalis* also contains a *luxS* homologue (48, 49), although the role of this QS system in virulence has not been investigated for this species.

Yersinia Species

Two *Yersinia* species, *Y. enterocolitica* and *Y. pseudotuberculosis,* can cause diarrhea in humans. A pair of LuxR-LuxI homologues, called YenR and YenI, were first found in *Y. enterocolitica* (65). Production of two AIs, 3-oxo-C_6-HSL and C_6-HSL, was attributed to this locus, but a specific phenotype controlled by this locus could not be established. Subsequently, *Y. pseudotuberculosis* was shown to contain two pairs of LuxIR homologues that control motility and clumping (3). The LuxIR pairs in *Y. pseudotuberculosis* are YpsI-YpsR and YtbI-YtbR, and three AI molecules, C_6-HSL, 3-oxo-C_6-HSL, and C_8-HSL, are produced from the two autoinducer synthases. YpsI is responsible for 3-oxo-C_6-HSL, and YtbI is responsible for C_8-HSL, while both YpsI and YtbI can synthesize C_6-HSL. Temperature appears to play a pivotal role in determining which autoinducer synthase is active, and various combinations of AI production were seen with *ypsI* or *ytbI* mutants at 22, 28, and 37°C. A mutation in *ypsR* results in overexpression of a major flagellin subunit and increased motility. The YpsIR and YtbIR systems appear to comprise a hierarchical QS cascade in which YpsR can help regulate YtbIR. Temperature plays an important role in the regulation of this cascade, which is particularly interesting in light of the temperature regulation of many *Yersinia* virulence genes.

Clostridium perfringens

Clostridium perfringens causes gas gangrene and is also capable of causing food-borne illness. Ohtani et al. (41) demonstrated that the *luxS*-mediated AI-2 QS system enhances the extracellular expression of alpha-, kappa-, and theta-toxins in a mechanism that appears to involve both transcriptional and posttranscriptional mechanisms. More than 20 years ago, *C. perfringens* was reported to produce an extracellular signaling molecule called substance A, which activated theta-toxin expression, but this appears to be an additional, as yet uncharacterized autoinducer that is distinct from AI-2 (41).

Shigella flexneri

Shigella species, the primary agents of bacillary dysentery, have a very low infectious dose and possess a T3SS that is essential for virulence. These two characteristics are shared with EHEC, and so Day and Maurelli (12) investigated QS in *S. flexneri* to determine if T3SS in this species was regulated by the LuxS/AI-2 system. They found that expression of the *ipa, mxi,* and *spa* loci of the T3SS that are responsible for invasion of host cells was enhanced by conditioned medium derived from stationary-phase cultures, suggesting the presence of an AI molecule in the medium. The AI-2 molecule was detected in culture supernatants, and mutation of the *luxS* locus resulted in decreased expression of VirB, a transcription factor that is essential for expression of these invasion loci. However, mutation of *luxS* did not affect invasion operon expression, and a *luxS* mutant was fully virulent in the tissue culture and guinea pig keratoconjunctivitis invasion assays used. The authors noted that in contrast to EPEC and EHEC, which persist in the intestinal lumen and are continuously exposed to high levels of AI-2 from members of the normal flora, *Shigella* efficiently invades host cells and is therefore likely to be exposed to luminal AI-2 for only a short time.

Campylobacter jejuni

Two studies have examined the role of *luxS* in *C. jejuni,* which has been shown in several studies to be the most frequently isolated bacterial agent of food-borne disease. A *luxS* mutant showed comparable growth rate, resistance to oxidative stress, and ability to invade Caco-2 cell monolayers relative to the parent strain but also showed decreased motility in semisolid medium (16). The effect on motility was confirmed by another group of investigators, who also showed that mutation of *luxS* reduced the transcription of *flaA*, the major flagellin gene in this species (26). These investigators also found reduced agglutination capability in a *luxS* mutant, suggesting that QS might be involved in the formation of surface structures in *C. jejuni*.

Vibrio vulnificus

V. vulnificus is the most frequent bacterial cause of death due to ingestion of seafood. Primary septicemia after ingestion of this species is particularly lethal for individuals with underlying hepatic diseases. Kim et al. (29) found that a *luxS V. vulnificus* mutant was significantly increased in 50% lethal dose and time required for death in a mouse model. Cytotoxicity for HeLa cells was also significantly decreased by the mutation. Mutation of *luxS* caused decreased protease activity and increased hemolysin activity, effects which were reversed by complementation with the wild-type *luxS* gene. These investigators found that *V. vulnificus* produced the AI-2 molecule but found no evidence for an AHL molecule similar to that produced by *V. harveyi*. However, a LuxR homologue was previously identified in this species by other investigators and named SmcR (50). Similar to a *luxS* mutant, a *smcR* mutant showed decreased protease and increased hemolytic activity; however, notably, virulence of the *smcR* mutant in mice was comparable to that of wild type, indicating that SmcR is not required for virulence in this model. The similar effects on protease and hemolysin suggest that an interaction or hierarchy involving these two QS systems is present in this species.

Vibrio parahaemolyticus

Although it is not as lethal as *V. vulnificus*, *V. parahaemolyticus* is a more frequently isolated bacterial agent of gastroenteritis due to ingestion of contaminated seafood. The genome sequence of *V. parahaemolyticus* revealed a T3SS, although the exact role of this system in the disease is not yet known. Henke and Bassler (23) recently showed that this system, as well as a similar system in *V. harveyi*, is regulated by QS. *V. parahaemolyticus* possesses all of the Lux regulators present in *V. harveyi* that comprise system 1 (LuxM and LuxR) and system 2 (LuxS). Mutation of the LuxR homologue of *V. parahaemolyticus*, called OpaR, had a striking effect on the T3SS. However, in contrast to the positive regulation by QS of the T3SS seen with EPEC and EHEC, QS acts to repress the T3SS in *V. parahaemolyticus*. These results suggest a scenario wherein *V. parahaemolyticus* secretes effector proteins at low cell density and terminates secretion at a high cell density. This scenario is similar to that seen with *V. cholerae* and may also play a role in preparing the transition of the vibrios from the intestine to the aquatic environment.

ADDITIONAL ASPECTS OF QUORUM SENSING IN THE INTESTINE

Prokaryotic-Eukaryotic Communication

The GI ecosystem has been called a "precarious alliance among epithelium, immunity, and microbiota" (35). Not only does QS affect the microbiota of the GI tract, but it may also affect the eukaryotic aspects of this alliance, specifically the epithelium and immunity. There are several examples of prokaryotic-eukaryotic communication in which bacterial signals can modulate the expression of eukaryotic genes or vice versa in cross-kingdom communication. One of the AIs of *P. aeruginosa* (3-oxo-C_{12}-HSL) has immunomodulatory activity and can downregulate tumor necrosis factor alpha and interleukin-12 production in leukocytes (64) and upregulate expression of the proinflammatory cytokine gamma interferon (55). This molecule also stimulates interleukin-8 production in human lung fibroblasts and epithelial cells through a mechanism involving the eukaryotic transcription factors NF-κB and activator protein 2 (54). Yet another example is the down-regulation of the nucleotide receptors P2Y2 and P2Y4 in tracheal gland serous cells of cystic fibrosis patients by this and other AHL signaling molecules (47). Although not yet linked to modulating the transcription of eukaryotic genes, the cytolysin of *E. faecalis*, which serves as an autoinducer of expression of the enterococcal *cyl* operon, also has toxic effects on erythrocytes, retinal tissues, intestinal epithelial cells, neutrophils, and macrophages (10).

Bacterial QS can even play a role in the development of normal host tissue. A symbiotic relationship between the squid *Euprymna scolopes* and *V. fischeri* has been well characterized by Visick et al., who showed that development of the normal crypt epithelium of the squid light organ depends on colonization with *V. fischeri* (67). Interestingly, isogenic *luxI* or *luxR* mutants were incapable of stimulating the normal crypt epithelium development seen with the QS-positive parent strain. Given the importance of the commensal flora in the normal development of the mammalian intestine (see chapter 13), it would not be surprising if QS among commensal bacteria is also important in mammalian intestinal development.

The converse situation, i.e., eukaryotic molecules affecting prokaryotic QS, has also been reported. As described above, the hormones epinephrine and norepinephrine can activate the expression of the LuxS/AI-3 bacterial QS system in EPEC and EHEC, resulting in transcription of the T3SS, which mediates the attaching and effacing intestinal histopathology that is characteristic of these pathogens. Chun et al. (9) recently reported that human airway epithelial cells produce a substance, as yet unidentified, that inactivates the 3-oxo-C_{12}-HSL signaling molecule produced by *P. aeruginosa.* In yet another kingdom, the red alga *Delisea pulchra* produces halogenated furanones that inhibit QS mechanisms of the plant pathogen *Erwinia carotovora* (32). No doubt other examples of eukaryotic-prokaryotic communication involving QS systems will be discovered as additional hosts, body niches, and signaling molecules are examined.

QS In Vivo

Evidence that QS actually occurs at mucosal surfaces has been gathered from studies of both the respiratory and GI tracts. Several signaling molecules of *P. aeruginosa,* including the two major AHL signaling molecules as well as AI-2 (15, 53), have been detected in sputum specimens from cystic fibrosis patients infected with this organism. Sperandio et al. (60) detected the presence of AI-2 in fecal specimens from the GI tracts of 10 of 12 healthy individuals examined. This study examined specimens from a continuous-flow anaerobic fecal culture system, which was inoculated with a fecal specimen from a healthy subject, using the *LEE1::lacZ* for detection of AI-3 and also found signaling activity in this simulated intestinal environment. As noted above, this continuous-flow culture system also had signaling activity that activated the expression of the major regulator of virulence gene expression in EAEC (46). There is also a report of multiple AHL AI molecules in the rumen contents of six of eight cattle examined, although, interestingly, no pure cultures of bacteria isolated from the rumen contents had AHL activity; only the rumen contents directly obtained from the cattle showed activity (17).

In the study of AI-2 activity in human fecal specimens, the concentrations of this signaling activity varied among the 10 subjects by as much as 1 log unit (60). This result suggests that different levels of signaling molecules in the intestines of different individuals may lead to different levels of QS activity and transcription of QS-regulated genes in the intestine. One speculation that arises from these results is that the course of disease may vary among different individuals as a result of variable levels of QS activity in their intestines.

QS-Based Therapy

The discovery of QS in human pathogens has led to considerable interest in developing new therapeutic interventions to interfere with these signaling molecules. Thus, instead of using an antibiotic to kill pathogenic bacteria, a compound that interferes with the QS mechanism would be used to repress the expression of the virulence genes responsible for the disease. Such an approach is particularly promising in light of increasing resistance to conventional antimicrobial agents, and several biotechnology companies have been formed to pursue this avenue of research. The topic of QS providing new targets for novel antibacterial drugs has been covered in several recent reviews (7, 24, 56, 69), and only a few brief points are discussed here.

There are at least three major strategies for the development of drugs that interrupt bacterial QS (25): (i) inhibition of QS signal synthesis, (ii) destruction or degradation of the signal, and (iii) inhibition of signal reception. As an example of the first approach, Parsek et al. used analogs of SAM to inhibit the synthesis of AHL by the *P. aeruginosa* RhlI protein (44). Several examples can be found for the second approach since several naturally occurring compounds have been found that can degrade signal molecules. One enzyme that catalyzes the hydrolysis of AHL signals, called AiiA, is produced by *Bacillus* species (14). Expression of AiiA in the plant pathogen *Erwinia carotovora* resulted in reduced levels of AHL molecules and attenuated pathogenicity in a variety of plant species. The application of the third approach, inhibition of signal reception, has yielded several promising results in animal models of disease due to human pathogens. Henzter et al. (25) developed a synthetic derivative of natural furanone compounds that can act as a potent antagonist of bacterial QS in *P. aeruginosa.* In a mouse pulmonary model of infection, addition of this compound 2 days after inoculation with *P. aeruginosa* resulted in the downregulation of bacterial genes regulated by QS in the bacteria present in the mouse. Application of this compound also resulted in substantial clearance of the organism from the lungs and rendered *P. aeruginosa* present in biofilms significantly more susceptible to tobramycin, an antibiotic routinely used to treat cystic fibrosis patients. Studies by other investigators have resulted in the development of compounds that inhibit QS-mediated virulence gene expression in *S. aureus,* thereby protecting mice against *S. aureus* infection (34).

The development of anti-QS therapy has been directed primarily toward nonintestinal pathogens, which do not have to deal with the huge numbers of commensal organisms present in the GI tract. The presence of this complex microbial flora and the variety of signaling molecules that might be produced by the microbial flora or even by the host itself (see above) greatly complicates the application of this approach to GI pathogens.

CONCLUSIONS

A fundamental property of QS is that the greater the density of bacteria, the greater the density of signaling molecules and the greater the opportunity for cell-to-cell communication. There is no environment in the human body with a greater density of bacteria and a greater potential for cell-to-cell signaling than the GI tract. The first report of QS regulating the virulence factors of GI pathogens appeared a mere 5 years ago, and numerous reports have since appeared documenting QS mechanisms in a variety of enteric pathogens. There is clearly much more to be learned about QS in the GI tract, but with our increased understanding of this phenomenon, new insights will emerge regarding the virulence mechanisms of enteric pathogens as well as the development and maintenance of our commensal intestinal microbial flora.

REFERENCES

1. **Ahmer, B. M.** 2004. Cell-to-cell signalling in *Escherichia coli* and *Salmonella enterica*. *Mol. Microbiol.* **52:**933–945.
2. **Ahmer, B. M., J. van Reeuwijk, C. D. Timmers, P. J. Valentine, and F. Heffron.** 1998. *Salmonella typhimurium* encodes an SdiA homolog, a putative quorum sensor of the LuxR family, that regulates genes on the virulence plasmid. *J. Bacteriol.* **180:**1185–1193.
3. **Atkinson, S., J. P. Throup, G. S. Stewart, and P. Williams.** 1999. A hierarchical quorum-sensing system in *Yersinia pseudotuberculosis* is involved in the regulation of motility and clumping. *Mol. Microbiol.* **33:**1267–1277.
4. **Bassler, B. L.** 2002. Small talk. Cell-to-cell communication in bacteria. *Cell* **109:**421–424.
5. **Berg, R. D.** 1996. The indigenous gastrointestinal microflora. *Trends Microbiol.* **4:**430–435.
6. **Borriello, S. P.** 2002. The normal flora of the gastrointestinal tract, p. 1–24. *In* A. L. Hart, A. J. Stagg, H. Graffner, H. Glise, P. G. Falk, and M. A. Kamm (ed.), *Gut Ecology.* Martin Dunitz, Ltd., London, United Kingdom.
7. **Camara, M., P. Williams, and A. Hardman.** 2002. Controlling infection by tuning in and turning down the volume of bacterial small-talk. *Lancet Infect. Dis.* **2:**667–676.
8. **Chen, X., S. Schauder, N. Potier, A. Van Dorsselaer, I. Pelczer, B. L. Bassler, and F. M. Hughson.** 2002. Structural identification of a bacterial quorum-sensing signal containing boron. *Nature* **415:**545–549.
9. **Chun, C. K., E. A. Ozer, M. J. Welsh, J. Zabner, and E. P. Greenberg.** 2004. Inactivation of a *Pseudomonas aeruginosa* quorum-sensing signal by human airway epithelia. *Proc. Natl. Acad. Sci. USA* **101:**3587–3590.
10. **Coburn, P. S., and M. S. Gilmore.** 2003. The *Enterococcus faecalis* cytolysin: a novel toxin active against eukaryotic and prokaryotic cells. *Cell. Microbiol.* **5:**661–669.
11. **Day, A. M., J. H. Cove, and M. K. Phillips-Jones.** 2003. Cytolysin gene expression in *Enterococcus faecalis* is regulated in response to aerobiosis conditions. *Mol. Genet. Genomics* **269:**31–39.
12. **Day, W. A., Jr., and A. T. Maurelli.** 2001. *Shigella flexneri* LuxS quorum-sensing system modulates *virB* expression but is not essential for virulence. *Infect. Immun.* **69:**15–23.
13. **De Kievit, T. R., R. Gillis, S. Marx, C. Brown, and B. H. Iglewski.** 2001. Quorum-sensing genes in *Pseudomonas aeruginosa* biofilms: their role and expression patterns. *Appl. Environ. Microbiol.* **67:**1865–1873.
14. **Dong, Y. H., J. L. Xu, X. Z. Li, and L. H. Zhang.** 2000. AiiA, an enzyme that inactivates the acylhomoserine lactone quorum-sensing signal and attenuates the virulence of *Erwinia carotovora*. *Proc. Natl. Acad. Sci. USA* **97:**3526–3531.
15. **Duan, K., C. Dammel, J. Stein, H. Rabin, and M. G. Surette.** 2003. Modulation of *Pseudomonas aeruginosa* gene expression by host microflora through interspecies communication. *Mol. Microbiol.* **50:**1477–1491.
16. **Elvers, K. T., and S. F. Park.** 2002. Quorum sensing in *Campylobacter jejuni:* detection of a *luxS* encoded signalling molecule. *Microbiology* **148:**1475–1481.
17. **Erickson, D. L., V. L. Nsereko, D. P. Morgavi, L. B. Selinger, L. M. Rode, and K. A. Beauchemin.** 2002. Evidence of quorum sensing in the rumen ecosystem: detection of *N*-acyl homoserine lactone autoinducers in ruminal contents. *Can. J. Microbiol.* **48:**374–378.
18. **Falcao, J. P., F. Sharp, and V. Sperandio.** 2004. Cell-to-cell signaling in intestinal pathogens. *Curr. Issues Intest. Microbiol.* **5:**9–17.
19. **Fuqua, C., and E. P. Greenberg.** 2002. Listening in on bacteria: acyl-homoserine lactone signalling. *Nat. Rev. Mol. Cell Biol.* **3:**685–695.
20. **Fuqua, C., M. R. Parsek, and E. P. Greenberg.** 2001. Regulation of gene expression by cell-to-cell communication: acyl-homoserine lactone quorum sensing. *Annu. Rev. Genet.* **35:**439–468.
21. **Haas, W., B. D. Shepard, and M. S. Gilmore.** 2002. Two-component regulator of *Enterococcus faecalis* cytolysin responds to quorum-sensing autoinduction. *Nature* **415:**84–87.
22. **Hammer, B. K., and B. L. Bassler.** 2003. Quorum sensing controls biofilm formation in *Vibrio cholerae*. *Mol. Microbiol.* **50:**101–104.
23. **Henke, J. M., and B. L. Bassler.** 2004. Quorum sensing regulates type III secretion in *Vibrio harveyi* and *Vibrio parahaemolyticus*. *J. Bacteriol.* **186:**3794–3805.
24. **Hentzer, M., and M. Givskov.** 2003. Pharmacological inhibition of quorum sensing for the treatment of chronic bacterial infections. *J. Clin. Investig.* **112:**1300–1307.
25. **Hentzer, M., H. Wu, J. B. Andersen, K. Riedel, T. B. Rasmussen, N. Bagge, N. Kumar, M. A. Schembri, Z. Song, P. Kristoffersen, M. Manefield, J. W. Costerton, S. Molin, L. Eberl, P. Steinberg, S. Kjelleberg, N. Hoiby, and M. Givskov.** 2003. Attenuation of *Pseudomonas aeruginosa* virulence by quorum sensing inhibitors. *EMBO J.* **22:**3803–3815.
26. **Jeon, B., K. Itoh, N. Misawa, and S. Ryu.** 2003. Effects of quorum sensing on *flaA* transcription and autoagglutination in *Campylobacter jejuni*. *Microbiol. Immunol.* **47:**833–839.
27. **Kanamaru, K., I. Tatsuno, T. Tobe, and C. Sasakawa.** 2000. SdiA, an *Escherichia coli* homologue of quorum-sensing regulators, controls the expression of virulence factors in enterohaemorrhagic *Escherichia coli* O157:H7. *Mol. Microbiol.* **38:**805–816.

28. **Kaper, J. B., and A. D. O'Brien (ed.).** 1998. Escherichia coli *and Other Shiga-Toxin Producing* E. coli *Strains*. ASM Press, Washington, D.C.
29. **Kim, S. Y., S. E. Lee, Y. R. Kim, C. M. Kim, P. Y. Ryu, H. E. Choy, S. S. Chung, and J. H. Rhee.** 2003. Regulation of *Vibrio vulnificus* virulence by the LuxS quorum-sensing system. *Mol. Microbiol.* **48:**1647–1664.
30. **Kovacikova, G., and K. Skorupski.** 2002. Regulation of virulence gene expression in *Vibrio cholerae* by quorum sensing: HapR functions at the *aphA* promoter. *Mol. Microbiol.* **46:**1135–1147.
31. **Macfarlane, G. T., and S. Macfarlane.** 1997. Human colonic microbiota: ecology, physiology and metabolic potential of intestinal bacteria. *Scand. J. Gastroenterol. Suppl.* **222:**3–9.
32. **Manefield, M., M. Welch, M. Givskov, G. P. Salmond, and S. Kjelleberg.** 2001. Halogenated furanones from the red alga, *Delisea pulchra,* inhibit carbapenem antibiotic synthesis and exoenzyme virulence factor production in the phytopathogen *Erwinia carotovora. FEMS Microbiol. Lett.* **205:**131–138.
33. **Manna, A. C., and A. L. Cheung.** 2003. *sarU,* a *sarA* homolog, is repressed by SarT and regulates virulence genes in *Staphylococcus aureus. Infect. Immun.* **71:**343–353.
34. **Mayville, P., G. Ji, R. Beavis, H. Yang, M. Goger, R. P. Novick, and T. W. Muir.** 1999. Structure-activity analysis of synthetic autoinducing thiolactone peptides from *Staphylococcus aureus* responsible for virulence. *Proc. Natl. Acad. Sci. USA* **96:**1218–1223.
35. **McCracken, V. J., and R. G. Lorenz.** 2001. The gastrointestinal ecosystem: a precarious alliance among epithelium, immunity and microbiota. *Cell. Microbiol.* **3:**1–11.
36. **McDaniel, T. K., K. G. Jarvis, M. S. Donnenberg, and J. B. Kaper.** 1995. A genetic locus of enterocyte effacement conserved among diverse enterobacterial pathogens. *Proc. Natl. Acad. Sci. USA* **92:**1664–1668.
37. **Michael, B., J. N. Smith, S. Swift, F. Heffron, and B. M. Ahmer.** 2001. SdiA of *Salmonella enterica* is a LuxR homolog that detects mixed microbial communities. *J. Bacteriol.* **183:**5733–5742.
38. **Miller, M. B., K. Skorupski, D. H. Lenz, R. K. Taylor, and B. L. Bassler.** 2002. Parallel quorum sensing systems converge to regulate virulence in *Vibrio cholerae. Cell* **110:**303–314.
39. **Mitsumori, M., L. Xu, H. Kajikawa, M. Kurihara, K. Tajima, J. Hai, and A. Takenaka.** 2003. Possible quorum sensing in the rumen microbial community: detection of quorum-sensing signal molecules from rumen bacteria. *FEMS Microbiol. Lett.* **219:**47–52.
40. **Novick, R. P.** 2003. Autoinduction and signal transduction in the regulation of staphylococcal virulence. *Mol. Microbiol.* **48:**1429–1449.
41. **Ohtani, K., H. Hayashi, and T. Shimizu.** 2002. The *luxS* gene is involved in cell-cell signalling for toxin production in *Clostridium perfringens. Mol. Microbiol.* **44:**171–179.
42. **Ozaki, E., H. Kato, H. Kita, T. Karasawa, T. Maegawa, Y. Koino, K. Matsumoto, T. Takada, K. Nomoto, R. Tanaka, and S. Nakamura.** 2004. *Clostridium difficile* colonization in healthy adults: transient colonization and correlation with enterococcal colonization. *J. Med. Microbiol.* **53:**167–172.
43. **Parsek, M. R., and E. P. Greenberg.** 2000. Acyl-homoserine lactone quorum sensing in gram-negative bacteria: a signaling mechanism involved in associations with higher organisms. *Proc. Natl. Acad. Sci. USA* **97:**8789–8793.
44. **Parsek, M. R., D. L. Val, B. L. Hanzelka, J. E. Cronan, Jr., and E. P. Greenberg.** 1999. Acyl homoserine-lactone quorum-sensing signal generation. *Proc. Natl. Acad. Sci. USA* **96:** 4360–4365.
45. **Qin, X., K. V. Singh, G. M. Weinstock, and B. E. Murray.** 2000. Effects of *Enterococcus faecalis fsr* genes on production of gelatinase and a serine protease and virulence. *Infect. Immun.* **68:**2579–2586.
46. **Ruiz-Perez, F., J. Sheikh, S. Davis, E. C. Boedeker, and J. P. Nataro.** 2004. Use of a continuous-flow anaerobic culture to characterize enteric virulence gene expression. *Infect. Immun.* **72:**3793–3802.
47. **Saleh, A., C. Figarella, W. Kammouni, S. Marchand-Pinatel, A. Lazdunski, A. Tubul, P. Brun, and M. D. Merten.** 1999. *Pseudomonas aeruginosa* quorum-sensing signal molecule *N*-(3-oxododecanoyl)-L-homoserine lactone inhibits expression of P2Y receptors in cystic fibrosis tracheal gland cells. *Infect. Immun.* **67:**5076–5082.
48. **Schauder, S., and B. L. Bassler.** 2001. The languages of bacteria. *Genes Dev.* **15:**1468–1480.
49. **Schauder, S., K. Shokat, M. G. Surette, and B. L. Bassler.** 2001. The LuxS family of bacterial autoinducers: biosynthesis of a novel quorum-sensing signal molecule. *Mol. Microbiol.* **41:**463–476.
50. **Shao, C. P., and L. I. Hor.** 2001. Regulation of metalloprotease gene expression in *Vibrio vulnificus* by a *Vibrio harveyi* LuxR homologue. *J. Bacteriol.* **183:**1369–1375.
51. **Sifri, C. D., E. Mylonakis, K. V. Singh, X. Qin, D. A. Garsin, B. E. Murray, F. M. Ausubel, and S. B. Calderwood.** 2002. Virulence effect of *Enterococcus faecalis* protease genes and the quorum-sensing locus fsr in *Caenorhabditis elegans* and mice. *Infect. Immun.* **70:**5647–5650.
52. **Singh, K. V., X. Qin, G. M. Weinstock, and B. E. Murray.** 1998. Generation and testing of mutants of *Enterococcus faecalis* in a mouse peritonitis model. *J. Infect. Dis.* **178:**1416–1420.
53. **Singh, P. K., A. L. Schaefer, M. R. Parsek, T. O. Moninger, M. J. Welsh, and E. P. Greenberg.** 2000. Quorum-sensing signals indicate that cystic fibrosis lungs are infected with bacterial biofilms. *Nature* **407:**762–764.
54. **Smith, R. S., E. R. Fedyk, T. A. Springer, N. Mukaida, B. H. Iglewski, and R. P. Phipps.** 2001. IL-8 production in human lung fibroblasts and epithelial cells activated by the *Pseudomonas* autoinducer *N*-3-oxododecanoyl homoserine lactone is transcriptionally regulated by NF-kappa B and activator protein-2. *J. Immunol.* **167:**366–374.
55. **Smith, R. S., S. G. Harris, R. Phipps, and B. Iglewski.** 2002. The *Pseudomonas aeruginosa* quorum-sensing molecule *N*-(3-oxododecanoyl)homoserine lactone contributes to virulence and induces inflammation in vivo. *J. Bacteriol.* **184:**1132–1139.
56. **Smith, R. S., and B. H. Iglewski.** 2003. *Pseudomonas aeruginosa* quorum sensing as a potential antimicrobial target. *J. Clin. Investig.* **112:**1460–1465.
57. **Sperandio, V., C. C. Li, and J. B. Kaper.** 2002. Quorum-sensing *Escherichia coli* regulator A: a regulator of the LysR family involved in the regulation of the locus of enterocyte effacement pathogenicity island in enterohemorrhagic *E. coli. Infect. Immun.* **70:**3085–3093.
58. **Sperandio, V., J. L. Mellies, W. Nguyen, S. Shin, and J. B. Kaper.** 1999. Quorum sensing controls expression of the type III secretion gene transcription and protein secretion in enterohemorrhagic and enteropathogenic *Escherichia coli. Proc. Natl. Acad. Sci. USA* **96:**15196–15201.
59. **Sperandio, V., A. G. Torres, J. A. Giron, and J. B. Kaper.** 2001. Quorum sensing is a global regulatory mechanism in enterohemorrhagic *Escherichia coli* O157:H7. *J. Bacteriol.* **183:** 5187–5197.
60. **Sperandio, V., A. G. Torres, B. Jarvis, J. P. Nataro, and J. B. Kaper.** 2003. Bacteria-host communication: the language of hormones. *Proc. Natl. Acad. Sci. USA* **100:**8951–8956.

61. **Sperandio, V., A. G. Torres, and J. B. Kaper.** 2002. Quorum sensing *Escherichia coli* regulators B and C (QseBC): a novel two-component regulatory system involved in the regulation of flagella and motility by quorum sensing in *E. coli. Mol. Microbiol.* **43:**809–821.

62. **Taga, M. E., and B. L. Bassler.** 2003. Chemical communication among bacteria. *Proc. Natl. Acad. Sci. USA* **100**(Suppl. 2)**:**14549–14554.

63. **Taga, M. E., S. T. Miller, and B. L. Bassler.** 2003. Lsr-mediated transport and processing of AI-2 in *Salmonella typhimurium. Mol. Microbiol.* **50:**1411–1427.

64. **Telford, G., D. Wheeler, P. Williams, P. T. Tomkins, P. Appleby, H. Sewell, G. S. Stewart, B. W. Bycroft, and D. I. Pritchard.** 1998. The *Pseudomonas aeruginosa* quorum-sensing signal molecule *N*-(3-oxododecanoyl)-L-homoserine lactone has immunomodulatory activity. *Infect. Immun.* **66:** 36–42.

65. **Throup, J. P., M. Camara, G. S. Briggs, M. K. Winson, S. R. Chhabra, B. W. Bycroft, P. Williams, and G. S. Stewart.** 1995. Characterisation of the yenI/yenR locus from *Yersinia enterocolitica* mediating the synthesis of two *N*-acylhomoserine lactone signal molecules. *Mol. Microbiol.* **17:**345–356.

66. **Venter, J. C., K. Remington, J. F. Heidelberg, A. L. Halpern, D. Rusch, J. A. Eisen, D. Wu, I. Paulsen, K. E. Nelson, W. Nelson, D. E. Fouts, S. Levy, A. H. Knap, M. W. Lomas, K. Nealson, O. White, J. Peterson, J. Hoffman, R. Parsons, H. Baden-Tillson, C. Pfannkoch, Y. H. Rogers, and H. O. Smith.** 2004. Environmental genome shotgun sequencing of the Sargasso Sea. *Science* **304:**66–74.

67. **Visick, K. L., J. Foster, J. Doino, M. McFall-Ngai, and E. G. Ruby.** 2000. *Vibrio fischeri lux* genes play an important role in colonization and development of the host light organ. *J. Bacteriol.* **182:**4578–4586.

68. **Wang, X. D., P. A. de Boer, and L. I. Rothfield.** 1991. A factor that positively regulates cell division by activating transcription of the major cluster of essential cell division genes of *Escherichia coli. EMBO J.* **10:**3363–3372.

69. **Williams, P.** 2002. Quorum sensing: an emerging target for antibacterial chemotherapy? *Expert Opin. Ther. Targets* **6:** 257–274.

70. **Winzer, K., K. R. Hardie, and P. Williams.** 2002. Bacterial cell-to-cell communication: sorry, can't talk now—gone to lunch! *Curr. Opin. Microbiol.* **5:**216–222.

71. **Xavier, K. B., and B. L. Bassler.** 2003. LuxS quorum sensing: more than just a numbers game. *Curr. Opin. Microbiol.* **6:**191–197.

72. **Xu, J., M. K. Bjursell, J. Himrod, S. Deng, L. K. Carmichael, H. C. Chiang, L. V. Hooper, and J. I. Gordon.** 2003. A genomic view of the human-*Bacteroides thetaiotaomicron* symbiosis. *Science* **299:**2074–2076.

73. **Yarwood, J. M., and P. M. Schlievert.** 2003. Quorum sensing in *Staphylococcus* infections. *J. Clin. Investig.* **112:**1620–1625.

74. **Zhu, J., and J. J. Mekalanos.** 2003. Quorum sensing-dependent biofilms enhance colonization in *Vibrio cholerae. Dev. Cell* **5:**647–656.

75. **Zhu, J., M. B. Miller, R. E. Vance, M. Dziejman, B. L. Bassler, and J. J. Mekalanos.** 2002. Quorum-sensing regulators control virulence gene expression in *Vibrio cholerae. Proc. Natl. Acad. Sci. USA* **99:**3129–3134.

Colonization of Mucosal Surfaces
Edited by James P. Nataro et al.

Chapter 15

Role of the Mucus Layer in Bacterial Colonization of the Intestine

DAVID C. LAUX, PAUL S. COHEN, AND TYRRELL CONWAY

Most interactions that occur between mammalian hosts and microorganisms present in the environment take place at mucosal surfaces. The nature of these interactions has been investigated for many years, and it is clear that mucosal surfaces have evolved a variety of innate and adaptive mechanisms to limit or prevent bacterial colonization of mucosal surfaces. Equally clear is the fact that many bacterial species are quite successful in colonizing mucosal surfaces. For the very large populations of microorganisms that constitute the normal microflora, colonization involves the establishment of relatively stable communities engaged in complex interactions with each other and the host. With pathogens, colonization may be temporary and the organisms may persist only long enough to cause disease.

Mucosal surfaces constitute a very large surface area and have a common feature in that they are covered by a layer of secreted mucus. While the mucus layer serves a variety of functions, one of the most important is its role in protecting the underlying epithelial cells by acting as a barrier or buffer zone between the epithelial cells and the external, usually luminal, environment (29). In spite of its prominence as a feature of mucosal surfaces, our knowledge about the role of the mucus layer in allowing or preventing bacterial colonization of mucosal surfaces is rudimentary. In this chapter we focus on the mucus layer of the intestinal tract and its role in colonization of the intestine by enteric bacteria.

OVERVIEW

The ability of enteric bacteria to colonize the intestinal tract is determined to a large extent by the properties of the microorganisms, the conditions encountered in the intestine, and the combined dynamics of the intestinal tract in general and the mucosal surface in particular. The human gastrointestinal (GI) tract, when viewed along the proximal-distal axis, consists of the duodenum, jejunum, ileum, and colon, while that of rodents, which are frequently used as model systems, is similar, with the notable exception of the cecum, which is quite prominent in rodents (125). In the small intestine, the flow of luminal contents is rapid (a transit time of 1 to 4 h in humans) and the conditions (low pH, bile secretion, digestive enzymes) do not favor colonization. Typically, the number of microorganisms recovered from the small intestine of a healthy human is small (10^3 to 10^4/ml of intestinal contents) and the indigenous flora lacks diversity, consisting primarily of *Streptococcus* spp. and *Lactobacillus* spp. (21). The exception occurs in the ileum, an area of transition between the small and large intestines, which may be colonized by significant numbers of microorganisms (10^8/ml of intestinal contents), probably as a result of reflux from the colon. In the lower intestine (the colon in humans and the cecum and colon in mice), the flow of luminal contents slows (10 h to several days for humans) and there is a tremendous increase in bacterial population density (10^{11} to 10^{12}/ml of intestinal contents) and diversity. It is thought that over 500 species of microorganisms, predominantly (>99.9%) anaerobes, inhabit the human large intestine. A representative list of colonic bacteria includes *Bacteroides* spp., *Bifidobacterium* spp., *Clostridium* spp., *Eubacterium* spp., *Peptococcus* spp., *Peptostreptococcus* spp., *Streptococcus* spp., *Enterococcus* spp., *Ruminococcus* spp., *Fusobacterium* spp., *Lactobacillus* spp., and enterobacteria (for a more detailed listing, see references 12, 21, 115, and 125). Together, these more or less permanent residents form the complex, functionally stable microbial communities that, in theory, occupy all of the available niches of the intestine.

David C. Laux and Paul S. Cohen • Department of Cell and Molecular Biology, University of Rhode Island, Kingston, RI 02881.
Tyrrell Conway • Department of Botany and Microbiology, The University of Oklahoma, Norman, OK 73019.

One of the benefits provided to the host by the indigenous flora relates to the phenomenon known as colonization resistance (32, 131). Colonization resistance, also described as bacterial antagonism (31), bacterial interference (23), barrier effect (24), and competitive exclusion (68), refers to the ability of the indigenous microbial populations to prevent or limit colonization by potentially harmful transient bacteria. Although the importance of the indigenous flora in defending the host against pathogenic invaders is well documented and the study of colonization resistance has a long history (32), the mechanism(s) by which the indigenous flora prevents colonization by newly introduced bacteria remains the subject of much research. While no one disputes the importance of colonization resistance, the complexity and sheer numbers of the microbial populations involved and the diversity of ecological niches in the intestine make systematic investigation of the basic underlying mechanisms a formidable task. Mechanisms implicated in colonization resistance include the production of inhibitory substances such as short-chain fatty acids (43, 106) and bacteriocins (13, 66), competition for binding sites at mucosal surfaces (9), and—the most prevalent theory—competition between the "invading" microorganisms and the established microflora for nutrients at mucosal surfaces (32–34).

INTESTINAL MUCOSAL SURFACES

Colonization, whether by newly transient microorganisms or by the indigenous microbial flora, is influenced to a large extent by the dynamic nature of the intestinal mucosal surface. The surface of the small intestine is characterized by the presence of a multitude (10 to 40/mm^2) of intestinal villi. At the base of each villus are the crypts of Lieberkühn, which contain stem cells that give rise to the basic epithelial cell types associated with the small intestine—columnar absorptive enterocytes (approximately 80% of all epithelial cells), mucus-secreting goblet cells, enteroendocrine cells, and Paneth cells, which secrete lysozyme and a variety of antimicrobial defensins (27, 91). One of the principal characteristics of the intestinal epithelial surface is the high rate of epithelial cell turnover. Epithelial cells are continuously produced by the stem cells located within the crypts, migrate up from the basal portion of the villus to the apical tip, and are shed through processes involving exfoliation and apoptosis. It is estimated that any individual cell takes approximately 2 to 3 days to complete this circuit. As a result of this process, the small intestine of humans produces an estimated 2×10^7 to 5×10^7 cells per min. The epithelial surface of the large intestine is structurally much less complex than that of the small intestine, in that villi are not present and the surface is relatively smooth. Here again, stem cells located in the crypts give rise to a continuous supply of epithelial cells, in this case absorptive enterocytes, numerous goblet cells, and widely scattered enteroendocrine cells, which migrate up from the crypts and are turned over at the luminal surface. It is estimated that the large intestine produces approximately 2×10^6 to 5×10^6 cells per min (27, 136).

An additional factor in the dynamics of the intestinal mucosal surface relates to the production of the mucus layer, which covers the epithelial cells. The mucus layer is thought to serve as a lubricant, facilitating the passage of luminal contents, a protective barrier separating the epithelial cells from the potentially hostile luminal environment, and a first line of defense against infection (1, 3, 61). In many locations, especially the large intestine, the mucus layer also serves as a habitat or niche, which is colonized by the indigenous microflora.

MUCUS LAYER

The gel-like mucus layer of the intestine is dynamic in that it is continuously being renewed by secretion of stored or newly synthesized components, sloughed or eroded by mechanical forces, and degraded by the indigenous flora. The components that give the mucus layer its gel-like properties are the very large, heavily glycosylated mucin proteins secreted by goblet cells. More specifically, secreted mucins are large filamentous glycoproteins characterized by complex O-linked glycosylated tandem repeats of proline, threonine, and serine peptide sequences (19, 120). In humans, the principal secreted GI gel-forming mucins (MUC2, MUC5AC, MUC5B, and MUC6) are sometimes given the designation 11p15 mucins on the basis of the chromosome location of the respective genes and their apparent sequence homologies (19). Other members of the so-called mucin family (19) are membrane glycoproteins, some of which form the membrane-glycocalyx present on the apical surface of the intestinal epithelium.

In addition to the major gel-forming mucins present, the mucus layer contains a variety of proteins, carbohydrates, lipids, nucleic acids, and other components that originate from a number of sources. These include components secreted as part of the digestive process (bile, enzymes, etc.); components secreted by the mucosal epithelial cells (defensins and lysozyme); cellular components released as a result of the continuous migration, exfoliation, and apoptosis of epithelial cells; partially digested and undigested dietary components; and end products and secondary metabolites produced by the indigenous microbial populations.

The actual structure of the mucus layer in vivo has proven to be somewhat difficult to elucidate, largely because of its inaccessibility and artifacts introduced in preparing samples for analysis. Recently, Atuma et al. (7) described the mucus layer of the rat intestine as consisting of a loosely adherent surface layer that can be removed by suction and a firmly adherent underlying layer closely associated with the epithelial surface. The overall thickness of the mucus layer and the thickness of the loose and firmly adherent layers vary in specific sections of the intestinal tract, being thinner (123 to 480 μm total thickness) and somewhat discontinuous in the small intestine and significantly thicker (830 μm total thickness) and more continuous in the colon.

BACTERIAL INTERACTIONS WITH THE MUCUS LAYER

In considering bacterial colonization of mucosal surfaces, the role of the mucus layer has been described in a number of ways. These include the mucus layer as a physical barrier that bacteria must enter or pass through to reach the epithelial surface, a biochemically complex molecular structure involved in trapping and facilitating the removal of bacteria, an initial point of contact for bacterial adhesion that favors the colonization of mucosal surfaces, a source of competitive inhibitors of bacterial adhesion to epithelial cells, a complex environmental niche in which bacterial communities replicate, and a source of nutrients for bacterial growth. All of these are valid considerations, whose relative importance will vary depending on the host, the region of the intestinal tract being colonized, and the properties of the colonizing microorganism.

Although the heterogeneous nature of the habitats of the intestinal mucosal surface and the diversity of microorganisms capable of populating those habitats make generalizations difficult, bacterial colonization of the GI tract can be separated to some extent on the basis of colonization as it occurs in the upper portions of the GI tract (duodenum, jejunum, and ileum) and colonization as it occurs in the lower portions (cecum and colon) of the GI tract. In the upper regions of the GI tract, where there is a relatively short transit time for luminal contents and a relatively low bacterial population density, the key elements for successful colonization appear to involve the ability of microorganisms to penetrate the mucus layer and adhere to epithelial cells (111-113). In the lower GI tract, the flow of luminal contents slows and the situation changes dramatically. In these regions, the mucus layer serves as a matrix and a source of nutrients supporting bacterial replication. Here the major obstacle faced by newly introduced bacteria is the need to establish a stable population in the highly competitive environment created by the presence of a large, very diverse, well-adapted normal flora (9, 32, 34). In the following sections, we consider various aspects of bacterial-mucosal interactions as they relate to the mucus layer.

Chemotaxis and Motility

The presence of the intestinal mucus layer as a viscous physical barrier and the need for bacteria to reach and maintain themselves in an appropriate environmental niche has led many investigators to propose chemotaxis and motility as important aspects of colonization for the microorganisms that exhibit these properties. As might be expected, the relative importance of chemotaxis and motility appears to vary with the organism and model system under study. In *Vibrio* spp. (36, 41), *Campylobacter* spp. (85, 87, 93, 124, 135), *Yersinia* spp. (137), *Helicobacter pylori* (25, 26), various anaerobes (33, 118), and the spirochete *Serpulina hyodysenteriae* (54), chemotaxis and motility appear to play an important role in pathogenesis and colonization. In each instance, nonmotile and/or nonchemotactic mutants were found to be less virulent or to colonize the intestine less well than the wild-type parent. It should perhaps be noted that in most studies of this type, the microorganisms involved were pathogens or potential pathogens and the models employed examined the relative ability of nonmotile or nonchemotactic microorganisms to initiate infection or colonization. Relatively little information is available on the role of motility and chemotaxis in maintaining colonization once these microorganisms have become established.

While chemotaxis and motility appear to be important for some species, they do not appear to play a significant role in colonization by other species. This is most apparent for the dominant members of the colonic anaerobic intestinal flora (i.e., *Bacteroides thetaiotaomicron* and *Bifidobacterium* spp.), which are nonmotile and colonize in very large numbers. Apparently, this is also the case for some species that usually exhibit motility or chemotaxis. For example, nonmotile (*mot* and *fla*) mutants of *Salmonella enterica* serovar Typhimurium (69) remain virulent in BALB/c mice. Similar findings were obtained with a competitive streptomycin-treated mouse model (see below for more details) that was used to evaluate the relative ability of motile and/or chemotactic and nonmotile and/or nonchemotactic *S. enterica* serovar Typhimurium and *Escherichia coli* to colonize the mouse large intestine. In those studies, motile and/or chemotactic avirulent *S. enterica* serovar Typhimurium and isogenic nonmotile (*fla*) or nonchemotactic (*cheA*) mutants were, when simultaneously fed to mice, found to colonize the large intestine equally well (80). In addition, a normally motile and/or chemotactic commensal *E. coli* fecal isolate (*E. coli* F-18) and isogenic nonchemotactic (*cheA*) or nonmotile (*flhD*) mutants colonized equally well (79). Indeed, in both of these studies, the wild-type *S. enterica* serovar Typhimurium and *E. coli* strains grown in vitro in mucus were shown to be flagellated but nonmotile.

While many of the studies conducted to date appear to have produced clear evidence for the importance of motility and chemotaxis in different model systems, deciphering the role of motility and chemotaxis per se in pathogenesis and colonization of the intestine is complicated by a number of factors. In some cases, assigning a direct role to motility is complicated by a growing number of reports indicating that flagella may also function as adhesins. Evidence for the action of flagella as adhesins has been reported for enteropathogenic *E. coli* (40), *E. coli* O78:K80 (a pathogen of poultry) (62), *S. enterica* serovars Typhimurium and Enteritidis (4, 5, 20, 38, 39, 104), *Yersinia enterocolitica* (137), *H. pylori* (26), *Vibrio cholerae* (37, 97), and *Clostridium difficile* (126). As a result of this apparent dual function, it is often difficult to distinguish between the role of flagella in motility and the role of flagella as adhesins.

Further complications arise as a result of the complex regulatory systems involved in flagellar synthesis and motility. For example, evidence has been presented linking the expression of genes involved in flagellar synthesis and assembly with genes involved in the expression of type III secretion systems in a number of enteropathogenic microorganisms (40, 49, 95, 96) and with the coordinated (17, 50, 60, 127), sometimes reciprocally regulated, expression of virulence factors, as is the case for *V. cholerae* (37). There is a general lack of information regarding the environmental signals that regulate motility and chemotaxis at the intestinal mucosal surface.

Overall, the picture that emerges is perhaps more complex than was originally thought. It is apparent that chemotaxis and motility involve a large number of genes (approximately 50 in *E. coli* and *Salmonella*), whose expression is highly regulated and carefully coordinated (73, 90, 119), often in response to environmental signals. Undoubtedly, appropriate regulation and coordinated expression of these genes represent important aspects of colonization and pathogenesis for many microorganisms (57). Future in situ studies designed to investigate gene expression and regulation of motility and chemotaxis at mucosal surfaces, particularly within the mucus layer, will be of interest. While the mucus layer has traditionally been viewed as a viscous barrier that must be penetrated, it also represents a complex environment likely to have a direct effect on the regulation of motility and chemotaxis.

Binding of Mucus Components

For many bacteria, particularly pathogens, the ability to interact with and adhere directly to the epithelial cell surface has long been recognized as a critical feature of colonization of mucosal surfaces (8, 112), and a large body of evidence relating to the bacterial surface structures (fimbriae, flagella, lipopolysaccharide [LPS], etc.) involved in adhesion to epithelial cells is available. In the case of the normal flora of the large intestine, association with the mucosal surface appears to be a critical aspect of colonization and, depending on the organism, may or may not involve direct adhesion to epithelial cells. In addition to examining bacterial adhesion to epithelial cells, numerous investigators have shown that bacteria bind in vitro to components present in mucus; examples include commensal *E. coli* strains (15, 16, 78), enteropathogenic *E. coli* (72, 117), pathogenic *E. coli* strains positive for K88 (16), K99 (63, 67), RDEC-1 (22, 72), 987P (18), *C. difficile* (52, 53, 126, 134), *Campylobacter* spp. (81, 82, 123), *Bifidobacterium* spp. (92), *Lactobacillus* spp. (51, 105, 107, 108), *Salmonella* serovars (132), *Yersinia* spp. (74–76), and *H. pylori* (130).

It is less clear how binding of mucus-associated components affects access to appropriate niches in vivo and facilitates or inhibits mucosal association and adhesion to epithelial cells. Most studies involve measuring the ability of isogenic strains to bind mucus in vitro and attempt to correlate in vitro binding with altered virulence or in vivo colonization. In theory, binding to components of the mucus layer could trap bacteria, limit their ability to penetrate the mucus layer, and facilitate their removal as the mucus layer is turned over. Binding to receptors in the mucus layer could also competitively inhibit adhesion to specific epithelial cell receptors, and numerous reports have demonstrated such inhibition in vitro (18, 22, 71, 72, 74, 117). In some cases, binding of mucus components appears to show a positive correlation with enhanced colonization (15, 132), although the mechanism by which such binding would enhance colonization is not clear; presumably, such binding favors mucosal associations, perhaps by providing an initial point of contact with the firmly adherent mucus layer.

In attempting to relate the binding of mucus components in vitro to events in vivo, it is difficult to draw conclusions, but several generalizations are apparent. It is likely that bacterial binding to mucus components in vitro reflects events associated with the initiation of colonization rather than the later stages of colonization, and, as with all aspects of colonization, the outcome in vivo may be affected by the presence or absence of competing organisms. For example, avirulent *S. enterica* serovar Typhimurium with wild-type LPS and isogenic LPS-deficient mutants were assayed for their ability to bind mucus components and to penetrate mucus layers in vitro and colonize the mouse large intestine (80). In all cases, compared to the wild-type parent, the LPS-deficient mutants showed increased binding to mucus in vitro and reduced ability to penetrate mucus in vitro and were poor colonizers of mice, presumably because the LPS-deficient mutants were less able to penetrate the mucus layer in vivo. Notably, the LPS-deficient mutants were poor colonizers only when placed in direct competition with the parent strain. When fed alone, each colonized well, and, if allowed to become established (i.e., were fed prior to the parent strain), none of the mutants were displaced by the parent strain. Thus, defective binding of mucus components and reduced ability to penetrate the mucus layer appeared to interfere with the initiation of colonization only when the more efficient competing parent strain was present.

An interesting but complicating factor in any investigation of the role of mucus binding in colonization relates to differences in expression of the bacterial surface structures involved (i.e., flagella, fimbriae, LPS, capsule, and outer membrane proteins). Such variation may exist within a given population as a result of phase variation, in which case binding to mucus could select for, or against, the subpopulation expressing particular surface structures at the time the microorganisms are introduced (10). Differences in expression could also occur as a result of normal molecular mechanisms regulating the expression of the bacterial surface components in response to the environment. In either case, expression at one site (e.g., within the mucus layer) may favor elimination while expression at a different site (e.g., the epithelial surface) may favor colonization. For example, *E. coli* F-18 can produce type 1 fimbriae and has been shown to do so in vivo (59, 133). In studies designed to investigate the role of type 1 fimbriae in colonization, a nonfimbriated *E. coli* F-18 *fimA* mutant was placed in direct competition with the isogenic wild-type *E. coli* F-18 strain and was found to be equally capable of colonizing the mouse large intestine, indicating that the expression of type 1 fimbriae was not necessary for colonization (77). However, when *E. coli* F-18 cells constitutively overexpressed type 1 fimbriae (i.e., were phase locked "on"), they avidly bound mucus components and were poor colonizers relative to the parent strain, apparently because they had difficulty in penetrating the mucus layer and becoming established in the large intestine (78). Although it did not appear to be required for colonization, expression of type 1 fimbriae and binding to mucus components appeared to initially select against the piliated portion of the population, but cells expressing type 1 fimbriae were subsequently able to colonize.

Biofilms in the Intestine

Much attention has been paid to the role played by biofilms in bacterial colonization of many environments, but surprisingly little information is available regarding biofilms in the intestine. On the one hand, the transit time of the intestinal contents is short compared to the timescale of biofilm development (100), and so it is hard to imagine how a stable biofilm might be maintained in the intestine. On the other hand, the mucus layer itself has many of the characteristics of a secreted biofilm matrix. When the concept of bacterial binding to mucus components is added to this scenario, it seems reasonable to at least consider the evidence for bacterial biofilms in the GI tract.

It is important to distinguish the physical location within the intestine of bacterial biofilms: those in association with host cells and those in association with food particles (101). Most studies have focused on biofilms that form around food particles in transit within the lumen (70). These particulate biofilms in the intestine appear to have qualities similar to those found on teeth in the oral cavity, and both seem to involve fusiform bacteria (56, 103).

There is also microscopic evidence that biofilms—defined as cells in close proximity on a surface—form on the mucosal surface. Type 1 fimbria-dependent biofilm formation by *S. enterica* serovar Typhimurium on HEp-2 cells has been documented (11). Similarly, *E. coli* biofilms on colonic pieces have been observed and are more resistant to antibiotics than planktonic cells, a common characteristic of biofilm cells (55). Colonic biofilms of enteroaggregative *E. coli* appear to require aggregative adherence fimbriae exclusively (114). Plasmid transfer, as it takes place in the intestine, is more similar to the way it occurs in biofilms than it is to the way it takes place in planktonic cells (65). Taken together, the available evidence suggests that *E. coli* may behave as a biofilm in the intestine, possibly by growing within the mucus layer (98).

To summarize, we now know that many bacteria express surface structures (fimbriae, flagella, LPS, and outer membrane proteins) that can bind components of the mucus layer. However, the role played by such interactions in selectively allowing or limiting the colonization of the intestinal tract remains largely speculative, and our knowledge is limited further by our lack of understanding of the factors affecting the expression of these structures at mucosal surfaces. The availability of recently developed procedures for in situ analysis of gene expression in specific habitats should help clarify many of these issues.

THE MUCUS LAYER AS HABITAT AND SOURCE OF NUTRIENTS

In addition to its role as a physical barrier limiting colonization—particularly in the upper GI tract—the mucus layer represents a habitat and source of nutrients for the bacterial communities that colonize mucosal surfaces. Freter (32), in discussing the colonization of the intestine by the normal flora, describes four basic microhabitats: the lumen, the epithelial cell surface, the mucus layer, and the rather specialized region represented by the deep mucus layer of the crypts (64). The lumen of the upper intestine may contain a number of microorganisms, but the rate of bacterial growth in the lumen in these regions does not does not appear to be high enough to compensate for the rapid flow of luminal contents (32). Thus, many of the microorganisms in the upper GI tract are transient. In the lower GI tract, where the flow rate diminishes, the lumen contains very large numbers of bacteria, but it is not clear whether these represent an accumulation of transient bacteria from the upper regions, populations that colonize the mucosal surfaces of the lower regions and are sloughed into the lumen, or distinct populations that are maintained without forming an association with the mucosal epithelium. However, we note that the ability to associate with the mucosal surface (i.e., growth within the mucus layer without specific adhesion to the epithelial surface) appears to be a common feature of many microorganisms that make up the microbial communities of the GI tract (33, 111–113).

Microscopic examination of mucosal tissue reveals the presence of a large and diverse population of microorganisms associated with the mucosal surface, including large numbers of bacteria in the mucus layer (109, 111). In some cases, because of artifacts introduced during sample preparation, it is difficult to distinguish between bacteria adhering directly to epithelial cells and bacteria free within the mucus layer. However, procedures designed to stabilize the mucus layer and reduce shear clearly demonstrate that large numbers of bacteria colonize the mucus layer without adhering to the underlying epithelial cells (109). Until recently, it has been difficult, if not impossible, to locate and identify specific strains of bacteria in mucosal tissue preparations. Today, with the development of specific 16S rRNA fluorescent oligonucleotide probes, it is possible to identify and visualize specific microorganisms associated with specific intestinal habitats (98). Such procedures should provide a valuable tool for elucidating the community structure of specific niches within the intestine.

In addition to being a physical niche occupied by large numbers of bacteria, the mucus layer serves as a source of nutrients for bacterial growth. The principal components of mucus include the large, complex mucin; a variety of smaller proteins and glycoproteins; lipids and glycolipids secreted by epithelial cells; and bacterial and epithelial cell debris. The ability of mucus to support the growth of bacteria is clearly evident from in vitro studies in which bacteria have been shown to grow readily in mucus preparations (30, 80, 99). For example, *E. coli* inoculated into mouse intestinal mucus grows to high densities (10^9 CFU/ml), with a generation time of approximately 30 min (86), roughly equivalent to the growth rate measured in vivo in the mouse intestine (99).

In vivo, the majority of intestinal bacteria require a fermentable carbohydrate, and this mode of metabolism is assumed to be the nutritional basis for colonization by most bacterial species (110). While the complexity of the mucus layer makes a detailed analysis of its composition difficult, it is estimated that the total carbohydrate content of mucus is approximately 50% (1, 2, 29, 120). Mucin, the dominant glycoprotein of the intestine, is about 80% polysaccharide and contains *N*-acetylglucosamine, *N*-acetylgalactosamine, galactose, fucose, sialic acid, and mannose. The low-molecular-weight glycoproteins in mucus contain smaller amounts of L-arabinose, mannose, and hexuronates (94). Ribose is also present (30, 94). The total content of both free and polymerized forms of the hexuronates, glucuronate and galacturonate, in mucus is about 0.6% by weight (122). Mouse cecal mucus also contains a small amount of gluconate as a free monosaccharide, which presumably arises from bacterial oxidation of glucose or dephosphorylation of 6-phosphogluconate released by lysed cells (94).

While the mucus layer appears to be rich in polysaccharides, the bulk of fermentable carbohydrate is in the form of complex polysaccharides, which are thought to be degraded by an impressive array of glycosidases produced by the consortium of anaerobes that dominate the colonic flora (136). The monosaccharides released by these glycosidases support the growth of the members of the intestinal flora, which for the most part are unable to degrade mucus-derived polysaccharides. This includes human fecal *E. coli* strains, which do not make any mucin-degrading enzymes (47).

Although it appears that colonization of mucosal tissues (mucus layer and/or epithelial surface) is important for maintaining stable populations of the indigenous microflora, the homeostatic mechanisms involved in maintaining such complex communities are poorly understood. Similarly, while the phenomenon of colonization resistance has been well documented (and is the basis for probiotics), the specific interactions that occur between the indigenous mucosal populations and "invading" microorganisms are complex and largely theoretical (3, 32–34).

One key aspect relating to the homeostatic mechanisms involved in the colonization of mucosal surfaces is based on the nutrient-niche hypothesis (33, 34). A basic tenet for colonization of mucosal surfaces is that for newly introduced bacteria to increase in numbers, they must initially grow at a rate that exceeds their rate of removal. Once this population has reached the maximum population density that can be sustained in that environment, maintaining a stable population density requires that the bacteria grow at a rate that equals the rate at which they are removed. The nutrient-niche hypothesis postulates that the growth rates and population densities achieved by specific microorganisms are limited by the quality and availability of particular nutrients and competition with other members of the microbiota for those nutrients.

In Vitro Models for Nutritional Studies

In vitro evidence supporting the nutrient-niche hypothesis has been reported by investigators who used continuous-flow chemostat culture systems designed to mimic conditions in the intestine (34, 64). The use of these systems has demonstrated the importance of microbial association with surfaces, the stability of the populations involved (at least with respect to the major genera), and the role of nutrient utilization in maintaining stable, mixed populations. Essentially, the evidence indicates that two microorganisms with a preference for the same growth-limiting nutrient cannot coexist in a chemostat; one will eventually outcompete the other (116). However, if two microorganisms utilize different growth-limiting nutrients, they can coexist and maintain stable populations (128, 129). If the analogy of a chemostat is applied to the intestinal tract, several hundred species of bacteria are in equilibrium, competing for resources from an extensive mixture of limiting nutrients, and the only way for a bacterial species to survive is to compete effectively for one or a few of the available nutrients.

In Vivo Models for Nutritional Studies

Although the idea that the intestine functions as a chemostat is appealing and the results provided by in vitro continuous-flow culture systems have provided very useful information, eventually the hypothesis must be tested in vivo. The animal models available for in vivo studies usually fall into three categories—conventional animals, gnotobiotic animals, and antibiotic-treated animals—each with its advantages and disadvantages. Conventional animals, usually mice, are those with an intact but usually undefined resident microbial flora. These animals, which are used most often in examining host-pathogen interactions, would most closely represent a "natural" situation, but colonizing such mice is often difficult due to colonization resistance. The diversity and poorly defined nature of the resident microbial populations often make it difficult to draw specific conclusions regarding the interactions that occur between the newly introduced bacteria and the resident flora. At the other extreme would be gnotobiotic animals, which have been colonized by one or more specific microorganisms. The lack of a resident flora makes colonization of these animals relatively easy, and studies involving gnotobiotic animals have contributed to our understanding of interactions among bacterial populations with a limited degree of diversity and, perhaps more importantly, have contributed a great deal to our understanding of the contributions made by the microbial flora to the normal development of the host mucosal immune system and the development and differentiation processes of the intestine itself (27, 45, 136). Particularly intriguing are recent studies of host-microbe interactions demonstrating what appears to be microbial coordination of nutrient production by the host, specifically, the ability of *Bacteroides thetaiotaomicron* to induce sustained production of fucosylated glycans, which can then be utilized by *Bacteroides* as a carbon source (46).

The Streptomycin-Treated Mouse Model

The third system for animal studies involves the use of conventional mice that have been subjected to antibiotic treatment. For many years it has been known that such treatment is capable of disrupting the normal flora and of lowering resistance to colonization. In other words, antibiotic treatment opens a previously unavailable niche, which can then be colonized by newly introduced microorganisms such as *E. coli* or minor populations of the existing flora. A frequently used model is the streptomycin-treated mouse (88). Streptomycin treatment reduces the facultative-anaerobe population but appears to leave the strict anaerobes largely intact (44). Following streptomycin treatment, the mice are simultaneously fed two or more competing strains of streptomycin-resistant *E. coli* and the relative ability of each strain to colonize the intestine is monitored, usually over several weeks. By genetically manipulating the strains and placing them in direct competition, it is possible to assess the role of specific genes and processes in the ability to colonize a specific intestinal niche. In these experiments, colonization is observed as occurring in two stages: initiation and maintenance. Initiation occurs between 5 h and 3 days, when the number of *E. coli* cells increases from 10^5 to approximately 10^7 to 10^8 CFU/g of feces. During the maintenance stage (3 to 15 days), the *E. coli* population density stabilizes and is maintained at the maximum level that can be sustained by the available nutrients.

Using this model, several basic patterns of competitive colonization are readily discernible. Mutant strains that do not compete effectively for nutrients involved in initiation will fail to reach the same large numbers as their wild-type parent. Mutants that do not compete effectively for nutrients involved in maintenance (whether or not they initiate efficiently) will decline in numbers relative to the wild-type parent during the maintenance stage. This results in the mutated strain being displaced and eventually eliminated or persisting at a lower level. Such persistence may be due to continued presence at a separate site or, more probably, may reflect a metabolic shift to utilization of an alternate, less preferred nutrient present at lower concentrations. In some instances, the mutated strain is at a competitive disadvantage during initiation and is either unable to colonize or colonizes at levels much lower than those of the competing strain. The streptomycin-treated mouse colonization assay thus provides a relative measure of fitness during both stages of colonization, revealing the nutrient(s) involved and indicating whether the nutrient plays a major, significant, or minor role.

In recent years, extensive use has been made of the streptomycin-treated mouse model to specifically define some of the parameters involved in colonization of mucosal surfaces, particularly since these relate to the metabolic pathways and nutrients utilized by *E. coli* strains colonizing the mucus layer (14, 58, 84, 121, 122). Although it is not a dominant member of the intestinal microbiota, *E. coli* was chosen because it is the predominant facultative anaerobe (28) and is arguably the best understood of all model organisms. Ironically, in spite of the vast amount of information available concerning *E. coli,* we still lack the most basic information about how it is able to colonize the GI tract (102). To begin to address these issues and provide a focus for in vivo studies, whole-genome expression profiling has been used. An overview of these studies is discussed below.

Gene Expression Profiling To Identify Pathways Involved in Colonization

There is no perfect in vivo model for investigating bacterial colonization of the large intestine. Previous attempts to obtain in vivo *E. coli* expression data with conventional and streptomycin-treated mice failed because of cross-hybridization by the other species sampled (T. Conway et al., unpublished data). The gnotobiotic-mouse model is subject to the criticism that the colonized bacteria do not face the intense competition encountered in conventional animals (83). Thus, the gene expression profile of commensal *E. coli* K-12 was determined under conditions designed to mimic nutrient availability in the intestine (14). Since *E. coli* is known to grow rapidly in vivo within the mucus layer of the intestine, presumably using nutrients derived from the mucus, mucus was prepared from streptomycin-treated mice (not colonized with *E. coli*) and standing cultures were grown in vitro in minimal salts medium containing lyophilized mucus as the sole source of carbon and energy.

The gene expression profiles of mucus-grown *E. coli* K-12 identified genes induced under these growth conditions (14). Significantly induced genes (>3 standard deviations) directed attention to pathways for catabolism of amino sugars: *N*-acetylglucosamine, *N*-acetylneuraminic (sialic) acid, and glucosamine; hexoses and pentoses: maltose, mannose, fucose, and ribose; sugar acids (glucuronate, galacturonate, and gluconate); and phosphatidylethanolamine. There was also evidence for catabolism of phospholipids and amino acids that require the tricarboxylic acid cycle and gluconeogenesis.

Competition for Nutrients and Colonization of the Mouse Intestine

On the basis of the data provided by whole-genome expression profiling, mutations were made in various genes likely to be involved in the colonization process, and each of the strains was assessed for the effect of the mutation on colonization in the streptomycin-treated mouse model. The experimental design involved simultaneously feeding small numbers (10^5 CFU/ml) of two strains, usually the wild-type parent and an isogenic mutant strain. Seven sugars were found to affect the colonization of the mouse intestine to a greater or lesser extent. Interestingly, the relative impact of these nutrients on colonization tends to parallel their preferred order of utilization in vitro (14). Gluconate, which represents a relatively minor component of mucus (94), was a major carbon source utilized by *E. coli* K-12 during both initiation and maintenance. *N*-Acetylglucosamine and sialic acid were involved in initiation but not maintenance, and glucuronate, mannose, fucose, and ribose were involved in maintenance but not initiation. It is also important to note that a large number of potential nutrients implicated by gene induction during in vitro growth in mucus did not affect colonization.

Global Regulators

Gene systems involved in related cellular processes are often controlled in blocks. It seems likely that some global regulatory circuits in the cell are important for growth in the intestinal environment while others are not. As a shortcut to identification of cellular processes that are important for *E. coli* to colonize, a series of mutants of *E. coli* MG1655, a K-12 strain, and *E. coli* EDL933, an O157:H7 strain, were constructed with defects in global regulatory networks and stress responses (Conway et al., unpublished). The fitness of these mutant strains for competition with their wild-type parents in streptomycin-treated mouse colonization assays is shown in Table 1. Not surprisingly, colonization was negatively affected by mutation of most, but not all, of the global regulators.

Induction of stationary-phase survival genes by RpoS (stationary-phase sigma factor) was not essential for competition with wild-type strains. Although each survival system controlled by RpoS responds directly to specific signals, apparently the integrated control of these systems by RpoS in stationary phase is not necessary. Mutation of global regulators involved in respiration (ArcA and FNR) and balancing carbon metabolism (CsrA and Cra) had major effects on colonization during initiation or maintenance. The ability to coordinate rapid growth and stasis (*relA*) also appeared to be important.

While the range of effects produced by mutations in these genes is broad, making it difficult to draw specific conclusions, the results are intriguing and warrant further investigation. For example, the colonization defects of the *fnr* and *arcA* mutants point to respiratory metabolism as being important. Since oxygen is available in the layer adjacent to the intestinal epithelium (42) and since enterohemorrhagic *E. coli* O157:H7 is closely associated with the epithelium (84), respiratory metabolism could play an important role in colonization. Therefore, it will be important to test the hypothesis that respiratory metabolism is important for colonization of the mouse intestine by *E. coli*.

Competition between *E. coli* Commensal and Pathogenic Strains

The systematic analysis of *E. coli* K-12 described above suggests the testable hypothesis that there is a nutritional basis for intestinal colonization which explains the steady progression of commensal strains in

Table 1. *E. coli* MG1655 and *E. coli* EDL933 global regulatory mutant colonizations

Mutation	Regulatory network	Function	Defect[a]	Stage[b]
MG1655 *arcA*	Aerobic respiratory control	Aerobic respiration	Major	I
MG1655 *crp*	Cyclic AMP receptor protein	Catabolite repression control	Major	M
MG1655 *csrA*	mRNA stability control	Gluconeogenesis and glycogen	Major	I
MG1655 *fnr*	Fumarate nitrate reductase	Anaerobic respiration and fermentation	Major	M
MG1655 *fruR* (*cra*)	Catabolite repressor-activator	Gluconeogenesis and Entner-Doudoroff pathway	Major	M
MG1655 *relA*	Relaxed stringent control	ppGpp-dependent stringent control	Major	M
MG1655 *rpoS*	Stationary-phase sigma factor	Survival in stationary phase	None	
MG1655 *uspA*	Universal stress protein	Oxidative stress	None	
MG1655 *gadXW*	Gad-acid tolerance activator	Glutamate-dependent acid resistance	None	
EDL933 *fnr*	Fumarate nitrate reductase	Anaerobic respiration and fermentation	Major	M
EDL933 *fruR* (*cra*)	Catabolite repressor-activator	Gluconeogenesis and Entner-Doudoroff pathway	Major	M
EDL933 *rpoS*	Stationary-phase sigma factor	Survival in stationary phase	None	

[a]Major, >3 log unit difference between wild type and mutant; significant, 2 to 3 log unit difference between wild type and mutant; minor, 1 to 2 log unit difference between wild type and mutant; none, < 1 log unit difference between wild type and mutant.
[b]I, initiation (5 h to day 3); M, maintenance (after day 7). The strain is essentially gone by the maintenance stage (day 7).

human hosts (6). Thus, there is a need to study the competition for resources by *E. coli* commensal and pathogenic strains. For example, it is known that certain *E. coli* strains can prevent *Shigella flexneri* (35) and *S. enterica* serovar Typhimurium (48) infections in mice. A reasonable hypothesis is that different enteric bacteria occupy distinct ecological niches and preferentially utilize different nutrients for colonization of the mouse intestine. The nutritional basis for cocolonization might involve one or more of the following possibilities: presence or absence of specific pathways, differences in the order of consumption of preferred nutrients, ability to switch to an alternative source of nutrition, and cometabolism of nutrients. Indeed, the progressive colonization of humans by diverse *E. coli* strains (6) may have a nutritional basis. Hence, individual differences in the *E. coli* flora could be the reason for differences in human susceptibility to gastroenteritis (89). If so, there is the opportunity to build a nutritional framework for understanding how commensal strains serve as the first line of defense against intestinal infections. The composition of the human commensal *E. coli* flora, and the consequent availability of certain nutrients, may be an important factor in susceptibility to infection by *E. coli* O157:H7, providing an open niche for infecting *E. coli* pathogens in some individuals and a barrier to infection in others.

CONCLUDING REMARKS

The types of experiments being used to elucidate the carbon nutrition for colonization of *E. coli* can and should be extended to include other members of the intestinal microflora. In this way, a framework will be established for understanding how competition for resources in the intestinal ecosystem results in the establishment of stable, complex communities of microorganisms. This body of data will allow us to understand the nutritional basis for colonization by various species, the nutrients that support their growth, the pathways that are important, etc. Adding onto this framework other physiological aspects of colonized bacterial cells (i.e., stress responses necessary for survival, production and resistance to antimicrobials, and attachment to the mucosal epithelium) will provide a picture of microbial interaction within the intestinal ecosystem at the cellular and molecular levels.

REFERENCES

1. **Allen, A.** 1989. Gastrointestinal mucus, p. 359–382. *In* J. G. Forte (ed.), *Handbook of Physiology. The Gastrointestinal System. Salivary, Gastric, Pancreatic, and Hepatobiliary Secretion,* vol. III. American Physiology Society, Bethesda, Md.
2. **Allen, A., A. Bell, M. Mantle, and J. P. Pearson.** 1982. The structure and physiology of gastrointestinal mucus. *Adv. Exp. Med. Biol.* **144:**115–133.
3. **Allen, A., and L. C. Hoskins.** 1988. Colonic mucus in health and disease, p. 65–94. *In* J. B. Kirsner and R. G. Shorter (ed.), *Diseases of the Rectum and Colon.* The Williams & Wilkins Co., Baltimore, Md.
4. **Allen-Vercoe, E., A. R. Sayers, and M. J. Woodward.** 1999. Virulence of *Salmonella enterica* serotype Enteritidis aflagellate and afimbriate mutants in a day-old chick model. *Epidemiol. Infect.* **122:**395–402.
5. **Allen-Vercoe, E., and M. J. Woodward.** 1999. The role of flagella, but not fimbriae, in the adherence of *Salmonella enterica* serotype Enteritidis to chick gut explant. *J. Med. Microbiol.* **48:**771–780.
6. **Apperloo-Renkema, H. Z., B. D. Van der Waaij, and D. Van der Waaij.** 1990. Determination of colonization resistance of the digestive tract by biotyping of *Enterobacteriaceae. Epidemiol. Infect.* **105:**355–361.
7. **Atuma, C., V. Strugala, A. Allen, and L. Holm.** 2001. The adherent gastrointestinal mucus gel layer: thickness and physical state *in vivo. Am. J. Physiol. Ser G.* **280:**G922–G929.
8. **Beachey, E. H.** 1981. Bacterial adherence: adhesin-receptor interactions mediating the attachment of bacteria to mucosal surface. *J. Infect. Dis.* **143:**325–345.
9. **Bernet, M. F., D. Brassart, J. R. Neeser, and A. L. Servin.** 1994. *Lactobacillus acidophilus* LA 1 binds to cultured human intestinal cell lines and inhibits cell attachment and cell invasion by enterovirulent bacteria. *Gut* **35:**483–489.
10. **Blomfield, I. C., and M. van der Woude.** 2002. Regulation and function of phase variation in *Escherichia coli,* p. 89–113. *In* M. Wilson (ed.), *Bacterial Adhesion to Host Tissues: Mechanisms and Consequences.* Cambridge University Press, Cambridge, United Kingdom.
11. **Boddicker, J. D., N. A. Ledeboer, J. Jagnow, B. D. Jones, and S. Clegg.** 2002. Differential binding to and biofilm formation on, HEp-2 cells by *Salmonella enterica* serovar Typhimurium is dependent upon allelic variation in the *fimH* gene of the *fim* gene cluster. *Mol. Microbiol.* **45:**1255–1265.
12. **Borrelio, S. P.** 1986. Microbial flora of the gastro-intestinal tract, p. 2–16. *In* M. J. Hill (ed.), *Microbial Metabolism in the Digestive Tract.* CRC Press, Inc., Boca Raton, Fla.
13. **Brook, I.** 1999. Bacterial interference. *Crit. Rev. Microbiol.* **25:**155–172.
14. **Chang, D. E., D. J. Smalley, D. L. Tucker, M. P. Leatham, W. E. Norris, S. J. Stevenson, A. B. Anderson, J. E. Grissom, D. C. Laux, P. S. Cohen, and T. Conway.** 2004. Carbon nutrition of *Escherichia coli* in the mouse intestine. *Proc. Nat. Acad. Sci. USA* **101:**7427–7432.
15. **Cohen, P. S., J. C. Arruda, T. J. Williams, and D. C. Laux.** 1985. Adhesion of a human fecal *Escherichia coli* strain to mouse colonic mucus. *Infect. Immun.* **48:**139–145.
16. **Cohen, P. S., R. Rossoll, V. J. Cabelli, S. L. Yang, and D. C. Laux.** 1983. Relationship between the mouse-colonizing ability of a human fecal *Escherichia coli* strain and its ability to bind a specific mouse colonic mucous gel protein. *Infect. Immun.* **40:**62–69.
17. **Correa, N. E., C. M. Lauriano, R. McGee, and K. E. Klose.** 2000. Phosphorylation of the flagellar regulatory protein FlrC is necessary for *Vibrio cholerae* motility and enhanced colonization. *Mol. Microbiol.* **35:**743–755.
18. **Dean, E. A.** 1990. Comparison of receptors for 987P pili of enterotoxigenic *Escherichia coli* in the small intestines of neonatal and older pig. *Infect. Immun.* **58:**4030–4035.
19. **Dekker, J., J. W. Rossen, H. A. Buller, and A. W. Einerhand.** 2002. The MUC family: an obituary. *Trends Biochem. Sci.* **27:**126–131.

20. **Dibb-Fuller, M. P., E. Allen-Vercoe, C. J. Thorns, and M. J. Woodward.** 1999. Fimbriae- and flagella-mediated association with and invasion of cultured epithelial cells by *Salmonella enteritidis*. *Microbiology* **145:**1023–1031.
21. **Drasar, B. S., and P. A. Barrow.** 1985. The bacterial flora of the normal intestine, p. 19–40, *Intestinal Microbiology,* vol. 10. American Society for Microbiology, Washington, D.C.
22. **Drumm, B., A. M. Roberton, and P. M. Sherman.** 1988. Inhibition of attachment of *Escherichia coli* RDEC-1 to intestinal microvillus membranes by rabbit ileal mucus and mucin in vitro. *Infect. Immun.* **56:**2437–2442.
23. **Dubos, R.** 1966. Staphylococci and infection immunity. *Am. J. Dis. Child.* **105:**643–645.
24. **Ducluzeau, R., M. Bellier, and P. Raibaud.** 1970. Transit through the digestive tract of the inocula of several bacterial strains introduced "per os" into axenic and "holoxenic" mice. The antagonistic effect of the microflora of the gastrointestinal tract. *Zentbl. Bakteriol. Orig.* **213:**533–548. (In German.)
25. **Eaton, K. A., D. R. Morgan, and S. Krakowka.** 1992. Motility as a factor in the colonisation of gnotobiotic piglets by *Helicobacter pylori*. *J. Med. Microbiol.* **37:**123–127.
26. **Eaton, K. A., S. Suerbaum, C. Josenhans, and S. Krakowka.** 1996. Colonization of gnotobiotic piglets by *Helicobacter pylori* deficient in two flagellin genes. *Infect. Immun.* **64:**2445–2448.
27. **Falk, P. G., L. V. Hooper, T. Midtvedt, and J. I. Gordon.** 1998. Creating and maintaining the gastrointestinal ecosystem: what we know and need to know from gnotobiology. *Microbiol. Mol. Biol. Rev.* **62:**1157–1170.
28. **Finegold, S. M., V. L. Sutter, and G. E. Mathisen.** 1983. Normal indigenous intestinal microflora, p. 3–31. *In* D. J. Hentges (ed.), *Human Intestinal Microflora in Health and Disease.* Academic Press, Inc., New York, N.Y.
29. **Forstner, J. F., and G. G. Forstner.** 1994. Gastrointestinal mucus, p. 1255–1283. *In* L. R. Johnson (ed.), *Physiology of the Gastrointestinal Tract,* 3rd ed., vol. 2. Raven Press, New York, N.Y.
30. **Franklin, D. P., D. C. Laux, T. J. Williams, M. C. Falk, and P. S. Cohen.** 1990. Growth of *Salmonella typhimurium* SL5319 and *Escherichia coli* F-18 in mouse cecal mucus: role of peptides and iron. *FEMS Micobiol. Ecol.* **74:**229–240.
31. **Freter, R.** 1956. Experimental enteric *Shigella* and *Vibrio* infections in mice and guinea pigs. *J. Exp. Med.* **104:**411–418.
32. **Freter, R.** 1992. Factors affecting the microecology of the gut, p. 111–144. *In* R. Fuller (ed.), *Probiotics: the Scientific Basis.* Chapman & Hall, London, United Kingdom.
33. **Freter, R.** 1988. Mechanisms of bacterial colonization of the mucosal surfaces of the gut, p. 45–60. *In* J. A. Roth (ed.), *Virulence Mechanisms of Bacterial Pathogens.* American Society for Microbiology, Washington, D.C.
34. **Freter, R.** 1983. Mechanisms that control the microflora in the large intestine, p. 33–54. *In* D. J. Hentges (ed.), *Human Intestinal Microflora in Health and Disease.* Academic Press, Inc., New York, N.Y.
35. **Freter, R.** 1972. Parameters affecting the association of vibrios with the intestinal surface in experimental cholera. *Infect. Immun.* **6:**134–141.
36. **Freter, R., P. C. O'Brien, and M. S. Macsai.** 1981. Role of chemotaxis in the association of motile bacteria with intestinal mucosa: in vivo studies. *Infect. Immun.* **34:**234–240.
37. **Gardel, C. L., and J. J. Mekalanos.** 1996. Alterations in *Vibrio cholerae* motility phenotypes correlate with changes in virulence factor expression. *Infect. Immun.* **64:**2246–2255.
38. **Gewirtz, A. T., T. A. Navas, S. Lyons, P. J. Godowski, and J. L. Madara.** 2001. Cutting edge: bacterial flagellin activates basolaterally expressed TLR5 to induce epithelial proinflammatory gene expression. *J. Immunol.* **167:**1882–1885.
39. **Gewirtz, A. T., P. O. Simon, Jr., C. K. Schmitt, L. J. Taylor, C. H. Hagedorn, A. D. O'Brien, A. S. Neish, and J. L. Madara.** 2001. *Salmonella typhimurium* translocates flagellin across intestinal epithelia, inducing a proinflammatory response. *J. Clin. Investig.* **107:**99–109.
40. **Giron, J. A., A. G. Torres, E. Freer, and J. B. Kaper.** 2002. The flagella of enteropathogenic *Escherichia coli* mediate adherence to epithelial cells. *Mol. Microbiol.* **44:**361–379.
41. **Guentzel, M. N., and L. J. Berry.** 1975. Motility as a virulence factor for *Vibrio cholerae*. *Infect. Immun.* **11:**890–897.
42. **He, G., R. A. Shankar, M. Chzhan, A. Samouilov, P. Kuppusamy, and J. L. Zweier.** 1999. Noninvasive measurement of anatomic structure and intraluminal oxygenation in the gastrointestinal tract of living mice with spatial and spectral EPR imaging. *Proc. Natl. Acad. Sci. USA* **96:**4586–4591.
43. **Hentges, D. J.** 1983. Role of the intestinal microflora in host defense against infection, p. 311–331. *In* D. J. Hentges (ed.), *Human Intestinal Microflora in Health and Disease.* Academic Press, Inc., New York, N.Y.
44. **Hentges, D. J., J. U. Que, S. W. Casey, and A. J. Stein.** 1984. The influence of streptomycin on colonization in mice. *Microecol. Theor.* **14:**53–62.
45. **Hooper, L. V., T. Midtvedt, and J. I. Gordon.** 2002. How host-microbial interactions shape the nutrient environment of the mammalian intestine. *Annu. Rev. Nutr.* **22:**283–307.
46. **Hooper, L. V., J. Xu, P. G. Falk, T. Midtvedt, and J. I. Gordon.** 1999. A molecular sensor that allows a gut commensal to control its nutrient foundation in a competitive ecosystem. *Proc. Natl. Acad. Sci. USA* **96:**9833–9838.
47. **Hoskins, L. C., M. Agustines, W. B. McKee, E. T. Boulding, M. Kriaris, and G. Niedermeyer.** 1985. Mucin degradation in human colon ecosystems. Isolation and properties of fecal strains that degrade ABH blood group antigens and oligosaccharides from mucin glycoproteins. *J. Clin. Investig.* **75:**944–953.
48. **Hudault, S., J. Guignot, and A. L. Servin.** 2001. *Escherichia coli* strains colonising the gastrointestinal tract protect germfree mice against *Salmonella typhimurium* infection. *Gut* **49:**47–55.
49. **Hueck, C. J.** 1998. Type III protein secretion systems in bacterial pathogens of animals and plants. *Microbiol. Mol. Biol. Rev.* **62:**379–433.
50. **Hughes, K. T., K. L. Gillen, M. J. Semon, and J. E. Karlinsey.** 1993. Sensing structural intermediates in bacterial flagellar assembly by export of a negative regulator. *Science* **262:**1277–1280.
51. **Jonsson, H., E. Strom, and S. Roos.** 2001. Addition of mucin to the growth medium triggers mucus-binding activity in different strains of *Lactobacillus reuteri in vitro*. *FEMS Microbiol. Lett.* **204:**19–22.
52. **Karjalainen, T., M. C. Barc, A. Collignon, S. Trolle, H. Boureau, J. Cotte-Laffitte, and P. Bourlioux.** 1994. Cloning of a genetic determinant from *Clostridium difficile* involved in adherence to tissue culture cells and mucus. *Infect. Immun.* **62:**4347–4355.
53. **Karjalainen, T., A. J. Waligora-Dupriet, M. Cerquetti, P. Spigaglia, A. Maggioni, P. Mauri, and P. Mastrantonio.** 2001. Molecular and genomic analysis of genes encoding surface-anchored proteins from *Clostridium difficile*. *Infect. Immun.* **69:**3442–3446.
54. **Kennedy, M. J., E. L. Rosey, and R. J. Yancey, Jr.** 1997. Characterization of *flaA*$^-$ and *flaB*$^-$ mutants of *Serpulina hyodysenteriae:* both flagellin subunits, FlaA and FlaB, are necessary for full motility and intestinal colonization. *FEMS Microbiol. Lett.* **153:**119–128.
55. **Ketyi, I.** 1994. Effectiveness of antibiotics on the autochthonous *Escherichia coli* of mice in the intestinal biofilm versus its

planktonic phase. *Acta Microbiol. Immunol. Hung.* **41:**189–195.

56. **Kolenbrander, P. E.** 2000. Oral microbial communities: biofilms, interactions, and genetic systems. *Annu. Rev. Microbiol.* **54:**413–437.
57. **Kresse, A. U., F. Ebel, and C. A. Guzman.** 2002. Functional modulation of pathogenic bacteria upon contact with host target cells, p. 203–220. *In* M. Wilson (ed.), *Bacterial Adhesion to Host Tissues: Mechanisms and Consequences.* Cambridge University Press, Cambridge, United Kingdom.
58. **Krivan, H. C., D. P. Franklin, W. Wang, D. C. Laux, and P. S. Cohen.** 1992. Phosphatidylserine found in intestinal mucus serves as a sole source of carbon and nitrogen for salmonellae and *Escherichia coli. Infect. Immun.* **60:**3943–3946.
59. **Krogfelt, K. A., B. A. McCormick, R. L. Burghoff, D. C. Laux, and P. S. Cohen.** 1991. Expression of *Escherichia coli* F-18 type 1 fimbriae in the streptomycin-treated mouse large intestine. *Infect. Immun.* **59:**1567–1568.
60. **Kutsukake, K.** 1994. Excretion of the anti-sigma factor through a flagellar substructure couples flagellar gene expression with flagellar assembly in *Salmonella typhimurium. Mol. Gen. Genet.* **243:**605–612.
61. **Lamont, J. T.** 1992. Mucus: the front line of intestinal mucosal defense. *Ann. N. Y. Acad. Sci.* **664:**190–201.
62. **La Ragione, R. M., A. R. Sayers, and M. J. Woodward.** 2000. The role of fimbriae and flagella in the colonization, invasion and persistence of *Escherichia coli* O78:K80 in the day-old-chick model. *Epidemiol. Infect.* **124:**351–363.
63. **Laux, D. C., E. F. McSweegan, and P. S. Cohen.** 1984. Adhesion of enterotoxigenic *Escherichia coli* to immobilized intestinal mucosal preparations: a model of adhesion to mucosal surface components. *J. Microbiol. Methods* **2:**27–39.
64. **Lee, A.** 1985. Neglected niches. The microbial ecology of the gastrointestinal tract. *Adv. Microb. Ecol.* **8:**115–162.
65. **Licht, T. R., B. B. Christensen, K. A. Krogfelt, and S. Molin.** 1999. Plasmid transfer in the animal intestine and other dynamic bacterial populations: the role of community structure and environment. *Microbiology* **145:**2615–2622.
66. **Lievin, V., I. Peiffer, S. Hudault, F. Rochat, D. Brassart, J. R. Neeser, and A. L. Servin.** 2000. *Bifidobacterium* strains from resident infant human gastrointestinal microflora exert antimicrobial activity. *Gut* **47:**646–652.
67. **Lindahl, M., and I. Carlstedt.** 1990. Binding of K99 fimbriae of enterotoxigenic *Escherichia coli* to pig small intestinal mucin glycopeptides. *J. Gen. Microbiol.* **136:**1609–1614.
68. **Lloyd, A. B., R. B. Cumming, and R. D. Kent.** 1977. Prevention of *Salmonella typhimurium* infection in poultry by pretreatment of chickens and poults with intestinal extracts. *Aust. Vet. J.* **53:**82–87.
69. **Lockman, H. A., and R. Curtiss, 3rd.** 1990. *Salmonella typhimurium* mutants lacking flagella or motility remain virulent in BALB/c mice. *Infect. Immun.* **58:**137–143.
70. **Macfarlane, S., A. J. McBain, and G. T. Macfarlane.** 1997. Consequences of biofilm and sessile growth in the large intestine. *Adv. Dent. Res.* **11:**59–68.
71. **Mack, D. R., and P. L. Blain-Nelson.** 1995. Disparate *in vitro* inhibition of adhesion of enteropathogenic *Escherichia coli* RDEC-1 by mucins isolated from various regions of the intestinal tract. *Pediatr. Res.* **37:**75–80.
72. **Mack, D. R., and P. M. Sherman.** 1991. Mucin isolated from rabbit colon inhibits in vitro binding of *Escherichia coli* RDEC-1. *Infect. Immun.* **59:**1015–1023.
73. **MacNab, R. M.** 1996. Flagella and motility, p. 123–145. *In* F. C. Neidhardt, R. Curtiss III, J. L. Ingraham, E. C. C. Lin, K. B. Low, B. Magasanik, W. S. Reznikoff, M. Riley, M. Schaechter, and H. E. Umbarger (ed.), Escherichia coli *and* Salmonella: *Cellular and Molecular Biology,* 2nd ed., vol. 1. ASM Press, Washington, D.C.
74. **Mantle, M., L. Basaraba, S. C. Peacock, and D. G. Gall.** 1989. Binding of *Yersinia enterocolitica* to rabbit intestinal brush border membranes, mucus, and mucin. *Infect. Immun.* **57:**3292–3299.
75. **Mantle, M., and S. D. Husar.** 1993. Adhesion of *Yersinia enterocolitica* to purified rabbit and human intestinal mucin. *Infect. Immun.* **61:**2340–2346.
76. **Mantle, M., and S. D. Husar.** 1994. Binding of *Yersinia enterocolitica* to purified, native small intestinal mucins from rabbits and humans involves interactions with the mucin carbohydrate moiety. *Infect. Immun.* **62:**1219–1227.
77. **McCormick, B. A., D. P. Franklin, D. C. Laux, and P. S. Cohen.** 1989. Type 1 pili are not necessary for colonization of the streptomycin-treated mouse large intestine by type 1-piliated *Escherichia coli* F-18 and *E. coli* K-12. *Infect. Immun.* **57:**3022–3029.
78. **McCormick, B. A., P. Klemm, K. A. Krogfelt, R. L. Burghoff, L. Pallesen, D. C. Laux, and P. S. Cohen.** 1993. *Escherichia coli* F-18 phase locked 'on' for expression of type 1 fimbriae is a poor colonizer of the streptomycin-treated mouse large intestine. *Microb. Pathog.* **14:**33–43.
79. **McCormick, B. A., D. C. Laux, and P. S. Cohen.** 1990. Neither motility nor chemotaxis plays a role in the ability of *Escherichia coli* F-18 to colonize the streptomycin-treated mouse large intestine. *Infect. Immun.* **58:**2957–2961.
80. **McCormick, B. A., B. A. Stocker, D. C. Laux, and P. S. Cohen.** 1988. Roles of motility, chemotaxis, and penetration through and growth in intestinal mucus in the ability of an avirulent strain of *Salmonella typhimurium* to colonize the large intestine of streptomycin-treated mice. *Infect. Immun.* **56:**2209–2217.
81. **McSweegan, E., D. H. Burr, and R. I. Walker.** 1987. Intestinal mucus gel and secretory antibody are barriers to *Campylobacter jejuni* adherence to INT 407 cells. *Infect. Immun.* **55:**1431–1435.
82. **McSweegan, E., and R. I. Walker.** 1986. Identification and characterization of two *Campylobacter jejuni* adhesins for cellular and mucous substrates. *Infect. Immun.* **53:**141–148.
83. **Midtvedt, T.** 1985. The influence of antibiotics upon microflora-associated characteristics in man and animals. *Prog. Clin. Biol. Res.* **181:**241–244.
84. **Miranda, R. L., T. Conway, M. P. Leatham, D. E. Chang, W. E. Norris, J. H. Allen, S. J. Stevenson, D. C. Laux, and P. S. Cohen.** 2004. Glycolytic and gluconeogenic growth of *Escherichia coli* O157:H7 (EDL933) and *E. coli* K-12 (MG1655) in the mouse intestine. *Infect. Immun.* **72:**1666–1676.
85. **Moens, S., and J. Vanderleyden.** 1996. Functions of bacterial flagella. *Crit. Rev. Microbiol.* **22:**67–100.
86. **Moller, A. K., M. P. Leatham, T. Conway, P. J. Nuijten, L. A. de Haan, K. A. Krogfelt, and P. S. Cohen.** 2003. An *Escherichia coli* MG1655 lipopolysaccharide deep-rough core mutant grows and survives in mouse cecal mucus but fails to colonize the mouse large intestine. *Infect. Immun.* **71:**2142–2152.
87. **Morooka, T., A. Umeda, and K. Amako.** 1985. Motility as an intestinal colonization factor for *Campylobacter jejuni. J. Gen. Microbiol.* **131:**1973–1980.
88. **Myhal, M. L., D. C. Laux, and P. S. Cohen.** 1982. Relative colonizing abilities of human fecal and K 12 strains of *Escherichia coli* in the large intestines of streptomycin-treated mice. *Eur. J. Clin. Microbiol.* **1:**186–192.
89. **Neill, M. A.** 1998. Treatment of disease due to Shiga toxin-producing *Escherichia coli:* Infectious disease management,

p. 357–363. *In* J. B. Kaper and A. D. O'Brien (ed.), Escherichia coli *O157:H7 and Other Shiga Toxin-Producing* E. coli *Strains.* ASM Press, Washington, D.C.

90. **Ottemann, K. M., and J. F. Miller.** 1997. Roles for motility in bacterial-host interactions. *Mol. Microbiol.* **24:**1109–1117.
91. **Ouellette, A. J., and M. E. Selsted.** 1996. Paneth cell defensins: endogenous peptide components of intestinal host defense. *FASEB J.* **10:**1280–1289.
92. **Ouwehand, A. C., E. Isolauri, P. V. Kirjavainen, S. Tolkko, and S. J. Salminen.** 2000. The mucus binding of *Bifidobacterium lactis* Bb12 is enhanced in the presence of *Lactobacillus* GG and *Lact. delbrueckii* subsp. bulgaricus. *Lett. Appl. Microbiol.* **30:**10–13.
93. **Pavlovskis, O. R., D. M. Rollins, R. L. Haberberger, Jr., A. E. Green, L. Habash, S. Strocko, and R. I. Walker.** 1991. Significance of flagella in colonization resistance of rabbits immunized with *Campylobacter* spp. *Infect. Immun.* **59:**2259–2264.
94. **Peekhaus, N., and T. Conway.** 1998. What's for dinner? Entner-Doudoroff metabolism in *Escherichia coli. J. Bacteriol.* **180:**3495–3502.
95. **Pierson, D. E.** 2002. Induction of protein secretion by *Yersinia enterocolitica* through contact with eukaryotic cells, p. 183–202. *In* M. Wilson (ed.), *Bacterial Adhesion to Host Tissues: Mechanisms and Consequences.* Cambridge University Press, Cambridge, United Kingdom.
96. **Plano, G. V., J. B. Day, and F. Ferracci.** 2001. Type III export: new uses for an old pathway. *Mol. Microbiol.* **40:**284–293.
97. **Postnova, T., O. G. Gomez-Duarte, and K. Richardson.** 1996. Motility mutants of *Vibrio cholerae* O1 have reduced adherence *in vitro* to human small intestinal epithelial cells as demonstrated by ELISA. *Microbiology* **142:**2767–2776.
98. **Poulsen, L. K., F. Lan, C. S. Kristensen, P. Hobolth, S. Molin, and K. A. Krogfelt.** 1994. Spatial distribution of *Escherichia coli* in the mouse large intestine inferred from rRNA in situ hybridization. *Infect. Immun.* **62:**5191–5194.
99. **Poulsen, L. K., T. R. Licht, C. Rang, K. A. Krogfelt, and S. Molin.** 1995. Physiological state of *Escherichia coli* BJ4 growing in the large intestines of streptomycin-treated mice. *J. Bacteriol.* **177:**5840–5845.
100. **Pratt, L. A., and R. Kolter.** 1998. Genetic analysis of *Escherichia coli* biofilm formation: roles of flagella, motility, chemotaxis and type I pili. *Mol. Microbiol.* **30:**285–293.
101. **Probert, H. M., and G. R. Gibson.** 2002. Bacterial biofilms in the human gastrointestinal tract. *Curr. Issues Intest. Microbiol.* **3:**23–27.
102. **Relman, D. A., and S. Falkow.** 2001. The meaning and impact of the human genome sequence for microbiology. *Trends Microbiol.* **9:**206–208.
103. **Rickard, A. H., P. Gilbert, N. J. High, P. E. Kolenbrander, and P. S. Handley.** 2003. Bacterial coaggregation: an integral process in the development of multi-species biofilms. *Trends Microbiol.* **11:**94–100.
104. **Robertson, J. M., G. Grant, E. Allen-Vercoe, M. J. Woodward, A. Pusztai, and H. J. Flint.** 2000. Adhesion of *Salmonella enterica* var Enteritidis strains lacking fimbriae and flagella to rat ileal explants cultured at the air interface or submerged in tissue culture medium. *J. Med. Microbiol.* **49:** 691–696.
105. **Rojas, M., F. Ascencio, and P. L. Conway.** 2002. Purification and characterization of a surface protein from *Lactobacillus fermentum* 104R that binds to porcine small intestinal mucus and gastric mucin. *Appl. Environ. Microbiol.* **68:**2330–2336.
106. **Rolfe, R. D.** 1984. Role of volatile fatty acids in colonization resistance to *Clostridium difficile. Infect. Immun.* **45:**185–191.
107. **Roos, S., and H. Jonsson.** 2002. A high-molecular-mass cell-surface protein from *Lactobacillus reuteri* 1063 adheres to mucus components. *Microbiology* **148:**433–442.
108. **Roos, S., F. Karner, L. Axelsson, and H. Jonsson.** 2000. *Lactobacillus mucosae* sp. nov., a new species with *in vitro* mucus-binding activity isolated from pig intestine. *Int. J. Syst. Evol. Microbiol.* **50:**251–258.
109. **Rozee, K. R., D. Cooper, K. Lam, and J. W. Costerton.** 1982. Microbial flora of the mouse ileum mucous layer and epithelial surface. *Appl. Environ. Microbiol.* **43:**1451–1463.
110. **Salyers, A. A., and J. A. Z. Leedle.** 1983. Carbohydrate metabolism in the human colon, p. 129–146. *In* D. J. Hentges (ed.), *Human Intestinal Microflora in Health and Disease.* Academic Press, Inc., New York, N.Y.
111. **Savage, D. C.** 1984. Adherence of the normal flora, p. 3–10. *In* E. C. Boedeker (ed.), *Attachment of Organisms to the Gut Mucosa,* vol. 1. CRC Press, Inc., Boca Raton, Fla.
112. **Savage, D. C.** 1983. Associations of indigenous microorganisms with gastrointestinal epithelial surfaces, p. 55–78. *In* D. H. Hentges (ed.), *Human Intestinal Microflora in Health and Disease.* Academic Press, Inc., New York, N.Y.
113. **Savage, D. C.** 1985. Effects on host animals of bacteria adhering to epithelial surfaces, p. 437–463. *In* D. C. Savage and M. Fletcher (ed.), *Bacterial Adhesion: Mechanisms and Physiological Significance.* Plenum Press, New York, N.Y.
114. **Sheikh, J., S. Hicks, M. Dall'Agnol, A. D. Phillips, and J. P. Nataro.** 2001. Roles for Fis and YafK in biofilm formation by enteroaggregative *Escherichia coli. Mol. Microbiol.* **41:**983–997.
115. **Simon, G. L., and S. L. Gorbach.** 1984. Intestinal flora in health and disease. *Gastroenterology* **86:**174–193.
116. **Slater, J. H.** 1988. Microbial populations and community dynamics, p. 51–74. *In* J. M. Lynch and J. E. Hobbie (ed.), *Microorganisms in Action: Concepts and Applications in Microbial Ecology.* Blackwell Scientific Publications Ltd., Oxford, United Kingdom.
117. **Smith, C. J., J. B. Kaper, and D. R. Mack.** 1995. Intestinal mucin inhibits adhesion of human enteropathogenic *Escherichia coli* to HEp-2 cells. *J. Pediatr. Gastroenterol. Nutr.* **21:**269–276.
118. **Stanton, T. B., and D. C. Savage.** 1984. Motility as a factor in bowel colonization by *Roseburia cecicola,* an obligately anaerobic bacterium from the mouse caecum. *J. Gen. Microbiol.* **130:**173–183.
119. **Stock, J. B., and M. G. Surette.** 1996. Chemotaxis, p. 1103–1129. *In* F. C. Neidhardt, R. Curtiss III, J. L. Ingraham, E. C. C. Lin, K. B. Low, B. Magasanik, W. S. Reznikoff, M. Riley, M. Schaechter, and H. E. Umbarger (ed.), Escherichia coli *and* Salmonella: 2nd ed., vol. 1. ASM Press, Washington, D.C.
120. **Strous, G. J., and J. Dekker.** 1992. Mucin-type glycoproteins. *Crit. Rev. Biochem. Mol. Biol.* **27:**57–92.
121. **Sweeney, N. J., P. Klemm, B. A. McCormick, E. Moller-Nielsen, M. Utley, M. A. Schembri, D. C. Laux, and P. S. Cohen.** 1996. The *Escherichia coli* K-12 *gntP* gene allows *E. coli* F-18 to occupy a distinct nutritional niche in the streptomycin-treated mouse large intestine. *Infect. Immun.* **64:**3497–3503.
122. **Sweeney, N. J., D. C. Laux, and P. S. Cohen.** 1996. *Escherichia coli* F-18 and *E. coli* K-12 *eda* mutants do not colonize the streptomycin-treated mouse large intestine. *Infect. Immun.* **64:**3504–3511.
123. **Sylvester, F. A., D. Philpott, B. Gold, A. Lastovica, and J. F. Forstner.** 1996. Adherence to lipids and intestinal mucin by a recently recognized human pathogen, *Campylobacter upsaliensis. Infect. Immun.* **64:**4060–4066.

124. **Takata, T., S. Fujimoto, and K. Amako.** 1992. Isolation of nonchemotactic mutants of *Campylobacter jejuni* and their colonization of the mouse intestinal tract. *Infect. Immun.* **60:**3596–3600.
125. **Tannock, G. W.** 1997. Normal microbiota of the gastrointestinal tract of rodents, p. 187–215. *In* R. I. Mackie, B. A. White, and R. E. Isaacson (ed.), *Gastrointestinal Microbiology,* vol. 2. Chapman & Hall, London, United Kingdom.
126. **Tasteyre, A., M. C. Barc, A. Collignon, H. Boureau, and T. Karjalainen.** 2001. Role of FliC and FliD flagellar proteins of *Clostridium difficile* in adherence and gut colonization. *Infect. Immun.* **69:**7937–7940.
127. **Taylor, C., and I. S. Roberts.** 2002. The regulation of capsule expression, p. 115–138. *In* M. Wilson (ed.), *Bacterial Adhesion to Host Tissues: Mechanisms and Consequences.* Cambridge University Press, Cambridge, United Kingdom.
128. **Taylor, P. A., and P. J. Williams.** 1975. Theoretical studies on the coexistence of competing species under continuous-flow conditions. *Can. J. Microbiol.* **21:**90–98.
129. **Tilman, D.** 1982. Resource competition and community structure. *Monogr. Popul. Biol.* **17:**1–296.
130. **Van de Bovenkamp, J. H., J. Mahdavi, A. M. Korteland-Van Male, H. A. Buller, A. W. Einerhand, T. Boren, and J. Dekker.** 2003. The MUC5AC glycoprotein is the primary receptor for *Helicobacter pylori* in the human stomach. *Helicobacter* **8:**521–532.
131. **van der Waaij, D., J. M. Berghuis-de Vries, and J. E. Lekkerkerk.** 1971. Colonization resistance of the digestive tract in conventional and antibiotic-treated mice. *J. Hyg.* (London) **69:**405–411.
132. **Vimal, D. B., M. Khullar, S. Gupta, and N. K. Ganguly.** 2000. Intestinal mucins: the binding sites for *Salmonella typhimurium. Mol. Cell. Biochem.* **204:**107–117.
133. **Wadolkowski, E. A., D. C. Laux, and P. S. Cohen.** 1988. Colonization of the streptomycin-treated mouse large intestine by a human fecal *Escherichia coli* strain: role of adhesion to mucosal receptors. *Infect. Immun.* **56:**1036–1043.
134. **Waligora, A. J., C. Hennequin, P. Mullany, P. Bourlioux, A. Collignon, and T. Karjalainen.** 2001. Characterization of a cell surface protein of *Clostridium difficile* with adhesive properties. *Infect. Immun.* **69:**2144–2153.
135. **Wassenaar, T. M., B. A. van der Zeijst, R. Ayling, and D. G. Newell.** 1993. Colonization of chicks by motility mutants of *Campylobacter jejuni* demonstrates the importance of flagellin A expression. *J. Gen. Microbiol.* **139:** 1171–1175.
136. **Xu, J., and J. I. Gordon.** 2003. Inaugural Article. Honor thy symbionts. *Proc. Natl. Acad. Sci. USA* **100:**10452–10459.
137. **Young, G. M., J. L. Badger, and V. L. Miller.** 2000. Motility is required to initiate host cell invasion by *Yersinia enterocolitica. Infect. Immun.* **68:**4323–4326.

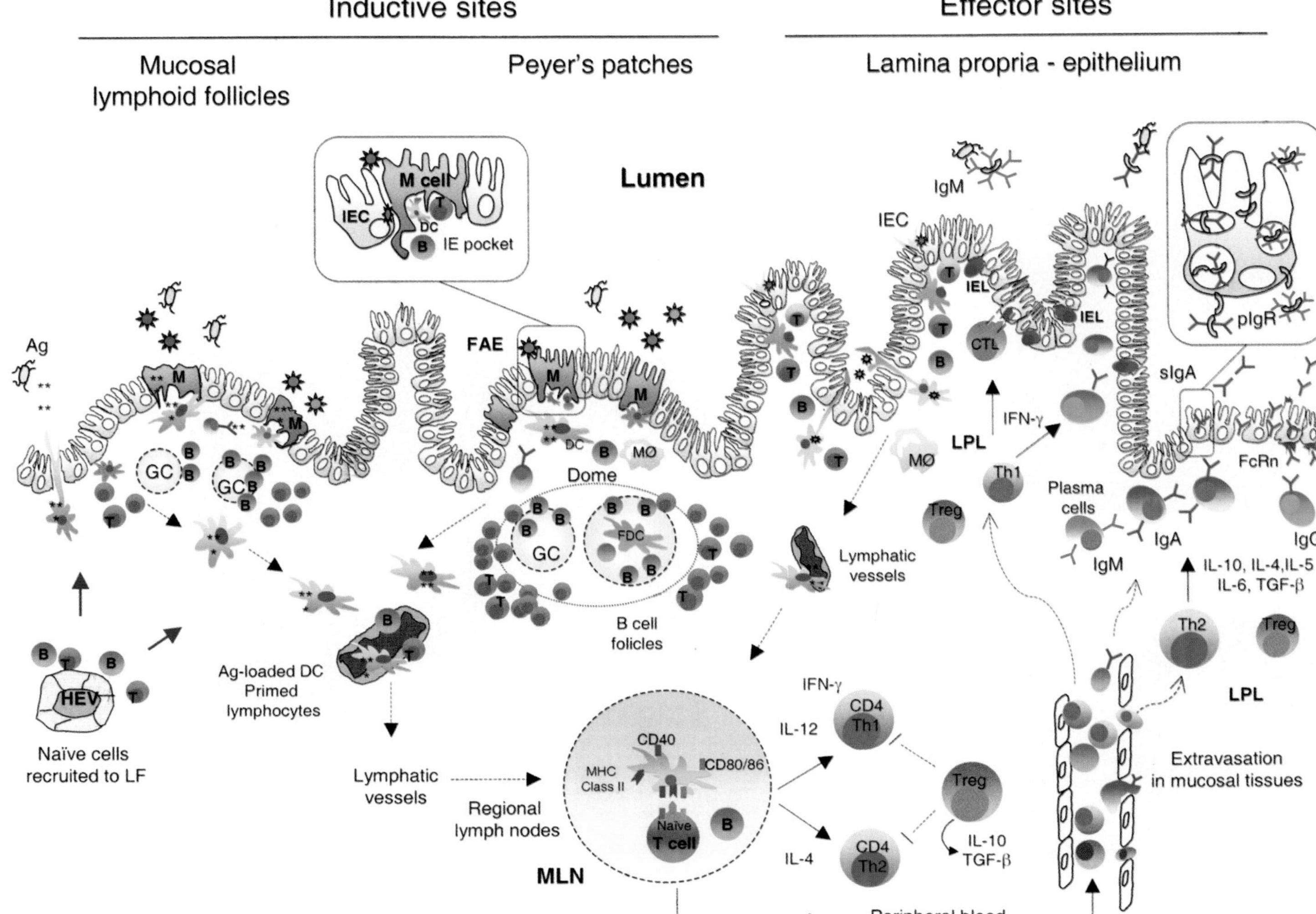

Color Plate 1. (Chapter 3) Schematic representation of inductive and effector sites of the GALT. Ag, antigen; B, B lymphocyte; CTL, cytotoxic T lymphocyte; DC, dendritic cell; FAE, follicle associated epithelium; FcRn, neonatal Fc receptor; FDC, follicular dendritic cells; GC, germinal center; IE, intraepithelial; IEC, intestinal epithelial cell; IEL, intraepithelial lymphocyte; IFN-γ, interferon-γ; IgA, immunoglobulin A; IgG, immunoglobulin G; IgM, immunoglobulin M; IL, interleukin; LF, lymphoid follicle; LPL, lamina propria lymphocyte; M, microfold-M cell; Mϕ, macrophage; MHC, major histocompatibility complex; MLN, mesenteric lymph node; pIgR, polymeric immunoglobulin receptor; sIgA, secretory IgA; T, T lymphocyte; TGF-β, transforming growth factor-β; Th1, T_{helper} type 1 lymphocyte; Th2, T_{helper} type 2 lymphocyte; T_{reg}, regulatory T cell. For other abbreviations, see chapter 3.

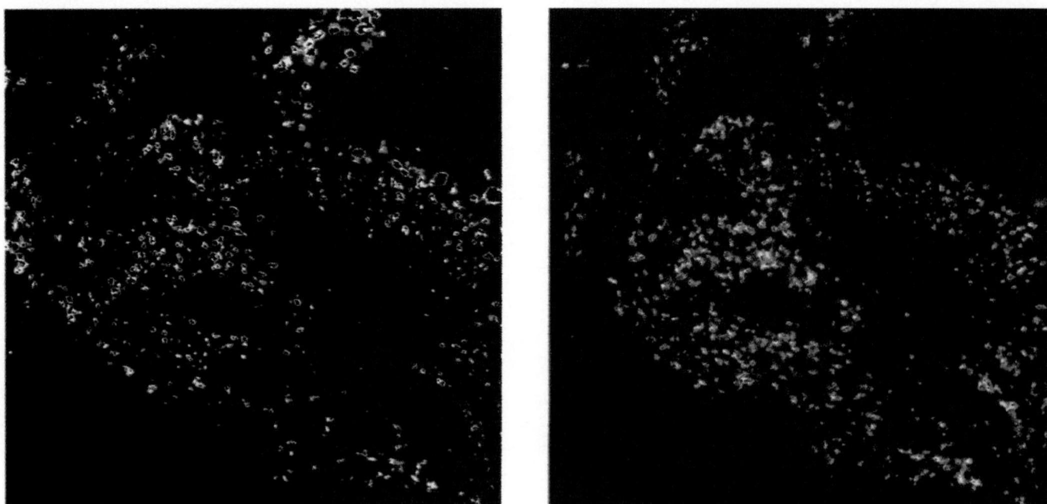

Color Plate 2. (Chapter 4) Confocal laser-scanning micrographs showing *P. aeruginosa* quorum sensing visualized in the lungs of mice. The cells were tagged with Rfp, and they expressed green fluorescence (Gfp ASV) in response to quorum-sensing signals received from their nearest neighbors (see references 31 and 32 in chapter 4 for details).

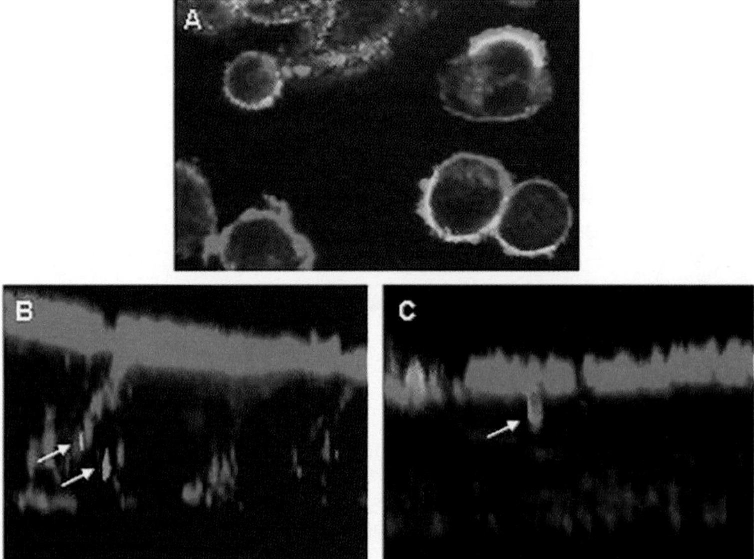

Color Plate 3. (Chapter 8) (A) CEACAM interactions, cytoskeletal reorganization and transmigration of *N. meningitidis* and *H. influenzae*. On interactions of *H. influenzae* with target CHO cells transfected with CEACAM1, cellular actin is reorganized and filamentous actin (stained with rhodamine-conjugated phalloidin) is colocalized beneath bacterially induced receptor caps. *H. influenzae* organisms are stained with fluorescein isothiocyanate-conjugated anti-*H. influenzae* antibodies (150). (B and C) *H. influenzae* and *N. meningitidis* transmigration across polarized human epithelial cells was studied using Caco-2 cells as a model system since they express high levels of CEACAMs. *x-z* sections obtained by confocal imaging show *N. meningitidis* located within (arrow in panel C) the epithelial cell whereas *H. influenzae* is often found in junctions between cells (arrows in panel B). Bacteria were labeled using antibodies raised against bacteria and fluorescein isothiocyanate-conjugated secondary antibodies. Filamentous actin was stained with rhodamine-conjugated phalloidin. Different modes of transmigration of *N. meningitidis* and *H. influenzae* are similar to those observed in organ culture studies (see Fig. 1 in chapter 8) (M. Soriani and M. Virji, unpublished data). Confocal images were obtained with the help of Mark Jepson in the MRC cell imaging facility at the School of Medical Sciences, University of Bristol.

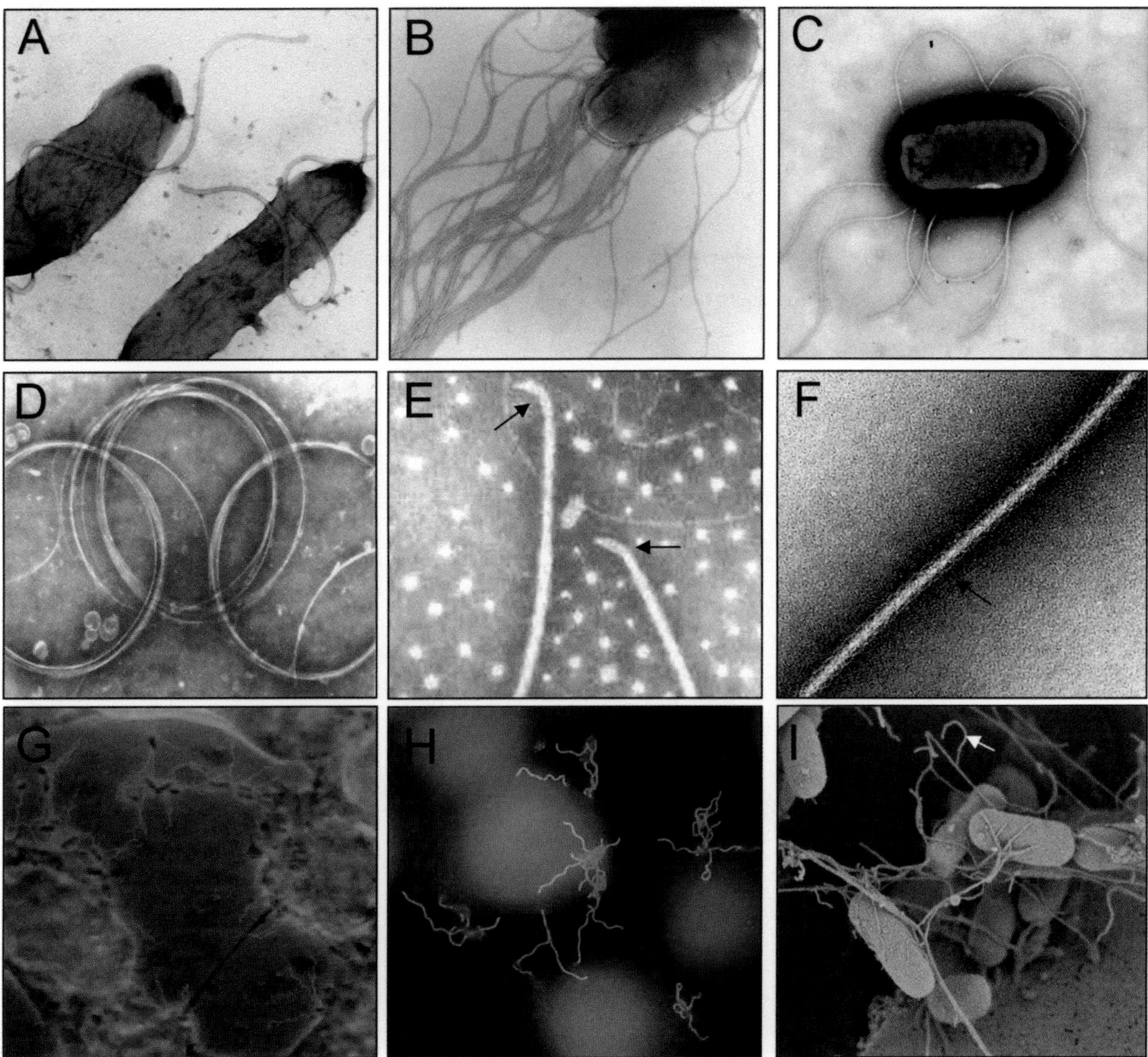

Color Plate 4. (Chapter 16) Micrographs of bacterial flagella. (A) *V. cholerae* O395 producing a sheathed polar flagellum. (B) *S. enterica* producing multiple peritrichous flagella. (C) EHEC EDL933 producing various flagella. (D) Flagellar structures isolated from an ETEC strain, demonstrating the flexible nature of the filaments. (E) Flagellar hook attached to the flagellum (arrows). (F) High magnification of a flagellum, showing the helical nature of the filament. (G and H) Flagella (green-fluorescent structures) produced by adhering EPEC to HeLa cells tethering bacteria to the inert substratum and to the cultured cells. (I) Scanning electron micrograph of EPEC adhering to HeLa cells and producing multiple interconnecting flagellar structures.

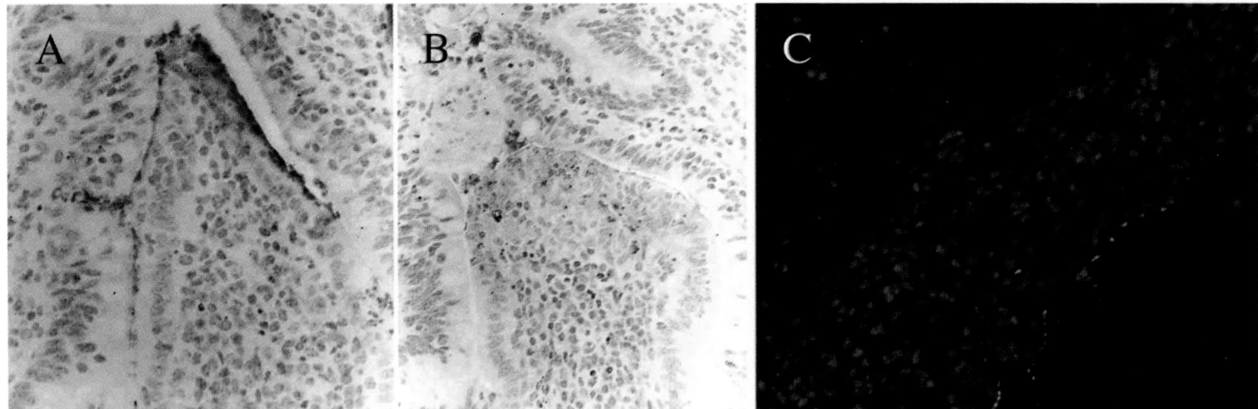

Color Plate 5. (Chapter 21) Colonization by *S. enterica* serotype Typhimurium of the bovine (A and B) and murine (C) mucosa. (A and B) Detection of serotype Typhimurium in sections of the bovine ileal mucosa collected at 15 min (A) and 2 h (B) after inoculation of bovine ligated loops by immunohistochemistry using rabbit anti-O4,5 antiserum (brown precipitate). The center of each microscopic field shows a domed villus of a Peyer's patch lymphoid follicle. The sections were counterstained with hematoxylin (blue signal). Reprinted from B. P. Reis, S. Zhang, R. M. Tsolis, A. J. Bäumler, L. G. Adams, and R. L. Santos, *Vet. Microbiol.* **97**:269–277, 2003, with permission from the publisher. (C) Section of the murine cecum collected 5 days after oral infection with a lethal dose of serotype Typhimurium (10^9 CFU/animal), showing bacterial microcolonies lining the epithelial surface. Serotype Typhimurium was detected by fluorescence microscopy using sheep anti-somatic (O4) antiserum and donkey anti-sheep immunoglobulin Alexa-Fluor 488 (green signal). Sections were counterstained with Hoechst nuclear stain (blue signal).

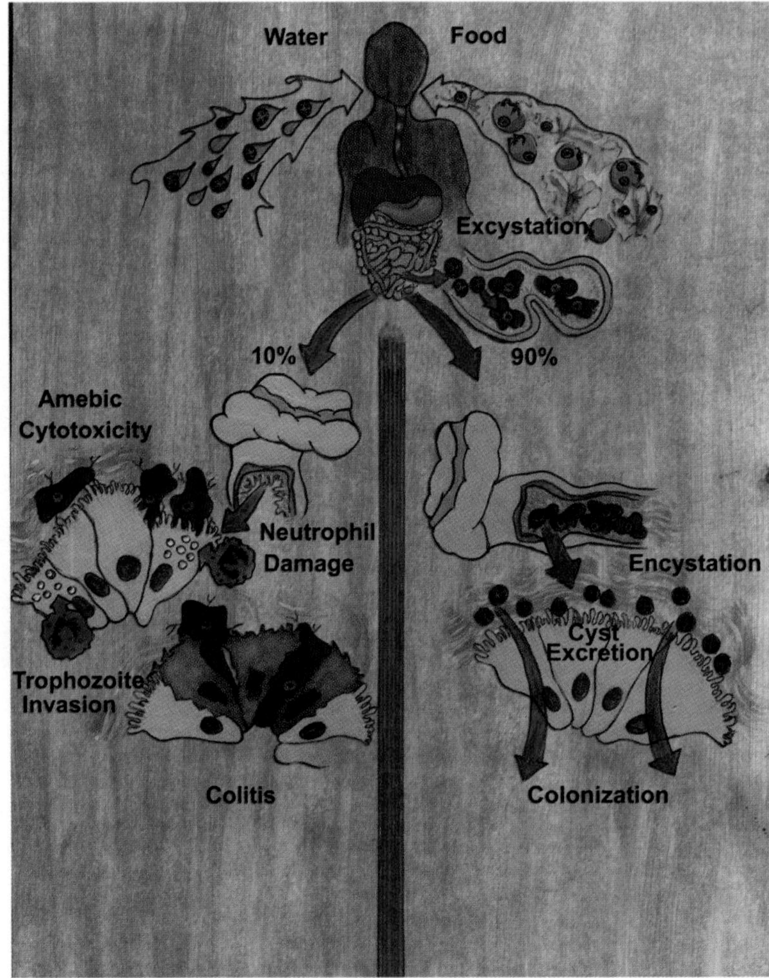

Color Plate 6. (Chapter 22) Life cycle of *E. histolytica*. Infection is initiated by ingestion of the cyst from fecally contaminated food or water. Excystation in the intestine to the trophozoite form leads to colonization in 90% or more of individuals and invasion in 10% or fewer. Colonization leads to the production of new cysts that are formed via a quorum-sensing-like interaction of amebae that requires the Gal/GalNAc lectin. Invasion is mediated by amebic contact-dependent cytotoxicity as well as via neutrophil-mediated damage.

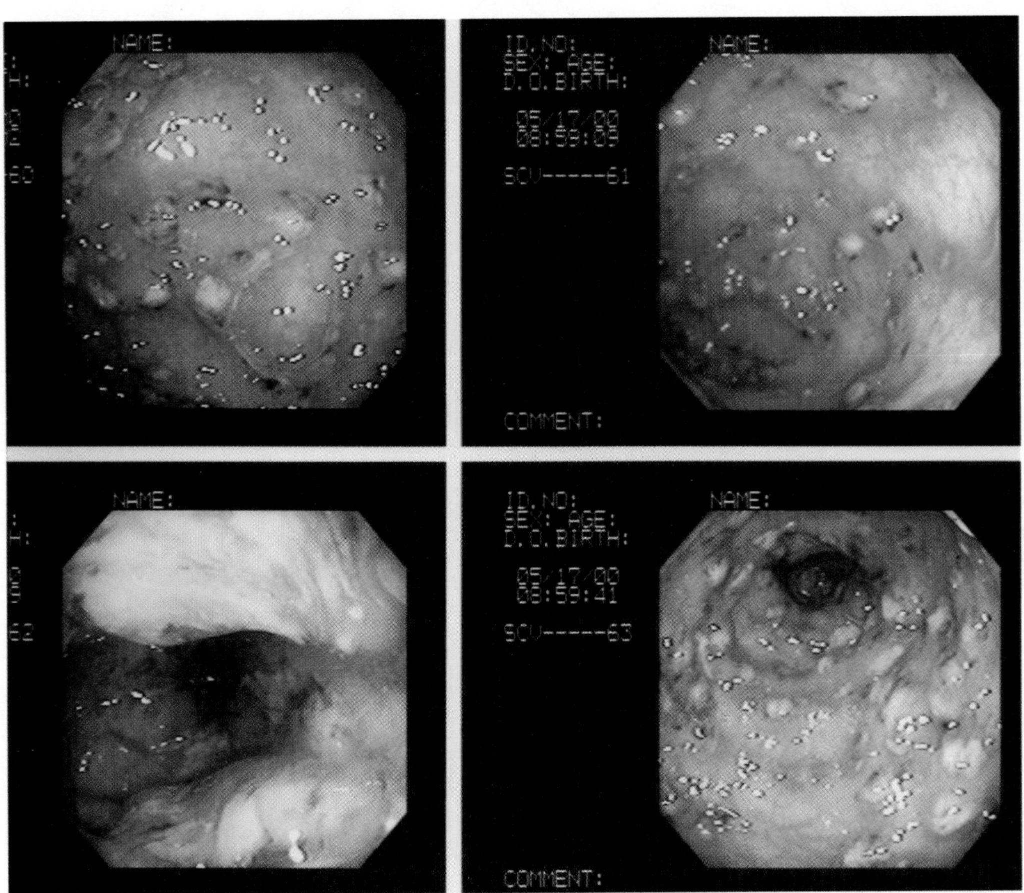

Color Plate 7. (Chapter 22) Amebic colitis in a human viewed by endoscopy. Note the raised white ulcerations in the intestinal epithelium. Reprinted from R. Haque, C. Huston, M. Hughes, E. Houpt, and W. A. Petri, *N. Engl. J. Med* **348:**1565–1573, 2003, with permission from the publisher.

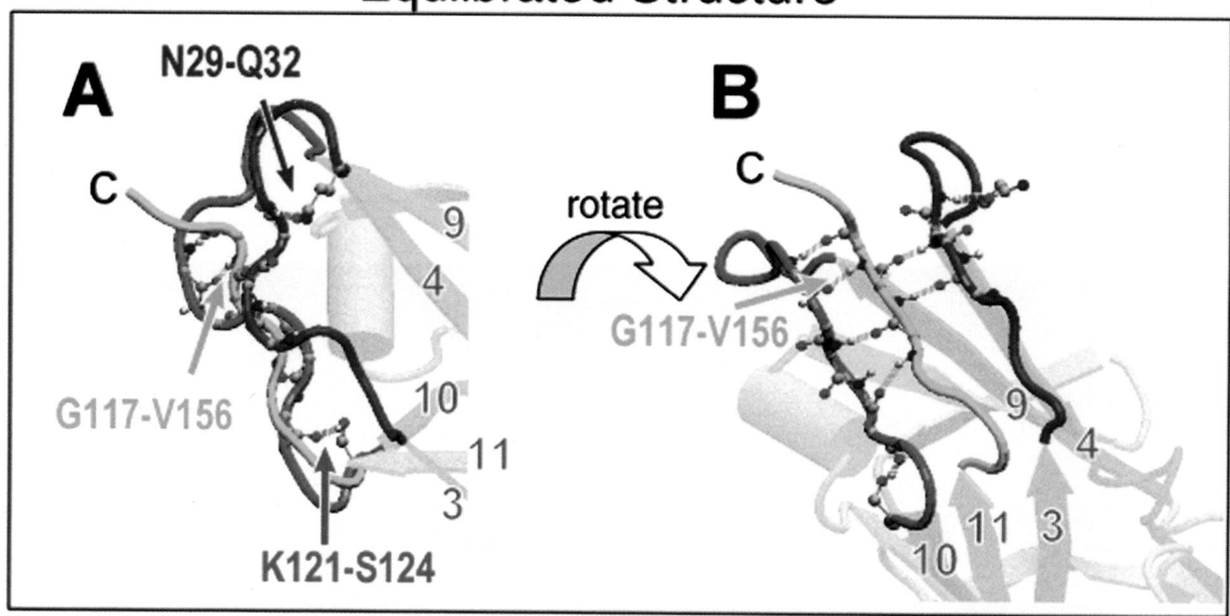

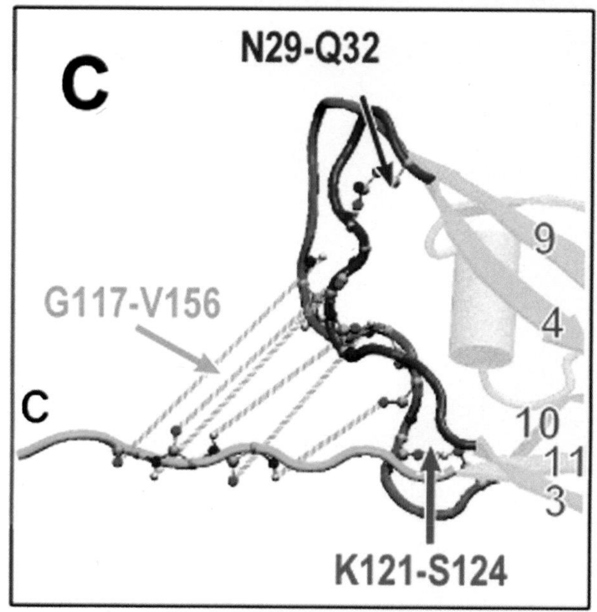

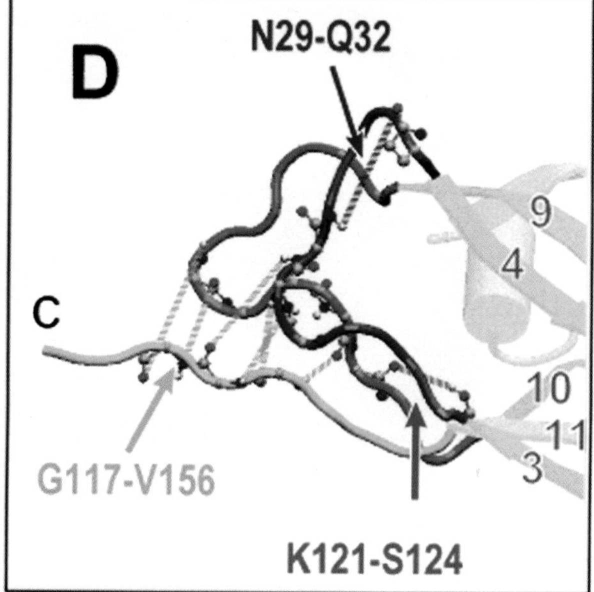

Color Plate 8. (Chapter 24) Structural changes occurring in the interdomain region of the FimH lectin domain during SMD simulations. Lateral (A) and end-on (B) views of residues A150 to T158 of the linker chain (light yellow) lying between the 3-4 loop (blue) and the 9-10 loop (red). These loops are identified by the β-strands that they connect, and the residue and strand numbers reflect the terminology published with the crystal structure (20). Six hydrogen bonds that anchor the linker chain to the 3–4 and 9–10 loops in the crystal structure are shown as dashed lines. (C and D) Lateral view similar to that shown in panel A, illustrating linker chain extension (C) and loop region deformation (D). Reprinted from W. E. Thomas, E. Trintchina, M. Forero, V. Vogel, and E. V. Sokurenko, *Cell* **109**:913–923, 2002, with permission from Elsevier.

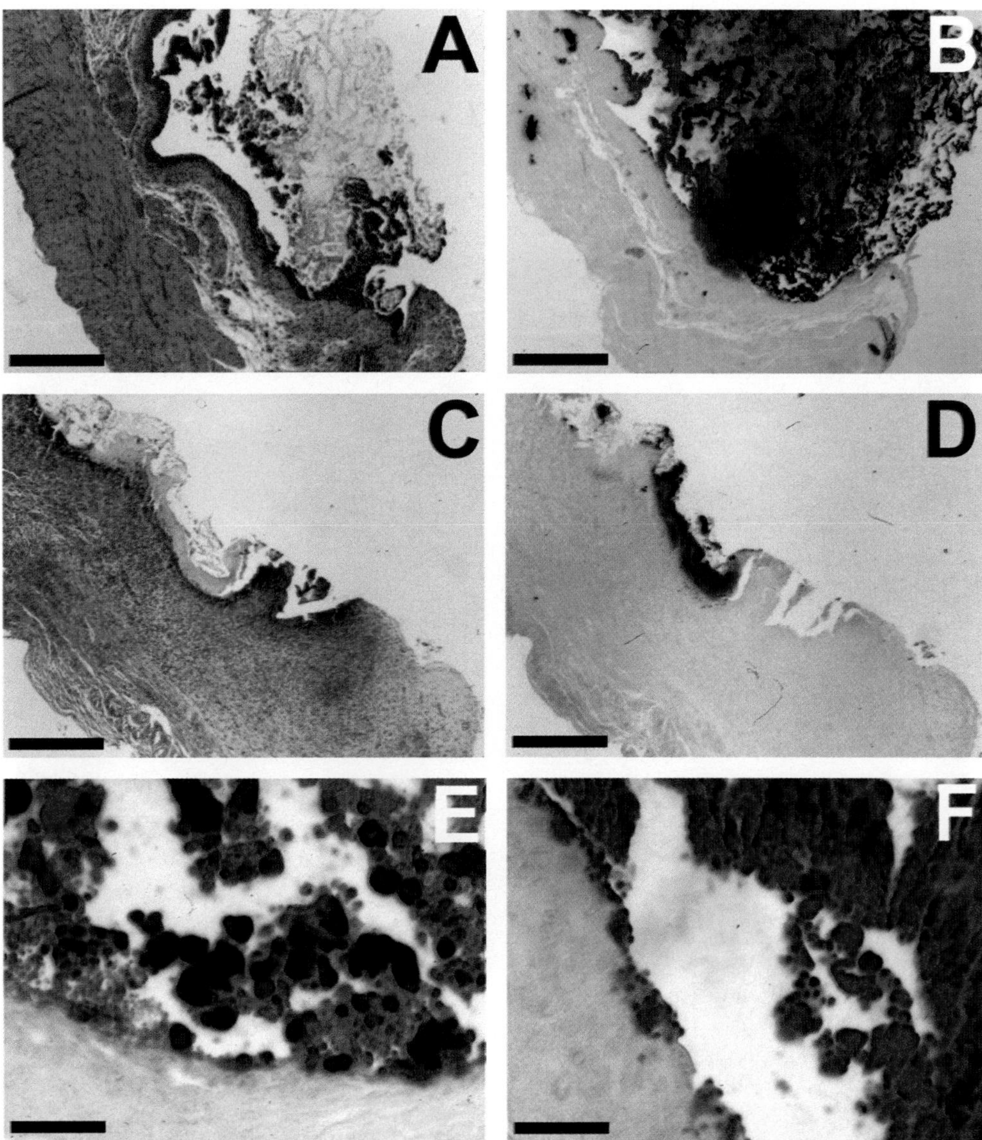

Color Plate 9. (Chapter 26) Hematoxylin-and-eosin- and alizarin red S-stained sections of *P. mirabilis*-infected mouse bladders. (A and B) Hematoxylin-and-eosin-stained (A) and alizarin red S-stained (B) consecutive sections of a mouse bladder that developed a macroscopically visible stone after infection by *P. mirabilis*. (C and D) Hematoxylin-and-eosin-stained (C) and alizarin red S-stained (D) consecutive sections of a mouse bladder that developed mild urolithiasis due to *P. mirabilis* infection. (E and F) Magnified views of areas in panel B to show alizarin red S-stained mineral deposits on the bladder epithelium. Scale bars, 400 μm (A to D) and 100 μm (E and F). Reprinted with permission from X. Li, H. Zhao, C. V. Lockatell, C. B. Drachenberg, D. E. Johnson, and H. L. Mobley, *Infect. Immun.* 70:389–394, 2002.

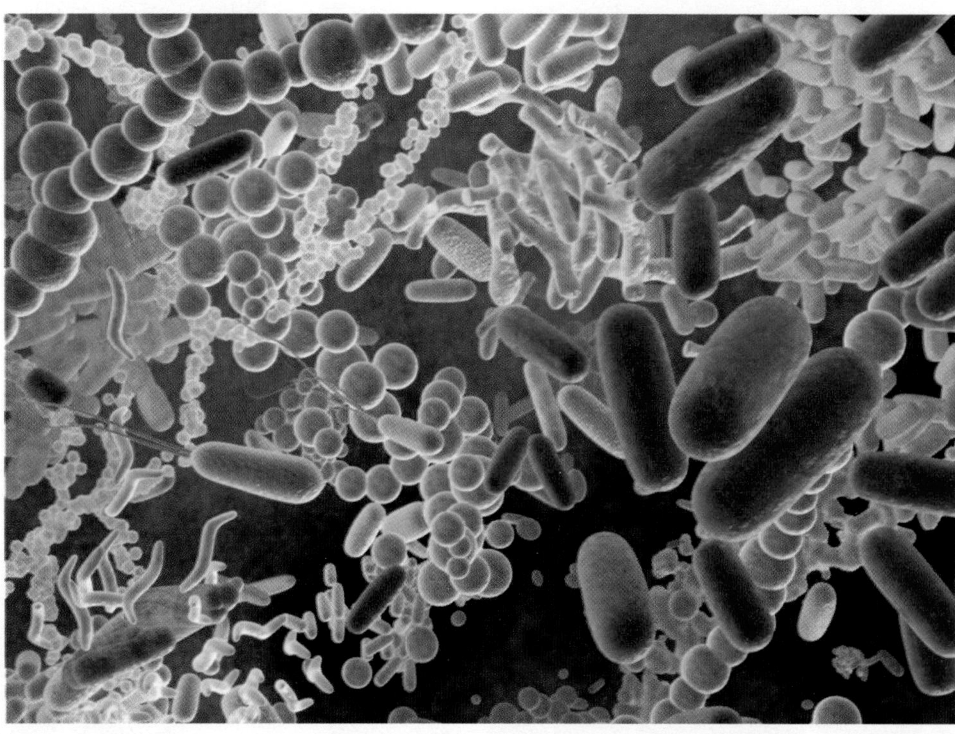

Color Plate 10. (Chapter 28) Within a short time after birth, a relatively diverse group of bacteria colonize the mucosal surfaces of the gut and perirurethra. Illustrated here are bifidobacteria (yellow), lactobacilli (blue rods), streptococci (light blue cocci), and coliforms (red, rod shaped).

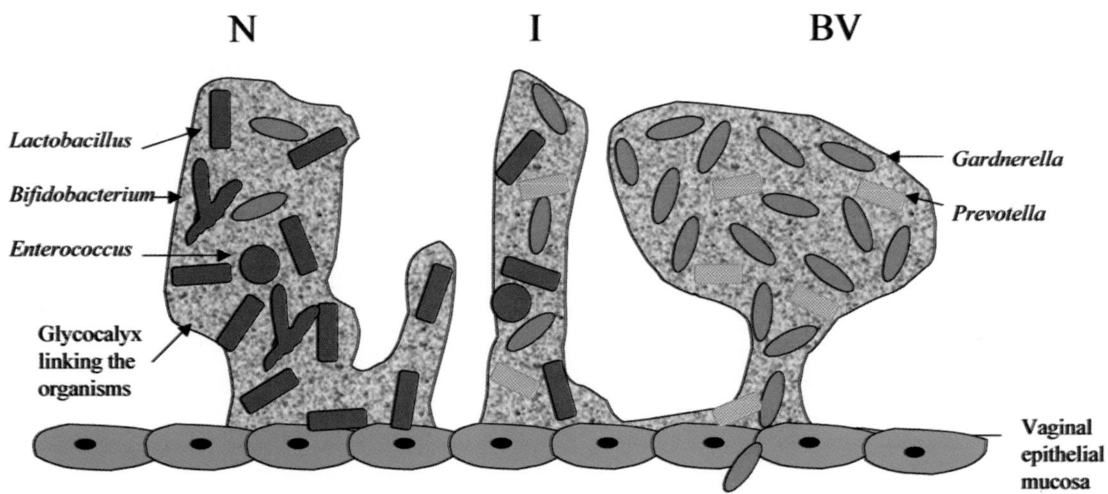

Color Plate 11. (Chapter 28) Three types of biofilms found on the vaginal epithelial mucosa. The so-called normal (N) state is when lactobacilli and often bifidobacteria dominate the microbiota. The intermediate (I) state is a transition between normal and BV, with more gram-negative anaerobes such as *Gardnerella* and *Prevotella* and enterococci; the BV state includes dense microcolonies of gram-negative anaerobes or other pathogens, with no or very few lactobacilli (105). The ability of the gram-negative pathogens to invade the epithelium has been shown for *E. coli* in the bladder (4) but has not been shown to date for vaginal pathogens.

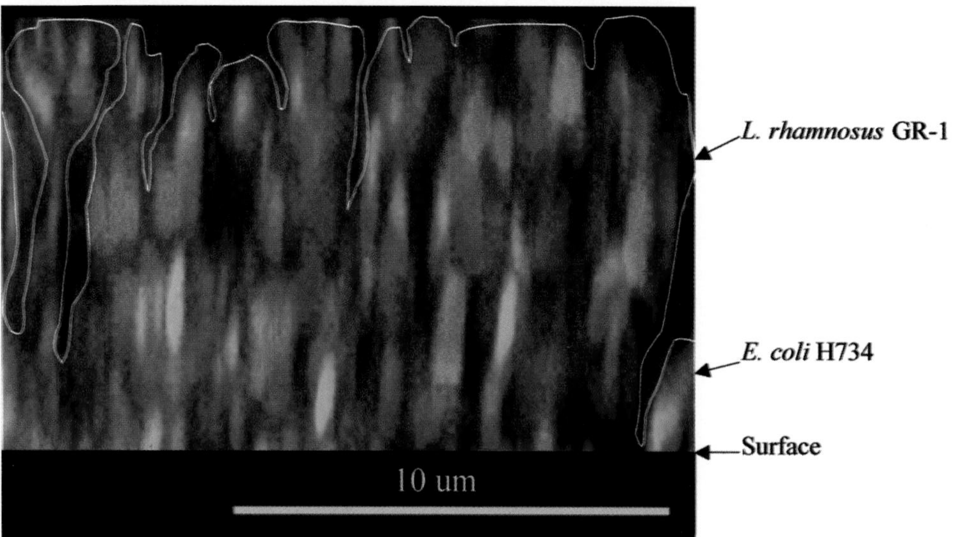

Color Plate 12. (Chapter 28) Deconvolution micrograph showing a 5-day *E. coli* biofilm protruding from a glass surface (DAPI live blue stain about five layers thick) into which *L. rhamnosus* GR-1 (Texas Red X stain) has penetrated after addition on day 2. The white line shows the outline of the biofilm.

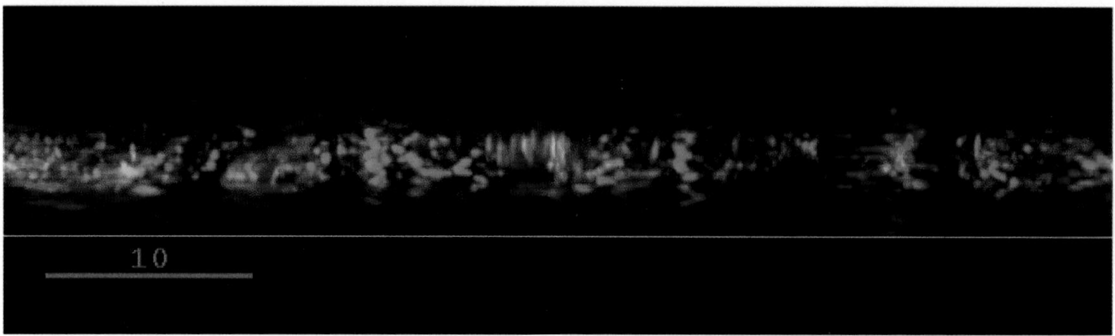

Color Plate 13. (Chapter 28) *E. coli* Hu734 after 96 h, grown in a four-chamber Lab-Tek coverglass system, stained with live/dead stain. Live cells are stained with SYTO 9 green-fluorescent nucleic acid stain, and dead cells are stained with propidium iodide red-fluorescent nucleic acid stain.

Colonization of Mucosal Surfaces
Edited by James P. Nataro et al.

Chapter 16

Role of Flagella in Mucosal Colonization

JORGE A. GIRÓN

Bacterial survival depends in part on the ability of organisms to respond to chemoattractants and to migrate toward beneficial environments or to escape from adverse ones. Bacteria achieve this through an orchestrated collection of complex molecular mechanisms and apparati that include flagella, motility, and chemotaxis (63, 92). These properties are viewed in some bacterial systems as virulence factors that may favor the host-pathogen interaction, resulting in the establishment of disease. In the human gut, enteric bacterial pathogens must penetrate the viscous mucus that naturally bathes epithelial surfaces before they can successfully reach their target niches. Thus, flagellar expression and motility are critical properties in translocating bacteria across the intestinal mucus, thereby facilitating interaction with specific target sites on the surface of enterocytes (14, 36, 40, 117, 147). A body of evidence reveals that the expression of flagella and virulence is often closely linked by complex regulatory networks involved in controlling virulence gene expression (40). Several studies have addressed the role of flagellum-mediated motility in virulence (for example, adherence, invasion, biofilm formation, and proinflammatory response) in several gram-negative pathogens (4, 5, 17, 27, 29, 41, 44, 111, 123, 128, 160). The remarkable similarities between the virulence-associated type III secretion systems (TTSS) (involved in secretion and translocation of effector molecules) and the flagellar type III secretion apparatus (involved in secretion of the flagellin protein and bioassembly of flagella) support the notion that these export systems have evolved to favor bacterial survival and host colonization (39, 56, 91). Several studies indicate that the flagella and flagellin of various bacterial species induce a proinflammatory response by inducing tumor necrosis factor alpha (TNF-α) and a cascade of interleukin production, most notably interleukin-8 (IL-8) (19, 45, 144, 157, 158, 170).

In recent years, different studies of bacterial flagella have unmasked novel features regarding their complex and sophisticated structure as well as their biological relevance beyond motility. This review focuses on these new structural and functional features of flagella, with emphasis on their ability to favor adherence, colonization, penetration, and translocation by bacterial pathogens and the resulting activation of innate immunity. How and what the implications of this knowledge will be in terms of vaccine development remains to be determined. It is clear that the interaction between bacterial flagellin and its eukaryotic recognition partners is complex, as demonstrated by the body of evidence accumulated in the literature, and it provides the basis for further studies to characterize the innate immune response to flagellin. The innate response may be critical to host resistance to mucosal colonization, since it is the first line of defense against pathogens before the adaptive immune system, which takes much longer to develop, comes into play to control infection.

THE FLAGELLUM APPENDAGE

Flagella and motility are bacterial features as ancient as the bacteria themselves. Bacteria may produce one or several polar or peritrichous flagella (Color Plate 4). Bacteria swim by rotating their helical flagellar filaments. Much of what is known about the structure and genetics of flagella come from studies of *Salmonella* and *Escherichia coli* (88, 92). Each flagellum is driven by a motor, which rotates extremely fast (up to 15,000 revolutions per min) and is driven by a flux of protons, ~1,000 protons per revolution (1, 90, 132, 161–163). The complex structure of the flagellum is generally simplified and described as consisting of three main components: (i) the basal body, (ii) the hook, and (iii) the filament (Fig. 1). Flagellum

Jorge A. Girón • Department of Microbiology and Immunology, Life Science North Building, Room 649, College of Medicine, University of Arizona, Tucson, AZ 85743.

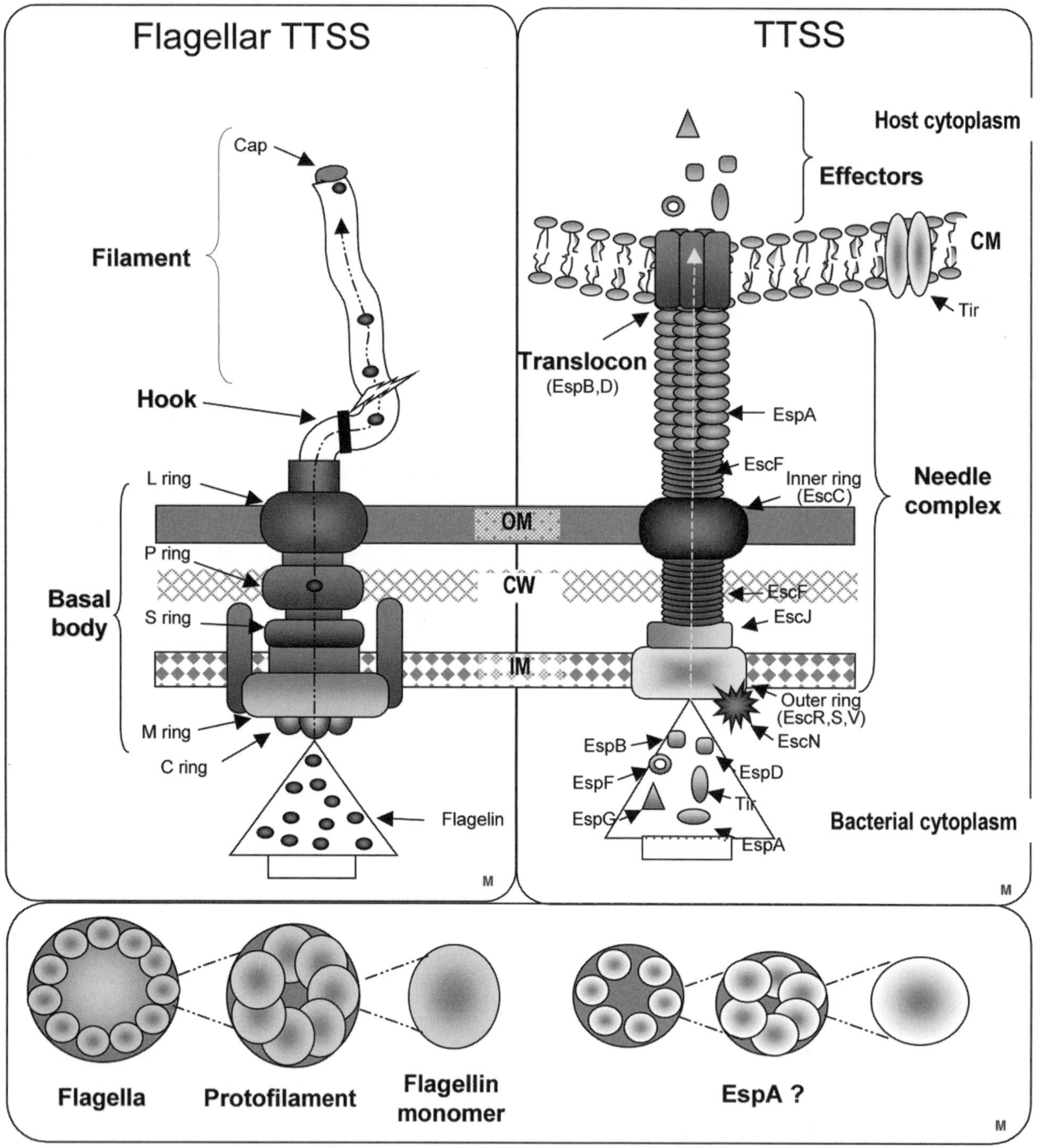

Figure 1. Schematic representation of the flagellar structure and the TTSS. The basal body is composed of a series of rings (L, P, S, M, and C), which span the inner and outer membranes and represent the motor that propels the flagellum filament. The filament is connected to the basal body through a hook structure. The flagellin subunits are exported across the cell envelope through the basal body to be assembled in a helical pattern at the tip of the growing filament. The tubular structure is formed by 11 strands of protofilaments of longitudinal helical arrays of flagellin subunits. The cap protein serves as a modulator of flagellum synthesis and secretion of proteins. The virulence-associated TTSS is structurally similar to the flagellar apparatus. Shown here is the TTSS of EPEC, which is composed of several Esc proteins and directs the secretion and translocation of secreted proteins (Esp and Tir) to the host cell cytoplasm. Like flagella, the presence of coiled-coil domains present in EspA suggests the possibility of formation of protofilaments yielding an EspA tubular helical structure. OM, outer membrane; IM, inner membrane; CW, cell wall; CM, cell membrane.

biogenesis is highly ordered such that the basal body is assembled first, followed by the hook and self-assembly of the filament. The basal body is the motor structure, composed of a series of rings (C, M, S, P, and L) that span the inner and outer membranes. Not only does the motor rotate, but also it rotates in either direction, clockwise and counterclockwise. In the presence of an attractant, it rotates almost exclusively counterclockwise; in the presence of a repellent, it rotates almost exclusively clockwise. Thus, in free-swimming bacteria, the chemotastatic signal is inexorably linked to the direction of flagellar rotation (1, 90, 132, 161–163).

The flagellum is composed of a single flagellin subunit, and its tubular structure is formed by 11 strands of protofilaments in two distinct conformations, L-type and R-type, which are nearly longitudinal helical arrays of flagellin subunits. The helical forms of the filament are caused by supercoiling, which is supposed to occur through a switching of conformation or packing interactions of the subunits between two distinct states and their non- and quasi-equivalent intersubunit interactions (132, 161–163). The presence of protofilaments in both states accounts for the flagellum helicity and is an essential feature for propulsion. For example, in the case of the form used for swimming, there are nine L and two R protofilaments. The protofilaments have an extraordinary capacity to switch between different states in a cooperative fashion. This manifests itself at the macroscopic level as conversion between different helical forms, notably the left-handed form used for swimming (when the motor rotates in a counterclockwise direction) and the right-handed form used for reorientation (tumbling, when the motor rotates clockwise). Details of the topological organization, interaction of the flagellin subunits, and the atomic structural features of the flagellin molecule revealed by cryoelectron microscopy, X-ray crystallography, and X-ray fiber diffraction have recently emerged (7, 99, 132, 162, 163).

The flagellum is about 20 nm in diameter, grows to a length of about 15 μm, and is made of as many as 20,000 to 30,000 flagellin subunits (7, 162, 163). The growth of the filament occurs at its distal end, where the cap protein is stably attached. The flagellum also functions as a sophisticated export and self-assembly apparatus. The flagellin subunits can travel through the hollow filament and self-assemble at the distal end in a helical array (99). The majority of the proteins which are located beyond the cytoplasmic membrane employ a specialized type III flagellar export pathway. The synthesis, assembly, and function of the flagellar system of *E. coli* and *Salmonella enterica* serovar Typhimurium are under the control of a flagellar regulon that comprises more than 50 genes divided among at least 17 operons. Details of the genetic organization, regulation, and bioassembly of flagella can be found elsewhere (16, 88, 90, 92).

FLAGELLIN

The flagellin molecule is the major structural subunit of the flagellum tubular filament, and depending on the bacterial species, it varies in molecular mass from 31 to 80 kDa (2, 27, 92, 103, 127). The flagellins of enterobacteria have extended sequence homology in the amino and carboxy termini, while considerable divergence exists within the middle region of the proteins (92, 126). This means that the middle hypervariable region of the flagellin protein is exposed on the flagellum filament while the conserved amino and carboxy termini remain hidden and embedded within the polymeric structure. As a consequence, it is this middle region of the flagellin which provides the serologic type specificity to the flagellar antigen, resulting in antigenic differences between divergent flagellins and forming the basis for H serotyping of *E. coli* (126).

Recent data reported by Samatey et al. and by Yonekura et al. have yielded further detail regarding the structural and functional domains of the flagellin molecule (132, 162, 163). Flagellin was crystallized, allowing for precise structural positioning of amino acid residues, thus identifying three globular domains (D1 to D3). The molecule is oriented with the D1 domain lying innermost in the filament while the D2 and D3 domains are exposed on the outermost of the filament. The D1 and D2 domains are discontinuous and are formed when residues in the amino and carboxyl termini are juxtaposed by the formation of a hairpin structure. The D1 domain consists of evolutionarily conserved α-helical regions of the amino and carboxyl termini and is strictly required for assembly of the monomeric subunits into the flagellum. The D2 and D3 domains are located within the central region of the extended flagellin molecule and are less well conserved in different bacteria. The middle hypervariable domain (D3) is exposed at the surface of the flagellar filament and is a common target for antibody responses. The D3 domain is not required for motility since it is not present in the flagellin molecules of some bacteria. Its high degree of variability indicates that it has evolved to permit considerable antigenic heterogeneity to evade adaptive immune responses. The D3 domain forms the molecular determinant of the H antigen that is widely used for subtyping bacterial strains (132, 162, 163).

FLAGELLAR TYPE III EXPORT APPARATUS

Some pathogenic bacteria, such as *Shigella, Salmonella, Pseudomonas, Yersinia,* enteropathogenic *E. coli* (EPEC), and enterohemorrhagic *E. coli* (EHEC), produce sophisticated secretion or export machineries, the TTSS (see above). TTSS function as microsyringes which deliver virulence factors into target sites in the host cell membrane or into the cytosolic compartment (39, 56, 67, 75, 119) (Fig. 1). In the cytosol, the injected effector molecules act on components of the cellular cytoskeleton and activate signaling pathways, leading to various manifestations such as intracellular penetration (e.g., for *Salmonella, Yersinia* and *Shigella*) or effacement of the cell membrane (e.g., for EPEC and EHEC) (39, 73). Soon after their discovery, it became quite obvious that there were striking structural and biochemical similarities between the TTSS and the flagellar type III export apparatus involved in flagellum assembly and motility. The body of experimental data supports the idea that the TTSS evolved from the flagellar apparatus. This suggests common functional features that have evolved for bacterial adaptation, survival, and virulence (1, 10, 22, 39, 56). Pallen et al. first noted that components of both the TTSS and the flagellum assembly machinery of several pathogens are predicted to have an α-helical coiled-coil property (115). The coiled-coil structure is a common and highly versatile assembly motif found in a wide range of structural and regulatory proteins with functions ranging from assembly of macromolecular complexes to molecular recognition, including the TTSS and flagellar components (26, 89, 115). Flagellins contain coiled coils involved in the assembly of flagella (99). Similarly, the EspA secreted protein, a monomeric component of the fiber connected to the TTSS syringe of EPEC, was noted to contain a putative coiled-coil α-helix at the carboxy end (residues 141 to 175), probably responsible for assembly of the EspA fiber (26, 115). Based on these structural analyses, it is reasonable to propose that the EspA fibers and other TTSS needle complexes of other pathogens have a similar structural organization to the flagellar filament in which the flagellin subunits are polymerized into protofilaments and that these are arranged into a highly organized flagellum filament (Fig. 1).

The translocation of virulence-associated nonflagellar proteins (e.g., phospholipase YlpA) by the type III flagellar export machinery of *Yersinia enterocolitica* has been reported (165). It remains to be determined whether other flagellated bacteria employ the type III flagellar apparatus to export virulence factors. It is reasonable to hypothesize that flagella could act to secrete effectors in bacteria lacking a dedicated TTSS.

ENVIRONMENTAL INDUCTION OF FLAGELLATION AND MOTILITY

It is clear that the production of flagella by many organisms appears to be modulated by environmental and nutritional cues. Recent data suggest that an in vivo environment (e.g., the chicken gut) induces motility in *S. enterica* serovars Gallinarum and Pullorum, serovars that are traditionally recognized as nonmotile (15). The effect of pH, temperature, and surface contact on the elaboration of flagella by *S. enterica* serovar Enteritidis has also been reported (152). Production of flagella in EPEC is enhanced by the presence of epithelial cells and by undefined signals derived from eukaryotic cells (48). Typical nonmotile enteroinvasive *E. coli* (EIEC) strains were shown to produce flagella under defined in vitro growth conditions (2), suggesting that many aflagellate/nonmotile organisms may in fact produce flagella under suitable growth conditions or on induction with host signals. The presence of bile significantly increases the motility of vibrios, as determined in in vitro assays (51, 72). It is speculated that in the intestine, the presence of bile would allow the bacteria to become highly motile in order to penetrate the mucus layer and gain access to the underlying epithelial cells. From this perspective, it is assumed that flagellum-driven motility would be critical in early stages of infection as the bacteria encounter bile secreted into the lumen of the duodenum from the gallbladder through the bile duct (72, 76).

ASSOCIATION OF BACTERIAL FLAGELLA WITH INFECTION

For most bacterial pathogens, flagella and flagellum-driven motility are recognized as essential elements in their virulence scheme. This notion is supported by a growing body of evidence derived from the use of isogenic mutants deficient in flagellar production, motility, and/or chemotaxis and their analysis in several in vitro, in vivo, and ex vivo models of infection. A number of gram-negative flagellated bacteria are associated with infections in the gastrointestinal, urinary, and respiratory tracts (5, 6, 23, 32, 40, 102, 103, 105, 123, 128, 137). *Borrelia burgdorferi* is a flagellate pathogen and is responsible for Lyme disease (131). The fact that animals and humans develop immunoglobulins directed against flagellins on infection with *Helicobacter pylori, Pseudomonas aeruginosa,* or *Campylobacter jejuni* demonstrates that flagella are expressed when these bacteria are inside the host. For many organisms, their natural nonflagellate/nonmotile variants and isogenic mutants deficient in flagellation are less virulent due to their inability to interact efficiently with host epithelial cells. On the other hand, bacterial organisms such as *Brucella, Klebsiella,* and *Shigella,* which are typically nonmotile, are nevertheless able to cause infections. These three organisms may produce flagella under certain modified in vitro growth conditions (47; Castaneda and J. A. Girón, unpublished data), but it remains unknown whether these organisms produce flagella during in vivo infections. Of note, commensal nonpathogenic bacteria may also possess flagella that allow them to be motile in the lumen of the gut and to persist on epithelial surfaces. In all, although flagella may be important for interaction with the host, flagella alone are not sufficient for the establishment of disease.

In the case of *Salmonella, Yersinia,* and *Bordetella,* motility seems to be important only for the initial phases of infection while colonization is being established. These organisms are also able to switch off motility in order to initiate their full virulence program. Other bacteria, such as *Legionella pneumophila,* require flagella and motility to establish and maintain infection in a human host as well as to survive during their life cycle in the external environment (e.g., in a protozoan host) (76, 122). During growth in the infected macrophage, these bacteria are nonflagellate, but once the macrophages are lysed the bacteria reinitiate their flagellar production, allowing them to search for new target sites (122). Numerous pathogens require continuous motility during colonization of a host and are less capable of surviving without being motile and competent for chemotaxis. Examples of organisms requiring in-host full-time motility include *B. burgdorferi, H. pylori, C. jejuni, P. aeruginosa,* and *Proteus mirabilis.* For *Vibrio cholerae* and other *Vibrio* spp., the role of motility during the course of infection is a matter of controversy, but the general feeling is that motility should assist the vibrios in penetrating the mucus layer overlying the gut epithelium before colonization. Further, the experimental data accumulated so far indicate that motility and virulence regulation in this bacterium are coupled (40). Similar regulatory relationships are likely to exist in other pathogens. Regarding the requirement of flagella for pathogenic *E. coli* strains in the gut, it would seem reasonable to propose that flagellum-driven motility would favor gut colonization, as in the vibrios. However, natural nonmotile strains of EIEC, EPEC, and EHEC can be isolated from patients with diarrhea (109), suggesting that, unless flagella are actually produced in vivo during infection, these pathogens can cause disease without possessing flagella.

Role of Flagella in *S. enterica* Invasiveness

A growing body of research strongly supports the involvement of flagella and motility in the pathogenesis of *S. enterica* serovars Typhimurium, Enteritidis, and Typhi (4, 5, 14, 15, 29, 65, 78–80, 87). Experimental *Salmonella* infection in animal models including mice, rats, and chickens, as well as in cultured human and avian epithelial cells, suggested the requirement of flagella for adherence, colonization, survival, persistence, and/or invasion. The general view from these studies is that aflagellate derivatives of *S. enterica* serovars Enteritidis and Typhimurium are attenuated in murine and avian models of infection compared to the wild-type strains.

Jones et al. reported that flagellum-driven motility was important for the initial interaction of *S. enterica* serovar Typhimurium with HeLa cells (61). Flagella were also implicated in assisting the survival of serovar Typhimurium within murine macrophages (156) and were required for full virulence in an immunocompetent C57BL/6J mouse model of infection (14). *S. enterica* serovar Typhimurium isogenic nonmotile mutants showed lower invasion rates in HeLa cells than did the parent strain, LT2; no significant differences were found between the invasion rate of nonflagellate and nonmotile mutants, suggesting the importance of these phenotypes in the invasive process. Of note, smooth-swimming nonchemotactic mutants exhibited 10-fold higher invasion rates than did the wild-type strain (65). Similarly, Jones et al. (62) found that *S. enterica* serovar Typhimurium SL1344 smooth-swimming *che* mutants possessed increased invasiveness for HEp-2 cells and suggested that the rotation and physical orientation of flagella around the bacteria affect the ability of the salmonellae to enter host cells. In a different study, a mutation in the master flagellar regulator gene *flhD* rendered *S. enterica* serovar Typhimurium more virulent than its parent strain in mice but less invasive in human and mouse cultured epithelial cells. These results suggest that the different in vivo and in vitro assays employed measure different aspects of virulence (135). An *S. enterica* serovar Typhimurium strain deficient in both mannose-sensitive hemagglutination and motility was reported to be greatly reduced in its ability to invade epithelial cells in vitro and to persist in the liver and spleen of orally challenged chicks. However, mutants defective in the production of flagella or fimbriae alone showed no significant effect in cell invasion or persistence in the chick model (80, 116). Other studies showed that aflagellate mutants of *S. enterica* serovar Typhimurium associated with and invaded fewer human INT-407 epithelial cells than did motile bacteria, although the addition of antiflagellum antibodies had no inhibitory effect (29). These authors speculated that flagella assist the colonization of epithelial cells by providing motility rather than acting as an adhesin for receptor recognition.

Salmonella possesses two antigenically distinct flagellins that are alternatively expressed. The reciprocal switch from FliC to FljB flagella, in a process called flagellar phase variation, contributes to *S. enterica* serovar Typhimurium virulence in a model of murine systemic infection but does not influence the invasion of epithelial cells in vitro (12, 58). An investigation of the role of flagella in the human pathogen *S. enterica* serovar Typhi found that all of the Tn*5* insertion motility mutants (Fla^-, Mot^-, or Che^-) analyzed were noninvasive, suggesting that flagella and motility are invasion-related factors of serovar Typhi (87).

Flagella also play a central role in the adherence and invasion of *S. enterica* serovar Enteritidis in various cultured epithelial cell lines and in explants from the intestine and trachea of chickens (4, 29, 151). Lillard suggested a role for flagella in the attachment of *S. enterica* serovar Typhimurium to poultry skin (85), and Allen-Vercoe and Woodward suggested that flagella might be considered an adhesin since flagellate, nonmotile *motAB* (motility genes) mutants of *S. enterica* serovar Enteritidis adhered well to cultured cells and chicken gut explant (4). In a chick infection model, fimbriate *S. enterica* serovar Enteritidis organisms unable to express flagella were recovered postmortem from livers and spleens in significantly reduced numbers compared to those of the parent strain, indicating that flagella, not motility or fimbriae, play a role in pathogenesis in in vivo infection (4, 116). Robertson et al. reported that active flagella appear to be an important factor in enabling salmonellae to penetrate the gastrointestinal mucus layer and attach specifically to epithelial cells in a rat model (129, 130). Isogenic flagellar isogenic mutants showed transiently reduced numbers in the liver and spleen, were less able to persist in the upper gastrointestinal tract, and did not trigger the severe inflammatory responses in the gut shown by the wild-type strain. These authors also showed that expression of SEF14, SEF17, SEF21, Pef, and Lpf fimbriae or chemotaxis was not a prerequisite for induction of *S. enterica* serovar Enteritidis infection in the rat. La Ragione et al. examined the participation of flagella in early stages of cell invasion and found that compared to the wild-type, aflagellate mutants of serovar Enteritidis showed a significant reduction of invasion, formation of ruffles, and cytoskeleton rearrangements in HEp-2 (human) and Div-1 (avian) cells (78, 79). They suggested that flagella play an active role in the early events of the invasive process. It has become clear from these studies that the adherence, invasive, and colonization capabilities of the *Salmonella* serovars is greatly enhanced by the production of flagella and motility. Whether the flagellum itself is an adhesin, as proposed in two studies (4, 85), is a matter that requires confirmation.

Role of Flagellar Activity in Virulence of *V. cholerae*

V. cholerae produces a potent cholera toxin which is primarily responsible for the diarrheal syndrome associated with *V. cholerae* infection. The toxin-coregulated pilus (Tcp) is an essential element for colonization by *V. cholerae,* as demonstrated in animal models and human volunteers (76). *V. cholerae* produces a mannose-sensitive hemagglutinin, which is also associated with intestinal colonization. *V. cholerae* strains are typically motile, and the role of their polar sheathed flagella (Color Plate 4), motility, and chemotaxis in intestinal colonization in various experimental animal models and in vitro assays has been explored (6, 36–38, 110, 120, 128, 148, 159). There appears to be a general consensus that nonmotile mutants of *V. cholerae* manifest reduced or no attachment to intestinal brush borders in vitro and in vivo. Nonmotile aflagellate mutants were less virulent in vivo and did not penetrate the crypts of Lierberkühn to the same extent in suckling mice as did the motile parent strain. The role of motility in fluid accumulation and adherence was also supported by analysis of nonmotile mutants in experimental cholera in adult rabbit ileal loops (159). Nonmotile aflagellate mutants required at least 100-fold-higher doses than did wild-type strains to produce comparable fluid accumulation responses. The decreased ability of nonmotile strains to produce fluid responses was attributed to their inability to associate with the intestinal mucosa. In a series of studies, Freter et al. investigated the role of motility and chemotaxis in *V. cholerae* colonization of the small intestine of rabbits and germfree mice (36–38). Employing motile nonchemotactic mutants and nonmotile mutants, they concluded that motility was beneficial to survival only when it was directed by chemotactic stimuli (36). In vitro, motile nonchemotactic vibrios penetrated the mucus gel of intestinal slices poorly compared to the penetration by chemotactic vibrios (37). Further, the nonchemotactic mutants were indistinguishable from the parent strain in their ability to attach to isolated intestinal brush border membranes, whereas the nonmotile mutant had lost this ability (38). In sum, these studies indicated that motile vibrios can efficiently invade the mucus gel barrier when their locomotion is driven by chemotactic stimuli.

In the infant-mouse cholera model, the flagellum was shown to be essential for attachment and colonization while the property of motility was neither necessary nor sufficient (6). Since it was possible to demonstrate independently both motility and binding capacities associated with the flagellum, it was concluded that the flagellum functions as the carrier of the moieties that promote adherence by facilitating the initial colonization of the small bowel (6). In contrast, Teppema et al. reported that flagella, motility, and the hemagglutinating activity of the vibrios are not essential for the pathogenesis of experimental cholera in ligated intestinal loops of rabbits (148). This conclusion is reasonable, considering that the ligated loop is a closed system and potential adherence factors associated with the flagellum may not be required. However, ultrastructural analysis by scanning electron microscopy of the association of the motile flagellated *V. cholerae* strain C5 with the rabbit intestinal epithelium revealed the presence of typical polar wavy flagellar structures sticking out to the lumen. Although not stated by the authors, the flagella in the published electron micrographs appeared to be promoting interbacterial interactions and mediating contact between the vibrios and the villi of the intestinal epithelium. However, it is clear from the results that both flagellate and nonflagellate vibrios were able to colonize the small bowel.

In an effort to further clarify the role of the flagellar binding capacity versus the role of motility in the colonization process, Richardson (128) studied three groups of El Tor and classical biotypes transposon mutants: (i) nonmotile and flagellate, (ii) nonmotile and aflagellate, and (iii) nonmotile and aflagellate but forming a sheath-like structure. These mutants were compared with the wild-type strains for their ability to colonize in three animal models including the rabbit ileal loop, the removable intestinal tie adult rabbit diarrhea (RITARD) model, and the suckling-mouse model. The comprehensive analysis clearly demonstrated a role for motility in *V. cholerae* pathogenicity, while reduced virulence was observed for the nonmotile mutants derived from the classical parent strain, strongly suggesting a role for flagellar structures during infection. In a different study, Postnova et al. (120) showed that *V. cholerae* mutants defective in motility, flagellar structure, or chemotaxis, previously shown to exhibit reduced colonization in animal models (128), exhibited decreased adherence to homogenized human small intestinal mucosal tissue. It is apparent from the above studies that variations in the results obtained by several authors are due to the nature of the experimental models employed. Klose and Mekalanos constructed an *rpoN* (encoding σ^{54})-null mutant of *V. cholerae* and found that this strain was defective in motility, flagellation, and colonization in the infant-mouse colonization assay (72). In this study, they also identified three flagellar regulatory genes (*flrABC*), among which *flrA* and *flrC* encode σ^{54}-activators; mutations in these two genes yielded mutants defective in colonization. A *flaA* strain, which is nonflagellate and hence nonmotile, is slightly defective in colonization in this assay. Because the colonization defect of the *rpoN* strain is greater than that of the *flaA* strain, it was suggested that the defect is not due to motility per se. Their results implicated FlrA and FlrC in the expression of genes that aid colonization but are not necessarily motility-related genes. In conclusion, RpoN is responsible for the proper balance between flagellar biogenesis and the expression of colonization factors during the infection. It is clear from several studies that *V. cholerae* motility and intestinal colonization are related in a complex and intimate way (40). Hase showed that inhibition of motility by increasing the medium viscosity or addition of ionophores that disrupt ion gradients used to energize motility results in increased *toxT* expression and that this induction is flagellum independent. She concluded that the flagellum of *V. cholerae* does not act as a mechanosensor, voltmeter, or signal transducer for environmental conditions (52).

For *Vibrio anguillarum,* it appears that flagella are not important for adherence, invasion, or virulence (112). For the most part, it is well accepted that flagellum-driven motility in *V. cholerae* is a key factor in gut colonization because it allows bacteria to travel across the intestinal mucus layer and permits efficient delivery of the cholera toxin. Whether the flagellum filament itself has adhesin properties is an issue that needs to be addressed further.

Flagella of Diarrheagenic *E. coli:* an Ancillary Adhesin and Interleukin-8 Activator

It is clear that the close interaction of EPEC with host epithelial cells involves the outer membrane protein intimin while the formation of microcolonies is attributed to the type IV bundle-forming pilus (Bfp) (Fig. 2). The EspA fiber associated with the TTSS may function as an adhesin as well (73). Recently, it was reported that flagella mediate the adherence of motile EPEC strains to cultured epithelial cells (Color Plate 4) (48). Insertional mutants with mutations in the flagellin structural gene *fliC,* constructed in two wild-type EPEC strains, were markedly impaired in adherence to and microcolony formation on cultured cells. Motility was apparently not required for adherence, since an isogenic *motB* (flagellar motor gene) mutant adhered to the same level as that of the wild-type strain. In support of these results, adhesion of several EPEC serotypes could be blocked by anti-flagellum antibodies (48). Further, purified H6 (EPEC) but not EHEC H7 flagella bound to cultured epithelial cells. Interestingly, flagellar production in EPEC was triggered by the presence of epithelial cells or by growth in culture medium preconditioned with epithelial cell monolayers. It was suggested that a molecule of eukaryotic origin acts as a signal to activate flagellar production and possibly the expression of other virulence factors in EPEC. Although flagella appear to be important for motile EPEC strains, this may not be the case for clinical nonmotile EPEC strains, which, although unable to produce flagella, still undergo localized adherence to epithelial cells. Quorum sensing was recently shown to influence the production of flagella in EPEC and EHEC (142). In all, there are compelling data suggesting that the flagella of EPEC mediate adherence to epithelial cells in vitro (Fig. 2); however, the precise mechanism of how this occurs deserves further investigation.

In favor of the role of *E. coli* flagella in adherence, Barnich et al. reported compelling data showing that flagella play a direct role in the adherence process via motility in an adherent/invasive *E. coli* strain isolated from a patient with Crohn's disease (9). Interestingly, an isogenic *fliC* mutant showed down-regulation of type 1 pili as a result of a preferential switch toward the off-position of the invertible DNA element located upstream of the type 1 pilus *fim* operon. The study suggested regulatory and functional cooperation of flagella and type 1 pili in this particular strain. More studies are required to further elucidate the role of flagella in the various *E. coli* pathogroups. Nevertheless, several authors have demonstrated that the flagella of pathogenic and nonpathogenic *E. coli* strains are potent stimulators of IL-8 induction by epithelial cells (144, 170). Details of these studies are described below.

Flagella of *Y. enterocolitica:* Gateway for Protein Secretion

Demonstration of a dual function for the flagellar export machinery in *Y. enterocolitica* comes from experimental data showing that it also functions to secrete nonflagellar virulence proteins (165). Among these proteins, YlpA, a phospholipase required for survival in Peyer's patches and for stimulation of the acute inflammatory response of the host to the infection, was shown to be secreted through the flagellar export apparatus. In fact, YlpA may be secreted by any of the three TTSS maintained in *Y. enterocolitica:* the two contact-dependent TTSS or the flagellar TTSS (167). Expression of the flagellar regulon is required for YlpA secretion, since conditions known to

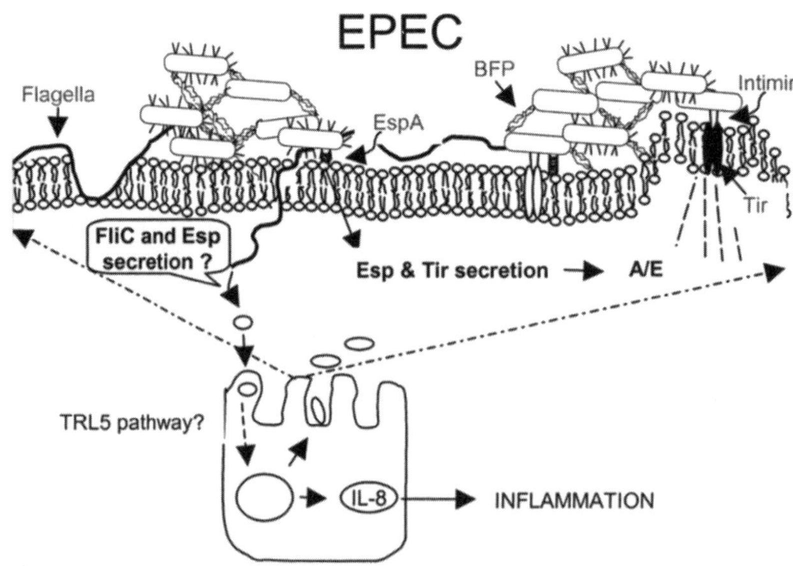

Figure 2. Schematic representation of the role of flagella of EPEC in adherence and IL-8 induction. Intimin, bundle-forming pilus (BFP), and EspA fiber are well-recognized EPEC adhesins. The EspA fibers connected to the TTSS secrete proteins (Esp and Tir) involved in attaching and effacing (A/E) lesion formation. The extracellular bacteria are tethered through the bundle-forming pilus and possibly via rod-like pili. The wavy flagellar filaments interconnect the bacteria and may mediate direct binding to a receptor on the cell membrane and hypothetically pierce the cell membrane and inject flagellins or other proteins to the cytosol. It is documented that the flagella may activate IL-8 and induce inflammation, but it is uncertain whether this activation employs the TLR5 signaling pathway.

repress motility were nonpermissive for YlpA activity (166). Further, a mutation in the master flagellar regulator, *flhDC*, yielded a mutant deficient in swimming motility, swarming motility, flagellin production, and secretion of YlpA (166, 167). Thus, the levels of YlpA activity were completely dependent on the level of *flhDC* expression. Consistent with the implication of the flagellar regulon in YlpA secretion, production of flagellin and YlpA activity was abolished in *Y. enterocolitica* carrying a mutation in *fliA*, a gene that encodes σ^{28} and is required for flagellar class II gene transcription. Further studies have revealed that secretion of YlpA requires the flagellum basal body and hook but does not require the filament (165).

Invasin-mediated invasion of host cells by *Y. enterocolitica* is affected by flagellum-dependent motility (134). Badger and Miller demonstrated that the expression of *inv* and flagellar genes is coordinated at the level of transcription and that they are part of an operon (8). Expression of the *inv* gene and motility is maximal during late exponential and early stationary phases at 23°C. Mutations in known regulatory genes of the flagellar regulon, *flhD* and *fliA*, resulted in low levels of invasion comparable to the reduced level of invasion exhibited by the *inv* mutant. In all, these experiments suggested that the flagellar regulon and motility are required for efficient host cell invasion by *Y. enterocolitica* (166). In contrast to these studies, Iriarte et al. reported that the *fliA* mutant showed the same rate of invasion as the wild type and concluded that at least in the mouse model of infection, motility seemed not to be required for *Y. enterocolitica* pathogenesis (59). Although important for protein secretion, the role of flagella in invasion of host cells by *Yersinia* spp. is still unclear.

Use of *H. pylori* Flagella for Successful Colonization of Gastric Cells

H. pylori is a human pathogen whose colonization of gastric tissue is facilitated by the presence of five sheathed flagella located at one pole of the bacterium. Each filament is composed of a major flagellin, FlaA, which is exposed on the outer regions, and a minor flagellin, FlaB, which localizes to the base of the flagellum (32). Elimination of either *flaA* or *flaB* resulted in mutants that were able to transiently colonize piglets for 4 days, but at 4 log units lower than did their wild-type parent. Mutation of both flagellins resulted in aflagellate nonmotile bacteria that weakly colonized animals and persisted only for 2 days as opposed to 10 days for the parent strain. Thus, full expression of motility is required for *H. pylori* to maintain high levels of infection in the piglet animal model. Contradictory results were reported by Clyne et al., who found that *H. pylori* and *H. mustalae flaA* and *flaB* mutants do not show a reduction in adherence to primary human or ferret gastric epithelial cells (21). Mutations in *fliD, fliQ,* and *flhB* flagellar genes eliminate flagellum production and the ability to colonize mouse stomachs (66). The product of the *flbA* gene regulates flagellum biosynthesis and secretion and represents a structural component of the flagellar secretion apparatus in *H. pylori*. A *flbA* mutant was also shown to be significantly reduced in its adherence to human gastric cells compared to the parent strain (21). In agreement, Ottemann and Lowenthal (114) found that the *flbA* mutant colonized only one out of six gerbils and caused no gastric lesions in the infected animal. This finding supports the requirement of *flbA*-mediated motility in colonization (94). *motB* mutants, although possessing a paralyzed flagellum, did not retain their ability to effectively infect mice. In sum, the data available strongly suggest that flagellum-driven motility of *H. pylori* is used to initially colonize the stomach and to attain full infection levels.

Requirement for Motility in Intestinal Colonization by *C. jejuni*

A role for motility in in vivo colonization by *C. jejuni* was demonstrated in the suckling-mouse model by comparing wild-type strains with aflagellate nonmotile mutants. All of the nonmotile strains were cleared from the mouse intestinal tract after 2 days of challenge, with no signs of colonization or illness, whereas the wild type colonized different regions of the intestine (105). In a later study, rabbits immunized orally with aflagellate mutants were not colonized or protected against challenge with the flagellate wild-type strain. This suggests that flagella are important in initiating colonization and eliciting protective immunity by *Campylobacter* spp. (117). Similarly, a study performed with a 3-day-old-chick model showed that intact flagella and motility are important colonization factors for *C. jejuni* in chicks (108). In contrast, Grant et al. suggested that flagella of *C. jejuni* are not involved in adherence to human epithelial INT 407 cells (50). However, nonmotile mutants were unable to travel across polarized Caco-2 monolayers, indicating that flagella do play a role in internalization. Two independent groups showed that mutants with defects in motility had lost their ability to adhere and invade INT 407 epithelial cells (49, 160). The swimming behavior of *C. jejuni* in a viscous environment was also suggested to be an important factor in the interaction of these organisms with host epithelial cells (145).

Down-Regulation of Flagella by *L. monocytogenes* To Evade Innate Immunity

The flagellin of *L. monocytogenes* induces inflammation through cell activation of Toll-like receptors (TLR) (55). Flagellin expression in *L. monocytogenes* is regulated by temperature and has been described as being shut off at 37°C (118), a reason to conclude that this is a mechanism whereby the bacteria evade TLR-mediated host innate immunity. Recently, it was shown that some laboratory-adapted strains and one-fifth of the clinical *L. monocytogenes* isolates studied do activate TLR5 and maintain flagellin expression at 37°C (155). A Δ*flaA* mutant derived from an isolate that expressed flagellin under all conditions examined was shown to be nonmotile. It was unable to activate TLR5-transfected HeLa cells but induced TNF-α production in vitro. However, when analyzed in a murine infection model, this Δ*flaA* mutant showed no significant alteration in virulence after intravenous or oral murine inoculation and did not show any difference in the generation of CD8 or CD4 T cells. From these studies, it is apparent that flagellin of *L. monocytogenes* is not essential for pathogenesis or for the induction of specific adaptive immune responses in normal mice.

Requirement of *P. aeruginosa* Flagella for Inflammation and Infection

The molecular interactions of *P. aeruginosa* with respiratory epithelial cells are complex and probably involve multiple ligands such as pili, flagella, and lipopolysaccharide (LPS) (103). The flagella of *P. aeruginosa* have been associated with induction of inflammation and adherence to the respiratory tract via recognition of mucin, GM_1 gangliosides, and asialogangliosides (3, 35, 124). FliD, a flagellar cap protein, was also shown to bind to mucin (3). The role of flagella in the initial stages of respiratory tract infection has been tested by comparing the virulence of *fliC* mutants in a neonatal-mouse model of pneumonia. No mortality was observed in the absence of *fliC* compared to the wild-type strain, and only 25% of the mice inoculated with the aflagellate mutants developed pneumonia. Further, they caused strictly focal inflammation and did not spread through the lungs, in contrast to the wild-type PAK strain. The data available suggest that the presence of flagella on the pseudomonads is a key factor in colonization and infection (35, 103).

Role of Flagella in Adherence and Virulence in Other Bacterial Species

Depending on the experimental model employed, the production of flagella in *Burkholderia pseudomallei* may or may not be considered a virulence factor (17, 28). In one study, no attenuation of virulence was observed between the *B. pseudomallei fliC* mutant and its parent strain in a diabetic infant-rat or Syrian hamster model (28). In contrast, a study with BALB/c mice showed that a *B. pseudomallei fliC* mutant was avirulent during intranasal infection (17). This is yet another example of how different experimental infection models can lead to different interpretations of the role of flagella in virulence. In the close relative *B. cepacia*, a nonmotile mutant was less invasive than the parent strain in lung epithelial cells, indicating that flagellum-mediated motility is required for full invasiveness since it facilitates the penetration of the host epithelial cells barriers and contributes to the onset of systemic spread of the organism (149).

It is well accepted that the mannose-resistant *Proteus*-like (MR/P) fimbriae and flagella are among the virulence factors of *P. mirabilis* that contribute to colonization in a murine model of ascending urinary tract infection. This notion is based on the finding that a *P. mirabilis* nonmotile mutant was significantly reduced in its ability to invade cultured human renal epithelial cells and to infect the murine urinary tract (102). Recent experimental data showed that a fimbrial *mrpJ* gene product represses transcription of the flagellar regulon, reducing flagellar synthesis when MR/P fimbriae are produced. The fact that an *mrpJ* null mutant was attenuated in the murine model strongly supports the role for flagella in colonization and infection of the urinary tract (82).

Clostridium difficile flagellate strains have a better capacity to associate with the murine cecal wall in vivo. The flagellin and flagellar cap protein FliD of *C. difficile* adhered to mucus isolated from axenic mice, and a nonflagellate strain was 10-fold less strongly adherent than a flagellate strain of the same serogroup (147). These results suggest that the production of flagella by *C. difficile* is central to the ability of this species to colonize the host intestinal epithelium. However, so far no genetic data have been provided to support this hypothesis.

Aeromonas spp. possess two distinct flagellar systems: a polar flagellum for swimming in liquid and lateral flagella for swarming over surfaces. The lateral flagella were proposed to be adhesins, to contribute to microcolony formation and efficient biofilm formation on surfaces, and to facilitate host invasion (123). It is possible, although not proven, that the ability of *Aeromonas* strains to cause persistent and dysenteric infections in the gastrointestinal tract is attributed to the expression of lateral flagella (41, 42, 68).

The association between loss of flagellar expression and loss of infectivity in *L. pneumophila* argues in favor of a coordinated regulation between virulence factors and flagellar expression (122). Insertional mutagenesis has linked the expression of flagella by *L. pneumophila* serogroup 1 and the ability to infect and multiply intracellularly within both *Hartmanella vermiformis* and the human macrophage-like (U937) cell line. Flagella were then proposed to be a positive predictor for virulence in *Legionella* (13).

Reduced invasion of human endothelial cells in vitro and in a mouse model has been noted for flagellar mutants of *B. burgdorferi* and *B. agrinii*, respectively (131, 136). The few reports on these organisms indicate that flagella might be important factors that promote the invasion of endothelial cells.

BINDING OF BACTERIAL FLAGELLA TO HOST PROTEINS

Bacterial flagella may possess adhesive properties per se. The flagella of *P. aeruginosa* and *C. difficile* promote bacterial adherence to mucus (3, 147). In the case of respiratory epithelial cells, five membrane-tethered mucins have been identified (reviewed in reference 86), among which Muc1 is the most abundantly expressed. Its strategic location in the upper airways permits interaction with inhaled microorganisms such as *P. aeruginosa*. Muc1 mucin on the epithelial cell surface was shown to be a flagellum-mediated adhesion site for *P. aeruginosa* (86). Adherence experiments performed with Chinese hamster ovarian (CHO) cells transfected with Muc1 DNA demonstrated that flagella, and not the flagellar cap protein FliD or the type IV pili, mediated *P. aeruginosa* adhesion to Muc1. In agreement, *P. aeruginosa* adhesion was blocked by pretreatment of bacteria with antibody to flagellin or pretreatment of CHO-Muc1 cells with purified flagellin (86). Contradictory results were presented by Arora et al., who found that *P. aeruginosa* FliD cap protein has the ability to bind to mucin (3). These differences could be attributed to the experimental conditions or strains employed. If FliD is in fact an adhesin, it would be equivalent to the PapG or FimH pilus tip adhesin proteins present in the uropathogenic *E. coli* Pap and type 1 pili, respectively (reviewed in reference 71). McNamara and Basbaum presented details of the pathway by which *Pseudomonas* flagellin stimulates transcription of the mucin gene *muc2* (95). The ability of *Pseudomonas* flagella to bind mucin may be a mechanism that facilitates colonization and persistence of the organism in the respiratory tract. The flagella of *C. difficile* were also shown to bind to axenic mouse cecal mucus and porcine stomach mucus (147).

A clinical *E. coli* strain (O25:H1) associated with bacteremia and meningitis was demonstrated to immobilize plasminogen on its flagella (77). In the presence of tissue-type plasminogen activator, this event leads to the formation of plasmin, a trypsin-like serine protease involved in a variety of physiological and pathological processes, including degradation of fibrin and mammalian extracellular matrices. This scenario is thought to promote bacterial penetration into host tissues and fluids. The flagellar H6 of EPEC was recently shown to bind directly to cultured epithelial cells (48) and to immobilized plasminogen (Avelino et al., unpublished results). Flagella of various EPEC serotypes (H2, H6, H34, and H40) were shown to specifically bind to extracellular matrix glycoproteins including fibronectin and laminin but not collagen type IV or vitronectin. In contrast, H7 flagella from EHEC O157 did not bind to any of these basement membrane proteins (Avelino et al., unpublished). In all, these data suggest that flagella from some bacterial species may possess the ability to bind to host proteins in such a way as to benefit bacterial colonization and survival.

WHAT IS THE ROLE OF FLAGELLA IN BIOFILM FORMATION AND DEVELOPMENT?

It has been suggested that the formation of a biofilm (defined as the association of bacteria into communities on abiotic surfaces) is influenced by flagella, motility, the presence of pili, and exopolysaccharide (46, 113). Alginate and colonic acid are essential components of the biofilm formed by *P. aeruginosa* and *E. coli* K-12, respectively (113, 121). In these organisms, the transcription of flagellar genes and biofilm-forming exopolysaccharides is inversely regulated. Once the bacteria associate with the surface, they shut off flagellar synthesis and initiate biosynthesis of exopolysaccharide, a key element in the formation, architecture, and stability of biofilms (Fig. 3).

In *E. coli* K-12, *V. cholerae* O1 El Tor, and *P. aeruginosa,* the initial attachment to a surface is promoted by functional flagella, suggesting that motility is needed to overcome the repulsive forces between the bacterium and the inert surface (113, 121, 154).

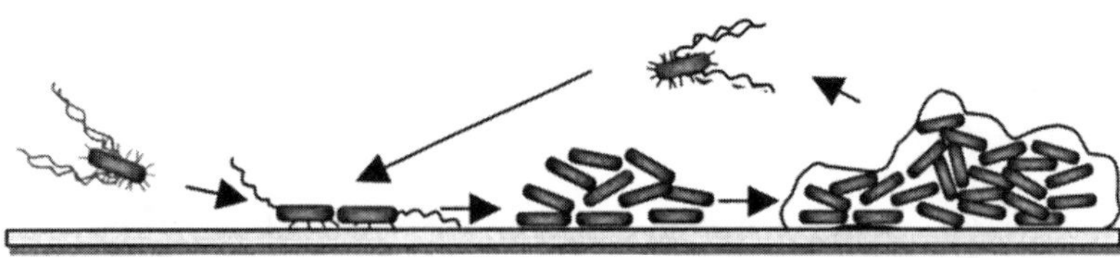

Figure 3. Role of flagella in biofilm formation. Shown is the model for biofilm formation by *P. aeruginosa*. The initial attachment of the bacteria to the abiotic surface is promoted by functional flagella and motility. Once attached, the bacteria employ type IV pilus-mediated twitching motility to spread out on the surface, with subsequent formation of aggregates that form three-dimensional domes or columns surrounded by exopolysaccharide material, which renders the bacteria resistant to many antimicrobials. The bacteria may dissociate from the columns to initiate a new community.

Flagellar and twitching motility (mediated by type IV pili) were described to be required for biofilm formation in *P. aeruginosa* PA14 (Fig. 3) (113). In contrast, a separate study with *P. aeruginosa* PAO1 and isogenic type IV pili and flagellar mutants found that type IV pili and flagella do not play a role in the attachment of *P. aeruginosa* to a glass surface in flow chambers or microtiter plate wells with citrate minimal medium (69, 70). In microtiter plate wells with glucose-Casamino Acids minimal medium, however, the flagella and type IV mutants of PAO1 were deficient in biofilm formation, in agreement with the findings of O'Toole and Kolter with strain PA14 (113). It is obvious that differences in the strains and the carbon source used in the growth medium are factors determining the involvement of flagella or type IV pili in biofilm formation and development (69, 70, 113). *Stenotrophomonas maltophilia* (formerly *P. maltophilia*) forms a biofilm whose architecture depends on the presence of flagella and fimbriae (27).

The developmental pathway leading to *V. cholerae* El Tor biofilm formation has been dissected into its constituent steps through genetic and microscopic analyses (153, 154). The bacteria use their polar flagella to establish cell-to-surface interactions, which are enhanced by the presence of the mannose-sensitive hemagglutinin type IV pili. These pili are then responsible for surface-associated motility, known as twitching motility. Flagellar mutants of *V. cholerae* O1 El Tor were delayed in biofilm formation compared to wild-type strains, suggesting a role for flagella in the early steps of biofilm formation (11). Gram-negative bacteria require exopolysaccharide synthesis for biofilm development. In *V. cholerae* O139, the expression of exopolysaccharide yields rugose colonies and the formation of a three-dimensional biofilm structure. Watnick et al. showed that *V. cholerae* O139 flagellar mutants produced inappropriate exopolysaccharide, which led to inefficient colonization of the infant-mouse intestinal epithelium (153). Thus, flagella play a role in the regulation of the rugose O139 polysaccharide biosynthesis in *V. cholerae*.

In contrast to the biofilm studies performed with *E. coli* K-12, enteroaggregative *E. coli* (EAEC) strain 042 does not require flagella/motility, curli, antigen 43, or type 1 pili to initiate biofilm formation (137). Instead, the aggregative adherence fimbriae (AAF/II) appear to be important for biofilm formation in EAEC strains producing these fimbriae. Controversy exists concerning the role of type 1 pili in aggregative adherence and biofilm formation by EAEC strains, since a recent report claimed that antibodies against type 1 pili largely blocked the aggregative adherence pattern of 042 and that a *fimD* (the type 1 pilus usher gene) mutant exhibited only 51% of the biofilm formation of the parent strain (104). It is possible that both AAF/II and type 1 pili are both important players in biofilm formation by EAEC strains.

INNATE IMMUNE RESPONSE TO BACTERIAL FLAGELLIN

The capacity of flagellin to induce an adaptive immune response has been known for a long time. More recently, a growing body of exciting data has made us aware of the ability of the innate immune system to selectively detect an ample range of microbial products. These include flagella, LPS, peptidoglycan, yeast cell wall, lipoteichoic acid, lipoproteins, heat shock proteins, and DNA. This selectivity is achieved via recognition of a family of TLR, which can be found in a wide spectrum of eukaryotic organisms ranging from insects to plants to humans (64, 98, 125, 139, 150). Epithelial cells from various tissues, including lung, intestine, and kidney, respond to flagellin (107). In addition to being expressed by phagocytic cells (monocytes and dendritic cells), TLR5 is expressed on the basolateral surface of epithelial cells (45, 101). The engagement of TLR with bacterial products activates host cells to initiate local and systemic inflammatory responses and drives the differentiation of antigen-presenting cells to trigger an adaptive immune response (93, 94, 96, 97, 100, 106, 111, 126, 133, 138, 140, 141). In particular,

TLR5 was shown in experiments with several intestinal epithelial cells lines to detect flagellin of *S. enterica* serovar Typhimurium and to induce the basolateral production of IL-8 (45, 55). It has been proposed that this system selectively discriminates between flagella produced by resident members of the normal flora and flagella of pathogenic bacteria. The flagellin-TLR5 signaling pathway is emerging as a novel concept in the interaction of motile bacteria with eukaryotic organisms and in inducing dendritic cell maturation and differentiation of T-helper cells (64, 98). Further discussion of the interaction between flagella and TLR5 is presented below.

PROINFLAMMATORY ROLE OF *S. ENTERICA* FLAGELLA

Salmonella infection in the gastrointestinal tract is characterized by an acute inflammatory response, which leads to luminal translocation and recruitment of neutrophils in intestinal epithelia. *Salmonella* species can stimulate TNF-α in primary human monocytes and in the promonocytic cell line U38 (18). *Salmonella* infection also results in activation of inflammatory mediators, most of which are typically controlled by the transcription of nuclear factor κB (NF-κB) and the mitogen-activated protein kinase (MAPK) signal transduction pathways (Fig. 4).

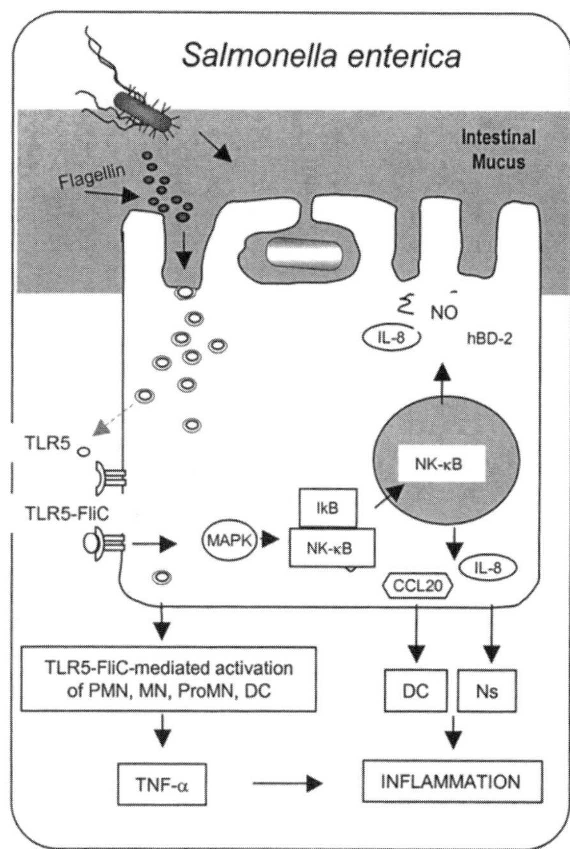

Figure 4. Schematic representation of the mechanism of flagella-mediated inflammation by *S. enterica*. The salmonellae utilize flagellum-driven motility to cross the intestinal mucus barrier. These bacteria secrete abundant flagellin to the extracellular milieu; flagellin then translocates to the basolateral surface of epithelial cells, where it interacts specifically with TLR5, leading to MAPK/NF-κB activation pathways, resulting in the induction of proinflammatory molecules (IL-8 and CCL20), synthesis of nitric oxide synthase and human β-defensin 2. (hBD-2). FliC may also induce TNF-α release from PMN, monocytes (MN), promonocytes (ProMN), and dendritic cells (DC) via activation of TLR5. CCL20 is a potent activator of DC, while IL-8 is a potent recruiter of neutrophils (Ns). The end result is inflammation.

Salmonella is able to secrete abundant amounts of flagellin into the medium (74). *Salmonella*-free growth supernatants are able to activate TNF-α release, and this activity has been linked to flagellin (19). A great deal of experimental data suggests that intact flagella and purified or recombinant flagellin (FliC or FljB) from *S. enterica* may directly activate host inflammatory signaling pathways (19, 20, 93). Thus, assembly of FliC into complete flagellar filaments is not strictly required for induction of TNF-α and IL-1 release. Human primary monocytes, monocyte-derived macrophages, and THP-1 cell lines produce TNF-α, IL-1β, IL-6, IL-10, and gamma interferon (IFN-γ) in response to bacterial flagellin (19, 20, 93, 157). Several research groups have independently shown that flagellins from diarrheagenic *E. coli*, *Y. enterocolitica*, and *P. aeruginosa* strains are responsible for eliciting TNF-α or IL-8 immune responses depending on whether they employed differentiated or undifferentiated macrophage or epithelial cell lines (19, 48, 144). This suggests that the activation or differentiation state of a monocyte may have a substantial effect on the responsiveness of a cell to flagellum-mediated stimulation of cytokine synthesis.

Bacterial flagellin is also a target of CD4 T cells during murine *S. enterica* serovar Typhimurium infection, suggesting that flagellin might also regulate cells of the adaptive immune system and function as adjuvants (96). Mutants defective in flagellum production or in the function of regulators controlling flagellum expression (e.g., *hns* and *flhDC*) generally lack the ability to induce TNF-α (19). *S. enterica* serovar Typhi flagella also have the ability to stimulate de novo synthesis of TNF-α and IL-1β, IL-6, and IL-10 proinflammatory cytokines and to alter in vitro T-cell proliferation by soluble antigens and mitogens (157, 158). Moreover, *S. enterica* serovar Typhi flagella decreased the percentage of $CD14^+$ cells, indicating that these flagella are potent monocyte activators (157, 158). In mice, intravenous bacterial flagellin is a potent activator of inflammatory responses, secretion of

TNF-α, IL-6, IL-12p40, IL-10, macrophage inflammatory protein 1α, and IFN-γ, and production of reactive nitrogen intermediates (33, 55, 83). While the magnitude of cytokine induction by flagellin is significantly lower than for LPS, flagellin may be an important modulator of gram-negative bacterium-mediated shock and of cytokine responses when LPS and flagellin are coadministered (84).

RESPONSES OF EPITHELIAL CELLS TO FLAGELLIN

Epithelial cells constitute a natural barrier to bacterial infections and respond to bacteria by secreting chemokines such as IL-8 and CCL20 and activating the synthesis of inducible nitric oxide synthase and β-defensin 2 (100, 111, 138, 144). A number of reports have shown that *E. coli* and *S. enterica* flagellins are responsible for proinflammatory IL-8 signals in T84, Caco-2, HT-29, and MCDK epithelial cell lines (111, 143, 144, 170). Notably, IL-8 is a potent polymorphonuclear (PMN) cell chemoattractant that is secreted basolaterally and is stimulated by the pathogen-host interaction at the apical surface.

Steiner et al. (144) showed that the H18 flagellin of EAEC strain 042 causes IL-8 release from Caco-2 cells. A *fliC* mutant of 042 did not induce IL-8 production. Not all of the EAEC strains studied showed increased IL-8 signaling, including a strain producing the same flagellar H18 type, suggesting that differences between H18 flagellins may exist and that this is not a universal property of EAEC strains. In this study, the flagella of EPEC and EHEC also induced IL-8 release from Caco-2 cells, while no IL-8 activation was seen with the flagella of *V. cholerae,* diffuse-adhering *E. coli* strain C1845, or nonpathogenic *E. coli* strains.

It was thought that the TTSS of EPEC was involved in triggering IL-8 production in epithelial cells by activating MAPK cascades (24, 25), but a recent study showed that the purified H6 and H34 flagella of EPEC, and not the TTSS, stimulate IL-8 production in T84 cells (Fig. 2) (170). In agreement, EPEC *fliC* isogenic mutants were reduced in their ability to trigger IL-8 production while TTSS mutants were unaffected in this activity. Flagella purified from EHEC and *E. coli* K-12 showed similar levels of IL-8 induction as those for H6 flagella, suggesting that this is a property of flagella of some pathogenic bacteria as well as some members of the normal flora. The differences in IL-8 induction by *E. coli* K-12 flagella observed by Steiner et al. and Zhou et al. may be attributed to the strains and epithelial cell lines employed (144, 170).

Recent evidence obtained on infection of T84 and HT-29 human colonic carcinoma cells with flagellate and nonflagellate *Salmonella* strains indicates that *S. enterica* serovar Typhimurium secretes and translocates flagellin monomers across intestinal epithelia to elicit a potent proinflammatory IL-8 response (Fig. 4) (44, 45). Flagellin translocation and promotion of PMN movement by the salmonellae is independent of their invasive capability, since mutants defective in invasion and function of the TTSS present in *Salmonella* pathogenicity island-1 (SPI-1) were still able to translocate flagellin and to fully activate proinflammatory signals (33, 44, 168).

Zeng et al. showed that the proinflammatory response by *S. enterica* serovar Typhimurium was contact dependent, since bacteria separated from the apical surface by a 0.2-μm-pore-size filter did not induce a proinflammatory response in T84 cells (168). In contrast, neither commensal *E. coli, S. enterica* serovar Typhi nor their flagellins activate proinflammatory signals even when applied apically. The differences in the cell types and strains employed may account for the differences between the observations of these independent researchers. Aflagellate *S. enterica* serovar Typhimurium strains with mutations in the flagellin structural genes *fliC*/*fljB* and *fliD* (encoding a capping protein) were nearly devoid of proinflammatory signaling. This results from their inability to activate the well-defined signaling pathways, MAPK and NF-κB, both of which can be triggered by TLR5. However, mutants with mutations in either *fliC* or *fljB* induced full proinflammatory responses similar to those induced by their parent strain (168). This result is consistent with the idea that the proinflammatory nature of flagellin is a property of the protein rather than a functional consequence of bacterial motility.

Contact-dependent interleukin stimulation by flagellin was also demonstrated during *P. aeruginosa* cell infection (57). A recent study investigated the role of epithelial polarity in the control of the innate immune response of airway epithelial cells (CF15 nasal cystic fibrosis cell line) to *P. aeruginosa* infection. Basal application of the bacteria caused significant changes in expression of 1,525 different genes, including those encoding IL-6, IL-8, IL-1β, and TNF-α, as well as genes associated with leukocyte adhesion, antibacterial factors, and NF-κB signaling, while apical stimulation elicited the expression of only 602 genes. These responses depended on the presence of flagellin but not of pili on the bacteria (57).

Defensins are antimicrobial peptides that play a key role in host defenses at mucosal surfaces. The flagellin of *S. enterica* serovar Enteritidis induces human β-defensin 2 mRNA expression in Caco-2 cells via NF-κB activation and plays an important role in up-regulation of the innate immune response (Fig. 4) (111). Further studies showed that *S. enterica* serovar Enteritidis FliC increased the concentration of intracellular Ca^{2+} via inositol-1,4,5-triphosphate, which was followed by up-regulation of human β-defensin 2 mRNA expression via the NF-κB-dependent pathway (146).

Sierro et al. showed that *S. enterica* serovars Enteritidis and Typhimurium and *L. monocytogenes* activated expression of the *ccl20* gene in Caco-2 cells (Fig. 4) (138). CCL20 is a chemokine that selectively attracts immature dendritic cells. In contrast, indigenous intestinal bacteria such as *E. coli, Bacillus bifidum,* and *Bacteroides vulgatus* were unable to induce CCL20 expression. Consistent with other studies, CCL20 stimulation does not require epithelial cell invasion or activity of the TTSS (33). Purified flagellin of *S. enterica* serovar Enteritidis and a *fliC* mutant complemented with *fliC* and *fljB* of *S. enterica* serovar Typhimurium induced CCL20 in Caco-2 and T84 cells. The coupling of CCL20 and IL-8 transcriptional activation could be crucial for the induction of protective immune responses in the gut. The CCL20-stimulating activity of flagellin could be essential to enhance the migration of fully activated dendritic cells into subepithelial areas of Peyer's patches and villi.

Another property associated with flagellin is its ability to induce self-tolerance. For example, prior exposure of human monocytes and THP-1, Jurkat, and Cos-1 cells to flagellin results in flagellin-induced self-tolerance, which is associated with a block in the release of IL-1R-associated kinase from the TLR5 complex in flagellin-tolerant cells (100). Further, flagella may act as an inducer of oxidative stress during sepsis, as shown by *S. enterica* serovar Muenchen infection in a rat model. Flagellin may circulate as free protein in the plasma of infected rats, which could contribute to generalized inflammatory responses, oxidative stress, and organ damage (33, 83, 84).

TOLL-LIKE RECEPTOR 5 ACTIVATION BY FLAGELLIN

It was reported that the immunostimulatory capacity of flagellin from gram-negative (*S. enterica* serovar Typhimurium) and gram-positive (*L. monocytogenes*) bacteria was mediated by recognition of TLR5 expressed by monocytes, immature dendritic cells, and epithelial cells (55). These authors suggested that the innate immune system utilizes TLR5 mainly to specifically detect a diverse set of flagellated bacterial pathogens. The flagellin-TLR5 interaction activates the transcription of inflammatory genes such as those encoding NF-κB, leading to TNF-α production. To support these notions, nonflagellate *E. coli* expressing *L. monocytogenes* flagellin was able to activate TLR5, and the deletion of the flagellin genes from *S. enterica* serovar Typhimurium abrogated TLR5-stimulating activity. Consistent with the emerging model of *S. enterica* serovar Typhimurium flagellin-induced inflammation, it was recently demonstrated that the basal body protein FliE of *S. enterica* serovar Typhimurium is essential for flagellin secretion, flagellar assembly, and IL-8 induction via basolateral TLR5 (126). The *fliE* mutant, although adherent and invasive, failed to activate the IκBα/NF-κB signaling pathway or induce the coordinated transepithelial migration of isolated human neutrophils. These authors showed that TLR5 signaling can be functionally down-regulated by preincubation with purified flagellin. Monolayers of MDCK or T84 cells that had been preincubated with basolateral flagellin did not secrete IL-8 in response to either wild-type *Salmonella* or basolateral flagellin, while the response of the monolayers to TNF-α was unaffected. Thus, functional down-regulation of TLR5 completely blocks the ability of the epithelial monolayer to mount a proinflammatory response to *Salmonella,* confirming again that the flagellin-TLR5 interaction is the sole mediator of the epithelial proinflammatory response to *Salmonella* (126). Although LPS is also proinflammatory, flagellin is the major molecular trigger by which *S. enterica* serovar Typhimurium specifically activates gut epithelial proinflammatory gene expression (168). It was recently shown that *S. enterica* serovar Typhimurium flagellin can directly stimulate human but not murine dendritic cell maturation, providing an additional mechanism by which motile bacteria can initiate an acquired immune response (97). Stimulation of immature dendritic cells with flagellin induced increased surface expression of CD80, CD83, CD86, major histocompatibility complex class II, and the chemokine receptor CCR7.

TLR5 signaling by flagellin was also demonstrated to occur in vivo. Intraperitoneal injection of wild-type mice with purified flagellin induced a systemic IL-6 release. Remarkably, stimulation of TLR5 does not require polymerization of the flagellin molecule since the monomer is sufficiently capable of activating TLR5 (45, 55).

TLR5 activation by epithelial cells in response to flagellin is topologically dependent. TLR5 is expressed exclusively on the basolateral surface of intestinal epithelia, which is consistent with the fact that flagellin can show proinflammatory activity through the basolateral, not the apical, surface (44, 45). For instance, *S. enterica* serovar Typhimurium flagellin does not activate an immune response when applied on the apical surface of intestinal epithelial cells (55). This non-proinflammatory scenario may mimic a condition that presumably exists when commensal flagellate bacteria colonize the mucosal surface of the gut. In agreement with this idea, commensal flagellate *E. coli* strains or their flagella failed to induce CCL20 to activate TLR5 or to elicit IL-8 release from epithelial cells (138, 143). Conversely, flagellin applied basolaterally elicits an expression profile of proinflammatory molecules largely similar to that observed with direct colonization with pathogenic *Salmonella*. Thus, it is the interaction of the bacteria with the apical surface of polarized epithelial cells that induces secreted flagellin to be translocated and transported to the TLR5-rich basolateral surface, triggering IL-8 production (44). As with TNF-α induction, TLR5 activation does not require bacterial invasion of epithelial cells or expression of the SPI-1-encoded TTSS (33, 45).

How do we explain the tolerance of the innate immune system to large numbers of commensal bacteria in the gut epithelia? On one hand, it was suggested that the existence of molecular tags on the flagellar structures and flagellins produced by bacterial pathogens allowed the innate system to discriminate between indigenous bacteria and microbial foes (98, 139). Further, it is possible that normal gut residents fail to break down or depolymerize flagellar filaments or to secrete or translocate flagellin across the epithelia to the basolateral region where TLR5 is located, and as a result, cannot elicit a proinflammatory response. Alternatively, the normal flora may shut off the program to induce flagella once they have established a niche. Several reports argue against these hypotheses. For instance, nonpathogenic flagellate *E. coli* strains and their purified flagella or flagellin were able to release IL-8 from T84 cells and nitric oxide from DLD-1 cells (33) to levels exhibited by flagella of EPEC (170). The differences in IL-8 stimulation obtained by several workers using fecal, laboratory, and pathogenic *E. coli* strains can be attributed to the intrinsic nature of their flagellins, to the origin of the strains, and to their ability or inability to secrete or translocate flagellin. More experimental data are required to define whether TLR5 recognition is a widespread property of bacterial flagellins and whether TLR5 serves as a general alarm system for a broad group of motile pathogens (55).

In a different model of bacterial infection, the two flagellins FlaA and FlaB of *H. pylori*, which exhibit extensive amino acid homology to stimulating flagellins of other bacterial species, did not show any immunostimulatory potential via TLR5 in human gastric epithelial cells (81). The authors demonstrated that different gastric cells lines transcribed and expressed TLR5 in the presence or absence of wild-type *H. pylori* or isogenic *flaA* and *flaB* mutants. They speculated that this low activating potential is a novel mechanism whereby *H. pylori* preserves the essential function of its flagella during chronic colonization of the stomach without inducing proinflammatory activity, thus evading the deleterious host immune response (81).

Hawn et al. reported that CHO and HEK 293 cells expressing TLR5 with a common stop codon polymorphism were unable to respond to *S. enterica* serovar Typhimurium flagellin (54). The presence of this stop codon in TLR5 was suggested to be associated with susceptibility to *L. pneumophila* infection in Legionnaires' disease in Caucasians. These authors demonstrated that *L. pneumophila* flagellin is the major stimulant of IL-8 production in lung A459 epithelial cells.

Diverse *P. aeruginosa* gene products stimulate respiratory epithelial cells to produce IL-8 (30). Previous studies found that *P. aeruginosa* flagellin stimulated IL-8 production in tracheal epithelial cells. These authors concluded that selective TLR5 deficiency can predispose an individual to an increased risk of pneumonia due to a flagellate organism. In a different study, *P. aeruginosa* flagellin was shown to induce inflammatory responses of corneal epithelium in a TLR5/NF-κB signaling pathway-dependent manner (169). It is apparent that the TLR5-flagellin interaction may lead to different activation patterns that depend on the organism and host tissue studied.

FliC REGIONS RESPONSIBLE FOR PROINFLAMMATORY ACTIVITY AND TOLL-LIKE RECEPTOR 5 BINDING

Several authors have attempted to determine whether inflammatory activity is a widespread feature of gram-negative bacterial flagellins and to identify regions of the flagellin protein responsible for activating TNF-α and IL-8 production in macrophage and epithelial cell lines, respectively. For this purpose, they have employed mutagenesis analysis, creating peptides derived from conserved and nonconserved regions of the FliC molecule and recombinant flagellin proteins lacking amino or carboxy regions of the molecule. Since flagellin of several bacterial species binds TLR5, it would seem reasonable to think that the conserved regions of the flagellin molecule are responsible for binding to TLR5. However, this would signify that most bacteria, both pathogenic and nonpathogenic, should be capable of activating proinflammatory signals via interaction of the conserved regions of their flagellins with TLR5. In fact, flagella from nonpathogenic *E. coli* strains have been shown to elicit an IL-8-stimulatory response (170). However, the experimental data obtained by independent research groups are somewhat conflicting in this respect. While some workers argue that the conserved region is crucial for proinflammatory activity, others favor the hypervariable region as the one responsible for stimulating an innate immune response. Although earlier studies suggested that the immunostimulatory activity of flagellin resided in the variable D3 domain (93, 144), more recent reports suggest that the conserved D1 and D2 domains are important for innate immune response (29, 33).

As discussed above, differences in proinflammatory potency manifested by different bacterial flagellins exist. Flagellar bioactivity even seems to be species specific, since *S. enterica* serovar Enteritidis flagella consistently induced higher levels of TNF-α in U38 cells than do flagella from *S. enterica* serovar Typhimurium. Variations in potency between flagellins of related *Salmonella* species may be due to differences in the structure or primary sequence of the hypervariable domain in these flagellin proteins (19). In agreement with these results, mutagenesis studies of *S. enterica* serovar Enteritidis *fliC* revealed that the central hypervariable region of the recombinant fusion-flagellin molecule is crucial for the TNF-α-inducing activity of the protein. Deletion of the variable domain of flagellin failed to cause proinflammatory activity. The central hypervariable region alone (containing residues 146 to 465) was quite active, albeit less so than the native protein, in inducing TNF-α production in monocytes and THP-1 cells. Flagellin lacking residues 151 to 289 and 296 to 421 of the hypervariable regions showed only 5% of the TNF-α production activity of the intact molecule, suggesting the presence of at least two independent TNF-α-inducing domains (93). Deletion of either the amino- or carboxy-terminal half of the variable domain reduced immunopotency but still yielded an inflammatory molecule. These authors concluded that for optimal activation, the presence of the conserved amino- and carboxy-terminal domains is required, although these regions alone did not stimulate TNF-α production. It is possible that the conserved regions play an important role in generating an optimal conformation of the hypervariable domain within the flagellin molecule and, in turn, on the flagellum filament in order to display proinflammatory epitopes effectively.

In contrast to these findings, Eaves-Pyles et al. (33) showed that a construct containing the amino- and carboxy-terminal constant regions of *S. enterica* serovar Dublin flagellin separated by an *E. coli* hinge element to provide proper folding was as active as wild-type flagellin. Unlike in previous studies, these workers generated a series of recombinant His_6 fusion proteins comprising different domains of the *S. enterica* serovar Dublin FliC protein, which were tested for bioactivities such as IκBα degradation, NF-κB activation, IL-8 secretion, and NO production in cultured intestinal (Caco-2 and DDl-1) cells and for TNF-α induction in human monocytes. The bioactivity was localized to a region in the stem that is dependent on the amino and carboxyl conserved sequences. Fusion proteins representing the D3, amino, or carboxyl regions alone did not induce proinflammatory mediators. The authors concluded from these data that the constant domains of flagella contain the proinflammatory structural epitopes and that the variable domains are necessary to keep the constant regions in proper apposition. The lack of bioactivity detected by McDermott et al. (93) in the conserved region of flagellin may be due to the inability of the fusion-FliC proteins to fold properly and form the domain responsible for bioactivity.

In a more detailed approach, Murthy et al. (106) deleted amino acids 95 to 108 in the N terminus amino acids and 441 to 449 in the C terminus of the flagellin of *S. enterica* serovar Muenchen and found that these deletions abolished flagellin proinflammatory activity completely. Site-directed mutagenesis of these regions provided further evidence for the importance of the amino and carboxy ends of the FliC molecule in proinflammatory induction.

Another study used the complementary hydropathy between TLR5 and flagellin to predict the binding regions within the complex formed by these two proteins. A region common to the flagellins of *S. enterica* serovar Typhimurium, *P. aeruginosa,* and *L. monocytogenes* was shown to be hydropathically complementary to the fragment from positions 552 to 561 (EILDISRNQL) of TLR5. The most stable structure obtained in this search showed an α-helical conformation, which forms the electrostatic interactions Glu552-Gln89, Asp555-Arg92, and Arg558-Glu93 with the predicted site of the flagellin of *S. enterica* serovar Typhimurium, formed by the chain fragment from positions 88 to 97 (LQRVRELAVQ), which is likewise α-helical (60). It is likely that α-helix conformations are important for flagellin-TLR5 interactions.

A separate study involving FliC of EAEC strain 042 suggested that the conserved D1 domain of the flagellin molecule is required for IL-8 release and TLR5 activation (31). These authors identified two regions in the conserved D1 domain of the FliC of EAEC 042 necessary for proinflammatory activity in epithelial cells. Large regions of the variable domain of EAEC FliC could be deleted or altered without disrupting inflammatory activity. However, mutations in one narrow region of the carboxy-terminal constant domain (residues 185 to 191) eliminated IL-8-releasing activity. Similarly, the amino-terminal conserved region (residues 57 to 190) was shown to contain an epitope required for proinflammatory activity in epithelial cells. Interestingly, these authors found epitopes of flagellin recognized by innate immune receptors in plants that are not required for flagellin signaling in mammalian cells.

It was recently reported that TLR5 recognizes a site on *Salmonella* flagellin composed of amino-terminal residues 79 to 117 and 408 to 439 that map to a discrete region of the FliC D1 domain, which is buried in the core of the flagellar filament. As discussed above, this site is also predicted to be involved in axial intermolecular contacts between individual flagellin monomers that form the protofilaments (141). Removal of the amino-terminal 99 amino acids of the FliC monomer prevented TLR5 recognition. Deletions that removed amino acids 416 to 444 within the C terminus of FliC were sufficient to abrogate TLR5 recognition. These sites fall within the D1-domain polypeptide regions, which are highly conserved among bacteria and are essential for motility. In contrast, no effect was observed when deletions in residues 444 to 492 of the C-terminal D1 domain or the D3-domain residues 185 to 306 within the hypervariable region were introduced. It was also shown that alanine residues within the D1 domain were crucial for TRL5 recognition and for motility.

There is no question that the variable region is responsible for the antigenic diversity of the flagellin molecule. Since flagella from pathogenic bacteria and members of the normal flora are able to induce an IL-8 proinflammatory response (either via TLR5 interaction or via an unknown mechanism), one would assume that the conserved domains of the flagellin account for this activity. The majority of the published data favor this hypothesis. It is tempting to speculate that bioactivities manifested by certain flagella of specific pathogens might be located in the variable region of the molecule. It remains to be determined which domain of the flagellin is responsible for the other bioactivities attributed to this fascinating molecule.

CONCLUDING REMARKS

Flagellar genes are highly conserved among gram-negative bacteria, and much similarity in structure and function exists. The close link between the flagellar TTSS and the TTSS implicated in virulence highlights the evolutionary relationship of these structures. Whether the flagellar TTSS is involved in secreting virulence factors other than flagellin in organisms lacking traditional virulence-dedicated TTSS is an interesting issue that deserves further investigation. For a number of bacterial pathogens, flagella and motility are important in mucosal colonization. Recently, other properties of flagella such as adherence, colonization, invasion of host cells, biofilm formation, and binding to host proteins have been unraveled. Thus, flagella are multipurpose structures composed of a multifunctional protein that has evolved to favor bacterial survival. It is clear that flagellins from residents of the normal flora and from pathogenic bacteria possess molecular signatures that are selectively recognized by both the innate and acquired immune systems resulting in tolerance to residents of the normal flora and in protection from pathogenic bacteria. The highly specific binding of basolateral TLR5 to flagellin is an important selective means of activating innate immune responses that appear to function to prevent host mucosal colonization, at least for *S. enterica*. It is still unclear how this system would work in vivo in pathogens that do not secrete flagellin or that do not depolymerize their flagella. Questions regarding intracellular trafficking of secreted flagellin to the basolateral surface of epithelial cells or across the epithelium to activate immune cells and induce inflammation still need to be addressed and clarified. The controversies observed between the results of some of the studies discussed here, regarding the role of flagella and motility in the host-cell interplay and regarding the domains of the flagellin molecule involved in TLR5 interaction and proinflammatory response, are a reflection of the various in vitro and in vivo models employed, which often give different results and interpretations. One would predict that the novel TLR5-flagellin signaling pathway recognized will stimulate further research to clarify some important discordant issues. The challenge will be to determine which models most faithfully reproduce natural infections and to explore the implications derived from these studies.

Acknowledgments. I thank James W. Moulder, Chris Alteri, and Monique Dall'Agnol for critical reading of the manuscript. Special thanks are due to María Rendón for the schematic illustrations.

REFERENCES

1. **Aizawa, S.** 2001. Bacterial flagella and type III secretion systems. *FEMS Microbiol. Lett.* **202:**157–164.
2. **Andrade, A., J. A. Girón, J. M. K. Amhaz, L. R. Trabulsi, and M. B. Martinez.** 2002. Expression and characterization of flagella in non-motile enteroinvasive *Escherichia coli* isolated from diarrhea cases. *Infect. Immun.* **70:**5882–5886.

3. **Arora, S. K., B. W. Ritchings, E. C. Almira, S. Lory and R. Ramphal.** 1998. The *Pseudomonas aeruginosa* flagellar cap protein, FliD, is responsible for mucin adhesion. *Infect. Immun.* **66:**1000–1007.
4. **Allen-Vercoe, E., and M. J. Woodward.** 1999. The role of flagella, but not fimbriae, in the adherence of *Salmonella enterica* serotype Enteriditis to chick gut explant. *J. Med. Microbiol.* **48:**771–780.
5. **Allen-Vercoe, E. A. R. Sayers, M. J. Woodward.** 1999. Virulence of *Salmonella enterica* serotype Enteritidis aflagellate and afimbriate mutants in a day-old chick model. *Epidemiol. Infect.* **122:**395–402.
6. **Attridge, S. R., and D. Rowley.** 1983. The role of flagellum in the adherence of *Vibrio cholerae*. *J. Infect. Dis.* **147:**864–872.
7. **Auvray, F., J. Thomas, G. M. Fraser, and C. Hughes.** 2001. Flagellin polymerization control by a cytosolic export chaperone. *J. Mol. Biol.* **308:**221–229.
8. **Badger, J. L., and V. L. Miller.** 1998. Expression of invasin and motility are coordinately regulated in *Yersinia enterocolitica*. *J. Bacteriol.* **180:**793–800.
9. **Barnich, N., J. Boudeau, L. Claret, and A. Darfeuille-Michaud.** 2003. Regulatory and functional co-operation of flagella and type 1 pili in adhesive and invasive abilities of AIEC strain LF82 isolated from a patient with Crohn's disease. *Mol. Microbiol.* **48:**781–794.
10. **Blocker, A., K. Komoriya, and S. Aizawa.** 2003. Type III secretion systems and bacterial flagella: insights into their function from structural similarities. *Proc. Natl. Acad. Sci. USA* **100:**3027–3030.
11. **Bomchil, N., P. Watnick, and R. Kolter.** 2003. Identification and characterization of a *Vibrio cholerae* gene, *mbaA*, involved in maintenance of biofilm architecture. *J. Bacteriol.* **185:**1384–1390.
12. **Bonifield, H. R., and K. T. Hughes.** 2003. Flagellar phase variation in *Salmonella enterica* is mediated by a posttranscriptional control mechanism. *J. Bacteriol.* **185:**3567–3574.
13. **Bosshardt, S. C., R. F. Benson, and B. S. Fields.** 1997. Flagella are a positive predictor for virulence in *Legionella*. *Microb. Pathog.* **23:**107–112.
14. **Carsiotis, M., D. L. Weionstein, H. Karch, I. A. Holder, A.D. O'Brien.** 1984. Flagella of *Salmonella typhimurium* are a virulence factor in infected C57BL/6J mice. *Infect. Immun.* **46:**814–818.
15. **Chaubal, L. H., and P. S. Holt.** 1999. Characterization of swimming motility and identification of flagellar proteins in *Salmonella pullorum* isolates. *Am. J. Vet. Res.* **60:**1322–1327.
16. **Chilcott, G. S., and K. Hughes.** 2000. Coupling of flagellar gene expression to flagellar assembly in *Salmonella enterica* serovar Typhimurium and *Escherichia coli*. *Microbiol. Mol. Biol. Rev.* **64:**694–708.
17. **Chua, K. L., Y. Y. Chan, and Y. H. Gan.** 2003. Flagella are virulence determinants of *Burkholderia pseudomallei*. *Infect. Immun.* **71:**1622–1629.
18. **Ciacci-Woolwine, F., L. S. Kucera, S. H. Richardson, N. P. Iyer, and S. B. Mizel.** 1997. Salmonellae activate tumor necrosis factor alpha production in a human promonocytic cell line via a released polypeptide. *Infect. Immun.* **65:**4624–4633.
19. **Ciacci-Woolwine, F., I. C. Blomfield, S. H. Richardson, and S. B. Mizel.** 1998. *Salmonella* flagellin induces tumor necrosis factor alpha in a human promonocytic cell line. *Infect. Immun.* **66:**127–1134.
20. **Ciacci-Woolwine, F., P. F. McDermott, and S. B. Mizel.** 1999. Induction of cytokine synthesis by flagella from gram-negative bacteria may be dependent on the activation or differentiation state of human monocytes. *Infect. Immun.* **67:**5176–5185.
21. **Clyne, M., T. Ocroinin, S. Suerbaum, C. Josenhans, and B. Drumms.** 2000. Adherence of isogenic flagellum negative mutants of *Helicobacter pylori* and *Helicobacter mustelae* to human and ferret gastric epithelial cells. *Infect. Immun.* **68:**4335–4339.
22. **Cordes, F. S., K. Komoriya, E. Larquet, S. Yang, E. D. Egelman, A. Blocker, and S. M. Lea.** 2003. Helical structure of the needle of the type III secretion system of *Shigella flexneri*. *J. Biol. Chem.* **278:**17103–17107.
23. **Correa, N. E., C. M. Lauriano, R. McGee, and K. E. Klose.** 2000. Phosphorylation of the flagellar regulatory protein FliC is necessary for *Vibrio cholerae* motility and enhanced colonization. *Mol. Microbiol.* **35:**743–755.
24. **Czerucka, D., S. Dahan, B. Mograbi, B. Rossi, and P. Rampal.** 2001. Implication of mitogen-activated protein kinases in T84 cell responses to enteropathogenic *Escherichia coli* infection. *Infect. Immun.* **69:**1298–1305.
25. **de Grado, M., C. M. Rosenberger, A. Gauthier, B. A. Vallance, and B. B. Finlay.** 2001. Enteropathogenic *Escherichia coli* infection induces expression of the early growth response factor by activating mitogen-activated protein kinase cascades in epithelial cells. *Infect. Immun.* **69:**6217–6224.
26. **Delahay, R. M., and G. Frankel.** 2002. Coiled-coil proteins associated with type III secretion systems: a versatile domain revisited. *Mol. Microbiol.* **45:**905–916.
27. **de Oliveira Garcia, D., M. Dall'Agnol, M. Rosales, A. C. Azzuz, M. B. Martinez, and J. A. Girón.** 2002. Characterization of flagella produced by clinical isolates of *Stenotrophomonas maltophilia*. *Emerg. Infect. Dis.* **8:**918–923.
28. **DeShazer, D., P. J. Brett, R. Carlton and D. E. Woods.** 1997. Mutagenesis of *Burkholderia pseudomallei* with Tn5-OT182: isolation of motility mutants and molecular characterization of the flagellin structural gene. *J. Bacteriol.* **179:**2116–2125.
29. **Dibb-Fuller, M. P., E. Allen-Vercoe, C. J. Thorns, and M. J. Woodward.** 1999. Fimbriae- and flagella-mediated association with and invasion of culture epithelial cells by *Salmonella enteritidis*. *Microbiology* **145:**1023–1031.
30. **DiMango, E, H. J. Zar, R. Bryan, and A. Prince.** 1995. Diverse *Pseudomonas aeruginosa* gene products stimulate respiratory epithelial cells to produce interleukin-8. *J. Clin. Investig.* **96:**2204–2210.
31. **Donnelly, M. A., and T. S. Steiner.** 2002. Two nonadjacent regions in enteroaggregative *Escherichia coli* flagellin are required for activation of toll-like receptor 5. *J. Biol. Chem.* **277:**40456–40461.
32. **Eaton, K. A., S. Suerbam, C. Josenhans, and S. Krakowka.** 1996. Colonization of gnotobiotic piglets by *Helicobacter pylori* deficient in two flagellin genes. *Infect. Immun.* **64:**2445–2448.
33. **Eaves-Pyles, T. D., H. R. Wong, K. Odoms, and R. B. Pyles.** 2001. *Salmonella* flagellin dependent proinflammatory responses are localized to the conserved amino and carboxyl regions of the protein. *J. Immunol.* **167:**7009–7016.
34. **Eubanks, E. R., M. N. Guentzel, and L. J. Berry.** 1977. Evaluation of surface components of *Vibrio cholerae* as protective immunogens. *Infect. Immun.* **15:**533–538.
35. **Feldman, M., R. Bryan, S. Rajan, L. Scheffler, S. Brunner, H. Tang, and A. Prince.** 1998. Role of flagella in pathogenesis of *Pseudomonas aeruginosa* pulmonary infection. *Infect. Immun.* **66:**43–51.
36. **Freter R., P. C. O'Brien, and M. S. Macsai.** 1981. Role of chemotaxis in the association of motile bacteria with intestinal mucosa: in vivo studies. *Infect. Immun.* **34:**234–240.

37. **Freter, R., B. Allweiss, P. C. O'Brien, S. A. Halstead, and M. S. Macsai.** 1981. Role of chemotaxis in the association of motile bacteria with intestinal mucosa: in vitro studies. *Infect. Immun.* **34:**241–249.
38. **Freter, R., and P. C. O'Brien.** 1981. Role of chemotaxis in the association of motile bacteria with intestinal mucosa: chemotactic responses of *Vibrio cholerae* and description of motile non-chemotactic mutants. *Infect. Immun.* **34:**215–221.
39. **Galán, J. E., and A. Collmer.** 1999. Type III secretion machines: bacterial devices for protein delivery into host cells. *Science* **284:**1322–1328.
40. **Gardel, C. L., and J. J. Mekalanos.** 1996. Alterations in *Vibrio cholerae* motility phenotypes correlate with changes in virulence factor expression. *Infect. Immun.* **64:**2246–2255.
41. **Gavin, R., S. Merino, M. Altarriba, R. Canals, J. G. Shaw, and J. M. Tomas.** 2003. Lateral flagella are required for increased cell adherence, invasion and biofilm formation by *Aeromonas* spp. *FEMS Microbiol. Lett.* **224:**77–83.
42. **Gavin, R., A. A. Rabaan, S. Merino, J. M. Tomas, I. Gryllos, and J. G. Shaw.** 2002. Lateral flagella of *Aeromonas* species are essential for epithelial cell adherence and biofilm formation. *Mol. Microbiol.* **43:**383–397.
43. **Gewirtz, A. T., A. M. Siber, J. L. Madara, and B. A. McCormick.** 1999. Orchestration of neutrophil movement by intestinal epithelial cells in response to *Salmonella typhimurium* can be uncoupled from bacterial internalization. *Infect. Immun.* **67:**608–617.
44. **Gewirtz, A. T., P. O. Simon, Jr., C. K. Schmitt, L. J. Taylor, C. H. Hagedorn, A. D. O'Brien, A. S. Neish, and J. L. Madara.** 2001. *Salmonella typhimurium* translocates flagellin across intestinal epithelia, inducing a proinflammatory response. *J. Clin. Investig.* **107:**99–109.
45. **Gewirtz, A. T., T. A. Navas, S. Lyons, P. J. Godowski, and J. L. Madara.** 2001. Cutting edge. Bacterial flagellin activates basolaterally expressed TLR5 to induce epithelial proinflammatory gene expression. *J. Immunol.* **167:**1882–1885.
46. **Ghigo, J. M.** 2003. Are there biofilm-specific physiological pathways beyond a reasonable doubt? *Res. Microbiol.* **154:** 1–8.
47. **Girón, J. A.** 1995. Expression of flagella and motility by *Shigella. Mol. Microbiol.* **18:**63–75.
48. **Girón, J. A., A. G. Torres, E. Freer, and J. B. Kaper.** 2002. The flagella of enteropathogenic *Escherichia coli* mediate adherence to epithelial cells. *Mol. Microbiol.* **44:**361–379.
49. **Golden, N. J., and D. W. K. Acheson.** 2002. Identification of motility and autoagglutination *Campylobacter jejuni* mutants by random transposon mutagenesis. *Infect. Immun.* **70:**1761–1771.
50. **Grant, C. C. R., M. E. Konkel, W. Cieplak, Jr., and L. S. Tompkins.** 1993. Role of flagella in adherence, internalization and translocation of *Campylobacter jejuni* in nonpolarized and polarized epithelial cell cultures. *Infect. Immun.* **61:**1764–1771.
51. **Gupta, S., and R. Chowdhruy.** 1997. Bile affects production of virulence factors and motility of *Vibrio cholerae. Infect. Immun.* **65:**1131–1134
52. **Hase, C. C.** 2001. Analysis of the role of flagella activity in virulence gene expression in *Vibrio cholerae. Microbiology* **147:**831–837.
53. **Harshey, R. M.** 2003. Bacterial motility on a surface. *Annu. Rev. Microbiol.* **57:**249–273.
54. **Hawn, T. R., A. Verbon, K. D. Lettinga, L. P. Zhao, S. S. Li, R. J. Laws, S. J. Skerrett, B. Beutler, L. Schroeder, A. Nachman, A. Ozinsky, K. D. Smith, and A. Aderem.** 2003. A common dominant *TLR5* stop codon polymorphism abolishes flagellin signaling and is associated with susceptibility to Legionnaires' disease. *J. Exp. Med.* **198:**1563–1572.
55. **Hayashi, F., K. D. Smith, A. Ozinsky, T. R. Hawn, E. C. Yi, D. R. Goodlett, J. K. Eng, S. Akira, D. M. Underhill, and A. Aderem.** 2001. The innate immune response to bacterial flagellin is mediated by Toll-like receptor 5. *Nature* **410:**1099–1103.
56. **Hueck, C. J.** 1998. Type III protein secretion systems in bacterial pathogens of animals and plants. *Microbiol. Mol. Biol. Rev.* **62:**379–433.
57. **Hybiske, K. J. K. Ichikawa, V. Huang, S. J. Lory and T. E. Machen.** 2004. Cystic fibrosis airway epithelial cell polarity and bacterial flagellin determine host response to *Pseudomonas aeruginosa. Cell. Microbiol.* **6:**49–63.
58. **Ikeda, J. S., C. K. Schmitt, S. C. Darnell, P. R. Watson, J. Bispham, T. S. Wallis, D. L. Weinstein, E. S. Metcalf, P. Adams, C. D. O'Connor, and A. D. O'Brien.** 2001. Flagellar phase variation of *Salmonella enterica* serovar Typhimurium contributes to virulence in the murine typhoid infection model but does not influence *Salmonella*-induced enteropathogenesis. *Infect. Immun.* **69:**3012–3030.
59. **Iriarte, M. I. Stainier, A. V. Mikulskis, and G. R. Conelis.** 1995. The *fliA* gene encoding σ-28 in *Yersinia enterocolitica. J. Bacteriol.* **177:**2299–2304.
60. **Jacchieri, S. G., R. Torquato, and R. R. Brentani.** 2003. Structural study of binding of flagellin by toll-like receptor 5. *J. Bacteriol.* **185:**4243–4247.
61. **Jones, G. W., L. A. Richardson, and D. Uhlman.** 1981. The invasion of HeLa cells by *Salmonella typhimurium:* reversible and irreversible bacterial attachment and the role of bacterial motility. *J. Gen. Microbiol.* **127:**351–360.
62. **Jones, B. D., C. A. Lee, and S. Falkow.** 1992. Invasion by *Salmonella typhimurium* is affected by the direction of flagellar rotation. *Infect. Immun.* **60:**2473–2480.
63. **Josenhans, C., and S. Suerbaum.** 2002. The role of motility as a virulence factor in bacteria. *Int. J. Med. Microbiol.* **291:**605–614.
64. **Kaisho, T., and S. Akira.** 2002. Toll-like receptors as adjuvant receptors. *Biochim. Biophys. Acta* **1589:**1–13.
65. **Khoramian-Falsafi, T., S. Harayama, K. Kutsukake, and J. C. Pechere.** 1990. Effect of motility and chemotaxis on the invasion of *Salmonella typhimurium* into HeLa cells. *Microb. Pathog.* **9:**47–53.
66. **Kim, J. S., J. H. Chang, S. I. Chung, and J. S. Yum.** 1999. Molecular cloning and characterization of the *Helicobacter pylori fliD* gene, an essential factor in flagellar structure and motility. *J. Bacteriol.* **181:**6969–6976.
67. **Kimbrough, T. G., and S. I. Miller.** 2002. Assembly of the type III secretion needle complex of *Salmonella typhimurium. Microb. Infect.* **4:**75–82.
68. **Kirov, S. M.** 2003. Bacteria that express lateral flagella enable dissection of the multifunctional roles of flagella in pathogenesis. *FEMS Microbiol. Lett.* **224:**151–159.
69. **Klausen M., A. Aaes-Jorgensen, S. Molin, and T. Nielsen.** 2003. Involvement of bacterial migration in the development of complex multicellular structures in *Pseudomonas aeruginosa* biofilms. *Mol. Microbiol.* **50:**61–68.
70. **Klausen M., A. Heydorn, P. Ragas, L. Lambertsen, A. Aaes-Jorgensen, S. Molin, and T. Nielsen.** 2003. Biofilm formation by *Pseudomonas aeruginosa* wild type, flagella and type IV pili mutants. *Mol. Microbiol.* **48:**1511–1524.
71. **Klemm, P., and M. A. Schembri.** 2000. Bacterial adhesins: function and structure. *Int. J. Med. Microbiol.* **290:**27–35.
72. **Klose, K. E., and J. J. Mekalanos.** 1998. Distinct roles of an alternative sigma factor during both free-swimming and colonizing phases of the *Vibrio cholerae* pathogenic cycle. *Mol. Microbiol.* **28:**501–520.
73. **Knutton, S., I. Rosenshine, M. J. Pallen, I. Nisan, B. C. Neves, C. Bain, C. Wolff, G. Dougan, and G. Frankel.** 1998.

A novel EspA-associated surface organelle of enteropathogenic *Escherichia coli* involved in protein translocation into epithelial cells. *EMBO J.* **17:**2166–2176.

74. **Komoriya, K., N. Shibano, T. Higano, N. Azuma, S. Yamaguchi, and S. Aizawa.** 1999. Flagellar proteins and type III-exported virulence factors are the predominant proteins secreted into the culture media of *Salmonella typhimurium*. *Mol. Microbiol.* **34:**767–779.
75. **Kubori, T., Y. Matsushima, D. Nakamura, J. Uralil, M. Lara-Tejero, A. Sukhan, J.E. Galán, and S. I. Aizawa.** 1998. Supramolecular structure of the *Salmonella typhimurium* type III protein secretion system. *Science* **280:**602–605.
76. **Krukonis, E. S., and V. J. DiRita.** 2003. From motility to virulence: sensing and responding to environmental signals in *Vibrio cholerae*. *Curr. Opin. Microbiol.* **6:**186–190.
77. **Lahteenmaki, K., B. Westerlund, P. Kuusela, and T. K. Korhonen.** 1993. Immobilization of plasminogen on *Escherichia coli* flagella. *FEMS Microbiol. Lett.* **106:**309–314.
78. **La Ragione, R. M., A. R. Sayers, and M. J. Woodward.** 2000. The role of fimbriae and flagella in the colonization, invasion and persistence of *Escherichia coli* O78:K80 in the day-old-chick model. *Epidemiol. Infect.* **124:**351–363.
79. **La Ragione, R. M., W. A. Cooley, P. Velge, M. A. Jepson, and M. J. Woodward.** 2003. Membrane ruffling and invasion of human avian cell lines is reduced for flagellate mutants of *Salmonella enterica* serotype Enteritiditis. *Int. J. Med. Microbiol.* **293:**261–272.
80. **Lee, M. D., R. Curtiss III, and T. Peay.** 1996. The effect of bacterial surface structures on the pathogenesis of *S. typhimurium* infection in chickens. *Avian Dis.* **40:**28–36.
81. **Lee, S. K., A. Stack, E. Katzowitsch, S. I. Aizawa, S. Suerbaum, and C. Josenhans.** 2003. *Helicobacter pylori* flagellins have very low intrinsic activity to stimulate human gastric epithelial cells via TLR5. *Microb. Infect.* **5:**1345–1356.
82. **Li, X., D. A. Rasko, C. V. Lockatell, D. E. Johnson, and H. L. T. Mobley.** 2001. Repression of bacterial motility by a novel fimbrial gene product. *EMBO J.* **20:**4854–4862.
83. **Liaudet, L., K, G. K. Murthy, J. G. Mabley, P. Pacher, F. G. Soriano, A. L. Salzman, and C. Szabo.** 2002. Comparison of inflammation, organ damage and oxidant stress induced by *Salmonella enterica* serovar Muenchen flagellin and serovar Enteritidis lipopolysaccharide. *Infect. Immun.* **70:** 192–198.
84. **Liaudet, L., C. Szabo, O. V. Evgenov, K. G. Murthy, P. Pacher, L. Virag, J. G. Mabley, A. Marton, F. G. Soriano, M. Y. Kirov, L. J. Bjertnaes, and A. L. Salzman.** 2003. Flagellin from gram-negative bacteria is a potent mediator of acute pulmonary inflammation in sepsis. *Shock* **19:**131–137.
85. **Lillard H. S.** 1986. Role of fimbriae and flagella in the attachment of *Salmonella typhimurium* to poultry skin. *J. Food Sci.* **51:**54–57.
86. **Lillehoj, E. P., B. T. Kim, and K. C. Kim.** 2002. Identification of *Pseudomonas aeruginosa* flagellin as an adhesin for Muc1 mucin. *Am. J. Physiol. Ser. L* **282:**L751–L756.
87. **Liu, S.-L., T. Ezaaki, H. Miura, K. Matsui, and E. Yabuuchi.** 1988. Intact motility as a *Salmonella typhi* invasion related factor. *Infect. Immun.* **56:**1967–1973.
88. **Liu, X., and P. Matsumara.** 1994. The FlhD/FlhC complex, a transcriptional activator of the *Escherichia coli* flagellar class II operons. *J. Bacteriol.* **176:**7345–7351.
89. **Lupas, A., M. Van Dyke, and J. Stock.** 1991. Predicting coiled coils from protein sequences. *Science* **252:**1162–1164.
90. **MacNab, R. M.** 1999. The bacterial flagellum: reversible rotary propeller and type III export apparatus. *J. Bacteriol.* **181:**7149–7153.
91. **MacNab, R. M.** 2000. Type III protein pathway exports *Salmonella* flagella. *ASM News* **66:**738–745.
92. **MacNab, R. M.** 2003. How bacteria assemble flagella. *Annu. Rev. Microbiol.* **57:**77–100.
93. **McDermott, P. E., F. Ciacci-Woolwine, J. A. Snipes, and S. B. Mizel.** 2000. High-affinity interaction between gram-negative flagellin and a cell surface polypeptide results in human monocyte activation. *Infect. Immun.* **10:**5525–5529.
94. **McGee, D. J., C. Coker, T. L. Testerman, J. M. Harro, S. V. Gibson, and H. L. T. Mobley.** 2002. The *Helicobacter pylori fibA*, flagellar biosynthesis and regulatory gene is required for motility and virulence and modulates urease of *H. pylori* and *Proteus mirabilis*. *J. Med. Microbiol.* **51:**958–970.
95. **McNamara, N., and C. Basbaum.** 2002. Mechanism by which bacterial flagellin stimulates host mucin production. *Adv. Exp. Med. Biol.* **506:**269–273.
96. **McSorley, S. J., B. D. Ehst, Y. Yu, and A. T. Gewirtz.** 2002. Bacterial flagellin is an effective adjuvant for $CD4^+$ T cells *in vivo*. *J. Immunol.* **169:**3914–3919.
97. **Means, T. K., F. Hayashi, K. D. Smith, A. Aderem, and A. D. Luster.** 2003. The toll-like receptor 5 stimulus: bacterial flagellin induces maturation and chemokine production in human dendritic cells. *J. Immunol.* **170:**5165–5175.
98. **Medzhitov, R.** 2001. Toll-like receptors and innate immunity. *Nat. Rev. Immunol.* **1:**135–145.
99. **Mimori-Kyouse, Y., F. Vonderviszt, and K. Namba.** Locations of terminal segments of flagellin in the filament structure and their roles in polymerization and polymorphism. *J. Mol. Biol.* **270:**222–237.
100. **Mizel, S. B., and J. A. Snipes.** 2002. Gram-negative flagellin-induced self-tolerance is associated with a block in interleukin-1 receptor-associated kinase release from Toll-like receptor 5. *J. Biol. Chem.* **277:**22414–22420.
101. **Mizel, S. B., A. P. West, and R. R. Hantgan.** 2003. Identification of a sequence in human toll-like receptor 5 required for the binding of gram-negative flagellin. *J. Biol. Chem.* **278:**23624–23629.
102. **Mobley, H. T., B. Belas, V. Lockatell, G. Chippendale, A. L. Trifillis, D. E. Johnson, and J. W. Warren.** 1996. Construction of a flagellum-negative mutant of *Proteus mirabilis:* effect on internalization by human renal epithelial cells and virulence in a mouse model of ascending urinary tract infections. *Infect. Immun.* **64:**5332–5340.
103. **Montie, T. C., D. Doyle-Huntzinger, R. C. Craven, and I. A. Holder.** 1982. Loss of virulence associated with absence of flagellum in an isogenic mutant of *Pseudomonas aeruginosa* in the burned-mouse model. *Infect. Immun.* **38:**1296–1298.
104. **Moreira, C. G., S. M. Carneiro, J. P. Nataro, L. R. Travulsi, and W. P. Elias.** 2003. Role of type 1 fimbriae in the aggregative adhesion pattern of enteroaggregative *Escherichia coli*. *FEMS Microbiol. Lett.* **226:**79–85.
105. **Morooka, T., A. Umeda, and K. Amako.** 1985. Motility as an intestinal colonization factor for *Campylobacter jejuni*. *J. Gen. Microbiol.* **131:**1973–1980.
106. **Murthy, K. B., A. Deb, S. Goonesekere, C. Szabo, and A. L. Salzman.** 2003. Identification of conserved domains in *Salmonella muenchen* flagellin that are essential for its ability to activate TLR5 and to induce an inflammatory response *in vitro*. *J. Biol. Chem.* **279:**5667–5675.
107. **Muzio, M., and A. Mantovani.** 2000. Toll-like receptors. *Microbes Infect.* **2:**251–255.
108. **Nachamkin, I, X.-H. Yang and N. J. Stern.** 1993. Role of *Campylobacter jejuni* flagella as colonization factors for three-day old chicks: analysis with flagellar mutants. *Appl. Environ. Microbiol.* **59:**1269–1273.
109. **Nataro, J. P., and J. B. Kaper.** 1998. Diarrheagenic *Escherichia coli*. *Clin. Microbiol. Rev.* **11:**142–201.
110. **Nelson, E. T., J. D. Clements, and R. A. Finkelstein.** 1976. *Vibrio cholerae* adherence and colonization in experimental

cholera: electron microscopic studies. *Infect. Immun.* **14:** 527–547.

111. **Ogushi, K., A. Wada, T. Niidome, N. Mori, K. Oishi, T. Nagatake, A. Takahashi, A. Asakura, S. Makino, H. Hojo, Y. Nakahara, M. Ohsaki, T. Hatakeyama, H. Aoyagi, H. Kurazono, J. Moss, and T. Hirayama.** 2001. *Salmonella enteritidis* FliC (flagella filament protein) induces human β-defensin-2 mRNA production by Caco-2 cells. *J. Biol. Chem.* **276:**30521–30526.
112. **Ormonde, P., P. Horstedt, R. O'Toole, and D. L. Milton.** 2000. Role of motility in adherence to and invasion of a fish cell line by *Vibrio anguillarum*. *J. Bacteriol.* **182:**2326–2328.
113. **O'Toole, G. A., and R. Kolter.** 1998. Flagellar and twitching motility are necessary for *Pseudomonas aeruginosa* biofilm development. *Mol. Microbiol.* **30:**295–304.
114. **Ottemann, K. M., and A. C. Lowenthal.** 2002. *Helicobacter pylori* uses motility for initial colonization and to attain robust infection. *Infect. Immun.* **70:**1984–1990.
115. **Pallen, M. J.** 1997. Coiled-coil domains in proteins secreted by type III secretion systems. *Mol. Microbiol.* **25:**423–425.
116. **Parker, C. T. and J. Guard-Petter.** 2001. Contribution of flagella and invasion proteins to pathogenesis of *Salmonella enterica* serovar Enteritidis in chicks. *FEMS Microbiol. Lett.* **204:**287–291.
117. **Pavlovskis, O. R., D. M. Rollins, R. L. Haberberger, Jr., A. E. Green, L. Habash, S. Strocko, and R. I. Walker.** 1991. Significance of flagella in colonization resistance of rabbits immunized with *Campylobacter* spp. *Infect. Immun.* **59:**2259–2264.
118. **Peel, M. W. Donachie, and A. Shaw.** 1988. Temperature-dependent expression of flagella of *Listeria monocytogenes* studied by electron microscopy, SDS-PAGE and western blotting. *J. Gen. Microbiol.* **134:**2171–2178.
119. **Plano, G. V., J. B. Day, and F. Ferracci.** 2001. Type III export: new uses for an old pathway. *Mol. Microbiol.* **40:**284–293.
120. **Postnova, T., O. G. Gomez-Duarte, and K. Richardson.** 1996. Motility mutants of *Vibrio cholerae* O1 have reduced adherence *in vitro* to human small intestinal epithelial cells as demonstrated by ELISA. *Microbiology* **142:**2767–2776.
121. **Pratt, L. A., and R. Kolter.** 1998. Genetic analysis of *Escherichia coli* biofilm formation: roles of flagella, motility, chemotaxis and type I pili. *Mol. Microbiol.* **30:**285–293.
122. **Pruckler, J. M., R. F. Benson, M. Moyenuddi, W. T. Martin, and B. S. Fields.** 1995. Association of flagellum expression and intracellular growth of *Legionella pneumophila*. *Infect. Immun.* **63:**4928–4932.
123. **Rabaan, A. A., I. Gryllos, J. M. Tomas, and J. G. Shaw.** 2001. Motility and the polar flagellum are required for *Aeromonas caviae* adherence to HEp-2 cells. *Infect. Immun.* **69:**4257–4267.
124. **Ramphal, R., S. K. Arora, and B. W. Ritchings.** 1996. Recognition of mucin by the adhesin-flagellar system of *Pseudomonas aeruginosa*. *Am. J. Respir. Crit. Care Med.* **154:**S170–S174.
125. **Read, R. C., and D. H. Wyllie.** 2001. Toll receptors and sepsis. *Curr. Opin. Crit. Care* **7:**371–375.
126. **Reed, K. A., M. E. Hobert, C. E. Kolenda, K. A. Sands, M. Rathman, M. O'Connor, S. Lyons, S. A. T. Gewirtz, P. J. Sansonetti, and J. L. Madara.** 2002. The *Salmonella typhimurium* flagellar basal body protein FliE is required for flagellin production and to induce a proinflammatory response in epithelial cells. *J. Biol. Chem.* **277:**13346–13353.
127. **Reid, S. D., R. Selander, and T. S. Whittam.** 1999. Sequence diversity of flagellin (FliC) alleles in pathogenic *Escherichia coli*. *J. Bacteriol.* **181:**153–160.
128. **Richardson, K.** 1991. Roles of motility and flagellar structure in pathogenicity of *Vibrio cholerae*: analysis of motility mutants in three animal models. *Infect. Immun.* **59:**2727–2736.
129. **Robertson, J. M., G. Grant, E. Allen-Vercoe, M. J. Woodward, A. Pusztai, and H. J. Flint.** 2000. Adhesion of *Salmonella enterica* var Enteritidis strains lacking fimbriae and flagella to rat ileal explants cultured at the air interface or submerged in tissue culture medium. *J. Med. Microbiol.* **49:**691–969.
130. **Robertson, J. M., N. H. McKenzie, M. Duncan, E. Allen-Vercoe, M. J. Woodward, and H. J. Flint.** 2003. Lack of flagella disadvantages *Salmonella enterica* serovar Enteritidis during the early stages of infection in the rat. *J. Med. Microbiol.* **52:**91–99.
131. **Sadziene, A., D. D. Thomas, V. G. Bundoc, S. C. Holt, and A. G. Barbour.** 1991. A flagella-less mutant of *Borrelia burgdorferi*. Structural, molecular and *in vitro* functional characterization. *J. Clin. Investig.* **88:**82–92.
132. **Samatey, F. A., K. Imada, S. Nagashima, F. Vendervisztt, T. Kumasaka, M. Yamamoto, and K. Namba.** 2001. Structure of the bacterial flagellar protofilament and implications for a switch for supercoiling. *Nature* **410:**331–337.
133. **Schilling, J. D., S. M. Matin, C. S. Hung, R. G. Lorenz, S. J. Hultgren.** 2003. Toll-like receptor 4 on stomal and hematopoietic cells mediate innate resistance to uropathogenic *Escherichia coli*. *Proc. Natl. Acad. Sci. USA* **100:**4203–4208.
134. **Schmiel, D. H., G. M. Young, and V. L. Miller.** 2000. The *Yersinia enterocolitica* phospholipase gene *yplA* is part of the flagellar regulon. *J. Bacteriol.* **182:**2314–2320.
135. **Schmitt, C. K., J. S. Ikeda, S. C. Darnell, P. R. Watson, J. Bispham, T. S. Wallis, D. L. Weinstein, E. S. Metcalf, and A. D. O'Brien.** 2001. Absence of all components of the flagellar export and synthesis machinery differentially alters virulence of *Salmonella enterica* serovar Typhimurium in models of typhoid fever, survival in macrophages, tissue culture invasiveness and calf enterocolitis. *Infect. Immun.* **69:**5619–5625.
136. **Sellek, R. E., R. Escudero, H. Gil, I. Rodriguez, E. Chaparro, E. Perez-Pastrana, A. Vivo, and P. Anda.** 2002. In vitro culture of *Borrelia garinii* results in loss of flagella and decreased invasiveness. *Infect. Immun.* **70:**4851–4858.
137. **Sheikh J, S. Hicks, M. Dall'Agnol, A. D. Phillips, and J. P. Nataro.** 2001 Roles for Fis and YafK in biofilm formation by enteroaggregative *Escherichia coli*. *Mol. Microbiol.* **5:**983–997.
138. **Sierro, F., B. Dubois, A. Coste, D. Kaiserlian, Kraehenbuhl, and J. Sirard.** 2001. Flagellin stimulation of intestinal epithelial cells triggers CCL20 mediated migration of dendritic cells. *Proc. Natl. Acad. Sci. USA* **98:**13722–13727.
139. **Smith, K. D., and A. Ozinsky.** 2002. Toll-like receptor-5 and the innate immune response to bacterial flagellin. *Curr. Top. Microbiol. Immunol.* **270:**93–108.
140. **Smith, M. F., A. Mitchell, G. Li, S. Ding, A. M. Fitzmaurice, K. Ryan, S. Crowe, and J. R. Goldberg.** 2003. Toll-like receptor (TLR) 2 and TLR5 but not TLR4, are required for *Helicobacter pylori*-induced NF-κB activation and chemokine expression by epithelial cells. *J. Biol. Chem.* **278:**32552–32560.
141. **Smith, K. D., E. Andersen-Nissen, F. Hayashi, K. Strobe, M. A. Bergman, S. L. Rassoulian Barrret, B. T. Cookson and A. Aderem.** 2003. Toll-like receptor 5 recognizes a conserved site on flagellin required for protofilament formation and bacterial motility. *Nat. Immunol.* **4:**1247–1253.
142. **Sperandio, V., A. G. Torres, J. A. Girón, and J. B. Kaper.** 2001. Quorum sensing is a global regulatory mechanism in enterohemorrhagic *Escherichia coli* O157:H7. *J. Bacteriol.* **183:**5187–5197.

143. **Steiner, T. S., A. A. Lima, J. P. Nataro, and R. L. Guerrant.** 1998. Enteroaggregative *Escherichia coli* produce intestinal inflammation and growth impairment and cause interleukin-8 release from intestinal epithelial cells. *J. Infect. Dis.* **177:**88–96.

144. **Steiner, T. S., J. P. Nataro, C. E. Poteet-Smith, J. A. Smith, and R. L. Guerrant.** 2000. Enteroaggregative *Escherichia coli* expresses a novel flagellin that causes IL-8 release from intestinal epithelial cells. *J. Clin. Investig.* **105:**1769–1777.

145. **Szymanski, C. M., M. King, M. Haardt, and G. D. Armstrong.** 1995. *Campylobacter jejuni* motility and invasion of Caco-2 cells. *Infect. Immun.* **3:**4295–4300.

146. **Takahashi, A., A. Wada, K. Ogushi, K. Maeda, T. Kawahara, K. Mawatari, H. Kurazono, J. Moss, T. Hirayama, and Y. Nakaya.** 2001. Production of β-defensin-2 by human colonic epithelial cells induced by *Salmonella enteritidis* flagella filament structural protein. *FEBS Lett.* **508:**484–488.

147. **Tasteyre, A., M. Barc, A. Collignon, H. Boureau, and T. Karjalainen.** 2001. Role of FliC and FliD flagellar proteins of *Clostridium difficile* in adherence and gut colonization. *Infect. Immun.* **69:**7937–7940.

148. **Teppema, J. S., P. A. M. Guinée, A. A. Abrahim, M. Páques, and E. J. Ruitenberg.** 1987. In vivo adherence and colonization of *Vibrio cholerae* strains that differ in hemagglutinating activity and motility. *Infect. Immun.* **55:**2093–2102.

149. **Tomich, M., C. A. Herfst, J. W. Golden, and C. D. Mohr.** 2002. Role of flagella in host cell invasion by *Burkholderia cepacia. Infect. Immun.* **70:**1799–1806.

150. **Underhill, D. M., and A. Ozinsky.** 2002. Toll-like receptors: key mediators of microbe detection. *Curr. Opin. Immunol.* **14:**103–110.

151. **Van Asten F. J. A. M., H. G. C. J. M. Hendriks, J. F. J. G. Koninkx, B. A. M. Van der Zeijst, and W. Gaastra.** 2000. Inactivation of the flagellin gene of *Salmonella enterica* serotype Enteritidis strongly reduces invasion into differentiated caco-2 cells. *FEMS Microbiol. Lett.* **185:**175–179.

152. **Walker, S. L., M. Sojka, M. Dibb-Fuller, and M. J. Woodward.** 1999. Effect of pH, temperature and surface contact on the elaboration of fimbriae and flagella by *Salmonella* serotype Enteritidis. *J. Med. Microbiol.* **48:**253–261.

153. **Watnick, P. I., C. M. Lauriano, K. E. Klose, L. Croal, and R. Kolter.** 2001. The absence of a flagellum leads to altered colony morphology, biofilm development and virulence in *Vibrio cholerae* O139. *Mol. Microbiol.* **39:**223–235.

154. **Watnick, P. I., and R. Kolter.** 1999. Steps in the development of a *Vibrio cholerae* El Tor biofilm. *Mol. Microbiol.* **34:**586–595.

155. **Way, S. S., L. J. Thompson, J. E. Lopes, A. M. Hajjar, T. R. Kollmann, N. E. Freitag, and C. B. Wilson.** 2004. Characterization of flagellin expression and its role in *Listeria monocytogenes* infection and immunity. *Cell. Microbiol.* **6:**235–242.

156. **Weinstein, D. L., M. Carsiotis, C. R. Lissner, and A. D. O'Brien.** 1984. Flagella help *Salmonella typhimurium* survive within murine macrophages. *Infect. Immun.* **46:**819–825.

157. **Wyant, T. L., M. K. Tanner, and M. B. Sztein.** 1999. *Salmonella typhi* flagella are potent inducers of proinflammatory cytokine secretion by human monocytes. *Infect. Immun.* **67:**3619–3624.

158. **Wyant, T. L., M. K. Tanner, and M. B. Sztein.** 1999. Potent immunoregulatory effects of *Salmonella typhi* flagella on antigenic stimulation of human peripheral blood mononuclear cells. *Infect. Immun.* **67:**1338–1346.

159. **Yancey, R. J., D. L. Willis, and L. J. Berry.** 1978. Role of motility in experimental cholera in adult rabbits. *Infect. Immun.* **22:**387–392.

160. **Yao, R., D. H. Burr, P. Doig, T. J. Trust, H. Niu, and P. Guerry.** Isolation of motile and non-motile insertional mutants of *Campylobacter jejuni:* the role of motility in adherence and invasion of eukaryotic cells. *Mol. Microbiol.* **14:**883–893.

161. **Yonekura, K., S. Maki, D. G. Morgan, D. J. DeRosier, F. Vonderviszt, K. Imada, and K. Namba.** 2000. The bacterial flagellar cap as the rotary promoter of flagellin self-assembly. *Science* **290:**2148–2152.

162. **Yonekura, K., S. Yonekura, and K. Namba.** 2002. Growth mechanism of the bacterial flagellar filament. *Res. Microbiol.* **153:**191–197.

163. **Yonekura, K., S. Yonekura, and K. Namba.** 2003. Complete atomic model of the bacterial flagellar filament by electron cryomicroscopy. *Nature* **424:**643–650.

164. **Young, G. M., M. J. Smith, S. A. Minnich, and V. L. Miller.** 1999. The *Yersinia enterocolitica* motility master regulatory operon, *flhDC,* is required for flagellin production, swimming motility and swarming motility. *J. Bacteriol.* **181:**2823–2833.

165. **Young, G. M., D. H. Schmiel, and V. L. Miller.** 1999. A new pathway for the secretion of virulence factors by bacteria: the flagellar export apparatus functions as a protein secretion system. *Proc. Natl. Acad. Sci. USA* **96:**6456–6461.

166. **Young, G. M., J. L. Badger, and V. L. Miller.** 2000. Motility is required to initiate host cell invasion by *Yersinia enterocolitica. Infect. Immun.* **68:**4323–4326.

167. **Young, B. M., and G. M. Young.** 2002. Yp1A is exported by the Ysc, Ysa and flagellar type III secretion systems of *Yersinia enterocolitica. J. Bacteriol.* **184:**1324–1334.

168. **Zeng, H., A. Q. Carlson, Y. Guo, Y. Yu, L. S. Collier-Hyams, J. L. Madara, A. T. Gewirtz, and A.S. Neish.** 2003. Flagellin is the major proinflammatory determinant of enteropathogenic *Salmonella. J. Immunol.* **171:**3668–3674.

169. **Zhang, J, K. Xu, B. Ambati, and F-S. X. Yu.** 2003. Toll-like receptor-5 mediated corneal epithelial inflammatory responses to *Pseudomonas aeruginosa* flagellin. *Investig. Ophthalmol. Visual Sci.* **44:**4247–4254.

170. **Zhou, X., J. A. Girón, A. G. Torres, J. A. Crawford, E. Negrete, S. N. Vogel, and J. B. Kaper.** 2003. Flagellin of enteropathogenic *Escherichia coli* stimulates interleukin-8 production in T84 cells. *Infect. Immun.* **71:**2120–2129.

Colonization of Mucosal Surfaces
Edited by James P. Nataro et al.
© 2005 ASM Press, Washington, D.C.

Chapter 17

Tissue Tropism in Intestinal Colonization

Elizabeth L. Hartland, Roy M. Robins-Browne, Alan D. Philips, and Gad Frankel

TISSUE TROPISM AND COLONIZATION OF THE HOST

The tropism of different bacterial pathogens for certain tissues and mucosal surfaces is a major feature of the host-pathogen interaction, and this restriction or predilection of pathogens for a particular host mucosal surface is likely to be determined by the ability of the organism to attach more efficiently to some epithelial cell types than to others. For most pathogens, initial bacterium-host contact is supported by a specific receptor-ligand interaction between a surface-located bacterial adhesin and an extracellular host cell receptor, and it is likely that this interaction plays an important role in tissue tropism Differences in the tissue tropism of bacterial pathogens may therefore depend on adhesins that are unique to particular pathogens or on adhesins that are shared by related pathogens but exhibit amino acid sequence variation. Even minor amino acid differences within adhesin-binding domains have the potential to alter receptor specificity and/or binding affinity, so that closely related pathogens may display a range of tropisms for different cell types, tissues, or hosts. For the most part, the bacterial factors that determine tissue and host specificity are poorly characterized, although for some pathogenic varieties of *Escherichia coli,* the determinants that contribute to tissue tropism have been genetically dissected to provide a framework for understanding the molecular basis and complexities of tissue tropism and the host-pathogen interaction. This chapter reviews our current understanding of the tissue tropism of bacterial pathogens, with a focus on virulent types of *E. coli.*

UROPATHOGENIC *E. COLI* AND PILUS-MEDIATED TISSUE TROPISM

As surface-expressed organelles involved in bacterial attachment to the host mucosa, bacterial pili are likely to contribute to the tissue tropism of many bacterial pathogens. An example is uropathogenic *E. coli* (UPEC), where type I pili expressly promote initial infection and colonization of the bladder epithelium of the urinary tract by UPEC (65). The sheath of the type I pilus is assembled via a chaperone/usher pathway from repeating FimA subunits that support a tip fibrillum composed of FimH (45). FimH comprises the adhesive domain of the pilus that typically binds to N-linked high-mannose carbohydrate in host cell glycoprotein receptors (40). The recognition and binding of mannose by FimH is essential for colonization of the host urinary tract and bladder by UPEC and makes a major contribution to the tissue tropism of this pathogen. The crystal structure of FimH has revealed a negatively charged mannose-binding pocket within an elongated 11-strand β-barrel receptor-binding domain (Fig. 1A) (40). Single-amino-acid changes of residues which interact with mannose in the binding pocket abolish mannose binding, hemagglutination, and adherence to bladder epithelial cells and bladder tissue (40). These amino acid residues are highly conserved among UPEC strains, demonstrating the importance of this invariant FimH mannose-binding pocket for function. Interestingly, the gastrointestinal pathogen enterohemorrhagic *E. coli* (EHEC) O157:H7 exhibits sequence variation within the mannose-binding pocket of FimH at residue 135, where lysine is present instead

Elizabeth L. Hartland • Department of Microbiology, School of Biomedical Sciences, Monash University, Victoria 3800, Australia. **Roy M. Robins-Browne** • Department of Microbiology and Immunology, University of Melbourne, Victoria 3010, Australia. **Alan D. Philips** • Centre for Paediatric Gastroenterology, Department of Paediatrics and Child Health, Royal Free Hospital, London NW3 2QG, United Kingdom. **Gad Frankel** • Centre for Molecular Microbiology and Infection, Department of Biological Sciences, Imperial College London, London SW7 2AZ, United Kingdom.

of asparagine. Mutation of UPEC FimH at Asn135 abolishes mannose-binding function and adherence to bladder epithelial cells (40). According to the crystal structure of FimH, the presence of Lys135 in FimH from EHEC O157:H7 would exclude mannose from entering the binding pocket, which suggests that FimH mannose-binding activity is not an essential feature of EHEC infection. Given this sequence variation, EHEC O157:H7 would not be expected to colonize the bladder epithelium efficiently, and, indeed, strains of EHEC O157:H7, unlike some other serotypes of EHEC, are not highly associated with urinary tract infections (88, 100). The molecular interaction between FimH and its host cell receptor is therefore highly conserved, where even a single amino acid change is sufficient to abolish function and alter tissue specificity and tropism.

Some strains of UPEC typically infect the kidneys and are termed pyelonephritic *E. coli.* Whereas type I pili are essential for colonization of the bladder, P pili are required for infection of the kidneys by pyelonephritic *E. coli.* Like type I pili, P pili are composite structures built from repeating PapA subunits that support an adhesive tip comprising the adhesin PapG, which is linked to the body of the pilus via the PapE, PapF, and PapK proteins (92). Instead of mannose, PapG recognizes the digalactoside [Galα(1-4)Gal] component of the globotriasylceramide family of glycolipid receptors, which are resident in the host cell membrane (62). PapG is more variable than FimH, and three different alleles exist that recognize slightly different glycolipid receptors with different affinities. These different binding specificities and the tissue distribution of the receptors account in part for the tissue and host tropism of pyelonephritic *E. coli.* Human pyelonephritic disease is associated predominantly with the class II PapG allele, which recognizes globoside (Gb4), whereas PapGI and PapGIII recognize globotriasylceramide (Gb3) and the Forssman antigen (Gb5), respectively (90). Similar to the structure of FimH, the PapG receptor-binding domain is mostly β-sheet protein flanking a sugar-binding pocket (Fig. 1B) (20). Structural determination and molecular modeling of the PapGII- and PapGIII-binding pockets has shown that the amino acid differences between the two alleles lie predominantly within the region of the PapGIII-binding pocket predicted to accommodate the extra GalNAcα residue of Gb5 (20). This predicted structural variation provides a molecular explanation for the tropism of PapGIII for the canine kidney, which is rich in Gb5 (90). Thus, as observed for FimH, only minor variations in the pilus adhesin may alter tissue specificity and, in this case, host specificity.

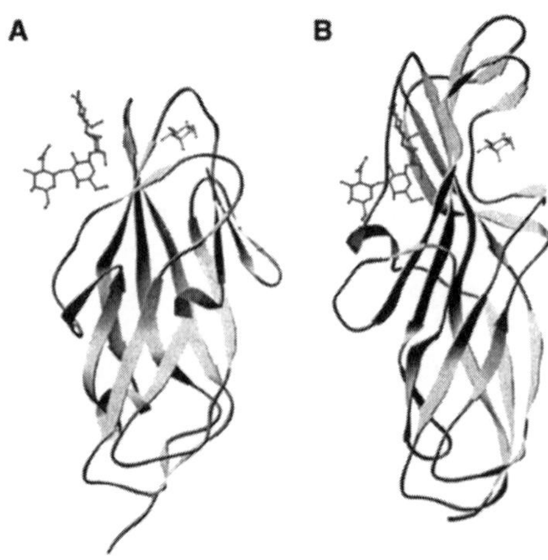

Figure 1. Structural representations of the receptor-binding domains of the pilus adhesins, FimH (A) and PapGII (B). D-Mannose and Gb4 are included in both structural illustrations for comparison of the binding pockets. Reprinted from reference 40 with permission from Blackwell Publishing.

ATTACHING AND EFFACING PATHOGENS

Adhesin-receptor interactions contribute to the tissue tropism of other *E. coli* pathogens, including EHEC and enteropathogenic *E. coli* (EPEC). Both EHEC and EPEC belong to a group of mucosal pathogens that induce characteristic histopathology on the host intestine described as the attaching and effacing (A/E) lesion (31). The lesions are characterized by intimate bacterial adherence to the host cell membrane, the destruction of microvilli at the site of adherence, and the rearrangement and accumulation of cytoskeletal proteins, in particular filamentous actin, in the host cell abutting the adherent bacteria (Fig. 2A) (53). The genes required for A/E lesion formation are located within a large pathogenicity island termed the locus for enterocyte effacement (LEE). Approximately half of the genes in the LEE code for a type III secretion system (TTSS) that directs the secretion of several LEE-encoded proteins, known as the *E. coli* secreted proteins or Esp proteins (23, 67). These include EspA, EspD, EspB, EspF, EspG, EspH, Tir/EspE, and Map, of which EspA, EspB, EspD, and Tir play a direct role in A/E lesion formation (21, 22, 49–51, 58, 68). Tir is transported to the host cell membrane by the LEE-encoded translocon, where it serves as a receptor for the bacterial outer membrane protein adhesin intimin, which is present on the bacterial cell surface and which is also encoded within the LEE (49). Tir contains two putative transmembrane domains and adopts a hairpin loop topology in the host cell membrane, with the middle region exposed extracellularly

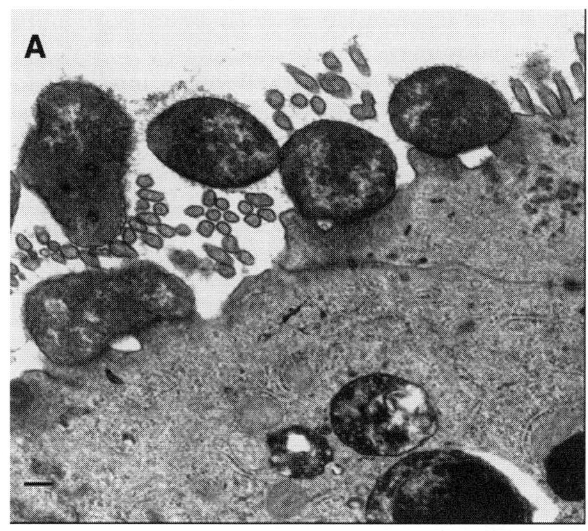

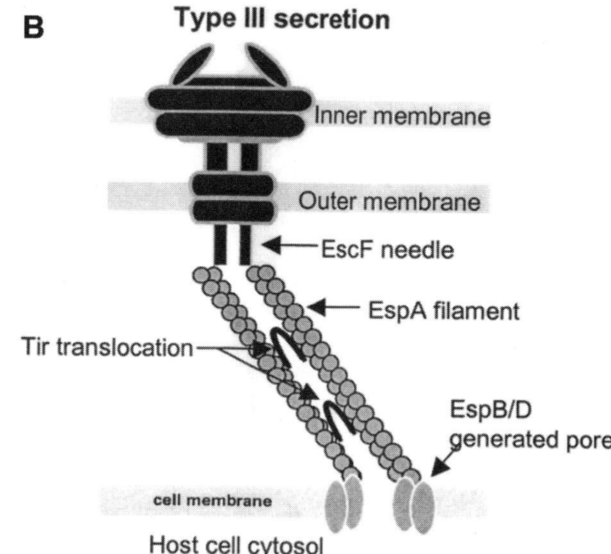

Figure 2. (A) Transmission electron micrograph of A/E lesions in vivo in a child with chronic diarrhea. *E. coli* O114 serogroup was identified in stool samples and on the intestinal mucosa. Bar, 175 nm. (B) Schematic representation of the LEE-encoded EPEC/EHEC type III translocon.

and the N- and C-terminal regions located intracellularly (37). It is the middle region of Tir that specifically interacts with intimin located in the bacterial outer membrane (37, 48).

The EPEC and EHEC translocon directs the delivery of selected proteins into the internal milieu of the host cell (Fig. 2B). A pore comprising EspD and possibly EspB is inserted into the host cell membrane by the LEE-encoded TTSS, which is linked to the translocon by a filamentous organelle made up of repeating EspA subunits (13, 15, 41, 54). Structural determination of the EspA filaments has shown that the organelle comprises a hollow tube made of helically arranged EspA subunits (14, 85). Thus, the EspA filament appears to act as a conduit for the delivery of translocated proteins such as Tir to the host cell cytosol and/or membrane through the EspD-EspB membrane pore.

INTIMIN FAMILY OF ADHESINS

Intimin is the product of the LEE-located *eae* gene, which was the first gene to be associated with A/E lesion formation (44). Structural analyses of the extracellular domains of intimin have shown that the protein comprises a series of four globular modules with a distinct organization (Fig. 3) (46, 63). The three domains (D1 to D3) closest to the bacterial cell surface comprise β-sheet sandwiches and structurally resemble immunoglobulin (Ig)-like folds. A fourth

Figure 3. Structural representation of the C-terminal region of intimin-α showing the Ig-like domains, D2 and D3, and the lectin-like module, D4.

domain (D4), located at the C-terminal tip of the molecule, is formed by a 76-amino-acid disulfide loop that shows structural similarity to members of C-type lectin family of carbohydrate-binding proteins (46). The N-terminal domains of intimin are highly conserved among A/E pathogens, but based on antigenic polymorphism within the C-terminal domain, several different types of intimin are recognized, the most common of which are α, β, γ, δ, and ϵ (2). Intimin type is associated with different clonal groups of A/E pathogens, and intimin-α and intimin-γ in particular are associated with different evolutionary lineages of EPEC and EHEC. The disulfide loop is essential for the function of the molecule and is highly conserved among different intimins and the related homolog, invasin, from enteropathogenic *Yersinia* species (28).

The interaction between intimin and Tir is central to the intimate attachment of bacteria to the host cell epithelium. Specific binding between intimin and Tir occurs between the C-terminal 190 amino acids of intimin and a central 55-amino-acid domain of Tir (37, 48). This protein-protein interaction involves the C-type lectin-like domain of intimin as well as amino acids from the most distal Ig-like domain. Together, the D3 and D4 domains constitute a superdomain of intimin that is essential for Tir binding and A/E lesion formation (3, 60). Extensive interactions occur between the central intimin-binding region of Tir and the superdomain of intimin, creating a largely hydrophobic binding interface between the bacterial cell surface and the host cell membrane (63). The crystal structure of the intimin-α/Tir complex from EPEC shows that the intimin-binding region of Tir functions as a dimer, which is stabilized by an antiparallel, four-helix bundle (63). The Tir-binding site of intimin is a surface-exposed region measuring 20 by 8 Å, located in the C-terminal 100 amino acids of the protein within D3 and D4 (3, 63). Alanine substitution of a conserved tryptophan residue in D4 (Trp899) that is predicted to lie just below the Tir-binding pocket of intimin abolishes Tir-binding activity and A/E lesion formation (3). Trp899 presumably provides structural support for the amino acid residues that interact directly with the intimin-binding region of Tir. Interestingly, mutagenesis of another conserved tryptophan residue, Trp795, results in a form of intimin that can still bind to Tir but is unable to mediate A/E lesion formation (78). Trp795 is located within the core of the D3 domain of intimin and is predicted to play a role in maintaining the structural integrity of the D3-D4 superdomain of intimin. Studies of intimin from EHEC O157:H7 confirm the location of the Tir-binding region in intimin-γ and support the correlation between Tir binding and the ability to focus actin during A/E lesion formation (61). For intimin-γ, the residues Ser890, Thr909, Ala916, and Ala927 are essential for Tir binding and actin nucleation. Modeling the positions of these amino acids on the structure of intimin-α shows that the residues lie within a binding pocket that interacts with Ile298 of Tir (61). Overall, these studies show that even minor structural changes in the binding domains of intimin can affect Tir-binding capacity and/or intimin function.

A/E lesion formation is a complex process requiring the coordinated expression, secretion, and translocation of several proteins. Most cell culture-based assays for A/E lesion formation support a model where EHEC and EPEC form an initial loose attachment with the host cell surface that is followed by intimate adherence and the development of A/E lesions. The identities of the adhesins responsible for the initial stage of attachment are still controversial, but they may include pili such as the bundle-forming pilus (BFP) of EPEC or other components encoded by the LEE pathogenicity island such as EspA or intimin itself. The timing and complexity of the process leading to A/E lesion formation suggest that the development of A/E lesions by EHEC and EPEC represents a later event in mucosal colonization. This aspect, together with the fact that EPEC and EHEC are able to induce A/E lesions on a wide range of cell types derived from different tissues and different hosts, suggests that A/E lesion formation per se is unlikely to contribute to the tissue tropism of EHEC and EPEC. Instead, a growing body of evidence suggests that intimin not only binds Tir but also interacts directly with a host mucosal surface receptor and that this interaction can influence tissue tropism and, in some cases, host specificity.

INTIMIN-MEDIATED TISSUE TROPISM AND THE PUTATIVE HOST CELL INTIMIN RECEPTOR

Although EPEC and EHEC are closely related human enteric pathogens, they nevertheless exhibit different tissue tropisms. Whereas EPEC is generally described as a small intestinal pathogen, colonization of the gut by strains of EHEC is widely thought to be restricted to the large intestine. The tropism of EHEC for the colon is assumed partly from the observed histological damage to the colon from hemorrhagic colitis, although it is unclear if this tissue destruction results from adherent bacteria or whether it is mediated only by the action of Shiga toxin, a potent, prophage-encoded A1B5 toxin that is released into the intestinal lumen during EHEC infection. Nevertheless, this assumption is supported by evidence derived from colonization studies performed with gnotobiotic piglets. In this infection model, EHEC O157:H7, which expresses intimin-γ, is associated with extensive colonization and destruction of the large intestinal epithelium while EPEC O127:H6, which expresses intimin-α, colonizes both the small and large intestines (95). In addition, complementation of an *eae* mutant of EHEC O157:H7 with *eae* from EPEC O127:H6 alters the pattern of colonization of the EHEC strain so that it colonizes the small and large intestines in a similar manner to EPEC (96). These studies show that not only is intimin an essential colonization factor in the gnotobiotic piglet model but that also the protein contributes to the differences in tissue tropism observed for the two pathogens.

The modification of colonization site via the mutation of residues in a single protein is worthy of additional comment. The complexities of colonization set forth in this volume, including sophisticated metabolic competition phenotypes, chemotaxis, and other biological feats, might suggest that the anatomic site of colonization could be the result of multiple finely orchestrated evolutionary adaptations. This may yet be so, but our data clearly indicate that adherence to the gut epithelium represents a primary prerequisite. Indeed, a degree of metabolic versatility of enteric organisms during the colonization process may itself be adaptive to perpetuation within the community. The benefit of asymptomatic (and presumably colonic) carriage of a small bowel pathogen, for example, could be such a phenomenon.

Recent work using human intestinal in vitro organ culture (IVOC) (39) to investigate the interactions of EPEC and EHEC with mucosal tissue supports the role of intimin in the tissue tropisms of EHEC and EPEC, although the tropisms observed in this system vary somewhat from those in the gnotobiotic piglet model. In human pediatric IVOCs, EPEC O127:H6 adheres equally well to biopsy specimens obtained from the proximal small intestine, the distal small intestine, and the follicle-associated epithelium of Peyer's patches, producing A/E lesions on all tissue types. EHEC O157:H7, on the other hand, adheres preferentially to the follicle-associated epithelium, where it induces A/E lesions, and not to biopsy specimens from other regions of the intestine, including the colon (Fig. 4) (77). The Peyer's patch epithelium has been implicated as a site of initial colonization for several intestinal pathogens, including the rabbit EPEC pathogen (REPEC) (98), and the predilection of EHEC O157:H7 for FAE on human IVOCs may therefore represent an early site of colonization that precedes dissemination to other intestinal tissues.

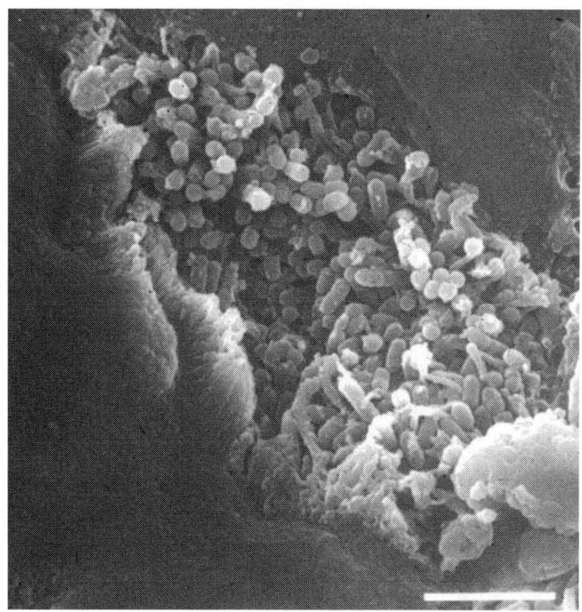

Figure 4. Scanning electron micrograph of EHEC O157:H7-induced A/E lesions on FAE of a Peyer's patch in a human intestinal organ culture. Bar, 5 μm.

The preference of EHEC O157:H7 for FAE depends on the expression of intimin-γ. As in the gnotobiotic piglet model, intimin exchanges show that the tissue tropism exhibited by both EHEC O157:H7 and EPEC O127:H6 on human IVOCs is controlled by intimin type. In these studies, an *eae* mutant of EHEC O157:H7 complemented with intimin-α from EPEC O127:H6 displayed an ability to colonize and induce A/E lesions on FAE and other intestinal sites in a manner similar to EPEC (25). In the same way, an *eae* mutant of EPEC O127:H6 complemented with intimin-γ was restricted to the FAE, as observed for EHEC O157:H7 (76). These differences in the biological activities of intimin-γ and intimin-α presumably depend on amino acid sequence differences in the variable C-terminal receptor-binding domain of the protein. While this region of intimin clearly contributes to Tir binding, as discussed above, it is also the part of the protein that contributes to tissue tropism and thus comprises a multifunctional region. Site-directed mutagenesis of intimin-α in EPEC O127:H6 has enabled the different biological properties of intimin to be distinguished from one another. Amino acid substitution of a hydrophobic valine residue predicted to lie exposed within the Tir-binding pocket (Val911 to Ala) results in a derivative of intimin-α that is still able to induce A/E lesions and interact with Tir in a conformation-dependent manner but is profoundly altered in tissue tropism (80). Adherence of EPEC O127:H6 expressing intimin-α V911A in human IVOCs is restricted to the FAE as if the strain expressed intimin-γ (80). Importantly, this intimin-mediated tissue tropism appears to be independent of intimin-Tir protein-protein interactions, suggesting that intimin also has a host cell receptor. In support of the existence of an intimin host cell receptor, site-directed mutagenesis to alanine of a cysteine residue (Cys937) that forms a conserved disulfide loop in the C-type lectin-like domain of intimin results in a protein that retains Tir-binding activity but cannot mediate A/E lesion formation or colonization of mucosal tissue (30, 37). Overall, this suggests that the interaction between intimin and a putative host cell receptor plays a critical role in determining not only tissue tropism but also the outcome of colonization of the intestinal mucosa.

The tissue tropism conferred by intimin-γ is borne out by the colonization patterns of EPEC strains of serogroup O55, the third most common EPEC serotype associated with infant diarrhea. Within the O55 serogroup, different clones associated with different flagellar types are also associated with different intimin types. Strains of EPEC and EHEC can be classified into different clonal lineages designated EPEC clones 1 and 2 and EHEC clones 1 and 2. These are characteristically recognized by flagellar type, where EPEC strains of clone 1 typically express H6 flagellar antigens and EPEC strains of clone 2 express H2 or are nonmotile. Hence, EPEC O55:H6 belongs to a different clonal group from EPEC O55:H7, which is more closely related to EHEC O157:H7. EPEC O55:H7 expresses intimin-γ and displays a corresponding restriction to the FAE of Peyer's patches of human IVOCs (26). Since EPEC O55:H7 is widely thought to be the evolutionary ancestor of EHEC O157:H7 and the two are closely related at a genetic level, it is not surprising that these strains share phenotypic properties. Interestingly, however, EPEC O55:H6 which expresses intimin-α also shows tropism for the FAE. This is the only example to date of an intimin-α-expressing strain that does not colonize the villous surface of the small intestinal mucosa. When the *eae* gene from EPEC O55:H6 was sequenced, one amino acid difference was identified compared with the sequence of intimin-α from EPEC O127:H6. The difference occurred at position 907, where this residue was valine in intimin-α from EPEC O55:H6 compared with alanine in EPEC O127:H6. This amino acid difference is not the reason for the change in tissue tropism, however, since mutation of alanine to valine (A907V) in intimin-α from EPEC O127:H6 does not alter the colonization pattern of this strain, suggesting that other adherence factors contribute to FAE restriction of O55 EPEC strains (26). Intimin-ϵ is associated with human and bovine strains of EHEC and, on the basis of the amino acid sequence, appears more closely related to intimin-γ than to the other intimin types. Similar to EHEC O157:H7, the intimin-ϵ-expressing O103:H2 and O103:H− EHEC serotypes show a tropism for the FAE of Peyer's patches in human IVOCs, where they form A/E lesions (27).

The significance of the different tropisms of intimin types in the IVOC model is difficult to assess without monitoring tissue colonization over a longer period. The IVOC infections performed to date proceed for only 8 h. It is therefore possible that adherence to the FAE may act as a starting point for further colonization of the intestinal tract, including the colon. Intimin types that do not show FAE restriction may bypass this initial step by adhering extensively to the mucosal epithelium early on. Nevertheless, the association of intimin type with colonization pattern strongly suggests that, in addition to binding Tir, intimin binds to a host cell receptor. Intimin-mediated tissue tropism may therefore result from different binding affinities of different intimin types for a host cell receptor or from different receptor types and availability. Recently, intimin-γ was shown to bind to surface-located host cell nucleolin, which is highly expressed in cells undergoing cell division (86). Nucleolin is shuttled between the cytoplasm, nucleus, and cell surface, where it can contribute to the regulation of cell proliferation (101). Purified intimin-γ specifically binds nucleolin from human epithelial cells, and during cell infection with EHEC O157:H7, intimin-γ colocalizes with surface-located nucleolin. In addition, nucleolin interacts specifically with the C-terminal region of intimin, which contains the host cell-binding domain (86). Despite this, the specificity of intimin-γ for Peyer's patch FAE has not yet been linked to binding of nucleolin. Since the distribution of surface located nucleolin on cells of the intestinal tract is not documented, it is difficult to explain the tropism of intimin-γ for FAE in this way. Nevertheless, binding of the extracellular matrix ligand laminin to the surface nucleolin of a rat intestinal cell line increases the translocation of intracellular nucleolin to the cell membrane. In this way, the binding of intimin-γ to nucleolin may also enhance surface translocation of nucleolin in human intestinal cells, making further receptor-binding sites available for attachment by EHEC O157:H7.

INTIMIN, INVASIN, AND INTEGRIN BINDING

Intimin has significant homology to the invasin protein from *Yersinia* species (36). Similar to the structure of intimin, invasin comprises a series of Ig-like domains that extend from the bacterial surface and culminate in a C-type lectin-like domain. Invasin mediates the internalization of the enteropathogenic *Yersinia* species into epithelial cells by binding to members of the β_1-integrin family (42, 43). This high-affinity interaction leads to cytoskeletal changes in the host cell that result in uptake of the bacteria through receptor-mediated endocytosis. The sequence and structural similarity between intimin and invasin implies that the molecules may also be functionally related. In solid-phase binding assays, intimin has been shown to bind $\alpha_4\beta_1$- and $\alpha_5\beta_1$-integrin, and this interaction can be inhibited by peptides containing the integrin-binding RGD motif (29). This interaction appears to depend on an intact C-type lectin-like domain, since the Cys937 mutant of intimin is unable to bind efficiently to β_1-integrins. The interaction between purified intimin and integrins does not competitively inhibit binding by invasin, however, indicating that despite their similarities, the two proteins exhibit different specificities for the same receptor (29). Although intimin has clear integrin-binding capacity, the importance of this interaction in vivo is under question. Since integrins are found mostly on the basolateral surface of the epithelium and EPEC can induce A/E lesions in epithelial cells that do not express β_1-integrins due to the translocation of Tir and subsequent intimin-Tir interaction, the interaction is thought to not be essential for colonization of the host (59). However, with increasing evidence from in vivo studies that colonization of the intestinal mucosa by A/E pathogens depends not only on the intimin-Tir interaction but also on additional functions of intimin, integrins may well represent a target host cell receptor for intimin during the early stages of infection with EPEC and EHEC. The possibility of binding to multiple host receptors may confer the ability to adhere efficiently to different tissues or cell types during the multiple stages of infection.

The interaction of invasin with β_1-integrins is also likely to be important for tissue tropism. The crystal structure of invasin shows that the molecule has some similarities to fibronectin, the natural ligand of β_1-integrins, including a similar exposed loop containing the critical integrin-binding residues, Asp911 for invasin and Asp1495 for fibronectin, within the RGD integrin-binding motif. In addition, recognition of both invasin and fibronectin by β_1-integrins involves a putative cation-binding site, suggesting that the interaction between invasin and β_1-integrins mimics binding by the natural ligand (56).

Like intimin-γ-expressing strains of EHEC and EPEC, the enteropathogenic *Yersinia* species target the FAE as an initial site of colonization and, in this case, entry into the host tissues. The specialized antigen-sampling cells (M cells) that are present in FAE appear to be the main target for invasin-mediated *Yersinia* invasion, and the basis of this tissue and cell-type tropism relies largely on the fact that the apical surface of M cells is rich in β_1-integrins compared with dome enterocytes or goblet cells (11, 84). Amino acid substitution of Asp911 in the C terminus of invasin results in a derivative that is unable to mediate uptake by M cells in vitro and that, when expressed in *Yersinia pseudotuberculosis,* is defective for colonization of Peyer's patches in the mouse model of yersiniosis (64). Indeed, loss of invasin-mediated integrin-binding activity leads to the colonization of other epithelial surfaces rich in mucus, suggesting that invasin is absolutely required for bacterial translocation into Peyer's patches (64). Hence, similar to intimin, invasin modulates the site of initial mucosal colonization by an enteric pathogen.

It is interesting that several pathogens have been observed to utilize the FAE as the first site of intestinal colonization. This may be due to the natural propensity of M cells to attract and sample intestinal contents (and the presence of a thinner glycocalyx), but this interaction may also be adaptive to the bacteria. It may be beneficial for enteric pathogens that suppress innate immunity to encounter FAE and its underlying lymphoid tissue in a "first-strike" effort to facilitate further colonization.

MUCOSAL COLONIZATION AND HOST SPECIFICITY

Like tissue tropism, the molecular basis of host specificity is for the most part not well defined. Examples of host specificity occur in almost every genus of bacteria that includes significant pathogens, including *Mycobacterium, Streptococcus, Staphylococcus, Brucella, Pasteurella, Haemophilus, Bordetella, Treponema, Chlamydia, Mycoplasma, Vibrio, Shigella,* and *Salmonella*. Recently, whole-genome comparison of highly related pathogens exhibiting different host restrictions has been used as a starting point to identify putative factors involved in host specificity. Although many of these factors remain to be characterized experimentally, this bioinformatic approach has provided an overall view of the sometimes subtle differences between organisms with different host preferences. For example, *Salmonella enterica* serovar Typhi is a highly invasive human-specific pathogen, whereas *S. enterica* serovar Typhimurium can colonize humans and other mammalian hosts but causes damage that is usually confined to the mucosal region. The genome sequences of *S. enterica* serovars Typhi and Typhimurium show that serovar Typhi has

601 serovar-specific genes while serovar Typhimurium has 479 genes that are absent from serovar Typhi (66, 73). Another notable difference between the two genomes is the large number of pseudogenes in *S. enterica* serovar Typhi (more than 140), which are intact in serovar Typhimurium (66). Many of the mutations are in genes coding for factors putatively involved in host-pathogen interactions, including seven chaperone/usher fimbrial operons. The genetic losses exhibited by *S. enterica* serovar Typhi may reflect its adaptation to a specialized host and niche (99). On the other hand, the fact that *S. enterica* serovar Typhimurium carries many more functional fimbrial operons than does serovar Typhi suggests that these contribute to its broader host range. However, recent work has shown that in a bovine model of invasive salmonellosis, the outcome of infection with different serovars of *S. enterica* depends on the ability of the bacteria to replicate and disseminate in deeper host tissues (74). Although the molecular basis of this phenomenon was not identified, persistence of the pathogen rather than adherence to and initial interaction with the host mucosal surface was crucial in determining virulence. Together, these data show that host specificity is likely to be multifactorial and complex.

In contrast to the salmonellae, a recent genomic comparison between the facultative intracellular pathogens *Brucella suis* and *B. melitensis* revealed only 74 different genes between these two species (17, 75). One reason for the presence of fewer genetic differences between *Brucella* species than between *Salmonella* serovars may be the lack of opportunity for lateral gene transfer in an intracellular environment. Both species of *Brucella* are well-known animal pathogens that can cause systemic infections in a variety of hosts but induce abortions in only a few host types. *B. suis* induces late-term abortions in swine, whereas *B. melitensis* affects only ruminant species. The basis of this host specificity is unclear but may depend on a niche specificity where interspecies differences in placental tissue affect the ability of the bacteria to survive and proliferate in this environment (94). Genes unique to each pathogen may play a role in this adaptation, although sequence variation among genes shared by related pathogens may also contribute to host specificity.

K88 AND K99 PILI: DETERMINANTS OF HOST SPECIFICITY

Similar to UPEC, the diarrheal pathogen enterotoxigenic *E. coli* (ETEC) produces fimbrial colonization factors that are essential for adherence to the host intestinal mucosa and for virulence. Strains of ETEC are a common cause of infantile and traveler's diarrhea in humans but also comprise a well-known group of animal pathogens. Development of vaccines against human ETEC has been impeded in part by the extreme heterogeneity of pili expressed by these strains. Animal strains of ETEC appear to be somewhat less diverse in terms of their fimbrial adhesins, and at least two of these, K88 (or F4) and K99 (or F5) pili, have been investigated as candidate vaccine antigens (70). K88 and K99 pili are classical mediators of species specificity, where strains expressing K99 are pathogens of pigs, calves, and lambs and K88-producing strains cause disease only in pigs (32). The genetic loci encoding K88 and K99 pili, *fae* and *fan*, respectively, are both present on large plasmids and share a similar genetic organization with the *fim* and *pap* operons of UPEC (69). Both K88 and K99 pili mediate the direct binding of ETEC to the intestinal mucosa via lectin-like activity, and both confer a tropism for small intestinal tissue in preference to the large bowel. The in vivo receptor for K99 is a ganglioside [Neu5Gc-α(2-3)Galp-β(1-4)Glcp-β(1-1)-ceramide] that has been found in the small intestinal mucosa of pigs, calves, and lambs but not humans (57). Adherence via K88 pili is more complex, since several antigenic variants of K88 exist (K88ab, K88ac, and K88ad), each with its own glycoconjugate receptor specificities (97). The K88 receptors characterized to date include intestinal mucin-type glycoproteins (K88ab and K88ac), porcine enterocyte transferrin (K88ab), and an intestinal glycosphingolipid (K88ad) (33, 34). Ultimately, the pilus variant, together with different host receptor combinations, determines the susceptibility of pigs to infection by K88-expressing strains of ETEC. Together with the pili of UPEC, these examples illustrate the importance of initial bacterial attachment to host cells in controlling tissue and host specificity.

HOST RESTRICTION AND THE PATHOGEN-SPECIFIC PILI OF EPEC, REPEC, AND *C. RODENTIUM*

In addition to the LEE-encoded factors discussed above, the plasmid-encoded type IV BFP of EPEC is essential for full colonization of the host and for virulence (5), possibly through its direct involvement in microcolony formation on the mucosal surface. BFP-negative strains are able to cause A/E lesions on the human intestine, indicating that BFP is not necessary for intestinal colonization. However, BFP can mediate cell-type-dependent adherence in vitro. On formalin-fixed tissue culture cells where A/E lesion formation is prevented, EPEC preferentially adheres to human epithelial cell lines over mouse intestinal cell lines. This species-specific adherence depends on BFP expression and correlates with preferential binding of the major fimbrial subunit, BfpA, to human cells. This suggests that the pilus can discriminate between cell lines of different origins. Interestingly, the binding of purified BfpA to human cells does not require disulfide bond formation, although the binding domain is thought to lie within the C-terminal intrachain disulfide loop that is conserved among type IV pili (93).

Other studies have shown that BFP binds to phosphatidylethanolamine (PE), which is a component of both eukaryotic and bacterial cell membranes (52). While the ability of BFP to bind PE from *E. coli* may explain the underlying basis of the interbacterial interactions and microcolony formation mediated by BFP, it appears to contradict a role for BFP in host cell-specific attachment, given the wide distribution of PE among eukaryotic cell types. Nevertheless, the levels of PE in different host cell types and the relative abundance of PE in the outer and inner leaflets of the host cell membrane are not known for cells of different origins. In addition, although BFP-dependent tissue specificity has not been demonstrated in vivo, BFP is found only in human EPEC strains and may contribute to tissue and host specificity, given the tropism of the pilus for cells of human origin in vitro.

Some A/E pathogens are genetically closely related to EPEC yet are natural pathogens of animals. REPEC, for example, causes severe diarrhea in susceptible animals, inducing the same clinicopathological features, age, and tissue specificity as the varieties of EPEC which infect humans (81). *Citrobacter rodentium* is an A/E pathogen of mice that is closely related to EPEC and REPEC (83). Instead of inducing diarrhea, however, *C. rodentium* is the causative agent of murine transmissible colonic hyperplasia, a disease affecting colonies of laboratory mice. As in other A/E pathogens, the genes required for the production of A/E lesions by REPEC and *C. rodentium* are contained within a LEE region that is generally conserved between these pathogens (18, 91, 102). By contrast, REPEC and *C. rodentium* differ considerably from EPEC in terms of their fimbrial adhesins. Strains of REPEC possess diverse fimbrial operons that vary among REPEC isolates and serotypes. Of the ones that have been characterized, all are required for the virulence of their corresponding strain. AF/R1 fimbriae (adherence factor/rabbit 1) are encoded by a 132-kb conjugative plasmid of REPEC O15:H− strain RDEC-1 (9). These prototypic REPEC fimbriae have a similar genetic organization and significant amino acid sequence similarity to components of the F18 fimbrial operon of *E. coli* strains responsible for pig edema (7). AF/R1 fimbriae promote the attachment of RDEC-1 to rabbit intestinal brush borders, and this binding is specific for rabbit small intestinal tissue, including tissue derived from the duodenum, jejunum, and ileum (4, 10). AF/R1 does not mediate attachment to large intestinal tissue, including tissue obtained from the cecum, proximal colon, or distal colon. Evidence obtained from brush border adherence assays suggests that AF/R1 fimbriae are associated with the tropism of AF/R1-expressing REPEC strains for the FAE of Peyer's patches in the rabbit intestinal mucosa (8, 98). Additional data, however, suggest that AF/R1 fimbriae recognize and bind to glycoprotein and glycolipid receptors in rabbit intestinal brush borders that are not specific for the FAE. These have been tentatively identified as the sucrase-isomaltase complex and galactosylceramide [Gal (1-1)Cer], respectively (82). The sucrase-isomaltase complex is expressed on the luminal surface of enterocytes in intestinal villi and is associated almost exclusively with small intestinal tissue. The identification of these receptors may explain the tropism of RDEC-1 for the small intestine but suggests that the tropism of AF/R1-expressing REPEC strains for FAE depends on other adherence factors. This putative factor is unlikely to be intimin, however, since RDEC expresses intimin-β, which exhibits a similar tropism to *E. coli* strains that express intimin-α (our unpublished observations).

Not all strains of REPEC carry the AF/R1 fimbrial operon, and these isolates express different fimbrial adhesins. Two of them have been characterized and sequenced, namely, the AF/R2 fimbriae identified in REPEC O103:H2 strain B10 and the Ral fimbriae identified in REPEC O15:H− strain 83/39 (1, 24). Although these two fimbrial operons are essential for virulence, they have not been as widely studied as AF/R1 from RDEC-1. The predicted products of both these fimbrial operons show significant similarity to each other and to components of the *fae* operon encoding K88 fimbriae, which may indicate some conservation of function between the two adhesins. Further work with the plasmid-encoded Ral fimbriae has shown that not only are the fimbriae essential for colonization of the rabbit intestine by REPEC, but also that they evidently mediate adherence to the rabbit intestinal mucosa prior to A/E lesion formation (55).

Overall, the REPEC fimbriae are undisputed colonization factors and do not appear to occur outside the REPEC group of pathogens. This could indicate that they play a role in host and tissue specificity, but since the receptors for Ral and AF/R2 have not been identified, it is difficult to comment on this possible contribution to host specificity. Experiments are under way in our laboratory to introduce Ral into strains of EHEC and EPEC to determine if this colonization factor is sufficient to allow the human pathogens to colonize a rabbit host and cause disease. This approach may help to elucidate the contribution of fimbriae to tissue and host specificity by using the REPEC/rabbit model.

The murine A/E pathogen *C. rodentium* utilizes pili to colonize the intestinal tracts of mice and produces a type IV pilus, colonization factor citrobacter (CFC), that is essential for colonization and virulence (71). It is tempting to speculate that since CFC belongs to the type IV family of pili, this colonization factor functions in a manner similar to BFP from EPEC. At this stage, the contribution of CFC to tissue tropism and host specificity is unknown, although since the pilus appears to be restricted to *C. rodentium,* CFC may represent a specific determinant for the interaction of *C. rodentium* with the murine intestinal mucosa. While the putative specificity of this interaction remains to be shown experimentally, the host specificity of both REPEC and *C. rodentium* and their relatedness to EPEC and EHEC provides us with small-animal models to investigate the molecular basis of host specificity.

INTIMIN AND HOST SPECIFICITY

Whereas EPEC is a human-specific pathogen, EHEC is a zoonotic pathogen that persists asymptomatically in the gastrointestinal tracts of ruminants, in particular cattle. Although the factors required by EHEC to colonize cattle have not been well defined, there is likely to be some overlap with the virulence determinants required for infection of humans. The shared virulence-associated genes of EHEC already identified as necessary for colonization of cattle include *eae* and a large 9663-bp open reading frame termed *efa1* (12, 16, 89). In keeping with the IVOC model employing human intestinal tissue, EHEC O157:H7 shows a tropism for the FAE of the bovine gut and in particular for the rectoanal junction, a region that is rich in lymphoid follicles (72). EHEC O157:H7 colonization of this site has implications for shedding into feces and subsequent transmission of EHEC from cattle to humans, and it is tempting (but somewhat premature) to speculate that intimin-γ may contribute to this tropism for bovine FAE.

In addition to the pathogen-specific fimbrial adhesins discussed above, LEE-encoded factors may play a role in the host specificity of REPEC and *C. rodentium.* The LEE of REPEC contains a ca. 35-kb region that is highly homologous to the LEE of other A/E pathogens containing the same genes in the same order and which confers A/E capacity on *E. coli* K-12 but not the ability to colonize rabbits (91). The LEE region of *C. rodentium* shares all 41 core open reading frames with the LEE regions of other A/E pathogens, and *C. rodentium* induces A/E lesions in vivo that are morphologically indistinguishable from those caused by REPEC and EPEC (18, 83). REPEC, *C. rodentium,* and most animal A/E pathogens express intimin-β. Nevertheless, intimin exchanges between human and animal A/E pathogens have shown that in addition to a role in tissue tropism, intimin may play some role in host specificity. For example, intimin-α from human EPEC is capable of restoring full virulence to an *eae* knockout mutant of *C. rodentium* and has 79% amino acid identity to intimin-β from *C. rodentium* (30). Intimin-γ from EHEC O157:H7, on the other hand, is unable to complement virulence in *C. rodentium* despite having 78% amino acid identity (38). Hence, the difference between a strain that is capable of colonizing one host and a strain that is apparently avirulent is only slight in this model of infection. This effect is apparently independent of Tir binding by intimin since in vitro binding assays show that intimin-α and intimin-γ bind Tir from *C. rodentium* with comparable affinity (38). Interestingly, when the FAE-restricted mutant of intimin-α, V911A, that exhibits an intimin-γ like tropism in human IVOCs was used to complement an *eae* mutant of *C. rodentium,* the resulting strain was avirulent (80). This further supports the observation that intimin-γ-like molecules are not functional when expressed by *C. rodentium* during infection of mice, even though they retain Tir-binding capacity.

The evaluation of intimin mutants in the *C. rodentium* model of infection carrying only minor changes in amino acid sequence strengthens a growing body of evidence supporting the existence of a host cell receptor(s) for intimin. Site-directed mutagenesis of the conserved tryptophan residues, Trp776 and Trp881, of intimin-α that function to stabilize a surface pocket formed within the D4 domain shows that these two residues are not required for intimin-Tir interactions or A/E lesion formation but are essential for the virulence of *C. rodentium* (78). In addition, specific substitution of the conserved tyrosine residues, Tyr779, Tyr887, and Tyr890, which are partially exposed and form an interface between D3 and D4 of the intimin superdomain, results in mutants of *C. rodentium* that retain the Tir-binding capacity and the ability to form A/E lesions but are avirulent in mice (79). While some colonization of the murine intestinal mucosa was observed for these site-directed mutants, the mice did not develop the colonic hyperplasia typically associated with infection by *C. rodentium.* On the other hand, substitution of Tyr889, which is present in the same conserved tyrosine-rich region of intimin as Tyr779, Tyr887, and Tyr890, had no effect on Tir binding, A/E lesion formation, colonization of mice by *C. rodentium,* or virulence (79). Together, these data suggest that a host mucosal receptor(s) for intimin may be responsible for the different tissue and host tropisms observed for different intimin types.

Nevertheless, intimin type is clearly not the only determinant of host specificity, since intimin-β expressing strains are common among human and animal A/E pathogens. In addition, unpublished data from our laboratory have shown that a human EPEC strain producing intimin-β is unable to cause disease or to colonize rabbits to any degree. However, the strain was capable of producing A/E lesions in ligated loops of rabbit ileum that were indistinguishable from those evoked by a intimin-β producing REPEC strain which is highly virulent (Table 1).

FUNCTIONAL PHENOTYPIC DIFFERENCES BETWEEN Tir MOLECULES

Although it appears unlikely that the interaction between intimin and Tir contributes to tissue tropism in a major way, some functional differences exist among Tir molecules from different A/E pathogens that can alter the A/E phenotype in vitro and influence the phenotype of *C. rodentium* in the mouse model of infection. On translocation into the host cell membrane, a tyrosine residue (Tyr474) located in the cytosolic C-terminal domain of Tir from EPEC undergoes tyrosine phosphorylation (48). This phosphorylation event is essential for the recruitment of the adapter protein Nck to the EPEC pedestal and subsequent induction of actin nucleation through N-WASP and the Arp2/3 complex (35). Although Tir from EHEC O157:H7 has more than 80% amino acid similarity to EPEC Tir, a serine residue is present in place of Tyr474. Hence, EHEC O157:H7 Tir does not undergo tyrosine phosphorylation and does not recruit Nck, suggesting that the EHEC O157:H7 protein has an alternative mechanism of actin nucleation (6). Tir exchange experiments have shown that when Tir of EPEC is expressed in EHEC O157:H7, the resulting strain can induce actin nucleation and A/E lesion formation with or without tyrosine phosphorylation (47). In contrast, Tir from EHEC O157:H7 is not functionally interchangeable when expressed in EPEC. The basis for this tyrosine phosphorylation-independent actin nucleation by Tir of EHEC O157:H7 is not clear, but the mechanism could involve other type III secreted and translocated proteins that are present in EHEC O157:H7 but are redundant, silent, or missing in EPEC.

Similar to Tir from EPEC, Tir from *C. rodentium* becomes tyrosine phosphorylated at Y471 on translocation into the host cell membrane (19). These two Tir molecules are functionally interchangeable for A/E lesion formation in tissue culture, where tyrosine phosphorylation is essential for actin nucleation. Tir from O157:H7, however, cannot induce actin nucleation in tissue culture cells when expressed by *C. rodentium,* further supporting the idea that two mechanisms of Tir-induced actin nucleation operate in EPEC/*C. rodentium* and EHEC O157:H7. Interestingly, Tir from *C. rodentium* is functional when expressed in either EPEC or EHEC O157:H7, although actin nucleation occurs independently of tyrosine phosphorylation in EHEC O157:H7 (19). This reinforces the suggestion that EHEC O157:H7 encodes additional factors which allow actin nucleation to occur without tyrosine phosphorylation.

The expression of these different Tir molecules in *C. rodentium* offers an opportunity to examine the contribution of Tir tyrosine phosphorylation to pathogenesis. In vivo, Tir proteins from both EPEC and EHEC O157:H7 can restore virulence to a *tir* mutant of *C. rodentium* (19). In addition, when Tyr471 of native *C. rodentium* Tir is replaced by phenylalanine, *C. rodentium* TirY471F can still cause disease and induce A/E lesions in vivo (19). This indicates that colonization and A/E lesion formation can proceed in vivo in the absence of Tir tyrosine phosphorylation. This disparity of phenotype in vitro and in vivo

Table 1. Number of bacteria in mucosal scrapings obtained from different regions of the intestinal tract of rabbits inoculated with an EPEC strain 4 days previously

Strain	Characteristics	CFU/g of mucosa[a] from:			
		Jejunum	Ileum	Cecum	Colon
E65/56	Human EPEC, intimin-β	$<10^2$	$<10^2$	$<10^2$	$<10^2$
JPN15	Human EPEC, intimin-α	$<10^2$	$<10^2$	1.5×10^5	3.9×10^5
83/39-23	Rabbit EPEC, intimin-β, *ral*	3.6×10^3	5.3×10^4	8.5×10^6	2.0×10^6

[a]Mean value from five rabbits. Limit of detection, 10^2 CFU/g.

is interesting and may signal an important difference between the behavior of A/E pathogens in tissue culture and in the complex environment of the gut. Perhaps specific host factors are required to induce the expression of the additional bacterial proteins needed for tyrosine phosphorylation-independent actin nucleation in *C. rodentium* and possibly in EPEC. If so, this additional layer of complexity could contribute to the tissue and host tropism of A/E pathogens. Interestingly, the gut hormone epinephrine was recently shown to play a role in regulating LEE gene expression in EHEC O157:H7 through a quorum-sensing mechanism, indicating that A/E pathogens can indeed respond to specific host signals (87).

SUMMARY

The tissue tropism of bacterial pathogens is the result of a complex encounter between the microorganism and the host. While initial attachment plays an early and crucial role in this process, especially for mucosal pathogens, other factors are undeniably also important for persistence within a particular niche. The molecular dissection of intimin and Tir from EPEC and the pili of UPEC have allowed us to gain significant insight into the complexity of bacterial tissue tropism and its importance for pathogenesis. Importantly, these *E. coli* pathogens have shown us that the specificity and affinity of bacterial surface attachment underlie the tissue tropism of mucosal pathogens and depend not only on timely expression of a specific bacterial adhesin but also on its binding affinity for a cognate host cell receptor and the tissue distribution and availability of the host receptor. While considerable work remains to be done before we have a true understanding of the basis of tissue tropism in a broad range of organisms, the *E. coli* pathogens described here provide us with invaluable models to explore the basis of tissue tropism for bacterial mucosal pathogens and the importance of this interaction for colonization of the host.

REFERENCES

1. **Adams, L. M., C. P. Simmons, L. Rezmann, R. A. Strugnell, and R. M. Robins-Browne.** 1997. Identification and characterization of a K88- and CS31A-like operon of a rabbit enteropathogenic *Escherichia coli* strain which encodes fimbriae involved in the colonization of rabbit intestine. *Infect. Immun.* **65:**5222–5230.
2. **Adu-Bobie, J., G. Frankel, C. Bain, A. G. Goncalves, L. R. Trabulsi, G. Douce, S. Knutton, and G. Dougan.** 1998. Detection of intimins alpha, beta, gamma, and delta, four intimin derivatives expressed by attaching and effacing microbial pathogens. *J. Clin. Microbiol.* **36:**662–668.
3. **Batchelor, M., S. Prasannan, S. Daniell, S. Reece, I. Connerton, G. Bloomberg, G. Dougan, G. Frankel, and S. Matthews.** 2000. Structural basis for recognition of the translocated intimin receptor (Tir) by intimin from enteropathogenic *Escherichia coli. EMBO J..* **19:**2452–2464.
4. **Berendson, R., C. P. Cheney, P. A. Schad, and E. C. Boedeker.** 1983. Species-specific binding of purified pili (AF/R1) from the *Escherichia coli* RDEC-1 to rabbit intestinal mucosa. *Gastroenterology* **85:**837–845.
5. **Bieber, D., S. W. Ramer, C. Y. Wu, W. J. Murray, T. Tobe, R. Fernandez, and G. K. Schoolnik.** 1998. Type IV pili, transient bacterial aggregates, and virulence of enteropathogenic *Escherichia coli. Science* **280:**2114–2118.
6. **Campellone, K. G., and J. M. Leong.** 2003. Tails of two Tirs: actin pedestal formation by enteropathogenic *E. coli* and enterohemorrhagic *E. coli* O157:H7. *Curr. Opin. Microbiol.* **6:** 82–90.
7. **Cantey, J. R., R. K. Blake, J. R. Williford, and S. L. Moseley.** 1999. Characterization of the *Escherichia coli* AF/R1 pilus operon: novel genes necessary for transcriptional regulation and for pilus-mediated adherence. *Infect. Immun.* **67:**2292–2298.
8. **Cantey, J. R., L. R. Inman, and R. K. Blake.** 1989. Production of diarrhea in the rabbit by a mutant of *Escherichia coli* (RDEC-1) that does not express adherence (AF/R1) pili. *J. Infect. Dis.* **160:**136–141.
9. **Cheney, C. P., S. B. Formal, P. A. Schad, and E. C. Boedeker.** 1983. Genetic transfer of a mucosal adherence factor (R1) from an enteropathogenic *Escherichia coli* strain into a *Shigella flexneri* strain and the phenotypic suppression of this adherence factor. *J. Infect. Dis.* **147:**711–723.
10. **Cheney, C. P., P. A. Schad, S. B. Formal, and E. C. Boedeker.** 1980. Species specificity of in vitro *Escherichia coli* adherence to host intestinal cell membranes and its correlation with in vivo colonization and infectivity. *Infect. Immun.* **28:**1019–1027.
11. **Clark, M. A., B. H. Hirst, and M. A. Jepson.** 1998. M-cell surface beta1 integrin expression and invasin-mediated targeting of *Yersinia pseudotuberculosis* to mouse Peyer's patch M cells. *Infect. Immun.* **66:**1237–1243.
12. **Cornick, N. A., S. L. Booher, and H. W. Moon.** 2002. Intimin facilitates colonization by *Escherichia coli* O157:H7 in adult ruminants. *Infect. Immun.* **70:**2704–2707.
13. **Daniell, S. J., R. M. Delahay, R. K. Shaw, E. L. Hartland, M. J. Pallen, F. Booy, F. Ebel, S. Knutton, and G. Frankel.** 2001. Coiled-coil domain of enteropathogenic *Escherichia coli* type III secreted protein EspD is involved in EspA filament-mediated cell attachment and hemolysis. *Infect. Immun.* **69:**4055–4064.
14. **Daniell, S. J., E. Kocsis, E. Morris, S. Knutton, F. P. Booy, and G. Frankel.** 2003. 3D structure of EspA filaments from enteropathogenic *Escherichia coli. Mol. Microbiol.* **49:**301–308.
15. **Daniell, S. J., N. Takahashi, R. Wilson, D. Friedberg, I. Rosenshine, F. P. Booy, R. K. Shaw, S. Knutton, G. Frankel, and S. Aizawa.** 2001. The filamentous type III secretion translocon of enteropathogenic *Escherichia coli. Cell. Microbiol.* **3:**865–871.
16. **Dean-Nystrom, E. A., B. T. Bosworth, H. W. Moon, and A. D. O'Brien.** 1998. *Escherichia coli* O157:H7 requires intimin for enteropathogenicity in calves. *Infect. Immun.* **66:**4560–4563.
17. **DelVecchio, V. G., V. Kapatral, R. J. Redkar, G. Patra, C. Mujer, T. Los, N. Ivanova, I. Anderson, A. Bhattacharyya, A. Lykidis, G. Reznik, L. Jablonski, N. Larsen, M. D'Souza, A. Bernal, M. Mazur, E. Goltsman, E. Selkov, P. H. Elzer,

S. Hagius, D. O'Callaghan, J. J. Letesson, R. Haselkorn, N. Kyrpides, and R. Overbeek. 2002. The genome sequence of the facultative intracellular pathogen *Brucella melitensis. Proc. Natl. Acad. Sci. USA* **99:**443–448.

18. **Deng, W., Y. Li, B. A. Vallance, and B. B. Finlay.** 2001. Locus of enterocyte effacement from *Citrobacter rodentium:* sequence analysis and evidence for horizontal transfer among attaching and effacing pathogens. *Infect. Immun.* **69:**6323–6335.
19. **Deng, W., B. A. Vallance, Y. Li, J. L. Puente, and B. B. Finlay.** 2003. *Citrobacter rodentium* translocated intimin receptor (Tir) is an essential virulence factor needed for actin condensation, intestinal colonization and colonic hyperplasia in mice. *Mol. Microbiol.* **48:**95–115.
20. **Dodson, K. W., J. S. Pinkner, T. Rose, G. Magnusson, S. J. Hultgren, and G. Waksman.** 2001. Structural basis of the interaction of the pyelonephritic *E. coli* adhesin to its human kidney receptor. *Cell* **105:**733–743.
21. **Donnenberg, M. S., J. Yu, and J. B. Kaper.** 1993. A second chromosomal gene necessary for intimate attachment of enteropathogenic *Escherichia coli* to epithelial cells. *J. Bacteriol.* **175:**4670–4680.
22. **Elliott, S. J., E. O. Krejany, J. L. Mellies, R. M. Robins-Browne, C. Sasakawa, and J. B. Kaper.** 2001. EspG, a novel type III system-secreted protein from enteropathogenic *Escherichia coli* with similarities to VirA of *Shigella flexneri. Infect. Immun.* **69:**4027–4033.
23. **Elliott, S. J., L. A. Wainwright, T. K. McDaniel, K. G. Jarvis, Y. K. Deng, L. C. Lai, B. P. McNamara, M. S. Donnenberg, and J. B. Kaper.** 1998. The complete sequence of the locus of enterocyte effacement (LEE) from enteropathogenic *Escherichia coli* E2348/69. *Mol. Microbiol.* **28:**1–4.
24. **Fiederling, F., M. Boury, C. Petit, and A. Milon.** 1997. Adhesive factor/rabbit 2, a new fimbrial adhesin and a virulence factor from *Escherichia coli* O103, a serogroup enteropathogenic for rabbits. *Infect. Immun.* **65:**847–851.
25. **Fitzhenry, R. J., D. J. Pickard, E. L. Hartland, S. Reece, G. Dougan, A. D. Phillips, and G. Frankel.** 2002. Intimin type influences the site of human intestinal mucosal colonisation by enterohaemorrhagic *Escherichia coli* O157:H7. *Gut* **50:**180–185.
26. **Fitzhenry, R. J., S. Reece, L. R. Trabulsi, R. Heuschkel, S. Murch, M. Thomson, G. Frankel, and A. D. Phillips.** 2002. Tissue tropism of enteropathogenic *Escherichia coli* strains belonging to the O55 serogroup. *Infect. Immun.* **70:**4362–4368.
27. **Fitzhenry, R. J., M. P. Stevens, C. Jenkins, T. S. Wallis, R. Heuschkel, S. Murch, M. Thomson, G. Frankel, and A. D. Phillips.** 2003. Human intestinal tissue tropism of intimin epsilon O103 *Escherichia coli. FEMS Microbiol. Lett.* **218:**311–316.
28. **Frankel, G., D. C. Candy, E. Fabiani, J. Adu-Bobie, S. Gil, M. Novakova, A. D. Phillips, and G. Dougan.** 1995. Molecular characterization of a carboxy-terminal eukaryotic-cell-binding domain of intimin from enteropathogenic *Escherichia coli. Infect. Immun.* **63:**4323–4328.
29. **Frankel, G., O. Lider, R. Hershkoviz, A. P. Mould, S. G. Kachalsky, D. C. A. Candy, L. Cahalon, M. J. Humphries, and G. Dougan.** 1996. The cell-binding domain of intimin from enteropathogenic *Escherichia coli* binds to beta1 integrins. *J. Biol. Chem.* **271:**20359–20364.
30. **Frankel, G., A. D. Phillips, M. Novakova, H. Field, D. C. Candy, D. B. Schauer, G. Douce, and G. Dougan.** 1996. Intimin from enteropathogenic *Escherichia coli* restores murine virulence to a *Citrobacter rodentium eaeA* mutant: induction of an immunoglobulin A response to intimin and EspB. *Infect. Immun.* **64:**5315–5325.
31. **Frankel, G., A. D. Phillips, I. Rosenshine, G. Dougan, J. B. Kaper, and S. Knutton.** 1998. Enteropathogenic and enterohaemorrhagic *Escherichia coli:* more subversive elements. *Mol. Microbiol.* **30:**911–921.
32. **Gaastra, W., and F. K. de Graaf.** 1982. Host-specific fimbrial adhesins of noninvasive enterotoxigenic *Escherichia coli* strains. *Microbiol. Rev.* **46:**129–161.
33. **Grange, P. A., A. K. Erickson, S. B. Levery, and D. H. Francis.** 1999. Identification of an intestinal neutral glycosphingolipid as a phenotype-specific receptor for the K88ad fimbrial adhesin of *Escherichia coli. Infect. Immun.* **67:**165–172.
34. **Grange, P. A., M. A. Mouricout, S. B. Levery, D. H. Francis, and A. K. Erickson.** 2002. Evaluation of receptor binding specificity of *Escherichia coli* K88 (F4) fimbrial adhesin variants using porcine serum transferrin and glycosphingolipids as model receptors. *Infect. Immun.* **70:**2336–2343.
35. **Gruenheid, S., R. DeVinney, F. Bladt, D. Goosney, S. Gelkop, G. D. Gish, T. Pawson, and B. B. Finlay.** 2001. Enteropathogenic *E. coli* Tir binds Nck to initiate actin pedestal formation in host cells. *Nat. Cell Biol.* **3:**856–859.
36. **Hamburger, Z. A., M. S. Brown, R. R. Isberg, and P. J. Bjorkman.** 1999. Crystal structure of invasin: a bacterial integrin-binding protein. *Science* **286:**291–295.
37. **Hartland, E. L., M. Batchelor, R. M. Delahay, C. Hale, S. Matthews, G. Dougan, S. Knutton, I. Connerton, and G. Frankel.** 1999. Binding of intimin from enteropathogenic *Escherichia coli* to Tir and to host cells. *Mol. Microbiol.* **32:**151–158.
38. **Hartland, E. L., V. Huter, L. M. Higgins, N. S. Goncalves, G. Dougan, A. D. Phillips, T. T. MacDonald, and G. Frankel.** 2000. Expression of intimin gamma from enterohemorrhagic *Escherichia coli* in *Citrobacter rodentium. Infect. Immun.* **68:**4637–4646.
39. **Hicks, S., G. Frankel, J. B. Kaper, G. Dougan, and A. D. Phillips.** 1998. Role of intimin and bundle-forming pili in enteropathogenic *Escherichia coli* adhesion to pediatric intestinal tissue in vitro. *Infect. Immun.* **66:**1570–1578.
40. **Hung, C. S., J. Bouckaert, D. Hung, J. Pinkner, C. Widberg, A. DeFusco, C. G. Auguste, R. Strouse, S. Langermann, G. Waksman, and S. J. Hultgren.** 2002. Structural basis of tropism of *Escherichia coli* to the bladder during urinary tract infection. *Mol. Microbiol.* **44:**903–915.
41. **Ide, T., S. Laarmann, L. Greune, H. Schillers, H. Oberleithner, and M. A. Schmidt.** 2001. Characterization of translocation pores inserted into plasma membranes by type III-secreted Esp proteins of enteropathogenic *Escherichia coli. Cell. Microbiol.* **3:**669–679.
42. **Isberg, R. R., and J. M. Leong.** 1990. Multiple beta 1 chain integrins are receptors for invasin, a protein that promotes bacterial penetration into mammalian cells. *Cell* **60:**861–871.
43. **Isberg, R. R., D. L. Voorhis, and S. Falkow.** 1987. Identification of invasin: a protein that allows enteric bacteria to penetrate cultured mammalian cells. *Cell* **50:**769–778.
44. **Jerse, A. E., J. Yu, B. D. Tall, and J. B. Kaper.** 1990. A genetic locus of enteropathogenic *Escherichia coli* necessary for the production of attaching and effacing lesions on tissue culture cells. *Proc. Natl. Acad. Sci. USA* **87:**7839–7843.
45. **Jones, C. H., J. S. Pinkner, R. Roth, J. Heuser, A. V. Nicholes, S. N. Abraham, and S. J. Hultgren.** 1995. FimH adhesin of type 1 pili is assembled into a fibrillar tip structure in the *Enterobacteriaceae. Proc. Natl. Acad. Sci. USA* **92:**2081–2085.
46. **Kelly, G., S. Prasannan, S. Daniell, K. Fleming, G. Frankel, G. Dougan, I. Connerton, and S. Matthews.** 1999. Structure

of the cell-adhesion fragment of intimin from enteropathogenic *Escherichia coli. Nat. Struct. Biol.* **6**:313–318.

47. **Kenny, B.** 2001. The enterohaemorrhagic *Escherichia coli* (serotype O157:H7) Tir molecule is not functionally interchangeable for its enteropathogenic *E. coli* (serotype O127:H6) homologue. *Cell. Microbiol.* **3**:499–510.
48. **Kenny, B.** 1999. Phosphorylation of tyrosine 474 of the enteropathogenic *Escherichia coli* (EPEC) Tir receptor molecule is essential for actin nucleating activity and is preceded by additional host modifications. *Mol. Microbiol.* **31**:1229–1241.
49. **Kenny, B., R. DeVinney, M. Stein, D. J. Reinscheid, E. A. Frey, and B. B. Finlay.** 1997. Enteropathogenic *E. coli* (EPEC) transfers its receptor for intimate adherence into mammalian cells. *Cell* **91**:511–520.
50. **Kenny, B., and M. Jepson.** 2000. Targeting of an enteropathogenic *Escherichia coli* (EPEC) effector protein to host mitochondria. *Cell. Microbiol.* **2**:579–590.
51. **Kenny, B., L. C. Lai, B. B. Finlay, and M. S. Donnenberg.** 1996. EspA, a protein secreted by enteropathogenic *Escherichia coli,* is required to induce signals in epithelial cells. *Mol. Microbiol.* **20**:313–323.
52. **Khursigara, C., M. Abul-Milh, B. Lau, J. A. Giron, C. A. Lingwood, and D. E. Foster.** 2001. Enteropathogenic *Escherichia coli* virulence factor bundle-forming pilus has a binding specificity for phosphatidylethanolamine. *Infect. Immun.* **69**:6573–6579.
53. **Knutton, S., T. Baldwin, P. H. Williams, and A. S. McNeish.** 1989. Actin accumulation at sites of bacterial adhesion to tissue culture cells: basis of a new diagnostic test for enteropathogenic and enterohemorrhagic *Escherichia coli. Infect. Immun.* **57**:1290–1298.
54. **Knutton, S., I. Rosenshine, M. J. Pallen, I. Nisan, B. C. Neves, C. Bain, C. Wolff, G. Dougan, and G. Frankel.** 1998. A novel EspA-associated surface organelle of enteropathogenic *Escherichia coli* involved in protein translocation into epithelial cells. *EMBO J.* **17**:2166–2176.
55. **Krejany, E. O., T. H. Grant, V. Bennett-Wood, L. M. Adams, and R. M. Robins-Browne.** 2000. Contribution of plasmid-encoded fimbriae and intimin to capacity of rabbit-specific enteropathogenic *Escherichia coli* to attach to and colonize rabbit intestine. *Infect. Immun.* **68**:6472–6477.
56. **Krukonis, E. S., and R. R. Isberg.** 2000. Integrin beta1-chain residues involved in substrate recognition and specificity of binding to invasin. *Cell. Microbiol.* **2**:219–230.
57. **Kyogashima, M., V. Ginsburg, and H. C. Krivan.** 1989. *Escherichia coli* K99 binds to *N*-glycolylsialoparagloboside and *N*-glycolyl-GM3 found in piglet small intestine. *Arch. Biochem. Biophys.* **270**:391–397.
58. **Lai, L. C., L. A. Wainwright, K. D. Stone, and M. S. Donnenberg.** 1997. A third secreted protein that is encoded by the enteropathogenic *Escherichia coli* pathogenicity island is required for transduction of signals and for attaching and effacing activities in host cells. *Infect. Immun.* **65**:2211–2217.
59. **Liu, H., L. Magoun, and J. M. Leong.** 1999. Beta1-chain integrins are not essential for intimin-mediated host cell attachment and enteropathogenic *Escherichia coli*-induced actin condensation. *Infect. Immun.* **67**:2045–2049.
60. **Liu, H., L. Magoun, S. Luperchio, D. B. Schauer, and J. M. Leong.** 1999. The Tir-binding region of enterohaemorrhagic *Escherichia coli* intimin is sufficient to trigger actin condensation after bacterial-induced host cell signalling. *Mol. Microbiol.* **34**:67–81.
61. **Liu, H., P. Radhakrishnan, L. Magoun, M. Prabu, K. G. Campellone, P. Savage, F. He, C. A. Schiffer, and J. M. Leong.** 2002. Point mutants of EHEC intimin that diminish Tir recognition and actin pedestal formation highlight a putative Tir binding pocket. *Mol. Microbiol.* **45**:1557–1573.
62. **Lund, B., F. Lindberg, B. I. Marklund, and S. Normark.** 1987. The PapG protein is the alpha-D-galactopyranosyl-(1-4)-beta-D-galactopyranose-binding adhesin of uropathogenic *Escherichia coli. Proc. Natl. Acad. Sci. USA* **84**:5898–5902.
63. **Luo, Y., E. A. Frey, R. A. Pfuetzner, A. L. Creagh, D. G. Knoechel, C. A. Haynes, B. B. Finlay, and N. C. Strynadka.** 2000. Crystal structure of enteropathogenic *Escherichia coli* intimin-receptor complex. *Nature* **405**:1073–1077.
64. **Marra, A., and R. R. Isberg.** 1997. Invasin-dependent and invasin-independent pathways for translocation of *Yersinia pseudotuberculosis* across the Peyer's patch intestinal epithelium. *Infect. Immun.* **65**:3412–3421.
65. **Martinez, J. J., M. A. Mulvey, J. D. Schilling, J. S. Pinkner, and S. J. Hultgren.** 2000. Type 1 pilus-mediated bacterial invasion of bladder epithelial cells. *EMBO J.* **19**:2803–2812.
66. **McClelland, M., K. E. Sanderson, J. Spieth, S. W. Clifton, P. Latreille, L. Courtney, S. Porwollik, J. Ali, M. Dante, F. Du, S. Hou, D. Layman, S. Leonard, C. Nguyen, K. Scott, A. Holmes, N. Grewal, E. Mulvaney, E. Ryan, H. Sun, L. Florea, W. Miller, T. Stoneking, M. Nhan, R. Waterston, and R. K. Wilson.** 2001. Complete genome sequence of *Salmonella enterica* serovar Typhimurium LT2. *Nature* **413**:852–856.
67. **McDaniel, T. K., K. G. Jarvis, M. S. Donnenberg, and J. B. Kaper.** 1995. A genetic locus of enterocyte effacement conserved among diverse enterobacterial pathogens. *Proc. Natl. Acad. Sci. USA* **92**:1664–1668.
68. **McNamara, B. P., and M. S. Donnenberg.** 1998. A novel proline-rich protein, EspF, is secreted from enteropathogenic *Escherichia coli* via the type III export pathway. *FEMS Microbiol. Lett.* **166**:71–78.
69. **Mol, O., and B. Oudega.** 1996. Molecular and structural aspects of fimbriae biosynthesis and assembly in *Escherichia coli. FEMS Microbiol. Rev.* **19**:25–52.
70. **Morona, R., J. K. Morona, A. Considine, J. A. Hackett, L. van den Bosch, L. Beyer, and S. R. Attridge.** 1994. Construction of K88- and K99-expressing clones of *Salmonella typhimurium* G30: immunogenicity following oral administration to pigs. *Vaccine* **12**:513–517.
71. **Mundy, R., D. Pickard, R. K. Wilson, C. P. Simmons, G. Dougan, and G. Frankel.** 2003. Identification of a novel type IV pilus gene cluster required for gastrointestinal colonization of *Citrobacter rodentium. Mol. Microbiol.* **48**:795–809.
72. **Naylor, S. W., J. C. Low, T. E. Besser, A. Mahajan, G. J. Gunn, M. C. Pearce, I. J. McKendrick, D. G. Smith, and D. L. Gally.** 2003. Lymphoid follicle-dense mucosa at the terminal rectum is the principal site of colonization of enterohemorrhagic *Escherichia coli* O157:H7 in the bovine host. *Infect. Immun.* **71**:1505–1512.
73. **Parkhill, J., G. Dougan, K. D. James, N. R. Thomson, D. Pickard, J. Wain, C. Churcher, K. L. Mungall, S. D. Bentley, M. T. Holden, M. Sebaihia, S. Baker, D. Basham, K. Brooks, T. Chillingworth, P. Connerton, A. Cronin, P. Davis, R. M. Davies, L. Dowd, N. White, J. Farrar, T. Feltwell, N. Hamlin, A. Haque, T. T. Hien, S. Holroyd, K. Jagels, A. Krogh, T. S. Larsen, S. Leather, S. Moule, P. O'Gaora, C. Parry, M. Quail, K. Rutherford, M. Simmonds, J. Skelton, K. Stevens, S. Whitehead, and B. G. Barrell.** 2001. Complete genome sequence of a multiple drug resistant *Salmonella enterica* serovar Typhi CT18. *Nature* **413**:848–852.
74. **Paulin, S. M., P. R. Watson, A. R. Benmore, M. P. Stevens, P. W. Jones, B. Villarreal-Ramos, and T. S. Wallis.** 2002.

Analysis of *Salmonella enterica* serotype-host specificity in calves: avirulence of *S. enterica* serotype Gallinarum correlates with bacterial dissemination from mesenteric lymph nodes and persistence in vivo. *Infect. Immun.* 70:6788–6797.

75. **Paulsen, I. T., R. Seshadri, K. E. Nelson, J. A. Eisen, J. F. Heidelberg, T. D. Read, R. J. Dodson, L. Umayam, L. M. Brinkac, M. J. Beanan, S. C. Daugherty, R. T. Deboy, A. S. Durkin, J. F. Kolonay, R. Madupu, W. C. Nelson, B. Ayodeji, M. Kraul, J. Shetty, J. Malek, S. E. Van Aken, S. Riedmuller, H. Tettelin, S. R. Gill, O. White, S. L. Salzberg, D. L. Hoover, L. E. Lindler, S. M. Halling, S. M. Boyle, and C. M. Fraser.** 2002. The *Brucella suis* genome reveals fundamental similarities between animal and plant pathogens and symbionts. *Proc. Natl. Acad. Sci. USA* **99:**13148–13153.
76. **Phillips, A. D., and G. Frankel.** 2000. Intimin-mediated tissue specificity in enteropathogenic *Escherichia coli* interaction with human intestinal organ cultures. *J. Infect. Dis.* **181:** 1496–1500.
77. **Phillips, A. D., S. Navabpour, S. Hicks, G. Dougan, T. Wallis, and G. Frankel.** 2000. Enterohaemorrhagic *Escherichia coli* O157:H7 target Peyer's patches in humans and cause attaching/effacing lesions in both human and bovine intestine. *Gut* **47:**377–381.
78. **Reece, S., C. P. Simmons, R. J. Fitzhenry, M. Batchelor, C. Hale, S. Matthews, A. D. Phillips, G. Dougan, and G. Frankel.** 2002. Mutagenesis of conserved tryptophan residues within the receptor-binding domain of intimin: influence on binding activity and virulence. *Microbiology* **148:** 657–665.
79. **Reece, S., C. P. Simmons, R. J. Fitzhenry, M. Ghaem-Maghami, R. Mundy, C. Hale, S. Matthews, G. Dougan, A. D. Phillips, and G. Frankel.** 2002. Tyrosine residues at the immunoglobulin-C-type lectin inter-domain boundary of intimin are not involved in Tir-binding but implicated in colonisation of the host. *Microbes Infect.* **4:**1389–1399.
80. **Reece, S., C. P. Simmons, R. J. Fitzhenry, S. Matthews, A. D. Phillips, G. Dougan, and G. Frankel.** 2001. Site-directed mutagenesis of intimin alpha modulates intimin-mediated tissue tropism and host specificity. *Mol. Microbiol.* **40:**86–98.
81. **Robins-Browne, R. M., A. M. Tokhi, L. M. Adams, V. Bennett-Wood, A. V. Moisidis, E. O. Krejany, and L. E. O'Gorman.** 1994. Adherence characteristics of attaching and effacing strains of *Escherichia coli* from rabbits. *Infect. Immun.* **62:**1584–1592.
82. **Ryu, H., Y. S. Kim, P. A. Grange, and F. J. Cassels.** 2001. *Escherichia coli* strain RDEC-1 AF/R1 endogenous fimbrial glycoconjugate receptor molecules in rabbit small intestine. *Infect. Immun.* **69:**640–649.
83. **Schauer, D. B., and S. Falkow.** 1993. Attaching and effacing locus of a *Citrobacter freundii* biotype that causes transmissible murine colonic hyperplasia. *Infect. Immun.* **61:**2486–2492.
84. **Schulte, R., G. A. Grassl, S. Preger, S. Fessele, C. A. Jacobi, M. Schaller, P. J. Nelson, and I. B. Autenrieth.** 2000. *Yersinia enterocolitica* invasin protein triggers IL-8 production in epithelial cells via activation of Rel p65-p65 homodimers. *FASEB J.* **14:**1471–1484.
85. **Sekiya, K., M. Ohishi, T. Ogino, K. Tamano, C. Sasakawa, and A. Abe.** 2001. Supermolecular structure of the enteropathogenic *Escherichia coli* type III secretion system and its direct interaction with the EspA-sheath-like structure. *Proc. Natl. Acad. Sci. USA* **98:**11638–11643.
86. **Sinclair, J. F., and A. D. O'Brien.** 2002. Cell surface-localized nucleolin is a eukaryotic receptor for the adhesin intimin-gamma of enterohemorrhagic *Escherichia coli* O157:H7. *J. Biol. Chem.* **277:**2876–2885.
87. **Sperandio, V., A. G. Torres, B. Jarvis, J. P. Nataro, and J. B. Kaper.** 2003. Bacteria-host communication: the language of hormones. *Proc. Natl. Acad. Sci. USA* **100:**8951–8956.
88. **Starr, M., V. Bennett-Wood, A. K. Bigham, T. F. de Koning-Ward, A. M. Bordun, D. Lightfoot, K. A. Bettelheim, C. L. Jones, and R. M. Robins-Browne.** 1998. Hemolytic-uremic syndrome following urinary tract infection with enterohemorrhagic *Escherichia coli:* case report and review. *Clin. Infect. Dis.* **27:**310–315.
89. **Stevens, M. P., P. M. van Diemen, G. Frankel, A. D. Phillips, and T. S. Wallis.** 2002. Efa1 influences colonization of the bovine intestine by shiga toxin-producing *Escherichia coli* serotypes O5 and O111. *Infect. Immun.* **70:**5158–5166.
90. **Stromberg, N., B. I. Marklund, B. Lund, D. Ilver, A. Hamers, W. Gaastra, K. A. Karlsson, and S. Normark.** 1990. Host-specificity of uropathogenic *Escherichia coli* depends on differences in binding specificity to Gal alpha 1-4Gal-containing isoreceptors. *EMBO J.* **9:**2001–2010.
91. **Tauschek, M., R. A. Strugnell, and R. M. Robins-Browne.** 2002. Characterization and evidence of mobilization of the LEE pathogenicity island of rabbit-specific strains of enteropathogenic *Escherichia coli. Mol. Microbiol.* **44:**1533–1550.
92. **Thanassi, D. G., E. T. Saulino, and S. J. Hultgren.** 1998. The chaperone/usher pathway: a major terminal branch of the general secretory pathway. *Curr. Opin. Microbiol.* **1:**223–231.
93. **Tobe, T., and C. Sasakawa.** 2002. Species-specific cell adhesion of enteropathogenic *Escherichia coli* is mediated by type IV bundle-forming pili. *Cell. Microbiol.* **4:**29–42.
94. **Tsolis, R. M.** 2002. Comparative genome analysis of the alpha-proteobacteria: relationships between plant and animal pathogens and host specificity. *Proc. Natl. Acad. Sci. USA* **99:**12503–12505.
95. **Tzipori, S., R. Gibson, and J. Montanaro.** 1989. Nature and distribution of mucosal lesions associated with enteropathogenic and enterohemorrhagic *Escherichia coli* in piglets and the role of plasmid-mediated factors. *Infect. Immun.* **57:**1142–1150.
96. **Tzipori, S., F. Gunzer, M. S. Donnenberg, L. de Montigny, J. B. Kaper, and A. Donohue-Rolfe.** 1995. The role of the *eaeA* gene in diarrhea and neurological complications in a gnotobiotic piglet model of enterohemorrhagic *Escherichia coli* infection. *Infect. Immun.* **63:**3621–3627.
97. **Van den Broeck, W., E. Cox, B. Oudega, and B. M. Goddeeris.** 2000. The F4 fimbrial antigen of *Escherichia coli* and its receptors. *Vet. Microbiol.* **71:**223–244.
98. **Von Moll, L. K., and J. R. Cantey.** 1997. Peyer's patch adherence of enteropathogenic *Escherichia coli* strains in rabbits. *Infect. Immun.* **65:**3788–3793.
99. **Wain, J., D. House, J. Parkhill, C. Parry, and G. Dougan.** 2002. Unlocking the genome of the human typhoid bacillus. *Lancet Infect. Dis.* **2:**163–170.
100. **Wilson, D., M. Tuohy, and G. W. Procop.** 2000. The low prevalence of Shiga-toxin production among sorbitol non-fermenting *Escherichia coli* urinary tract isolates does not warrant routine screening. *Clin. Infect. Dis.* **31:**1313.
101. **Yu, D., M. Z. Schwartz, and R. Petryshyn.** 1998. Effect of laminin on the nuclear localization of nucleolin in rat intestinal epithelial IEC-6 cells. *Biochem. Biophys. Res. Commun.* **247:**186–192.
102. **Zhu, C., T. S. Agin, S. J. Elliott, L. A. Johnson, T. E. Thate, J. B. Kaper, and E. C. Boedeker.** 2001. Complete nucleotide sequence and analysis of the locus of enterocyte effacement from rabbit diarrheagenic *Escherichia coli* RDEC-1. *Infect. Immun.* **69:**2107–2115.

Colonization of Mucosal Surfaces
Edited by James P. Nataro et al.

Chapter 18

Aggregation and Dispersal on Mucosal Surfaces

JAMES P. NATARO AND ANGELA JANSEN

Studies of mucosal colonization have long focused on the requirement for microorganisms to adhere to surfaces. Adherence clearly adapts the microorganism to resist the shearing forces of peristalsis or urine or the action of the mucociliary escalator. In this chapter, we consider two additional facets of adherence: (i) that in addition to adhering to a surface, bacteria often adhere to each other; and (ii) that colonization is sometimes promoted by relinquishing adherence. For the purposes of this discussion, we use the terms "adhesin" to describe factors that mediate binding to a surface, "aggregin" for factors that mediate interbacterial attachment, and "antiaggregin" and "antiadhesin" to describe factors that counter such phenotypes.

BACTERIAL AGGREGATION

Role of Aggregation

Anchoring of bacteria to the mucosa can be strengthened by adherence not only to epithelial cells but also to each other. From a mathematical viewpoint, multiple points of adhesion obviously increase the strength of binding. Unsurprisingly, therefore, bacteria adhering to mucosal surfaces commonly exhibit interbacterial aggregation or agglutination (the terms are used here interchangeably), perhaps with relatively few points of anchorage to the substratum (16). On the mucosa, interbacterial aggregation presumably plays the same biophysical role as it does on abiotic substrata (adherence to which is more thoroughly studied), although data to this effect are few.

Microbial biofilms have been defined as "microbially derived sessile communities characterized by cells that are irreversibly attached to a substratum or interface or to each other, are embedded in a matrix of extracellular polymeric substances that they have produced, and exhibit an altered phenotype with respect to growth rate and gene transcription" (20). Biofilms may be found on any environmental surface where sufficient moisture and nutrients are present (14, 17, 18). This includes surfaces such as substrata in contact with moisture, liquid-air interfaces, soft tissue surfaces in living organisms, and medical devices. Sources of biofilm-related infections can include the surfaces of catheters, medical implants, wound dressings, and other types of medical devices. The contribution of biofilms may be partly to confer resistance to host-mediated killing, yet established biofilms can also be extremely difficult to eradicate in abiotic systems, such as pipelines. This suggests that an important contribution of the biofilm is resistance to physical stressors. Life in a biofilm may also provide resistance to phagocytosis or other host factors. Interestingly, Mah et al. have recently reported that resistance to antibacterial agents by bacteria within a biofilm can be unlinked from the phenotype of physical association (40). Whereas early hypotheses regarding the mechanism of biofilm resistance focused on resistance to penetration of antibacterial compounds through the biofilm matrix, more modern concepts postulate that the high bacterial density within the biofilm mass may promote interbacterial communication via quorum-sensing systems and that the response to a bacterial quorum is the induction of genes that independently encode resistance phenotypes (4). The dramatic evolutionary convergence resulting in the near universality of biofilm forms among bacterial species suggests that the advantages to biofilm formation are many, complex, and versatile. Research in this area is probably in its infancy.

It is difficult for phagocytes to engulf bacteria adhering in a firmly anchored mass. However, bacterial aggregates may be resistant to phagocytes even

James P. Nataro • Departments of Pediatrics, Medicine, and Microbiology & Immunology, University of Maryland School of Medicine, Baltimore, MD 21201. **Angela Jansen** • Department of Microbiology and Immunology, University of Maryland School of Medicine, Baltimore, MD 21201.

when the aggregates are not fixed to a surface. Ochiai et al. reported that multicell aggregates of either *Actinomyces viscosus* or *Streptococcus mitis* produced a greater number of abscesses than did single-cell suspensions when injected into mice (48). Moreover, these investigators showed that aggregated bacteria were more resistant to phagocytosis and killing by neutrophils both in vitro and in vivo. *A. viscosus* in aggregates was resistant to killing after engulfment by neutrophils. The similarity of these authors' findings when studying such distinct pathogens suggests that these observations could be generally applicable. It is interesting to consider that quorum-sensing systems may contribute to these phenotypes, in addition to the physical challenges posed by engulfment of a very large particle.

Genetic studies have begun to address the contributions of aggregation to pathogenesis. Frick et al. demonstrated that *Streptococcus pyogenes* mutants with mutations in the autoaggregative domain of protein H were deficient in colonization, demonstrating less adherence to epithelial cells, lower resistance to phagocytosis, and overall lower virulence than wild-type *S. pyogenes* in a mouse skin model (23). Whereas aggregated *S. pyogenes* bacteria grew well in human blood, nonaggregating mutants were rapidly killed. Aggregates of the *S. pyogenes* strain adhered abundantly to epithelial cells in vitro, but disruption of aggregates by gentle sonication prior to application to the cells resulted in dramatically less adherence.

The ability to aggregate plays a role in the virulence of gram-negative bacteria as well. Recent studies demonstrated that *Escherichia coli* aggregation mediated by antigen 43 (Ag43) protects bacteria from H_2O_2-dependent killing (52).

Factors Conferring Aggregation

Staphylococcus biofilm formation

Staphylococcal species are abundant biofilm producers, which complicates management of patients with indwelling medical devices. Coagulase-negative staphylococci commonly produce a polysaccharide intercellular adhesin (PIA), which provides (or contributes to) the anchoring matrix and which has been extensively studied. PIA is also produced by *Staphylococcus aureus* strains, although, interestingly, the expression of this locus appears to be more limited to the in vivo state. Despite early reports to the contrary, consensus is building that the PIA, composed of poly-*N*-acetylglucosamine, is the major and perhaps the only significant polysaccharide interbacterial adhesin among infecting staphylococcal strains (41). In addition to its role in promoting adherence to indwelling medical devices, the PIA may promote the infection of host tissues. Shiro et al. showed that a given inoculum of an *S. epidermidis* strain producing PIA was 42 times more likely to cause endocarditis in rabbits than an isogenic strain mutated at the PIA-encoding locus. Moreover, when the PS/A-positive strain was adherent to a catheter surface in the rabbit, the organism persisted in rabbit blood, whereas under the same conditions approximately 90% of the PIA-negative strain was killed in 1 h. The study of aggregation and biofilm formation by staphylococcal species continues to represent the paradigm for bacterial infections. However, information regarding a role for staphylococcal biofilm formation in colonization of mucosal surfaces is thus far lacking.

Antigen 43

Ag43 of *E. coli* is probably the best characterized of the gram-negative agglutinins. Ag43 mediates agglutination of *E. coli* in the liquid phase and also promotes *E. coli* biofilm formation. No function other than agglutination is known for this protein. Ag43 is translocated across the outer membrane of *E. coli* by the autotransporter pathway; the 1,039-amino-acid Ag43 preprotein possesses the three domains typical of proteins exported via this strategy. An N-terminal signal peptide directs the preprotein across the cytoplasmic membrane into the periplasm. The resulting periplasmic species possesses a passenger (α) domain, which is ultimately secreted to the cell surface and represents the final mature species, and the transporter (β) domain, which forms a β-barrel in the outer membrane. During the outer membrane translocation process, the passenger is thought to thread through the β-barrel and to undergo final processing by proteolytic cleavage, although the passenger domain remains noncovalently attached to the cell surface (30).

Ag43 is encoded by the 3-kb *flu* gene, named for the fluffy colony type it confers. Expression of *flu* is phase variable, with a switching rate of $\sim 10^{-3}$ per cell per generation under normal laboratory growth conditions (27). Dam methylase effects positive regulation on this system, and OxyR, the cellular redox sensor, exerts negative regulation (28, 29). Mutations in *dam* (the gene encoding Dam methyl transferase) result in bacteria being locked in the "off" state, whereas disruption of OxyR results in strong constitutive production (28, 29). Expression of the *flu* gene is controlled by a sigma-70 promoter that contains three tightly spaced GATC sites overlapping the OxyR binding sites (29). Bound OxyR prevents methylation of these GATC sites, leading to repression of transcription, while methylation of any two of the GATC sites in the promoter region blocks binding of the repressor (60).

There is reciprocal interaction between fimbriae and Ag43 at both the protein and transcriptional levels (27). The presence of fimbriae sterically blocks Ag43-mediated aggregation of bacteria; in addition, synthesis of type 1 fimbriae abrogates Ag43 expression. Type 1 fimbriae expression is also phase variable, with switching controlled by products of two genes lying within the fimbrial operon: FimB and FimE. Type 1 fimbria production is thought to interfere with Ag43 production via the necessity of a cysteine bridge in FimA, the major structural subunit of type 1 fimbriae. During fimbrial biosynthesis, disulfide bridge formation is occurring at high levels, due to the production of large amounts of FimA. Theoretically, this could affect the thiol-disulfide status of the cell, thus conveying a signal to the cell via the oxidative-reductive state of OxyR (52). Evidence supporting this hypothesis include the lack of repression of *flu* in strains in which *oxyR* is mutated and the reduced expression of *flu* in cells when dithiothreitol, a simple reducing agent, is added. Thus, there is hierarchical regulation—expression of type 1 fimbriae allows attachment of bacteria to mannose residues during infection and prohibits the production of Ag43, so that bacteria do not aggregate. When the fimbriae are in the "off" phase, Ag43 is expressed and produced on the surface of the bacteria, allowing microcolony and biofilm formation.

Bacterial aggregation and microcolony formation are considered to be an initial stage of biofilm formation (reviewed in reference 18). Several groups have shown that *flu* expression enhances *E. coli* biofilm formation on a variety of abiotic substrata (19, 38). Schembri and colleagues used an *E. coli* microarray to determine that *flu* expression is upregulated in sessile-growing bacteria specifically, in contrast to both stationary and exponential planktonic growth (52).

Torres et al. have described a pathogen-specific Ag43 homolog called Cah, which promotes the aggregation of enterohemorrhagic *E. coli* (59). It is possible that forms of Ag43 have been adapted in response to the severe challenge to colonization faced by enteric pathogens. Nataro has also identified a similar Ag43 homolog in the genome of enteroaggregative *E. coli* (E. Dudley and J. Nataro, unpublished data).

Type 1 fimbriae

Type 1 fimbriae are 7-nm-diameter adhesins expressed by the majority of isolates of the *Enterobacteriaceae*. They are subject to high-frequency phase variation, permitting the presence of rapid transition of a bacterial population from mostly off to mostly on (and vice versa) states. Many phenotypes have been attributed to type 1 fimbriae, and, to some extent, different alleles of the tip adhesin (FimH) provide the versatility to adapt the host bacterium to different mucosal niches. FimH mediates a ligand-receptor interaction of high affinity. However, in addition to a receptor-specific interaction, type 1 fimbriae have been implicated as strong bacterial aggregins. Their role in this process is not entirely clear. McCormick et al. performed a seminal experiment in which they incubated type 1 fimbria locked-on mutants with epithelial cells coated with an artificial mucus layer (44). Compared with wild-type bacteria, which penetrated the mucus layer efficiently, locked-on mutants were arrested at the top of the layer. Interestingly, wild-type bacteria were found to turn on fimbrial expression once they were in the mucus. This raises the intriguing possibility that one function of type 1 fimbria-mediated adherence may be to anchor *E. coli* within the mucus layer of the intestine. This could be mediated simply via interbacterial aggregation, resulting in the generation of larger particles that are retarded in entry into or movement through the viscous layer. If this hypothesis is true, it could be promoted by other nonspecific aggregation factors, including Ag43 and the curli adhesins.

If both type 1 fimbriae and Ag43 mediate aggregation, why should they be reciprocally regulated? One obvious possibility is that the two aggregins may not adapt the bacteria to the same niche. Indeed, Ag43 produces dramatic biofilms on abiotic surfaces, and this may yet turn out to be its principal role. The two adhesins may also be providing aggregation phenotypes at different stages of the colonization process. Lastly, it is possible that aggregation by type 1 fimbriae, although dramatic at the air-liquid interface in broth culture, may be superseded by adherence to mucosal surfaces in vivo. Dissecting the various contributions of specific aggregins and adhesins in the in vivo state will require focused experimentation.

Other aggregins

A growing number of other factors have been implicated in the aggregation phenotype, yet it is not known whether this is their principal contribution. Curli are thin, coiled aggregative fimbriae produced by *E. coli*, *Salmonella*, and other gram-negative enteric pathogens (13, 51, 65). Curli have been implicated in aggregation and biofilm formation (3), as well as in adherence to epithelial cells (58). However, curli expression is maximal at temperatures prevailing outside the host (11), raising questions about the in vivo role of these interesting structures.

DISPERSAL OF AGGREGATED BACTERIA

Dispersal of Biofilms

The dynamic interaction of bacteria with living surfaces can be difficult to study in real time and impossible to model accurately. Work with bacterial biofilms, initially discovered on abiotic surfaces, has suggested that bacterial adherence to surfaces is an extraordinarily complex phenomenon. Moreover, recent evidence has suggested that bacterial growth in association with mucosal surfaces may have many of the elements described for biofilms on abiotic surfaces. Thus, the study of colonization can benefit from a reading of the abundant and rapidly growing biofilm literature.

One of the truly astonishing observations made in the course of biofilm research is the description of a complex organization and morphology. The typical biofilm comprises mushrooms with water channels or void spaces interspersed (15, 20). The development of this organization is itself a complex phenomenon, which includes the requirement for interbacterial communication via quorum-sensing signals and probably physiological signals as well (15).

Biofilms generally exist in the face of forces that should promote bacterial detachment. Spontaneous detachment of bacteria from substrata has been divided into three distinct processes: erosion, sloughing, and abrasion. Erosion is the continual loss of single bacteria or small groups of bacteria from the surface of the biofilm. Stoodley et al. used digital time-lapse microscopy to characterize the detachment of small bacterial clusters from mixed-species biofilms in glass tubing flow cells (57). These experiments demonstrated that the erosion of large clusters represented a disproportionately large number of bacteria leaving the biofilm. Charaklis reported that the rate of erosion increases with biofilm thickness and fluid shear at the biofilm/bulk-liquid interface. Sloughing of the full-thickness biofilm is a more random, episodic process than is erosion. It may occur as a result of nutrient depletion within the biofilm and is most common in thick biofilms (10). Abrasion occurs when the biofilm is bombarded by particles in the fluid phase.

Detachment with colonization of noncontiguous sites appears to be a fundamental feature of bacterial biofilms and occurs at predictable rates depending on species. *Pseudomonas fluorescens* disperses and colonizes a surface over approximately 5 h, whereas *Vibrio harveyi* may accomplish this feat in as little as 2 h.

There is substantial evidence that the propensity to detach is an evolved feature whose adaptive characteristics may be to promote spread across the substratum, ensure the most appropriate biofilm thickness given the environmental conditions, and provide optimal biofilm architecture. Gilbert et al. and Allison et al. (1, 24) reported that the surface hydrophobicity of daughter cells formed at the biofilm-fluid interface was significantly lower than that of established sessile bacteria, predisposing the daughter to leave the biofilm. This feature may be a fundamental characteristic of many biofilms, permitting young to "leave the nest" and establish new distant sessile colonies.

Biofilms of *Pseudomonas aeruginosa* represent bacteria embedded in an alginate polysaccharide matrix (8, 9). Boyd and Chakrabarty overexpressed the enzyme alginate lyase from *P. aeruginosa* biofilms and demonstrated an increased rate of detachment (9), suggesting that control of the rate of detachment is naturally exercised via modulation of alginate lyase activity. In a parallel phenomenon, Dow et al. identified an enzyme in *Xanthomonas campestris* that mediates the detachment of the bacteria from biofilms and aggregates (21). Using an enzyme purification strategy, these investigators identified this enzyme as endo-β-1,4-mannanase and showed that it can digest *X. campestris* aggregates. A mutant with a mutation in the mannanase-encoding gene was deficient in its ability to spread along the vascular system of infected plants.

Several oral bacteria exhibit elegant biofilm dispersal phenotypes. Kaplan and coworkers have shown that biofilms of *Actinobacillus actinomycetemcomitans, Neisseria subflava, Haemophilus aphrophilus,* and two *Streptococcus* spp. (35–37, 53) release streamers of colonies from the biofilm in the direction of higher temperature (Fig. 1). These investigators showed that biofilms of *A. actinomycetemcomitans* feature internal lacunae, which harbor unaggregated cells destined for release from the biofilm. Once free, these released planktonic cells bind to distant sites, establishing new biofilm colonies (36). Using transposon mutagenesis, Kaplan et al. identified a β-hexosaminidase enzyme which mediated the detachment of *A. actinomycetemcomitans* from biofilms. When this purified enzyme was applied to preformed biofilms, an increased number of released bacteria was observed.

Hap Detachase of *V. cholerae*

Vibrio cholerae is perhaps the most intensively studied enteric pathogen. As early as 1947, Burnet and Stone described a secreted mucinase of *V. chol-*

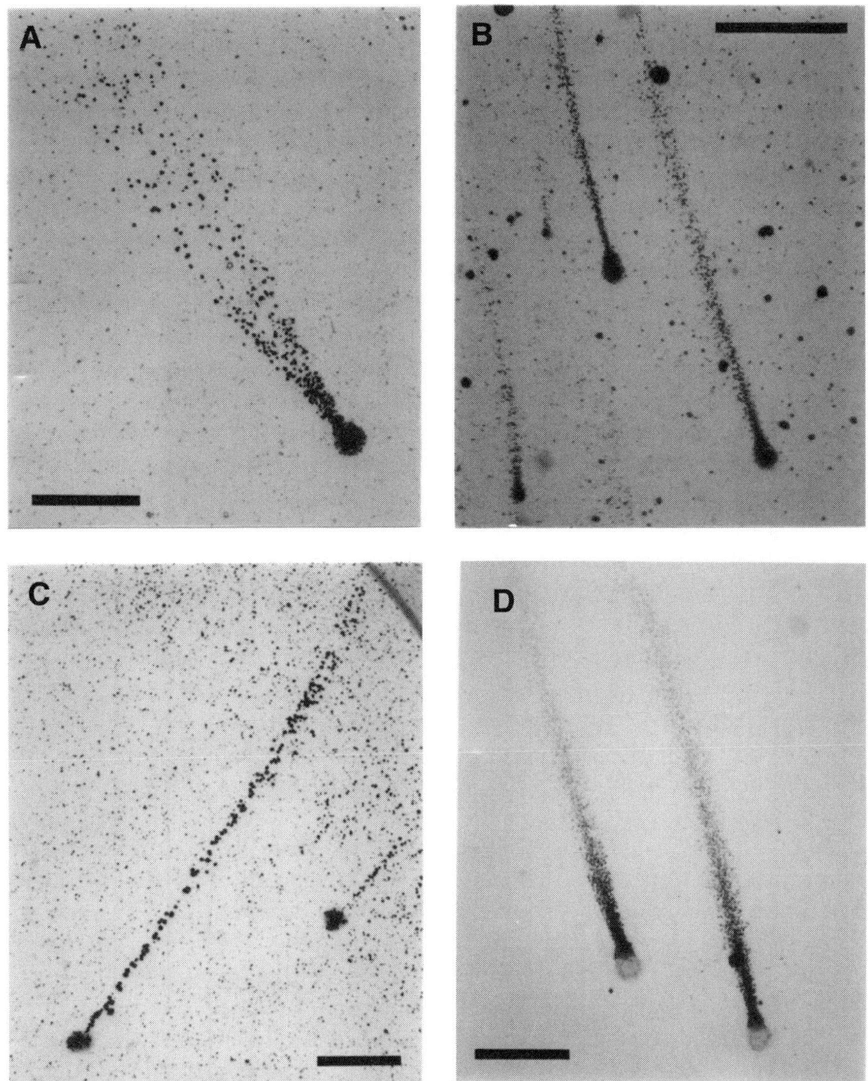

Figure 1. Dispersal of biofilm colonies of various oral bacteria growing attached to polystyrene petri dishes. The growing colonies release cells into the medium, and the released cells attach to the surface of the petri dish and form new colonies, enabling the biofilms to spread. The bacteria were stained with crystal violet. (A) *N. subflava;* (B) *H. aphrophilus;* (C) *A. actinomycetemcomitans;* (D) *Streptococcus mutans.* Bar, 0.5 μm in panel A and 2 μm in others. Panels A and C from reference 35. Courtesy of Dr. Jeffery Kaplan.

erae that caused disruption of the intestinal epithelium of guinea pigs. Although interest in this protein waned with the characterization of cholera toxin, Finkelstein was later to propose a role for this protease in cholera pathogenicity. In a classic experiment, the Finkelstein laboratory purified a 32-kDa secreted "hemagglutinin/lectin/protease," corresponding to the mucinase originally described by Burnet. These investigators showed that the protease cleaved fibronectin, lactoferrin, and ovomucin, in addition to cholera toxin A subunit and albumin. In addition, the protein was found to mediate hemagglutination of "responder" chicken erythrocytes. These investigators cited preliminary and inconsistent data (22) suggesting that this soluble protease and hemagglutinin could inhibit the attachment of vibrios to rabbit ileum. This protein, subsequently termed the cholera hemagglutinin protease (Hap), was shown by the same investigators to be a zinc metalloprotease (7).

Hase and Finkelstein cloned, sequenced, and mutated the *hap* gene from *V. cholerae* El Tor Ogawa strain 3083. The *hap* gene encodes a precursor protein of 609 amino acids; this protein undergoes cleavage of a signal sequence and propeptide during secretion to yield the mature zinc metalloprotease of 32 kDa (25). Interestingly, the predicted Hap gene product demonstrates 70% amino acid similarity to the elastase of *P. aeruginosa,* placing Hap within a family of bacterial proteases that includes enzymes from *Bacillus, Serratia,* and *Legionella* spp. (26).

With the mutation in hand, the Finkelstein laboratory set out to assess the role of Hap in cholera. Although constrained by imperfect models, these investigators hypothesized that the protease could promote the colonization of the gastrointestinal tract via roles in adherence to and/or detachment from the mucosa. In an influential study from 1992, Finkelstein showed that strain 3083*hap* was as virulent as its parent in the 6-day-old rabbit model. However, in in vitro assays of adherence to the human intestinal epithelial cell line Intestine-407, assessed by microscopic counting of bacteria attached to cells, these investigators observed a novel phenomenon. Over the first 15 min after addition of strain 3083 to cell monolayers, the number of adherent bacteria per cell increased rapidly, to approximately 50 per cell. After 15 min, however, there was a significant diminution in the number of adherent bacteria: to 25 per cell by 30 min and less than 20 per cell by 1 h. Interestingly, the numbers of adherent bacteria per cell continued to increase in the Hap mutant at 30 and 60 min, reaching a plateau of >80 vibrios per cell after 60 min of incubation. Strain 3083*hap* complemented with the *hap* gene expressed in the low-copy-number plasmid pACYC184 displayed the wild-type phenotype. In addition, pretreatment of cells with the purified Hap protease prior to addition of the bacteria resulted in diminution of the number of adherent bacteria per cell in a dose-dependent fashion, compared with that in untreated epithelial cells. The authors concluded from these studies that the Hap protease may cleave a surface receptor for *V. cholerae* and that the role of this phenotype may be to promote the detachment of adherent bacteria from the mucosa, which could in turn promote dispersal across the mucosa and/or shedding in the stool, thereby facilitating transmission to another host. Importantly, the observation of adherence followed by detachment of vibrios from the epithelium had been previously suspected as a result of experiments with animal models (32, 46).

Although the *hap* mutant had previously been shown not to possess reduced virulence in the rabbit model, a *hap* mutant of El Tor Ogawa (Δ*ctx zot ace orfU cep*) colonized the infant (3- to 5-day-old) CD1 mouse approximately fivefold more abundantly than did its parent by 16 h after inoculation (50). When inoculated simultaneously, the *hap* mutant was 1.5 $\pm$ 0.28 times more abundant than its parent in an intestinal homogenate (which represents both luminal and adherent organisms). Benitez et al. subsequently studied the behavior of *hap* mutants in an intestinal goblet cell monolayer (5). In this model, the *hap* mutants adhered better and grew faster than their parents. After 2 h of incubation with the goblet cells, 27% of adherent vibrios were reported to have detached in the parent strain 3083, compared with only 11% in strain 3083*hap*.

The data from models involving cells in culture suggest that Hap mediates the detachment of *V. cholerae* from its receptor and perhaps from the intestinal mucus layer. Silva et al. subsequently showed that *V. cholerae* strains expressing Hap are able to penetrate a column of mucus-agarose significantly faster than are *hap* mutants. These investigators showed that *hap* expression was activated by bile and repressed by glucose; penetration of the mucus column paralleled expression of the *hap* gene.

Data regarding the Hap protease of *V. cholerae* suggest that this factor could play multiple roles in virulence, yet data from animal models suggest that its role is probably related to colonization and dissemination. Several roles can be hypothesized. Studies with goblet cell lines and the mucus column suggest that the mucinase could facilitate penetration of the mucus layer and thereby accelerate the approach of the organism to the intestinal mucosa. However, if this role were to prevail, one would expect less efficient colonization by the *hap* mutant, which is not observed. Alternatively, Hap could facilitate the release of adhered vibrios from their receptors and penetration of the mucus blanket during escape from the mucosa. Released vibrios could spread laterally through the mucus gel, thereby promoting dispersal across the mucosa, or they could ascend out of the mucus layer and into the lumen. Once again, increased shedding with decreased colonization by the wild type is more consistent with the latter role. Also consistent with this model is recent evidence regarding the control of virulence genes in *V. cholerae.* Zhu et al. reported that at high cell density, as would prevail among vibrios adherent to the mucosa, the LuxO system induces repression of the toxin-coregulated pilus colonization factor and activation of Hap expression (63, 64). This sequence could result in simultaneous release of the bacterium from the mucosa (perhaps from bacterial aggregates) and acceleration of its release from the mucus layer. Testing of this model will require sophisticated observation of in vivo systems. An alternative function for the Hap protease is also suggested by experiments with biofilms. Since in the natural state *V. cholerae* may be ingested as biofilm-coated plankton, Hap could facilitate the release from this biofilm early in the infection process.

Dispersal by Type IV Fimbriae

Type IV fimbriae (Tfp) are surface organelles of gram-negative bacteria that mediate complex functions in colonization and adherence. They have been implicated in colonization processes in *V. cholerae*, enteropathogenic *E. coli* (EPEC), enterotoxigenic *E. coli*, *Neisseria gonorrhoeae*, and *P. aeruginosa*, in addition to several nonpathogenic species.

Genetic studies of Tfp suggest that expression of these adhesins requires at least 11 distinct genes. This observation suggests an obvious question: if adherence to cells can be mediated by simple fimbrial systems requiring only four or five genes, what additional benefit is provided by systems as complex as the type IV adhesins?

What are now termed type IV fimbriae were first characterized in the 1970s as polar organelles linked to gliding motility in *Myxococcus xanthus* (34). Myxobacteria move by gliding at liquid-solid or solid-air interfaces (reviewed in references 61 and 62). Bacterial motion is smooth and directed along the long axis of the cell. Tfp mediate the so-called social motility of *Myxococcus*, so designated because it requires bacterial contact. Motile cell clusters swarm in spearhead-like clusters of >50 cells; the bacteria are highly aligned, with their long axes in the direction of motion, and *Myxococcus* Tfp are localized at one pole of the bacterium. Many genes have been implicated in surface translocation by *Myxococcus*, including at least 10 required for Tfp assembly or function (62). Absence of the PilT protein resulted in loss of motility despite an apparently normal pilus.

Tfp have been detected in a large number of gram-negative bacteria, including several mucosal pathogens. In contrast to the gliding motility of *Myxococcus*, other bearers of Tfp demonstrate twitching motility, so called because the bacteria move in small intermittent jerks, often changing direction (42). The extent of motion can vary from extensive swarming behavior to small, barely perceptible jerking movements. Like gliding, twitching motility also requires a minimum quorum of bacteria. In the 1970s, Henrichsen published a series of articles correlating twitching motility with the presence of polar fimbriae in several species of bacteria (reviewed in reference 31); later experiments demonstrated Tfp in most of these strains. Moreover, twitching motility in several systems has been correlated with the phenotype of pilus retraction (33, 43, 45, 55), an energy-dependent process that is mediated by homologs of the *P. aeruginosa* PilT protein.

In *Pseudomonas*, the presence of twitching motility correlates with the ability of the organism to spread within an infected host. Zolfaghar et al. reported that *P. aeruginosa* strains deficient in Tfp production displayed reduced virulence in a mouse corneal-infection model (66). In that study, twitching motility contributed to the role of pili in corneal disease but was not involved in adherence to or invasion of corneal epithelial cells.

EPEC bundle-forming pilus

EPEC is a diarrheal pathogen whose pathogenetic scheme involves adherence to and colonization of the small bowel mucosa, followed by the development of a complex signal transduction cascade mediated largely via a chromosomally encoded type III secretion system (47). Signal transduction is accompanied by intimate adherence of the bacteria to the eukaryotic plasma membrane, followed by disruption of the cytoskeleton and induction of a net secretory state. Nearly all EPEC strains express the bundle-forming pilus (BFP), a member of the Tfp family (12). BFP is encoded by a cluster of 14 genes residing on a ca. 60-MDa virulence plasmid called EAF (56). BFP intertwine into structures that extend from the bacterium and confer attachment to other EPEC strains and to epithelial cells.

Several investigators have suggested that BFP mediate aggregation (interbacterial clumping in the liquid phase), agglutination (interbacterial adherence on the surface of cells), and anchoring of the aggregates to the cells, although this is mediated principally by the chromosomally encoded intimin protein (47). Anantha et al. reported that the *bfpF* gene within the BFP biogenesis operon was a homolog of *P. aeruginosa pilT;* these investigators hypothesized by analogy that *pilT*-induced retraction could be observed in EPEC and was associated with a functional *bfpF* gene product (2). The results of these experiments revealed that the *bfpF* mutant of strain E2348/69 formed larger bacterial aggregates in the liquid phase and adhered to epithelial cells in larger numbers. However, inconsistent with a role for *bfpF* in pilus retraction, the *bfpF* mutant did not exhibit increased pilus length or greater distance between bacteria and epithelial cells.

Bieber et al. extended these studies by constructing a *bfpF* mutant in EPEC strain B171 (6). In agreement with results obtained with strain E2348/69, these mutants were found to adhere in greater numbers to epithelial cells. In addition, these investigators reported a novel phenomenon: wild-type EPEC strains tended to form aggregates in the late logarithmic phase, followed by dispersal of the aggregates in stationary phase. In contrast, their *bfpF* mutant formed normal aggregates but failed to subsequently disperse. Microscopic observation of these aggregates revealed an increased number of interbacterial filaments in the *bfpF* mutant. Importantly, these investigators showed that the *bfpF* mutant was significantly attenuated in virulence in an adult volunteer model (11 of 13 volunteers developed diarrhea on wild-type challenge versus 4 of 13 who ingested the mutant; $P < 0.003$). Colonization of these volunteers was not rigorously quantitated, but subjects fed with B171*bfpF* were reported to be colonized with the challenge strain.

Knutton et al. extended these observations still further by describing the behavior of strain E2348/69 in the presence of HEp-2 epithelial cells (39). After 3 h of coincubation, wild-type E2348/69 adhered in localized microcolonies. After 6 h, however, microcolonies had dispersed and bacteria were distributed more diffusely over the cell surface. By immunofluorescence microscopy, BFP appeared in 40-nm-diameter pilus bundles, which mediated interbacterial agglutination and microcolony formation. After 3 h, however, BFP underwent a striking morphological alteration: the bundles became longer and thicker (approximately 100 nm in diameter) and were frequently aggregated laterally to form even thicker bundles arranged in a loose three-dimensional network. Notably, dispersal of EPEC from the microcolonies was associated with this BFP transformation from thin to thick bundles. Dispersal and transformation of BFP from thin to thick bundles did not occur with E2348/69*bfpF*.

Knutton et al. thus corroborated the observations of Beiber et al. and also explained the findings of Anantha et al. BfpF may mediate not pilus retraction but, rather, a pilus morphologic transformation that results in dispersal. Interestingly, Knutton et al. suggested a specific mechanism to explain the discrepancy: retraction of individual filaments could result in thickening of the bundle in the same way as pulling on a single fiber results in thickening of a twine. Regardless of the mechanism (which remains ambiguous), BFP is suggested to be involved in the dispersal of EPEC from the initial microcolony. The dispersal phase presumably facilitates colonization of distant sites on the intestinal mucosa. Proof of this model may await sophisticated human studies, since animal models do not require BFP for efficient colonization.

Neisseria meningitidis Tfp

Dispersal with an additional level of complexity has been reported for meningococci (MC), which also express a Tfp (49). MC interaction with epithelial cells comprises two fairly discrete steps. In the first phase, the bacteria form microcolonies on the cell surface; this phase features distinct interbacterial aggregates and is dependent on the Tfp. In the second phase, MC disperse and shed their pili, leading to a more diffuse pattern of adherence. In this second stage, microvilli have disappeared and the MC interact intimately with cells. MC can be observed to be mounted on pedestals extending from the cell surface, and condensed actin is prominent in these structures. This lesion is remarkably similar to the attaching-and-effacing phenotype described for EPEC, and thus it was predicted that the MC Tfp may play a role similar to that of BFP. Pujol et al. engineered a mutation in the cytoplasmic nucleotide binding protein PilT, a homolog of BfpF. The MC *pilT* mutant displayed increased piliation and adherence, similar to the phenotype observed for the EPEC *bfpF* mutant (49). Also similar to the BFP mutant, the PilT mutant did not disperse from the microcolony. Interestingly, however, PilT mutants did not progress to the stage of intimate attachment and attaching-and-effacing lesion formation. Rather, they remained adherent as piliated clumps at all times. Instead of a shortening and thickening of the fimbriae as seen with EPEC, MC shed their fimbriae completely and initiate a complex signal transduction cascade within the cell. The differences between the contributions and effects of BfpF and PilT are not clear. Both may exert their effects primarily via pilus retraction, leading to shortening or loss, respectively. However, PilT may exert an additional effect, perhaps directly participating in the initial stages of cell signaling. Beyond a potential direct role in signaling, PilT-mediated dispersal may free the bacteria from the initial aggregate so that they are free to interact directly with the cells rather than each other. Thus, one could envision an early aggregation phase promoting early colonization, followed by a second phase of cellular damage in which aggregation is more an impediment than an advantage.

EAEC Dispersin

Enteroaggregative *E. coli* (EAEC) is a diarrheal pathogen defined by its characteristic aggregative adherence to HEp-2 cells in culture; such adherence is usually accompanied by aggregation on the glass substratum. EAEC adheres to the intestinal mucosa in interbacterial aggregates, much like those of EPEC and MC described above. However, scanning electron microscopy of infected epithelium also demonstrates the presence of single bacteria scattered across the intestinal surface. A second phase of dispersal has

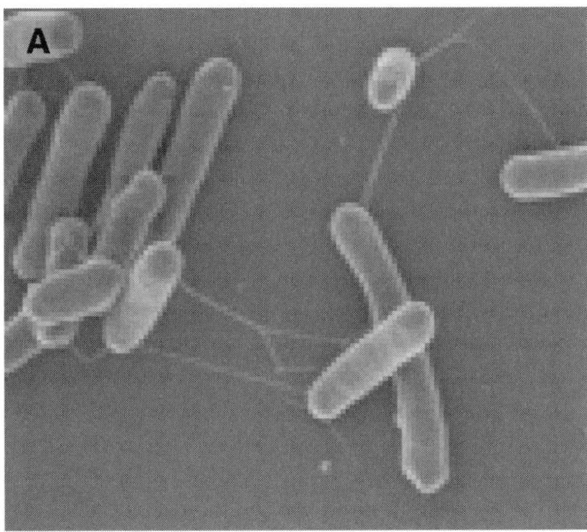

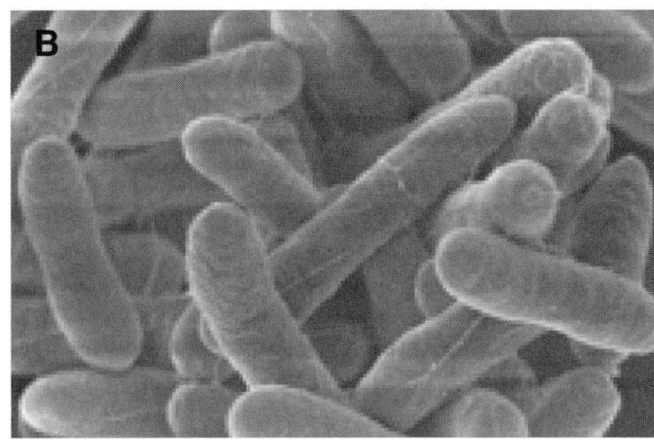

Figure 2. Mutation in the dispersin-encoding gene *aap* results in collapse of AAF/II fimbriae onto the surface of the bacterium, precluding interbacterial adherence at a distance. (A) 042. (B) 042*aap*.

not been reported. One of us (J.P.N.) has described a 10-kDa protein encoded by a gene (called *aap*) lying immediately upstream of that encoding the AggR transcriptional activator and under control of AggR (54). The product of *aap* has a typical signal sequence and is secreted to the extracellular milieu, where it remains noncovalently attached to the surface of the bacterium. *aap* mutants aggregate more intensely than the wild-type parent in a number of assays, forming larger aggregates and fewer individual bacteria in broth culture, on glass, and on the surface of cultured epithelial cells (Fig. 2). Infection of colonic biopsy specimens with wild-type EAEC strain 042 and its *aap* mutant revealed more dramatic autoagglutination of the mutant than of the wild-type parent. Our data suggest that the *aap* gene product participates in the formation of a surface coat which acts to disperse the bacteria, thus partially counteracting AAF-mediated aggregation. We have accordingly named the *aap* gene product "dispersin." The mechanism of the function of dispersin in EAEC is unclear. Dispersin mutants are deficient in colonization of the streptomycin-treated mouse model, suggesting that EAEC mutants exhibiting collapsed fimbriae and hyperaggretation are not able to establish colonization of the intestine.

One possible explanation for the role of dispersin is suggested from biophysical studies. AAF/II fimbriae assume a net positive charge at physiologic pH (the pI of the fimbrial subunit is 10.0). Presumably, expression of this flexible surface organelle in the vicinity of the negatively charged lipopolysaccharide would result in electrostatic attraction of the fimbriae to the surface of the bacterium. Dispersin, expressed before or concurrently with AAF/II, could neutralize the cell surface and permit the AAF to avoid electrostatic attraction. We have recently solved the solution structure of dispersin (Velarde and J. P. Nataro, unpublished data); the structure has a slight net positive surface charge conferred by a number of exposed lysine side chains. We envision a model in which dispersin is translocated to the bacterial cell surface and subsequently binds noncovalently to the lipopolysaccharide, partially masking or neutralizing the negative surface charge; the fimbriae are then free to splay outward from the cell surface.

Other Dispersin Mechanisms

The ability of bacteria to modulate their adhesion and agglutination phenotypes may be more widespread than described above. Preliminary evidence suggests that *Bordetella pertussis* filamentous hemagglutinin may execute this function after release from the surface of the bacterium and subsequent readherence to other bacteria in the colonizing community (F. Jacob-Dubuison, personal communication).

CONCLUSIONS AND FUTURE DIRECTIONS

Although each bacterium interacting with a mucosal surface must modify its colonization approach to fit its particular life-style and survival strategy, comparative biology continues to indentify new conserved general functions such as aggregation and its counterpoint, dispersal. The search for these activities among other microorganisms at mucosal surfaces is likely to yield new virulence factors and functions.

Acknowledgments. Work in the Nataro laboratory is funded by U.S. Public Health Service grants AI033096 and AI043615.

REFERENCES

1. **Allison, D. G., D. J. Evans, M. R. Brown, and P. Gilbert.** 1990. Possible involvement of the division cycle in dispersal of *Escherichia coli* from biofilms. *J. Bacteriol.* **172:**1667–1669.
2. **Anantha, R. P., K. D. Stone, and M. S. Donnenberg.** 1998. Role of BfpF, a member of the PilT family of putative nucleotide-binding proteins, in type IV pilus biogenesis and in interactions between enteropathogenic *Escherichia coli* and host cells. *Infect. Immun.* **66:**122–131.
3. **Austin, J. W., G. Sanders, W. W. Kay, and S. K. Collinson.** 1998. Thin aggregative fimbriae enhance *Salmonella enteritidis* biofilm formation. *FEMS Microbiol. Lett.* **162:**295–301.
4. **Bassler, B. L.** 2002. Small talk. Cell-to-cell communication in bacteria. *Cell* **109:**421–424.
5. **Benitez, J. A., R. G. Spelbrink, A. Silva, T. E. Phillips, C. M. Stanley, M. Boesman-Finkelstein, and R. A. Finkelstein.** 1997. Adherence of *Vibrio cholerae* to cultured differentiated human intestinal cells: an in vitro colonization model. *Infect. Immun.* **65:**3474–3477.
6. **Bieber, D., S. W. Ramer, C. Y. Wu, W. J. Murray, T. Tobe, R. Fernandez, and G. K. Schoolnik.** 1998. Type IV pili, transient bacterial aggregates, and virulence of enteropathogenic *Escherichia coli. Science* **280:**2114–2118.
7. **Booth, B. A., M. Boesman-Finkelstein, and R. A. Finkelstein.** 1983. *Vibrio cholerae* soluble hemagglutinin/protease is a metalloenzyme. *Infect. Immun.* **42:**639–644.
8. **Boyd, A., and A. M. Chakrabarty.** 1995. *Pseudomonas aeruginosa* biofilms: role of the alginate exopolysaccharide. *J. Ind. Microbiol.* **15:**162–168.
9. **Boyd, A., and A. M. Chakrabarty.** 1994. Role of alginate lyase in cell detachment of *Pseudomonas aeruginosa. Appl. Environ. Microbiol.* **60:**2355–2359.
10. **Brading, M. G., J. Jass, and H. M. Lappin-Scott.** 1995. Dynamics of bacterial biofilm formation, p. 46–63. *In* H. M. Lappin-Scott and J. W. Costerton (ed.), *Microbial Biofilms.* Cambridge University Press, Cambridge, United Kingdom.
11. **Brown, P. K., C. M. Dozois, C. A. Nickerson, A. Zuppardo, J. Terlonge, and R. Curtiss, 3rd.** 2001. MlrA, a novel regulator of curli (AgF) and extracellular matrix synthesis by *Escherichia coli* and *Salmonella enterica* serovar Typhimurium. *Mol. Microbiol.* **41:**349–363.
12. **Cleary, J., L. C. Lai, R. K. Shaw, A. Straatman-Iwanowska, M. S. Donnenberg, G. Frankel, and S. Knutton.** 2004. Enteropathogenic *Escherichia coli* (EPEC) adhesion to intestinal epithelial cells: role of bundle-forming pili (BFP), EspA filaments and intimin. *Microbiology* **150:**527–538.
13. **Collinson, S. K., S. C. Clouthier, J. L. Doran, P. A. Banser, and W. W. Kay.** 1996. *Salmonella enteritidis agfBAC* operon encoding thin, aggregative fimbriae. *J. Bacteriol.* **178:**662–667.
14. **Costerton, J. W., and P. S. Stewart.** 2000. Biofilms and device-related infections, p. 423–439. *In* J. P. Nataro, M. J. Blaser, and S. Cunningham-Rundles (ed.), *Persistent Bacterial Infections.* American Society for Microbiology, Washington, D.C.
15. **Costerton, J. W.** 1999. Introduction to biofilm. *Int. J. Antimicrob. Agents* **11:**217–221, 237–239.
16. **Costerton, J. W., Z. Lewandowski, D. E. Caldwell, D. R. Korber, and H. M. Lappin-Scott.** 1995. Microbial biofilms. *Annu. Rev. Microbiol.* **49:**711–745.
17. **Costerton, J. W., P. S. Stewart, and E. P. Greenberg.** 1999. Bacterial biofilms: a common cause of persistent infections. *Science* **284:**1318–1322.
18. **Costerton, W., R. Veeh, M. Shirtliff, M. Pasmore, C. Post, and G. Ehrlich.** 2003. The application of biofilm science to the study and control of chronic bacterial infections. *J. Clin. Investig.* **112:**1466–1477.
19. **Danese, P. N., L. A. Pratt, S. L. Dove, and R. Kolter.** 2000. The outer membrane protein, antigen 43, mediates cell-to-cell interactions within *Escherichia coli* biofilms. *Mol. Microbiol.* **37:**424–432.
20. **Donlan, R. M., and J. W. Costerton.** 2002. Biofilms: survival mechanisms of clinically relevant microorganisms. *Clin. Microbiol. Rev.* **15:**167–193.
21. **Dow, J. M., L. Crossman, K. Findlay, Y. Q. He, J. X. Feng, and J. L. Tang.** 2003. Biofilm dispersal in *Xanthomonas campestris* is controlled by cell-cell signaling and is required for full virulence to plants. *Proc. Natl. Acad. Sci. USA* **100:**10995–11000.
22. **Finkelstein, R. A., and L. F. Hanne.** 1982. Purification and characterization of the soluble hemagglutinin (cholera lectin) produced by *Vibrio cholerae. Infect. Immun.* **36:**1199–1208.
23. **Frick, I. M., M. Morgelin, and L. Bjorck.** 2000. Virulent aggregates of *Streptococcus pyogenes* are generated by homophilic protein-protein interactions. *Mol. Microbiol.* **37:**1232–1247.
24. **Gilbert, P., D. J. Evans, and M. R. Brown.** 1993. Formation and dispersal of bacterial biofilms in vivo and in situ. *J. Appl. Bacteriol.* **74**(Suppl):67S–78S.
25. **Hase, C. C., and R. A. Finkelstein.** 1991. Cloning and nucleotide sequence of the *Vibrio cholerae* hemagglutinin/protease (HA/protease) gene and construction of an HA/protease-negative strain. *J. Bacteriol.* **173:**3311–3317.
26. **Hase, C. C., and R. A. Finkelstein.** 1990. Comparison of the *Vibrio cholerae* hemagglutinin/protease and the *Pseudomonas aeruginosa* elastase. *Infect. Immun.* **58:**4011–4015.
27. **Hasman, H., T. Chakraborty, and P. Klemm.** 1999. Antigen-43-mediated autoaggregation of *Escherichia coli* is blocked by fimbriation. *J. Bacteriol.* **181:**4834–4841.
28. **Henderson, I. R., M. Meehan, and P. Owen.** 1997. Antigen 43, a phase-variable bipartite outer membrane protein, determines colony morphology and autoaggregation in *Escherichia coli* K-12. *FEMS Microbiol. Lett.* **149:**115–120.
29. **Henderson, I. R., M. Meehan, and P. Owen.** 1997. A novel regulatory mechanism for a novel phase-variable outer membrane protein of *Escherichia coli. Adv. Exp. Med. Biol.* **412:**349–355.
30. **Henderson, I. R., and P. Owen.** 1999. The major phase-variable outer membrane protein of *Escherichia coli* structurally resembles the immunoglobulin A1 protease class of exported protein and is regulated by a novel mechanism involving Dam and *oxyR. J. Bacteriol.* **181:**2132–2141.
31. **Henrichsen, J.** 1983. Twitching motility. *Annu. Rev. Microbiol.* **37:**81–93.
32. **Jones, G. W., G. D. Abrams, and R. Freter.** 1976. Adhesive properties of *Vibrio cholerae:* adhesion to isolated rabbit brush border membranes and hemagglutinating activity. *Infect. Immun.* **14:**232–239.
33. **Kaiser, D.** 2000. Bacterial motility: how do pili pull? *Curr. Biol.* **10:**R777–R780.
34. **Kaiser, D.** 1979. Social gliding is correlated with the presence of pili in *Myxococcus xanthus. Proc. Natl. Acad. Sci. USA* **76:**5952–5956.
35. **Kaplan, J. B., and D. H. Fine.** 2002. Biofilm dispersal of *Neisseria subflava* and other phylogenetically diverse oral bacteria. *Appl. Environ. Microbiol.* **68:**4943–4950.
36. **Kaplan, J. B., M. F. Meyenhofer, and D. H. Fine.** 2003. Biofilm growth and detachment of *Actinobacillus actinomycetemcomitans. J. Bacteriol.* **185:**1399–1404.
37. **Kaplan, J. B., C. Ragunath, N. Ramasubbu, and D. H. Fine.** 2003. Detachment of *Actinobacillus actinomycetemcomitans* biofilm cells by an endogenous beta-hexosaminidase activity. *J. Bacteriol.* **185:**4693–4698.

38. **Kjaergaard, K., M. A. Schembri, H. Hasman, and P. Klemm.** 2000. Antigen 43 from *Escherichia coli* induces inter- and intraspecies cell aggregation and changes in colony morphology of *Pseudomonas fluorescens*. *J. Bacteriol.* **182**:4789–4796.
39. **Knutton, S., R. K. Shaw, R. P. Anantha, M. S. Donnenberg, and A. A. Zorgani.** 1999. The type IV bundle-forming pilus of enteropathogenic *Escherichia coli* undergoes dramatic alterations in structure associated with bacterial adherence, aggregation and dispersal. *Mol. Microbiol.* **33**:499–509.
40. **Mah, T. F., B. Pitts, B. Pellock, G. C. Walker, P. S. Stewart, and G. A. O'Toole.** 2003. A genetic basis for *Pseudomonas aeruginosa* biofilm antibiotic resistance. *Nature* **426**:306–310.
41. **Maira-Litran, T., A. Kropec, D. Goldmann, and G. B. Pier.** 2004. Biologic properties and vaccine potential of the staphylococcal poly-*N*-acetyl glucosamine surface polysaccharide. *Vaccine* **22**:872–879.
42. **Mattick, J. S.** 2002. Type IV pili and twitching motility. *Annu. Rev. Microbiol.* **56**:289–314.
43. **McBride, M. J.** 2001. Bacterial gliding motility: multiple mechanisms for cell movement over surfaces. *Annu. Rev. Microbiol.* **55**:49–75.
44. **McCormick, B. A., P. Klemm, K. A. Krogfelt, R. L. Burghoff, L. Pallesen, D. C. Laux, and P. S. Cohen.** 1993. Escherichia coli F-18 phase locked 'on' for expression of type 1 fimbriae is a poor colonizer of the streptomycin-treated mouse large intestine. *Microb. Pathog.* **14**:33–43.
45. **Merz, A. J., M. So, and M. P. Sheetz.** 2000. Pilus retraction powers bacterial twitching motility. *Nature* **407**:98–102.
46. **Nelson, E. T., J. D. Clements, and R. A. Finkelstein.** 1976. *Vibrio cholerae* adherence and colonization in experimental cholera: electron microscopic studies. *Infect. Immun.* **14**:527–547.
47. **Nougayrede, J. P., P. J. Fernandes, and M. S. Donnenberg.** 2003. Adhesion of enteropathogenic *Escherichia coli* to host cells. *Cell. Microbiol.* **5**:359–372.
48. **Ochiai, K., T. Kurita-Ochiai, Y. Kamino, and T. Ikeda.** 1993. Effect of co-aggregation on the pathogenicity of oral bacteria. *J. Med. Microbiol.* **39**:183–190.
49. **Pujol, C., E. Eugene, M. Marceau, and X. Nassif.** 1999. The meningococcal PilT protein is required for induction of intimate attachment to epithelial cells following pilus-mediated adhesion. *Proc. Natl. Acad. Sci. USA* **96**:4017–4022.
50. **Robert, A., A. Silva, J. A. Benitez, B. L. Rodriguez, R. Fando, J. Campos, D. K. Sengupta, M. Boesman-Finkelstein, and R. A. Finkelstein.** 1996. Tagging a *Vibrio cholerae* El Tor candidate vaccine strain by disruption of its hemagglutinin/protease gene using a novel reporter enzyme: *Clostridium thermocellum* endoglucanase A. *Vaccine* **14**:1517–1522.
51. **Romling, U., Z. Bian, M. Hammar, W. D. Sierralta, and S. Normark.** 1998. Curli fibers are highly conserved between *Salmonella typhimurium* and *Escherichia coli* with respect to operon structure and regulation. *J. Bacteriol.* **180**:722–731.
52. **Schembri, M. A., L. Hjerrild, M. Gjermansen, and P. Klemm.** 2003. Differential expression of the *Escherichia coli* autoaggregation factor antigen 43. *J. Bacteriol.* **185**:2236–2242.
53. **Schreiner, H. C., K. Sinatra, J. B. Kaplan, D. Furgang, S. C. Kachlany, P. J. Planet, B. A. Perez, D. H. Figurski, and D. H. Fine.** 2003. Tight-adherence genes of *Actinobacillus actinomycetemcomitans* are required for virulence in a rat model. *Proc. Natl. Acad. Sci. USA* **100**:7295–7300.
54. **Sheikh, J., J. R. Czeczulin, S. Harrington, S. Hicks, I. R. Henderson, C. Le Bouguenec, P. Gounon, A. Phillips, and J. P. Nataro.** 2002. A novel dispersin protein in enteroaggregative *Escherichia coli*. *J. Clin. Investig.* **110**:1329–1337.
55. **Skerker, J. M., and H. C. Berg.** 2001. Direct observation of extension and retraction of type IV pili. *Proc. Natl. Acad. Sci. USA* **98**:6901–6904.
56. **Stone, K. D., H. Z. Zhang, L. K. Carlson, and M. S. Donnenberg.** 1996. A cluster of fourteen genes from enteropathogenic *Escherichia coli* is sufficient for the biogenesis of a type IV pilus. *Mol. Microbiol.* **20**:325–337.
57. **Stoodley, P., Z. Lewandowski, J. D. Boyle, and H. M. Lappin-Scott.** 1999. The formation of migratory ripples in a mixed species bacterial biofilm growing in turbulent flow. *Environ. Microbiol.* **1**:447–455.
58. **Sukupolvi, S., R. G. Lorenz, J. I. Gordon, Z. Bian, J. D. Pfeifer, S. J. Normark, and M. Rhen.** 1997. Expression of thin aggregative fimbriae promotes interaction of *Salmonella typhimurium* SR-11 with mouse small intestinal epithelial cells. *Infect. Immun.* **65**:5320–5325.
59. **Torres, A. G., N. T. Perna, V. Burland, A. Ruknudin, F. R. Blattner, and J. B. Kaper.** 2002. Characterization of Cah, a calcium-binding and heat-extractable autotransporter protein of enterohaemorrhagic *Escherichia coli*. *Mol. Microbiol.* **45**:951–966.
60. **Waldron, D. E., P. Owen, and C. J. Dorman.** 2002. Competitive interaction of the OxyR DNA-binding protein and the Dam methylase at the antigen 43 gene regulatory region in *Escherichia coli*. *Mol. Microbiol.* **44**:509–520.
61. **Wall, D., and D. Kaiser.** 1999. Type IV pili and cell motility. *Mol. Microbiol.* **32**:1–10.
62. **Wolgemuth, C., E. Hoiczyk, D. Kaiser, and G. Oster.** 2002. How myxobacteria glide. *Curr. Biol.* **12**:369–377.
63. **Zhu, J., and J. J. Mekalanos.** 2003. Quorum sensing-dependent biofilms enhance colonization in *Vibrio cholerae*. *Dev. Cell* **5**:647–656.
64. **Zhu, J., M. B. Miller, R. E. Vance, M. Dziejman, B. L. Bassler, and J. J. Mekalanos.** 2002. Quorum-sensing regulators control virulence gene expression in *Vibrio cholerae*. *Proc. Natl. Acad. Sci. USA* **99**:3129–3134.
65. **Zogaj, X., W. Bokranz, M. Nimtz, and U. Romling.** 2003. Production of cellulose and curli fimbriae by members of the family *Enterobacteriaceae* isolated from the human gastrointestinal tract. *Infect. Immun.* **71**:4151–4158.
66. **Zolfaghar, I., D. J. Evans, and S. M. Fleiszig.** 2003. Twitching motility contributes to the role of pili in corneal infection caused by *Pseudomonas aeruginosa*. *Infect. Immun.* **71**:5389–5393.

Colonization of Mucosal Surfaces
Edited by James P. Nataro et al.

Chapter 19

Signal Transduction in the Intestinal Mucosa

BETH A. MCCORMICK

The epithelial cells that line the gastrointestinal tract are the front line of defense against the diverse population of commensal and potentially pathogenic bacteria that flourish within the lumen of the intestine. Not surprisingly, there exists considerable communication between cells of the intestinal epithelium and the underlying immune system. The microbes of the healthy human intestine provide some benefit, especially by generating important metabolites and by reducing the colonization efficiency of dietary pathogens by occupying their potential niches. However, such commensal microbes have the potential to cause disease, principally on gaining systemic access to the host. There are also communication networks between host cells and pathogens that seek to breach the intestinal mucosa. In fact, many microbial pathogens have evolved the capacity to engage their host cells in very complex interactions commonly involving the exchange of biochemical signals. On detection of such pathogens, the intestinal epithelium coordinates a complex mucosal inflammatory response designed to eliminate all potentially dangerous foreign entities. Although much of this bacterium-epithelial and immune cell cross talk maintains the well-being of the intestinal mucosa, many pathogens have developed strategies to exploit mucosal immune responses to their own benefit. This chapter considers the remarkable signal transduction networks that have evolved between intestinal microbes and their host in trying to maintain the balance of health and disease.

BENEFICIAL ROLE OF COMMENSAL BACTERIA IN THE INTESTINAL MUCOSA

The microflora of the intestinal microenvironment as a unit provides important protective, metabolic, and trophic functions. Resident bacteria serve a central line of resistance to colonization by exogenous microbes and thus assist in preventing the potential invasion of the intestinal mucosa by an incoming pathogen. This protective function is known as the barrier effect or colonization resistance and plays a number of important roles. In one role, adherent nonpathogenic bacteria can often prevent attachment and subsequent entry of suspected pathogens into epithelial cells (11). In another, commensal bacteria compete for nutrient availability in ecological niches and, as such, maintain their collective microenvironment by administering and consuming all resources. In this case, the host actively provides nutrients needed by the microflora, and, reciprocally, the microflora actively indicates how much it needs to the host (49). This mutual and beneficial relationship helps to dampen unwanted overproduction of nutrients, which could potentially support the intrusion of microbial competitors with a potential pathogenic outcome for the host.

Perhaps the major metabolic function of the colonic microflora is the fermentation of nondigestible carbohydrates, which is a key source of energy in the colon (86). Such nondigestible carbohydrates include large polysaccharides (resistant starches, pectins, and cellulose), some oligosaccharides that escape digestion, and unabsorbed sugars and alcohols. The primary metabolic end point of such fermentation is the generation of short-chain fatty acids. A fundamental role of short-chain fatty acids on colonic physiology is their trophic effect on the intestinal epithelium. All three major short-chain fatty acids (proprionate, acetate, and butyrate) stimulate epithelial cell proliferation and differentiation in the colon in vivo, whereas butyrate inhibits cell proliferation and stimulates cell differentiation in vitro (31). Therefore, short-chain fatty acids appear to play an essential role in the control of epithelial cell proliferation and differentiation

Beth A. McCormick • Department of Pediatric Gastroenterology and Nutrition, Massachusetts General Hospital, and Department of Microbiology and Molecular Genetics, Harvard Medical School, Boston, MA 02129.

in the colon. Another important trophic factor is the communication network that exists between the host and the resident microflora at the mucosal interface, which channels the development of a competent immune system. There is a clear indication that microbial colonization of the gastrointestinal tract affects the composition of the gut-associated lymphoid tissue (GALT). As one example, mice bred under germfree conditions have low densities of lymphoid cells in the gut mucosa, possess smaller specialized follicles, and have considerably lower circulating concentrations of immunoglobulins in the blood (101). In contrast, immediately following exposure to luminal microbes, the number of intraepithelial lymphocytes is significantly increased (102), germinal centers with immunoglobulin-producing cells rapidly appear in follicles as well as in the lamina propria (15), and concentrations of immunoglobulins increase substantially in the serum (13). Not surprisingly, the interplay between the GALT and the microflora early in life is also crucial for the appropriate development of the complex mucosal and systemic immunoregulator circuitry (98).

Furthermore, the intimate relationship between the mucosal microflora residents, the intestinal epithelium, and the GALT is involved in sculpting the memory mechanisms of systemic immunity, such as oral tolerance. This was initially recognized by the discovery that the systemic response to a specific pathogen can be abrogated after ingestion of the antigen; this effect continues for several months in conventionally colonized mice, whereas systemic unresponsiveness persists for only a few days in germfree mice (76). Therefore, the innate immune system must be able to discriminate between potential pathogens and commensal bacteria. One way that this is achieved is by the mammalian cell expression of Toll-like receptors (TLR), which recognize conserved motifs on bacteria that are not present in higher eukaryotes (1). As described below, this innate strategy allows for the immediate recognition of bacteria to rapidly respond to a potential threat. Therefore, the unique dialogue that has developed between the bacteria of the normal flora and the epithelium with its associated GALT is critical to promoting not only development but also homeostasis of the immune system.

ROLE OF COMMENSAL BACTERIA IN INDUCING INTESTINAL INFLAMMATION

Resident members of the bacterial flora have been implicated as an essential factor in driving the inflammatory response which underlies human inflammatory bowel disease. The most compelling evidence that normal luminal bacteria induce inflammatory bowel disease is based on the collective findings from at least 11 distinct models of induced or spontaneous intestinal inflammation; in these models, no disease occurs in the absence of bacteria (i.e., in a germfree or sterile state) (95). One of the first animal models developed, which demonstrated the link between normal flora residence and the ability to promote colitis was in the interleukin-10 (IL-10) knockout mouse (63). In this particular model system, IL-10 knockout mice in a germfree environment fail to produce intestinal inflammation by clinical, histological, or immunological criteria but do develop cecum-predominant colitis within 1 week of colonization with specific-pathogen-free fecal bacteria. By 5 weeks after bacterial colonization, this disease progresses to aggressive, transmural inflammation characterized by mucosal ulcers and crypt abscesses (90). What is more, the inflammation is confined to the colon, which is the site of the highest level of bacterial colonization.

While the animal models of mucosal inflammation reflect a wide array of causes, the resulting inflammation that develops is almost always directed into a common pathway of inflammation, mediated by either an excessive Th-1 T-cell receptor response associated with an increase in IL-12 gamma interferon, and tumor necrosis factor alpha (TNF-α) secretion or an up-regulation of the Th-2 T-cell response associated with an increase in IL-4 and IL-5 secretion (96). However, the explicit role played by the resident microflora in the induction of mucosal inflammation remains an important unanswered question. Several lines of evidence suggest that the antigen in the microflora, at least in part, drives such mucosal inflammation, presumably because with the onset of the colitis, tolerance to many antigens of the microflora is lost. Substantiating this concept is the observation that mouse lamina propria T cells are usually unresponsive to their own microbial flora but are responsive to the microflora of other individuals even if the other individual mouse is of the same genetic background (22). A related observation is that mice with trinitrobenzene sulfonic acid (TNBS)-induced colitis lose their nonresponsiveness to their own flora and regain it when the colitis resolves (22). Furthermore, in patients with Crohn's disease, intestinal T cells are hyperreactive, suggesting that local tolerance has been broken (107). This patient population also has increased intestinal mucosal secretion of immunoglobulin G antibodies against a broad spectrum of commensal bacteria (97) and has larger numbers of bacteria attached to epithelial cell surfaces compared to healthy individuals (101).

Another way in which the antigens in the normal microbial flora may direct mucosal inflammation is by provoking a loss of balance between induction of effector and regulatory cells. To illustrate this point, studies performed with mice bearing a STAT4 transgene show that in vitro exposure of T cells from mice with an increased tendency to undergo Th1 T-cell differentiation in response to autologous microflora antigens induces in these T cells the capacity to cause a Th1 colitis in SCID recipients (108). Although this example reflects a direct demonstration that T cells specific for the mucosal microflora act as effector cells, it should be noted that regulatory cells could also be stimulated (108). In sum, the mucosal microflora plays a critical role in mediating experimental mucosal inflammatory disease by providing the major stimulus for the induction of effector T cells that cause the inflammation. At present, no single bacterial antigen has been shown to be responsible for this stimulation, although obviously some bacteria may be more important than others. A continued search for a particular organism and/or antigen that triggers inflammatory bowel disease remains an important objective of many investigators.

ACTIVATION OF NF-κB

One of the primary means by which gut epithelial cells activate expression of the genes involved in host defense mechanisms is via the activation of the proinflammatory transcription factor NF-κB. A number of enteric pathogens and some opportunistic commensal bacteria possess the means to provoke NF-κB activation and, subsequently, intestinal inflammation. The different mechanism by which microbes stimulate epithelial cell NF-κB activation are discussed and include (i) receptor-mediated detection of soluble extracellular ligands, (ii) direct activation of host signaling proteins by bacterial products translocated into the host cytoplasm, and (iii) detection of intracellular organisms by intracellular receptors (Fig. 1).

NF-κB is composed of two subunits, both heterodimers and homodimers, of various members of the Rel family of DNA-binding transcription factors [e.g., p50/RelA(p65)] (62). Such NF-κB dimers are held in an inactivated state by an inhibitor of κB (IKB), which binds NF-κB and sequesters it in the cytosol. On generation of the appropriate set of signals, IKB is phosphorylated; this triggers their polyubiquitination and proteasomal degradation, an event that results in NF-κB translocation to the nucleus and activation of NF-κB-responsive genes. IKB degradation is preceded by a phosphorylation step mediated by IKK (IκB kinase) (20), a large-subunit complex that is generally viewed as the key rate-limiting step in the activation of this pathway. A wide variety of signaling pathways converge into IKK activation, and for this reason IKK serves as a central integrator of many signaling pathways (Fig. 1).

The IKK complex is composed of multiple subunits, and its exact composition has yet to be defined. The main catalytic subunits of the complex are IKKα and IKKβ, which recognize a conserved 6-amino-acid motif present on all IKB isoforms and phosphorylate two serine residues within the motif (20, 58). Activation of the NF-κB signaling via IKKβ has two major physiological functions: the activation of genes whose products are important for signaling and mediating innate immune defense and inflammation, and the activation of genes whose products are critical for the suppression of apoptosis. Ultimately, however, degradation of IKB results in exposure of NF-κB nuclear localization signals, leading to translocation across the nuclear membrane, DNA binding to NF-κB elements, and eventual transcription of the genes (20, 58). Therefore, although enteric pathogens use different strategies and mechanisms to interact with intestinal epithelial cells, NF-κB appears to a be central regulator of intestinal epithelial innate defense. In this way, the diverse signals that are activated by enteric pathogens can be integrated into a common signaling pathway, culminating in the activation of a conserved set of proinflammatory genes in infected host cells. While NF-κB is required for activation of the host innate defense, other signal transduction pathways are also important, most notably mitogen-activated protein (MAP) kinase pathways (47).

PATHOGEN RECOGNITION BY INTESTINAL EPITHELIAL CELLS: EXTRACELLULAR SIGNALING MECHANISMS

Innate epithelial defense mechanisms provide a rapid response whereby microbial pathogens in the host are quickly detected and signals are generated that activate mucosal antimicrobial defense mechanisms. In essence, the intestinal epithelium functions as a central dispatch system that receives signals from pathogenic microbes and their products and conveys important messages to the cells in the underlying lamina propria. Orchestration of this biological detector involves a number of sophisticated strategies that have coevolved among the microbes and the host epithelial cells, which are the primary targets of infection. The surfaces of microbes are composed of complex macromolecules, often with a number of repeated structural motifs. Although microbes can

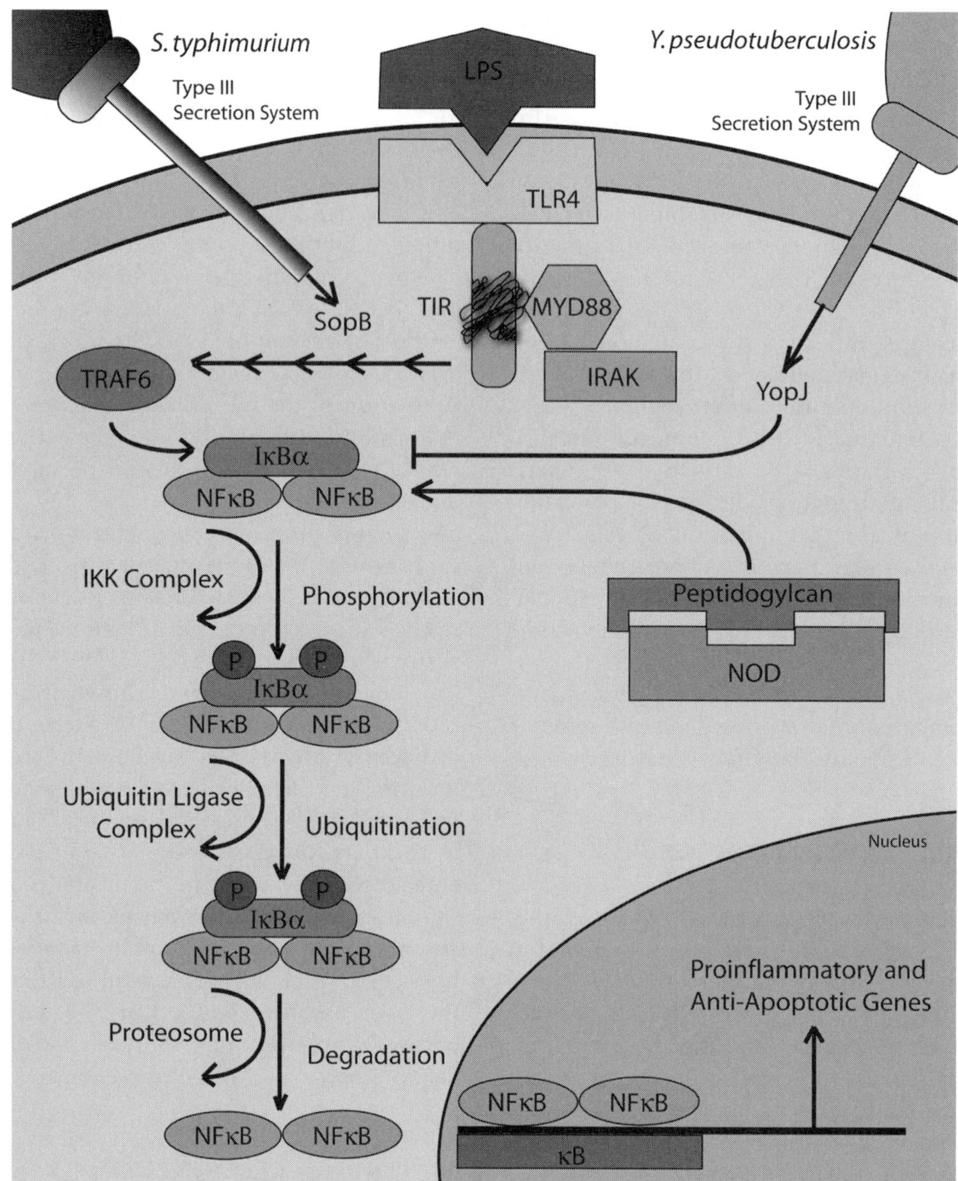

Figure 1. Activation of the NF-κB pathway can be induced by a variety of bacterial constituents. In unstimulated cells, NF-κB is sequestered in the cytoplasm by IκB. Bacterial components such as LPS selectively bind to TLR4, while components of the peptidoglycan selectively interact with intracytoplasmic Nods (Nod1 or Nod2) to initiate signaling that sets in motion a series of enzymatic modifications of IκB, such as phosphorylation, ubiquitination, and degradation. Loss of IκB allows NF-κB to translocate to the nucleus, bind to the promoters of many proinflammatory effector genes, and activate the proinflammatory program. Many of these cellular events are also caused by bacterial effector proteins, which are delivered into the intestinal cells (by type III secretion systems) and directly modulate the activities of host cell proteins (e.g., SopB from *S. enterica* serovar Typhimurium). Perturbation of any of these enzymatic steps (such as with YopJ from *Yersinia*) could inhibit the entire pathway.

demonstrate tremendous variability, certain structural features permit relatively little variation in their structure to maintain their function. Such critical microbial structural motifs have been termed pathogen-associated molecular patterns of (PAMPs) (74). Examples of PAMPs include lipopolysaccharide (LPS), peptidoglycans and lipoteichoic acid (74), bacterial flagellin (38), and nonmethylated bacterial DNA (45) and double-stranded RNA (2).

PAMPs are recognized by receptors called pattern recognition receptors. One family of PRR, TLR, is expressed by intestinal epithelial cells (14). Essentially, TLR are involved in the extracellular detection of microbes. Biochemically, TLR are transmembrane receptors defined by the presence of leucine-rich repeats (LRR) in the extracellular portion of the molecule and a TIR (Toll-interleukin 1R) cytoplasmic domain. The extracellular LRR domain is essential for ligand recognition, whereas the TIR domain functions in signaling (48). Recent progress has been made in identifying ligands for TLR. Principally, TLR2 is involved in recognizing a broad range of PAMPs such as peptidoglycans and bacterial lipoproteins from gram-positive bacteria, mycobacterial cell wall components, atypical LPS from certain gram-negative bacteria, and yeast cell wall components (4, 103), whereas TLR4 is involved mostly in recognition of LPS (100). TLR3, which is also expressed by intestinal epithelial cell lines, mediates responses to double-stranded RNA (2). Other TLR family members include TLR5, which is involved in recognition of bacterial flagellin (38), and TLR9, which mediates responses to nonmethylated, CpG-containing bacterial DNA (45). Imparting yet another tier of sophistication, TLR can also exist as dimers, and recent evidence indicates that responses to some ligands (e.g., soluble tuberculosis factor and *Borrelia burgdorferi* outer surface protein A lipoprotein) require a specific TLR combination, thus greatly increasing both the specificity and potential ligands for TLR activation (12).

Signaling through TLR activates genes whose products play a key role in further activating or mediating innate defense mechanisms. TLR signaling ultimately leads to the activation of NF-κB and MAP kinase. Binding of an appropriate ligand to a TLR results in the sequential recruitment of the adaptor protein MyD88 and the protein kinase IL-1 receptor-associated kinase (IRAK) to the TIR cytoplasmic domain on the TLR (Fig. 1). IRAK, a serine kinase, then activates the cytoplasmic intermediate TRAF6, with eventual downstream activation of NF-κB and the MAP kinase cascades (Fig. 1) (19). The consequences of cellular activation through TLR have been studied most extensively by using nonepithelial cell types such as macrophages and dendritic cells. However, engagement of TLR5 by bacterial flagellin may perhaps play a key functional role in innate immune defense by activating NF-κB (34, 36, 37).

TLR must not be routinely activated by the resident microflora, since to do so would result in a constant state of gut inflammation. How, then, is the host able to tolerate the presence of the mucosal microflora while at the same time activating an innate immune and inflammatory response to enteric pathogens? One mechanism by which epithelial cells may discriminate between pathogens and nonpathogens is by restricting the bacterial sensor to the basolateral membrane of epithelial cells. As an example, intestinal epithelial cells express TLR5 only on their basolateral surfaces (38). In this manner, presentation of a flagellin receptor restricted to the basolateral membrane of epithelial cells would maintain the apical pole of the epithelial cells, which is in contact with the normal flora, in a refractory state. Therefore, only bacteria that have successfully breached the epithelia (see below) or translocated flagellin across the epithelia will activate this receptor. Whether other TLR share a similar mechanism of restricting activation awaits further investigation. Another means by which the epithelial cells might retain the epithelium in a state of equilibrium is through an auxiliary mechanism designed to desensitize TLR. This concept is based largely on the fact that intestinal epithelial cells are refractory to LPS (36). A likely explanation, which is still the subject of much debate, is that intestinal epithelial cells may not express certain TLR or may not effectively signal through some of these receptors (i.e., intestinal epithelial cells do not appear to express CD14 and MD2, molecules important for signaling through TLR4).

DIRECT ACTIVATION OF HOST SIGNALING PROTEINS BY BACTERIAL PRODUCTS TRANSLOCATED INTO THE HOST CYTOPLASM

Many pathogens possess a specific set of virulence factors that affect the host, in addition to having the ability to invade. As a consequence, the host develops specialized strategies to resist such infection. It is this cross talk between enteric pathogens and their intestinal host that permits the induction of inflammation and development of disease, and it has been widely investigated. Many of these cellular events are caused by bacterial effector proteins, which are delivered into intestinal cells and directly modulate the activities of host cell proteins. The effector proteins are secreted and translocated into host cells via bacterial type III secretion systems (Fig. 1). This type of bacterial communication network has been most extensively studied for enteropathogenic *Escherichia coli* (EPEC), enterohemorrhagic *E. coli* (EHEC), *Salmonella enterica* serovars, *Shigella* species, and *Yersinia* species.

By way of example, the *S. enterica* serovar Typhimurium effector protein, SopB, activates small GTP-binding proteins (112) that presumably lead to the activation of MAP kinase cascades (47). While a direct signaling link between these bacterial effector proteins and the NF-κB pathway has not been demonstrated under physiologic conditions, it is likely that mechanisms of activation of host signals exist and play a role in the activation of this important proinflammatory transcription factor. By a different means, EPEC uses the type III secretion system to activate tyrosine kinase pathways and to deliver its own receptor into host cells (61). The intimate interaction between EPEC and the intestinal epithelium causes the induction of phosphate fluxes within the host cells, as well as the activation of protein kinase C (PKC), phospholipase C, and NF-κB (104). Although the activation of NF-κB correlated with increased IL-8 production, it remains unclear whether these responses are specifically triggered by bacterial effector proteins or are a nonspecific consequence of EPEC infection resulting in pedestal formation.

PATHOGEN RECOGNITION BY INTESTINAL EPITHELIAL CELLS: INTRACELLULAR SIGNALING

The ability of epithelial cells to discriminate between pathogens and nonpathogens may also involve the subcellular localization of the recognition system. A newly discovered family of proteins, the nucleotide-binding site (NBS)-LLR proteins, is involved in intracellular recognition of microbes and their products (41). NBS-LRR proteins are characterized by three structural domains: a C-terminal LRR domain able to sense a microbial motif, an intermediary NBS essential for the oligomerization of the molecule that is necessary for the signal transduction induced by different N-terminal motifs (such as a pyrin domain), and a caspase-activating and recruitment domain (CARD) (41).

Two of these family members, Nod1 and Nod2, play a role in the regulation of proinflammatory pathways through NF-κB induced by bacterial ligands (52), and the bacterial motifs detected by both Nod1 and Nod2 have been identified (39). These molecules both sense peptidoglycan through the detection of different *N*-acetylglucosamine (GlcNAc) and Nacetylmuramic (MurNAc) acid peptide fragments (i.e., muropeptides). However, Nod1 detects the naturally occurring peptidoglycan degradation product GlcNAc-MurNAc-L-Ala-γ-D-Glu-*meso*-DAP (GM-Tri-$_{DAP}$) (39), whereas Nod2 senses the muramyl dipeptide MurNAc-L-Ala-D-iso-Gln, also known as MDP (40). As a consequence, Nod2 is considered a general sensor of peptidoglycans from gram-negative and gram-positive bacteria because MDP corresponds to the minimal motif found in all peptidoglycans. On the other hand, the peptidogylcan motif detected by Nod1, GM-Tri-$_{DAP}$, is found primarily in peptidoglycans from gram-negative bacteria (39), suggesting a specific role for Nod1 in innate immune responses to these bacteria. Biochemical investigation into the signaling pathways induced downstream of Nod1 showed that oligomerization of this molecule is sufficient to induce the recruitment of Rip2 (an adaptor molecule with homology to IRAK) through homophilic CARD-CARD interactions (53). The subsequent interaction between Nod1 and Rip2 ultimately leads to activation of the NF-κB pathway (Fig. 1).

Several human genetic disorders have been linked to mutations in the genes encoding some of the NBS-LRR family members and associated proteins. The CARD15/Nod2 gene is located on chromosome 16q12, which has been identified by linkage analysis as the susceptibility locus *IBD1* for Crohn's disease (51). Many patients with Crohn's genes inherit germ line mutations in the CARD15/Nod2 gene that result in an altered protein (81). More than 90% of the mutated alleles associated with Crohn's disease affect the LRR region, suggesting a crucial role of this domain in the development of inflammatory bowel disease. The most common disease-associated variant is insertion of cytosine at nucleotide 3020, resulting in a mutant CARD15/Nod2 with truncated LRRs. This variant is nonresponsive to a preparation containing bacterial LPS and peptidoglycan (51), implying that the impairment of regulation of the host-microorganism interaction may contribute to the development of Crohn's disease.

REGULATION OF EPITHELIAL TIGHT JUNCTIONS BY ENTERIC PATHOGENS

Epithelial cells that line the intestine are uniquely positioned to serve as a direct line of communication between the immune system and the external environment. In their normal state, mucosal surfaces of the alimentary tract are exposed on the luminal surface to high concentrations of foreign antigens and, at the same time, are intimately associated with the immune system via subepithelial lymphoid tissue (9). Thus, the epithelium forms an important barrier, preventing the free mixing of luminal antigenic material with the lamina propria, which harbors the mucosal immune system. This ability is attributable to intercellular tight junctions, present at gaskets, which circumferentially join epithelial cells at their apices. Epithelial cell tight junctions are dynamic structures that control the passive permeation of hydrophilic solutes through the paracellular space via a highly regulated process (66).

The intercellular tight junction is a complex multiprotein structure that is composed of both intracellular and membrane-spanning proteins (Fig. 2). The intracellular complex of the tight junction-associated proteins includes zonula oocludens protein 1 (ZO-1), ZO-2, ZO-3, cingulin, 7H6, symplekin, and ZA-1TJ,

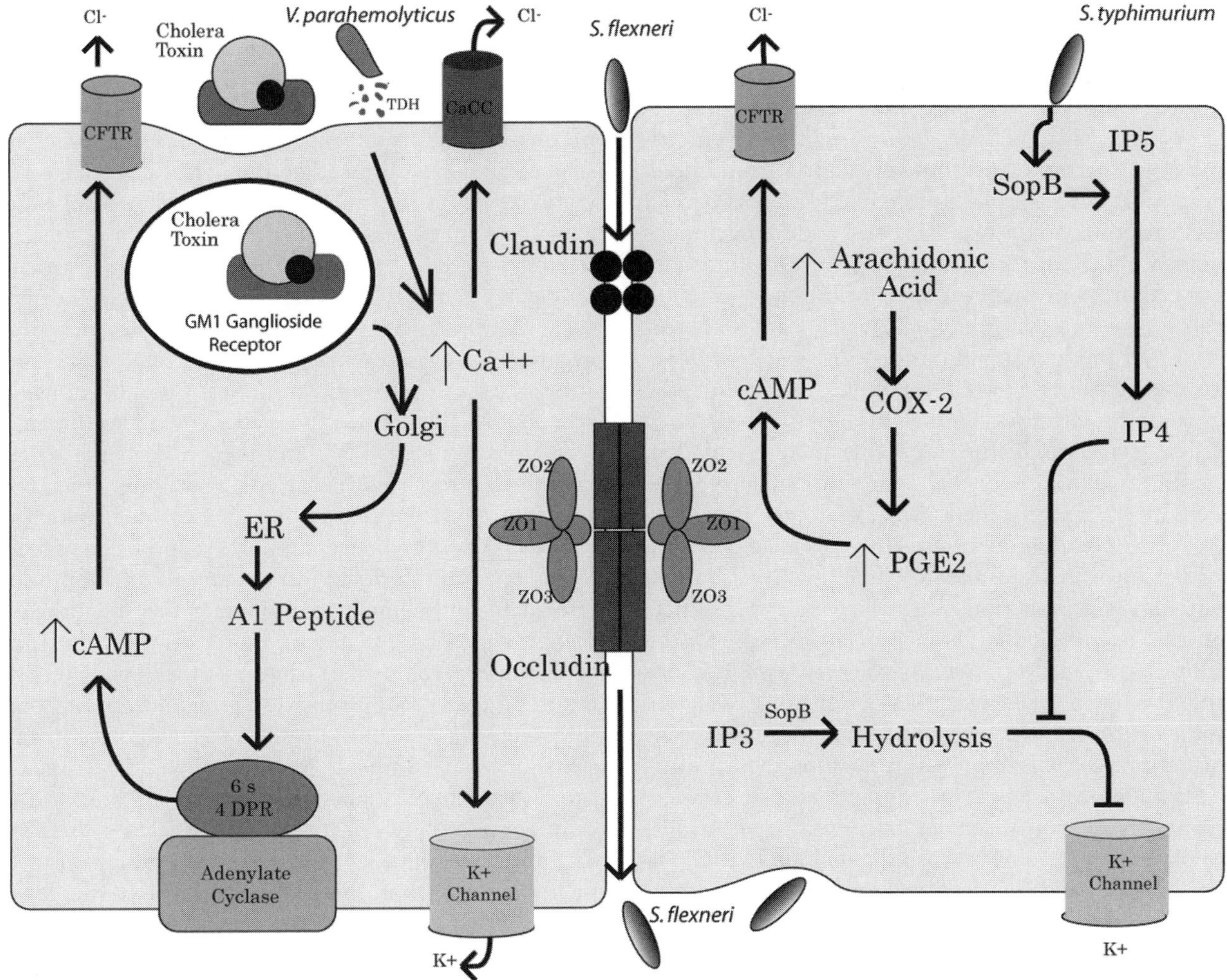

Figure 2. Intestinal bacteria have evolved different strategies to induce chloride secretion and regulate the tight-junction complex. Cholera toxin binds to the ganglioside receptor, GM_1, and enters epithelial cells as an AB_5 complex by retrograde trafficking through the Golgi and endoplasmic reticulum (ER). Dissociation and cleavage of the A subunit results in the A1 peptide-mediated ADP-ribosylation of $G_s\alpha$. This results in sustained activation of adenylate cyclase and elevation of the cAMP concentration, which in turn increase electrogenic chloride secretion. Through a different pathway, TDH of *V. parahemolyticus* elevates the intracellular Ca^{2+} concentration, resulting in the activation of CaCC. Phosphorylated inositol derivatives are also involved in regulating Ca^{2+}-mediated chloride secretion and have stimulatory or inhibitory effects. As shown here, the *S. enterica* serovar Typhimurium intracellular SopB protein affects inositol phosphate signaling events. One such event is the transient increase in the concentration of Ins(1,4,5,6)P_4 (IP4), which antagonizes the closure of chloride channels, influencing net electrolyte transport and hence fluid secretion. Infection of epithelial cells also results in the production of PGs such as PGE_2; this elevates cAMP levels, which can lead to further Cl^- secretion. The epithelial tight junction is a macromolecular structure consisting of both transmembrane-spanning proteins, such as occludin, and a number of claudin isoforms. This complex provides a barrier to the paracellular space, preventing free access of bacteria or their products to the underlying compartment. Pathogens, however, have developed strategies to disrupt the tight-junction barrier. *S. flexneri,* for example, can modulate the function of the tight-junction components in a manner which allows passage of the organism through the paracellular space, which is particularly relevant for the ability of *Shigella* to infect the colon.

in addition to several signaling molecules (7, 110). At present, four integral tight junction membrane proteins have been identified: occludin, claudins, junctional adhesion molecule, and coxsackievirus and adenovirus receptor. Occludin and the claudins interact with the PDZ domains of ZO-1 and ZO-2 (33, 54). The latter are membrane-associated guanylate kinase-like homologues, which may play a general role in creating and maintaining specialized membrane domains by cross-linking multiple integral membrane proteins at the cytoplasmic surface of plasma membranes (6, 56, 109). ZO-1 and ZO-2 bind directly to actin filaments at their COOH-terminal regions, enabling them to function as cross-linkers between TJ strands and actin filaments (29, 54, 80, 109).

A limited number of pathogens and virulence factors have been reported to be able to affect the paracellular permeability of intestinal epithelial cells. Breakdown of the tight junction is caused by bacterial products such as *Clostridium perfringens* enterotoxin (94), cytotoxic necrotizing factor 1 from *E. coli* (84), *Clostridium difficile* toxins (42), *Bordetella* dermonecrotic toxin (50, 91), and zonula occludens toxin from *Vibrio cholerae* (30). Several microbes also elaborate proteases to disrupt the tight-junction seal. In this category, the *Bacteroides fragilis* enterotoxin and the hemagglutinin protease of *V. cholerae* are the best-characterized examples. Moreover, the invasion of intestinal epithelial cells by *S. enterica* serovar Typhi can disturb the tight-junction seal (16), and EPEC can also dephosphorylate and dissociate occludin from tight junctions (92).

The outcome most frequently associated with a perturbation in tight-junction function (i.e., barrier function) is a reduction in transepithelial resistance, but this occurs in the absence of an increase in the chloride current. At present, of the organisms demonstrated to decrease transepithelial resistance, only the studies with EPEC have lead to some mechanistic information. EPEC employs three major mechanisms to stimulate a decrease in transepithelial resistance: (i) an increase in intracellular calcium concentrations, (ii) phosphorylation of myosin light chain (MLC) via MLC kinase, and (iii) dephosphorylation of occludin (89, 92). It is proposed that the increased calcium concentration acts, in part, by activation of MLC kinase. This kinase phosphorylates MLC and subsequently prompts the contraction of the perijuctional actomyosin ring, located beneath the tight junction, and in this manner increases the tension on the tight junction to reduce transepithelial resistance. The specific phosphatase that mediates the dephosphorylation of occludin, leading to its dissociation from the tight junction, has yet to be identified. However, the EPEC-induced decrease in transepithelial resistance has been linked to the expression of the EPEC-secreted protein, EspF (27). In fact, a direct relationship between the amount of EspF produced and the extent of the reduction in transepithelial resistance has been demonstrated. While the mechanism by which EspF exerts its effect on host cell function is unknown, this secreted protein contains three proline-rich repeats, suggesting that there is high potential for interacting with host cell proteins.

The reason why enteric pathogens have evolved the capacity to disrupt the highly regulated tight-junction complex is coming to light. An emerging concept is that such pathogens have engineered different strategies to gain access to basolateral membrane receptors since the intestinal epithelial receptors for many enteric pathogens are restricted to the basolateral membrane. Thus, impairment of tight-junction proteins can either allow microbes to traverse the paracellular pathway or disrupt the polar distribution of membrane proteins such that basolaterally restricted molecules relocalize the apical domain; each tactic results in the ability of pathogenic microbes to gain access to cognate host receptors (73).

As a specific example, *Shigella flexneri* is an enteric pathogen that lacks the ability to bind to the apical surface of host epithelial cells. After ingestion, the organism progresses along the intestinal tract and colonizes the colon and rectum. It is generally considered that shigellae cross the epithelial barrier through microfold (M) cells (77), and then once at the basolateral pole, the bacteria are either available for invasion into epithelial cells or ingested by macrophages. However, new evidence supports a model in which *S. flexneri* can interfere with the apical pole of model intestinal epithelium and modulate the function of components of tight junctions, resulting in the specific temporal removal of claudin 1, the dephosphorylation of occludin, and the down-regulation of ZO-1 (88) (Fig. 2). Concomitant with these changes in the tight-junction complex, *S. flexneri* rapidly translocates through the tight-junction seal into the paracellular space between intestinal epithelial cells (88). The ability of shigellae to disrupt the epithelial tight-junction complex in this manner may be particularly important for the infection of the colon, in which the number of M cells is markedly reduced in comparison with the number in the small intestine.

ACTIVATION OF CHLORIDE SECRETORY PATHWAYS BY ENTERIC PATHOGENS

Another important function of the intestinal epithelium is the maintenance of mucosal hydration. Mucosal surfaces, such as the lining of the intestine, maintain hydration in part as a result of the ion transport process known as electrogenic chloride secretion. With chloride secretion, paracellular movement of sodium follows, and the resulting accumulation of luminal sodium choride provides an osmotic gradient for the diffusion of water. The secretion of water and electrolytes is a highly regulated system and is a key physiologic process of mucosal surfaces (10). Important human diseases result when the ability to mount chloride secretion is either impaired (as in cystic fibrosis) or enhanced (as in secretory diarrhea). Chloride secretion in the intestinal mucosa involves the collaborative effort of several transporters (Fig. 2). Chloride enters the basolateral membrane via the Na/K/2Cl cotransporter (10). Energy for this transport mechanism is provided principally by an inward-directed sodium gradient; potassium channels, also situated at the basolateral membrane, provide the exit route for transported K^+. The apically located cyclic AMP (cAMP)-dependent cystic fibrosis transmembrane conductance regulator (CFTR) is responsible for the majority of apical chloride secretion (26). There are also calcium-activated chloride channels (CaCC), which secrete chloride in response to increased intracellular calcium levels (60). Various enteric pathogens elicit a chloride secretory response by stimulating either one of these two apical secretory channels.

Numerous pathogens, such as *V. cholerae, E. coli, Salmonella, Campylobacter jejuni,* and *Shigella dysenteriae,* mediate chloride secretion in a cAMP-dependent manner. This is commonly accomplished through the elaboration of enterotoxins of the AB_5 family (75). These toxins consist of an A subunit noncovalently bound to five identical B polypeptides (46). Cholera toxin, elaborated by *V. cholerae,* is the prototype and best characterized of these toxins. To induce disease, cholera toxin is released into the intestinal lumen and enters the intestinal epithelial cell at the apical membrane, where it eventually activates epithelial adenylyl cyclase at the cytoplasmic surface of the basolateral membrane (Fig. 2). This event leads to the massive intestinal salt and water secretory response characteristic of cholera, as discussed below. Interestingly, the microbe *V. cholerae* does not invade the intestinal mucosa or directly assist the delivery of toxin into the cell cytoplasm by other mechanisms (such as by type III secretion) (64). Thus, the molecular determinants that drive the entry of cholera toxin into the host intestinal epithelial cell are encoded entirely within the structure of the fully assembled and folded protein toxin itself.

The cholera toxin holotoxin initially binds to ganglioside GM_1 in the apical membrane; the binding specificity is determined by the B subunit. After endocytosis, the cholera toxin-GM_1 complex traffics retrograde through Golgi cisternae into the lumen of the endoplasmic reticulum, where the A1 peptide, the catalytically active portion, is unfolded and dissociated from the B pentamer. The unfolded A1 peptide is probably dislocated to the cytosol through the Sec61p complex. Once in the cytosol, the A1 peptide is transported to the basolateral membrane, where the Gs/adenylate cyclase complex is located. ADP-ribosylation of the heterotrimeric GTPase Gsα by the toxin produces an increase in the amount of intracellular AMP. The resulting increase in the amount of intracellular cAMP fosters the activation of cAMP-dependent protein kinase (Fig. 2) and subsequently CFTR, the outcome of which is the stimulation of electrogenic chloride secretion and massive diarrhea (87).

The CaCC apically located channels are activated by increases in cytosolic calcium levels. The underlying mechanism is initiated by the binding of a ligand to a receptor, which results in the activation of the membrane-associated phosphilipase C (32). This enzyme hydrolyses phosphatidylinositol-4,5-biphosphate releasing inositol-1,4,5-triphosphate and diacylglycerol (DAG). Inositol-1,4,5-triphosphate acts to increase intracellular calcium levels, whereas DAG activates PKC. Intracellular calcium levels are generally very low and are under tight physiological control. In fact, changes in calcium concentration are transient even in the continued presence of agonist. Several neurohormonal substances can increase the intracellular calcium level, but at present only a few microorganisms (*Vibrio parahaemolyticus* and rotavirus) with this capacity have been identified. *V. parahaemolyticus* is the leading cause of gastroenteritis due to the consumption of seafoods worldwide. A protein secreted by *V. parahemolyticus,* known as thermostable direct hemolysin (TDH) toxin, is not only the main pathogenic determinant associated with this organism but also the factor responsible for promoting an increase in intracellular calcium levels (99) (Fig. 2). In particular, TDH has been shown to cause intestinal fluid secretion, cytotoxicity, and a dose-dependent increase in intracellular calcium levels when applied to the luminal surface of colonic epithelial cells. Regulation of these channels is a complex process that appears to involve phosphorylation by the calmodulin-dependent protein kinase CaMKII and possibly PKC (99).

HOST-DERIVED FACTORS THAT INCREASE CHLORIDE SECRETION IN RESPONSE TO ENTERIC INFECTION

In addition to bacterial toxins that function directly as secretagogues, bacterial infection of host intestinal epithelial cells, through paracrine or autocrine effects, can stimulate chloride secretion (e.g., galanin, nitric oxide, and cyclooxygenase-2). Galanin is a neuropeptide that is widely distributed in enteric nerve terminals lining the gastrointestinal tract (44). This neuropeptide modulates intestinal motility and induces chloride secretion via the activation of the galanin-1 receptor (59, 60). Most recently, galanin has been distinguished as a mediator of intestinal ion secretion in response to infection by enteric bacterial pathogens. For instance, the enteric pathogens *S. enterica* serovar Typhimurium, *S. flexneri,* EHEC, EPEC, and enterotoxigenic *E. coli* (ETEC) all increase the expression of colonocyte galanin-1 receptor, whose activation is largely responsible for the augmented colonic fluid secretion. By contrast, normal commensal organisms failed to cause an increase in galanin-1 receptor expression (68). Therefore, galanin-1 receptor up-regulation may represent a novel mechanism accounting for the increased colonic fluid secretion noted in infectious diarrhea due to several different pathogens.

Another host-derived bacterial induced secretory agent is nitric oxide (NO). NO serves several important functions in the intestinal epithelium, including regulation of barrier function and antimicrobial activity (3, 28). Recently it has been shown that NO is capable of stimulating intestinal epithelial chloride secretion by increasing intracellular cGMP levels. Increased intracellular cGMP levels can lead to phosphorylation and activation of CFTR primarily by cGMP-dependent protein kinase II, resulting in chloride and bicarbonate secretion. Perhaps not surprisingly, certain enteric pathogens have evolved the ability to regulate chloride secretion in this manner. Bacterial infection, principally by *Salmonella, Shigella,* and enteroinvasive *E. coli* (EIEC), of human colon epithelial cells also rapidly up-regulates inducible nitric oxide synthase and NO production (85). This suggests that NO and/or its redox products can be important components of the intestinal epithelial cell response to microbial infection. Indeed, infection by these pathogens was shown to increase intracellular cGMP levels and electrogenic chloride secretion. However, the ability to induce this response appears to be dependent on the invasive capacity of the organism, since nonpathogenic as well as noninvasive enteric pathogens do not share this property (85).

Increased intestinal fluid secretion is an important host innate defense mechanism, and, coupled with intestinal motility, it plays an important role in flushing the enteric bacteria from the intestinal tract. Infection of the intestinal epithelial cells with *Salmonella* spp. results in increased epithelial cell chloride secretion, generated presumably through at least two different pathways. One pathway involves the up-regulated expression of cyclooxygenase-2 with a subsequent increase in production of the prostanoid prostaglandin E_2 (PGE_2) (Fig. 2). PG are formed from free arachidonic acid through the conversion of arachidonic acid to PGH, which is catalyzed by the enzyme PGH synthase (PGHS). PGH is thereafter converted by specific synthases to PGE, PGF, thromboxanes, or prostacyclins (23, 93). Several studies have implicated prostaglandins in *Salmonella*-induced enteropathogenic responses that may influence chloride secretory pathways. One study determined that pretreatment of rabbits with indomethacin, an inhibitor of prostaglandin biosynthesis, abolishes fluid secretion in ligated loops (106). Further, PGE_1 stimulates chloride secretion in polarized cells and PGE analogues were found to induce diarrhea in vivo. Another study evaluated the role of the intestinal epithelium in the secretory response after infection with *Salmonella* (25). In this investigation, infection of intestinal epithelial cells with *Salmonella* resulted in a rapid up-regulation in PGHS expression that subsequently stimulated the production of PGE_2 and PGF_{2a} and consequently led to an increase in Cl^- secretion (25). Therefore, PGE_2 released by intestinal epithelial cells in response to *Salmonella* infection acts in an autocrine/paracrine manner and stimulates chloride secretion by exploiting the cAMP-mediated pathway.

A second avenue leading to increased chloride secretion occurs through the Ca^{2+}-mediated pathway. Underlying this mechanism, *Salmonella* induces a rapid increase in D-myo-inositol-1,4,5,6-tetrakisphosphate [Ins(1,4,5,6)P_4] levels in model intestinal epithelial cells. This molecule antagonizes the epidermal growth factor-induced inhibition of calcium-mediated chloride secretion. Fundamental to this process is the *Salmonella* protein SopB, a protein secreted into host cells by the type III secretion apparatus. The SopB protein is an inositol polyphosphate phosphatase that is capable of hydrolyzing phosphates from both phosphoinositol-containing phospholipids and phosphorylated inositol sugars (79), resulting in the increased cellular levels of Ins(1,4,5,6)P_4 (Fig. 2). This, in turn, prevents inhibition by epidermal growth factor of a basolateral potassium channel that is essential for Ca^{2+}-activated chloride secretion. Hence, this is an example of a bacterial protein that can directly counteract a host cell inhibitory pathway, with the net result being increased chloride secretion.

PATHOGEN-INDUCED EPITHELIAL ORCHESTRATION OF THE INFLAMMATORY RESPONSE

Pathogen-induced changes in epithelial gene expression encompass a remarkable assortment of effects. Arguably, the central feature of this response is the induction of NF-κB genes that directly result in the recruitment of inflammatory cells, the militia of the innate immune system. Such immune cell recruitment, especially that of neutrophils, is mediated by epithelial expression of adhesion molecules and chemokines, both of which are NF-κB regulated. Chemokine secretion is absolutely critical to the recruitment of neutrophils, which migrate from the vascular endothelium across the intestinal mucosa to the intestinal lumen. A consequence of such neutrophil infiltration is the formation of an intestinal crypt abscess, a histopathologic finding characteristic of acute inflammation of the intestine. While the histology of this response has long been observed, only relatively recently has the mechanism by which enteric pathogens elicit this acute response begun to be appreciated. The molecular and cellular events involved in this complex phenomenon have been most extensively studied vis-à-vis the intestinal epithelial response to *S. enterica* serovar Typhimurium infection.

The current paradigm postulates that intestinal epithelial cells respond to *S. enterica* serovar Typhimurium by the polarized release of distinct proinflammatory chemoattractants, which sequentially orchestrate neutrophil movement across the intestinal epithelium (Fig. 3). In particular, interactions between *S. enterica* serovar Typhimurium and intestinal epithelial cells induce the epithelial synthesis and

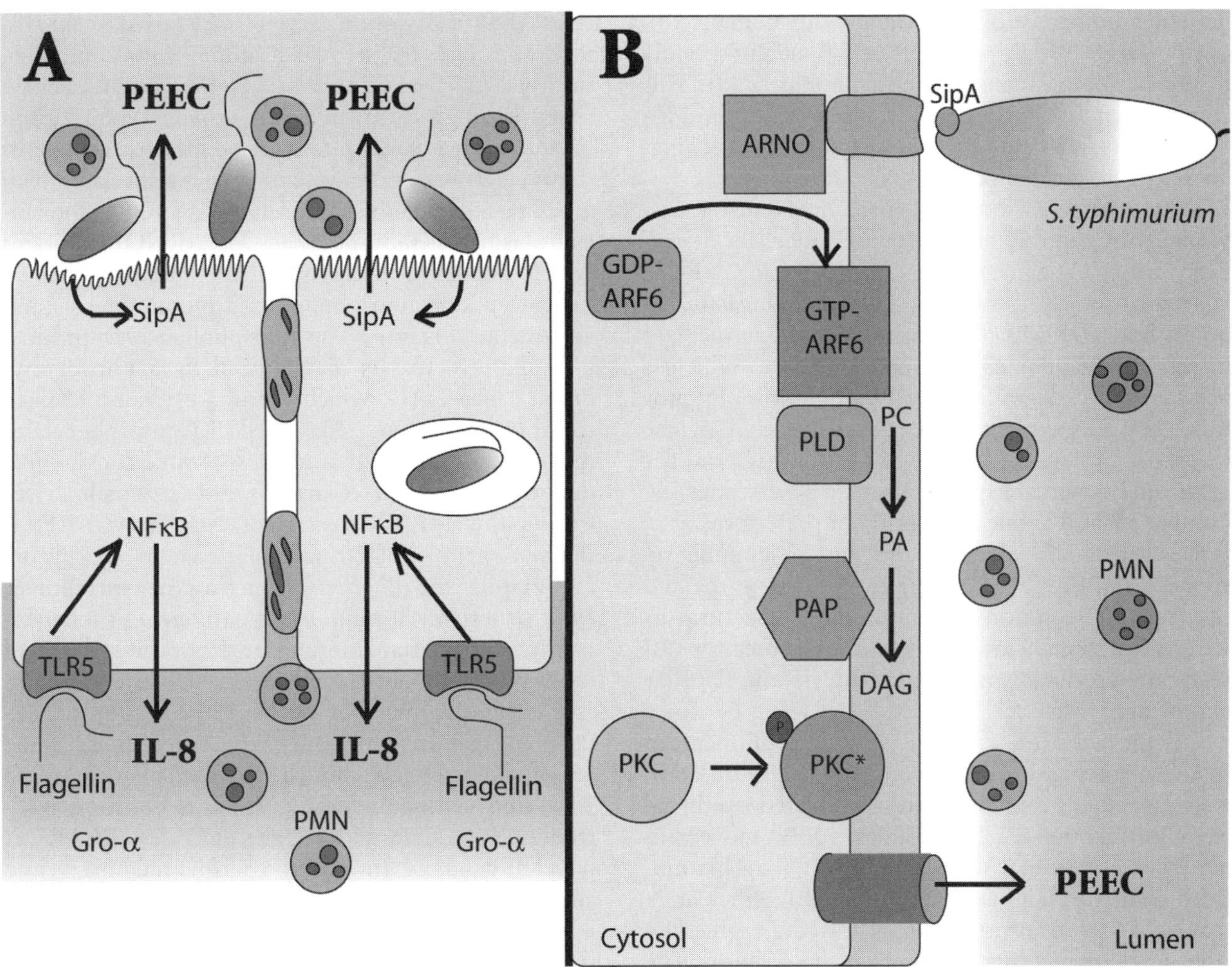

Figure 3. (A) Model of proposed events affecting *Salmonella*-induced PMN transmigration across the intestinal epithelium. *Salmonella* evokes a potent inflammatory response in the host, the hallmark of which is the migration of PMN across the intestinal mucosa. This process includes extravasation of circulating PMN from the microvasculature, passage of PMN across the lamina propria, and paracellular movement of PMN across the epithelium. PMN recruitment is coordinated by the release of proinflammatory cytokines, among which are IL-8 and PEEC. By an unknown mechanism, *Salmonella* invasion also causes the transcellular transport of flagellin to the basolateral membrane domain, where it promotes the release of IL-8 by interacting with TLR-5 and activates the NF-κB pathway. Concurrently, the *Salmonella* type III secretion product, SipA, is necessary and sufficient for induction of PMN transmigration across model intestinal epithelia in a PKC-dependent manner, which leads to the apical secretion of PEEC. (B) Model of *S. enterica* serovar Typhimurium-induced signaling in epithelial cells by the *Salmonella* secreted protein SipA. Interaction of SipA with the apical domain of polarized epithelial cells leads to activation of ARF6 (GTP-ARF6) at the apical membrane, most probably through the mammalian guanine exchange factor ARNO. This leads to an increase in PLD activity and local production of phosphatidic acid (PA), which is metabolized to DAG by phosphatidic acid phosphohydrolase (PAP). Generation of DAG recruits PKC to the apical membrane. Activation of PKC at this site (PKC*) is necessary for the apical release of the chemokine PEEC and subsequent basolateral-to-apical PMN transepithelial migration.

polarized basolateral release of the potent polymorphonuclear leukocyte [PMN] chemokine IL-8 (35, 70–72). Such basolateral release of IL-8 imprints subepithelial matrices with long-lived haptotactic gradients that function to guide PMN through the lamina propria and to a subepithelial position (71). Epithelial cell-derived IL-8 exhibits many properties that would make this chemokine well suited for this purpose. One important property is that IL-8 is extremely resistant to inactivation. Thus, once present in the inflamed tissue, it is likely to retain its biological activity for several hours, as shown by local intradermal administration to animals and humans (8). Further, IL-8 is highly cationic and thus binds avidly to glycosaminoglycans of the tissue matrix (8). This combination of characteristics makes such bound IL-8 gradients particularly resistant to the sweeping-away effects of fluid flow.

Activation of the IL-8 promoter during *Salmonella* invasion of cultured epithelial cells is dependent primarily on the transcription factor NF-κB. Up-regulation of IL-8 during *Salmonella* invasion requires both p38 MAP kinase activity and an increase in the intracellular calcium concentration, which is necessary for degradation of the NF-κB inhibitor IκBα and consequent nuclear translocation of the transcription factor (36, 47). Moreover, bacterial flagellin may activate proinflammatory cascades by binding to TLR5, one of a family of pattern recognition receptors involved in the innate immune response (38, 43). The flagellar export apparatus and the type III secretion apparatus have structural as well as functional similarities (67), and some type III secretion products can be secreted via the flagellar export apparatus (55, 111).

Although IL-8 is required for recruitment of PMN to the basal aspect of epithelial cells, an additional chemoattractant, pathogen-elicited epithelial chemoattractant (PEEC), guides PMN movement across the intestinal epithelium (69, 70). This apically secreted chemoattractant is highly efficient in driving PMN transmigration across the epithelium (the final step of transepithelial migration). The PEEC bioactivity possesses properties that appear to set it apart from other known peptide- or lipid-based PMN chemoattractants. The distinguishing features of PEEC reveal that it is relatively small (<1 kDa), stable, and not highly hydrophobic and does not appear to signal via the *n*-formyl peptide or the IL-8 receptors. Like most neutrophil chemoattractants, PEEC induces a PMN signal transduction cascade involving a GTP-binding protein ($G_{\alpha i}$) that also elicits a rise in $[Ca^{2+}]_i$. In addition, bioassays reveal that PEEC directly signals PMN, but, unlike known chemoattractants, PMN respond to PEEC with essentially a purely chemotactic response in which degranulation or superoxide generation is virtually not detectable even at saturating concentrations of PEEC bioactivity. Thus, PEEC appears to exhibit characteristics that are even distinct from "pure" PMN chemoattractants. Therefore, in the final step of transepithelial migration, subepithelial PMN would be ideally positioned to traffic across the monolayer in response to secreted PEEC activity.

Recent evidence reveals that the regulation of *S. enterica* serovar Typhimurium-PMN transmigration is governed by a novel mechanism involving the GTPase ADP-ribosylation factor 6 (ARF6) (18). In this model, *S. enterica* serovar Typhimurium contacting the apical surface of polarized epithelial cells elicits a signal through the *S. enterica* serovar Typhimurium bacterial effector, SipA (a secreted effector protein of the type III secretion system) that recruits an ARF6 guanine exchange factor (such as ARNO) to the apical plasma membrane (Fig. 3). ARNO facilitates ARF6 activation at the apical membrane, which in turn stimulates phospholipase D recruitment to and activity at this site. The phospholipase D product phosphatidic acid is metabolized by a phosphohydrolase into DAG, which recruits cytosolic PKC to the apical membrane. Activated PKC phosphorylates downstream targets that are responsible for the production and apical release of PEEC, which drives transepithelial PMN movement. At present, ARF6 is the first example of a "molecular switch" specifically modulating the PMN transmigration aspect of *S. enterica* serovar Typhimurium pathogenesis independently of bacterial internalization or the release of other proinflammatory mediators such as IL-8.

ARF6 is a member of the Arf subgroup of GTPases of the Ras superfamily, which were first defined as cofactors necessary for the cholera toxin-catalyzed ADP-ribosylation of the α_s subunit of the heterotrimeric G proteins. ARF6, the only class III ARF, is novel in its localization to the plasma membrane and endosomal structures in nonpolarized cells (21) and is highly expressed in polarized cells, where it localizes primarily to the apical brush border and apical early endosomes (5, 65). Although ARF6 was initially identified as a modulator of vesicular traffic and of cortical actin cytoarchitecture, recent evidence suggests that it is also a regulator of signal transduction cascades initiated by a variety of external stimuli and appears to be a key signal transducer downstream of G-protein-coupled receptors (105). Even though the identity of the cellular "receptor" for *Salmonella* and/or its effector SipA is not known, these observations suggest that bacterial adherence to the apical membrane of polarized epithelial cells promotes ARF6 in an analogous manner.

Furthermore, other studies have also shown that stimulation of human intestinal epithelial cells with *S. enterica* serovar Typhimurium evokes an increased expression and secretion of a number of other cytokines with chemoattractant and proinflammatory functions. For instance, stimulated epithelial cells express and secrete the chemokines GRO-α, GRO-β, GRO-γ, and ENA-78 (24, 57). Like IL-8, these cytokines belong to the CXC family of chemokines and are characterized by their ability to chemoattract and activate PMN, suggesting an important redundant function of intestinal epithelial cells to initiate the mucosal influx of PMN. In addition, *Salmonella*-infected human intestinal epithelial cells express and secrete other proinflammatory cytokines, including TNF-α, granulocyte-macrophage colony-stimulating factor, IL-1α, and IL-1β), although expression of these cytokines is generally much lower (orders of magnitude) than that observed for chemokines (24, 57). Further, intestinal epithelial cells do not appear to express a number of cytokines, such as IL-2, IL-4, IL-5, IL-12, or gamma interferon, that are more commonly associated with antigen-specific acquired immune responses (24, 57). These findings indicate that such cytokines secreted by intestinal epithelial cells are likely to play a more important role in initiating and regulating the innate mucosal inflammatory response rather than the antigen-specific mucosal immune response.

ACTIVE SUPPRESSION OF INFLAMMATORY PATHWAYS

An emerging concept is that certain microbes may take a proactive approach and actually suppress host responses. In fact, direct microbial inhibition of NF-κB activation is being reported with increasing frequency. Perhaps the best-described example is exhibited by *Y. enterocolitica* and *Y. pseudotuberculosis,* which are enteric pathogens known for their ability to inhibit TNF-α production and induce apoptosis in infected macrophages; this capability is controlled by the type III-secreted effector molecule YopJ (82). YopJ blocks the innate immune response by inhibiting cytokine production and inducing apoptosis in infected macrophages (83). Consequently, the infected host cell cannot respond to the *Yersinia* infection because YopJ inhibits MAP kinase and the NF-κB pathways by preventing the activation, via phosphorylation, of MAP kinase kinases and IKB (Fig. 1). Abrogation of IKB phosphorylation prevents sequential ubiquitination, degradation, and NF-κB nuclear translocation. The derived secondary structure of YopJ is highly similar to that of the known secondary structure of a cysteine protease from the eukaryotic ubiquitin-like protein protease, and YopJ is thought to mimic this type of hydrolase activity. A YopJ homologue expressed by the enteric pathogen *Salmonella,* AvrA, inhibits only the NF-κB pathway but not the MAP kinase pathways (17). Thus, as with *Yersinia,* inhibition of NF-κB is potentially proapoptotic, since this organism has evolved a unique mechanism to eliminate the threat of phagocytic cells by obstructing their proinflammatory responses and ultimately inducing apoptotic destruction.

Teleologically, the blockage of the NF-κB pathway may be an underlying mechanism for commensalisms. Recent observations of the anti-inflammatory activity of *Salmonella* spp. that are intestinal commensals in young birds and reptiles lend support to this view. For example, epithelial cells colonized by several strains of nonpathogenic *Salmonella* fail to induce the up-regulation of inflammatory effector molecules or the activation of NF-κB (78). Cells colonized with such nonpathogens were refractory to activation by known proinflammatory strains, as well as other potent pharmacologic activators of NF-κB. Blockade of the proinflammatory gene activation occurred pretranscriptionally by inhibiting the activation of NF-κB and preventing translocation to the nucleus. Unlike the situation with *Yersina,* IκBα was phosphorylated in cells colonized by nonpathogens, but subsequent ubiquitination of IκBα was completely abolished. Speculatively, microorganisms intimately associated with the intestinal mucosa may have evolved such mechanisms to dampen the host proinflammatory and immune responses without provoking apoptotic death. Hence, this mechanism may play a role in the distinctive tolerance that the human intestinal epithelium displays toward the enteric microflora.

SUMMARY

It has become evident that the intestinal epithelium is not a passive barrier but, rather, is the central mediator in the complex interactions between the commensal microflora and the immune cells that underlie the intestinal epithelium. Moreover, the intestinal epithelium has evolved an innate immune response that is armed with the capacity to sense pathogenic bacteria among this complex environment and to respond appropriately, even if the ultimate outcome results in unpleasant consequences for the host (e.g., diarrhea). A more complete understanding of the signal transduction cascades that exist between the intestinal bacteria (pathogenic or otherwise) and the human host in the intestinal mucosa may uncover new insights into human diseases and reveal novel approaches to treating them.

REFERENCES

1. **Aderem, A., and R. J. Ulevitch.** 2000. Toll-like receptors in the induction of the innate immune response. *Nature* **406:** 782–787.
2. **Alexopoulou, L., A. C. Holt, R. Medzhitov, and R. A. Flavell.** 2001. Recognition of double stranded RNA and activation of NF-κB by toll-like receptor 3. *Nature* **413:**732–738.
3. **Alicon, I., and P. Kubes.** 1996. A critical role for nitric oxide in intestinal barrier function and dysfunction. *Am. J. Physiol. Ser. G* **270:**G225–G237.
4. **Aliprantis, A., R.-B. Yang, M. Mark, S. Suggett, B, Devaux, J, Radolf, G. Klimple, P. Godowski, and A. Zychlinsky.** 1999. Cell activation and apoptosis by bacterial lipoproteins through toll-like receptor-2. *Science* **285:**736–739.
5. **Altschuler, Y., S. Liu, L. Katz, K. Tang, S. Hardy, F. Brodsky, G. Apodacca, and K. Mostov.** 1999. ADP-ribosylation factor 6 and endocytosis at the apical surface of Madin-Darby canine kidney cells. *J. Cell Biol.* **147:**7–12.
6. **Anderson, J. M.** 1996. Cell signalling: MAGUK magic. *Curr. Biol.* **6:**382–384.
7. **Anderson, J. M.** 2001. Molecular structure of tight junctions and their role in epithelial transport. *News Physiol. Sci.* **16:**126–130.
8. **Baggiolini, M., B. Dewald, and A. Walz.** 1992. Interleukin-8 and related cytokines, p. 247–263. *In* J. I. Gallin, I. M. Golstein, and R. Snyderman (ed.), *Inflammation: Basic Principles and Clinical Correlates,* 2nd ed. Raven Press, New York, N.Y.
9. **Beagley, K. W., and A. J. Husband.** 1998. Intraepithelial lymphocytes: origins, distributions, and function. *Crit. Rev. Immunol.* **18:**237–254.
10. **Berkes, J., V. K. Viswanathan, S. D. Savkovic, and G. Hecht.** 2003. Intestinal epithelial responses to enteric pathogens: effects on the tight junction barrier, ion transport, and inflammation. *Gut* **52:**439–451.
11. **Bernet, M. F., D. Brassart, J. R. Nesser, and A. L. Servin.** 1994. *Lactobacillus* LA binds to cultured human intestinal cell lines and inhibits cell attachment and cell invasion by enterovirulent bacteria. *Gut* **35:**483–489.
12. **Bulut, Y., E. Faure, L. Thomas, O. Equils, and M. Arditi.** 2001. Cooperation of Toll-like receptor 2 and 6 for cellular activation by soluble tuberculosis factor and *Borrelia burgdorferi* outer protein A lipoprotein: role of Toll-interacting protein and IL-1 receptor signaling molecules in Toll-like receptor-2 signaling. *J. Immunol.* **167:**987–994.
13. **Butler, J. E., J. Sun, P. Weber, P. Navarro, and D. Francis.** 2000. Antibody repertoire development in fetal and newborn piglets. III. Colonization of the gastrointestinal tract selectively diversifies the preimmune repertoire in mucosal lymphoid tissue. *Immunology* **100:**119–130.
14. **Cario, E., and D. K. Podolsky.** 2000. Differential alteration in intestinal epithelial cell expression of toll-like receptor 3 (TLR3) and TLR4 in inflammatory bowel disease. *Infect. Immun.* **68:**7010–7017.
15. **Cebra, J. J., S. B. Periwal, G. Lee, F. Lee, and K. E. Shroff.** 1998. Development and maintenance of the gut-associated lymphoid tissue (GALT): the roles of enteric bacteria and viruses. *Dev. Immunol.* **6:**13–18.
16. **Chen, L. M., S. Hobbie, and J. E. Galan.** 1996. Requirement of CDC42 for *Salmonella*-induced cytoskeletal and nuclear responses. *Science* **274:**2115–2118.
17. **Collier-Hyams, L. S., H. Zeng, J. Sun, A. D. Tomlinson, Z. O. Bao, H. Chen, J. L. Madara, K. Orth, and A. S. Neish.** 2002. Cutting edge: *Salmonella* AvrA effector inhibits the key proinflammatory, anti-apoptotic NF-κB pathway. *J. Immunol.* **169:**2846–2850.
18. **Criss, A. K., M. Silva, J. E. Casanova, and B. A. McCormick.** 2001. Regulation of *Salmonella*-induced neutrophil transmigration by epithelial ADP-ribosylation factor 6. *J. Biol. Chem.* **276:**48431–48439.
19. **Deng, L., C. Wang, E. Spencer, L. Yang, A. Brqun, X. You, C. Slaughter, C. Pickart, and Z. Chen.** 2000. Activation of the IκB kinase complex by TRAF6 requires a dimeric ubiquitin-conjugating enzyme complex and a unique polyubiquitin chain. *Cell* **103:**351–361.
20. **Didonato, J. A., M. Hayakawa, D. M. Rothwarf, E. Zandi, and M. Karin.** 1997. A cytokine-responsive IkappaB kinase that activates the transcription factor NF-kappa-B. *Nature* **388:**548–554.
21. **D'Souza-Shorey, C, G. Li, M. I. Colombo, and P. D. Stahl.** 1995. A regulatory role for ARF6 in receptor-mediated endocytosis. *Science* **267:**1175–1178.
22. **Duchmann, R., E. Scmitt, P. Knolle, K. H. Meyer zum Buschenfelde, and M. Neurath.** 1996. Tolerance towards resident intestinal flora in mice is abrogated in experimental colitis and restored by treatment with interleukin 10 or antibodies to interleukin 12. *Eur. J. Immunol.* **26:**934–938.
23. **Eberhart, C. E., and R. N. DuBois.** 1995. Eicosanoids and the gastrointestinal tract. *Gastroenterology* **109:**285–301.
24. **Eckmann, L., H.-C. Jung, C.-C. Schuerer-Maly, A. Panja, E. Morzycka-Wroblewska, and M. F. Kagnoff.** 1993. Differential cytokine expression by human intestinal epithelial cell lines: regulated expression of interleukin-8. *Gastroenterology* **105:**1689–1697.
25. **Eckmann, L., W. F. Stenson, T. C. Savidge, D. C. Lowe, K. E. Barrett, J. Fierer, J. R. Smith, and M. F. Kagnoff.** 1997. Role of intestinal epithelial cells in the host secretory response to infection by invasive bacteria. *J. Clin. Investig.* **100:**296–309.
26. **Eggermont, E.** 1996. Gastrointestinal manifestations in cystic fibrosis. *Eur. J. Gastrointerol. Hepatol.* **8:**731–738.
27. **Elliot, S. J., C. B. O'Connell, A. Koutsouris, C. Brinkley, M. S. Donnenberg, G. Hecht, and J. B. Kaper.** 2002. A gene from the locus of enterocyte effacement that is required for enteropathogenic *Escherichia coli* to increase tight-junction permeability encodes a chaperone for EspF. *Infect. Immun.* **70:**2271–2277.
28. **Fang, F.** 1997. Mechanisms of nitric oxide-related antimicrobial activity. *J. Clin. Investig.* **99:**2818–2825.
29. **Fanning, A. S., B. J. Jameson, L. A. Jesaitis, and J. M. Anderson.** 1998. The tight junction protein ZO-1 establishes a link between the transmembrane protein occludin and the actin cytoskeleton. *J. Biol. Chem.* **273:**29745–29753.
30. **Fasano, A., T. Not, W. Wang, S. Uzzau, I. Berti, A. Tommasini, and S. E. Goldblum.** 2000. Zonulin, a newly discovered modulator of intestinal permeability, and its expression in coeliac disease. *Lancet* **355:**1518–1519.
31. **Frankel, W. L., W. Zhang, A. Singh, D. M. Kurfield, S. Don, T. Sakata, I. Modlin, and J. L. Rombeau.** 1994. Mediation of trophic effects of short chain fatty acids on the rat jejunum and colon. *Gastroenterology* **106:**375–380.
32. **Fuller, C. M., I. I. Ismailov, D. A. Keeton, and D. J. Benos.** 1994. Phosphorylation and activation of a bovine tracheal anion channel by $Ca2^{+}$/calmodulin-dependent protein kinase II. *J. Biol. Chem.* **269:**26642–26650.
33. **Furuse, M., M. Itoh, T. Hirase, A. Nagafuchi, S. Yonemura, and S. Tsukita.** 1994. Direct association of occludin with

ZO-1 and its possible involvement in the localization of occludin at tight junctions. *J. Cell Biol.* **127**:1617–1626.

34. **Gewirtz, A. T., A. T. Navas, S. Lyons, P. J. Godowski, and J. L. Madara.** 2001. Bacterial flagella activates basolaterally expressed TLR5 to induce epithelial proinflammatory gene expression. *J. Immunol.* **167**:1882–1885.
35. **Gewirtz, A. T., A. M. Siber, J. L. Madara, and B. A. McCormick.** 1999. Orchestration of neutrophil movement by intestinal epithelial cells in response to *Salmonella typhimurium* can be uncoupled from bacterial internalization. *Infect. Immun.* **67**:608–617.
36. **Gewirtz, A. T., A. S. Rao, P. O. Simon, D. Merlin, D. Carnes, J. L. Madara, and A. S. Neish.** 2000. *Salmonella typhimurium* induces epithelial IL-8 expression via $Ca2^+$-mediated activation of the NF-κB pathway. *J. Clin. Investig.* **105**:79–92.
37. **Gewirtz, A. T., P. O. Simon, C. K. Schmidt, L. J. Taylor, C. H. Hagedor, A. D. O'Brien, A. S. Neish, and J. L. Madara.** 2001. *Salmonella typhimurium* translocates flagellin across the intestinal epithelia inducing a pro-inflammatory response. *J. Clin. Investig.* **107**:99–109.
38. **Gewirtz, A. T., T. A. Navas, S. Lyons, P. J. Godowski, and J. L. Madara.** 2001. Cutting edge: bacterial flagellin activates basolaterally expressed TLR5 to induce epithelial pro-inflammatory genes expression. *J. Immunol.* **167**:1882–1885.
39. **Girardin, S. E., I. G. Boneca, L. A. M. Carnairo, A. Antignac, M. Jehanno, J. Viala, P. J. Sansonetti, and D. J. Phillpot.** 2003. Nod1 detects a unique muropeptide from Gram-negative bacterial peptidoglycan. *Science* **300**:1584–1587.
40. **Girardin, S. E., I. G. Boneca, J. Viala, M. Chamaillard, A. Labigne, G. Thomas, D. J. Phillpot, and P. J. Sansonetti.** 2003. Nod2 is a general sensor of peptidoglycan through muramyl dipeptide (MDP) detection. *J. Biol. Chem.* **278**:8869–8872.
41. **Girardin, S. E., P. J. Sansonetti, and D. J. Phillpot.** 2002. Intracellular vs. extracellular recognition of pathogens-common concepts in mammals and flies. *Trends Microbiol.* **10**:193–199.
42. **Gopalakrishnan, S., N. Raman, S. J. Atkinson, and J. A. Marrs.** 1998. Rho GTPase signaling regulates tight junction assembly and protects tight junctions during ATP depletion. *Am. J. Physiol. Ser. C* **275**:C798–C809.
43. **Hayashi, F., K. D. Smith, A. Ozinsky, T. R. Hawn, E. C. Yi, D. R. Goodlett, J. K. Eng, S. Akira, D. M. Underhill, and A. Aderem.** 2001. The innate immune response to bacterial flagellin is mediated by Toll-like receptor 5. *Nature* **410**:1099–1103.
44. **Hayle, C., and G. Burnstock.** 1989. Galanin-like immunoreactivity in enteric neurons of the human colon. *J. Anat.* **166**:23–33.
45. **Hemmi, H. O., T. Takeuchi, T. Kawai, S. Kaisho, H. Sato, M. Sanjo, K. Matsumoto, H. Hoshino, K. Wahner, K, Takeda, and S. Akira.** 2000. A Toll-like receptor recognizes bacterial DNA. *Nature* **408**:740–745.
46. **Hirst, T. R.** 1995. Biogenesis of cholera and related oligomeric enterotoxins, p. 123–184. *In* J. Moss, M. Vaughan, B. Iglewski, and A. T. Tu (ed.), *Bacterial Toxins and Virulence Factors in Disease,* vol 8. Marcel Dekker, Inc., New York, N.Y.
47. **Hobbie, S., L. Chen, R. Davis, and J. E. Galan.** 1997. Involvement of mitogen-activated protein kinase pathways in the nuclear responses and cytokine production induced by *Salmonella typhimurium* in cultured intestinal epithelial cells. *J. Immunol.* **159**:5550–5559.
48. **Hoffman, J. F. Kafatos, C. Janeway, and R. Ezekowitz.** 1999. Phylogenic perspectives in innate immunity. *Science* **284**: 1313–1318.
49. **Hooper, L. V., J. Xu, P. G. Falk, T. Midvedt, and J. I. Gordan.** 1999. A molecular sensor that allows a gut commensal to control its nutrient foundation in a competitive ecosystem. *Proc. Natl. Acad. Sci. USA* **96**:9833–9838.
50. **Horiguchi, Y., T. Senda, N. Sugimoto, J. Katahira, and M. Matsuda.** 1995. *Bordetella bronchiseptica* dermonecrotizing toxin stimulates assembly of actin stress fibers and focal adhesions by modifying the small GTP- binding protein rho. *J. Cell. Sci.* **108**:3243–3251.
51. **Hugot, J. P., P. Laurent-Puig, C. Gower-Rousseau, J. M. Olson, J. C. Lee, L. Beaugerie, I. Naom, J. L. Dupas, A. Van Gossum, M. Orholm, S. Bonaiti-Pellie, J. Weissenbach, C. G. Mathew, J. E. Lennard-Jones, A. Cortot, J. F. Colombel, and G. Thomas.** 1996. Mapping of a susceptibility locus for Crohn's disease. *Nature* **379**:821–823.
52. **Inohara, N., and G. Nunez.** 2003. Nods. Intracellular proteins involved in inflammation and apoptosis. *Nat. Rev. Immunol.* **3**:371–382.
53. **Inohora, N., T. Koseki, J. Lin, L. del Paso, P. C. Lucas, F. F. Chen, Y. Ogura, and G. Nunez.** 2000. An induced proximity model for NF-kappaB activation in the Nod1/Rick and RIP signaling pathways. *J. Biol. Chem.* **275**:27823–27831.
54. **Itoh, M., K. Morita, and S. Tsukita.** 1999. Characterization of ZO-2 as a MAGUK family member associated with tight as well as adherens junctions with a binding affinity to occludin and alpha catenin. *J. Biol. Chem.* **274**:5981–5986.
55. **Iyoda, S., T. Kamidoi, K. Hirose, K. Kutsukake, and H. Watanabe.** 2001. A flagellar gene *fliz* regulates the expression of invasion genes and virulence phenotype in *Salmonella enterica* serovar Typhimurium. *Microb. Pathog.* **30**:81–90.
56. **Jesaitis, L. A., and D. A. Goodenough.** 1994. Molecular characterization and tissue distribution of ZO-2, a tight junction protein homologous to ZO-1 and the *Drosophila* discs-large tumor suppressor protein. *J. Cell Biol.* **124**:949–961.
57. **Jung, H. C., L. Eckmann, S. K. Yang, A. Panja, J. Fierer, E. Morzycka-Wroblewska, and M. F. Kagnoff.** 1995. A distinct array of proinflammatory cytokines is expressed in human colon epithelial cells in response to bacterial invasion. *J. Clin. Investig.* **95**:55–65.
58. **Karin, M.** 1999. The beginning of the end: I(B kinase (IKK) and NF-κB activation. *J. Biol. Chem.* **274**:27339–27342.
59. **Katsoulis, S., A. Clemens, C. Morys-Wortmann, H. Schworer, H. Schaube, H. J. Klomp, U. R. Folsch, and W. E. Schmidt.** 1996. Human galanin modulates colonic motility in vitro. Characterization of structural requirements. *Scand. J. Gastroenterol.* **31**:446–451.
60. **Keely, S., and K. Barrett.** 2000. Regulation of chloride secretion: novel pathways and messengers. *Ann. N. Y. Acad. Sci.* **915**:67–76.
61. **Kenny, B., R. DeVinney, M. Stein, D. J. Reinscheid, G. A. Frey, and B. B. Finlay.** 1997. Enteropathogenic *E. coli* (EPEC) transfers its receptor for intimate adherence into mammalian cells. *Cell* **91**:511–520.
62. **Kopp, E., and S. Ghosh.** 1995. NF-κB and Rel proteins in innate immunity. *Adv. Immunol.* **58**:1–12.
63. **Kuhn, R., J. Lohler, D. Rennick, K. Rajewsky, and W. Muller.** 1993. Interleukin-10-deficient mice develop chronic enterocolitis. *Cell* **75**:263–274.
64. **Lencer, W. I.** 2001. Microbes and microbial toxins: paradigms for microbial mucosal interactions. V. Cholera: invasion of the intestinal epithelial barrier by a stably folded protein toxin. *Am. J. Physiol. Ser. G* **280**:G781–G786.
65. **Londono, I., V. Marshansky, S. Bourgoin, P. Vinay, and M. Bendayan.** 1999. Expression and distribution of adeno-

sine diphosphate ribosylation factorsin the rat kidney. *Kidney Int.* **55:**1407–1416.

66. **Madara, J. L.** 2000. Modulation of tight junction permeability. *Adv. Drug Deliv. Rev.* **41:**251–253.
67. **Makishima, S., K. Komoriya, S. Yamaguchi, and S. I. Aizawa.** 2001. Length of the flagella hook and the capacity of the type III export apparatus. *Science* **291:**2411–2413.
68. **Matkowskyj, K. A., A. Danilkovich, J. Marrero, S. D. Savkovic, G. Hecht, and R. V. Benya.** 2000. Galanin-1 receptor up-regulation mediates the excess colonic fliud production caused by infection with enteric pathogens. *Nat. Med.* **6:**1048–1051.
69. **McCormick, B. A., C. A. Parkos, S. P. Colgan, D. K. Carnes, and J. L. Madara.** 1998. Apical secretion of a pathogen-elicited epithelial chemoattractant (PEEC) activity in response to surface colonization of intestinal epithelia by *Salmonella typhimurium. J. Immunol.* **160:**455–466.
70. **McCormick, B. A., S. P. Colgan, C. D. Archer, S. I. Miller, and J. L. Madara.** 1993. *Salmonella typhimurium* attachment to human intestinal epithelial monolayers: transcellular signalling to subepithelial neutrophils. *J. Cell. Biol.* **123:**895–907.
71. **McCormick, B. A., P. M. Hofman, J. Kim, D. K. Carnes, S. I. Miller, and J. L. Madara.** 1995. Surface attachment of *Salmonella typhimurium* to intestinal epithelia imprints the subepithelial matrix with gradients chemotactic for neutrophils. *J. Cell. Biol.* **131:**1599–1608.
72. **McCormick, B. A., S. I. Miller, D. Carnes, and J. L. Madara.** 1995. Transepithelial signaling to neutrophils by salmonellae: a novel virulence mechanism for gastroenteritis. *Infect. Immun.* **63:**2302–2309.
73. **McCormick, B. A., A. Nusrat, L. D'Andrea, C. A. Parkos, P. M. Hofman, D. K. Carnes, and J. L. Madara.** 1997. Unmasking of intestinal epithelial lateral membrane β_1-integrin consequent to transepithelial neutrophil migration in vitro facilitates *inv*-mediated invasion by *Yersinia. Infect. Immun.* **65:**1414–1421.
74. **Medzhitov, R., and C. Janeway, Jr.** 2000. Innate immunity. *N. Engl. J. Med.* **343:**338–344.
75. **Molina, N., and J. Petterson.** 1980. Cholera toxin-like toxin released by *Salmonella* species in the presence of mitomycin C. *Infect. Immun.* **30:**224–230.
76. **Moreau, M. C., and V. Gaboriau-Routhiau.** 1996. The absence of the gut flora, the doses of antigen ingested and aging affect the long-term peripheral tolerance induced by ovalbumin feeding in mice. *Res. Immunol.* **147:**49–59.
77. **Mounier, J., T. Vasselon, R. Hellio, M. Lesourd and P. J. Sansonetti.** 1992. *Shigella flexneri* enters human colonic Caco-2 epithelial cells through the basolateral pole. *Infect. Immun.* **60:**237–248.
78. **Neish, A., A. Gewirtz, H. Zeng, A. Young, M. Hobert, V. Karmali, A. Rao, and J. Madara.** 2000. Prokaryotic regulation of epithelial responses by inhibition of IκB-α ubiquitination. *Science* **289:**1563.
79. **Norris, F. A., M. R. Wilson, T. S. Wallis, E. E. Galyov, and P. W. Majerus.** 1998. SopB a protein required for virulence of *Salmonella* Dublin is an inositol phosphate phosphatase. *Proc. Natl. Acad. Sci. USA* **95:**14057–14059.
80. **Nusrat, A., M. Giry, J. R. Turner, S. P. Colgan, C. A. Parkos, D. Carnes, E. Lemichez, P. Boquet, and J. L. Madara.** 1995. Rho protein regulates tight junctions and perijunctional actin organization in polarized epithelia. *Proc. Natl. Acad. Sci. USA* **92:**10629–10633.
81. **Ogura, Y., D. K. Bonen, N. Inohara, D. L. Nicolae, F. F. Chen, R. Rames, H. Britton, T. Moran, R. Karaliuskas, R. H. Duerr, J. P. Achkar, S. R. Brant, P. M. Bayless, B. S. Kirschner, J. B. Hanaver, G. Nunez, and J. H. Cho.** 2001. A frameshift mutation in Nod2 associated with susceptibility to Crohn's disease. *Nature* **411:**603–608.
82. **Orth, K., L. Palmer, Z. Bao, S. Stewart, A. Rudolph, J. Bliska, and J. Dixon.** 1999. Inhibition of the mitogen-activated protein kinase kinase superfamily by a *Yersinia* effector. *Science* **285:**1920–1923.
83. **Orth, K., Z. Xu, M. Mudgett, Z. Bao, L. Palmer, J. Bliska, W. Mangel, B. Staskawicz, and J. Dixon.** 2000. Disruption of signaling by *Yersinia* effector YopJ, a ubiquitinin-like protein protease. *Science* **290:**1594–1597.
84. **Oswald, E., M. Sugai, A. Labigne, H. C. Wu, C. Fiorentini, P. Boquet, and A. D. O'Brien.** 1994. Cytotoxic necrotizing factor type 2 produced by virulent *Escherichia coli* modifies the small GTP-binding proteins Rho involved in assembly of actin stress fibers. *Proc. Natl. Acad. Sci. USA* **91:**3814–3818.
85. **Resta-Lenert, S., and K. Barrett.** 2002. Enteroinvasive bacteria alter barrier and transport properties of human intestinal epithelium: role of iNOS and COX-2.*Gastroenterology* **122:** 1070–1087.
86. **Roberfroid, M. B., F. Bornet, C. Bouley, and J. H. Cummings.** 1995. Colonic microflora: nutrition and health. Summary of conclusions of an Internattional Life Sciences Institute [(ILSI) Europe] Workshop held in Barcelona, Spain. *Nutr. Rev.* **53:**127–130.
87. **Rodighiero, C., and W. I. Lencer.** 2003. Trafficking of cholera toxin and related bacterial enterotoxins: pathways and endpoints, p. 385–422. *In* G. Hecht (ed.) *Microbial Pathogenesis and the Intestinal Epithelial Cell.* ASM Press, Washington, D.C.
88. **Sakaguchi, T., H. Kohler, X. Gu, B. A. McCormick, and H. C. Reinecker.** 2002. *Shigella flexneri* regulates tight junction-associated proteins in human intestinal epithelial cells. *Cell. Microbiol.* **6:**367–381.
89. **Sears, C. L., and J. B. Kaper.** 1996. Enteric bacterial toxins: mechanisms of action and linkage to intestinal secretion. *Microbiol. Rev.* **60:**167–215.
90. **Sellon, R. K., S. Tonkonogy, M. Shultz, L. A. Dieleman, W. Grenther, E. Balish, D. M. Rennick, and R. B. Sartor.** 1998. Resident enteric bacteria are necessary for the development of spontaneous colitis and immune system activation in interleukin-10-deficient mice. *Infect. Immun.* **66:**5224–5231.
91. **Senda, T., Y. Horiguchi, M. Umemoto, N. Sugimoto, and M. Matsuda.** 1997. *Bordetella bronchiseptica* dermonecrotizing toxin, which activates a small GTP-binding protein rho, induces membrane organelle proliferation and caveolae formation. *Exp. Cell Res.* **230:**163–168.
92. **Simonovic, I., J. Rosenberg, A. Koutsouris, and G. Hecht.** 2000. Enteropathogenic Escherichia coli dephosphorylates and dissociates occluding from intestinal epithelial tight junctions. *Cell. Microbiol.* **4:**305–315.
93. **Smith, W. L., and D. L. DeWitt.** 1996. Prostaglandin endoperoxidae H synthase-1 and -2. *Adv. Immunol.* **62:**167–215.
94. **Sonoda, N., M. Furuse, H. Sasaki, S. Yonemura, J. Katahira, Y. Horiguchi, and S. Tsukita.** 1999. *Clostridium perfringens* enterotoxin fragment removes specific claudins from tight junction strands. Evidence for direct involvement of claudins in tight junction barrier. *J. Cell Biol.* **147:**195–204.
95. **Strober, W., I. J. Fuss, and R. S. Blumberg.** 2002. The immunology of mucosal models of inflammation. *Annu. Rev. Immunol.* **20:**495–549.
96. **Strober, W., I. J. Fuss, R. O. Ehrhardt, M. Neurath M. Boirivant, and B. R. Ludviksson.** 1998. Mucosal immunoregulation and inflammatory bowel disease: new insights from murine models of inflammation. *Scand. J. Immunol.* **30:**2101–2111.

97. **Suau, A., R. Bonnet, M. Sutren, J. J. Gordan, G. R. Gibson, M. D. Collin, and J. Dore.** 1999. Direct rDNA community analysis reveals a myriad of novel bacterial lineages within the human gut. *Appl. Environ. Microbiol.* **65:**4799–4807.

98. **Sudo, N., S. Sawamura, K. Tanaka, Y. Aiba, C. Kubo, and Y. Koga.** 1997. The requirement of intestinal bacteria flora for the development of an IgE production system fully susceptible to oral tolerance induction. *J. Immunol.* **159:**1739–1745.

99. **Takahashi, A., N. Kenjyo, K. Imura, Y. Myosum, and T. Honda.** 2000. Cl^- secretion in colonic epithelial cells induced by *Vibrio parahaemolyticus* hemolytic toxin related to thermostable direct hemolysin. *Infect. Immun.* **68:**5435–5438.

100. **Takeuchi, O., K. Hoshino, T. Kawai, H. Sanjo, H. Takaada, T. Ogawa, K. Takeda, and S. Akira.** 1999. Differential roles of TLR-2 and TLR4 in recognition of Gram-negative and Gram-positive bacterial cell wall components. *Immunity* **11:**443–451.

101. **Tannock, G. W.** 2001. Molecular assessment of intestinal microflora. *Am. J. Clin. Nutr.* **73:**(Suppl.):54799–54807.

102. **Umesaki, Y., H. Setoyama, S. Matsumoto, and Y. Okada.** 1993. Expression of alpha beta T-cell receptor-bearing intestinal intraepithelial lymphocytes after microbial colonization in germ-free mice and its independence from the thymus. *Immunology* **79:**32–37.

103. **Underhill, D. M., A. Ozinsky, K. D. Smith, and A. Aderem.** 1999. Toll-like receptor-2 mediates mycobacteria-induced proinflammatory signaling in macrophage. *Proc. Natl. Acad. Sci. USA* **96:**14459–14463.

104. **Vallance, B. A., and B. B. Finlay.** 2000. Exploitation of host cells by enteropathogenic *Escherichia coli. Proc. Natl. Acad. Sci. USA* **97:**8799–8806.

105. **Venkateswarlu, K., and P. J. Cullen.** 2000. Signalling via ADP-ribosylation factor 6 lies downstream of phosphatidylinositide 3-kinase. *Biochem. J.* **345:**719–724.

106. **Wallis, T. S., A. T. M. Vaughan, G. J. Clarke, G.-M. Qi, K. J. Woron, D. C. A. Candy, M. P. Osborne, and J. Stephen.** 1990. The role of leucocytes in the induction of fluid secretion by *Salmonella typhimurium. J. Med. Microbiol.* **31:**27–35.

107. **Wilson, K. H., and R. B. Blitchington.** 1996. Human colonic biota studies by ribosomal DNA sequence analysis. *Appl. Environ. Microbiol.* **62:**2273–2278.

108. **Wirtz, S., S. Finotto, A. W. Lohse, M. Blessing, H. A. Lehr, P. R. Galle, and M. F. Neurath.** 1999. Chronic intestinal inflammation in STAT-4 transgenic mice: characterization of disease and adoptive transfer by TNF-plus IFN gamma-producing $CD4^+$ T cells that respond to bacterial antigens. *J. Immunol.* **162:**1884–1888.

109. **Wittchen, E. S., J. Haskins, and B. R. Stevenson.** 1999. Protein interactions at the tight junction. Actin has multiple binding partners, and ZO-1 forms independent complexes with ZO-2 and ZO-3. *J. Biol. Chem.* **274:**35179–35185.

110. **Woods, D. F., and P. J. Bryant.** 1993. ZO-1, DlgA and PSD-95/SAP90: homologous proteins in tight, septate and synaptic cell junctions. *Mech. Dev.* **44:**85–89.

111. **Young, G. M., D. H. Schiel, and V. L. Miller.** 1999. A new pathway for the secretion of virulence factors by bacteria: the flagellar export apparatus functions as a protein secretion system. *Proc. Natl. Acad. Sci. USA* **96:**6456–6461.

112. **Zhou, D., L. M. Chen, L. Hernandez, S. B. Shears, and J. E. Galan.** 2001. A *Salmonella* inositol phosphatase acts in conjunction with other bacterial effectors to promote host cell actin cytoskeleton rearrangements and bacterial internalization. *Mol. Microbiol.* **39:**248–259.

Colonization of Mucosal Surfaces
Edited by James P. Nataro et al.

Chapter 20

Pathogen Gene Expression during Intestinal Infection

Susan M. Butler, Anna D. Tischler, and Andrew Camilli

According to the World Health Organization *Global Water Supply and Sanitation Assessment, 2000 Report* (www.who.int/docstore/water_sanitation_health/Globalassessment/GlobalTOC.htm), diarrheal diseases cause over 2.2 million deaths annually worldwide, mostly in children younger than 5 years. Although intestinal parasites contribute to this figure, viruses and bacterial intestinal pathogens are the predominant causes. In the United States, diarrhea due to infection by known pathogens results in a reported 60,000 hospitalizations and 1,800 deaths annually, with over one billion dollars worth of lost wages and medical costs due to salmonellosis alone (http://www.cdc.gov). Examining the patterns of virulence factor expression during intestinal infection by these pathogens, in addition to providing insights into how pathogens cause disease, is important in order to obtain information that may be used in the design of improved or novel vaccines and therapeutic agents.

In bacteria, regulation of the expression of virulence factors and other cellular products is mediated primarily at the level of transcription. Therefore, the patterns of virulence factor expression during infection can be conveniently ascertained using one of several available research tools for measuring transcriptional induction or steady-state transcript levels of genes. In this chapter, we discuss what is known about bacterial gene expression during infection of the intestinal tract in animal models of disease. Despite the publication of numerous studies investigating gene expression by pathogens that colonize the intestinal tract, many of these studies were performed in vitro or in cell culture models of infection. While the importance of these investigations cannot be disputed, these systems do not fully represent the complex milieu of the intestinal tract, on which this chapter is focused. In addition, since many intestinal pathogens have the ability to cause serious systemic disease, bacterial gene expression has been examined at sites of systemic infection. This chapter, however, deals only with bacterial gene expression within the intestine; as a result, a number of intestinal pathogens will not be covered. We therefore apologize to the authors of the many important papers involving bacterial gene expression that are not discussed here. We refer the reader interested in gene expression at systemic sites of infection or in tissue culture to the following selection of articles: *Listeria monocytogenes* (28, 36, 41, 42, 104), *Shigella flexneri* (5, 76), *Escherichia coli* (74, 79), and *Campylobacter jejuni* (105).

METHODS FOR EXAMINING BACTERIAL GENE EXPRESSION IN VIVO

This section briefly discusses the techniques used to study bacterial gene expression during infection (in vivo). It is not intended to provide detailed technical descriptions but, rather, to present an overview of the various methods for studying bacterial gene expression in vivo (for more technically detailed reviews, see reference 56). All the techniques described in this section (with the exception of in vivo-induced antigen technology [IVIAT]) involve the examination of gene expression at the transcriptional level. Traditionally, these techniques have been used individually; there is, however, an increasing trend toward the simultaneous use of multiple techniques to circumvent the disadvantages associated with each method.

In Vivo Expression Technology

In vivo expression technology (IVET) was one of the first techniques applied to the identification of in vivo-induced (*ivi*) genes (reviewed in reference 1). In

Susan M. Butler, Anna D. Tischler, and Andrew Camilli • Department of Molecular Biology & Microbiology, Tufts University School of Medicine, 150 Harrison Ave., Boston, MA 02111.

IVET, transcriptional fusions are generated within the genome of the pathogen by random insertion of a promoterless reporter gene. One advantage of IVET is that even genes essential for survival in vivo can be identified if the fusions are generated by insertion-duplication of a nonreplicating vector harboring random genomic fragments; the resulting merodiploids often contain an undisturbed (wild-type) copy of the gene in addition to the reporter fusion. In the first two variations of IVET developed, expression of the reporter gene is required for survival of the bacteria in vivo. This was achieved either by complementing a Δ*purA* strain that is otherwise rapidly cleared from the host due to purine auxotrophy, using fusions to a promoterless *purA* gene (92, 122), or by using fusions to a promoterless antibiotic resistance gene (*cat*) and administering the antibiotic during infection (93). Due to the stringent conditions, only bacteria harboring fusions that are well expressed throughout the infection period will survive. Therefore, one disadvantage of these methods is that genes that are transiently expressed or expressed at low levels in vivo will not be identified.

A variation of IVET known as recombinase-based IVET (RIVET) circumvented this drawback by generating transcriptional fusions to a promoterless resolvase gene (*tnpR*). Such fusions are generated in a strain background that harbors an antibiotic resistance marker (e.g., *tet*) flanked by two DNA sequences, *res,* which constitute the substrate for the resolvase (12). Expression of the fusion results in the irreversible excision (resolution) of the *res-tet-res* antibiotic resistance cassette. Conversion from antibiotic resistance to antibiotic sensitivity is therefore a direct result of gene induction. The irreversible nature of the resolution reaction has allowed RIVET to be used to examine the kinetics of bacterial gene induction in vivo (84–86). RIVET cannot give information about later changes in expression, however, due to the irreversible nature of the resolution, and the levels of gene induction observed using RIVET are only qualitative (99). Quantitative gene expression data can be generated using quantitative PCR on cDNA from infected tissue. This technique has been used to examine *Helicobacter pylori* gene expression in animal and human gastric mucosa (112), but has yet to be successfully applied to bacteria in intestinal tissue.

Use of GFP

A technique known as differential fluorescence induction, which uses random transcriptional fusions of genomic DNA to a promoterless gene encoding the enhanced green fluorescent protein (GFP) (GFPmut3) of *Aequorea victoria* (20), has been extensively used to identify *ivi* genes in cell culture models of infection (131, 132). Expression of the GFP fusion results in green fluorescence that can be detected by flow cytometry, and fluorescing bacteria can be isolated using fluorescence-activated cell sorting. The temporal kinetics of gene expression can also be monitored by fusing a promoter of interest to the promoterless *gfpmut3* gene and performing flow cytometry analysis on bacteria recovered from the host. One advantage of this technique is that it can be used to monitor transient gene expression, although the relatively long half-life of GFP presents a limitation to this application. Until recently, however, GFP fusions were unsuitable for monitoring gene expression in intestinal homogenates due to the intrinsic fluorescence of host particles that could not be distinguished from bacterial fluorescence (10). Two-color flow cytometry, which discriminates between host cell autofluorescence (yellow) and the green fluorescence of GFP, has allowed this distinction to be made and has been used to study temporal and spatial induction of gene expression following oral infection of mice (10).

Microarrays

Since the advent of genome sequencing, the use of DNA microarrays, which allow the investigator to monitor gene expression across the entire bacterial genome under a particular experimental condition (transcriptional profiling), has become widespread. Although microarrays have been most commonly used to monitor host expression in response to bacterial infection, in a few specialized cases they have been applied to studying infections from the pathogen's perspective. This technique involves isolating total RNA from an infected tissue, performing cDNA synthesis using a fluorescent deoxyribonucleotide tag, and hybridizing these cDNAs to arrayed oligonucleotides or PCR products of open reading frames encompassing the genome. Array experiments can be performed in several ways. Either cDNAs from two conditions (e.g., infected tissue and in vitro-grown bacteria) can be directly compared to each other by labeling the cDNAs with two different fluorescent tags, or cDNAs from each condition can be compared to a common internal reference (either cDNA or genomic DNA). In the latter method, the cDNA to internal reference ratios are compared between conditions to detect differences in gene expression. Although the expression ratios are a more qualitative than quantitative measure of relative gene expression, the power to simultaneously monitor the expression of so many genes is unparalleled. In most experimental infection situations, however, transcriptional profiling of the pathogen is made impossible by the presence of large amounts of contaminating host RNA. For this reason, this method has been limited to a handful of special situations where large numbers of bacteria can be recovered from infected tissue or fluid in relatively pure form.

In Vivo-Induced Antigen Technology

One caveat to using animal models is that they often only partially reproduce the disease observed in human infections. On the other hand, in many cases there is little knowledge about what actually happens in the human host during infection. IVIAT, one method that seeks to circumvent this problem, involves using convalescent-phase serum from patients who were infected with a particular organism to identify antigenic proteins expressed during infection (53). Pooled sera are repeatedly absorbed against in vitro-grown bacteria to remove antibodies to proteins expressed in vitro. The sera are then incubated against an expression library to identify proteins expressed during infection. One major advantage of IVIAT is that it provides information about the set of antigenic proteins that are most highly expressed during an actual human infection—potentially an excellent method for the identification of vaccine candidate antigens. However, one complicating factor for the purposes of this chapter is the inability to distinguish between the antigenic proteins expressed during infection of the intestine and those expressed extraintestinally by pathogens that disseminate from the intestine.

Signature-Tagged Mutagenesis

In contrast to IVET and DFI, signature-tagged mutagenesis (STM) is a negative selection technique that looks for bacterial mutants that grow normally in vitro but do not survive and/or multiply during an infection (62). This technique is therefore best suited for discovering factors required for infection; it only indirectly indicates in vivo expression. STM is therefore not discussed further in this chapter, but we refer the interested reader to the following references: *Salmonella* (62, 120), *Yersinia* (22, 95); *Vibrio cholerae* (18, 100), *L. monocytogenes* (2), and *E. coli* (94).

VIBRIO CHOLERAE

In contrast to the other bacterial pathogens discussed in this chapter, *V. cholerae,* which causes the acute intestinal disease cholera, colonizes solely at the intestinal mucosal surface and does not disseminate into deeper tissue (25). The profuse secretory diarrhea characteristic of cholera ("rice water" stool) results from the action of cholera toxin (CT). CT is an ADP-ribosylating toxin (47), encoded by the *ctxAB* genes on the lysogenic CTXϕ (133). The receptor for this phage is the type IV bundle-forming toxin coregulated pilus (TCP) (133), which itself is required for *V. cholerae* infection of humans (64) and infant mice (129). Although rabbit ligated ileal loops are also used as a model for *V. cholerae* infection, the susceptibility of suckling mammals to *V. cholerae* colonization makes infant mice the most commonly used animal model (81).

V. cholerae Intestinal *ivi* Genes

V. cholerae was the first organism studied using RIVET, and its characterization has provided important insights into mucosal colonization (13, 98; C. G. Osorio, J. A. Crawford, J. Michalsky, H. Martinez-Wilson, J. B. Kaper, and A. Camilli, submitted for publication; D. Merrell, M. Angelichio, and A. Camilli, unpublished data). A number of metabolic and nutrient-scavenging genes are induced during infection of infant mice (Table 1), suggesting that the infant-mouse small intestine is a nutrient-limiting environment. Genes involved in motility and chemotaxis were also identified. The phenotypes of virulence gene expression and motility are thought to be linked at the level of expression (45), and the presence of bile was shown to result in increased motility and decreased virulence gene expression (51, 118). It has therefore been suggested that *V. cholerae* is motile on entering the small intestine but that after it enters the crypt spaces, the reduced concentration of bile results in induction of virulence gene expression and inhibition of motility (83).

Although RIVET has been applied to two animal models of *V. cholerae* infection (Table 1), of the 16 genes that were induced in the rabbit model only 6 were also induced in infant mice. One of these genes, *cadA,* which codes for lysine decarboxylase (98), is known to be involved in the acid tolerance response of *V. cholerae* and *Salmonella enterica* serovar Typhimurium (108). Although *cadA* was subsequently shown to be dispensable for *V. cholerae* colonization of the infant mouse small intestine, acid-adapted *V. cholerae* has a tremendous competitive advantage over nonadapted bacteria during infection (98). Whether this competitive advantage is due specifically to acid tolerance or to induction of a general stress response as a result of exposure to low pH is unknown.

DNA microarrays have been recently used to examine *V. cholerae* gene expression in rabbit ligated

Table 1. Genes induced during intestinal infection by bacterial pathogens

Organism and gene designation[a]	Gene name	Protein function	Role in pathogenesis	Animal model	Method	Reference
V. cholerae						
VC2316	*argA*	*N*-Acetylglutamate synthase	Arginine/polyamine synthesis	Infant mouse	RIVET	13
VC2641	*argH*	Arginosuccinate lyase	Arginine/polyamine synthesis	Infant mouse	RIVET	Osorio et al., submitted
VC0385	*cysI*	Sulfite reductase	Sulfur metabolism	Infant mouse	RIVET	13
VC0384	*cysJ*	Sulfite reductase	Sulfur metabolism	M9 and infant mouse	RIVET	M. J. Angelichio and A. Camilli; unpublished
VC1137	*hisA*	Ribotide isomerase	Histidine biosynthesis	Infant mouse	RIVET[b]	Osorio et al., submitted
VC2373	*gltB-1*	Glutamate synthase	Glutamate biosynthesis	Infant mouse	RIVET[b]	Osorio et al., submitted
VC1173	*trpG*	Anthranilate synthase	Aromatic amino acid synthesis	Infant mouse	RIVET	Osorio et al., submitted
VC0063	*thiF*	Thiazole synthase	Thiamine biosythesis	Infant mouse	RIVET	Osorio et al., submitted
VC1111	*bioA*	Aminotransferase	Biotin synthesis	Infant mouse	RIVET[b]	Osorio et al., submitted
VC1658	*sdaC-2*	L-Serine transporter	Amino acid transport	Infant mouse	RIVET	Osorio et al., submitted
VC0541	*cysA*	Sulfate ABC, ATP binding	Sulfate transport	Infant mouse	RIVET	13
VC0207		Phosphotransferase system: IIC (sucrose)	Carbohydrate transport	Infant mouse	RIVET	Osorio et al., submitted
VC0721		Phosphate-binding protein	Phosphate transport	Infant mouse	RIVET	Osorio et al., submitted
VC2705		Sodium/solute symporter	Unknown	Infant mouse	RIVET	Osorio et al., submitted
VCA0529	*kup*	K uptake protein	Potassium transport	Infant mouse	RIVET	Osorio et al., submitted
VCA1008		Outer membrane protein	Unknown	Infant mouse	RIVET	Osorio et al., submitted
VC0201		Iron(III) ABC, ATP binding	Ferric iron uptake	Infant mouse	RIVET	Osorio et al., submitted
VC0202		Iron(III)ABC, iron binding	Ferric iron uptake	Infant mouse	RIVET	Osorio et al., submitted
VC0203		Iron(III) ABC, permease	Ferric iron uptake	Infant mouse	RIVET	Osorio et al., submitted
VCA0687		Iron(III) ABC, ATP binding	Ferric iron uptake	Infant mouse	RIVET[b]	Osorio et al., submitted
VC2087	*sucA*	Succinyl coenzyme A biosynthesis	Energy metabolism: tricarboxylic acid cycle	Infant mouse	RIVET	13
VC1338	*acnA*	Aconitase	Energy metabolism: tricarboxylic acid cycle	Infant mouse	RIVET	Osorio et al., submitted
VC2646	*ppc*	Phosphoenolpyruvate carboxylase	Energy metabolism	Infant mouse	RIVET	Osorio et al., submitted
VCA1057		Short-chain oxidoreductase	Central metabolism	Infant mouse	RIVET	Osorio et al., submitted
VC1693	*torC* (*nirT*)	Cytochrome *c*	Electron transport, anaerobic growth	Infant mouse	RIVET	13
VCA0752	*trxC*	Thioredoxin 2	Electron transport	Infant mouse	RIVET	Osorio et al., submitted
VCA0221	*hlyC*	Triacylglyceride lipase	Fatty acid metabolism	Infant mouse	RIVET	13
VCA0242	*sgbH*	Hexulose-6-phosphate synthesis	Sugar metabolism	Infant mouse	RIVET[b]	Osorio et al., submitted
VCA0014	*malQ*	4-α-Glucanotransferase	Polysaccharide synthesis	Infant mouse	RIVET	Osorio et al., submitted
VCA0016	*glgB*	1,4-α-Glucan branching	Polysaccharide synthesis	Infant mouse	RIVET	Osorio et al., submitted
VC0924	*capK*	Exopolysaccharide synthesis	Unknown	Human serum	IVIAT	54
VC2392	*mutT*	8-oxo-dGTP hydrolysis	DNA repair	Infant mouse	RIVET	Osorio et al., submitted
VC2419	*xerD*	Recombinase/integrase	DNA replication, CTXφ integration/excision	Infant mouse	RIVET	Osorio et al., submitted

VC2621		Extracellular nuclease	Unknown	Infant mouse and rabbit ileal loop	RIVET	D. Merrell and A. Camilli, unpublished; Osorio et al., submitted
VC2742	*rbn*	RNase BN	RNA processing	Infant mouse	RIVET	Osorio et al., submitted
VC1034	*udp-1*	Uridine phosphorylase	Nucleotide metabolism	Infant mouse	RIVET	Osorio et al., submitted
VC1651	*vieB*	Response regulator	Gene regulation	Infant mouse	RIVET	13
VC0622	*baeS*	Sensor kinase	Gene regulation	Infant mouse	RIVET	Osorio et al., submitted
VC0130		GGDEF family protein	Cyclic-di-GMP signaling	Infant mouse	RIVET	Osorio et al., submitted
VC0280	*cadA*	Lysine decarboxylase	Acid tolerance	Rabbit ileal loop and infant mouse	RIVET	98
VC0828	*tcpA*	Toxin-coregulated pilin	Colonization	Human serum	IVIAT	54
VC0412/3	*mshOP*	Mannose-sensitive hemagglutinin pilus biogenesis	Unknown	Human serum	IVIAT[b]	54
VC2423	*pilA*	Type IV pilin	Unknown	Human serum	IVIAT	54
VCA0068	*iviIV*	MCP	Chemotaxis	Infant mouse	RIVET	13
VCA0176		MCP	Chemotaxis	Human serum	IVIAT[b]	54
VCA1056		MCP	Chemotaxis	Human serum	IVIAT	54
VC0216		MCP	Chemotaxis	Human serum	IVIAT	54
VCA0773		MCP	Chemotaxis	Infant mouse	RIVET[b]	Osorio et al., submitted
VC1535		MCP	Chemotaxis	Infant mouse	RIVET	Osorio et al., submitted
VC1397	*cheA-1*	Histidine kinase	Chemotaxis	Human serum	IVIAT	54
VC1399	*cheR*	Methyltransferase	Chemotaxis	Human serum	IVIAT	54
VC2130	*fliI*	Flagellum ATP synthase	Motility and chemotaxis	Infant mouse	RIVET	Osorio et al., submitted
VC1619.1		Cadherin-RTX	Toxin	Infant mouse	RIVET	Osorio et al., submitted
VC2209	*vibF*	Peptide synthetase	Antibiotic synthesis	M9 and infant mouse	RIVET[b]	M. J. Angelichio and A. Camilli, unpublished
S. enterica serovar Typhimurium						
STM0191	*fhuA*	Ferric hydroxamate transport	Ferric iron uptake	Mouse small intestine	IVET	59
STM2557	*cadC*	Cadaverine synthesis	Acid tolerance	Mouse small intestine	IVET	59
STM0405	*vacC (tgt)*	tRNA processing	Posttranscriptional regulation	Mouse small intestine	IVET	59
STM2526	*ndk*	Nucleotide balance	Alarmone synthesis	Mouse small intestine	IVET	59
STM1427	*cfa*	Membrane modification	Stationary-phase survival	Mouse small intestine	IVET	59
STM1928	*otsA*	Trehalose synthesis	Thermo/osmotic protection	Mouse small intestine	IVET	59
STM1281	*phoP*	Response regulator	Gene regulation	Mouse spleen (intragastric route)	IVET	59
STM2599	*gipA*	DNA-binding protein	Unknown	Mouse PP	IVET	125
STM0543	*fimA*	Fimbrial subunit	Colonization	Bovine ileal loop	IF/FC[c]	69
STM0021	*bcfA*	Fimbrial subunit	Colonization	Bovine ileal loop	IF/FC	69
STM3640	*lpfA*	Fimbrial subunit	Colonization	Bovine ileal loop	IF/FC	69
PSLT018	*pefA*	Fimbrial subunit	Colonization	Bovine ileal loop	IF/FC	69
STM0340	*stbA*	Fimbrial subunit	Colonization	Bovine ileal loop	IF/FC	69
STM2152	*stcA*	Fimbrial subunit	Colonization	Bovine ileal loop	IF/FC	69
STM2039	*stdA*	Fimbrial subunit	Colonization	Bovine ileal loop	IF/FC	69
STM0195	*stfA*	Fimbrial subunit	Colonization	Bovine ileal loop	IF/FC	69
STM0177	*stiA*	Fimbrial subunit	Colonization	Bovine ileal loop	IF/FC	69

Continued on following page

Table 1. *Continued*

Organism and gene designation[a]	Gene name	Protein function	Role in pathogenesis	Animal model	Method	Reference
Y. enterocolitica[d]						
	mtpS	DNA methylase	DNA repair	Mouse PP	IVET	137
	mutL	Mismatch repair	DNA repair	Mouse PP	IVET	137
	recB	Exodeoxyribonuclease	DNA repair	Mouse PP	IVET	137
	mdoH	Periplasmic glucan synthesis	Bile resistance/osmoprotection	Mouse PP	IVET	137
	lpxA	Acyltransferase	Lipid A modification	Mouse PP	IVET	137
	irp2	Yersiniabactin synthesis	Fe^{3+} transport	Mouse PP	IVET	137
	fyuA	Yersiniabactin receptor	Fe^{3+} transport	Mouse PP	IVET	137
	foxA	Siderophore receptor	Fe^{3+} transport	Mouse PP	IVET	137
	yfuB	Fe transporter	Fe^{3+} transport	Mouse PP	IVET	137
	prcA	Protease	Unknown	Mouse PP	IVET	137
	tnp	Transposase	Unknown	Mouse PP	IVET	137
	vacC (*tgt*)	tRNA processing	Posttranscriptional regulation	Mouse PP	IVET	137
	gsh	Glutathione synthesis	Oxidative stress response	Mouse PP	IVET	137
	ydhD	Glutaredoxin	Oxidative stress response	Mouse PP	IVET	137
	hemD	Heme synthesis	Peroxide resistance	Mouse PP	IVET	137
	kpyI	Pyruvate kinase	Unknown	Mouse PP	IVET	137
	yeiY	Transcriptional regulator	Gene regulation	Mouse PP	IVET	137
	hoxQ	Nickel tranporter	Nickel transport	Mouse PP	IVET	137
	cpdP	Cyclic AMP phosphodiesterase	Unknown	Mouse PP	IVET	137
	aceB	Malate synthase	Unknown	Mouse PP	IVET	137

[a]Gene designations provided by The Institute for Genomic Research (http://www.tigr.org/tigr-scripts/CMR2/CMRHomePage.spl).
[b]Greater than twofold induction of gene expression in transcriptional profiles comparing *V. cholerae* grown in rabbit ligated ilial loops to mid-exponential-phase in vitro-grown bacteria (136).
[c]IF/FC, immunofluorescence and flow cytometry.
[d]Gene designations have not yet been assigned to the *Y. enterocolitica* genome.

ileal loops (136). Growth of *V. cholerae* to high cell densities in the lumen of the ligated loops allowed sufficiently pure *V. cholerae* RNA to be isolated. In these experiments, bacteria in the loops were growing at an exponential rate, and transcriptional profiles were compared with in vitro patterns of gene expression in *V. cholerae* cells grown aerobically to mid-exponential phase in rich medium. Transcriptional profiling indicated that *V. cholerae* experiences anaerobiosis as well as iron and nutrient limitation during infection of the ligated loops (136). The *V. cholerae* genome is composed of two chromosomes (58); interestingly, a shift toward increased expression of genes located on chromosome II during infection of rabbit ligated loops was observed (136).

Surprisingly, *ctxA* and *tcpA*, which codes for the pilin subunit of TCP (129a), were not expressed appreciably in rabbit ligated ileal loops (7, 136). Although it is unknown which animal model best represents the human host (83), these data raise questions about the validity of the rabbit ligated ileal loop model for studying virulence gene expression. We anticipate, however, that microarray technology will also be used to study gene expression during infection of infant mice, perhaps aided by the use of gene-specific primers to increase the relative ratio of bacterial cDNA to host cDNA (128). In addition, the use of microarrays for temporal analyses during infection is likely to lead to new discoveries regarding gene expression patterns in vivo.

The expression of *V. cholerae* genes in human diarrhea (rice water stool) has also been examined using microarrays (7, 97). Whereas one group found many genes to be induced in stool relative to in vitro-grown *V. cholerae* (7), the other found that the majority of differentially expressed genes were repressed, including those involved in chemotaxis (97). Expression of *ctxA* and *tcpA* was detected at intermediate levels in one study (7) but was not detected in the other study (97). However, these studies each used different internal reference standards and different stages of growth in vitro (exponential phase [7] versus stationary phase [97]), and thus a direct comparison between these sets of results cannot easily be made (7). Both groups, however, saw induction of genes characteristic of the presence of low iron levels and anaerobiosis. These are conditions that were previously shown to be present in rice water stool (38).

Unlike RIVET and DNA microarrays, which monitor transcription induction and steady-state mRNA levels, respectively, IVIAT was used to examine the protein expression of *V. cholerae* during human infection. Although TCP is known to be required for *V. cholerae* infection, this bacterium also encodes two additional type IV pili: PilA and mannose-sensitive hemagglutinin (58). All three pili were determined to be expressed using IVIAT (54) (Table 1), but, in contrast to TCP, mannose-sensitive hemagglutinin was dispensable for infection of humans (127) and PilA was not required for infection of infant mice (40). In addition to pili, the second major class of expressed antigenic proteins identified by IVIAT are involved in chemotaxis (Table 1). This class includes three methyl-accepting chemotaxis proteins (MCPs) as well as the histidine kinase CheA (VC1397) and the methyltransferase CheR (VC1399). In contrast to the well-studied *E. coli* chemotaxis paradigm, widespread sequencing of bacterial genomes has revealed that many bacterial species encode multiple paralogs of chemotaxis genes (8). *V. cholerae* is no exception, having 43 genes encoding MCPs and three paralogs each of CheA and CheR (58). The CheA paralogue expressed during human infection was previously shown not to be required for chemotaxis in vitro (48). Moreover, we have found that the CheR identified using IVIAT is dispensable for chemotaxis of *V. cholerae* in vitro and that neither CheA and CheR paralogues are required for infection of infant mice (S. Butler and A. Camilli, unpublished data). The role of these chemotaxis paralogues has not yet been uncovered. Since neither CheA nor CheR is predicted to be exported or membrane localized, this suggests that bacterial lysis occurs during infection before presentation to the immune system.

Spatial and Temporal Analysis of *V. cholerae* Intestinal Gene Expression

Unlike other IVET-based methods, RIVET can be used to examine the spatial and temporal expression of genes in vivo (99) (Table 2). This was first used to monitor the expression of the *ivi* gene *vieB*, which encodes a response regulator (13). By infecting infant mice with a *vieB*::*tnpR* fusion strain, the expression of *vieB* could be monitored as a function of both time and location within the small intestine. Expression of this fusion was induced in the duodenum specifically at 3 to 4 h postinfection (84). Despite knowledge of the spatial and temporal expression of *vieB* during infection, the precise role of *vieB* remains to be determined. However, a role in the positive regulation of CT expression was recently assigned to the response regulator *vieA*, a part of the three-component regulatory system that includes *vieB* and the sensor kinase *vieS* (130).

Due to the sensitivity of RIVET to low levels of *tnpR* transcription, this technique was initially suitable only for studying the in vivo expression of genes that are transcriptionally silent during growth in vitro. Therefore, a "tunable" RIVET system was

Table 2. Genes whose spatial and temporal expression patterns in the intestine have been studied

Organism and gene designation[a]	Gene name	Expression patterns	Regulators of observed expression pattern	Method	Reference(s)
V. cholerae					
VC1651	*vieB*	Induced 4 h postinfection requires TcpA-mediated colonization; expressed in duodenum	Unknown	RIVET	84
VC0828	*tcpA*	Biphasic pattern of induction (low-level expression at 2 h, second phase of induction at 4–5 h)	ToxT absolutely required; ToxR needed at 2 h; TcpPH not required	RIVET	86
VC1457	*ctxA*	Induced 3–4 h postinfection; requires TcpA-mediated colonization	ToxR and ToxT required; TcpPH not required; VieS required for maximum expression	RIVET	86, 130
VC0838	*toxT*	Induced 3 h postinfection	Nonchemotactic mutants exhibit delayed induction	RIVET	85, 86
S. enterica serovar Typhimurium					
STM2886	*sicA* (SPI-1)	Induced in lumen 1 h postinfection; expressed in PP 5 h postinfection; undetectable after 24 h in PP; undetectable in liver and spleen	Unknown	GFP	10
STM1407	*ssaH* (SPI-2)	Not expressed in intestinal lumen; induced in PP at 5 h postinfection; expressed through 4 days in PP, liver, and spleen	Unknown	GFP	10
Y. enterocolitica	*fhuA* and *hemR*	Induced in PP at 24 h; not expressed in lumen	Unknown	GFP	72
	yopE	Expressed in PP at 24 h; not induced in lumen	Unknown	GFP	73

[a]See Table 1, footnote *a*.

developed that is less sensitive and can be used to study the expression of genes with low to moderate basal levels of transcription in vitro (86). Using this tunable RIVET, important differences between the regulation of *V. cholerae* virulence factor expression in vitro and in vivo were observed (86). Although *tcpA* and *ctxA* are coregulated under in vitro-inducing conditions, the expression of *tcpA* was found to occur before that of *ctxA* during infection of infant mice (Table 2). In addition, the induction of *ctxA* was dependent on the presence of *tcpA* in vivo but not in vitro. This led the authors to propose a spatiotemporal model for coordinated CT and TCP expression in vivo, whereby TCP is expressed at the earliest stage of infection to aid in microcolony formation on the intestinal epithelium followed by expression of CT in response to unknown signals present at the epithelial surface (86).

ToxR and TcpP are inner membrane-associated transcriptional regulators that induce the expression of *toxT* cooperatively and in response to particular environmental signals (15, 55, 65, 103). ToxT is an AraC family transcriptional regulator that directly activates the transcription of the CT and TCP structural genes as well as other virulence factors (26, 66). Interesting differences with respect to the regulation of *ctxA* and *tcpA* expression were revealed using the tunable RIVET system (Table 2). Specifically, although in vitro induction of *tcpA* required the expression of both *toxR* and *tcpP*, individually these regulators were not required for expression in vivo. In contrast, although *ctxA* did not require *tcpP* for induction in vivo, *toxR* was required. These studies revealed facets of *V. cholerae* in vivo virulence gene regulation that could never have been realized using in vitro conditions alone. These studies demonstrate the importance of studying the expression of virulence genes during infection.

Additionally, RIVET was applied in a genetic selection to identify in vivo-specific positive regulators of the *V. cholerae* virulence genes *ctxA* and *toxT* (85). RIVET reporter strains (e.g., *ctxA*::*tnpR* and *res-tet-res*) were mutagenized with a transposable element and passaged through infant mice, and mutant strains that failed to induce the *tnpR* fusion in vivo were selected by growth ex vivo in the presence of tetracycline. A number of genes disrupted by the transposon were identified as regulators, including several chemotaxis genes (Table 3). RIVET was then used to examine the temporal in vivo induction of

Table 3. In vivo regulators of *V. cholerae ctxA* and *toxT* genes identified using RIVET[a]

Organism and gene designation[b]	Gene name	Regulated *tnpR* fusion	Putative protein function
VC2063	*cheA-2*	*toxT*	Chemotaxis sensor kinase
VC2065	*cheY-3*	*toxT*	Chemotaxis response regulator
VC2064	*cheZ*	*ctxA*	CheY phosphatase
VC2161	*mcpX*	*ctxA*	Methyl-accepting chemotaxis protein
VC1653	*vieS*	*ctxA*	Three-component sensor kinase
VC0866	*sltA*	*toxT*	Soluble lytic transglycosylase
VC1336	*prpB*	*toxT*	Carboxy-phosphoenolpyruvate phosphomutase
VC0924	*capK*	*toxT*	Biofilm-associated exopolysaccharide synthesis
VC1448	*rtxB*	*toxT*	ABC transporter for RTX toxin

[a]Data from reference 85.
[b]See Table 1, footnote *a*.

ctxA and *toxT* expression in strains with mutations in several of the regulators identified by the selection. Although chemotaxis mutants had only modest defects in the extent of induction of these virulence genes (with the exception of the *cheZ* mutant, which had a severe defect), defects with respect to the timing of induction were observed. We think the primary reason why mutations in chemotaxis genes were isolated in the selection, however, is that nonchemotactic strains of *V. cholerae* outcompete the wild type in the infant-mouse small intestine (37, 85) and that these mutants were therefore selected for in vivo. Whereas wild-type *V. cholerae* mainly colonizes the ileum in the early stages of infection, a nonchemotactic *cheY-3D60N* mutant colonizes all sections of the infant-mouse small intestine (85). Using RIVET, *ctxA* and *tcpA* expression was found to be induced to the same extent throughout the small intestine in the *cheY-3D60N* mutant (11). This implies that the signals required for *V. cholerae* virulence gene induction are present throughout the small intestine, not just in the ileum, where wild-type *V. cholerae* predominantly colonizes. In addition, it suggests that detection of these signals by *V. cholerae* does not require chemotaxis.

We observe a complex contribution of chemotaxis to *V. cholerae* pathogenicity. Our data suggest that the increased colonization efficiency of chemotaxis mutants requires the expression of counterclockwise-biased flagellar rotation, since clockwise-biased flagellar mutants are attenuated in colonization. Analysis of net movement of these biased flagellar rotation mutants reveals that counterclockwise mutants and their wild-type parents travel in straight runs, which are likely to encounter the mucosa, whereas clockwise mutants undergo frequent changes in direction and may fail to reach the epithelium. The discovery of a complex role for chemotaxis, which remains partially obscure, represents another important contribution of in vivo expression technology.

S. ENTERICA SEROVAR TYPHIMURIUM

Salmonella species are responsible for a number of diseases ranging from enterocolitis to enteric fever (23). Of the serotypes of *Salmonella* that have been described, much work has focused on *S. enterica* serovar Typhimurium. Since serovar Typhimurium is capable of infecting a wide range of warm-blooded animals, several models have been developed for serovar Typhimurium infections. As is the case with many bacterial pathogens, however, mice have become the most commonly used animal model. Results from various animal models, including mice, show that serovar Typhimurium attaches to and subsequently invades the intestinal epithelium. This invasion leads to colonization of the mesenteric lymph nodes and more distal sites such as the liver and spleen (23, 78, 124, 134, 135).

S. enterica Serovar Typhimurium Intestinal *ivi* Genes

Both auxotrophic- and antibiotic-based IVET strategies have been used to identify *S. enterica* serovar Typhimurium intestinal *ivi* genes during infection in mice (59, 92, 93, 122, 123). Since 25% of serovar Typhimurium cells within the murine small intestine are located in the intestinal lymphoid aggregates known as Peyer's patches (PP) (67), bacteria have been isolated from both the PP and small intestine as a whole. Similar to other IVET screens, the major classes of genes identified were involved in metabolism and stress response (Table 1). One such gene is *fhuA* (ferric hydroxamate uptake), which encodes a siderophore uptake system (59). The expression of *fhuA* is induced fivefold under iron-limiting conditions in vitro. The identification of *fhuA* as an intestinal *ivi* gene is consistent with the small intestine being an iron-limiting environment (60).

Two of the intestinal *ivi* genes identified, *vacC* and *ndk,* are involved in nucleotide metabolism (Table 1) (59). *vacC* encodes a tRNA transglycosylase (*tgt*) (30), an enzyme required for introducing the precursor of a modified nucleoside called queuosine into a subset of tRNAs (106). In addition to improving the survival of *E. coli* during stationary phase (106), the presence of queuosine influences codon choice (96) and prevents reading of the stop codon UAG by $tRNA^{Tyr}$ (6, 39). Tgt does not, however, have a general effect on the translation of mRNAs; instead, it influences the translation of specific mRNAs (29). For example, Tgt is required for the translation of *virF* mRNA (29, 30), which encodes the major virulence regulator of *Shigella flexneri* (27). It is unknown whether Tgt is required for *S. enterica* serovar Typhimurium virulence or the expression of specific serovar Typhimurium virulence factors in vivo.

Nucleotide diphosphate kinase, an enzyme important for maintaining levels of nucleoside triphosphates and their deoxy derivatives within the cell (111), as well as for synthesis of GTP and the alarmone ppGpp (80), was also identified as an intestinal *ivi* gene (59). Levels of ppGpp rise dramatically in *E. coli* in response to amino acid, carbon, or inorganic phosphate starvation, and this results in activation of the stationary-phase sigma factor RpoS (46). RpoS controls the expression of many genes that enable cells to survive in the stationary phase (89) and is required for the full virulence of *S. enterica* serovar Typhimurium (33). Interestingly, two additional intestinal *ivi* genes identified (*cfa* and *otsA*) are regulated by RpoS (59). Thus, nucleotide diphosphate kinase may play a role in the activation of RpoS during infection, enabling serovar Typhimurium to survive nutrient-limiting conditions present in the small intestine. Although both *tgt* and *ndk* are involved in nucleotide metabolism, given the importance of these genes in other processes, it is likely that they also play roles in the serovar Typhimurium stress response during intestinal infection.

Genes encoding transcriptional regulatory factors, including *phoP,* were absent from the set of *S. enterica* serovar Typhimurium *ivi* genes induced specifically in the small intestine. PhoP is the transcriptional activator of the two-component PhoPQ system (49) that has been previously demonstrated to be required for virulence in systemic infection models (34, 101), as well as for survival within the macrophage phagosome (102). The *phoP* gene was identified as a spleen *ivi* gene following both oral and intraperitoneal infections (59), suggesting that *phoP* might be expressed both at early stages of intestinal colonization and at later, systemic stages of disease. It is therefore likely that *phoP* is also induced in the intestinal tract, although this has not been demonstrated experimentally. In addition to low Mg^{2+} levels and low pH, conditions that activate the PhoPQ regulatory system in vitro (44), cationic antimicrobial peptides were recently shown to induce in vitro expression of *phoP* (3). Since PhoPQ is required for the resistance of serovar Typhimurium to antimicrobial peptides (32) expressed both in phagocytic vacuoles and on mucosal surfaces (52, 87), the presence of antimicrobial peptides in the intestine may help to induce *phoP* expression. Mice infected with wild-type serovar Typhimurium exhibit reduced expression of antimicrobial peptides, but *phoP* mutant strains do not elicit this effect (117). It is interesting to speculate that secretion of antimicrobial peptides by Paneth cells within the small intestine induces *phoP* expression by serovar Typhimurium, resulting not only in resistance to these compounds but also in a decrease in their production, enabling increased survival of the bacteria. Whether a similar phenomenon occurs for virulence regulators of other intestinal pathogens is unknown.

In contrast to PhoP, which is thought to be expressed at both early and late stages of infection, the *gipA* gene is expressed specifically in the PP but not at systemic sites (125). This gene, which is carried on a lysogenic bacteriophage called Gifsy-1, is required for survival and/or growth in the PP. The presence of virulence genes on bacteriophage is not uncommon; for example, the structural genes for *V. cholerae* CT are located on CTXϕ (133). The authors propose that GipA may be involved in the regulation of virulence factors due to the presence of a zinc finger DNA-binding motif within the protein. The precise function of this *ivi* gene remains unknown.

Spatial and Temporal Analysis of *S. enterica* Serovar Typhimurium Intestinal *ivi* Gene Expression

Although numerous *S. enterica* serovar Typhimurium intestinal *ivi* genes have been identified, there is a dearth of information regarding the spatial and temporal expression of these and other virulence factors within the intestine. GFP fusions have proven to be an invaluable tool for monitoring the temporal expression of serovar Typhimurium genes in cell culture and in vitro (19, 131, 132). Now, with the advent of two-color flow cytometry, the expression of GFP fusions can be examined during infection as well. The induction of two promoters previously implicated in serovar Typhimurium pathogenesis has been studied using this technique (Table 2) (10). The *PsicA* promoter drives the expression of the *Salmonella* pathogenicity island 1 (SPI-1) *sicA sipBCDA iacP* operon (23). SPI-1 encodes a type III secretion apparatus (68, 90) that is required for invasion of epithelial cells (23, 43). The second promoter examined, *PssaH*, is located within another pathogenicity island (SPI-2) and drives the expression of the *ssaHIJKLMV* operon (19). SPI-2 is expressed on entry into cultured mammalian cells and is required for intracellular growth in host cells and systemic infection (61, 63). In addition to induction in phagocytes, SPI-2 was induced in infected epithelial cells (109).

Using two-color flow cytometry, *PsicA* was determined to be activated in the intestinal lumen 1 h after infection and in the PP 5 h after infection. However, by 24 h, activation of *PsicA* was no longer observed in the PP. This promoter was inactive during growth in Luria broth (LB). Despite the presence of heavy colonization by *S. enterica* serovar Typhimurium, no induction of the *PsicA* promoter was observed in the MLN, liver, or spleen. It was concluded that the *PsicA* promoter is active very early during infection of the small intestine but inactive during later stages of infection. SPI-1 expression is induced in vitro by oxygen limitation and high osmolarity, conditions that are considered to be present in the intestinal lumen (4). Thus, the in vivo data (10) are consistent with the in vitro data regarding the induction of SPI-1. Furthermore, Bajaj et al. found that induction of SPI-1 requires *phoPQ* in vitro (4). These data are consistent with the induction of *phoPQ* expression during intestinal infection, as discussed above. It would therefore be most interesting to determine the temporal and spatial induction of *phoPQ* during infection of mice by using two-color flow cytometry.

In contrast to *PsicA*, the *PssaH* promoter was inactive in LB and in the intestinal lumen. However, this promoter was active in PP 5 h after infection and remained active in PP during the first 4 days of infection (10). Induction was also observed in infected systemic organs. Therefore, this promoter is active in both the early and late stages of infection. The spatial and temporal analyses of the *PsicA* and *PssaH* promoters within the intestine show that both SPI-1 and SPI-2 are induced in the PP at 5 h. The early induction of SPI-1 at 5 h in the PP is consistent with its requirement for invasion of the PP. Furthermore, the induction of SPI-2 is consistent with intracellular bacterial growth within the PP. The fact that SPI-1 is no longer expressed in the PP at 24 h but that SPI-2 remains induced for up to 4 days after infection implies that nearly all the bacteria in the PP are growing intracellularly at this time point.

In addition to monitoring the transcriptional induction of GFP fusions in vitro, flow cytometry has been used to study *S. enterica* serovar Typhimurium protein expression during intestinal infection (69). The serovar Typhimurium genome has 13 operons with homology to fimbrial gene sequences, and there is thought to be functional redundancy with respect to the requirements for fimbriae during infection (69). Humphries et al. examined fimbrial expression in serovar Typhimurium isolated from bovine ligated ileal loops following 8 h of infection (69). The recovered bacteria were incubated with individual highly specific antisera against each of the predicted fimbrial subunit proteins and incubated with fluorescently labeled secondary antibody. Flow cytometry was then used to measure the fluorescence for each fimbrial subunit. Although FimA was the only fimbrial subunit detected following static growth in LB, nine of the fimbrial subunits (FimA, BcfA, LpfA, PefA, StbA, StcA, StdA, StfA, and StiA) were detected following growth in vivo (Table 1). This technique represents a novel approach to examining the expression of bacterial surface proteins during infection.

YERSINIA

Of the three *Yersinia* species that infect humans, *Y. enterocolitica* and *Y. pseudotuberculosis* are enteric pathogens. Both species are associated with gastroenteritis, produce an enterotoxin that causes diarrhea (24), are lymphotropic, and can cause systemic infections in rare instances (9). Mice are the most commonly used animal model for *Yersinia* infection, and like *Salmonella, Y. enterocolitica* can colonize the PP of mice (16). Invasion of *Yersinia* into the M cells of the PP is promoted by binding of the bacterial protein invasin to β_1-integrins (70, 71), which are expressed on the apical side of M cells (119). Like *Salmonella, Yersinia* ultimately disseminates from the PP to systemic sites of infection. In contrast to *Salmonella,* however, after the initial invasion of M cells *Yersinia* is predominantly an extracellular pathogen for the remainder of the infection (21, 121).

Y. enterocolitica Intestinal *ivi* Genes

Y. enterocolitica harbors a virulence plasmid, known as pYV, that carries many genes required for virulence (88). Using a strain lacking this plasmid, *Y. enterocolitica* chromosomal intestinal *ivi* genes were identified in the PP of mice 24 h after oral infection (137). An antibiotic-based IVET selection strategy was employed wherein fusions that resulted in strains that were able to survive in vivo in the presence of antibiotic but that were unable to grow in either LB or a minimal medium plus antibiotic were characterized. The in vitro growth parameters were included in the selection strategy so that genes required for biosynthesis or nutrient uptake under nutrient-limiting conditions would not be identified. Therefore, in contrast to the results of other IVET screens, in which these two classes make up the bulk of identified genes, relatively few genes in these two categories were identified (Table 1).

The predominant class of *ivi* genes in *Y. enterocolitica* included genes involved in the stress response (Table 1); three of these genes are involved in DNA repair (*mtpS, mutL,* and *recB*). In addition, two genes involved in cell envelope maintenance were identified and may be expressed to counteract stresses that occur during infection. The first, *mdoH,* was required for optimal survival and/or growth in PP (137). In *E. coli,* MdoH is involved in the synthesis of membrane-derived oligosaccharides, a class of glucose polymers located in the periplasmic space that are thought to contribute to bile resistance and osmoprotection (77, 107). Membrane-derived oligosaccharides are required for virulence in several plant pathogens and are thought to play a role in host recognition and survival under low-osmolarity conditions (77, 91, 107). The second gene involved in cell envelope maintenance, *lpxA,* was essential for full virulence of *Y. enterocolitica* (137). *lpxA* encodes an acyltransferase and is predicted to modify lipid A, a major component of the outer leaflet of the outer membrane (137).

The members of the second major class of *Yersinia ivi* genes are involved in the iron starvation response (Table 1). Yersiniabactin (Ybt) is a siderophore that is essential for virulence (57, 110) as well as for growth in medium depleted of iron (14). The *irp2* and *fyuA* genes, which are required for Ybt synthesis and uptake, respectively (110), were identified as intestinal *ivi* genes. In addition to Ybt-associated genes, those involved in uptake of the TonB-dependent siderophore ferrioxamine (*foxA*) (31) and the TonB-independent, siderophore-independent iron transport system (*yfuB*) (116) were identified. Although only Ybt is required for infection (57), it appears that other iron uptake systems are also expressed during intestinal infection. The reason for the requirement of Ybt but not the other iron uptake systems during infection is unclear.

Another *ivi* gene encoding a protease with homology to PrcA was found to be required for the survival of *Yersinia* and its spread to sites of systemic infection (137). The exact role of this protease during infection remains unclear. Finally, *Y. enterocolitica tgt* was identified as an *ivi* gene. It is interesting that *tgt* was identified as an *ivi* gene in both *Yersinia* and *Salmonella,* suggesting that *tgt* acts as a posttranscriptional regulator of in vivo gene expression in multiple enteric pathogens.

Spatial and Temporal Analysis of *Y. enterocolitica* Intestinal Gene Expression

The expression of *fyuA* and *hemR,* which encode the receptors for Ybt and heme uptake, respectively (126), were examined during infection of mice by constructing translational fusions of these genes to *gfp* (72). Neither the *fyuA* nor the *hemR* fusion was induced in the lumen of the small intestine, but induction was seen in PP after a 24-h infection. Since the intestinal lumen is anaerobic, iron is predicted to be present in the more soluble ferrous form and to be transported via the Fe(II) active-transport Feo system (75). The authors propose that there is an increase in the concentration of Fe(III) in PP, thus necessitating the need for Ybt synthesis and uptake.

Once the *Yersinia* enter PP, they secrete a battery of proteins (Yops) via a contact-dependent TTS apparatus (9, 17) that collectively acts to prevent phagocytosis by macrophages. YopE is a cytotoxin that induces disruption of the actin microfilament structure of the host cell (113, 114). YopH is a protein tyrosine kinase (50) that inhibits phagocytosis (115). The in vivo expression of *yopE* and *yopH* was originally examined in *Y. pseudotuberculosis* by using transcriptional fusions to a promoterless *luxAB* operon (35). Both genes were expressed in PP, and expression was highest on day 1, the early stage of colonization. This was confirmed for *Y. enterocolitica* by a subsequent study using a *yopE::gfp* fusion (73). Expression of *yopE::gfp* was observed at 24 h in PP, although expression was not monitored past this time point. In addition, *yopE::gfp* was not expressed in the intestinal lumen. The expression data for *yopE* and *yopH* within PP are consistent with a role for these proteins in phagocytosis inhibition, enabling the proliferation of *Y. enterocolitica* at this site as extracellular microcolonies (9).

CONCLUSIONS

The examination of gene expression in diverse bacterial pathogens during intestinal infection has revealed much about the conditions experienced during this process. It is evident that these pathogens are adapting to nutrient-deprived, chemically stressful conditions in similar ways, and it appears that pathogens have evolved redundant mechanisms for certain important functions needed for survival and growth in the host intestine. Although a great deal of work has been done to investigate gene expression at the transcriptional level, almost nothing is known about posttranscriptional or posttranslational regulation of virulence factors that may be occurring during infection of the intestinal tract. Discoveries such as the Tgt-mediated posttranscriptional control of virulence in several pathogens suggest that posttranscriptional and perhaps posttranslational regulation may be widespread and important for precisely controlling virulence factor expression during infection. Thus, studies characterizing the proteome of intestinal pathogens in a spatiotemporal manner during infection will be of particular importance in the future. In vivo induction studies point up the exquisite complexity of spatiotemporal gene expression during colonization and disease, illuminating a multistep, finely tuned process, which we are only beginning to understand.

REFERENCES

1. **Angelichio, M. J., and A. Camilli.** 2002. In vivo expression technology. *Infect. Immun.* **70:**6518–6523.
2. **Autret, N., I. Dubail, P. Trieu-Cuot, P. Berche, and A. Charbit.** 2001. Identification of new genes involved in the virulence of *Listeria monocytogenes* by signature-tagged transposon mutagensis. *Infect. Immun.* **69:**2054–2065.
3. **Bader, M. W., W. W. Navarre, W. Shiau, H. Nikaido, J. G. Frye, M. McClelland, F. C. Fang, and S. I. Miller.** 2003. Regulation of *Salmonella typhimurium* virulence gene expression by cationic antimicrobial peptides. *Mol. Microbiol.* **50:**219–230.
4. **Bajaj, V., R. L. Lucas, C. Hwang, and C. A. Lee.** 1996. Coordinate regulation of *Salmonella typhimurium* invasion genes by environmental and regulatory factors is mediated by control of *hilA* expression. *Mol. Microbiol.* **22:**703–714.
5. **Bartoleschi, C., M. C. Pardini, C. Scaringi, M. C. Martino, C. Pazzani, and M. L. Bernardini.** 2002. Selection of *Shigella flexneri* candidate virulence genes specifically induced in bacteria resident in host cell cytoplasm. *Cell. Microbiol.* **4:**613–626.
6. **Bienz, M., and E. Kubli.** 1981. Wild-type $tRNA_G^{Tyr}$ reads the TMV RNA stop codon, but Q base-modified $tRNA_Q^{Tyr}$ does not. *Nature* **294:**188–190.
7. **Bina, J., J. Zhu, M. Dziejman, S. Faruque, S. Calderwood, and J. Mekalanos.** 2003. ToxR regulon of *Vibrio cholerae* and its expression in vibrios shed by cholera patients. *Proc. Natl. Acad. Sci. USA* **100:**2801–2806.
8. **Bourret, R. B., N. W. Charon, A. M. Stock, and A. H. West.** 2002. Bright lights, abundant operons—fluorescence and genomic technologies advance studies of bacterial locomotion and signal transduction: review of the BLAST meeting, Cuernavaca, Mexico, 14 to 19 January 2001. *J. Bacteriol.* **184:**1–17.
9. **Boyd, A. P., and G. P. Cornelis.** 2001. *Yersinia*, p. 227–264. *In* E. A. Groisman (ed.), *Principles of Bacterial Pathogenesis.* Academic Press, Ltd., London, United Kingdom.
10. **Bumann, D.** 2002. Examination of *Salmonella* gene expression in an infected mammalian host using the green fluorescent protein and two-colour flow cytometry. *Mol. Microbiol.* **43:**1269–1283.
11. **Butler, S. M., and A. Camilli.** 2004. Both chemotaxis and net motility greatly influence the infectivity of *Vibrio cholerae. Proc. Natl. Acad. Sci. USA* **101:**5018–5023.
12. **Camilli, A., D. Beattie, and J. Mekalanos.** 1994. Use of genetic recombination as a reporter of gene expression. *Proc. Natl. Acad. Sci. USA* **91:**2634–2638.
13. **Camilli, A., and J. J. Mekalanos.** 1995. Use of recombinase gene fusions to identify *Vibrio cholerae* genes induced during infection. *Mol. Microbiol.* **18:**671–683.
14. **Carniel, E., A. Guiyoule, I. Guilvout, and O. Mercereau-Puijalon.** 1992. Molecular cloning, iron-regulation and mutagenesis of the *irp2* gene encoding HMWP2, a protein specific for the highly pathogenic *Yersinia. Mol. Microbiol.* **6:**379–388.
15. **Carroll, P. A., K. T. Tashima, M. B. Rogers, V. J. DiRita, and S. B. Calderwood.** 1997. Phase variation in *tcpH* modulates expression of the ToxR regulon in *Vibrio cholerae. Mol. Microbiol.* **25:**1099–1111.
16. **Carter, P. B.** 1975. Pathogenicity of *Yersinia enterocolitica* for mice. *Infect. Immun.* **11:**164–170.
17. **Cheng, L. W., D. M. Anderson, and O. Schneewind.** 1997. Two independent type II secretion mechanisms for YopE in *Yersinia enterocolitica. Mol. Microbiol.* **24:**757–765.
18. **Chiang, S. L., and J. J. Mekalanos.** 1998. Use of signature-tagged transposon mutagenesis to identify *Vibrio cholerae* genes critical for colonization. *Mol. Microbiol.* **27:**797–805.
19. **Cirillo, D. M., R. H. Valdivia, D. M. Monack, and S. Falkow.** 1998. Macrophage-dependent induction of the *Salmonella* pathogenicity island 2 type III secretion system and its role in intracellular survival. *Mol. Microbiol.* **30:**175–188.
20. **Cormack, B., R. H. Valdivia, and S. Falkow.** 1996. FACS-optimized mutants of green fluorescent protein (GFP). *Gene* **173:**33–38.
21. **Cornelis, G., Y. Laroche, G. Balligand, M.-P. Sory, and G. Wauters.** 1987. *Yersinia enterocolitica*, a primary model for bacterial invasiveness. *Rev. Infect. Dis.* **9:**64–87.
22. **Darwin, A. J., and V. L. Miller.** 1999. Identification of *Yersinia enterocolitica* genes affecting survival in an animal host using signature-tagged transposon mutagenesis. *Mol. Microbiol.* **32:**51–62.
23. **Darwin, K. H., and V. L. Miller.** 1999. Molecular basis of the interaction of *Salmonella* with the intestinal mucosa. *Clin. Microbiol. Rev.* **12:**405–428.
24. **Delor, I., and G. R. Cornelis.** 1992. Role of *Yersinia enterocolitica* Yst toxin in experimental infection of young rabbits. *Infect. Immun.* **60:**4269–4277.
25. **DiRita, V. J.** 2001. Molecular basis of *Vibrio cholerae* pathogenesis, p. 457–508. *In* E. A. Groisman (ed.), *Principles of Bacterial Pathogenesis.* Academic Press, Ltd., London, United Kingdom.
26. **DiRita, V. J., and J. J. Mekalanos.** 1991. Periplasmic interaction between two membrane regulatory proteins, ToxR and ToxS, results in signal transduction and transcriptional activation. *Cell* **64:**29–37.
27. **Dorman, C. J., and M. E. Porter.** 1998. The *Shigella* virulence gene regulatory cascade: a paradigm of bacterial gene control mechanisms. *Mol. Microbiol.* **29:**677–684.
28. **Dubail, I., P. Berche, and A. Charbit.** 2000. Listeriolysin O as a reporter to identify constitutive and in vivo-inducible promoters in the pathogen *Listeria monocytogenes. Infect. Immun.* **68:**3242–3250.
29. **Durand, J. M., B. Dagberg, B. E. Uhlin, and G. R. Bjork.** 2000. Transfer RNA modification, temperature and DNA superhelicity have a common target in the regulatory network of the virulence of *Shigella flexneri:* the expression of the *virF* gene. *Mol. Microbiol.* **35:**924–935.
30. **Durand, J. M., N. Okada, T. Tobe, M. Watarai, I. Fukuda, T. Suzuki, N. Nakata, K. Komatsu, M. Yoshikawa, and C. Sasakawa.** 1994. *vacC*, a virulence-associated chromosomal locus of *Shigella flexneri*, is homologous to *tgt*, a gene encoding tRNA-guanine transglycosylase (Tgt) of *Escherichia coli* K-12. *J. Bacteriol.* **176:**4627–4634.

31. **Earhart, C. F.** 1996. *Uptake and Metabolism of Iron and Molybdenum.* ASM Press, Washington, D.C.
32. **Ernst, R. K., T. Guina, and S. I. Miller.** 2001. *Salmonella typhimurium* outer membrane remodeling: role in resistance to host innate immunity. *Microbes Infect.* **3:**1327–1334.
33. **Fang, F. C., S. J. Libby, N. A. Buchmeier, P. C. Loewen, J. Switala, J. Harwood, and D. G. Guiney.** 1992. The alternative sigma factor *katF* (*rpoS*) regulates *Salmonella* virulence. *Proc. Natl. Acad. Sci. USA* **89:**11978–11982.
34. **Fields, P. I., E. A. Groisman, and F. Heffron.** 1989. A *Salmonella* locus that controls resistance to microbicidal proteins from phagocytic cells. *Science* **243:**1059–1062.
35. **Forsberg, A., and R. Rosqvist.** 1993. *In vivo* expression of virulence genes of *Yersinia pseudotuberculosis. Infect. Agents Dis.* **2:**275–278.
36. **Freitag, N. E., and K. E. Jacobs.** 1999. Examination of *Listeria monocytogenes* intracellular gene expression using the green fluorescent protein of *Aequorea victoria. Infect. Immun.* **67:**1844–1852.
37. **Freter, R., and P. C. O'Brien.** 1981. Role of chemotaxis in the association of motile bacteria with intestinal mucosa: fitness and virulence of nonchemotactic *Vibrio cholerae* mutants in infant mice. *Infect. Immun.* **34:**222–233.
38. **Freter, R., H. L. Smith, and F. J. Sweeney.** 1961. An evaluation of intestinal fluids in the pathogenesis of cholera. *J. Infect. Dis.* **109:**35–42.
39. **Frey, B., G. Janel, U. Michelsen, and H. Kersten.** 1989. Mutation in the *Escherichia coli fnr* and *tgt* genes: control of molybdate reductase activity and the cytochrome *d* complex by *fnr. J. Bacteriol.* **171:**1524–1530.
40. **Fullner, K. J., and J. J. Mekalanos.** 1999. Genetic characterization of a new type IV-A pilus gene cluster found in both classical and El Tor biotypes of *Vibrio cholerae. Infect. Immun.* **67:**1393–1404.
41. **Gahan, C. G. M., and C. Hill.** 2001. Characterization of the *groESL* operon in *Listeria monocytogenes:* utilization of two reporter systems (*gfp* and *hly*) for evaluating in vivo expression. *Infect. Immun.* **69:**3924–3932.
42. **Gahan, C. G. M., and C. Hill.** 2000. The use of listeriolysin to identify *in vivo* induced genes in the gram-positive intracellular pathogen *Listeria monocytogenes. Mol. Microbiol.* **36:**498–507.
43. **Galan, J. E., C. Ginocchio, and P. Costeas.** 1992. Molecular and functional characterization of the *Salmonella* invasion gene *invA:* homology of InvA to members of a new protein family. *J. Bacteriol.* **174:**4338–4349.
44. **Garcia Vescovi, E., F. C. Soncini, and E. A. Groisman.** 1996. Mg^{2+} as an extracellular signal: environmental regulation of *Salmonella* virulence. *Cell* **84:**165–174.
45. **Gardel, C. L., and J. J. Mekalanos.** 1996. Alterations in *Vibrio cholerae* motility phenotypes correlate with changes in virulence factor expression. *Infect. Immun.* **64:**2246–2255.
46. **Gentry, D. R., V. J. Hernandez, L. H. Nguyen, D. B. Jensen, and M. Cashel.** 1993. Synthesis of the stationary-phase sigma factor sigma s is positively regulated by ppGpp. *J. Bacteriol.* **175:**7982–7989.
47. **Gill, D. M.** 1977. Mechanism of action of cholera toxin. *Adv. Cyclic Nucleotide Res.* **8:**85–118.
48. **Gosink, K. K., R. Kobayashi, I. Kawagishi, and C. C. Hase.** 2002. Analyses of the roles of the three *cheA* homologs in chemotaxis of *Vibrio cholerae. J. Bacteriol.* **184:**1767–1771.
49. **Groisman, E. A.** 2001. The pleiotropic two-component regulatory system PhoP-PhoQ. *J. Bacteriol.* **183:**1835–1842.
50. **Guan, K. L., and J. E. Dixon.** 1990. Protein tyrosine phosphatase activity is an essential virulence determinant of *Yersinia. Science* **249:**553–556.
51. **Gupta, S., and R. Chowdhury.** 1997. Bile affects production of virulence factors and motility of *Vibrio cholerae. Infect. Immun.* **65:**1131–1134.
52. **Hancock, R. E., and M. G. Scott.** 2000. The role of antimicrobial peptides in animal defenses. *Proc. Natl. Acad. Sci. USA* **97:**8856–8861.
53. **Handfield, M., L. J. Brady, A. Progulske-Fox, and J. D. Hillman.** 2000. IVIAT: a novel method to identify microbial genes expressed specifically during human infections. *Trends Microbiol.* **8:**336–339.
54. **Hang, L., M. John, M. Asaduzzaman, E. A. Bridges, C. Vanderspurt, T. J. Kirn, R. K. Taylor, J. D. Hillman, A. Progulske-Fox, M. Handfield, E. T. Ryan, and S. B. Calderwood.** 2003. Use of *in vivo*-induced antigen technology (IVIAT) to identify genes uniquely expressed during human infection with *Vibrio cholerae. Proc. Natl. Acad. Sci. USA* **100:**8508–8513.
55. **Hase, C. C., and J. J. Mekalanos.** 1998. TcpP protein is a positive regulator of virulence gene expression in *Vibrio cholerae. Proc. Natl. Acad. Sci. USA* **95:**730–734.
56. **Hautefort, I., and J. C. Hinton.** 2000. Measurement of bacterial gene expression *in vivo. Philos. Trans. R. Soc. Lond. Ser. B* **355:**601–611.
57. **Heesemann, J., K. Hantke, T. Vocke, E. Saken, A. Rakin, I. Stojilijkovic, and R. Berner.** 1993. Virulence of *Yersinia enterocolitica* is closely associated with siderophore production, expression of an iron-repressible outer membrane polypeptide of 65,000Da and pesticin sensitivity. *Mol. Microbiol.* **8:**397–408.
58. **Heidelberg, J. F., J. A. Eisen, W. C. Nelson, R. A. Clayton, M. L. Gwinn, R. J. Dodson, D. H. Haft, E. K. Hickey, J. D. Peterson, L. Umayam, S. R. Gill, K. E. Nelson, T. D. Read, H. Tettelin, D. Richardson, M. D. Ermolaeva, J. Vamathevan, S. Bass, H. Qin, I. Dragoi, P. Sellers, L. McDonald, T. Utterback, R. D. Fleishmann, W. C. Nierman, and O. White.** 2000. DNA sequence of both chromosomes of the cholera pathogen *Vibrio cholerae. Nature* **406:**477–483.
59. **Heithoff, D. M., C. P. Conner, P. C. Hanna, S. M. Julio, U. Hentschel, and M. J. Mahan.** 1997. Bacterial infection as assessed by *in vivo* gene expression. *Proc. Natl. Acad. Sci. USA* **94:**934–939.
60. **Heithoff, D. M., C. P. Conner, U. Hentschel, F. Govantes, P. C. Hanna, and M. J. Mahan.** 1999. Coordinate intracellular expression of *Salmonella* genes induced during infection. *J. Bacteriol.* **181:**799–807.
61. **Hensel, M.** 2000. *Salmonella* pathogenicity island 2. *Mol. Microbiol.* **36:**1015–1023.
62. **Hensel, M., J. E. Shea, C. Gleeson, M. D. Jones, E. Dalton, and D. W. Holden.** 1995. Simultaneous identification of bacterial virulence genes by negative selection. *Science* **269:**400–403.
63. **Hensel, M., J. E. Shea, S. R. Waterman, R. Mundy, T. Nikolaus, G. Banks, A. Vazquez-Torres, C. Gleeson, F. C. Fang, and D. W. Holden.** 1998. Genes encoding putative effector proteins of the type III secretion system of *Salmonella* pathogenicity island 2 are required for bacterial virulence and proliferation in macrophages. *Mol. Microbiol.* **30:**163–174.
64. **Herrington, D. A., R. H. Hall, G. Losonsky, J. J. Mekalanos, R. K. Taylor, and M. M. Levine.** 1988. Toxin, toxin-coregulated pili, and the *toxR* regulon are essential for *Vibrio cholerae* pathogenesis in humans. *J. Exp. Med.* **168:**1487–1492.
65. **Higgins, D. E., and V. J. DiRita.** 1994. Transcriptional control of *toxT,* a regulatory gene in the *ToxR* regulon of *Vibrio cholerae. Mol. Microbiol.* **14:**17–29.

66. **Higgins, D. E., E. Nazareno, and V. J. DiRita.** 1992. The virulence gene activator ToxT from *Vibrio cholerae* is a member of the AraC family of transcriptional activators. *J. Bacteriol.* **174:**6874–6980.
67. **Hohmann, A. W., G. Schmidt, and D. Rowley.** 1978. Intestinal colonization and virulence of *Salmonella* in mice. *Infect. Immun.* **22:**763–770.
68. **Hueck, C. J.** 1998. Type III protein secretion systems in bacterial pathogens of animals and plants. *Microbiol. Mol. Biol. Rev.* **62:**379–433.
69. **Humphries, A. D., M. Raffatellu, S. Winter, E. H. Weening, R. A. Kingsley, R. Droleskey, S. Zhang, J. Figueiredo, S. Khare, J. Nunes, L. G. Adams, R. M. Tsolis, and A. J. Baumler.** 2003. The use of flow cytometry to detect expression of subunits encoded by 11 *Salmonella enterica* serotype Typhimurium fimbrial operons. *Mol. Microbiol.* **48:**1357–1376.
70. **Isberg, R. R., and J. M. Leong.** 1990. Multiple beta 1 chain integrins are receptors for invasin, a protein that promotes bacterial penetration into mammalian cells. *Cell* **60:**861–871.
71. **Isberg, R. R., D. L. Voorhis, and S. Falkow.** 1987. Identification of invasin: a protein that allows enteric bacteria to penetrate cultured mammalian cells. *Cell* **50:**769–778.
72. **Jacobi, C. A., S. Gregor, A. Rakin, and J. Heesemann.** 2001. Expression analysis of the yersiniabactin receptor gene *fyuA* and the heme receptor *hemR* of *Yersinia enterocolitica* in vitro and in vivo using the reporter genes for green fluorescent protein and luciferase. *Infect. Immun.* **69:**7772–7782.
73. **Jacobi, C. A., A. Roggenkamp, A. Rakin, R. Zumbihl, L. Leitritz, and J. Heesemann.** 1998. *In vitro* and *in vivo* expression studies of *yopE* from *Yersinia enterocolitica* using the *gfp* reporter gene. *Mol. Microbiol.* **30:**865–882.
74. **Jarvis, K. G., J. A. Giron, A. E. Jerse, T. K. McDaniel, M. S. Donnenberg, and J. B. Kaper.** 1995. Enteropathogenic *Escherichia coli* contains a putative type III secretion system necessary for the export of proteins involved in attaching and effacing lesion formation. *Proc. Natl. Acad. Sci. USA* **92:**7996–8000.
75. **Kammler, M., C. Schon, and K. Hantke.** 1993. Characterization of the ferrous iron uptake system of *Escherichia coli. J. Bacteriol.* **175:**6212–6219.
76. **Kane, C. D., R. Schuch, W. A. Day, Jr., and A. T. Maurelli.** 2002. MxiE regulates intracellular expression of factors secreted by the *Shigella flexneri* 2a type III secretion system. *J. Bacteriol.* **184:**4409–4419.
77. **Kennedy, E. P.** 1996. Membrane-derived oligosaccharides (periplasmic β-D-glucans) of *Escherichia coli,* p. 1064–1071. *In* F. C. Neidhardt, R. Curtiss III, J. L. Ingraham, E. C. C. Lin, K. B. Low, B. Magasanik, W. S. Reznikoff, M. Riley, M. Schaechter, and H. E. Umbarger (ed.), Escherichia coli *and* Salmonella: *Cellular and Molecular Biology,* 2nd ed. ASM Press, Washington, D.C.
78. **Kent, T. H., S. B. Formal, and E. H. Labrec.** 1966. *Salmonella* gastroenteritis in rhesus monkeys. *Arch. Pathol.* **82:**272–279.
79. **Khan, M. A., and R. E. Isaacson.** 2002. Identification of *Escherichia coli* genes that are specifically expressed in a murine model of septicemic infection. *Infect. Immun.* **70:**3404–3412.
80. **Kim, H. Y., D. Schlictman, S. Shankar, Z. Xie, A. M. Chakrabarty, and A. Kornberg.** 1998. Alginate, inorganic polyphosphate, GTP and ppGpp synthesis co-regulated in *Pseudomonas aeruginosa:* implications for stationary phase survival and synthesis of RNA/DNA precursors. *Mol. Microbiol.* **27:**717–725.
81. **Klose, K. E.** 2000. The suckling mouse model of cholera. *Trends Microbiol.* **8:**189–191.
82. Reference deleted.
83. **Krukonis, E. S., and V. J. DiRita.** 2003. From motility to virulence: sensing and responding to environmental signals in *Vibrio cholerae. Curr. Opin. Microbiol.* **6:**186–190.
84. **Lee, S. H., M. J. Angelichio, J. J. Mekalanos, and A. Camilli.** 1998. Nucleotide sequence and spatiotemporal expression of the *Vibrio cholerae vieSAB* genes during infection. *J. Bacteriol.* **180:**2298–2305.
85. **Lee, S. H., S. M. Butler, and A. Camilli.** 2001. Selection for *in vivo* regulators of bacterial virulence. *Proc. Natl. Acad. Sci. USA* **98:**6889–6894.
86. **Lee, S. H., D. L. Hava, M. K. Waldor, and A. Camilli.** 1999. Regulation and temporal expression patterns of *Vibrio cholerae* virulence genes during infection. *Cell* **99:**625–634.
87. **Lehrer, R. I., A. K. Lichtenstein, and T. Ganz.** 1993. Defensins: antimicrobial and cytotoxic peptides of mammalian cells. *Annu. Rev. Immunol.* **11:**105–128.
88. **Lian, C. J., and C. H. Pai.** 1987. Plasmid-mediated resistance to phagocytosis in *Yersinia enterocolitica. Infect. Immun.* **55:**1176–1183.
89. **Loewen, P. C., and R. Hengge-Aronis.** 1994. The role of the sigma factor sigma S (KatF) in bacterial global regulation. *Annu. Rev. Microbiol.* **48:**53–80.
90. **Lostroh, C. P., and C. A. Lee.** 2001. The *Salmonella* pathogenicity island-1 type III secretion system. *Microbes Infect.* **3:**1281–1291.
91. **Loubens, L., L. Debarbieux, A. Bohin, J.-M. Lacroix, and J. P. Bohin.** 1993. Homology between a genetic locus (*mdoA*) involved in the osmoregulated biosynthesis of periplasmic glucans in *Escherichia coli* and a genetic locus (*hrpM*) controlling the pathogenicity of *Pseudomonas syringae. Mol. Microbiol.* **10:**329–340.
92. **Mahan, M. J., J. M. Slauch, and J. J. Mekalanos.** 1993. Selection of bacterial virulence genes that are specifically induced in host tissues. *Science* **259:**686–688.
93. **Mahan, M. J., J. W. Tobias, J. M. Slauch, P. C. Hanna, R. J. Collier, and J. J. Mekalanos.** 1995. Antibiotic-based selection for bacterial genes that are specifically induced during infection of a host. *Proc. Natl. Acad. Sci. USA* **92:**669–673.
94. **Martindale, J., D. Stroud, E. R. Moxon, and C. M. Tang.** 2000. Genetic analysis of *Escherichia coli* K1 gastrointestinal colonization. *Mol. Microbiol.* **37:**1293–1305.
95. **Mecsas, J., I. Bilis, and S. Falkow.** 2001. Identification of attenuated *Yersinia pseudotuberculosis* strains and characterization of an orogastric infection in BALB/c mice on day 5 postinfection by signature-tagged mutagenesis. *Infect. Immun.* **69:**2779–2787.
96. **Meier, F., B. Suter, H. Grosjean, G. Keith, and E. Kubli.** 1985. Queuosine modification of the wobble base in $tRNA^{His}$ influences '*in vivo*' decoding properties. *EMBO J.* **4:**823–827.
97. **Merrell, D. S., S. M. Butler, F. Qadri, N. A. Dolganov, A. Alam, M. B. Cohen, S. B. Calderwood, G. K. Schoolnik, and A. Camilli.** 2002. Host-induced epidemic spread of the cholera bacterium. *Nature* **417:**642–645.
98. **Merrell, D. S., and A. Camilli.** 1999. The *cadA* gene of *Vibrio cholerae* is induced during infection and plays a role in acid tolerance. *Mol. Microbiol.* **34:**836–849.
99. **Merrell, D. S., and A. Camilli.** 2000. Detection and analysis of gene expression during infection by *in vivo* expression technology. *Philos. Trans. R. Soc. Lond. Ser. B* **355:**587–599.
100. **Merrell, D. S., D. L. Hava, and A. Camilli.** 2002. Identification of novel factors involved in colonization and acid tolerance of *Vibrio cholerae. Mol. Microbiol.* **43:**1471–1491.
101. **Miller, S. I., A. M. Kukral, and J. J. Mekalanos.** 1989. A two-component regulatory system (*phoP phoQ*) controls *Salmo-*

nella typhimurium virulence. *Proc. Natl. Acad. Sci. USA* **86:** 5054–5058.

102. **Miller, S. I., and J. J. Mekalanos.** 1990. Constitutive expression of the *phoP* regulon attenuates *Salmonella* virulence and survival within macrophages. *J. Bacteriol.* **172:**2485–2490.
103. **Miller, V. L., R. K. Taylor, and J. J. Mekalanos.** 1987. Cholera toxin transcriptional activator ToxR is a transmembrane DNA binding protein. *Cell* **48:**271–279.
104. **Moors, M. A., B. Levitt, P. Youngman, and D. A. Portnoy.** 1999. Expression of listeriolysin O and ActA by intracellular and extracellular *Listeria monocytogenes. Infect. Immun.* **67:** 131–139.
105. **Nachamkin, I., and A. M. Hart.** 1985. Western blot analysis of the human antibody response to *Campylobacter jejuni* cellular antigens during gastrointestinal infection. *J. Clin. Microbiol.* **21:**33–38.
106. **Noguchi, S., Y. Nishimura, Y. Hirota, and S. Nishimura.** 1982. Isolation and characterization of an *Escherichia coli* mutant lacking tRNA-guanine transglycosylase. Function and biosynthesis of queuosine in tRNA. *J. Biol. Chem.* **257:** 6544–6550.
107. **Page, F., S. Altabe, N. Hugouvieux-Cotte-Pattat, J.-M. Lacroix, J. Robert-Baudouy, and J.-P. Bohin.** 2001. Osmoregulated periplasmic glucan synthesis is required for *Erwinia chrysanthemi* pathogenicity. *J. Bacteriol.* **183:**3134–3141.
108. **Park, Y. K., B. Bearson, S. H. Bang, I. S. Bang, and J. W. Foster.** 1996. Internal pH crisis, lysine decarboxylase and the acid tolerance response of *Salmonella typhimurium. Mol. Microbiol.* **20:**605–611.
109. **Pfeifer, C. G., S. L. Marcus, O. Steele-Mortimer, L. A. Knodler, and B. B. Finlay.** 1999. *Salmonella typhimurium* virulence genes are induced upon bacterial invasion into phagocytic and nonphagocytic cells. *Infect. Immun.* **67:**5690–5698.
110. **Rakin, A., E. Saken, D. Harmsen, and J. Heesemann.** 1994. The pesticin receptor of *Yersinia enterocolitica:* a novel virulence factor with dual function. *Mol. Microbiol.* **13:**253–262.
111. **Ray, N. B., and C. K. Mathews.** 1992. Nucleoside diphosphokinase: a functional link between intermediary metabolism and nucleic acid synthesis. *Curr. Top. Cell Regul.* **33:** 343–357.
112. **Rokbi, B., D. Seguin, B. Guy, V. Mazarin, E. Vidor, F. Mion, M. Cadoz, and M. J. Quentin-Millet.** 2001. Assessment of *Helicobacter pylori* gene expression within mouse and human gastric mucosa. *Infect. Immun.* **69:**4759–4766.
113. **Rosqvist, R., A. Forsberg, M. Rimpilainen, T. Bergman, and W. H. Wolf.** 1990. The cytotoxic protein YopE of *Yersinia* obstructs the primary host defence. *Mol. Microbiol.* **4:**657–667.
114. **Rosqvist, R., A. Forsberg, and W. H. Wolf.** 1991. Intracellular targeting of the *Yersinia* YopE cytotoxin in mammalian cells induces actin microfilament disruption. *Infect. Immun.* **59:**4562–4569.
115. **Ruckdeschel, K., A. Roggenkamp, S. Schubert, and J. Heesemann.** 1996. Differential contribution of *Yersinia enterocolitica* virulence factors to evasion of microbicidal action of neutrophils. *Infect. Immun.* **64:**724–733.
116. **Saken, E., A. Rakin, and J. Heesemann.** 2000. Molecular characterization of a novel siderophore-independent iron transport system in *Yersinia. Int. J. Med. Microbiol.* **290:**51–60.
117. **Salzman, N. H., M. M. Chou, H. de Jong, L. Liu, E. M. Porter, and Y. Paterson.** 2003. Enteric *Salmonella* infection inhibits Paneth cell antimicrobial peptide expression. *Infect. Immun.* **71:**1109–1115.
118. **Schuhmacher, D. A., and K. E. Klose.** 1999. Environmental signals modulate ToxT-dependent virulence factor expression in *Vibrio cholerae. J. Bacteriol.* **181:**1508–1514.
119. **Schulte, R., S. Kerneis, S. Klinke, H. Bartels, S. Preger, J.-P. Kraehenbuhl, E. Pringault, and I. B. Autenrieth.** 2000. Translocation of *Yersina enterocolitica* across reconsituted intestinal epithelial monolayers is triggered by *Yersina* invasin binding to β-1 integrins apically expressed on M-like cells. *Cell. Microbiol.* **2:**173–185.
120. **Shea, J. E., M. Hensel, C. Gleeson, and D. W. Holden.** 1996. Identification of a virulence locus encoding a second type III secretion system in *Salmonella typhimurium. Proc. Natl. Acad. Sci. USA* **93:**2593–2597.
121. **Simonet, M., S. Richard, and P. Berche.** 1990. Electron microscopic evidence for in vivo extracellular localization of *Yersinia pseudotuberculosis* harboring the pYV plasmid. *Infect. Immun.* **58:**841–845.
122. **Slauch, J. M., and A. Camilli.** 2000. IVET and RIVET: use of gene fusions to identify bacterial virulence factors specifically induced in host tissues. *Methods Enzymol.* **326:**73–96.
123. **Slauch, J. M., M. J. Mahan, and J. J. Mekalanos.** 1994. *In vivo* expression technology for selection of bacterial genes specifically induced in host tissues. *Methods Enzymol.* **235:** 481–492.
124. **Smith, H. W., and J. E. Jones.** 1967. Observations on experimental oral infection with *Salmonella Dublin* in calves and *Salmonella choleraesuis* in pigs. *J. Pathol. Bacteriol.* **93:**141–156.
125. **Stanley, T. L., C. D. Ellermeier, and J. M. Slauch.** 2000. Tissue-specific gene expression identifies a gene in the lysogenic phage Gifsy-1 that affects *Salmonella enterica* serovar Typhimurium survival in Peyer's patches. *J. Bacteriol.* **182:** 4406–4413.
126. **Stojiljkovic, I., and K. Hantke.** 1992. Hemin uptake system of *Yersinia enterocolitica:* similarities with other TonB-dependent systems in gram-negative bacteria. *EMBO J.* **12:**4359–4367.
127. **Tacket, C. O., R. K. Taylor, G. Losonsky, Y. Lim, J. P. Nataro, J. B. Kaper, and M. M. Levine.** 1998. Investigation of the roles of toxin-coregulated pili and mannose-sensitive hemagglutinin pili in the pathogenesis of *Vibrio cholerae* O139 infection. *Infect. Immun.* **66:**692–695.
128. **Talaat, A. M., Hunter, P., and S. A. Johnson.** 2000. Genome-directed primers for selective labeling of bacterial transcripts for DNA microarray analysis. *Nat. Biotechnol.* **18:**679–682.
129. **Taylor, R. K., V. L. Miller, D. B. Furlong, and J. J. Mekalanos.** 1986. Identification of a pilus colonization factor that is coordinately regulated with cholera toxin. *Ann. Sclavo Collana Monogr.* **3:**51–61.

129a. **Taylor, R. K., V. L. Miller, D. B. Furlong, and J. J. Mekalanos.** 1987. Use of *phoA* gene fusions to identify a pilus colonization factor coordinately regulated with cholera toxin. *Proc. Natl. Acad. Sci. USA* **84:**2833–2837.

130. **Tischler, A. D., S. H. Lee, and A. Camilli.** 2002. The *Vibrio cholerae vieSAB* locus encodes a pathway contributing to cholera toxin production. *J. Bacteriol.* **184:**4104–4113.
131. **Valdivia, R. H., and S. Falkow.** 1996. Bacterial genetics by flow cytometry: rapid isolation of *Salmonella typhimurium* acid-inducible promoters by differential fluorescence induction. *Mol. Microbiol.* **22:**367–378.
132. **Valdivia, R. H., and S. Falkow.** 1997. Fluorescence-based isolation of bacterial genes expressed within host cells. *Science* **277:**2007–2011.

133. **Waldor, M. K., and J. J. Mekalanos.** 1996. Lysogenic conversion by a filamentous phage encoding cholera toxin. *Science* **272**:1910–1914.
134. **Wallis, T. S., S. M. Paulin, J. S. Plested, P. R. Watson, and P. W. Jones.** 1995. The *Salmonella Dublin* virulence plasmid mediates systemic but not enteric phases of salmonellosis in cattle. *Infect. Immun.* **63**:2755–2761.
135. **Wallis, T. S., W. G. Starkey, J. Stephen, S. J. Haddon, M. P. Osborne, and D. C. Candy.** 1986. The nature and role of mucosal damage in relation to *Salmonella typhimurium*-induced fluid secretion in the rabbit ileum. *J. Med. Microbiol.* **22**:39–49.
136. **Xu, Q., M. Dziejman, and J. J. Mekalanos.** 2003. Determination of the transcriptome of *Vibrio cholerae* during intraintestinal growth and midexponential phase *in vitro*. *Proc. Natl. Acad. Sci. USA* **100**:1286–1291.
137. **Young, G. M., and V. L. Miller.** 1997. Identification of novel chromosomal loci affecting *Yersinia enterocolitica* pathogenesis. *Mol. Microbiol.* **25**:319–328.

Colonization of Mucosal Surfaces
Edited by James P. Nataro et al.

Chapter 21

Mechanisms of *Salmonella enterica* Serotype Typhimurium Intestinal Colonization

Caleb W. Dorsey, Manuela Raffatellu, Robert A. Kingsley, and Andreas J. Bäumler

Salmonella-induced enterocolitis is the single most common cause of death from food-borne illnesses associated with viruses, parasites, or bacteria in the United States (72). The serotype associated most frequently with this diarrheal disease is *Salmonella enterica* serotype Typhimurium (26% of total *Salmonella* isolates reported to the Centers for Disease Control and Prevention in 1998) (15). Volunteers infected with serotype Typhimurium develop an illness within 12 to 72 h after infection that is characterized by diarrhea, vomiting, and abdominal pain (10). Rectal biopsy specimens from patients infected with serotype Typhimurium reveal an acute enteritis characterized by a severe inflammatory infiltrate that is composed primarily of neutrophils (20, 70). Serotype Typhimurium infection in calves is an excellent model for the intestinal pathology, host response, and disease syndrome observed in humans (107, 113, 122). Bovine ligated ileal loops can be used to study fluid accumulation and host responses characterizing the acute phase of serotype Typhimurium infection (31, 102, 103).

Diarrhea and intestinal inflammation subside within 10 days, but the organism can be isolated from the feces of patients for a median duration of 5 weeks (12). Similarly, the acute phase of serotype Typhimurium infection in cattle is followed by a period of 1 to several months during which the organism is shed with the feces (108). Intestinal persistence of serotype Typhimurium can be modeled experimentally by using genetically resistant mice (58, 77, 88). Fecal contamination of the environment allows serotype Typhimurium to persist in reservoirs of livestock and domestic fowl (53, 71, 121, 123). Furthermore, intestinal persistence of serotype Typhimurium in apparently healthy livestock and domestic fowl leads to its subsequent introduction into the derived food products, thereby resulting in animal-to-human transmission. This route of infection is responsible for most of the estimated 1.4 million annual cases of *Salmonella*-induced enterocolitis in the United States (72).

BACTERIAL INVASION AND THE ACUTE PHASE OF INFECTION

The interaction of *S. enterica* serotype Typhimurium with the intestinal mucosa in calves and humans results in the recruitment of neutrophils whose presence is the histopathologic hallmark for the acute phase of *Salmonella*-induced enterocolitis (20, 31, 70, 103). This neutrophilic infiltrate is associated with necrosis of the upper mucosa in large areas of the terminal ileum and colon (113). Necrosis presumably occurs because neutrophils in particular release substances (e.g., proteases, myeloperoxidase, and NADPH oxidase) that result in tissue injury. The injury to the intestinal epithelium and the resulting loss of epithelial barrier function lead to leakage of extravascular fluids, thereby contributing to diarrhea by causing liquid to flow from the blood to the intestinal lumen (127). Neutrophils have also been implicated in contributing to diarrhea by stimulating chloride secretion in intestinal epithelial cells (67). Although the neutrophil influx in calves and humans may be potentially harmful to the host (since the resulting diarrhea may lead to severe dehydration), this response is probably necessary to clear the infection and prevent systemic spread of the pathogen. A neutrophil-mediated arrest of bacterial translocation into the lamina propria has been demonstrated for

Caleb W. Dorsey, Manuela Raffatellu, Robert A. Kingsley, and Andreas J. Bäumler • Department of Medical Microbiology and Immunology, College of Medicine, Texas A&M University System Health Science Center, 407 Reynolds Medical Bldg, College Station, TX 77843-1114.

Shigella flexneri infection in a rabbit ligated loop model (101).

Recent work suggests that the factors triggering the massive neutrophil influx during serotype Typhimurium infection are part of the innate immune system. The innate immune system can distinguish between self and microbial intruders by recognizing molecular patterns found exclusively in microorganisms. These molecular patterns are conserved among groups of microbes, thereby allowing the detection of infectious agents with a limited number of receptors. Molecular patterns are recognized by host cells through pathogen recognition receptors (PRR), which include the membrane-localized Toll-like receptors (TLR1 through TLR-10) (105) and the cytosolic Nod receptors (Nod1 and Nod2) (54). Although molecular patterns recognized by PRR have been termed pathogen-associated molecular patterns (PAMPs), they are clearly also present in nonpathogenic environmental and commensal microorganisms. Commensal microbes permanently colonize mucosal surfaces in the terminal ileum and the large intestine, with densities between 10^8 bacteria/ml (ileum) and 10^{11} bacteria/ml (colon). One challenge for the innate immune recognition in the intestinal mucosa is thus to distinguish between harmless commensal bacteria present in the intestinal lumen and invasive infections (i.e., invasion by the normal flora after traumatic injury or infection with an invasive enteric pathogen). Therefore, PAMP recognition of microbes present in the intestinal lumen needs to be controlled to avoid an excessive immune response and uncontrolled inflammation. At the same time, the intestinal mucosa needs to be able to detect invasion by microbes through PAMP recognition to trigger an appropriate inflammatory response. Although the molecular mechanisms of distinguishing luminal from invasive microbes are just beginning to be worked out, it is becoming increasingly clear that the intestinal epithelium is a key player during this process (4).

Several properties of epithelial cells appear to be essential for distinguishing between the presence of commensals and the presence of pathogenic bacteria. First, the intestinal epithelium serves as a barrier to luminal microorganisms. Disruption of this epithelial barrier by expressing a dominant-negative N-cadherin transgene that prevents intraepithelial-cell adhesion results in the induction of inflammatory bowel disease in mice (46). These data demonstrate convincingly the importance of the intestinal epithelial barrier function in preventing the induction of an inflammatory response by microbes present in the intestinal lumen. Second, recent work provides evidence for a functional compartmentalization of PRR expression in intestinal epithelial cells that prevents direct contact between PRRs and their PAMP ligands in the intestinal lumen (4, 47). TLR5 appears to be preferentially expressed at the basolateral pole of intestinal epithelial cells in vitro (37) and in vivo (13), while TLR4 is found in vesicular structures in the cytoplasm of these cells (50, 51). Furthermore, invasive pathogens that manage to escape from a vacuole into the cytosol of an epithelial cell may be recognized through Nod receptors, as recently shown for the Nod1 recognition of *Shigella flexneri* (40). Thus, functional compartmentalization of PRR expression may enable intestinal epithelial cells to recognize invasive pathogens while at the same time preventing PRR stimulation through microbes present in the intestinal lumen. Third, TLR and Nod recognition of PAMPs by intestinal epithelial cells triggers the expression and release of proinflammatory cytokines in vitro (37, 40, 56, 124, 125), which probably results in the initiation of an inflammatory response in vivo. Their barrier function, functional compartmentalization of PRR expression, and ability to produce cytokines allows intestinal epithelial cells to play a key role in orchestrating a host response to invasive enteric pathogens such as serotype Typhimurium.

Serotype Typhimurium initiates interaction with epithelial cells by causing the formation of membrane ruffles, a process that results in bacterial internalization (29). This invasion process is mediated by a type III secretion system (T3SS-1) encoded by genes located on *Salmonella* pathogenicity island 1 (35, 75). The main function of the T3SS-1 is to translocate effector proteins into the cytosol of a host cell (34). Effector proteins are transported across the inner and outer membranes of the bacterial cell by the T3SS-1. Subsequently, effector proteins are delivered into the host cell cytoplasm via the action of three secreted proteins, SipB (SspB), SipC (SspC), and SipD (SspD), which are thought to assemble into a translocation complex in the eukaryotic membrane (18, 32, 36, 44, 120). A subset of effector proteins, including SopB (SigD), SopE, SopE2, and SipA (SspA), induce rearrangements of the host cell cytoskeleton which result in the formation of membrane ruffles and bacterial internalization (43, 55, 76, 109, 129–131). Invasion of epithelial cells is evident between 15 min and 1 h after infection of bovine ligated ileal loops by serotype Typhimurium as detected by electron microscopy (103). After 1 h postinfection, most of the bacteria within epithelial cells are located at the basal side of the cells as opposed to the apical location seen at earlier time points. Detection of serotype Typhimurium in ligated loops by immunohistochemistry shows that bacteria preferentially colonize epithelial cells in the follicle-associated epithelium (Color Plate 5A) and the tips of absorptive villi (within 10 to 15 min after infection) and then rapidly enter the underlying tissue (Color Plate 5B). Epithelial transmigration takes up to 4 h to complete in vivo (96). Bacteria detected in the lamina propria always have an intracellular location, either within mononuclear phagocytic cells or within neutrophils (103).

Once serotype Typhimurium has invaded the epithelial barrier, it is recognized by PRRs in epithelial cells. In vitro studies using a polarized human colon carcinoma cell line (T84 cells) suggest that serotype Typhimurium flagella are recognized by TLR5, which is expressed at the basolateral pole of the host cells (37). Recognition of flagellin through TLR5 triggers the expression of proinflammatory genes in T84 cells (125), including a family of genes encoding neutrophil chemoattractants known as CXC chemokines. CXC chemokine gene expression is also observed in response to serotype Typhimurium infection of bovine ligated ileal loops (103, 126). The picture emerging from these studies is that invasion of the intestinal mucosa by serotype Typhimurium is detected by host epithelial cells via TLR proteins that are not accessible from the intestinal lumen. TLR stimulation leads to the production of CXC chemokines (e.g., interleukin-8) by epithelial cells, which in turn leads to the massive neutrophil influx characteristic of enterocolitis.

The host response to serotype Typhimurium infection has been extensively analyzed by using cultured macrophages (82, 83, 98) and epithelial cell lines (22, 125) with cDNA microarrays. Microarray expression profiling of cultured cells stimulated with purified serotype Typhimurium or *Escherichia coli* PAMPs, including lipopolysaccharide (LPS) (TLR4/MD-2/CD14 ligand), Braun lipoprotein (TLR2 ligand), and flagella (TLR5 ligand), reveals both shared and unique responses (82, 85, 98, 125). It is likely that host cells use several different TLRs to detect several features (i.e., PAMPs) of serotype Typhimurium simultaneously. This TLR recognition transmits information about the nature of the invading bacterium into the cell, thereby providing the innate immune system with the information needed to elicit a response (i.e., neutrophil influx) that is appropriate to the threat (i.e., serotype Typhimurium mucosal invasion). A comparative analysis of responses elicited by serotype Typhimurium and gram-positive bacteria in human macrophages reveals that in addition to a common expression profile induced by all bacteria, serotype Typhimurium (and other gram-negative organisms such as *E. coli*) induced the expression of a unique set of genes (83). This unique set of genes could also be induced by stimulating macrophages with purified LPS, suggesting that gene expression changes in macrophages exposed to serotype Typhimurium encompass those induced by gram-positive bacteria plus a distinct TLR4 response (83). These data show that the host response to serotype Typhimurium infection is dominated by the accumulated input from multiple TLRs.

Although the available data can be assembled into a rough draft of the series of events leading to neutrophil influx and diarrhea, some major discrepancies remain. First, invasion mediated by T3SS-1 is essential for eliciting CXC chemokine expression and neutrophil influx in bovine ligated ileal loops (2, 113, 114, 118, 119, 126, 128). However, this well-documented contribution of T3SS-1 to CXC chemokine production in vivo is not apparent in colon carcinoma cell lines in vitro (38, 125). A second apparent contradiction between in vivo and in vitro data relates to the relative contribution of flagellin to the recognition of serotype Typhimurium by epithelial cells. In vitro data obtained with colon carcinoma cell lines suggest that all changes in host cell gene expression can be attributed to TLR5 recognition of flagellin because a serotype Typhimurium strain carrying mutations in both flagellin genes (a *fljB fliC* mutant) does not induce a proinflammatory gene expression program (39, 125). In bovine ligated ileal loops infected with a *fljB fliC* mutant, the neutrophil influx is reduced to approximately 80% of the levels seen in loops infected with wild-type serotype Typhimurium, thereby supporting a role for flagella in triggering inflammation. However, in loops infected with strains that are noninvasive (due to inactivation of T3SS-1), the neutrophil influx is reduced to approximately 20% of the levels seen in loops infected with wild-type serotype Typhimurium (104). These data show that stimulation of TLR5 by flagellin cannot be the only mechanism contributing to inflammation in vivo. A PAMP that may contribute to inflammation in vivo is the lipid A domain of the serotype Typhimurium LPS because a mutation in *msbB* (which results in reduced acylation of lipid A, thereby abrogating its PAMP activity) has recently been shown to reduce fluid accumulation in a ligated ileal loop model (25).

The apparent contradictions between in vivo and in vitro data may be explained by some of the limitations inherent to tissue culture studies. First, polarized colon carcinoma cell lines need to be viewed as a gnotobiotic system, a fact that probably alters their responsiveness to microbial exposure compared to that of epithelial cells colonized by members of the normal flora in vivo. Second, it is not clear how accurately colon carcinoma cell lines model the repertoire and the location of TLR expression found in the epithelium of the small and large intestines. The colon carcinoma cell lines commonly used to study serotype Typhimurium pathogenesis (T84, HT-29, and Caco-2) do not express identical repertoires of TLRs. For example, TLR2 is expressed in Caco-2 and T84 cells but not by the HT-29 line (14, 80). Evaluation of cell culture models is further complicated by the fact that it is not clear how well the TLR repertoire of T84, HT-29, or Caco-2 cells matches that seen in the normal human small or large intestine. This limitation is significant, because the primary source of differential signaling through TLRs in vivo is thought to arise not from different intracellular signaling cascades but from expression of different TLRs in different cell types (116). Relevant information is limited to studies investigating the tissue distribution of TLR2 and TLR4 in the human fetal small intestine (33) and TLR2, TLR3, TLR4, and TLR5 expression in the human colon (13). Interestingly, the latter study reported a marked upregulation of TLR4 expression in epithelial cells during inflammation. LPS is recognized by the innate immune system through a complex composed of TLR4, CD14, and MD-2 (4). The colon carcinoma cell lines T84, Caco-2, and HT-29 do not express MD-2, while T84 and HT-29 cells do not produce CD14, which explains their nonresponsiveness to LPS (1, 14, 80). However, LPS responsiveness is seen with human fetal enterocytes (81) and a murine small intestinal cell line (50). The recent suggestion that LPS stimulation of epithelial cells may require bacterial invasion since TLR4 is located in a vesicular compartment in the cytoplasm further complicates an evaluation of published work on the responsiveness of epithelial cells to LPS (4). Thus, while commonly used human colon carcinoma cell lines are clearly LPS nonresponsive, it is far from certain whether this accurately models the situation found in the healthy human small and large intestine during invasion with gram-negative pathogens.

BACTERIAL ATTACHMENT TO AND PERSISTENT COLONIZATION OF INTESTINAL TISSUES

With the onset of adaptive immunity, *S. enterica* serotype Typhimurium infection in calves and humans is controlled and the acute signs of disease (i.e., diarrhea and severe acute neutrophil influx) subside. However, despite the development of adaptive immunity, serotype Typhimurium persists for extended periods in some intestinal and extraintestinal reservoirs. In genetically resistant mice (*Nramp1*$^{+/+}$), serotype Typhimurium can be isolated from mesenteric lymph nodes for up to 1 year after infection in approximately 60% of animals (77). Persistence of serotype Typhimurium in the mesenteric lymph nodes has also been described for apparently healthy cattle (78, 99, 100). In mesenteric lymph nodes of chronically infected mice, serotype Typhimurium can be found within macrophages, which may be a mechanism for immune evasion. Neutralization of gamma interferon leads to reactivation of the infection, with rapid bacterial multiplication, suggesting that chronic persistence in mesenteric lymph nodes represents a standoff between host defense mechanisms and bacterial multiplication (77).

A second important reservoir for serotype Typhimurium persistence is the intestinal lumen. Our knowledge about mechanisms contributing to intestinal persistence is limited to studies using rodent models. Serotype Typhimurium is excreted with the feces of genetically resistant mice (*Nramp1*$^{+/+}$) for 2 to 3 months after inoculation (58, 88). The main reservoir for bacteria that are shed in the feces of mice appears to be the cecum (58). Serotype Typhimurium LPS-deficient mutants have a reduced capacity to colonize the murine large intestine (86), presumably because they have a reduced ability to penetrate the intestinal mucus layer (65). Serotype Typhimurium is able to colonize the epithelial lining of the cecal mucosa in mice (Color Plate 5C) (60), which may contribute to intestinal persistence. Whole-genome sequencing has revealed the presence of a large number of putative fimbrial adhesins in the serotype Typhimurium genome that may be involved in intestinal colonization (69). The 13 operons containing fimbrial gene sequences that are present in the genome are *pefBACD*, *bcfABCDEFG*, *stiABCD*, *stfACDEFG*, *safABCD*, *stbABCDE*, *fimAICDHF*, *agfBACagfDEFG*, *stcABCD*, *stdABCD*, *lpfABCDE*, *stjABCDE* and *sthABCDE* (17, 24, 28, 30, 69, 79, 97, 112, 115). A genomic comparison of serotype Typhimurium with other *S. enterica* serotypes and *E. coli* shows that while some of these operons are conserved among members of both species (e.g., *agf*, *fim*, and *stc*), others have a scattered phylogenetic distribution (Fig. 1) (6, 112). The remarkable heterogeneity of some DNA regions containing fimbrial gene sequences provides evidence for their frequent horizontal gene transfer and/or loss by deletion during divergence of the *S. enterica* and *E. coli* lineages. As a result, most *Salmonella* serotypes possess unique repertoires of fimbrial gene sequences (6, 112). Figure 1 further illustrates that the majority of putative fimbrial proteins are well conserved among different *Salmonella* serotypes. However, the deduced amino acid sequences of *S. enterica* serotype Typhi *safA* and *S. enterica* serotype Enteritidis *safA*, *stcABCD*, and *pefI* are highly divergent from those of their serotype Typhimurium orthologues. The high degree of se-

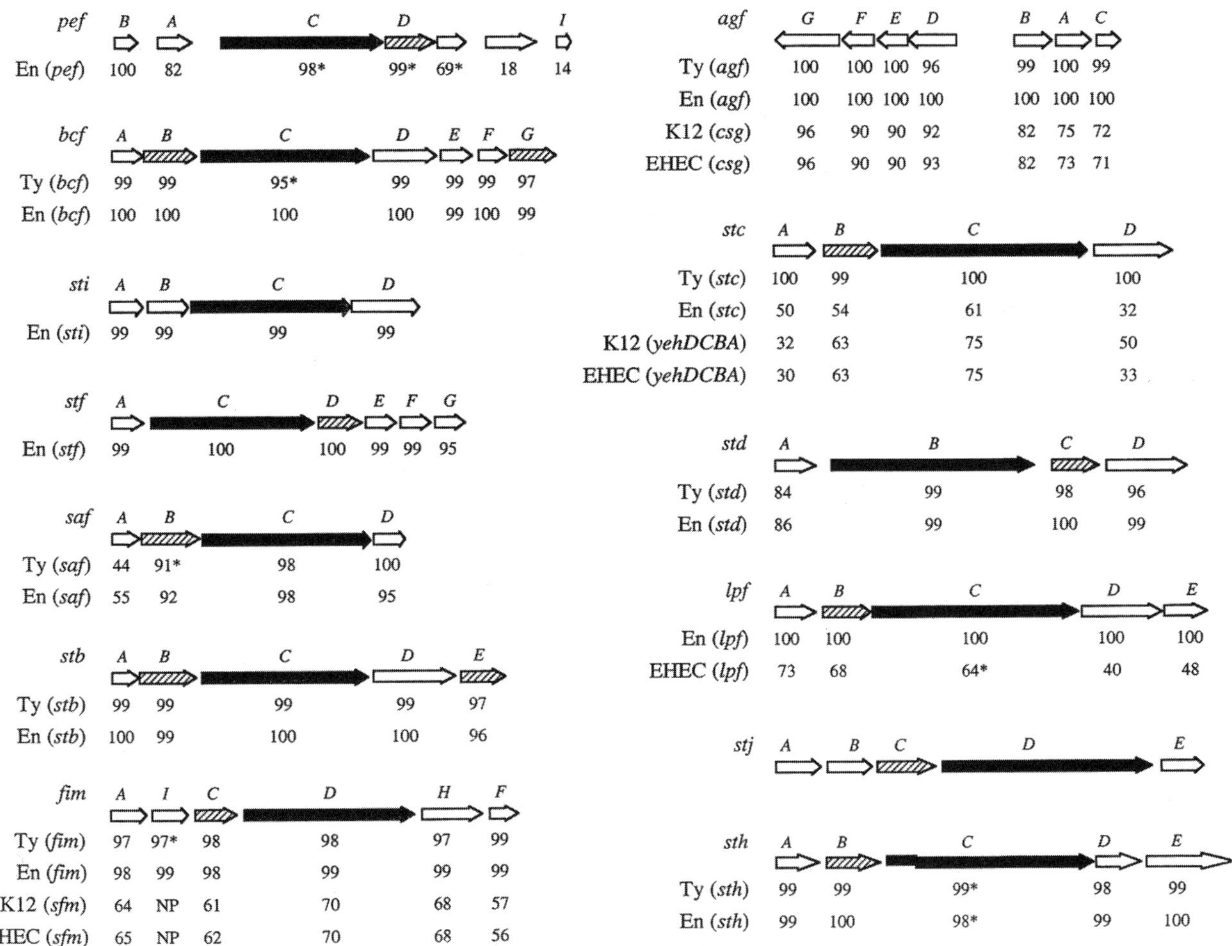

Figure 1. Comparison of the amino acid sequences deduced from fimbrial genes of *S. enterica* serotype Typhimurium strain LT-2 (69) with orthologues present in serotype Typhi strain CT-18 (Ty) (92), serotype Enteritidis phage type 4 (En) (http://www.sanger.ac.uk/Projects/Salmonella/ and http://www.salmonella.org/), *E. coli* K-12 strain MG1655 (K12) (11), and *E. coli* O157:H7 strain EDL933 (EHEC) (93). Serotype Typhimurium genes are indicated by arrows. Genes whose deduced amino acid sequences show homology to fimbrial chaperones and outer membrane fimbrial usher proteins are indicated as hatched and black arrows, respectively. The percent identity of deduced serotype Typhimurium amino acid sequences to orthologues (defined by the same location on the genome, as indicated by homology of flanking DNA regions) present in serotype Typhi, serotype Enteritidis, *E. coli* K-12, and *E. coli* O157:H7 are indicated below each arrow. Asterisks indicate that sequences were modified prior to the alignment to correct for truncations, frameshift mutations, or stop codons. NP, gene not present.

quence divergence of the serotype Enteritidis *stc* operon may explain why its presence was detected with serotype Typhimurium- or serotype Typhi-specific DNA probes only in some studies (16, 94) while others scored it as absent (112). The presence of a large number of fimbrial gene sequences that differ in their distribution among *Salmonella* serotypes and whose deduced amino acid sequences exhibit different degrees of sequence identity to orthologous proteins in related serotypes are features of a genetic design that is currently poorly understood.

Three serotype Typhimurium fimbrial operons (*agf*, *pef*, and *lpf*) have been implicated in colonization of the murine intestine (7, 8, 110), while one operon (*bcf*) is thought to contribute to colonization of bovine intestinal tissues (115). Characterization of strains carrying mutations in multiple fimbrial operons provides indirect evidence for a partial functional redundancy of these colonization factors (117). However, receptors for the encoded fimbriae on the surface of epithelial cells have not yet been identified. Furthermore, the possible role in intestinal colonization of the majority of fimbrial gene sequences, including *stiABCD*, *stfACDEFG*, *safABCD*, *stbABCDE*, *stcABCD*, *stdABCD*, *stjABCDE*, and *sthABCDE*, has not yet been explored. One obstacle to a functional analysis of fimbrial gene sequences in vitro appears to be the fact that, with the exception of type

1 fimbriae (encoded by the *fim* operon) (21) and thin aggregative fimbriae (encoded by the *agf* operon) (41, 42), the encoded fimbrial proteins are poorly expressed after growth of serotype Typhimurium in standard laboratory medium (52).

In addition to fimbrial adhesins, the serotype Typhimurium genome contains numerous genes with homology to genes encoding nonfimbrial adhesins. These include genes encoding members of the autotransporter family of outer membrane proteins (ShdA, MisL, and BigA) and a protein with homology to the *Yersinia enterocolitica* invasin (termed SivH) (9, 58, 61, 69). Two of these nonfimbrial adhesins, ShdA and SivH, have been implicated in colonization of murine intestinal tissues (58, 61). ShdA is an outer membrane protein which binds to the extracellular matrix proteins collagen and fibronectin (59, 60). Deletion of the *shdA* gene results in a reduced ability of serotype Typhimurium to colonize the murine cecum and a reduction in the time during which bacteria are shed in the feces of mice (58, 61). Colonization of the murine cecum by serotype Typhimurium results in epithelial erosion in a few isolated areas of the mucosal surface, which exposes the extracellular matrix to the intestinal lumen. Immunohistochemical analysis of cecal tissue from infected mice demonstrates that in addition to colonization of epithelial surfaces, serotype Typhimurium colonizes the cecal mucosa at these areas of epithelial erosion (60). These data suggest that ShdA-mediated binding of extracellular matrix proteins in areas of epithelial erosion may be a mechanism for persistent intestinal carriage of serotype Typhimurium. Additional adhesins that may contribute to extracellular matrix binding are thin aggregative fimbriae (curli), encoded by the *agf* operon, which bind fibronectin (19), and type 1 fimbriae, encoded by the *fim* operon, which bind laminin (64). However, no in vivo role during infection has so far been ascribed to type 1 fimbriae (66, 84).

The reasons for the presence in serotype Typhimurium of numerous fimbrial and nonfimbrial adhesins with partially redundant functions are not fully understood. The genetic design of intestinal colonization factors in serotype Typhimurium is further complicated by the finding that expression of several fimbrial structures is regulated by phase variation (87, 90, 111). As a result of phase variation, fimbrial filaments are expressed in only a fraction of cells in a bacterial population, whereas all cells express some invariant surface structures such as LPS (52). The benefit of regulating fimbrial expression by phase variation while at the same time displaying invariant antigens (i.e., LPS) on the surface of all bacterial cells is not immediately apparent. The LPS of serotype Typhimurium carries terminal structures formed by repeating oligosaccharide units (O antigen) that are highly immunogenic (57). A monoclonal immunoglobulin A antibody directed against the serotype Typhimurium O antigen can prevent attachment to and invasion of epithelial cells in vitro (74) and can protect mice from oral challenge by immune exclusion at the mucosal surface (73). These data suggest that bacteria residing in a niche in which they are exposed to adaptive immunity will become cleared by responses directed against their O antigen, regardless of whether they are able to evade responses directed against fimbrial proteins by phase variation. It is therefore unlikely that fimbrial phase variation represents an effective mechanism for serotype Typhimurium to evade adaptive immune responses during an infection.

EVASION OF CROSS-IMMUNITY AND TRANSMISSION

The selective advantage of fimbrial phase variation becomes apparent only when the properties that allow *S. enterica* serotype Typhimurium to persist in its host reservoirs are considered. During persistence of serotype Typhimurium in an animal reservoir, direct competition arises as a result of the presence of related pathogens that share antigens important for protective immunity (immunodominant antigens) (57). The O antigen of serotype Typhimurium is one of the major targets for the adaptive immune response generated during infection. The immunodominant epitope of the serotype Typhimurium O antigen is an abequose branch (O4 antigen) on the trisaccharide backbone of its O-antigen repeat unit (49, 106). Serotype Typhimurium shares its immunodominant O4 antigen with several hundred other *Salmonella* serotypes, which together form a serologically related group termed serogroup B. Coexistence of serotype Typhimurium with another *Salmonella* serotype of serogroup B in an animal reservoir results in direct competition because the host immune response to the O4 antigen will effectively limit the number of susceptible individuals suitable for transmission (57). Theory indicates that this between-serotype competition will eventually be won by the serotype with the higher transmission success by excluding the other serotype expressing the O4 antigen from the host population through the generation of herd immunity (3). The theoretical prediction that one member of serogroup B will exclude other serotypes expressing the O4 antigen from the host population provides an attractive explanation for the observation that in each population of livestock only one serotype expressing the O4 antigen is frequently isolated from animals with disease (57).

The above considerations explain why a particular *Salmonella* serotype expressing the O4 antigen is not able to persist in a host population which is already occupied by another well-adapted member of serogroup B (i.e., one with higher transmissibility). If this particular *Salmonella* serotype is not able to infect other host species, the consequence of competitive exclusion by the well-adapted member of serogroup B will be its eradication. One way to avoid eradication and to coexist in the host population with the well-adapted member of serogroup B is for the particular *Salmonella* serotype to evade cross-immunity against the O4 antigen (3). A possible mechanism for this evasion of cross-immunity would be the expression of an O antigen that does not cross-react with the O4 antigen. Comparison of O-antigen biosynthesis gene clusters (*rfb* locus) between different *Salmonella* serotypes provides compelling evidence that acquisition of new genetic material by horizontal transfer has occurred much more frequently in this region than in other areas of the genome, often generating closely related strains expressing different immunodominant O antigens (95). Evasion of cross-immunity to the immunodominant O antigen of a competitor may have provided the advantage necessary for selecting variants that acquired new O-antigen biosynthesis genes by horizontal gene transfer, thereby generating the observed hypervariability of the *rfb* locus (57).

While mathematical models combining epidemiology with population biology provide a theoretical basis for predicting the selective advantage of evading cross-immunity (3), review of the available literature on protective immunity in animal models provides compelling experimental evidence that O-antigen polymorphism is a mechanism that allows *Salmonella* serotypes to evade cross-immunity (57). However, in reviewing the available literature, one has to make the important distinction between vaccination studies in which protection was conferred solely by a specific memory response and those in which protection was due at least in part to nonspecific immune defenses. Nonspecific immunity to serotype Typhimurium is thought to be due to nonspecific macrophage activation that may be triggered during an infection with related or unrelated pathogens. For instance, mice infected with mouse hepatitis virus become resistant to serotype Typhimurium infection (26, 27). This immunity to serotype Typhimurium induced in animals that are currently infected with a different pathogen is presumably mediated by responses that are not antigen specific. Nonspecific immunity to serotype Typhimurium can be induced experimentally by injecting mice with gamma interferon or PAMPs prior to challenge, thereby supporting the idea that the underlying mechanism is nonspecific macrophage activation (62, 68, 91). Immunization with a live attenuated serotype Typhimurium vaccine strain elicits nonspecific immunity that begins to subside only when the vaccine strain has been cleared from the animal. Most serotype Typhimurium live attenuated vaccines are cleared from internal organs within 4 to 6 weeks after immunization of mice. Thus, when challenge is performed at 5 or 6 weeks after immunization, immunity is not antigen specific and vaccination with a live attenuated serotype Typhimurium strain can cross-protect mice against challenge with serotypes that possess unrelated O antigens (45, 48). In contrast, if challenge is performed at 10 or 11 weeks after vaccination, when nonspecific immune responses have returned to their normal level, immunity becomes highly specific for the immunodominant O antigen of the vaccine strain and is conferred solely by an adaptive memory response.

The importance of immunodominant O antigens in eliciting a protective memory response to *Salmonella* serotypes has been convincingly demonstrated in a seminal study by Hormaeche et al., using serotype Typhimurium (O4 antigen) and Enteritidis strains (49). The LPS of serotype Enteritidis does not contain the O4 antigen but instead carries a tyvelose branch (O9 antigen) as the immunodominant epitope on the trisaccharide backbone of its O-antigen repeat unit (5, 49). When mice immunized with a serotype Enteritidis *aroA* vaccine strain are challenged 3 months later with virulent serotype Enteritidis, a 10,000-fold protection is observed compared to that in unvaccinated control animals. In contrast, mice vaccinated with serotype Enteritidis *aroA* vaccine are only 200- to 500-fold protected when challenge is performed with a serotype Enteritidis strain genetically engineered to express the O4 antigen instead of its O9 antigen. Similarly, 3 months after vaccination with a serotype Enteritidis *aroA* vaccine, mice are only fivefold protected against challenge with virulent serotype Typhimurium. However, 200-fold protection is observed when challenge is performed with a fully virulent serotype Typhimurium strain genetically engineered to express the O9 antigen instead of the O4 antigen (49). These data show that O-antigen polymorphism is a mechanism that contributes significantly to the ability of serotypes Typhimurium and Enteritidis to evade cross-immunity.

Serotypes Typhimurium and Enteritidis are the *Salmonella* serotypes most commonly isolated from rodent animal reservoirs (23, 63). As outlined above, evasion of cross-immunity by O-antigen polymorphism is one of the mechanisms which allows these organisms to evade between-serotype competition during coexistence in their rodent animal reservoir.

While serotypes Typhimurium and Enteritidis can evade cross-immunity against LPS by O-antigen polymorphism, many fimbrial antigens (e.g., LpfA) are highly conserved between the two organisms (Fig. 1). Recent studies suggest that fimbrial phase variation is a mechanism to evade cross-immunity between serotypes Enteritidis and Typhimurium that is elicited by immune responses to their fimbrial antigens (88, 89). When mice immunized with a serotype Typhimurium *aroA* vaccine are challenged 10 weeks later with serotype Enteritidis, no protection is observed but recovery of the challenge strain reveals a selection against bacteria expressing the *lpf* operon in the phase-on expression state compared to the situation for unvaccinated controls (89). However, a serotype Enteritidis strain genetically engineered to express the *lpf* operon constitutively is recovered in smaller numbers from mice immunized with a serotype Typhimurium *aroA* vaccine than from unvaccinated controls (88). Thus, serotype Enteritidis is not able to evade cross-immunity elicited by exposure to serotype Typhimurium once *lpf* fimbrial phase variation has been eliminated.

These data support the idea that fimbrial phase variation is a mechanism to evade cross-immunity between serotypes Typhimurium and Enteritidis. While it is difficult to envision how fimbrial phase variation could mediate continued evasion of adaptive immune responses developing during an infection, a simple on-off switch is clearly sufficient to evade a memory response encountered in a host with previous exposure to a related pathogen. Control of fimbrial expression by phase variation has important consequences for the genetic design of attachment factors in serotype Typhimurium. The finding that *lpf* phase-off variants of serotype Enteritidis can evade cross-immunity against serotype Typhimurium suggests that they retain considerable virulence because they express alternate attachment factors. Consequently, a partial functional redundancy of the 13 fimbrial operons present in serotype Typhimurium may be a prerequisite for effectively using phase variation as a mechanism of immune evasion. Furthermore, while genomic comparisons of *Salmonella* serotypes reveals that many virulence factors are shared between different serotypes, the repertoires of fimbrial operons are highly polymorphic, and this genetic design probably facilitates evasion of cross-immunity against fimbrial proteins between different serotypes (16, 94, 112). Their role in evasion of cross-immunity provides an attractive explanation of why genomes of *S. enterica* serotypes contain large but distinct repertoires of fimbrial operons that are regulated by phase variation and exhibit partial functional redundancy.

Acknowledgments. We thank the Wellcome Trust Sanger Institute, Hinxton, United Kingdom, for communication of DNA sequence data prior to publication. Work in the laboratory of A.J.B. is supported by USDA/NRICGP grant 2002-35204-12247 and Public Health Service grants AI40124 and AI44170.

REFERENCES

1. **Abreu, M. T., P. Vora, E. Faure, L. S. Thomas, E. T. Arnold, and M. Arditi.** 2001. Decreased expression of Toll-like receptor-4 and MD-2 correlates with intestinal epithelial cell protection against dysregulated proinflammatory gene expression in response to bacterial lipopolysaccharide. *J. Immunol.* **167:**1609–1616.
2. **Ahmer, B. M., J. van Reeuwijk, P. R. Watson, T. S. Wallis, and F. Heffron.** 1999. *Salmonella* SirA is a global regulator of genes mediating enteropathogenesis. *Mol. Microbiol.* **31:**971–982.
3. **Anderson, R. M.** 1995. Evolutionary pressures in the spread and persistence of infectious agents in vertebrate populations. *Parasitology* **111:**S15–S31.
4. **Bäckhed, F., and M. Hornef.** 2003. Toll-like receptor 4-mediated signaling by epithelial surfaces: necessity or threat? *Microbes Infect.* **5:**951–959.
5. **Barrow, P. A., A. Berchieri, Jr., and O. al-Haddad.** 1992. Serological response of chickens to infection with *Salmonella gallinarum-S. pullorum* detected by enzyme-linked immunosorbent assay. *Avian Dis.* **36:**227–236.
6. **Bäumler, A. J., A. J. Gilde, R. M. Tsolis, A. W. M. van der Velden, B. M. M. Ahmer, and F. Heffron.** 1997. Contribution of horizontal gene transfer and deletion events to the development of distinctive patterns of fimbrial operons during evolution of *Salmonella* serotypes. *J. Bacteriol.* **179:**317–322.
7. **Bäumler, A. J., R. M. Tsolis, F. Bowe, J. G. Kusters, S. Hoffmann, and F. Heffron.** 1996. The *pef* fimbrial operon of *Salmonella typhimurium* mediates adhesion to murine small intestine and is necessary for fluid accumulation in the infant mouse. *Infect. Immun.* **64:**61–68.
8. **Bäumler, A. J., R. M. Tsolis, and F. Heffron.** 1996. The *lpf* fimbrial operon mediates adhesion of *Salmonella typhimurium* to murine Peyer's patches. *Proc. Natl. Acad. Sci. USA* **93:**279–283.
9. **Blanc-Potard, A. B., F. Solomon, J. Kayser, and E. A. Groisman.** 1999. The SPI-3 pathogenicity island of salmonella enterica. *J. Bacteriol.* **181:**998–1004.
10. **Blaser, M. J., and L. S. Newman.** 1982. A review of human salmonellosis. I. Infective dose. *Rev. Infect. Dis.* **4:**1096–1106.
11. **Blattner, F. R., G. Plunkett III, C. A. Bloch, N. T. Perna, V. Burland, M. Riley, J. Colladovides, J. D. Glasner, C. K. Rode, G. F. Mayhew, J. Gregor, N. W. Davis, H. A. Kirkpatrick, M. A. Goeden, D. J. Rose, B. Mau, and Y. Shao.** 1997. The complete genome sequence of *Escherichia coli* K-12. *Science* **277:**1453–1474.
12. **Buchwald, D. S., and M. J. Blaser.** 1984. A review of human salmonellosis. II. Duration of excretion following infection with nontyphi *Salmonella*. *Rev. Infect. Dis.* **6:**345–356.
13. **Cario, E., and D. K. Podolsky.** 2000. Differential alteration in intestinal epithelial cell expression of Toll-like receptor 3 (TLR3) and TLR4 in inflammatory bowel disease. *Infect. Immun.* **68:**7010–7017.
14. **Cario, E., I. M. Rosenberg, S. L. Brandwein, P. L. Beck, H. C. Reinecker, and D. K. Podolsky.** 2000. Lipopolysaccharide activates distinct signaling pathways in intestinal epithelial cell lines expressing Toll-like receptors. *J. Immunol.* **164:**966–972.

15. **Centers for Disease Control and Prevention.** 1999. Salmonella *Surveillance: Annual Tabulation Summary, 1998.* U.S. Department of Health and Human Services, Centers for Disease Control and Prevention, Atlanta, Ga.
16. **Chan, K., S. Baker, C. C. Kim, C. S. Detweiler, G. Dougan, and S. Falkow.** 2003. Genomic comparison of *Salmonella enterica* serovars and *Salmonella bongori* by use of an *S. enterica* serovar Typhimurium DNA microarray. *J. Bacteriol.* **185:** 553–563.
17. **Clegg, S., B. K. Purcell, and J. Pruckler.** 1987. Characterization of genes encoding type 1 fimbriae of *Klebsiella pneumoniae, Salmonella typhimurium,* and *Serratia marcescens. Infect. Immun.* **55:**281–287.
18. **Collazo, C. M., and J. E. Galan.** 1997. The invasion-associated type III system of *Salmonella typhimurium* directs the translocation of Sip proteins into the host cell. *Mol. Microbiol.* **24:** 747–756.
19. **Collinson, S. K., P. S. Doig, J. L. Doran, S. Clouthier, T. J. Trust, and W. W. Kay.** 1993. Thin, aggregative fimbriae mediate binding of *Salmonella enteritidis* to fibronectin. *J. Bacteriol.* **175:**12–18.
20. **Day, D. W., B. K. Mandal, and B. C. Morson.** 1978. The rectal biopsy appearances in *Salmonella* colitis. *Histopathology* **2:**117–131.
21. **Duguid, J. P., E. S. Anderson, and I. Campbell.** 1966. Fimbriae and adhesive properties in salmonellae. *J. Pathol. Bacteriol.* **92:**107–137.
22. **Eckmann, L., J. R. Smith, M. P. Housley, M. B. Dwinell, and M. F. Kagnoff.** 2000. Analysis by high density cDNA arrays of altered gene expression in human intestinal epithelial cells in response to infection with the invasive enteric bacteria *Salmonella. J. Biol. Chem.* **275:**14084–14094.
23. **Edwards, P. R., and D. W. Bruner.** 1943. The occurrence and distribution of *Salmonella* types in the United States. *J. Infect. Dis.* **72:**58–67.
24. **Emmerth, M., W. Goebel, S. I. Miller, and C. J. Hueck.** 1999. Genomic subtraction identifies *Salmonella typhimurium* prophages, F-related plasmid sequences, and a novel fimbrial operon, *stf,* which are absent in *Salmonella typhi. J. Bacteriol.* **181:**5652–5661.
25. **Everest, P., J. Ketley, S. Hardy, G. Douce, S. Khan, J. Shea, D. Holden, D. Maskell, and G. Dougan.** 1999. Evaluation of *Salmonella typhimurium* mutants in a model of experimental gastroenteritis. *Infect. Immun.* **67:**2815–2821.
26. **Fallon, M. T., W. H. Benjamin, Jr., T. R. Schoeb, and D. E. Briles.** 1991. Mouse hepatitis virus strain UAB infection enhances resistance to *Salmonella typhimurium* in mice by inducing suppression of bacterial growth. *Infect. Immun.* **59:** 852–856.
27. **Fallon, M. T., T. R. Schoeb, W. H. Benjamin, Jr., J. R. Lindsey, and D. E. Briles.** 1989. Modulation of resistance to *Salmonella typhimurium* infection in mice by mouse hepatitis virus (MHV). *Microb. Pathog.* **6:**81–91.
28. **Folkesson, A., A. Advani, S. Sukupolvi, J. D. Pfeifer, S. Normark, and S. Lofdahl.** 1999. Multiple insertions of fimbrial operons correlate with the evolution of *Salmonella* serovars responsible for human disease. *Mol. Microbiol.* **33:**612–622.
29. **Frances, C. L., T. A. Ryan, B. D. Jones, S. J. Smith, and S. Falkow.** 1993. Ruffles induced by *Salmonella* and other stimuli direct macropinocytosis of bacteria. *Nature* **364:**639–642.
30. **Friedrich, M. J., N. E. Kinsey, J. Vila, and R. J. Kadner.** 1993. Nucleotide sequence of a 13.9 kb segment of the 90 kb virulence plasmid of *Salmonella typhimurium:* the presence of fimbrial biosynthetic genes. *Mol. Microbiol.* **8:**543–558.
31. **Frost, A. J., A. P. Bland, and T. S. Wallis.** 1997. The early dynamic response of the calf ileal epithelium to *Salmonella typhimurium. Vet. Pathol.* **34:**369–386.
32. **Fu, Y., and J. E. Galan.** 1998. The *Salmonella typhimurium* tyrosine phosphatase SptP is translocated into host cells and disrupts the actin cytoskeleton. *Mol. Microbiol.* **27:**359–368.
33. **Fusunyan, R. D., N. N. Nanthakumar, M. E. Baldeon, and W. A. Walker.** 2001. Evidence for an innate immune response in the immature human intestine: Toll-like receptors on fetal enterocytes. *Pediatr. Res.* **49:**589–593.
34. **Galán, J. E.** 1999. Interaction of *Salmonella* with host cells through the centisome 63 type III secretion system. *Curr. Opin. Microbiol.* **2:**46–50.
35. **Galán, J. E., and R. Curtiss III.** 1989. Cloning and molecular characterization of genes whose products allow *Salmonella typhimurium* to penetrate tissue culture cells. *Proc. Natl. Acad. Sci. USA* **86:**6383–6387.
36. **Galyov, E. E., M. W. Wood, R. Rosqvist, P. B. Mullan, P. R. Watson, S. Hedges, and T. S. Wallis.** 1997. A secreted effector protein of *Salmonella dublin* is translocated into eukaryotic cells and mediates inflammation and fluid secretion in infected ileal mucosa. *Mol. Microbiol.* **25:**903–912.
37. **Gewirtz, A. T., T. A. Navas, S. Lyons, P. J. Godowski, and J. L. Madara.** 2001. Cutting edge: bacterial flagellin activates basolaterally expressed TLR5 to induce epithelial proinflammatory gene expression. *J. Immunol.* **167:**1882–1885.
38. **Gewirtz, A. T., A. M. Siber, J. M. Madara, and B. A. McCormick.** 1999. Orchestration of neutrophil movement by intestinal epithelial cells in response to *Salmonella typhimurium* can be uncoupled from bacterial internalization. *Infect. Immun.* **67:**608–617.
39. **Gewirtz, A. T., P. O. Simon, Jr., C. K. Schmitt, L. J. Taylor, C. H. Hagedorn, A. D. O'Brien, A. S. Neish, and J. L. Madara.** 2001. *Salmonella typhimurium* translocates flagellin across intestinal epithelia, inducing a proinflammatory response. *J. Clin. Investig.* **107:**99–109.
40. **Girardin, S. E., I. G. Boneca, L. A. Carneiro, A. Antignac, M. Jehanno, J. Viala, K. Tedin, M. K. Taha, A. Labigne, U. Zahringer, A. J. Coyle, P. S. DiStefano, J. Bertin, P. J. Sansonetti, and D. J. Philpott.** 2003. Nod1 detects a unique muropeptide from gram-negative bacterial peptidoglycan. *Science* **300:**1584–1587.
41. **Grund, S., and R. Meyer.** 1987. Fimbriae formation, disease form and vaccination in infection with *Salmonella typhimurium* variatio copenhagen (STMVC). *Berl. Münch. Tierärztl. Wochenschr.* **100:**367–373. (In German.)
42. **Grund, S., and A. Seiler.** 1993. Electron microscopic studies of fimbriae and lectin phagocytosis of *Salmonella typhimurium* variety copenhagen (STMVC). *Zentbl. Veterinaermed. Reihe B* **40:**105–112. (In German.)
43. **Hardt, W. D., L. M. Chen, K. E. Schuebel, X. R. Bustelo, and J. E. Galan.** 1998. *S. typhimurium* encodes an activator of Rho GTPases that induces membrane ruffling and nuclear responses in host cells. *Cell* **93:**815–826.
44. **Hardt, W. D., and J. E. Galan.** 1997. A secreted *Salmonella* protein with homology to an avirulence determinant of plant pathogenic bacteria. *Proc. Natl. Acad. Sci. USA* **94:**9887–9892.
45. **Heithoff, D. M., E. Y. Enioutina, R. A. Daynes, R. L. Sinsheimer, D. A. Low, and M. J. Mahan.** 2001. *Salmonella* DNA adenine methylase mutants confer cross-protective immunity. *Infect. Immun.* **69:**6725–6730.
46. **Hermiston, M. L., and J. I. Gordon.** 1995. Inflammatory bowel disease and adenomas in mice expressing a dominant negative N-cadherin. *Science* **270:**1203–1207.

47. **Hershberg, R. M.** 2002. The epithelial cell cytoskeleton and intracellular trafficking. V. Polarized compartmentalization of antigen processing and Toll-like receptor signaling in intestinal epithelial cells. *Am. J. Physiol. Ser. G* **283:**G833–G839.
48. **Hormaeche, C. E., H. S. Joysey, L. Desilva, M. Izhar, and B. A. Stocker.** 1991. Immunity conferred by Aro$^-$ *Salmonella* live vaccines. *Microb. Pathog.* **10:**149–158.
49. **Hormaeche, C. E., P. Mastroeni, J. A. Harrison, R. Demarco de Hormaeche, S. Svenson, and B. A. Stocker.** 1996. Protection against oral challenge three months after i.v. immunization of BALB/c mice with live Aro *Salmonella typhimurium* and *Salmonella enteritidis* vaccines is serotype (species)-dependent and only partially determined by the main LPS O antigen. *Vaccine* **14:**251–259.
50. **Hornef, M. W., T. Frisan, A. Vandewalle, S. Normark, and A. Richter-Dahlfors.** 2002. Toll-like receptor 4 resides in the Golgi apparatus and colocalizes with internalized lipopolysaccharide in intestinal epithelial cells. *J. Exp. Med.* **195:**559–570.
51. **Hornef, M. W., B. H. Normark, A. Vandewalle, and S. Normark.** 2003. Intracellular recognition of lipopolysaccharide by toll-like receptor 4 in intestinal epithelial cells. *J. Exp. Med.* **198:**1225–1235.
52. **Humphries, A. D., M. Raffatellu, S. Winter, E. H. Weening, R. A. Kingsley, R. Droleskey, S. Zhang, J. Figueiredo, S. Khare, J. Nunes, L. G. Adams, R. M. Tsolis, and A. J. Baumler.** 2003. The use of flow cytometry to detect expression of subunits encoded by 11 *Salmonella enterica* serotype Typhimurium fimbrial operons. *Mol. Microbiol.* **48:**1357–1376.
53. **Hurd, H. S., J. K. Gailey, J. D. McKean, and M. H. Rostagno.** 2001. Rapid infection in market-weight swine following exposure to a *Salmonella typhimurium*-contaminated environment. *Am. J. Vet. Res.* **62:**1194–1197.
54. **Inohara, N., Y. Ogura, and G. Nunez.** 2002. Nods: a family of cytosolic proteins that regulate the host response to pathogens. *Curr. Opin. Microbiol.* **5:**76–80.
55. **Jepson, M. A., B. Kenny, and A. D. Leard.** 2001. Role of *sipA* in the early stages of *Salmonella typhimurium* entry into epithelial cells. *Cell. Microbiol.* **3:**417–426.
56. **Jung, H. C., L. Eckmann, S. K. Yang, A. Panja, J. Fierer, E. Morzycka-Wroblewska, and M. F. Kagnoff.** 1995. A distinct array of proinflammatory cytokines is expressed in human colon epithelial cells in response to bacterial invasion. *J. Clin. Investig.* **95:**55–65.
57. **Kingsley, R. A., and A. J. Bäumler.** 2000. Host adaptation and the emergence of infectious disease: the salmonella paradigm. *Mol. Microbiol.* **36:**1006–1014.
58. **Kingsley, R. A., A. D. Humphries, E. H. Weening, M. R. De Zoete, S. Winter, A. Papaconstantinopoulou, G. Dougan, and A. J. Bäumler.** 2003. Molecular and phenotypic analysis of the CS54 island of *Salmonella enterica* serotype Typhimurium: identification of intestinal colonization and persistence determinants. *Infect. Immun.* **71:**629–640.
59. **Kingsley, R. A., A. M. Keestra, M. R. de Zoete, and A. J. Bäumler.** 2004. The Shad adhesin binds to the cationic cradle of the fibronectin 13FnIII repeat module: evidence for molecular mimicry of heparin binding. *Mol. Microbiol.* **52:**345–355.
60. **Kingsley, R. A., R. L. Santos, A. M. Keestra, L. G. Adams, and A. J. Bäumler.** 2002. *Salmonella enterica* serotype Typhimurium ShdA is an outer membrane fibronectin-binding protein that is expressed in the intestine. *Mol. Microbiol.* **43:**895–905.
61. **Kingsley, R. A., K. van Amsterdam, N. Kramer, and A. J. Bäumler.** 2000. The *shdA* gene is restricted to serotypes of *Salmonella enterica* subspecies I and contributes to efficient and prolonged fecal shedding. *Infect. Immun.* **68:**2720–2727.
62. **Kita, E., M. Emote, D. Oku, F. Nishikawa, A. Hamburg, N. Kamikaidou, and S. Kashiba.** 1992. Contribution of interferon gamma and membrane-associated interleukin 1 to the resistance to murine typhoid of Ityr mice. *J. Leukoc. Biol.* **51:**244–250.
63. **Krabisch, P., and P. Dorn.** 1980. Epidemiologic significance of live vectors in the transmission of *Salmonella* infections in broiler flocks. *Berl. Muench. Tieraerztl. Wochenschr.* **93:**232–235.
64. **Kukkonen, M., T. Raunio, R. Virkola, K. Lahteenmaki, P. H. Makela, P. Klemm, S. Clegg, and T. K. Korhonen.** 1993. Basement membrane carbohydrate as a target for bacterial adhesion: binding of type I fimbriae of *Salmonella enterica* and *Escherichia coli* to laminin. *Mol. Microbiol.* **7:**229–237.
65. **Licht, T. R., K. A. Krogfelt, P. S. Cohen, L. K. Poulsen, J. Urbance, and S. Molin.** 1996. Role of lipopolysaccharide in colonization of the mouse intestine by *Salmonella typhimurium* studied by in situ hybridization. *Infect. Immun.* **64:**3811–3817.
66. **Lockman, H. A., and R. Curtiss III.** 1992. Virulence of non-type 1-fimbriated and nonfimbriated nonflagellated *Salmonella typhimurium* mutants in murine typhoid fever. *Infect. Immun.* **60:**491–496.
67. **Madara, J. L., T. W. Patapoff, B. Gillece-Castro, S. P. Colgan, C. A. Parkos, C. Delp, and R. J. Mrsny.** 1993. 5′-Adenosine monophosphate is the neutrophil-derived paracrine factor that elicits chloride secretion from T84 intestinal epithelial cell monolayers. *J. Clin. Investig.* **91:**2320–2325.
68. **Matsumura, H., K. Onozuka, Y. Terada, Y. Nakano, and M. Nakano.** 1990. Effect of murine recombinant interferon-gamma in the protection of mice against *Salmonella. Int. J. Immunopharmacol.* **12:**49–56.
69. **McClelland, M., K. E. Sanderson, J. Spieth, S. W. Clifton, P. Latreille, L. Courtney, S. Porwollik, J. Ali, M. Dante, F. Du, S. Hou, D. Layman, S. Leonard, C. Nguyen, K. Scott, A. Holmes, N. Grewal, E. Mulvaney, E. Ryan, H. Sun, L. Florea, W. Miller, T. Stoneking, M. Nhan, R. Waterston, and R. K. Wilson.** 2001. Complete genome sequence of *Salmonella enterica* serovar Typhimurium LT2. *Nature* **413:**852–856.
70. **McGovern, V. J., and L. J. Slavutin.** 1979. Pathology of *Salmonella* colitis. *Am. J. Surg. Pathol.* **3:**483–490.
71. **McLaren, I. M., and C. Wray.** 1991. Epidemiology of *Salmonella typhimurium* infection in calves: persistence of salmonellae on calf units. *Vet. Rec.* **129:**461–462.
72. **Mead, P. S., L. Slutsker, V. Dietz, L. F. McCaig, J. S. Bresee, C. Shapiro, P. M. Griffin, and R. V. Tauxe.** 1999. Food-related illness and death in the United States. *Emerg. Infect. Dis.* **5:**607–625.
73. **Michetti, P., M. J. Mahan, J. M. Slauch, J. J. Mekalanos, and M. R. Neutra.** 1992. Monoclonal secretory immunoglobulin A protects mice against oral challenge with the invasive pathogen *Salmonella typhimurium. Infect. Immun.* **60:**1786–1792.
74. **Michetti, P., N. Porta, M. J. Mahan, J. M. Slauch, J. J. Mekalanos, A. L. Blum, J. P. Kraehenbuhl, and M. R. Neutra.** 1994. Monoclonal immunoglobulin A prevents adherence and invasion of polarized epithelial cell monolayers by *Salmonella typhimurium. Gastroenterology* **107:**915–923.
75. **Mills, D. M., V. Bajaj, and C. A. Lee.** 1995. A 40kb chromosomal fragment encoding *Salmonella typhimurium* invasion genes is absent from the corresponding region of the *Escherichia coli* K-12 chromosome. *Mol. Microbiol.* **15:**749–759.

76. **Mirold, S., K. Ehrbar, A. Weissmuller, R. Prager, H. Tschape, H. Russmann, and W. D. Hardt.** 2001. *Salmonella* host cell invasion emerged by acquisition of a mosaic of separate genetic elements, including *Salmonella* pathogenicity island 1 (SPI1), SPI5, and *sopE2*. *J. Bacteriol.* **183:**2348–2358.
77. **Monack, D. M., D. M. Bouley, and S. Falkow.** 2004. *Salmonella typhimurium* persists within macrophages in the mesenteric lymph nodes of chronically infected Nramp1$^{+/+}$ mice and can be reactivated by IFN-γ neutralization. *J. Exp. Med.* **199:**231–241.
78. **Moo, D., D. O'Boyle, W. Mathers, and A. J. Frost.** 1980. The isolation of *Salmonella* from jejunal and caecal lymph nodes of slaughtered animals. *Aust. Vet. J.* **56:**181–183.
79. **Morrow, B. J., J. E. Graham, and R. Curtiss III.** 1999. Genomic subtractive hybridization and selective capture of transcribed sequences identify a novel *Salmonella typhimurium* fimbrial operon and putative transcriptional regulator that are absent from the *Salmonella typhi* genome. *Infect. Immun.* **67:**5106–5116.
80. **Naik, S., E. J. Kelly, L. Meijer, S. Pettersson, and I. R. Sanderson.** 2001. Absence of Toll-like receptor 4 explains endotoxin hyporesponsiveness in human intestinal epithelium. *J. Pediatr. Gastroenterol. Nutr.* **32:**449–453.
81. **Nanthakumar, N. N., R. D. Fusunyan, I. Sanderson, and W. A. Walker.** 2000. Inflammation in the developing human intestine: a possible pathophysiologic contribution to necrotizing enterocolitis. *Proc. Natl. Acad. Sci. USA* **97:**6043–6048.
82. **Nau, G. J., J. F. Richmond, A. Schlesinger, E. G. Jennings, E. S. Lander, and R. A. Young.** 2002. Human macrophage activation programs induced by bacterial pathogens. *Proc. Natl. Acad. Sci. USA* **99:**1503–1508.
83. **Nau, G. J., A. Schlesinger, J. F. Richmond, and R. A. Young.** 2003. Cumulative Toll-like receptor activation in human macrophages treated with whole bacteria. *J. Immunol.* **170:** 5203–5209.
84. **Naughton, P. J., G. Grant, S. Bardocz, E. Allen-Vercoe, M. J. Woodward, and A. Pusztai.** 2001. Expression of type 1 fimbriae (SEF 21) of *Salmonella enterica* serotype Enteritidis in the early colonisation of the rat intestine. *J. Med. Microbiol.* **50:**191–197.
85. **Neilsen, P. O., G. A. Zimmerman, and T. M. McIntyre.** 2001. *Escherichia coli* Braun lipoprotein induces a lipopolysaccharide-like endotoxic response from primary human endothelial cells. *J. Immunol.* **167:**5231–5239.
86. **Nevola, J. J., B. A. Stocker, D. C. Laux, and P. S. Cohen.** 1985. Colonization of the mouse intestine by an avirulent *Salmonella typhimurium* strain and its lipopolysaccharide-defective mutants. *Infect. Immun.* **50:**152–159.
87. **Nicholson, B., and D. Low.** 2000. DNA methylation-dependent regulation of pef expression in *Salmonella typhimurium*. *Mol. Microbiol.* **35:**728–742.
88. **Nicholson, T. L., and A. J. Bäumler.** 2001. *Salmonella enterica* serotype Typhimurium elicits cross-immunity against a *Salmonella enterica* serotype Enteritidis strain expressing LP fimbriae from the lack promoter. *Infect. Immun.* **69:**204–212.
89. **Norris, T. L., and A. J. Bäumler.** 1999. Phase variation of the *lpf* fimbrial operon is a mechanism to evade cross immunity between *Salmonella* serotypes. *Proc. Natl. Acad. Sci. USA* **96:**13393–13398.
90. **Norris, T. L., R. A. Kingsley, and A. J. Bäumler.** 1998. Expression and transcriptional control of the *Salmonella typhimurium lpf* fimbrial operon by phase variation. *Mol. Microbiol.* 29:311–320.
91. **Onozuka, K., S. Shimada, H. Yamasu, Y. Osada, and M. Nakano.** 1993. Non-specific resistance induced by a low-toxic lipid A analogue, DT-5461, in murine salmonellosis. *Int. J. Immunopharmacol.* **15:**657–664.
92. **Parkhill, J., G. Dougan, K. D. James, N. R. Thomson, D. Pickard, J. Wain, C. Churcher, K. L. Mungall, S. D. Bentley, M. T. Holden, M. Sebaihia, S. Baker, D. Basham, K. Brooks, T. Chillingworth, P. Connerton, A. Cronin, P. Davis, R. M. Davies, L. Dowd, N. White, J. Farrar, T. Feltwell, N. Hamlin, A. Haque, T. T. Hien, S. Holroyd, K. Jagels, A. Krogh, T. S. Larsen, S. Leather, S. Moule, P. O'Gaora, C. Parry, M. Quail, K. Rutherford, M. Simmonds, J. Skelton, K. Stevens, S. Whitehead, and B. G. Barrell.** 2001. Complete genome sequence of a multiple drug resistant *Salmonella enterica* serovar Typhi CT18. *Nature* **413:** 848–852.
93. **Perna, N. T., G. Plunkett III, V. Burland, B. Mau, J. D. Glasner, D. J. Rose, G. F. Mayhew, P. S. Evans, J. Gregor, H. A. Kirkpatrick, G. Posfai, J. Hackett, S. Klink, A. Boutin, Y. Shao, L. Miller, E. J. Grotbeck, N. W. Davis, A. Lim, E. T. Dimalanta, K. D. Potamousis, J. Apodaca, T. S. Anantharaman, J. Lin, G. Yen, D. C. Schwartz, R. A. Welch, and F. R. Blattner.** 2001. Genome sequence of enterohaemorrhagic *Escherichia coli* O157:H7. *Nature* **409:**529–533.
94. **Porwollik, S., R. M. Wong, and M. McClelland.** 2002. Evolutionary genomics of *Salmonella:* gene acquisitions revealed by microarray analysis. *Proc. Natl. Acad. Sci. USA* **99:**8956–8961.
95. **Reeves, P.** 1993. Evolution of *Salmonella* O antigen variation by interspecific gene transfer on a large scale. *Trends Genet.* **9:**17–22.
96. **Reis, B. P., S. Zhang, R. M. Tsolis, A. J. Bäumler, L. G. Adams, and R. L. Santos.** 2003. The attenuated *sopB* mutant of *Salmonella enterica* serovar Typhimurium has the same tissue distribution and host chemokine response as the wild type in bovine Peyer's patches. *Vet. Microbiol.* **97:**269–277.
97. **Romling, U., Z. Bian, M. Hammar, W. D. Sierralta, and S. Normark.** 1998. Curli fibers are highly conserved between *Salmonella typhimurium* and *Escherichia coli* with respect to operon structure and regulation. *J. Bacteriol.* **180:**722–731.
98. **Rosenberger, C. M., M. G. Scott, M. R. Gold, R. E. Hancock, and B. B. Finlay.** 2000. *Salmonella typhimurium* infection and lipopolysaccharide stimulation induce similar changes in macrophage gene expression. *J. Immunol.* **164:**5894–5904.
99. **Samuel, J. L., J. A. Eccles, and J. Francis.** 1981. *Salmonella* in the intestinal tract and associated lymph nodes of sheep and cattle. *J. Hyg.* **87:**225–232.
100. **Samuel, J. L., D. A. O'Boyle, W. J. Mathers, and A. J. Frost.** 1979. Isolation of *Salmonella* from mesenteric lymph nodes of healthy cattle at slaughter. *Res. Vet. Sci.* **28:**238–241.
101. **Sansonetti, P. J., J. Arondel, M. Huerre, A. Harada, and K. Matsushima.** 1999. Interleukin-8 controls bacterial transepithelial translocation at the cost of epithelial destruction in experimental shigellosis. *Infect. Immun.* **67:**1471–1480.
102. **Santos, R. L., R. M. Tsolis, S. Zhang, T. A. Ficht, A. J. Bäumler, and L. G. Adams.** 2001. *Salmonella*-induced cell death is not required for enteritis in calves. *Infect. Immun.* **69:**4610–4617.
103. **Santos, R. L., S. Zhang, R. M. Tsolis, A. J. Bäumler, and L. G. Adams.** 2002. Morphologic and molecular characterization of *Salmonella typhimurium* infection in neonatal calves. *Vet. Pathol.* 39:200–215.
104. **Schmitt, C. K., J. S. Ikeda, S. C. Darnell, P. R. Watson, J. Bispham, T. S. Wallis, D. L. Weinstein, E. S. Metcalf, and A. D. O'Brien.** 2001. Absence of all components of the flagellar export and synthesis machinery differentially alters virulence of *Salmonella enterica* serovar Typhimurium in models of

typhoid fever, survival in macrophages, tissue culture invasiveness, and calf enterocolitis. *Infect. Immun.* **69**:5619–5625.

105. **Sieling, P. A., and R. L. Modlin.** 2002. Toll-like receptors: mammalian "taste receptors" for a smorgasbord of microbial invaders. *Curr. Opin. Microbiol.* **5**:70–75.
106. **Smith, B. P., G. W. Dilling, J. K. House, H. Konrad, and N. Moore.** 1995. Enzyme-linked immunosorbent assay for *Salmonella* serology using lipopolysaccharide antigen. *J. Vet. Diagn. Investig.* **7**:481–487.
107. **Smith, B. P., F. Habasha, M. Reina-Guerra, and A. J. Hardy.** 1979. Bovine salmonellosis: experimental production and characterization of the disease in calves, using oral challenge with *Salmonella typhimurium. Am. J. Vet. Res.* **40**:1510–1513.
108. **Sojka, W. J., and H. I. Field.** 1970. Salmonellosis in England and Wales 1958–1967. *Vet. Bull.* **40**:515–531.
109. **Stender, S., A. Friebel, S. Linder, M. Rohde, S. Mirold, and W. D. Hardt.** 2000. Identification of SopE2 from *Salmonella typhimurium,* a conserved guanine nucleotide exchange factor for Cdc42 of the host cell. *Mol. Microbiol.* **36**:1206–1221.
110. **Sukupolvi, S., R. G. Lorenz, J. I. Gordon, Z. Bian, J. D. Pfeifer, S. J. Normark, and M. Rhen.** 1997. Expression of thin aggregative fimbriae promotes interaction of *Salmonella typhimurium* SR-11 with mouse small intestinal epithelial cells. *Infect. Immun.* **65**:5320–5325.
111. **Swenson, D. L., and S. Clegg.** 1992. Identification of ancillary *fim* genes affecting *fimA* expression in *Salmonella typhimurium. J. Bacteriol.* **174**:7697–7704.
112. **Townsend, S. M., N. E. Kramer, R. Edwards, S. Baker, N. Hamlin, M. Simmonds, K. Stevens, S. Maloy, J. Parkhill, G. Dougan, and A. J. Bäumler.** 2001. *Salmonella enterica* serotype Typhi possesses a unique repertoire of fimbrial gene sequences. *Infect. Immun.* **69**:2894–2901.
113. **Tsolis, R. M., L. G. Adams, T. A. Ficht, and A. J. Bäumler.** 1999. Contribution of *Salmonella typhimurium* virulence factors to diarrheal disease in calves. *Infect. Immun.* **67**:4879–4885.
114. **Tsolis, R. M., L. G. Adams, M. J. Hantman, C. A. Scherer, T. Kimborough, R. A. Kingsley, T. A. Ficht, S. I. Miller, and A. J. Bäumler.** 2000. SspA is required for lethal *Salmonella typhimurium* infections in calves but is not essential for diarrhea. *Infect. Immun.* **68**:3158–3163.
115. **Tsolis, R. M., S. M. Townsend, E. A. Miao, S. I. Miller, T. A. Ficht, L. G. Adams, and A. J. Bäumler.** 1999. Identification of a putative *Salmonella enterica* serotype typhimurium host range factor with homology to IpaH and YopM by signature-tagged mutagenesis. *Infect. Immun.* **67**:6385–6393.
116. **Underhill, D. M., and A. Ozinsky.** 2002. Toll-like receptors: key mediators of microbe detection. *Curr. Opin. Immunol.* **14**:103–110.
117. **van der Velden, A. W. M., A. J. Bäumler, R. M. Tsolis, and F. Heffron.** 1998. Multiple fimbrial adhesins are required for full virulence of *Salmonella typhimurium* in mice. *Infect. Immun.* **66**:2803–2808.
118. **Wallis, T. S., M. Wood, P. Watson, S. Paulin, M. Jones, and E. Galyov.** 1999. Sips, Sops, and SPIs but not *stn* influence *Salmonella* enteropathogenesis. *Adv. Exp. Med. Biol.* **473**:275–280.
119. **Watson, P. R., E. E. Galyov, S. M. Paulin, P. W. Jones, and T. S. Wallis.** 1998. Mutation of *invH,* but not *stn,* reduces *Salmonella*-induced enteritis in cattle. *Infect. Immun.* **66**:1432–1438.
120. **Wood, M. W., R. Rosqvist, P. B. Mullan, M. H. Edwards, and E. E. Galyov.** 1996. SopE, a secreted protein of *Salmonella dublin,* is translocated into the target eukaryotic cell via a *sip*-dependent mechanism and promotes bacterial entry. *Mol. Microbiol.* **22**:327–338.
121. **Wray, C., Y. E. Beedell, and I. M. McLaren.** 1991. A survey of antimicrobial resistance in salmonellae isolated from animals in England and Wales during 1984–1987. *Br. Vet. J.* **147**:356–369.
122. **Wray, C., and W. J. Sojka.** 1978. Experimental *Salmonella typhimurium* infection in calves. *Res. Vet. Sci.* **25**:139–143.
123. **Wray, C., J. N. Todd, and M. Hinton.** 1987. Epidemiology of *Salmonella typhimurium* infection in calves: excretion of *S. typhimurium* in the faeces of calves in different management systems. *Vet. Rec.* **121**:293–296.
124. **Yang, S. K., L. Eckmann, A. Panja, and M. F. Kagnoff.** 1997. Differential and regulated expression of C-X-C, C-C, and C-chemokines by human colon epithelial cells. *Gastroenterology* **113**:1214–1223.
125. **Zeng, H., A. Q. Carlson, Y. Guo, Y. Yu, L. S. Collier-Hyams, J. L. Madara, A. T. Gewirtz, and A. S. Neish.** 2003. Flagellin is the major proinflammatory determinant of enteropathogenic *Salmonella. J. Immunol.* **171**:3668–3674.
126. **Zhang, S., L. G. Adams, J. Nunes, S. Khare, R. M. Tsolis, and A. J. Bäumler.** 2003. Secreted effector proteins of *Salmonella enterica* serotype Typhimurium elicit host-specific chemokine profiles in animal models of typhoid fever and enterocolitis. *Infect. Immun.* **71**:4795–4803.
127. **Zhang, S., R. A. Kingsley, R. L. Santos, H. Andrews-Polymenis, M. Raffatellu, J. Figueiredo, J. Nunes, R. M. Tsolis, L. G. Adams, and A. J. Bäumler.** 2003. Molecular pathogenesis of *Salmonella enterica* serotype Typhimurium-induced diarrhea. *Infect. Immun.* **71**:1–12.
128. **Zhang, S., R. L. Santos, R. M. Tsolis, S. Stender, W.-D. Hardt, A. J. Bäumler, and L. G. Adams.** 2002. SipA, SopA, SopB, SopD, and SopE2 act in concert to induce diarrhea in calves infected with *Salmonella enterica* serotype Typhimurium. *Infect. Immun.* **70**:3843–3855.
129. **Zhou, D., L. M. Chen, L. Hernandez, S. B. Shears, and J. E. Galan.** 2001. A *Salmonella* inositol polyphosphatase acts in conjunction with other bacterial effectors to promote host cell actin cytoskeleton rearrangements and bacterial internalization. *Mol. Microbiol.* **39**:248–259.
130. **Zhou, D., M. S. Mooseker, and J. E. Galan.** 1999. An invasion-associated *Salmonella* protein modulates the actin-bundling activity of plastin. *Proc. Natl. Acad. Sci. USA* **96**:10176–10181.
131. **Zhou, D., M. S. Mooseker, and J. E. Galan.** 1999. Role of the *S. typhimurium* actin-binding protein SipA in bacterial internalization. *Science* **283**:2092–2095.

Colonization of Mucosal Surfaces
Edited by James P. Nataro et al.

Chapter 22

Colonization and Invasion of Humans by *Entamoeba histolytica*

KRISTINE M. PETERSON AND WILLIAM A. PETRI, JR.

ENTAMOEBA SPECIES AND PREVALENCE

Amoebiasis is defined as infection with the protozoan parasite *Entamoeba histolytica.* This amoeba most often resides asymptomatically in the large bowel, but occasionally it penetrates the intestinal mucosa to cause invasive diseases, including amoebic colitis and liver abscess (48). It is estimated that *E. histolytica* infects hundreds of millions of people each year (46), but it was not until 1993 that enough biochemical, immunological, and genetic studies had accumulated to separate this organism into two morphologically identical but genetically distinct species: *E. dispar* and *E. histolytica.* All invasive disease is caused by *E. histolytica,* whereas *E. dispar* is nonpathogenic (48).

Since most prevalence and incidence data were collected prior to this reclassification, the true incidence of *E. histolytica* can only be estimated. Studies have shown that colonization with *E. dispar* is much more frequent than is colonization with *E. histolytica.* One of the earliest studies which distinguished *E. histolytica* from *E. dispar* was a community cross-sectional survey of asymptomatic individuals in South Africa. The overall prevalence of *E. histolytica* or *E. dispar* infection was 10%, and it was determined that 90% of these individuals were colonized with the nonpathogenic *E. dispar* while 10% were asymptomatically infected with *E. histolytica* (17). A study at an urban refugee camp in Dhaka, Bangladesh, revealed that infections with *E. histolytica* and *E. dispar* were present in 5 and 13% of asymptomatic children between 2 and 5 years of age, respectively (20). Another study conducted in a slum community of northeastern Brazil revealed that 20% of the population sampled were colonized with *E. dispar* or *E. histolytica* and 10.6% were colonized with *E. histolytica* alone (10).

A third morphologically identical species, *Entamoeba moshkovskii,* has subsequently been recognized and may have a high prevalence as well. In a study of preschool children in Bangladesh, PCR was used to detect *E. moshkovskii* in stool samples. Of 109 stool samples tested, 15.6% were positive for *E. histolytica,* 35.8% were positive for *E. dispar,* and 21.1% were positive for *E. moshkovskii* infection. Of the stools positive for *E. moshkovskii,* 73.9% had a mixed infection with *E. histolytica* or *E dispar. E. moshkovskii* has not been associated with invasive disease and is considered a nonpathogenic organism (1).

COLONIZATION

Entamoeba parasites have a two-stage life cycle consisting of a disease-causing trophozoite stage and an infectious cyst stage (Fig. 1 and Color Plate 6). Ingestion of *E. histolytica* cysts from fecally contaminated food or water initiates infection in humans. This occurs regularly among impoverished people in developing countries, but it also has the potential to occur in developed countries, as the outbreak of amoebiasis in Tbilisi demonstrates. *E. histolytica* infections were rarely diagnosed prior to June 1998 in Tbilisi. However, from July through September 1998, 177 cases of suspected amoebiasis were identified. The epidemic was linked to a contaminated water supply (3).

The ingested cyst is quadrinucleate and is resistant to chlorination, gastric acidity, and desiccation. Excystation occurs in the intestinal lumen, producing eight trophozoites by nuclear and cytoplasmic division. There is no sexual reproductive cycle. The trophozoites adhere to colonic mucin glycoproteins via a galactose- and *N*-acetyl-D-galactosamine (Gal/GalNac)-specific lectin (Fig. 2) (37). Encystation may

Kristine M. Peterson and William A. Petri, Jr. • Division of Infectious Diseases and International Health, Department of Medicine, University of Virginia School of Medicine, MR4 Bldg. Rm. 2115, Lane Rd., P.O. Box 801340, Charlottesville, VA 22908-1340.

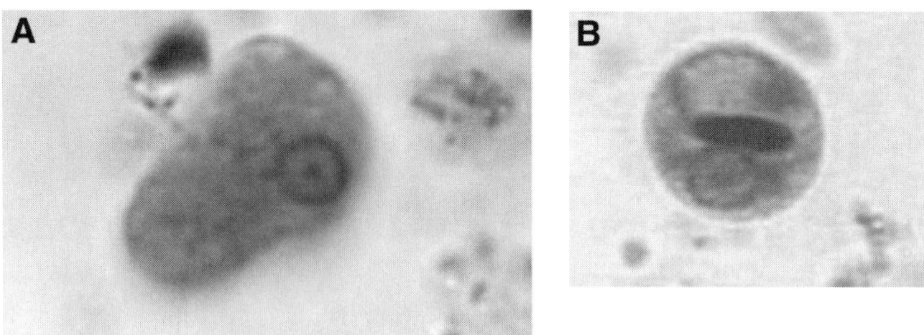

Figure 1. (A) Trophozoite of *E. histolytica*. Trophozoites are 10 to 30 μm in diameter. They are the invasive form of the parasite and are the only form seen in invasive lesions due to *E. histolytica*. (B) Cyst of *E. histolytica*. Cysts are approximately 12 μm in diameter. They are the infectious form of the parasite and are environmentally stable, being resistant to desiccation and chlorination. Reprinted from reference 23 with permission from the publisher. © 2003 Massachusetts Medical Society. All rights reserved.

then occur in the mucin layer, followed by excretion of cysts in the stool to perpetuate the life cycle by further fecal-oral spread. If encystation does not occur, invasion rather than simply colonization can ensue.

The mechanism of formation of the cyst has been elucidated by studies of the distantly related reptilian parasite *Entamoeba invadens*. It was initially found that the kinetoplastid *Crithidia fasciculata* induced encystation of *E. invadens* by using a mechanism dependent on galactose, because removal of galactose from the surface of *C. fasciculata* prevented the flagellate-induced amoebic encystation.

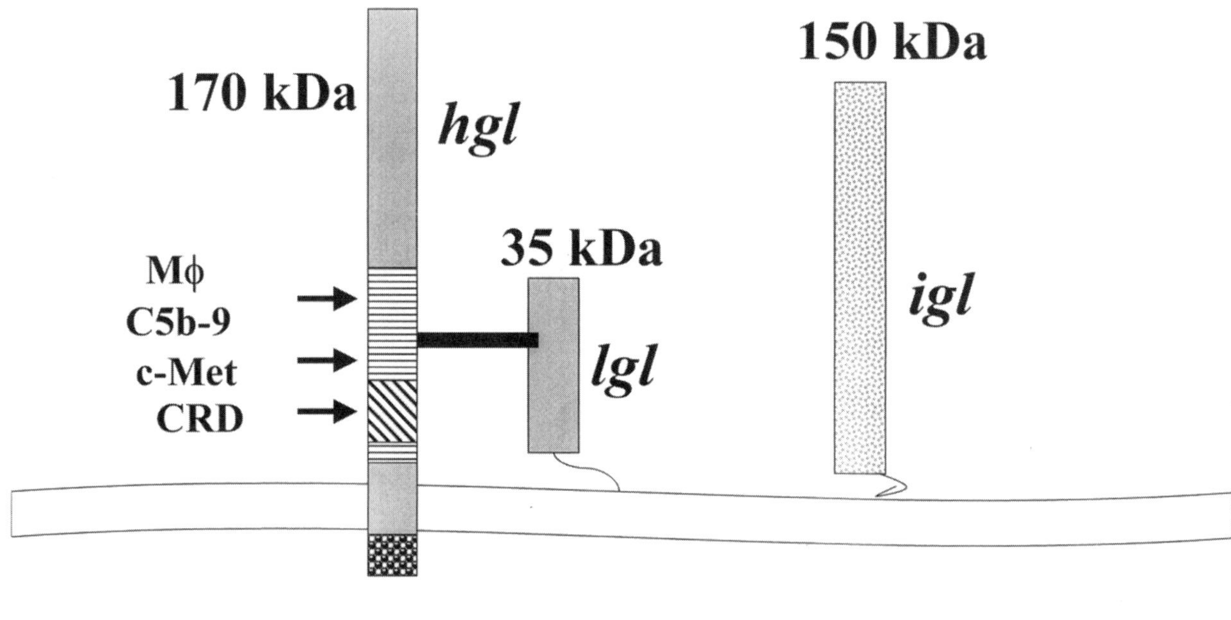

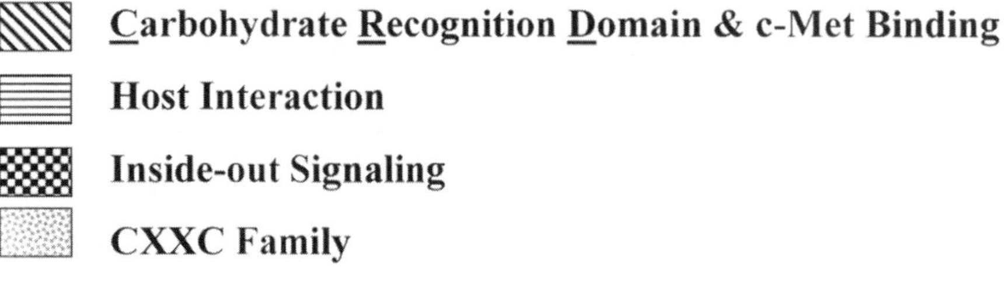

Figure 2. The Gal/GalNAc lectin of *E. histolytica*. The lectin mediates the adherence of *E. histolytica* to host immune cells, intestinal epithelium, and colonic mucins and is required for contact-dependent cytolysis and serum resistance. It is composed of three major subunits, each of which is encoded by gene families. By a combination of monoclonal antibody epitope mapping and in vitro expression studies, domains of the lectin involved in carbohydrate binding, host recognition, serum resistance, and cell signaling have been identified. Mϕ, macrophage.

Adding free galactose to medium containing *C. fasciculata* and *E. invadens* also prevented encystation, presumably by competing for the same galactose-binding sites on the amoeba surface to which *C. fasiculata* bound (12). Subsequently, highly efficient encystation of *E. invadens* was found to require a precise concentration of galactose-terminated molecules such as asialofetuin, galactose-bovine serum albumin, and mucin. In vitro, stress stimuli such as osmotic shock and glucose deprivation are capable of stimulating encystment, but the presence of a galactose-ligand source is also needed (13). The data suggest that *E. invadens* contains a galactose lectin that mediates the encystment process, and *E. invadens* has been found to express a Gal lectin, which has a heterodimeric structure similar to that of the *E. histolytica* Gal/GalNac lectin (16).

ADHERENCE AND INVASION

Instead of forming cysts, some trophozoites of *E. histolytica* occasionally invade the tissue of the large intestine. The amoebae can then enter the portal system draining the colon and be carried to the liver or can enter the general circulatory system and disseminate to other organs. The first line of defense against *E. histolytica* invasion is provided by the intestinal mucins, which are high-molecular-weight glycoproteins lining the mucosal epithelium. It has been shown that colonic mucins inhibit amoebic adherence and cytolysis of target cells and therefore protect against amoebic invasion. Mucins are secreted by intestinal goblet cells and form viscous gels that trap microorganisms, limiting their diffusion to the intestinal epithelium (4). The adherent mucus layer of the colon averages 150 μm in thickness, a considerable barrier for *Entamoeba* trophozoites, which average 20 μm in diameter (16). After penetrating this mucus layer, trophozoites must contact the colonic epithelial cells by using a Gal/GalNac adherence lectin prior to cytolysis (37).

The surface of *E. histolytica* contains the Gal/GalNac adherence lectin, which consists of a 170-kDa heavy subunit linked by disulfide bonds to a glycosylphosphatidylinositol-anchored 35- or 31-kDa light subunit. These subunits are noncovalently associated with a 150-kDa intermediate subunit. Amoebae can attach to mucins and colonic epithelial cells via the 170-kDa heavy subunit of the Gal/GalNac adherence lectin (Fig. 2) (37). After attachment to epithelial cells, one of the initial steps in amoebic invasion involves an increase in the permeability of enterocytes. Studies of *E. histolytica* with human enteric T84-cell layers have shown that this increase in permeability is due to the disruption of tight-junction proteins, including degradation of ZO-1 and dephosphorylation of ZO-2 (32).

Pathogenicity is related in part to the association of amoebae with certain bacterial species. *E. histolytica* trophozoites grown with bacteria actively phagocytize the bacteria of select species. In vitro, bacteria with mannose-binding lectin can attach to *E. histolytica* trophozoites and certain other bacteria without mannose-binding capacity are able to attach to trophozoites after they have been opsonized. Binding of opsonized bacteria to trophozoites is mediated by the trophozoite Gal/GalNac lectin. The virulence of trophozoites is stimulated by this interaction. *E. histolytica* trophozoites gradually lose their virulence on prolonged axenic cultivation, but the virulence can be regained after reassociation with certain bacteria. Increased trophozoite virulence is not due to bacterial plasmid or virulence factors, because bacteria chemically cured of virulence factors or devoid of plasmids can stimulate virulence as long as they possess recognition mechanisms (lectin or receptor for amoebic lectin) and attach to amoebae. The intestinal bacterial flora probably affects whether *E. histolytica* causes an asymptomatic infection or becomes invasive (36).

IMMUNE RESPONSE

Interaction of the parasite *E. histolytica* with the intestinal epithelium causes an inflammatory response, resulting in the attraction of activated neutrophils and macrophages to the site of pathogen invasion. Studies using a severe combined immunodeficiency phenotype mouse-human intestinal xenograft (SCID-HU-INT) model of amoebic colitis have helped to establish the cytokines and chemokines produced by epithelial cells in response to *E. histolytica*, as well as to study the resultant intestinal inflammation. In this model, human fetal intestinal tissue is engrafted into the subcutaneous space of mice with SCID. Because the intestinal xenograft is of human origin and the inflammatory cells are of murine origin, the contribution of intestinal epithelial cells to the production of inflammatory mediators can be determined by using assays specific for human cytokines and chemokines. A study using enzyme-linked immunosorbent assays (ELISAs) specific for human interleukin-1β (IL-1β) and IL-8 revealed that human intestinal xenografts infected with *E. histolytica* produced 50-fold more human IL-1β and six fold more IL-8 than did uninfected human intestinal xenografts. Immunofluorescence studies confirmed that human intestinal epithelial cells were the source of IL-8 in infected xenografts and that this cytokine may be produced in intestinal epithelial cells not in direct contact with amoebae. Neutrophils were the predominant inflammatory cells initially present at the areas of tissue damage in the intestinal xenografts after infection with *E. histolytica* (43).

Subsequent studies established a link between this inflammatory response induced by epithelial cells infected with amoebae and the resultant tissue damage seen in amoebic colitis. The transcription factor NF-κB controls the expression of a number of genes, including those encoding IL-1β and IL-8 (19). Activation of NF-κB in human intestinal cells is required for initiation of the inflammatory response. When the antisense oligonucleotide to the p65 subunit of human NF-κB was introduced into the lumen of intestinal xenografts, the increase in IL-1β and IL-8 production seen in response to *E. histolytica* infection was significantly reduced. The degree of *E. histolytica*-induced intestinal inflammation and tissue damage was significantly reduced by this blockade as well. In addition, neutrophils were found to play a key role in inducing this tissue damage, and depletion of neutrophils from SCID-HU-INT mice reduced the intestinal tissue damage as well (44).

The role of macrophages in host defense against *E. histolytica* has also been elucidated in a number of in vitro studies. It was observed that unstimulated murine peritoneal macrophages or bone marrow-derived macrophages were inefficient at killing *E. histolytica*. However, gamma interferon (IFN-γ) and tumor necrosis factor alpha (TNF-α) activate macrophages for cytotoxicity against *E. histolytica*. Furthermore, it has been shown that the combination of IFN-γ and TNF-α or of IFN-γ and colony-stimulating factor 1 can synergize with one another to enhance killing. These cytokines alone, without macrophages, are unable to induce the killing of *E. histolytica,* and studies suggest the requirement for contact between the amoeba and macrophage for killing to take place. It was also shown that the macrophages exert their effects by a combination of oxidative and nonoxidative mechanisms (14). A subsequent study found that IFN-γ and TNF-α activate macrophages for cytotoxicity against *E. histolytica* by inducing nitric oxide (NO) production and that NO is the major effector molecule produced by activated macrophages for in vitro cytotoxicity against *E. histolytica* trophozoites. The radioactive oxygen intermediates O_2^- and H_2O_2 may act as cofactors for the NO effector molecule (33).

While a protective role for IFN-γ, TNF-α, and NO in phagocytic killing of *E. histolytica* is demonstrable in vitro, it is important to establish the significance of this response in vivo. To better understand the contribution of the host immune response to amoebiasis, a study was done using a murine animal model (Fig. 3). Intracecal inoculation with *E. histolytica* trophozoites into C3H mice led to an established infection in 60% of these mice, although C57BL/6 and BALB/c mice were relatively resistant to infection. When infection did occur (4 of 27 C57BL/6 mice), however, the disease severity was similar to that in C3H mice. Interestingly, this resistance persisted when mice genetically deficient for IL-12, IFN-γ, or inducible NO synthase were used, suggesting that resistance cannot be due only to robust innate macrophage activity in these mice. The study suggested that a combination of genetics, environmental factors such as different bacteria in the intestine, and parasite factors may contribute to this resistance (26).

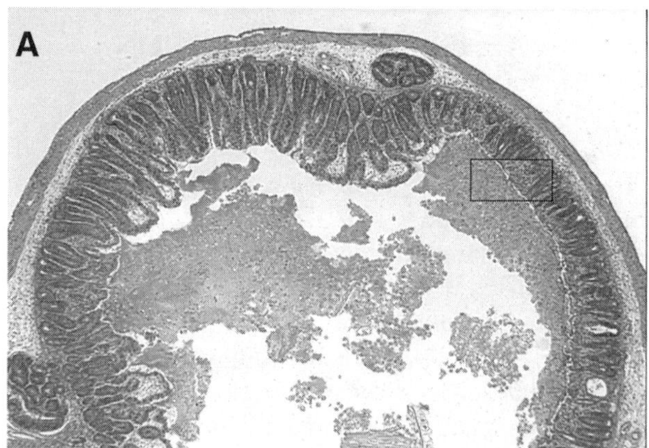

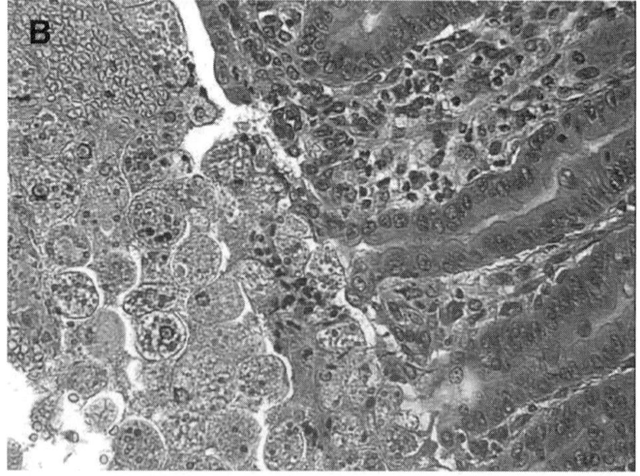

Figure 3. Murine model of amebic colitis. A section of the cecum from a C3H mouse infected with *E. histolytica* is shown at magnification of ×40 (A) and ×1,000 (B) with hematoxylin and eosin stain. The large amebic ulceration seen in panel A is shown under higher power in panel B, where individual amebic trophozoites are demonstrated invading the intestine. Reprinted from reference 26 with permission.

This study was also the first to conclude that the $CD4^+$ T-cell response in murine amoebic colitis may actually increase the parasite burden and worsen the disease. Chronic amoebic colitis was found to be associated with increased levels of IL-4 and IL-13 in the mesenteric lymph nodes and ceca of C3H mice. Since $CD4^+$ T cells are a primary source of these cytokines, mice depleted of $CD4^+$ T cells were infected to detect the impact of these cells on the course of disease. Depletion of $CD4^+$ T cells resulted in a decreased production of IL-4 and IL-13 and in a significant decrease in both the parasite burden and cecal inflammation (26). The contribution of the cell-mediated immune responses to human amoebiasis remains unclear. Invasive amoebiasis is increased in incidence and severity by factors that depress the cell-mediated immune response, such as steroid use. In contrast to this observation, there is a striking lack of invasive amoebic infection in symptomatic human immunodeficiency virus-infected individuals, despite an increased incidence of carriage in these individuals (34).

VIRULENCE FACTORS

As discussed above, the Gal/GalNac lectin of the amoebic trophozoite is used to adhere to colonic mucin and thereby colonize the large intestine. The amoebae may then aggregate in the mucin layer and trigger encystation by means of the Gal/GalNac-specific lectin. Cysts can then be excreted in the stool to continue the life cycle by fecal-oral spread. Alternatively, colitis may occur if the trophozoites do not encyst but instead penetrate the mucin layer. Invasion occurs when the trophozoites adhere to and subsequently kill intestinal epithelial cells, leading to the release of cytokines and the attraction of neutrophils and macrophages.

Secretion by the amoeba of amoebapores may contribute to the epithelial cell killing. Amoebapores are a family of at least three small peptides capable of forming pores in lipid bilayers. Three amoebapore isoforms (A, B, and C) have been isolated and are 77 amino acids long with six cysteines. They are present in cytoplasmic granules of the amoeba trophozoite and are released after the amoeba Gal/GalNac lectin recognizes and adheres to its target cell. The amoebapore is then inserted into its target membrane and forms ion channels, leading to lysis of the cell. Further evidence to support the identity of the amoebapore as a virulence factor comes from the finding that inhibition of amoebapore A synthesis by the production of specific antisense RNA in amoebae suppressed the killing of host cells (8).

Another established virulence factor for *E. histolytica* is cysteine proteinase, which can be released by *E. histolytica* trophozoites and plays a key role in intestinal invasion and inflammation. At least seven distinct genes encoding pre-pro forms of papain family proteinases have been identified in this species. Most of the proteinase activity seen in *E. histolytica* lysates can be attributed to four proteinases: amoebic cysteine proteinase 1 (ACP1), ACP2, ACP3, and *E. histolytica* cysteine proteinase 5. *E. histolytica* trophozoites can secrete 10 to 1,000 times more cysteine proteinases than the nonpathogenic *E. dispar* does (39).

Cysteine proteinases have diverse functions. They digest extracellular matrix proteins such as collagen, elastin, fibrinogen, and laminin, which may facilitate invasion of and dispersal within the submucosal tissues (39). Cysteine proteinases also have a cytopathic effect in vitro on tissue monolayers, causing the detachment of cells from their underlying substrate. This cytopathic effect correlates with the amount of cysteine proteinase activity released into the medium and is blocked by specific cysteine proteinase inhibitors (29, 39).

To assess the role of *E. histolytica* cysteine proteinases in amoebic colitis, human intestinal xenografts in SCID-HU-INT mice were infected with a plasmid that permitted the expression of an antisense mRNA to *E. histolytica* cysteine proteinase 5. This significantly reduced the total cysteine proteinase expression. The xenografts infected with proteinase-deficient amoebae failed to stimulate the production of human IL-1β or IL-8 and showed significantly less mucosal damage and neutrophil influx into the intestine. The cysteine proteinase-deficient amoebae were also defective in their ability to invade through the mucosa and were rarely detected in the submucosal tissues. It was also found that cysteine proteinases have IL-1-converting enzyme (ICE or caspase-1) activity and that they proteolytically cleave the inactive pIL-1β released from injured epithelial cells to the biologically active IL-1β (50).

Cysteine proteinases also help *E. histolytica* evade the host immune system through the degradation of immunoglobulin A (IgA) and IgG (30, 47). Although the cysteine proteinases of *E. histolytica* trophozoites activate the complement system by cleaving C3 (42), they also inactivate the subsequent anaphylotoxins, C3a and C5a (41) *E. histolytica* is also resistant to being killed by the membrane attack complex C5b-9. The Gal/GalNac lectin can bind C8 and C9 and therefore confers C5b-9 resistance. The lectin shares cross-reactivity with CD59, a membrane inhibitor of the membrane attack complex in human red blood cells (11).

E. HISTOLYTICA-INDUCED CELL DEATH

DNA fragmentation characteristic of apoptosis is observed in host cells upon intestinal invasion by *E. histolytica* (Fig. 4). A number of studies have been conducted to elucidate the mechanism of cell death by amoebae. It was initially found that contact of amoebae with target cells is mediated by the Gal/GalNac lectin and is required for cytolytic activity, although the pathway of the contact-dependent cytolysis was unknown. An initial in vitro study found that coculture of murine myeloid cells with amoeba trophozoites led to an apoptotic death of the myeloid cells, as evidenced by a characteristic laddering pattern of DNA (40). In contrast, the results of another in vitro study using Jurkat and HL-60 cell lines concluded that amoebae caused target cell death via a necrotic mechanism. The isolated amoebapore also led to necrotic cell death in vitro (5). It is likely that both apoptotic and necrotic mechanisms of cell death are utilized in vivo by *E. histolytica.*

A murine model of amoebic liver abscess subsequently revealed that *E. histolytica* does induce apoptosis among hepatocytes in liver abscesses. In addition, two common pathways of apoptosis were found to not be utilized by *E. histolytica* when causing apoptotic target cell death. Fas and its ligands are surface cell markers with single transmembrane domains, and ligation of Fas led to apoptotic death under a variety of conditions such as immune responses, autoimmunity, and hepatitis. However, apoptosis still occurred within liver abscesses in mice that failed to produce Fas protein and in mice that expressed nonfunctional Fas ligand. TNF-α with TNF receptor 1 (TNFR1) is an alternative pathway for apoptosis, but mice genetically engineered to lack TNFR1 were still found to experience apoptotic hepatocyte death (45). In addition, overexpression of the *bcl-2* gene protects cells from apoptosis under a variety of stressful conditions, but it does not protect myeloid cells from amoebic killing (40). Therefore, a nonclassical pathway for apoptosis is suspected.

Apoptosis occurs when a signal activates effectors such as caspases (cysteine proteinases with specificity for aspartate residues) to degrade cell components (49). It has been found that the inhibition of caspase-8 (an initiator of apoptosis following ligation of cell surface death receptors) and caspase-9 (the farthest upstream member of the apoptotic protease cascade) did not prevent *E. histolytica*-induced apoptosis in vitro. However, caspase-3, a distal effector caspase, was found to be activated in Jurkat cells killed by *E. his-*

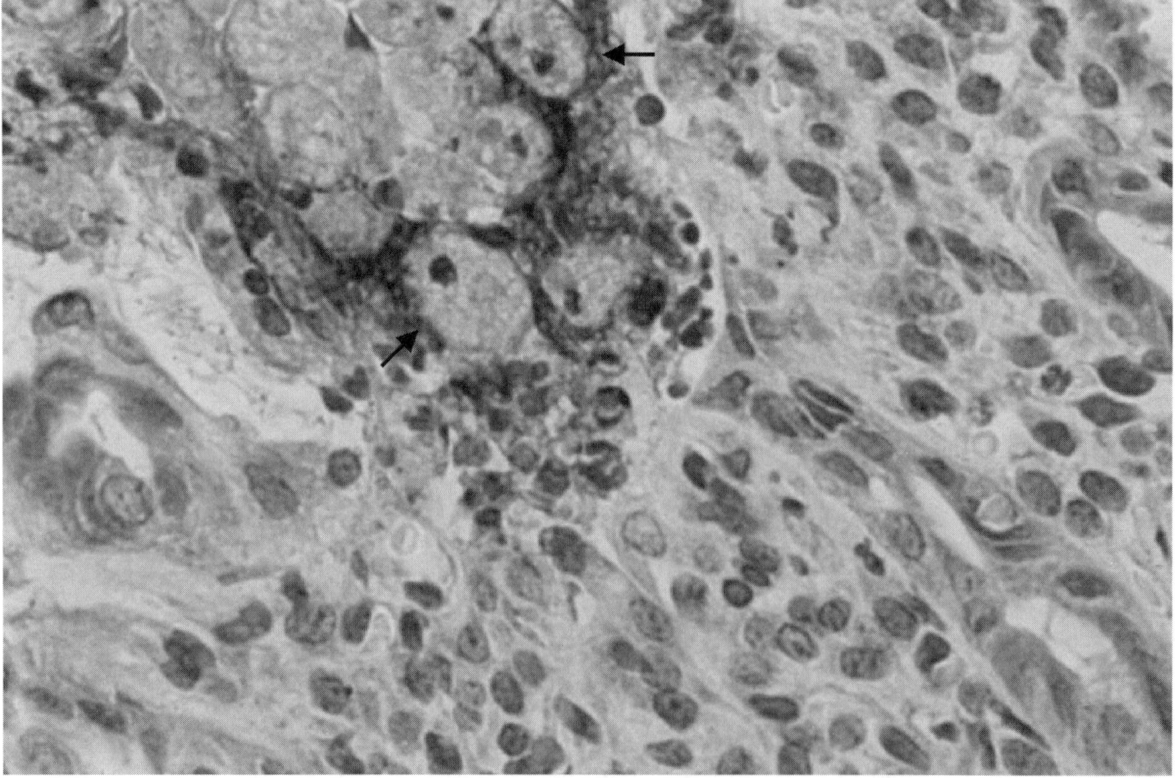

Figure 4. DNA fragmentation of murine intestinal cells at the site of colonic invasion by *E. histolytica.* Magnification, ×1,000. Reprinted from reference 28 with permission from Blackwell Publishing.

tolytica, and caspase-3/caspase-3-like activity was required for contact-dependent cell killing via a caspase-8- and caspase-9-independent mechanism (28).

The role of caspases in vivo on amoebic liver abscesses has been studied by using a mouse model of infection. Amoebic liver abscesses were significantly smaller in mice receiving a caspase inhibitor. The abscesses from control mice without a caspase inhibitor had large areas of dead hepatocytes, with diffuse terminal deoxyribonucleotidyltransferase-mediated dUMP-biotin nick end labeling (TUNEL) staining consistent with apoptosis. Amoebic liver abscesses from mice receiving a caspase inhibitor showed smaller areas of dead hepatocytes, with some cytoplasmic uptake of TUNEL staining. The mice receiving caspase inhibitor showed little to no TUNEL activity in nuclei from hepatocytes adjacent to *E. histolytica* trophozoites or in the nuclei of surrounding inflammatory cells. It was hypothesized from these results that direct contact with amoebic trophozoites may cause either hepatocyte necrosis or apoptosis but that hepatocyte death distal to areas of immediate contact between *E. histolytica* trophozoites and hepatocytes may be due primarily to apoptosis (49).

COLONIZATION AND INVASION IN HUMANS

Asymptomatic infection with *E. histolytica* is defined as the presence of *E. histolytica* in the stool in the absence of colitis or extraintestinal infection. As discussed above, colonization with the morphologically identical *E. dispar* is at least three times more common than asymptomatic colonization with *E. histolytica.* Infection with *E. histolytica* usually results in asymptomatic colonization, but there is an approximately 10% risk of developing invasive disease. A study in Bangladesh of children 2 to 5 years old colonized with *E. histolytica* revealed that there was a low risk of developing invasive amoebiasis with *E. histolytica* colonization: 2 of 17 colonized children developed dysentery during a 1-year follow-up (20). A subsequent study in Bangladesh revealed a fivefold risk of developing *E. histolytica*-associated diarrhea and a threefold risk of *E. histolytica*-associated dysentery among individuals asymptomatically colonized with *E. histolytica* (10). A study in South Africa found that 10% of individuals colonized with *E. histolytica* developed invasive disease within 1 year (17).

Patients who develop amoebic colitis typically present with a history of cramping abdominal pain, weight loss, and watery or bloody diarrhea for several weeks. Colonic lesions can vary from mucosal thickening to flask-shaped ulcerations to necrosis of the intestinal wall (Color Plate 7) (38). Acute necrotizing colitis is an unusual manifestation of amoebic colitis and occurs in fewer than 0.5% of patients with colitis; however, it is associated with a mortality of 40% (23, 38). An ameboma is another unusual manifestation of amoebic colitis. It results from the formation of annular colonic granulation tissue and may mimic carcinoma. Perianal ulcerations may also result and have the potential to form fistulas (23).

Amoebic liver abscesses (ALAs) can also result from invasive infection with *E. histolytica* (Fig. 5). It is thought that *E. histolytica* trophozoites cause ALAs by ascending the portal venous system (27). The incidence of ALA is highly variable between different amoebiasis-endemic areas. Most amoebic liver abscesses reported are from Mexico or Southeast Asia.

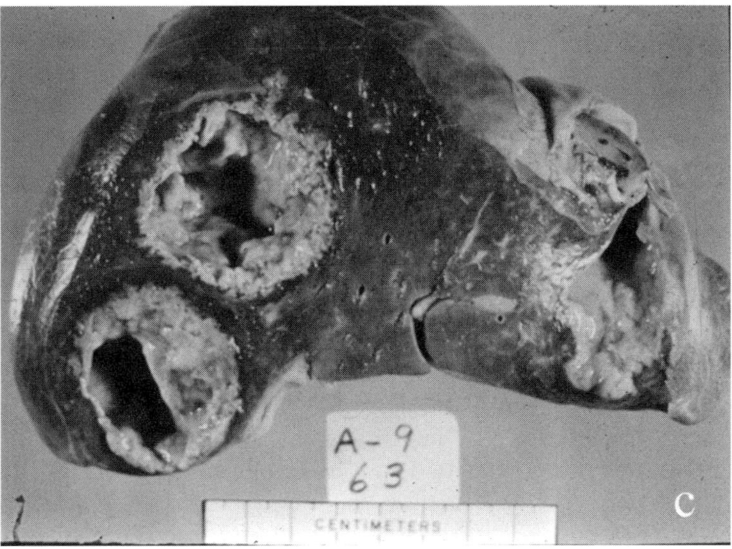

Figure 5. Gross appearance of amebic liver abscess. Reprinted from reference 23.

A very large number of cases was recently reported in Hue City, Vietnam, with an estimated frequency of ALA as high as 21 cases per 100,000 inhabitants (7). The risk of ALA appears to be age and sex dependent, with more than 95% of ALA cases occurring in adults, of whom more than 80% were males (9).

The diversity and organ tropisms of infection by *E. histolytica* may be the result of amoebae and/or host factors. A recent study evaluated the role of parasitic genetic diversity in *E. histolytica* virulence. Of 54 *E. histolytica* isolates analyzed by nested PCR of DNA extracted from stool and liver aspirate pus, half had unique polymorphisms. Most of the polymorphisms from the liver isolates were not present in the intestinal isolates, consistent with there being genetic differences in the *E. histolytica* strains that infect the colon and those that cause liver disease (2).

Host genetic variability may also affect whether an individual develops *E. histolytica* colonization and/or invasive disease. $CD4^+$ T-helper lymphocytes recognize antigens in association with class II major histocompatibility complex gene products. The presence or lack of a particular class II allele therefore has the potential to alter the repertoire of proteins presented to $CD4^+$ T cells and affect the response of the host to infection. A study of Bangladeshi children recently found a protective association of the HLA class II allele DQB1*0601 and the heterozygous haplotype DQB1*0601/DRB1*1501 with *E. histolytica* infection and disease. These findings suggest that class II-restricted immune responses play a role in protecting children from *E. histolytica* infections (15).

The duration of *E. histolytica* colonization has been evaluated in a few studies. In a series of 20 patients in South Africa, 18 of whom were younger than 20 years, the initial infecting *E. histolytica* zymodeme (isoenzyme pattern) had disappeared in all patients by 6 months after infection. Four patients did acquire a new zymodeme, but their infections all resolved within the subsequent 6 months (17). Another study of Bangladeshi children aged 2 to 5 years also suggested a clearance rate of *E. histolytica* colonization within months (22).

In contrast, a recent study evaluated asymptomatic adult carriers of *E. histolytica* living in an area of Vietnam where infection is endemic and reported a much longer duration of *E. histolytica* colonization. Using genetic fingerprints to assess persistence of infection versus reinfection, it was found that the mean half-life of colonization was 12.9 months. The finding that the average half-life of infection is about 13 months in adults is consistent with the finding that travelers returning from areas of endemic infection can develop amoebic liver abscesses months or even years later (6). A study in Germany concluded that the latent period between time of *E. histolytica* infection and ALA ranges between 8 and 20 weeks. However, there are case reports of the onset of ALA more than 20 years after exposure to an area of endemic infection, suggesting that colonization can persist for an extended period (31).

Immunity in Humans

Reinfection with *E. histolytica* is possible, and seropositivity to *E. histolytica* antigen does not confer protection against colonization. A study conducted in northeastern Brazil identified cases of *E. histolytica,* and the index case and household contacts were examined. ELISA for detection of Gal/GalNac lectin in stool and serologic evaluation for the detection of serum anti-lectin antibodies were performed. Of seropositive people who were initially stool negative for *E. histolytica,* 63% converted to stool positive during the course of the study. Of 84 people who were stool positive for *E. histolytica* at the start of the study, 66 (85%) became stool negative within 30 to 45 days. However, four people were thought to be reinfected, since they were stool positive on day 1, stool negative on day 15, and stool positive again on day 45 (24). A study of diarrheal illness in Dhaka, Bangladesh, also showed recurrence of *E. histolytica* infection. Of 43 children with a first episode of *E. histolytica*-associated diarrhea, 14 had a subsequent episode (10). Furthermore, a study of ALA in Vietnam concluded that the risk that a man will develop a second ALA within 1 year was 0.33%, which is more than five times higher than the risk that a man will develop the first ALA (0.06% per year) (9).

The finding that reinfection with *E. histolytica* does occur could suggest that acquired immunity is nonexistent. However, there is evidence that acquired immunity does exist, although it is incomplete. A study of preschool children (2 to 5 years old) in Dhaka was undertaken to look for evidence of acquired immunity. The children were monitored every other day by health care workers, and stool for antigen and culture were obtained when diarrheal disease was detected. A monthly surveillance stool sample was also obtained for evidence of *E. histolytica* infection, and ELISA was used to detect anti-GalNac lectin IgG and IgA antibodies in serum and stool samples, respectively. The study found that there was evidence for immunity linked to a mucosal anti-adherence lectin IgA response. In a cross-sectional analysis, none of the 64 children with stool anti-lectin IgA were colonized with *E. histolytica.* Children with stool anti-lectin IgA also acquired fewer new *E. histolytica* infections over a prospective period of observation. They had 64% fewer new *E. histolytica* infections by 5 months. In addition, of the four children who developed *E. histolytica*-associated dysentery, all were IgA negative. Finally, the appearance of a stool IgA anti-lectin response coincided with resolution of infection. Lectin-specific IgA was detected in the month during which the infection resolved in 67% of cases and lasted for an average of 17 days. Although the rate of infections was decreased in children with IgA anti-lectin antibodies, the immunity was not complete. The limited duration of stool IgA may be the reason for the incomplete immunity. The *E. histolytica* trophozoites also produce cysteine proteinases, which can degrade IgA, and genetically distinct strains of *E. histolytica* may play a role in reinfections as well (21).

Interestingly, there was actually an increased rate of new *E. histolytica* infections in children with serum IgG antibodies to the lectin: children with serum anti-lectin IgG antibodies had 53% more new infections. A follow-up study found that siblings of children with anti-trophozoite IgG have 4.8-fold-higher odds of having anti-trophozoite IgG themselves. Also, most children who are anti-trophozoite IgG negative do not become anti-trophozoite IgG positive after contracting a new infection. These findings may suggest that the anti-trophozoite trait is inherited, and they indicate the individuals at an increased risk for *E. histolytica* infections (22).

This study also examined the association between protection from infection and IgA antibody response to the active site of the lectin, the carbohydrate recognition domain (CRD). It found that the children with stool anti-CRD IgA had a 7% incidence of new *E. histolytica* infection at 1 year compared to a 43% incidence in children without stool anti-CRD IgA at the onset of the study (Fig. 6). There was also a lower incidence of amoebic diarrhea and colitis in children with stool anti-CRD IgA (22).

The Gal/GalNac lectin mediates adherence of the *E. histolytica* trophozoite to the colonic mucin glycoproteins and the colonic epithelium. This adherence is required in vivo for virulence. Mucosal IgA functions primarily to prevent adherence of microorganisms to the gut epithelium, and the mucosal anti-adherence lectin IgA response has been associated with acquired immunity to *E. histolytica*. Therefore, the Gal/GalNac lectin is a potential vaccine candidate. A recent study was done to determine if vaccination with the *E. histolytica* Gal/GalNac lectin could prevent cecal infection in a C3H mouse model of amoebic colitis. Two trials using a purified lectin and two trials using a recombinant fragment of the lectin heavy subunit that contains the CRD (LecA) were performed. Earlier trials had shown that mice must be boosted intraperitoneally with Freund's adjuvant prior to intranasal immunization with lectin and cholera toxin to produce a measurable fecal IgA response. Therefore, intranasal and intraperitoneal vaccines were used in each trial. Overall, there was significant protection after administration of native lectin or LecA vaccination, although native lectin was more efficacious than LecA (94 and 55%, respectively). A significant correlation existed between prechallenge fecal anti-native lectin IgA and subsequent resistance to infection. A total of 96% of mice with detectable fecal anti-lectin IgA were resistant to

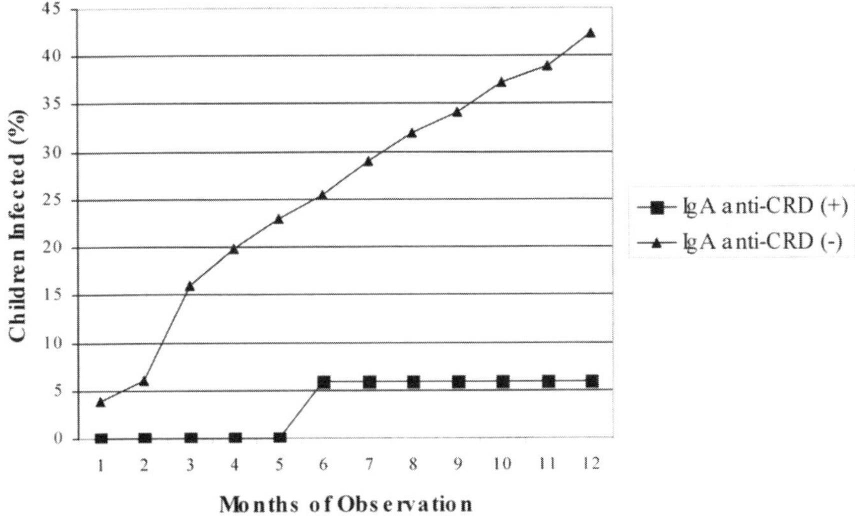

Figure 6. Acquired immunity of children to amebiasis. Children who developed stool IgA anti-CRD antibodies in the first year of the study (n = 81) had a lower incidence of new *E. histolytica* infections in the second year compared to children who remained IgA anti-CRD negative (n = 149). The two groups are statistically significantly different ($P < 0.04$) at every time point. Reprinted from reference 22 with permission from University of Chicago Press. © 2002 by the Infectious Disease Society of America. All rights reserved.

infection, compared to 68% of mice without anti-lectin IgA. The presence of fecal anti-LecA IgA antibody was not a significant marker of protection. Other factors may also play a role in the immunity seen in this study. The highly protective native-lectin trials utilized complete Freund's adjuvent, and this can induce IFN-γ production and stimulate macrophage activity in vitro (25).

Several other potential vaccine candidates are being evaluated. Oral immunization of gerbils with live-attenuated *Salmonella* vector expressing a portion of the Gal/GalNac lectin was shown to protect the animals from developing amoebic liver abscesses (35). It has also been found through an in vivo study of mice that an injectable codon-optimized DNA vaccine, encoding a portion of the galactose-lectin heavy subunit (including the CDR), stimulated a Th1-type galactose-lectin-specific cellular immune response. It also induced the development of serum antibodies that recognized a recombinant portion of the heavy subunit. Further studies to assess this vaccine via an oral delivery route are in progress (18).

Diarrheal illness is a major cause of morbidity and mortality among children in the developing world, and amoebiasis is a significant contributor to these illnesses. An effective vaccine would substantially improve the health and well-being of children in developing countries. Although significant advances are being made, extensive work remains to be done in order to create an effective vaccine. Further knowledge is needed to understand what leads to asymptomatic amoebic colonization versus invasive disease. The discovery that humans naturally acquire partial immunity gives hope that an effective vaccine can be made, and the Gal/GalNac lectin, which mediates adherence, is an intriguing vaccine candidate. Importantly, since humans are the only significant reservoir for *E. histolytica* infections, a vaccine that blocks colonization could lead to the elimination of amoebiasis.

REFERENCES

1. **Ali, I. K. M., M. B. Hassain, S. Roy, P. F. Ayeh-Kumi, W. A. Petri, Jr., R. Haque, and C. Clark.** 2003. *Entamoeba moshkovskii* infection in childen in Bangladesh. *Emerg. Infect. Dis.* **9:**580–584.
2. **Ayeh-Kumi, P. F., I. M. Ali, L. A. Lockhart, C. A. Gilchrist, W. A. Petri, and R. Haque.** 2001. *Entamoeba histolytica:* genetic diversity of clinical isolates from Bangladesh as demonstrated by polymorphisms in the serine-rich gene. *Exp. Parasitol.* **99:**80–88.
3. **Barwick, R., A. Uzicanin, S. Lareau, N. Malakmadze, P. Imnadze, M. Iosava, N. Ninashvili, M. Wilson, A. Hightower, S. Johnston, H. Bishop, W. A. Petri, and D. Juranek.** 2002. Outbreak of amoebiasis in Tbilisi, Republic of Georgia, 1998. *Am. J. Trop. Med. Hyg.* **67:**623–631.
4. **Belley, A., K. Keller, M. Goettke, and K. Chadee.** 1999. Intestinal mucins in colonization and host defense against pathogens. *Am. J. Trop. Med. Hyg.* **60:**10–15.
5. **Berninghausen, O., and M. Leippe.** 1997. Necrosis versus apoptosis as the mechanism of target cell death induced by *Entamoeba histolytica. Infect. Immun.* **65:**3615–3621.
6. **Blessman, J., I. K. M. Ali, P. A. T. Nu, B. T. Dinh, T. Q. N. Viet, A. L. Van, C. G. Clark, and E. Tannich.** 2003. Longitudinal study of intestinal *Entamoeba histolytica* infections in asymptomatic adult carriers. *J. Clin. Microbiol.* **41:**4745–4750.
7. **Blessmann, J., P. Van Linh, P. A. T. Nu, H. D. Thi, B. Muller-Myshsok, H. Buss, and E. Tannich.** 2002. Epidemiology of amebiasis in a region of high incidence of amebic liver abscess in central Vietnam. *Am. J. Trop. Med. Hyg.* **66:**578–583.
8. **Bracha, R., Y. Nuchamowitz, M. Leippe, and D. Mirelman.** 1999. Antisense inhibition of amoebapore expression in *Entamoeba histolytica* causes a decrease in amoebic virulence. *Mol. Microbiol.* **34:**463–472.
9. **Braga, L. L., M. L. Gomes, M. W. Da Silva, F. E. Facanha, L. Fiuza, and B. J. Mann.** 2001. Household epidemiology of *Entamoeba Histolytica* infection in an urban community in northeastern Brazil. *Am. J. Trop. Med. Hyg.* **65:**268–271.
10. **Braga, L. L., Y. Mednonca, C. A. Paiva, A. Sales, A. L. M. Cavalcante, and B. J. Mann.** 1998. Seropositivity and intestinal colonization with *Entamoeba histolytica* and *Entamoeba dispar* in indviduals in northeastern Brazil. *J. Clin. Microbiol.* **36:**3044–3045.
11. **Braga, L. L., H. Ninomiya, J. J. McCoy, S. Eacker, T. Wiedmer, C. Pham, S. Wood, P. J. Sims, and W. A. Petri.** 1992. Inhibition of the complement membrane attack complex by the galactose-specific adhesin of *Entamoeba histolytica. J. Clin. Investig.* **90:**1131–1137.
12. **Cho, J., and D. Eichinger.** 1998. *Crithidia fasciculata* induces encystation of *Entamoeba invadens* in a galactose-dependent manner. *J. Parasitol.* **84:**705–710.
13. **Coppi, A., and D. Eichinger.** 1999. Regulation of *Entamoeba invadens* encystation and gene expression with galactose and *N*-acetylglucosamine. *Mol. Biochem. Parasitol.* **102:**67–77.
14. **Denis, M., and K. Chadee.** 1989. Cytokine activation of murine macrophages for in vitro killing of *Entamoeba histolytica* trophozoites. *Infect. Immun.* **57:**1750–1756.
15. **Duggal, P., R. Haque, R. Shantanu, D. Mondal, R. B. Sack, B. M. Farr, T. H. Beaty, and W. A. Petri.** 2004. Influence of human leukocyte antigen class II alleles on susceptibility to *Entamoeba histolytica* infection in Bangladeshi children. *J. Infect. Dis.* **189:**520–526.
16. **Eichinger, D.** 2001. A role for a galactose lectin and its ligands during encystment of *Entamoeba. J. Eukaryot. Microbiol.* **48:**17–21.
17. **Gathiram, V., and T. F. Jackson.** 1987. A longitudinal study of asymptomatic carriers of pathogenic zymodemes of *Entamoeba histolytica. S. Afr. Med. J.* **72:**669–672.
18. **Gaucher, D., and K. Chadee.** 2002. Construction and immunogenicity of a codon-optimized *Entamoeba histolytica* Gal-lectin based DNA vaccine. *Vaccine* **20:**3244–3253.
19. **Ghosh, S., M. J. May, and E. B. Kopp.** 1998. NF-κB and Rel proteins: evolutionarily conserved mediators of immune responses. *Annu. Rev. Immunol.* **16:**225–260.
20. **Haque, R., I. K. M. Ali, and W.A. Petri.** 1999. Prevalence and immune response to *Entameoba histolytica* in preschool children in an urban slum of Dhaka, Bangladesh. *Am. J. Trop. Med. Hyg.* **60:**1031–1034.
21. **Haque, R., I. M. Ali, B. Sack, B. M. Farr, G. Ramakrishnan, and W. A. Petri.** 2001. Amoebiasis and mucosal IgA antibody against the *Entamoeba Histolytica* adherence lectin in Bangladeshi children. *J. Infect. Dis.* **183:**1787–1793.

22. **Haque, R., P. Duggal, I. M. Ali, M. B. Hossain, D. Mondal, R. B. Sack, B. M. Farr, T. M. Beaty, and W. A. Petri.** 2002. Innate and acquired resistance to amoebiasis in Bangladeshi children. *J. Infect. Dis.* **186:**547–552.
23. **Haque, R., C. Huston, M. Hughes, E. Houpt, and W. A. Petri.** 2003. Amoebiasis. *N. Engl. J. Med.* **348:**1565–1573.
24. **Haque, R., D. Mondal, B. Kirkpatrick, S. Akther, B. Farr, R. Sack, and W. A. Petri.** 2003. Epidemiologic and clinical characteristics of acute diarrhea with emphasis on *Entamoeba histolytica* infections in preschool children in an urban slum of Dhaka, Bangladesh. *Am. J. Trop. Med. Hyg.* **69:**398–405.
25. **Houpt, E., L. Barroso, L. Lockhart, R. Wright, C. Cramer, D. Lyerly, and W. A. Petri.** 2004. Prevention of intestinal amoebiasis by vaccination with the *Entamoeba histolytica* Gal/GalNac lectin. *Vaccine* **22:**611–617.
26. **Houpt, E., D. Glembocki, T. Obrig, C. Moskaluk, L. Lockhart, R. Wright, R. Seaner, T. Keepers, T. Wilkins, and W. Petri.** 2002. The mouse model of amebic colitis reveals mouse strain susceptibility to infection and exacerbation of disease by $CD4^+$ T cells. *J. Immunol.* **169:**4496–4503.
27. **Hughes, M., and W. A. Petri.** 2000. Amebic liver abscess. *Infect. Liver* **14:**565–582.
28. **Huston, C., E. Houpt, B. Mann, C. Hahn, and W. A. Petri.** 2000. Caspase 3-dependent killing of host cells by the parasite *Entamoeba histolytica. Cell. Microbiol.* **2:**617–625.
29. **Keene, W.E., M.G. Pettit, A. Sand, and J. H. McKerrow.** 1986. The major neutral proteinase of *Entamoeba histolytica. J. Exp. Med.* **163:**536–549.
30. **Kelsall, B. L., and J. I. Ravdin.** 1993. Degradation of human immunoglobulin A by *Entamoeba histolytica. J. Infect. Dis.* **168:**1319–1322.
31. **Knobloch, J., and E. Mannweiler.** 1983. Development and persistence of antibodies to *Entamoeba histolytica* in patients with amebic liver abscess: analysis of 216 cases. *Am. J. Trop. Med. Hyg.* **32:**727–732.
32. **Leroy, A., T. Lauwaet, G. De Bruyne, M. Cornelissen, and M. Mareel.** 2000. *Entamoeba histolytica* disturbs the tight junction complex in human enteric T84 cell layers. *FASEB J.* **14:**1139–1146.
33. **Lin, J., and K. Chadee.** 1992. Macrophage cytotoxicity against *Entamoeba histolytica* trophozoites is mediated by nitric oxide from L-arginine. *J. Immunol.* **148:**3999–4005.
34. **Lucas, S. B.** 1990. Missing infections in AIDS. *Trans. R. Soc. Trop. Med. Hyg.* **84**(Suppl. 1)**:**34–38.
35. **Miller-Sims, V. C., and W. A. Petri.** 2002. Opportunities and obstacles in developing a vaccine for *Entamoeba histolytica. Curr. Opin. Immunol.* **14:**549–552.
36. **Mirelman, D.** 1997. Ameba-bacterium relationship in amebiasis. *Microbiol. Rev.* **51:**272–284.
37. **Petri, W. A., B. J. Mann, and R. Haque.** 2002. The bittersweet interface of parasite and host: lectin-carbohydrate interactions during human invasion by the parasite *Entamoeba histolytica. Annu. Rev. Microbiol.* **56:**39–64.
38. **Petri, W. A., and U. Singh.** 1999. Diagnosis and management of amoebiasis. *Clin. Infect. Dis.* **29:**1117–1125.
39. **Que, X., and S. Reed.** 2000. Cysteine proteinases and the pathogenesis of amoebiasis. *Clin. Microbiol. Rev.* **13:**196–206.
40. **Ragland, B., L. Ashley, D. Vaux, and W. Petri.** 1994. *Entamoeba histolytica:* target cells killed by trophozoites undergo DNA fragmentation which is not blocked by Bcl-2. *Exp. Parasitol.* **79:**460–467.
41. **Reed, S. L., J. A. Ember, D. S. Herdman, R. G. Discipio, T. E. Hugli, and I. Gigli.** 1995. The extracellular neutral cysteine proteinase of *Entamoeba histolytica* degrades anaphylotoxins C3a and C5a. *J. Immunol.* **155:**266–274.
42. **Reed, S. L., W. E. Keene, J. H. McKerrow, and I. Gigli.** 1989. Cleavage of C3 by a neutral cysteine proteinase of *Entamoeba histolytica. J. Immunol.* **143:**189–195.
43. **Seydel, K. B., E. Li, P. E. Swanson, and S. L. Stanley.** 1997. Human intestinal epithelial cells produce proinflammatory cytokines in response to infection in a SCID mouse-human intestinal xenograft model of amoebiasis. *Infect. Immun.* **65:** 1631–1639.
44. **Seydel, K. B., E. Li, Z. Zhang, and S. L. Stanley.** 1998. Epithelial cell-initiated inflammation plays a crucial role in early tissue damage in amoebic infection of human intestine. *Gastroenterology* **115:**1446–1453.
45. **Seydel, K. B., and S. Stanley.** 1998. *Entamoeba histolytica* induces host cell death in amebic liver abscess by a non-Fas-dependent, non-tumor necrosis factor alpha-dependent pathway of apoptosis. *Infect. Immun.* **66:**2980–2983.
46. **Stauffer, W., and J. Ravdin.** 2003. *Entamoeba histolytica:* an update. *Curr. Opin. Infect. Dis.* **16:**479–485.
47. **Tran, V. Q., D. S. Herdman, B. E. Torian, and S. L. Reed.** 1998. The neutral cysteine proteinase of *Entamoeba histolytica* degrades IgG and prevents its binding. *J. Infect. Dis.* **177:**508–511.
48. **World Health Organization.** 1997. Amoebiasis. *Wkly. Epidemiol. Rec.* **72:**97–99.
49. **Yan, L., and S. L. Stanley.** 2001. Blockade of caspases inhibits amebic liver abscess formation in a mouse model of disease. *Infect. Immun.* **69:**7911–7914.
50. **Zhang, Z., L. Wang, K. B. Seydel, E. Li, S. Ankri, D. Mirelman, and S. L. Stanley.** 2000. *Entamoeba histolytica* cysteine proteinases with interleukin-1 beta converting enzyme (ICE) activity cause intestinal inflammation and tissue damage in amoebiasis. *Mol. Microbiol.* **37:**542–548.

IV. COLONIZATION OF THE GENITOURINARY TRACT

Colonization of Mucosal Surfaces
Edited by James P. Nataro et al.

Chapter 23

Role of Phase and Antigenic Variation in *Neisseria gonorrhoeae* Colonization

AMY N. SIMMS AND ANN E. JERSE

OVERVIEW OF GONOCOCCAL PATHOGENESIS

Colonization of Mucosal Surfaces

Neisseria gonorrhoeae is responsible for a large number of uncomplicated lower urogenital tract infections and serious upper reproductive tract infections in the United States and the developing world. The most common site of gonococcal colonization is the urethra of males and the endocervix of females. The female urethra is often also infected. *N. gonorrhoeae* colonizes the vaginas of prepubescent girls and postmenopausal women who are not receiving hormonal replacement therapy. Rectal and pharyngeal colonization is common in both genders. Between 10 and 20% of endocervical infections ascend to the upper genital tract to cause pelvic inflammatory disease (PID) and the associated sequelae of ectopic pregnancy, involuntary infertility, and chronic pelvic pain. In acute gonorrhea, colonization induces a purulent exudate consisting of polymorphonuclear leukocytes (PMNs), exfoliated epithelial cells, and intracellular and extracellular gonococci (107). Mucosal infections are often asymptomatic, however, with 50% of genital tract infections in women being inapparent (reported range, 19 to 80%) (23, 155, 184). Most pharyngeal infections (35, 272, 280) and a high percentage of rectal infections (59, 140, 191) also occur without symptoms.

N. gonorrhoeae is a human-specific pathogen with no outside reservoir. As attested by its existence in the human population since biblical times (235), the success of this pathogen undoubtedly stems from the evolution of highly effective adaptation mechanisms that promote transmission and survival on mucosal surfaces. A diverse set of colonization factors provides the gonococcus with multiple avenues for establishing infection. In addition to pili, several other adhesins have been identified, some of which lead to uptake by epithelial cells (60, 68, 99, 159, 167). Once anchored to the mucosal surface, *N. gonorrhoeae* strengthens its foothold by evading numerous nonspecific physiological and immunological factors that constitute the host innate defense. For example, *N. gonorrhoeae* utilizes one or more active efflux systems to protect itself from toxic hydrophobic antimicrobial substances present in the rectum (95, 226, 284) and the female lower genital tract (116). It circumvents host iron sequestration by expressing specific receptors for human transferrin, lactoferrin, and siderophores of members of the commensal flora (236). Gonococci evade complement-mediated defenses by down-regulating complement activation (194), and, in addition to reducing opsonophagocytic uptake (79, 130, 195, 197), the gonococcus may defend itself from phagocytic killing with enzymes that neutralize toxic oxygen species (11, 102, 119, 260) or repair oxidatively damaged proteins (232). A nonenzymatic quenching mechanism may protect against phagocyte-derived superoxide anion (259). Mechanisms by which gonococci reduce the respiratory burst have also been described (34, 147, 196). *N. gonorrhoeae* also effectively evades the adaptive immune response, as evidenced by a high incidence of repeated infections with the same serovar or strain (235) despite the presence of specific antibodies in previously infected individuals (reviewed in reference 36). Mechanisms for avoiding gonococcus-specific antibodies include the expression of blocking antigens, specific cleavage of immunoglobulin A1, molecular mimicry, retreat into

Amy N. Simms and Ann E. Jerse • Department of Microbiology and Immunology, Uniformed Services University of the Health Sciences, F. Edward Hèbert School of Medicine, Bethesda, MD 20814.

epithelial cells, and blebbing of membranes to create a decoy (as reviewed by others [53, 236]), as well as antigenic variation of surface molecules, as discussed below.

Phase and Antigenic Variation of Surface Molecules

The first reported evidence of phase variation in *N. gonorrhoeae* was a careful description of changes in colony morphology during in vitro passage of clinical isolates in 1963 by Douglas Kellogg and colleagues at the Center for Communicable Diseases (129). These researchers described four morphologies based on size, color, photo-opacity, shape, and consistency, which were subsequently referred to as Kellogg types T1, T2, T3, and T4 (39, 128, 129). Freshly isolated clinical isolates produced colonies of the T1 type. In contrast, laboratory isolates were mainly of the T4 type. T1 colony morphology was maintained by selective passage, with occasional T2-type offspring observed; loss of the T1-type colony and isolation of T3- and T4-type colonies followed nonselective passage. Notably, intraurethral inoculation of male volunteers with non-T1-type colonies resulted in reversion to the T1 type (129). The demonstration that T1-type colonies remained infectious after 440 passages (performed over 3.5 years) (128), in contrast to T4-type colonies, which were infective after 38 passages but not after 69 passages (129), further established a link between colony morphology and virulence. The factor(s) responsible for the difference in infectivity of the T1- and T4-type colonies used in these landmark studies was not identified.

Eight years after Kellogg's initial description, it was discovered that gonococci within T1- and T2-type colonies were piliated (112, 250). Recognition that gonococcal opacity (Opa) proteins (111, 137, 206, 207, 245) and lipooligosaccharide (LOS) (25) also contribute to differences in colony phenotypes suggested the presence of other phase-variable factors in *N. gonorrhoeae*. Fueled by the emerging power of molecular biology, the race to define the genetic basis for these observations led to an early recognition of *N. gonorrhoeae* as a leading paradigm of a pathogen that uses phase and antigenic variation to maintain a reservoir of antigenically and functionally different phenotypes. Here we define phase variation as the reversible expression of a gene at a rate higher than that of spontaneous mutation (218). Most phase-variable genes in the pathogenic *Neisseria* spp. turn expression on and off via a frameshift mutation caused by the insertion or deletion of one or more repeated units during replication. This mechanism is consistent with that of slipped-strand mispairing, in which mutational events occur during DNA replication due to local denaturation of the duplex followed by mispairing of bases in the repeat region (142, 261). If the repeated elements are within the structural gene (and are not composed of nucleotides in multiples of three), expression will be affected at the level of translation due to the production of transcripts that are in or out of frame. If the repeated elements are within a promoter region, expression is affected at the level of transcription as a result of altered spacing of promoter sequences (96). As many as 83 gonococcal genes may be phase variable via this frameshift mechanism, based on a survey of the *N. gonorrhoeae* genome sequence. Potential phase-variable genes include those that encode outer membrane proteins or enzymes used in the biosynthesis of surface molecules, genes predicted to encode inner membrane and cytoplasmic proteins, and genes of unknown function (234).

Phase-variable expression of different versions of the same gene, as in the case of *opa* genes, or of genes that contribute to the structure of the same macromolecule, as occurs with LOS biosynthesis genes, results in reversible changes in the antigenic makeup of the bacterial surface. *N. gonorrhoeae* also demonstrates true antigenic variation, defined here as the expression of different antigenic versions of a molecule as a direct result of reversible changes in the structural gene. Pilin antigenic variation, the result of new genetic information recombining into the pilin gene, is perhaps the most fascinating example of true antigenic variation in *N. gonorrhoeae*. A second, less extensively explored form of antigenic variation is similar to the frameshift mechanism of phase variation, except that the number of nucleotides within the repeated unit is in multiples of three and the repeated element is always within the structural gene. The end result is the addition or loss of amino acids, which may alter function or antigenicity, the latter of which depends on surface exposure of the altered region. As many as 27 proteins may be structurally variable by this mechanism; 12 of these are likely to be surface exposed based on an analysis of the *N. gonorrhoeae* genome sequence (124).

Challenges and Limitations of Pathogenesis Studies

Research on phase-variable and antigenically variable phenotypes in *N. gonorrhoeae* can be highly frustrating due to the challenge of obtaining sufficiently pure populations of bacteria that do or do not express the gene of interest. While subcultures of *N. gonorrhoeae* are still routinely inspected under a stereomicroscope in the research laboratory to monitor the expression of surface phenotypes, reliance on colony morphology can now be replaced by molecular tools to estimate the purity of a population and to assess phase and/or antigenic variation in response to test conditions. These methods differ in accuracy and suitability for screening large numbers of isolates. The importance of reliable technical methods for detecting antigenic and phase variants of *N. gonorrhoeae* cannot be overstated. The rate of variation is relatively high for most genes described (10^{-2} to 10^{-5}) (46, 154, 214, 224, 247, 248); therefore, most populations of gonococci contain variants that do or do not express the molecule of interest, whether it be a colony, a broth culture, or a bacterial suspension inoculated into a tissue culture well. Genetic manipulations that stabilize expression to a locked "on" or locked "off" state have been helpful in this regard. In some cases, however, phase variation of unrelated but functionally redundant genes may still confound the interpretation of data. For example, studies of pilus-mediated adherence may be compromised by changes in Opa protein expression (153). Similarly, studies of Opa-mediated serum resistance may be confounded by LOS phase variation (27).

A second obstacle in studying the impact of phase and antigenic variation on gonococcal pathogenesis is the inability to test predictions made from in vitro systems in an animal model that closely mimics natural infection. Our current understanding of how variable expression of gonococcal factors contributes to adherence to and invasion of host cells is based on tissue culture cell (26, 48, 85, 267), primary cell (99), and organ culture (81, 82) models. Similarly, in vitro assays using human serum, antibody or isolated complement components have helped define the potential role of antigenic and phase variation in evasion of host immune defenses (27, 86, 195, 198). A human male urethritis model, similar to that used in Kellogg's day, continues to be a valuable tool for characterizing the kinetics of phase and antigenic variation during early urethritis (51, 52, 211, 223, 246, 251). The volunteer model is highly relevant to natural infection; however, ethical restrictions against manipulating the host immune system or maintaining long-term infection limit its power to identify host factors that select for variable phenotypes. Also, questions concerning the adaptation of *N. gonorrhoeae* to the female genital tract, which differs substantially from the male urethra, cannot be addressed by studies of female volunteers due to the higher risk of complications. The development of a female-mouse model of genital tract infection (113, 115), together with the increasing availability of transgenic animals to circumvent issues of host restriction (117), has broadened research possibilities in this area. The quest for a surrogate animal model that fully mimics human infection may never be realized, however, due to the many host restrictions that exist.

Despite the experimental challenges inherent in studying this human-specific pathogen, evidence that variable expression of surface molecules plays a critical role in gonococcal pathogenesis is strong. Here we focus on four well-described phase-variable or antigenically variable factors for which there is evidence of a role in adherence to specific host receptors or evasion of host defenses. Collectively, these functions promote the survival of *N. gonorrhoeae* in vivo and therefore are important for successful colonization. For each factor, we review the genetic mechanism of variation, the tools used to monitor or stabilize expression, and the manner in which expression of each factor, or lack thereof, contributes to pathogenesis as hypothesized from existing clinical and experimental data.

COLONIZATION PILI

Structure and Function

The colonization pili of *N. gonorrhoeae,* which are also produced by *N. meningitidis,* are members of the type IV family of pili (242). These macromolecules mediate adherence to host cells (121, 158, 167, 181, 190, 209, 244) and also play a role in natural transformation (77), bacterial aggregation (179, 254), twitching motility (38, 160), and host cell signaling (12, 126, 141). The pilus fiber is a filamentous structure composed of an ordered array of a pilin (PilE), the major subunit (74, 178), and the minor components PilC (202) and PilV (281). An interesting feature of neisserial pilin is the covalent modification of certain serine residues, including phosphorylation (73) and the addition of galactose α-1,3-*N*-acetylglucosamine (13, 188), α-glycerol-phosphate (73, 240), and phosphorylcholine (134, 278).

Unlike pilus genes in many other bacteria, the genes that encode the structural components and biosynthesis machinery of gonococcal pili are unlinked (reviewed in reference 257). Among the components involved in pilus structure and synthesis, PilE undergoes antigenic variation, as discussed extensively below. The secretin PilQ, which polymerizes to form the pore through which pili are assembled, may undergo structural variation via the insertion or deletion of a 24-nucleotide tandemly repeated element in the *pilQ* N-terminus-encoding region. Changes in this region did not alter PilQ function in *N. meningitidis;* whether insertion or deletion of this repeated element confers antigenic differences is not known (256). This aspect of PilQ has not been studied in *N. gonorrhoeae,* and fewer copies of this element are found in the gonococcal *pilQ* gene (124, 256). PilC, a 110-kDa protein found on the pilus tip (202) and also on the outer membrane (192), undergoes phase variation (122). PilC is required for pilus assembly (122, 192), although piliated PilC$^-$ gonococci have been isolated (200, 203). PilC is probably the primary pilus adhesin (164, 201, 202, 209), whose putative receptor is the human membrane cofactor protein (CD46) (125, 127). PilC requires assembled pili to mediate adherence (121, 268). Both PilC and PilE are required for transformation competence (77). Expression of pilus glycosyltransferase A (*pgtA*), which glycosylates pilin at the serine-63 residue (14), is also phase variable. Other glycosylation loci exist in *N. gonorrhoeae,* although the modifications catalyzed by these enzymes are still not defined (13). Among these, *pgtE* and *pgtG* are likely to undergo phase variation based on the presence of homopolymeric regions that cause reversible frameshift mutations in the meningococcal homologues of these genes (189). The phosphorylcholine (ChoP) epitope on gonococcal pilin also appears to be phase variable (278). The genetic basis for this observation is unclear, however, since the repeated region that confers phase variation of the pilin ChoP transferase A (*pptA*) gene in *N. meningitidis* is considered too short in the gonococcal *pptA* homologue for strand slippage to occur. Instead, ChoP phase variation in *N. gonorrhoeae* may stem from variable expression of the target or of genes needed for ChoP biosynthesis or uptake (274).

Mechanism of Pilin Antigenic Variation

Kellogg's seminal volunteer studies generated much enthusiasm for the notion of using pilin as a gonorrhea vaccine (33, 216). Research in this area was closely shadowed by a growing appreciation of pilin antigenic variation, however (32, 40), and after several years of intense effort, development of a gonorrhea pilin vaccine was abandoned. Clues to the basis of pilin antigenic variation were first revealed in a series of cloning and mapping studies in which two expressed *pilE* genes and several silent pilus (*pilS*) loci were detected in gonococcal strain MS11 (161, 162). The silent loci appeared to lack promoters, based on the inability to express pilin in *E. coli* from a clone carrying the *pilS1* locus (161); a cloned *pilE* gene, in contrast, conferred expression (162). More than one pilin variable region was detected in a nonexpressed locus, and although the picture was not yet clear, the mechanism appeared similar to that of immunoglobulin gene rearrangement (161). Determination of the predicted amino acid sequence of a silent locus (91) and of different pilin mRNA transcripts (94) and cloned *pilE* genes (174) helped demarcate regions of variation within the pilin monomer. These regions, called minicassettes, were separated by strictly conserved sequences (91, 94). Working in parallel, Schoolnik et al. (216) determined the primary amino acid sequence of serologically distinct pilins. Together with antibody binding studies, this information defined variable regions in the central and C-terminal regions of pilin (69, 199), whose locations corresponded to those identified by nucleotide sequencing (Fig. 1A). We now know the source of pilin antigenic variation to be the introduction of one or more minicassettes within a partial gene copy, within a *pilS* locus, into the *pilE* gene via recombination (222). Recombination is unidirectional and may occur via intragenic recombination or with exogenous DNA taken up from lysed gonococci (reviewed in references 219 and 239) (Fig. 1B). As predicted, recombination occurs within conserved regions that flank the minicassettes. However, points of recombination have also been traced to variable regions (97, 157, 270, 271), an event that expands the potential repertoire of new *pilE* sequences beyond that originally estimated.

The depth of variability created by the size of the pilin repertoire and the seemingly random manner by which cassettes are inserted make *Neisseria* pilus antigenic variation one of the most fascinating stories of genetic diversity in bacterial pathogenesis. The reserve of pilin sequences is completely (97) or partially (93) known for two strains and consists of one or two expression loci (*pilE*) and five *pilS* loci, which collectively contain 17 or 19 copies (91, 93, 97). None of the silent copies are identical between the two strains, and variability between the minicassettes is high (97). The theoretical number of pilin variants that can be produced by a single strain through these recombination events is well over 10^6 variants per strain. The actual number, however, is much smaller due to missense and nonsense mutations and to structural constraints (93, 97, 247, 248).

Mechanisms of Pilus Phase Variation

Pilus phase variation occurs at a high rate in vitro, with nonpiliated (P$^-$) colony variants arising from piliated (P$^+$) colony variants at a rate of 10^{-2} to 10^{-5} per cell (250). Piliation can be lost via several mechanisms, only a few of which are true phase variation (reviewed in reference 219). A high percentage of gonococci cultured in vitro become P$^-$ due to deletion of the 5′ end of the *pilE* gene (P$^-$n; pilus-nonreverting), an event that is not reversible (20, 247). Gonococci also secrete an unassembled form of pilin (S pilin) that results from posttranslational

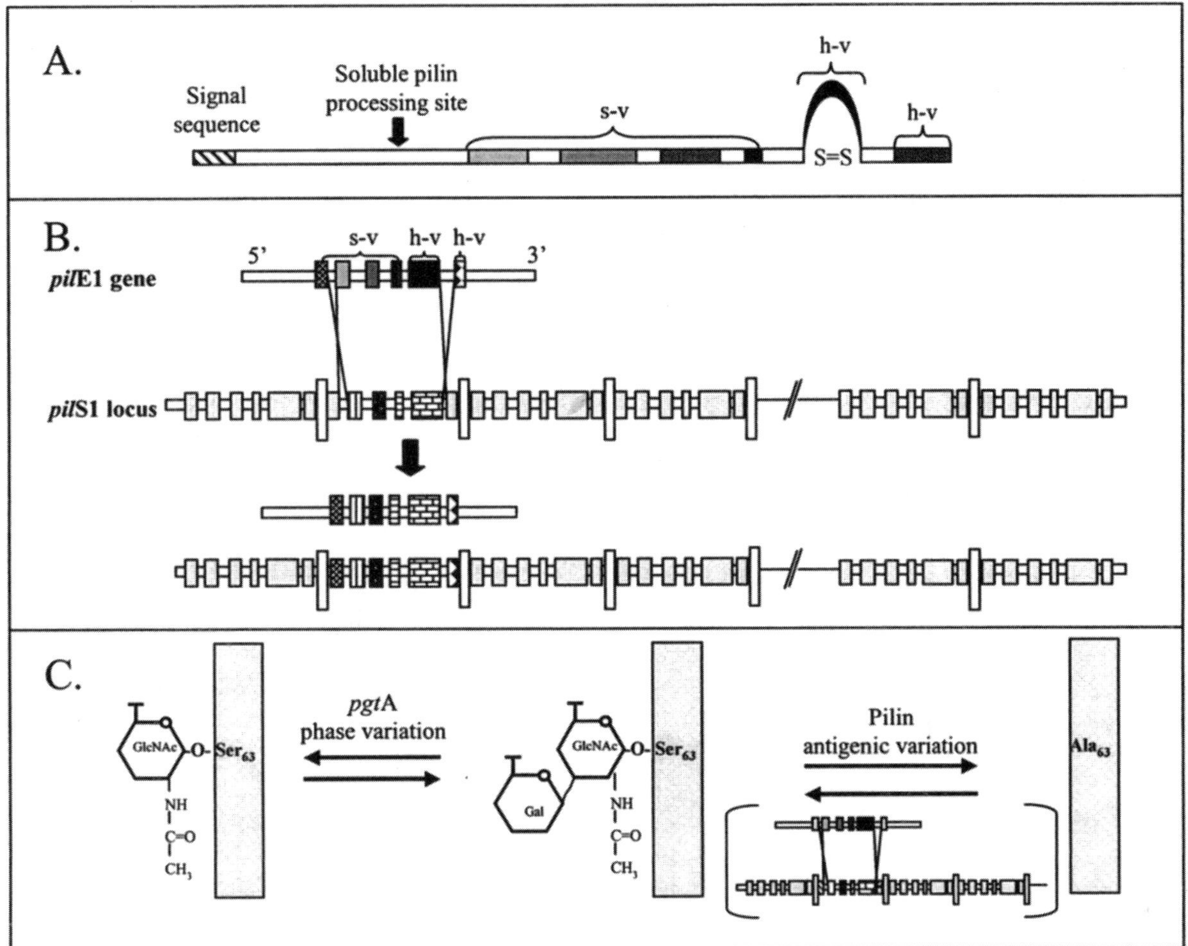

Figure 1. Mechanisms of pilin antigenic variation and variation of pilus glycosylation. (A) Diagram of the pilin monomer, showing complete conservation at the N terminus and variability in the central and the C-terminal regions. Hypervariable (h-v) and semivariable (s-v) regions as defined by Hagblom et al. (94) are shaded; conserved regions are represented in white. The presence of a hypervariable region flanked by two cystines, which form a disulfide bond, was predicted from both the primary amino acid (216) and nucleotide (161) sequences. This hypervariable loop is the immunodominant region of the pilin (216), which suggests that immune pressure may have selected for the high degree of variation in this region. Posttranslational cleavage at amino acid 39 from the N terminus results in secreted (soluble) pilin (S pilin). (B) Schematic of the *pilE* gene and a *pilS* locus in *N. gonorrhoeae* and generation of a new *pilE* gene via nonreciprocal recombination. The number of copies per silent locus varies, with six copies being present in *pilS1* of strain MS11 (93). Each copy contains six variable regions (minicassettes), which, when translated, correspond to the semivariable and hypervariable regions in the pilin protein. The *pilE* gene may receive a complete or partial *pilS* copy (as shown here). Therefore, over time, repeated recombination events into the *pilE* gene create a chimeric *pilE* gene composed of sequences from multiple *pilS* loci. The minicassettes are depicted here by shading, except for the ones within the *pilE* gene and the silent copy shown to be undergoing recombination, which are patterned. (C) Phase variation of pilus glycosylation in *N. gonorrhoeae.* Phase-variable expression of pilus glycosyltransferase (*pgtA*) can result in the presence or absence of galactose bound to an O-linked galactose *N*-acetylglucosamine molecule linked to a surface-exposed serine at position 63. The glycosylation state is also dependent on the presence of a serine residue at this position and therefore is also controlled by pilin antigenic variation. Only 11 of 17 pilin copies in the *pilS* loci of strain FA1019 encode a serine residue at this position (97). This finding suggests that antigenic variation may be a significant source of changes in pilus glycosylation.

cleavage at the N terminus (92). Originally, S pilin was associated with reduced piliation (92); however, S-pilin-producing gonococci that express numerous pili have been isolated (145). True pilus phase variation, as in high-frequency, spontaneous, reversible change in pilus production, occurs by one of three known mechanisms. (i) Recombination of silent copies into the *pilE* gene may create pilin sequences that contain missense or nonsense mutations, which results in no pilin being expressed or pilin that cannot be assembled. In either case, pili are not produced (20, 248, 249). (ii) Inaccurate recombination can lead to more than one "copy" of information in the *pilE* gene, with the result being the production of

L pilin (20 to 32 kDa), which is larger than the normal 18-kDa pilin monomer and is not assembled (92, 152). (iii) Phase variation of *pilC,* which is required for pilus biosynthesis, can also cause loss of piliation. There are two *pilC* alleles in *N. gonorrhoeae;* reversible frameshift mutations in a poly(G) region within the signal sequence-encoding region of each gene confers phase variation (122). Interestingly, loss of PilC expression seems to select for secreted pilin variants, presumably because the buildup of unassembled pilin in PilC$^-$ bacteria may be toxic (123).

Tools for Studying Pilus Phase and Antigenic Variation

Variations in colony shape and edges are generally attributed to differences in piliation (247, 250). To the trained eye, these differences are easy to see; therefore, simple inspection of colony morphology is often used to assess piliation. This technique is adequate for maintaining a P$^+$ (domed with defined edges) or P$^-$ (flat with irregular edges) colony phenotype during subculture. However, colony morphology is not always an accurate predictor of piliation, and, in fact, piliated gonococci can produce colonies with the classic P$^-$ morphology (92, 203, 271). Reliance on changes in colony morphology as a surrogate for pilus phase variation is especially risky, since gonococci that produce a P$^-$ or intermediate P$^{+/-}$ colony morphology often do not represent true pilus phase variants (145). Another indirect measure of piliation in *N. gonorrhoeae* is DNA transformation efficiency (221, 250). However, extended pili are actually not required for competence (146, 201), and although the degree of piliation can correlate with transformation efficiency when expression is stabilized to a single antigenic type (144), this functional phenotype does not always correlate with piliation when gonococci that express different pilin molecules are compared. The latter observation is particularly true when partially and highly piliated variants (145), or variants that express different levels of pilin (146), are compared. Immunodetection of PilC (173), in conjunction with detection of normal-size pilin (as opposed to S pilin or L pilin), with cross-reactive antibodies against pilin is a good method for confirming piliation (123, 145). Immunodetection methods are not quantitative, however, since the degree of antibody binding may differ among pilin variants (145). Transmission electron microscopy remains the "gold standard" for assessing piliation in *N. gonorrhoeae.*

Accurate detection of pilin antigenic variation within a population of gonococci is complicated by the sheer magnitude of variants that can be expressed. Serological reagents, therefore, are not useful for this purpose, and detection at the genetic level is impeded by the seemingly unrestricted number of potential recombination sites within numerous pilin copies. Direct sequencing of the *pilE* gene ranks high in accuracy; however, the number of isolates that can be examined by this approach is too limited to assess the rate of variation or the frequency of variants within a population. Historically, colony morphology phase variation has been used to measure the rate of pilin antigenic variation since recombination of new sequences into the *pilE* gene frequently results in missense or nonsense mutations or structurally defective molecules (247, 249). This approach is validated by the fact that *recA* mutants, which are unable to undergo pilin antigenic variation, have a significantly lower rate of transition from P$^+$ to P$^-$ colonies (123, 135). Loss of piliation due to *pilC* phase variation or the P$^-$n phenotype occurs at a lower frequency than antigenic variation; therefore, these events do not falsely affect the antigenic-variation rate at a detectable level when assessed by colony morphology phase variation. To address the need for a more accurate method of measuring antigenic variation, Wainwright et al. (271) developed a reverse transcription-PCR recombination assay that detects recombination between *pilE* and one or two copies within a single *pilS* locus. Modification of this assay to a competitive reverse transcription-PCR method enhanced the quantitative power of this approach; sensitivity was also increased by using primers with sufficient homology to enable detection of sequences within several silent copies (224).

The use of gonococci that are locked on or off for pilus expression or stabilized to express a certain antigenic pilin type is often desirable for pathogenesis studies. P$^-$n gonococci, which are not true phase variants, can be easily isolated in vitro and are frequently used to study the effect of nonpiliation in tissue culture assays or to provide a less adherent background for studying other gonococcal adhesins. Pilus phase variation can also be significantly reduced by *recA* mutation, which cripples recombination with the *pilE* gene (135). A caveat to this approach is that *recA* mutants can turn off piliation via *pilC* phase variation (123). PilC expression can be stabilized by altering the poly(G) region responsible for *pilC* phase variation (122). Stabilized expression of a single pilin antigenic type through *recA* mutation has been helpful in studies of functional differences among pilins (123, 144, 145). An inducible *recA* system constructed by Seifert (220) has the advantage of allowing one to start with a uniform population of a piliated or nonpiliated variant before initiating test conditions; induction of *recA* expression when desired ensures that the majority of variants detected do not represent subpopulations present in the initial population.

Consequences of Pilus Antigenic and Phase Variation

Much evidence suggests that gonococcal pili mediate the initial adherence of *N. gonorrhoeae* to epithelial cells (121, 158, 167, 181, 190, 209, 244), a step that presumably enables the gonococcus to withstand the flushing force of urine and the constant shedding of cervical mucus. The importance of gonococcal pili in colonization is supported by volunteer studies in which immunization to pilin blocked infection by strains that expressed the homologous pilin type (32, 41, 42). The fact that clinical isolates exhibit the P^+ colony morphology further supports a strong functional role for pili during infection. It is reasonable to assume that enhanced colonization provides the selective advantage for piliation in vivo; this assumption may underestimate the selective factors involved, however, due to the multifunctional nature of these molecules. Also, while pili undoubtedly mediate colonization of mucosal surfaces, they are not essential for urethral infection in men (51), despite years of consensus to the contrary based on the early studies by Kellogg et al. (128, 129).

The most likely advantage conferred by pilin antigenic variation is evasion of a specific antibody response. This hypothesis is strongly supported by the demonstration that pilin-specific antibodies do not inhibit the adherence of gonococci that express a heterologous pilin type (40, 181, 265) and the failure of pilin vaccines to protect volunteers challenged with heterologous strains (29, 33). In addition to blocking adherence, which presumably is most important early in infection, pilin-specific antibodies promote opsonophagocytosis by neutrophils (190, 266) and macrophages (120), and some are weakly bactericidal (42). All these reported antibody activities were pilin specific. The rate of pilin antigenic variation during infection is not known but may be higher in the low-iron environment of mucosal surfaces than under standard in vitro conditions, based on evidence that iron limitation significantly increases the frequency of pilin antigenic variation (225). Importantly, new pilin variants emerged within 1 to 2 days following intraurethral inoculation of volunteers with gonococci that expressed primarily a single pilin type (97, 223, 251). This early time point suggests that antigenic variation of pilin is not driven by a developing specific immune response. Rather, the most important advantage of pilin antigenic variation may be enhanced transmission, since the likelihood that the next host will be immunologically naive is high (223).

In addition to immune evasion, there is evidence that pilin antigenic variation may promote functional diversity. As well as having a structural function, PilE mediates hemagglutination (203, 209), adherence to epithelial cells (121, 175, 209, 216), and bacterial aggregation (179, 254). Antigenically different pilins mediate different degrees of adherence (121, 138, 203). This finding cannot be clearly interpreted, however, since the PilE domain that mediates epithelial cell adherence has not been identified. Other pilin-specific attributes might be responsible for differences in adherence. For example, the amount of pilin and the number of pili produced per cell differs among pilin variants (145), which can affect adherence (144). Also, the net adherence of pilin variants that form pilus-mediated aggregates is higher than that of variants which have been specifically mutated to lose the aggregative phenotype (179). Studies with *N. meningitidis* suggest that changes in the glycosylation state may be responsible for differences in PilE-mediated adherence (269), a result consistent with reduced adherence of gonococci following treatment with galactosidase (88) and the predicted surface exposure of this glycosylation site (74). Pilin glycosylation state can change due to *pgtA* phase variation or as a consequence of pilin antigenic variation, since many silent pilin copies lack the serine-63 residue that undergoes glycosylation (97) (Fig. 1C). The extent to which *pgtA* phase variation plays a role in pilus glycosylation during genital tract infection is questioned by the work of Banerjee et al. (14), who showed that gonococci carry one of two different alleles of the *pgtA* gene, one of which is phase variable due to the presence of a poly(G) tract in the structural gene and one of which lacks this region and is constitutively expressed. Interestingly, all of 24 strains isolated from patients with disseminated gonococcal infection (DGI) possessed a *pgtA* allele with the poly(G) tract. In contrast, the majority of isolates from patients with uncomplicated mucosal infections did not contain this poly(G) region. Whether gonococci isolated from systemic sites are glycosylated has not been reported, and how phase variable pilus glycosylation might be advantageous during dissemination is not known; avoidance of natural anti-Gal antibodies is one explanation (14). Marceau and Nassif (153) showed that substitution of the serine-63 residue was associated with the production of high levels of S pilin. S pilin may act as an immunological decoy (92) or serve to release a toxic accumulation of pilin monomers (123). S pilin also signals via the CD46 receptor in the absence of assembled pili and may act to amplify signaling beyond that which occurs through intact pili (204).

The purpose of pilus phase variation in bacterial pathogenesis is less intuitive than that of antigenic variation. Loss of piliation may promote invasion and traversal through cells, as suggested by a report that nonpiliated variants were more invasive than piliated variants when invasion was expressed relative to the number of adherent bacteria (148). Also, mostly nonpiliated gonococci were isolated from the basal side of polarized endometrial cells following inoculation with piliated variants (109). Evasion of the specific immune response might also be enhanced by loss of piliation in that mostly nonpiliated variants were recovered after incubation with bactericidal pilin-specific antibodies (42). Testing of these theories is handicapped by the infrequent isolation of nonpiliated gonococci from humans. For similar reasons, it is difficult to determine which of the several mechanisms of pilus phase variation occur during infection. Loss of piliation due to the formation of structurally defective pilin molecules or *pilE* genes does not appear to be simply an unavoidable casualty of pilin antigenic variation. For example, although piliated wild-type and *recA* mutant gonococci invaded polarized endometrial tissue culture cells to an equal degree, only wild-type gonococci were released from the basal side, and all of them were nonpiliated due to recombination into the *pilE* gene, causing a loss of pilus production (109). The existence of a seemingly more deliberate mechanism of pilus phase variation, namely, *pilC* phase variation, also supports the likelihood that turning off piliation during infection can be advantageous.

OPACITY PROTEINS

Structure and Function

Differences in colony photo-opacity in *N. gonorrhoeae* are caused by the expression of a family of outer membrane proteins called Opa proteins. Originally referred to as P.II proteins in *N. gonorrhoeae* and class 5 proteins in *N. meningitidis,* Opa proteins range from 25 to 30 kDa in molecular mass and are encoded on separate unlinked chromosomal genes. *N. gonorrhoeae* strains possess 8 to 11 *opa* alleles. The repertoires of *N. meningitidis* (2, 105, 165, 283) and some commensal *Neisseria* spp. (255) are more limited, with only three or four *opa* alleles present in these species. Strains carry different *opa* genes, and the complete Opa repertoire has been fully defined for only two gonococcal strains (22, 54). Three regions are responsible for the differences in molecular mass and antigenicity, namely, the semivariable region (SV) and two hypervariable regions (HV_1 and HV_2) (Fig. 2A). These regions form surface-exposed loops in the mature Opa protein, as predicted by two-dimensional models (149) and antibody binding studies (63, 149). A fourth conserved loop is also surface exposed. *opa* gene phase variation occurs via a reversible frameshift mechanism by which one or more copies of a pentameric sequence (CTCTT) within the signal sequence-encoding region is inserted or deleted (19, 170, 238). Individual *opa* genes phase vary independently of each other, and gonococci can express no Opa proteins, a single Opa protein, or multiple Opa proteins simultaneously.

Opa proteins mediate attachment to and invasion of a variety of cultured epithelial cells, including cervical, endometrial, Chang human conjunctival, and endothelial cells (21, 136, 148, 230, 264, 275). Some Opa proteins also mediate uptake by phagocytes in the absence of opsonization (17, 26, 49, 72, 136) (Fig. 2B). The failure to identify a conserved receptor binding region presented a paradox as to how such antigenically different proteins could mediate the same function. This dilemma was resolved by the discovery that several Opa receptors exist, including different members of the carcinoembryonic antigen (CEACAM) family, specifically CEACAM1, CEACAM3, CEACAM5, and CEACAM6, as well as heparin sulfate proteoglycans (HSPG) (26, 50, 85, 263, 267; N. B. Guyer, M. C. Piriou, M. M. Hobbs, and J. G. Cannon, *Abstr. 101st Gen. Meet. Am. Soc. Microbiol. 2001,* p. 303–304, 2001). Serum factors, specifically vitronectin, enhance invasion via the HSPG binding pathway (66, 80). Binding of Opa proteins to these specific cellular receptors is dependent on conformation. For example, both HV loops are critical for CEACAM binding among Opa proteins that bind these receptors, and specific combinations of HV regions confer tropism for different CEACAM receptors (28, 61, 62, 84). Structure-function analysis of an HSPG-binding Opa protein revealed that the HV_1 loop was absolutely necessary for invasion via HSPG receptors, with positively charged residues implicated in this interaction. The SV or HV_2 loops were also important, in that deletion of these regions decreased recognition of the receptor (84). The SV loop, in contrast, is completely dispensable for Opa-mediated binding to CEACAM receptors (28).

Tools for Studying *opa* Gene Phase Variation

Different Opa proteins confer different degrees of photo-opacity to a colony, with colony phenotypes ranging from transparent to deeply opaque. Unfortunately, many early studies of Opa protein expression are invalidated by the fact that some Opa proteins do

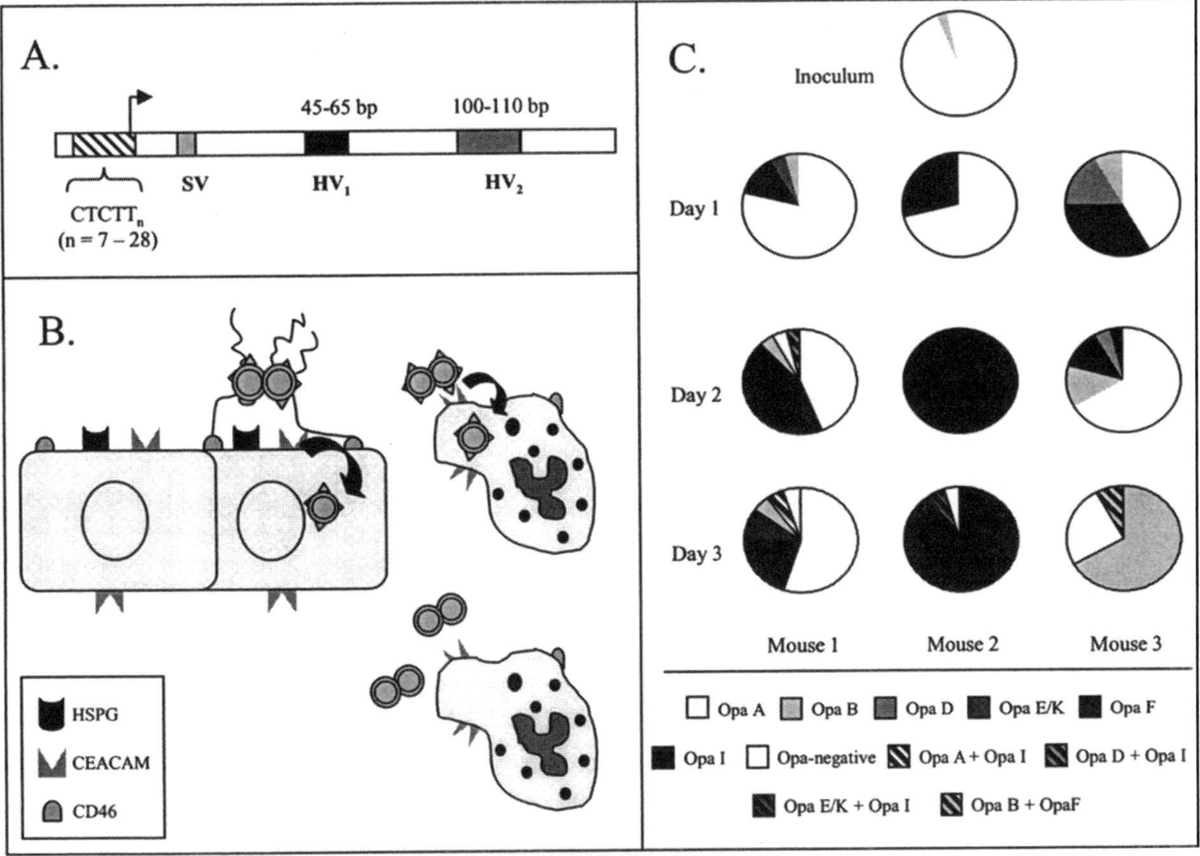

Figure 2. Opa protein structure and function, and selection or induction of Opa protein expression during experimental murine genital tract infection. (A) Diagram of an *opa* gene showing the three regions of variability (SV, semivariable; HV_1, and HV_2, hypervariable regions 1 and 2) that define different *opa* alleles. The pentameric repeat responsible for phase variation is in the signal sequence-encoding region, with 7 to 28 copies present in different *opa* genes (54, 238). (B) Cartoon depicting Opa-mediated adherence to and invasion of epithelial cells and nonopsonic uptake by phagocytes. (C) Opa phenotypes of vaginal isolates from three mice that demonstrated an increased percentage of Opa-positive variants among vaginal isolates following inoculation with a predominantly Opa-negative population of strain FA1090. More than 50% of the vaginal isolates expressed at least one Opa protein within 24 h following inoculation into the lower genital tract. Different Opa proteins predominated in different mice, with OpaB being most highly represented in mouse 3 and OpaI being mostly highly represented in mice 1 and 2. OpaI of this strain is known to bind HSPG, which is likely to be present in mice. OpaB does not bind HSPG but does bind to human CEACAM receptors (Guyer et al., *Abstr. 101st Gen. Meet. Am. Soc. Microbiol. 2001*). Gonococci that expressed more than one Opa protein simultaneously are represented by stripes.

not confer detectable opacity to a colony (and thus the colony appears to be composed of Opa-negative gonococci). Monitoring Opa protein expression by direct nucleotide sequencing of *opa* genes to identify those that are in or out of frame is impractical due to the large number of *opa* alleles within a strain. Also, because *opa* gene phase variation effectively occurs at the level of translation, detection of mRNA is not useful since all *opa* genes are constitutively transcribed. Therefore, the Opa phenotype must be determined phenotypically. Gel electrophoresis alone or in conjunction with immunoblotting using broadly reactive Opa-specific antibodies (1) is often used to determine the predominant Opa phenotype produced by a population of gonococci. High-resolution gels and a familiarity with the relative migration of all the Opa proteins expressed by the strain being studied are required to reliably detect the minor differences in molecular weight between Opa proteins (24). Also, due to the cumbersome nature of gel electrophoresis as a means of screening large numbers of variants, this method is limited in its ability to accurately quantitate the frequency of different Opa phenotypes within a population. The development of HV-specific monoclonal antibodies (MAbs) (16, 24) or polyclonal antisera (A. N. Simms and A. E. Jerse, unpublished data) has aided accurate determination of the Opa phenotype. Opa-specific antibodies can be used in immunoblot assays to probe suspensions of individual colonies isolated from test conditions (114); this strategy minimizes in vitro passage. HV-specific antibodies can also be used in immunostaining protocols to directly determine the Opa phenotype of gonococci in exudates from experimentally infected volunteers or laboratory animals without in vitro passage. Unfortunately, these serological reagents are useful only for the strain for which they were created, since different strains possess different Opa repertoires.

Translational gene fusions containing a reporter gene downstream of the pentameric repeat region of an *opa* gene have been useful to study *opa* gene phase variation. For example, a rate of *opa* gene phase variation of $\sim10^{-3}$ to 10^{-4} per cell per generation was measured in *Escherichia coli* by using a translational fusion driven from a single- or low-copy-number plasmid backbone (19, 170). The effect of promoter strength on *opa* gene phase variation was examined using a similar approach (18). To date, the rate of *opa* gene phase variation has not been appropriately determined in a *Neisseria* background; hence, information is based solely on colony morphology (154). *opa* gene expression can be locked on for functional studies by replacing the signal sequence that carries the CTCTT repeat responsible for frameshifts with a signal sequence from another protein (175) or by altering the repeat in a way that does not change the amino acid sequence (136). Even when expression is stabilized, however, functional studies are usually complicated by background expression of other Opa proteins. This problem can be overcome by ectopic expression of *opa* genes in a different host; in fact, the first evidence that Opa proteins mediate the invasion of epithelial cells was based on expression of a locked-on *opa* gene in *E. coli* (230). The tropism of one particular Opa protein for human conjunctival tissue culture cells was demonstrated by complementation of *N. gonorrhoeae* mutated in this specific *opa* gene in *trans* by using a plasmid carrying a locked-on copy of the wild-type *opa* gene (136). Finally, the presence of numerous *opa* genes per strain challenges the construction of an *opa* knockout mutant.

Consequences of *opa* Gene Phase Variation

Interest in the role of *opa* gene phase variation in gonococcal pathogenesis originated with two seminal studies by James and Swanson (110, 111), in which the photo-opacity of over 200 clinical isolates was carefully documented with respect to gender and body site. Urethral isolates from men were predominantly opaque. In contrast, endocervical isolates varied in colony opacity, with opaque colonies being isolated more often during the proliferative stage of the menstrual cycle and transparent colonies being isolated during or shortly after menses. The majority of rectal and pharyngeal isolates from either gender were opaque. Others later reported that fallopian tube isolates produced colonies that were more transparent than colonies produced by endocervical isolates from the same individual. These workers hypothesized that Opa-negative variants were therefore more invasive (65), a hypothesis that is consistent with the clinical observation that PID generally occurs during or after menstruation, when Opa-negative variants predominate (106). Differences in the relative sensitivity of Opa-positive and Opa-negative variants to proteases (110) and progesterone (205), both of which show fluctuating levels depending on the stage of the menstrual cycle, have been reported. Unfortunately, the results of these studies are compromised by the assumption that transparent colonies are always composed of Opa-negative variants.

The link between hormonal state and Opa expression in the female genital tract is not yet resolved and, in fact, should be revisited with more accurate tools to assess the Opa phenotype. In contrast, data to support the hypothesis that Opa protein expression is selected for or induced during urethritis in males are substantial. In volunteer studies using two different strains, a clear predominance of Opa-positive variants was isolated from a majority of volunteers after inoculation with primarily Opa-negative gonococci (114, 210, 246). Different Opa variants predominated among volunteers, a result suggestive of random selection or the involvement of host factors. Interestingly, as infection progressed, a high percentage of reisolates expressed two to five Opa proteins simultaneously (114), a phenotype that is uncommon in vitro.

Selection of Opa-expressing gonococci due to the adherence function of Opa proteins may explain the results of volunteer studies. This hypothesis is potentially foiled by reports that relevant CEACAM receptors are not expressed by all primary urethral cells (99), all primary endocervical or fallopian tube cells, or all immortalized cell lines of the female genital tract (252). The inability to detect these receptors may be explained by evidence that CEACAM expression is induced during infection (168, 169). Although some Opa proteins bind to HSPGs, which are more widespread in tissues, only one Opa protein in each of three gonococcal strains examined binds to this receptor (48, 263; Guyer et al., *Abstr. 101st Gen. Meet. Am. Soc. Microbiol. 2001*) and a majority of urethral isolates that expressed non-HSPG-binding Opa proteins were recovered from several volunteers (114). The existence of other Opa receptors, distinct from those already reported, may also explain these results. Alternatively, components of the innate response may play a selective role. Bos et al. (27)

showed that Opa-positive variants were more resistant to the bactericidal activity of normal human serum and LOS-specific MAbs. Critical to the success of this study was the matching of Opa-positive and Opa-negative variants for LOS phenotype, since the LOS phenotype more dramatically affected serum resistance than did Opa expression. LOS sialylation did not mask the difference in sensitivity between Opa-positive and Opa-negative variants, which supports the potential significance of Opa-mediated serum resistance during infection. The molecular basis for increased serum resistance of Opa-expressing variants is not known. Experimental infection of mice may be a useful tool for investigating the kinetics of Opa expression in vivo. Recovery of Opa-positive variants occurs following vaginal inoculation of mice with a predominantly Opa-negative inoculum (113) (Fig. 2C). Data from our laboratory, obtained by using a marked strain to monitor the recovery of particular Opa variants from mice, suggest that selection is responsible for the increased isolation of both HSPG binding and CEACAM binding variants rather than induction of *opa* gene expression in vivo (Simms and Jerse, *Abstr. 14th Int. Path. Neisseria Conf. 2004*). This observation is perhaps consistent with the existence of selective forces other than adherence, since with the possible exception of CEACAM-1 (98, 231, 264), murine and human CEACAM molecules are not highly related.

An intriguing and highly elegant aspect of Opa protein biology is the possibility that different Opa variants within a strain confer different tissue tropisms in vitro (136). Consistent with the known signal transduction pathways triggered by this broad group of receptors, interactions with HSPG or different CEACAM receptors induces different responses in host cells, which leads to distinct mechanisms of uptake (reviewed in reference 103). Phase variation of Opa expression may also play a critical role in avoiding phagocytic killing. PMNs bind and ingest most Opa-positive variants in the absence of opsonins via binding to CEACAM-3 and CEACAM-6 receptors (49, 85, 267). Opa-negative gonococci and some Opa-positive variants, in contrast, do not associate with and are not killed by PMNs (17, 71, 72, 136, 171) or monocytes (133) in the absence of serum.

Transcriptional factors, not yet defined, may further contribute to Opa protein expression during infection, as indicated by the evidence obtained by Belland et al. (18) that differences in promoter strength correlate with rates of *opa* gene phase variation. It is therefore conceivable that stimuli within the host may drive phase variation of certain *opa* genes rather than *opa* genes being turned on or off spontaneously. As yet, no specific environmental factor(s) has been identified that influences the rate of *opa* gene phase variation. In one study, rates of colony opacity phase variation were measured on exposure to 194 culture conditions, none of which had a detectable effect (154). However, as discussed, the use of colony photo-opacity as a marker of Opa protein expression may have reduced the sensitivity of the results.

Finally, although the Opa repertoire expressed by a single strain of *N. gonorrhoeae* pales in comparison to that of pilin variants, phase variation of Opa protein expression may also facilitate immune evasion. This hypothesis is supported by the demonstration that Opa-specific antibodies block the adherence of homologous Opa variants in vitro (243) and by a reported correlation between Opa-specific antibody and a reduced risk of gonococcal salpingitis (186). Some new and exciting data demonstrate that Opa-mediated binding to CEACAM1 on $CD4^+$ T cells can down-regulate the activation and proliferation of these cells (30). It is not known to what extent these in vitro experiments reflect the situation during infection. However, this finding has led some workers to hypothesize that Opa-mediated immunosuppression increases susceptibility to repeated infection or to coinfection with other pathogens (30).

LIPOOLIGOSACCHARIDE

Structure, Mechanism of Phase Variation, and Tools for Study

Early biochemical investigations gave conflicting results about LOS differences in gonococci within the virulent T1-type versus the T4-type colonies, as well as whether a repeating polysaccharide chain was present (182, 237, 282). We now know that LOS plays a role in colony morphology through its interactions with Opa proteins to cause intragonococcal adhesion (25) and that the LOS of *Neisseria* spp. differs from the lipopolysaccharide of enteric gram-negative bacteria in that the repeating O antigen is absent. Instead, gonococcal LOS consists of a lipid A moiety attached to a branched carbohydrate core that varies in length and sugar composition within and between strains (87) (Fig. 3). The structural heterogeneity of this major glycolipid complicated the development of a serotyping system for *N. gonorrhoeae* (5, 6, 8). Certain technical advances were key in unraveling the complexity of LOS phenotypes in *N. gonorrhoeae*. An improved staining technique (258) and methods for extraction and resolution of LOS by gel electrophoresis (104), together with the development of MAbs against gonococcal and meningococcal LOS (7, 288), helped clearly determine that the LOS

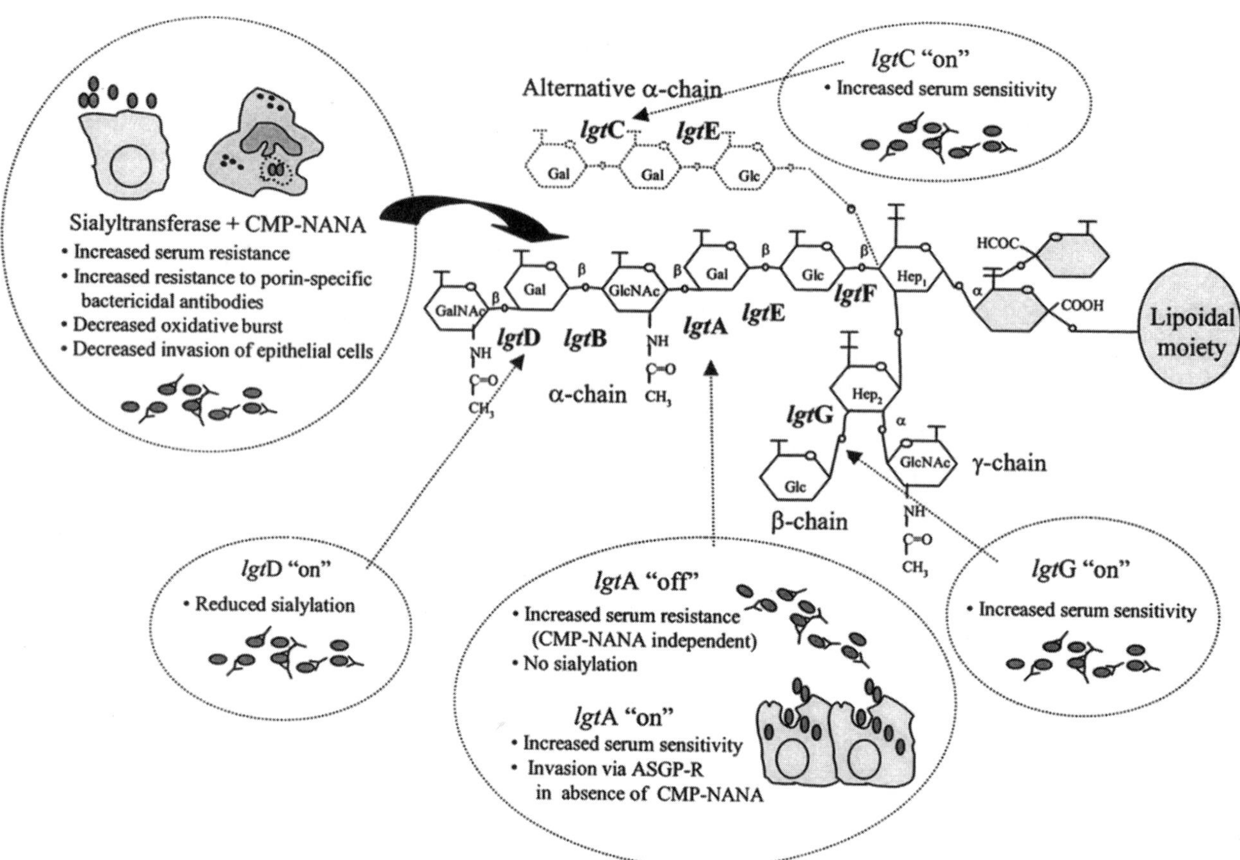

Figure 3. Prototype *N. gonorrhoeae* LOS molecule, showing points of structural variation that lead to diversity in function. The basic structure and corresponding glycosyltransferase genes are based on references 15 and 83. Phase variation of LOS structure occurs due to frameshifts in a poly(G) region within *lgtA* (also called *lsi2* [43, 58]), *lgtC*, or *lgtD* (83, 286) or a poly(C) region within *lgtG* (15). The synthesis of an alternative α chain, in which *lgtC* participates, occurs in only a minor population of gonococcal strains. Slippage of *lgtA* to an "off" position causes the production of a short-chain LOS species that confers high levels of serum resistance independent of growth in CMP-NANA (83, 227). Expression of *lgtA* or *lgtC* increases serum sensitivity (227). In contrast, slippage of *lgtG* to an "on" position results in the formation of a highly bactericidal epitope on the β-chain (15). The addition of CMP-NANA via the action of sialyltransferase to the terminal galactose residue of the lacto-*N*-tetraose moiety is shown, a modification that results in several adaptive advantages, as described in the text. Phase variation of *lgtA* or *lgtD* can influence LOS sialylation by controlling the presence of the target species or by blocking the target residue, respectively (83, 262). Sialylation blocks epithelial cell invasion mediated by interactions between the lacto-*N*-neotetraose moiety and the ASGP-R (99, 262). Lectin-like interactions with gonococcal Opa proteins also rely on this tetrasaccharide species and are blocked by sialic acid (25).

species produced by different gonococcal strains were antigenically heterogeneous and also differed in size, ranging from 3.2 to 7.2 kDa (150, 213). Even more intriguing was the observation that some MAbs bound LOS species of different size and that some LOS species bound to more than one MAb (150). A clue to this puzzle was the discovery of pilus and Opa protein phase variation; accordingly, in 1987, Apicella et al. (10) hypothesized that phase variation of genes that function at different points in the biosynthesis of LOS might cause the production of different LOS species that share some epitopes but not others. In this important study, researchers departed from the use of electrophoresis to analyze LOS phenotypes, since the sample loaded represents LOS extracted from a relatively large population of gonococci. Instead, individual gonococci derived from a single colony were stained with two LOS-specific MAbs, both of which recognize a 4.8-kDa LOS species and one of which also recognizes a larger (5.9-kDa) species. Using immunoelectron and fluorescence microscopy, gonococci that bound to one or the other MAb, or to both MAbs, were visualized on each slide, a result consistent with variable expression of LOS biosynthesis genes within a strain. This work was followed by a study by Schneider et al. (214) showing that changes in LOS reactivity with MAbs occurred at high frequency (10^{-2} to 10^{-3}) and was reversible. This finding was demonstrated by carefully subculturing MAb-reactive sectors of individual colonies and documenting the occurrence of reversion on serial passage (214). Detection of LOS variants continues to rely on gel electrophoresis and the use of well-characterized MAbs against specific LOS species.

Subsequent advances in LOS structure were aided by the use of pyocin, a bacteriocin that binds to gonococci that express certain LOS molecules. Biochemical characterization of the LOS structure of a set of pyocin-resistant mutants (118) confirmed the predicted epitopes of serotyping MAbs against LOS (7, 55, 150, 166) and facilitated the cloning of two genes involved in early steps of LOS biosynthesis (183, 208, 217, 287). A major breakthrough in understanding the variable nature of LOS was the work of Gotschlich (83), who identified a five-gene operon (*lgtA* to *lgtE*) important in synthesis of the α chain, the longest branch of the LOS molecule. Three of these genes (*lgtA*, *lgtC*, and *lgtD*) had poly(G) tracts, suggestive of phase variation, which was later verified experimentally (286). Identification of other glycosyltransferase genes followed, including the phase-variable gene, *lgtG*, which initiates the synthesis of the β chain (15) (Fig. 3). A related discovery to that of LOS phase variation was the identification of a sialic acid modification to the lacto-*N*-tetraose moiety of neisserial LOS (172, 180). This important modification is catalyzed by gonococcal α-2,3-sialyltransferase (Lst) (78, 151, 229). Unlike *N. meningitidis*, *N. gonorrhoeae* is unable to synthesize the substrate for this modification and must therefore use host CMP-*N*-acetylneuraminic acid (CMP-NANA). Therefore, sialylation is lost on routine subculture of primary gonococcal isolates (273).

Consequences of LOS Phase Variation

LOS phase variation generates diversity within a population; this may facilitate evasion of immune responses acquired during infection or cross-reactive antibodies in normal human serum that recognize gonococcal LOS. The penta-, tetra-, tri-, and disaccharide components of the α chain are present in human glycolipids and therefore may participate in molecular mimicry (118; reviewed in reference 101). The β chain, which is synthesized in ca. 80% of gonococcal strains, carries a bactericidal epitope that induces a strong immune response during natural infection (89). This epitope, which is considered a promising vaccine target (89, 90), is lost if *lgtG* is out of frame (15). Slippage of *lgtA* to an out-of-frame position produces a short-chain species of 3.6 kDa (83) that confers high-level resistance to the bactericidal activity of normal human serum (212). Expression of *lgtA* or *lgtC* increases serum-sensitivity (227). The addition of galactosamine (GalNAc) via the action of *lgtD* creates a highly serum sensitive species due to the creation of an epitope for bactericidal immunoglobulin M in normal human serum (86). If sialic acid is added instead, the serum sensitivity is dramatically reduced (172, 180, 262). Therefore, although sialyltransferase expression is constitutive (78), phase variation of *lgtA* or *lgtD* can control LOS sialylation. *lgtA* must be in frame to build, with the help of *lgtB*, the tetrasaccharide species to which sialic acid is added. An in-frame *lgtD* gene reduces sialylation by conferring the ability to add the terminal GalNAc, which blocks the addition of sialic acid to the terminal Gal of the lacto-*N*-neotetraose chain (83, 262, 285). The functional consequences of *lgt* phase variation are illustrated in Fig. 3.

The lacto-*N*-neotetraose species of the α chain, which confers an intermediate level of serum resistance (27), appears to be selected for early in infection of volunteers inoculated with gonococci that express primarily the 3.6-kDa LOS species (211). Oxygen tension (76) and other transcriptional factors (31) may also influence which LOS species predominates. The vast majority of gonococci are sialylated in urethral exudates (9). The consequence of LOS sialylation is impressive in terms of evasion of both innate and acquired defenses (reviewed in reference 233). Sialylated gonococci are more resistant to the bactericidal activity of serum; this CMP-NANA-dependent form of serum resistance, also called unstable serum resistance, differs from stable serum resistance, which is porin mediated. Strains that are serum sensitive in the absence of CMP-NANA express porin molecules that do not down-regulate the classical pathway via binding to C4 binding protein (193). Serum-sensitive strains are associated with mucosal infection but seldom with DGI, and LOS sialylation is proposed to be the main mechanism by which these strains evade complement-mediated killing in the genital tract (reviewed in reference 198). The mechanism of CMP-NANA-dependent serum resistance was identified by Ram et al. (195) as involving the binding of factor H (fH) to sialic acid on the LOS. fH is a regulatory protein of the alternative pathway (176, 177, 276). Binding of fH to the gonococcal surface down-regulates the alternative pathway, which causes increased surface deposition of C3bi and blocks the complement cascade at the level of C3 (176). Sialylated gonococci are more resistant to opsonophagocytic killing by human PMNs

(79, 197), since uptake via the C3b receptor of phagocytes is reduced if the majority of C3b on the bacterial surface is inactivated. Sialylated gonococci also exhibit reduced surface deposition of C4, a step that is needed in the activation of the classical pathway (156). Interestingly, LOS sialylation may also protect against PMN killing by causing a reduced oxidative respiratory burst (197). Finally, in addition to evasion of innate defenses, LOS sialylation may protect gonococci from antibodies induced by previous infection, as evidenced by increased resistance of sialylated gonococci to the bactericidal activity of porin-specific antibodies in vitro (64, 70, 279).

There also seems to be an advantage for gonococci to have nonsialylated LOS. In a classic paper that weighs the functional advantages conferred by LOS sialylation and phase variation of components of the LOS α chain, van Putten (262) reported a dramatic reduction in gonococcal invasion of epithelial cells when certain LOS phase variants were cultured in the presence of CMP-NANA; invasion of LOS phase variants that were unable to be sialylated was not affected by CMP-NANA. Apicella and colleagues identified the reason for this observation as being the binding of the lacto-*N*-neotetraose LOS moiety to the asialoglycoprotein receptor (ASGP-R) followed by receptor-mediated uptake, an event that is inhibited by sialic acid (99, 187). These investigators propose this interaction to be the main pathway by which gonococci invade urethral cells (99). The hypothesis that nonsialylated gonococci that express the lacto-*N*-tetraose LOS species have a functional advantage early in infection is weakly supported by a volunteer study in which sialylated gonococci appeared to be less infectious (215). This result was not statistically significant, however, possibly because of the small number of subjects tested. LOS-mediated binding of gonococci to the ASGP-R present on sperm cells suggests that the lacto-*N*-tetraose moiety may also play a role in transmission (100).

HEMOGLOBIN RECEPTOR

Structure and Mechanism of Phase Variation

Acquisition of iron for growth and as a cofactor of several key enzymes in the low-iron environment of the host is important for successful colonization by most microbes (277). *N. gonorrhoeae* acquires iron from extracellular stores by expressing receptors specific for human transferrin (Tf) and lactoferrin (Lf) (reviewed in reference 57), hemoglobin (Hg), and Hg-haptoglobin complexes (46, 47). A putative siderophore receptor, FetA, is also expressed by *N. gonorrhoeae* (44, 45), although its role in iron acquisition is only speculated on at this time. Transcription of each of these receptors is repressed by iron; expression of the Hg receptor and FetA is further controlled by phase variation based on reversible frameshift mutations (45).

Identification of the gonococcal Hg receptor was initially delayed by a lack of evidence that all strains of *N. gonorrhoeae* could utilize Hg as an iron source (67, 163). While working with a strain that appeared unable to utilize Hg, Chen et al. (46), investigated the appearance of rare colonies on agar containing Hg as a sole iron source and determined the reason to be the expression of a phase-variable Hg receptor. Like the Tf and Lf receptors, the gonococcal Hg receptor is composed of an outer membrane channel and an accessory lipoprotein (47). The corresponding genes (*hpuA* and *hpuB*) form an operon that is repressed on iron-supplemented media via regulation by a ferric uptake regulator (Fur) homologue (46, 143). A poly(G) region in the *hpuA* structural gene is the source of phase variation (Fig. 4A). Measured frequencies of Hg^- to Hg^+ transition range from 10^{-3} to 10^{-4}, depending on the strain (46).

Tools for Studying Hg Receptor Phase Variation

Monitoring the expression of the Hg receptor is easy compared to monitoring that of other phase-variable genes in *N. gonorrhoeae*. Phase variants that are "on" for the Hg receptor (Hg^+) can simply be isolated on media containing Hg as the sole iron source, and from this, the frequency of Hg^+ variants within a population can be calculated. Measurement of the Hg^+ to Hg^-, transition, however, is problematic due to the obstacle of selecting for a negative phenotype. The phenotype can be confirmed by Western blotting (47) or by determining the nucleotide sequence of the *hpuA* gene amplified from single colonies by PCR to identify genes that are in frame (115). Phase-locked mutants can be constructed by mutation of the poly(G) region.

Consequence of Phase Variation

The importance of the Tf and Lf receptors during urethral colonization is strongly supported by the results of volunteer studies using genetically defined mutants. From these studies, we know that Tf and Lf are critical sources of iron in the urethra and that Lf can substitute for loss of Tf utilization (3, 56). The importance of the gonococcal Hg receptor in urethral colonization has not been similarly tested in volunteers. However, no evidence of an increased frequency of Hg^+ variants was found among urethral

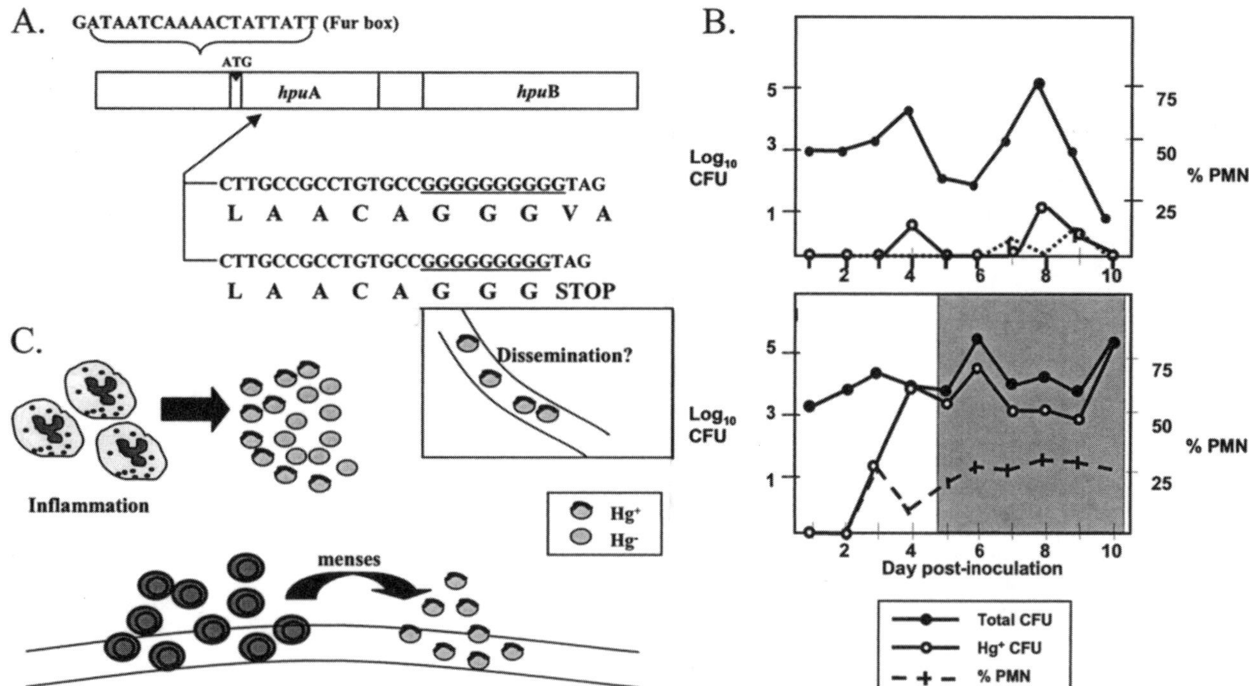

Figure 4. Genetic control of the gonococcal Hg receptor and selection of "on" phase variants under certain circumstances in vivo. (A) Diagram of the Hg receptor (HpuAB) operon as described by Chen et al. (46), showing the poly(G) region in *hpuA* that is responsible for strand slippage during replication. The upper transcript is in frame; the lower transcript has lost a guanidine in the poly(G) tract and is out of frame. A putative Fur box is present upstream of the start site, which is likely to play a role in iron repression of transcription. (B) Representative graphs showing total recovery of gonococci versus recovery of Hg^+ variants over time in a mouse with no influx of vaginal PMNs (top) versus a mouse that developed a PMN response during infection (bottom). The shaded area in the bottom graph corresponds to the period during which the frequency of Hg^+ variants among vaginal isolates was significantly elevated over that of the inoculum. Numbers of vaginal PMNs were elevated during this period. Hg was detected on day 5 in vaginal washes from the mouse with inflammation; none was detected at any time point in mice without inflammation. (C) Cartoon depicting the circumstances during which Hg^+ receptor expression may be advantageous. Selection for Hg^+ variants by menstrual blood is supported by an analysis of endocervical isolates from women (4); whether the presence of Hg as an additional iron source in the female genital tract leads to increased virulence is not known. Selection for Hg^+ variants on the development of a PMN influx in the lower genital tract of mice suggests that gonococci may capitalize on the introduction of Hg, which exudes into the lumen with other serum components, during inflammation (115). It is possible that Hg^+ variants are also selected for during the bloodstream stage of disseminated infection; however, no evidence for this hypothesis has been reported.

isolates from men, a result which suggests that Hg is not a critical iron source in this body site. In contrast, phase variation of the Hg receptor appears to play an adaptive role in infection of the female genital tract. Anderson et al. (4) detected significantly more Hg^+ isolates from women within 12 days of the onset of menses than from women at later time points in their cycle. These investigators therefore concluded that Hg^+ variants are selected for by menstrual blood (Fig. 4C). Based on the greater occurrence of PID (185, 253) and DGI (106) within a week after menses, it is tempting to speculate that the increased proliferation of variants capable of using Hg and Hg-haptoglobin complexes may promote ascension and dissemination. No growth advantage was detected for Hg^+ variants in media containing Hg and Tf. However, limiting concentrations of Tf were not tested, and the relative concentration of Hg and Tf during the menstrual cycle is not known (4).

Although female mice do not menstruate and therefore cannot be used to test the consequence of Hg use by *N. gonorrhoeae* during menses, mouse infection studies suggest an additional role for Hg receptor phase variation. An increased frequency of Hg^+ variants was detected among vaginal isolates from mice that developed a PMN response during infection. The increased frequency was temporally associated with PMN influx into the vagina and detection of Hg in vaginal washes (Fig. 4B). These results suggest that Hg or Hg-haptoglobin complexes exude onto mucosal surfaces during inflammation and select for Hg^+ variants. Unlike Lf and Tf, the use of Hg by *N. gonorrhoeae* is not host restricted (241). No obvious growth advantage accompanied the increase in the number of Hg^+ variants during infection, however (115). The lack of correlation between isolation of Hg^+ variants and an increased colonization load may reflect a balance between a growth advantage conferred by Hg and killing by components of the inflammatory response. A caveat to this conclusion is that such an advantage may be insignificant during natural infection, where Tf and Lf are plentiful.

FUTURE DIRECTIONS

Unraveling the molecular basis for changes in colony morphology was the first glimpse into the sophisticated biology of *N. gonorrhoeae*. Research by many investigators since that time collectively supports a role for phase and antigenic variation of gonococcal surface molecules in colonization. This area remains fertile ground for continued investigation. Definition of the genetic mechanisms (132) and host factors (225) involved in pilus phase and antigenic variation is an exciting area of study. The functional domains of PilE and the molecular basis for the dual functions of PilC are also unresolved (257). Studies of the covalent modifications of pilin may clarify the functional properties of pili and illuminate the stealth with which this organism evades the immune response (14, 73, 278). The use of transgenic mice may help test hypotheses about Opa-mediated tissue tropism and the consequence of signaling through these receptors; mouse models may also help elucidate the selection of Opa phenotypes by factors of the innate response. Although a better understanding of how LOS phase variation and sialylation protect gonococci from humoral defenses frustrates the hope of active immunization against gonorrhea, strategies to subvert these evasion mechanisms may follow (90). Finally, the genomic era has revealed numerous other gonococcal genes that may undergo phase (234) or structural (124) variation. Tapping this new source of information should yield more fascinating insights into how this finely tuned pathogen maintains itself within a host and within communities.

REFERENCES

1. **Achtman, M., M. Neibert, B. A. Crowe, W. Strittmatter, B. Kusecek, E. Weyse, M. J. Walsh, B. Slawig, G. Morelli, A. Moll, et al.** 1988. Purification and characterization of eight class 5 outer membrane protein variants from a clone of *Neisseria meningitidis* serogroup A. *J. Exp. Med.* **168:**507–525.
2. **Aho, E. L., J. A. Dempsey, M. M. Hobbs, D. G. Klapper, and J. G. Cannon.** 1991. Characterization of the opa (class 5) gene family of *Neisseria meningitidis. Mol. Microbiol.* **5:**1429–1437.
3. **Anderson, J. E., M. M. Hobbs, G. D. Biswas, and P. F. Sparling.** 2003. Opposing selective forces for expression of the gonococcal lactoferrin receptor. *Mol. Microbiol.* **48:**1325–1337.
4. **Anderson, J. E., P. A. Leone, W. C. Miller, C. Chen, M. M. Hobbs, and P. F. Sparling.** 2001. Selection for expression of the gonococcal hemoglobin receptor during menses. *J. Infect. Dis.* **184:**1621–1623.
5. **Apicella, M. A.** 1974. Antigenically distinct populations of *Neisseria gonorrhoeae:* isolation and characterization of the responsible determinants. *J. Infect. Dis.* **130:**619–625.
6. **Apicella, M. A.** 1976. Serogrouping of *Neisseria gonorrhoeae:* identification of four immunologically distinct acidic polysaccharides. *J. Infect. Dis.* **134:**377–383.
7. **Apicella, M. A., K. M. Bennett, C. A. Hermerath, and D. E. Roberts.** 1981. Monoclonal antibody analysis of lipopolysaccharide from *Neisseria gonorrhoeae* and *Neisseria meningitidis. Infect. Immun.* **34:**751–756.
8. **Apicella, M. A., and N. C. Gagliardi.** 1979. Antigenic heterogeneity of the non-serogroup antigen structure of *Neisseria gonorrhoeae* lipopolysaccharides. *Infect. Immun.* **26:**870–874.
9. **Apicella, M. A., R. E. Mandrell, M. Shero, M. E. Wilson, J. M. Griffiss, G. F. Brooks, C. Lammel, J. F. Breen, and P. A. Rice.** 1990. Modification by sialic acid of *Neisseria gonorrhoeae* lipooligosaccharide epitope expression in human urethral exudates: an immunoelectron microscopic analysis. *J. Infect. Dis.* **162:**506–512.
10. **Apicella, M. A., M. Shero, G. A. Jarvis, J. M. Griffiss, R. E. Mandrell, and H. Schneider.** 1987. Phenotypic variation in epitope expression of the *Neisseria gonorrhoeae* lipooligosaccharide. *Infect. Immun.* **55:**1755–1761.
11. **Archibald, F. S., and M. N. Duong.** 1986. Superoxide dismutase and oxygen toxicity defenses in the genus *Neisseria. Infect. Immun.* **51:**631–641.
12. **Ayala, B. P., B. Vasquez, S. Clary, J. A. Tainer, K. Rodland, and M. So.** 2001. The pilus-induced $Ca2^+$ flux triggers lysosome exocytosis and increases the amount of Lamp1 accessible to *Neisseria* IgA1 protease. *Cell. Microbiol.* **3:**265–275.
13. **Banerjee, A., and S. K. Ghosh.** 2003. The role of pilin glycan in neisserial pathogenesis. *Mol. Cell. Biochem.* **253:**179–190.
14. **Banerjee, A., R. Wang, S. L. Supernavage, S. K. Ghosh, J. Parker, N. F. Ganesh, P. G. Wang, S. Gulati, and P. A. Rice.** 2002. Implications of phase variation of a gene (*pgtA*) encoding a pilin galactosyl transferase in gonococcal pathogenesis. *J. Exp. Med.* **196:**147–162.
15. **Banerjee, A., R. Wang, S. N. Uljon, P. A. Rice, E. C. Gotschlich, and D. C. Stein.** 1998. Identification of the gene (*lgtG*) encoding the lipooligosaccharide beta chain synthesizing glucosyl transferase from *Neisseria gonorrhoeae. Proc. Natl. Acad. Sci. USA* **95:**10872–10877.
16. **Barritt, D. S., R. S. Schwalbe, D. G. Klapper, and J. G. Cannon.** 1987. Antigenic and structural differences among six proteins II expressed by a single strain of *Neisseria gonorrhoeae. Infect. Immun.* **55:**2026–2031.
17. **Belland, R. J., T. Chen, J. Swanson, and S. H. Fischer.** 1992. Human neutrophil response to recombinant neisserial Opa proteins. *Mol. Microbiol.* **6:**1729–1737.
18. **Belland, R. J., S. G. Morrison, J. H. Carlson, and D. M. Hogan.** 1997. Promoter strength influences phase variation of neisserial *opa* genes. *Mol. Microbiol.* **23:**123–135.
19. **Belland, R. J., S. G. Morrison, P. van der Ley, and J. Swanson.** 1989. Expression and phase variation of gonococcal P.II genes in *Escherichia coli* involves ribosomal frameshifting and slipped-strand mispairing. *Mol. Microbiol.* **3:**777–786.

20. **Bergstrom, S., K. Robbins, J. M. Koomey, and J. Swanson.** 1986. Piliation control mechanisms in *Neisseria gonorrhoeae. Proc. Natl. Acad. Sci. USA* **83:**3890–3894.
21. **Bessen, D., and E. C. Gotschlich.** 1986. Interactions of gonococci with HeLa cells: attachment, detachment, replication, penetration, and the role of protein II. *Infect. Immun.* **54:** 154–160.
22. **Bhat, K. S., C. P. Gibbs, O. Barrera, S. G. Morrison, F. Jahnig, A. Stern, E. M. Kupsch, T. F. Meyer, and J. Swanson.** 1992. The opacity proteins of *Neisseria gonorrhoeae* strain MS11 are encoded by a family of 11 complete genes. *Mol. Microbiol.* **6:**1073–1076.
23. **Biro, F. M., S. L. Rosenthal, and M. Kiniyalocts.** 1995. Gonococcal and chlamydial genitourinary infections in symptomatic and asymptomatic adolescent women. *Clin. Pediatr.* **34:**419–423.
24. **Black, W. J., R. S. Schwalbe, I. Nachamkin, and J. G. Cannon.** 1984. Characterization of *Neisseria gonorrhoeae* protein II phase variation by use of monoclonal antibodies. *Infect. Immun.* **45:**453–457.
25. **Blake, M. S., C. M. Blake, M. A. Apicella, and R. E. Mandrell.** 1995. Gonococcal opacity: lectin-like interactions between Opa proteins and lipooligosaccharide. *Infect. Immun.* **63:**1434–1439.
26. **Bos, M. P., F. Grunert, and R. J. Belland.** 1997. Differential recognition of members of the carcinoembryonic antigen family by Opa variants of *Neisseria gonorrhoeae. Infect. Immun.* **65:**2353–2361.
27. **Bos, M. P., D. Hogan, and R. J. Belland.** 1997. Selection of Opa$^+$ *Neisseria gonorrhoeae* by limited availability of normal human serum. *Infect. Immun.* **65:**645–650.
28. **Bos, M. P., D. Kao, D. M. Hogan, C. C. Grant, and R. J. Belland.** 2002. Carcinoembryonic antigen family receptor recognition by gonococcal Opa proteins requires distinct combinations of hypervariable Opa protein domains. *Infect. Immun.* **70:**1715–1723.
29. **Boslego, J. W., E. C. Tramont, R. C. Chung, D. G. McChesney, J. Ciak, J. C. Sadoff, M. V. Piziak, J. D. Brown, C. C. Brinton, Jr., S. W. Wood, et al.** 1991. Efficacy trial of a parenteral gonococcal pilus vaccine in men. *Vaccine* **9:**154–162.
30. **Boulton, I. C., and S. D. Gray-Owen.** 2002. Neisserial binding to CEACAM1 arrests the activation and proliferation of CD4$^+$ T lymphocytes. *Nat. Immunol.* **3:**229–236.
31. **Braun, D. C., and D. C. Stein.** 2004. The *lgtABCDE* gene cluster, involved in lipooligosaccharide biosynthesis in *Neisseria gonorrhoeae,* contains multiple promoter sequences. *J. Bacteriol.* **186:**1038–1049.
32. **Brinton, C. C., J. Bryan, J.-A. Dillon, N. Guernia, L. J. Jackobson, A. Labik, S. Lee, A. Levine, S. Lim, J. McMichael, S. A. Polen, K. Rogers, A. C.-C. To, and S. C.-M. To.** 1978. Use of pili in gonorrhoea control: role of bacterial pilin in disease, purification, and properties of gonococcal pili and progress in the development of a gonococcal pilus vaccine for gonorrhoea, p. 155–178. *In* G. F. Brooks, E. C. Gotschlich, K. K. Holmes, W. D. Sawyer, and F. E. Young (ed.), *Immunobiology of* Neisseria gonorrhoeae. American Society for Microbiology, Washington, D.C.
33. **Brinton, C. C., S. W. Wood, and A. Brown.** 1982. The development of a neisserial pilus vaccine for gonorrhea adn meningococcal meningitis. *Semin. Infect. Dis.* **1982:**140–159.
34. **Bridgman, B. E., D. Klapper, T. Svendsen, and M. S. Cohen.** 1988. Phagocyte-derived lactate stimulates oxygen consumption by *Neisseria gonorrhoeae.* An unrecognized aspect of the oxygen metabolism of phagocytosis. *J. Clin. Investig.* **81:** 318–324.
35. **Bro-Jorgensen, A., and T. Jensen.** 1973. Gonococcal pharyngeal infections. Report of 110 cases. *Br. J. Vener. Dis.* **49:** 491–499.
36. **Brooks, G. F., and C. J. Lammel.** 1989. Humoral immune response to gonococcal infections. *Clin. Microbiol. Rev.* **2**(Suppl.):S5–S10.
37. **Brooks, G. F., L. Linger, C. J. Lammel, K. S. Bhat, C. A. Colville, M. L. Palmer, J. S. Knapp, and R. S. Stephens.** 1991. Prevalence of gene sequences coding for hypervariable regions of Opa (protein II) in *Neisseria gonorrhoeae. Mol. Microbiol.* **5:**3063–3072.
38. **Brossay, L., G. Paradis, R. Fox, M. Koomey, and J. Hebert.** 1994. Identification, localization, and distribution of the PilT protein in *Neisseria gonorrhoeae. Infect. Immun.* **62:**2302–2308.
39. **Brown, W. J., and S. J. Kraus.** 1974. Gonococcal colony types. *JAMA* **228:**862.
40. **Buchanan, T. M.** 1975. Antigenic heterogeneity of gonococcal pili. *J. Exp. Med.* **141:**1470–1475.
41. **Buchanan, T. M., and R. J. Arko.** 1977. Immunity to gonococcal infection induced by vaccination with isolated outer membranes of *Neisseria gonorrhoeae* in guinea pigs. *J. Infect. Dis.* **135:**879–887.
42. **Buchanan, T. M., W. A. Pearce, G. K. Schoolnik, and R. J. Arko.** 1977. Protection against infection with *Neisseria gonorrhoeae* by immunization with outer membrane protein complex and purified pili. *J. Infect. Dis.* **136**(Suppl.):S132–S137.
43. **Burch, C. L., R. J. Danaher, and D. C. Stein.** 1997. Antigenic variation in *Neisseria gonorrhoeae:* production of multiple lipooligosaccharides. *J. Bacteriol.* **179:**982–986.
44. **Carson, S. D., P. E. Klebba, S. M. Newton, and P. F. Sparling.** 1999. Ferric enterobactin binding and utilization by *Neisseria gonorrhoeae. J. Bacteriol.* **181:**2895–2901.
45. **Carson, S. D., B. Stone, M. Beucher, J. Fu, and P. F. Sparling.** 2000. Phase variation of the gonococcal siderophore receptor FetA. *Mol. Microbiol.* **36:**585–593.
46. **Chen, C. J., C. Elkins, and P. F. Sparling.** 1998. Phase variation of hemoglobin utilization in *Neisseria gonorrhoeae. Infect. Immun.* **66:**987–993.
47. **Chen, C. J., P. F. Sparling, L. A. Lewis, D. W. Dyer, and C. Elkins.** 1996. Identification and purification of a hemoglobin-binding outer membrane protein from *Neisseria gonorrhoeae. Infect. Immun.* **64:**5008–5014.
48. **Chen, T., R. J. Belland, J. Wilson, and J. Swanson.** 1995. Adherence of pilus$^-$ Opa$^+$ gonococci to epithelial cells in vitro involves heparan sulfate. *J. Exp. Med.* **182:**511–517.
49. **Chen, T., and E. C. Gotschlich.** 1996. CGM1a antigen of neutrophils, a receptor of gonococcal opacity proteins. *Proc. Natl. Acad. Sci. USA* **93:**14851–14856.
50. **Chen, T., F. Grunert, A. Medina-Marino, and E. C. Gotschlich.** 1997. Several carcinoembryonic antigens (CD66) serve as receptors for gonococcal opacity proteins. *J. Exp. Med.* **185:**1557–1564.
51. **Cohen, M. S., and J. G. Cannon.** 1999. Human experimentation with *Neisseria gonorrhoeae:* progress and goals. *J. Infect. Dis.* **179**(Suppl. 2):S375–S379.
52. **Cohen, M. S., J. G. Cannon, A. E. Jerse, L. M. Charniga, S. F. Isbey, and L. G. Whicker.** 1994. Human experimentation with *Neisseria gonorrhoeae:* rationale, methods, and implications for the biology of infection and vaccine development. *J. Infect. Dis.* **169:**532–537.
53. **Cohen, M. S., and P. F. Sparling.** 1992. Mucosal infection with *Neisseria gonorrhoeae.* Bacterial adaptation and mucosal defenses. *J. Clin. Investig.* **89:**1699–1705.
54. **Connell, T. D., D. Shaffer, and J. G. Cannon.** 1990. Characterization of the repertoire of hypervariable regions in the

Protein II (opa) gene family of *Neisseria gonorrhoeae. Mol. Microbiol.* **4:**439–449.

55. **Connelly, M. C., and P. Z. Allen.** 1983. Antigenic specificity and heterogeneity of lipopolysaccharides from pyocin-sensitive and -resistant strains of *Neisseria gonorrhoeae. Infect. Immun.* **41:**1046–1055.
56. **Cornelissen, C. N., M. Kelley, M. M. Hobbs, J. E. Anderson, J. G. Cannon, M. S. Cohen, and P. F. Sparling.** 1998. The transferrin receptor expressed by gonococcal strain FA1090 is required for the experimental infection of human male volunteers. *Mol. Microbiol.* **27:**611–616.
57. **Cornelissen, C. N., and P. F. Sparling.** 1994. Iron piracy: acquisition of transferrin-bound iron by bacterial pathogens. *Mol. Microbiol.* **14:**843–850.
58. **Danaher, R. J., J. C. Levin, D. Arking, C. L. Burch, R. Sandlin, and D. C. Stein.** 1995. Genetic basis of *Neisseria gonorrhoeae* lipooligosaccharide antigenic variation. *J. Bacteriol.* **177:**7275–7279.
59. **Deheragoda, P.** 1977. Diagnosis of rectal gonorrhoea by blind anorectal swabs compared with direct vision swabs taken via a proctoscope. *Br. J. Vener. Dis.* **53:**311–313.
60. **Dehio, C., S. D. Gray-Owen, and T. F. Meyer.** 1998. The role of neisserial Opa proteins in interactions with host cells. *Trends Microbiol.* **6:**489–495.
61. **de Jonge, M. I., M. P. Bos, H. J. Hamstra, W. Jiskoot, P. van Ulsen, J. Tommassen, L. van Alphen, and P. van der Ley.** 2002. Conformational analysis of opacity proteins from *Neisseria meningitidis. Eur. J. Biochem.* **269:**5215–5223.
62. **de Jonge, M. I., H. J. Hamstra, L. van Alphen, J. Dankert, and P. van der Ley.** 2003. Mapping the binding domains on meningococcal Opa proteins for CEACAM1 and CEA receptors. *Mol. Microbiol.* **50:**1005–1015.
63. **de Jonge, M. I., G. Vidarsson, H. H. van Dijken, P. Hoogerhout, L. van Alphen, J. Dankert, and P. van der Ley.** 2003. Functional activity of antibodies against the recombinant OpaJ protein from *Neisseria meningitidis. Infect. Immun.* **71:**2331–2340.
64. **de la Paz, H., S. J. Cooke, and J. E. Heckels.** 1995. Effect of sialylation of lipopolysaccharide of *Neisseria gonorrhoeae* on recognition and complement-mediated killing by monoclonal antibodies directed against different outer-membrane antigens. *Microbiology* **141:**913–920.
65. **Draper, D. L., J. F. James, G. F. Brooks, and R. L. Sweet.** 1980. Comparison of virulence markers of peritoneal and fallopian tube isolates with endocervical *Neisseria gonorrhoeae* isolates from women with acute salpingitis. *Infect. Immun.* **27:**882–888.
66. **Duensing, T. D., and J. P. van Putten.** 1997. Vitronectin mediates internalization of *Neisseria gonorrhoeae* by Chinese hamster ovary cells. *Infect. Immun.* **65:**964–970.
67. **Dyer, D. W., E. P. West, and P. F. Sparling.** 1987. Effects of serum carrier proteins on the growth of pathogenic neisseriae with heme-bound iron. *Infect. Immun.* **55:**2171–2175.
68. **Edwards, J. L., E. J. Brown, K. A. Ault, and M. A. Apicella.** 2001. The role of complement receptor 3 (CR3) in *Neisseria gonorrhoeae* infection of human cervical epithelia. *Cell. Microbiol.* **3:**611–622.
69. **Edwards, M., R. L. McDade, G. Schoolnik, J. B. Rothbard, and E. C. Gotschlich.** 1984. Antigenic analysis of gonococcal pili using monoclonal antibodies. *J. Exp. Med.* **160:**1782–1791.
70. **Elkins, C., N. H. Carbonetti, V. A. Varela, D. Stirewalt, D. G. Klapper, and P. F. Sparling.** 1992. Antibodies to N-terminal peptides of gonococcal porin are bactericidal when gonococcal lipopolysaccharide is not sialylated. *Mol. Microbiol.* **6:**2617–2628.
71. **Elkins, C., and R. F. Rest.** 1990. Monoclonal antibodies to outer membrane protein P.II block interactions of *Neisseria gonorrhoeae* with human neutrophils. *Infect. Immun.* **58:**1078–1084.
72. **Fischer, S. H., and R. F. Rest.** 1988. Gonococci possessing only certain P.II outer membrane proteins interact with human neutrophils. *Infect. Immun.* **56:**1574–1579.
73. **Forest, K. T., S. A. Dunham, M. Koomey, and J. A. Tainer.** 1999. Crystallographic structure reveals phosphorylated pilin from *Neisseria:* phosphoserine sites modify type IV pilus surface chemistry and fibre morphology. *Mol. Microbiol.* **31:**743–752.
74. **Forest, K. T., and J. A. Tainer.** 1997. Type-4 pilus-structure: outside to inside and top to bottom—a minireview. *Gene* **192:**165–169.
75. **Francioli, P., H. Shio, R. B. Roberts, and M. Muller.** 1983. Phagocytosis and killing of *Neisseria gonorrhoeae* by *Trichomonas vaginalis. J. Infect. Dis.* **147:**87–94.
76. **Frangipane, J. V., and R. F. Rest.** 1993. Anaerobic growth and cytidine 5′-monophospho-*N*-acetylneuraminic acid act synergistically to induce high-level serum resistance in *Neisseria gonorrhoeae. Infect. Immun.* **61:**1657–1666.
77. **Fussenegger, M., T. Rudel, R. Barten, R. Ryll, and T. F. Meyer.** 1997. Transformation competence and type-4 pilus biogenesis in Neisseria gonorrhoeae—a review. *Gene* **192:**125–134.
78. **Gilbert, M., D. C. Watson, A. M. Cunningham, M. P. Jennings, N. M. Young, and W. W. Wakarchuk.** 1996. Cloning of the lipooligosaccharide alpha-2,3-sialyltransferase from the bacterial pathogens *Neisseria meningitidis* and *Neisseria gonorrhoeae. J. Biol. Chem.* **271:**28271–28276.
79. **Gill, M. J., D. P. McQuillen, J. P. van Putten, L. M. Wetzler, J. Bramley, H. Crooke, N. J. Parsons, J. A. Cole, and H. Smith.** 1996. Functional characterization of a sialyltransferase-deficient mutant of *Neisseria gonorrhoeae. Infect. Immun.* **64:**3374–3378.
80. **Gomez-Duarte, O. G., M. Dehio, C. A. Guzman, G. S. Chhatwal, C. Dehio, and T. F. Meyer.** 1997. Binding of vitronectin to *opa*-expressing *Neisseria gonorrhoeae* mediates invasion of HeLa cells. *Infect. Immun.* **65:**3857–3866.
81. **Gorby, G. L., A. F. Ehrhardt, M. A. Apicella, and C. Elkins.** 2001. Invasion of human fallopian tube epithelium by *Escherichia coli* expressing combinations of a gonococcal porin, opacity-associated protein, and chimeric lipo-oligosaccharide. *J. Infect. Dis.* **184:**460–472.
82. **Gorby, G. L., and G. B. Schaefer.** 1992. Effect of attachment factors (pili plus Opa) on *Neisseria gonorrhoeae* invasion of human fallopian tube tissue in vitro: quantitation by computerized image analysis. *Microb. Pathog.* **13:**93–108.
83. **Gotschlich, E. C.** 1994. Genetic locus for the biosynthesis of the variable portion of *Neisseria gonorrhoeae* lipooligosaccharide. *J. Exp. Med.* **180:**2181–2190.
84. **Grant, C. C., M. P. Bos, and R. J. Belland.** 1999. Proteoglycan receptor binding by *Neisseria gonorrhoeae* MS11 is determined by the HV-1 region of OpaA. *Mol. Microbiol.* **32:**233–242.
85. **Gray-Owen, S. D., C. Dehio, A. Haude, F. Grunert, and T. F. Meyer.** 1997. CD66 carcinoembryonic antigens mediate interactions between Opa-expressing *Neisseria gonorrhoeae* and human polymorphonuclear phagocytes. *EMBO J.* **16:**3435–3445.
86. **Griffiss, J. M., G. A. Jarvis, J. P. O'Brien, M. M. Eads, and H. Schneider.** 1991. Lysis of *Neisseria gonorrhoeae* initiated by binding of normal human IgM to a hexosamine-containing lipooligosaccharide epitope(s) is augmented by strain-specific, properdin-binding-dependent alternative complement pathway activation. *J. Immunol.* **147:**298–305.

87. **Griffiss, J. M., H. Schneider, R. E. Mandrell, R. Yamasaki, G. A. Jarvis, J. J. Kim, B. W. Gibson, R. Hamadeh, and M. A. Apicella.** 1988. Lipooligosaccharides: the principal glycolipids of the neisserial outer membrane. *Rev. Infect. Dis.* **10**(Suppl. 2):S287–S295.

88. **Gubish, E. R., Jr., K. C. Chen, and T. M. Buchanan.** 1982. Attachment of gonococcal pili to lectin-resistant clones of Chinese hamster ovary cells. *Infect. Immun.* **37:**189–194.

89. **Gulati, S., D. P. McQuillen, R. E. Mandrell, D. B. Jani, and P. A. Rice.** 1996. Immunogenicity of *Neisseria gonorrhoeae* lipooligosaccharide epitope 2C7, widely expressed in vivo with no immunochemical similarity to human glycosphingolipids. *J. Infect. Dis.* **174:**1223–1237.

90. **Gulati, S., D. P. McQuillen, J. Sharon, and P. A. Rice.** 1996. Experimental immunization with a monoclonal anti-idiotope antibody that mimics the *Neisseria gonorrhoeae* lipooligosaccharide epitope 2C7. *J. Infect. Dis.* **174:**1238–1248.

91. **Haas, R., and T. F. Meyer.** 1986. The repertoire of silent pilus genes in *Neisseria gonorrhoeae:* evidence for gene conversion. *Cell* **44:**107–115.

92. **Haas, R., H. Schwarz, and T. F. Meyer.** 1987. Release of soluble pilin antigen coupled with gene conversion in *Neisseria gonorrhoeae. Proc. Natl. Acad. Sci. USA* **84:**9079–9083.

93. **Haas, R., S. Veit, and T. F. Meyer.** 1992. Silent pilin genes of *Neisseria gonorrhoeae* MS11 and the occurrence of related hypervariant sequences among other gonococcal isolates. *Mol. Microbiol.* **6:**197–208.

94. **Hagblom, P., E. Segal, E. Billyard, and M. So.** 1985. Intragenic recombination leads to pilus antigenic variation in *Neisseria gonorrhoeae. Nature* **315:**156–158.

95. **Hagman, K. E., W. Pan, B. G. Spratt, J. T. Balthazar, R. C. Judd, and W. M. Shafer.** 1995. Resistance of *Neisseria gonorrhoeae* to antimicrobial hydrophobic agents is modulated by the *mtrRCDE* efflux system. *Microbiology* **141:**611–622.

96. **Hallet, B.** 2001. Playing Dr Jekyll, and Mr Hyde: combined mechanisms of phase variation in bacteria. *Curr. Opin. Microbiol.* **4:**570–581.

97. **Hamrick, T. S., J. A. Dempsey, M. S. Cohen, and J. G. Cannon.** 2001. Antigenic variation of gonococcal pilin expression in vivo: analysis of the strain FA1090 pilin repertoire and identification of the *pilS* gene copies recombining with *pilE* during experimental human infection. *Microbiology* **147:**839–849.

98. **Han, E., D. Phan, P. Lo, M. N. Poy, R. Behringer, S. M. Najjar, and S. H. Lin.** 2001. Differences in tissue-specific and embryonic expression of mouse Ceacam1 and Ceacam2 genes. *Biochem. J.* **355:**417–423.

99. **Harvey, H. A., M. P. Jennings, C. A. Campbell, R. Williams, and M. A. Apicella.** 2001. Receptor-mediated endocytosis of *Neisseria gonorrhoeae* into primary human urethral epithelial cells: the role of the asialoglycoprotein receptor. *Mol. Microbiol.* **42:**659–672.

100. **Harvey, H. A., N. Porat, C. A. Campbell, M. Jennings, B. W. Gibson, N. J. Phillips, M. A. Apicella, and M. S. Blake.** 2000. Gonococcal lipooligosaccharide is a ligand for the asialoglycoprotein receptor on human sperm. *Mol. Microbiol.* **36:** 1059–1070.

101. **Harvey, H. A., W. E. Swords, and M. A. Apicella.** 2001. The mimicry of human glycolipids and glycosphingolipids by the lipooligosaccharides of pathogenic *Neisseria* and *Haemophilus. J. Autoimmun.* **16:**257–262.

102. **Hassett, D. J., and M. S. Cohen.** 1989. Bacterial adaptation to oxidative stress: implications for pathogenesis and interaction with phagocytic cells. *FASEB J.* **3:**2574–2582.

103. **Hauck, C. R., and T. F. Meyer.** 2003. 'Small' talk: Opa proteins as mediators of *Neisseria*–host-cell communication. *Curr. Opin. Microbiol.* **6:**43–49.

104. **Hitchcock, P. J., and T. M. Brown.** 1983. Morphological heterogeneity among *Salmonella* lipopolysaccharide chemotypes in silver-stained polyacrylamide gels. *J. Bacteriol.* **154:**269–277.

105. **Hobbs, M. M., A. Seiler, M. Achtman, and J. G. Cannon.** 1994. Microevolution within a clonal population of pathogenic bacteria: recombination, gene duplication and horizontal genetic exchange in the opa gene family of *Neisseria meningitidis. Mol. Microbiol.* **12:**171–180.

106. **Holmes, K. K., G. W. Counts, and H. N. Beaty.** 1971. Disseminated gonococcal infection. *Ann. Intern. Med.* **74:**979–993.

107. **Hook, E., and H. Handsfield.** 1999. Gonococcal infections in the adult, p. 451–466. *In* (ed.), *Sexually Transmitted Diseases,* 3rd ed. McGraw-Hill Co. Inc., New York, N.Y.

108. **Householder, T. C., W. A. Belli, S. Lissenden, J. A. Cole, and V. L. Clark.** 1999. *cis-* and *trans*-acting elements involved in regulation of *aniA,* the gene encoding the major anaerobically induced outer membrane protein in *Neisseria gonorrhoeae. J. Bacteriol.* **181:**541–551.

109. **Ilver, D., H. Kallstrom, S. Normark, and A. B. Jonsson.** 1998. Transcellular passage of *Neisseria gonorrhoeae* involves pilus phase variation. *Infect. Immun.* **66:**469–473.

110. **James, J. F., and J. Swanson.** 1978. Color/opacity colonial variants of *Neisseria gonorrhoeae* and their relationship to the menstrual cycle, p. 338–343. *In* G. F. Brooks, E. C. Gotschlich, K. K. Holmes, W. D. Sawyer and F. E. Young (ed.), *Immunobiology of* Neisseria gonorrhoeae. American Society for Microbiology, Washington, D.C.

111. **James, J. F., and J. Swanson.** 1978. Studies on gonococcus infection. XIII. Occurrence of color/opacity colonial variants in clinical cultures. *Infect. Immun.* **19:**332–340.

112. **Jephcott, A. E., A. Reyn, and A. Birch-Andersen.** 1971. *Neisseria gonorrhoeae.* 3. Demonstration of presumed appendages to cells from different colony types. *Acta Pathol. Microbiol. Scand. Ser. B* **79:**437–439.

113. **Jerse, A. E.** 1999. Experimental gonococcal genital tract infection and opacity protein expression in estradiol-treated mice. *Infect. Immun.* **67:**5699–5708.

114. **Jerse, A. E., M. S. Cohen, P. M. Drown, L. G. Whicker, S. F. Isbey, H. S. Seifert, and J. G. Cannon.** 1994. Multiple gonococcal opacity proteins are expressed during experimental urethral infection in the male. *J. Exp. Med.* **179:**911–920.

115. **Jerse, A. E., E. T. Crow, A. N. Bordner, I. Rahman, C. N. Cornelissen, T. R. Moench, and K. Mehrazar.** 2002. Growth of *Neisseria gonorrhoeae* in the female mouse genital tract does not require the gonococcal transferrin or hemoglobin receptors and may be enhanced by commensal lactobacilli. *Infect. Immun.* **70:**2549–2558.

116. **Jerse, A. E., N. D. Sharma, A. N. Simms, E. T. Crow, L. A. Snyder, and W. M. Shafer.** 2003. A gonococcal efflux pump system enhances bacterial survival in a female mouse model of genital tract infection. *Infect. Immun.* **71:**5576– 5582.

117. **Johansson, L., A. Rytkonen, P. Bergman, B. Albiger, H. Kallstrom, T. Hokfelt, B. Agerberth, R. Cattaneo, and A. B. Jonsson.** 2003. CD46 in meningococcal disease. *Science* **301:** 373–375.

118. **John, C. M., J. M. Griffiss, M. A. Apicella, R. E. Mandrell, and B. W. Gibson.** 1991. The structural basis for pyocin resistance in *Neisseria gonorrhoeae* lipooligosaccharides. *J. Biol. Chem.* **266:**19303–19311.

119. **Johnson, S. R., B. M. Steiner, D. D. Cruce, G. H. Perkins, and R. J. Arko.** 1993. Characterization of a catalase-deficient strain of *Neisseria gonorrhoeae:* evidence for the significance of catalase in the biology of *N. gonorrhoeae. Infect. Immun.* **61:**1232–1238.

120. **Jones, R. B., J. C. Newland, D. A. Olsen, and T. M. Buchanan.** 1980. Immune-enhanced phagocytosis of *Neisseria gonorrhoeae* by macrophages: characterization of the major antigens to which opsonins are directed. *J. Gen. Microbiol.* **121:**365–372.
121. **Jonsson, A. B., D. Ilver, P. Falk, J. Pepose, and S. Normark.** 1994. Sequence changes in the pilus subunit lead to tropism variation of *Neisseria gonorrhoeae* to human tissue. *Mol. Microbiol.* **13:**403–416.
122. **Jonsson, A. B., G. Nyberg, and S. Normark.** 1991. Phase variation of gonococcal pili by frameshift mutation in *pilC*, a novel gene for pilus assembly. *EMBO J.* **10:**477–488.
123. **Jonsson, A. B., J. Pfeifer, and S. Normark.** 1992. *Neisseria gonorrhoeae* PilC expression provides a selective mechanism for structural diversity of pili. *Proc. Natl. Acad. Sci. USA* **89:**3204–3208.
124. **Jordan, P., L. A. Snyder, and N. J. Saunders.** 2003. Diversity in coding tandem repeats in related *Neisseria* spp. *BMC Microbiol.* **3:**23.
125. **Kallstrom, H., D. Blackmer Gill, B. Albiger, M. K. Liszewski, J. P. Atkinson, and A. B. Jonsson.** 2001. Attachment of *Neisseria gonorrhoeae* to the cellular pilus receptor CD46:identification of domains important for bacterial adherence. *Cell. Microbiol.* **3:**133–143.
126. **Kallstrom, H., M. S. Islam, P. O. Berggren, and A. B. Jonsson.** 1998. Cell signaling by the type IV pili of pathogenic *Neisseria. J. Biol. Chem.* **273:**21777–21782.
127. **Kallstrom, H., M. K. Liszewski, J. P. Atkinson, and A. B. Jonsson.** 1997. Membrane cofactor protein (MCP or CD46) is a cellular pilus receptor for pathogenic *Neisseria. Mol. Microbiol.* **25:**639–647.
128. **Kellogg, D. S., Jr., I. R. Cohen, L. C. Norins, A. L. Schroeter, and G. Reising.** 1968. *Neisseria gonorrhoeae.* II. Colonial variation and pathogenicity during 35 months in vitro. *J. Bacteriol.* **96:**596–605.
129. **Kellogg, D. S., Jr., W. L. Peacock, Jr., W. E. Deacon, L. Brown, and D. I. Pirkle.** 1963. *Neisseria gonorrhoeae.* I. Virulence genetically linked to clonal variation. *J. Bacteriol.* **85:**1274–1279.
130. **Kim, J. J., D. Zhou, R. E. Mandrell, and J. M. Griffiss.** 1992. Effect of exogenous sialylation of the lipooligosaccharide of *Neisseria gonorrhoeae* on opsonophagocytosis. *Infect. Immun.* **60:**4439–4442.
131. **Klimpel, K. W., S. A. Lesley, and V. L. Clark.** 1989. Identification of subunits of gonococcal RNA polymerase by immunoblot analysis: evidence for multiple sigma factors. *J. Bacteriol.* **171:**3713–3718.
132. **Kline, K. A., E. V. Sechman, E. P. Skaar, and H. S. Seifert.** 2003. Recombination, repair and replication in the pathogenic neisseriae: the 3 R's of molecular genetics of two human-specific bacterial pathogens. *Mol. Microbiol.* **50:**3–13.
133. **Knepper, B., I. Heuer, T. F. Meyer, and J. P. van Putten.** 1997. Differential response of human monocytes to *Neisseria gonorrhoeae* variants expressing pili and opacity proteins. *Infect. Immun.* **65:**4122–4129.
134. **Kolberg, J., E. A. Hoiby, and E. Jantzen.** 1997. Detection of the phosphorylcholine epitope in streptococci, *Haemophilus* and pathogenic neisseriae by immunoblotting. *Microb. Pathog.* **22:**321–329.
135. **Koomey, M., E. C. Gotschlich, K. Robbins, S. Bergstrom, and J. Swanson.** 1987. Effects of *recA* mutations on pilus antigenic variation and phase transitions in *Neisseria gonorrhoeae. Genetics* **117:**391–398.
136. **Kupsch, E. M., B. Knepper, T. Kuroki, I. Heuer, and T. F. Meyer.** 1993. Variable opacity (Opa) outer membrane proteins account for the cell tropisms displayed by *Neisseria gonorrhoeae* for human leukocytes and epithelial cells. *EMBO J.* **12:**641–650.
137. **Lambden, P. R., J. E. Heckels, L. T. James, and P. J. Watt.** 1979. Variations in surface protein composition associated with virulence properties in opacity types of *Neisseria gonorrhoeae. J. Gen. Microbiol.* **114:**305–312.
138. **Lambden, P. R., J. N. Robertson, and P. J. Watt.** 1980. Biological properties of two distinct pilus types produced by isogenic variants of *Neisseria gonorrhoeae* P9. *J. Bacteriol.* **141:**393–396.
139. **Laskos, L., J. P. Dillard, H. S. Seifert, J. A. Fyfe, and J. K. Davies.** 1998. The pathogenic neisseriae contain an inactive *rpoN* gene and do not utilize the *pilE* sigma54 promoter. *Gene* **208:**95–102.
140. **Lebedeff, D. A., and E. B. Hochman.** 1980. Rectal gonorrhea in men: diagnosis and treatment. *Ann. Intern. Med.* **92:**463–466.
141. **Lee, S. W., R. A. Bonnah, D. L. Higashi, J. P. Atkinson, S. L. Milgram, and M. So.** 2002. CD46 is phosphorylated at tyrosine 354 upon infection of epithelial cells by *Neisseria gonorrhoeae. J. Cell Biol.* **156:**951–957.
142. **Levinson, G., and G. A. Gutman.** 1987. Slipped-strand mispairing: a major mechanism for DNA sequence evolution. *Mol. Biol. Evol.* **4:**203–221.
143. **Lewis, L. A., E. Gray, Y. P. Wang, B. A. Roe, and D. W. Dyer.** 1997. Molecular characterization of *hpuAB*, the haemoglobin-haptoglobin-utilization operon of *Neisseria meningitidis. Mol. Microbiol.* **23:**737–749.
144. **Long, C. D., S. F. Hayes, J. P. van Putten, H. A. Harvey, M. A. Apicella, and H. S. Seifert.** 2001. Modulation of gonococcal piliation by regulatable transcription of *pilE. J. Bacteriol.* **183:**1600–1609.
145. **Long, C. D., R. N. Madraswala, and H. S. Seifert.** 1998. Comparisons between colony phase variation of *Neisseria gonorrhoeae* FA1090 and pilus, pilin, and S-pilin expression. *Infect. Immun.* **66:**1918–1927.
146. **Long, C. D., D. M. Tobiason, M. P. Lazio, K. A. Kline, and H. S. Seifert.** 2003. Low-level pilin expression allows for substantial DNA transformation competence in *Neisseria gonorrhoeae. Infect. Immun.* **71:**6279–6291.
147. **Lorenzen, D. R., D. Gunther, J. Pandit, T. Rudel, E. Brandt, and T. F. Meyer.** 2000. *Neisseria gonorrhoeae* porin modifies the oxidative burst of human professional phagocytes. *Infect. Immun.* **68:**6215–6222.
148. **Makino, S., J. P. van Putten, and T. F. Meyer.** 1991. Phase variation of the opacity outer membrane protein controls invasion by *Neisseria gonorrhoeae* into human epithelial cells. *EMBO J.* **10:**1307–1315.
149. **Malorny, B., G. Morelli, B. Kusecek, J. Kolberg, and M. Achtman.** 1998. Sequence diversity, predicted two-dimensional protein structure, and epitope mapping of neisserial Opa proteins. *J. Bacteriol.* **180:**1323–1330.
150. **Mandrell, R., H. Schneider, M. Apicella, W. Zollinger, P. A. Rice, and J. M. Griffiss.** 1986. Antigenic and physical diversity of *Neisseria gonorrhoeae* lipooligosaccharides. *Infect. Immun.* **54:**63–69.
151. **Mandrell, R. E., H. Smith, G. A. Jarvis, J. M. Griffiss, and J. A. Cole.** 1993. Detection and some properties of the sialyltransferase implicated in the sialylation of lipopolysaccharide of *Neisseria gonorrhoeae. Microb. Pathog.* **14:**307–313.
152. **Manning, P. A., A. Kaufmann, U. Roll, J. Pohlner, T. F. Meyer, and R. Haas.** 1991. L-pilin variants of *Neisseria gonorrhoeae* MS11. *Mol. Microbiol.* **5:**917–926.
153. **Marceau, M., and X. Nassif.** 1999. Role of glycosylation at Ser63 in production of soluble pilin in pathogenic *Neisseria. J. Bacteriol.* **181:**656–661.

154. **Mayer, L. W.** 1982. Rates in vitro changes of gonococcal colony opacity phenotypes. *Infect. Immun.* **37:**481–485.

155. **McCormack, W. M., R. J. Stumacher, K. Johnson, and A. Donner.** 1977. Clinical spectrum of gonococcal infection in women. *Lancet* **i:**1182–1185.

156. **McQuillen, D. P., S. Gulati, S. Ram, A. K. Turner, D. B. Jani, T. C. Heeren, and P. A. Rice.** 1999. Complement processing and immunoglobulin binding to *Neisseria gonorrhoeae* determined in vitro simulates in vivo effects. *J. Infect. Dis.* **179:**124–135.

157. **Mehr, I. J., and H. S. Seifert.** 1998. Differential roles of homologous recombination pathways in *Neisseria gonorrhoeae* pilin antigenic variation, DNA transformation and DNA repair. *Mol. Microbiol.* **30:**697–710.

158. **Merz, A. J., C. A. Enns, and M. So.** 1999. Type IV pili of pathogenic neisseriae elicit cortical plaque formation in epithelial cells. *Mol. Microbiol.* **32:**1316–1332.

159. **Merz, A. J., and M. So.** 2000. Interactions of pathogenic neisseriae with epithelial cell membranes. *Annu. Rev. Cell Dev. Biol.* **16:**423–457.

160. **Merz, A. J., M. So, and M. P. Sheetz.** 2000. Pilus retraction powers bacterial twitching motility. *Nature* **407:**98–102.

161. **Meyer, T. F., E. Billyard, R. Haas, S. Storzbach, and M. So.** 1984. Pilus genes of *Neisseria gonorrheae:* chromosomal organization and DNA sequence. *Proc. Natl. Acad. Sci. USA* **81:**6110–6114.

162. **Meyer, T. F., N. Mlawer, and M. So.** 1982. Pilus expression in *Neisseria gonorrhoeae* involves chromosomal rearrangement. *Cell* **30:**45–52.

163. **Mickelsen, P. A., and P. F. Sparling.** 1981. Ability of *Neisseria gonorrhoeae, Neisseria meningitidis,* and commensal *Neisseria* species to obtain iron from transferrin and iron compounds. *Infect. Immun.* **33:**555–564.

164. **Morand, P. C., P. Tattevin, E. Eugene, J. L. Beretti, and X. Nassif.** 2001. The adhesive property of the type IV pilus-associated component PilC1 of pathogenic *Neisseria* is supported by the conformational structure of the N-terminal part of the molecule. *Mol. Microbiol.* **40:**846–856.

165. **Morelli, G., B. Malorny, K. Muller, A. Seiler, J. F. Wang, J. del Valle, and M. Achtman.** 1997. Clonal descent and microevolution of *Neisseria meningitidis* during 30 years of epidemic spread. *Mol. Microbiol.* **25:**1047–1064.

166. **Morse, S. A., and M. A. Apicella.** 1982. Isolation of a lipopolysaccharide mutant of *Neisseria gonorrhoeae:* an analysis of the antigenic and biologic difference. *J. Infect. Dis.* **145:**206–216.

167. **Mosleh, I. M., H. J. Boxberger, M. J. Sessler, and T. F. Meyer.** 1997. Experimental infection of native human ureteral tissue with *Neisseria gonorrhoeae:* adhesion, invasion, intracellular fate, exocytosis, and passage through a stratified epithelium. *Infect. Immun.* **65:**3391–3398.

168. **Muenzner, P., O. Billker, T. F. Meyer, and M. Naumann.** 2002. Nuclear factor-kappa B directs carcinoembryonic antigen-related cellular adhesion molecule 1 receptor expression in Neisseria gonorrhoeae-infected epithelial cells. *J. Biol. Chem.* **277:**7438–7446.

169. **Muenzner, P., M. Naumann, T. F. Meyer, and S. D. Gray-Owen.** 2001. Pathogenic *Neisseria* trigger expression of their carcinoembryonic antigen-related cellular adhesion molecule 1 (CEACAM1; previously CD66a) receptor on primary endothelial cells by activating the immediate early response transcription factor, nuclear factor-kappaB. *J. Biol. Chem.* **276:**24331–24340.

170. **Murphy, G. L., T. D. Connell, D. S. Barritt, M. Koomey, and J. G. Cannon.** 1989. Phase variation of gonococcal protein II: regulation of gene expression by slipped-strand mispairing of a repetitive DNA sequence. *Cell* **56:**539–547.

171. **Naids, F. L., and R. F. Rest.** 1991. Stimulation of human neutrophil oxidative metabolism by nonopsonized *Neisseria gonorrhoeae. Infect. Immun.* **59:**4383–4390.

172. **Nairn, C. A., J. A. Cole, P. V. Patel, N. J. Parsons, J. E. Fox, and H. Smith.** 1988. Cytidine 5′-monophospho-*N*-acetylneuraminic acid or a related compound is the low M_r factor from human red blood cells which induces gonococcal resistance to killing by human serum. *J. Gen. Microbiol.* **134:**3295–3306.

173. **Nassif, X., J. L. Beretti, J. Lowy, P. Stenberg, P. O'Gaora, J. Pfeifer, S. Normark, and M. So.** 1994. Roles of pilin and PilC in adhesion of *Neisseria meningitidis* to human epithelial and endothelial cells. *Proc. Natl. Acad. Sci. USA* **91:**3769–3773.

174. **Nicolson, I. J., A. C. Perry, M. Virji, J. E. Heckels, and J. R. Saunders.** 1987. Localization of antibody-binding sites by sequence analysis of cloned pilin genes from *Neisseria gonorrhoeae. J. Gen. Microbiol.* **133:**825–833.

175. **Palmer, L., G. F. Brooks, and S. Falkow.** 1989. Expression of gonococcal protein II in *Escherichia coli* by translational fusion. *Mol. Microbiol.* **3:**663–671.

176. **Pangburn, M. K., and H. J. Muller-Eberhard.** 1978. Complement C3 convertase: cell surface restriction of β1H control and generation of restriction on neuraminidase-treated cells. *Proc. Natl. Acad. Sci. USA* **75:**2416–2420.

177. **Pangburn, M. K., R. D. Schreiber, and H. J. Muller-Eberhard.** 1977. Human complement C3b inactivator: isolation, characterization, and demonstration of an absolute requirement for the serum protein β1H for cleavage of C3b and C4b in solution. *J. Exp. Med.* **146:**257–270.

178. **Parge, H. E., K. T. Forest, M. J. Hickey, D. A. Christensen, E. D. Getzoff, and J. A. Tainer.** 1995. Structure of the fibre-forming protein pilin at 2.6 A resolution. *Nature* **378:**32–38.

179. **Park, H. S., M. Wolfgang, J. P. van Putten, D. Dorward, S. F. Hayes, and M. Koomey.** 2001. Structural alterations in a type IV pilus subunit protein result in concurrent defects in multicellular behaviour and adherence to host tissue. *Mol. Microbiol.* **42:**293–307.

180. **Parsons, N. J., J. R. Andrade, P. V. Patel, J. A. Cole, and H. Smith.** 1989. Sialylation of lipopolysaccharide and loss of absorption of bactericidal antibody during conversion of gonococci to serum resistance by cytidine 5′-monophospho-*N*-acetyl neuraminic acid. *Microb. Pathog.* **7:**63–72.

181. **Pearce, W. A., and T. M. Buchanan.** 1978. Attachment role of gonococcal pili. Optimum conditions and quantitation of adherence of isolated pili to human cells in vitro. *J. Clin. Investig.* **61:**931–943.

182. **Perry, M. B., and V. Daoust.** 1975. The lipopolysaccharides of *Neisseria gonorrhoeae* colony types 1 and 4. *Can. J. Biochem.* **53:**623–629.

183. **Petricoin, E. F., III, R. J. Danaher, and D. C. Stein.** 1991. Analysis of the *lsi* region involved in lipooligosaccharide biosynthesis in *Neisseria gonorrhoeae. J. Bacteriol.* **173:**7896–7902.

184. **Phillips, R. S., P. A. Hanff, A. Wertheimer, and M. D. Aronson.** 1988. Gonorrhea in women seen for routine gynecologic care: criteria for testing. *Am. J. Med.* **85:**177–182.

185. **Platt, R., P. A. Rice, and W. M. McCormack.** 1983. Risk of acquiring gonorrhea and prevalence of abnormal adnexal findings among women recently exposed to gonorrhea. *JAMA* **250:**3205–3209.

186. **Plummer, F. A., H. Chubb, J. N. Simonsen, M. Bosire, L. Slaney, N. J. Nagelkerke, I. Maclean, J. O. Ndinya-Achola, P. Waiyaki, and R. C. Brunham.** 1994. Antibodies to opacity

proteins (Opa) correlate with a reduced risk of gonococcal salpingitis. *J. Clin. Investig.* **93:**1748–1755.

187. **Porat, N., M. A. Apicella, and M. S. Blake.** 1995. *Neisseria gonorrhoeae* utilizes and enhances the biosynthesis of the asialoglycoprotein receptor expressed on the surface of the hepatic HepG2 cell line. *Infect. Immun.* **63:**1498–1506.
188. **Power, P. M., and M. P. Jennings.** 2003. The genetics of glycosylation in gram-negative bacteria. *FEMS Microbiol. Lett.* **218:**211–222.
189. **Power, P. M., L. F. Roddam, K. Rutter, S. Z. Fitzpatrick, Y. N. Srikhanta, and M. P. Jennings.** 2003. Genetic characterization of pilin glycosylation and phase variation in *Neisseria meningitidis. Mol. Microbiol.* **49:**833–847.
190. **Punsalang, A. P., Jr., and W. D. Sawyer.** 1973. Role of pili in the virulence of *Neisseria gonorrhoeae. Infect. Immun.* **8:**255–263.
191. **Quinn, T. C., W. E. Stamm, S. E. Goodell, E. Mkrtichian, J. Benedetti, L. Corey, M. D. Schuffler, and K. K. Holmes.** 1983. The polymicrobial origin of intestinal infections in homosexual men. *N. Engl. J. Med.* **309:**576–582.
192. **Rahman, M., H. Kallstrom, S. Normark, and A. B. Jonsson.** 1997. PilC of pathogenic *Neisseria* is associated with the bacterial cell surface. *Mol. Microbiol.* **25:**11–25.
193. **Ram, S., M. Cullinane, A. M. Blom, S. Gulati, D. P. McQuillen, B. G. Monks, C. O'Connell, R. Boden, C. Elkins, M. K. Pangburn, B. Dahlback, and P. A. Rice.** 2001. Binding of C4b-binding protein to porin: a molecular mechanism of serum resistance of *Neisseria gonorrhoeae. J. Exp. Med.* **193:**281–295.
194. **Ram, S., F. G. Mackinnon, S. Gulati, D. P. McQuillen, U. Vogel, M. Frosch, C. Elkins, H. K. Guttormsen, L. M. Wetzler, M. Oppermann, M. K. Pangburn, and P. A. Rice.** 1999. The contrasting mechanisms of serum resistance of *Neisseria gonorrhoeae* and group B *Neisseria meningitidis. Mol. Immunol.* **36:**915–928.
195. **Ram, S., A. K. Sharma, S. D. Simpson, S. Gulati, D. P. McQuillen, M. K. Pangburn, and P. A. Rice.** 1998. A novel sialic acid binding site on factor H mediates serum resistance of sialylated *Neisseria gonorrhoeae. J. Exp. Med.* **187:**743–752.
196. **Rest, R. F., S. H. Fischer, Z. Z. Ingham, and J. F. Jones.** 1982. Interactions of *Neisseria gonorrhoeae* with human neutrophils: effects of serum and gonococcal opacity on phagocyte killing and chemiluminescence. *Infect. Immun.* **36:**737–744.
197. **Rest, R. F., and J. V. Frangipane.** 1992. Growth of *Neisseria gonorrhoeae* in CMP-*N*-acetylneuraminic acid inhibits nonopsonic (opacity-associated outer membrane protein-mediated) interactions with human neutrophils. *Infect. Immun.* **60:**989–997.
198. **Rice, P. A.** 1989. Molecular basis for serum resistance in *Neisseria gonorrhoeae. Clin. Microbiol. Rev.* **2**(Suppl.):S112–S117.
199. **Rothbard, J. B., R. Fernandez, L. Wang, N. N. Teng, and G. K. Schoolnik.** 1985. Antibodies to peptides corresponding to a conserved sequence of gonococcal pilins block bacterial adhesion. *Proc. Natl. Acad. Sci. USA* **82:**915–919.
200. **Rudel, T., H. J. Boxberger, and T. F. Meyer.** 1995. Pilus biogenesis and epithelial cell adherence of *Neisseria gonorrhoeae pilC* double knock-out mutants. *Mol. Microbiol.* **17:**1057–1071.
201. **Rudel, T., D. Facius, R. Barten, I. Scheuerpflug, E. Nonnenmacher, and T. F. Meyer.** 1995. Role of pili and the phase-variable PilC protein in natural competence for transformation of *Neisseria gonorrhoeae. Proc. Natl. Acad. Sci. USA* **92:**7986–7990.
202. **Rudel, T., I. Scheurerpflug, and T. F. Meyer.** 1995. *Neisseria* PilC protein identified as type-4 pilus tip-located adhesin. *Nature* **373:**357–359.
203. **Rudel, T., J. P. van Putten, C. P. Gibbs, R. Haas, and T. F. Meyer.** 1992. Interaction of two variable proteins (PilE and PilC) required for pilus-mediated adherence of *Neisseria gonorrhoeae* to human epithelial cells. *Mol. Microbiol.* **6:**3439–3450.
204. **Rytkonen, A., L. Johansson, V. Asp, B. Albiger, and A. B. Jonsson.** 2001. Soluble pilin of *Neisseria gonorrhoeae* interacts with human target cells and tissue. *Infect. Immun.* **69:**6419–6426.
205. **Salit, I. E.** 1982. The differential susceptibility of gonococcal opacity variants to sex hormones. *Can. J. Microbiol.* **28:**301–306.
206. **Salit, I. E., M. Blake, and E. C. Gotschlich.** 1980. Intra-strain heterogeneity of gonococcal pili is related to opacity colony variance. *J. Exp. Med.* **151:**716–725.
207. **Salit, I. E., and E. C. Gotschlich.** 1978. Gonococcal color and opacity variants: virulence for chicken embryos. *Infect. Immun.* **22:**359–364.
208. **Sandlin, R. C., and D. C. Stein.** 1994. Role of phosphoglucomutase in lipooligosaccharide biosynthesis in *Neisseria gonorrhoeae. J. Bacteriol.* **176:**2930–2937.
209. **Scheuerpflug, I., T. Rudel, R. Ryll, J. Pandit, and T. F. Meyer.** 1999. Roles of PilC and PilE proteins in pilus-mediated adherence of *Neisseria gonorrhoeae* and *Neisseria meningitidis* to human erythrocytes and endothelial and epithelial cells. *Infect. Immun.* **67:**834–843.
210. **Schmidt, K. A., C. D. Deal, M. Kwan, E. Thattassery, and H. Schneider.** 2000. *Neisseria gonorrhoeae* MS11mkC opacity protein expression in vitro and during human volunteer infectivity studies. *Sex. Transm. Dis.* **27:**278–283.
211. **Schneider, H., J. M. Griffiss, J. W. Boslego, P. J. Hitchcock, K. M. Zahos, and M. A. Apicella.** 1991. Expression of paragloboside-like lipooligosaccharides may be a necessary component of gonococcal pathogenesis in men. *J. Exp. Med.* **174:**1601–1605.
212. **Schneider, H., J. M. Griffiss, R. E. Mandrell, and G. A. Jarvis.** 1985. Elaboration of a 3.6-kilodalton lipooligosaccharide, antibody against which is absent from human sera, is associated with serum resistance of *Neisseria gonorrhoeae. Infect. Immun.* **50:**672–677.
213. **Schneider, H., T. L. Hale, W. D. Zollinger, R. C. Seid, Jr., C. A. Hammack, and J. M. Griffiss.** 1984. Heterogeneity of molecular size and antigenic expression within lipooligosaccharides of individual strains of *Neisseria gonorrhoeae* and *Neisseria meningitidis. Infect. Immun.* **45:**544–549.
214. **Schneider, H., C. A. Hammack, M. A. Apicella, and J. M. Griffiss.** 1988. Instability of expression of lipooligosaccharides and their epitopes in *Neisseria gonorrhoeae. Infect. Immun.* **56:**942–946.
215. **Schneider, H., K. A. Schmidt, D. R. Skillman, L. Van De Verg, R. L. Warren, H. J. Wylie, J. C. Sadoff, C. D. Deal, and A. S. Cross.** 1996. Sialylation lessens the infectivity of Neisseria gonorrhoeae MS11mkC. *J. Infect. Dis.* **173:**1422–1427.
216. **Schoolnik, G. K., R. Fernandez, J. Y. Tai, J. Rothbard, and E. C. Gotschlich.** 1984. Gonococcal pili. Primary structure and receptor binding domain. *J. Exp. Med.* **159:**1351–1370.
217. **Schwan, E. T., B. D. Robertson, H. Brade, and J. P. van Putten.** 1995. Gonococcal *rfaF* mutants express Rd2 chemotype LPS and do not enter epithelial host cells. *Mol. Microbiol.* **15:**267–275.
218. **Seifert, H. S.** 1992. Molecular mechanisms of antigenic variation in *Neisseria gonorrhoeae. Mol. Cell. Biol. Hum. Dis. Ser.* **1:**1–22.

219. **Seifert, H. S.** 1996. Questions about gonococcal pilus phase and antigenic variation. *Mol. Microbiol.* **21:**433–440.
220. **Seifert, H. S.** 1997. Insertionally inactivated and inducible *recA* alleles for use in *Neisseria. Gene* **188:**215–220.
221. **Seifert, H. S., R. S. Ajioka, D. Paruchuri, F. Heffron, and M. So.** 1990. Shuttle mutagenesis of *Neisseria gonorrhoeae:* pilin null mutations lower DNA transformation competence. *J. Bacteriol.* **172:**40–46.
222. **Seifert, H. S., and M. So.** 1988. Genetic mechanisms of bacterial antigenic variation. *Microbiol. Rev.* **52:**327–336.
223. **Seifert, H. S., C. J. Wright, A. E. Jerse, M. S. Cohen, and J. G. Cannon.** 1994. Multiple gonococcal pilin antigenic variants are produced during experimental human infections. *J. Clin. Investig.* **93:**2744–2749.
224. **Serkin, C. D., and H. S. Seifert.** 1998. Frequency of pilin antigenic variation in *Neisseria gonorrhoeae. J. Bacteriol.* **180:**1955–1958.
225. **Serkin, C. D., and H. S. Seifert.** 2000. Iron availability regulates DNA recombination in *Neisseria gonorrhoeae. Mol. Microbiol.* **37:**1075–1086.
226. **Shafer, W. M., J. T. Balthazar, K. E. Hagman, and S. A. Morse.** 1995. Missense mutations that alter the DNA-binding domain of the MtrR protein occur frequently in rectal isolates of *Neisseria gonorrhoeae* that are resistant to faecal lipids. *Microbiology* **141:**907–911.
227. **Shafer, W. M., A. Datta, V. S. Kolli, M. M. Rahman, J. T. Balthazar, L. E. Martin, W. L. Veal, D. S. Stephens, and R. Carlson.** 2002. Phase variable changes in genes *lgtA* and *lgtC* within the *lgtABCDE* operon of *Neisseria gonorrhoeae* can modulate gonococcal susceptibility to normal human serum. *J. Endotoxin. Res.* **8:**47–58.
228. **Shaw, J. H., and S. Falkow.** 1988. Model for invasion of human tissue culture cells by *Neisseria gonorrhoeae. Infect. Immun.* **56:**1625–1632.
229. **Shell, D. M., L. Chiles, R. C. Judd, S. Seal, and R. F. Rest.** 2002. The *Neisseria* lipooligosaccharide-specific alpha-2,3-sialyltransferase is a surface-exposed outer membrane protein. *Infect. Immun.* **70:**3744–3751.
230. **Simon, D., and R. F. Rest.** 1992. *Escherichia coli* expressing a *Neisseria gonorrhoeae* opacity-associated outer membrane protein invade human cervical and endometrial epithelial cell lines. *Proc. Natl. Acad. Sci. USA* **89:**5512–5516.
231. **Singer, B. B., I. Scheffrahn, R. Heymann, K. Sigmundsson, R. Kammerer, and B. Obrink.** 2002. Carcinoembryonic antigen-related cell adhesion molecule 1 expression and signaling in human, mouse, and rat leukocytes: evidence for replacement of the short cytoplasmic domain isoform by glycosylphosphatidylinositol-linked proteins in human leukocytes. *J. Immunol.* **168:**5139–5146.
232. **Skaar, E. P., D. M. Tobiason, J. Quick, R. C. Judd, H. Weissbach, F. Etienne, N. Brot, and H. S. Seifert.** 2002. The outer membrane localization of the *Neisseria gonorrhoeae* MsrA/B is involved in survival against reactive oxygen species. *Proc. Natl. Acad. Sci. USA* **99:**10108–10113.
233. **Smith, H., J. A. Cole, and N. J. Parsons.** 1992. The sialylation of gonococcal lipopolysaccharide by host factors: a major impact on pathogenicity. *FEMS Microbiol. Lett.* **79:**287–292.
234. **Snyder, L. A., S. A. Butcher, and N. J. Saunders.** 2001. Comparative whole-genome analyses reveal over 100 putative phase-variable genes in the pathogenic *Neisseria* spp. *Microbiology* **147:**2321–2332.
235. **Sparling, P. F.** 1999. Biology of *Neisseria gonorrhoeae,* p. 433–449. *In* K. H. Holmes, P.-A. Mardh, P. F. Sparling, S. M. Lemon, W. E. Stamm, P. Piot, and J. N. Wasserheit (ed.), *Sexually Transmitted Diseases,* 3rd ed. McGraw-Hill, Inc., New York, N.Y.
236. **Sparling, P. F., J. Tsai, and C. N. Cornelissen.** 1990. Gonococci are survivors. *Scand. J. Infect. Dis. Suppl.* **69:**125–136.
237. **Stead, A., J. S. Main, M. E. Ward, and P. J. Watt.** 1975. Studies on lipopolysaccharides isolated from strains of *Neisseria gonorrhoeae. J. Gen. Microbiol.* **88:**123–131.
238. **Stern, A., M. Brown, P. Nickel, and T. F. Meyer.** 1986. Opacity genes in *Neisseria gonorrhoeae:* control of phase and antigenic variation. *Cell* **47:**61–71.
239. **Stern, A., and T. F. Meyer.** 1987. Common mechanism controlling phase and antigenic variation in pathogenic neisseriae. *Mol. Microbiol.* **1:**5–12.
240. **Stimson, E., M. Virji, S. Barker, M. Panico, I. Blench, J. Saunders, G. Payne, E. R. Moxon, A. Dell, and H. R. Morris.** 1996. Discovery of a novel protein modification: alpha-glycerophosphate is a substituent of meningococcal pilin. *Biochem. J.* **316:**29–33.
241. **Stojiljkovic, I., J. Larson, V. Hwa, S. Anic, and M. So.** 1996. HmbR outer membrane receptors of pathogenic *Neisseria* spp.: iron-regulated, hemoglobin-binding proteins with a high level of primary structure conservation. *J. Bacteriol.* **178:**4670–4678.
242. **Strom, M. S., D. N. Nunn, and S. Lory.** 1994. Posttranslational processing of type IV prepilin and homologs by PilD of *Pseudomonas aeruginosa. Methods Enzymol.* **235:**527–540.
243. **Sugasawara, R. J., J. G. Cannon, W. J. Black, I. Nachamkin, R. L. Sweet, and G. F. Brooks.** 1983. Inhibition of *Neisseria gonorrhoeae* attachment to HeLa cells with monoclonal antibody directed against a protein II. *Infect. Immun.* **42:**980–985.
244. **Swanson, J.** 1977. Surface components affecting interactions between *Neisseria gonorrhoeae* and eucaryotic cells. *J. Infect. Dis.* **136**(Suppl.):S138–S143.
245. **Swanson, J.** 1978. Studies on gonococcus infection. XII. Colony color and opacity varienats of gonococci. *Infect. Immun.* **19:**320–331.
246. **Swanson, J., O. Barrera, J. Sola, and J. Boslego.** 1988. Expression of outer membrane protein II by gonococci in experimental gonorrhea. *J. Exp. Med.* **168:**2121–2129.
247. **Swanson, J., S. Bergstrom, O. Barrera, K. Robbins, and D. Corwin.** 1985. Pilus-gonococcal variants. Evidence for multiple forms of piliation control. *J. Exp. Med.* **162:**729–744.
248. **Swanson, J., S. Bergstrom, J. Boslego, and M. Koomey.** 1987. Gene conversion accounts for pilin structural changes and for reversible piliation "phase" changes in gonococci. *Antonie Leeuwenhoek* **53:**441–446.
249. **Swanson, J., S. Bergstrom, K. Robbins, O. Barrera, D. Corwin, and J. M. Koomey.** 1986. Gene conversion involving the pilin structural gene correlates with pilus$^+$ in equilibrium with pilus$^-$ changes in *Neisseria gonorrhoeae. Cell* **47:**267–276.
250. **Swanson, J., S. J. Kraus, and E. C. Gotschlich.** 1971. Studies on gonococcus infection. I. Pili and zones of adhesion: their relation to gonococcal growth patterns. *J. Exp. Med.* **134:**886–906.
251. **Swanson, J., K. Robbins, O. Barrera, D. Corwin, J. Boslego, J. Ciak, M. Blake, and J. M. Koomey.** 1987. Gonococcal pilin variants in experimental gonorrhea. *J. Exp. Med.* **165:**1344–1357.
252. **Swanson, K. V., G. A. Jarvis, G. F. Brooks, B. J. Barham, M. D. Cooper, and J. M. Griffiss.** 2001. CEACAM is not necessary for *Neisseria gonorrhoeae* to adhere to and invade female genital epithelial cells. *Cell. Microbiol.* **3:**681–691.
253. **Sweet, R. L., M. Blankfort-Doyle, M. O. Robbie, and J. Schacter.** 1986. The occurrence of chlamydial and gonococcal salpingitis during the menstrual cycle. *JAMA* **255:**2062–2064.
254. **Todd, W. J., G. P. Wray, and P. J. Hitchcock.** 1984. Arrangement of pili in colonies of *Neisseria gonorrhoeae. J. Bacteriol.* **159:**312–320.

255. **Toleman, M., E. Aho, and M. Virji.** 2001. Expression of pathogen-like Opa adhesins in commensal *Neisseria:* genetic and functional analysis. *Cell. Microbiol.* **3:**33–44.
256. **Tonjum, T., D. A. Caugant, S. A. Dunham, and M. Koomey.** 1998. Structure and function of repetitive sequence elements associated with a highly polymorphic domain of the *Neisseria meningitidis* PilQ protein. *Mol. Microbiol.* **29:**111–124.
257. **Tonjum, T., and M. Koomey.** 1997. The pilus colonization factor of pathogenic neisserial species: organelle biogenesis and structure/function relationships—a review. *Gene* **192:** 155–163.
258. **Tsai, C. M., and C. E. Frasch.** 1982. A sensitive silver stain for detecting lipopolysaccharides in polyacrylamide gels. *Anal. Biochem.* **119:**115–119.
259. **Tseng, H. J., Y. Srikhanta, A. G. McEwan, and M. P. Jennings.** 2001. Accumulation of manganese in *Neisseria gonorrhoeae* correlates with resistance to oxidative killing by superoxide anion and is independent of superoxide dismutase activity. *Mol. Microbiol.* **40:**1175–1186.
260. **Turner, S., E. Reid, H. Smith, and J. Cole.** 2003. A novel cytochrome *c* peroxidase from *Neisseria gonorrhoeae:* a lipoprotein from a Gram-negative bacterium. *Biochem. J.* **373:**865–873.
261. **van Belkum, A., S. Scherer, L. van Alphen, and H. Verbrugh.** 1998. Short-sequence DNA repeats in prokaryotic genomes. *Microbiol. Mol. Biol. Rev.* **62:**275–293.
262. **van Putten, J. P.** 1993. Phase variation of lipopolysaccharide directs interconversion of invasive and immuno-resistant phenotypes of *Neisseria gonorrhoeae. EMBO J.* **12:**4043–4051.
263. **van Putten, J. P., and S. M. Paul.** 1995. Binding of syndecan-like cell surface proteoglycan receptors is required for *Neisseria gonorrhoeae* entry into human mucosal cells. *EMBO J.* **14:**2144–2154.
264. **Virji, M., D. Evans, A. Hadfield, F. Grunert, A. M. Teixeira, and S. M. Watt.** 1999. Critical determinants of host receptor targeting by *Neisseria meningitidis* and *Neisseria gonorrhoeae:* identification of Opa adhesiotopes on the N-domain of CD66 molecules. *Mol. Microbiol.* **34:**538–551.
265. **Virji, M., and J. E. Heckels.** 1984. The role of common and type-specific pilus antigenic domains in adhesion and virulence of gonococci for human epithelial cells. *J. Gen. Microbiol.* **130:**1089–1095.
266. **Virji, M., and J. E. Heckels.** 1985. Role of anti-pilus antibodies in host defense against gonococcal infection studied with monoclonal anti-pilus antibodies. *Infect. Immun.* **49:**621–628.
267. **Virji, M., K. Makepeace, D. J. Ferguson, and S. M. Watt.** 1996. Carcinoembryonic antigens (CD66) on epithelial cells and neutrophils are receptors for Opa proteins of pathogenic neisseriae. *Mol. Microbiol.* **22:**941–950.
268. **Virji, M., K. Makepeace, I. Peak, G. Payne, J. R. Saunders, D. J. Ferguson, and E. R. Moxon.** 1995. Functional implications of the expression of PilC proteins in meningococci. *Mol. Microbiol.* **16:**1087–1097.
269. **Virji, M., J. R. Saunders, G. Sims, K. Makepeace, D. Maskell, and D. J. Ferguson.** 1993. Pilus-facilitated adherence of *Neisseria meningitidis* to human epithelial and endothelial cells: modulation of adherence phenotype occurs concurrently with changes in primary amino acid sequence and the glycosylation status of pilin. *Mol. Microbiol.* **10:**1013–1028.
270. **Wainwright, L. A., J. V. Frangipane, and H. S. Seifert.** 1997. Analysis of protein binding to the Sma/Cla DNA repeat in pathogenic neisseriae. *Nucleic Acids Res.* **25:**1362–1368.
271. **Wainwright, L. A., K. H. Pritchard, and H. S. Seifert.** 1994. A conserved DNA sequence is required for efficient gonococcal pilin antigenic variation. *Mol. Microbiol.* **13:**75–87.
272. **Wallin, J., and M. S. Siegel.** 1979. Pharyngeal *Neisseria gonorrhoeae:* coloniser or pathogen? *Br. Med. J.* **1:**1462–1463.
273. **Ward, M. E., P. J. Watt, and A. A. Glynn.** 1970. Gonococci in urethral exudates possess a virulence factor lost on subculture. *Nature* **227:**382–384.
274. **Warren, M. J., and M. P. Jennings.** 2003. Identification and characterization of *pptA:* a gene involved in the phase-variable expression of phosphorylcholine on pili of *Neisseria meningitidis. Infect. Immun.* **71:**6892–6898.
275. **Weel, J. F., C. T. Hopman, and J. P. van Putten.** 1991. In situ expression and localization of *Neisseria gonorrhoeae* opacity proteins in infected epithelial cells: apparent role of Opa proteins in cellular invasion. *J. Exp. Med.* **173:**1395–1405.
276. **Weiler, J. M., M. R. Daha, K. F. Austen, and D. T. Fearon.** 1976. Control of the amplification convertase of complement by the plasma protein β1H. *Proc. Natl. Acad. Sci. USA* **73:** 3268–3272.
277. **Weinberg, E. D.** 1978. Iron, and infection. *Microbiol. Rev.* **42:**45–66.
278. **Weiser, J. N., J. B. Goldberg, N. Pan, L. Wilson, and M. Virji.** 1998. The phosphorylcholine epitope undergoes phase variation on a 43-kilodalton protein in *Pseudomonas aeruginosa* and on pili of *Neisseria meningitidis* and *Neisseria gonorrhoeae. Infect. Immun.* **66:**4263–4267.
279. **Wetzler, L. M., K. Barry, M. S. Blake, and E. C. Gotschlich.** 1992. Gonococcal lipooligosaccharide sialylation prevents complement-dependent killing by immune sera. *Infect. Immun.* **60:**39–43.
280. **Wiesner, P. J., E. Tronca, P. Bonin, A. H. Pedersen, and K. K. Holmes.** 1973. Clinical spectrum of pharyngeal gonococcal infection. *N. Engl. J. Med.* **288:**181–185.
281. **Winther-Larsen, H. C., F. T. Hegge, M. Wolfgang, S. F. Hayes, J. P. van Putten, and M. Koomey.** 2001. *Neisseria gonorrhoeae* PilV, a type IV pilus-associated protein essential to human epithelial cell adherence. *Proc. Natl. Acad. Sci. USA* **98:**15276–15281.
282. **Wiseman, G. M., and J. D. Caird.** 1977. Composition of the lipopolysaccharide of *Neisseria gonorrhoeae. Infect. Immun.* **16:**550–556.
283. **Woods, J. P., and J. G. Cannon.** 1990. Variation in expression of class 1 and class 5 outer membrane proteins during nasopharyngeal carriage of *Neisseria meningitidis. Infect. Immun.* **58:**569–572.
284. **Xia, M., W. L. Whittington, W. M. Shafer, and K. K. Holmes.** 2000. Gonorrhea among men who have sex with men: outbreak caused by a single genotype of erythromycin-resistant *Neisseria gonorrhoeae* with a single-base pair deletion in the *mtrR* promoter region. *J. Infect. Dis.* **181:**2080–2082.
285. **Yamasaki, R., J. M. Griffiss, K. P. Quinn, and R. E. Mandrell.** 1993. Neuraminic acid is alpha 2→3 linked in the lipooligosaccharide of *Neisseria meningitidis* serogroup B strain 6275. *J. Bacteriol.* **175:**4565–4568.
286. **Yang, Q. L., and E. C. Gotschlich.** 1996. Variation of gonococcal lipooligosaccharide structure is due to alterations in poly-G tracts in *lgt* genes encoding glycosyl transferases. *J. Exp. Med.* **183:**323–327.
287. **Zhou, D., D. S. Stephens, B. W. Gibson, J. J. Engstrom, C. F. McAllister, F. K. Lee, and M. A. Apicella.** 1994. Lipooligosaccharide biosynthesis in pathogenic *Neisseria.* Cloning, identification, and characterization of the phosphoglucomutase gene. *J. Biol. Chem.* **269:**11162–11169.
288. **Zollinger, W. D., and R. E. Mandrell.** 1983. Importance of complement source in bactericidal activity of human antibody and murine monoclonal antibody to meningococcal group B polysaccharide. *Infect. Immun.* **40:**257–264.

Colonization of Mucosal Surfaces
Edited by James P. Nataro et al.

Chapter 24

Allelic Variation of the FimH Lectin of *Escherichia coli* Type 1 Fimbriae and Uropathogenesis

David L. Hasty, Xue-Ru Wu, Daniel E. Dykuizen, and Evgeni V. Sokurenko

Escherichia coli is a typical member of the family *Enterobacteriaceae* that principally inhabits (i.e., asymptomatically infects) the large intestine of humans and animals as commensal microbiota (143). *E. coli* is also able to cause symptomatic infections in a variety of host niches, including the gastrointestinal tract and the central nervous system. One of the more common sites infected by *E. coli,* and a host niche in which *E. coli* is by far the most common pathogen, is the urinary tract. Roughly half of adult women have at least one urinary tract infection (UTI) during their lifetime (28). In fact, UTIs are considered to be the most common bacterial infection, accounting for 7 million physician office visits, 1 million emergency department visits, and 100,000 hospitalizations and costing approximately $1.5 billion per year in the United States alone (41). *E. coli* accounts for 80% or more of all UTIs.

It is clear that certain strains of *E. coli* are better equipped than others to cause UTIs. Determining the genetic makeup that may define uropathogenic *E. coli* has been of considerable interest to epidemiologists and basic scientists for many years (67, 68, 151). The occurrence of dramatic outbreaks of UTI caused by a single clone of *E. coli* (93) highlights the fact that even among the various uropathogenic *E. coli* lineages, some genotypes are more likely than others to cause UTI.

Most of the virulence determinants that are considered to be important for survival in the urinary tract reside on specialized regions of the chromosome called pathogenicity islands (PAIs). Examples of virulence factors found on PAIs include P fimbriae (51), P-related fimbriae (14), hemolysin (14), cytotoxic necrotizing factor 1 (14), and S fimbriae (105). A great many PAIs have now been identified in a variety of *E. coli* pathotypes, as well as in other genera that contain both commensal and pathogenic lineages (73). It has come to be accepted that these "islands" of virulence factor genes have been acquired by horizontal transfer from foreign sources. This type of acquisition of additional genes that encode specific virulence factors constitutes a "gain-of-function" mechanism of evolution of a bacterial species toward a more pathogenic phenotype.

Another mechanism for evolution toward a more pathogenic phenotype is one that we have called pathoadaptive mutation. This mechanism does not require the horizontal transfer of specific virulence factor genes. Rather, it involves a "change-of-function" mutation (i.e., loss or modification of function) in a preexisting commensal gene that enables the clone to move into a new host niche and cause a symptomatic infection.

One of the clearest examples of pathoadaptive mutation can be found in the allelic variation of the FimH lectin adhesin of type 1 fimbriae. Type 1 fimbriae are arguably the most common of all bacterial adhesins. These hair-like structures are expressed by virtually all *E. coli* strains and by many other members of the *Enterobacteriaceae.* Twenty years or so ago, very few investigators considered type 1 fimbriae to be an important virulence factor. Indeed, the purpose of these common fimbriae was not clear. Epidemiologists would not assign to them any particular role in virulence because they were equally represented among nonpathogen and pathogen alike. The experimental data for a potential role of type 1 fimbriae in

David L. Hasty • Department of Anatomy & Neurobiology, University of Tennessee Health Science Center, and Research Service (151), Veterans Affairs Medical Center, Memphis, TN 38104. **Xue-Ru Wu** • Departments of Urology and Microbiology, Kaplan Comprehensive Cancer Center, New York University School of Medicine, New York, NY 10016, and Veterans Affairs Medical Center in Manhattan, New York, NY 10010. **Daniel E. Dykuizen** • Department of Ecology and Evolution, State University of New York, Stony Brook, NY 11794. **Evgeni V. Sokurenko** • Department of Microbiology, University of Washington, Seattle, WA 98195.

UTIs were all but dismissed because of the lack of epidemiological data to implicate them in disease. Theories that they must be important in the ability of *E. coli* to colonize its primary habitat, the gastrointestinal tract (121), were never solidified, because while some investigators found them to play a role, others found them to be nonessential. Therefore, the controversy over the precise role of type 1 fimbriae in the natural habitat of *E. coli* continued. Over the last decade, however, gradually improving genetic technologies have facilitated the demonstration that type 1 fimbriae can indeed serve as a virulence factor. In fact, a recent publication has proclaimed type 1 fimbriae (and extracellular polysaccharides) to be a ". . . preeminent uropathogenic *Escherichia coli* virulence determinant . . ." (6). In this chapter, we review evidence for the role of type 1 fimbriae as urovirulence factors. While our focus is on the FimH lectin and the occurrence of mutations that cause some *fimH* alleles to be pathoadaptive, the story of allelic variation of FimH is presented within the broader context of type 1 fimbrial biology.

ADHESION AS AN ESSENTIAL EVOLUTIONARY TRAIT

Over the last several decades, it has become a general tenet of bacterial ecology that the ability of bacteria to adhere to surfaces is essential for the survival of bacterial species (116). Indeed, very few free-floating (planktonic) bacteria can be found in natural situations outside the laboratory. Most bacteria are found attached to the surface of an organic or inorganic surface, often within a complex biofilm (23, 24). Because the survival of microorganisms is so dependent on their adhesion to surfaces, the bacterial surface structures that accomplish this function (i.e., adhesins) must have been basic traits of the very first microbial cells that emerged on Earth. As new hosts and new surfaces appeared, adhesins had to evolve to accommodate adhesion in these new niches, with a strong selective pressure toward best-fit adhesins. In the evolutionary process, a sufficient number of adhesins with distinct receptor specificity have evolved to ensure that most bacterial clones can normally colonize a defined range of substrata, but one that is sufficiently broad to ensure species survival.

Macroevolution occurs over a millennial time frame, but more rapid changes in receptor specificity can come about through microevolution of adhesins within a single host. Microevolution of the adhesion process is the result of intraspecies diversification of adhesins to become complementary to new receptors (153). These two processes take place, in pathogen and commensal alike, by different mechanisms, which generally fall into three basic categories, two of which have already been mentioned: (i) transfer of genes from one strain or species to another via mobile elements (e.g., plasmids and transposons [73]), (ii) gene rearrangement (e.g., *Neisseria gonorrhoeae* pili [49]), and (iii) allelic variation via mutation of a parental gene (e.g., FimH [139, 140]). In the last case, variation of receptor specificity may be directly due to conformational changes in the adhesin itself or the conformation may be indirectly affected by interactions with other proteins involved in presenting the adhesin on the bacterial surface.

TYPE 1 FIMBRIAE

Over the millennia, many different classes of adhesive structures have evolved to provide for adhesion (26, 65, 80, 116). One of the best-studied classes of adhesive structures is commonly referred to by the general term "fimbriae" (31), which are hair-like appendages of the bacterial surface, also sometimes called pili (16, 17). The primary function of these organelles, as far as is currently known, is to mediate adhesion. Fimbriae are particularly common among members of the family *Enterobacteriaceae* (65, 80, 116).

Because the adhesive function of many fimbriae can be inhibited by simple saccharides, fimbrae are considered lectins. For certain types of fimbriae, there is one primary structural subunit that also serves as the adhesin (27). There are also heteropolymeric fimbriae in which the lectin subunit is present in one or a very small number of copies and located only (or at least primarily) at the distal tips of a polymeric shaft composed of the primary structural subunit. The focus of this chapter, *E. coli* type 1 fimbriae, is one of the heteropolymeric types of fimbriae. The FimH lectin is only a minor component of the overall type 1 fimbrial superstructure, but it plays a prominent role in initiating fimbrial biogenesis (7, 82) and is almost completely responsible for the binding activity (138). Nevertheless, because in most normal instances the ability of FimH to function in adhesion is dependent on the ability of the bacteria to express intact fimbriae, consideration of FimH structure and function must be made within the context of both its genetic operon and its structural scaffold.

Type 1 fimbriae bearing the FimH lectin are expressed on the surfaces of virtually all *E. coli* strains and most other members of the family *Enterobacteriaceae* (e.g., *Klebsiella, Enterobacter,* and *Salmonella* spp.) (2, 53). The ubiquity of type 1 fimbriae alone should have signified their importance for enterobacterial populations; instead, this fact long presented more of a problem than a solution, particularly in epidemiologic analyses of the contribution of the various fimbrial types to *E. coli* infections (67). The fact that type 1 fimbriae are produced by almost all normal and pathogenic isolates alike led many investigators in the field to the long-held assumption that these fimbriae could not be important factors in *E. coli* pathogenesis. In addition, the role(s) of type 1 fimbriae in normal *E. coli* ecology was also long debated.

The early idea that type 1 fimbriae must contribute in a direct way to the maintenance of *E. coli* in the primary colonic niche was seriously questioned following studies indicating that type 1 fimbriae were not required for intestinal colonization (99, 100). An inability to produce type 1 fimbriae did not impair the ability of recombinant *E. coli* to colonize the large intestine when animals were inoculated with large numbers of bacteria (10, 99). However, when small numbers of *E. coli* were instilled into the oropharyngeal cavity of mice, lack of type 1 fimbrial expression had a very deleterious effect on the transient colonization of the oropharyngeal cavity and, apparently thereby, on successful passage through the acid barrier of the stomach in route to the intestine. Whether transient colonization of the oral cavity was important because it allowed the concentration of bacteria to reach a critical level, because it induced the expression of specific genes, or because a continuous supply of bacteria over a period of hours or days enabled their passage through the stomach at a time when acid production was low is not yet known. Nevertheless, these studies demonstrated a way in which type 1 fimbriae might be important in gaining access to the primary colonic niche (10, 45). It remains to be confirmed, however, that this is the ancestral function of type 1 fimbriae.

Numerous studies over the last 20 years suggested that type 1 fimbriae might be important in UTIs, particularly cystitis, although this role was often questioned. Over the last 5 to 10 years, several studies have much more clearly documented that type 1 fimbriae are important in the colonization of the urinary bladder by *E. coli*, as reviewed in greater detail below.

HISTORICAL PERSPECTIVE

Type 1 fimbriae were the first fimbriae to be described, but they should probably also be designated type 1 because they are the most common type of fimbriae among *E. coli* strains and, indeed, among most other members of the family *Enterobacteriaceae*. In fact, they are sometimes referred to simply as common fimbriae. The ability of *E. coli* to agglutinate erythrocytes was first described by Guyot at the turn of the 20th century (48). In 1991, Hultgren et al. (62) noted that this discovery coincided with the last time the Chicago Cubs won the baseball World Series. To the exasperation of baseball fans everywhere (even fans of other teams, e.g., the Boston Red Sox), this statement is still accurate a dozen years later. In the early 1940s, Rosenthal observed that *E. coli* also agglutinated many other types of cells (127). Interestingly, in that report he compared the agglutinating activity of *E. coli* to that of the phytoagglutinins and found them to be similar in several respects, even though the lectin nature of the phytoagglutinins had not yet been discovered.

The introduction of the electron microscope enabled researchers to get the first good look at bacterial surface structures that might be involved in the agglutination reactions described by Guyot and by Rosenthal. One of the earliest images of fimbriae appeared in a 1950 electron microscopic study of flagellation (60), but these new organelles were not mentioned until the article's Appendix. It is a testimony to the superior observational skills of Houwink and van Iterson that, even in what almost amounted to an afterthought, the authors speculated correctly that these thin, straight filaments emanating from the surface of *E. coli* were probably organelles of attachment. It is not unreasonable to think that the filaments shown in this 1950 study were type 1 fimbriae, because of their structural resemblance to type 1 fimbriae and because type 1 is the single most common fimbrial type of *E. coli*.

In 1955, Collier and de Miranda (21) were the first to note the ability of D-mannose to inhibit the agglutinating properties of *E. coli*, but it was primarily Duguid's and Brinton's research groups that initiated the characterization of bacterial agglutination by type 1 fimbriae (17, 18, 30–33).

In 1977, Salit and Gotschlich showed that hemagglutination caused by purified type 1 fimbriae could be inhibited by D-mannose (128). In the same year, Ofek et al. (111) showed that adherent fractions of *E. coli* were better fimbriated than were nonadherent fractions and that adhesion of *E. coli* to human epithelial cells was also inhibited by D-mannose. In 1979, Aronson et al. (5) were able to prevent urinary tract colonization by *E. coli* by instilling the bacteria in the presence of D-mannose. In the mid-1980s, Duncan and coworkers (61, 130) showed that *E. coli* phase variants selected in broth to produce type 1 fimbriae were better able to colonize mouse bladders than were nonfimbriated phase variants and that, when nonfimbriated cells were able to colonize the bladder, they had shifted to a fimbriated phenotype. Keith et al. (75) created mutants with mutations in the fimbrial operon and found that they were not able to colonize mouse bladders. Using a much better characterized set of strains, Connell et al. (22) confirmed, by complementation, that type 1 fimbriae, more specifically the FimH lectin, were important for bladder colonization. In spite of these strongly suggestive studies, however, epidemiological studies could not implicate type 1 fimbriae in disease because of their ubiquity.

E. COLI TYPE 1 FIMBRIA STRUCTURE AND FUNCTION

fim Genetics

The cluster of nine genes directly responsible for the expression of type 1 fimbriae (Fig. 1) is located on a single 9-kb fragment of DNA in the *E. coli* K-12 chromosome (8, 81, 84, 119, 120). This is in contrast to some other fimbrial genes, which are located on plasmids or present in multiple copies on the chromosome. The production of several hundred of these relatively large surface appendages (Fig. 2) consumes a significant fraction of cellular resources, and so their expression is carefully regulated. One essential

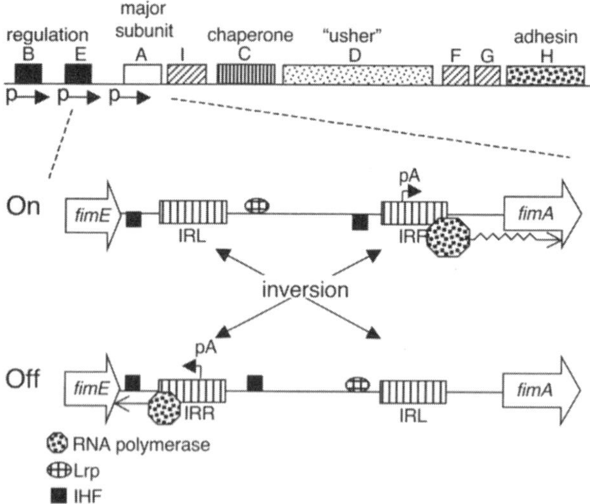

Figure 1. Schematic illustration of the genetic organization of the *fim* genes. The location of the genes within the *fim* gene cluster and their roles in regulation or biogenesis are indicated. Switching from ON to OFF phases is controlled by the inversion of a segment of DNA located between the *fimE* and *fimA* genes that contains the promoter for the *fimA* gene. Inversion is affected by FimB and FimE, as described in the text. In the ON orientation, *fimAICDFGH* is successfully transcribed, but in the OFF orientation, the message is aborted. Integration host factor (IHF) and Lrp bind to elements within the switch and also affect rates of inversion. Modified from reference 116 with permission.

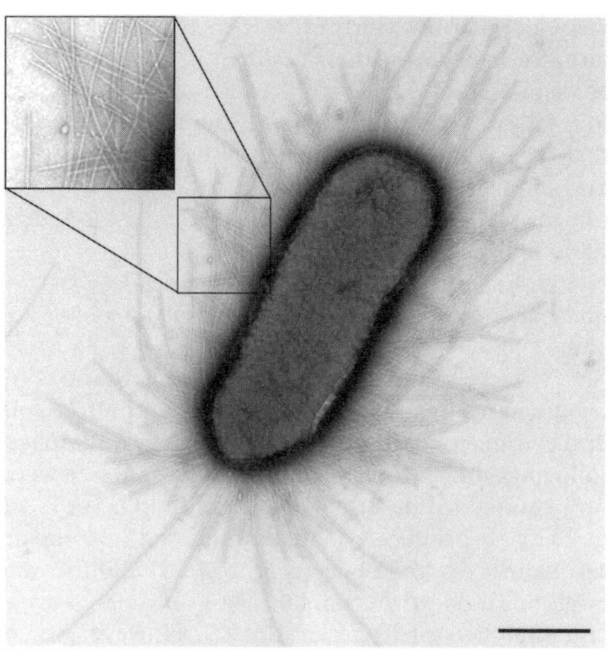

Figure 2. Electron micrograph of type 1 fimbriated *E. coli* strain illustrating the typical numbers, lengths, and general morphology of the hair-like surface appendages. Type 1 fimbriae, and others of its class, are peritrichously arranged, can number in the hundreds per cell, and are rigid-appearing, straight structures. There are 1,000 or more FimA subunits polymerized into a 7-nm-diameter helical structure. This subunit makes up the majority of the fimbrial structure. The mannose-binding lectin, FimH, is located at the distal tips of the fimbriae (see Fig. 4 and 5). Reprinted from reference 52 with permission from Elsevier.

characteristic of type 1 fimbrial expression is the ability to vary between ON and OFF phases. Interestingly, the ability to turn the expression of fimbriae OFF is at least as important as being able to turn them ON, such as when the organisms need to evade phagocytes or invade the interstices of a mucous gel (97, 100, 114).

Variation from a nonfimbriated to a fimbriated phase is controlled primarily by a phase "switch," a 314-bp invertible segment of DNA immediately upstream of the *fimA* gene encoding the major subunit (1) (Fig. 1). This element includes a promoter that directs the transcription of *fimAICDFGH* when it is in the ON orientation. Two regulatory genes at the 5′ end of the cluster, *fimB* and *fimE,* encode proteins that resemble the lambda integrase family of site-specific recombinases (29, 78) and affect the rate of DNA inversion. FimE promotes inversion preferentially from ON to OFF, while FimB promotes inversion in both directions, with minimal preference for OFF to ON (44, 98). Certain wild-type *E. coli* strains possess a variety of mutations within or adjacent to the *fim* switch that affect phase variation to different degrees (90).

There is an interesting cross talk between certain members of the family of regulatory proteins to which FimB belongs (58, 150). PapB, a regulator of the expression of P fimbriae, as well as its homologs SfaB (regulator of S fimbriae) and PefB (regulator of *Salmonella enterica* serovar Typhimurium Pef fimbriae), was shown to repress FimB-promoted switching of type 1 fimbrial expression. PapB was the only one of the three to be completely effective, while other members of the family, DaaA (of F1845 fimbriae), FaeB (of K88 fimbriae), FanA and FanB (of K99 fimbriae), and ClpB (of CS31A fimbriae), had no effect on the *fim* switch. Such cross talk could be tremendously important in switching from one fimbrial type to another in different segments of the urinary tract.

Phase variation of the *fim* switch is also affected by multiple *cis*-active sequences upstream of the start site for *fimB* (35). The *fimB* gene is separated from the divergently transcribed *yjhATS* operon by a 1.4-kb intergenic region whose function is unknown. Two regions in this intergenic segment appear to be involved in affecting *fimB* expression by an antirepression mechanism. This mechanism of control is in some way related to the sensing of *N*-acetylneuraminic acid, which suppresses type 1 fimbriae expression. The authors suggest that this is a mechanism whereby *E. coli* may be able to detect the activation of a host defense reaction.

Several other accessory proteins bind to the *fim* switch region (Fig. 1), resulting in reorientation of the DNA into a conformation more favorable for inversion. These accessory proteins include integration host factor (13, 29, 34), the leucine-responsive regulatory protein (Lrp) (12, 43) and the histone-like protein, HNS (74, 118).

The presence of an asymmetric restriction site within the *fim* switch enabled the development of a relatively simple means of determining ON versus OFF orientation in isolates from experimental infections or from clinical specimens. Isolates from experimental murine UTI and from women with UTI showed that type 1 fimbrial genes were turned ON in both the bladder and the kidneys (91). Interestingly, genes from the more virulent strains (i.e., those causing pyelonephritis) showed a resistance to switch from OFF to ON when they were grown in liquid medium, while those from normal fecal isolates readily switched from OFF to ON. Cystitis isolates were more likely to maintain the switch in the ON orientation, while pyelonephritis isolates were more likely to be in the OFF orientation. A subsequent study of experimental murine UTI showed that the percentage of the *E. coli* population with the switch in the ON orientation correlated with the CFU per gram of bladder tissue but not with the CFU per gram of kidney tissue (46). When the phase switch was genetically locked in the OFF position, bacteria were recovered from experimental infections in much smaller numbers than when the phase switch was locked in the ON orientation, at least during the first 24 h (47). Gunther et al. (46, 47) have suggested that the phase switch itself is a urovirulence determinant.

Although the *fim* gene cluster is not located on a PAI, expression of type 1 fimbriae can still be affected by them, at least in certain strains of *E. coli*. One facet of the mobility of PAIs is that they sometimes spontaneously excise (50). Since they are frequently inserted within or next to tRNA genes, PAI insertion or excision can have effects on the translation of any genes utilizing codons specific for the affected tRNA. PAI excision affected the *fim* switch in this manner. PAI II of uropathogenic *E. coli* strain 536 is inserted at the leucine-specific tRNA locus, *leuX* (110). Excision of this PAI II destroys the *leuX* locus, and so genes containing the rare leucine codon TTG, which is recognized specifically by the *leuX* product, $tRNA_5^{Leu}$, are not transcribed. The *fimB* gene has five such codons, and *fimE* has two, so that the interruption of *leuX* has a dramatic effect on type 1 fimbria production in uropathogenic *E. coli* strain 536 (126). Undoubtedly, there are other phenomena, yet to be described, that will add to the complexities of fimbrial regulation.

Fimbrial Biogenesis

E. coli typically produces on the order of several hundred type 1 fimbriae per cell (Fig. 2). Fimbrial lengths vary for unknown reasons, but typically they are 1 to 2 μm. Using electron microscopy and X-ray diffraction, Brinton showed in the mid-1960s that type 1 fimbriae consisted of subunits polymerized into a right-handed helix, with 3.125 subunits per turn, having a pitch distance of 2.38 nm, a screw angle of 115.2°, a diameter of 7 nm, and an axial hole of 2 nm (17). More recent studies of type 1 fimbriae (52, 104) or P fimbriae, which follow a similar structural plan (19), have refined Brinton's early work (Fig. 3).

The functions of all but one of the proteins encoded by the *fimAICDFGH* region of the operon are known (Fig. 1 and Fig. 4) (83). FimA is the primary structural subunit and makes up 98% or more of the fimbrial mass (36, 54, 77). Data on the possible function of FimI have appeared only very recently. The results suggest that FimI is required for fimbriation (146), but its precise role is still far from clear. FimC is a periplasmic chaperone which is essential for proper folding of the nascent polypeptide chains of *fim* gene products, docking at the FimD outer membrane

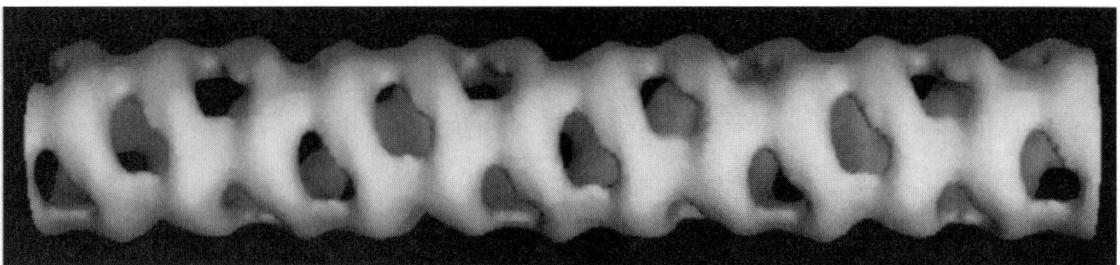

Figure 3. A 3-D reconstruction of a type 1 fimbria. The segment displayed in this model comprises 40 FimA subunits and covers 1.5 helical repeats. It has been surface rendered to include 100% of the nominal mass. The model shows the type 1 fimbria to be a hollow tube with walls that are formed by a helical string of elongated subunits associated in a head-to-tail orientation. Adjacent turns of the helix are connected via three binding sites, making the fimbriae relatively rigid structures. Reprinted from reference 52 with permission from Elsevier.

assembly complex and initiating the incorporation of subunits into fimbriae (69, 70, 79). As each fimbria develops, it extends toward the periphery by adding subunits at the base (92). FimF, FimG, and FimH, whose genes are at the 3′ end of the *fim* cluster, are incorporated into the fimbrial structure in extremely small amounts (3, 54, 86, 87). These minor subunits are now known to be presented as a tip fibrillum that is roughly half the diameter of the main fimbrial shaft (Fig. 5). Maurer and Orndorff (95) and Minion et al. (103) were the first to show evidence that the type 1 fimbrial adhesin was distinct from the primary structural subunit. In the next year, Klemm and Christiansen (82) as well as Maurer and Orndorff (96) published much more convincing sequences of the entire *fim* gene cluster, along with somewhat more in-

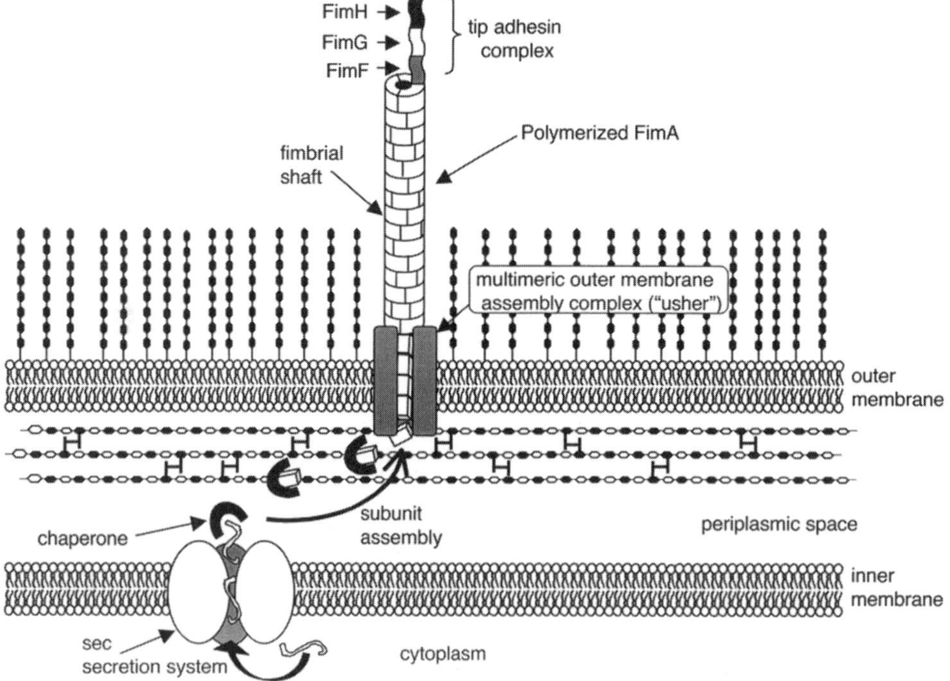

Figure 4. Schematic model for biogenesis of type 1 fimbriae. Nascent polypeptides of fimbrial subunits are transported across the inner membrane by the general secretion pathway. The periplasmic chaperone, FimC, binds to the polypeptides as they are being transported through the inner membrane. Polypeptides not protected by the chaperone are thought to be degraded by periplasmic proteases. The FimH subunit is held by FimC in a mannose-binding, nonpolymerizing form in the periplasm until delivered to the FimD assembly complex, or usher. The chaperone-subunit complexes arrive at the outer membrane usher, where they bind to previously delivered subunits, traverse the outer membrane as a ~2-nm-diameter linear filament, and then coil into a helical form at the external face of the usher. In this pathway, the translocation of subunits is highly ordered, with translocation of a FimH subunit being followed by that of FimF, FimG, and, finally, hundreds to thousands of copies of FimA. The precise number of FimF, FimG, and FimH subunits in this illustration is not known with absolute certainty. Reprinted from reference 116, with permission.

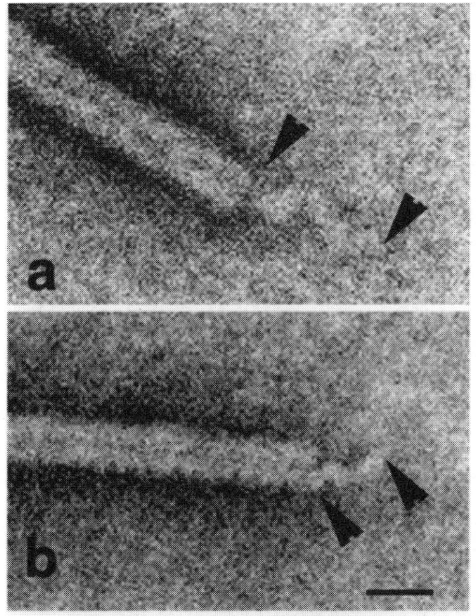

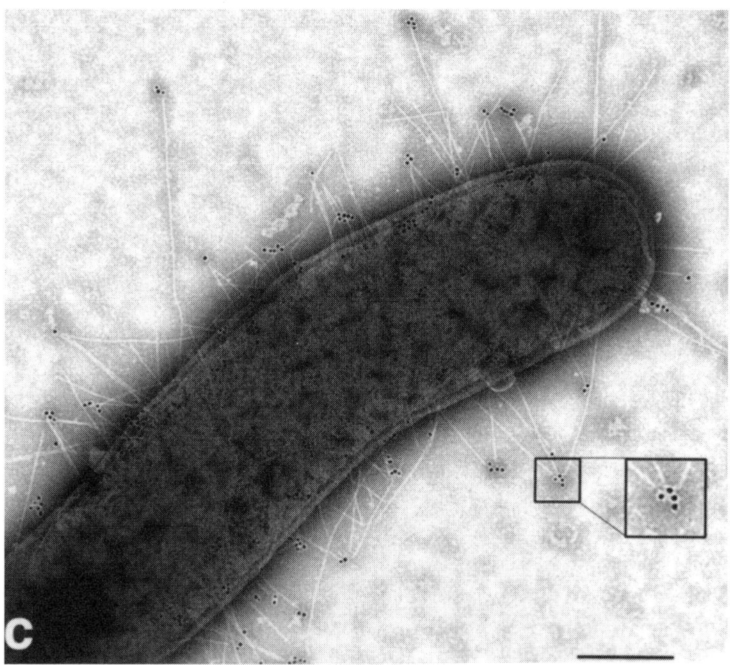

Figure 5. Electron micrographs of type 1 fimbriae. (a and b) Isolated type 1 fimbriae negatively stained with uranyl formate. The main fimbrial shaft appears to be rather rigid and contains a central cavity, indicated by the dark thread of stain running parallel to the fimbrial axis. The 7-nm shafts end in a flexible, loosely coiled tip fibrillum (arrowheads) roughly half the diameter of the main shaft. (c) Immunolocalization of FimH at the fimbrial tips. Colloidal gold particles 8 nm in diameter were coated with polyclonal rabbit antibody against FimH. Gold particles are found exclusively at the fimbrial tips. Reprinted from reference 52 with permission from Elsevier.

clusive functional analyses. In this same year, Hanson and Brinton (53) published chemical analyses of purified fimbriae that confirmed several of the observations of those taking a more genetic approach. Because FimF, FimG, and FimH make up a tip fibrillum (52, 64, 71) that is the initial fimbrial structure to polymerize, their presence or absence has significant effects on fimbrial biogenesis. In fact, in the absence of FimH, the bacteria usually form no fimbriae or a single, unusually long, nonfunctional fimbria.

FimH RECEPTOR SPECIFICITY PRIOR TO 1992

We owe much of our understanding of the fine sugar specificity of the FimH adhesin to the combined expertise of a bacteriologist, Itzhak Ofek, and a lectinologist, Nathan Sharon and their colleagues (38–40, 111–113), with additional key experiments being performed by Neeser et al. (108). Up until 1992, it was assumed that the primary, if not exclusive, receptor for the FimH adhesin of *E. coli* type 1 fimbriae was a Man_5 unit of N-linked glycans (Fig. 6, structure 1). The FimH saccharide-binding pocket was thought to be in the form of a trisaccharide, and because hydrophobic derivatives of mannosides were much more potent inhibitors, the ligand-binding pocket was thought to have a closely associated hydrophobic group. Because hybrid-type N-linked glycans had an exposed trimannose unit (Fig. 6, structure 3), these types of N-linked glycans were also thought to be

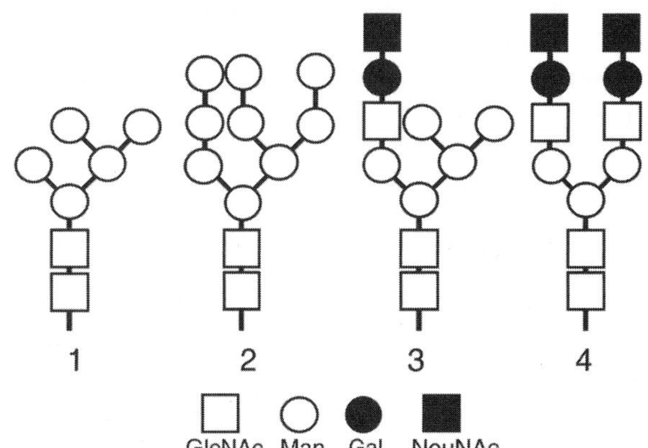

Figure 6. Schematic diagram of typical N-linked glycans. Saccharide structures 1 and 2 are examples of high-mannose oligosaccharide chains. Saccharide 3 is an example of a typical hybrid-type glycan unit, one arm of which bears a trisaccharide. Saccharide 4 is an example of a complex-type glycan unit, both arms of which are terminally substituted with saccharides other than mannose (i.e., there is no terminal mannose). Prior to 1992, the primary type 1 fimbrial receptor was expected to have the structure of saccharide 1 or 3. Modified from reference 55 with permission.

potential receptors. Terminal substitution of the trimannose units with other sugars, even mannose (Fig. 6, structures 2 and 4), dramatically reduced the effectiveness of inhibitors. There was certainly no indication at the time that FimH would bind single terminal mannose residues. Indeed, monosaccharide receptors are not typical for lectins.

E. COLI FimH PHENOTYPES

E. coli Binds to Human Plasma Fibronectin

The first indication that there were important functional differences among the FimH adhesin subunits of *E. coli* isolates was the observation that the CSH-50 strain, a derivative of *E. coli* K-12 which expresses only type 1 fimbriae, was able to adhere to human plasma fibronectin (Fn). Fn does not possess high-mannose oligosaccharides. It possesses only O-linked glycans that do not contain mannose and complex-type N-linked glycans with no terminal mannose. Therefore, it was quite surprising to find that Fn was an excellent substratum for attachment of *E. coli* CSH-50 and that it was sensitive to inhibition by D-mannose (135). It was even more surprising to find that CSH-50 bound in a mannose-sensitive manner to Fn domains completely devoid of oligosaccharide moieties. Because a recombinant strain bearing the *fim* gene cluster of *E. coli* K-12 strain PC31 [HB101(pPKL4)] produced type 1 fimbriae that did not confer either Fn- or protein-binding activities, we initially speculated that the mutagenesis experiments to which *E. coli* CSH-50 had been subjected at Cold Spring Harbor Laboratory (101) had created an aberrant protein. However, a survey of *E. coli* strains isolated from human urine showed that the unusual binding activities were also found in wild-type strains.

Mannose-sensitive, FimH-mediated binding to a protein moiety has also been found in a meningitis-associated *E. coli* isolate, strain IHE 3034 (123). IHE 3034, an O18ac:K1:H7 serotype strain, binds to types I and IV collagen but not to type III collagen. None of these extracellular matrix components possess terminal mannose residues, yet adhesion was blocked by α-methyl mannopyranoside. The possibility that adhesion to collagens was mediated by S fimbriae (another meningitis-associated virulence factor) was ruled out. The IHE 3034 FimH varied from the PC31 FimH at five residues, but a single A62S substitution was shown to be responsible for conferring the ability to bind to collagen. A similar mutation was found in two other O18ac:K1:H7 strains. The precise role of the A62S substitution in the pathogenesis of meningitis remains to be determined.

Three Distinct Phenotypes of FimH

Because the unusual binding activities described above had never been observed during the almost 40 years of studying adhesion of type 1-fimbriated *E. coli* strains, we were initially skeptical of the validity of the observations. Therefore, we constructed a series of recombinant strains that differed only by the *fimH* genes they contained (136). In initial studies, the *fimH* genes were cloned by PCR from CSH-50, PC31, and four human UTI isolates. Plasmids bearing the *fimH* genes were transformed into *E. coli* strain AAEC191A, a Δ*fimBEAICDFGH* derivative of *E. coli* K-12 strain MG1655 (11), along with plasmid pPKL114 that bears the entire *fim* cluster but with a translational stop-linker inserted into the 5′ end of the *fimH* gene (Fig. 7). These isogenic strains then expressed type 1 fimbriae that differed only by the FimH protein that was incorporated into the fimbriae. The ability of these isogenic parent/mutant

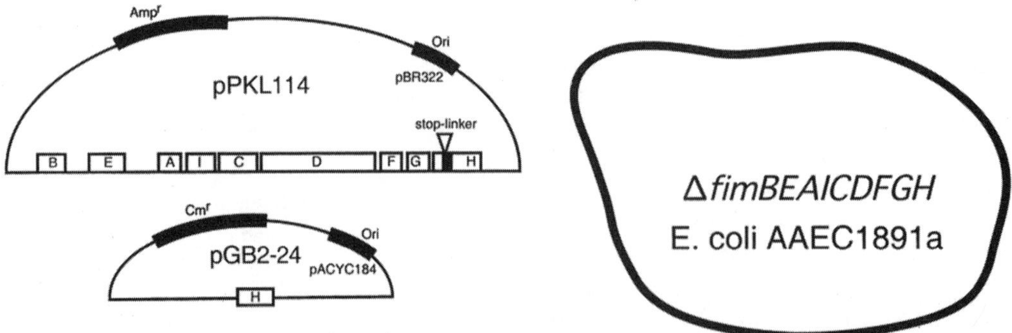

Figure 7. Schematic diagram of the recombinant strains constructed to test the phenotypes conferred by different alleles of *fimH*. The host strain used was the Δ*fimBEAICDFGH E. coli* K-12 strain AAEC191A (11). Plasmid pPKL114 contains the entire *fim* gene cluster in a pBR322 replicon, with a stop linker inserted into the *fimH* gene. Because *fimH* is the last gene in the *fim* cluster, no polar effects would be expected. Plasmid pGB2-24 and subsequent derivatives contain *fimH* genes in a pACYC184 replicon, and these plasmids complement the *fimH* defect of pPKL114.

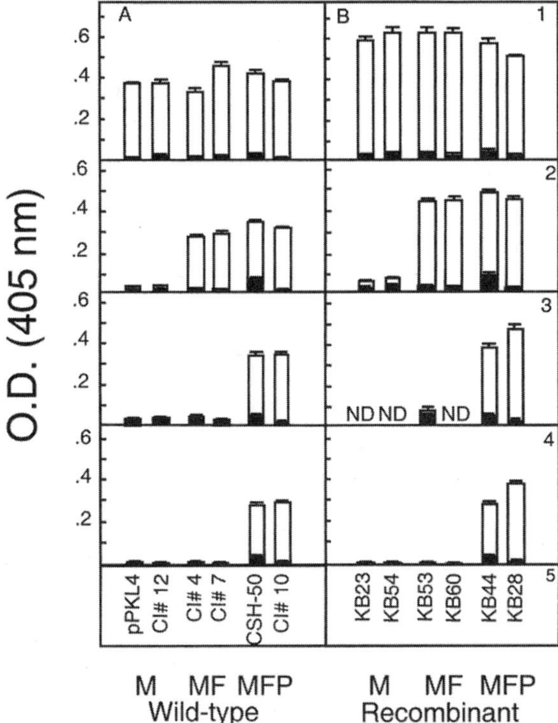

Figure 8. Adhesion of representative wild-type (A) and recombinant (B) M, MF, and MFP class strains to mannan (panel 1), Fn (panel 2), periodate-treated Fn (panel 3), and a synthetic peptide (panel 4). Strain designations are given in panels 5. The recombinant strains are constructed as indicated in the legend to Fig. 7 and the text. Open columns indicate bacteria incubated without D-mannose; solid columns indicate bacteria incubated with D-mannose. Values are the means and standard errors of the mean (n = 4) for each column. ND, not determined. O.D., optical density. Reprinted from reference 136 with permission.

pairs to adhere to four different substrates was evaluated. The test substrata were (i) yeast mannan, (ii) human plasma Fn, (iii) periodate- or endogylcosidase-treated human plasma Fn, and (iv) a synthetic peptide of Fn (Fig. 8). Three different adhesive phenotypes were observed. One, designated M, bound only to mannan out of the four substrata tested. Another, designated MF, could bind to both mannan and Fn. A third, designated MFP, could bind to mannan, Fn, and the protein/peptide substrata. The *fimH* genes of these and six other strains exhibiting M, MF, or MFP phenotypes were sequenced (Fig. 9) (137). The deduced FimH sequences showed variations in structure, as logically followed from the phenotypic variation observed, but there was no clear motif belonging to any of the phenotypic classes.

Quantitative Variation within the M Phenotype

As additional wild-type *E. coli* strains were analyzed in the above manner (137), we found that the majority of the isolates were of the M class, meaning that they could bind to mannan but not to Fn or the peptide substrata. Within the M phenotype, adhesion to mannan varied by up to 15-fold (Fig. 10). In fact, the level of adhesion of many of these strains was so low that one wondered why the organism would devote the tremendous resources required by fimbrial expression, simply to generate such an extremely poor adhesin. This observation suggested that a great many *E. coli* strains that have been characterized in epidemiological studies as non-type-1 fimbriated may simply have been strains that expressed the low-binding FimH and were therefore missed in phenotypic assays, especially those relying on mannan binding. Again, the recombinant strains varying only by the *fimH* genes showed quantitative differences comparable to the parent strains from which the *fimH* genes were obtained (Fig. 11). No striking differences could be found between the highest binding and lowest binding of the strains in terms of fimbrial number, fimbrial length, and relative amounts of FimH protein incorporated into fimbriae. These results suggested that conformational differences in the FimH subunit alone were responsible for the differences in *E. coli* adhesion.

Further efforts to understand the adhesive properties of FimH variants were focused primarily on the high- and low-binding variants of the M phenotype, since they are predominant among wild strains (137). A survey of 42 UTI isolates and 43 isolates from the feces of healthy adults indicated that a majority of the normal fecal isolates exhibited a low-mannan-binding phenotype while the UTI isolates had predominantly a high-mannan-binding phenotype (Fig. 12). The ability of seven isogenic strains constructed as described above to bind to mannan correlated directly with their ability to bind either to J82 human bladder epithelial cells (Fig. 13) or to A498 human kidney epithelial cells (138). Thus, the quantitative differences observed in the ability of these variants to bind to mannan extended to their ability to bind to these two human cells.

In an attempt to explain the nature of these variations in terms of different receptor specificity or affinity, the adhesion of one high-mannan-binding strain (KB54) and one low-mannan-binding strain (KB91) was examined in more detail (138). Scatchard plot analyses of equilibrium measurements showed that differential binding to mannan could be explained by differences in the numbers of high- and low-affinity sites on the mannan substratum for the two different FimH lectins (Fig. 14).

Because these quantitative studies showed that the low-mannan-binding strain did exhibit a small number of high-affinity binding sites, we predicted that the low-binding strain might serve as an effective

Residue Number	Wild Strain	Recombinant Strain	Adhesive Phenotype	Source
--+++++++++++++++++				
111111122				
21 23567779011116607				
16173860381767893619				
||||||||||||||||||||				
MTFVNLGNGSPLGVAIVRHQ	PC31	KB	M[a]	K-12
-----**R**--------------	CSH50	KB44	MFP	K-12
---**A**----------------	- -	KB21	M	chimera
---**A**--------**ΔΔΔΔ**----	CI#10	KB28	MFP	UTI, Memphis, TN
-N-**A**----------**V**-----	CI#3	KB59	M	UTI, Memphis, TN
-N-**AH**-----L---------	CI#7	KB60	MF	UTI, Memphis, TN
---**A**-------------**H**--	KS54	KB92	M	UTI, Sweden
---**A**-------------**H**--	MJ9-3	KB97	M	UTI, Boston, MA
---**A**---S-N----------	F18	KB91	M	Feces, Kingston, RI
---**A**---S-N------**A**---	MJ11-2	KB99	M	UTI, Boston, MA
---**A**--**D**S-N------**A**---	MJ2-2	KB96	M	UTI, Nairobi, Kenya
---**A**---S**E**N--------D-	CI#4	KB53	MF	UTI, Memphis, TN
---**A**---S-N-------**C**--	CI#12	KB54	M	UTI, Memphis, TN

[a] The adhesive subclass of PC31 is based on data using recombinant strain only.

Figure 9. Deduced amino acid sequences of several FimH variants. The polymorphic sites (sites in which there has been a nonsynonymous mutation in the codon) within the 300-residue FimH sequence are indicated. The positions are numbered vertically above each polymorphic amino acid, compared to the original FimH sequence published by Klemm and Christiansen (82). Positions that do not vary among the FimH alelles sequenced thus far are not present in the figure. Δ indicates a deleted residue. Substitutions that affect the adhesion phenotype are indicated in boldface type. The sequences are divided into two groups that differ from each other at residues 70 and 78, where Asn-to-Ser and Ser-to-Asn substitutions occur. Reprinted from reference 137 with permission.

adhesin if we could identify the correct substratum (138). This hypothesis was confirmed when a series of 16 glycoproteins with different glycosylation patterns were surveyed for their ability to bind type 1-fimbriated *E. coli.* Neither of the M-phenotype strains bound to glycoproteins that possessed only complex-type N-linked glycans. Among the 13 other glycoproteins, which possess variable fractions of high-mannose and complex-type N-linked glycans, there was a variable binding of KB54 and KB91 (Fig.

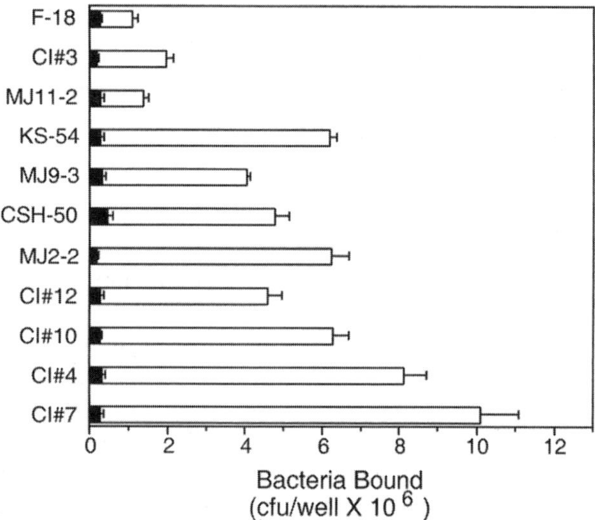

Figure 10. Adhesion of *fimB*-transformed wild-type strains to mannan. Extra copies of *fimB* result in a more uniform expression of type 1 fimbriae, eliminating variable expression levels as one explanation for differences in adhesion. Open columns indicate bacteria incubated without α-methylmannoside; solid columns indicate bacteria incubated with 50 mM α-methylmannoside. Values are the means and standard errors of the mean ($n = 4$). Reprinted from reference 137 with permission.

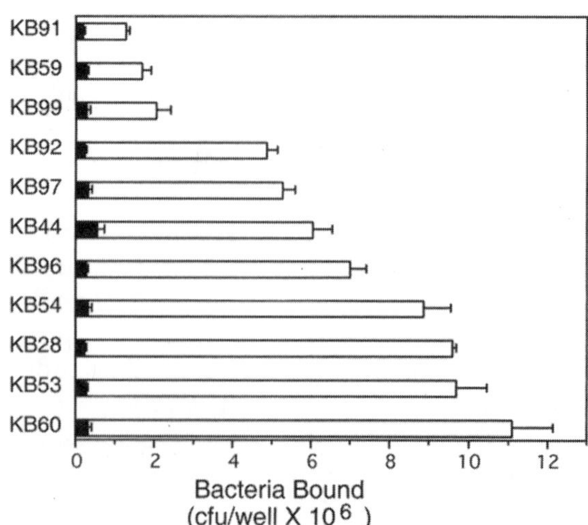

Figure 11. Adhesion of recombinant strains constructed using *fimH* genes cloned from the wild-type strains shown in Fig. 10. Columns and values are as in Fig. 10. Reprinted from reference 137 with permission.

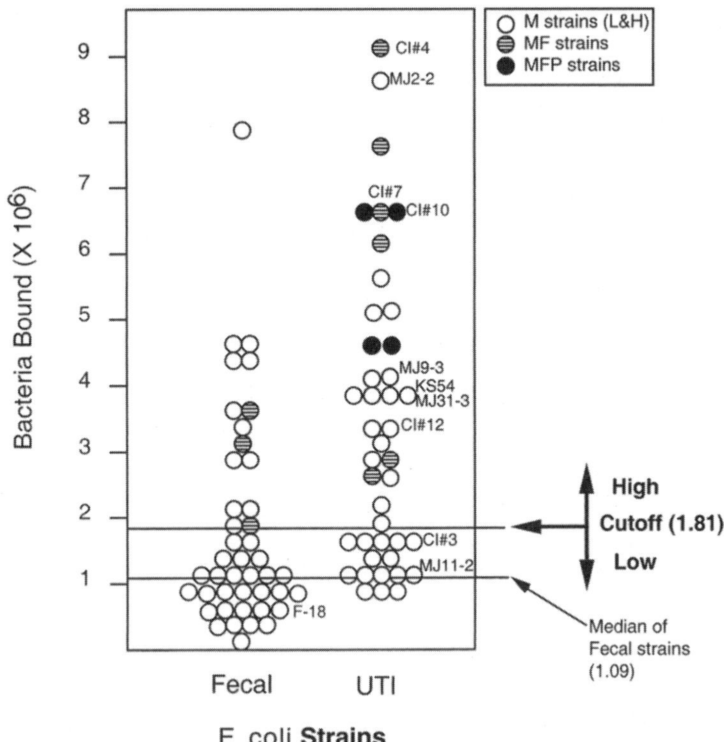

Figure 12. Adhesion of wild-type fecal isolates and isolates from patients with UTIs to mannan. All binding was inhibited by >80% by α-methylmannoside. To simplify the graphic presentation, data are arranged in groups of 0.25×10^6 bacteria bound per well. Since the data are plotted in this way, the actual numbers for circles placed behind the two reference lines fall below the line values, whereas those placed in front of the lines fall above the line values. Reprinted from reference 137 with permission.

15). Of the 13 substrata, 3 (i.e., human immunoglobulin A, lactoferrin, and bovine RNase B) bound the two strains equally. Of these three, bovine RNase B was selected for further study because it has only a single N-linked glycan per molecule and the structure is relatively uniform, being 57 mol% Man_5-$GlcNAc_2$ units and 31 mol% Man_6-$GlcNAc_2$ units (42) (Fig. 16). When adhesion of the low-mannan-binding and high-mannan-binding strains to RNase B was studied in equilibrium-binding experiments, the curves in a

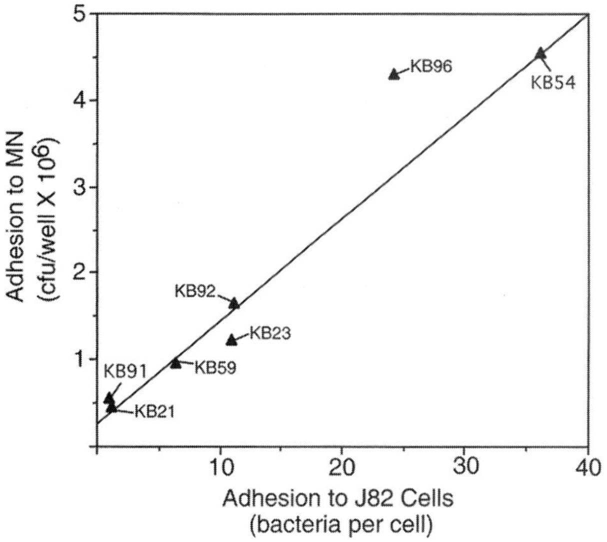

Figure 13. Correlation of the abilities of seven recombinant strains to bind to mannan (MN) with their abilities to adhere to the J82 bladder epithelial cell line. Strain numbers are shown. Statistical analysis is given in the text. Reprinted from reference 138 with permission from the publisher.

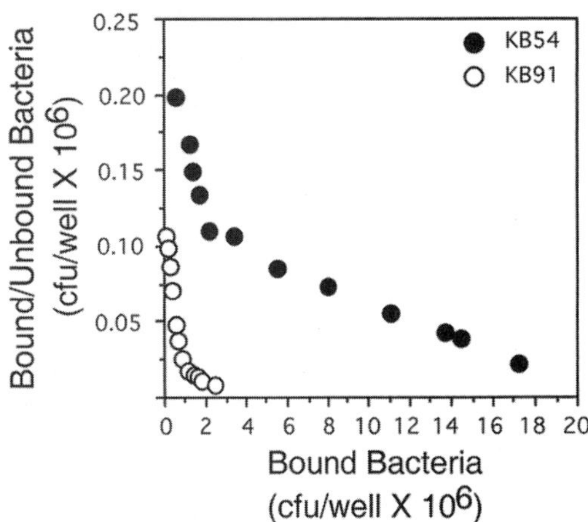

Figure 14. Scatchard plot analyses of the binding of strains KB54 and KB91 to mannan at equilibrium. Data from a single experiment are presented, but the experiment was repeated several times and the results were essentially the same. Reprinted from reference 138 with permission from the publisher.

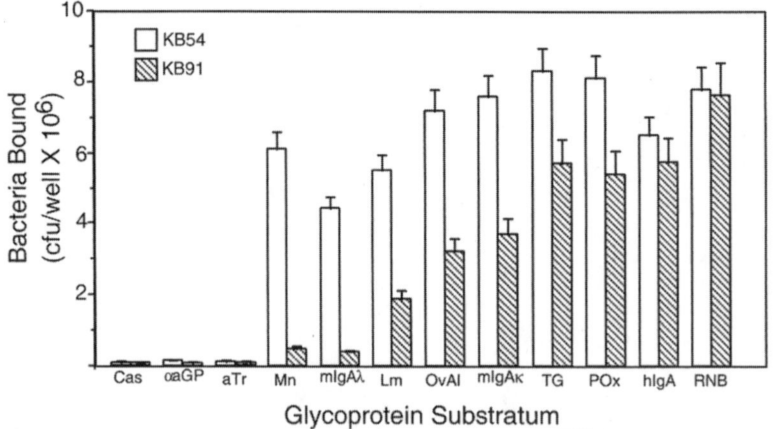

Figure 15. Adhesion of strains KB54 and KB91 to various glycoproteins. Abbreviations: Cas, bovine milk casein; αaGP, human α-acid glycoprotein; aTr, human serum *apo*-transferrin; Mn, yeast mannan; mIgAλ, mouse immunoglobulin Aλ; Lm, human laminin; OvAl, chicken egg albumin; mIgAκ, mouse immunoglobulin Aκ; TG, porcine thyroglobulin; POx, horseradish peroxidase; hIgA human immunoglobulin A; RNB, bovine RNase B. Values are means and standard errors of the mean (n = 3). Reprinted from reference 138 with permission from the publisher.

Scatchard plot were indistinguishable (Fig. 17). These experiments demonstrated that the low-mannan-binding FimH subunit would serve as a very effective adhesin in the presence of an appropriate substratum, and they also clearly showed that the two types of FimH subunit utilized different mechanisms of ligand-receptor interaction.

Quantitative Differences Are Due to Variable Ability to Bind to Single Mannosyl (Man_1) Residues

Simpler mannosides were used in an effort to further examine the ligand-receptor interactions exhibited by these two FimH subunits. From earlier work, the trimannosyl core of the Man_5 units could be predicted to be a major receptor, and so we compared the ability of *E. coli* to bind to trimannosyl (Man_3) groups coupled to bovine serum albumin (BSA) with their ability to bind to single mannosyl (Man_1) groups conjugated to BSA. Again, the same seven isogenic strains used above were compared in their ability to bind to Man_1-BSA and Man_3-BSA. All of the strains bound relatively well to Man_3-BSA, and there was a positive correlation of their ability to bind to Man_3-BSA and their ability to bind to bovine RNase B (Fig. 18). Some of the strains bound well to Man_1-BSA, and others bound very poorly. There was a very strong positive

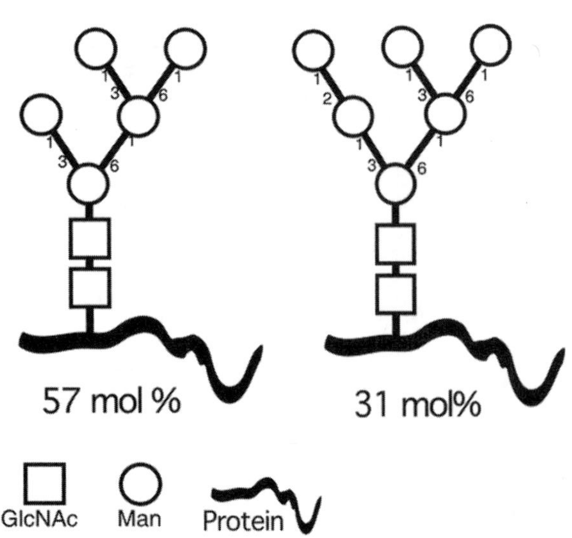

Figure 16. Schematic diagram of the N-linked glycan units of bovine RNase B. Data from reference 42. Reprinted from reference 55 with permission.

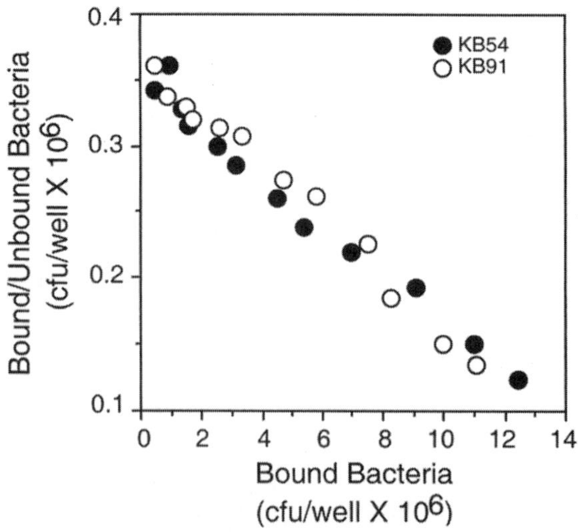

Figure 17. Scatchard plot analyses of binding of strains KB54 and KB91 to bovine RNase B at equilibrium. Data from a single experiment are presented, but the experiment was repeated several times and the results were essentially the same. Reprinted from reference 138 with permission from the publisher.

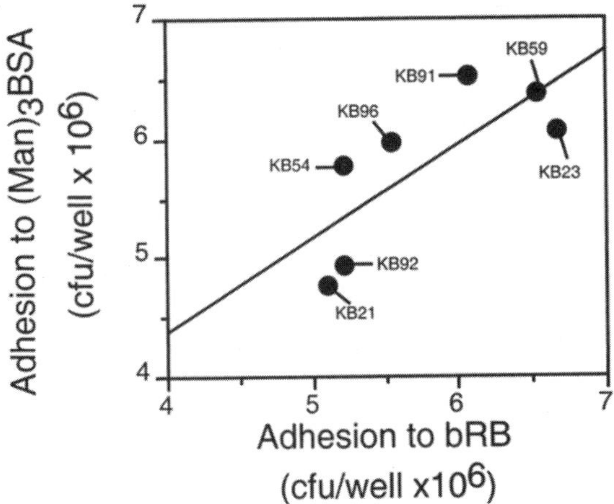

Figure 18. Correlation of the adhesion of seven recombinant strains (Fig. 13) to Man_3-BSA with their adhesion to bovine RNase B (bRB). Strain numbers are shown. Reprinted from reference 138 with permission from the publisher.

correlation between the ability of the strains to bind to Man_1-BSA and their ability to bind to mannan (Fig. 19). Additional experiments with wild-type strains provided further evidence that all type 1 fimbriated *E. coli* strains bound in relatively similar numbers to Man_3-BSA but there were dramatic differences in their abilities to bind Man_1-BSA (Fig. 20). Because FimH binding to Man_3 residues is conserved while binding to Man_1 residues varies considerably, the two phenotypes are now referred to as low-Man_1-binding and high-Man_1-binding FimH alleles. The critical importance of Man_1 binding in the tropism of type 1 fimbriated *E. coli* to the urinary tract has recently been confirmed by another laboratory (63).

Publication of the crystal structure of FimH (20, 63) provided insights into how the various mutations could affect the FimH phenotype. To crystalize FimH, it was necessary to begin with a soluble molecule. Overexpression of FimH alone led to its degradation in the periplasm, but soluble FimH could be obtained by overexpressing FimH in a complex with its periplasmic chaperone, FimC. The three-dimensional (3-D) structure showed that FimH is folded into two domains that are connected by a short, extended linker arm (Fig. 21). The NH_2-terminal domain (residues 1 to 156) is the lectin domain. The COOH-terminal domain (residues 160 to 279) is the pilin domain. The lectin domain is an 11-strand elongated β-barrel with the mannose-binding pocket located at the distal tip of the lectin domain. The pilin domain has an immunoglobulin-like topology with a missing strand, which is temporarily completed by a strand contributed by the FimC chaperone. During fimbrial biogenesis, the donor strand of the chaperone is eventually exchanged for a similar strand from another subunit (7), presumably FimG. All of the naturally occurring mutations that changed the low-Man_1-binding phenotype to a high-Man_1-binding phenotype were not located within the mannose-binding site. Most mutations were located within the lectin domain, but opposite from the mannose-binding site, at the "bottom" of the lectin domain beta-barrels. The *fimH* gene of PC31 confers a high-Man_1-binding phenotype. When this gene was randomly mutagenized by nucleotide misincorporation during suboptimal PCR, all mutations that changed the phenotype toward low-Man_1-binding also tended to be distant from the mannose-binding pocket (132, 141)

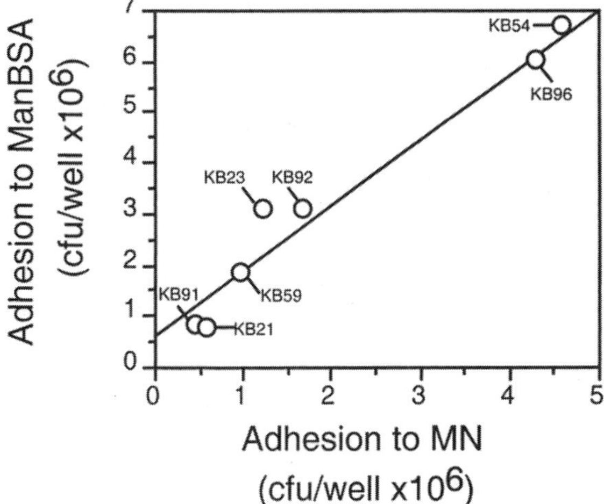

Figure 19. Correlation of the levels of adhesion of the same recombinant strains shown in Fig. 18 to Man_3-BSA with their adhesion to mannan (MN). Strain numbers are shown. Reprinted from reference 138 with permission from the publisher.

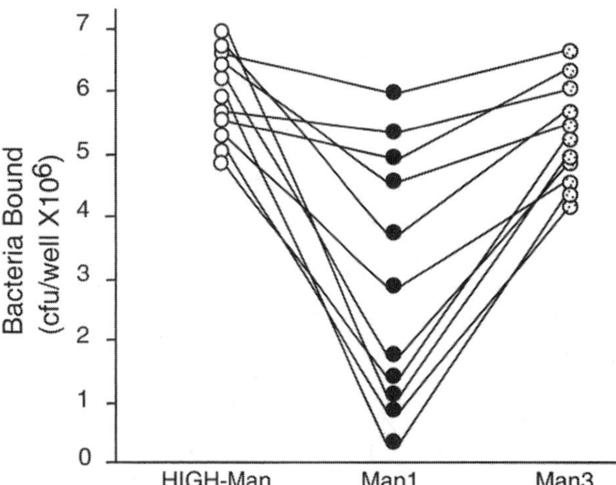

Figure 20. Adhesion of 11 wild-type strains to high-Man moieties of bovine RNase B (HIGH-Man), monomannosylated BSA (Man1) and trimannosylated BSA (Man3). Reprinted from reference 55 with permission from the publisher.

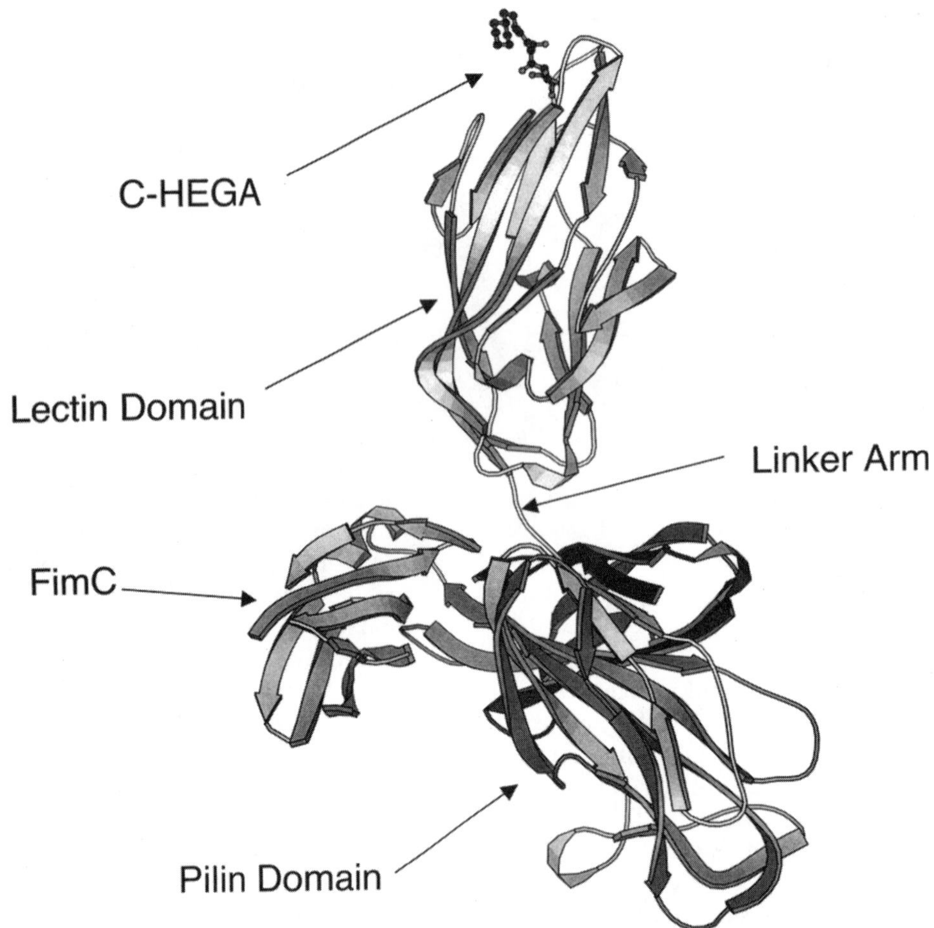

Figure 21. Ribbon diagram illustrating the three-dimensional structure of the FimH lectin subunit complexed with the FimC chaperone and cocrystallized with cyclohexylbutanoyl-*N*-hydroxyethyl-D-glucamide (C-HEGA). FimH is folded into two domains connected by a short linker arm. The NH_2-terminal lectin domain binds to mannosylated receptors, and the COOH-terminal pilin domain anchors FimH to the proximal subunits of the fimbrial superstructure. The lectin domain is an 11-strand elongated β-barrel that exhibits a mannose-size pocket at the tip of the domain distal from the linker arm connecting the two domains. The figure was generously provided by Stefan Knight.

(Fig. 22). In fact, mutations near or within the binding pocket tend to abolish the ability of FimH to mediate binding. The 3-D structure of a single FimH did not provide clear insights into low-Man_1-binding and high-Man_1-binding phenotypes.

Shear Force Affects the FimH Phenotype

Detailed studies of the effects of increasing shear force on the interaction of type 1 fimbriated *E. coli* with erythrocytes provided additional insights into mechanisms of FimH function (145). *E. coli* bearing the low-Man_1-binding FimH-*f18* allele bound erythrocytes weakly. However, at moderate shear forces, much firmer binding was observed. Reciprocating levels of binding were observed when shear forces were alternately increased or decreased, demonstrating the reversibility of shear enhancement of adhesion. *E. coli* bearing the high-Man_1-binding FimH-*j96* allele bound erythrocytes firmly even under low shear forces. It was hypothesized that the changes in adhesion were due to effects of tensile shear force on the tertiary structure of receptor-bound FimH. A computational method called steered molecular dynamics (SMD) simulations was used to predict stretch-induced conformational changes with angstrom precision.

The FimH-*j96* variant of FimH that has been crystallized (20) was used as the model structure in SMD simulations. The C terminus of the linker arm which connects the lectin domain and the pilin domain was pulled in one direction, while the 13 residues taking part in the receptor-binding site were pulled with an equivalent sum force in the opposite

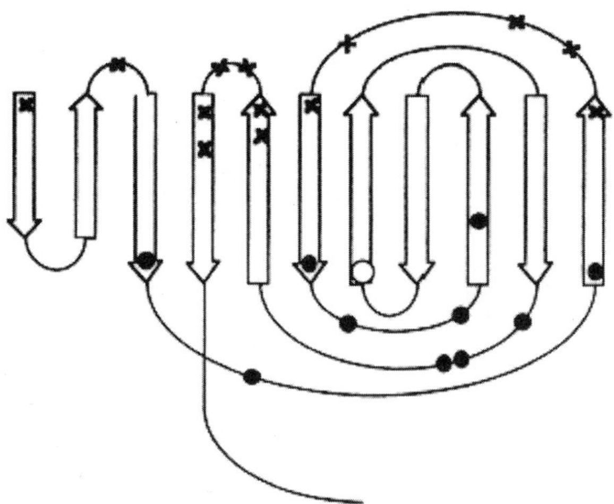

Figure 22. β-Sheet topology diagram of the lectin domain of FimH. The lengths of the sheets and loops do not relect the actual size, but the relative positions of the labeled residues are accurately indicated. The filled circles indicate the positions of point mutations that induce a dual low-Man_1-binding/high-Man_1-binding phenotype (for details, see reference 132). The crosses indicate residues interacting with the receptor analog, C-HEGA. The open circle indicates the position of an E89K mutation (see reference 141). Reprinted from reference 141 with permission from Blackwell Publishing.

direction. This was intended to simulate tension across the lectin domain between the cell-bound mannosylated receptor and the bacterium via the linkage of the pilin domain to the tips of the fimbriae. The greatest structural change predicted was between residues 154 and 156 (VVV) immediately upstream from the linker arm connecting two domains in the crystal structure and residues in two loops of the lectin domain that incorporate the position 154 to 156 residues into the bottom of lectin domain (residues 117 to 120 [GVAI] of the 9-10 loop and residues 26 to 28 [PVV] of the 3-4 loop). The external force applied was predicted to cause hydrogen bonds between these residues to break, causing the linker arm, which is folded into the lectin domain, to pull away, thus lengthening the distance between the lectin and pilin domains. This predicted effect was called "linker extension" (Color Plate 8). This phenomenon, which would eliminate contacts between the lectin and pilin domains, must then lead to other structural events in the FimH molecule. The functional importance of linker extension was then tested by engineering changes in the FimH protein that would make linker extension either easier or more difficult.

Mutations destabilizing bonds between the 154 and 156 (VVV) linker chain anchor and the surrounding loops of the lectin domain were engineered in the low-Man_1-binding FimH-*f18* allele. It was predicted by SMD simulations that a V156P point mutation would make the linker extension easier. In the V156P mutant, the ability to bind to Man_3 residues was conserved, indicating that this mutation did not cause major structural changes in FimH. However, the V156P FimH mutant was able to mediate a dramatically stronger adhesion under static conditions than was the wild-type FimH. That is, the mutation had a functional effect similar to that of the naturally occurring Man_1-enhancing mutations. It is interesting that many of the mutations enhancing Man_1-binding ability previously identified in natural isolates are located within or immediately adjacent to the region of linker arm-stabilizing bonds.

Mutations suppressing linker extension were created in the high-Man_1-binding FimH-*j96* allele by introducing Q32L and S124A substitutions. These mutations make the bonds between the linker arm and surrounding loops less rigid, causing the latter to extend under the tensile force together with the linker arm and thus preventing it from being zipped out (i.e., extended) from the loops. The Q32L/S124A FimH mutant had an unaltered Man_3-binding ability, a significantly reduced ability to bind Man_1 under static conditions, and strong binding only under high shear forces. Thus, the mutations suppressing linker extension converted the high-Man_1-binding variant into rather low-Man_1-binding variants.

Taken together, the FimH mutagenesis results suggest that the change in affinity toward Man_1 residues and the shear force-dependent linker extension of FimH are concurrent processes. Further work is required to determine the precise molecular mechanism of shear enhancement of adhesion and conversion to high-Man_1 binding. One possible advantage of shear-dependent binding is that bacteria could attach to surfaces under moderate to high shear forces but retain the ability to detach and spread when shear forces were reduced. Another advantage is that shear-enhanced binding would tend to favor attachment of the lectin to surface-bound receptors over interaction with soluble inhibitors. Indeed, the low-Man_1-binding FimH-*f18* allele was less susceptible to inhibition than was the V156P mutant.

CONTRIBUTION OF FimH VARIANTS TO TISSUE TROPISM IN COMMENSAL AND PATHOGENIC NICHES

FimH Variants Differ Quantitatively in Binding to AUMs and Bladder Epithelium In Situ

It was logical to hypothesize, on the basis of the in vitro studies, that the ability to bind effectively to Man_1 receptors was a key factor in the pathogenesis of cystitis. There was a highly significant correlation between the level of bacterial adhesion to J82 and A498 bladder and kidney cells and their ability to bind to Man_1 receptors (see "Quantitative variation within the M phenotype" above). It could be, however, that the phenotypic differences in FimH-mediated binding were a purely in vitro phenomenon. The epithelial cells studied were originally derived from poorly differentiated bladder tumor tissues that bear little resemblance to transitional epithelium lining the urinary tract. The stability of the so-called "umbrella" cells at the surface of transitional epithelium is remarkable in itself, with their turnover being on the order of several months instead of a few days, as is typical of many epithelia (57). The umbrella cells also exhibit a very unusual specialization of their luminal membrane that is not observed in urothelial cell lines. The membranes possess rigid-appearing plaques connected by more flexible hinge regions, such that the apical plasmalemma has an accordion-like appearance in transmission electron micrographs.

The plaque areas have been called asymmetric unit membranes (AUMs) owing to the increased electron density of the luminal leaflet. These portions of the membrane have been purified, and their integral membrane protein components have been analyzed in considerable detail (147, 148). The proteins are called uroplakins (UPs), and four primary protein components have been described, UPIa, UPIb, UPII, and UPIII. These proteins are the primary, if not exclusive, components of the hexagonally arrayed 16-nm AUM particles that can be seen in electron micrographs of negative stains of purified plaques or quick-freeze/deep-etch surface replicas of intact tissue (72) (Fig. 23). All of the UPs, except for mature UPII, are glycosylated, and both high-mannose-type and complex-type oligosaccharide chains are present within the 16-nm AUM particle. Wu et al. (149) were the first to provide evidence, using in vitro studies, that UPIa and/or UPIb can serve as a receptor for *E. coli.* It has been shown more recently that the FimH receptor of the mouse AUM particles is UPIa and that its single potential N-glycosylation site is occupied by a high-mannose saccharide (102, 152) (Fig. 24). The UPs are highly conserved across a variety of species, including bovine and murine tissues. Uroepithelial cell lines derived from poorly differentiated human bladder tumors, such as J82 and T24, however, do not exhibit AUMs and do not appear to express UPs in detectable amounts (X.-R. Wu, unpublished observations).

Therefore, to continue testing the hypothesis regarding the tropism of the high-Man_1-binding phenotype for urinary tract epithelium, the ability of *E. coli* strains KB54 and KB91 to bind to AUMs purified from the transitional epithelium of bovine bladder was tested. The high-Man_1-binding *E. coli* strain KB54 adhered in significantly greater numbers to AUMs than did the low-Man_1-binding strain KB91 (Fig. 25). Adhesion of KB54 and KB91 to intact mouse bladder gave similar results. *E. coli* KB91 could be seen to bind to the hexagonal "umbrella cells" in very small numbers, while *E. coli* KB54 adhered in large numbers (Fig. 26). Surprisingly, the adhesion of KB54 occurred in a very striking mosaic pattern, in which cells that bound up to 500 *E. coli* organisms were surrounded by cells which bound no bacteria (Fig. 26). The *E. coli*-binding cells appeared to make up approximately 10 to 20% of the total. These results are consistent with the hypothesis that the high-Man_1-binding FimH phenotype is important in the ability of certain clones to adhere well to urothelium.

FimH Variants Differ in Their Ability To Colonize the Murine Bladder

To determine the relative abilities of low-Man_1-binding and high-Man_1-binding FimH phenotypes to facilitate colonization of the urinary tract, we created isogenic strains in a UTI isolate background, *E. coli* strain CI #10. This strain is genotypically negative for common urovirulence factors, such as P and P-related fimbriae, S/F1C fimbriae, Dr adhesin, aerobactin, group II and III capsules, hemolysin, cytotoxic necrotizing factor I, and outer membrane protein T (139). The *fimH* gene of CI #10 was insertionally inactivated using the pCH103 suicide plasmid, as described previously (22, 131), creating CI #10-9. When strain CI #10-9 was complemented with *fimH* genes that encode either a low-Man_1-binding or high-Man_1-binding FimH (such as those of KB91 and KB54, respectively), binding patterns similar to those of the K-12 derivatives were seen in vitro. At 24 h following transurethral instillation of equal numbers of these recombinant strains into the bladders of mice, the recombinant strain expressing the high-Man_1-binding FimH was recovered from bladders in 15-fold-larger numbers than was the strain expressing a low-Man_1-binding FimH (Fig. 27). Thus, in a uropathogen background, high-Man_1-binding FimH confers a significant advantage for colonization of the bladder. These data provide a rationale for the predominance of the high-Man_1-binding phenotype among UTI isolates and are consistent with our hypothesis regarding the tropism of FimH phenotypes in cystitis.

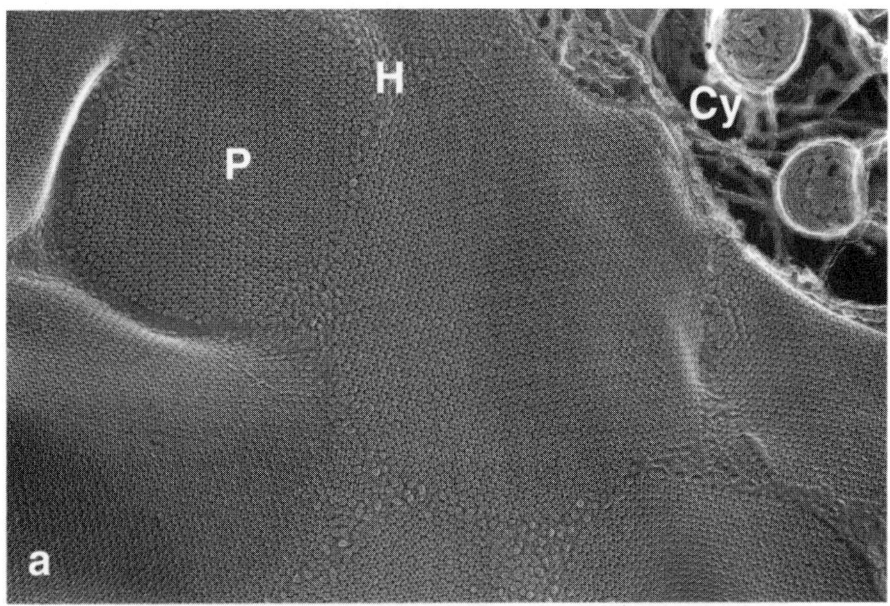

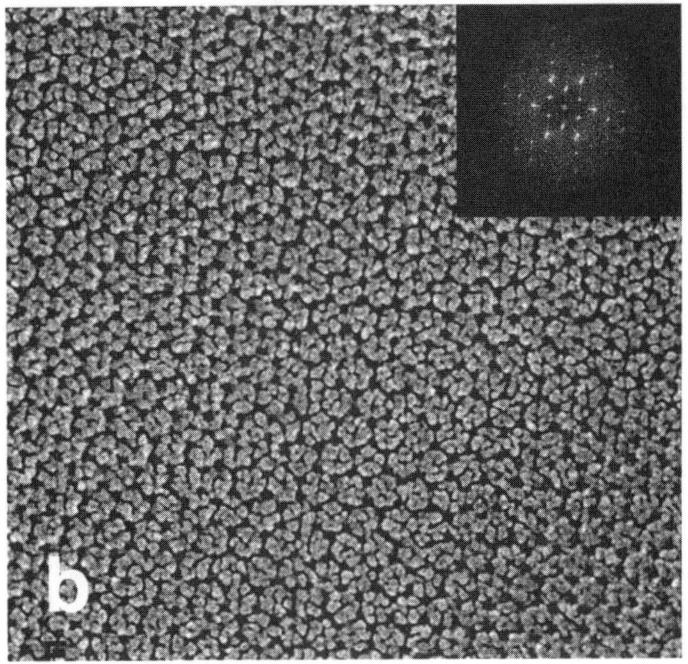

Figure 23. Luminal surface topography of the mouse urothelium. Quick-freeze, deep-etch, and rotary shadowed images of mouse bladder urothelium are shown. (a) An overview showing several crystalline plaques (P) interrupted by hinge areas (H). The upper right corner is an area where the apical surface membrane has been cross-fractured and the cytosol has been water etched, thus exposing the underlying cytoskeleton (Cy). (b) Higher-magnification image of the urothelial plaque and its fast Fourier transform (inset), showing the hexagonal symmetry of the packing and the twisted hexagonal symmetry of individual particles. Reprinted from reference 72 with permission from Elsevier.

Mutations of FimH That Increase Urovirulence Are Detrimental for Adhesion to Oropharyngeal Cells

The high-Man_1-binding FimH phenotype would appear at first thought to be a superior adhesin in every way; the high-Man_1-binding FimH confers greater binding than the low-Man_1-binding FimH phenotype to virtually every model substratum and cell type tested and leads to greatly increased colonization of mouse bladders. However, it is the low-Man_1-binding FimH phenotype that predominates among intestinal isolates, the primary population of *E. coli*.

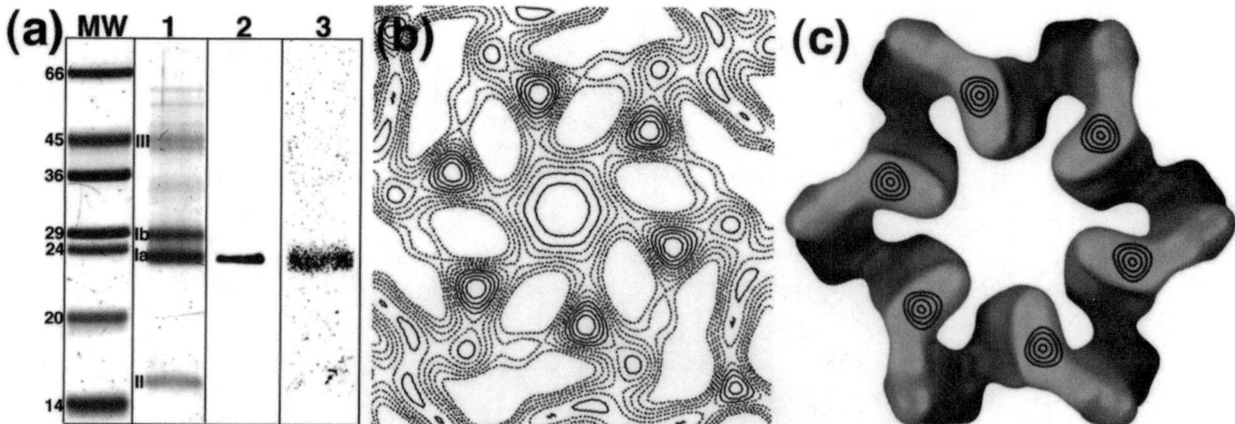

Figure 24. Localization of UPIa receptor for FimH on the six inner domains of the mouse 16-nm AUM particle. (a) UPIa specificity of the FimH-FimC complex. The FimH-FimC complex was biotinylated and incubated with proteins from purified mouse urothelial plaques that had been resolved by sodium dodecyl sulfate-polyacrylamide gel electrophoresis and transferred to a nitrocellulose membrane. Lane 1 shows total proteins of mouse urothelial plaques visualized by Coomassie blue staining, showing the separation of UPII (15 kDa), UPIa (24 kDa), UPIb (27 kDa), and UPIII 947 kDa). There is excellent separation between UPIa and UPIb in this figure. Lane 2 shows the selective binding of biotinylated FimH/FimC to UPIa. Lane 3 shows the selective binding of [^{35}S] methionine-labeled type 1 fimbriated *E. coli* to UPIa. MW, molecular mass standards. (b) A 2-D difference map of the mouse urothelial plaque images collected in the presence and absence of FimH and FimC. (c) Localization of the FimH-binding site on the six inner domains of the 16-nm AUM particle when projected onto a 3-D model of the particle. Reprinted from reference 102 with permission from Elsevier.

To understand why low-Man$_1$-binding variants of FimH would be more fit than high-Man$_1$-binding variants for the intestinal *E. coli* population, we compared the ability of FimH variants to bind oropharyngeal cells. As mentioned earlier in this chapter, there are data suggesting that type 1 fimbriae are required for transient colonization of the oropharyngeal portal during transmission between hosts (9, 10).

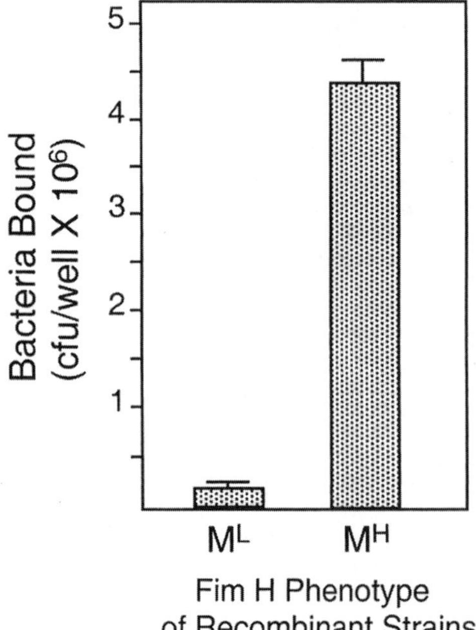

Figure 25. Adhesion of low-Man$_1$-binding and high-Man$_1$-binding strains to AUMs. Reprinted from reference 55 with permission from the publisher.

Although bacteria bearing the high-Man$_1$-binding FimH had been found to bind in much larger numbers to virtually every other cell or model receptor tested than did the low-Man$_1$-binding FimH, the high-Man$_1$-binding and low-Man$_1$-binding strains interacted with buccal epithelial cells in equivalent numbers (Fig. 28). Because the number of bacteria bound to mucosal cells is a function not only of the affinity of the adhesin for the epithelial cell receptor but also of the interference of inhibitors present in body fluids bathing the mucosal surfaces, it was hypothesized that any advantage provided by the low-Man$_1$-binding FimH in the oropharyngeal cavity might be due, at least in part, to differential effects of inhibitors, since the shear-dependent low-Man$_1$-binding FimH variants are expected to be more difficult to inhibit than the high-Man$_1$-binding variants by mannose-like inhibitors. Indeed, although the two variants bound buccal cells in equivalent numbers, the low-Man$_1$-binding variant was much less sensitive to mannoside and at intermediate levels of the inhibitor it provided for greater levels of attachment (Fig. 28). Free mannose is not usually found in the natural environment, and so we tested the effects of several mannose-containing glycoproteins; in each instance, the low-Man$_1$-binding FimH was the superior adhesin. Since inhibitors that are most likely to affect the binding of *E. coli* to buccal epithelial cells in situ should be present in saliva, we also tested the

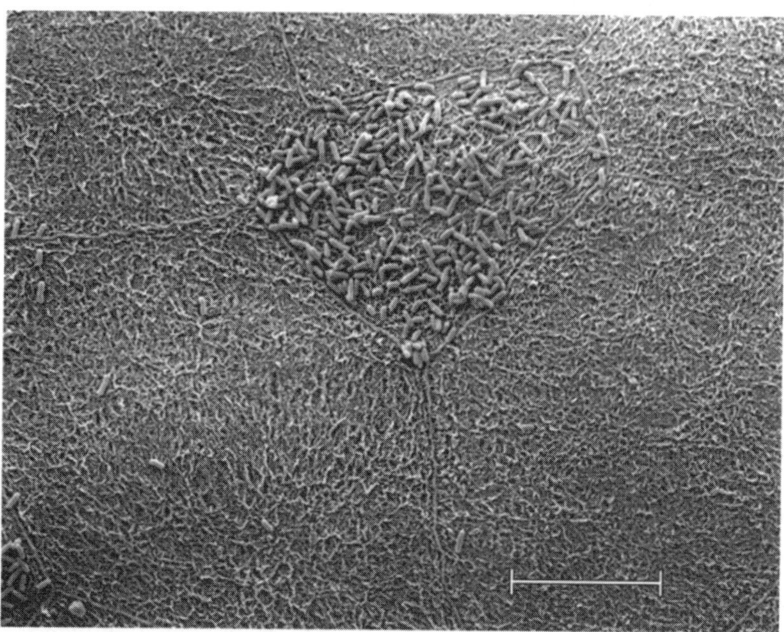

Figure 26. Scanning electron micrograph of a high-Man$_1$-binding phenotype recombinant *E. coli* strain binding to the surface of mouse bladder epithelial cells. The mosaic pattern of adhesion of this *E. coli* strain is striking. Cells bearing hundreds of bound *E. coli* bacteria are intermingled with cells bearing essentially none. Bar, 20 μm.

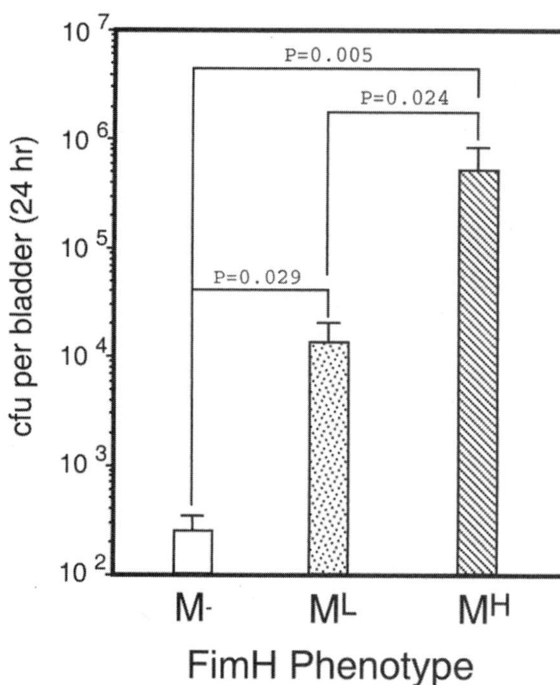

Figure 27. Colonization of mouse bladders by isogenic *E. coli* expressing nonfunctional (M^-) FimH, low-Man$_1$-binding (M^L) FimH, or high-Man$_1$-binding (M^H) Fim H subunits. Bars indicate mean CFU per bladder; error bars indicate standard error of the mean. *P* values indicating level of significance between different groups are indicated. Reprinted from reference 139 with permission from the publisher.

ability of clarified, whole human saliva to inhibit adhesion (Fig. 29). In this case, as well, the low-Man$_1$-binding Fim H phenotype was the superior adhesin. These data provide a strong suggestion that the reduced sensitivity of the strains bearing a low-Man$_1$-binding FimH to inhibitors confers on *E. coli* a greater ability to bind to buccal epithelial cells in the presence of inhibitors. Thus, in a physiological environment, they would more effectively accomplish the transient colonization of the saliva-bathed oropharyngeal cavity and therefore would ensure host-to-host transmission at a higher rate.

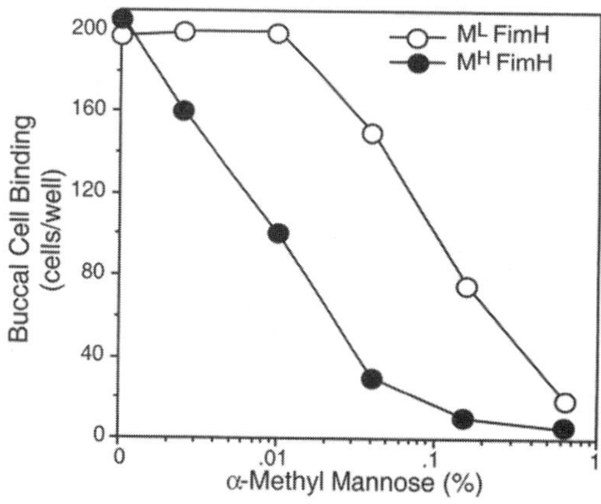

Figure 28. Essentially equivalent binding of *E. coli* to buccal epithelial cells (0% α-methylmannoside) and inhibition of this interaction by increasing levels of α-methyl-D-mannopyrannoside. Reprinted from reference 139 with permission from the publisher.

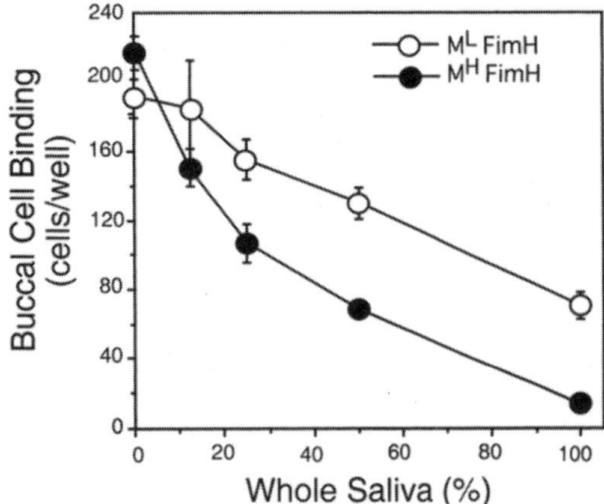

Figure 29. Inhibition of the interaction of *E. coli* and buccal cells by whole, stimulated human saliva. Reprinted from reference 139 with permission from the publisher.

FimH VARIANTS AND THE EVOLUTION OF VIRULENCE

It is now clear that type 1 fimbriae, due in large part to specific high-Man$_1$-binding alleles of FimH, are important virulence factors in the urinary tract. In addition to the role of FimH and type 1 fimbriae in adhesion to urothelial cells, these organelles are required for internalization of selected *E. coli* strains by urothelial cells (94, 106, 107). Internalization leads to persistence of bacteria within a space protected from antibiotics and the host immune response that, at least in mice, facilitates a recurrence of infection. Internalization is not mediated by P fimbriae (94) but is dependent on a functional high-Man$_1$-binding FimH allele. In fact, FimH immobilized on inert beads is sufficient to activate signaling cascades leading to internalization (94). Type 1 fimbriae and FimH are also known to be important in the formation of biofilms in vitro (122, 124, 133, 134). Specific alleles of the *fimH* gene of *Salmonella* were also found to influence the efficiency of biofilm formation (15). It has recently been suggested that the *E. coli* bacteria growing within urothelial cells mature into bulging, biofilm-like structures that have been labeled "pods" (4). All of these interactions between *E. coli* and urothelial cells appear to be dependent, at least in large part, on specific, high-Man$_1$-binding alleles of FimH. If these alleles are adaptive, how have they evolved and how stable are they within the *E. coli* population?

Pathoadaptive Mutations

A prevailing view of the evolution of bacterial pathogens is that adaptation to the pathologic habitat is due to the presence in pathogenic strains of virulence factor genes that are absent from commensal strains, i.e., that virulence has evolved by quantum leaps through the horizontal transfer of large DNA insertions (25, 37, 129). While it is absolutely clear that the genetic transfer of, for instance, PAIs contributes to pathogen evolution, we introduced the concept of pathogenicity-adaptive (pathodaptive) mutations (140) to call attention to the idea that pathogen potential can be increased without horizontal gene transfer by the adaptive modification of existing genes through the effects of selective pressures on randomly mutated traits that are shared with and probably originated in commensal bacterial lineages. The adaptive mutations of the *fimH* gene that enhance Man$_1$-binding activity and increase the fitness of *E. coli* for the urinary tract are some of the best and most direct examples of pathoadaptive mutations. The pathoadaptive nature of at least some FimH alleles has also been recently shown by Hommais et al. (59), who sequenced *fimH* genes from 44 commensal and 69 pathogenic *E. coli* strains and measured the lethality of the strains in mice inoculated intraperitoneally. Fifteen different FimH protein variants were found in the collection of 113 strains. Among B2 strains, Hommais et al. found a highly significant correlation between the A27V substitution and virulence in mice. It is noteworthy that this mutation confers a high-Man$_1$-binding phenotype and is found in one of the loops surrounding the interdomain linker chain. It was also suggested that the A27V mutation has arisen several distinct times during evolution and possibly has been transferred horizontally between different strains.

If evolution of the FimH adhesin reflects an ongoing pathoadaptive niche expansion by *E. coli*, it should leave a detectable "footprint" in the phylogeny of *fimH* alleles that might clarify the stage of this expansion and its evolutionary dynamics. To determine whether there is a specific "selection footprint" for the pathoadaptive evolution of FimH, we recently analyzed DNA variation patterns in *fimH* alleles of *E. coli* isolates of commensal and extraintestinal infection origin. In parallel, we analyzed variations in the gene encoding the periplasmic chaperone, FimC, as a control, since it is not expressed on the bacterial surface and so is unlikely to be subjected to selective pressures similar to those affecting receptor-binding properties of the adhesin subunit.

We were unable to detect any clear selective footprints by using traditional molecular evolution analyses for neutrality of mutations (i.e., ratio of the rate of nonsynonymous and synonymous mutations), such as Tajima's D statistic or Fu and Li's D^* statistic (109, 144). However, it was not surprising that these methods failed to uncover evidence for the influence of positive selection on structural mutations in the FimH adhesin. These methods are highly conservative, and the pattern of DNA variation among randomly sampled *fimH* alleles is not particularly different from that expected to occur with selectively neutral evolution (76), because functional mutations in FimH are relatively few (usually single-point mutations), arise independently in different allelic backgrounds, and are scattered across the protein structure.

In contrast to the more traditional methods, a newly developed zonal analysis of gene sequences provided strong evidence for the effects of positive selection on the FimH adhesin (142). The method is based on separation of evolutionarily ancient nodes containing silent variations on the protein tree (i.e., the primary zone) from nodes representing subsequent single-replacement evolution (i.e., the secondary zone) and also from nodes representing multiple-replacement evolution (i.e., the extended zone) (Fig. 30). This separation then permits an estimation of the combined number of replacement and silent substitutions along the branches that connect the different zones. Collective analysis of all connecting branches (instead of analyzing individual branches or individual codons) makes zonal analysis very sensitive in the detection of selective footprints. Zonal analysis led to the identification of a dramatically higher rate of replacement mutations relative to silent mutations along branches connecting the extended-zone nodes with one another or with the rest of the FimH tree. The hypothesis that FimH variants from the extended zone, and at least some variants from the secondary zone, were created by adaptive replacment was further supported by the fact that most of the mutations that formed these zones occurred in a limited number of positions in FimH (i.e., "hot spots"). Furthermore, FimH alleles of uropathogenic strains were found more often on the

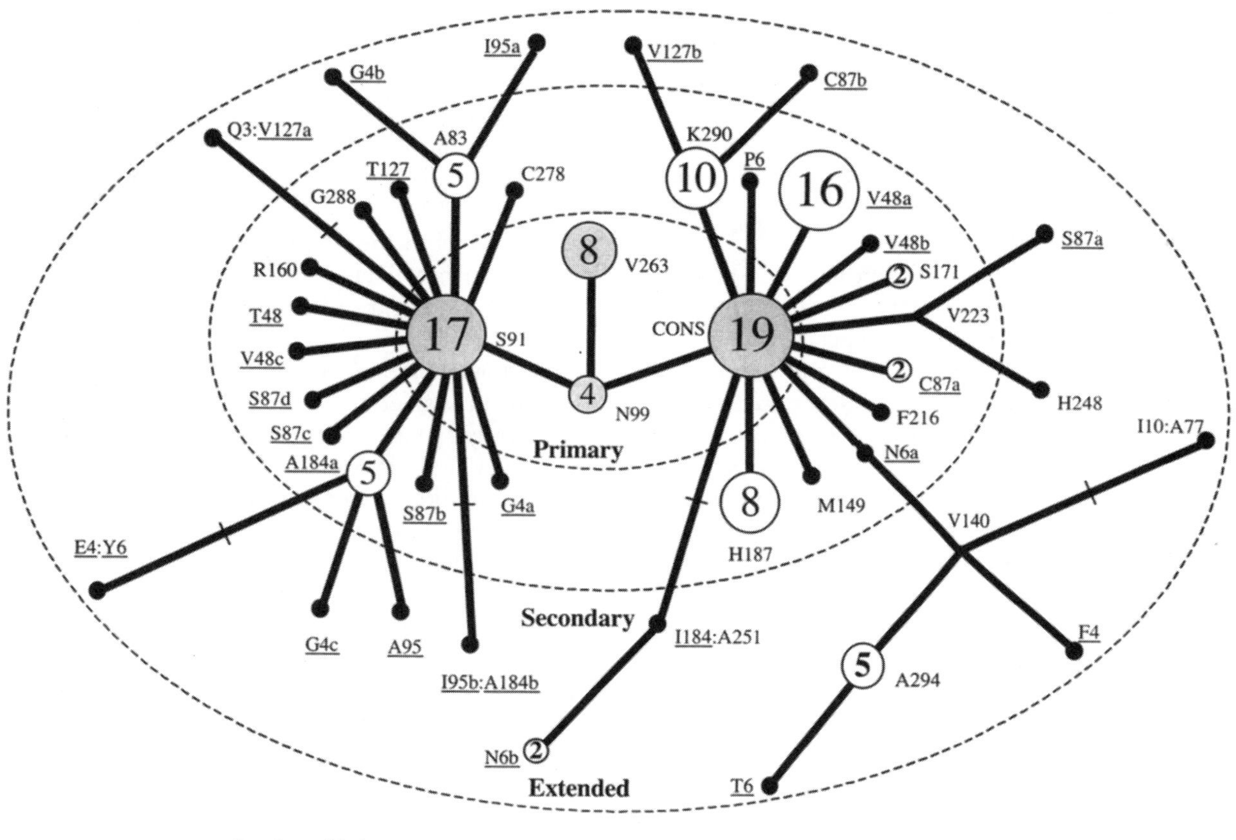

Figure 30. Phylogenetic tree (unrooted phylogram) of FimH protein variants. The protein tree was built by collapsing branch regions containing silent mutations, using a maximum-likelihood *fimH* gene tree. The nodes are separated into primary, secondary, and extended zones (see the text). The node labeled CONS corresponds to the consensus structure FimH. All other nodes are labeled with the replacement mutation by which they vary from the immediate ancestral node. Where the same replacement was acquired independently, they are distinguished by a lowercase letter. The grey circles represent FimH variants with intranodal synonymous variations, i.e., encoded by multiple gene alleles. The numbers within circles give the total numbers of strains with the indicated mutation. The small solid circles represent FimH variants found in a single strain in the collection. Reprinted from reference 142 with permission from the publisher.

extended-zone nodes than were alleles of commensal intestinal strains. Alleles of uropathogenic strains also exhibited more multiple-strain and hot-spot nodes in the secondary zone than did alleles of commensal strains. Importantly, zonal analysis of *fimC* alleles from the same strains did not reveal any similar signs of adaptive selection. Finally, the structural evolution of FimH variants from the secondary and extended zones is accompanied by the acquisition of a high-Man_1-binding phenotype. Thus, we think that the structural evolution of FimH reflects pathoadaptive niche expansion of *E. coli.*

We should note, however, that the adaptive effects of FimH variation may not be limited to increased Man_1-binding. For example, positions Val4 and Thr6 are hot-spot mutations within the leader sequence of the nascent protein (residues 1 to 21). These mutations may affect fimbrial length or number, rather than receptor specificity, and this is currently under investigation. Furthermore, it is necessary to note that, although extended-zone FimH variants are associated with uropathogenic strains and are pathoadaptive (i.e., enhance the urovirulence of *E. coli*), it is unclear at present whether FimH evolution is driven by the uropathogenicity itself or by a different, possibly nonpathogenic type of adaptive niche expansion by *E. coli.* Zonal analysis of *fimH* evolution also provides insight into the evolutionary dynamics of adaptive niche expansion of *E. coli.* Silent mutations are generally considered to be selectively neutral and to accumulate in a random fashion and at a constant rate for a given gene (76). The absence of silent mutations along the FimH tree branches that connect secondary and extended nodes suggests that diversifying FimH evolution has happened too recently for silent mutations to accumulate. By similar reasoning, the formation of large, multistrain, multiclone nodes within the secondary zone is also evolutionarily recent. Thus, it appears that the adaptive evolution of FimH in the course of *E. coli* niche differentiation has been occurring over a relatively brief period. The short-term nature of diversifying FimH evolution argues against the occurrence of "balanced-polymorphism" dynamics in *E. coli* adaptation, because balancing selection would be expected to allow the accumulation of at least some silent mutations along the evolutionary branches (85). The evolutionary dynamics observed by zonal analysis also do not fit the "population replacement" mode of niche expansion characterized by selective sweeps in the course of adaptive evolution (85). In fact, in our sample the extended-zone FimH variants coexist with a pool of primary variants.

We think that the FimH variation footprint identified here fits best with yet a third pattern of niche expansion, termed "source-sink" habitatat dynamics. Source-sink habitats have been proposed as an ecological model with two fundamental requirements: (i) a stable primary niche (or "source" habitat), to which the species is well adapted and in which it maintains populations over a long period of evolutionary time, and (ii) the existence of alternative niches (or "sink" habitats), into which the organisms can spread but in which, for one reason or another, they do not maintain a stable population (125). Thus, genes involved in this source-sink adaptation are constantly adapting to the alternative niches, but the adapted forms also constantly become extinct. As a result, primary alleles well adapted to the main niche contain silent variation, while those adapted to the alternative niches do not (56). This pattern of niche expansion is consistent with our data, where commensal *E. coli* strains from the principal, intestinal niche commonly express evolutionarily stable, primary forms of FimH while extraintestinal uropathogenic *E. coli* strains primarily express recently evolved FimH variants that are clearly unstable over the long term.

As discussed above, the increased Man_1 binding of FimH is accompanied by its increased susceptibility to inhibition by soluble mannosylated compounds, including salivary glycoproteins and intestinal mucin (137, 139). Thus, FimH mutations that are advantageous in the urinary tract (or an alternative sink niche) are likely to be selected against in the course of oral transmission and/or intestinal colonization, resulting in the evolutionary instability of uropathogenic *E. coli* clones. On the other hand, one must also consider an alternative explanation for the recent origin of adapted FimH variants. It is possible that novel habitats providing selective conditions for FimH evolution have become available to *E. coli* clones only relatively recently and the end-point dynamics of niche expansion have not yet emerged.

In any case, genes that adapt according to source-sink or other models of niche differentiation may occur quite commonly throughout many biological species, possibly among opportunistic microbial pathogens in particular, and may exhibit an evolutionary footprint similar to that of *E. coli fimH* alleles. Further studies will also show whether and how the pathoadaptive evolution of *fimH* and other genes is linked with acquisition of larger segments of novel genetic material through horizontal gene transfer.

SUMMARY AND SPECULATIONS

Over the past several years, it has become clear that type 1 fimbriae, the most common adhesins of enterobacteria, are much more complicated and interesting organelles than has been heretofore appreciated, from the perspective of both structure-function and virulence. Adaptive mutations of the FimH lectin, an originally commensal trait, have a dramatic effect on the tropism of *E. coli* and shift the organism to a more virulent phenotype. The studies summarized here provide a rationale for the contribution of type 1 fimbriae to colonization of both the primary commensal niche and a pathologic niche. The studies also suggest that future epidemiologic studies would benefit from taking into consideration the variation of FimH phenotypes. A better understanding of FimH structural and functional properties could also aid in the development of an anti-FimH vaccine, the prospects for which have been supported by a variety of animal studies (88, 89). It is also very possible that, no matter how strongly FimH might be implicated as a primary virulence factor for UTI, the ability of *E. coli* to produce multiple adhesins will require a much more complicated formulation for the successful development of antiadhesin vaccines or other antiadhesion therapies (115, 117).

REFERENCES

1. **Abraham, J. M., C. S. Freitag, J. R. Clements, and B. I. Eisenstein.** 1985. An invertible element of DNA controls phase variation of type 1 fimbriae of *Escherichia coli. Proc. Natl. Acad. Sci. USA* **82:**5724–5727.
2. **Abraham, S. N, D. Sun, J. B. Dale, and E. H. Beachey.** 1988. Conservation of the D-mannose-adhesion protein among type 1 fimbriated members of the family *Enterobacteriaceae. Nature* **336:**682–684.
3. **Abraham, S. N., J. D. Goguen, D. Sun, P. Klemm, and E. H. Beachey.** 1987. Identification of two ancillary subunits of *Escherichia coli* type 1 fimbriae by using antibodies against synthetic oligopeptides of *fim* gene products. *J. Bacteriol.* **169:**5530–5536.
4. **Anderson, G. G., J. J. Palermo, J. D. Schilling, R. Roth, J. Heuser, and S. J. Hultgren.** 2003. Intracellular bacterial biofilm-like pods in urinary tract infections. *Science* **301:** 105–107.
5. **Aronson, M., O. Medalia, L. Schori, D. Mirelman, N. Sharon, and I Ofek.** 1979. Prevention of colonization of the urinary tract of mice with *Escherichia coli* by blocking of bacterial adherence with methyl alpha-D-mannopyranoside. *J. Infect. Dis.* **139:**329–332.
6. **Bahrani-Mougeot, F.K., E. L. Buckles, C. V. Lockatell, J. R. Hebel, D. E Johnson, C. M. Tang, and M. S. Donnenberg.** 2002. Type 1 fimbriae and extracellular polysaccharides are preeminent uropathogenic *Escherichia coli* virulence determinants in the murine urinary tract. *Mol. Microbiol.* **45:**1079–1093.
7. **Barnhart, M. M., F. G. Sauer, J. S. Pinkner, and S. J. Hultgren.** 2003. Chaperone-subunit-usher interactions required for donor strand exchange during bacterial pilus assembly. *J. Bacteriol.* **185:**2723–2730.
8. **Blattner, F. R., G. Plunkett III, C. A. Bloch, N. T. Perna, V. Burland, M. Riley, J. Collado-Vides, J. D. Glasner, C. K. Rode, G. F. Mayhew, J. Gregor, N. W. Davis, H. A. Kirkpatrick, M. A. Goeden, D. J. Rose, B. Mau, and Y. Shao.** 1997. The complete genome sequence of *Escherichia coli* K-12. *Science* **277:**1453–1474.
9. **Bloch, C. A., and P. E. Orndorff.** 1990. Impaired colonization by and full invasiveness of *Escherichia coli* K1 bearing a site-directed mutation in the type 1 pilin gene. *Infect. Immun.* **58:**275–278.
10. **Bloch, C. A., B. A. D. Stocker, and P. Orndorff.** 1992. A key role for type 1 fimbriae in enterobacterial communicability. *Mol. Microbiol.* **6:**697–701.
11. **Blomfield, I. C., M. S. McClain, and B. I. Eisenstein.** 1991. Type 1 fimbriae mutants of *Escherichia coli* K12: characterization of recognized afimbriate strains and construction of new *fim* deletion mutants. *Mol. Microbiol.* **5:**1439–1445.
12. **Blomfield, I. C., P. J. Calie, K. J. Eberhardt, M. S. McClain, and B. I. Eisenstein.** 1993. Lrp stimulates phase variation of type 1 fimbriation in *Escherichia coli* K-12. *J. Bacteriol.* **175:** 27–36.
13. **Blomfield, I. C., D. H. Kulasekara, and B. I. Eisenstein.** 1997. Integration host factor stimulates both FimB- and FimE-mediated site-specific DNA inversion that controls phase variation of type 1 fimbriae expression in *Escherichia coli. Mol. Microbiol.* **23:**705–717.
14. **Blum, G., V. Falbo, A. Caprioli, and J. Hacker.** 1995. Gene clusters encoding the cytotoxic necrotizing factor type 1, Prs-fimbriae and α-hemolysin form the pathogenicity island II of the uropathogenic *Escherichia coli* strain J96. *FEMS Microbiol. Lett.* **126:**189–196.
15. **Boddicker, J. D., N. A. Ledeboer, J. Jagnow, B. D. Jones, and S. Clegg.** 2002. Differential binding to and biofilm formation on HEp-2 cells by *Salmonella enterica* serovar Typhimurium is dependent upon allelic variation in the *fimH* gene of the *fim* gene cluster. *Mol. Microbiol.* **45:**1255–1265.
16. **Brinton, C. C., Jr.** 1959. Non-flagellar appendages of bacteria. *Nature* **183:**782–786.
17. **Brinton, C. C., Jr.** 1965. The structure, function, synthesis and genetic control of bacterial pili and a molecular model for DNA and RNA transport in gram negative bacteria. *Trans. N. Y. Acad. Sci.* **27:**1003–1054.
18. **Brinton, C. C., Jr., A. Buzzell, and M. A. Lauffer.** 1954. Electrophoresis and phage susceptibility studies on a filament-producing variant of the *E. coli* B bacterium. *Biochim. Biophys. Acta* **15:**533–542.
19. **Bullitt, E., and L. Makowski.** 1995. Structural polymorphism of bacterial adhesion pili. *Nature* **373:**164–167.
20. **Choudhury, D., A. Thompson, V. Stojanoff, S. Langermann, J. Pinkner, S. J. Hultgren, and S. D. Knight.** 1999. X-ray structure of the FimC-FimH chaperone-adhesin complex from uropathogenic *Escherichia coli. Science* **285:**1061–1066.
21. **Collier, W. A., and J. C. de Miranda.** 1955. Bacterien-Haemagglutination. III. Die Hemmung der Coli-Haemagglutination durch Mannose. *Antonie Leeuwenhoek* **21:**133–140.
22. **Connell, H., W. Agace, P. Klemm, M. Schembri, S. Mårild, and C. Svanborg.** 1996, Type 1 fimbrial expression enhances *Escherichia coli* virulence for the urinary tract. *Proc. Natl. Acad. Sci. USA* **93:**9827–9832.
23. **Costerton, J. W., Z. Lewandowski, D. E. Caldwell, D. R. Korber, and H. M. Lappin-Scott.** 1995. Microbial biofilms. *Annu. Rev. Microbiol.* **49:**711–745.
24. **Costerton, J. W., P. S. Stewart, and E. P. Greenberg.** 1999. Bacterial biofilms: a common cause of persistent infections. *Science* **284:**2137–2142.
25. **Covacci, A., S. Falkow, D. E. Berg, and R. Rappuoli.** 1997. Did the inheritance of a pathognicity island modify the

virulence of *Helicobacter pylori? Trends Microbiol.* **5**:205–208.
26. **De Graaf, F. K., and R. R. Mooi.** 1986. The fimbrial adhesins of *Escherichia coli. Adv. Microb. Physiol.* **28**:65–143.
27. **De Graaf, F. K., and W. Gaastra.** 1994. Fimbriae of enterotoxigenic *Escherichia coli,* p. 53–83. *In* P. Klemm (ed.), *Fimbriae: Adhesion, Genetics, Biogenesis, and Vaccines.* CRC Press, Inc., Boca Raton, Fla.
28. **Donnenberg, M. S., and R. A. Welch.** 1996. Virulence determinants of uropathogenic *Escherichia coli,* p. 135–174. *In* H. L. T. Mobley and J. W. Warren (ed.), *Urinary Tract Infections: Molecular Pathogenesis and Clinical Management.* ASM Press, Washington, D.C.
29. **Dorman, C. J., and C. F. Higgins.** 1987. Fimbrial phase variation in *Escherichia coli:* dependence on integration host factor and homologies with other site-specific recombinases. *J. Bacteriol.* **169**:3840–3843.
30. **Duguid, J. P.** 1964. Functional anatomy of *Escherichia coli* with special reference to enteropathogenic *E. coli. Rev. Latinam. Microbiol.* **7**(Suppl. 13-4):1–16.
31. **Duguid, J. P., I. W. Smith, G. Dempster, and P. N. Edmunds.** 1955. Non-flagellar filamentous appendages ("fimbriae") and haemagglutinating activity in *Bacterium coli. J. Pathol. Bacteriol.* **70**:335–348.
32. **Duguid, J. P., and R. R. Gillies.** 1957. Fimbriae and adhesive properties in dysentery bacilli. *J. Pathol.* **74**:397–411.
33. **Duguid, J. P., and D. C. Old.** 1994. Introduction: A historical perspective, p. 1–7. *In* P. Klemm (ed.), *Fimbriae: Adhesion, Genetics, Biogenesis, and Vaccines.* CRC Press, Inc., Boca Raton, Fla.
34. **Eisenstein, B. I., D. S. Sweet, V. Vaughn, and D. I. Friedman.** 1987. Integration host factor is required for the DNA inversion that controls phase variation in *Escherichia coli. Proc. Natl. Acad. Sci. USA* **84**:6506–6510.
35. **El-Labany, S., B. K. Sohanpal, M. Lahooti, R. Akerman and I. C. Blomfield.** 2003. Distant *cis*-active sequences and sialic acid control the expression of *fimB* in *Escherichia coli* K-12. *Mol. Microbiol.* **49**:1109–1118.
36. **Eshdat, Y., F. J. Silverblatt, and N. Sharon.** 1981. Dissociation and reassembly of *Escherichia coli* type 1 pili. *J. Bacteriol.* **148**:308–314.
37. **Falkow, S.** 1997. What is a pathogen? *ASM News* **63**:359–365.
38. **Firon, N., I. Ofek, and N. Sharon.** 1983. Carbohydrate specificity of the surface lectins of *Escherichia coli, Klebsiella pneumoniae,* and *Salmonella typhimurium. Carbohydr. Res.* **120**:235–249.
39. **Firon, N., I. Ofek, and N. Sharon.** 1984. Carbohydrate-binding sites of the mannose-specific fimbrial lectins of enterobacteria. *Infect. Immun.* **43**:1088–1090.
40. **Firon, N., S. Ashkenazi, D. Mirelman, I. Ofek, and N. Sharon.** 1987. Aromatic alpha-glycosides of mannose are powerful inhibitors of the adherence of type 1 fimbriated *Escherichia coli* to yeast and intestinal epithelial cells. *Infect. Immun.* **55**:472–476.
41. **Foxman, B.** 2002. Epidemiology of urinary tract infections: incidence, morbidity, and economic costs. *Am. J. Med.* **113** (Suppl. 1):5S–13S.
42. **Fu, D., L. Chen, and R. A. O'Neill.** 1994. A detailed structural characterization of ribonuclease B oligosaccharides by ^{1}H NMR spectroscopy and mass spectrometry. *Carbohydr. Res.* **261**:173–186.
43. **Gally, D. L, T. J. Rucker, and I. C. Blomfield.** 1994. The leucine-responsive regulatory protein binds to the *fim* switch to control phase variation of type 1 fimbrial expression in *Escherichia coli* K-12. *J. Bacteriol.* **176**:5665–5672.
44. **Gally, D. L., J. Leathart, and I. C. Blomfield.** 1996. Interaction of *fimB* and *fimE* with the *fim* switch that controls the phase variation of type 1 fimbriae in *Escherichia coli* K-12. *Mol. Microbiol.* **21**:725–738.
45. **Guerina, N. G., T. W. Kessler, V. J. Guerina, M. R. Neutra, H. W. Clegg, S. Langermann, F. A. Scannapieco, and D. A. Goldman.** 1983. The role of pili and capsule in the pathogenesis of neonatal infection with *Escherichia coli* K1. *J. Infect. Dis.* **148**:395–405.
46. **Gunther, N. W., IV, V. Lockatell, D. E. Johnson, and H. L. T. Mobley.** 2001. In vivo dynamics of type 1 fimbria regulation in uropathogenic *Escherichia coli* during experimental urinary tract infection. *Infect. Immun.* **69**:2838–2846.
47. **Gunther, N. W., IV, J. A. Snyder, V. Lockatell, I. Blomfield, D. E. Johnson, and H. L. T. Mobley.** 2002. Assessment of virulence of uropathogenic *Escherichia coli* type 1 fimbrial mutants in which the invertible element is phase-locked on or off. *Infect. Immun.* **70**:3344–3354.
48. **Guyot, G.** 1908. Uber die bakterielle Haemagglutination. *Zentbl. Bakteriol. Abt. I Orig.* **47**:640–653.
49. **Haas, R., and T. F. Meyer.** 1992. Silent pilin genes of *Neisseria gonorrhoeae* MS11 and the occurrence of related hypervariant sequences among gonocococcal isolates. *Mol. Microbiol.* **6**:197–208.
50. **Hacker, J., G. Blum-Oehler, I. Muhldorfer, and H. Tschape.** 1997. Pathogenicity islands of virulent bacteria: structure, function and impact on microbial evolution. *Mol. Microbiol.* **23**:1089–1097.
51. **Hacker, J., G. Blum-Oehler, B. Janke, G. Nagy, and W. Goebel.** 1999. Pathogenicity islands of extraintestinal *Escherichia coli,* p. 59–76. *In* J. B. Kaper and J. Hacker (ed.), *Pathogenicity Islands and Other Mobile Virulence Elements.* ASM Press, Washington, D.C.
52. **Hahn, E., P. Wild, U. Hermanns, P. Sebbel, R. Glockshuber, M. Häner, N. Taschner, P. Burkhard, U. Aebi and S. A. Müller.** 2002. Exploring the 3D molecular architecture of *Escherichia coli* type 1 pili. *J. Mol. Biol.* **323**:845–857.
53. **Hanson, M. S., and C. C. Brinton, Jr.** 1988. Identification and characterization of *E. coli* type-1 pilus tip adhesion protein. *Nature* **332**:265–268.
54. **Hanson, M. S., J. Hempel, and C. C. Brinton, Jr.** 1988. Purification of the *Escherichia coli* type 1 pilin and minor pilus proteins and partial characterization of the adhesin protein. *J. Bacteriol.* **170**:3350–3358.
55. **Hasty, D. L., and E. V. Sokurenko.** 2000. The FimH lectin of *Escherichia coli* type 1 fimbriae. An adaptive adhesin, p. 481–515. *In* R. J. Doyle (ed.), *Glycomicrobiology.* Kluwer Academic/Plenum Publishers, New York, N.Y.
56. **Hedrick, P. W.** 1986. Genetic polymorphism in heterogeneous environments—a decade later. *Annu. Rev. Ecol. Syst.* **17**:535–566.
57. **Hicks, R. M.** 1975. The mammalian urinary bladder: an accommodating organ. *Biol. Rev. Camb. Philos. Soc.* **50**:215–246.
58. **Holden, N. J., B. E. Uhlin, and D. L. Gally.** 2001. PapB paralogues and their effect on the phase variation of type 1 fimbriae in *Escherichia coli. Mol. Microbiol.* **42**:319–330.
59. **Hommais, F., S. Gouriou, C. Amorin, H. Bui, M. C. Rahimy, B. Picard, and E. Denamur.** 2003. The FimH A27V mutation is pathoadaptive for urovirulence in *Escherichia coli* B2 phylogenetic group isolates. *Infect. Immun.* **71**:3619–3622.
60. **Houwink, A. L., and W. van Iterson.** 1950. Electron microscopical observations on bacterial cytology. II. A study on flagellation. *Biochim. Biophys. Acta* **5**:10–44.
61. **Hultgren, S. J., T. N. Porter, A. J. Schaeffer, and J. L. Duncan.** 1985. Role of type 1 pili and effects of phase variation on

lower urinary tract infections produced by *Escherichia coli. Infect. Immun.* 50:370–377.

62. **Hultgren, S.J., S. Normark, and S. N. Abraham.** 1991. Chaperone-assisted assembly and molecular architecture of adhesive pili. *Annu. Rev. Microbiol.* 45:383–415.
63. **Hung, C.-S., J. Bouckaert, D. Hung, J. Pinkner, C. Widberg, A. DeFusco, C. G. Auguste, R. Strouse, S. Langermann, G. Waksman, and S. J. Hultgren.** 2002. Structural basis of tropism of *Escherichia coli* to the bladder during urinary tract infection. *Mol. Microbiol.* 44:903–915.
64. **Jacob-Dubuisson, F., J. Heuser, K. Dodson, S. Normark, and S. J. Hultgren.** 1993. Initiation of assembly and association of the structural elements of a bacterial pilus depend on two specialized tip proteins. *EMBO J.* 12:837–847.
65. **Jann, K., and B. Jann (ed.).** 1990. *Bacterial Adhesins.* Springer-Verlag, KG, Berlin, Germany.
66. Reference deleted.
67. **Johnson, J. R.** 1991. Virulence factors in *Escherichia coli* urinary tract infection. *Clin. Microbiol. Rev.* 4:80–128.
68. **Johnson, J. R., T. A. Russo, F. Scheutz, J. J. Brown, L. X. Zhang, K. Palin, C. Rode, C. Bloch, C. F. Marrs, and B. Foxman.** 1997. Discovery of disseminated L96-like strains of uropathogenic *Escherichia coli* O4:H5 containing genes for both papG$_{J96}$ (class I) and prsG$_{J96}$ (class III) Gal(α1-4)Gal-binding adhesins. *J. Infect. Dis.* 175:983–988.
69. **Jones, H. C., F. Jacob-Dubuisson, K. Dodson, M. Kuehn, L. Slonim, R. Striker, and S. J. Hultgren.** 1992. Adhesin presentation in bacteria requires molecular chaperones and ushers. *Infect. Immun.* 60:4445–4451.
70. **Jones, C. H., J. S. Pinkner, A. V. Nicholes, L. N. Slonim, S. N. Abraham, and S. J. Hultgren.** 1993. FimC is a periplasmic PapD-like chaperone that directs assembly of type 1 pili in bacteria. *Proc. Natl. Acad. Sci. USA* 90:8397–8401.
71. **Jones, C. H., J. S. Pinkner, R. Roth, J. Heuser, A. V. Nicholes, S. N. Abraham, and S. J. Hultgren.** 1995. FimH adhesin of type 1 pili is assembled into a fibrillar tip structure in the *Enterobacteriaceae. Proc. Natl. Acad. Sci. USA* 92:2081–2085.
72. **Kachar, B., F. Liang, U. Lins, M. Ding, X.-R. Wu, D. Stoffler, U. Aebi, and T.-T. Sun.** 1999. Three-dimensional analysis of the 16 nm urothelial plaque particle: luminal surface exposure, preferential head-to-head interaction, and hinge formation. *J. Mol. Biol.* 285:595–608.
73. **Kaper, J. B., and J. Hacker.** 1999. *Pathogenicity Islands and Other Mobile Virulence Elements.* ASM Press, Washington, D.C.
74. **Kawula, T. H., and P. E. Orndorff.** 1991. Rapid site-specific DNA inversion in *Escherichia coli* mutants lacking the histonelike protein H-NS. *J. Bacteriol.* 173:4116–4123.
75. **Keith, B. R., L. Maurer, P. A. Spears, and P. E. Orndorff.** 1986. Receptor-binding function of type 1 pili effects bladder colonization by a clinical isolate of *Escherichia coli. Infect. Immun.* 53:693–696.
76. **Kimura, M.** 1983. *The Neutral Theory of Molecular Evolution.* Cambridge University Press, Cambridge, United Kingdom.
77. **Klemm, P.** 1984. The *fimA* gene encoding the type-1 fimbrial subunit of *Escherichia coli.* Nucleotide sequence and primary structure of the proteins. *Eur. J. Biochem.* 143:395–399.
78. **Klemm, P.** 1986. Two regulatory *fim* genes, *fimB* and *fimE,* control the phase variation of type 1 fimbriae in *Escherichia coli. EMBO J.* 5:1389–1393.
79. **Klemm, P.** 1992. FimC, a chaperone-like periplasmic protein of *Escherichia coli* involved in biogenesis of type 1 fimbriae. *Res. Microbiol.* 143:831–838.
80. **Klemm, P. (ed.).** 1994. *Fimbriae. Adhesion, Genetics, Biogenesis, and Vaccines,* CRC Press, Inc., Boca Raton, Fla.
81. **Klemm, P., B. J. Jorgensen, I. van Die, H. de Ree, and H. Bergmans.** 1985. The *fim* genes responsible for synthesis of type 1 fimbriae in *Escherichia coli:* cloning and genetic organization. *Mol. Gen. Genet.* 199:410–414.
82. **Klemm, P., and G. Christiansen.** 1987. Three *fim* genes required for the regulation of length and mediation of adhesion of *Escherichia coli* type 1 fimbriae. *Mol. Gen. Genet.* 208: 439–445.
83. **Klemm, P., and K. A. Krogfelt.** 1994. Type 1 fimbriae of *Escherichia coli,* p. 9–26. *In* P. Klemm (ed.), *Fimbriae: Adhesion, Genetics, Biogenesis, and Vaccines.* CRC Press, Inc., Boca Raton, Fla.
84. **Krallmann-Wenzel, U., M. Ott, J. Hacker, and G. Schmidt.** 1989. Chromosomal mapping of genes encoding mannose-sensitive (type I) and mannose-resistant F8 (P) fimbriae of *Escherichia coli* O18:K5:H5. *FEMS Microbiol. Lett.* 58: 315–322.
85. **Kreitman, M., and R. R. Hudson.** 1991. Inferring the evolutionary histories of the Adh and Adh-dup loci in *Drosophila melanogaster* from patterns of polymorphism and divergence. *Genetics* 127:565–582.
86. **Krogfelt, K. A., and P. Klemm.** 1988. Investigation of minor components of *E. coli* type 1 fimbriae: protein chemical and immunological aspects. *Microb. Pathog.* 4:231–238.
87. **Krogfelt, K. A., H. Bergmans, and P. Klemm.** 1990. Direct evidence that the FimH protein is the mannose-specific adhesin of *Escherichia coli* type 1 fimbriae. *Infect. Immun.* 58: 1995–1998.
88. **Langermann, S., S. Palaszynski, M. Barnhart, G. Auguste, J. S. Pinkner, J. Burlein, P. Barren, S. Koenig, S. Leath, C. H. Jones, and S. J. Hultgren.** 1997. Prevention of mucosal *Escherichia coli* infection by FimH-adhesin-based systemic vaccination. *Science* 276:607–611.
89. **Langermann, S., R. Möllby, J. E. Burlein, S. R. Palaszynski, C. G. Auguste, A. DeFusco, R. Strouse, M. A. Schenerman, S. J. Hultgren, J. S. Pinkner, J. Winberg, L. Guldevall, M. Söderhäll, K. Ishikawa, S. Normark, and S. Koenig.** 2000. Vaccination with FimH adhesin protects *Cynomolgus* monkeys from colonization and infection by uropathogenic *Escherichia coli. J. Infect. Dis.* 181:774–778.
90. **Leathart, J. B. S., and D. L. Gally.** 1998. Regulation of type 1 fimbrial expression in uropathogenic *Escherichia coli:* heterogeneity of expression through sequence changes in the *fim* switch region. *Mol. Microbiol.* 28:371–381.
91. **Lim, J. K., N. W. Gunther IV, H. Zhao, D. E. Johnson, S. K. Keay, and H. L. T. Mobley.** 1998. In vivo phase variation of *Escherichia coli* type 1 fimbrial genes in women with urinary tract infection. *Infect. Immun.* 66:3303–3310.
92. **Lowe, M. A., S. C. Holt, and B. I. Eisenstein.** 1987. Immunoelectron microscopic analysis of elongation of type 1 fimbriae in *Escherichia coli. J. Bacteriol.* 169:157–163.
93. **Manges, A. R., J. R. Johnson, B. Foxman, T. T. O'Bryan, K. E. Fullerton, and L. W. Riley.** 2001. Widespread distribution of urinary tract infections caused by a multidrug-resistant *Escherichia coli* clonal group. *N. Engl. J. Med.* 345:1007–1013.
94. **Martinez, J. J., M. A. Mulvey, J. D. Schilling, J. S. Pinkner, and S. J. Hultgren.** 2000. Type 1 pilus-mediated bacterial invasion of bladder epithelial cells. *EMBO J.* 19:28803–2812.
95. **Maurer, L., and P. E. Orndorff.** 1985. A new locus, *pilE,* required for the binding of type 1 piliated *Escherichia coli* to erythrocytes. *FEMS Microbiol. Lett.* 30:59–66.
96. **Maurer, L., and P. E. Orndorff.** 1987. Identification and characterization of genes determining receptor binding and pilus length of *Escherichia coli* type 1 pili. *J. Bacteriol.* 169:640–645.

97. **May, A. K., C. A. Bloch, R. G. Sawyer, M. D. Spengler, and T. L. Pruett.** 1993. Enhanced virulence of *Escherichia coli* bearing a site-targeted mutation in the major structural subunit of type 1 fimbriae. *Infect. Immun.* **61:**1667–1673.
98. **McClain, M. S., I. C. Blomfield, and B. I. Eisenstein.** 1991. Roles of *fimB* and *fimE* in site-specific DNA inversion associated with phase variation of type 1 fimbriae in *Escherichia coli. J. Bacteriol.* **173:**5308–5314.
99. **McCormick, B. A., D. P. Franklin, D. C. Laux, and P. S. Cohen.** 1989. Type 1 pili are not necessary for colonization of the streptomycin-treated mouse large intestine by type 1-piliated *Escherichia coli* F-18 and *E. coli* K-12. *Infect. Immun.* **57:**3022–3029.
100. **McCormick, B. A., P. Klemm, K. A. Krogfelt, R. L. Burghoff, L. Pallesen, D. C. Laux, and P. S. Cohen.** 1993. *Escherichia coli* F-18 phase locked 'on' for expression of type 1 fimbriae is a poor colonizer of the streptomycin-treated mouse large intestine. *Microb. Pathog.* **14:**33–43.
101. **Miller, J. H.** 1972. *Experiments in Molecular Genetics.* Cold Spring Harbor Laboratory, Cold Spring Harbor, N.Y.
102. **Min, G., M. Stolz, G. Zhou, F. Liang, P. Sebbel, D. Stoffler, R. Glockshuber, T.-T. Sun, U. Aebi, and X.-P. Kong.** 2002. Localization of Uroplakin Ia, the urothelial receptor for bacterial adhesin FimH, on the six inner domains of the 16 nm urothelial plaque particle. *J. Mol. Biol.* **317:**697–706.
103. **Minion, F. C., S. N. Abraham, E. H. Beachey, and J. D. Goguen.** 1986. The genetic determinant of adhesive function in type 1 fimbriae of *Escherichia coli* is distinct from the gene encoding the fimbrial subunit. *J. Bacteriol.* **165:**1033–1036.
104. **Mitsui, Y., F. P. Dyer, and R. Langridge.** 1973. X-ray diffraction studies on bacterial pili. *J. Mol. Biol.* **79:**57–64.
105. **Morschhäuser, J., V. Vetter, L. Emödy, and J. Hacker.** 1994. Adhesin regulatory genes within large, unstable DNA regions of pathogenic *Escherichia coli:* cross-talk between different adhesin gene clusters. *Mol. Microbiol.* **11:**555–566.
106. **Mulvey, M. A., Y. S. Lopez-Boado, C. L. Wilson, R. Roth, W. C. Parks, J. Heuser, and S. J. Hultgren.** 1998. Induction and evasion of host defenses by type 1-piliated uropathogenic *Escherichia coli. Science* **282:**1494–1497.
107. **Mulvey, M. A., J. D. Schilling, and S. J. Hultgren.** 2001. Establishment of a persistent *Escherichia coli* reservoir during the acute phase of a bladder infection. *Infect. Immun.* **69:**4572–4579.
108. **Neeser, J.-R., B. Koellreutter, and P. Wuersch.** 1986. Oligomannoside-type glycopeptides inhibiting adhesion of *Escherichia coli* strains mediated by type 1 pili: preparation of potent inhibitors from plant glycoproteins. *Infect. Immun.* **52:**428–436.
109. **Nei, M., and T. Gojobori.** 1986. Simple methods for estimating the numbers of synonymous and nonsynonymous nucleotide substitutions. *Mol. Biol. Evol.* **3:**418–426.
110. **Newman, J. V., R. L. Burghoff, L. Pallesen, K. A. Krogfelt, C. S. Kristensen, D. C. Laux, and P. S. Cohen.** 1994. Stimulation of *Escherichia coli* F-18Col$^-$ type-1 fimbriae synthesis by *leuX. FEMS Microbiol. Lett.* **122:**281–287.
111. **Ofek, I., D. Mirelman, and N. Sharon.** 1977. Adherence of *Escherichia coli* to human mucosal cells mediated by mannose receptors. *Nature* **265:**623–625.
112. **Ofek, I., and E. H. Beachey.** 1978. Mannose binding and epithelial cell adherence of *Escherichia coli. Infect. Immun.* **22:** 247–254.
113. **Ofek, I., and N. Sharon.** 1986. Mannose specific bacterial surface lectins, p. 55–81. *In* D. Mirelman (ed.), *Microbial Lectins and Agglutinins. Properties and Biological Activity.* John Wiley & Sons, Inc., New York, N.Y.
114. **Ofek, I., and N. Sharon.** 1988. Lectinophagocytosis: a molecular mechanism of recognition between cell surface sugars and lecting in the phagocytosis of bacteria. *Infect. Immun.* **56:**539–547.
115. **Ofek, I., D. L. Hasty, and N. Sharon.** 2003. Anti-adhesion therapy of bacterial infections: achievements, problems and prospects. *FEMS Immunol. Med. Microbiol.* **8:**181–191.
116. **Ofek, I., D. L. Hasty, and R. J. Doyle.** 2003. *Bacterial Adhesion to Animal Cells and Tissues.* ASM Press, Washington, D.C.
117. **Ofek, I., D. L. Hasty and R. J. Doyle.** 2003. Antiadhesion therapy, p. 157–176. *In* I. Ofek, D. L. Hasty, and R. J. Doyle (ed.), *Bacterial Adhesion to Animal Cells and Tissues.* ASM Press, Washington, D.C.
118. **Olsen, P. B., and P. Klemm.** 1994. Localization of promoters in the *fim* gene cluster and the effect of H-NS on the transcription of *fimB* and *fimE. FEMS Microbiol. Lett.* **116:**95–100.
119. **Orndorff, P. E., and S. Falkow.** 1984. Identification and characterization of a gene product that regulates type 1 piliation in *Escherichia coli. J. Bacteriol.* **160:**61–66.
120. **Orndorff, P. E., and S. Falkow.** 1984, Organization and expression of genes responsible for type 1 piliation in *Escherichia coli. J. Bacteriol.* **159:**736–744.
121. **Ørskov, I., F. Ørskov, and A. Birch-Andersen.** 1980. Comparison of *Escherichia coli* fimbrial antigen F7 with type 1 fimbriae. *Infect. Immun.* **27:**657–666.
122. **O'Toole, G. A., H. B. Kaplan and R. Kolter.** 2000. Biofilm formation as microbial development. *Annu. Rev. Microbiol.* **54:**49–79.
123. **Pouttu, R., T. Puustinen, R. Virkola, J. Hacker, P. Klemm, and T. K. Korhonen.** 1999. Amino acid residue Ala-62 in the FimH fimbrial adhesin is critical for the adhesiveness of meningitis-associated *Escherichia coli* to collagens. *Mol. Microbiol.* **31:**1747–1757.
124. **Pratt, L. A., and R. Kolter.** 1998. Genetic analysis of *Escherichia coli* biofilm formation: roles of flagella, motility, chemotaxis and type 1 pili. *Mol. Microbiol.* **30:**285–293.
125. **Pulliam, H. R.** 1988. Sources, sinks, and population regulation. *Am. Nat.* **132:**652–661.
126. **Ritter, A., D. L. Gally, P. B. Olsen, U. Dobrindt, A. Friedrich, P. Klemm, and J. Hacker.** 1997. The Pai-associated *leuX* specific $tRNA_5^{Leu}$ affects type 1 fimbriation in pathogenic *Escherichia coli* by control of FimB recombinase expression. *Mol. Microbiol.* **25:**871–882.
127. **Rosenthal, L.** 1943. Agglutinating properties of *Escherichia coli.* Agglutination of erythrocytes, leucocytes, thrombochtes, speermatozoa, spores of molds, and pollen by strains of *E. coli. J. Bacteriol.* **45:**545–550.
128. **Salit, I. E., and E. C. Gotschlich.** 1977. Haemagglutination by purified *Escherichia coli* pili. *J. Exp. Med.* **146:**1169–1181.
129. **Salyers, A. A., and D. D. Whitt.** 1994. *Bacterial Pathogenesis. A Molecular Approach.* ASM Press, Washington, D.C.
130. **Schaeffer, A. J., W. R. Schwan, S. J. Hultgren, and J. L. Duncan.** 1987. Relationship of type 1 pilus expression in *Escherichia coli* to ascending urinary tract infections in mice. *Infect. Immun.* **55:**373–380.
131. **Schembri, M. A., L. Pallesen, H. Connell, D. L. Hasty, and P. Klemm.** 1996. Linker insertion analysis of the FimH adhesin of type 1 fimbriae in an *Escherichia coli fimH*-null background. *FEMS Microbiol. Lett.* **137:**257–263.

132. **Schembri, M. A., E. V. Sokurenko, and P. Klemm.** 2000. Functional flexibility of the FimH adhesin: insights from a random mutant library. *Infect. Immun.* **68:**2638–2646.
133. **Schembri, M. A., and P. Klemm.** 2001. Biofilm formation in a hydrodynamic environment by novel FimH variants and ramifications for virulence. *Infect. Immun.* **69:**1322–1328.
134. **Schembri, M. A., K. Kjaegaard and P. Klemm.** 2003. Global gene expression in *Escherichia coli* biofilms. *Mol. Microbiol.* **48:**253–267.
135. **Sokurenko, E. V., H. S. Courtney, S. N. Abraham, P. Klemm, and D. L. Hasty.** 1992. Functional heterogeneity of type 1 fimbriae of *Escherichia coli. Infect. Immun.* **60:**4709–4719.
136. **Sokurenko, E. V., H. S. Courtney, D. E. Ohman, P. Klemm, and D. L. Hasty.** 1994. FimH family of type 1 fimbrial adhesins: functional heterogeneity due to minor sequence variations among *fimH* genes. *J. Bacteriol.* **176:**748–755.
137. **Sokurenko, E. V., H. S. Courtney, J. Maslow, A. Siitonen, and D. L. Hasty.** 1995. Quantitative differences in adhesiveness of type 1 fimbriated *Escherichia coli* due to structural differences in *fimH* genes. *J. Bacteriol.* **177:**3680–3686.
138. **Sokurenko, E. V., V. Chesnokova, R. J. Doyle, and D. L. Hasty.** 1997. Diversity of the *Escherichia coli* type 1 fimbrial lectin. Differential binding to mannosides and uroepithelial cells. *J. Biol. Chem.* **272:**17880–17886.
139. **Sokurenko, E. V., V. Chesnokova, D. E. Dykhuizen, I. Ofek, X.-R. Wu, K. A. Krogfelt, C. Struve, M. A. Shembri, and D. L. Hasty.** 1998. Pathogenic adaptation of *Escherichia coli* by natural variation of the FimH adhesin. *Proc. Natl. Acad. Sci. USA* **95:**8922–8926.
140. **Sokurenko, E. V., D. L. Hasty, and D. E. Dykhuizen.** 1999. Pathoadaptive mutations: gene loss and vation in bacterial pathogens. *Trends Microbiol.* **7:**191–195.
141. **Sokurenko, E. V., M. A. Schembri, E. Trintchina, K. Kjaergaard, D. L. Hasty, and P. Klemm.** 2001. Valency conversion in the type 1 fimbrial adhesin of *Escherichia coli. Mol. Microbiol.* **41:**675–686.
142. **Sokurenko, E. V., M. Feldgarden, E. Trintchina, S. J. Weissman, S. Avagyan, J. Johnson, and D. E. Dykhuizen.** 2004. Selection footprint in the FimH adhesin shows pathogenicity-adaptive niche differentiation in *Escherichia coli. Mol. Biol. Evol.* **21:**1373–1383.
143. **Sussman, M.** 1997. Escherichia coli. *Mechanisms of Virulence.* Cambridge University Press, Cambridge, United Kingdom.
144. **Tajima, F.** 1989. Statistical method for testing the neutral mutation hypothesis by DNA polymorphisms. *Genetics* **123:**585–595.
145. **Thomas, W. E., E. Trintchina, M. Forero, V. Vogel, and E. V. Sokurenko.** 2002. Bacterial adhesion to target cells enhanced by shear force. *Cell* **109:**913–923.
146. **Valenski, M.L., S.L. Harris, P. A. Spears, J. R. Horton and P. E. Orndorff.** 2003. The product of the fimI gene is necessary for *Escherichia coli* type 1 pilus biosynthesis. *J. Bacteriol.* **185:**5007–5011.
147. **Wu, X.-R., M. Manabe, J. Yu, and T.-T. Sun.** 1990. Large scale purification and immunolocalization of bovine uroplakins I, II, and III. Molecular markers of urothelial differentiation. *J. Biol. Chem.* **265:**19170–19179.
148. **Wu, X.-R., L.-H. Lin, T. Walz, M. Häner, J. Yu, U. Aebi, and T.-T. Sun.** 1994. Mammalian uroplakins: a group of highly conserved urothelial differentiation-related membrane proteins. *J. Biol. Chem.* **269:**13716–13724.
149. **Wu, X.-R., T.-T. Sun, and J. J. Medina.** 1996. *In vitro* binding of type 1 fimbriated *Escherichia coli* to uroplakins Ia and Ib: relation to urinary tract infections. *Proc. Natl. Acad. Sci. USA* **93:**9630–9635.
150. **Xia, Y., D. Gally, K. Forsman-Semb, and B. E. Uhlin.** 2000. Regulatory cross-talk between adhesin operons in *Escherichia coli:* inhibition of type 1 fimbriae expression by the PapB protein. *EMBO J.* **19:**1450–1457.
151. **Zhang, L. X., B. Foxman, P. Tallman, E. Claderar, C. Le Bouguenec, and C. F. Marrs.** 1997. Distribution of *drb* genes coding for Dr binding adhesins among uropathogenic and fecal *Escherichia coli* isolates and identification of new subtypes. *Infect. Immun.* **65:**2011–2018.
152. **Zhou, G., W.-J. Mo, P. Sebbel, G. Min, T. A. Neubert, R. Glockshuber, X.-R. Wu, T.-T. Sun, and X.-P. Kong.** 2001. Uroplakin Ia is the urothelial receptor for uropathogenic *Escherichia coli:* evidence from in vitro FimH binding. *J. Cell Sci.* **114:**4095–4103.
153. **Ziebuhr, W., K. Ohlsen, H. Karch, T. Korhonen, and J. Hacker.** 1999. Evolution of bacterial pathogenesis. *Cell. Mol. Life Sci.* **56:**719–728.

Colonization of Mucosal Surfaces
Edited by James P. Nataro et al.

Chapter 25

Fimbriae, Signaling, and Host Response to Urinary Tract Infection

NIAMH ROCHE, GÖRAN BERGSTEN, HANS FISCHER, GABRIELA GODALY, HEIKKE IRJALA, ANN CHARLOTTE LUNDSTEDT, PATRIK SAMUELSSON, MAJLIS SVENSSON, BRYNDIS RAGNARSDOTTIR, AND CATHARINA SVANBORG

The urinary tract should be sterile, yet urinary tract infections (UTIs) are common (73). At least half of all adult women experience one symptomatic UTI episode (73, 122), and an estimated 150 million people are diagnosed annually with recurrent symptomatic UTI (123). These frequency estimates illustrate the massive morbidity caused by UTIs but do not take into consideration the different forms of disease that arise in response to infection. UTIs may be symptomatic or asymptomatic, sporadic or recurrent, acute or chronic, and a cause of end-stage renal disease. Acute pyelonephritis may be accompanied by bacteremia and may be life-threatening, while acute cystitis causes severe discomfort but no systemic disease. Paradoxically, asymptomatic bacteriuria (ABU) is the most common form of UTI, and ABU patients may carry $>10^5$ CFU/ml of urine for months or years without developing symptoms or sequelae (67, 80, 149). ABU occurs in about 1% of girls, 2 to 11% of pregnant women, and about 20% of elderly men and women (73, 90).

THE "TWO-STEP" MODEL OF THE HOST RESPONSE TO URINARY TRACT INFECTION

The difference in disease severity reflects the virulence of the infecting strain and the propensity of the host to respond to infection (125). In asymptomatic carriers, the mucosa remains inert, despite large bacterial numbers in the lumen. In patients with acute pyelonephritis, both local and systemic inflammatory response pathways are activated. If the innate defense is efficient, infection may be cleared from the urinary tract; however, if it is not, chronic sequelae may ensue.

The "two-step" model of mucosal activation and disease induction is shown in Fig. 1. Pathogenic strains trigger step 1 and breach the integrity of the mucosal barrier through specific adherence and the activation of signaling pathways in the host cells. The pathogens utilize fimbriae to bind their receptors, and, as a result, Toll-like receptors (TLRs) signal to the innate host defense (Fig. 2). Attachment triggers an epithelial response, including the production of chemokines and expression of chemokine receptors, and thus mucosal inflammation is established (25, 36, 52, 75). The carrier strains may fail to elicit this response due to the lack of critical virulence factors. Alternatively, the host may be unresponsive due to the lack of primary receptors or aborted TLR4 signaling (42, 104, 133).

The second critical step is the effector phase of the anti-bacterial host defense (step 2, Fig. 1 and 3). The healthy urinary tract rapidly regains sterility as bacteriuria is cleared within hours or days. Neutrophilic granulocytes, the main effectors of the innate defense, are guided across the mucosa by interleukin-8 (IL-8) and by IL-8 receptors on neutrophils and infected epithelial cells (4, 37, 38). In the murine model, macrophage inflammatory protein 2 (MIP-2), the mouse homologue of IL-8, has the same function as IL-8 (45). As the neutrophils migrate across the mucosa, the infecting bacteria are killed. However, in susceptible individuals, infection may persist and cause exaggerated inflammation with tissue destruction (46). There is evidence that in both the murine UTI model and human patients, IL-8 receptor (CXCR1) mutations may disarm the neutrophils, disturb their migration, and cause severe disease (31).

Steps 1 and 2 are controlled through specific signaling pathways. Step 1 signaling involves bacterial

Niamh Roche, Göran Bergsten, Hans Fischer, Gabriela Godaly, Heikke Irjala, Ann Charlotte Lundstedt, Patrik Samuelsson, Majlis Svensson, Bryndis Ragnarsdottir, and Catharina Svanborg • Institute for Laboratory Medicine, Section for Microbiology, Immunology and Glycobiology, Lund University, Lund, Sweden.

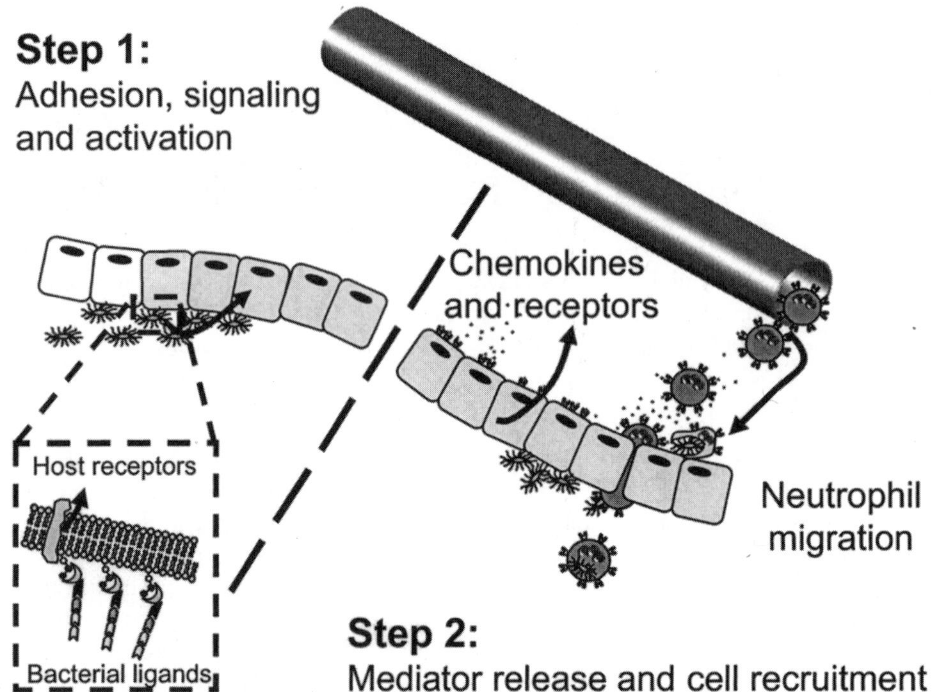

Figure 1. The two steps of the host response to uropathogenic *E. coli.* Step 1 involves triggering of the epithelial cells, and step 2 involves the neutrophil-dependent clearance of infection.

fimbriae and the TLR4-host cell receptor complex (Fig. 1 and 2), while step 2 is controlled by chemokines and chemokine receptors (Fig. 1 and 3). This review describes these signaling pathways in the urinary tract and their relevance to asymptomatic carriage, acute symptomatic disease, and chronic infection with tissue damage.

STEP 1: INDUCTION OF THE MUCOSAL HOST RESPONSE IN THE URINARY TRACT

Escherichia coli strains usually gain access to the urinary tract after having established a population at an adjacent site in the same host, for example, the large intestine or the vaginal introitus. Virulence fac-

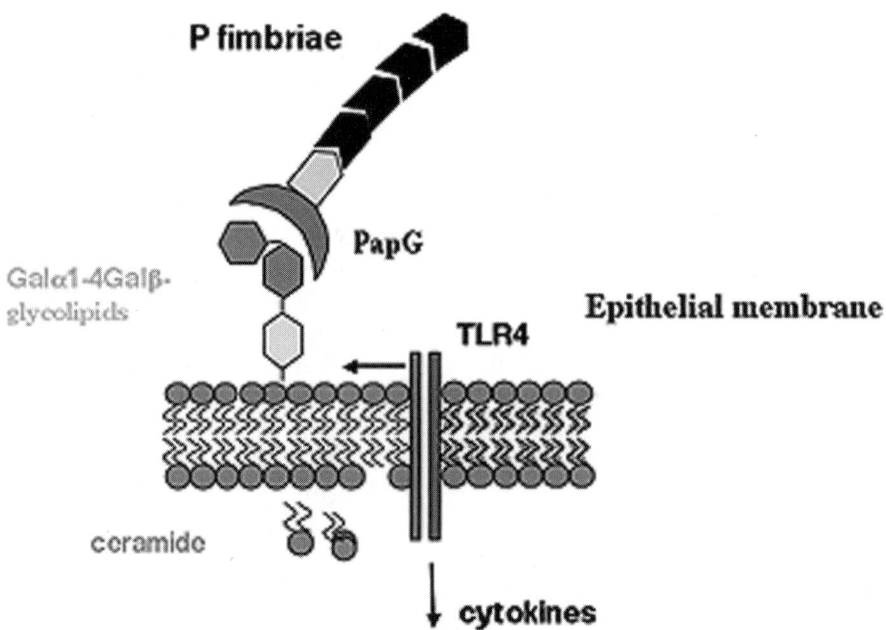

Figure 2. Step 1: host response induction. The PapG adhesin of the P fimbriae adheres to GLS receptors bearing the Gal(α1-4)Galβ epitope. Ceramide is released, and TLR4 is recruited. Ultimately, the cells produce cytokines such as IL-6 and IL-8.

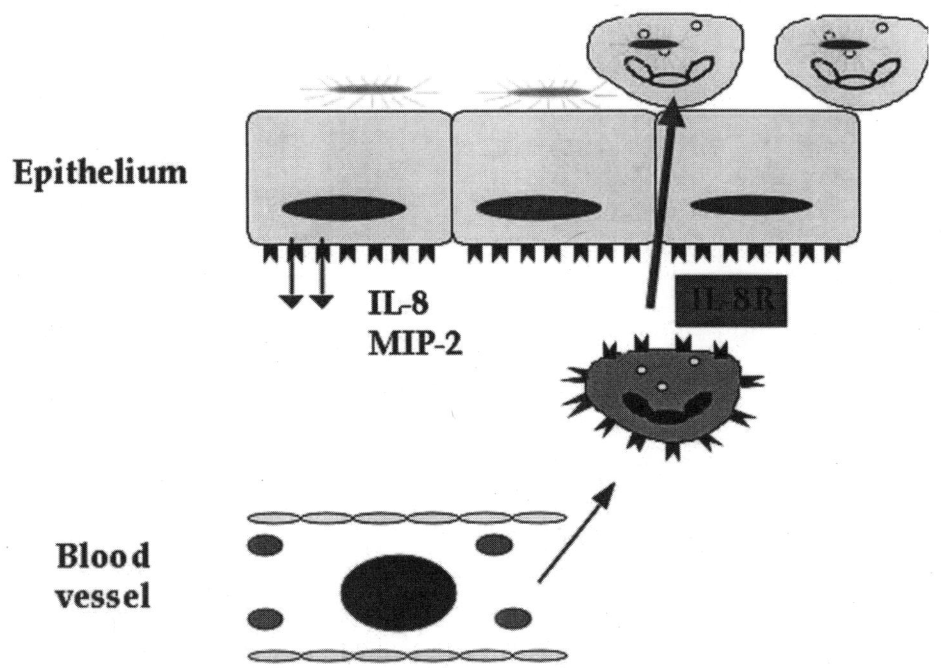

Figure 3. Step 2: effector phase of the innate host defense. Infected epithelial cells express IL-8 receptors on their surface and produce IL-8. Both IL-8 and the IL-8 receptors guide neutrophils across the epithelial barrier into the urine. In the process, infection is cleared.

tors that promote persistence in the urinary tract may also enhance survival at these sites (78). However, step 1 signaling appears to be activated only in the urinary tract, suggesting that qualitative differences exist in microbial recognition by the different mucosal cells and in the propensity to engage specific signaling pathways. For example, P-fimbriated bacteria colonize the large intestines of UTI-prone individuals and ascend into the urinary tract more efficiently than do strains lacking the *pap* operon, but they appear to cause inflammation and disease only in the urinary tract (102).

The pathogenic *E. coli* strains belong to a restricted set of clones expressing adherence factors, iron-sequestering molecules, exo- and endotoxins, and capsules (64, 83, 98, 124, 127, 145). Virulence genes can be located on chromosomal gene clusters called pathogenicity islands (40), and their expression can be regulated by the host and by environmental signals (41). ABU strains have been regarded as "avirulent" compared to the uropathogenic clones. Early studies showed that many carrier strains lacked a defined O antigen (80), were unencapsulated (66), and failed to adhere to uroepithelial cells in vitro (127). More recently, the ABU strains have been shown to carry virulence genes even though they fail to express the phenotype during long-term carriage (26, 102, 103). It is quite probable that ABU strains express adherence factors and other critical adaptive properties during the establishment of bacteriuria but then lose the phenotype during long-term persistence in the urinary tract. The difference between uropathogenic and carrier strains reflects their ability to trigger the innate host response and to cause inflammation. Thus, pathogens engage productive host signaling pathways while ABU strains avoid these activation steps.

Bacterial Adherence

The host response is remarkably selective for strains. Bacterial adherence is a key event and is guided by exquisite molecular specificity. The tissue attack starts as bacterial adhesive surface ligands identify host cell receptors at the site of infection. Early studies showed that the virulent clones bind more avidly to uroepithelial cells than do the carrier strains and thus identified adherence as a virulence factor in the human urinary tract (127). Subsequently, attachment was shown to be the first critical step in tissue attack by the pathogenic strains (for a review, see reference 126). Attachment may act indirectly by facilitating the delivery of toxins or the invasion into mucosal cells, but attachment per se may also activate specific signaling pathways.

The uropathogenic clones are fimbriated (128) and may express several adhesive surface organelles, such as Dr, afa, S, P, and type 1 fimbriae (75, 91, 141, 144). P fimbriae show the strongest association with acute-disease severity, with at least 90% of acute pyelonephritis but fewer than 20% of ABU strains expressing this phenotype (76, 102, 139). In contrast, type 1 fimbriae are expressed by more than 90% of both commensal and virulent *E. coli* strains (43, 64, 152).

P Fimbriae Enhance Bacterial Establishment and Trigger the Host Response in the Human Urinary Tract

The pathogenic strains use P-fimbria-mediated attachment to perturb the mucosal barrier. There is evidence that P fimbriae enhance bacterial virulence by promoting both intestinal colonization and spread to the urinary tract (103, 145), by promoting the establishment of bacteriuria (15, 150), by facilitating the establishment of bacteremia (143), by activating the innate host response (15, 23, 25, 52, 151), and by resisting neutrophil killing (129, 136). Although ABU strains carry toxic lipopolysaccharide (LPS), they are not recognized by the host. This suggests that their lack of P fimbriae is responsible for the lack of clearance of these strains.

The role of P fimbriae in bacterial persistence and virulence in the urinary tract has been debated (28, 86). Early studies with the mouse UTI model showed that P fimbriae enhance bacterial persistence (44, 74, 92), but later studies have produced contradictory results (12, 86). The importance of fimbrial expression in the human urinary tract environment has also been unclear (6). Recent studies with the human colonization model have conclusively shown that P fimbriae fulfill molecular Koch's postulates as independent virulence factors in the human urinary tract.

The ABU strain *E. coli* 83972 was used in these studies as a representative of the carrier strains that cause persistent bacteriuria without breaking the integrity of the mucosal barrier. P fimbria-expressing transformants of *E. coli* 83972 were constructed and were compared to the nonfimbriated parent strain after intravesical inoculation. Expression of P fimbriae was shown to enhance bacterial persistence. The P-fimbriated transformants became established faster and required fewer inoculations to reach 10^5 CFU/ml than did *E. coli* 83972 ($P < 0.0001$). Intraindividual analysis of patients colonized with both strains, but on different occasions, showed higher bacterial counts with the P-fimbriated transformants. The P-fimbriated strain triggered a local inflammatory response in all patients, but the response to the ABU strain was minor or absent. The host response was found to vary with the expression of P fimbriae, since urine samples that contained fimbriated bacteria also contained larger numbers of neutrophils and cytokines than did samples containing the negative phenotype (151).

PapG-Dependent Adhesion Breaks Mucosal Inertia and Triggers the Innate Host Response

P fimbriae are encoded by the *pap* (for "pyelonephritis associated pili") chromosomal gene cluster (60, 79). Of the 11 genes, 6 encode structural proteins: PapA, PapH, PapK, PapE, PapF, and PapG (61, 71, 79). P fimbriae are composite fibers consisting of a thin tip fibril (the PapE-PapF-PapG complex) bound to the distal end of the pilus rod (79). The PapG adhesin is at the tip of the fibril, and the proximal end is joined to the rod by the PapK adaptor (62, 71). The PapG adhesin at the fimbrial tip recognizes glycosphingolipid receptors (GSLs) on epithelial cells, and this specific interaction activates the innate host response in the murine and human urinary tracts.

To study the effect of adhesion in vivo, a *papG* deletion mutant was constructed and the host response to the full-length and the nonadhesive fimbriae was compared (15). The patients were subjected to intravesical inoculation with the *E. coli* pap^+ transformant or a *papG* deletion mutant and analyzed for epithelial cell adherence by using the *gfp* reporter. The *E. coli* pap^+ strain adhered to uroepithelial cells in vivo and triggered a significant host response, but the *papG* deletion mutant failed to adhere and was unable to break the mucosal inertia. These findings confirmed the importance of PapG-mediated host cell interactions for the host response. P fimbriation thus converted a carrier strain to a host response inducer in the human urinary tract. This effect was PapG dependent.

Epithelial Cell Response to P Fimbriae

The epithelial cells arbitrate the signaling between microbes in the lumen and underlying tissues. In early studies, P-fimbriated *E. coli* bacteria were shown to trigger an epithelial cell response both in isolated cells and in the murine urinary tract epithelium (50, 53). Cytokines such as IL-6 and IL-8 were secreted, and de novo synthesis of these cytokines was detected. More recently, the cytokine response in human biopsy specimens was investigated (109). The results confirmed the epithelial cell response to bacterial challenge in the human host and showed that fimbria-mediated adherence enhances this response. Stimulation of the biopsy specimens with recombinant *E. coli* strains expressing P fimbriae resulted in a significant increase in the amounts of IL-6 and IL-8 compared to those in the nonfimbriated control strains. Thus, isolated epithelial cells offer a suitable model to dissect the signaling pathways involved in host response induction.

Recognition Receptors and TLR4 in P-Fimbrial Signaling

E. coli P fimbriae use GSLs as primary receptors to adhere to the host cells and use TLR4 as coreceptors in transmembrane signaling and cell activation. The specificity for the GSL receptors is defined by the extracellular oligosaccharide sequence. The PapG adhesin specifically recognizes a Gal(α1-4)Galβ disaccharide motif present in GSLs along the urinary tract and in the renal tubuli (19, 75).

The GSL-specific binding is critical for cell attachment and cell activation, as shown by several experimental approaches. For example, attachment and host response induction are inhibited by the selective removal of receptor GSLs from the host cell surface. This is achieved by treatment with the ceramide glycosylation inhibitor NB-DNJ, which reduces the expression of Gal(α1-4)Galβ-containing oligosaccharides. In addition, the GSL receptors are P blood group antigens (P_1, P_2, P^k), and the expression varies between individuals of blood group P_1 or P_2. Individuals of the P blood group lack these receptor structures. P-fimbriated *E. coli* bacteria do not bind to epithelial cells from P individuals (84).

P Fimbriae Recruit TLR4 as Coreceptors in Cell Signaling

The Toll receptor was first discovered in *Drosophila melanogaster* (7, 51), where it determines dorsoventral patterning and the immune responses to microbes and fungi (77, 108). Today, more than 10 TLRs have been found in humans. TLRs are type 1 transmembrane receptors characterized by an extracellular domain containing leucine-rich repeats (107) and an intracellular Toll–IL-1 receptor (TIR) domain, which generates the signal by interacting with intracellular adaptor proteins (134). An overview of TLR signaling is given in Fig. 4.

TLR4 was first characterized in the mouse as an LPS response gene. A point mutation in the TIR domain of TLR4 was shown by Beutler and coworkers to abrogate the LPS response (104, 106). Subsequently, TLR4 was also identified as an LPS response element in humans. In the membrane, TLR4 is recruited to the CD14/MD-2 complex (116), and MD-2 is involved in the targeting of TLR4 to the cell surface by associating with the extracellular domain of TLR4 in the endoplasmic reticulum and *cis*-Golgi (89, 142). TLR4 is expressed in a wide range of cells and tissues, but its function has been studied mainly in myeloid cells (59, 88, 153). Besides LPS, some other ligands have tentatively been assigned to TLR4. These include heat shock proteins HSP60 (93, 137) and HSP70 (138) and fragments of host glycoproteins such as fibronectin (95), fibrinogen (117), hyaluronic acid (135), and heparan sulfate (63).

TLR4 Expression in the Human Urinary Tract

Epithelial cells express TLR4 in a different molecular context from that of myeloid cells (8, 32, 109). Cell lines from the intestinal epithelium show low TLR4 expression (3). In the mouse intestine, TLR4/MD-2 and CD14 complexes were detected only in the epithelial cells at the bottom of the crypts, and, more specifically, TLR4 was localized to the Golgi compartment of these cells (58, 100). This is consistent with the barrier function of the epithelium, since LPS responses would cause chronic inflammation at mucosal sites that are populated by a gram-negative flora (53).

TLR4 and CD14 expression in the human urinary tract has been examined by using biopsy specimens from the urinary bladder, ureter, renal pelvis, and cortex. TLR4 was detected in the epithelium along the entire urinary tract, and some TLR4 staining was observed in subepithelial tissues and in the connective tissue surrounding the tubuli. Epithelial TLR4 expression was confirmed by flow cytometry of exfoliated cells in the urine of healthy individuals. CD14, on the other hand, was not detected in epithelial cells from any part of the urinary tract, but occasional cells in the subepithelial compartment were CD14 positive.

Early studies showed that kidney epithelial cells lack CD14 (9, 54, 55), but more recently, CD14 was detected in a number of bladder epithelial cell lines, which responded to LPS (112). Bäckhed et al. attributed this to the presence of soluble CD14 (sCD14) (9). Results with the human biopsy specimens confirmed the lack of CD14 on human urinary tract epithelial cells. CD14 may still play a role in the LPS response, however, if secreted CD14 (sCD14) is secreted into the urine on infection. An LPS-sCD14 complex has been purified from the urine of proteinuric patients and shown to stimulate primary tubular epithelial cells (21). When recombinant proteins were used, both LBP and sCD14 were required for LPS to activate the cells. We have quantified LBP in the urine from healthy controls and found very low concentrations to be present. A more extensive study is required, however, to understand the role of LBP and sCD14 in the different phases of the response to *E. coli* infection.

The lack of CD14 may offer a mechanism to avoid a mucosal LPS response to avirulent strains and to reserve TLR4 signaling for the pathogens.

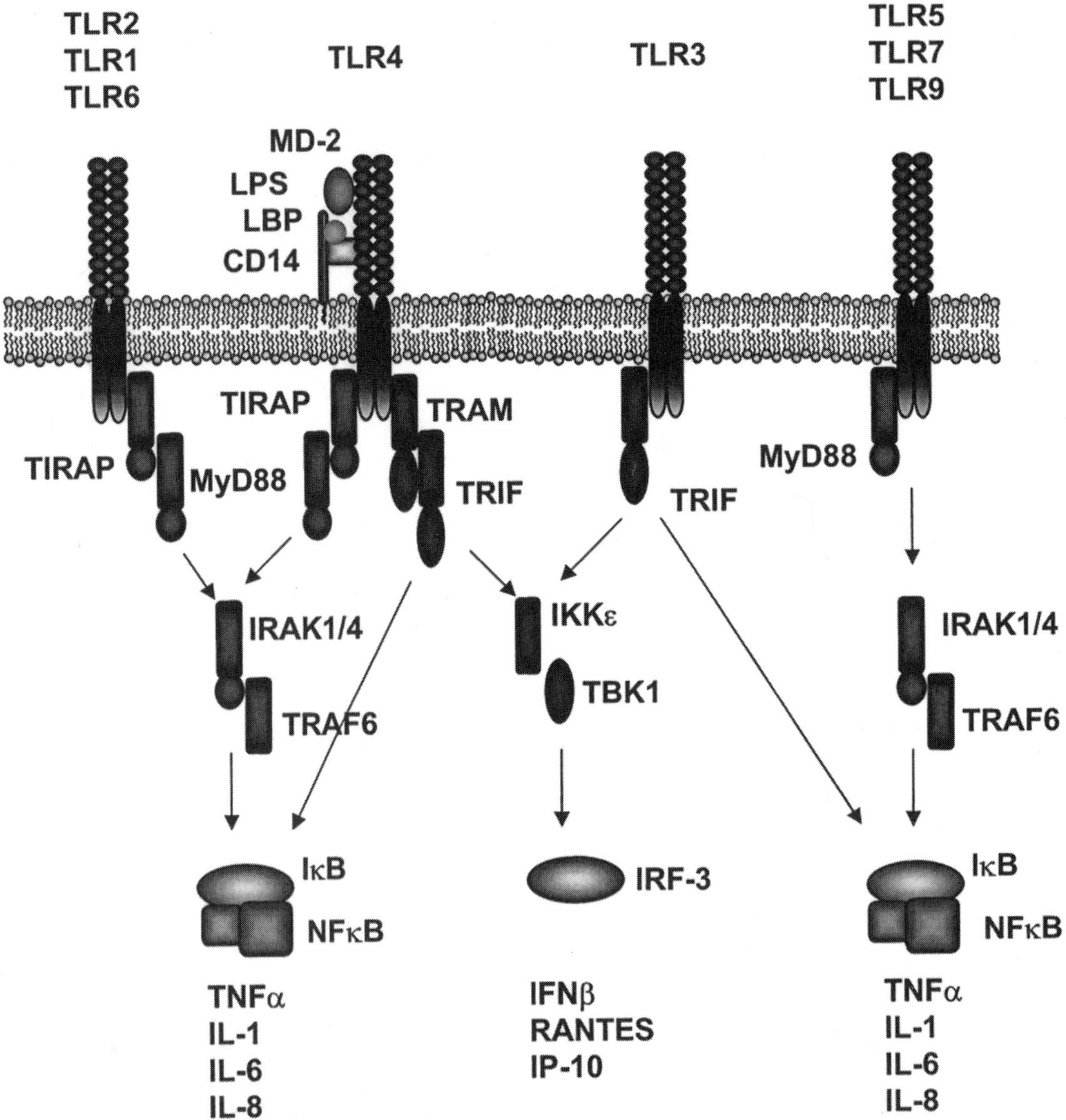

Figure 4. TLR signaling. The adaptor protein MyD88 is used by most TLRs, the probable exception being TLR3. Signaling through TLR4 can occur via MyD88-dependent or MyD88-independent pathways, and TIRAP is needed for MyD88-dependent signaling by TLR2 and TLR4. TLR4 signaling also involves the TRIF/TRAM-dependent but MyD88-independent pathway, which can activate the transcription factors NF-κB interferon regulating factor 3 (IRF-3). Activation of NF-κB or IRF-3 results in the production of IL-6 and IL-8. The TLRs are depicted as dimers, although dimer formation following stimulation has not been confirmed for all TLRs.

TLR4 and the Urinary Tract Response to Infection

TLR4 controls the mucosal response in the urinary tract. This was first observed following experimental UTI in C3H/HeJ mice (42, 130) and was attributed to the Lps nonresponder phenotype and *lps* gene defect. Later, the *lps* gene was identified as the gene encoding Tlr4, and C3H/HeJ mice were shown to carry a point mutation in the TIR domain of Tlr4 (104). The importance of Tlr4 has subsequently been confirmed by studies with naturally Tlr4-deficient mice and Tlr4 knockout mice. The role of Tlr4 in the response to P-fimbriated *E. coli* in vivo was studied by using C3H mice with an intact or dysfunctional Tlr4 gene. Infection was established with recombinant *E. coli* strains differing in P-fimbrial expression. The P-fimbriated bacteria were shown to trigger a mucosal response in the $Tlr4^{+/+}$ mice, as measured by the secretion of murine MIP-2 (mMIP-2) and neutrophil recruitment into the urine. However, the $Tlr4^{-/-}$ mice did not respond to infection. The nonfimbriated *E. coli* strain did not induce a host response in either $Tlr4^{+/+}$ or $Tlr4^{-/-}$ mice, demonstrating that P fimbriae and Tlr4 both were needed to trigger the innate host response.

The TLR repertoire of human uroepithelial cells was examined (32). Prior to infection, the cells expressed TLR1, TLR5, and TLR6 and, to a lesser extent, TLR2, TLR3, and TLR4 mRNA. Infection with P-fimbriated *E. coli* induced a rapid increase in the level of TLR4 mRNA and a slower increase in the levels of TLR1 and TLR6 mRNA. TLR3 and TLR5 did not respond. Infection with the nonfimbriated strain had no effect on TLR4 mRNA levels. TLR4 protein was detected in the cytoplasm and on the cell surface by flow cytometry and confocal microscopy using polyclonal antibodies to TLR4.

Studies with adaptor protein mutant mice have shown that P fimbriae trigger Tlr4 signaling through the TRIF/TRAM rather than the LPS-related MyD88/TIRAP pathway (H. Fischer, P. Samuelsson, M. Yamamoto, K. Hoebe, M. Svensson, K. Ekström, L. Gustafsson, S. Akira, B. Beutler, and C. Svanborg, submitted for publication). This is consistent with the febrile UTI patients being infected predominantly with P-fimbriated *E. coli* and producing mainly monocyte chemotactic protein 1 (MCP-1), which is a product of the TRIF/TRAM pathway (G. Otto, G. Godaly, M. Burdick, R. Strieter, and C. Svanborg, submitted for publication).

We conclude that P-fimbriated *E. coli* strains activate TLR4 signaling in epithelial cells by mechanisms that are CD14 independent, suggesting that signaling in this case does not involve the conventional LPS-CD14 /MD-2 signaling mechanism.

Role of LPS

The endotoxicity of LPS is determined by the structure of lipid A. Removal of one or more acyl chains leads to inactivation of the endotoxic effect or even to antagonistic effects (55, 120). In particular, bacteria carrying penta-acylated LPS antagonize the effects of *E. coli* LPS in TLR4 signaling. The intestinal flora contains several species carrying penta-acylated LPS, and it has therefore been speculated that this antagonistic effect of bacteria could be a mechanism of regulating the response to LPS (10).

The role of LPS was examined in vitro and in vivo by using *E. coli* expressing nonstimulatory LPS due to defective lipid A myristoylation (*msbB* mutant) (32, 55). The human kidney epithelial cell line responded to the P-fimbriated bacteria in both the wild-type and *msbB* mutant backgrounds but not to the nonfimbriated recombinants. TLR4-proficient or -deficient mice were infected with P-fimbriated strains in the *msbB* mutant or wild-type *E. coli* background. There was no significant difference in the host response to the P-fimbriated strains, regardless of the LPS genotype. The $Tlr4^{-/-}$ mice did not respond to any of the strains. Furthermore, inhibitors of LPS were used. P-fimbriated and nonfimbriated control bacteria were preincubated with polymyxin B or bactericidal permeability-increasing protein, but no inhibitory effect was observed. It may be speculated that P fimbriae activate TLR4 via their GSL receptors rather than via CD14 and overcome the mucosal LPS barrier that normally controls inflammation.

The results illustrate how the receptor specificity of the fimbriae and the nature of their cell surface receptors may influence the signaling pathways that activate innate immunity. The nonadhesive bacteria fail to trigger the innate host response, even though they carry LPS and other surface molecules that can activate TLR4-specific signaling in other cell types. While TLR4 regulates signaling and the host response, primary, ligand-binding receptors are needed for selective targeting of the pathogens to mucosal sites.

Ceramide as a Potential Signaling Intermediate between the GSLs and TLR4

How can TLR4 signaling in epithelial cells be activated by adhesive ligands that bind to membrane glycolipids? We have proposed that ceramide acts as a signaling intermediate between the GSLs and TLR4 (55, 56). The receptor GSLs lack a transmembrane domain but are bound to ceramide in the outer leaflet of the lipid bilayer (75). P-fimbriated *E. coli* stimulates an increase in intracellular ceramide levels, with a simultaneous decrease in surface-expressed receptor-active GSLs, suggesting that ceramide is released by hydrolysis of the receptor itself (56). Ceramide release and phosphorylation of ceramide are detected within minutes after exposure to P-fimbriated *E. coli*. Signaling involved Ser/Thr protein kinases, since chemokine responses to P-fimbriated *E. coli* were blocked in the presence of Ser/Thr protein kinase inhibitors. This signaling pathway was selectively activated by P fimbriae, since isogenic *E. coli* strains expressing type 1 fimbriae with mannose specificity failed to release ceramide.

Recent studies have suggested that the release of ceramide may activate the TLR4 signaling pathway. Sphingomyelinase (SMase) specifically releases ceramide from sphingomyelin (for a review, see reference 48) and has been used as a prototype activator of the "ceramide-signaling pathway." SMase treatment stimulates a cytokine response in many cell types, including uroepithelial cells (57). We used SMase to investigate if ceramide may act as a signaling intermediate and activate the TLR4 signaling pathway, and we found evidence that this is the case. It may be speculated that ceramide release offers a general mechanism of TLR4 coreceptor recruitment.

Type 1 Fimbriae

Type 1 fimbriae are encoded by the *fim* gene cluster, which is found in both commensal and pathogenic *E. coli* strains (20, 29, 43, 64, 152). Five of the nine genes in the *fim* gene cluster encode the structural proteins FimA, FimI, FimF, FimG, and FimH (65). Typically, a type 1-fimbriated bacterium has 200 to 500 peritrichously arranged fimbriae on its surface (110). Each organelle is a 7-nm-wide and ca. 1-μm long rod-shaped structure (110). The bulk of the fimbriae consists of around 1,000 copies of the major subunit protein, FimA, which serves as a scaffold for the receptor-recognizing adhesin, FimH. FimH is located at the tip and perhaps also interspersed along the fimbrial shaft (49, 69, 70).

Type 1 fimbriae recognize terminally located D-mannose moieties on cell-bound and secreted glycoproteins and peptide epitopes with similar structures (35, 114, 118, 147). Uroplakins 1a and 1b were shown to be receptors for FimH on the bladder epithelium (148). These cellular receptors contain a single N-linked carbohydrate of the high-mannose type, common to most glycoproteins recognized by FimH. Other reported FimH receptors include the Tamm-Horsfall protein (99, 101), secretory immunoglobulin A (sIgA) (147), CD48 (13), laminin (72), collagen (105), fibronectin (119), and complement receptors (115).

Type 1 fimbrial expression is controlled by site-specific recombination (1, 16). *fimA* transcription initiates at a promoter in a short (314-bp) invertible DNA element. Fimbriae are expressed when the element is in the ON but not the OFF orientation (96). The FimB and FimE recombinases mediate inversion of the *fim* invertible element (34, 68). Both are members of the tyrosine recombinase family (30). FimB recombinase promotes OFF-to-ON switching while FimE primarily catalyzes ON-to-OFF inversion (34, 68). A number of global regulators, including integration host factor (17), the leucine-responsive regulatory protein (18), and the histone-like nonstructural protein (H-NS) (96, 97, 111, 121) also affect switching of the invertible element.

Type 1 Fimbriae and Virulence for the Urinary Tract

Type 1 fimbriae are expressed by most enterobacteria, but their function remains something of an enigma. Type 1 fimbriae have been assigned a large number of different functions that may enhance virulence. They may promote bacterial persistence in the large intestine by binding to mannosylated glycoprotein receptors including secretory IgA (146, 147). Type 1-fimbriated *E. coli* strains invade bladder dome epithelial cells, and intracellular replication has been suggested to result in the formation of "bacterial factories" (85, 87). Type 1 fimbriae bind to neutrophils and macrophages and stimulate phagocytosis and the oxidative burst (14, 94), while the FimH adhesin of type 1 fimbriae binds CD14 expressed by mast cells (2).

The fimbriae have been identified as virulence factors in the murine experimental model of UTI and as colonization factors of the large intestine, but a role in virulence is potentially difficult to reconcile with the occurrence of type 1 fimbriae in both virulent and commensal strains. Since most *E. coli* isolates carry the *fim* gene regardless of their source (44), the expression of type 1 fimbriae, or the possession of the *fim* gene cluster, does not correlate with uropathogenicity in humans (43, 76, 131, 140). However, type 1 fimbriae promote virulence in the murine UTI model by enhancing bacterial survival and mucosal inflammation, suggesting that they can act as independent virulence factors (24). Type 1 fimbriae also contribute to virulence in the background of a fully virulent strain. Deletion of the *fim* gene from an *E. coli* O1:K1:H7 strain was shown to reduce virulence in the mouse model. Finally, the *fim* gene was shown by sequence-tagged mutagenesis to be up-regulated in vivo in the mouse urinary tract (24).

LPS and TLR4 Dependence of Responses to Type 1-Fimbriated Bacteria

Type 1 fimbriae stimulate an epithelial cell response in vitro. After binding to the mannosylated glycoprotein receptors, they trigger a cytokine response (54, 112). The response differs from that to P fimbriae in that it is partly LPS and TLR4 dependent. Thus, type 1-fimbriated recombinant strains with intact LPS (wild-type *msbB*) triggered a significantly higher cytokine response in A498 cells than did the isogenic type 1-fimbriated *msbB* mutant strain; in the mouse model, there was a higher response to the type 1-fimbriated wild-type strain than to the *msbB* mutant strain. TLR4 controlled the LPS-dependent response, as shown by infection in *tlr4*$^{+/+}$ and *tlr4*$^{-/-}$ mice. The type 1-fimbriated strain triggered a high and rapid neutrophil response in the *tlr4*$^{+/+}$ mice and a low but significant response in the *tlr4*$^{-/-}$ mice. Similar results were obtained using the *msbB* mutant strain. After detoxification of the lipid A, type 1-fimbriated strains triggered a lower response (54), suggesting that the type 1-fimbriated bacteria trigger both an LPS- and *tlr4*-dependent response and a fimbria-dependent response.

Type 1 Fimbriae in the Human Urinary Tract

Despite the extensive results from the mouse model, the in vivo function of type 1 fimbriae in the human urinary tract remains unclear. Patients have been inoculated with *E. coli* 83972 *FIM*$^+$ *GFP*$^+$, and postinoculation urine samples have been cultured and analyzed for fimbrial expression in vivo by hemagglutination and flow cytometry. Epithelial cell adherence was detected by fluorescence microscopy. The IL-8 and IL-6 concentrations and the polymorphonuclear leukocyte numbers in urine were used to monitor the innate host response to intravesical inoculation with type 1-fimbriated *E. coli* 83972. Surprisingly, we did not observe adherent *GFP*$^+$ bacteria in the urine samples. This might reflect the inhibitory activity of soluble mannosylated glycoproteins such as the Tamm-Horsfall glycoprotein and IgA in urine (99, 127). Secretory IgA antibodies may also block adhesion if they are directed against relevant surface epitopes, including the FimH adhesin. The *FIM*$^+$ strain did not trigger a significant IL-8 or an IL-6 response, and no polymorphonuclear leukocyte response was observed.

To address the issue of adaptor protein usage by type 1-fimbriated *E. coli,* studies were carried out with mutant mice. The results have shown that type 1 fimbriae trigger a partially MyD88-dependent response (Fig. 4). This is in line with the IL-8 response elicited by type 1-fimbriated *E. coli* strains, since the MyD88 pathway favors chemokines such as IL-8.

These limited results with type 1 fimbriae suggest that there are functional differences between P fimbriae and type 1 fimbriae in the human urinary tract. P fimbriae clearly promoted bacterial persistence and enhanced the mucosal host response, but type 1 fimbriae appeared not to influence the establishment of bacteriuria or trigger a mucosal host response to the same extent. Furthermore, the results emphasize the difference in the mechanisms of cell activation and the signaling pathways engaged by P or type 1 fimbriae.

STEP 2: NEUTROPHILS AS EFFECTORS OF THE INNATE DEFENSE

Signals between the activated epithelium and the neutrophils determine the efficiency of bacterial clearance from the urinary tract. The activated epithelial cells secrete a variety of cytokines, including the CXC chemokines involved in neutophil recruitment. Neutrophils dominate the acute inflammatory infiltrate in the urinary tract and are essential effector cells of the innate host defense (79). Normal mice have a transient neutrophil response, during which bacteria are eliminated from the tissues, but neutrophil-deficient mice fail to clear the infection. The importance of neutrophils has been demonstrated by antibody depletion experiments (36, 47) and by infection of mice with genetically determined aberrations in neutrophil migration or activation (31, 113, 132).

Molecular Mechanisms of Neutrophil Recruitment and "Pyuria"

The urinary tract epithelium directs the neutrophil response by secreting chemoattractants. The mediators that recruit the neutrophils from the blood have not been identified, but IL-8 directs the egress of neutrophils across the epithelial cell barrier and into the urine, causing "pyuria" (4). In the Transwell model, recombinant IL-8 was able to replace the bacterial stimulus as the force driving transepithelial neutrophil migration and anti-IL-8 antibodies blocked the migration in response to the bacteria (37).

These in vitro observations have been corroborated in vivo. Epithelial cells in the human urinary tract mucosa synthesize IL-8 (5), and intravesical infection causes a rapid IL-8 response (15, 52, 151). MIP-2 is an important IL-8 equivalent in the murine urinary tract (33, 36), and MIP-2 depletion blocks neutrophil egress across the epithelial barrier in vivo. Anti-MIP-2 antibody treatment caused the neutrophils to accumulate under the epithelium, demonstrating that chemokines such as IL-8 direct the exodus of neutrophils from the tissues and into the urine (45).

Fimbriae influence the chemokine response repertoire. Type 1-fimbriated strains mainly elicit neutrophil-activating chemokines (e.g., IL-8 and GRO-α), but P-fimbriated strains trigger a chemokine repertoire favoring the recruitment of many different cell types, including lymphocytes and monocytes, in addition to neutrophils (e.g., MCP-1 and MIP-1α) (G. Godaly, unpublished results). For example, the early MCP-1 response may activate mast cells, macrophages, monocytes, and even neutrophils, since receptors for MCP-1 are found on these different cell types.

In addition, infection enhances the expression of the CXC chemokine receptors on both epithelial cells and neutrophils. In humans, IL-8 mediates its biological activity through the G-protein-coupled receptors CXCR1 and CXCR2, and CXCR1 controls the neutrophil chemotaxis (11, 39). UTI stimulates CXCR1 and CXCR2 expression in the human urinary tract epithelium. In vitro studies have identified CXCR1 as the crucial receptor required to support neutrophil exit across infected human uroepithelial cell layers.

IL-8Rh$^{-/-}$ Mice Develop Acute Pyelonephritis with Bacteremia and Chronic Tissue Damage

Mice have a single functional receptor for the IL-8 CXC chemokine family, and IL-8 receptor knockout mice (mIL-8Rh KO) were constructed by insertion of a neomycin cassette in the murine IL-8 receptor gene (22). Their neutrophils fail to respond to the IL-8 homologues in general but retain responses to other chemotactic signals. Since all of the murine IL-8 homologues converge on one murine IL-8 receptor, inactivation of this gene should cause a peripheral neutrophil migration deficiency. We have used the mIL-8Rh KO mouse to study the contribution of chemokines and chemokine receptors to the defense against UTI (31). The mIL-8Rh KO mice developed severe acute pyelonephritis with bacteremia and symptoms during the first week following intravesical infection, and about half of the mice succumbed to lethal infection. Surviving mice then developed chronic inflammation and kidney damage. In control mice, the neutrophil influx was transient and the successful exit of neutrophils across the epithelial barrier into the lumen was detected as "pyuria." Infection was cleared in a few days, and there was no evidence of tissue damage.

The acute and chronic phases of infection were attributed to the dysfunctional neutrophil response in the mIL-8Rh KO mice. Neutrophils were recruited more slowly into infected kidneys and were unable to escape from the tissues by crossing the mucosal barrier (38). As a consequence, a massive buildup of neutrophils occurred in the kidney tissue. After 7 days, mIL-8Rh KO mice had swollen kidneys with neutrophil abscesses, and after 35 days, they had developed kidney pathology and renal scarring (46). The accumulation of neutrophils thus resulted in tissue destruction, resembling renal scarring in humans. In addition, the lack of the mIL-8Rh affects the neutrophil function. The receptor is needed for CXC chemokines to activate neutrophils, and this is a crucial step in phagocytosis and killing of the bacteria. The lack of phagocytosis and killing due to deficient neutrophil activation may contribute to the acute and fulminant disease in the mIL-8Rh KO mice.

We conclude that a single gene encoding the mIL-8Rh receptor homologue determines if mice will be resistant to experimental UTI, as defined by both acute disease severity and chronic tissue damage.

GENETIC CONTROL OF SIGNALING PATHWAYS AND RESISTANCE TO URINARY TRACT INFECTION

The results obtained with the murine model suggest at least three ways in which host receptor expression might influence the inflammatory response. Variations in step 1 may be caused by TLR4 and/or GSL receptor expression. The GSL receptors should influence whether fimbriae find their primary receptor, and TLR4 expression should influence whether cell signaling occurs, and both these mechanisms influence the functionality of the epithelial cell response. Variations in step 2 may be caused by chemokine receptor signaling and may play a crucial role in the development of kidney scarring after pyelonephritis.

Genetics of Glycolipid Receptor Variation as Deduced from the P Blood Group

The GSL recognition receptors are essential for P fimbriae to adhere and to recruit TLR4 for signaling. The expression of receptors for P fimbriae reflects the P blood group, since the receptor structures also act as the P blood group of the host. Patients prone to UTI show a higher density of epithelial cell receptors, and individuals of blood group P_1 have an 11-fold-increased risk of developing recurrent pyelonephritis (82). The receptor repertoire may also influence which fimbrial type can cause infection. Individuals of blood group A_1P_1 express the globoA receptor and become infected with bacteria recognizing this receptor (81). In theory, individual lacking receptors would be protected against infection with P-fimbriated *E. coli,* but there are too few receptor-negative individuals to investigate this hypothesis in studies of humans. This hypothesis was tested in studies of mice by pharmacological inhibition of epithelial receptor expression in vivo. The glucose analogue *N*-butyldeoxynojirimycin (NB-DNJ) is a metabolic inhibitor of the ceramide-specific glycosyltransferase involved in receptor biosynthesis (133). When infected with P-fimbriated *E. coli,* NB-DNJ-treated mice were protected against colonization and inflammation, but this effect was less pronounced when type 1-fimbriated bacteria were used. The results confirmed the importance of the primary receptor for the host response in vivo.

Disruptions of TLR4 Signaling May Cause Asymptomatic Bacteriuria

TLR4 mutant mice failed to respond to infection and were unable to clear bacteria from the urinary tract (32, 130), but they developed an asymptomatic carrier state. C3H/HeJ mice remained asymptomatic for months, and there was no evidence of tissue damage. The failure to activate the host response appeared to be protective, in that the animals could avoid developing asymptomatic disease. Furthermore, the lack of antibacterial defense activation permitted bacterial persistence, but apparently without causing side effects or tissue damage.

This carrier state resembles ABU in patients who maintain high bacterial counts for months and years if left untreated. Thus, mutations in TLR4 or downstream signaling pathways may underlie the unresponsiveness in ABU patients. Prospective clinical studies of the TLR4 genotype in patients with a history of primary ABU are currently being evaluated.

Genetics of IL-8 Receptor Expression in Pyelonephritis-Prone Children

In a prospective clinical study, CXCR expression in children with at least one episode of acute pyelonephritis was compared with that in age-matched controls with no history of UTI. CXCR1 expression was dramatically lower in the patients than in the controls (31; A.-C. Lundstedt, S. McCarthy, G. Godaly, D. Karpman, I. Leijonhufvud, M. Samuelsson, M. Svensson, B. Andersson, and C. Svanborg, submitted for publication). Subsequently, the low expression of CXCR1 was confirmed in studies of other pyelonephritis-prone patients. We conclude that pyelonephritis-prone children have low chemokine receptor expression.

Subsequent studies have shown that the CXCR1 gene is polymorphic and that mutations and/or insertions may underlie differences in the susceptibility to acute pyelonephritis and renal scarring. The single-nucleotide polymorphisms were observed in the intron, the coding region, and the 3′ untranslated region. No mutations were found in the age-matched children without a history of UTI.

SUMMARY

The urinary tract is a normally sterile site, yet UTIs are common. The severity of acute infection and the long-term sequelae are determined by the infecting strain and the virulence factors that it expresses. In addition, however, there is a strong influence of host genetics, and the genes involved are those that determine the response to bacterial attack. A specific molecular interaction between bacteria and host is needed to initiate the disease process, and a similar degree of molecular specificity is needed to clear the infection. Specific signaling events determine the activation at each step in the process. Step 1, which activates the host response, is controlled by recognition receptors for bacterial fimbriae and by TLR4 signaling. Thus, severe symptomatic infections are caused mostly by P-fimbriated bacteria, which bind GSL receptors and recruit TLR4 for transmembrane signaling. Type 1 fimbriae also activate TLR4 signaling but do so through a different mechanism. Step 2, which controls bacterial clearance, is controlled by CXC chemokines and their signaling receptors. Genetic dysfunctions in these signaling pathways may either be protective or cause the pathology. These new insights provide the first molecular background for host susceptibility to the well-known variation in UTI in human patients.

Acknowledgments. We thank Lennart Philipson for critically reviewing the manuscript and C. Tallqvist for technical assistance. This study was supported by grants from The Swedish Medical Research Council (K97-06X-07934); The Crafoord, Wallenberg (97.123); The Swedish-Japan Foundation; Österlund Foundation; and The Royal Physiografic Society. CS is the recipient of an Unrestricted Grant from Bristol-Meyers Squibb. PS was supported by the program "Glycoconjugates in Biological Systems" sponsored by the Swedish Foundation for Strategic Research.

REFERENCES

1. **Abraham, J. M., C. S. Freitag, J. R. M. Clements, and B. I. Eisentein.** 1985. An invertible element of DNA controls phase variation of type 1 fimbriae of *Escherichia coli. Proc. Natl. Acad. Sci. USA* **82:**5724–5727.
2. **Abraham, S. N., and A. Michael.** 1998. Mast cells and basophils in innate immunity. *Semin. Immunol.* **10:**373–381.
3. **Abreu, M. T., P. Vora, E. Faure, L. S. Thomas, E. T. Arnold, and M. Arditi.** 2001. Decreased espression of Toll-like receptor-4 and MD-2 correlates with intestinal epithelial cell protection against dysregulated proinflammatory gene expression in response to bacterial lipopolysaccharide. *J. Immunol.* **167:**1609–1616.
4. **Agace, W., S. Hedges, U. Andersson, J. Andersson, M. Ceska, and C. Svanborg.** 1993. Selective cytokine production by epithelial cells following exposure to *Escherichia coli. Infect. Immun.* **61:**602–609.
5. **Agace, W. W., S. R. Hedges, M. Ceska, and C. Svanborg.** 1993. Interleukin-8 and the neutrophil response to mucosal gram-negative infection. *J. Clin. Investig.* **92:**780–785.
6. **Anderson, P., I. Engberg, G. Lidin-Janson, K. Lincoln, R. Hull, S. Hull, and C. Svanborg.** 1991 Persistence of *Escherichia coli* bacteriuria is not determined by bacterial adherence. *Infect. Immun.* **59:**2915–2921.
7. **Anderson, K. V., L. Bokla and C. Nusslein-Volhard.** 1985. Establishment of the dorsal-ventral polarity in the *Drosophilia* embryo: the induction of polarity by the Toll gene product. *Cell* **42:**791–798.
8. **Bäckhed, F., M. Soderhall, P. Ekman, S. Normark, and A. Richter-Dahlfors.** 2001. Induction of the innate immune responses by *Escherichia coli* and purified lipopolysaccharide correlate with organ- and cell-specific expression of Toll-like receptors within the human urinary tract. *Cell. Microbiol.* **3:**153–158.
9. **Bäckhed, F., L. Meijer, S. Normark, and A. Richter-Dahlfors.** 2002. TLR4-dependent recognition of lipopolysaccharide by epithelial cells requires sCD14. *Cell. Microbiol.* **4:**493–501.
10. **Bäckhed, F., S. Normark, E. K. H. Schweda, S. Oscarsson, and A. Richter-Dahlfors.** 2003. Structural requirements for TLR4-mediated LPS signaling: A biological role for LPS modifications. *Microbes Infect.* **5:**1057–1063.
11. **Baggiolini, M., B. Dewald, and B. Moser.** 1994. Interleukin-8 and related chemotactic cytokines—CXC and CC chemokines. *Adv. Immunol.* **55:**97–179.
12. **Bahrani-Mougeot, F. K., E. L. Buckles, C. V. Lockatell, J. R. Hebel, D. E. Johnson, C. M. Tang, and M. S. Donnenberg.** 2002. Type 1 fimbriae and extracellular polysaccharides are

preeminent *Escherichia coli* virulence determinants in the murine urinary tract. *Mol. Microbiol.* **45**:1079–1093.

13. **Baorto, D. M., Z. Gao, R. Malaviya, M. L. Dustin, A. van der Merwe, D. M. Lublin, and S. N. Abraham.** 1997. Survival of FimH-expressing enterobacteria in macrophages relies on glycolipid traffic. *Nature* **389**:636–639.
14. **Bar-Shavit, Z., I. Ofek, G. Goldman, Mirelman D, and N. Sharon.** 1977. Mannose residues on phagocytes as receptors for the attachment of *Escherichia coli* and *Salmonella typhi. Biochem. Biophys. Res. Commun.* **78**:455–460.
15. **Bergsten, G., M. Samuelsson, B. Wullt, I. Leijonhufvud, H. Fischer, and C. Svanborg.** 2004. PapG dependent adhesion breaks mucosal inertia and triggers the innate host response. *J. Infect. Dis.* **189**:1734–1742.
16. **Blomfield, I., and M. van der Woude.** 2002. Regulation and function of phase variation in *Escherichia coli,* p 89–113. *In* M. Wilson (ed.), *Bacterial Adhesion to Host Tissues: Mechanisms and Consequences,* vol. 1. Cambridge University Press, Cambridge, United Kingdom.
17. **Blomfield, I. C., D. H. Kulasekara, and B. I. Eisenstein.** 1997. Integration host factor stimulates both FimB- and FimE-mediated site-specific DNA inversion that controls phase variation of type 1 fimbriae expression in *Escherichia coli. Mol. Microbiol.* **23**:705–717.
18. **Blomfield, I. C., P. J. Calie, K. J. Eberhardt, M. S. McClain, and B. I. Eisenstein.** 1993. lrp stimulates phase variation of type 1 fimbriation in *Escherichia coli* K-12. *J. Bacteriol.* **175**:27–36.
19. **Bock, K., M. E. Breimer, A. Brignole, G. C. Hansson, K.-A. Karlsson, G. Larson, H. Leffler, B. E. Samuelsson, N. Strömberg, C. Svanborg-Edén, and J. Thurin.** 1985. Specificity of binding of a strain of *Escherichia coli* to Gal alpha 1-4 Gal containing glycosphingolipids. *J. Biol. Chem.* **260**:8545–8551.
20. **Brinton, C.** 1965. The structure, function, synthesis, and genetic control of bacterial pili, and a molecular model of DNA and RNA transport in gram-negative bacteria. *Trans. N.Y. Acad. Sci.* **27**:1003–1054.
21. **Bussolati, B., S. David, V. Cambi, P. S. Tobias, and G. Camussi.** 2002. Urinary soluble CD14 mediates human proximal tubular epithelial cell injury induced by LPS. *Int. J. Mol. Med.* **10**:441–449.
22. **Cacalano, G., J. Lee, K. Kikly, A. M. Ryan, S. Pitts-Meek, B. Hultgren, W. I. Wood, and K. W. Moore.** 1994. Neutrophil and B cell expansion in mice that lack the murine IL-8 receptor homolog. *Science* **265**:682–684.
23. **Condron, C., D. Toomey, R. G. Casey, M. Shaffii, T. Creagh, and D. Bouchier-Hayes.** 2003. Neutrophil bacterial function is defective in patients with recurrent urinary tract infections. *Urol. Res.* **31**:329–334.
24. **Connell, I., W. Agace, P. Klemm, M. Schembri, S. Marild, and C. Svanborg.** 1996. Type 1 fimbrial expression enhances *Escherichia coli* virulence for the urinary tract. *Proc. Natl. Acad. Sci. USA* **93**:9827–9832.
25. **de Man, P., C. van Kooten, L. Aarden, I. Engberg, H. Linder, and C. Svanborg Edén.** 1989. Interleukin-6 induced at mucosal surfaces by gram-negative bacterial infection. *Infect. Immun.* **57**:3383–3388.
26. **Dobrindt, U., F. Agerer, K. Michaelis, A. Janka, C. Buchrieser, M. Samuelsson, C. Svanborg, G. Gottschalk, H. Karch, and J. Hacker.** 2003. Analysis of genome plasticity in pathogenic and commensal *Escherichia coli* isolates by use of DNA arrays. *J. Bacteriol.* **185**:1831–1840.
27. **Dodson, K. W., J. S. Pinkner, T. Rose, G. Magnusson, S. J., Hultgren, and G. Waksman.** 2001. Structural basis of the interaction of the pyelonephritic *E. coli* adhesin to its kidney receptor. *Cell* **105**:733–743.
28. **Donnenberg, M. S., and R. A. Welsh.** 1996. Virulence determinants of uropathogenic *Escherichia coli,* p. 135–174. *In* H. L. T. Mobley and J. W. Warren (ed.), *Urinary Tract Infections: Molecular Pathogenesis and Clinical Management.* ASM Press, Washington, D.C.
29. **Duguid, J., I. Smith, G. Dempster, and P. Edmunds.** 1955. Non-flagellar filamentous appendages ("fimbriae") and haemagglutinating activity in *Bacterium coli. J. Pathol. Bacteriol.* **70**:335–348.
30. **Esposito, D., and J. J. Scocca.** 1997. The integrase family of tyrosine recombinases: evolution of a conserved active site domain. *Nucleic Acids Res.* **25**:3605–3614.
31. **Frendéus, B., G. Godaly, L. Hang, D. Karpman, A. C. Lundstedt, and C. Svanborg.** 2000. Interleukin 8 receptor deficiency confers susceptibility to acute experimental pyelonephritis and may have a human counterpart. *J. Exp. Med.* **192**:881–890.
32. **Frendéus, B., C. Wachtler, M. Hedlund, H. Fischer, P. Samuelsson, M. Svensson, and C. Svanborg.** 2001. *Escherichia coli* P fimbriae utilize the Toll-like receptor 4 pathway for cell activation. *Mol. Microbiol.* **40**:37–51.
33. **Frendéus, B., G. Godaly, L. Hang, D. Karpman, and C. Svanborg.** 2001. Interleukin-8 receptor deficiency confers susceptibility to acute pyelonephritis. *J. Infect. Dis.* **183**(Suppl. 1):S56–S60.
34. **Gally, D. L., J. Leathart, and I. C. Blomfield.** 1996. Interaction of FimB and FimE with the fim switch that controls the phase variation of type 1 fimbriae in *Escherichia coli* K-12. *Mol. Microbiol.* **21**:725–738.
35. **Giampapa, C. S., S. N. Abraham, T. M. Chiang, and E. H. Beachey.** 1988. Isolation and characterization of a receptor for type 1 fimbriae of *Escherichia coli* from guinea pig erythrocytes. *J. Biol. Chem.* **263**:5362–5367.
36. **Godaly, G., B. Frendéus, A. Proudfoot, M. Svensson, P. Klemm, and C. Svanborg.** 1998. Role of fimbriae-mediated adherence for neutrophil migration across *Escherichia coli*-infected epithelial cell layers. *Mol. Microbiol.* **30**:725–735.
37. **Godaly, G., A. E. Proudfoot, R. E. Offord, C. Svanborg, and W. W. Agace.** 1997. Role of epithelial interleukin-8 (IL-8) and neutrophil IL-8 receptor A in *Escherichia coli*-induced transuroepithelial neutrophil migration. *Infect. Immun.* **65**:3451–3456.
38. **Godaly, G., L. Hang, B. Frendéus, and C. Svanborg.** 2000. Transepithelial neutrophil migration is CXCR1 dependent *in vitro* and is defective in IL-8 receptor knockout mice. *J. Immunol.* **165**:5287–5294.
39. **Godaly, G., G. Bergsten, L. Hang, H. Fischer, B. Frendéus, A.-C. Lundstedt, M. Samuelsson, P. Samuelsson, and C. Svanborg.** 2001. Neutrophil recruitment, chemokine receptors, and resistance to mucosal infection. *J. Leukoc. Biol.* **69**:899–906.
40. **Hacker, J., L. Bender, M. Ott, J. Wingender, B. Lund, R. Marre, and W. Goebel.** 1990. Deletions of chromosomal regions coding for fimbriae and hemolysins occur *in vitro* and *in vivo* in various extraintestinal *Escherichia coli* isolates. *Microb. Pathog.* **8**:213–225.
41. **Hacker, J., G. Blum-Oehler, I. Muhldorfer, and H. Tschape.** 1997. Pathogenicity islands of virulent bacteria: structure, function and impact on microbial evolution. *Mol. Microbiol.* **23**:1089–1097.
42. **Hagberg, L., R. Hull, S. Hull, J. R. McGhee, S. M. Michalek, and C. Svanborg-Edén.** 1984. Differences in the susceptibility to gram-negative urinary tract infection between C3H/HeJ and C3H/HeN mice. *Infect. Immun.* **46**:839–844.
43. **Hagberg, L., U. Jodal, T. K. Korhonen, G. Lidin-Janson, U. Lindberg, and C. Svanborg-Edén.** 1981. Adhesion,

hemagglutination, and virulence of *Escherichia coli* causing urinary tract infections. *Infect. Immun.* **31**:564–570.

44. **Hagberg, L., R. Hull, S. Falkow, R. Freter, and C. Svanborg-Edén.** 1983. Contribution of adhesion to bacterial persistence in the mouse urinary tract. *Infect. Immun.* **40**:265–272.

45. **Hang, L., M. Haraoka, W. W. Agace, H. Leffler, M. Burdick, R. Strieter, and C. Svanborg.** 1999. Macrophage inflammatory protein-2 is required for neutrophil passage across the epithelial barrier of the infected urinary tract. *J. Immunol.* **162**:3037–3044.

46. **Hang, L., B. Frendéus, G. Godaly, and C. Svanborg.** 2000. Interleukin-8 receptor knockout mice have subepithelial entrapment and renal scarring following acute pyelonephritis. *J. Infect. Dis.* **182**:1738–1748.

47. **Hang, L., B. Wullt, Z. Shen, D. Karpman, and C. Svanborg.** 1998. Cytokine repertoire of epithelial cells lining the human urinary tract. *J. Urol.* **159**:2185–2192.

48. **Hannun, Y. A.** 1994. The sphingomyelin cycle and the second messenger function of ceramide. *J. Biol. Chem.* **269**: 3125–3128.

49. **Hanson, M. S., and C. C. Brinton, Jr.** 1988. Identification and characterization of *E. coli* type 1 pilus tip adhesion protein. *Nature* **332**:265–268.

50. **Haraoka, M., L. Hang, W. Agace, M. Burdick, R. Strieter, and C. Svanborg.** 1999. Neutrophil recruitment and resistance to mucosal bacterial infection. *J. Infect. Dis.* **180**:1220–1229.

51. **Hashimoto, C., K. L. Hudson and K. V. Anderson.** 1988. The Toll gene of *Drosophilia*, required for dorsal-ventral embryonic polarity, appears to encode a transmembrane protein. *Cell* **52**:269–279.

52. **Hedges, S., P. Anderson, G. Lidin-Janson, P. de Man, and C. Svanborg.** 1991. Interleukin-6 response to deliberate colonization of the human urinary tract with gram-negative bacteria. *Infect. Immun.* **59**:421–427.

53. **Hedges, S., M. Svensson, and C. Svanborg.** 1992. Interleukin-6 response of epithelial cell lines to bacterial stimulation in vitro. *Infect. Immun.* **60**:1295–1301.

54. **Hedlund, M., B. Frendéus, C. Wachtler, L. Hang, H. Fischer, and C. Svanborg.** 2001. Type 1 fimbriae deliver an LPS- and TLR4-dependent activation signal to CD14-negative cells. *Mol. Microbiol.* **39**:542–552.

55. **Hedlund, M., C. Wachtler, E. Johansson, L. Hang, J. E. Sommerville, R. P. Darveau, and C. Svanborg.** 1999. P fimbriae-dependent, lipopolysaccharide-independent activation of epithelial cytokine responses. *Mol. Microbiol.* **33**:693–703.

56. **Hedlund, M., M. Svensson, Å. Nilsson, R.-D. Duan, and C. Svanborg.** 1996. Role of the ceramide signaling pathway in cells exposed to P fimbriated *Escherichia coli*. *J. Exp. Med.* **183**:1037–1044.

57. **Hedlund, M., Å. Nilsson, R.-D. Duan, and C. Svanborg.** 1998. Sphingomyelin, glycosphingolipids and ceramide signaling in cells exposed to P-fimbriated *Escherichia coli*. *Mol. Microbiol.* **29**:1297–1306.

58. **Hornef, M. W., T. Frisan, A. Vandewalle, S. Normark, and A. Richter-Dahlfors.** 2002. Toll-like receptor 4 resides in the Golgi apparatus and colocalizes with internalized lipopolysaccharide in the intestinal epithelial cells. *J. Exp. Med.* **195**:559–570.

59. **Hornung, V., S. Rothenfusser, S. Britsch, A. Krug, B. Jahrsdorfer, T. Giese, S. Endres, and G. Hartman.** 2002. Quantitive expression of Toll-like receptor 1-10 mRNA in cellular subsets of human peripheral blood mononuclear cells and sensitivity to CpG oligodeoxynucleotides. *J. Immunol.* **168**:4531–4537.

60. **Hull, R. A., R. E. Gill, P. Hsu, B. H. Minshew, and S. Falkow.** 1981. Construction and expression of recombinant plasmids encoding type 1 or D-mannose-resistant pili from a urinary tract infection. *Infect. Immun.* **33**:933–938.

61. **Hultgren S. J., S. Abraham, M. Caparon, P. Falk, J. W. St Geme III, and S. Normark.** 1993. Pilus and nonpilus bacterial adhesins; assembley and function in cell recognition. *Cell* **73**:887–901.

62. **Jacob-Dubuisson, F., J. Heuser, K. Dodson, S. Normark, and S. Hultgren.** 1993. Initiation of assembly and association of the structural elements of a bacterial pilus depend on two specialized tip proteins. *EMBO J.* **12**:837–847.

63. **Johnson, G. B., G. J. Brunn, Y. Kodaira, and J. L Platt.** 2002. Receptor-mediated monitoring of tissue well-being via detection of soluble heparan sulfate by Toll-like receptor 4. *J. Immunol.* **168**:5233–5239.

64. **Johnson, J.** 1991. Virulence factors in *Escherichia coli* urinary tract infection. *Clin. Microbiol Rev.* **4**:80–128.

65. **Jones, C. H., K. Dodson, and S. Hultgren.** 1996. Structure, function, and assembly of adhesive P pili, p. 175–219. *In* L. T. H. Mobley and J. Warren (ed.), *Urinary Tract Infections: Molecular Pathogenesis and Clinical Management.* ASM Press, Washington, D.C.

66. **Kaijser, B., L. A. Hanson, U. Jodal, G. Lidin-Janson, and J. B. Robbins.** 1977. Frequency of *E. coli* K antigens in urinary-tract infections in children. *Lancet* **i**:663–666.

67. **Kass, E. H.** 1956. Asymptomatic infections in the urinary tract. *Trans. Assoc. Am. Physicians* **69**:56–64.

68. **Klemm. P.** 1986. Two regulatory *fim* genes, *fimB* and *fimE*, control the phase variation of type 1 fimbriae of *Escherichia coli*. *EMBO J.* **5**:1389–1393.

69. **Krogfelt, K. A., and P. Klem.** 1988. Investigation of minor components of *Escherichia coli* type 1 fimbria: protein, chemical and immunological aspects. *Microb. Pathog.* **4**:231–238.

70. **Krogfelt, K. A., H. Bergmans, and P. Klemm.** 1990. Direct evidence that the FimH protein is the mannose-specific adhesin of *Escherichia coli* type 1 fimbriae. *Infect. Immun.* **58**:1995–1998.

71. **Kuehn, M. J., J. Heuser, S. Normark, and S. J. Hultgren.** 1992. P pili in *Escherichia coli* are composite fibers with distinct fibrillar adhesive tips. *Nature* **356**:252–255.

72. **Kukkonen, M., T. Raunio, R. Virkola, K. Lahteenmaki, P. H. Makela, P. Klemm, S. Clegg, and T. K. Korhonen.** 1993. Basement membrane carbohydrate as a target for bacterial adhesion: binding of type 1 fimbriae of *Salmonella enterica* and *Escherichia coli* to laminin. *Mol. Microbiol.* **7**:229–237.

73. **Kunin, C.** 1987. *Urinary Tract Infections. Detection, Prevention and Management*, 5th ed. The Williams & Wilkins Co., Baltimore, Md.

74. **Lanne, B., B. M. Olsson, P. A. Jovall, J. Ångström, H. Linder, B. I. Marklund, J. Bergström, and K.-A. Karlsson.** 1995. Glycoconjugate receptors for P-fimbriated *Escherichia coli* in the mouse: an animal model of urinary tract infection. *J. Biol. Chem.* **270**:9017–9025.

75. **Leffler, H., and C. Svanborg-Edén.** 1980. Chemical identification of a glycosphingolipid receptor for *Escherichia coli* attaching to human urinary tract epithelial cells and agglutinating human erythrocytes. *FEMS Microbiol. Lett.* **8**:127–134.

76. **Leffler, H., and C. Svanborg-Edén.** 1981. Glycolipid receptors for *Escherichia coli* on human erythrocytes and uroepithelial cells. *Infect. Immun.* **34**:920–929.

77. **Lemaitre, B., E. Nicolas, L. Michaut, J. M. Reichhart, and J. A. Hoffmann.** 1996. The dorso-ventral regulatory gene cassette spätzle/Toll/cactus controls the potent antifungal response in *Drosophila* adults. *Cell* **86**:973–983.

78. **Levin, B. R., and C. Svanborg.** 1990. Selection and evolution of virulence in bacteria: an ecumenical excursion and modest suggestion. *Parasitology* **100**(Suppl.):S103–S115.

79. **Lindberg, F., B. Lund, L. Johansson, and S. Normark.** 1987. Localization of the receptor binding protein adhesin at the tip of the bacterial pilus. *Nature* **328:**84–87.
80. **Lindberg, U.** 1975. Asymptomatic bacteriuria in school girls. V. The clinical course and response to treatment. *Acta Paediatr. Scand.* **64:**718–724.
81. **Lindstedt, R., G. Larson, P. Falk, U. Jodal, H. Leffler, and C. Svanborg-Edén.** 1991. The receptor repertoire defines the host range for attaching *Escherichia coli* recognizing globo-A. *Infect. Immun.* **59:**1086–1092.
82. **Lomberg, H., U. Jodal, C. Svanborg Eden, H. Leffler, and B. Samuelsson.** 1981. P_1 blood group and urinary tract infection. *Lancet* **i:**551–552.
83. **Mabeck, C. E., F. Orskov, and I. Orskov.** 1971. *Escherichia coli* serotypes and renal involvement in urinary-tract infection. *Lancet* **i:**1312–1314.
84. **Marcus, D. M., M. Naiki, and S. K. Kundu.** 1976. Abnormalities in the glycosphingolipid content of human P^k and p erythrocytes. *Proc. Natl. Acad. Sci. USA* **73:**3263–3267.
85. **Martinez, J. J., M. A. Mulvey, J. D. Schilling, J. S. Pinkner, and J. S. Hultgren.** 2000. Type 1 pilus-mediated bacterial invasion of the bladder epithelial cells. *EMBO J.* **19:**2803–2812.
86. **Mobley, H. L., K. G. Jarvis, J. P. Elwood, D. I. Whittle, C. V. Lockatell, R. G. Russell, D. E. Johson, M. S. Donnenberg, and J. Warren.** 1993. Isogenic P-fimbrial deletion mutation of pyelonephritogenic *Escherichia coli:* the role of alpha Gal(1-4) beta Gal binding in virulence of a wild type strain. *Mol. Microbiol.* **10:**143–155.
87. **Mulvey, M. A., J. D. Schilling, and S. J. Hultgren.** 2001. Establishment of a persistent *Escherichia coli* reservoir during the acute phase of a bladder infection. *Infect. Immun.* **69:** 4572–4579.
88. **Muzio, M., D. Bosisio, N. Polentarutti, G. D'Amico, A. Stoppacciaro, R. Mancinelli, C. van't Veer, G. Penton-Rol, L. P. Ruco, P. Allavena, and A. Mantovani.** 2000. Differential expression and regulation of Toll-like receptors (TLR) in human leukocytes: selective expression of TLR3 in dendritic cells. *J. Immunol.* **164:**5998–6004.
89. **Nagai, Y., S. Akashi, M. Nagafuku, M. Ogata, Y. Iwakura, S. Akira, T. Kitamura, A. Kosugi, M. Kimoto, and K. Miyake.** 2002. Essential role of MD-2 in LPS responsiveness and TLR4 distribution. *Nat. Immunol.* **3:**667–672.
90. **Nordenstam, G. R., C. A. Brandenberg, A. S. Oden, C. M. Svanborg-Edén, and A. Svanborg.** 1986. Bacteriuria and mortality in an elderly population. *N. Engl. J. Med.* **314:** 1152–1156.
91. **Nowicki, B, C. Svanborg-Edén, R. Hull, and S. Hull.** 1989. Molecular analysis and epidemiology of the Dr hemagglutinin of *Escherichia coli. Infect. Immun.* **57:**446–451.
92. **O'Hanley, P., D. Lark, S. Falkow, and G. Schoolnik.** 1985. Molecular basis of *Escherichia coli* colonization of the upper urinary tract in BALB/c Mice. *J. Clin. Investig.* **75:**347–360.
93. **Ohashi, K., V. Burkart, S. Flohe, and H. Kolb.** 2000. Cutting edge. Heat shock protein 60 is a putative endogenous ligand of the Toll-like receptor-4 complex. *J. Immunol.* **164:**558–561.
94. **Ohman, L., K. E. Magnusson, and O. Stendahl.** 1985. Mannose-specific and hydrophobic interaction between *Escherichia coli* and polymorphonuclear leukocytes—influence of bacterial culture period. *Acta Pathol. Microbiol. Immunol. Scand Ser. B* **93:**125–131.
95. **Okamura, Y., M. Watari, E. S. Jerud, D. W. Young, S. T. Ishizaka, J. Rose, J. C. Chow, and J. F. Strauss III.** 2001. The extra domain A of fibronectin activates Toll-like receptor 4. *J. Biol. Chem.* **276:**10229–10233.
96. **Olsen, P. B., and P. Klemm.** 1994. Localisation of promoters in the *fim* gene cluster and the effect of H-NS on the transcription of *fimB* and *fimE. FEMS Microbiol. Lett.* **116:**95–100.
97. **Olsen, P. B., M. A. Schembri, D. L. Gally, and P. Klemm.** 1998. Differential temperature modulation by H-NS of the *fimB* and *fimE* recombinase genes, which control the orientation of the type 1 fimbrial switch. *FEMS Microbiol. Lett.* **162:**17–23.
98. **Orskov, I., C. Svanborg Edén, and F. Orskov.** 1988. Aerobactin production of serotyped *Escherichia coli* from urinary tract infections. *Med. Microbiol. Immunol.* **177:**9–14.
99. **Orskov, I., A. Ferencz, and F. Orskov.** 1980. Tamm-Horsfall protein or uromucoid is the normal urinary slime that traps type 1 fimbriated *Escherichia coli. Lancet* **i:**887.
100. **Otega-Cava, C. F., S. Ishihara, M. A. Rumi, K. Kawashima, N. Ishimura, H. Kazumori, J. Udagawa, Y. Kadowaki, and Y. Kinoshita.** 2003. Strategic compartmentalization of Toll-like receptor 4 in the mouse gut. *J. Immunol.* **170:**3977–3985.
101. **Pak, J., Y. Pu, Z. T. Zhang, D. L. Hasty, and X. R. Wu.** 2001. Tamm-Horsfall protein binds to type 1 fimbriated *Escherichia coli* and prevents *E. coli* from binding to uroplakin 1a and 1b receptors. *J. Biol. Chem.* **276:**9924–9930.
102. **Plos, K, H. Connell, U. Jodal, B. I. Markund, S. Marild, B. Wettergren, and C. Svanborg.** 1995. Intestinal carriage of P fimbriated *Escherichia coli* and the susceptibility to urinary tract infection in young children. *J. Infect. Dis.* **171:**625–631.
103. **Plos, K., T. Carter, S. Hull, and C. Svanborg-Edén.** 1990. Frequency and organization of pap homologous DNA in relation to clinical origin of *Escherichia coli. J. Infect. Dis.* **161:** 518–524.
104. **Poltorak, A., X. He, I. Smirnova, M. Y. Liu, C. V. Huffel, X. Du, D. Birdwell, E. Alejos, M. Silva, C. Galanos, M. Freudenberg, P. Ricciardi-Castagnoli, B. Layton, and B. Beutler.** 1998. Defective LPS signaling in C3H/HeJ and C57BL/10ScCr mice: mutations in the *tlr4* gene. *Science* **282:** 2085–2088.
105. **Pouttu, R., T. Puustinen, R. Virkola, J. Hacker, P. Klemm, and T. K. Korhonen.** 1999. Amino acid residue Ala-62 in the FimH fimbrial adhesin is critical for the adhesiveness of meningitis associated *Escherichia coli* to collagens. *Mol. Microbiol.* **31:**1747–1757.
106. **Qureshi, S.T., L. Lariviere, G. Leveque, K. J. Moore, P. Gros, and D. Malo.** 1999. Endotoxin-tolerant mice have mutations in Toll-like receptor 4 (Tlr4). *J. Exp. Med.* **189:**615–625.
107. **Rock, F. L., G. Hardiman, J. C. Timans, R. A. Kastelein, and J. F. Bazan.** 1998. A family of human receptors structurally related to *Drosophila* Toll. *Proc. Natl. Acad. Sci. USA* **95:**588–593.
108. **Rosetto, M., Y. Engstrom, C. T. Baldari, J. L. Telford, and D. Hultmark.** 1995. Signals from the IL-1 receptor homolog, Toll, can activate an immune response in a *Drosophila* hemocyte cell line. *Biochem. Biophys. Res. Commun.* **209:**111–116.
109. **Samuelsson, P., L. Hang, B. Wullt, H. Irjala, and C. Svanborg.** 2004. Expression and cytokine response in the human urinary tract mucosa. *Infect. Immun.* **72:**3179–3186.
110. **Schembri, M. A., D. W. Ussery, C. Workman, H. Hasman, and P. Klemm.** 2002. DNA microarray analysis of *fim* mutations in *Escherichia coli. Mol. Genet. Genomics* **267:**721–729.
111. **Schembri, M. A., P. B. Olsen, and P. Klemm.** 1998. Orientation-dependent enhancement by H-NS of the activity of the type 1 fimbrial phase switch promoter in *Escherichia coli. Mol. Gen. Genet.* **259:**336–344.

112. **Schilling, J. D., S. M. Martin, D. A. Hunstad, K. P. Patel, M. A. Mulvey, S. S. Justice, R. G. Lorenz, and S. J. Hultgren.** 2003. CD14- and Toll-like receptor-dependent activation of bladder epithelial cells by lipopolysaccharide and type 1 piliated *Escherichia coli*. *Infect. Immun.* **71**:1470–1480.
113. **Shahin, R. D., I. Engberg, L. Hagberg, and C. S. Svanborg.** 1987. Neutrophil recruitment and bacterial clearance correlated with LPS responsiveness in local gram-negative infection. *J. Immunol.* **138**:3475–3480.
114. **Sharon, N., Y. Eshdat, J. Silverblatt, and I. Ofek.** 1981. Bacterial adherence to cell surface sugars. *Ciba Found. Symp.* **80**:119–141.
115. **Sharon, N.** 1987. Bacterial lectins, cell-cell recognition and infectious disease. *FEBS Lett.* **217**:145–157.
116. **Shimazu, R., S. Akashi, H. Ogata, Y. Nagai, K. Fukudome, K. Miyake, and M. Kimoto.** 1999. MD-2, a molecule that confers lipopolysaccharide responsiveness on Toll-like receptor 4. *J. Exp. Med.* **189**:1777–1782.
117. **Smiley, S. T., J. A. King, and W. W. Hancoock.** 2001. Fibrinogen stimulates macrophage chemokine secretion through toll-like receptor 4. *J. Immunol.* **167**:2887–2894.
118. **Sokurenko, E. V., H. S. Courtney, J. Maslow, A. Siitonen, and D. L Hasty.** 1995. Quantitative differences in adhesiveness of type 1 fimbriated *Escherichia coli* due to structural differences in *fimH* genes. *J. Bacteriol.* **177**:3680–3686.
119. **Sokurenko, E. V., H. S. Courtney, D. E. Ohman, P. Klemm, and D. L. Hasty.** 1994. FimH family of type 1 fimbrial adhesins: functional heterogeneity due to minor sequence variations among *fimH* genes. *J. Bacteriol.* **176**:748–755.
120. **Sommerville, J. E. J., L. Cassiano, B. Bainbridge, M. D. Cunningham, and R. P. Darveau.** 1996. A novel *Escherichia coli* lipid A mutant that produces an anti-inflammatory lipopolysaccharide. *J. Clin. Investig.* **97**:359–365.
121. **Spears, P. A., D. Schauer, and P. E. Orndorff.** 1986. Metastable regulation of type 1 piliation in *Escherichia coli* and isolation and characterization of a phenotypically stable mutant. *J. Bacteriol.* **168**:179–185.
122. **Stamm, W. E., M. McKevitt, P. L. Roberts, and N. J. White.** 1991. Natural history of recurrent urinary tract infections in women. *Rev. Infect. Dis.* **13**:77–84.
123. **Stamm, W. E., and S. R. Norrby.** 2001. Urinary tract infections: disease panorama and challenges. *J. Infect. Dis.* **183**(Suppl. 1):S1–S4.
124. **Stenqvist, K., T. Sandberg, G. Lidin-Janson, F. Orskov, I. Orskov, and C. Svanborg-Edén.** 1987. Virulence factors of *Escherichia coli* in urinary isolates from pregnant women. *J. Infect. Dis.* **156**:870–877.
125. **Svanborg, C., G. Bergsten, H. Fischer, B. Frendéus, G. Godaly, E. Gustafsson, L. Hang, M. Hedlund, D. Karpman, A. C. Lundstedt, M. Samuelsson, P. Samuelsson, M. Svensson, and B. Wullt.** 2001. The "innate" host response protects and damages the infected urinary tract. *Ann. Med.* **33**:563–570.
126. **Svanborg, C., G. Bergsten, H. Fischer, B. Frendéus, G. Godaly, E. Gustafsson, L. Hang, M. Hedlund, A. C. Lundstedt, M. Samuelsson, P. Samuelsson, M. Svensson, and B. Wullt.** 2002. Adhesion, signal transduction and mucosal inflammation, p. 223–240. *In* M. Wilson (ed.), *Bacterial Adhesion to Host Tissues*, vol. 1. Cambridge University Press, Cambridge, United Kingdom.
127. **Svanborg-Edén, C., L. A. Hanson, U. Jodal, U. Lindberg, and A. S. Åkerlund.** 1976. Variable adherence to normal human urinary-tract epithelial cells of *Escherichia coli* strains associated with various forms of urinary-tract infection. *Lancet* **i**:490–492.
128. **Svanborg-Edén, C., B. Eriksson, and L. A. Hanson.** 1977. Adhesion of *Escherichia coli* to human uroepithelial cells in vitro. *Infect. Immun.* **18**:767–774.
129. **Svanborg-Edén, C., L.-M. Bjursten, R. Hull, S. Hull, K.-E. Magnusson, Z. Moldovano, and H. Leffler.** 1984. Influence of adhesins on the interaction of *Escherichia coli* with human phagocytes. *Infect. Immun.* **44**:672–680.
130. **Svanborg-Edén, C., L. Hagberg, L. A. Hanson, S. Hull, R. Hull, U. Jodal, H. Leffler, H. Lomberg, and E. Strobe.** 1983. Bacterial adherence—a pathogenic mechanism in urinary tract infections caused by *Escherichia coli*. *Prog. Allergy* **33**:175–188.
131. **Svanborg-Edén, C., R. Freter, L. Hagberg, R. Hull, H. Leffler, and G. Schoolnik.** 1982. Inhibition of experimental ascending urinary tract infection by epithelial cell-surface receptor analogue. *Nature* **298**:560–562.
132. **Svanborg-Edén, C., L. Hagberg, R. Hull, K.-E. Magnusson, and L. Öhman.** 1987. Bacterial virulence verus host resistance in the urinary tracts of mice. *Infect. Immun.* **55**:1224–1232.
133. **Svensson, M., B. Frendéus, T. Butters, F. Platt, R. Dwek, and C. Svanborg.** 2003. Glycolipid depletion in antimicrobial therapy. *Mol. Microbiol.* **47**:453–461.
134. **Takeda, K., T. Kaisho, and S. Akira.** 2003. Toll-like receptors. *Annu. Rev. Immunol.* **21**:335–376.
135. **Termeer, C., F. Benedix, J. Sleeman, C. Fieber, U. Voith, T. Ahrens, K. Miyake, M. Freudenberg, C. Galanos, and J. C. Simon.** 2002. Oligosaccharides of hyaluronan activate dendritic cells via Toll-like receptor 4. *J. Exp. Med.* **195**:99–111.
136. **Tewari, R., T. Ikeda, R. Malaviya, J. I. MacGregor, J. R. Little, S. J. Hultgren, and S. N. Abraham.** 1994. The *papG* tip adhesin of P fimbriae protects *Escherichia coli* from neutrophil bacterial activity. *Infect. Immun.* **62**:5296–5304.
137. **Vabulas, R. M., P. Ahmad-Nejad, C. de Costa, T. Miethke, C. J. Kirschning, H. Hacker, and H. Wagner.** 2001. Endocytosed HSP60s use Toll-like receptor 2 (TLR2) and TLR4 to activate the Toll/interleukin-1 receptor signaling pathway in innate immune cells. *J. Biol. Chem.* **276**:31332–31339.
138. **Vabulas, R.M., P. Ahmad-Nejad, S. Ghose, C. J. Kirschning, R. D. Issels, and H. Wagner.** 2002. HSP70 as endogenous stimulus of the Toll/interleukin-1 receptor signal pathway. *J. Biol. Chem.* **277**:15107–15112.
139. **Väisanen, V., J. Elo, L. G. Tallgren, A. Siitonen, P. H. Mäkelä, C. Svanborg-Edén, G. Källenius, S. B. Svensson, H. Hultberg, and T. Korhonen.** 1981. Mannose-resistant haemagglutination and P antigen recognition are characteristic of *Escherichia coli* causing primary pyelonephritis. *Lancet* **i**:1366–1369.
140. **Vaisanen-Rhen, V., J. Elo, E. Vaisanen, A. Siitonen, I. Orskov, F. Orskov, S. B., Svensson, P. H. Makela, and T. K. Korhonen.** 1984. P-fimbriated clones among *Escherichia coli* strains. *Infect. Immun.* **43**:149–155.
141. **Virkola, R., B. Westerlund, H. Holthofer, J. Parkkinen, M. Kekomaki, and T. K. Korhonen.** 1988. Binding characteristics of *Escherichia coli* adhesins in human urinary bladder. *Infect. Immun.* **56**:2615–2622.
142. **Visintin, A., A. Mazzoni, J. A. Spitzer, and D. M. Segal.** 2001. Secreted MD-2 is a large polymeric protein that efficiently confers lipopolysaccharide sensitivity to Toll-like receptor 4. *Proc. Natl. Acad. Sci. USA* **98**:12156–12161.
143. **Warren, J. W., H. L. Mobley, and A. L. Trifillis.** 1988. Internalization of *Escherichia coli* into human renal tubular epithelial cells. *J. Infect. Dis.* **158**:221–223.
144. **Walz, W., M. A. Schmidt, A. F. Labigne-Roussel, S. Falkow, and G. Schoolnik.** 1985. AFA-1, a cloned afimbrial X-type

adhesin from a human pyelonephritic *Escherichia coli* strains. Purification and chemical, functional and serological characterization. *Eur. J. Biochem.* **152**:315–321.

145. **Welsh, R. A., P. Shahaireen, D. Robbins, W. F. Keene, G. Gekker, and P. R. Peterson.** 1989. Epidiemiologic observations involving the *Escherichia coli* hemolysin p 136–143. *In* E. H. Kass and C. Svanborg-Edén (ed.), *Host-Parasite Interactions in the Urinary Tract.* The University of Chicago Press Ltd., London, United Kingdom.

146. **Wold, A. E., M. Thorssén, S. Hull, and C. Svanborg-Edén.** 1988. Attachment of *Escherichia coli* via mannose or Galα1-4Galβ-containing receptors to human colonic epithelial cells. *Infect. Immun.* **56**:2531–2537.

147. **Wold, A. E., J. Mestecky, M. Tomana, A. Kobata, H. Ohbayashi, T. Endo and C. Svanborg-Edén.** 1990. Secretory immunoglobulin A carries oligosaccharide receptors for *Escherichia coli* type 1 fimbrial lectin. *Infect. Immun.* **58**:3073–3077.

148. **Wu, X., T. T. Sun, and J. J. Medina.** 1996. *In vitro* binding of type 1-fimbriated *Escherichia coli* to uroplakins 1a and 1b: relation to urinary tract infections. *Proc. Natl. Acad. Sci. USA* **93**:9630–9635.

149. **Wullt, B., G. Bergsren, H. Fischer, G. Godaly, D. Karpman, I. Leijonhufvud, A. C. Lundstedt, P. Samuelsson, M. Samuelsson, M. Svensson, and C. Svanborg.** 2003. The host response to urinary tract infection. *Infect. Dis. Clin. North Am.* **17**:279–301.

150. **Wullt, B., G. Bergsten, H. Connell, P. Röllano, N. Gebretsadik, R. Hull, and C. Svanborg.** 2000. P fimbriae enhance the early establishment of *Escherichia coli* in the human urinary tract. *Mol. Microbiol.* **38**:456–464.

151. **Wullt, B., G. Bergsten, H. Connell, P. Röllano, N. Gebratsedik, L. Hang, and C. Svanborg.** 2001. P-fimbriae trigger mucosal responses to *Escherichia coli* in the human urinary tract. *Cell. Microbiol.* **3**:255–264.

152. **Yamamoto, S., T. Tsukamoto, A. Terai, H. Kurazono, Y. Takeda, and O. Yoshida.** 1995. Distribution of virulence factors in *Escherichia coli* isolated from urine of cystitis patients. *Microbiol. Immunol.* **39**:401–404.

153. **Zarember, K. A., and P. J. Godowski.** 2002. Tissue expression of human Toll-like receptors and differential regulation of Toll-like receptor mRNA in leukocytes in response to microbes, their products and cytokines. *J. Immunol.* **168**:554–561.

Colonization of Mucosal Surfaces
Edited by James P. Nataro et al.

Chapter 26

Urease, Urolithiasis, and Colonization of the Urinary Tract

HARRY L. T. MOBLEY

INTRODUCTION

Urinary Tract Infection

The urinary tract is among the most common sites of bacterial infection, and *Escherichia coli* is by far the most common infecting agent at this site (37). Individuals at high risk for symptomatic urinary tract infection (UTI) include neonates, preschool girls, sexually active women, and elderly women and men. In 1997, the last year for which pertinent information is available, UTIs were the cause of 8.318 million physician visits (56) and were noted in 1.658 million hospital discharges (55) in the United States. These frequencies placed UTIs first among kidney and urologic diseases in terms of total cost, above that of chronic renal failure with its attendant dialysis and transplant expenses.

E. coli is the predominant cause of uncomplicated cystitis. Of women going to physicians for a UTI, 95% do so for symptoms of cystitis (16). It is estimated that 40% of adult women will suffer symptoms of cystitis during their lifetime, and that *E. coli* will be identified as the etiologic agent in 75 to 80% of these cases. Although less frequent, the serious clinical syndrome of acute pyelonephritis is also most commonly caused by strains of *E. coli* (87).

Studies suggest that up to 95% of all UTIs develop by an ascending route (1), meaning that infection begins by colonizing the periurethral area, followed by an upward progression to infect the bladder and, in some cases, continued progression of the bacteria through the ureters to infect the kidneys, if conditions of infection allow. The *E. coli* strains that colonize the periurethral area typically arise from the host intestinal flora. Such colonizing bacteria, which are normally confined to the colon of the host, thus gain access to a new environment (1).

Uncomplicated UTIs

Uncomplicated UTIs include cystitis infections in adult women who are not pregnant and do not suffer from structural or neurological dysfunction (1). Cystitis is a clinical diagnosis presumed to represent an infection of the bladder. It is often defined by the presence of $\geq 10^3$ bacteria/ml in a midstream, clean-catch urine sample from a patient with symptoms including dysuria, urinary urgency, and frequency (15, 89).

Complicated UTIs

In urinary tracts with functional or anatomical abnormalities in which the normal flow of urine is disrupted or in which residual urine cannot be expressed from the bladder, a distinct group of bacterial species may cause infection (91). These infections are most frequently polymicrobial (91). The placement of a long-term indwelling catheter ensures that a UTI will develop (91). *Proteus mirabilis* is among the most common organisms causing complicated UTI. Other common etiologic agents of catheter-associated bacteriuria are *Providencia stuartii, Morganella morganii, Klebsiella pneumoniae, Pseudomonas aeruginosa* (91), and enterococci. *E. coli* less commonly causes these infections.

Uropathogenic *P. mirabilis*

P. mirabilis, a dimorphic gram-negative bacterium, commonly causes urinary tract infection in individuals with structural abnormalities or long-term catheterization (91), i.e., complicated UTI. The hallmark of UTI with *P. mirabilis* is the development of bladder and kidney stones due to urease

Harry L. T. Mobley • Department of Microbiology & Immunology, University of Michigan, 5641 Medical Science Bldg II, 1150 West Medical Center Dr., Ann Arbor, MI 48409-0620.

production (52). Flagella and the mannose-resistant, *Proteus*-like (MR/P) fimbriae and *P. mirabilis* fimbriae (PMF) have been identified as virulence factors of *P. mirabilis* that contribute to its colonization of the urinary tract in a murine model (41, 46). Flagellum synthesis in *P. mirabilis* is also an integral part of swarmer cell differentiation, during which, on agar, single vegetative cells with polar flagella develop into multicellular swarm cells expressing thousands of peritrichous flagella, allowing the organism to move rapidly across surfaces in a coordinated manner known as swarming (18, 27). The MR/P fimbria belongs to a family of bacterial fimbriae, modeled after the P fimbria of *E. coli,* which is assembled through the chaperone-usher pathway (79).

A current model of pathogenesis holds *P. mirabilis* as a virulent opportunistic pathogen that is well adapted for life in the urinary tract. *P. mirabilis,* which contaminates the periurethral area, can colonize the bladder epithelium via MR/P and PMF fimbriae that may bind to specific (but unknown) receptors on the bladder epithelium (94). We postulate that, once established, motile flagellated bacteria ascend the ureters to the kidney. MR/P fimbriae are then able to bind specifically to the renal epithelium (3, 75, 82, 83). As a consequence of binding, a biofilm microenvironment can be established (42). During this process, urease hydrolyzes urea in the urine, causing a rise in local pH, which initiates precipitation of supersaturated polyvalent cations and anions in the form of struvite or apatite stones (i.e., urolithiasis); epithelial cell necrosis may follow. Hemolysin, a potent in vitro cytotoxin, appears to play a role once the organism has reached the kidney parenchyma. In addition, bacteria may cross the renal tubule epithelial cell barrier and enter the bloodstream.

Urease and Urolithiasis

Besides swarming motility, *P. mirabilis* is probably best known for production of urease. This multimeric nickel metalloenzyme is localized to the cytosol of the bacterium and catalyzes the hydrolysis of urea, an abundant nitrogenous waste product excreted in the urine of mammals, to ammonia and carbamate. Ammonia can act as a nitrogen source for the bacterium but also alkalinizes the urine, causing normally soluble ions to precipitate in the form of bladder and kidney stones; this process is known as urolithiasis. Bacteria are trapped within the matrix of the stone and can be protected from antibiotic treatment.

EPIDEMIOLOGY

Urease-Positive Organisms and Urinary Tract Infections

Long-term urethral catheterization, a common management technique for intractable urinary incontinence or urinary retention, is universally complicated by polymicrobial bacteriuria (10, 24, 47, 88, 91). The sequelae of this bacteriuria may include fevers, bacteremia, acute pyelonephritis, periurinary tract infections, urinary stones, chronic renal inflammation, renal failure, and death (24, 76, 90). A large proportion of the bacteriuric species include *P. mirabilis, M. morganii,* and *Providencia stuartii* (47, 91). These species commonly produce urease, an enzyme which catalyzes the hydrolysis of urea in the urine, resulting in a rapid rise in the pH (23). The alkaline environment prompts the precipitation of polyvalent cations and anions, including struvite (magnesium ammonium phosphate, $MgNH_4PO_4 \cdot 6H_2O$) and carbonate-apatite [carbonate—calcium phosphate, $Ca_{10}(PO_4CO_3OH)_6(OH)_2$] crystals (2, 8, 17, 23, 76), structural components of renal calculi called infections stones. Catheter encrustations are also composed of struvite and carbonate-apatite (8).

Urease-Positive Organisms and Stone Formation

Bacterial urease is recognized a virulence factor principally because of its role in kidney and bladder stone formation. The earliest reported urinary stone from an individual unearthed at an Egyptian excavation was dated about 4800 B.C. (73). Somewhat later (460 to 370 B.C.), Hippocrates correlated infected ("putrefying") urine with renal calculus formation. However, it was not until the late 1800s and early 1900s that the relationship between urinary tract infection and urinary tract stone formation was realized (21). Alkalination of the urine by urease-catalyzed urea hydrolysis facilitates the precipitation of polyvalent cations and anions (Fig. 1). Struvite and carbonate-apatite are the typical in vivo or in vitro insolubility products in urine with a pH of 8.5 or greater (23). The proportion of stones due to infection has been estimated to be between 20 and 40% (20, 67) and is particularly high in chronically catheterized patients with spinal cord injury (62). Indeed, electron micrographs of stone cross sections have demonstrated bacteria as "seeds" for crystal formation (84).

Urease-Positive Organisms and Catheter Encrustation

In a study of 1,135 weekly urine specimens from 32 long-term-catheterized patients, 86% of the speci-

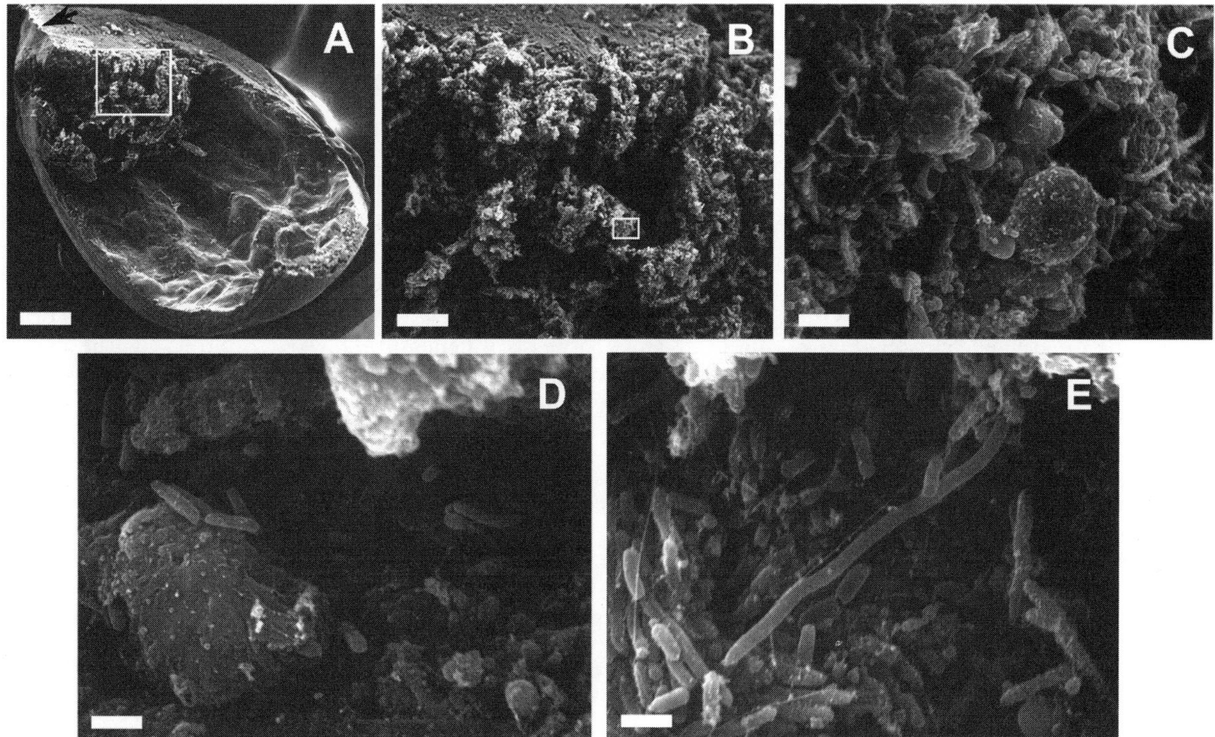

Figure 1. Scanning electron micrographs of a *P. mirabilis* urease-induced bladder stone. (A) One-quarter of the bladder viewed at low magnification (bar, 500 μm). The orientation of the bladder is indicated by an arrowhead pointing to the bottom of the bladder (the end leading to the urethra). (B) Higher magnification (bar, 100 μm) of the boxed area in panel A. (C) Higher magnification (bar, 5 μm) of the boxed area in panel B. (D and E) Representative views of bladder stone (bars, 2 μm). Reprinted from reference 43.

mens contained urease-positive bacterial species at $>10^5$ CFU/ml (51). The most common species were *P. mirabilis* and *M. morganii,* each found in over half the specimens. *P. mirabilis,* but no other urease-positive species, was significantly associated with the 67 obstructions observed in 23 patients. Specimens from patients with obstructions more frequently contained *P. mirabilis* (539 [65%] of 831 specimens) than did those from unobstructed patients (121 [40%] of 304 specimens) ($P < 0.001$). A progressive association of *P. mirabilis* and catheter obstruction was seen: from patients with obstructions, 73% of obstructed specimens and 64% of unobstructed specimens contained *P. mirabilis,* whereas from patients without obstructions, 40% of specimens contained *P. mirabilis. M. morganii* had a more complex association and in some way may protect the catheter from obstruction. The topic of catheter encrustation is covered more thoroughly in chapter 27.

STRUCTURE OF THE UREASE PROTEIN

The urease of *P. mirabilis* is a cytoplasmic nickel metalloenzyme of 281 kDa, as determined by gel filtration (256 kDa is the mass predicted from the nucleotide sequence); it is composed of three distinct subunits of 11.0, 12.2, and 61.0 kDa in a stoichiometry of $(11\text{ kDa-}12\text{ kDa-}61\text{kDa})_3$ (50). Because of the close apparent evolutionary relationship to the urease of *Klebsiella aerogenes,* for which the three-dimensional structure has been elucidated (29), the protein is probably also a trimer of trimers (Fig. 2). That is, each of three catalytic units present in each urease protein is composed of one copy of each subunit, 11, 12, and 61 kDa. In addition, two nickel ions are coordinated into each of three active sites by residues of the UreC subunit via amino acids His-134, His-163, Lys-217, His-246, His-272, and Asp-360 (29).

The catalytically inactive apoenzyme is assembled even in the absence of nickel or the accessory proteins that deliver the ions to the active site. The accessory polypeptides, UreD, UreE, UreF, and UreG, interact with nickel, each other, and the apoenzyme to activate urease by proper insertion of two nickel ions (Ni^{2+}) into each of the three nickel metallocenters of the enzyme (see "Metalloenzyme formation" below).

UREASE GENETICS

Genetic Organization

The *P. mirabilis* urease gene cluster is encoded within a 6.45-kb DNA sequence (GenBank accession numbers M31834, Z18752, and Z21940). This was the first urease gene cluster to be sequenced. There are now several completely sequenced urease gene clusters; many of these sequences were revealed by whole-genome sequencing. Eight contiguous genes comprise the urease gene cluster (Fig. 2) (36, 60, 81). Structural genes, *ureABC*, which encode subunits of the enzyme, are flanked immediately upstream by *ureD* and downstream by *ureEFG* genes. These four accessory genes play a role in the insertion of nickel ions into the apoenzyme, resulting in formation of a catalytically active urease protein.

Metalloenzyme Formation

It is known that in addition to the three genes encoding the structural subunits of the enzyme, there are four accessory genes, critical for incorporation of nickel ions into the apoenzyme, which activates the protein (Fig. 2) (28, 35, 39, 50, 54). Clones that lack accessory genes *ureD, ureF,* or *ureG* synthesize and assemble an inactive apoenzyme that, with the exception of lacking catalytic activity, is identical to the active native enzyme (39). *ureE* mutants are still capable of activating urease, but at a much reduced level.

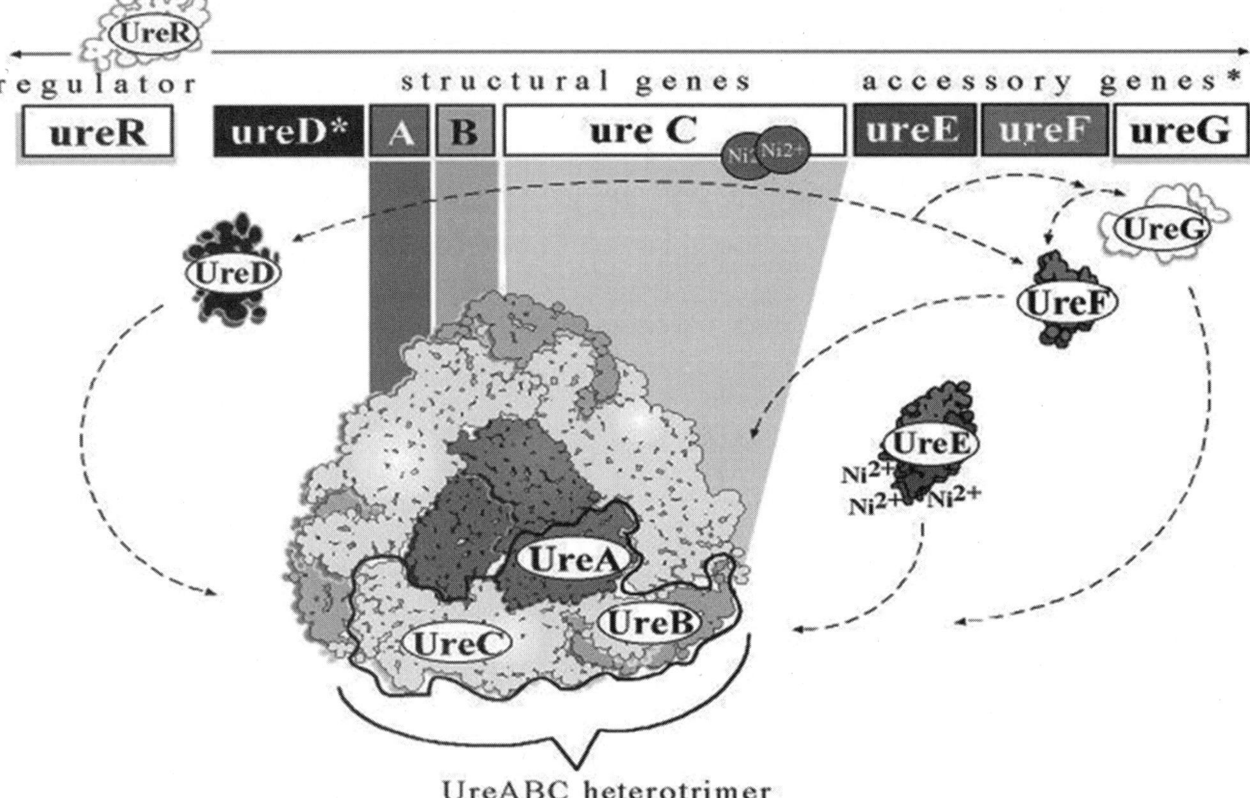

Figure 2. Hypothetical models of *P. mirabilis* urease structural and accessory protein interactions. (Top) The 6,500-bp *P. mirabilis* urease gene cluster encodes eight proteins that comprise, regulate, and assemble the urease holoenzyme. Previously described UreA and UreE homomultimeric interactions were confirmed in vivo (26). Likewise, UreA and UreC structural interactions were also confirmed in vivo (26). UreD associates with UreC in the context of the apo-urease independently of the UreA structural protein; UreD was arbitrarily drawn contacting the apo-enzyme face opposite to UreA; there is no direct evidence for this structure. UreD may interact with coaccessory protein UreF prior to associating with the apo-urease. Although UreD and UreF interact in the absence of structural proteins, UreD is still capable of associating with the apo-urease without coaccessory proteins such as UreF. (Bottom) Data reported in reference 26 suggest that UreD is capable of homomultimeric interactions in vivo. Based on the homotrimeric nature of the apo-urease, one explanation for our observation is that a single molecule of UreD associated with UreABC may interact with additional UreD molecules bound to adjacent UreABC heterotrimers. These interactions could stabilize overall the accessory protein interactions with the apo-urease and hypothetically coordinate nickel uptake among the three active sites of urease. A similar hypothesis could be applied to UreF; homomultimeric UreF interactions in vivo could occur between individual UreF molecules bound through UreD to adjacent UreABC heterotrimers. The three-dimensional structure of urease is inferred from that of the closely related urease of *K. aerogenes* (29). Reprinted from reference 26.

Culture in the presence of high $NiCl_2$ concentrations partially complements this defect (80).

Studies with the highly homologous *K. aerogenes* urease have revealed the formation of complexes including UreABCD (63, 65), UreABCDF (53), and UreABCDFG (64, 78). The formation of such complexes is required for urease activation. Presumably, transient association of the UreE dimer carrying Ni^{2+} ions occurs to complete the activation processes. However, this complex has not been detected. This activation also requires GTP, which is hydrolyzed in a UreG-dependent reaction. UreE, a dimer, possesses a histidine-rich motif at the carboxyl terminus (40, 77, 80) and binds approximately six nickel ions per dimer (40), a fact consistent with a possible nickel donor role. Beyond nucleotide sequence information, not much is known about the UreF and UreG polypeptides (50).

In genetic studies with *P. mirabilis* urease genes, using yeast two-hybrid technology and immunoprecipitation, Heimer and Mobley (26) provided evidence for interactions of UreD and UreF as well as homopolymeric interactions of UreD with itself and UreF with itself. UreE dimer formation was also confirmed.

Regulation of *P. mirabilis* Urease Gene Expression

For all urease gene clusters examined thus far, only one true regulatory gene, *ureR*, has been identified; this gene is present only in gene clusters that are inducible by urea (Fig. 2). *ureR* of *P. mirabilis* and its close homolog (also designated *ureR*) in the plasmid-encoded urease gene clusters of *Providencia stuartii* (47) and the rare urease-positive *E. coli* (12), act as an AraC-like positive activator of gene expression in the presence of urea. The *ureR* regulatory gene of *P. mirabilis* (60), *P. stuartii, E. coli* (13), and *Corynebacterium glutamicum* (70) does not have a homolog in the urease gene clusters of *Bacillus* spp., *Helicobacter pylori, K. aerogenes, Yersinia enterocolitica,* or other bacterial species examined to date (50).

UreR Transcriptional Activator

In contrast to *Klebsiella,* nitrogen control plays no role in regulation of the *P. mirabilis* urease gene cluster (60). Rather, ureases of *Proteus* and *Providencia* spp. are induced by the substrate urea (71, 72) to levels 5- to 25-fold over those in uninduced cultures (34, 48). Proper regulation of cloned urease genes requires the presence of closely linked sequences encoding UreR. The induction is highly specific for urea since no induction was observed by closely related structural analogs of urea (61).

For *P. mirabilis,* a urea-regulated, *ureR*-dependent promoter was found immediately upstream of *ureD* (28). Because *ureA* is regulated in a similar manner and has no independent urea-regulated promoter immediately upstream of *ureA,* this promoter apparently also controls the expression of *ureA*. D'Orazio and Collins (13) have provided evidence that *ureR* is transcribed under the control of UreR itself and is inducible by urea.

Primer extension analysis of the *P. mirabilis* urea-inducible *ureR* and *ureD* promoters was performed with *E. coli* strains harboring cloned *P. mirabilis* urease genes (14). Two transcriptional start sites were determined for *ureD,* and one major transcriptional start site was determined for *ureR*. Interestingly, three transcriptional start sites (P1, P2, and P3) for *ureR* were detected in primer extension analysis from wild-type *P. mirabilis.* Using a plasmid-encoded hybrid UreR-His_6-tagged protein (rUreR) in DNA footprinting analyses, a consensus UreR DNA binding site consisting of nucleotides T(A/G)(T/C)(A/T)(T/G)(C/T)T(A/T)(T/A)ATTG was described (85).

UREASE CATALYTIC ACTIVITY

Enzymatic Reaction

Urease (urea amidohydrolase, EC 3.5.1.5), the most commonly cited virulence factor of *P. mirabilis,* is central to the pathogenesis of *Proteus* UTIs. The enzyme catalyzes the hydrolysis of urea to yield ammonia and carbamate. The latter compound spontaneously decomposes to yield a second molecule of ammonia and carbonic acid. Thus, while urease cleaves only one ammonia group from urea, the net result is liberation of two ammonium ions and a single molecule of carbon dioxide. The precise catalytic mechanism is not finalized but is discussed in detail by Hausinger and Karplus (25).

Kinetics

The *P. mirabilis* urease appears to exhibit simple Michaelis-Menten-type kinetic behavior, with no evidence for substrate inhibition or allosteric behavior. The K_ms for the ureases of different *P. mirabilis* strains range from 13 to 60 mM (7, 34). Despite the relatively low affinity of urease for its substrate, the K_m value is nevertheless appropriate for its niche within the urinary tract, in that urea is present in very high concentrations in urine (0.4 to 0.5 M) (23) and therefore the enzyme is always saturated and working at the V_{max}. Urease has been unambiguously localized to the cytoplasm of the bacterium by comparison to control enzymes that partition with the cytoplasm, periplasmic space, and membrane fractions (35).

Urease Inhibitors

The catalytic activity of urease can be specifically inhibited by various compounds (49). Because of the high substrate specificity of urease for urea, the enzyme binds only a relatively few inhibitors. Substrate analogs such as methylurea and thiourea, while closely related in structure to urea, are actually poor inhibitors, with K_is much greater than 25 mM. Of 16 N-substituted hydroxyureas tested by Uesato et al. (86), *m*-methyl- and *m*-methoxyphenyl-substituted hydroxyurea showed the most potent inhibitory activities against the enzyme. Hydroxamic acids, such as acetohydroxamic acid, are potent inhibitors of the enzyme. They are good metal chelators; thus, the mechanism of inhibition has been generally assumed to involve binding to the active-site nickel ion. Indeed, the crystal structure of the *Bacillus pasteurii* urease complexed with the acetohydroxamate anion was solved to 1.55 Å. The structure clearly shows the binding mode of the inhibitor anion, symmetrically bridging the two Ni^{2+} ions in the active site through the hydroxamate oxygen and chelating one Ni^{2+} ion through the carbonyl oxygen (4). Phosphoramide compounds, also extremely potent inhibitors of the enzyme, possess a tetrahedral geometry that may mimic an intermediate state in enzymatic catalysis, thus acting as transition-state analogs.

Urease inhibitors have been used to treat patients with urolithiasis. Clinical trials have demonstrated efficacy for these inhibitors in therapeutic use for treating urolithiasis in individuals colonized by urease-producing bacteria (38), but the relative toxicity of these compounds has reduced the use of this regimen as a common treatment option (22, 92).

UROLITHIASIS

Impact

The percentage of adults aged 20 to 74 years who reported ever having kidney stones (1988 to 1994) was 5.2% (6.3% of men and 4.1% of women) (Social and Scientific Systems, Inc., 2003, NHANES, unpublished data). In 2001, there were 177,509 hospital dischares among adults aged 20 years or older with "calculus of kidney and ureters" as the primary diagnosis (Social and Scientific Systems, Inc., 2003, NHDS, unpublished data). In addition, urolithiasis was estimated to be responsible for 2.2 million doctor office visits (Social and Scientific Systems, Inc., 2003, NAMCS, unpublished data). The total cost of urolithiasis in the United States in 1993 was $1.83 billion for evaluation and treatment (9). These statistics are for all causes of urolithiasis. It is estimated that 20 to 40% of stones are due to infection (23).

Stone Formation

Stone formation, a hallmark of infection with *P. mirabilis*, is caused by the expression of a highly active urease that hydrolyzes urea to ammonia, causing the local pH to rise, with subsequent precipitation of struvite and apatite crystals (17, 23, 49, 51). Such stones can be seen in the bladder of a mouse infected with *P. mirabilis* (Fig. 1 and Color Plate 9). The stones resulting from aggregation of such crystals complicate infection for three reasons. First, the *P. mirabilis* organisms caught within the interstices of the forming stones are very difficult to clear with antibiotics alone. Second, the stone is a nidus for non-*P. mirabilis* bacteria to establish UTI which also are difficult to eradicate. Third, the stone can obstruct urine flow; pelvic and renal stones are often associated with pyonephrosis and/or chronic pyelonephritis.

Bacteria within the Stone

Bacteria reside within the matrix of bladder and kidney stones in the urinary tract infected with *P. mirabilis*. Using confocal laser-scanning microscopy and scanning electron microscopy, we have identified bacteria residing within the matrix of stones formed as a result of infection (Fig. 3). We have observed both short vegetative cells and long swarmer cells in the stone matrix. However, the significance of the presence of the swarmer cells in the stone is yet to be determined. Is swarmer cell differentiation in a stone merely a surface-induced phenomenon? Do swarmer cells play an active role in the bacterial community living in the stone? Answers to these questions await further investigation. Electron micrographs showed that most bacteria appeared to reside in the peripheral layers of the stone matrix, suggesting that during the development of the stone, bacteria may move out toward the surface of the stone to gain more access to the nutrients. The deep crevices of the stone may shield bacteria from clearance by host immune cells while allowing access to nutrients (Fig. 1). Takeuchi et al. reported the presence of *P. mirabilis* in areas from the nuclei to the peripheral layers in struvite stones (84). Bacteria encased in stones are protected from antibiotic-mediated killing and thus cause persistence and recurrence of the infection. On the other hand, the surface-localized bacteria, while gaining more access to nutrients, seemed also to be exposing themselves to killing by antibiotics. It is possible that bacteria living on the stone surface may exist in a biofilm form and thus be relatively resistant to antibiotic treatment.

Stones appear to form first in association with the surface of the epithelium in both the bladder and the kidneys, as revealed by alizarin red S stain. Ali-

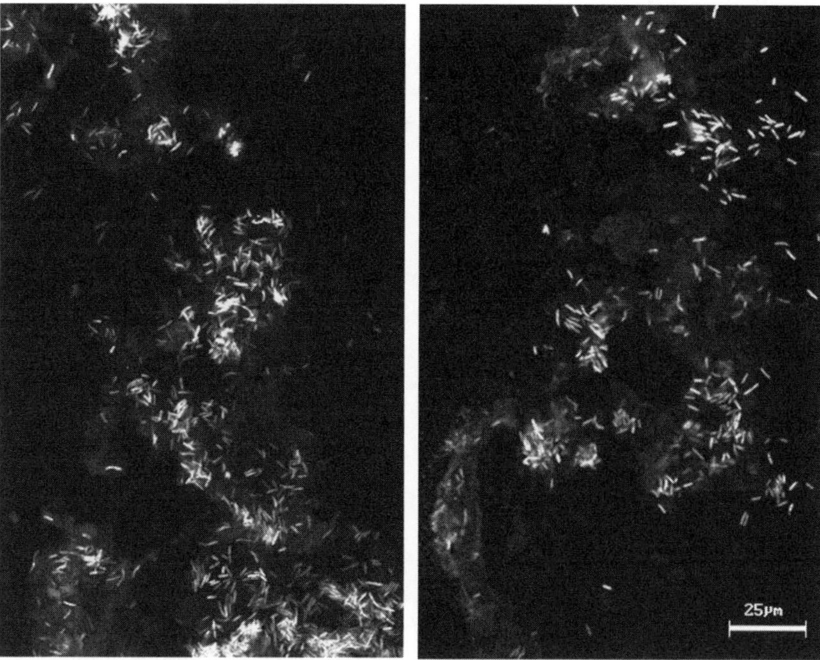

Figure 3. Confocal images of GFP-expressing *P. mirabilis* in a bladder stone. GFP-expressing *P. mirabilis* organisms in a mouse bladder stone were revealed using confocal laser-scanning microscopy. Two representative pictures are shown (scale bar, 25 µm). Reprinted from reference 43.

zarin red S has been used to identify calcium oxalate crystals in the kidneys of patients with familial hyperoxaluria, ethylene glycol poisoning, or chronic renal failure (68, 93). It should be noted that alizarin red S is not strictly specific for calcium and that other cations such as magnesium, manganese, and iron may interfere. Urease stones are composed of magnesium ammonium phosphate (struvite) and calcium phosphate (apatite) (21). Alizarin red S stains bladder stones to a brick-red color and stains the tissue to a very light pink color. Using alizarin red S stain, we were able to observe the early stages of bladder stone development, which indicated that stones appear to form first in close association with the bladder epithelium. Alizarin red S staining also revealed calcification of kidney tubules in *P. mirabilis*-infected mice. To correlate alizarin red S staining with infection, we stained the consecutive section of the sample with hematoxylin and eosin. Alizarin red S-stained calcium deposits in the bladder correlated well with the infection, as indicated by necrosis of the bladder epithelium underneath the stone. However, it became a rather difficult task to pinpoint the exact location of the calcified tubule in the hematoxylin-and-eosin stained section of the kidneys due to weak staining of the tissue in the alizarin red S-stained section. Nonetheless, the calcification of kidney tubules also correlated well with severe infection in the same area. Further investigation is needed to determine whether the observed calcification of kidney tubules is urease induced.

Visualization of GFP-expressing *P. mirabilis* in a bladder stone by confocal laser-scanning microscopy

Urease-induced stone formation in patients infected with *P. mirabilis* is detrimental in that it often causes persistence and recurrence of the infection. It is hypothesized that bacteria within stone matrix are protected from antibiotic treatment and clearance by the host immune system. To identify the presence of bacteria in a bladder stone, we constructed green fluorescent protein (GFP)-expressing *P. mirabilis*. A bladder stone that had formed in a mouse, challenged with 10^9 CFU of the GFP-expressing *P. mirabilis* and sacrificed on day 4, was processed for confocal laser-scanning microscopy. As indicated in two representative micrographs of the bladder stone, abundant GFP-expressing bacteria were present (Fig. 3).

Visualization of *P. mirabilis* and host immune cells in bladder stones by scanning electron microscopy

Two mouse bladders that developed severe urolithiasis were processed for scanning electron microscopy. In most cases of urolithiasis, the bladder stone settles at the bottom of the bladder, blocking the urethral opening; this was also the case for the two bladders analyzed by scanning electron microscopy (Fig. 1A). The infected bladder displayed gross pathological effects, including thickening of the bladder wall and distension of the epithelium. The structure of the bladder stone is compact in the core and relatively loose in the peripheral layers (Fig. 1B). No significant numbers of bacteria were observed on the bladder epithelium or in the core of bladder stone exposed after sectioning. Most bacteria resided in the loose peripheral layers of the stone (Fig. 1C to E). The surface of the stone was covered with fragments of host cells and an enormous number of bacteria. The majority of the bacteria were short rod-shaped vegetative cells; however, some appeared to represent elongated swarmer cells.

Visualization of bladder stones and calcium deposits in kidney tissue with alizarin red S stain

Histochemical staining of bladder stones and calcium deposits in kidney tissue was investigated by using alizarin red S. Alizarin red S stain has been used for diagnosis of calcium deposits in tissue (19, 43, 66, 69). Bladder and kidney tissues of mice, infected with wild-type *P. mirabilis,* were processed for histochemical staining with alizarin red S. Hematoxylin-and-eosin staining of the consecutive sections was performed to assess the histological damage due to infection. As shown in Color Plate 9, alizarin red S stained a macroscopically visible bladder stone to a brick-red color, creating a drastic contrast with respect to the light pink color of the bladder tissue. Stained mineral deposits were also observed on the bladder epithelium (Color Plate 9F). Color Plates 9D and E show a smaller bladder stone in development, which is still in close contact with the bladder epithelium. The bladder epithelium in close contact with the stone was strongly stained with hematoxylin, indicating severe necrosis. Similarly, necrotic areas in the kidneys were often stained red with alizarin red S.

ASSESSMENT OF UREASE AS A VIRULENCE DETERMINANT

Early Virulence Studies Using Undefined Urease Mutants

Studies using animal models of infection have implicated urease as a contributing factor in the severity of disease. Braude et al. (5) used a rat model of pyelonephritis to demonstrate that particular species (*P. mirabilis, M. morganii, P. vulgaris,* and *P. rettgeri,* all of which are urease positive) colonized the kidney epithelium more avidly than *E. coli* and *Pseudomonas aeruginosa* did and produced more severe histological damage than enterococci did. A subsequent report by Braude and Siemienski (6) further implicated urease in nephrotoxicity by creating alkalinity (pH 8.2) in the kidneys, leading to necrosis of the renal tubular epithelium, and by favoring intracellular infection of the tubular epithelium. In addition, it was noted the 22% of the *Proteus*-infected animals had kidney stones. MacLaren (44) used ethylmethane sulfonate (EMS)-generated urease mutants of *P. mirabilis* in a rat model of pyelonephritis and found that the ability to infect the kidneys, as judged by the abscess rate, was equal for the urease-deficient mutant and the parent strain. However, the urease-negative mutant produced much smaller renal abscesses. Renal failure and death were caused by the parent strain but only rarely by the urease-negative mutant. A subsequent report by MacLaren (45) described pyelonephritis including large microabscesses, as well as tissue and the cortical necrosis, caused by the parent strain, which was not generated by the urease-deficient mutant. Although these reports were suggestive that urease contributed to pathogenesis, undetected genetic lesions introduced by EMS in this mutant cannot be ruled out, and complementation was not practised in this era.

Virulence Studies with Defined Urease Mutants

Previously, we constructed an isogenic urease-negative mutant from wild-type *P. mirabilis* strain HI4320 by disrupting *ureC* through homologous recombination (33). Virulence studies with the murine model of ascending UTI showed that this urease-negative mutant is severely attenuated in its ability to colonize and persist in bladders and kidneys (32). More importantly, no bladder stone (urolithiasis) or death occurred in any mouse challenged with the urease-negative mutant. In a later study, the urease-negative mutant was further characterized and examined by using a CBA mouse model of ascending UTI involving a long-term indwelling catheter.

Catheter material (4-mm-long catheter segments) was surgically implanted in the bladders of CBA mice (30, 31). Two weeks later, the bladders were histologically normal and the mice were transurethrally challenged with either the urease-negative mutant or the wild-type strain. Compared to uncatheterized mice, the catheterized mice were more susceptible to ascending UTI by the wild-type *P. mirabilis* strain. At doses of 10^4 and 10^6 CFU/mouse, *P. mirabilis* wild-type strain HI4320 colonized the catheterized mice at significantly higher levels than it colonized the uncatheterized mice: the levels in the bladder were >5,000-fold higher, and the levels in the kidney were >100-fold higher. Increased colonization in the bladder also led to a higher incidence of urolithiasis.

When challenged with the urease-negative mutants at a dose of 10^9 CFU/mouse, the catheterized mice were more susceptible than the uncatheterized mice to bladder colonization but not kidney colonization. The level of bladder colonization by the urease-negative mutant was >200-fold higher in the catheterized mice than in the uncatheterized mice. Despite the successful colonization of the bladders of the catheterized mice by the urease-negative mutant, urolithiasis or death was never observed.

ureR Mutants

P. mirabilis, a cause of complicated UTI, produces urease, an essential virulence factor for this species. UreR, a member of the AraC/XylS family of transcriptional regulators, positively activates expression of the *ure* gene cluster in the presence of urea. To specifically evaluate the contribution of UreR to urease activity and virulence in the urinary tract, a *ureR* mutation was introduced into *P. mirabilis* HI4320 by homologous recombination (11). The isogenic *ureR::aphA* mutant, deficient in UreR production, lacked measurable urease activity. By Western blotting with monoclonal antibodies raised against UreD, expression was not detected in the UreR-deficient strain. Urease activity and UreD expression were restored by complementation of the mutant strain with *ureR* expressed from a low-copy-number plasmid. Virulence was assessed by transurethral cochallenge of CBA mice with wild-type and mutant strains. The isogenic *ureR::aphA* mutant of HI4320 was outcompeted in the urine ($P = 0.004$), bladder ($P = 0.016$), and kidneys ($P \leq 0.001$) 7 days after inoculation. Thus, UreR is required for basal urease activity in the absence of urea, induction of urease by urea, and virulence of *P. mirabilis* in the urinary tract.

Urease Mutants in the Catheterized Urinary Tract

In the catheterized urinary tract, the urease-negative defect in the bladder can be overcome by using a high inoculum. Unlike *E. coli, P. mirabilis* is not a common cause of UTI in normal host but frequently infects individuals with long-term urinary catheters in place or those with complicated urinary tracts. In this study, we tested the urease-negative (*ureC*) mutant in a clinically more relevant model, the CBA mouse model of ascending UTI modified to include catheter material in the bladder (33). Johnson et al. (unpublished data) have shown that in the long-term-catheterized mice, *P. mirabilis* colonizes better and persists longer at low infection doses (10^4 to 10^6) and causes more deaths at high infectious doses (10^8 to 10^9). Compared to its parent, the urease-negative (*ureC*) mutant was significantly attenuated in the catheterized mice, more so in the kidneys than in the bladder. However, with a 1,000-fold increase in infection dose, the mutant was able to colonize the bladder of catheterized mice at similar levels to the wild type. This raises a couple of interesting points.

First, urease, rather than the level of colonization, is required for stone formation in bladders of mice infected by *P. mirabilis.* We have shown previously that the urease-negative mutant is severely attenuated in its ability to colonize the urinary tract or cause urolithiasis in CBA mice (43). However, we were unable to attribute the loss of urolithiasis directly to the lack of urease because it could be due simply to the absence of infection. In the catheterized mice, at an inoculum size 1,000-fold higher than that of the wild type, the mutant colonized the bladder to a similar level. Even so, no bladder stones were ever detected.

Second, it raised some concern about the effectiveness of urease-inhibitory drugs in preventing bladder colonization by *P. mirabilis.* As mentioned above, *P. mirabilis* frequently causes UTI in long-term-catheterized patients. Urease-inhibitory drugs may be very effective in preventing the infection from ascending to the kidneys and eliminating urolithiasis, as indicated by our data. However, treatment with such drugs will not completely eradicate bladder colonization by *P. mirabilis* in catheterized patients.

The *ureC* mutant is urease negative due to the loss of UreC, the largest of the three structural subunits of the urease apoenzyme. Although the polar effect of this mutation on the expression of the downstream genes (*ureEFG*) was not determined in this study, it was indicated by our data that this mutation did not alter the urea-inducible expression of an upstream gene, *ureD.* The low-level urease activity detected in the uninduced culture of the wild type is consistent with a previous finding that the expression of UreD is not as tightly regulated in *P. mirabilis* as it is in *E. coli* expressing cloned urease genes (35). Zhao et al. showed low-level expression of the UreD-GFP fusion protein in *P. mirabilis,* but not in *E. coli* expressing UreR, in the absence of urea (95). Western blot analysis also indicated low-level expression of UreD in the uninduced cultures. Since urease activity absolutely requires UreC, it appears that low-level expression of UreC was also present in the uninduced cultures but was not detectable by the rabbit polyclonal antiserum against UreC. The molecular mechanism and the significance of this low-level expression of urease activity in the absence of urea remain unclear.

UREA AND THE URINARY TRACT

Kidney Disease

Individuals in the United States with kidney conditions, including infection, kidney stones, cancer, missing kidney, or other related disorders, numbered 2.55 million in 1996 (57), the last year for which a comprehensive database is available. In 1999, 424,179 people had end-stage renal disease (58). Of the individuals who were treated for this condition, 66,964 (15.8%) died, despite an expenditure of $17.87 billion (58). There were 13,483 kidney transplants in 1999, similar to the previous 2 years (58). In the same year, 243,320 people underwent dialysis, with a 1-year survival rate of 78.4%; however, only 31.3% of these patients survived for 5 years.

Urea Synthesis

Excretion of nitrogenous wastes is central to kidney function, and, as such, urea metabolism is the best indicator of kidney function. Urea $[(NH_2)_2CO]$ is synthesized in the mitochondrial matrix and cytosol of epithelial cells in the liver as a product of the urea cycle and as a mechanism to eliminate excess amino acid nitrogen (74). Arginine is cleaved by arginase to form ornithine and urea; the latter is the only product of the urea cycle, since the other components are recycled. Urea enters the bloodstream and diffuses via the glomerulus into the proximal and distal tubules of the kidneys. Although about half of the urea is reabsorbed, the other half is eliminated in the urine, thus maintaining nitrogen homeostasis. The urea concentration in urine is typically 0.4 to 0.5 M (23).

Blood Urea Nitrogen Measurements

The blood urea nitrogen test is a routine but critical clinical laboratory test used to evaluate renal function. The test quantifies serum urea that is produced in the liver as an end point of protein degradation. The majority of renal diseases affect the serum urea concentration, and so this test is a sensitive indicator of such disease. Normal levels of urea in serum range from 7 to 20 mg/dl (1.2 to 3.3 mM). Higher than normal levels (azotemia) may indicate congestive heart failure, excessive protein catabolism or ingestion, gastrointestinal bleeding, hypovolemia (as in burns or dehydration), myocardial infarction, renal diseases (including glomerulonephritis, acute pyelonephritis, and acute tubular necrosis), renal failure, shock, and urinary tract obstruction (such as from stones, tumor, or prostatic hypertrophy) (58, 59, 74). In contrast, lower than normal urea concentrations may signal liver failure, low protein diet, malnutrition, or overhydration. Numerous pharmaceutical agents and antibiotics can also affect blood urea levels.

SUMMARY

Urea, a nitrogenous waste product of mammals, provides an opportunity for urease-producing bacteria to colonize the urinary tract. The bacterial species, prominently represented by *P. mirabilis,* can hydrolyze urea by using urease. The resulting elevation in pH, coupled with the release of ammonia, initiates urolithiasis by precipitating normally soluble ions in the form of struvite, carbonate-apatite, or other minerals. Thus, these organisms, by forming stones, complicate the urinary tract, transforming this niche into one that is more hospitable to these opportunistic bacteria. Urolithiasis and catheter encrustation are the result of these bacteria using urea, an abundant nutrient source in urine, as their preferred nitrogen source, ammonia.

REFERENCES

1. **Bacheller, C. D., and J. M. Bernstein.** 1997. Urinary tract infections. *Med. Clin. North Am.* **81:**719–730.
2. **Barnhouse, D. H.** 1968. In vitro formation of precipitates in sterile and infected urines. *Investig. Urol.* **5:**342–347.
3. **Belas, R., and D. Flaherty.** 1994. Sequence and genetic analysis of multiple flagellin-encoding genes from *Proteus mirabilis. Gene* **148:**33–41.
4. **Benini, S., W. R. Rypniewski, K. S. Wilson, S. Miletti, S. Ciurli, and S. Mangani.** 2000. The complex of *Bacillus pasteurii* urease with acetohydroxamate anion from X-ray data at 1.55 A resolution. *J. Biol. Inorg. Chem.* **5:**110–118.
5. **Braude, A. I., A. P. Shapiro, and J. Siemienski.** 1959. Hematogenous pyelonephritis in rats. III. Relationship of bacterial species to the pathogenesis of acute pyelonephritis. *J. Bacteriol.* 77:270–280.
6. **Braude, A. I., and J. Siemienski.** 1960. Role of Bacterial urease in experimental pyelonephritis. *J. Bacteriol.* **80:**171–179.
7. **Breitenbach, J. M., and R. P. Hausinger.** 1988. *Proteus mirabilis* urease. Partial purification and inhibition by boric acid and boronic acids. *Biochem. J.* **250:**917–920.
8. **Bruce, A. W., S. S. Sira, A. F. Clark, and S. A. Awad.** 1974. The problem of catheter encrustation. *Can. Med. Assoc. J.* **111:**238–239.
9. **Clark, J. Y., I. M. Thompson, and S. A. Optenberg.** 1995. Economic impact of urolithiasis in the United States. *J. Urol.* **154:**2020–2024.
10. **Damron, D. J., J. W. Warren, G. R. Chippendale, and J. H. Tenney.** 1986. Do clinical microbiology laboratories report complete bacteriology in urine from patients with long-term urinary catheters? *J. Clin. Microbiol.* **24:**400–404.
11. **Dattelbaum, J. D., C. V. Lockatell, D. E. Johnson, and H. L. Mobley.** 2003. UreR, the transcriptional activator of the *Proteus mirabilis* urease gene cluster, is required for urease activity and virulence in experimental urinary tract infections. *Infect. Immun.* **71:**1026–1030.

12. **D'Orazio, S. E., and C. M. Collins.** 1993. Characterization of a plasmid-encoded urease gene cluster found in members of the family *Enterobacteriaceae. J. Bacteriol.* **175:**1860–1864.
13. **D'Orazio, S. E., and C. M. Collins.** 1993. The plasmid-encoded urease gene cluster of the family *Enterobacteriaceae* is positively regulated by UreR, a member of the AraC family of transcriptional activators. *J. Bacteriol.* **175:**3459–3467.
14. **D'Orazio, S. E., V. Thomas, and C. M. Collins.** 1996. Activation of transcription at divergent urea-dependent promoters by the urease gene regulator UreR. *Mol. Microbiol.* **21:**643–655.
15. **Faro, S., and D. E. Fenner.** 1998. Urinary tract infections. *Clin. Obstet. Gynecol.* **41:**744–754.
16. **Ferry, S., L. G. Burman, and S. E. Holm.** 1988. Clinical and bacteriological effects of therapy of urinary tract infection in primary health care: relation to in vitro sensitivity testing. *Scand. J. Infect. Dis.* **20:**535–544.
17. **Fowler, J. E., Jr.** 1984. Bacteriology of branched renal calculi and accompanying urinary tract infection. *J. Urol.* **131:**213–215.
18. **Fraser, G. M., and C. Hughes.** 1999. Swarming motility. *Curr. Opin. Microbiol.* **2:**630–635.
19. **Gilmore, S. K., S. W. Whitson, and D. E. Bowers, Jr.** 1986. A simple method using alizarin red S for the detection of calcium in epoxy resin embedded tissue. *Stain Technol.* **61:**89–92.
20. **Griffith, D. P.** 1978. Struvite stones. *Kidney Int.* **13:**372–382.
21. **Griffith, D. P.** 1979. Urease stones. *Urol. Res.* **7:**215–221.
22. **Griffith, D. P., M. J. Gleeson, H. Lee, R. Longuet, E. Deman, and N. Earle.** 1991. Randomized, double-blind trial of Lithostat (acetohydroxamic acid) in the palliative treatment of infection-induced urinary calculi. *Eur. Urol.* **20:**243–247.
23. **Griffith, D. P., D. M. Musher, and C. Itin.** 1976. Urease. The primary cause of infection-induced urinary stones. *Investig. Urol.* **13:**346–350.
24. **Gross, P. A., M. Flower, and G. Barden.** 1976. Polymicrobic bacteriuria: significant association with bacteremia. *J. Clin. Microbiol.* **3:**246–250.
25. **Hausinger, R. P., and P. A. Karplus.** 2001. Urease, p. 867–879. *In* A. Messerschmidt, R. Huber, T. Poulos, and K. Wieghardt (ed.), *Handbook of Metalloproteins.* John Wiley & Sons, Inc., Chichester, United Kingdom.
26. **Heimer, S. R., and H. L. Mobley.** 2001. Interaction of *Proteus mirabilis* urease apoenzyme and accessory proteins identified with yeast two-hybrid technology. *J. Bacteriol.* **183:**1423–1433.
27. **Henrichsen, J.** 1972. Bacterial surface translocation: a survey and a classification. *Bacteriol. Rev.* **36:**478–503.
28. **Island, M. D., and H. L. Mobley.** 1995. *Proteus mirabilis* urease: operon fusion and linker insertion analysis of *ure* gene organization, regulation, and function. *J. Bacteriol.* **177:**5653–5660.
29. **Jabri, E., M. B. Carr, R. P. Hausinger, and P. A. Karplus.** 1995. The crystal structure of urease from *Klebsiella aerogenes. Science* **268:**998–1004.
30. **Johnson, D. E., and C. V. Lockatell.** 1999. Mouse model of ascending UTI involving short- and long-term catheters, p. 441–445. *In* M. Sande (ed.), *Handbook of Animal Models of Infection.* Academic Press, Ltd., London, United Kingdom.
31. **Johnson, D. E., C. V. Lockatell, M. Hall-Craggs, and J. W. Warren.** 1991. Mouse models of short- and long-term foreign body in the urinary bladder: analogies to the bladder segment of urinary catheters. *Lab. Anim. Sci.* **41:**451–455.
32. **Johnson, D. E., R. G. Russell, C. V. Lockatell, J. C. Zulty, J. W. Warren, and H. L. Mobley.** 1993. Contribution of *Proteus mirabilis* urease to persistence, urolithiasis, and acute pyelonephritis in a mouse model of ascending urinary tract infection. *Infect. Immun.* **61:**2748–2754.
33. **Jones, B. D., C. V. Lockatell, D. E. Johnson, J. W. Warren, and H. L. Mobley.** 1990. Construction of a urease-negative mutant of *Proteus mirabilis:* analysis of virulence in a mouse model of ascending urinary tract infection. *Infect. Immun.* **58:**1120–1123.
34. **Jones, B. D., and H. L. Mobley.** 1987. Genetic and biochemical diversity of ureases of *Proteus, Providencia,* and *Morganella* species isolated from urinary tract infection. *Infect. Immun.* **55:**2198–2203.
35. **Jones, B. D., and H. L. Mobley.** 1988. *Proteus mirabilis* urease: genetic organization, regulation, and expression of structural genes. *J. Bacteriol.* **170:**3342–3349.
36. **Jones, B. D., and H. L. Mobley.** 1989. *Proteus mirabilis* urease: nucleotide sequence determination and comparison with jack bean urease. *J. Bacteriol.* **171:**6414–6422.
37. **Kunin, C. M.** 1987. *Detection, Prevention and Management of Urinary Tract Infections,* 4th ed., p. 195–201. Lea & Febiger, Philadelphia, Pa.
38. **Lake, K. D., and D. C. Brown.** 1985. New drug therapy for kidney stones: a review of cellulose sodium phosphate, acetohydroxamic acid, and potassium citrate. *Drug. Intell. Clin. Pharm.* **19:**530–539.
39. **Lee, M. H., S. B. Mulrooney, M. J. Renner, Y. Markowicz, and R. P. Hausinger.** 1992. *Klebsiella aerogenes* urease gene cluster: sequence of *ureD* and demonstration that four accessory genes (*ureD, ureE, ureF,* and *ureG*) are involved in nickel metallocenter biosynthesis. *J. Bacteriol.* **174:**4324–4330.
40. **Lee, M. H., H. S. Pankratz, S. Wang, R. A. Scott, M. G. Finnegan, M. K. Johnson, J. A. Ippolito, D. W. Christianson, and R. P. Hausinger.** 1993. Purification and characterization of *Klebsiella aerogenes* UreE protein: a nickel-binding protein that functions in urease metallocenter assembly. *Protein Sci.* **2:**1042–1052.
41. **Li, X., D. E. Johnson, and H. L. Mobley.** 1999. Requirement of MrpH for mannose-resistant *Proteus*-like fimbria-mediated hemagglutination by *Proteus mirabilis. Infect. Immun.* **67:**2822–2833.
42. **Li, X., H. Zhao, L. Geymonat, F. Bahrani, D. E. Johnson, and H. L. Mobley.** 1997. Proteus mirabilis mannose-resistant, *Proteus*-like fimbriae: MrpG is located at the fimbrial tip and is required for fimbrial assembly. *Infect. Immun.* **65:**1327–1334.
43. **Li, X., H. Zhao, C. V. Lockatell, C. B. Drachenberg, D. E. Johnson, and H. L. Mobley.** 2002. Visualization of *Proteus mirabilis* within the matrix of urease-induced bladder stones during experimental urinary tract infection. *Infect. Immun.* **70:**389–394.
44. **MacLaren, D. M.** 1968. The significance of urease in proteus pyelonephritis: a bacteriological study. *J. Pathol. Bacteriol.* **96:**45–56.
45. **MacLaren, D. M.** 1969. The significance of urease in proteus pyelonephritis: a histological and biochemical study. *J. Pathol.* **97:**43–49.
46. **Mobley, H. L., R. Belas, V. Lockatell, G. Chippendale, A. L. Trifillis, D. E. Johnson, and J. W. Warren.** 1996. Construction of a flagellum-negative mutant of *Proteus mirabilis:* effect on internalization by human renal epithelial cells and virulence in a mouse model of ascending urinary tract infection. *Infect. Immun.* **64:**5332–5340.
47. **Mobley, H. L., G. R. Chippendale, M. H. Fraiman, J. H. Tenney, and J. W. Warren.** 1985. Variable phenotypes of *Providencia stuartii* due to plasmid-encoded traits. *J. Clin. Microbiol.* **22:**851–853.
48. **Mobley, H. L., G. R. Chippendale, K. G. Swihart, and R. A. Welch.** 1991. Cytotoxicity of the HpmA hemolysin and urease of *Proteus mirabilis* and *Proteus vulgaris* against cultured

human renal proximal tubular epithelial cells. *Infect. Immun.* **59**:2036–2042.

49. **Mobley, H. L., and R. P. Hausinger.** 1989. Microbial ureases: significance, regulation, and molecular characterization. *Microbiol. Rev.* **53**:85–108.
50. **Mobley, H. L., M. D. Island, and R. P. Hausinger.** 1995. Molecular biology of microbial ureases. *Microbiol. Rev.* **59**:451–480.
51. **Mobley, H. L., and J. W. Warren.** 1987. Urease-positive bacteriuria and obstruction of long-term urinary catheters. *J. Clin. Microbiol.* **25**:2216–2217.
52. **Mobley, H. L. T.** 2000. Virulence of the two primary uropathogens. *ASM News* **66**:403–410.
53. **Moncrief, M. B., and R. P. Hausinger.** 1996. Purification and activation properties of UreD-UreF-urease apoprotein complexes. *J. Bacteriol.* **178**:5417–5421.
54. **Mulrooney, S. B., and R. P. Hausinger.** 1990. Sequence of the *Klebsiella aerogenes* urease genes and evidence for accessory proteins facilitating nickel incorporation. *J. Bacteriol.* **172**: 5837–5843.
55. **National Center for Health Statistics.** 2001. 1999 *National Hospital Discharge Survey: Annual Summary with Detailed Diagnoses and Procedures Data. Vital and Health Statistics,* series 13, no. 151. National Center for Health Statistics, Centers for Disease Control and Prevention, Atlanta, Ga.
56. **National Center for Health Statistics.** 1999. *Ambulatory Care Visits to Physician Offices, Hospital Outpatient Departments, and Emergency Departments: United States, 1997. Vital and Health Statistics,* series 13, no. 143. National Center for Health Statistics, Centers for Disease Control and Prevention, Atlanta, Ga.
57. **National Center for Health Statistics.** 1999. *Current Estimates from National Health Interview Survey. Vital and Health Statistics,* series 10, no. 200. National Center for Health Statistics, Centers for Disease Control and Prevention, Atlanta, Ga.
58. **National Institutes of Health.** 2001. *U.S. Renal Data System. Annual Data Report.* National Institute of Diabetes and Digestive and Kidney Diseases, National Institutes of Health, Bethesda, Md. [Online.] http://www.usrds.org.
59. **National Institutes of Health.** 1990. *The National Kidney and Urologic Diseases Advisory Board 1990 Long-Range Plan—Window on the 21st Century.* NIH publication 90-583. National Institutes of Health, Bethesda, Md.
60. **Nicholson, E. B., E. A. Concaugh, P. A. Foxall, M. D. Island, and H. L. Mobley.** 1993. *Proteus mirabilis* urease: transcriptional regulation by UreR. *J. Bacteriol.* **175**:465–473.
61. **Nicholson, E. B., E. A. Concaugh, and H. L. Mobley.** 1991. *Proteus mirabilis* urease: use of a *ureA-lacZ* fusion demonstrates that induction is highly specific for urea. *Infect. Immun.* **59**:3360–3365.
62. **Nikakhtar, B., N. D. Vaziri, F. Khonsari, S. Gordon, and M. D. Mirahmadi.** 1981. Urolithiasis in patients with spinal cord injury. *Paraplegia* **19**:363–366.
63. **Park, I. S., M. B. Carr, and R. P. Hausinger.** 1994. In vitro activation of urease apoprotein and role of UreD as a chaperone required for nickel metallocenter assembly. *Proc. Natl. Acad. Sci. USA* **91**:3233–3237.
64. **Park, I. S., and R. P. Hausinger.** 1995. Evidence for the presence of urease apoprotein complexes containing UreD, UreF, and UreG in cells that are competent for in vivo enzyme activation. *J. Bacteriol.* **177**:1947–1951.
65. **Park, I. S., and R. P. Hausinger.** 1996. Metal ion interaction with urease and UreD-urease apoproteins. *Biochemistry* **35**: 5345–5352.
66. **Paul, H., A. J. Reginato, and H. R. Schumacher.** 1983. Alizarin red S staining as a screening test to detect calcium compounds in synovial fluid. *Arthritis Rheum.* **26**:191–200.
67. **Peacock, M., and W. G. Robertson.** 1980. Treatment of urinary tract stone disease. *Drugs* **20**:225–232.
68. **Proia, A. D., and N. T. Brinn.** 1985. Identification of calcium oxalate crystals using alizarin red S stain. *Arch. Pathol. Lab. Med.* **109**:186–189.
69. **Puchtler, H., S. N. Meloan, and M. S. Terry.** 1969. On the history and mechanism of alizarin and alizarin red S stains for calcium. *J. Histochem. Cytochem.* **17**:110–124.
70. **Puskas, L. G., M. Inui, and H. Yukawa.** 2000. Structure of the urease operon of *Corynebacterium glutamicum. DNA Seq.* **11**: 383–394, 467.
71. **Rosenstein, I., J. M. Hamilton-Miller, and W. Brumfitt.** 1980. The effect of acetohydroxamic acid on the induction of bacterial ureases. *Investig. Urol.* **18**:112–114.
72. **Rosenstein, I. J., J. M. Hamilton-Miller, and W. Brumfitt.** 1981. Role of urease in the formation of infection stones: comparison of ureases from different sources. *Infect. Immun.* **32**:32–37.
73. **Rosenstein, I. J. M., and J. M. T. Hamilton-Miller.** 1985. Inhibitors of urease as chemotherapeutic agents. *Crit. Rev. Microbiol.* **11**:1–12.
74. **Rubin, R. H., N. E. Tolkoff-Rubin, and R. S. Cotran.** 1986. Urinary tract infection, pyelonephritis, and reflux nephropathy, p. 1085–1141. *In* B. M. Brenner and F. C. Rector (ed.), *The Kidney.* The W. B. Saunders Co., Philadelphia, Pa.
75. **Sareneva, T., H. Holthofer, and T. K. Korhonen.** 1990. Tissue-binding affinity of *Proteus mirabilis* fimbriae in the human urinary tract. *Infect. Immun.* **58**:3330–3336.
76. **Smith, P. H., J. B. Cook, and W. G. Robertson.** 1969. Stone formation in paraplegia. *Paraplegia* **7**:77–85.
77. **Song, H. K., S. B. Mulrooney, R. Huber, and R. P. Hausinger.** 2001. Crystal structure of *Klebsiella aerogenes* UreE, a nickel-binding metallochaperone for urease activation. *J. Biol. Chem.* **276**:49359–49364.
78. **Soriano, A., and R. P. Hausinger.** 1999. GTP-dependent activation of urease apoprotein in complex with the UreD, UreF, and UreG accessory proteins. *Proc. Natl. Acad. Sci. USA* **96**:11140–11144.
79. **Soto, G. E., and S. J. Hultgren.** 1999. Bacterial adhesins: common themes and variations in architecture and assembly. *J. Bacteriol.* **181**:1059–1071.
80. **Sriwanthana, B., M. D. Island, D. Maneval, and H. L. Mobley.** 1994. Single-step purification of *Proteus mirabilis* urease accessory protein UreE, a protein with a naturally occurring histidine tail, by nickel chelate affinity chromatography. *J. Bacteriol.* **176**:6836–6841.
81. **Sriwanthana, B., M. D. Island, and H. L. Mobley.** 1993. Sequence of the Proteus mirabilis urease accessory gene *ureG. Gene* **129**:103–106.
82. **Stamm, W. E., M. McKevitt, P. L. Roberts, and N. J. White.** 1991. Natural history of recurrent urinary tract infections in women. *Rev. Infect. Dis.* **13**:77–84.
83. **Stapleton, A. E., M. R. Stroud, S. I. Hakomori, and W. E. Stamm.** 1998. The globoseries glycosphingolipid sialosyl galactosyl globoside is found in urinary tract tissues and is a preferred binding receptor In vitro for uropathogenic *Escherichia coli* expressing *pap*-encoded adhesins. *Infect. Immun.* **66**:3856–3861.
84. **Takeuchi, H., H. Takayama, T. Konishi, and T. Tomoyoshi.** 1984. Scanning electron microscopy detects bacteria within infection stones. *J. Urol.* **132**:67–69.

85. **Thomas, V. J., and C. M. Collins.** 1999. Identification of UreR binding sites in the *Enterobacteriaceae* plasmid-encoded and *Proteus mirabilis* urease gene operons. *Mol. Microbiol.* **31:**1417–1428.
86. **Uesato, S., Y. Hashimoto, M. Nishino, Y. Nagaoka, and H. Kuwajima.** 2002. N-substituted hydroxyureas as urease inhibitors. *Chem. Pharm. Bull.* (Tokyo) **50:**1280–1282.
87. **Warren, J. W.** 1996. Clinical presentations and epidemiology of urinary tract infections, p. 3–27. *In* J. W. Warren (ed.), *Urinary Tract Infections—Molecular Pathogenesis and Clinical Management.* ASM Press, Washington, D.C.
88. **Warren, J. W.** 1983. Nosocomial urinary tract infections, p. 283–318. *In* R. A. Gleckman and M. N. Gantz (ed.), *Infections in the Elderly.* Little, Brown, & Co., Boston, Mass.
89. **Warren, J. W., E. Abrutyn, J. R. Hebel, J. R. Johnson, and A. J. Schaeffer for the Infectious Diseases Society of America (IDSA). and W. E. Stamm.** 1999. Guidelines for antimicrobial treatment of uncomplicated acute bacterial cystitis and acute pyelonephritis in women. *Clin. Infect. Dis.* **29:**745–758.
90. **Warren, J. W., D. Damron, J. H. Tenney, J. M. Hoopes, B. Deforge, and H. L. Muncie, Jr.** 1987. Fever, bacteremia, and death as complications of bacteriuria in women with long-term urethral catheters. *J. Infect. Dis.* **155:**1151–1158.
91. **Warren, J. W., J. H. Tenney, J. M. Hoopes, H. L. Muncie, and W. C. Anthony.** 1982. A prospective microbiologic study of bacteriuria in patients with chronic indwelling urethral catheters. *J. Infect. Dis.* **146:**719–723.
92. **Williams, J. J., J. S. Rodman, and C. M. Peterson.** 1984. A randomized double-blind study of acetohydroxamic acid in struvite nephrolithiasis. *N. Engl. J. Med.* **311:**760–764.
93. **Yasue, T.** 1969. Histochemical identification of calcium oxalate. *Acta Histochem. Cytochem.* **2:**83–95.
94. **Zhao, H., X. Li, D. E. Johnson, I. Blomfield, and H. L. Mobley.** 1997. In vivo phase variation of MR/P fimbrial gene expression in *Proteus mirabilis* infecting the urinary tract. *Mol. Microbiol.* **23:**1009–1019.
95. **Zhao, H., R. B. Thompson, V. Lockatell, D. E. Johnson, and H. L. Mobley.** 1998. Use of green fluorescent protein to assess urease gene expression by uropathogenic *Proteus mirabilis* during experimental ascending urinary tract infection. *Infect. Immun.* **66:**330–335.

Colonization of Mucosal Surfaces
Edited by James P. Nataro et al.
© 2005 ASM Press, Washington, D.C.

Chapter 27

Polymicrobial Bacteriuria: Biofilm Formation on Foreign Bodies and Colonization of the Urinary Tract

DAVID J. STICKLER

Prosthetic devices such as catheters and stents are used to manage many of the pathological conditions that affect the urinary tract. While they are generally effective at restoring the lost physiological functions, these foreign bodies are vulnerable to bacterial contamination. They become responsible for infection-associated complications which can severely undermine their utility and compromise the health and welfare of the patient. Indwelling catheters are used to control the release of urine from the bladder. Various types of stents maintain the flow of urine through the ureters and the urethra (Fig. 1). Polymers such as silicone, polyurethane or latex that has been coated with silicone, Teflon, and hydrogels are used in the manufacture of these devices. These materials were chosen because their mechanical properties allow them to perform the required physiological function without causing trauma to the surrounding tissues. They are not inert, however, and their surfaces are capable of physical, chemical, and biological interactions. In addition, they have none of the defense mechanisms which protect mucosal tissue surfaces from bacterial colonization. The surface irregularities common on these devices also induce the passive entrapment of bacterial cells. These foreign bodies are thus extremely vulnerable to colonization by any contaminating microbes that might be in their vicinity.

THE INDWELLING BLADDER CATHETER

By far the most common of the foreign bodies is the indwelling bladder catheter. It has been estimated that each year about 100 million of these catheters are used worldwide (81). Several prevalence surveys have indicated that about 10% of hospital patients and 5% of patients being cared for in nursing homes will at any given time have indwelling suprapubic or urethral catheters (31, 107, 114). The catheter provides a convenient way to drain urine from the bladder and relieve urinary retention caused by anatomical or neurophysiological blockage of the urethra. Its use also facilitates the repair of the urethra after prostatic or gynecological surgery. In addition, enormous numbers of these catheters are used for the long-term management of urinary incontinence in the elderly and in patients disabled by strokes, spinal injury, or major neurological disease. Unfortunately, urinary catheters also provide easy access for bacteria from a contaminated external environment into a vulnerable body cavity.

The ways in which bacteria infect the catheterized bladder are well established. Most organisms causing catheter-associated urinary tract infection derive from the patient's own colonic and perineal flora or from the hands of health care personnel during catheter insertion or manipulation of the collection system. The mere insertion of the catheter can carry urethral organisms into the bladder. Bacteria colonizing the periurethral epithelia and skin can also migrate into the bladder through the mucous film on the external surface of the catheter. By applying *Serratia marcescens* to the periurethral skin of three catheterized patients and subsequently culturing their urine, Kass and Schneiderman (41) demonstrated that these motile bacilli took just 2 to 4 days to ascend from the urethral meatus to the bladder. Several clinical studies since then have demonstrated that the appearance of an organism in the urine of catheterized patients is often preceded by its colonization of the periurethral skin (13, 24, 30, 84). Bacteria contaminating the drainage system can also gain access to the bladder through the lumen of the catheter. The regular emptying of the urine drainage bag provides opportunities for bacterial contamination. Multiplication of the bacteria in the bag urine produces dense

David J. Stickler • Cardiff School of Biosciences, Cardiff University, Cardiff CF10 3TL, Wales, United Kingdom.

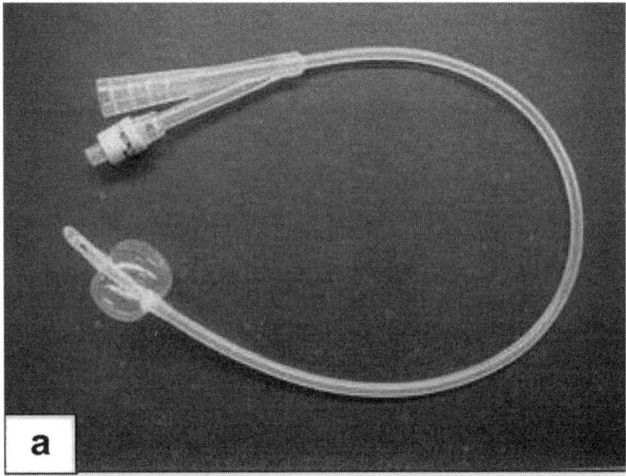

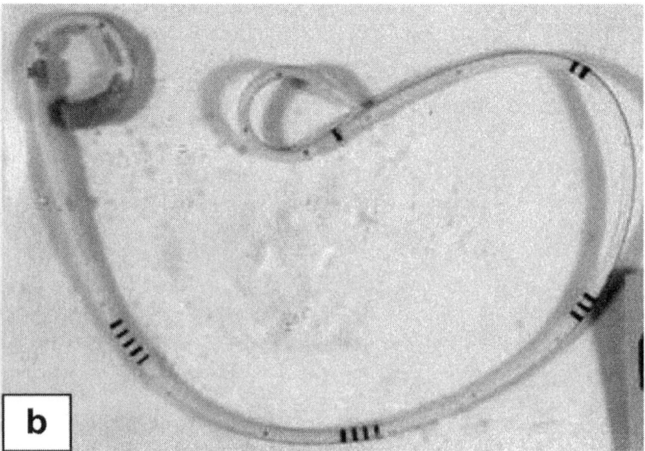

Figure 1. Foreign bodies that are placed in the urinary tract. (a) An unused all-silicone catheter. The retention balloon near to the tip of the catheter has been inflated, and the eyehole through which urine flows is visible. (b) A ureteral stent that had been lying in situ between a patient's kidney and bladder for 12 weeks. Encrustation is visible on the end of the stent that was lying in the bladder. Modified from reference 89 and published with permission from Elsevier BV and the International Society of Chemotherapy.

populations of contaminating organisms, and from these reservoirs of infection, bacteria can migrate to the drainage tube, the catheter, and the bladder (65). Disconnection of the catheter from the drainage tube can also lead to contamination (86). Overall, the evidence suggests that the extraluminal route is more frequent but that both routes are important (100).

In the healthy urinary tract, the bladder contracts when the internal pressure exerted by the expanding volume of urine reaches a critical level. This contraction is synchronized with the relaxation of the external sphincter. The resulting flow of urine distends the normally collapsed urethra, and the bladder empties. This process is a major mechanical defense mechanism against infection. It ensures that bacteria colonizing the distal urethra rarely manage to migrate into the bladder and that, if they do so, they are washed out as the full bladder empties. The indwelling catheter undermines this important host defense mechanism. With the catheter in situ, the cycle of filling, expansion, and emptying of the bladder is replaced by a continuous dribble of urine from the bladder through the catheter. The bladder cannot empty completely since the position of the eyehole is above the retention balloon. A residual sump of urine forms, in which any contaminating bacteria can flourish.

CATHETER-ASSOCIATED URINARY TRACT INFECTION

Generally, the development of urinary tract infection in the healthy urinary tract results from the ability of certain strains of bacteria to express virulence factors that allow them to overcome host defenses. When the host defenses are compromised, however, as in the catheterized tract, many species can establish themselves in the bladder urine and produce bacteriuria. A prospective study in which urine specimens from catheterized patients were cultured daily showed that isolation of any bacteria from such a specimen (even in numbers as small as 3 to 4 CFU/ml) is highly predictive of subsequent infection. In patients not receiving intercurrent antibiotics, the level of bacteriuria increased uniformly to $>10^5$ CFU/ml within 24 to 48 h, demonstrating the vulnerability of the catheterized urinary tract to infection once any microorganism gains access to the lumen of the catheter or bladder (87).

The likelihood of acquiring infection is related to the length of time the catheter remains in situ and to the standard of catheter care (29, 32). In practice, some 10 to 50% of patients undergoing short-term catheterization (1 to 7 days) will develop bacteriuria. Patients undergoing long-term catheterization (>28 days), even with modern closed-drainage systems and meticulous nursing care, will inevitably develop bacteriuria. The numbers of patients undergoing this form of bladder management is so large that catheter-associated urinary tract infection is the most common of the infections acquired in hospitals and other health care facilities (34, 58, 107).

These infections are generally asymptomatic, but patients are vulnerable to severe complications which can put their health at risk. Platt et al. (70), for instance, in a prospective study of 1,485 hospitalized patients, reported that the acquisition of urinary tract infection (defined as at least 10^5 organisms/ml) in patients undergoing long-term catheterization was associated with a threefold increase in mortality rate. A similar study of a large population of nursing home residents (50) established that irrespective of their age and general medical condition, catheterized patients were three times more likely to be dead at the end of a year than noncatheterized patients were. The specific complications that can afflict these patients include urolithiaisis, pyelonephritis, septicemia, and bladder cancer (98).

If infection occurs in the first week of catheterization, it is usually caused by a single species such as *Staphylococcus epidermidis, Enterococcus faecalis,* or *Escherichia coli.* The longer the catheter remains in place, the greater the variety of organisms that accumulate in the bladder (7, 47). Polymicrobial bacteriuria is thus characteristic of patients enduring long-term bladder management by indwelling catheter. Typically, their urines become infected by complex, mixed-species communities of nosocomial organisms. A review of polymicrobial bacteriuria and colonization of foreign bodies in the urinary tract must consider long-term catheterization in settings where most patients are resident in rehabilitation units, nursing homes, residential facilities, and their own homes, rather than the short-term catheterization more typical of the acutely ill hospital patient.

Several studies have shown that these infections are not only polymicrobial, they are also dynamic. Warren et al. (108), for example, analyzed 605 consecutive weekly specimens from 20 chronically catheterized patients and found that 98% contained bacteria at high concentrations and 76% were polymicrobial. The duration of the bacteriuric episodes varied greatly by species. Of the episodes caused by gram-positive cocci (other than enterococci), >75% lasted less than 1 week, whereas the mean durations of episodes due to *E. coli, Proteus mirabilis,* and *Pseudomonas aeruginosa* were 4 to 6 weeks, and the mean duration of those due to *Providencia stuartii* was 10 weeks.

A similar study by Clayton et al. (9) of the urinary flora of patients with spinal injuries undergoing long-term indwelling catheterization in a rehabilitation unit found complex mixed urinary communities of up to seven species. The mean number of species per sample was 3.6. Observations were also made on how these populations changed with time and in response to antibiotic therapy and bladder washouts with antiseptics. It was clear that some species were stable and others were transient inhabitants of the catheterized tract. Persistence factors were calculated for each species by using the following simple relationship: persistence factor = mean period of consecutive isolation/mean period of observation. During periods when patients were not receiving antibacterials, *E. coli, P. stuartii, Citrobacter diversus, P. aeruginosa, P. mirabilis, Morganella morganii,* and *Klebsiella pneumoniae* were stable inhabitants; once they had been recovered from the urine, they were subsequently recovered throughout the period of observation until specific chemotherapy was applied. *Acinetobacter* spp. *S. epidermidis,* and *E. faecalis* proved to be transient components of the mixed communities. The remarkable stability of *P. mirabilis* in the catheterized urinary tract was recently demonstrated by Sabbuba et al. (79). The same genotype of *P. mirabilis* was isolated from catheter urine specimens from a patient over a period of 121 days, during which eight changes of catheter occurred.

It is interesting that Anderson et al. (2) had previously presented data indicating that the relative growth rates of bacteria in urine were major factors in determining survival in mixed populations. They determined the generation times of various species in urine and then used a physical model of the bladder (not a catheterized bladder) to show that mixed urine-grown cultures containing equal numbers of the rapidly growing *E. coli,* paired with one of a variety of slower-growing species such as *P. stuartii, P. mirabilis,* or *P. aeruginosa,* became virtually pure cultures of *E. coli* within 24 h. This clearly does not happen in the catheterized urinary tract.

Urine collected from catheters that had been in place for over 4 weeks contains dense bacterial populations, commonly exceeding 10^8 CFU/ml (9, 101). Maki and Tambyah (55) have pointed out that the urine of patients undergoing long-term catheterization constitutes the largest institutional reservoir of nosocomial antibiotic-resistant pathogens.

Tenney and Warren (101) also investigated the stability of the bacterial flora of urine in the catheterized bladder. In a study of 62 elderly women, urine specimens were aspirated from the port of the indwelling catheter. The old catheter was removed, a new sterile silicone catheter was inserted, and then a second urine specimen was obtained from the replacement catheter. Analysis of these paired urine specimens produced some extremely interesting information. The geometric mean of the bacterial numbers in urine from the indwelling catheter was >10-fold higher than that in the urine from the replacement catheters. The authors concluded that the indwelling catheter provided an "environment" that allowed larger bacterial populations to develop than were present in the bladder urine. Further analysis revealed that the catheter provided a niche for the development of specific communities of bacteria. The mean populations of *P. mirabilis, P. stuartii, M. morganii, P. aeruginosa,* and group D streptococci were 10-fold higher in urine from the indwelling catheters than in urine from the replacement catheters. It was further reported that in a group of patients who had their catheters removed to undergo a trial of intermittent catheterization, these species usually disappeared from the urine within a few days to a week. The indwelling catheter thus seems to facilitate the persistence of the *Proteaee, P. aeruginosa,* and enterococci in the urinary tract. However, the mean population densities of *E. coli* and *K. pneumoniae,* were similar in specimens from the indwelling and replacement catheters. These species were also present in the largest numbers in urine from the replacement catheters. It appears that while these organisms persist in the catheterized urinary tract, they do not use the catheter as their niche. Gram-positive organisms including streptococci (other than enterococci), staphylococci, and *Corynebacterium* spp. rarely persisted during catheter replacement. They had short half-lives in the catheterized bladder. While these organisms are frequently introduced into the catheterized bladder, they cannot persist on the catheter or in the bladder urine.

Mobley et al. (59) presented evidence that the ability of *E. coli* to persist in the urine of patients with long-term catheters is associated with mannose sensitive type 1 pili. These structures mediate adherence to uromucoid or uroepithelial cells and are thought to be involved in the pathogenesis of bladder colonization. Benton et al. (6) examined the strains of *E. coli* isolated from the urine of catheterized spine-injured patients and found that 76% of them expressed type 1 pili but only a minority were able to produce virulence factors (P-pili, hemolysin, and aerobactin) normally associated with pyelonephritic strains.

CATHETER BIOFILMS

Subsequent observations using scanning electron microscopy revealed the extent to which catheters become colonized by bacteria. Substantial communities of organisms have been found growing as bacterial biofilms on catheters. Nickel et al. (64) first observed that both the external and luminal surfaces of a catheter removed from a patient infected with *P. aeruginosa* were heavily colonized by bacteria embedded in an extensive polysaccharide matrix. Ohkawa et al. (67) examined 57 catheters that had been in place for 1 to 14 days and found that bacteria could be recovered from the surface of 54% of the devices and that in over half of the cases, two or more organisms were isolated. Some 21 species were recovered from the catheters, with the most commonly isolated being *E. faecalis, P. aeruginosa,* and *S. epidermidis.* Scanning electron microscopy detected biofilms on 75% of catheters that had been in place for 7 days or longer but not on any of the ones that had been in place for less than a week.

Subsequent studies found that long-term catheters are commonly colonized, particularly on the luminal surface, by extensive bacterial biofilms. Ganderton et al. (28), for example, found biofilms on 44 of 50 catheters that had been indwelling from 3 to 83 days. When a freeze-fracturing technique was used to produce cross sections of the catheters, scanning electron microscopy revealed that the biofilms varied in thickness from patchy monolayers of cells to extensive films some 400 cells deep. Bacteriological analysis showed that mixed communities containing up to four species were isolated from 75% of the catheters. It was also recorded that *E. coli,* group D streptococci, *P. aeruginosa,* and *P. mirabilis* were the organisms most often recovered. It was interesting that *P. aeruginosa,* in particular, was recovered from nine of the catheters but that only in four of these cases was it found in the urine samples taken at the time of catheter replacement, suggesting that this species has a particular preference for the sessile, biofilm mode of growth.

Many of the biofilms on catheters from long-term-catheterized patients contain crystalline material embedded in the matrix (12, 28). This is especially true of catheters colonized by *P. mirabilis* in single or mixed-species biofilms. The organisms commonly found in association with *P. mirabilis* were *E. faecalis, E. coli, Enterobacter aerogenes, P. aeruginosa,* and *P. stuartii. P. mirabilis* was not isolated from catheters that showed no sign of crystalline biofilms, and crystals were rarely seen on catheters that were colonized by organisms other than urease-producing species (90).

A laboratory model of the catheterized bladder was used to investigate the ability of various urease-producing urinary tract pathogens to encrust catheters with crystalline biofilms (95). Models were inoculated with various test organisms, and the extent of encrustation produced over 24 h was assessed by measuring the amounts of calcium and magnesium deposited on the catheters. It was surprising that strains of *M. morganii, K. pneumoniae,* and *P. aeruginosa* shown to be capable of producing urease in conventional bacterial identification tests failed to raise the pH of the urine in the bladder above 6.2 or produce a significant crystalline biofilm on the all-silicone catheters. In contrast, the growth of strains of *P. mirabilis, P. vulgaris,* and *Providencia rettgeri* produced urinary pHs ranging from 8.3 to 8.6 and generated extensive catheter encrustation. Since these latter two species are not commonly found on encrusted catheters removed from patients, the evidence points to *P. mirabilis* as the prime cause of crystalline catheter biofilms.

The encrustations that form on the catheters resemble infection-induced bladder and kidney stones in their crystalline structure, being composed of a mixture of struvite (magnesium ammonium phosphate) and hydroxyapatite (a poorly crystalline form of calcium phosphate) (11). As in stone formation, it is the activity of the urease enzyme that causes the formation of these crystals. The hydrolysis of urea produces the alkaline conditions in which the calcium and magnesium phosphates are relatively insoluble. While urease is the main driving force of the process, it appears that the bacterial exopolysaccharide also plays a role. In vitro studies have demonstrated that the gel formed by these exopolymers facilitates the rapid formation of crystals (8). There is also evidence that the purified exopolysaccharide from *P. mirabilis* is uniquely capable of binding magnesium and accelerating struvite formation (20).

McLean et al. (56) studied the early stages in the formation of crystalline *P. mirabilis* biofilms on glass surfaces in a flow cell system. They reported that the first stage was the deposition of organic material from the urine growth medium onto the test surface. Crystals of struvite then started to form on this conditioning film. Aggregates of cells were seen to attach to the surface, and microcolonies developed from the multiplication of these cells. A complete biofilm layer formed by the spreading growth of the microcolonies, and crystals of struvite then formed rapidly in the biofilm matrix.

The colonization of catheter materials might of course be different. Winters et al. (112) used their simple physical model of the catheterized bladder to study the formation of encrustation on silicone catheters. Transmission electron micrographs (Fig. 2) of sections through the crystalline biofilms revealed elongated struvite crystals lying in the matrix of the biofilm and in contact with the catheter surface, where they had obviously become heavily colonized by bacteria. The amorphous calcium phosphate particles were distributed throughout the biofilm, and many seemed to have formed around the bacterial cells.

CLINICAL SIGNIFICANCE OF CATHETER BIOFILMS

There is evidence from clinical observations that bacteria within the protective biofilm matrix survive in the presence of concentrations of antibiotics generated in urine by conventional treatment regimes, even though they register as antibiotic sensitive in conventional laboratory sensitivity tests (28, 65). In vitro tests have also shown that strains of *P. aeruginosa* in biofilms on catheter materials acquired resistance to tobramycin, amikacin, ceftazidime, ciprofloxacin, and meropenem (63). Many studies with other model systems have demonstrated that the biofilm mode of growth protects cells against the lethal action of antibiotics and disinfectants (10). Although systematic studies of the antibiotic sensitivities of the complex mixed-biofilm communities colonizing catheters have not been performed, the evidence suggests that the innate resistance of biofilms contributes to the difficulties of treating these infections. A common observation is that soon after the completion of a course of antibiotic treatment for a symptomatic catheter-associated urinary tract infection, the organisms will rapidly recolonize the urine (9). It is likely that organisms survive on the catheter and then, on completion of therapy, simply reinoculate the urine. These ideas suggest that changing the

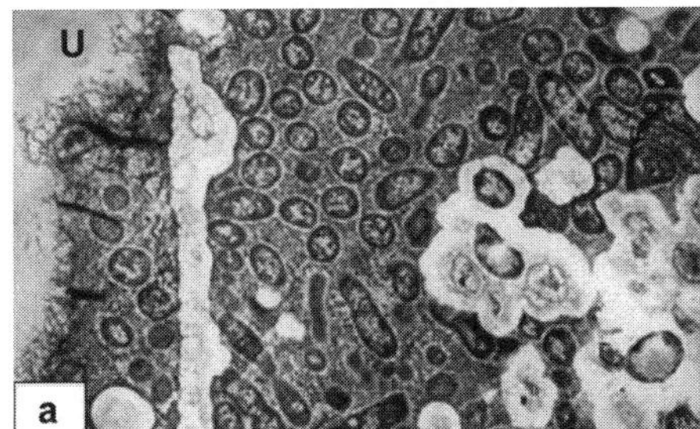

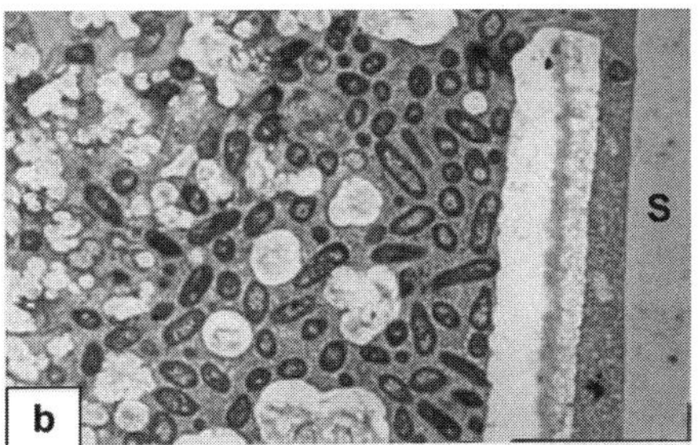

Figure 2. Transmission electron micrographs showing sections through a crystalline biofilm (120 μm thick) that had developed on an all-silicone catheter over 24 h in a model of the catheterized bladder. The model had been inoculated with a mixed urine culture of *P. mirabilis* and *E. coli.* (a) Boundary of the biofilm over which urine (U) has been flowing. (b) Edge of the biofilm adjacent to the silicone surface (S) of the catheter. Bacterial cells can be seen distributed throughout the ruthenium red-stained polysaccharide matrix. The unstained areas are struvite and apatite crystals. Bacteria are visible within the apatite formations. From reference 88a with permission from the European Board of Urology.

catheter during the course of antibiotic treatment might delay recurrence of the infection (61).

The crystalline biofilms produced by urease-positive bacilli cause particular problems for long-term-catheterized patients. They can form on the balloon and external surface of the catheter tip and can traumatize the bladder mucosa and the urethra. The mineralized deposits can also build up rapidly on the luminal surfaces and obstruct the flow of urine from the bladder. This can result in urine by-passing the catheter, leaking around the outside of the device, and causing the patient to become incontinent. Alternatively, urine may be retained, producing painful distension of the bladder. If this problem is not detected and dealt with, it can induce vesicoureteral reflux, triggering episodes of pyelonephritis, bacteremia, and shock. About half of the many patients enduring long-term catheterization will suffer from this complication (46, 49, 60).

STENTS

Various types of metallic or polymeric stents have been used to manage obstruction of the bladder outlet due to benign prostatic enlargement and recurrent urethral strictures. While bacteriuria and encrustation of the stents have been found in patients fitted with these devices, there is very little information in the literature on the nature of these infections (40, 51). For example, Nielsen et al. (66) reported that 4 of 29 patients who received a ProstaKath stent (a gold-plated spiral of steel) developed bacteriuria, but no details were given about these infections. Rosenkilde et al. (77) observed 29 patients fitted with this urethral stent for a mean period of 22 weeks. They reported that severe calcification of the stents and chronic urinary tract infections resistant to antibiotic treatment occurred in five patients. Encrustation by crystalline biofilms may well have occurred on these stents, but no information about the nature of the urinary tract infections was given.

The indwelling self-retaining double-J ureteral stent is used to relieve obstruction of the ureter and to ensure the clearance of kidney stone fragments after extracorporeal shock wave lithotripsy. As the use of these devices has increased in recent years, data on the incidence of ureteral stent-associated infection has emerged. The recent study by Paick et al. (69) reported that 21% of patients had significant bacteriuria (>10^5 CFU/ml) on the day of stent removal and 44% of stents were colonized by bacteria. In all the patients with significant bacteriuria, the same organisms were isolated from the urine and stents. Bacteriuria was not found in patients with sterile stents. None of the stents that were in place for only 2 weeks were contaminated, but 75% of those removed after 84 days or longer were colonized. *Enterococcus* spp. and *E. coli* were the organisms most commonly isolated from the stents. In general, single species were isolated from each specimen. In only two cases were two organisms recovered from a urine sample.

Previous studies had reported rather higher rates of bacteriuria and stent colonization. Reid et al. (73) examined 30 stents that had been inserted for 5 to 128 days following extracorporeal shock wave lithotripsy. They reported the presence of adherent pathogens on 90% of the stents, with mixed organisms being recovered from 44%. Positive urine cultures were reported for 27% of the patients. Of the organisms isolated, 77% were gram-positive cocci, 15% were gram-negative bacilli, and 8% were *Candida* spp. Farsi et al. (23) reported their experience with 237 stents that had been in place for a median duration of 4 months. Bacteriuria was detected in 30% of urine cultures taken just before stent removal, and 68% of the stents were colonized. *P. aeruginosa* was the predominant organism. *S. epidermidis*, *Streptococcus* spp., and *E. coli* were also commonly isolated. Riedl et al. (75) examined 27 "permanent" stents that had been placed due to malignant ureteral obstruction and 66 "temporary" stents and found rates of bacteriuria of 45% in the patients with temporary stents and 100% in the patients with permanent stents. The rates of colonization of the stents were 69 and 100%, respectively. *Enterococcus* spp. and *S. epidermidis* dominated the floras of the temporary devices. Enterococci were also predominant on the long-term stents, but in these cases gram-negative bacilli were also prevalent (*Proteus* and *Klebsiella* were isolated from 26 and 22% of these stents, respectively). Stickler et al. (89) found that biofilms were visible on 46% of 72 stents, most of which had been in place for 12 weeks. Enterococci and coagulase-negative staphylococci were again reported as the organisms commonly colonizing the stents. Mixed communities (two species) were recovered from only eight of the stents. *P. aeruginosa* was the only gram-negative bacillus to be recovered to any extent, being found on four of the stents. In this study, extensive macroscopic encrustation was seen on 8 (11%) of the stents, particularly on the end of the stent located in the bladder (Fig. 3). It was surprising that urease-producing organisms were only rarely found in the crystalline material. Overall, it seems that while the presence of stents in the urinary tract can lead to infection and colonization of the device, mixed-community biofilms and polymicrobial bacteriuria are not commonly found. The cycle of bladder filling and emptying is still in operation in

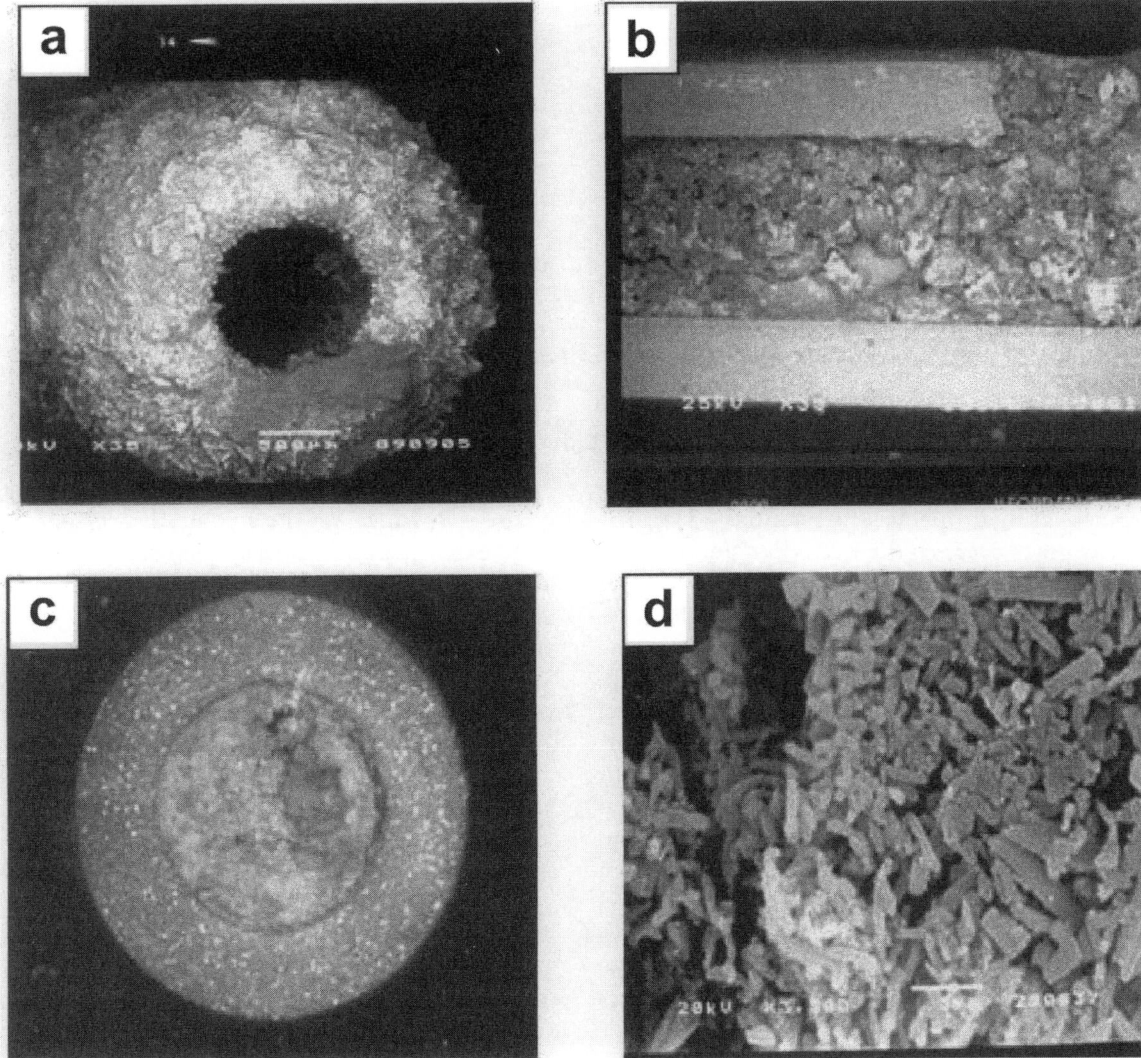

Figure 3. Scanning electron micrographs of a ureteral stent that had been removed from a patient after 12 weeks. (a to c) Crystalline material can be seen on the outer surface (a) and completely occluding an eyehole (b) and the central channel (c). (d) Bacteria in this crystalline material. Modified from reference 89 and published with permission from Elsevier BV and the International Society of Chemotherapy.

these patients, and the stents, unlike indwelling bladder catheters, do not form bridges to the contaminated external surface of the body. There is probably less opportunity for the mixed flora of the perineum to gain access to and colonize the bladder urine.

DEVELOPMENT OF BIOFILMS ON ABIOTIC SURFACES

When a device such as a catheter is placed in the urinary tract, it seems that organic molecules adsorb rapidly onto its surface, forming a conditioning film. Little is known about this aspect of biofilm formation (53). Precisely which molecules bind to the surfaces? Do they form complete layers covering the whole surface, or do they produce a heterogeneous patchy covering? Is the formation of the conditioning film a prerequisite for bacterial adhesion or just an incidental process? Unequivocal answers to these questions do not appear to be available. Ohkawa et al. (67) observed the presence of a fine fibrillar material on catheters that had been in place for less than 3 days (where no biofilm had formed). It was identified as fibrin by an immunofluorescent-staining method. On the surface of catheters that had been in place for longer than 7 days, bacteria and crystals were often found associated with the fibrin. The authors suggested that this protein might assist the adherence of crystals and microbes to the catheter surfaces. Santin et al. (82) reported finding albumin and Tamm-Horsfall protein on ureteral stents recovered from patients. Wassall et al. (109) exposed glass and polystyrene surfaces to pooled human urine and were then able to elute a 29-kDa protein, which was identified as α_1-microglobulin, from these materials. They also presented evidence that the adhesion of a clinical isolate of *P. aeruginosa* onto these surfaces was mediated through this protein. Whether this would occur on the silicone, Teflon, or hydrogel surfaces of catheters is unknown. It seems that the characterization of conditioning films by modern physicochemical techniques has yet to be achieved.

The adhesion of bacteria to living tissues generally occurs through specific mechanisms of molecular "docking" to particular ligands or receptors. For abiotic surfaces such as catheters and stents, primary adhesion is thought to occur by nonspecific mechanisms such as hydrophobic interactions (21). The primary contact between the bacterium and the surface is accidentally brought about either by passive transport, as contaminated urine flows over the catheter, or by active motility of the organism, causing the cell to bump into the catheter surface. What happens after the initial collision depends on the physicochemical properties of the cell and the polymer surface. Interactions with the biomaterial may occur directly via cell surface components such as lipopolysaccharides or exopolysaccarides or by using appendages such as flagella and pili that extend from the bacterial envelope. This second stage in the process results in the cells being firmly anchored to the surface. Lejeune (53) compiled a useful list of the bacterial appendages and envelope surface structures that have been implicated in attachment to abiotic surfaces. Overall, the evidence suggests that bacteria have a range of options they can use to attach to the variety of surfaces they might meet. Faced with hydrophobic materials such as silicone, they use nonpolar surface structures such as pili to secure the attachment. With hydrophilic surfaces such as hydrogels, they exploit their exopolysaccharides or lipopolysaccharides.

Pratt and Kolter (71) used transposon insertion mutagenesis to produce mutants of *E. coli* that were unable to colonize polyvinyl chloride; such mutants lacked either type 1 pili or flagella. They concluded that motility is important to overcome surface forces that repel bacteria. Once the surface is reached, the type 1 pili are required to achieve stable cell-to-surface attachment. A similar study by O'Toole and Kolter (68) showed that for *P. aeruginosa,* flagellum-mediated motility was required for adhesion to polyvinyl chloride surfaces. Bacteria thus seem to have evolved multiple mechanisms which allow them to attach to a great variety of surfaces under different environmental conditions (18).

Once firmly attached to the surface, the bacteria can multiply to produce microcolonies. This occurs rapidly on the surface of urinary catheters because the cells are sitting in a gentle flow of a good growth medium (97). The cells up-regulate their production of exopolysaccharide to form a matrix which encases the developing and spreading microcolonies (15). The spreading and continued growth of these microcolonies leads to formation of the mature biofilm.

CONSENSUS MODEL OF MATURE-BIOFILM STRUCTURE

The biofilms produced by *Pseudomonas* spp. under oligotrophic conditions in which the aqueous phase is flowing over the colonized surface have been used extensively as model systems for the study of the development and structure of mature biofilms (10). The optical sectioning of these biofilms by confocal laser-scanning micrcoscopy has revealed that they are certainly not just homogeneous layers of cells within an exopolymer matrix. Rather, they have a sophisticated structure in which the microcolonies grow up from the basal layer of the biofilm and are interspersed by open channels into which the surrounding fluid can flow (19, 52).

The current consensus view of the architecture of mature natural biofilms portrays them as highly structured communities. The bacterial cells are located in matrix-enclosed microcolonies which develop vertically into tower or mushroom-shaped structures. The open water channels, which permeate the biofilms, act as primitive circulatory systems delivering nutrients to and removing waste products from the populations of cells in the microcolonies (99).

An important question addresses how the adhered cells coordinate their behavior to build the mature biofilms. Presumably they have to communicate in some way to aggregate, stack themselves into the mushroom-like microcolonies, and ensure that the water channels do not become blocked. In many gram-negative bacteria, cell communication can be achieved through the activity of acylated homoserine lactones (AHLs) (26). These small signal molecules are excreted by cells and accumulate in bacterial communities as a function of cell density. When the population density increases to a certain level (the quorum), the accumulated AHLs in the community reach a critical concentration which allows them to interact with receptors on the surfaces of cells that control gene expression. Cells in the biofilm mode express genes in a pattern that differs profoundly from that of their planktonic counterparts (83, 111).

Such a coordinated expression of sets of genes can thus be achieved in response to local cell density. Since biofilms contain high population densities, there was speculation that AHLs might have functions to perform in these communities, and AHLs were duly detected in natural biofilms (57).

Davies et al. (16) presented evidence suggesting that AHLs do indeed play crucial roles in the development of mature biofilms. Mutants of *P. aeruginosa* that were unable to make one of the AHLs were found to produce thin, undifferentiated layers of cells on glass surfaces. The addition of the AHL to the medium flowing through the reaction chamber restored the ability of the mutant to produce mature wild-type biofilm structures. Subsequent studies have resulted in the consensus among most workers in the field that the development of biofilms is a stepwise process involving adhesion of cells to a surface, production of exopolysaccharide, multiplication, and migration of the cells into the mushroom-like microcolonies. In this process, cells communicate via quorum-sensing signals that orchestrate the development and differentiation of the mature biofilm (99).

There is, however, another school of thought about the factors that control biofilm development. By using mathematical models and considering the effects of changing nutrient concentrations, Wimpenny and Colasanti (113) and Van Loosdrecht et al. (104) independently concluded that the development and structure of biofilms is simply a predictable consequence of physicochemical conditions in the biofilm environment. In this model, the cells do not sense that they are on a surface in high cell densities and consequently invoke a quorum-sensing differentiation process, they simply respond to nutrient gradients and stresses in the same way as planktonic cells. Kjelleberg and Molin (43) reviewed the merits of these contrasting ideas and concluded that the quorum-sensing circuit probably controls biofilm development for several organisms under certain sets of conditions. They proposed, however, that it is an oversimplification to consider that this is the exclusive mechanism controlling biofilm development. They pointed out that control systems such as those regulating catabolite repression, stationary-phase gene expression, and motility have been implicated in determining the formation of biofilms. Their conclusion was that biofilms need more than one control system to develop stable communities. They need to invoke a range of regulatory systems if they are to flourish in the diverse environmental conditions they have to face in nature.

Recent studies of *P. aeruginosa* in flowthrough cells have shown that the development of mature biofilms is dependent on the carbon source in the growth medium (44). When glucose was used as the carbon source, *P. aeruginosa* produced the typical mushroom-shaped muticellular structures separated by water-filled channels. In contrast, when citrate was provided as the carbon source, simple flat films of mobile cells, which failed to differentiate into mushroom-like structures, were formed. In a follow-up paper (45), these authors presented a detailed description of the way in which the mushroom-shaped muticellular structures are formed. They performed time-lapse confocal laser-scanning microscopy on biofilms formed by various combinations of a wild-type *P. aeruginosa* strain (tagged with yellow fluorescent protein) and a mutant unable to produce type IV pili (tagged with a cyan blue fluorescent protein). They observed that after the attachment phase, the population of cells consisted of some motile cells moving by the extension-grip-retraction twitching motility mediated by type IV pili and some nonmotile cells in which twitching had been down-regulated. At certain foci, the nonmotile cells multiply, pile up on top of each other, and form the stalks of the mushrooms. In the meantime, the motile cells spread out over the surface and then climb up on the stalks by using the type IV pili in a coordinated process which leads to the formation of the caps of the mushrooms. The cause of the down-regulation of the twitching in a subpopulation of the cells is unknown. It was suggested that it might be due to local environmental cues. Perhaps at certain foci, the cells adhere to each other strongly and create a matrix in which twitching is arrested. It is interesting that twitching motility is not regulated by any of the known cell-to-cell signaling systems in *P. aeruginosa*. The process seems to be dependent on nutritional conditions since down-regulation does not occur and stalks are not formed when cells are grown on citrate as the carbon source.

STRUCTURE OF MATURE BIOFILMS ON URINARY CATHETERS

While there is discussion about the nature of the regulatory mechanisms controlling biofilm development, most workers in the field subscribe to the *Pseudomonas* model for the structure of mature biofilm. What relevance does this consensus model of biofilm structure have for those of us who are trying to understand the formation of biofilms on long-term catheters? Little is known about what happens when a foreign body such as a silicone or hydrogel-coated latex catheter is introduced to complex communities of planktonic organisms present in polymicrobial bacteriuria. Is there competition between the species for the surface? Once one species has bound to a surface, does it promote or inhibit the adhesion of other members of the urine community? Is coaggregation between species a factor? *P. aeruginosa* is a common member of the urinary and biofilm communities; does it, in this context, form stacks of nontwitching cells and mushroom caps of the motile cells? Perhaps the adherence of nonmotile species such as enterococci, staphylococci, and *Klebsiella* spp. and their multiplication to form colony-like aggregations could be the base over which species capable of surface migration clamber and form the caps of the mushroom-like structures. Perhaps some species use their swarmer-cell-forming ability to ascend these stalks.

The precise structures of mature biofilms vary with location, nature of the surface, constituent organisms, and availability of nutrients. The cell density in the biofilm increases in more nutritious media (113). Biofilms on urinary catheters have been found to contain up to 5×10^9 CFU/cm^2 of surface area (Fig. 4) and can form along the entire length of the catheter (96). Scanning electron microscopy has revealed that despite the shrinkage during sample preparation, many of these biofilms can be >500 μm in depth. Instances have been recorded (93) in which the biofilm completely fills the catheter lumen and can be removed from the catheter as a solid "worm-like" structure (Fig. 5). Such a biofilm was found to be a mixed community composed of *P. aeruginosa, E. coli, E. faecalis,* and *P. mirabilis.* It blocked a patient's catheters regularly at 3- to 4-day intervals, with 15 of the "worms" being produced in a 10-week period.

The images presented in Fig. 4 and 5 are difficult to reconcile with the structure of the oligotrophic pseudomonad model. While they have certainly been distorted during sample preparation, it is hard to imagine how a series of discrete mushroom-shaped structures surrounded by open water channels could be transformed by dehydration into the blocks of material visible in these micrographs. The catheter biofilms are certainly permeated by channels, but these seem to produce a spongiform structure in which the matrix-encased cells are not individual towers but form part of a linked three-dimensional network. The biofilms produced by *P. mirabilis* in urine offer a rather special opportunity to study mature biofilm structure by electron microscopy. As the pH of the biofilm and surrounding medium becomes alkaline, hydroxyapatite starts to form around the cells in the matrix. This cements the components of the biofilm in place and creates a "plaster cast" of the structure, which is not so subject to distortion by the dehydration involved in sample preparation. The images of these calcified biofilms (Fig. 6) again suggest that urinary catheter biofilms are somewhat different in structure from the pseudomonad model. They indicate that, rather than consisting of a series of mushrooms dispersed in a fluid bed, the catheter biofilms have a sponge-like structure permeated by urine-filled pores.

WHY ARE CATHETERS SO VULNERABLE TO ENCRUSTATION AND BLOCKAGE BY CRYSTALLINE BIOFILM?

Stickler et al. (97) used their simple model of the catheterized bladder to observe the early stages in the formation of a *P. mirabilis* biofilm. The idea was to understand why catheters are so vulnerable to blockage and to establish whether there are sites on the catheter surface that are particularly susceptible to encrustation. Models were assembled, supplied with

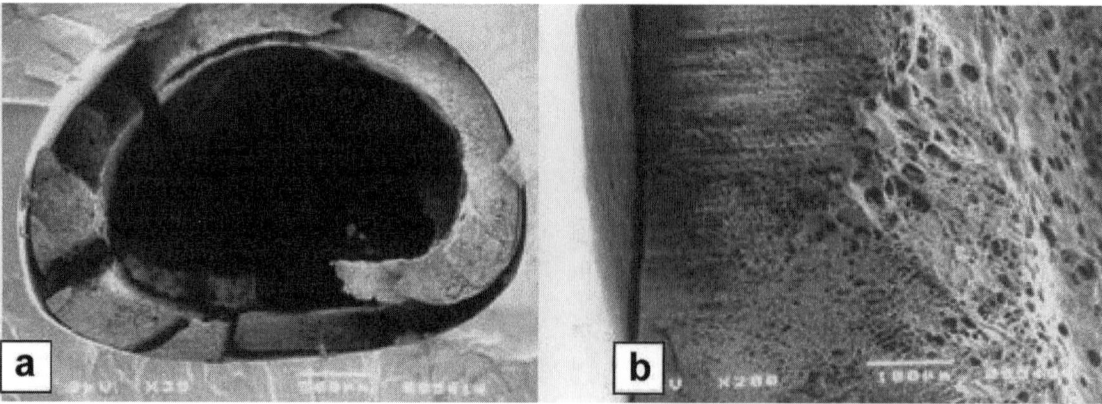

Figure 4. Scanning electron micrographs of a *P. aeruginosa* biofilm on a catheter that had been removed from a patient after 6 weeks. (a) Freeze-dried preparation of a freeze-fractured cross section of the catheter lumen. (b) Spongiform structure of the biofilm. Modified from reference 88a with permission from the European Board of Urology.

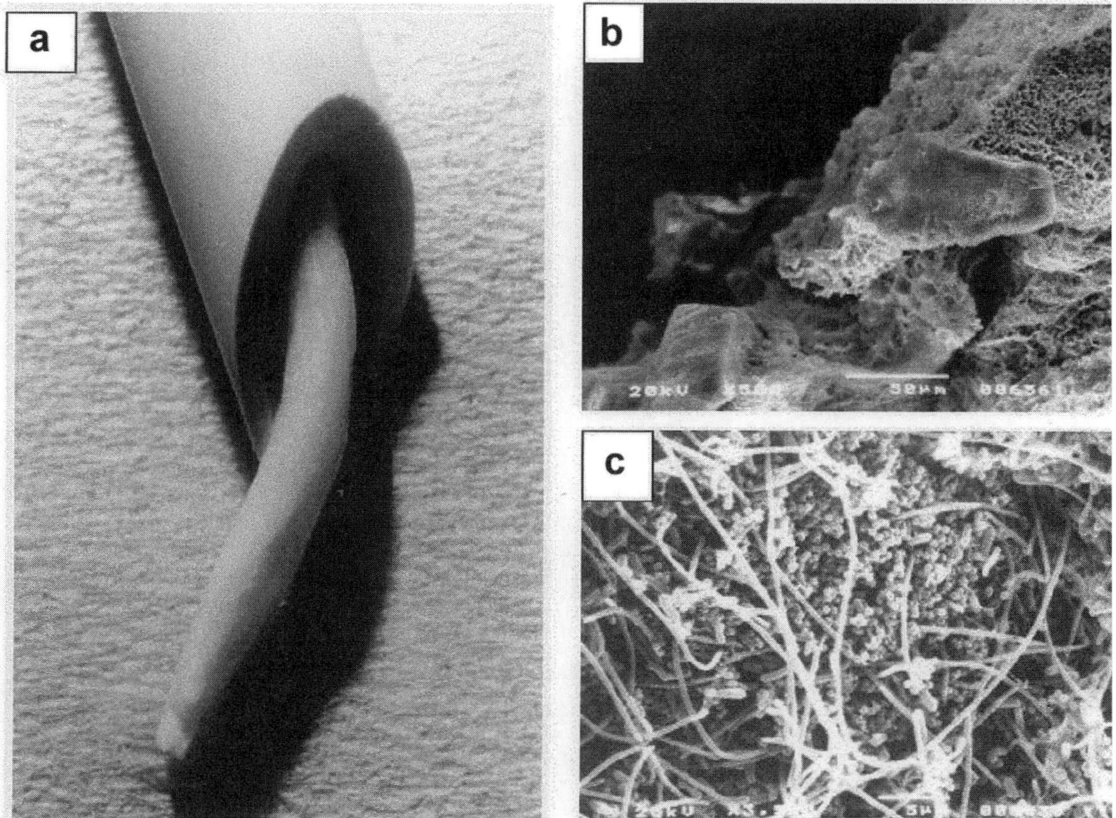

Figure 5. Images of a worm-like structure that blocked a patient's catheter regularly at 4-day intervals. (a) Photograph of a cut section of the catheter, revealing the "worm" that blocked the lumen. (b) Scanning electron micrograph of a freeze-dried preparation of the "worm," showing the crystal formations of the outer surface. (c) Scanning electron micrograph of a fixed, critical-point-dried specimen, showing that beneath the crystalline coat the "worm" is composed of masses of cocci, short rods, and a tangle of very long bacilli. Modified from reference 93 with permission.

Figure 6. Micrograph of a *P. mirabilis* biofilm that had developed on a hydrogel-coated latex catheter after 20 h in a model of the catheterized bladder. Pores are visible penetrating the calcified "plaster cast" of the biofilm. Kindly provided by Rob Young, Cardiff University.

urine, and infected with *P. mirabilis*. The catheters were removed from the models after various times and examined by electron microscopy for encrustation. It can be seen from this experiment the catheter eyehole is a narrow slit about 1 mm wide (Fig. 7). The surface of its rim is extremely rough and irregular, and after just 2 h, cells can be seen trapped in crevices in the hydrogel-coated latex surface. At 4 h, microcolonies became visible at these sites. The clumps of cells seemed to have developed in depressions in the irregular surface. At this stage there was little sign of biofim in the central channel. At 6 h, the microcolonies on the eyeholes had developed further and amorphous crystalline material, typical of calcium apatite, formed in association with these bacterial aggregates. There were also signs that the biofilm had started to move down the catheter lumen. At 20 h, the eyehole was extensively encrusted and the biofilm had spread along the luminal surface. While the encrustation eventually formed along the full length of the catheter, it developed most extensively and generally blocked the catheter lumen at or just below the eyehole.

The micrographs presented in Fig. 7 demonstrate that the techniques used in catheter manufacture produce particularly rough irregular surfaces on

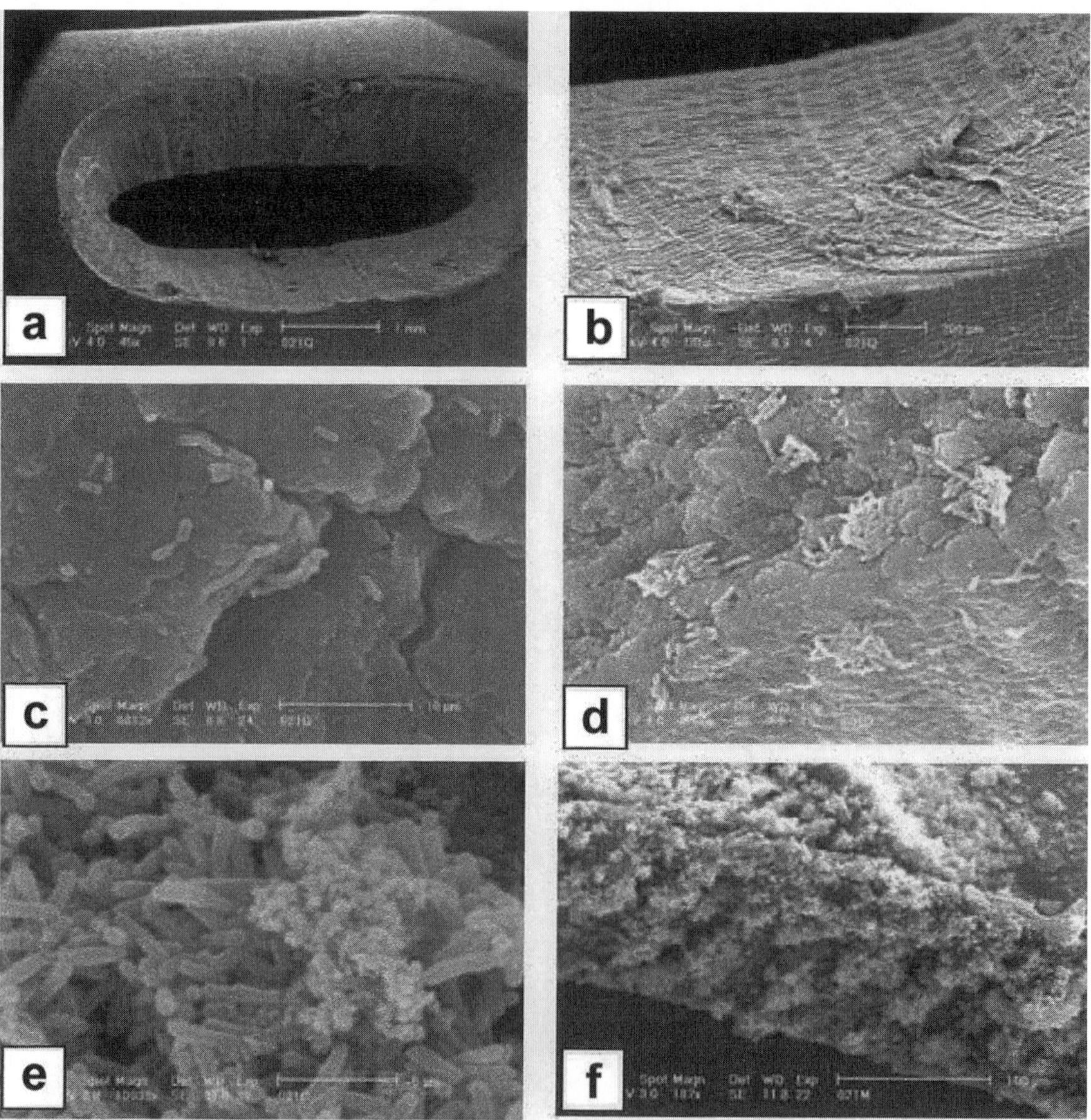

Figure 7. Micrographs illustrating the formation of a crystalline *P. mirabilis* biofilm around the eyehole of a hydrogel-coated latex catheter over a period of 20 h in a laboratory model. (a and b) Rough, irregular surface of the catheter. (c) After 2 h in the model, bacteria have adhered to the crevices in the surface. (d) After 4 h, microcolonies have developed in depressions in the surface. (e) After 6 h, amorphous crystalline material typical of apatite is associated with the cells. (f) After 20 h, extensive crystalline biofilm has formed around the eyehole. Modified from reference 97 with permission.

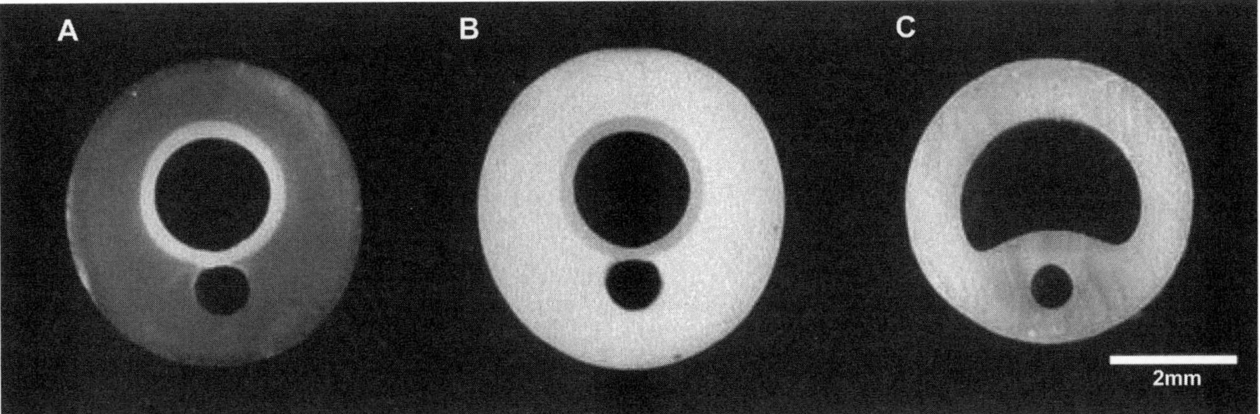

Figure 8. Photographs of unused catheters illustrating the narrow central channels through which the urine has to flow. (A) Silicone-coated latex catheter. (B) Hydrogel-coated latex catheter. (C) All-silicone catheter. From reference 97 with permission.

the rims surrounding the eye-holes. Figure 8 illustrates that while size 14 catheters, for example, have an external diameter of 4.7 mm, the walls are so thick that the central channel has an internal diameter of only 1.8 mm. These narrow channels are easily blocked by developing biofilm and debris from the infected bladder. All-silicone catheters have thinner walls and slightly larger internal diameters (around 2.5 mm for a size 14 catheter). However, the central channels of silicone catheters are crescent shaped, and the corners of the crescents are sites which often trap crystalline debris and initiate encrustation. Silicone surfaces are smoother than latex and not so crudely engineered. The eyeholes tend to be larger and smoother, but surface striations and irregularities were frequently observed both around the eyeholes and along the length of the luminal surfaces. The pattern of development of crystalline biofilm on the silicone catheters was similar to that observed on the latex-based devices. These observations indicate that there is ample opportunity for catheter design to be improved.

OBSERVATIONS ON THE MIGRATION OF URINARY PATHOGENS OVER CATHETERS

P. mirabilis is of course well known for its ability to swarm over solid surfaces. When the small planktonic bacilli come to rest on an agar surface, for example, they have the ability to differentiate into elongated, highly flagellated forms that organize themselves into rafts of parallel cells and migrate rapidly over the agar. Belas (5) suggested that this swarming ability might be involved in the colonization of urethral catheters. Using a simple laboratory model, test organisms were shown to be able to migrate across sections of hydrogel-coated, silver/hydrogel-coated, and silicone-coated latex and all-silicone catheters. Migration was shown to be significantly more rapid over the hydrogel than the silicone surfaces. Scanning electron micrographs showed that the swarming fronts on the catheters were composed of discrete rafts of tightly packed masses of typically elongated swarmer cells (Fig. 9). It was suggested that swarming facilitates both the spread of crystalline biofilm over the catheters and the migration of *P. mirabilis* along the external surface of the catheter from the urethral meatus to the bladder (91).

A subsequent study (80) examined the ability of a range of urinary tract pathogens to migrate over catheter sections. Species capable of swarming (*P. mirabilis, P. vulgaris,* and *S. marcescens*) were able to migrate over hydrogel and silicone surfaces. *P. aeruginosa* was the most mobile of the species that were motile by normal flagellum-mediated swimming. Nonmotile species (*E. faecalis, K. pneumoniae,* and *S. aureus*) generally failed to migrate over all types of catheter. It was also clear that the hydrogel coatings facilitated the migration of urinary tract pathogens over catheter surfaces.

Since the periurethral skin is commonly colonized by mixed populations of organisms (24), the mobility of *P. mirabilis* was also examined in the presence of other species. *P. mirabilis* was capable of swarming through dense populations of *E. coli, K. pneumoniae, S. aureus,* and *E. faecalis* and then migrating successfully over hydrogel-coated latex catheters. Evidence was also presented that the swarmers were also consistently capable of transporting *K. pneumoniae* and *S. aureus* (but not the other species) over the hydrogel surface. Populations of *P. aeruginosa* and *S. marcescens,* however, had an

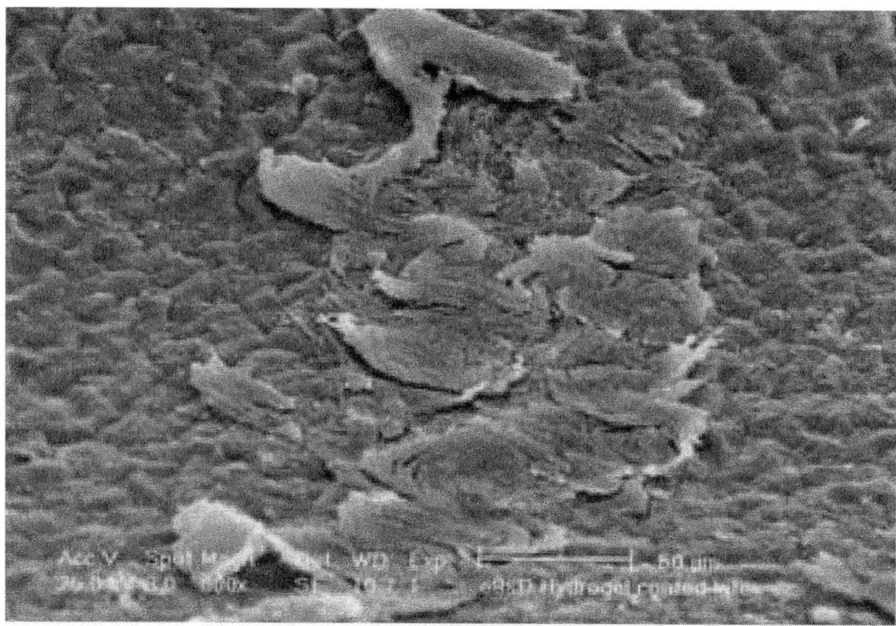

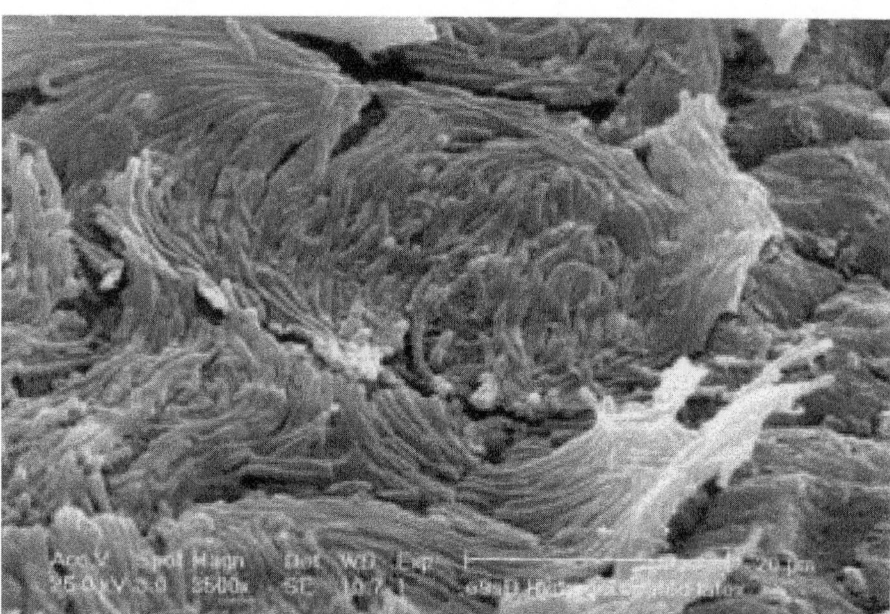

Figure 9. Scanning electron micrographs showing a section of hydrogel-coated latex catheter over which *P. mirabilis* swarmers are migrating. The section has been removed from a laboratory model. Multicellular rafts of elongated swarmer cells can be seen on the rough, irregular luminal surface of the catheter. The swarmers are migrating from left to right over the catheter. Kindly provided by Rob Broomfield, Cardiff University.

inhibitory effect on the migration of the *Proteus* swarmers.

These observations raise a number of interesting questions. When *P. mirabilis* is a component of the periurethral skin flora, is it capable of swarming over the catheter surface, taking other members of the community with it and thus inducing polymicrobial bacteriuria? Also, what is the basis of the antagonism between these strains of *P. mirabilis* and those of *P. aeruginosa* and *S. marcescens*? Does it lead to incompatibility between strains in a biofilm?

PREVENTION OF BACTERIURIA AND DEVICE COLONIZATION

Many strategies have been deployed to try to prevent bacteriuria in patients undergoing long-term catheterization. Attempts to block the routes of infection by including antibacterial agents in the drainage bags or applying topical antibacterial agents to the urethra-catheter junction have all failed (48, 105). Catheters have been coated or impregnated with antimicrobial agents in efforts to prevent bacteriuria and biofilm formation. There is some evidence that catheters impregnated with nitrofurazone or coated with a silver-containing hydrogel might delay bacteriuria in the short term and might have applications in acute-care hospitals (55). There is no evidence, however, that they are effective in preventing infections in long-term-catheterized patients. It has also become clear that crystalline *P. mirabilis* biofilms readily form on these catheters (62, 85).

Nitrofurazone, of course, has little activity against *P. mirabilis* (110). The antiseptic triclosan might be a more appropriate agent for the control of crystalline biofilms since strains of *P. mirabilis* isolated from encrusted catheters are extremely sensitive to this agent (MIC, 0.5 μg/ml) (88). Recently, a novel method for introducing triclosan into the bladder has been described (92). The retention balloons of silicone catheters were inflated with triclosan solutions rather than water. In laboratory models infected with *P. mirabilis,* it was demonstrated that triclosan diffused through the balloon into the urine, cleared the culture of test organisms from the bladder, and prevented the rise in urinary pH that triggers crystal formation. Whereas the control catheters inflated in the normal way with water became blocked after just 25 h, the test catheters drained freely for 7 days, the stage at which the experiment was stopped. Little sign of crystalline biofilm was found on these triclosan-inflated catheters after the 7 days in the model. In addition, it was found that the catheters had become impregnated with the antibacterial agent. These in vitro observations suggest an approach that could have practical applications in controlling catheter encrustation. The method does not disturb the integrity of the drainage system, and it might be possible to deliver other agents, including antibiotics, through the retention balloons. The treatment of catheter-associated infections by delivering antibacterials directly to the bladder could avoid the selection of an antibiotic-resistant gut flora as a result of oral administration of drugs.

An alternative approach to the use of antibacterials is to develop materials with physicochemical properties that prevent the adherence of bacterial cells. Phosphorylcholine (PC) has been incorporated into synthetic polymers to mimic the naturally occurring erythrocyte membrane lipid dipalmitoylphosphatidylcholine. Surface coatings of this material are extremely hydrophilic, and it seems that water molecules bind tightly to the PC head groups and make it difficult for proteins and other materials to react with the surface. Ishihara and Iwasaki (38) found that proteins exposed to PC-coated surfaces remained in their native state, whereas exposure to the base polymers induced conformational changes in protein structure which allowed them to bind irreversibly as conditioning film. In view of these properties, it was of interest to test whether PC-coated catheters resist colonization by biofilms such as those produced by *P. mirabilis.*

It was disappointing to find that, in laboratory tests, PC coating did not reduce the rate at which a crystalline *P. mirabilis* biofilm encrusted both latex- and silicone-based catheters. However, a clinical study reported more encouraging results for PC-coated ureteral stents. Both biofilm formation and encrustation were less extensive on coated stents that had been in situ for 12 weeks than on control stents that had been previously removed from the same group of patients (89).

Heparin, a strongly electronegative polysaccharide that might be expected to repel bacteria, has also been proposed as a coating for urological devices (78). A recent clinical study reported that in contrast to uncoated polyurethane ureteral stents, heparin-coated devices did not show any degree of colonization by biofilms or crystals for up to 6 weeks after placement in patients (76). A number of other surface coatings have been reported to reduce bacterial colonization of ureteral stents in laboratory tests (4). For example, a hydrophilic polyvinylpyrrolidine coating applied to polyurethane reduced the adherence of *E. faecalis* (103).

Recent advances in our understanding of the involvement of chemical signals in the development of biofilms have suggested some very subtle approaches to biofilm control. Quorum-sensing signal molecules have been detected in *P. aeruginosa* biofilms that developed on silicone catheters in a laboratory model. In addition, sections of silicone catheters taken directly from patients, cleaned of biofilm by sonication, and then autoclaved also produced strong positive reactions for AHLs in an *Agrobacterium tumefaciens* cross-feeding test. These results show that AHLs are produced by biofilms while the catheter is in situ in the patient's bladder and that the quorum-sensing signals adsorb to silicone (94). If these signals are indeed crucial for the development of catheter biofilms, there is an opportunity for a control strategy based on scrambling these cell communication systems. Natural and synthetic quorum-sensing inhibitors have been identified (35). Recently, it has been demonstrated that in the presence of such an inhibitor, a halogenated furanone, *P. aeruginosa* could produce only flat, undifferentiated, and unstable biofilms (36). We look forward to further studies on these inhibitors.

It seems, therefore, that while the only sure way at present to control bacteriuria in patients undergoing long-term catheterization is to remove the catheter, there are some exciting new strategies on the horizon for controlling biofilm formation on catheters and other urological devices.

INTERACTIONS BETWEEN SPECIES IN MIXED-COMMUNITY BIOFILMS

Reviews by Stoodley et al. (99) and Donlan (18) have emphasized that natural biofilms are mixed communities of microbes demonstrating a level of structural differentiation that requires cell-to-cell communication and cellular specialization. Furthermore, the proximity and high densities of cells within the matrix-enclosed microcolonies provide highly suitable conditions for physiological cooperation, exchange of genetic material, and communication by messages such as quorum-sensing signals. We are only just beginning to understand the environmental cues and phenotypic responses that produce these physiologically integrated multispecies communities. For the example of urinary catheter biofilms, the question arises whether the communities that develop are predetermined in some way by interactions between species, so that particular clusters of species commonly form effective communities, or whether the whole process is serendipitous.

Rickard et al. (74) have reviewed the role of bacterial coaggregation in the development of multispecies biofilms. Coaggregation occurs between pairs of genetically distinct bacteria. It is brought about by protein adhesins on one cell binding with complementary saccharide receptors on the other. These interactions may contribute to the development of biofilms in two ways. Single cells in suspension might recognize partner cells that have already adhered to the surface. Prior coaggregation of the cells might also occur in suspension, and the pairs of organisms then coadhere to the surface. This process is thought to be particularly important in the mouth, where it ensures the mutually beneficial pairing of species within dental plaque. The subsequent multiplication of the cells produces clusters of the various species (termed clonal mosaics) within the multispecies biofilm. Coaggregation has recently been observed in bacteria isolated from multispecies biofilms occurring in the mammalian gut, the human urogenital tract, and potable water supply systems (33). Perhaps it is involved in determining the structure and composition of catheter biofilms. Has a clonal mosaic formed in the biofilm shown in Fig. 5?

The crystalline biofilms that form on catheters usually contain *P. mirabilis* and several other species (28). It would be useful to know whether any of these other species interact with *Proteus* spp. and modulate its activity. It is well known that in some patients, catheters become blocked with crystalline biofilm regularly at 2- to 3-day intervals, whereas that in others, blockage can take up to 6 weeks to develop. Do some of these other organisms influence the urease activity of the community, the pH of the urine, and the rate at which crystals form in the biofilm? It would be equally interesting to establish whether other biofilm communities, for example those that form on the catheters of patients who do not suffer from this complication, could exclude *P. mirabilis*.

The early study by Mobley and Warren (60) presented data that give a tantalizing insight into these questions. Over the period of a year, these authors collected urine samples weekly from 32 elderly women who had been catheterized for more than 100 consecutive days. Bacteriological analysis of 1,135 of these specimens showed that 972 (86%) contained urease-producing bacteria. *P. mirabilis* and *M. morganii* were the most common urease-positive species, each being recovered in over half of the specimens. During the study period, 67 catheter obstructions were recorded in 23 of the 32 patients. Of the specimens taken immediately prior to an obstruction episode, 73% contained *P. mirabilis*. Only 40% of specimens from patients who did not have an obstructed catheter during the year contained this species. *P. mirabilis* (but not the other urease-producing species [*M. morganii, P. stuartii, K. pneumoniae, P. rettgeri,* and *P. vulgaris*]) was strongly correlated with the obstruction of long-term catheters. *M. morganii* was particularly interesting since it was identified in urine from unobstructed catheters more frequently than from obstructed specimens (59 and 33%, respectively). The authors made the interesting suggestion that perhaps *M. morganii* may in some way protect against catheter obstruction.

These analyses were, of course, all performed with urine samples, and it would be interesting to examine the catheter biofilm flora in this way. It could also prove fruitful to use statistical techniques such as cluster analysis (25) to test for significant associations between biofilm species. Any signs of associations or antagonisms detected by this ecological approach could then be tested experimentally in laboratory models. If bacterial factors prove to be capable of modulating catheter encrustation, it might be possible to exploit them in interference strategies (14) to control this common complication.

TREATMENT OF POLYMICROBIAL BACTERIURIA

Many studies have shown that conventional antibiotic treatment of asymptomatic polymicrobial bacteriura in long-term catheterized patients is futile. Warren et al. (106), for example, conducted a prospective clinical study to examine whether antibiotic treatment of catheter-associated bacteruria prevented symptomatic infections. Cephalexin was administered to patients whenever susceptible organisms appeared in the urine. The authors reported no decrease in the number of bacterial species per urine specimen or in the incidence of febrile episodes. The only change was an increase in the proportion of antibiotic-resistant organisms.

It is common practice not to treat the polymicrobial bacteriuria in the long-term-catheterized patients unless the clinical symptoms suggest that the kidneys or the bloodstream has become involved in the infection. While it is recognised that many patients will suffer from symptomatic infections (98), surprisingly few studies have investigated the appropriate treatment for these conditions.

Raz et al. (72) suggested that, prior to treatment, catheters should be replaced and urine specimens should be obtained for culture from the fresh catheter before initiating antibiotic therapy. They argued that catheter replacement provided a more valid specimen, allowing the culture of the organisms present in the urine rather than those colonizing the old catheter, and that it would also remove the biofilm-laden catheters, with their dense populations of organisms, from the infected urinary tract. These authors went on to conduct a prospective randomized trial with elderly nursing-home residents to determine whether routine replacement of a chronic indwelling catheter, before instituting antimicrobial therapy for symptomatic urinary tract infection, leads to improved outcomes. A total of 54 patients with symptomatic catheter-associated urinary tract infection were randomized to either catheter replacement or no catheter replacement before antimicrobial therapy was given. Analysis of urine specimens obtained 7 days after therapy had been completed showed that polymicrobial bacteriuria had decreased in 18 of the patients in the catheter replacement group and in only 9 of those whose catheters were not replaced ($P = 0.01$). Catheter replacement was also associated with a shorter time to return to afebrile status, improved clinical condition at 72 h after initiation of therapy, and a significantly lower rate of symptomatic clinical relapse 28 days after therapy ($P = 0.015$). There is a clear lesson here.

Raz et al. (72) gave oral ciprofloxacin or ofloxacin to patients who had been febrile for 24 h. If the pretherapy urine sample showed the presence of quinolone-resistant organisms, treatment was changed appropriately. A complication here is that the resistance profiles of the mixed urinary community might well differ and multidrug resistance might be common. Under these circumstances, an option is to take a blood culture and base treatment on the sensitivity of the organisms responsible for the bloodstream infection. It would be reassuring to have more information about which organisms in patients with polymicrobial bacteriuria are more likely to cause pyelonephritis or bacteremia.

Vaccines have been developed against uropathogenic *E. coli* and shown to induce effective prophylaxis against infections in patients with "normal" urinary tracts (i.e., who were not undergoing catheterization [3]). Li and Mobley (54) have shown that a surface antigen of *P. mirabilis* is a promising vaccine candidate. It will be extremely interesting to discover whether these vaccines are capable of preventing the colonization of the catheter and the urine in the catheterized urinary tract.

ARE THERE LONG-TERM CONSEQUENCES TO ASYMPTOMATIC POLYMICROBIAL BACTERIURIA?

Current dogma accepts that polymicrobial bacteriuria is inevitably associated with long-term catheterization and that, in the absence of symptoms, attempts should not be made to clear organisms from the urine with drugs. It is reasonable to question whether this policy has any undesirable long-term sequelae for these patients. Galloway et al. (27) pointed out that clinical signs are unreliable indicators that urinary organisms have invaded the upper urinary tract in catheterized patients. In an attempt to detect invasive infection, they determined serial concentrations of C-reactive protein (CRP) in urine from paraplegic patients undergoing long-term catheterization. While they confirmed that symptomatic urinary tract infection was associated with elevated CRP levels in urine, they found six patients with abnormally high levels of CRP in whom the only known site of infection was the urinary tract and who were clinically well. These observations raise the question whether asymptomatic pyelonephritis is causing chronic renal damage in these patients.

Jewes et al. (39) conducted a survey of bacteremia in 115 domiciliary patients with long-term catheters. Regular blood samples were taken, and asymptomatic bacteremia was often found (10% of the blood cultures were positive). The potential dangers of catheter insertion and removal were highlighted by their finding that bacteremia followed 20 of the 197 catheter changes that were monitored. During the study, only two patients displayed symptoms of bacteremia. The number of bacteremic patients with symptoms is thus a very small proportion of the total. The authors suggested that frequent asymptomatic bacteremia could put these patients at risk of endocarditis.

Several studies of patients with spine injuries have indicated that catheterization for longer than 10 years is associated with an increased risk of carcinoma of the bladder. Davies (17) examined the death certificates of men and women who had died of bladder cancer in England and Wales and concluded that there was an excess mortality from this cause among paraplegics and other groups of patients likely to suffer chronic urinary tract infections. Kaufman et al. (42) performed random bladder biopsies on 62 patients with spine injuries and found that 6 had diffuse squamous cell bladder carcinoma. El-Masri and Fellows (22) reviewed the histological records of 6744 paraplegic and tetraplegic patients at Stoke Mandeville Hospital in England and found 25 cases of bladder cancer. The mean interval between spinal injury and diagnosis of cancer was 23 years. The cancer was registered as the cause of death in 20 of the patients. The expected number of deaths from bladder cancer in this group based on national statistics was calculated as 1.1. Chronic inflammation of the bladder mucosa brought about by the presence of the catheter and the concomitant infections could be the carcinogenic stimulus in these patients (1). Another possible explanation is the production of carcinogenic nitrosamines in infected urine. Hill & Hawksworth (37) suggested that since normal urine contains nitrates and secondary amines from dietary and endogenous sources, infection of the urinary tract by nitrate-reducing bacteria sets up conditions under which *N*-nitrosamines will be produced in the bladder. A study of patients with spine injuries who were undergoing various forms of bladder management, including indwelling catheterization, detected *N*-nitrosamines in the infected urine of 32 of 33 individuals. These carcinogens were not found in uninfected urine from control subjects in the same hospital (102).

CONCLUDING THOUGHTS

It appears that at present we can do little to deal effectively with the polymicrobial bacteriuria that is commonly associated with the long-term placement of foreign bodies, particularly catheters, in the urinary tract. We know that these mixed bacterial populations are dynamic and have stable and transient components, but we need to understand more about their ecology. We are largely ignorant about the interactions between the species in these communities and the ways in which they respond to environmental challenges. The catheter drainage system, with its enormous load of nosocomial pathogens in the urine bag and colonizing biofilms, comprises a major reservoir of antibiotic-resistant organisms. Surely we should be making more strenuous efforts to deal with this issue in health care facilities such as rehabilitation units and nursing homes. We need to resolve concerns about the long-term consequences of not treating device-associated asymptomatic bacteriuria.

In an era when significant advances are being made in the development of prosthetic devices in other areas of health care, it is difficult to understand why we are still failing to solve the relatively simple problem of draining urine from the bladder without producing infections and all the associated complications. The morbidity associated with the long-term use of the currently available devices undermines the quality of life for many individuals and is no longer acceptable. Developing alternatives to the catheter is surely possible. The medical device industry must be encouraged to take up the challenge.

REFERENCES

1. **Akaza, H., W. M. Murphy, and M. S. Soloway.** 1984. Bladder cancer induced by noncarcinogenic substances. *J. Urol.* **131:**152–155.
2. **Anderson, J. D., F. Eftekhar, M. Y. Aird, and J. Hammond.** 1979. The role of bacterial growth rates in the epidemiology and pathogenesis of urinary infections in women. *J. Clin. Microbiol.* **10:**766–771.
3. **Bauer, H. W., V. W. Rahlfs, P. A. Lauener, and S. S. Blessmann.** 2002. Prevention of recurrent urinary tract infection with immuno-active *E. coli* fractions: a meta-analysis of five placebo-controlled double-blind studies. *Int. J. Antimicrob. Agents* **19:**451–456.
4. **Beiko, D. T., B. E. Knudsen, J. D. Watterson, and J. D. Denstedt.** 2003. Biomaterials in urology. *Curr. Urol. Rep.* **4:** 51–55.
5. **Belas, R.** 1996. *Proteus mirabilis* swarmer cell differentiation and urinary tract infection, p. 271–298. *In* H. T. L. Mobley and J. W. Warren (ed.), *Urinary Tract Infections: Molecular Pathogenesis and Clinical Management.* ASM Press, Washington, D.C.
6. **Benton, J., J. C. Chawla, S. Parry, and D. J. Stickler.** 1992. Virulence factors in *Escherichia coli* from urinary tract infections in patients with spinal injuries. *J. Hosp. Infect.* **22:**117–127.
7. **Bultitude, M. J., and S. Eykyn.** 1973. The relationship between the urethral flora and urinary infection in the catheterized male. *Br. J. Urol.* **45:**678–683.
8. **Clapham, L., R. J. C. McLean, J. C. Nickel, J. Downey, and W. J. Costerton.** 1990. The influence of bacteria on struvite crystal habit and its importance in urinary stone formation. *J. Crystal Growth* **104:**475–484.
9. **Clayton, C. L., J. C. Chawla, and D. J. Stickler.** 1982. Some observations on urinary tract infections in patients undergoing long-term bladder catheterization. *J. Hosp. Infect.* **3:** 39–47.
10. **Costerton, J. W., Z. Lewandowski, D. E. Caldwell, D. R. Korber, and H. M. Lappin-Scott.** 1995. Microbial biofilms. *Annu. Rev. Microbiol.* **49:**711–745.
11. **Cox, A. J., and D. W. L. Hukins.** 1989. Morphology of mineral deposits on encrusted urinary catheters investigated by scanning electron microscopy. *J. Urol.* **142:**1347–1350.

12. **Cox, A. J., D. W. L. Hukins, and T. M. Sutton.** 1989. Infection of catheterized patients: bacterial colonization of encrusted Foley catheters shown by scanning electron microscopy. *Urol. Res.* **17:**349–352.
13. **Daifuku, R., and W. E. Stamm.** 1984. Association of rectal and urethral colonization with urinary tract infection in patients with indwelling catheters. *JAMA* **252:**2028–2030.
14. **Darouiche, R. O., and R. A. Hull.** 2000. Bacterial interference for prevention of urinary tract infection. *J. Spinal Cord Med.* **23:**136–141.
15. **Davies, D. G., and G. G. Geesey.** 1995. Regulation of the alginate biosynthesis gene *algC* in *Pseudomonas aeruginosa* during biofilm development in continous culture. *Appl. Environ. Microbiol.* **61:**860–867.
16. **Davies, D. G., M. R. Parsek, J. P. Pearson, B. H. Iglewski, J. W. Costerton, and E. P. Greenberg.** 1998. The involvement of cell-to-cell signals in the development of a bacterial biofilm. *Science* **280:**295–298.
17. **Davies, J. M.** 1977. Two aspects of the epidemiology of bladder cancer in England and Wales. *Proc. R. Soc. Med.* **70:** 411–413.
18. **Donlan, R. M.** 2002. Biofilms: microbial life on surfaces. *Emerg. Infect. Dis.* **8:**881–890.
19. **Donlan, R. M., and J. W. Costerton.** 2002. Biofilms: survival mechanisms of clinically relevant microorganisms. *Clin. Microbiol. Rev.* **15:**167–193.
20. **Dumanski, A. J. H. Hedelin, A. Edin-Lijegren, D. Beauchemin, and R. J. C. McLean.** 1994. Unique ability of *Proteus mirabilis* capsule to enhance mineral growth in infectious urinary calculi. *Infect. Immun.* **62:**2998–3003.
21. **Dunne, W. M.** 2002. Bacterial adhesion: seen any good biofilms lately? *Clin. Microbiol. Rev.* **15:**155–166.
22. **El-Masri, W. S., and G. Fellows.** 1981. Bladder cancer after spinal cord injury. *Paraplegia* **19:**265–270.
23. **Farsi, H. M., H. A. Mosli, M. F. Al-Zemaity, A. A. Bahnassy, and M. Alvarez.** 1995. Bacteriuria and colonization of double-pigtail ureteral stents: long-term experience with 237 patients. *J. Endourol.* **9:**469–472.
24. **Fawcett, C., J. C. Chawla, A. Quoraishi, and D. J. Stickler.** 1986. A study of the skin flora of spinal cord injured patients. *J. Hosp. Infect.* **8:**149–158.
25. **Fry, J. C.** 1993. *Biological Data Analysis: a Practical Approach.* IRL Press, Oxford, United Kingdom.
26. **Fuqua, W. C., S. C. Winnans, and E. P. Greenberg.** 1996. Census and consensus in bacterial ecosystems: the LuxR-LuxI family of quorum sensing transcriptional regulators. *Annu. Rev. Microbiol.* **50:**727–751.
27. **Galloway, A., H. T. Green, J. J. Windsor, K. K. Mennon, and B. D. Gardner.** 1986. Serial concentrations of C-reactive protein as an indicator of urinary tract infection in patients with spinal injury. *J. Clin. Pathol.* **39:**851–855.
28. **Ganderton, L., J. C. Chawla, C. Winters, J. Wimpenny, and D. J. Stickler.** 1992. Scanning electron microscopy of bacterial biofilms on indwelling bladder catheters. *Eur. J. Clin. Microbiol. Infect. Dis.* **11:**789–797.
29. **Garibaldi, R. A., J. P. Burke, M. R. Britt, W. A. Miller, and C. B. Smith.** 1974. Factors predisposing to bacteriuria during indwelling urethral catheterization. *N. Engl. J. Med.* **291:** 215–219.
30. **Garibaldi, R. A., J. P. Burke, M. R. Britt, W. A. Miller, and C. B. Smith.** 1980. Meatal colonization and catheter-associated bacteriuria. *N. Engl. J. Med.* **303:**316–318.
31. **Getliffe, K., and A. Mulhall.** 1991. The encrustation of indwelling catheters. *Br. J. Urol.* **67:**337–341.
32. **Gillespie, W. A., J. E. Jones, C. Teasdale, R. A. Simpson, L. Nashef, and D. A. Speller.** 1983. Does the addition of disinfectant to urine drainage bags prevent infection in catheterized patients? *Lancet* **i:**1037–1039.
33. **Handley, P. S., A. H. Rickard, N. J. High, and S. A. Leach.** 2001. Coaggregation—is it a universal phenomenon? p. 1–10. *In* P. Gilbert, D. Allison, M. Brading, J. Verran, and J. Walker (ed.), *Biofilm Community Interactions: Chance or Necessity?* Bioline Press, Cardiff, United Kingdom.
34. **Hayley, R. W., D. H. Culver, J. W. White, W. M. Morgan, and T. G. Emori.** 1985. The nationwide nosocomial infection rate. *Am. J. Epidemiol.* **121:**159–167.
35. **Hentzer, M., L. Eberl, J. Nielsen, and M. Givskov.** 2003. Quorum sensing: a novel target for the treatment of biofilm infections. *Biodrugs* **17:**241–250.
36. **Hentzer, M., K. Riedel, T. B. Rasmussen, A. Heydorn, J. B. Andersen, M. R. Parsek, S. A. Rice, L. Eberl, S. Molin, N. Høiby, S. Kjelleberg, and M. Givskov.** 2002. Inhibition of quorum sensing in *Pseudomonas aeruginosa* biofilm bacteria by a halogenated furanone compound. *Microbiology* **148:** 87–102.
37. **Hill, M. J., and G. Hawksworth.** 1972. Bacterial production of nitrosamines *in vitro* and *in vivo*, p. 116–121. *In* P. Bogouski, R. Preussmann, and E. A. Walker (ed.), *N-Nitroso Compounds: Analysis and Formation.* IARC Publications, Lyons, France.
38. **Ishihara, K., and Y. Iwasaki.** 1998. Reduced protein adsorption on novel phospholipids polymers. *J. Biomater. Appl.* **13:** 111–127.
39. **Jewes, L. A., W. A. Gillespie, A. Leadbetter, B. Myers, R. A. Simpson, M. J. Stower, and A. C. Viant.** 1988. Bacteriuria and bacteremia in patients with long-term indwelling catheters—a domiciliary study. *J. Med. Microbiol.* **26:**61–65.
40. **Kapoor, R., E. N. Liatsikos, and G. Badlani.** 2000. Endoprostatic stents for management of benign prostatic hyperplasia. *Curr. Opin. Urol.* **10:**19–22.
41. **Kass, E. H., and L. J. Schneiderman.** 1957. Entry of bacteria into the urinary tracts of patients with inlying catheters. *N. Engl. J. Med.* **256:**566–567.
42. **Kaufman, J. M., B. Fam, S. C. Jacobs, F. Gabilondo, S. Yalla, J. P. Kane, and A. B. Rossier.** 1977. Bladder cancer and squamous metaplasia in spinal cord injured patients. *J. Urol.* **118:** 967–971.
43. **Kjelleberg, S., and S. Molin.** 2002. Is there a role for quorum sensing signals in bacterial biofilms? *Curr. Opin. Microbiol.* **5:**254–258.
44. **Klausen, M. A., A. Heydorn, P. Ragas, L. Lambertsen, A., Aaes-Jorgensen, S. Molin, and T. Tolker-Nielsen.** 2003. Biofilm formation by *Pseudomonas aeruginosa* wild-type, flagella and type IV pili mutants. *Mol. Microbiol.* **48:**1511–1524.
45. **Klausen, M., A. Aaes-Jorgensen, S. Molin, and T. Tolker-Nielsen.** 2003. Involvement of bacterial migration in the development of complex multicellular structures in *Pseudomonas aeruginosa* biofilms. *Mol. Microbiol.* **50:**61–68.
46. **Kohler-Ockmore, J., and R. C. L. Feneley.** 1996. Long-term catheterization of the bladder: prevalence and morbidity. *Br. J. Urol.* **77:**347–351.
47. **Kunin, C. M.** 1987. *Detection, Prevention and Management of Urinary Tract Infections,* 4th ed., p. 245–298. Lea & Febiger, Philadelphia, Pa.
48. **Kunin, C. M.** 1997. *Urinary Tract Infections: Detection, Prevention and Management,* 5th ed., p. 226–278. The Williams & Wilkins, Co., Baltimore, Md.
49. **Kunin, C. M., Q. F. Chin, and S. Chambers.** 1987. Formation of encrustations on indwelling catheters in the elderly: a comparison of blockers and non-blockers. *J. Urol.* **138:** 899–902.

50. **Kunin, C. M., S. Douthitt, J. Dancing, J. Anderson, and M. Moeschberger.** 1992. The association between the use of urinary catheters and morbidity and mortality among elderly patients in nursing homes. *Am. J. Epidemiol.* **135:** 291–301.
51. **Laaksovirta, S., T. Valimaa, T. Isotalo, P. Tormala, M, Talja, and T. L. Tammela.** 2003. Encrustation and strength retention properties of the self-expandable, biodegradable, self-reinforced L-lactide-glycolic acid copolymer 80:20 spiral urethral stent in vitro. *J. Urol.* **170:**468–471.
52. **Lawrence, J. R., D. R. Korber, B. D. Hoyle, J. W. Costerton, and D. E. Caldwell.** 1991. Optical sectioning of microbial biofilms. *J. Bacteriol.* **173:**6558–6567.
53. **Lejeune, P.** 2003. Contamination of abiotic surfaces: what a colonizing bacterium sees and how to blur it. *Trends Microbiol.* **11:**179–184.
54. **Li, X., and H. L. T. Mobley.** 2002. Vaccines for *Proteus mirabilis* in urinary tract infection. *Int. J. Antimicrob. Agents* **19:**461–465.
55. **Maki, D. G., and P. A. Tambyah.** 2001. Engineering out the risk of infection with urinary catheters. *Emerg. Infect. Dis.* **7:** 342–347.
56. **McLean, R. J. C., J. R. Lawrence, D. K. Korber, and D. E. Caldwell.** 1991. *Proteus mirabilis* biofilm protection against struvite crystal dissolution and its implications for struvite urolithiasis. *J. Urol.* **146:**1138–1142.
57. **McLean, R. J. C., M. Whiteley, D.J. Stickler, and W. C. Fuqua.** 1997. Evidence of autoinducer activity in naturally occurring biofilms. *FEMS Microbiol. Lett.* **154:**259–263.
58. **Meers, P. D., G. A. J. Ayliffe, and A. M. Emmerson.** 1981. National survey of infections in hospitals 1980. Part 2. Urinary tract infection. *J. Hosp. Infect.* **2**(Suppl.):23–28.
59. **Mobley, H. L. T., G. R. Chippendale, J. H. Tenney, R. A. Hull, and J. W. Warren.** 1987. Expression of type 1 fimbrae may be required for persistence of *Escherichia coli* in the catheterized urinary tract. *J. Clin. Microbiol.* **25:**2253–2257.
60. **Mobley, H. L. T., and J. W. Warren.** 1987. Urease-positive bacteriuria and obstruction of long-term catheters. *J. Clin. Microbiol.* **25:**2216–2217.
61. **Morris, N. S., D. J. Stickler, and R. J. C. McLean.** 1999. The development of bacterial biofilms on urethral catheters. *World J. Urol.* **17:**345–350.
62. **Morris, N. S., D. J. Stickler, and C. Winters.** 1997. Which indwelling urethral catheters resist encrustation by *Proteus mirabilis* biofilms? *Br. J. Urol.* **80:**58–63.
63. **Nickel, J. C., J. W. Costerton, R. J. C. McLean, and M. Olsen.** 1994. Bacterial biofilms: influence on the pathogenesis, diagnosis and treatment of urinary tract infections. *J. Antimicrob. Chemother.* **33**(Suppl. A):31–41.
64. **Nickel, J. C., J. A. Downey, and W. J. Costerton.** 1989. Ultrastructural study of microbiological colonization of urinary catheters. *Urology* **34:**284–291.
65. **Nickel, J. C., S. K. Grant, and W. J. Costerton.** 1985. Catheter-associated bacteriuria: an experimental study. *Urology* **26:**369–375.
66. **Nielsen, K. K., P. Klarskov, J. Nordling, J. T. Andersen, and H. H. Holm.** 1990. The intraprostaticspiral. New treatment for urinary retention. *Br. J. Urol.* **65:**500–503.
67. **Ohkawa, M., T. Sugata, M. Sawaki, T. Nakashima, H. Fuse, and H. Hisazumi.** 1990. Bacterial and crystal adherence to the surfaces of indwelling urethral catheters. *J. Urol.* **143:** 717–721.
68. **O'Toole, G. A., and R. Kolter.** 1998. Flagellar and twitching motility are necessary for *Pseudomonas aeruginosa* biofilm development. *Mol. Microbiol.* **30:**295–305.
69. **Paick, S. H., H. K. Park, S. Oh, and H. H. Kim.** 2003. Characteristics of bacterial colonization and urinary tract infection after indwelling of double-J ureteral stents. *Urology* **62:** 214–217.
70. **Platt, R., B. F. Polk, B. Murdoch, and B. Rosner.** 1982. Mortality associated with nosocomial urinary tract infection. *N. Engl. J. Med.* **307:**637–642.
71. **Pratt, L. A., and R. Kolter.** 1998. Genetic analysis of *Escherichia coli* biofilm formation: roles of flagella, motility, chemotaxis and type 1 pili. *Mol. Microbiol.* **30:**285–293.
72. **Raz, R., D. Schiller, and L. E. Nicolle.** 2000. Chronic indwelling catheter replacement before antimicrobial therapy for symptomatic urinary tract infection. *J. Urol.* **164:**1254–1253.
73. **Reid, G., J. D. Denstedt, Y. S. Kang, D. Lam, and C. Nause.** 1992. Microbial adhesion and biofilm formation on ureteral stents in vitro and in vivo. *J. Urol.* **148:**1592–1594.
74. **Rickard, A. H., P. Gilbert, N. J. High, P. E. Kolenbrander, and P. S. Handley.** 2003. Bacterial coaggregation: an integral process in the development of multi-species biofilms. *Trends Microbiol.* **11:**94–100.
75. **Riedl, C. R., E. Plas, W. A. Hubner, H. Zimmerl, W. Ulrich, and H. Pfluger.** 1999. Bacterial colonization of ureteral stents. *Eur. Urol.* **36:**53–59.
76. **Riedl, C. R., M. Witkowski, E. Plas, and H. Pfluger.** 2002. Heparin coating reduces encrustation on ureteral stents: a preliminary report. *Int. J. Antimicrob. Agents* **19:**507–510.
77. **Rosenkilde, P., J. F. Pedersen, and H. H. Meyerhoff.** 1991. Late complications of Prostakath treatment for benign prostatic hypertrophy. *Br. J. Urol.* **68:**387–389.
78. **Ruggieri, M. R., P. M. Hanno, and R. M. Levin.** 1987. Reduction of bacterial adherence to catheter surface with heparin. *J. Urol.* **138:**423–426.
79. **Sabbuba, N. A., E. Mahenthiralingam, and D. J. Stickler.** 2003. Molecular epidemiology of *Proteus mirabilis* infections of the catheterized urinary tract. *J. Clin. Microbiol.* **41:**4961–4965.
80. **Sabbuba, N. A., G. Hughes, and D. J. Stickler.** 2002. The migration of *Proteus mirabilis* and other urinary tract pathogens over Foley catheters. *Br. J. Urol. Int.* **89:**55–60.
81. **Saint, S., and C. E. Chenoweth.** 2003. Biofilm and catheter-associated urinary tract infections. *Infect. Dis. Clin. North Am.* **17:**411–432.
82. **Santin, M., A. Motta, S. P. Denyer, and M. Cannas.** 1999. Effect of urine conditioning film on ureteral stent encrustation and characterization of its protein composition. *Biomaterials* **20:**1245–1251.
83. **Sauer, K., A. K. Camper, G. D. Ehrlich, J. W. Costerton, and D. G. Davies.** 2002. *Pseudomonas aeruginosa* displays multiple phenotypes during development as a biofilm. *J. Bacteriol.* **184:**1140–1154.
84. **Schaeffer, A. J., and J. Chmiel.** 1983. Urethral meatal colonization in the pathogenesis of catheter-associated bacteriuria. *J. Urol.* **130:**1096–1099.
85. **Schierholz, J. M., D. P. Konig, J. Beuth, and G. Pulverer.** 1999. The myth of encrustation inhibiting materials. *J. Hosp. Infect.* **42:**162–163.
86. **Stamm, W. E.** 1991. Catheter-associated urinary tract infections: pathogenesis and prevention. *Am. J. Med.* **91**(Suppl. 3B):65S–72S.
87. **Stark, R. P., and D. G. Maki.** 1984. Bacteriuria in the catheterized patient: what quantitative level of bacteriuria is relevant? *N. Engl. J. Med.* **311:**560–564.
88. **Stickler, D. J.** 2002. Susceptibility of antibiotic-resistant gram-negative bacteria to biocides: a perspective from the

study of catheter biofilms. *J. Appl. Microbiol. Symp. Suppl.* **92:**163S–170S.

88a. **Stickler, D. J.** 1996. Biofilms, catheters and urinary tract infections. *Urol. Update* **5:**1–8.

89. **Stickler, D. J., A. Evans, N. S. Morris, and G. Hughes.** 2002. Strategies for the control of catheter encrustation. *Int. J. Antimicrob. Agents* **19:**499–506.

90. **Stickler, D. J., L. Ganderton, J. B. King, J. Nettleton, and C. Winters.** 1993. *Proteus mirabilis* biofilms and the encrustation of urethral catheters. *Urol. Res.* **21:**407–411.

91. **Stickler, D. J., and G. Hughes.** 1999. Ability of *Proteus mirabilis* to swarm over urethral catheters. *Eur. J. Clin. Microbiol. Infect. Dis.* **18:**206–208.

92. **Stickler, D. J., G. L. Jones, and A. D. Russell.** 2003. Control of encrustation and blockage of Foley catheters. *Lancet* **361:** 1435–1437.

93. **Stickler, D. J., J. B. King, C. Winters, and S. L. Morris.** 1993. Blockage of urethral catheters by bacterial biofilms. *J. Infect.* **27:**133–135.

94. **Stickler, D. J., N. S. Morris, R. J. C. McLean, and C. Fuqua.** 1998. Biofilms on indwelling catheters produce quorum-sensing signal molecules in situ and in vitro. *Appl. Environ. Microbiol.* **64:**3486–3490.

95. **Stickler, D. J., N. Morris, M. Moreno, and N. A. Sabbuba.** 1998. Studies on the formation of crystalline bacterial biofilms on urethral catheters. *Eur. J. Clin. Microbiol. Infect. Dis.* **17:**1–4.

96. **Stickler, D. J., and C. Winters.** 1994. Biofilms and urethral catheters, p. 97–104. *In* J. Wimpenny, W. Nichols, D. Stickler, and H. Lappin-Scott (ed.), *Bacterial Biofilms and Their Control in Medicine and Industry*. Bioline Press, Cardiff, United Kingdom.

97. **Stickler, D. J., R. Young, G. Jones, N. A. Sabbuba, and N. S. Morris.** 2003. Why are Foley catheters so vulnerable to encrustation and blockage by crystalline bacterial biofilm? *Urol. Res.* **31:**306–311.

98. **Stickler, D. J., and J. Zimakoff.** 1994. Complications of urinary tract infections associated with devices used for long-term bladder management. *J. Hosp. Infect.* **28:**177–194.

99. **Stoodley, P., K. Sauer, D. G. Davies, and J. W. Costerton.** 2002. Biofilms as complex differentiated communities. *Annu. Rev. Microbiol.* **56:**187–209.

100. **Tambyah, P. A., K. T. Halvorson, and D. G. Maki.** 1999. A prospective study of pathogenesis of catheter-associated urinary tract infections. *Mayo Clin. Proc.* **74:**131–136.

101. **Tenney, J. H., and J. W. Warren.** 1988. Bacteriuria in women with long-term catheters: paired comparison of indwelling and replacement catheters. *J. Infect. Dis.* **157:**199–202.

102. **Tricker, A. R., D. J. Stickler, J. C. Chawla, and R. Preussmann.** 1991. Increased urinary nitrosamine excretion in paraplegic patients. *Carcinogenesis* **12:**934–936.

103. **Tunney, M. M., and S. P. Gorman.** 2002. Evaluation of a poly(vinly pyrollidone)-coated biomaterial for urological use. *Biomaterials* **23:**4601–4608.

104. **Van Loosdrecht, M. C. M., C. Picioreanu, and J. J. Heinen.** 1997. A more unifying hypothesis for biofilm structures. *FEMS Microbiol. Ecol.* **24:**181–183.

105. **Warren, J. W.** 2001. Catheter-associated urinary tract infections. *Int. J. Antimicrob. Agents* **17:**299–303.

106. **Warren, J. W., W. C. Anthony, J. M. Hoopes, and H. L. Muncie.** 1982. Cephalexin for susceptible bacteriuria in afebrile, long-term catheterized patients. *JAMA* **248:**454–458.

107. **Warren, J. W., L. Steinberg, R. Hebel, and J. H. Tenney.** 1989. The prevalence of urethral catheterization in Maryland nursing homes. *Arch. Intern. Med.* **149:**1535–1537.

108. **Warren, J. W., J. H. Tenney, J. M. Hoopes, H. L. Muncie, and W. C. Anthony.** 1982. A prospective microbiological study of bacteriuria in patients with chronic indwelling urethral catheters. *J. Infect. Dis.* **146:**719–723.

109. **Wassall, M. A., M. Santin, G. Peluso, and S. P. Denyer.** 1998. Possible role of alpha-1-microglobulin in mediating bacterial attachment to model surfaces. *J. Biomed. Mater. Res.* **40:**365–370.

110. **Wazait, H. D., H. R. H. Patel, V. Veer, M. Kelsey, J. H. P. Van der Meulen, R. A. Miller, and M. Emberton.** 2003. Catheter-associated urinary tract infections: prevalence of uropathogens and patterns of antimicrobial resistance in a UK hospital (1996–2001). *Br. J. Urol. Int.* **91:**806–809.

111. **Whiteley, M., M. G. Bangera, R. E. Bumgarner, M. R. Parsek, G. M. Teitzel, S. Lory, and E. P. Greenberg.** 2001. Gene expression in *Pseudomonas aeruginosa* biofilms. *Nature* **413:** 860–864.

112. **Winters, C., D. J. Stickler, N. S. Howe, T. J. Williams, N. Wilkinson, and C. J. Buckley.** 1995. Some observations on the structure of encrusting biofilms of *Proteus mirabilis* on urethral catheters. *Cells Mat.* **5:**245–253.

113. **Wimpenny, J. W. T., and R. Colasanti.** 1997. A unifying hypothesis for the structure of microbial biofilms based on cellular automaton models. *FEMS Microbiol. Ecol.* **22:**1–16.

114. **Zimakoff, J., B. Pontoppidan, S. O. Larsen, and D. J. Stickler.** 1993. Management of urinary bladder function in Danish hospitals, nursing homes and home care. *J. Hosp. Infect.* **24:** 183–199.

Colonization of Mucosal Surfaces
Edited by James P. Nataro et al.

Chapter 28

Colonization of the Vaginal and Urethral Mucosa

Gregor Reid

This chapter attempts to answer six key questions as a means of examining current knowledge of the urogenital bacterial microbiota in women and exploring methods to manipulate the organisms to restore and maintain health.

The urogenital tract includes the perineal skin close to the anus, the vulva, vagina, cervix, uterus, urethra, bladder, ureters, and kidneys. For the most part, the uterus, bladder, ureters, and kidneys are sterile while the other regions are colonized by microorganisms. The reasons why the proximal urethra and upper cervix remain free of organisms is not well understood and is worthy of further investigation, given the importance of infections caused by organisms that break through this barrier.

The importance of the urogenital tract in the health of women cannot be understated. Given its proximity to potential pathogens emerging from the rectum, exposure to sexually transmitted organisms, and the extent to which hormones, tampons, douches, and the birthing process can influence the microbial content, it is remarkable that more infections do not occur.

Nevertheless, an estimated 1 billion urogenital infections do occur each year. The risk of 1 of the 3 billion females in the world getting an infection during 1 week of any given year is around 1:156 on average per person. This is extremely high, yet few studies are funded in this area of research and little government or media attention is paid to improving the situation. This is partly because most urogenital infections are not lethal and because no lobby group has undertaken to make this a political issue. Nevertheless, there are two serious outcomes which can be lethal: infections leading to preterm delivery, and acquisition of human immunodeficiency virus (HIV). These and other topics are discussed in this chapter.

WHEN DOES UROGENITAL COLONIZATION BEGIN?

The first few hours and days of life provide an opportunity for bacteria to colonize the urogenital tract. For 75% of Americans, the primary colonizers come from the vagina and rectum of the mother at birth. For the others, born by caesarian section, the organisms enter through handling and feeding. In healthy mothers, the vaginal microflora consists of lactobacilli, and while these organisms are early entrants into the newborn gut, they are not necessarily the most important in terms of conferring health benefits or in dominating the intestinal and urogenital mucosae. The first impact of bacterial entry into the newborn could be from organisms such as *Bacteroides thetaiotaomicron,* found to stimulate angiogenesis in the intestines of mice (177). These bacteria, along with lactobacilli, bifidobacteria, and other primary colonizers at birth, cause antigen-presenting cells to respond in a manner that could be important for the maturation of the immune system (87).

The one study that effectively monitored colonization of two newborns from birth showed that the primary colonizers in the first week of life in vaginally born, breast-fed babies are *Escherichia coli, Clostridium, Enterobacter, Streptococcus,* and *Enterococcus,* with bifidobacteria not appearing until the second week (46). The absence of lactobacilli and the later emergence of bifidobacteria is surprising, given the exposure of the newborn to lactobacilli during birthing, but the findings from only two babies may not be typical of a wider population. Nevertheless, that study illustrates the rapidity and multispecies nature of the gut flora (Color Plate 10). In terms of urogenital colonization, most studies have focused on the transfer of pathogens, such as group B streptococci, from mother to newborn. No molecular

Gregor Reid • Canadian Research and Development Centre for Probiotics, Lawson Health Research Institute, and Departments of Microbiology & Immunology and of Surgery, University of Western Ontario, London, Ontario, Canada N6A 4V2.

studies have plotted the acquisition of the urogenital microbiota at birth, but early culture studies showed that *E. coli,* enterococci, and streptococci were the dominant primary colonizers (12).

With the cloning of human and bacterial genomes, it will soon be possible to study selective factors involved in primary and secondary colonization of the host. It is known that there are selective proteins in the oral cavity that act as receptors for certain organisms, and so the host is choosing the strains it "wants" as colonizers (60). If similar selectivity occurs in the gut and urogenital tract and if receptors are discovered for organisms with limited or no pathogenic potential, it could be possible to "program" the host to receive a more beneficial microbiota at birth. This would be achieved by administration of probiotic organisms ("probiotic" is defined as "live microorganisms which when administered in adequate amounts confer a health benefit on the host") (52), possibly accompanied by prebiotic nutrients such as fructose (145) ("prebiotics" are defined as "non-digestible substances that provide a beneficial physiological effect on the host by selectively stimulating the favorable growth or activity of a limited number of indigenous bacteria") (139) that encourage the growth of bifidobacteria, lactobacilli, or other beneficial organisms. The programming process will not be easy, since not all administered strains may be able to colonize the host in the long term. For example, even although *Lactobacillus rhamnosus* GG was administered to expectant mothers and newborns in a study performed in Finland (86), the organism did not colonize the subjects in the long term. Likewise, in a New Zealand study, lactobacilli found in the maternal vagina were not detected in the newborn gut (161).

The programming concept will have to be considered in ethical terms, but the benefits could far outnumber the risks, especially if microbiota constituents reduce the risk of cancers (107, 114), diabetes (142), arthritic diseases (159), infections and inflammation (68, 80, 115, 116, 120, 158), renal calculi (21, 90), caries (101), cardiovascular disease (1, 100), and other serious ailments.

Once a child is born, intestinal infections are common (21 million to 37 million diarrheal disease episodes in 16.5 million U.S. children each year) and deaths can occur (around 200 annually in the United States and Canada). While these cases are mostly due to ingestion of contaminated food or water, they pose the question of why the indigenous microbiota is not protective. The flip side of this is that, given the frequency of pathogenic colonization in the urogenital tract, the fact that infections do not occur every day or week in every woman is something of a miracle. The reason that the rate of urinary tract infection (UTI) is so low in girls (1%) and boys (2.5%) (175) is unclear and may involve innate immunity or host antibacterial peptides that have not yet been identified. The finding that the flora of healthy girls is dominated by anaerobic gram-positive cocci and gram-positive rods suggests that there could be a role for bacterial competition in protecting the host.

Later in life, the higher incidence of group B *Streptococcus* colonization (162) and bacterial vaginosis (BV) among African-American women (102, 141) implies that racial differences exist which influence bacterial colonization of the vagina. Having stated that, similar *Lactobacillus* species have been recovered from women in four continents (S. L. Hillier, personal communication), and so the racial influences may be host derived or due to cultural practices such as douching or to socioeconomic or other similar factors. The interesting part about the finding by Hillier (now confirmed by our group) is that even when diets and environmental living conditions are dissimilar, the same species of bacteria are present in women, perhaps reflecting the fact that bacterial colonization of the planet occurred long before humans arrived. For now, the differences between races in terms of which organisms colonize and how receptor sites differ remain to be investigated effectively.

WHAT IS THE BACTERIAL COMPOSITION OF THE VAGINA, CERVIX, AND URETHRA?

To obtain a better understanding of the composition of the urogenital microbiota, new molecular-biology techniques have been utilized. The combined use of chromotyping, ribotyping, and plasmid typing was found to be more accurate in identifying urogenital lactobacilli than were the biochemical characteristics (180) that were used up to the 1990s (61). Twelve different plasmid profiles were found among 100 vaginal isolates (136), but this method was not sufficiently specific to differentiate species. Chromosome typing (restriction endonuclease fingerprinting of chromosomal DNA) is more specific and reproducible than plasmid content analysis; however, it is not easy to differentiate electrophoretic patterns with up to 100 bands. Ribotyping combines Southern hybridization of chromosomal DNA fingerprints with the use of *E. coli* rRNA probes, thereby discriminating between species and individual strains of lactobacilli. The rRNA genes are separated and identified, making it possible to delineate species based on differences in the restriction fragment length polymorphisms of the rRNA genes. Specific regions of the rRNA genes have remained well conserved because of their functional importance, thus allowing the detection of a broad range of bacteria with 16S and 23S rRNA of *E. coli* as probes.

A random-primed-DNA-labeling kit has been used to prepare whole-chromosome probes for 20 *Lactobacillus* species, and a resultant analysis of vaginal samples from 215 healthy women showed *L. crispatus* to be the most commonly isolated species (5). However, in this method the bacteria are first cultured and then identified. In a more recent study, *L. iners,* an organisms that does not grow on traditionally used *Lactobacillus* media (Rogosa's and MRS media), was found to be the predominant colonizer of the vagina, as determined by PCR-denaturing gradient gel electrophoresis and sequencing of the V2–V3 region of the 16S rRNA gene (17, 17a).

No recent studies have examined the microbiota of the cervix or urethra, with the exception of a recent restriction fragment length polymorphism analysis of the amplicons cloned from the mixtures of PCR products, which showed that the healthy male urethra was colonized by 71 clones, including *Pseudomonas, Streptococcus, Burkholderia, Ralstonia, Sphingomonas,* and various unidentified organisms (140). Earlier studies of the distal urethra showed *Corynebacterium* to be the most common in girls, lactobacilli to predominate after puberty, and *Bacteroides melaninogenicus* to be dominant after menopause (93). The mean number of species per physiological age group was 6.5 (premenarcheal), 7.7 (reproductive), and 10.3 (postmenopausal), with, surprisingly, no aerobic gram-negative rods isolated from any of the premenarcheal or reproductive-age subjects. A repeat of this study using molecular typing and large sample sizes would probably show a more diverse microbiota.

An approximately 600-bp region of the chaperonin 60 gene (*cpn60*), amplified by PCR with a single pair of degenerate primers, has utility as a universal target for bacterial identification (62). In this method *cpn60* gene segments from complex DNA templates extracted from complex microbial communities are amplified, cloned, and sequenced, which helps to identify the organisms from which they were derived (71a, 72). Combined with high-throughput sequencing, this genomic approach, though a labor-intensive one, enables the characterization of the structures of even complex microbial communities. To date, a database of *cpn60* gene sequences (*cpnDB*) from reference organisms covering a wide phylogenetic range has been generated, with 3,000 distinct sequence entries. Analysis of 30,000 cloned *cpn60* gene sequences from libraries representing the microflora from various sources, including the human vagina, has been completed, but the findings have yet to be published. These and other molecular biology-based methods will soon provide an accurate microbial map of the urogenital tracts of males and females, thereby permitting studies to be undertaken that will determine the factors most strongly associated with health.

WHAT FACTORS INFLUENCE THE FLORA?

It has long been recognized that hormones influence the female urogenital microbiota. This includes in vitro data showing that receptivity to uropathogenic *E. coli* (121, 143) and lactobacilli (27) follows the pattern of estrogen over the menstrual cycle, peaking at midcycle, while *E. coli* adhesion is highest in the first trimester of pregnancy (121). The relative absence of lactobacilli in the vagina before puberty is thought to be due to low estrogen and glycogen concentrations, although relatively little is known about the microbiota at this life stage. The most compelling evidence for hormonal effects on the vaginal flora come from clinical studies by Raz and coworkers, which have shown that lactobacilli return after the start of administration of vaginal estriol (118) and, to a lesser extent, oral estrogen therapy (117). Studies with oral premarin have now confirmed these findings and indeed have shown 100% presence of lactobacilli in the vagina compared to around 30% in normal postmenopausal subjects (37). This recovery of lactobacilli correlates well with a significant reduction in susceptibility to UTI (117) and recovery from asymptomatic BV, further illustrating the protective nature of the microbiota.

Less is known about the influence of growth factors and antibacterial peptides on the microbiota. Injection of keratinocyte growth factor, a member of the fibroblast growth factor family, into neonatal mice induced long-term changes of the uterine and vaginal epithelia such that the epithelial surface area significantly increased (74). The impact of such abnormal changes on microbial colonization are not known, but it might be worthy of investigation in women who suffer chronic infections. The growth of vaginal epithelial cells appears to be enhanced by extracellular collagen gel (79), which could potentially increase receptivity for organisms, such as *Lactobacillus reuteri* (formerly *L. fermentum*) RC-14 and *Staphylococcus aureus,* which are known to colonize via collagen binding proteins (50, 71). Vitamins also affect the vaginal epithelia, as shown by the response to addition of vitamin A (topical retinyl palmitate), which corrected squamous metaplastic vaginal epithelia into a normal phenotype (11). Ascorbic acid can modulate catecholaminergic activity, increase vascular function and oxytocin release, and increase intercourse frequency (15), while a mixed nutritional supplement containing ginseng, ginkgo, damiana, L-arginine, and other multivitamins and minerals has been found to enhance sexual desire and reduce vaginal dryness (82). Insulin use appears to affect bacterial colonization and increase UTI rates in postmenopausal women (14).

Little is known about the extent to which host antimicrobial peptides affect the urogenital microbiota. The vaginal fluid and cervical mucus contain a range of antimicrobial compounds, such as calprotectin, lysozyme, β-defensin 1, cathelicidin, lactoferrin, cationic polypeptide histone (possibly released from lysed vaginal cells), and secretory leukocyte protease inhibitor, with activity against pathogens such as *E. coli* attributed to low-molecular-weight (protein-free) fractions (70, 168). One human peptide, FALL-39, which may or may not be in the vagina, has been shown to kill urinary pathogens but not to affect lactobacilli, thereby potentially being an ideal antimicrobial agent for women at risk of BV and UTI (151).

The effect of diet on microbes in the human body has not been well studied, and thus there is little or no information about how this affects the urogenital flora. In an effort to understand how nutrients might alter the proportion of microbes, studies have been carried out in vitro using various compounds such as vitamins, minerals, glycogen, and estrogen (59, 124). It was discovered that some combinations (pyridoxal, *p*-aminobenzoic acid, nicotinic acid, glycogen, and lactose) supported the growth of lactobacilli but not uropathogens, and vice versa (124). The origin of these and other nutrients includes the epithelium, bloodstream, urine, and mucus, but their concentrations over time have never been studied in relation to providing a substratum for the microbiota. Foods such as cranberry juice interfere with fimbrial adhesion of *E. coli* to uroepithelial cells (78), reduce the risk of UTI, and lower the extent of biofilm formation on the bladder surface (138). Therefore, presumably, the presence of cranberry metabolites can also affect the urethral, perineal, and vaginal microbiota. Oxygen levels and pH have been measured in the vagina, but not in terms of supporting microbial survival and growth. The diet can play a role in the resurgence of lactobacilli, as shown by experiments in which intake of *Lactobacillus* GR-1 and RC-14 in milk resulted in ascension of the organisms from the rectum to the vagina and restoration of the flora to normal (123). A suggestion for future studies would be to identify nutrients used by bacteria and yeast in the vagina and examine how these and microenvironmental changes affect microbial proliferation, hibernation, and eradication.

Biofilm studies clearly demonstrate that nutrients affect the composition of microbial communities. The biofilm itself can prioritize nutrient transport with respect to its need for mechanical pliability to resist shear stress, such as is caused by the flow of urinary micturition (10). Nutrient availability changes the structure of biofilms, for example by quickening the shift toward compact and smooth biofilms (111). In the vagina and urethra, there is an ample supply of molecules [for example NaCl, KOH, $Ca(OH)_2$, serum albumin, lactic acid, acetic acid, glycerol, urea, glucose, glycogen, and estrogen] that can act as nutrients for microbes (109, 124), but the role of these molecules in creating tissue-adherent *Lactobacillus* biofilms and loosely bound streptococcal and staphylococcal biofilms (38) remains to be determined. At any rate, these components, especially estrogen, play a significant role in producing changes to the microbiota in the vagina. However, we do not yet know how and why this environment shifts from "normal" to BV, two totally different states which confer different health or disease risk outcomes on the host (Color Plate 11).

Antibiotic therapy, depending on the drug used, can significantly disrupt the urogenital microbiotia. One study showed that the normal inhabitants did not become reestablished until 6 weeks after cessation of ampicillin therapy (122), while a more recent study showed a trend toward increased rates of BV following the use of various 10-day antibiotic regimens (132). Antibiotics such as tetracyclines, sulfonamides, ciprofloxacin, macrolides, streptomycin, clopidol, ethopabate, and nitromide, as well as persistent organic pollutants present in the food chain (113, 144), may influence the microbial population of the intestine and vagina, but to date this has not been investigated. It was once thought that an inability of antibiotics to penetrate biofilms was the reason for their failure to eradicate these microbial communities. However, a study of *Klebsiella* biofilms has shown that ciprofloxacin can diffuse into the biofilm but is not able to eradicate the bacteria completely (3), thereby suggesting that other mechanisms are responsible for treatment failure.

Clearly, sexual intercourse and oral sex affect the urogenital microbiota. In addition to transferring sexually transmitted pathogens to the site, sexual practices can increase the number of BV pathogens (92). Indeed, BV has itself been described as a sexually transmitted disease (9). The direct introduction of oral bacteria into the vagina is one way for the microbiota to alter, but environmental influences such as swimming pools can also cause changes (17). Semen can carry microbes into the vagina from the partner (30), and its alkalinity produces an environment that is no longer virucidal to organisms such as HIV.

Stress per se is also a likely participant in microbiota changes. One study with rats showed that *E. coli* could translocate to intestinal mesenteric lymph nodes during stress (103), while another suggested that psychological stress might contribute to disruption of the vaginal flora and poor pregnancy outcome (85). A further study has shown that stress causes parasympathetic activation via a mechanism involving corticotropin-releasing factor, ultimately affecting mucosal mast cells and induction of inflammation, as well as increased pathogen adhesion and depletion of lactobacilli (67).

Proteomic studies have shown that the protein composition of the urine fluctuates, with stone formers having higher concentrations of albumin (19). One study showed that Tamm-Horsfall protein, localized in the thick ascending limb of the loop of Henle and early distal convoluted tubule, can alter the conditions for bacterial colonization (150).

Little is known about microbial and host communication in the vagina, but some data are emerging and information from other sites can be extrapolated. More than 15 years ago, acyl-homoserine lactone was first discovered as a system by which bacteria mediate signaling to themselves and to other species of bacteria. Some gram-negative bacteria can make acyl-homoserine lactones, and genes encoding homologs of its receptor are present in pathogens (63). Pneumococci and some other gram-positive organisms use a small-peptide quorum-sensing signal. These messages appear to control virulence, growth, nutrient utilization, motion and other factors which affect the ability of the bacteria to colonize and infect the host. Cell-cell communication is often mediated by secreted inducer peptide pheromones. When they reach a threshold concentration in the environment, this activates a cognate membrane-localized histidine protein kinase.

One study of *L. plantarum* has shown that the determinants for pheromone binding and specificity are contained within the transmembrane domain (83). This, and other studies of lactobacilli, could be relevant to the urogenital tract. Production of bacteriocins by lactobacilli, although not yet proven to occur within urogenital biofilms in vivo, is controlled in a cell-density dependent manner by a secreted peptide-pheromone (42). Two studies have shown that lactobacilli can signal host cell responses, including mucin gene production in gut cells (91, 94). This is only the tip of the iceberg, with additional messages to the immune response, mucosal epithelia, and other host sites soon to be uncovered.

While many questions about the factors affecting the urogenital microbiota remain unanswered, new molecular-biology and nanotechnology tools offer promise that swift progress can be made. In the latter case, the ability to detect and measure nanoforces operating on the outer surfaces of a bacterium (Fig. 1) (176) and the host will make it possible to gain insight into the balance between health and dis-

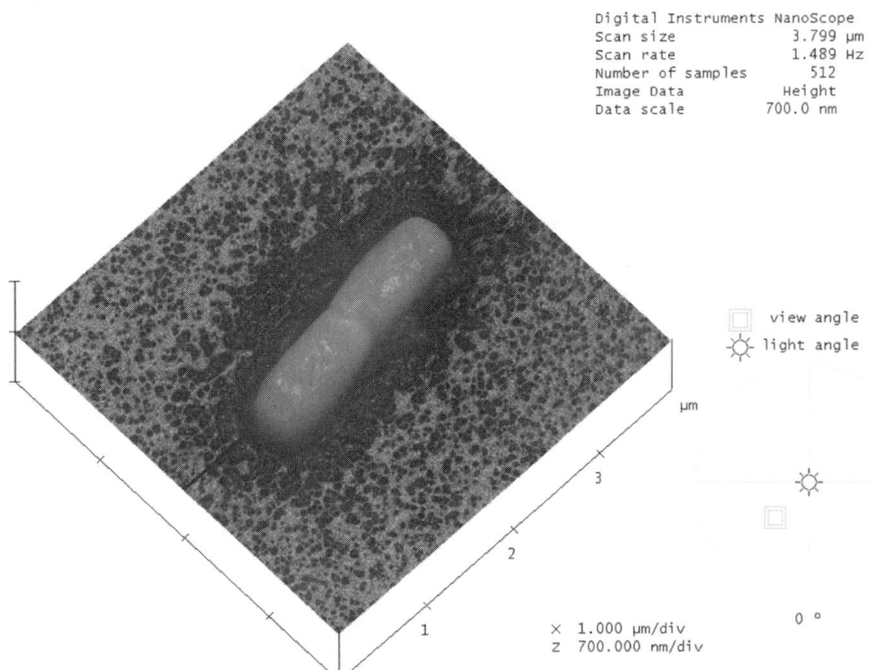

Figure 1. Atomic force microscopy image of *Lactobacillus rhamnosus* GR-1, illustrating the ability to examine nanoscale forces at the bacterial surface interface. Courtesy of Jonathan Hui and Jana Jass.

ease. This is particularly important in attempts to determine the forces required to detach pathogens and retain commensals.

HOW DOES THE BODY REACT TO THE UROGENITAL MICROBIOTA?

The host response to the presence of indigenous vaginal microbes is not well understood. There is considerable knowledge about how sexually transmitted disease pathogens such as gonococci, herpes simplex virus, and HIV interact with the host, but the focus of this chapter is on bacterial and yeast populations on the vaginal and urethral mucosa. While fewer studies have been performed to investigate nonsexually transmitted pathogens, there is still some understanding of the host response to *Candida albicans*, uropathogenic *E. coli*, and BV agents.

There may be immunoregulation and tolerance responses to members of the vaginal microbiota. This was suggested by a study in which commercial *Candida* skin test antigen introduced into the vagina did not invoke local immune stimulation, including changes in Th and proinflammatory cytokines, immunoglobulin E (IgE), histamine, and prostaglandin, despite a natural modulation of vaginal cytokines over the course of the menstrual cycle (48). In contrast, a mannoprotein extract and secreted aspartyl proteinases from *Candida albicans* administered vaginally or nasally induced anti-mannoprotein extract and anti-secreted aspartyl proteinases vaginal antibodies and conferred a high degree of protection against vaginitis (34), thereby suggesting an intolerant capacity. Other studies confirm this immunostimulatory potential against *Candida* (23, 97, 112) and demonstrate the complexity of the reactions. Indeed, proinflammatory cytokines, such as interleukin-1α (IL-1α) and tumor necrosis factor alpha but not IL-6, can be produced in the vagina in response to *Candida*, while Th1-type (IL-12 and gamma interferon) and Th2-type immunoregulatory (IL-10 and transforming growth factor β) cytokines and the chemokines monocyte chemoattractant protein 1 and IL-8 are produced in low to undetectable concentrations (153).

Given that *Candida* colonizes up to 60% of women deemed to have a healthy vaginal mucosa, more studies are needed to understand which hosts react in an inflammatory and symptomatic manner and to determine the triggers for this response (numbers of yeast cells present, hormonal changes, copresence of certain bacteria, etc.). One study with rats showed that an oral habitant, *Streptococcus gordonii,* engineered to secrete or display a microbicidal single-chain antibody (H6) could colonize the rat vagina and treat an experimental candidiasis (8). Two problems arise in applying this finding, namely, the ethical issue of administering recombinant bacteria and the fact that although oral bacteria can be found in the vagina, *S. gordonii* would certainly not be regarded as a normal commensal inhabitant.

Elevated production of IL-6 has been found in the bladder (156) and vagina in response to infection by *E. coli* and a range of anaerobic and aerobic (39) BV pathogens, respectively. The innate immune response to uropathogenic *E. coli* has been linked to lipopolysaccharide (LPS) and type 1, S, and P fimbriae (22, 76, 89, 152). The LPS recognition by the bladder epithelial mucosa is critical, with the primary receptor being Toll-like receptor 4 (TLR4) (146). The binding of P-fimbriated *E. coli* to uroepithelial cells causes the release of ceramide and activation of the ceramide-signaling pathway, whereby agonists such as tumor necrosis factor alpha and IL-1β release ceramide from sphingomyelin by the activation of endogenous sphingomyelinases and hydrolysis of sphingomyelin, resulting in an IL-6 response in the epithelial cells (69). When the uropathogens ascend into the kidneys, infection and inflammation can be induced by various factors, the most recently identified being *E. coli* induction of a constant, low-frequency oscillatory $[Ca^{2+}]_i$ response caused by secreted alpha-hemolysin RTX (repeats-in-toxin), stimulating the production of IL-6 and IL-8 (166). Such inflammatory processes can lead to renal scarring, especially in children (35), or, indeed, to clearance of the pathogens (157).

The urogenital mucosal response to the presence of BV-associated bacteria such as *Gardnerella* and *Prevotella* is only now being uncovered. *Gardnerella,* like *Candida,* is often found in the vaginas of healthy women, yet vaginal inflammatory IL-1β levels are nearly 13-fold higher in women with BV and have been associated with an anti-*Gardnerella* hemolysin (Gyh) IgA response (24). Concentrations of sialidase, prolidase, and anti-Gyh IgA were higher in vaginal fluids of 75 fertile women with BV than in vaginal fluids of 85 healthy control subjects (25). The inflammatory process, including IL-8 production in BV and degradation of IgA by sialidase (26), contributes to the cascade of reactions leading to preterm labor (120) and increased risk of HIV infection (28) in cases where there is reduced natural killer cell cytotoxicity (66) but an enhanced microflora-driven clonal expansion of CD4 T cells (178), particularly those expressing C chemokine receptor 5 (CCR5) (169). The presence of vaginal secretions, such as cathepsin and estrogen-regulated protease D, may increase HIV-1 transmission to women by increasing its infectivity for $CD4^+$ cells and allowing its entrance into some $CD4^+$ epithelial cells (44). More studies are required to determine how BV and other urogenital pathogens affect HIV-1 interactions with immature dendritic cells, including binding to the CD4 antigen, dendritic cell-specific ICAM-3-grabbing nonintegrin, mannose binding C-type lectin receptors, and heparan sulfate proteoglycans (65).

WHAT IS THE OUTCOME FOR THE HOST OF AN ABNORMAL VAGINAL MICROBIOTA?

Extensive evidence shows the negative impact of a pathogen-dominated urogenital microbiota on human health. It is clear that an "abnormal" microbiota, even though the definition of this condition is evolving and depends on host factors including age, race, hormonal status, and symptoms at the time of sampling, is not a preferred state for a woman. Still, the key factors which trigger the process leading to disease versus a return to a healthy microbiota remain to be determined.

As Fig. 2 illustrates and as studies have shown (88, 148), the vaginal microbiota is often not "normal," in that it is dominated by pathogenic and potentially pathogenic microbes such as *E. coli,* BV organisms, yeast, and gram-positive cocci. Clearly, there are factors which cause this flora to revert to a normal *Lactobacillus* colonization while others cause it to progress to an infectious, symptomatic state. The focus of research has been on how uropathogens infect the bladder mucosa, for example, rather than how they are able to ascend from the intestine, survive on the perineum and vagina, outcompete the indigenous flora, and then ascend into the bladder. Of more interest to us, and surely a line of inquiry that is important for therapeutic purposes, is the nature of the mechanisms used by pathogens to overcome these barriers and, likewise, the key factors used by the commensal organisms to prevent infection.

The net outcome of an abnormal microbiota is a steady rise in the number of vaginal infections, increased acquisition of sexually transmitted diseases, and persistently high rates of UTI, BV, and yeast vaginitis. In short, there is a silent epidemic associated with the vaginal microbiota.

Bacterial Vaginosis

Odor and vaginal discharge are the most common symptoms of BV, along with vulvovaginal irritation, which is found in half the cases. Classically, BV

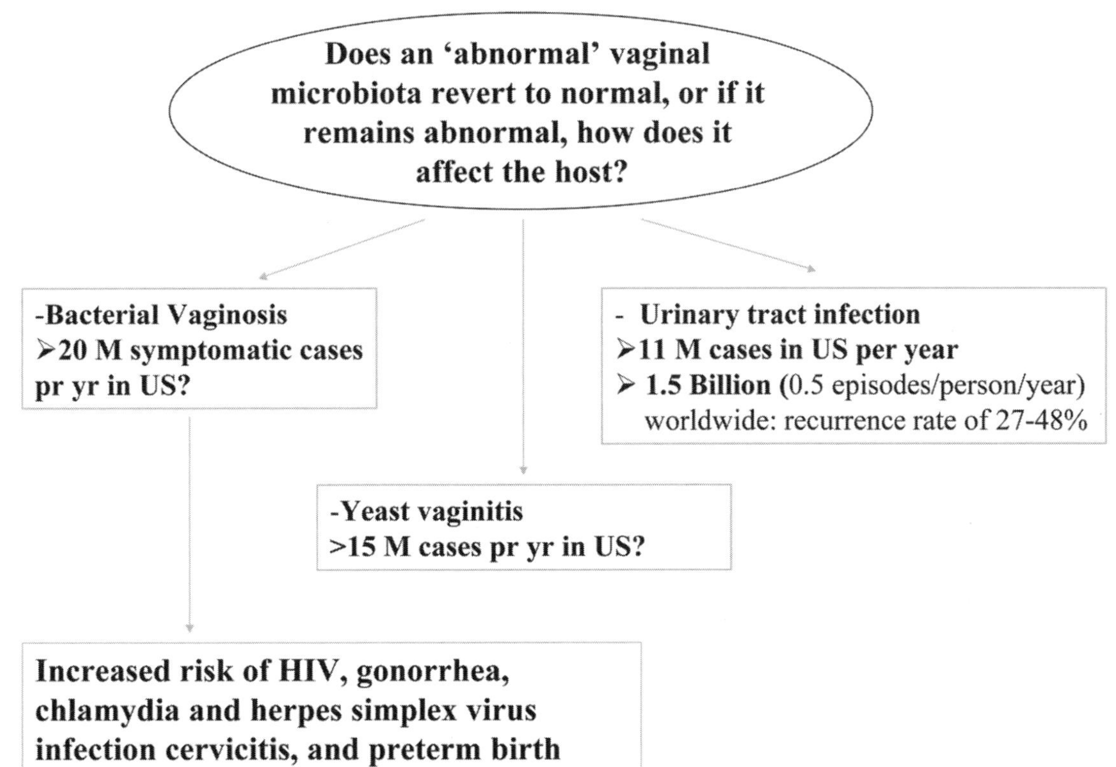

Figure 2. Vaginal health versus disease: what sways the balance?

has been diagnosed by finding the presence of at least 20% of "clue" cells (engulfed by gram-negative organisms) in the squamous cell population on microscopic examination of a saline suspension of vaginal discharge, associated with two of the following (147): (i) anterior fornix vaginal pH equal or greater than 4.7; (ii) release of a fishy odor on addition of 10% KOH to the vaginal discharge (positive "whiff test"), and (iii) presence of an increased thin, homogeneous, white vaginal discharge.

Since culture is not very effective in diagnosing BV, the series of tests outlined represents the "gold standard." However, in practical terms, these tests are rarely used in family practice settings, where many women with BV are seen. Plus, in many cases, odor or discharge are absent (88a). A simple Gram stain of a vaginal smear can identify clue cells engulfed by gram-negative organisms or can establish a Nugent score, in which the presence of mainly gram-positive rods (indicative of lactobacilli) is scored "normal" and the presence of clue cells and gram-negative rods and the absence of lactobacilli is scored as BV (105). When this method was used, 36% of women reporting to a clinic with vaginal symptoms were diagnosed with yeast infection and 24% were diagnosed with BV (31). In a study of healthy women reporting no symptoms or signs of vaginitis, 25% were found to have asymptomatic BV by Nugent scoring (127), again emphasizing the prevalence of this condition. In a study of over 200 women presenting with signs of preterm delivery, one-third were found to have BV (A. Bocking et al., unpublished data), and while BV per se did not predict birth within the next 24 h, it has been shown to be a risk factor in preterm delivery.

New methods are being tested for BV detection, not only for clinic and family medicine offices but also for women to self-diagnose. The BVBlue test (Gryphus Diagnostics) is a chromogenic system that detects elevated sialidase levels in vaginal fluid (98). Compared to the Amsel and Nugent methods, the BVBlue system showed good sensitivity, specificity, and positive predictive values, especially when patients had a vaginal pH of >4.5, a positive amine test, or clue cells. The FemExam card is another test being developed. It involves swabbing a vaginal sample onto a card designed to detect elevated pH and amine levels and onto a second card for measuring proline iminopeptidase activity. Studies of 230 women in The Gambia showed the FemExam cards to have a sensitivity of 91.0% and specificity of 61.5% (174), but its cost was significantly higher than that of Gram staining (U.S. $18.4 versus $1.54) and was prohibitive for use in developing countries and clinics in developed countries where patients pay their own fees. Of more concern, a study in our clinic showed the FemExam kit to be very poor at detecting BV (125a), with specificity being 42% (49 of 118) and sensitivity being 30% (7 of 23), well below acceptable levels.

Urinary Tract Infection

Lower UTI in women is one of the three most common reasons for patient visits to family physicians. In the United States alone, the estimated annual cost for treatment of the 11 million cases of UTI is around $2 billion (55), not including the cost of managing 250,000 patients with pyelonephritis. It has been estimated that uncomplicated UTI occurs as 0.5 episode/woman/year, with a recurrence rate of between 27 and 48% (75). Although UTI is not often fatal, it causes dysuria, frequency of micturition, occasional haematuria, suprapubic pain, and significant loss of quality of life (43).

In terms of bacterial colonization, the intestine and vagina are reservoirs for the organisms that eventually infect the urinary tract. This was shown most recently in an analysis of virulence factors of extraintestinal strains, in which the prevalence of the virulence factors *papC* (pilus associated with pyelonephritis), *hlyA* (hemolysin), *cnf1* (cytotoxic necrotizing factor), PAI (pathogenicity-associated island marker), *ibeA* (invasion brain endothelium), and K1 antigen was 45, 22, 19, 78, 32, and 44%, respectively, similar to that in fecal isolates (106). Thus, uropathogens must be able to colonize different cell types (intestinal, squamous vaginal, and transitional bladder) in order to infect the host.

Not all UTI episodes are symptomatic, and occasionally even the symptomatic ones self-resolve. Chronic or persistent bacteriuria is common in patients with neurogenic bladder disease or those who have lost functionality such as patients who have suffered spinal cord injury (SCI). Although intermittent catheterization somewhat mimics the shear force of micturition, the bladders of these patients tend to remain densely colonized irrespective of the use of antibiotic therapy. Indeed, biofilm formation is prevalent on bladder cells recovered from SCI patients (128, 138). The host responds to the presence of pathogens in urine by production of IL-6 and IL-8 locally and systemically, polymorphonuclear neutrophils, and $\gamma\delta$ T lymphocytes, as well as B lymphocytes and antibodies including local secretory IgA (165). Thus, even though SCI patients do not often perceive an infectious condition, the bacteria and host continue to interact in ways that could have long-term detrimental outcomes.

Most studies report that chronic bladder colonization causes a 30 to 50% reduction in survival

(40, 45), but one analysis refuted the claim that bacteriuria was the main cause of the high death rates (104). The chronic presence of bacteria in the bladder also has the potential to increase the risk of cancer at that site (33) and elsewhere (64, 99) due to production of carcinogens by the organisms. Thus, lack of symptoms and signs of infection do not always mean that other disease processes are not under way because of this microbial colonization.

Although there have been great improvements in the understanding of the etiology, diagnosis, and treatment of UTI, there remains a need for alternative methods to retain and restore health. The use of long-term, low-dose antibiotics, while reasonably effective, still leads to one or two breakthrough infections per year, increases in antibiotic resistance, and a range of side effects (134).

Yeast-Mediated Vaginitis

Yeast vaginitis, caused mostly by *C. albicans,* affects about 1 in 5 African American women and close to 1 in 10 Caucasian women during any given 2-month period (54), with 1 in 12 reporting four or more episodes per year. Patients generally present with a white vaginal discharge characterized by a malodourous, nonhomogeneous, caseous appearance, accompanied by vaginal and introital itch and irritation and evidence of vaginal inflammatory reaction. A study of various anatomical areas of the urogenital tract has shown that on day 3 of the menstrual cycle, yeast were detected at the hymen only, while enterococci and gram-negative enterobacteria were found in the perineum, below and above the hymen, and in the posterior fornix (73).

The presence of *Candida* in healthy women or those also colonized by lactobacilli (36) demonstrates a difference between the UTI and BV situations. The infecting organisms appear to come in part from the intestine, but more commonly they emerge from the patient's own vagina (51). The three stages of yeast colonization involve adhesion, blastopore germination, and epithelium invasion (47). *C. albicans* has four structurally related adhesins, Hwp1, Ala1p, Als5p, and Als1p, and *C. glabrata* has one, Epa1p (155). These are members of a class of proteins termed glycosylphosphatidylinositol-dependent cell wall proteins, which have N-terminal signal peptides and C-terminal features that mediate glycosylphosphatidylinositol membrane anchor addition, plus other determinants that aid attachment to cell wall glucan. They can adhere to basement membrane collagen N-terminal cross-linking domain (7S) and can be inhibited from adhering by several sugars known to be part of the N-linked oligosaccharide chains of collagen IV, namely, *N*-acetylglucosamine, L-fucose, methylmannoside, and, particularly, *N*-acetyllactosamine, but not glucose, galactose, lactose, or heparan sulfate (2). The organism synthesizes unique sequences of β-1,2-oligomannosides that act as adhesins, induce cytokine production, and generate protective antibodies (164). The yeast are particularly intriguing since they not only express adhesins but also can perform morphogenesis (the reversible transition between unicellular yeast cells and filamentous, growth forms) and produce secreted aspartyl proteases and phospholipases (20).

A study of 13,914 pregnant women showed that the prevalence of moderate to heavy *Candida* colonization at midgestation was 10%; 83% of these women carried *C. albicans* (29). Most of the colonized women were African American or Hispanic, unmarried, and previous oral-contraceptive users. *Candida* colonization was positively associated with *Trichomonas vaginalis,* group B streptococci, and aerobic *Lactobacillus.* The ability of yeast to coexist with bacteria is clear, but it is still not known what triggers their emergence as dominant, infection-causing agents. Investigations that shed light on this issue will be important if new approaches to prevention and treatment are to be found.

Cell-mediated immunity may not be the predominant host defense mechanism against *C. albicans;* rather, locally acquired mucosal immunity appears to be a more important defense at the vaginal mucosa (49). However, while candidiasis is more common in HIV-seropositive than seronegative women, cell-mediated immunodeficiency ($CD4^+$ lymphocyte count of <500 cells/ml) was not found to be associated with increased odds of vaginal colonization (108). The finding that vulvovaginal candidiasis occurs with higher incidence and greater persistence, but not greater severity, among HIV-infected women (41) emphasizes the multidimensional and complex pathogenesis of this organism and the need to better understand how it becomes infectious.

WHAT IS THE OUTCOME WHEN PROBIOTIC LACTOBACILLI ARE ADMINISTERED TO THE UROGENITAL TRACT?

The ability of probiotic strains of lactobacilli to colonize the host and interfere with pathogenesis has been the primary area of interest for our group for over 23 years. While the concept of probiotics dates back 100 years—longer if we consider the practice of consuming fermented foods by paleolithic humans (7)—it has only been within the past 10 years that significant scientific interest has emerged.

The process of selecting *Lactobacillus* strains for urogenital application has evolved from in vitro assessment of adhesion, competitive exclusion, production of inhibitory substances, and ability to coaggregate (27, 84, 129, 135), along with animal studies (126, 130) to the extent that unless proven health benefits are conferred on the host, these selection criteria only document characteristics but are essentially meaningless in proving probiotic capacity. To date, there has been no evidence to prove that, for example, the production of hydrogen peroxide or lactic acid by lactobacilli kills or inhibits pathogens in the vagina. In preliminary studies, we identified an anti-infective protein present on a human vaginal cell as a component of a biosurfactant from *L. reuteri* RC-14. The intent was to prove that the p29 protein was actually expressed in vivo. However, the antibody used for immunostaining was not sufficiently specific to conclusively prove the point. Thus, in vitro experiments provide useful insight into the functions of organisms but not necessarily how well they will perform in humans. The same is true for bacteriocins and antibiotics such as reuterin, which may not even be biologically active (172a).

With that in mind, four particularly interesting mechanistic attributes of vaginal lactobacilli have been studied in vitro: ability to kill pathogenic bacteria and viruses, biosurfactant activity, cell-cell signaling, and biofilm penetration.

The production of H_2O_2 is known to inhibit bacterial growth (167), possibly also with coexpression of catalase inhibitors (179). One study suggested that agitation and pH 6.5. were needed for H_2O_2 production, which, if true, makes it less likely that significant quantities are produced in the relatively static, acidic mucosa of the vagina (163). Peroxidase converts H_2O_2 into hypochlorous acid, thereby creating a microbicidal vaginal milieu by maintaining a balanced, nontoxic, steady-state level of the H_2O_2 microbicides and HOCl. This reaction may drive superoxide anion-producing transformed cells into apoptosis, thereby potentially preventing vaginal cancer (6). The production of H_2O_2 can act to self-inhibit *lactobacillus* growth and may be one factor in stopping the vaginal *Lactobacillus* concentration from exceeding around 10^8 viable organisms per ml of fluid.

Other *Lactobacillus*-produced compounds, for example bacteriocins and lactic acid (95, 96, 135, 154, 168), can inhibit and kill pathogens. The acidic environment created by the growth of lactobacilli is also able to kill viruses (18). The production of an acidic vaginal tract has been attributed to lactobacilli (13), making it feasible to apply probiotic strains to women with BV as a means of reducing the risk of HIV infection, particularly in women engaging in high-risk sexual practices.

The discovery of biosurfactant activity by lactobacilli (171) has led to several new avenues of investigation. The crude mixture contains a number of proteins and carbohydrates (172), including ones responsible for lowering fluid surface tension (surfactant activity) and others with anti-infective properties. In vitro studies have shown that the biosurfactant significantly reduces the adhesion of a range of uropathogens (170). The biosurfactant mixture, along with a p29 collagen binding protein, prevents infectious *S. aureus* sepsis in an animal surgical implant model (56). The ability of p29 to inhibit the binding of gram-positive cocci to surfaces (71, 77) could be one mechanism of action, but there appear to be two other possibilities at least: cell-cell signaling between the two organisms (*L. reuteri* RC-14 and *S. aureus*) and modulation of the host immune response by viable RC-14 and by p29. Novel amino acid sequences have been identified within the biosurfactant by using ProteinChip MS/MS technology (131), but their function remains to be determined.

By using a two-chamber system that separates lactobacilli from a second organism (such as *S. aureus* or *E. coli*), and reverse transcription-PCR and two-dimensional gel electrophoresis, cell signaling effects have now been discovered that down regulate three virulence factors in *S. aureus* and another three in uropathogenic *E. coli* (unpublished data). Insertion of a green fluorescent protein reporter system into the staphylococci has confirmed the shutting down of virulence genes. Attempts are under way to identify the signaling molecule(s) produced by *L. reuteri* RC-14.

The potential exists that cell-cell signaling occurs within urogenital biofilms. The ability or lactobacilli (both strains RC-14 and GR-1) to penetrate biofilms formed by uropathogenic *E. coli* has now been demonstrated in vitro (Color Plate 12) (133). This is intriguing in that it potentially provides a means of treating a recalcitrant infected biofilm by using other bacteria rather than antibiotics (which invariably fail to eradicate infection), especially if the added organisms are nonpathogenic and if they are able to down regulate the virulence of the infecting bacteria. An approach similar to this has involved the use of avirulent *E. coli* 83972 to reduce symptomatic UTI in SCI patients (32).

Another useful microscopic method of monitoring changes in biofilms exposed to bacteria such as lactobacilli is to use viability staining. As shown in Color Plate 13, the areas of viability and dead cells can be seen, providing insight into the structure and dynamics of the biofilm.

Probiotic lactobacilli can be administered directly into the vagina in douche or capsule forms (16, 18, 110, 125). The net result is a displacement of pathogens (or down regulation of virulence, as noted above but not yet proven in vivo) encountered on the surface (Fig. 3) and colonization by the lactobacilli of the vaginal mucosa, albeit only for several weeks or months (137). It is unclear why the strains do not appear to colonize for longer, but this may be due to host defenses (immune and commensal flora) displacing or killing them or to failure to adhere sufficiently well and thrive. Long-term colonization is perhaps not necessary for retention of health, as long as the woman's own *Lactobacillus* flora returns in sufficient numbers to help with defense. Alternatively, the probiotics can be administered weekly, monthly, or at other intervals designed to retain a normal microbiota as opposed to one dominated by BV and UTI pathogens.

Oral administration of lactobacilli can also lead to "normalization" of the vaginal microbiota through ascension of the ingested strains from the rectal skin to the vagina. It can also possibly lead to enhancement of generalized mucosal immunity, reduction of the pathogen load from the rectum, and creation of an environment in the vagina that allows the commensal lactobacilli to thrive. Studies of women prone to urogenital infection (119, 123) and healthy women (125a, 127) have shown that even asymptomatic BV can be cleared in some cases, while in others the flora maintains a normal Nugent score on more occasions than in placebo-treated women.

It is important to be able to track probiotic strains after administration. This can be achieved by a number of techniques such as randomly amplified polymorphic DNA, which generates DNA fingerprints of organisms recovered from swabs plated onto agar by using a method in which low-stringency PCR of genome DNA is performed with single short primers with arbitrary sequences (173). The assumption is that there are areas of similar DNA to anneal with primers in a random fashion. The nucleic acid between the primer sites is amplified, and the fragments are separated by electrophoresis. Pulsed-field gel electrophoresis is even more accurate than randomly amplified polymorphic DNA, albeit more time-consuming since it requires a comparison of the entire chromosome without the complicated patterns obtained by frequently cutting restriction enzymes (173). Using this method, *L. reuteri* RC-14 has been recovered from the stool and vagina (57, 58) and shown to be different from other commercial and type strains of this species (Fig. 4). The molecular tracking methods involve amplification by PCR of 16S rRNA genes (16S rDNA) from microbial DNA extracted from samples. The bacterial species are differentiated on the basis of the 500-bp spacer region located between the 16S and 23S rRNA genes. The nucleotide base sequence of this region can be obtained by sequencing the amplified DNA; therefore, the amplified 16S rDNA sequences are cloned and aligned with databank sequences to identify the organism (160). It was the recent use of this PCR-16S rRNA method which showed that *L. fermentum* RC-14 was classified as *L. reuteri* RC-14. This exemplifies why FAO/WHO (53) produced guidelines for probiotics, recognizing that many so-called probiotic commercial strains are actually mislabeled as

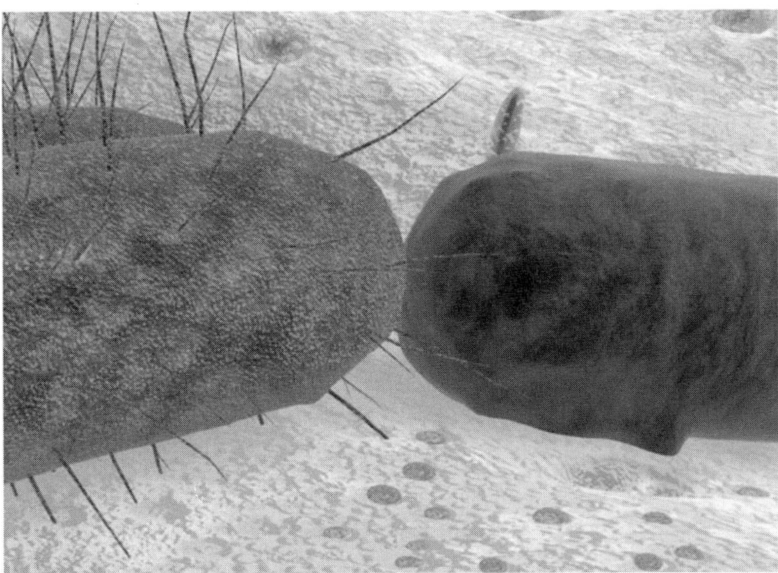

Figure 3. *Lactobacillus reuteri* (formerly *L. fermentum*) RC-14 (depicted on the right) colonizes the vagina, displaces pathogens, and sends signals to fimbriated uropathogenic *E. coli* (left) that down regulate some virulence factor activity.

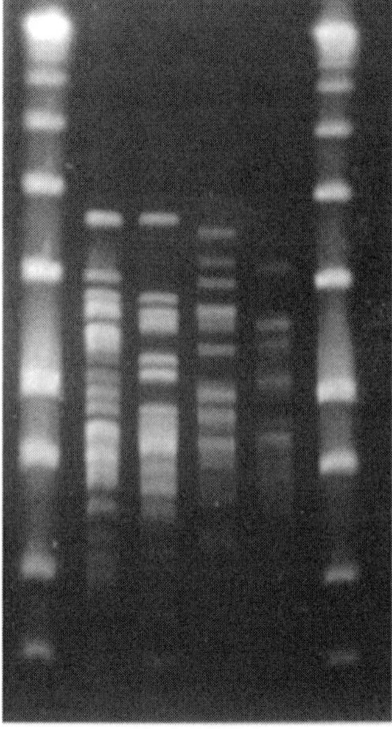

Figure 4. Pulsed-field gel electrophoresis of *Lactobacillus* strains by using ApaI. Lanes: 1 and 6, low-range pulsed-field gel electrophoresis marker; 2, *L. reuteri* RC-14, 3, *L. reuteri* ATCC 23272 (DSM 20016); 4, *L. reuteri* NCIMB 702656; 5, *L. reuteri* NCIMB 701359. Note the significant difference between strain RC-14 and the other *L. reuteri* strains.

L. acidophilus or the nonexistent species *L. sporogenes,* possibly to avoid legislative bans on the product, when in fact they are other species, such as *L. crispatus* or *L. reuteri.*

The final test of the impact of probiotic administration, apart from the clinical outcome for the host, is to understand the mucosal immune response to the strains and how the treatment affects the microbiota of the host. No data are available yet with respect to the vaginal mucosal response to probiotic use, but based on studies of the intestine, the response is likely to vary with the host, her own microbiota, and the strain of probiotic used (81). One study has examined the vaginal microbiota before and after probiotic instillation of *Lactobacillus* strains GR-1 and RC-14 and has found that in some women there is no major alteration while in others commensal *Lactobacillus* spp. recolonize or environmental (soil, freshwater, and feces) and oral organisms appear (17).

SUMMARY

The microbiota of the vagina and urethra is both complex and simple: complex in the number and types of organisms (a range of aerobic and anaerobic bacteria, protozoa, chlamydias, and viruses) that can colonize and in the fluctuations that occur through life (prepuberty, reproductive years and postmenopause) and even on an hourly basis (149); and simple in that usually only 1 to 10 species are present at any given time, even though each day the region is exposed to many organisms with adequate nutrient supplies from traces of urine, feces, and vaginal fluids. The ability of commensal organisms and host defenses to preserve the area from continual, or at least frequent, attack by intruders from the host herself and from sexual partners is quite remarkable. Nevertheless, the mucosal environment is in a constant state of change, due to hormones, diet, micturition and defecation, menstruation, hygienic cleansing, etc., and infections do occur, affecting the woman's quality of life.

Vaccine development such as the recent use of *L. jensenii* recombined to express the CD4 receptor for HIV (27a) will continue for some pathogens, as will the search for new remedies for viral infections. Still, the potential to retain and restore health through probiotic use is one novel approach that is worthy of further consideration and support. Although techniques are being developed to enable babies to be created without the need for a female host, until those days of "science fiction" appear, the health of the woman's reproductive tract is paramount to the survival of humankind. Given the 50% rate of HIV infection in pregnant women in countries such as Botswana and the growing number of cases of genital herpes, BV, UTI, and yeast vaginitis, even in an era where condoms are plentiful, the need to better understand the urogenital microbiota and generate new methods of preventing and effectively treating abnormal colonization has never been greater.

Acknowledgments. I am grateful to Mark Neysmith for the graphics, Gillian Gardiner for the pulsed-field gel electrophoresis gel, Mickey Zeller and Melissa Dell for their work on biofilms and the deconvolution photos, Jana Jass and Jonathan Hui for the atomic force microscopy photo, Dominique Lam for the Gram stain photo, and my many colleagues and collaborators, especially Andrew Bruce, for their continued support of our research. Funding from NSERC is appreciated.

REFERENCES

1. **Agerholm-Larsen, L., A. Raben, N. Haulrik, A. S. Hansen, M. Manders, and A. Astrup.** 2000. Effect of 8 week intake of probiotic milk products on risk factors for cardiovascular diseases. *Eur. J. Clin. Nutr.* **54:**288–297.
2. **Alonso, R., I. Llopis, C. Flores, A. Murgui, and J. Timoneda.** 2001. Different adhesins for type IV collagen on *Candida albicans:* identification of a lectin-like adhesin recognizing the 7S(IV) domain. *Microbiology* **147:**1971–1981.
3. **Anderl, J. N., M. J. Franklin, and P. S. Stewart.** 2000. Role of antibiotic penetration limitation in *Klebsiella pneumoniae* biofilm resistance to ampicillin and ciprofloxacin. *Antimicrob. Agents Chemother.* **44:**1818–1824.

4. **Anderson, G. G., J. J. Palermo, J. D. Schilling, R. Roth, J. Heuser, and S. J. Hultgren.** 2003. Intracellular bacterial biofilm-like pods in urinary tract infections. *Science* **301:** 105–107.
5. **Antonio, M. A., S. E. Hawes, and S. L. Hillier.** 1999. The identification of vaginal *Lactobacillus* species and the demographic and microbiologic characteristics of women colonized by these species. *J. Infect. Dis.* **180:**1950–1956.
6. **Bauer, G.** 2001. Lactobacilli-mediated control of vaginal cancer through specific reactive oxygen species interaction. *Med. Hypotheses* **57:**252–257.
7. **Bengmark S.** 2001. Use of prebiotics, probiotics and synbiotics in clinical immunonutrition. *In Int. Symp. Food Nutr. Health 21st Century,* p. 187–213. Korean Society of Food Science and Nutrition, Seoul.
8. **Beninati, C., M. R. Oggioni, M. Boccanera, M. R. Spinosa, T. Maggi, S. Conti, W. Magliani, F. De Bernardis, G. Teti, A. Cassone, G. Pozzi, and L. Polonelli.** 2000. Therapy of mucosal candidiasis by expression of an anti-idiotype in human commensal bacteria. *Nat. Biotechnol.* **18:**1060–1064.
9. **Berger, B. J., S. Kolton, J. M. Zenilman, M. C. Cummings, J. Feldman, and W. M. McCormack.** 1995. Bacterial vaginosis in lesbians: a sexually transmitted disease. *Clin. Infect. Dis.* **21:**1402–1405.
10. **Beyenal, H., and Z. Lewandowski.** 2002. Internal and external mass transfer in biofilms grown at various flow velocities. *Biotechnol. Prog.* **18:**55–61.
11. **Biesalski, H. K., U. Sobeck, and H. Weiser.** 2001. Topical application of vitamin A reverses metaplasia of rat vaginal epithelium: a rapid and efficient approach to improve mucosal barrier function. *Eur. J. Med. Res.* **6:**391–398.
12. **Bollgren, I., and J. Winberg.** 1976. The periurethral aerobic bacterial flora in healthy boys and girls. *Acta Paediatr. Scand.* **65:**74–80.
13. **Boskey, E. R., R. A. Cone, K. J. Whaley, and T. R. Moench.** 2001. Origins of vaginal acidity: high D/L-lactate ratio is consistent with bacteria being the primary source. *Hum. Reprod.* **16:**1809–1813.
14. **Boyko, E. J., S. D. Fihn, D. Scholes, C. L. Chen, E. H. Normand, and P. Yarbro.** 2002. Diabetes and the risk of acute urinary tract infection among postmenopausal women. *Diabetes Care* **25:**1778–1783.
15. **Brody, S.** 2002. High-dose ascorbic acid increases intercourse frequency and improves mood: a randomized controlled clinical trial. *Biol. Psychiatry* **52:**371–374.
16. **Bruce, A. W., and G. Reid.** 1988. Intravaginal instillation of lactobacilli for prevention of recurrent urinary tract infections. *Can. J. Microbiol.* **34:**339–343.
17. **Burton, J. P., P. Cadieux, and G. Reid.** 2003. Improved understanding of the bacterial vaginal microbiota of women before and after probiotic instillation. *Appl. Environ. Microbiol.* **69:**97–101.
17a. **Burton, J. P., and G. Reid.** 2003. Evaluation of the bacterial vaginal flora of twenty postmenopausal women by direct (Nugent Score) and molecular (polymerase chain reaction and denaturing gradient gel electrophoresis) techniques. *J. Infect. Dis.* **186:**1777–1780.
18. **Cadieux, P., J. Burton, C. Y. Kang, G. Gardiner, I. Braunstein, A. W. Bruce, and G. Reid.** 2002. *Lactobacillus* strains and vaginal ecology. *JAMA* **287:**1940–1941.
19. **Cadieux, P. A., D. T. Beiko, J. D. Watterson, J. P. Burton, J. Howard, B. E. Knudsen, B. S. Gan, J. McCormick, A. F. Chambers, J. D. Denstedt, and G. Reid.** 2004. Surface-enhanced laser desorption/ionization time-of-flight mass spectrometry (SELDI-TOF-MS): a new proteomic urinary test for patients with urolithiasis. *J. Clin. Lab. Anal.* **18:**170–175.
20. **Calderone, R. A., and W. A. Fonzi.** 2001. Virulence factors of *Candida albicans. Trends Microbiol.* **9:**327–335.
21. **Campieri, C., M. Campieri, V. Bertuzzi, E. Swennen, D. Matteuzzi, S. Stefoni, F. Pirovano, C. Centi, S. Ulisse, G. Famularo and C. DeSimone.** 2001. Reduction of oxaluria after an oral course of lactic acid bacteria at high concentration. *Kidney Int.* **60:**1097–1105.
22. **Carbone, M., D. L. Hasty, K. C. Yi, J. Rue, M. T. Fera, F. La Torre, M. Giannone, and E. Losi.** 2002. Cytokine induction in murine bladder tissue by type 1 fimbriated *Escherichia coli. Ann. N.Y. Acad. Sci.* **963:**332–335.
23. **Cardenas-Freytag, L., C. Steele, F. L. Wormley Jr, E. Cheng, J. D. Clements, and P. L. Fidel Jr.** 2002. Partial protection against experimental vaginal candidiasis after mucosal vaccination with heat-killed *Candida albicans* and the mucosal adjuvant LT(R192G). *Med. Mycol.* **40:**291–299.
24. **Cauci, S., S. Driussi, S. Guaschino, M. Isola, and F. Quadrifoglio.** 2002. Correlation of local interleukin-1beta levels with specific IgA response against *Gardnerella vaginalis* cytolysin in women with bacterial vaginosis. *Am. J. Reprod. Immunol.* **47:**257–264.
25. **Cauci, S., S. Guaschino, S. Driussi, D. De Santo, P. Lanzafame, and F. Quadrifoglio.** 2002. Correlation of local interleukin-8 with immunoglobulin A against *Gardnerella vaginalis* hemolysin and with prolidase and sialidase levels in women with bacterial vaginosis. *J. Infect. Dis.* **185:**1614–1620.
26. **Cauci, S., R. Monte, S. Driussi, P. Lanzafame, and F. Quadrifoglio.** 1998. Impairment of the mucosal immune system: IgA and IgM cleavage detected in vaginal washings of a subgroup of patients with bacterial vaginosis. *J. Infect. Dis.* **178:**1698–1706.
27. **Chan, R. C. Y., A. W. Bruce, and G. Reid.** 1984. Adherence of cervical, vaginal and distal urethral normal microbial flora to human uroepithelial cells and the inhibition of adherence of uropathogens by competitive exclusion. *J. Urol.* **131:**596–601.
27a. **Chang, T. L., C. H. Chang, D. A. Simpson, Q. Xu, P. K. Martin, L. A. Lagenaur, G. K. Schoolnik, D. D. Ho, S. L. Hillier, M. Holodniy, J. A. Lewicki, and P. P. Lee.** 2003. Inhibition of HIV infectivity by a natural human isolate of *Lactobacillus jensenii* engineered to express functional two-domain CD4. *Proc. Natl. Acad. Sci. USA* **100:**11672–11677.
28. **Cohen, C. R., F. A. Plummer, N. Mugo, I. Maclean, C. Shen, F. A. Bukusi, E. Irungu, S. Sinei, J. Bwayo, and R. C. Brunham.** 1999. Increased interleukin-10 in the the endocervical secretions of women with non-ulcerative sexually transmitted diseases: a mechanism for enhanced HIV-1 transmission? *AIDS* **13:**327–332.
29. **Cotch, M. F., S. L. Hillier, R. S. Gibbs, D. A. Eschenbach, and the Vaginal Infections and Prematurity Study Group.** 1998. Epidemiology and outcomes associated with moderate to heavy *Candida* colonization during pregnancy. *Am. J. Obstet. Gynecol.* **178:**374–380.
30. **Cottell, E., R. F. Harrison, M. McCaffrey, T. Walsh, E. Mallon, and C. Barry-Kinsella.** 2000. Are seminal fluid microorganisms of significance or merely contaminants? *Fertil. Steril.* **74:**465–470.
31. **Dan, M., N. Kaneti, D. Levin, F. Poch, and Z. Samra.** 2003. Vaginitis in a gynecologic practice in Israel: causes and risk factors. *Isr. Med. Assoc. J.* **5:**629–632.
32. **Darouiche, R. O., W. H. Donovan, M. Del Terzo, J. I. Thornby, D. C. Rudy, and R. A. Hull.** 2001. Pilot trial of bacterial interference for preventing urinary tract infection. *Urology* **58:**339–344.

33. **Davis, C. P., M. S. Cohen, R. L. Hackett, M. D. Anderson, and M. M. Warren.** 1991. Urothelial hyperplasia and neoplasia. III. Detection of nitrosamine production with different bacterial genera in chronic urinary tract infections of rats. *J. Urol.* **145:**875–880.

34. **De Bernardis, F., M. Boccanera, D. Adriani, A. Girolamo, and A. Cassone.** 2002. Intravaginal and intranasal immunizations are equally effective in inducing vaginal antibodies and conferring protection against vaginal candidiasis. *Infect. Immun.* **70:**2725–2729.

35. **de Man, P.** 1991. Bacterial attachment, inflammation and renal scarring in urinary tract infection. *Wien. Med. Wochenschr.* **141:**537–540.

36. **Demirezen, S.** 2002. The lactobacilli–*Candida* relationship in cervico-vaginal smears. *Cent. Eur. J. Public Health* **10:**97–99.

37. **Devillard, E., J. P. Burton, J.-A. Hammond, D. Lam, and G. Reid.** 2004. Novel insight into the vaginal microflora in postmenopausal women under hormone replacement therapy as analyzed by PCR-denaturing gradient gel electrophoresis. *Eur. J. Obstet. Gynecol.* **117:**76–81.

38. **Domingue, P. A., K. Sadhu, J. W. Costerton, K. Bartlett, and A. W. Chow.** 1991. The human vagina: normal flora considered as an in situ tissue-associated, adherent biofilm. *Genitourin. Med.* **67:**226–231.

39. **Donder, G. G., A. Vereecken, E. Bosmans, A. Dekeersmaecker, G. Salembier, and B. Spitz.** 2002. Definition of a type of abnormal vaginal flora that is distinct from bacterial vaginosis: aerobic vaginitis. *BJOG* **109:**34–43.

40. **Dontas, A. S., P. Kasviki-Charvati, P. C. Papanayiotou, and S. G. Marketos.** 1981. Bacteriuria and survival in old age. *N. Engl. J. Med.* **304:**939–943.

41. **Duerr, A., C. M. Heilig, S. F. Meikle, S. Cu-Uvin, R. S. Klein, A. Rompalo, and J. D. Sobel for the HER Study Group.** 2003. Incident and persistent vulvovaginal candidiasis among human immunodeficiency virus-infected women: risk factors and severity. *Obstet. Gynecol.* **101:**548–556.

42. **Eijsink, V. G., L. Axelsson, D. B. Diep, L. S. Havarstein, H. Holo, and I. F. Nes.** 2002. Production of class II bacteriocins by lactic acid bacteria; an example of biological warfare and communication. *Antonie Leeuwenhoek* **81:**639–654.

43. **Ellis, A. K., and S. Verma.** 2002. Quality of life in women with urinary tract infections: is benign disease a misnomer? *J. Am. Board Fam. Pract.* **13:**392–397.

44. **El Messaoudi, K., L. Thiry, N. Van Tieghem, C. Liesnard, Y. Englert, N. Moguilevsky, and A. Bollen,** 1999. HIV-1 infectivity and host range modification by cathepsin D present in human vaginal secretions. *AIDS* **13:**333–339.

45. **Evans, D. A., E. H. Kass, C. H. Hennekens, B. Rosner, L. Miao, M. I. Kendrick, W. E. Miall, and K. L. Stuart.** 1982. Bacteriuria and subsequent mortality in women. *Lancet* **i:** 156–158.

46. **Favier, C. F., E. E. Vaughan, W. M. De Vos, and A. D. Akkermans.** 2002. Molecular monitoring of succession of bacterial communities in human neonates. *Appl. Environ. Microbiol.* **68:**219–226.

47. **Ferrer, J.** 2000. Vaginal candidosis: epidemiological and etiological factors. *Int. J. Gynaecol. Obstet.* **71**(Suppl. 1):S21–S27.

48. **Fidel, P. L., Jr, M. Barousse, V. Lounev, T. Espinosa, R. R. Chesson, and K. Dunlap.** 2003. Local immune responsiveness following intravaginal challenge with *Candida* antigen in adult women at different stages of the menstrual cycle. *Med. Mycol.* **41:**97–109.

49. **Fidel, P. L., Jr., and J. D. Sobel.** 1996. Immunopathogenesis of recurrent vulvovaginal candidiasis. *Clin. Microbiol. Rev.* **9:**335–348.

50. **Flock, J.-L.** 1999. Extracellular-matrix-binding proteins as targets for the prevention of *Staphylococcus aureus* infections. *Mol. Med. Today* **5:**532–537.

51. **Fong, I. W.** 1994. The rectal carriage of yeast in patients with vaginal candidiasis. *Clin. Investig. Med.* **17:**426–431.

52. **Food and Agriculture Organization.** 2001. *Evaluation of Health and Nutritional Properties of Powder Milk and Live Lactic Acid Bacteria.* Food and Agriculture Organization of the United Nations and World Health Organization expert consultation report. http://www.fao.org/es/ESN/Probio/probio.htm.

53. **Food and Agriculture Organization.** 2002. *Guidelines for the Evaluation of Probiotics in Food.* Food and Agriculture Organization of the United Nations and World Health Organization working group report. ftp://ftp.fao.org/es/esn/food/wgreport2.pdf.

54. **Foxman, B., R. Barlow, H. D'Arcy, B. Gillespie, and J. D. Sobel.** 2000. *Candida* vaginitis: self-reported incidence and associated costs. *Sex. Transm. Dis.* **27:**230–252.

55. **Foxman, B., R. Barlow, H. D'Arcy, B. Gillespie, and J. D. Sobel.** 2000. Urinary tract infection: self-reported incidence and associated costs. *Ann. Epidemiol.* **10:**509–515.

56. **Gan, B. S., J. Kim, G. Reid, P. Cadieux, and J. C. Howard.** 2002. *Lactobacillus fermentum* RC-14 inhibits *Staphylococcus aureus* infection of surgical implants in rats. *J. Infect. Dis.* **185:**1369–1372.

57. **Gardiner, G., C. Heinemann, M. L. Baroja, A. W. Bruce, D. Beuerman, J. Madrenas, and G. Reid.** 2002. Oral administration of the probiotic combination *Lactobacillus rhamnosus* GR-1 and *L. fermentum* RC-14 for human intestinal applications. *Int. Dairy J.* **12:**191–196.

58. **Gardiner, G., C. Heinemann, D. Beuerman, A. W. Bruce, and G. Reid.** 2002. Persistence of *Lactobacillus fermentum* RC-14 and *L. rhamnosus* GR-1, but not *L. rhamnosus* GG, in the human vagina as demonstrated by randomly amplified polymorphic DNA (RAPD). *Clin. Diagn. Lab. Immunol.* **9:** 92–96.

59. **Geshnizgani, A. M., and A. B. Onderdonk.** 1992. Defined medium simulating genital tract secretions for growth of vaginal microflora. *J. Clin. Microbiol.* **30:**1323–1326.

60. **Gibbons, R. J., D. I. Hay, W. C. Childs III, and G. Davis.** 1990. Role of cryptic receptors (cryptitopes) in bacterial adhesion to oral surfaces. *Arch. Oral Biol.* **35**(Suppl.):107S–114S.

61. **Gilliland, S. E., M. L. Speck, and C. G. Morgan.** 1975. Detection of *Lactobacillus acidophilus* in feces of humans, pigs, and chickens. *Appl. Microbiol.* **30:**541–545.

62. **Goh, S. H., R. R. Facklam, M. Chang, J. E. Hill, G. J. Tyrrell, E. C. Burns, D. Chan, C. He, T. Rahim, C. Shaw, and S. M. Hemmingsen.** 2000. Identification of *Enterococcus* species and phenotypically similar *Lactococcus* and *Vagococcus* species by reverse checkerboard hybridization to chaperonin 60 gene sequences. *J. Clin. Microbiol.* **38:**3953–3959.

63. **Greenberg, E. P.** 2003. Bacterial communication and group behavior. *J. Clin. Investig.* **112:**1288–1290.

64. **Guarner, F., and J. R. Malagelada.** 2003. Gut flora in health and disease. *Lancet* **361:**512–519.

65. **Gummuluru, S., M. Rogel, L. Stamatatos, and M. Emerman.** 2003. Binding of human immunodeficiency virus type 1 to immature dendritic cells can occur independently of DC-SIGN and mannose binding c-type lectin receptors via a cholesterol-dependent pathway. *J. Virol.* **77:**12865–12874.

66. **Hafez, E. S., G. Merino, R. Bailon, and C. Moran.** 1992. HIV/STD interactions immunosuppression and future research development. *Arch. AIDS Res.* **6:**221–246.

67. **Hart, A., and M. A. Kamm.** 2002. Review article: mechanisms of initiation and perpetuation of gut inflammation by stress. *Aliment. Pharmacol. Ther.* **16:**2017–2028.
68. **Hatakka, K., E. Savilahti, A. Ponka, J. H. Meurman, T. Poussa, L. Nase, M. Saxelin, and R. Korpela.** 2001. Effect of long term consumption of probiotic milk on infections in children attending day care centres: double blind, randomised trial. *Br. Med. J.* **322:**1327.
69. **Hedlund, M., R. D. Duan, A. Nilsson, and C. Svanborg.** 1998. Sphingomyelin, glycosphingolipids and ceramide signalling in cells exposed to P-fimbriated *Escherichia coli. Mol. Microbiol.* **29:**1297–1306.
70. **Hein, M., E. V. Valore, R. B. Helmig, N. Uldbjerg, and T. Ganz.** 2002. Antimicrobial factors in the cervical mucus plug. *Am. J. Obstet. Gynecol.* **187:**137–144.
71. **Heinemann, C., J. E. T. Van Hylckama Vlieg, D. B. Janssen, H. J. Busscher, H. C. van der Mei, and G. Reid.** 2000. Purification and characterization of a surface-binding protein from *Lactobacillus fermentum* RC-14 inhibiting *Enterococcus faecalis* 1131 adhesion. *FEMS Microbiol. Lett.* **190:**177–180.
71a. **Hill, J. E., S. L. Penny, K. G. Crowell, S. H. Goh, and S. M. Hemmingsen.** 2004. cpnDB: a chaperonin sequence database. *Genome Res.* **14:**1669–1675.
72. **Hill, J. E., R. P. Seipp, M. Betts, L. Hawkins, A. G. Van Kessel, W. L. Crosby, and S. M. Hemmingsen.** 2002. Extensive profiling of a complex microbial community by high-throughput sequencing. *Appl. Environ. Microbiol.* **68:**3055–3066.
73. **Hochwalt, A. E., R. W. Berg, S. J. Meyer, and R. Eusebio.** 2002. Site-specific prevalence and cell densities of selected microbes in the lower reproductive tract of menstruating tampon users. *Infect. Dis. Obstet. Gynecol.* **10:**141–151.
74. **Hom, Y. K., P. Young, A. A. Thomson, and G. R. Cunha.** 1998. Keratinocyte growth factor injected into female mouse neonates stimulates uterine and vaginal epithelial growth. *Endocrinology* **139:**3772–3779.
75. **Hooton, T. M., D. Scholes, J. P. Hughes, C. Winter, P. L. Roberts, A. E. Stapleton, A. Stergachis, and W. E. Stamm.** 1996. A prospective study of risk factors for symptomatic urinary tract infection in young women. *N. Engl. J. Med.* **335:**468–474.
76. **Hopkins, W., A. Gendron-Fitzpatrick, D. O. McCarthy, J. E. Haine, and D. T. Uehling.** 1996. Lipopolysaccharide-responder and nonresponder C3H mouse strains are equally susceptible to an induced *Escherichia coli* urinary tract infection. *Infect. Immun.* **64:**1369–1372.
77. **Howard, J., C. Heinemann, B. J. Thatcher, B. Martin, B. S. Gan, and G. Reid.** 2000. Identification of collagen-binding proteins in *Lactobacillus* spp. with surface-enhanced laser desorption/ionization-time of flight ProteinChip technology. *Appl. Environ. Microbiol.* **66:**4396–4400.
78. **Howell, A. B.** 2002. Cranberry proanthocyanidins and the maintenance of urinary tract health. *Crit. Rev. Food Sci. Nutr.* **42**(3 Suppl.)**:**273–278.
79. **Iguchi, T., F. D. Uchima, P. L. Ostrander, and H. A. Bern.** 1983. Growth of normal mouse vaginal epithelial cells in and on collagen gels. *Proc. Natl. Acad. Sci. USA* **80:**3743–3747.
80. **Isolauri, E., P. V. Kirjavainen, and S. Salminen.** 2002. Probiotics: a role in the treatment of intestinal infection and inflammation? *Gut* **50**(Suppl. 3)**:**III54–III59.
81. **Isolauri, E., Y. Sutas, P. Kankaanpaa, H. Arvilommi, and S. Salminen.** 2001. Probiotics: effects on immunity. *Am. J. Clin. Nutr.* 73(2 Suppl.):444S–450S.
82. **Ito, T. Y., A. S. Trant, and M. L. Polan.** 2001. A double-blind placebo-controlled study of ArginMax, a nutritional supplement for enhancement of female sexual function. *J. Sex Marital Ther.* **27:**541–549.
83. **Johnsborg, O., D. B. Diep, and I. F. Nes.** 2003. Structural analysis of the peptide pheromone receptor PlnB, a histidine protein kinase from *Lactobacillus plantarum. J. Bacteriol.* **185:**6913–6920.
84. **Juarez Tomas, M. S., V. S. Ocana, B. Wiese, and M. E. Nader-Macias.** 2003. Growth and lactic acid production by vaginal *Lactobacillus acidophilus* CRL 1259, and inhibition of uropathogenic *Escherichia coli. J. Med. Microbiol.* **52:**1117–1124.
85. **Kalinka, J., T. Laudanski, W. Hanke, and M. Wasiela.** 2003. Do microbiological factors account for poor pregnancy outcome among unmarried pregnant women in Poland? *Fetal Diagn. Ther.* **18:**345–352.
86. **Kalliomaki, M., S. Salminen, T. Poussa, H. Arvilommi, and E. Isolauri.** 2003. Probiotics and prevention of atopic disease: 4-year follow-up of a randomised placebo-controlled trial. *Lancet* **361:**1869–1871.
87. **Karlsson, H., C. Hessle, and A. Rudin.** 2002. Innate immune responses of human neonatal cells to bacteria from the normal gastrointestinal flora. *Infect. Immun.* **70:**6688–6696.
88. **Keane, F. E., C. A. Ison, and D. Taylor-Robinson.** 1997. A longitudinal study of the vaginal flora over a menstrual cycle. *Int. J. STD AIDS* **8:**489–494.
88a. **Klebanoff, M. A., J. R. Schwebke, J. Zhang, T. R. Nansel, K. F. Yu, and W. W. Andrews.** 2004. Vulvovaginal symptoms in women with bacterial vaginosis. *Obstet. Gynecol.* **104:** 267–272.
89. **Kreft, B., S. Bohnet, O. Carstensen, J. Hacker, and R. Marre.** 1993. Differential expression of interleukin-6, intracellular adhesion molecule 1, and major histocompatibility complex class II molecules in renal carcinoma cells stimulated with S fimbriae of uropathogenic *Escherichia coli. Infect. Immun.* **61:**3060–3063.
90. **Kwak, C., H. K. Kim, E. C. Kim, M. S. Choi, and H. M. Kim.** 2003. Urinary oxalate levels and the enteric bacterium *Oxalobacter formigenes* in patients with calcium oxalate urolithiasis. *Eur. Urol.* **44:**475–481.
91. **Mack, D. R., S. Michail, S. Wei, L. McDougall, and M. A. Hollingsworth.** 1999. Probiotics inhibit enteropathogenic *E. coli* adherence in vitro by inducing intestinal mucin gene expression. *Am. J. Physiol. Ser. G* **276:**G941–G950.
92. **Marrazzo, J. M., L. A. Koutsky, D. A. Eschenbach, K. Agnew, K. Stine, and S. L. Hillier.** 2002. Characterization of vaginal flora and bacterial vaginosis in women who have sex with women. *J. Infect. Dis.* **185:**1307–1313.
93. **Marrie, T. J., C. A. Swantee, and M. Hartlen.** 1980. Aerobic and anaerobic urethral flora of healthy females in various physiological age groups and of females with urinary tract infections. *J. Clin. Microbiol.* **11:**654–659.
94. **Mattar, A. F., D. H. Teitelbaum, R. A. Drongowski, F. Yongyi, C. M. Harmon, and A. G. Coran.** 2002. Probiotics up-regulate MUC-2 mucin gene expression in a Caco-2 cell-culture model. *Pediatr. Surg. Int.* **18:**586–590.
95. **McGroarty, J. A., and G. Reid.** 1988. Detection of a *lactobacillus* substance which inhibits *Escherichia coli Can. J. Microbiol.* **34:**974–978.
96. **McGroarty, J. A., and G. Reid.** 1988. Inhibition of enterococci by *Lactobacillus* species *in vitro. Microb. Ecol. Health Dis.* **1:**215–219.
97. **Mulero-Marchese, R. D., K. J. Blank, and T. G. Sieck.** 1999. Strain-dependent migration of lymphocytes to the vaginal mucosa after peripheral immunization. *Immunogenetics* **49:** 973–980.

98. **Myziuk, L., B. Romanowski, and S. C. Johnson.** 2003. BVBlue test for diagnosis of bacterial vaginosis. *J. Clin. Microbiol.* **41:**1925–1928.
99. **Nagase, S., S. Shumiya, T. Emori, and H. Tanaka.** 1983. High incidence of renal tumors induced by *N*-dimethylnitrosamine in analbuminemic rats. *Gann* **74:**317–318.
100. **Naruszewicz, M., M. L. Johansson, D. Zapolska-Downar, and H. Bukowska.** 2002. Effect of *Lactobacillus plantarum* 299v on cardiovascular disease risk factors in smokers. *Am. J. Clin. Nutr.* **76:**1249–1255.
101. **Nase, L., K. Hatakka, E. Savilahti, M. Saxelin, A. Ponka, T. Poussa, R. Korpela, and J. H. Meurman.** 2001. Effect of long-term consumption of a probiotic bacterium, *Lactobacillus rhamnosus* GG, in milk on dental caries and caries risk in children. *Caries Res.* **35:**412–420.
102. **Ness, R. B., S. Hillier, H. E. Richter, D. E. Soper, C. Stamm, D. C. Bass, R. L. Sweet, and P. Rice.** 2003. Can known risk factors explain racial differences in the occurrence of bacterial vaginosis? *J. Natl. Med. Assoc.* **95:**201–212.
103. **Nettelbladt, C. G., M. Katouli, T. Bark, T. Svenberg, R. Mollby, and O. Ljungqvist.** 2003. Orally inoculated *Escherichia coli* strains colonize the gut and increase bacterial translocation after stress in rats. *Shock* **20:**251–256.
104. **Nordenstam, G. R., C. A. Brandberg, A. S. Oden, C. M. Svanborg Eden, and A. Svanborg.** 1996. Bacteriuria and mortality in an elderly population. *N. Engl. J. Med.* **314:** 1152–1156.
105. **Nugent, R. P., M. A. Krohn, and S. L. Hillier.** 1991. Reliability of diagnosing bacterial vaginosis is improved by a standardization method of Gram stain interpretation. *J. Clin. Microbiol.* **29:**297–301.
106. **Obata-Yasuoka, M., W. Ba-Thein, T. Tsukamoto, H. Yoshikawa, and H. Hayashi.** 2002. Vaginal *Escherichia coli* share common virulence factor profiles, serotypes and phylogeny with other extraintestinal *E. coli. Microbiology* **148:**2745–2752.
107. **Ohashi, Y., S. Nakai, T. Tsukamoto, N. Masumori, H. Akaza, N. Miyanaga, T. Kitamura, K. Kawabe, T. Kotake, M. Kuroda, S. Naito, H. Koga, Y. Saito, K. Nomata, M. Kitagawa, and Y. Aso.** 2002. Habitual intake of lactic acid bacteria and risk reduction of bladder cancer. *Urol. Int.* **68:**273–280.
108. **Ohmit, S. E., J. D. Sobel, P. Schuman, A. Duerr, K. Mayer, A. Rompalo, and R. S. Klein for the HIV Epidemiology Research Study (HERS) Group.** 2003. Longitudinal study of mucosal *Candida* species colonization and candidiasis among human immunodeficiency virus (HIV)-seropositive and at-risk HIV-seronegative women. *J. Infect. Dis.* **188:**118–127.
109. **Owen, D. H., and D. F. Katz.** 1999. A vaginal fluid simulant. *Contraception* **59:**91–95.
110. **Patton, D. L., Y. T. Cosgrove Sweeney, M. A. Antonio, L. K. Rabe, and S. L. Hillier.** 2003. *Lactobacillus crispatus* capsules: single-use safety study in the Macaca nemestrina model. *Sex. Transm. Dis.* **30:**568–570.
111. **Picioreanu, C., M. C. Van Loosdrecht, and J. J. Heijnen.** 2000. Effect of diffusive and convective substrate transport on biofilm structure formation: a two-dimensional modeling study. *Biotechnol. Bioeng.* **69:**504–515.
112. **Polonelli, L., F. De Bernadis, S. Conti, M. Boccanera, W. Magliani, M. Gerloni, C. Cantelli, and A. Cassone.** 1996. Human natural yeast killer toxin-like candidacidal antibodies. *J. Immunol.* **156:**1880–1885.
113. **Quon, D. J.** 2000. Monitoring of domestic and imported eggs for veterinary drug residues by the Canadian Food Inspection Agency. *J. Agric. Food Chem.* **48:**6421–6427.
114. **Rafter, J.** 2002. Lactic acid bacteria and cancer: mechanistic perspective. *Br. J. Nutr.* **88**(Suppl. 1):S89–S94.
115. **Rayes, N., S. Hansen, D. Seehofer, A. R. Muller, S. Serke, S. Bengmark, and P. Neuhaus.** 2002. Early enteral supply of fiber and lactobacilli versus conventional nutrition: a controlled trial in patients with major abdominal surgery. *Nutrition* **18:**609–615.
116. **Rayes, N., D. Seehofer, S. Hansen, K. Boucsein, A. R. Muller, S. Serke, S. Bengmark, and P. Neuhaus.** 2000. Early enteral supply of *Lactobacillus* and fiber versus selective bowel decontamination: a controlled trial in liver transplant recipients. *Transplantation* **74:**123–127.
117. **Raz, R., R. Colodner, Y. Rohana, S. Battino, E. Rottensterich, I. Wasser, and W. Stamm 2003.** Effectiveness of estriol-containing vaginal pessaries and nitrofurantoin macrocrystal therapy in the prevention of recurrent urinary tract infection in postmenopausal women. *Clin. Infect. Dis.* **36:**1362–1368.
118. **Raz, R., and W. E. Stamm.** 1993. A controlled trial of intravaginal estriol in postmenopausal women with recurrent urinary tract infections. *N. Engl. J. Med.* **329:**753–756.
119. **Reid, G., D. Beuerman, C. Heinemann, and A. W. Bruce.** 2001. Probiotic *Lactobacillus* dose required to restore and maintain a normal vaginal flora. *FEMS Immunol. Med. Microbiol.* **32:**37–41.
120. **Reid, G., and A. Bocking.** 2003. The potential for probiotics to prevent bacterial vaginosis and preterm labor. *Am. J. Obstet. Gynecol.* **189:**1202–1208.
121. **Reid, G., H. J. L. Brooks, and D. F. Bacon.** 1983. *In vitro* attachment of *Escherichia coli* to human uroepithelial cells. Variation in receptivity during the menstrual cycle and pregnancy. *J. Infect. Dis.* **148:**412–421.
122. **Reid, G., A. W. Bruce, R. L. Cook, and M. Llano.** 1990. Effect on the urogenital flora of antibiotic therapy for urinary tract infection. *Scand. J. Infect. Dis.* **22:**43–47.
123. **Reid, G., A. W. Bruce, N. Fraser, C. Heinemann, J. Owen, and B. Henning.** 2001. Oral probiotics can resolve urogenital infections. *FEMS Microbiol. Immunol.* **30:**49–52.
124. **Reid, G., A. W. Bruce, F. Soboh, and M. Mittelman.** 1998. Effect of nutrient composition on the *in vitro* growth of urogenital *Lactobacillus* and uropathogens. *Can. J. Microbiol.* **44:**1–6.
125. **Reid, G., A. W. Bruce, and M. Taylor.** 1995. Instillation of *Lactobacillus* and stimulation of indigenous organisms to prevent recurrence of urinary tract infections. *Microecol. Ther.* **23:**32–45.
125a. **Reid, G., J. Burton, J.-A. Hammond, and A. W. Bruce.** 2004. Nucleic acid-based diagnosis of bacterial vaginosis and improved management using probiotic lactobacilli. *J. Med. Food* **7:**223–228.
126. **Reid, G., R. C. Y. Chan, A. W. Bruce, and J. W. Costerton.** 1985. Prevention of urinary tract infection in rats with an indigenous *Lactobacillus casei* strain. *Infect. Immun.* **49:**320–324.
127. **Reid, G., D. Charbonneau, J. Erb, B. Kochanowski, D. Beuerman, R. Poehner, and A. W. Bruce.** 2003. Oral use of *Lactobacillus rhamnosus* GR-1 and *L. fermentum* RC-14 significantly alters vaginal flora: randomized, placebo-controlled trial in 64 healthy women. *FEMS Immunol. Med. Microbiol.* **35:**131–134.
128. **Reid, G., R. Charbonneau-Smith, D. Lam, M. Lacerte, Y. S. Kang and K. C. Hayes.** 1992. Bacterial biofilm formation in the urinary bladder of spinal cord injured patients. *Paraplegia* **30:**711–717.
129. **Reid, G., R. L. Cook, and A. W. Bruce.** 1987. Examination of strains of lactobacilli for properties which may influence bacterial interference in the urinary tract. *J. Urol.* **138:**330–335.
130. **Reid, G., R. L. Cook, L. Hagberg, and A. W. Bruce.** 1989. Lactobacilli as competitive colonizers of the urinary tract,

p. 390–396. *In* E. H. Kass and C. Svanborg Eden (ed.), *Host-Parasite Interactions in Urinary Tract Infections.* University of Chicago Press, Chicago, Ill.

131. **Reid, G., B. S. Gan, Y.-M. She, W. Ens, S. Weinberger, and J. C. Howard.** 2002. Rapid identification of probiotic lactobacilli biosurfactant proteins by ProteinChip MS/MS tryptic peptide sequencing. *Appl. Environ. Microbiol.* **68:** 977–980.
132. **Reid, G., J.-A. Hammond, and A. W. Bruce.** 2003. Effect of lactobacilli oral supplement on the vaginal microflora of antibiotic treated patients: randomized, placebo-controlled study. *Nutraceut. Foods* **8:**145–148.
133. **Reid, G., C. Heinemann, J. Howard, G. Gardiner, and B. S. Gan.** 2001. Understanding urogenital biofilms and the potential impact of probiotics. *Methods Enzymol.* **336:**403–410.
134. **Reid, G., J. Jass, T. Sebulsky and J. McCormick.** 2003. Probiotics in clinical practice. *Clin. Microbiol. Rev.* **16:**658–672.
135. **Reid, G., J. A. McGroarty, R. Angotti, and R. L. Cook.** 1988. *Lactobacillus* inhibitor production against *E. coli* and coaggregation ability with uropathogens. *Can. J. Microbiol.* **34:**344–351.
136. **Reid, G., J. A. McGroarty, L. Tomeczek, and A. W. Bruce.** 1996. Identification and plasmid profiles of *Lactobacillus* species from the vagina of 100 healthy women. *FEMS Immunol. Med. Microbiol.* **15:**23–26.
137. **Reid, G., K. Millsap, and A. W. Bruce.** 1994. Implantation of *Lactobacillus casei* var *rhamnosus* into the vagina. *Lancet* **344:**1229.
138. **Reid, G., P. Potter, D. Lam, D. Warren, M. Borrie, and K Hayes.** 2003. Cranberry juice to reduce bladder biofilms and infection in geriatric and spinal cord injured patients with dysfunctional bladders. *Nutraceut. Foods* **8:**24–28.
139. **Reid, G., M. E. Sanders, R. Gaskins, G. Gibson, A. Mercenier, B. Rastall, M. Roberfroid, I. Rowland, C. Cherbut, and T. Klaenhammer.** 2003. New scientific paradigms for probiotics and prebiotics. *J. Clin. Gastroenterol.* **37:**105–118.
140. **Riemersma, W. A., C. J. van der Schee, W. I. van der Meijden, H. A. Verbrugh, and A. van Belkum.** 2003. Microbial population diversity in the urethras of healthy males and males suffering from nonchlamydial, nongonococcal urethritis. *J. Clin. Microbiol.* **41:**1977–1986.
141. **Royce, R. A., T. P. Jackson, J. M. Thorp Jr, S. L. Hillier, L. K. Rabe, L. M. Pastore, and D. A. Savitz.** 1999. Race/ethnicity, vaginal flora patterns, and pH during pregnancy. *Sex. Transm. Dis.* **26:**96–102.
142. **Rozanova, G. N., D. A. Voevodin, M. A. Stenina, and M. V. Kushnareva.** 2002. Pathogenetic role of dysbacteriosis in the development of complications of type 1 diabetes mellitus in children. *Bull. Exp. Biol. Med.* **133:**164–166.
143. **Schaeffer, A. J., S. K. Amundsen, and L. N. Schmidt.** 1979. Adherence of *Escherichia coli* to human urinary tract epithelial cells. *Infect. Immun.* **24:**753–759.
144. **Schafer, K. S., and S. E. Kegley.** 2002. Persistent toxic chemicals in the US food supply. *J. Epidemiol. Community Health* **56:**813–817.
145. **Schell, M. A., M. Karmirantzou, B. Snel, D. Vilanova, B. Berger, G. Pessi, M. C. Zwahlen, F. Desiere, P. Bork, M. Delley, R. D. Pridmore, and F. Arigoni.** 2002. The genome sequence of *Bifidobacterium longum* reflects its adaptation to the human gastrointestinal tract. *Proc. Natl. Acad. Sci. USA* **99:**14422–14427.
146. **Schilling, J. D., S. M. Martin, C. S. Hung, R. G. Lorenz, and S. J. Hultgren.** 2003. Toll-like receptor 4 on stromal and hematopoietic cells mediates innate resistance to uropathogenic *Escherichia coli. Proc. Natl. Acad. Sci. USA* **100:**4203–4208.
147. **Schwebke, J. R., S. L. Hillier, J. D. Sobel, J. A. McGregor, and R. L. Sweet.** 1996. Validity of the vaginal Gram stain for the diagnosis of bacterial vaginosis. *Obstet. Gynecol.* **88:** 573–576.
148. **Schwebke, J. R., C. M. Richey, and H. L. Weiss.** 1999. Correlation of behaviors with microbiological changes in vaginal flora. *J. Infect. Dis.* **180:**1632–1636.
149. **Seddon, J. M., A. W. Bruce, P. Chadwick, and P. Carter.** 1976. Introital bacterial flora—effect of increased frequency of micturition. *Br. J. Urol.* **48:**211–218.
150. **Serafini-Cessi, F., N. Malagolini, and D. Cavallone.** 2003. Tamm-Horsfall glycoprotein: biology and clinical relevance. *Am. J. Kidney Dis.* **42:**658–676.
151. **Smeianov, V., K. Scott, and G. Reid.** 2000. Activity of Fall-39 and Cecropin P1 against urogenital microflora. *Microbes Infect.* **2:**773–777.
152. **Soderhall, M., S. Normark, K. Ishikawa, K. Karlsson, S. Teneberg, J. Winberg, and R. Mollby.** 1997. Induction of protective immunity after *Escherichia coli* bladder infection in primates. Dependence of the globoside-specific P-fimbrial tip adhesin and its cognate receptor. *J. Clin. Investig.* **100:**364–372.
153. **Steele, C., and P. L. Fidel, Jr.** 2002. Cytokine and chemokine production by human oral and vaginal epithelial cells in response to *Candida albicans. Infect. Immun.* **70:**577–583.
154. **Strus, M., L. Pakosz, H. Gosciniak, A. Przondo-Mordarska, E. Rozynek, H. Pituch, F. Meisel-Mikolajczyk, and P. B. Heczko.** 2001. Antagonistic activity of *Lactobacillus* bacteria strains against anaerobic gastrointestinal tract pathogens (*Helicobacter pylori, Campylobacter coli, Campylobacter jejuni, Clostridium difficile*). *Med. Dosw. Mikrobiol.* **53:**133–142. (In Polish.)
155. **Sundstrom, P.** 2002. Adhesion in *Candida* spp. *Cell. Microbiol.* **4:**461–469.
156. **Svanborg, C., W. Agace, S. Hedges, R. Lindstedt, and M. L. Svensson.** 1994. Bacterial adherence and mucosal cytokine production. *Ann. N.Y. Acad. Sci.* **730:**162–181.
157. **Svanborg, C., G. Bergsten, H. Fischer, B. Frendeus, G. Godaly, E. Gustafsson, L. Hang, M. Hedlund, D. Karpman, A. C. Lundstedt, M. Samuelsson, P. Samuelsson, M. Svensson, and B. Wullt.** 2001. The 'innate' host response protects and damages the infected urinary tract. *Ann. Med.* **33:**563–570.
158. **Tamboli, C. P., C. Caucheteux, A. Cortot, J. F. Colombel, and P. Desreumaux.** 2003. Probiotics in inflammatory bowel disease: a critical review. *Best Pract. Res. Clin. Gastroenterol.* **17:**805–820.
159. **Tannock, G. W.** 2002. Exploring the relationships between intestinal microflora and inflammatory conditions of the human bowel and spine. *Antonie Leeuwenhoek* **81:**529–535.
160. **Tannock, G. W.** 2001. Molecular assessment of intestinal microflora. *Am. J. Clin. Nutr.* 73(2 Suppl.):410S–414S.
161. **Tannock, G. W., R. Fuller, S. L. Smith, and M. A. Hall.** 1990. Plasmid profiling of members of the family *Enterobacteriaceae*, lactobacilli, and bifidobacteria to study the transmission of bacteria from mother to infant. *J. Clin. Microbiol.* **28:**1225–1228.
162. **Taylor, J. K., R. W. Hall, and A. R. Dupre.** 2002. The incidence of group B streptococcus in the vaginal tracts of pregnant women in central Alabama. *Clin. Lab. Sci.* **15:**16–17.
163. **Tomas, M. S., E. Bru, and M. E. Nader-Macias.** 2003. Comparison of the growth and hydrogen peroxide production by vaginal probiotic lactobacilli under different culture conditions. *Am. J. Obstet. Gynecol.* **188:**35–44.
164. **Trinel, P. A., E. Maes, J. P. Zanetta, F. Delplace, B. Coddeville, T. Jouault, G. Strecker, and D. Poulain.** 2002.

Candida albicans phospholipomannan, a new member of the fungal mannose inositol phosphoceramide family. *J. Biol. Chem.* **277**:37260–37271.

165. **Uehling, D. T., D. B. Johnson, and W. J. Hopkins.** 1999. The urinary tract response to entry of pathogens. *World J. Urol.* **17**:351–358.
166. **Uhlen, P., A. Laestadius, T. Jahnukainen, T. Soderblom, F. Backhed, G. Celsi, H. Brismar, S. Normark, A. Aperia, and A. Richter-Dahlfors.** 2000. Alpha-haemolysin of uropathogenic *E. coli* induces Ca^{2+} oscillations in renal epithelial cells. *Nature* **405**:694–697.
167. **Vallor, A. C., M. A. Antonio, S. E. Hawes, and S. L. Hillier.** 2001. Factors associated with acquisition of, or persistent colonization by, vaginal lactobacilli: role of hydrogen peroxide production. *J. Infect. Dis.* **184**:1431–1436.
168. **Valore, E. V., C. H. Park, S. L. Igreti, and T. Ganz.** 2002. Antimicrobial components of vaginal fluid. *Am. J. Obstet. Gynecol.* **187**:561–568.
169. **Veazey, R. S., P. A. Marx, and A. A. Lackner.** 2003. Vaginal $CD4^+$ T cells express high levels of CCR5 and are rapidly depleted in simian immunodeficiency virus infection. *J. Infect. Dis.* **187**:769–776.
170. **Velraeds, M. C., B. van der Belt, H. C. van der Mei, G. Reid, and H. J. Busscher.** 1998. Interference in initial adhesion of uropathogenic bacteria and yeasts silicone rubber by a *Lactobacillus acidophilus* biosurfactant. *J. Med. Microbiol.* **49**: 790–794.
171. **Velraeds, M. C., H. C. van der Mei, G. Reid, and H. J. Busscher.** 1996. Inhibition of initial adhesion of uropathogenic *Enterococcus faecalis* by biosurfactants from *Lactobacillus* isolates. *Appl. Environ. Microbiol.* **62**:1958–1963.
172. **Velraeds, M. C., H. C. van der Mei, G. Reid, and H. J. Busscher.** 1996. Physicochemical and biochemical characterization of biosurfactants released from *Lactobacillus* strains. *Colloids Surf. Ser. B.* **8**:51–61.

172a. **Vollenweider, S., G. Grassi, I. Konig, and Z. Puhan.** 2003. Purification and structural characterization of 3-hydroxypropionaldehyde and its derivatives. *J. Agric. Food Chem.* **51**:3287–3293.

173. **Weller, T. M.** 2000. Methicillin-resistant *Staphylococcus aureus* typing methods: which should be the international standard? *J. Hosp. Infect.* **44**:160–172.
174. **West, B., L. Morison, M. S. van der Loeff, E. Gooding, A. A. Awasana, E. Demba, and P. Mayaud.** 2003. Evaluation of a new rapid diagnostic kit (FemExam) for bacterial vaginosis in patients with vaginal discharge syndrome in The Gambia. *Sex. Transm. Dis.* **30**:483–489.
175. **Wettergren, B., U. Jodal, and G. Jonasson.** 1985. Epidemiology of bacteriuria during the first year of life. *Acta Paediatr. Scand.* **74**:925–933.
176. **Xia, Y., K. Forsman, J. Jass, and B. E. Uhlin.** 1998. Oligomeric interaction of the PapB transcriptional regulator with the upstream activating region of pili adhesin gene promoters in *Escherichia coli*. *Mol. Microbiol.* **30**:513–523.
177. **Xu, J., M. K. Bjursell, J. Himrod, S. Deng, L. K. Carmichael, H. C. Chiang, L. V. Hooper, and H. I. Gordon.** 2003. A genomic view of the human-*Bacteroides thetaiotaomicron* symbiosis. *Science* **299**:2074–2076.
178. **Yoshida, M., T. Watanabe, T. Usui, Y. Matsunaga, Y. Shirai, M. Yamori, T. Itoh, S. Habu, T. Chiba, T. Kita, and Y. Wakatsuki.** 2001. CD4 T cells monospecific to ovalbumin produced by *Escherichia coli* can induce colitis upon transfer to BALB/c and SCID mice. *Int. Immunol.* **13**:1561–1570.
179. **Zheng, H., J. Cao, H. Jin, and J. Wang.** 1999. Preliminary examination of a LB-H2O2 substance that inhibit *Neisseria gonorrhea* growth. *Zhongguo Yi Xue Ke Xue Yuan Xue Bao* **21**:379–383. (In Chinese.)
180. **Zhong, W., K. Millsap, H. Bialkowska-Hobrzanska, and G. Reid.** 1998. Differentiation of *Lactobacillus* species by molecular typing. *Appl. Environ. Microbiol.* **64**:2418–2423.

INDEX